								2 HELIUM **He** 4.00
			5 BORON **B** 10.81	6 CARBON **C** 12.01	7 NITROGEN **N** 14.01	8 OXYGEN **O** 15.999	9 FLUORINE **F** 18.998	10 NEON **Ne** 20.18
			13 ALUMINUM **Al** 26.98	14 SILICON **Si** 28.08	15 PHOSPHORUS **P** 30.97	16 SULFUR **S** 32.06	17 CHLORINE **Cl** 35.45	18 ARGON **Ar** 39.95
28 NICKEL **Ni** 58.69	29 COPPER **Cu** 63.55	30 ZINC **Zn** 65.38	31 GALLIUM **Ga** 69.72	32 GERMANIUM **Ge** 72.59	33 ARSENIC **As** 74.92	34 SELENIUM **Se** 78.96	35 BROMINE **Br** 79.90	36 KRYPTON **Kr** 83.80
46 PALLADIUM **Pd** 106.42	47 SILVER **Ag** 107.87	48 CADMIUM **Cd** 112.41	49 INDIUM **In** 114.82	50 TIN **Sn** 118.69	51 ANTIMONY **Sb** 121.75	52 TELLURIUM **Te** 127.60	53 IODINE **I** 126.90	54 XENON **Xe** 131.29
78 PLATINUM **Pt** 195.08	79 GOLD **Au** 196.97	80 MERCURY **Hg** 200.59	81 THALLIUM **Tl** 204.38	82 LEAD **Pb** 207.2	83 BISMUTH **Bi** 208.98	84 POLONIUM **Po** (209)	85 ASTATINE **At** (210)	86 RADON **Rn** (222)

63 EUROPIUM **Eu** 151.96	64 GADOLINIUM **Gd** 157.25	65 TERBIUM **Tb** 158.92	66 DYSPROSIUM **Dy** 162.50	67 HOLMIUM **Ho** 164.93	68 ERBIUM **Er** 167.26	69 THULIUM **Tm** 168.93	70 YTTERBIUM **Yb** 173.04	71 LUTETIUM **Lu** 174.97
95 AMERICIUM **Am** (243)	96 CURIUM **Cm** (247)	97 BERKELIUM **Bk** (247)	98 CALIFORNIUM **Cf** (251)	99 EINSTEINIUM **Es** (252)	100 FERMIUM **Fm** (257)	101 MENDELEVIUM **Md** (258)	102 NOBELIUM **No** (259)	103 LAWRENCIUM **Lr** (260)

GENERAL, ORGANIC, AND BIOLOGICAL

CHEMISTRY

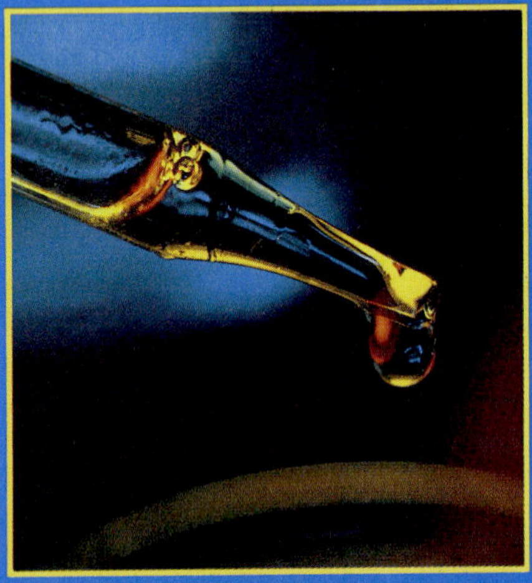

GENERAL, ORGANIC, AND BIOLOGICAL CHEMISTRY

John R. Amend
Montana State University

Bradford P. Mundy
Montana State University

Melvin T. Armold
Adams State College

SAUNDERS COLLEGE PUBLISHING
Philadelphia New York Chicago
San Francisco Montreal Toronto
London Sydney Tokyo

Illustration Credits appear on page A–53, which constitutes a continuation of the copyright page.

Printed in the United States of America

GENERAL, ORGANIC, AND BIOLOGICAL CHEMISTRY

0-03-046988-0

Library of Congress Catalog Card Number: 89-85711

9012 041 987654321

Preface

This textbook can be used for a range of introductory courses in chemistry. It is intended primarily for use in a first course in what is often called allied health chemistry. Students in such a course are preparing for careers in such areas as nursing and the health sciences, agriculture and its related fields, nutrition, home economics, medical technology, physical therapy, and all those areas of teaching that involve some applications of chemistry. These occupations require a broad-based conceptual understanding of the principles of general, organic, and biological chemistry, as well as mastery of some basic chemical arithmetic.

Students who work their way carefully through this book should be able to achieve two basic results. First, they will have mastered a relatively small number of core concepts through which most of chemistry can be understood—the structure of the atom, the nature of chemical bonding, the factors that influence reactions between molecules, and the quantitative tools used to describe chemical behavior. Second, they will have come to see how great a role chemistry plays in everyday life—how advances in medicine, material sciences, agriculture, and the generation of energy are dependent upon chemical principles.

This textbook is divided into two parts. The first, *General Chemistry*, is made up of 13 chapters that introduce the fundamental concepts of chemistry. It is intended to establish a foundation of chemical principles and to provide students with an opportunity to come to terms with the kinds of calculations that chemists and other allied health practitioners must make. The second part of the book, *Organic and Biological Chemistry*, consists of 17 chapters. This material builds on the already established framework in order to explore those topics that drew most students to this course—topics that involve the chemistry of living systems.

KEY FEATURES

The class-tested features of this book are designed to guide the introductory student into the complexities of chemistry. Each chapter begins with a list of principal learning *Objectives*. Definitions of all new, important key terms and concepts are listed at the end of each major section in a *New Terms* list, which functions as a running glossary. Accompanying most New Terms is a feature unique to this book called *Testing Yourself*. This is a brief self-test (with answers) that covers chapter objectives introduced in that section. The location of these two features—New Terms and Testing Yourself—within the chapter immediately following discussion of the concepts is especially helpful because it allows students to review and check their knowledge before going on to a new concept.

Placed throughout each chapter are several useful features. *Marginal Notes* highlight key points, making review easier. Numbered *Examples*, especially in the general and organic chemistry chapters of the book, present problems that are solved in a clear, step-by-step manner. Numerous *Application Boxes*, mostly on allied health-related issues, cover such diverse topics as Nuclear Waste, Artificial Blood, and "Crack."

Each chapter ends with a *Summary* and a list of *Terms* given in the same order as their presentation in the chapter. The *Chapter Review* that follows consists of three elements. First, a set of about ten diagnostic *Questions* provides a review of key chapter concepts. Second, a *Diagnostic Chart*, another unique feature of this book, identifies problem areas and directs the student to specific topics for additional study. Third, a set of *Exercises* organized by chapter topics provides a thorough problem-solving experience. These exercises are augmented by *Supplemental Exercises* in Appendix I; answers to odd-numbered exercises (both end-of-chapter and supplemental) are given in Appendix II.

SUPPLEMENTS

In recognition of the fact that both instructors and students might profit from some supplemental information, this textbook is accompanied by a number of ancillary publications:

Laboratory Manual
Instructor's Manual with Tests
Computerized Test Bank
Transparencies
Study Guide

ACKNOWLEDGMENTS

This book has been a team effort in the best sense of the word. The authors have been privileged to work with a fine development team at Harcourt Brace Jovanovich—Tom Thompson and Jeff Holtmeier, Acquisitions Editors; Kathy Walker, Manuscript Editor; Joan Harlan, Production Editor; Lynne Bush, Production Manager; Stacy Simpson, Art Editor; and Ann Smith, Designer.

The authors also express their appreciation to the reviewers who worked with us through several revisions of the manuscript and who contributed considerably to the organization and content of this book.

Margaret Asirrathan
University of Colorado, Boulder

Muriel Bishop
Clemson University

Pete Gardner
University of Utah

Isidore Goodman
Los Angeles Pierce College

Robert Harris
University of Nebraska

Alvin Herman
California State University, Fullerton

James Hoobler
Chemeketa Community College

Eileen Johann
Miami–Dade Community College

Joseph Lehmann
Colorado State University

Kevin Mayo
Temple University

Margaret Merrifield
California State University, Long Beach

Michael Mikita
University of Colorado, Denver

Russell Petter
University of Pittsburgh

Thomas Pynadth
Kent State University

John Richardson
Indiana University

Abraham Shina
San Diego City College

James Sodetz
University of South Carolina

Rich White
Sam Houston State University

ALLIED HEALTH-RELATED TOPICS

The following topics, examples, and applications in this book will be of special interest to students in health-related career programs. (Note: Section numbers are given unless otherwise indicated.)

Medicine and Nursing

Nutrition and Food Science

Health Risks

Medical Laboratory and Radiology Technology

Drugs

Contents

PART I

General Chemistry

CHAPTER 1

CHAPTER 2

CHAPTER 3
The Structure of Atoms 70

CHAPTER 4
Electrons and Chemical Periodicity 88

CHAPTER 5
The Chemical Bond 110

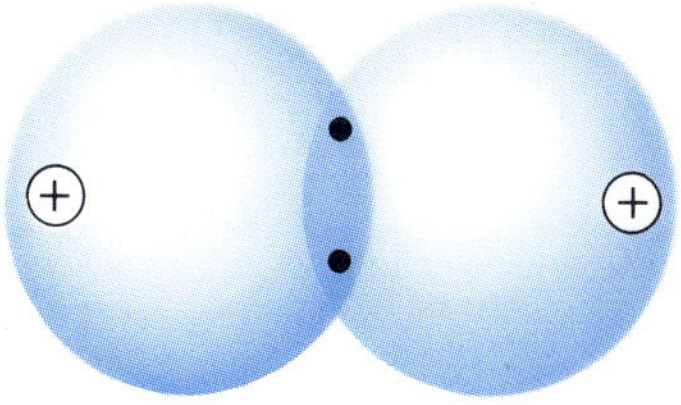

+
+

CHAPTER 6
Shapes and Polarities of Covalent Molecules 137

CHAPTER 7
Chemical Formulas and Equations 157

CHAPTER 8
Chemical Arithmetic: Weight Relationships in Chemical Reactions 198

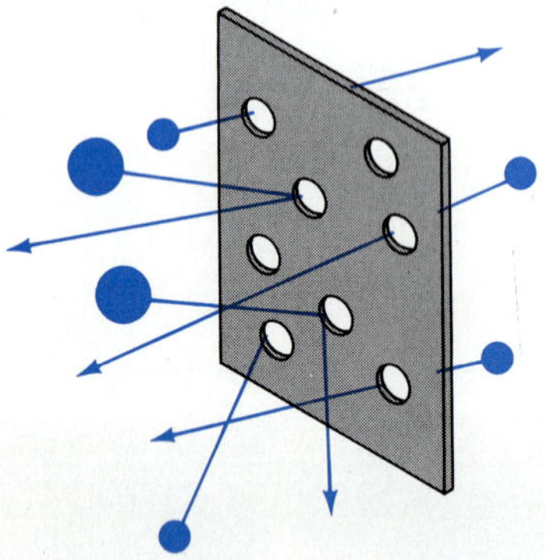

CHAPTER 9
Solutions, Colloids, and Suspensions 221

CHAPTER 10
Energy: The Driving Force in Chemical Reactions 243

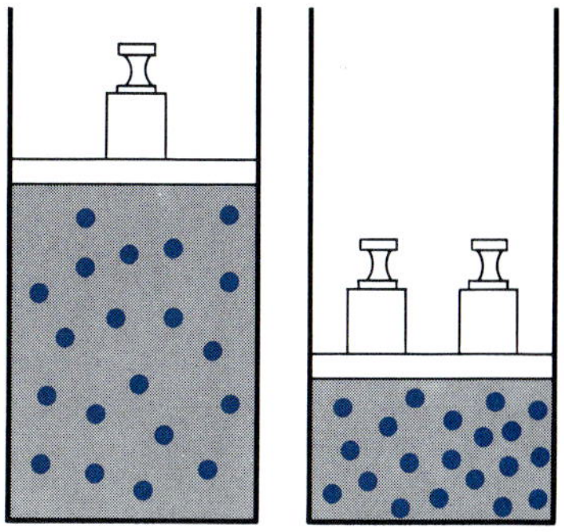

CHAPTER 11
Acids and Bases 268

CHAPTER 12
The Behavior of Gases 305

CHAPTER 13
Nuclear Chemistry 322

PART II
Organic and Biological Chemistry
355

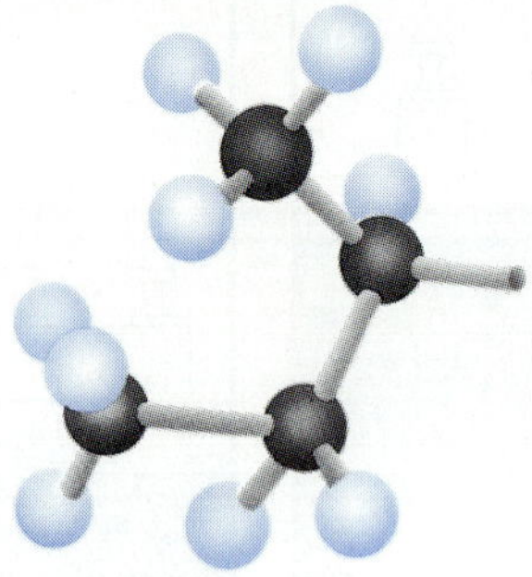

CHAPTER 14
An Introduction to Organic Chemistry 359

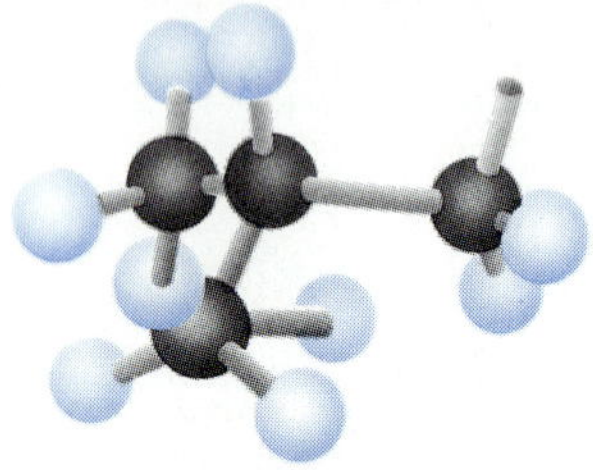

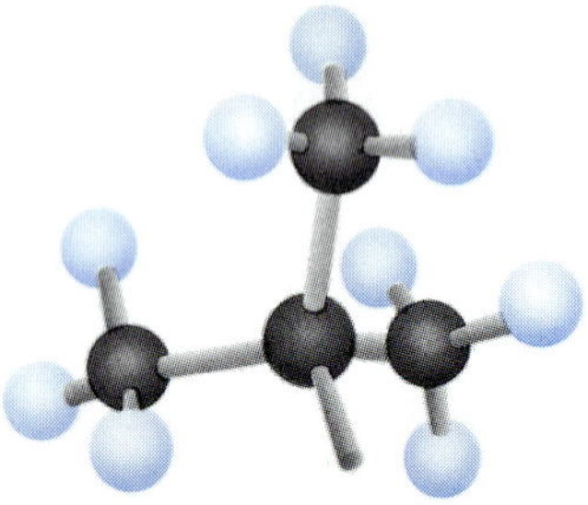

CHAPTER 15
Alkanes and Cycloalkanes: Single-Bonded Hydrocarbons 377

CHAPTER 16
Unsaturated Hydrocarbons 421

CHAPTER 17
Aromatic Hydrocarbons 456

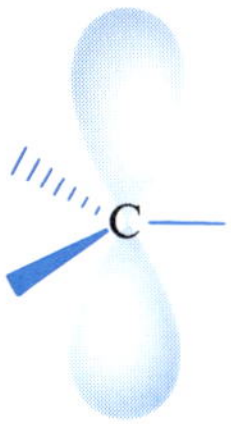
C

CHAPTER 18
Alcohols, Phenols, Ethers, and Thiols 476

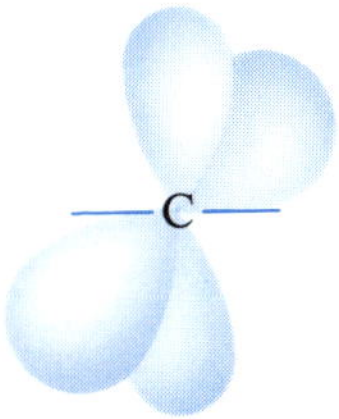
C

CHAPTER 19
Carbonyl Group and Its Compounds: Aldehydes and Ketones 515

CHAPTER 20
Carboxylic Acids and Their Derivatives 549

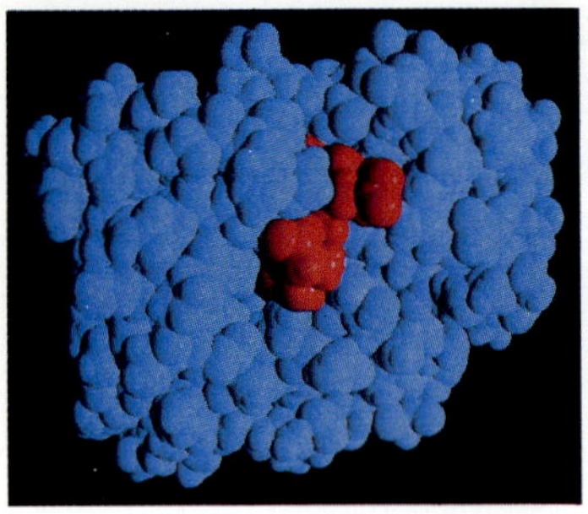

CHAPTER 24
Lipids 657

CHAPTER 25
Amino Acids, Peptides, and Proteins 689

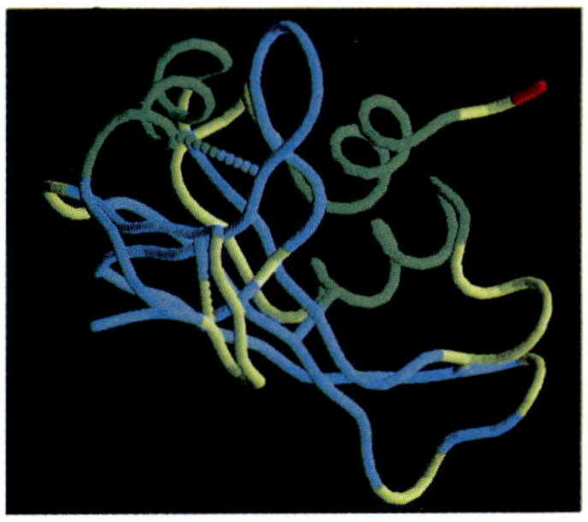

CHAPTER 26
Molecular Basis of Heredity 719

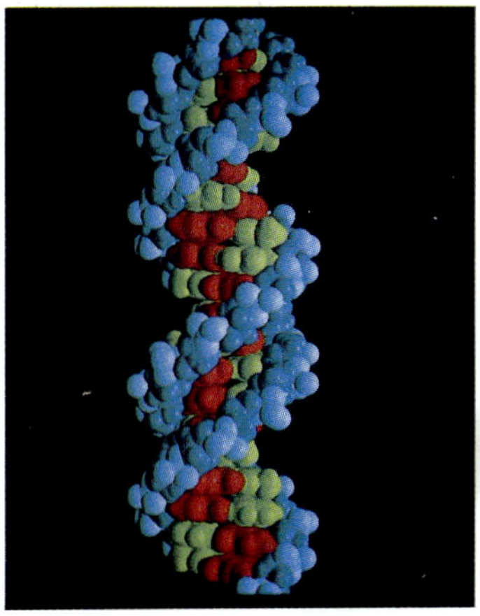

CHAPTER 27
Metabolism, Enzymes, and Bioenergetics 759

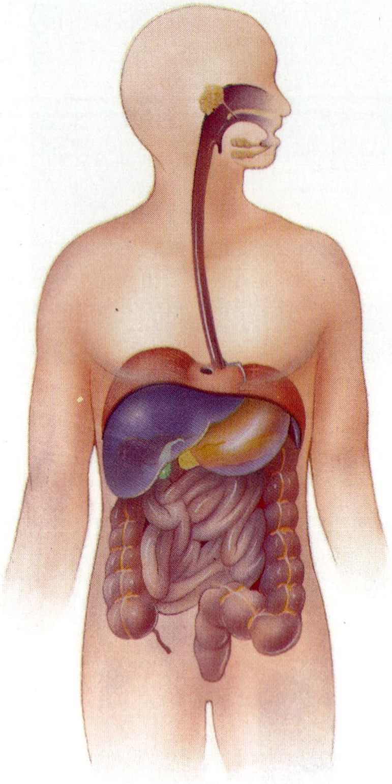

CHAPTER 28
Catabolism 787

CHAPTER 29
Anabolism 820

CHAPTER 30
Blood: The Constant Internal Environment 845

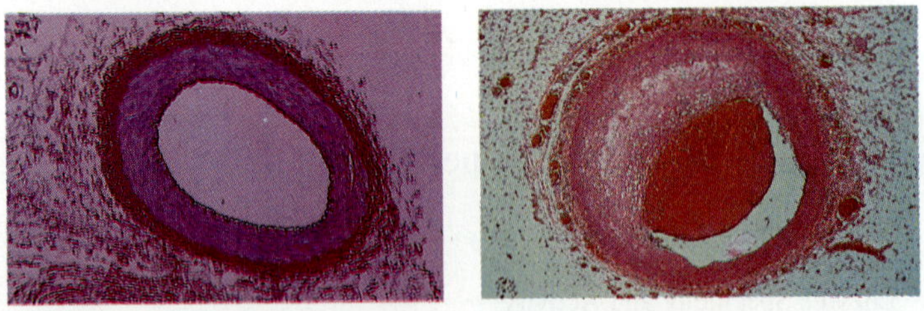

GENERAL, ORGANIC, AND BIOLOGICAL CHEMISTRY

PART

I

General Chemistry

(*Opposite page*)
This NASA photograph of a view of the earth from an Apollo mission to the moon illustrates the limited set of natural resources available to the planet's inhabitants. Except for light from the sun and dust from meteorites, every resource that will ever be available to mankind is here today. The importance of water and air are emphasized by the stark landscape of the moon in the foreground, devoid of water, air, and life.

Methods for recycling most of these resources exist in nature—the hydrologic cycle for recycling water, the nitrogen and carbon cycles which permit us to use relatively scarce nitrogen and carbon atoms over and over again, and systems for reusing the limited numbers of enzyme molecules in biological systems. These recycling mechanisms are essential to both the planet and to mankind.

> There is no science so intimately connected with the very life of man as Chemistry. Almost all kinds of cooking depend on chemical principles; as also the preparation of medicine, the detection of poisons, the arts of bleaching, and pottery, of making glass, ink, and leather, of dyeing, burning lime, working metals, &c. In most of the mechanical arts, and in some of the professions, those who understand chemistry have a great advantage over those who do not. This science bears an important relation to housekeeping in a variety of ways, as in the making of gravies, soups, jellies, and preserves, bread, butter, and cheese, in the washing of clothes, making soap, and the economy of heat in cooking, and in warming rooms . . . .

So began an introductory chemistry book written in 1838. Although our understanding of the atoms and molecules that make up our universe has increased enormously in the past few generations, the author's thought is still true today. The text quoted above was written only twelve years after atoms were proven to have measurable atomic weights. Fifty-nine years would pass before the discovery of the electron. Ernest Rutherford's discovery of the nucleus of the atom (1911) and the development of bonding theory are almost as close to us today as they were distant from the author of this 1838 text.

As our understanding of the behavior of atoms and molecules has increased, so has our ability to use these resources for the good of mankind. Improvements in sanitation and health care made possible by advances in chemistry have significantly increased the average human life span (Figure 1). Most striking is the decrease in infant death rate. In 1900 the U.S. death rate for children under one year of age was about 160 per 1000 population. Almost one child in six could be expected to die before its first birthday.

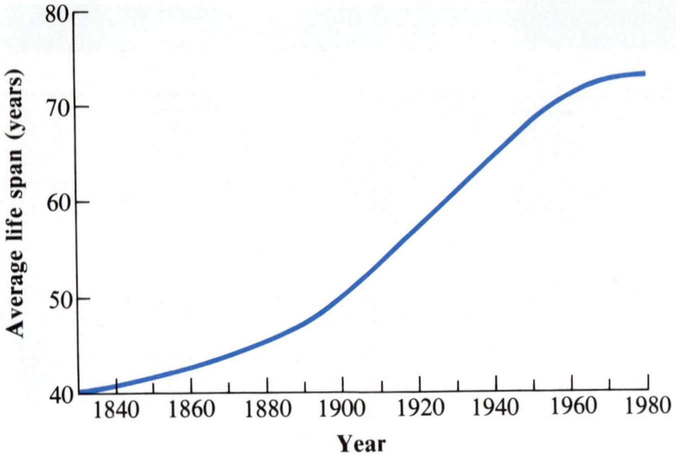
80
70
60
50
40
Average life span (years)
1840
1860
1880
1900
1920
1940
1960
1980
Year

FIGURE 1
This graph shows the increase in life span in the world's population from 1840 to 1980. This increase is due primarily to eradication of many childhood and communicable diseases.

Few among the grandparents of today's students have not lost a child. The current death rate for the children under 1 year of age is less than 25 per 1000. In 1900 the communicable diseases influenza, pneumonia, and tuberculosis were the principal causes of death. Tuberculosis has now practically been eliminated, and influenza and pneumonia have been controlled. Today's killers—heart disease and cancer—are principally enemies of the aged.

The advent of better health care has itself created problems, particularly in the areas of agriculture and energy production. There are many more of us attempting to find food, warmth, and a comfortable place to live on this planet. In 1838 the world population was about 800 million. Today it is more than 5 billion, an increase of more than six times (Figure 2).

The insecticide DDT (Figure 3) is credited with saving more human lives than any other single substance. According to the World Health Organization, the role of DDT in controlling the malaria mosquito saved about 5 million lives and prevented about 100 million cases of malaria between 1944 and 1953. Unfortunately, DDT does not break down easily and remains in the environment long after it is applied, which has caused very severe and undesirable effects on many forms of wildlife. Although it is controversial whether DDT causes harm to humans as normally applied, it kills useful as well as harmful insects. Today, considerable effort is being directed toward the development of species-selective, rapidly degradable insecticides.

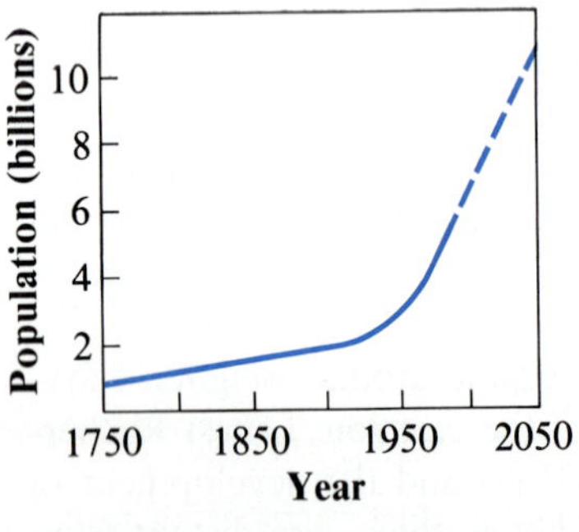
10
8
6
4
2
Population (billions)
1750
1850
1950
2050
Year

FIGURE 2
The world's population has increased more than six times in the last 200 years. The large upswing that began shortly before 1950 was largely due to the development of effective insecticides, which dramatically reduced death caused by insect-carried disease.

The dramatic increase in agricultural production necessary to feed an increased population has in part been made possible by the development of chemical fertilizers, insecticides, and herbicides. In 1900 more than 1/4 of the grain crop was eaten by insects. Today, with effective insect control, less than 1/20 is lost in this manner. Land that in 1900 produced 15 bushels of wheat per acre today produces 50 bushels with applications of chemical fertilizers.

As our population has expanded, so has our demand for energy. In 1838 most families in the United States owned a horse and buggy. In 1938, most families owned one automobile. Today a two-car family is common. The discovery of coal fueled the industrial revolution in Europe in the 1600s. In fact, until that time, the size of European cities was limited by the distance one could carry or haul wood for fuel. Heat was obtained from wood and coal in the United States until the late 1800s when petroleum was discovered

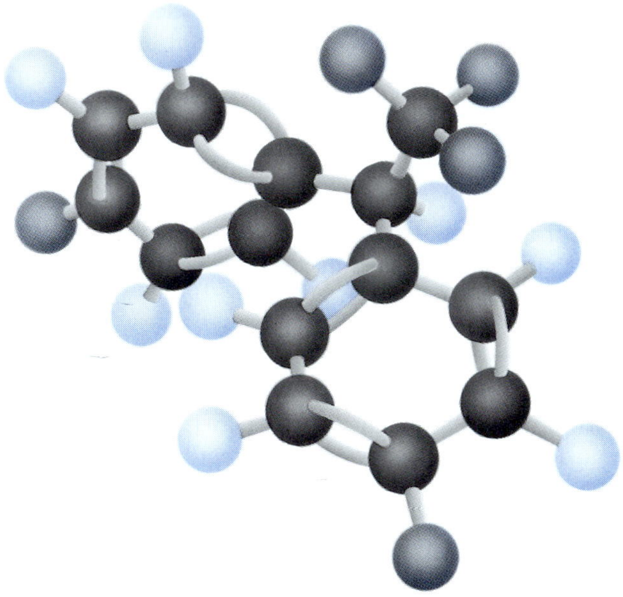

FIGURE 3
Chlorinated hydrocarbons are widely used as insecticides. The best known of this group, *di*chloro*di*phenyl*tri*chloroethane (DDT), is one of the most effective insecticides ever developed. A ball-and-stick model of DDT is shown in this figure. Chemical bonds, each consisting of a shared pair of electrons, are shown as sticks between the atoms. In some cases, atoms share two pairs of electrons to form a "double bond." Note that each carbon atom (black) participates in four bonds, while each hydrogen (light blue) and each chlorine (dark blue) only participate in one bond.

in Pennsylvania. Today, the combustion of petroleum provides about 43% of the energy used in the United States. Natural gas provides about 26% and coal about 22%. The remaining 9% of our energy comes from hydroelectric and nuclear installations. The petrochemical industry spawned by the discovery of this fossil resource has produced plastics, medicines, fabrics, and a wealth of materials unknown three generations ago.

Our demand for energy is dictated by two factors: an increasing population and an increasing material standard of living. Figure 4 illustrates this increasing demand and the mix of energy resourses used to meet it over the last 80 years. Easily visible is the shift from coal and wood to petroleum,

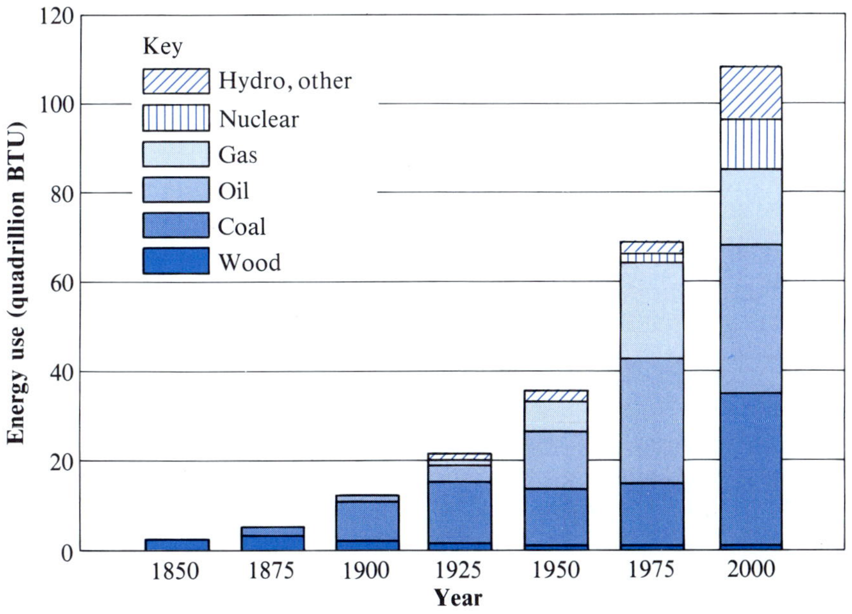

Energy use (quadrillion BTU)
120
100
80
60
40
20
0
Key
Hydro, other
Nuclear
Gas
Oil
Coal
Wood
1850
1875
1900
1925
1950
1975
2000
Year

FIGURE 4
This graph shows a breakdown of U.S. energy use from 1850 projected to the year 2000. It is evident that the United States has experienced considerable growth in energy demand as well as a significant shift in its source of energy over the past 100 years. The resources in greatest use today are also those in shortest supply.

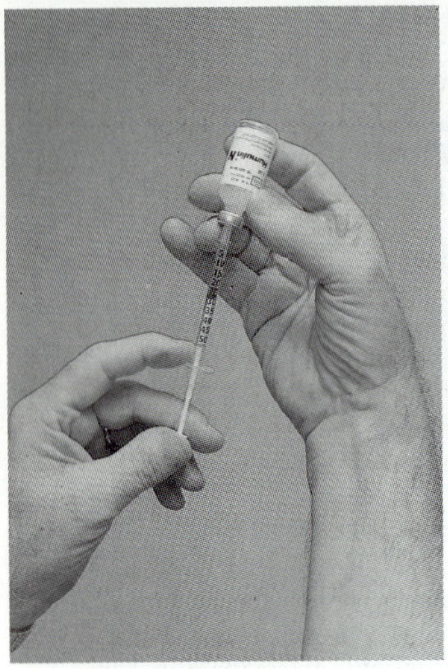

FIGURE 5
An understanding of the molecules involved in life processes have made possible the development of drugs to fight disease.

which became significant in the early 1900s. Nuclear power became a significant contributor about 1970. Most important, however, is something the graph does not show: what lies beyond the year 2000. Except for coal and nuclear energy, the other resources are near their limit of production and may soon begin to decline in availability. If our life-style is to be preserved, new resources must be found or developed to meet this demand.

Since it became a science at about the beginning of this century, chemistry has made four major contributions that have changed the lives of those of us who inhabit this planet. The first of these contributions provided the opportunity for adequate food through better fertilizers and particularly through pest control. The second has changed our way of transportation and the way we heat our homes through the development of high energy fuels. The third has been to provide a new set of synthetic materials that were unknown to our grandparents. These materials are used to clothe us, to package our foods, and to provide our communication and information processing. The fourth contribution involves chemical intervention in disease. An understanding of the molecular basis of life, and of the molecules involved in disease, has made possible vaccines to prevent communicable disease and drugs to treat thousands of diseases of both mind and body (Figure 5).

Nuclear energy, born in wartime and viewed as a power for destruction, may prove to be a useful source of peaceful energy. The application of artificial radioisotopes to medicine has resulted in significant advances in diagnostic and therapeutic techniques (Figure 6). Several years ago the balance shifted to the side where more lives had been saved by nuclear medicine than were taken at the end of World War II.

Throughout our lifetimes, most of us will use principles of chemistry daily in our work: in the health sciences, agriculture, home economics, and other fields. This book is designed to help you understand some of the most important concepts of chemistry—concepts that will explain the behavior of molecules as you encounter them in your work and in every aspect of your life.

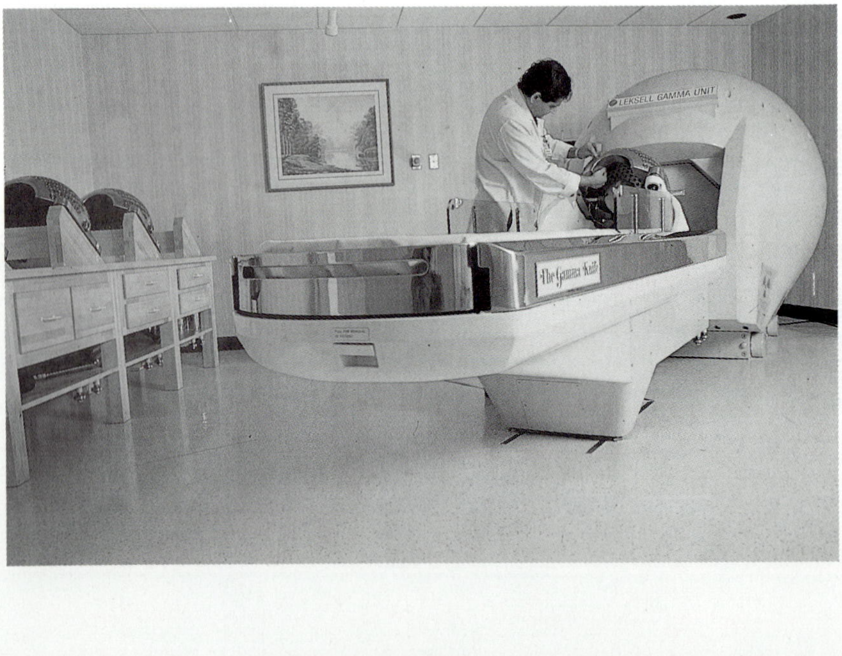
LEKSELL GAMMA UNIT
The Gamma Knife

FIGURE 6
Radioactivity is used for both medical diagnosis and treatment. This figure shows a machine used to treat cancer by applying a beam of penetrating radiation from cobalt-60 to selectively destroy diseased tissue.

CHAPTER

1

Measurement: The Basic Science

Chemistry has much in common with woodworking and art. They each have some basic tools we must master before we begin to work: the hammer and saw or the pen and brush. The basic tools of chemistry are the units we use for measurement. This chapter will introduce the Système Internationale (SI), an easy-to-use and internationally accepted measurement system.

OBJECTIVES

After completing this chapter, you should be able to

- Define the common SI prefixes, and relate SI measurement units to the physical standards from which they were derived.
- Use dimensional analysis to convert from one unit of measure to another.
- Differentiate between mass and weight, and describe the effect of gravitational field strength on these two quantities.
- Define the term *density*, and given mass and volume data, compute the density of a material.
- Express uncertainty in measurement by use of the proper number of significant figures.
- Identify the reference points for Fahrenheit and Celsius temperature measurement.
- Differentiate between heat and temperature, and define the terms *calorie* and *specific heat*.

1.1 Measurement Systems

Today's version of the metric system is called the Système Internationale (SI).

To understand chemistry, one must understand its language. During the last century, a language common to all nationalities was developed for communication of technical measurements. This system, called the **Système International d'Unités**, or **SI**, had its beginning during the French revolution. Until that time, measurement systems were regional in their use, and most were based on biological standards.

The *foot* (*ft*) for example, originated from the length of the average English carpenter's foot, convenient for him since he had his "foot" with him at all times. The *inch* (*in.*), formally defined as the distance occupied by four barley corns placed end to end, was 1/12 of a foot. A *yard*, or stride, was three feet. The *cubit*, the length from the elbow to the end of the index finger, is surprisingly constant between people of similar build, and it has been a standard for length from Old Testament times. As long as a craftsman worked by himself, his cubit, foot, or yard was consistent and useful as a unit of measure. As commerce between areas increased, however, it became important that a unit of measure be consistent throughout the trading area.

This chapter will show how a measurement system based on the circumference of the Earth has evolved into an international standard. It will also introduce a powerful but easy-to-use method of conversion between units of measure called dimensional analysis (see Section 1.3).

NEW TERM

Système International d'Unites (SI) An internationally accepted measurement system based on physical rather than biological standards. The system is unique in its use of decimal prefixes to modify its basic units.

1.2 Units of Length

The first successful attempt at international standardization of measurement occurred during the French revolution. In an attempt to find a nonbiological standard, the French scientists turned to the earth itself. A new standard of length, the **meter (m)**, was defined as one 10-millionth the distance from the equator to the north pole (Figure 1.1). The best estimate of this distance was scribed on a platinum-irridium bar in the Bureau of Standards in Paris, which then became the standard meter. The meter is about 10% larger than the English yard and is equal to 39.37 inches. The meter has since been more accurately defined as equal to the distance traveled by light in a vacuum during an interval of 1/299,792,458 second (sec).

1 meter = 1/10,000,000 of earth's quadrant

FIGURE 1.1
The meter was originally defined as one ten-millionth of the distance from the equator to the earth's pole.

Besides its nonbiological base, the French **metric system** has another advantage: there is a simple relationship between its units. No rational relationship exists between the English units of length. To convert from yards to feet, for example, one multiplies by 3. To convert from feet to inches, the conversion factor is 12. To convert from miles to feet, one multiplies by 5280. The metric system, on the other hand, uses Latin *decimal prefixes* to modify its basic units. The most common of these are listed in Table 1.1. These prefixes are related by powers of ten. For example, one **kilometer (km)** is

TABLE 1.1
Some Common Metric Prefixes

Prefix	Multiplier	Examples
mega	1,000,000 or 10^6	
kilo*	1,000 or 10^3	1 km = 1000 m
hecto	100 or 10^2	
deka	10 or 10^1	
deci	1/10 or 10^{-1}	
centi	1/100 or 10^{-2}	1 cm = 1/100 m
milli	1/1000 or 10^{-3}	1 mm = 1/1000 m
micro	1/1,000,000 or 10^{-6}	
nano	1/1,000,000,000 or 10^{-9}	

* The most common prefixes are in boldface.

equal to 1000 meters. One **centimeter (cm)** is equal to 1/100 meter, and ten **millimeters (mm)** equal one centimeter. As we will see later in this chapter, these decimal prefixes may be used with equal ease with other metric units. The metric system's units for distance, volume, and mass have carried over into the Système Internationale, as have its decimal prefixes.

NEW TERMS

meter (m) The basic unit of length in the metric and SI measurement systems, originally defined as one ten-millionth of the distance from the earth's equator to the north pole. It is now defined as the distance traveled by light in a vacuum during an interval of 1/299,792,458 sec.

metric system The French measurement system from which evolved the SI. The metric system was the first measurement system to use the earth as a standard.

kilometer (km) A metric unit of length equal to 1000 m (the decimal prefix *kilo-* means times 1000).

centimeter (cm) A metric unit of length equal to 1/100 m (the decimal prefix *centi-* means times 1/100).

millimeter (mm) A metric unit of length equal to 1/1000 m (the decimal prefix *milli-* means times 1/1000).

1.3 Dimensional Analysis

Conversion from one unit of measure to another is easily accomplished by a simple mathematical process called **dimensional analysis**. This approach treats units as numbers, subjecting them to the same arithmetic operations as the numbers that accompany the units. In dimensional analysis, conversions between units are performed by using conversion factors obtained from the relationships between these units. Dimensional analysis problems

Dimensional analysis is an easy way to convert from one unit of measure to another.

are always set up so that the measured unit is multiplied by a conversion factor. If done carefully, it is foolproof.

Consider, for example, the problem of converting inches to feet. We know that 12 in. = 1 ft. Because of this equality, the fraction

$$\frac{1\text{ ft}}{12\text{ in.}} = 1$$

Likewise, the fraction

$$\frac{12\text{ in.}}{1\text{ ft}} = 1$$

Conversion factors provide a mathematical relationship between two systems of units.

These fractions are called **conversion factors**. Since both are mathematically exactly equal to 1, we can multiply any quantity by either conversion factor without changing its true value. The trick is to choose a conversion factor with its units arranged so that the unit we wish to *change* is canceled out during the multiplication, and the unit we wish to *keep* remains when the problem is complete. Conversion factors will be highlighted with a blue screen throughout the text and examples in this chapter.

$$(\cancel{\text{Measured}}\text{ unit})\left(\frac{\text{Desired unit}}{\cancel{\text{Measured unit}}}\right) = \text{Desired unit}$$

Suppose that we wanted to convert a distance of 24 in. to feet. Let's try multiplying by our first conversion factor just using the units:

$$(\cancel{\text{in.}}) \times \left(\frac{\text{ft}}{\cancel{\text{in.}}}\right) = \text{ft}$$

This works out with the correct units, so let's insert the numbers:

$$(24\,\cancel{\text{in.}})\left(\frac{1.0\text{ ft}}{12\,\cancel{\text{in.}}}\right) = 2.0\text{ ft}$$

Suppose that an incorrect conversion factor was chosen. If this occurs, the units will *not* cancel to produce the desired result:

$$(\text{in.})\left(\frac{\text{in.}}{\text{ft}}\right) = \text{in.}^2/\text{ft}$$

If you know the units of the desired answer, the dimensional analysis method will always tell you if your problem is set up properly, before you begin to work the arithmetic.

Conversion factors may by definition be exact or approximate.

Conversion factors are of two types—exact and approximate. For example, the relationship 12 in. = 1 ft is exact, as is 100 cm = 1 m. Conversion between *systems* of measurement, however, is usually approximate. For example, 1 ft = about 30.5 cm. More exactly, 1 ft = 30.48 cm; even more exactly, 1 ft = 30.48005 cm. When approximate conversion factors are used, one must be aware of the accuracy desired. As we will see in Section 1.7, the accuracy of any computed result is limited by the least accurate measurement used in the computation.

Let's demonstrate this method with several examples; one in the English system, two in the SI, and two converting between the two systems.

EXAMPLE 1.1

A sheet of plywood is 4.0 ft in width. Express this value as inches.

The conversion factor is 1.0 ft = 12 in. (exactly). We have our choice of the conversion factors shown above. First try the problem just with units:

$$(\text{ft}) \times \left(\frac{\text{in.}}{\text{ft}}\right) = \text{in.}$$

The conversion factor is set up so that the unit we wish to keep is in the numerator, and the unit we wish to cancel is in the denominator. Since the unit analysis gave the proper result, let's insert the numbers:

$$(4.0\ \text{ft})\left(\frac{12\ \text{in.}}{1.0\ \text{ft}}\right) = 48\ \text{in.}$$

When multiplication is performed, the units "feet" cancel out, leaving 48 as the number and inches as the unit. Thus, 4.0 ft = 48 in.

Note again that if the conversion factor is set up upside down, the problem will not produce a correct unit:

$$(4.0\ \text{ft})\left(\frac{1.0\ \text{ft}}{12\ \text{in.}}\right) = 0.33\ \text{ft}^2/\text{in.}$$

An incorrect set-up always produces an incorrect unit, even though the arithmetic step will produce a number. This number, of course, will not be the correct answer to the problem.

EXAMPLE 1.2

Express a distance of 750 cm in meters.

The conversion factor is 100 cm = 1.00 m (exactly). Since we want to complete the problem with the unit in meters, the conversion factor will be set up with meters in the numerator:

$$\text{Conversion factor} = \frac{1.00\ \text{m}}{100\ \text{cm}} = 1$$

So, checking the units, we have

$$(\text{cm})\left(\frac{\text{m}}{\text{cm}}\right) = \text{m}$$

Since the unit analysis works, we are ready to insert numbers:

$$(750\ \text{cm})\left(\frac{1.00\ \text{m}}{100\ \text{cm}}\right) = 7.50\ \text{m}$$

Note that the arithmetic involved in this metric conversion is division by 100, which is done by simply moving the decimal point two places to the left. All metric conversions involve multiplication or division by powers of ten, and thus require only that the decimal point be moved.

EXAMPLE 1.3

Express a distance of 350 mm in centimeters.

Since we know that 10.0 mm = 1.00 cm (exactly), the conversion factor is

$$\frac{1.00\text{ cm}}{10.0\text{ mm}} = 1$$

Checking the units, we have

$$(\cancel{\text{mm}})\left(\frac{\text{cm}}{\cancel{\text{mm}}}\right) = \text{cm}$$

Since the unit analysis works, we are ready to insert numbers:

$$(350\ \cancel{\text{mm}})\left(\frac{1.00\text{ cm}}{10.0\ \cancel{\text{mm}}}\right) = 35.0\text{ cm}$$

TABLE 1.2
Some Common English to SI Conversion Factors for Length

English	SI
1.000 in. =	2.540 cm
1.000 ft =	30.48 cm
39.37 in. =	1.000 m
1.000 mile =	1.609 km

Dimensional analysis may also be used to convert from one system of measure to another, given appropriate conversion factors. Table 1.2 gives several English to SI conversions for length.

EXAMPLE 1.4

One English yard is a distance of 36.0 in. Express this distance as centimeters.

Since we know that 1.00 in. = 2.54 cm (approximately), the conversion factor is

$$\frac{2.54\text{ cm}}{1.00\text{ in.}} = 1$$

Checking the units, we have

$$(\cancel{\text{in.}})\left(\frac{\text{cm}}{\cancel{\text{in.}}}\right) = \text{cm}$$

Since the unit analysis works, we can insert numbers:

$$(36.0\ \cancel{\text{in.}})\left(\frac{2.54\text{ cm}}{1.00\ \cancel{\text{in.}}}\right) = 91.4\text{ cm}$$

A logical extension is the use of multiple conversion factors in a single problem, such as expressing a distance of feet in millimeters.

EXAMPLE 1.5

An English yard is also defined as a distance of 3.0 ft. Express this distance in millimeters.

We will need to use three conversion factors in this problem:

$$\frac{12.0 \text{ in.}}{1.00 \text{ ft}} = 1, \quad \frac{2.54 \text{ cm}}{1.00 \text{ in.}} = 1, \quad \text{and} \quad \frac{10.0 \text{ mm}}{1.00 \text{ cm}} = 1$$

First check the units:

$$(\cancel{\text{ft}})\left(\frac{\cancel{\text{in.}}}{\cancel{\text{ft}}}\right)\left(\frac{\cancel{\text{cm}}}{\cancel{\text{in.}}}\right)\left(\frac{\text{mm}}{\cancel{\text{cm}}}\right) = \text{mm}$$

Now insert the numbers:

$$(3.00 \cancel{\text{ft}})\left(\frac{12.0 \cancel{\text{in.}}}{1.00 \cancel{\text{ft}}}\right)\left(\frac{2.54 \cancel{\text{cm}}}{1.00 \cancel{\text{in.}}}\right)\left(\frac{10.0 \text{ mm}}{1.00 \cancel{\text{cm}}}\right) = 914 \text{ mm}$$

NEW TERMS

dimensional analysis A method of converting a measurement from one unit to another by multiplying by an appropriate conversion factor. Conversion factors are always set up so that the desired unit is in the numerator, and the unit you wish to convert *from* is in the denominator. The unwanted unit then cancels as the conversion factor and the known quantities are multiplied. Since the conversion factor is equal to 1, the mathematic effect is that of multiplication by 1. The measurement does not change, only its unit.

conversion factor A fraction in which numerator and denominator are in different units, but which are equal to the same quantity. The algebraic value of the conversion factor is always 1.

TESTING YOURSELF

Common Metric Prefixes and Dimensional Analysis

1. 3.00 m equals how many centimeters?
2. 500 mm equals how many centimeters?
3. 600 m equals how many kilometers?
4. 75 cm equals how many millimeters?
5. A person is 182.9 cm tall.
   a. She is how many meters tall?
   b. Given that 1 in. = 2.54 cm, the person is how many inches tall?
   c. Given that 1 ft = 12 in., the person is how many feet tall?
6. A football field is 100 yards long. Given that 1 yard = 3 ft, 1 ft = 12 in., and 1 in. = 2.54 cm, a football field is how many meters in length?

Answers **1.** 300 cm **2.** 50 cm **3.** 0.6 km **4.** 750 mm **5. a.** 1.829 m **b.** 72.00 in. **c.** 6.000 ft **6.** 91.44 m

1.4 Units of Volume

In contrast to the English system, the Système Internationale has an internal relationship among units that is refreshingly logical and simple to use once one discovers it.

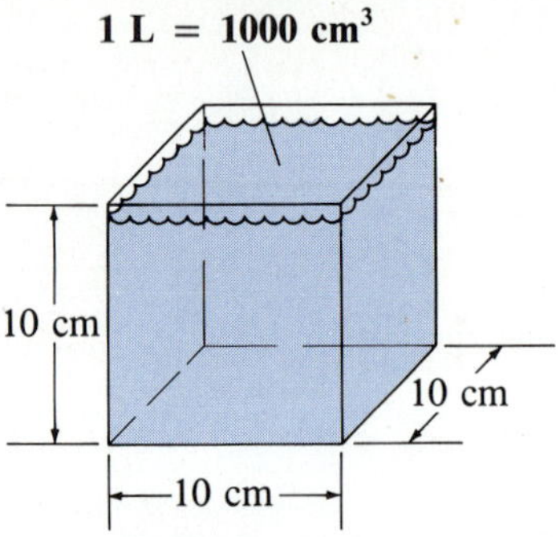
1 L = 1000 cm³
10 cm
10 cm
10 cm

FIGURE 1.2
The SI volume units are derived from SI length units. The liter is equal to 1000 cubic centimeters.

FIGURE 1.3
Several types of laboratory glassware commonly used to measure volume are illustrated in this figure.

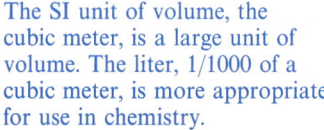

The SI unit of volume, the cubic meter, is a large unit of volume. The liter, 1/1000 of a cubic meter, is more appropriate for use in chemistry.

In the English measurement system, volume can be expressed in gallons, quarts, pints, cups, fluid ounces, and cubic feet or cubic inches. The relationships among all these units are far from straightforward. One quart, for example, is equal to 1/4 gallon, 2 pints, or 57.749 cubic in. In contrast, the SI unit of volume, the liter, is directly tied to the metric unit of length. A **liter (L)** is defined as the volume of a cube 10 cm on a side; thus, 10 cm × 10 cm × 10 cm = 1000 **cubic centimeters** (**cm^3** or **cc**). Applying metric prefixes to this unit shows us that 1 L equals 1000 milliliters; thus, 1 **milliliter (mL)** is equal in volume to 1 cm^3 (Figure 1.2). In most scientific work, volume is expressed in liters or milliliters, and these metric units are marked on all laboratory glassware (Figure 1.3). In medicine, small values are indicated in cubic centimeters instead of milliliters, although the units are equivalent. Several conversion factors for common volume units are listed in Table 1.3. Note that only the SI units have a rational relationship among themselves.

Volume calculations are accomplished by the dimensional analysis procedures introduced earlier in this chapter. The following examples illustrate calculation of volume and conversion between SI and English volume units.

TABLE 1.3
Some Common Units of Volume

English Units	SI Units
1 pint = 2 cups	1.000 L = 1000 mL
1 quart = 2 pints	1.000 mL = 1.000 cm^3
1 gallon = 4 quarts	
32 fluid ounces = 1 quart	
1 quart = 57.749 cubic inches	
1 cord = 128 cubic feet	

1 quart = 946 mL = 0.946 L
1.06 quarts = 1 L

EXAMPLE 1.6

Compute the volume of a cube 2.00 cm × 3.00 cm × 50.0 cm. Give the answer in milliliters.

First, calculate the volume in cubic centimeters:

$$\text{Volume} = \text{length} \times \text{width} \times \text{height}$$
$$= (2.00\ \text{cm})(3.00\ \text{cm})(50.0\ \text{cm}) = 300\ \text{cm}^3$$

Since 1.000 liter = 1000 cm^3 (exactly), the volume may be expressed in liters using this conversion factor:

$$\frac{1.000\ \text{L}}{1000\ \text{cm}^3} = 1$$

Now check the units:

$$(\text{cm}^3)\left(\frac{\text{L}}{\text{cm}^3}\right) = \text{L}$$

Now solve the problem for liters:

$$(300\ \text{cm}^3)\left(\frac{1.000\ \text{L}}{1000\ \text{cm}^3}\right) = 0.300\ \text{L}$$

To convert this measurement to milliliters, use this conversion factor:

$$\frac{1.000\ \text{L}}{1000\ \text{mL}} = 1$$

And complete the solution:

$$(0.300\ \text{L})\left(\frac{1000\ \text{mL}}{1.000\ \text{L}}\right) = 300\ \text{mL}$$

EXAMPLE 1.7

Express a volume of 300 mL in quarts.

Since we know from Table 1.3 that 1.00 quart = 946 mL and since we want to complete the problem with the answer in quarts, the conversion factor should be set up with quarts in the numerator:

$$\frac{1.00\ \text{quart}}{946\ \text{mL}} = 1$$

First check the units:

$$(\text{mL})\left(\frac{\text{quart}}{\text{mL}}\right) = \text{quart}$$

Now solve the problem:

$$(300\ \text{mL})\left(\frac{1.00\ \text{quart}}{946\ \text{mL}}\right) = 0.317\ \text{quart}$$

EXAMPLE 1.8

Express a volume of 16.0 L in gallons.

We will need to use three conversion factors in this problem:

$$\frac{1000\ \text{mL}}{1.000\ \text{L}} = 1 \qquad \frac{1.00\ \text{quart}}{946\ \text{mL}} = 1, \quad \text{and} \quad \frac{1.00\ \text{gallon}}{4.00\ \text{quart}} = 1$$

First check the units:

$$(\text{L})\left(\frac{\text{mL}}{\text{L}}\right)\left(\frac{\text{quart}}{\text{mL}}\right)\left(\frac{\text{gallon}}{\text{quart}}\right) = \text{gallons}$$

Now insert the numbers and solve the problem:

$$(16.0\ \text{L})\left(\frac{1000\ \text{mL}}{1.000\ \text{L}}\right)\left(\frac{1.00\ \text{quart}}{946\ \text{mL}}\right)\left(\frac{1.00\ \text{gallon}}{4.00\ \text{quart}}\right) = 4.23\ \text{gallons}$$

The mass unit in SI (the kilogram) was initially obtained by weighing a liter volume filled with water. The next section of this chapter will consider the development of SI units for mass.

NEW TERMS

liter (L) The SI volume unit equal to 1000 cm^3.

cubic centimeter (cm^3 or cc) A unit of volume equal to 1 mL; the preferred unit in medicine (for which the abbreviation "cc" is used).

milliliter (mL) A common unit of volume used in science and medicine because of its convenient, small size equal to 1/1000 L.

TESTING YOURSELF

Volume Units

1. An automobile engine has a piston displacement of 1600 cc. How many liters is this?
2. Since 1 quart = 946 mL, then 1 gallon equals how many liters?
3. A soft drink can has a volume of 12.0 fluid ounces. Using the conversion factors in Table 1.3, express this volume in milliliters.

Answers **1.** 1.6 L **2.** 3.784 L **3.** 355 mL

1.5 Mass and Weight

Weight is a force unit. Its value is dependent upon the gravitational field.

Although they mean different things, the terms *mass* and *weight* are often used interchangeably. They are similar in that they both reflect an *amount* of material. However, they differ in that **weight** is a measure of the *force* the object exerts down on the earth, while **mass** is an absolute measure of the amount of material present, regardless of the gravitational field.

Imagine measuring the weight of a 5-pound (lb) fish with a spring scale (Figure 1.4). The exact weight would depend on one's location, since the earth's gravitational field becomes weaker as one moves farther from the center of the earth. The earth is slightly more elongate at the equator than at the poles, thus an object weighed in Alaska would be closer to the center of the earth and would exert more force on the spring scale than an object weighed in Panama (Figure 1.5). The difference in gravitational field between these two locations is about 0.4%. Thus, the 5.00-lb fish weighed in Alaska would weigh 4.98 lb in Panama. This can cause some problems in commerce. Exactly 1 ounce of gold as weighed in Alaska, for example, would weigh about 0.9986 ounces in Panama, a significant difference for such an expensive material.

FIGURE 1.4
Spring scales are often used to measure weight. The stretch of the spring is directly proportional to the force with which the sample is attracted toward the earth.

To carry this to an extreme, consider weighing a 5.00-lb object on the moon. The gravitational force exerted by an object on the moon is about 1/6 of what it would exert on the earth. A spring scale would indicate a weight of $(1/6)(5 \text{ lb}) = 0.83$ lb for the object. The amount of material present, however, would be the same.

Mass is a more logical measure of the *amount* of a material because it is independent of the local gravitational field. The SI unit of mass is the **gram (g)**, or **kilogram (kg)**, which was originally derived from the metric volume unit using water as the standard. Exactly 1 L (1000 mL) of water has a mass of 1000 g. Thus, 1 mL (or 1 cm^3) of water has a mass of 1 g. Since water expands with increasing temperature, this definition is made at the temperature at which water is most dense, 3.4 degrees Celsius (°C) (see Section 1.8). The measurement is not related to gravitational field strength. The kilogram is now formally defined as the mass of a cylinder of platinum-irridium alloy kept by the International Bureau of Weights and Measures in Paris. This is the only basic unit still defined by an artifact. Other common mass units include the *milligram* (*mg*), equal to 1/1000 g, and the *microgram* (μg), equal to 1/1,000,000 g.

The kilogram is the SI unit of mass. One liter of water has a mass of 1 kg.

Measurement of mass is made by a comparison technique, as shown in Figure 1.6, which illustrates a simple equal arm balance. If two objects are

Location	Latitude	Elevation (m)	Gravity* (cm/sec^2)
Fort Egbert, Alaska	65°	269	982.183
Calgary, Alberta	51°	1044	980.823
Seattle, Washington	47°	58	980.733
Minneapolis, Minnesota	45°	256	980.597
New York, New York	40°	38	980.267
Denver, Colorado	39°	1638	979.609
San Francisco, California	37°	114	979.965
Austin, Texas	30°	189	979.283
Panama City, Canal Zone	9°	6	978.243

* Gravitational field strength is usually expressed in terms of how fast an object will pick up speed when it is dropped. In Alaska, a falling object will pick up speed at the rate of 982 cm/sec each second it falls.

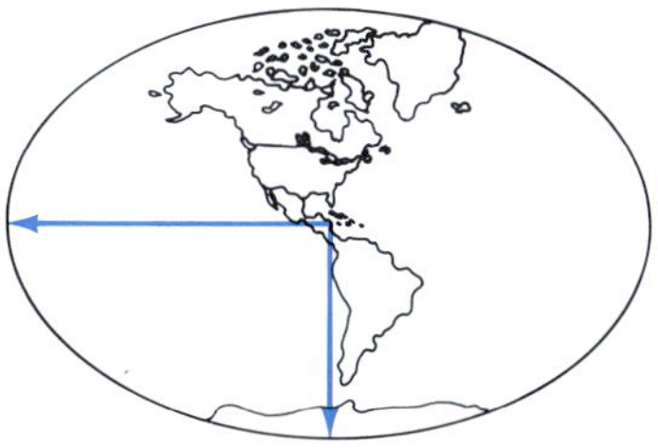

FIGURE 1.5
The diameter of the earth is greater at the equator than at the poles. As an object is moved to a higher latitude, it moves closer to the center of the earth and, as a result, exhibits greater weight.

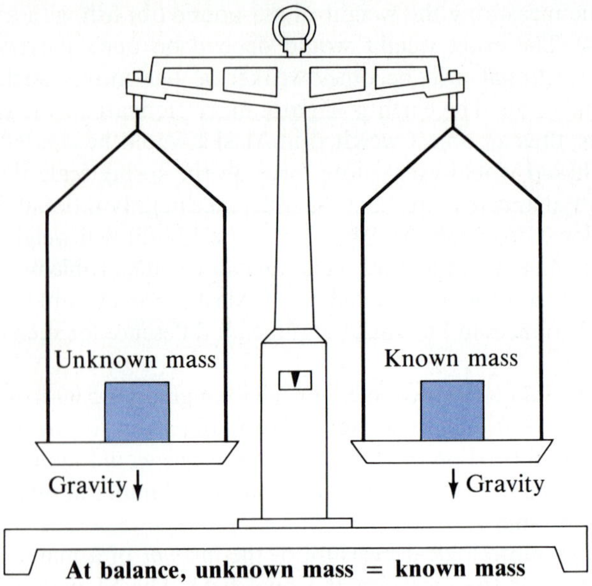
Unknown mass
Known mass
Gravity
Gravity
At balance, unknown mass = known mass

FIGURE 1.6
A simple equal arm balance used to measure mass by a comparison technique. Modern laboratory balances contain a very precise set of known factory-calibrated masses that can be placed on the balance arm with a dial to balance the unknown mass.

Mass is measured by a comparison technique. This measurement does not depend on gravitational field strength.

placed on the two pans of the balance and if the gravitational attraction for each object is the same, the pans will "balance." If the object on the right pan were a known mass of 1.000 g, the object on left pan would have the same mass if balanced. This measurement technique is independent of the gravitational field as long as the field is equal at both pans of the balance. Since the distance separating the pans is seldom more than a few centimeters, for practical purposes the gravitational field is the same. A 10-g mass would be measured as 10 g everywhere on the earth, as well as on the moon.

Conversion between the English *pound* (a force unit) and the SI kilogram (a mass unit) is accomplished by assuming an "average" gravitational field for the earth. Conversion factors resulting from this assumption are

$$1.00 \text{ kg} = 2.20 \text{ lb}$$

$$1.000 \text{ lb} = 453.6 \text{ g}$$

EXAMPLE 1.9

A soft drink can has a volume of 355 mL. What is the mass of water that can be placed in this can, expressed in grams?

The conversion factor is 1.00 mL water = 1.00 g water.

First check the units:

$$(\cancel{\text{mL}} \text{ water})\left(\frac{\text{g water}}{\cancel{\text{mL}} \text{ water}}\right) = \text{g water}$$

Now solve the problem:

$$(355 \cancel{\text{mL}})\left(\frac{1.00 \text{ g}}{1.00 \cancel{\text{mL}}}\right) = 355 \text{ g}$$

EXAMPLE 1.10

What is the weight of 355 g of water, expressed in ounces (oz)?

The conversion factors are 1.000 lb = 16.00 oz and 1.000 lb = 453.6 g.

First check the units:

$$(\text{g})\left(\frac{\text{lb}}{\text{g}}\right)\left(\frac{\text{oz}}{\text{lb}}\right) = \text{oz}$$

Now solve the problem:

$$(355\text{ g})\left(\frac{1.000\text{ lb}}{453.6\text{ g}}\right)\left(\frac{16.00\text{ oz}}{1.000\text{ lb}}\right) = 12.5\text{ oz}$$

NEW TERMS

weight The force a mass exerts downward on the earth, which varies with the gravitational field strength.

mass A measure of the *amount* of material, independent of gravitational field strength.

gram (g) The mass of 1 cm^3 or 1 mL of water.

kilogram (kg) The basic SI unit of mass; 1000 mL of water have a mass of 1.000 kg.

TESTING YOURSELF

Mass, Weight, and Gravity

1. Would astronaut in orbit in the space shuttle have the same mass or the same weight as if he or she were on earth?
2. Where would an object weigh slightly more: in New York or Atlanta?
3. A candy bar has a mass of 39.72 g.
   a. This is _______ milligrams (mg).
   b. Given that 1.000 lb = 453.6 g and 1.000 lb = 16.00 oz, the candy bar weighs _______ oz.

Answers **1.** Mass **2.** New York **3. a.** 39,720 mg **b.** 1.401 oz

1.6 Density and Specific Gravity

Density is a characteristic property of matter that is formally defined as mass per unit of volume. It is expressed as grams per milliliter, grams per cubic centimeter, or kilograms per liter. The density of water, for example, is 1.000 grams per milliliter (g/mL), or 1.000 g/cc at 4 °C. Densities of a number of common substances are listed in Table 1.4.

$$\text{Density} = \frac{\text{Mass}}{\text{Volume}}$$

TABLE 1.4
Densities of Some Common Substances

Substance	Density (g/mL)
Balsa wood	0.12–0.20
White pine	0.37
Douglas fir	0.51
Walnut	0.56
Isopropanol	0.78
Water	**1.00**
Chloroform	1.50
Carbon tetrachloride	1.59
Quartz	2.6
Aluminum	2.70
Hematite	5.1
Iron	7.8
Copper	8.9
Lead	11.3
Mercury	13.6
Uranium	18.7

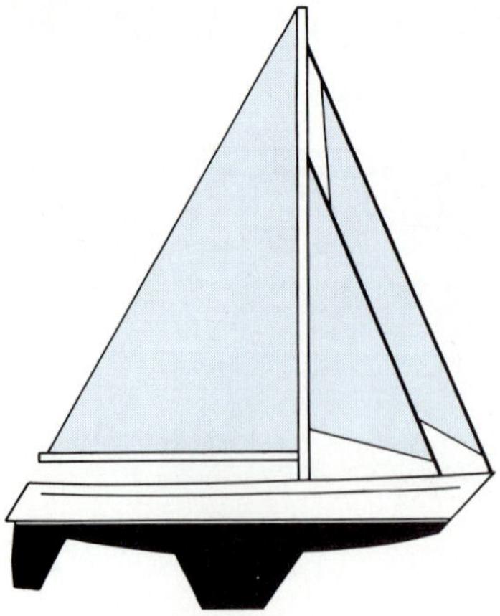

FIGURE 1.7
The keel of a sailboat is often filled with heavy ballast to help keep the boat upright against the force of the wind on the sails.

EXAMPLE 1.11

A volume of 25.0 mL of ethyl alcohol has a mass of 19.73 g. What is the density of this substance?

$$\text{Density} = \frac{\text{Mass}}{\text{Volume}} = \frac{19.73\ \text{g}}{25.0\ \text{mL}} = 0.789\ \text{g/mL}$$

EXAMPLE 1.12

A sailing enthusiast wishes to fill the keel of a small sailboat with lead to help keep it upright in strong winds (Figure 1.7). The keel is hollow and will hold 20 L of liquid (measured by pouring water in it). If poured full of molten lead, how many kilograms of ballast would be present? The density of lead is 11.3 g/mL (Table 1.4).

Since

$$\text{Density} = \frac{\text{Mass}}{\text{Volume}}$$

then

$$\begin{aligned}\text{Mass} &= (\text{Volume})(\text{Density})\\ &= (20\ \cancel{\text{L}})\left(\frac{1000\ \cancel{\text{mL}}}{1\ \cancel{\text{L}}}\right)\left(11.3\ \frac{\text{g}}{\cancel{\text{mL}}}\right)\\ &= 226{,}000\ \text{g} = 226\ \text{kg}\end{aligned}$$

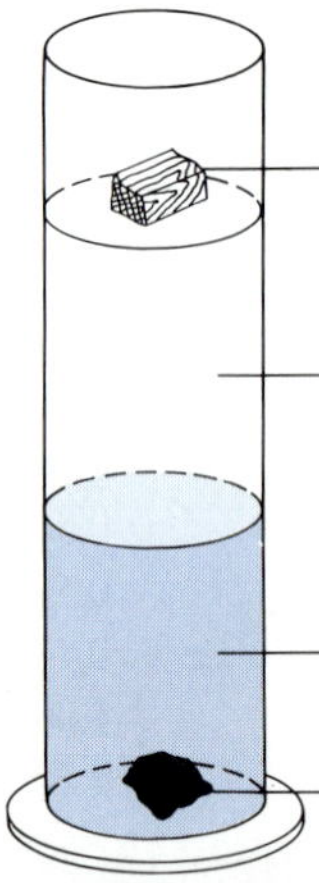

Wood
Water
Carbon tetrachloride
Aluminum

FIGURE 1.8
Specific gravity can be used to determine if one material will float on another. This illustration shows wood (specific gravity 0.5) floating on water (specific gravity 1.0), water floating on carbon tetrachloride (specific gravity 1.59), and aluminum (specific gravity 2.7) sinking through the carbon tetrachloride.

Another term often used to express the density of a material is **specific gravity**, which is the ratio of the mass of a given volume of material to the mass of the same volume of water. Note that specific gravity is a *ratio* and thus has no units. For example, 10.0 mL of carbon tetrachloride has a mass of 15.9 g, 1.59 times that of the same amount of water (Table 1.4). Its specific gravity is therefore 1.59. Specific gravity is used when one wishes to compare the density of a material with that of water (Figure 1.8). Since the density of water is 1.000 g/mL, the specific gravity of a material is numerically equal to its density.

Specific gravity is commonly used in medicine. Urine, for example, carries dissolved solids from the body, and it thus has a specific gravity greater than pure water. The concentration of these dissolved solids may be determined by measuring the specific gravity of the urine. When the kidneys are functioning properly, the specific gravity of urine is in the range of 1.018 to 1.025 at 25 °C. If the kidneys are not working properly, the urine will be more or less concentrated and its specific gravity will be greater or smaller. The specific gravity of urine can thus be used as a diagnostic tool.

EXAMPLE 1.13

A 200-mL sample of urine has a mass of 210 g. Does the specific gravity of this sample fall within the normal range?

To find the specific gravity, first find the density by dividing the mass by the volume:

$$\frac{210 \text{ g}}{200 \text{ mL}} = 1.05 \text{ g/mL}$$

Thus, the specific gravity is 1.05, slightly higher than normal.

Rapid and accurate measurement of the specific gravity of a liquid can be made with an instrument called a **hydrometer** (Figure 1.9). This is simply a float with a calibrated stem. The more dense the liquid, the higher the hydrometer will float. The specific gravity is read at the point where the liquid intersects the stem.

NEW TERMS

density Mass per unit volume, commonly expressed as grams per cubic centimeter or grams per milliliter.

$$\text{Density} = \frac{\text{Mass}}{\text{Volume}}$$

specific gravity The ratio of the mass of a sample of material to the mass of the same volume of water.

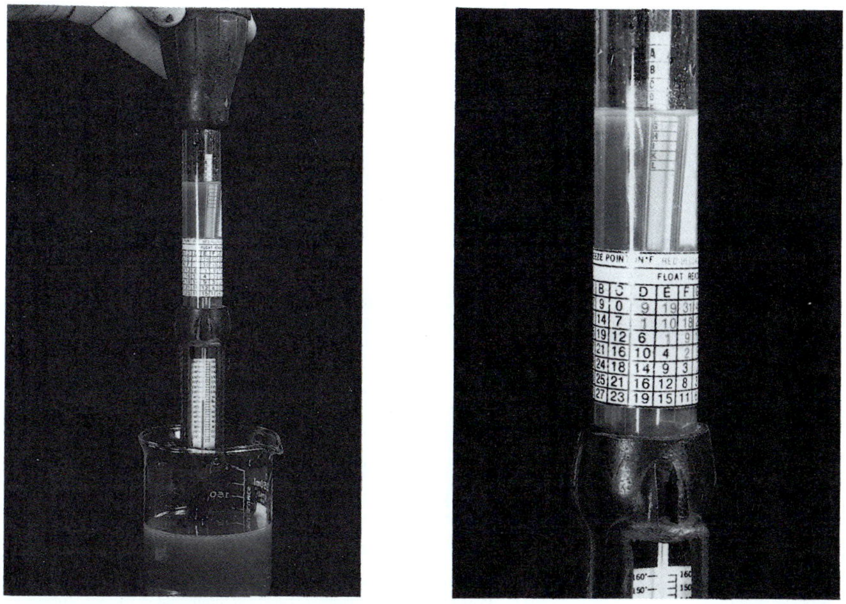

(a) (b)

FIGURE 1.9
A hydrometer is used to measure the specific gravity of a liquid. (a) The greater the specific gravity of the liquid, the higher the hydrometer will float. (b) The scale is read where the liquid surface crosses the stem of the instrument. This hydrometer is used to measure the specific gravity of antifreeze.

1.7 Calculators, Significant Figures, and Scientific Notation

Calculators and Significant Figures

Accuracy and precision mean different things.

Experimental observations or data may be characterized in terms of both **accuracy**, which is how close they are to the true value, and **precision**, which is how well they agree with each other. A measurement made with a meter stick is limited to an accuracy of plus or minus 1 mm (written ± 1 mm). A volume measurement made with a 100-mL graduated cylinder is limited to an accuracy of about ± 0.1 mL (Figure 1.10). All measurements produce numbers of which we are sure and numbers which are estimates. One therefore needs to keep in mind the effect of the measuring instrument on the degree of accuracy. However, the real danger lies in the ability of arithmetic operations to generate more figures, and therefore more apparent precision or accuracy, than is justified by the experimental data.

EXAMPLE 1.14

Consider the density of a sample of acetic acid. A volume of 45.0 mL of this material (measured with a graduated cylinder) has been determined to have a mass of 47.206 g (measured with a good analytical balance).

$$\text{Density} = \frac{\text{Mass}}{\text{Volume}} = \frac{47.206 \text{ g}}{45.0 \text{ mL}} = 1.049022222 \text{ g/mL}$$

Is this result correct?

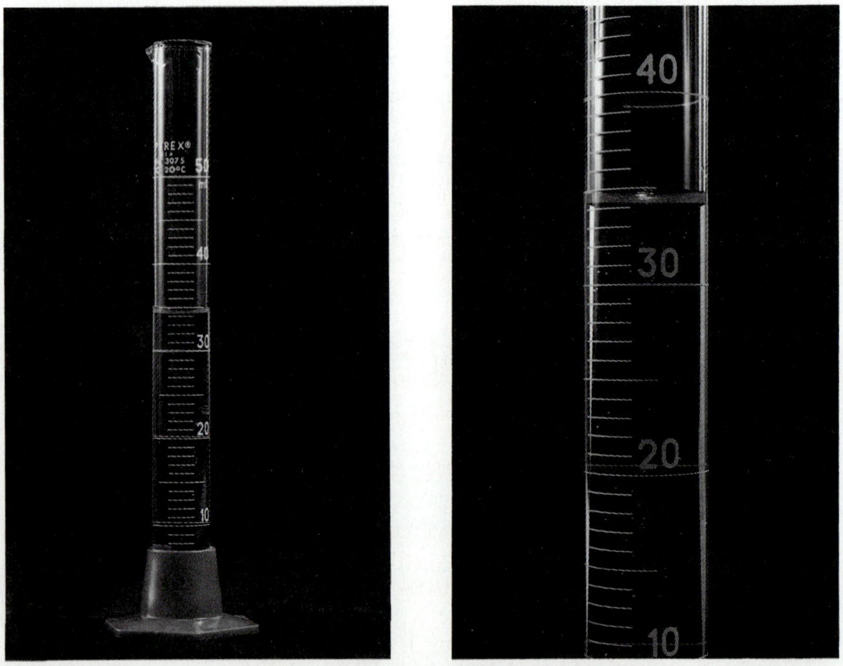
50
40
30
20
10
40
30
20
10

(a) (b)

FIGURE 1.10
(a) The graduated cylinder illustrated in this figure has 1.0-mL graduations. Volume may be estimated to 1/10 of each graduation or to an accuracy of about ± 0.1 mL. (b) The measured volume in this example is closer to 34.2 mL than to 34.1 mL or 34.3 mL. The numbers 3 and 4 are certain, and the 2 is the best estimate. The result, 34.2 mL, has three significant figures.

Most calculators display eight to ten digits; an answer of this sort is common (Figure 1.11). However, is it meaningful? Consider the accuracy of the numbers that were used in the computation. The mass of the sample was measured as 47.206 g. This means that the measured mass was closer to 47.206 g than it was to 47.205 g or 47.207 g. Five of the digits are meaningful or significant; the number is said to have five **significant figures**. Sample volume, on the other hand, was reported as 45.0 mL. This indicates that the measurement was closer to 45.0 mL than to 44.9 mL or 45.1 mL. This number has three significant figures. The division step on the calculator, however, produced a density value with ten (or more) digits. Are all of these significant? No. The least certain number used in this computation was volume, which had only three significant figures. The computed density must then be rounded to three significant figures.

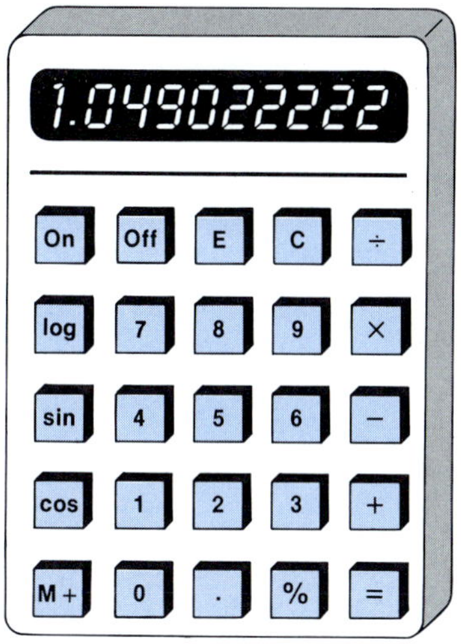
1.049022222
On
Off
E
C
÷
log
7
8
9
×
sin
4
5
6
−
cos
1
2
3
+
M+
0
.
%
=

FIGURE 1.11
Arithmetic calculations can produce answers that appear to be far more accurate than the data can justify. Although ten figures are shown in this example, only the first three are significant.

EXAMPLE 1.15

Express the density of the acetic acid sample in Example 1.14 to the correct number of significant figures.

$$\text{Density} = \frac{\text{Mass}}{\text{Volume}} = \frac{47.206\ \text{g}}{45.0\ \text{mL}} = 1.049022222\ \text{g/mL}$$
(Calculated result)

The least accurate data is 45.0 mL with three significant figures. The result must be rounded to three significant figures:

$$1.049022222\ \text{g/mL} \approx 1.05\ \text{g/mL}$$
(Rounded result)

An important general rule to remember about significant figures is that the result of any arithmetic operation may not have any more significant figures than the least certain number used in the operation.

The number of significant figures in an arithmetic result should equal the least number used in the operation.

Consider again the problem of the sailor filling his sailboat keel with molten lead that we discussed earlier in the chapter (Section 1.6). Recall that he determined the internal volume of the keel to be 20 L by pouring water into it until it was full. The reported "20 L" indicates that it was closer to 20 L than to 19 L or 21 L. There are two significant figures in this number. Consider the following calculation in Example 1.16.

EXAMPLE 1.16

Compute the weight (in pounds) of 20.0 L of lead.

$$\begin{aligned}\text{Mass} &= \text{Volume} \times \text{Density}\\ &= (20.0\ \text{L})\left(\frac{1000\ \text{mL}}{1\ \text{L}}\right)\left(11.3\ \frac{\text{g}}{\text{mL}}\right)\\ &= 226{,}000\ \text{g} = 220\ \text{kg}\end{aligned}$$

At 453.6 g per pound, this is

$$226{,}000\ \text{g}\left(\frac{1\ \text{lb}}{453.6\ \text{g}}\right) = 498.2363316\ \text{lb}$$
(Calculated result)

Since the original data only justifies three significant figures, we must round the answer to 498 lb.

$$498.2363316 \text{ lb} \approx 498 \text{ lb}$$

(Rounded result)

Rounding is the process of limiting the number of significant figures by dropping the last place digit and increasing the next higher digit by 1 if the last place digit was 5 or higher.

EXAMPLE 1.17

Round these numbers to three significant figures.

$$2257 \approx 2260$$

$$5743 \approx 5740$$

$$2.496 \approx 2.50$$

$$11.534 \approx 11.5$$

(Calculated results) (Rounded results)

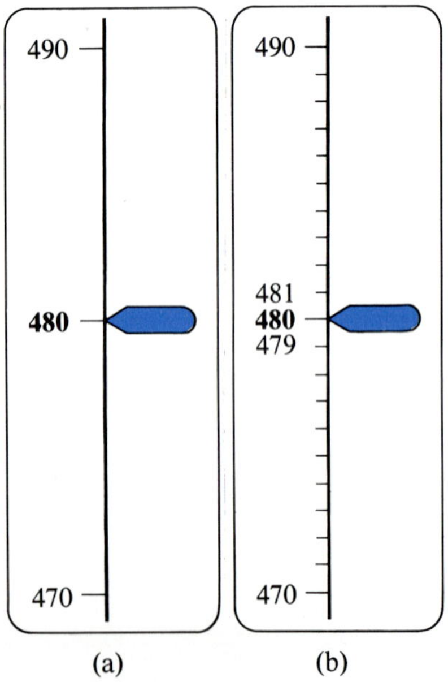

490
480
470
490
481
480
479
470
(a)
(b)

FIGURE 1.12
A crate of potatoes are reported to have a weight of 480 1b. (a) Does this mean that the weight is closer to 480 1b than to 470 1b or 490 1b, or (b) does it mean that it is closer to 480 1b than to 479 1b or 481 1b? One cannot tell by reading the reported value of 480 1b.

EXAMPLE 1.18

A student is 64.0 in. tall. Convert this number to feet, rounding to the proper number of significant figures.

$$(64.0 \text{ in.})\left(\frac{1 \text{ ft}}{12 \text{ in.}}\right) = 5.333333333 \text{ ft}$$

(Calculated result)

Since the measured value (64.0) has three significant figures, the correct result would also have three:

$$\approx 5.33 \text{ ft}$$

(Rounded result)

Scientific Notation

Both large and small numbers carry an additional hazard in that one cannot tell if a following zero is significant or not. Does 480 lb mean that the value is closer to 480 than to 479 or 481? Or does it mean that it is closer to 480 than to 470 or 490 (Figure 1.12)? A tool called scientific notation is used to deal with this problem. **Scientific notation** is the process of writing every number as a number between 1 and 10 multiplied times a power of ten. For example, 480 is equal to 4.8×100, so it would be written as 4.8×10^2. When written in scientific notation, it is clear that this number has two significant figures. It is closer to 4.8×10^2 than to 4.7×10^2 or 4.9×10^2.

TABLE 1.5
Examples of the Use of Significant Figures and Scientific Notation

Decimal Notation	Scientific Notation	Number of Significant Figures
121	1.21×10^2	3
420	4.2×10^2	2
423	4.23×10^2	3
0.001257	1.257×10^{-3}	4
46,200	4.62×10^4	3
46,200	4.6200×10^4	5

It is easy to tell the number of significant figures if a measurement is expressed in scientific notation.

All zeros in numbers expressed in scientific notation are considered to be measured and thus significant, such as in the last two rows of Table 1.5.

EXAMPLE 1.19

Express the following numbers in scientific notation.

(a) 525,000

First, move the decimal to the left until a number between 1 and 10 results.

525000.

Then add an exponential multiplier equal to the number of places you moved the decimal point. Since the decimal point was moved to make a smaller number out of a larger number, the exponential multiplier will be positive.

$$5.25 \times 10^5$$

(b) 0.0000015

First, move the decimal to the right until a number between 1 and 10 results.

0.0000015

Then add an exponential multiplier equal to the number of places you moved the decimal point. Since the decimal point was moved to make a larger number out of a small number, the exponential multiplier will be negative.

$$1.5 \times 10^{-6}$$

EXAMPLE 1.20

A 1-oz load of number 6 shotgun pellets contains 225 pellets. Reloaders purchase shot in 25.0-lb bags. How many pellets are in a 25.0-lb bag? Express the result in scientific notation with the proper number of significant figures.

$$(25.0\ \text{lb})\left(\frac{16\ \text{oz}}{1\ \text{lb}}\right)\left(\frac{225\ \text{pellets}}{1\ \text{oz}}\right) = 90{,}000\ \text{pellets}$$
(Calculated result)

Since both the number of pellets and the weight of the bag are known to three significant figures, we express the total number of pellets in scientific notation to the proper number of significant figures as

$$= 9.00 \times 10{,}000 = 9.00 \times 10^4\ \text{pellets}$$
(Rounded result)

Determining the Number of Significant Figures

There are several easy rules for determining the number of significant figures.

1. Zeros that follow nonzero digits after the decimal point are significant. For example,

   27.0230 (6 significant figures)

   2.30 (3 significant figures)

2. Zeros that precede nonzero digits in decimal numbers are *not* significant. For example,

   0.012 (2 significant figures)

   0.000386 (3 significant figures)

3. When multiplying or dividing, the answer should be rounded to the same number of significant figures as the *least* accurate number in the calculation.

EXAMPLE 1.21

Round off the results to the correct number of significant figures.
(a) Multiplication

0.0218	(**3** significant figures)
× 4200	(4 significant figures)
= 91.56	(Calculated result)
≈ 91.6	(Rounded result, **3** significant figures)

(b) Division

$$\frac{387\ \text{(3 significant figures)}}{0.003\ \text{(\textbf{1} significant figure)}} = 12{,}900$$
(Calculated result)

$= 1.29 \times 10^4$

$\approx 1 \times 10^4$
(Rounded result, **1** significant figure)

4. When carrying out addition or subtraction operations, carry all figures until the computation is complete, then round to the *number of decimal places* in the least certain number used in the operation.

EXAMPLE 1.22

Round off the result of this addition problem.

382.93	(2 decimal places)
82.01	(2 decimal places)
+ 13.2	(**1** decimal place)
= 478.14	(Calculated result)
≈ 478.1	(Rounded result, **1** decimal place)

NEW TERMS

accuracy How close the measured value is to the true value.

precision How close a set of measured values are to each other.

significant figures The digits in a measured or computed number that are meaningful. Arithmetic operations sometimes artificially create additional digits. The result of multiplication or division operations may never have more significant figures than the least certain number used in the operation. The result of an addition or subtraction operation may never have more numbers after the decimal place than the least certain number used in the operation.

scientific notation The process of writing a measurement or computed result as a number between 1 and 10 times a power of ten. It is particularly useful in rounding off and presenting results to the appropriate number of significant figures.

rounding The process by which the appropriate number of significant figures are maintained after an arithmetic operation.

TESTING YOURSELF

Density, Significant Figures, and Scientific Notation

1. Anhydrous ammonia is used as a fertilizer. This material will boil at −33.35 °C and must be kept in a pressurized tank. If the density of liquid ammonia is 0.770 g/mL, what would be the mass of ammonia in a 2000-L tank?
2. Express the answer to question (1) in scientific notation to the proper number of significant figures. Assume that the 2000-L tank is closer to 2000 L than to 1999 L or 2001 L. (Hint: which item of data is least certain?)
3. An investor paid a large price for a chunk of gold that he was told was pure. The gold bar had a mass of 440.00 g, but was slightly irregular so an exact volume could not be calculated. The investor filled a large graduated cylinder with water, immersed the chunk of gold, and observed an increase in the apparent volume of material in the graduated cylinder of 25.0 mL. Pure gold has a density of 19.3 g/mL. Did the investor get his money's worth?

4. A 50.0-mL sample of acetone was found to have a mass of 39.61 g. What is the density of this material expressed to the proper number of significant figures?
5. Consider a sample of pure uranium about the size of a large candy bar: 10.1 cm long, 1.50 cm high, and 2.0 cm wide. By reference to the density data in Table 1.4, compute the mass of this sample in grams and report your result to the proper number of significant figures.

Answers **1.** 1,540,000 g **2.** 1.54×10^6 g **3.** No; the density was 17.6 g/mL **4.** 0.792 g/mL **5.** Calculated result 566.61 g; rounded result 570 g

1.8 Energy, Heat, and Temperature

Energy

Energy comes in many forms.

We all have an intuitive definition of the term **energy**. We say that a person who has a lot of "energy" can get a lot of work done. A physicist would say more exactly that energy is the ability to do work. A chemist might say that energy is the ability to cause change. Energy comes in many forms: heat, light, sound, and the motion of objects such as people and automobiles.

Stored energy is called potential energy.

Energy can be divided into two classes, which might be called "stored energy" and "energy in action." Stored energy is called *potential energy*. Water stored behind a dam has the *potential* to do work when it is released. Food contains stored chemical energy. Gasoline in an automobile's gas tank has the potential to move the automobile when called upon.

Objects in motion have kinetic energy.

When stored energy is released, it often shows itself as a change in the motion of large or small objects. An automobile rolling down a hill gains energy of motion, or *kinetic energy*, as it accelerates. The kinetic energy of a moving object is related to its mass (m) and the square of its velocity (v):

$$\text{Kinetic energy} = \tfrac{1}{2}mv^2$$

When the velocity of an object is doubled, its kinetic energy increases by a factor of four. An increase in speed from 55 miles per hour (mph) to 65 mph will increase the kinetic energy of an automobile (and its occupants) by about 40%.

The type of energy expressed as molecular motion is called heat.

Individual molecules are constantly in motion and thus possess kinetic energy. This type of energy is called *heat*. Energy released by chemical reactions often appears as an increase in energy of motion of nearby molecules.

Heat and Temperature

Heat and energy are closely related. Many of us have huddled around a warm campfire on a cold day (Figure 1.13). We feel cold because the molecules in our bodies are moving slowly. Energy is released by the fire as the fuel molecules react with oxygen molecules from the air. Some of this energy appears as a type of light called *infrared radiation*, which is absorbed by our hands and clothes and causes the molecules located there to move faster. Some of the energy appears as motion of air molecules near the fire. These rapidly moving air molecules also strike us and transfer some of their energy

FIGURE 1.13
Heat from one chemical system such as campfire can be transferred to another chemical system such as a cold camper.

to our molecules, again causing them to move more rapidly. **Heat** is a form of energy of molecular motion. It may be generated by chemical reactions and can be transferred from one system to another. The amount of heat available is determined by two factors: how fast the molecules are moving and how many molecules are moving.

Temperature, in contrast to heat, is a measure *only* of how fast the molecules are moving and is not concerned with how many molecules are present. Consider two bricks, one small and one large, both at the same temperature. The heat energy contained in the large brick is more than the heat energy of the small brick because it has more moving molecules. But because the two bricks are at the same temperature, the *average energy of motion* of their molecules is identical.

Two temperature scales are commonly in use in the United States. These are the Fahrenheit and Celsius scales.

Fahrenheit Scale

The **Fahrenheit scale**, which uses **Fahrenheit degrees (°F)**, was developed by the German physicist Gabriel Fahrenheit in 1724. Fahrenheit's 0 °F was the coldest mixture of ice, water, and salt that he could produce in his laboratory. For 100 °F he chose what he believed to be normal body temperature. On this scale, water freezes at 32 °F and boils (at sea level atmospheric pressure) at 212 °F, a span of 180 °F.

The Fahrenheit temperature scale has arbitrary reference points.

Celsius Scale

The **Celsius scale**, which uses Celsius degrees (°C), was developed by the Swedish scientist Anders Celsius in 1742. This scale is based on the properties of water: the freezing point of pure water is defined as 0 °C, and the boiling point of water at sea level is 100 °C.

The Celsius scale is based on the freezing and boiling points of water.

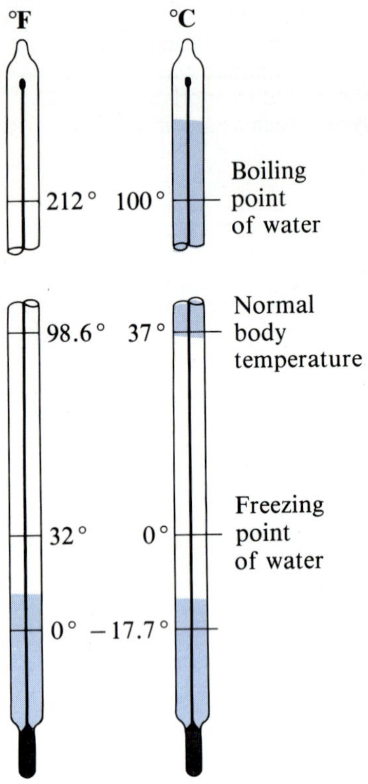
°F
°C
212°
100°
Boiling point of water
98.6°
37°
Normal body temperature
32°
0°
Freezing point of water
0°
−17.7°

FIGURE 1.14
The Fahrenheit and Celsius temperature scales are compared in this figure. The Celsius scale uses the freezing and boiling points of water as its reference values. Note that 180 °F occupy the same range as 100 °C.

Originally called the *centigrade scale* because there were 100° between the freezing and boiling points of water, this scale was renamed the Celsius scale with the advent of the Système Internationale (Figure 1.14). The Celsius temperature scale is used in science throughout the world and for general measurement in all major countries except the United States.

Fahrenheit to Celsius Temperature Conversion

There is a simple relationship between Fahrenheit and Celsius temperature.

A relatively simple conversion is possible between the Celsius and Fahrenheit temperature scales. The two scales differ in two ways: (a) in the position of their zero and (b) in the size of their degree. Celsius degrees are larger increments than Fahrenheit degrees. While 100 °C span the interval from the freezing point to the boiling point of water, 180 °F span the same interval in the Fahrenheit scale. A Celsius degree is thus 180/100 = 1.8 times as large as a Fahrenheit degree.

To convert from one temperature scale to another, one first corrects for the size of the degree and then changes the zero position. Consider room temperature, 68 °F. Let's compute its value in degrees Celsius. We have one point in common to the two scales: the freezing point of water.

0°
32°
36°
68°
Fahrenheit
0°
?
?
Celsius
Water freezes

First we subtract 32° from 68 °F to get the span, which is 36 °F. Now find this same span on the Celsius scale. Our conversion factor is

$$\frac{1.0\,^\circ\text{C}}{1.8\,^\circ\text{F}} = 1$$

Then multiply this conversion factor by the span of degrees:

$$^\circ\text{C} = (36\,\cancel{^\circ\text{F}})\left(\frac{1.0\,^\circ\text{C}}{1.8\,\cancel{^\circ\text{F}}}\right) = 20\,^\circ\text{C}$$

Thus, a temperature of 68 °F is equal to 20 °C.

This process can be generalized into the following equation:

$$^\circ\text{C} = \frac{(^\circ\text{F} - 32)}{1.8}$$

EXAMPLE 1.23

Find the Celsius temperature corresponding to 180 °F.

$$
\begin{aligned}
^\circ\text{C} &= \frac{(180\,^\circ\text{F} - 32)}{1.8} \\
&= \frac{148}{1.8} \\
&= 82.2222\,^\circ\text{C} \qquad \text{(Calculated result)} \\
&= 82.2\,^\circ\text{C} \qquad \text{(Rounded result with 3 significant figures)}
\end{aligned}
$$

Note that the numbers 1.8 and 32 are by definition *exact* numbers and can be assumed to have an infinite number of significant figures.

Conversion from Celsius to Fahrenheit degrees uses the opposite procedure. Let's find the Fahrenheit equivalent of 37.0 °C.

0°
32°
?
?
Fahrenheit
0°
37.0°
37.0°
Celsius
Water freezes

First, we know we are 37 °C above the reference point (the freezing point of water). How many degrees Fahrenheit is that? Our conversion factor is

$$\frac{1.8\ ^\circ\text{F}}{1\ ^\circ\text{C}} = 1$$

which we then multiply times the span:

$$37.0\ ^\circ\text{C}\left(\frac{1.8\ ^\circ\text{F}}{1\ ^\circ\text{C}}\right) = 66.6\ ^\circ\text{F}$$

So the value in question is 66.6 °F above the freezing point:

$$32\ ^\circ\text{F} + 66.6\ ^\circ\text{F} = 98.6\ ^\circ\text{F}$$

Thus, 37 °C is equal to normal body temperature, 98.6 °F.

This process can also be generalized into an equation:

$$^\circ\text{F} = (^\circ\text{C})(1.8) + 32$$

EXAMPLE 1.24

Find the Fahrenheit temperature corresponding to 10 °C.

$$\begin{aligned} ^\circ\text{F} &= (10\ ^\circ\text{C})(1.8) + 32 \\ &= 18 + 32 \\ &= 50\ ^\circ\text{F} \end{aligned}$$

Kelvin Scale

A third temperature scale of importance in chemistry is the **Kelvin scale** or *absolute scale.* The Kelvin scale is based on the behavior of gases and is best explained by a simple experiment.

Consider a small glass tube, sealed at its lower end and plugged with a small movable plug about half way up (Figure 1.15). This tube is immersed in a beaker of liquid that can be heated to 100 °C and cooled to −40 °C. As the temperature is changed, the molecules moving in the air trapped in the tube move faster or slower, and the volume changes accordingly. This occurs in the same way that a hot air balloon expands when heated and contracts when cooled.

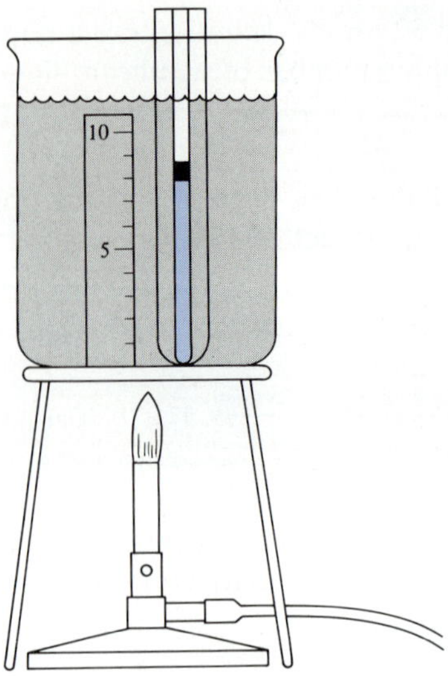
10
5

Temperature (°C)	Volume (mL)
−40	5.8
0	6.8
15	7.2
30	7.6
40	7.8
60	8.3
77	8.7
100	9.3

FIGURE 1.15
A gas expands in a regular manner when heated. This property is used as the base for the Kelvin temperature scale.

As the temperature is lowered, the volume of gas decreases in a very regular manner. For example, for a temperature change of 0 °C to −40 °C (a change of 40 °C), the volume change is 6.8 − 5.8 = 1.0 mL. For a 40 °C change from 40 °C to 0 °C the volume change is 7.8 − 6.8 = 1.0 mL. For the 40 °C change from 100 °C to 60 °C, the volume change is again 1.0 mL (Figure 1.16).

Gases occupy space because their molecules are moving rapidly, bumping into the walls of their container. If the rate of molecular motion is decreased, the bumping will be slower and less energetic, and the volume required to contain the gas will be less. Thus, the gas sample contracts as it cools.

The Kelvin temperature scale is based on the motion of molecules in a gas.

Two important concepts can be drawn from this graph. First, gas will expand or contract in a regular manner when heated or cooled. At reasonable temperatures, the volume of a gas sample gradually decreases as the temperature is reduced. Since the data are so regular, one can predict that this decrease will continue as the temperature is further reduced. The dotted line in the graph shows this projection. The second point is both important and unusual. The graph predicts that, if the gas is cooled to about −273 °C, its volume will drop to zero and the gas will vanish. Common sense, however, tells us that this is impossible.

The only way that the volume of gas could drop to zero would be if (1) its molecules were *infinitely* small, and (2) the molecules ceased to move. Although we believe that molecules may cease to move if cooled enough, we know that molecules are not infinitely small.

Zero degrees Kelvin is the temperature at which molecular motion ceases.

Although a gas will liquify before its molecules stop moving, the temperature at which molecular motion ceases can be predicted by projection of this graph. This temperature is called **absolute zero** and is −273.18 °C, usually rounded to −273 °C. The temperature scale based on absolute zero is used in science and engineering to describe the rate of molecular motion in all substances. The Kelvin scale has degrees called kelvins (K) that are the same size as those in the Celsius scale. Zero degrees in the Kelvin scale is absolute zero. To convert from Celsius to Kelvin one simply adds 273

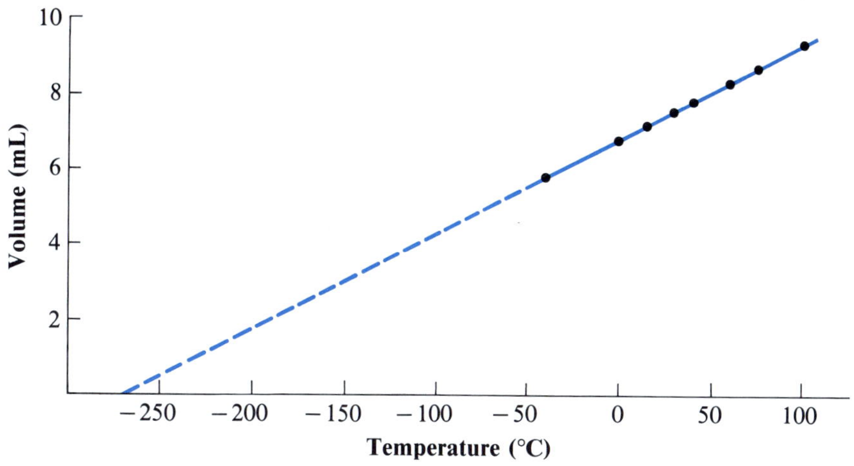
10
8
6
4
2
Volume (mL)
−250
−200
−150
−100
−50
0
50
100
Temperature (°C)

FIGURE 1.16
A plot of experimental data relating the volume of a sample of gas to its temperature. As the temperature is increased, the volume increases. As the temperature is reduced, the volume decreases. The graph theoretically predicts that the gas would vanish at about −273 °C.

(Figure 1.17). Note that the degree sign (°) is not used with Kelvin temperatures.

Kelvin temperature units are called kelvins rather than degrees.

EXAMPLE 1.25

Room temperature of 20 °C corresponds to what Kelvin temperature?

$$K = °C + 273$$
$$= 20 + 273 = 293 \text{ K}$$

NEW TERMS

energy The ability to do work. Energy is found in many forms, including heat.

temperature A measure of the average energy of motion of individual molecules or atoms.

heat A form of energy exhibited as molecular motion.

Fahrenheit scale A temperature scale used in the United States in which water freezes at 32 °F and boils at 212 °F. The conversion of Celcius to Fahrenheit is

$$°F = (1.8 \times °C) + 32$$

Celsius scale A temperature scale used worldwide in which water freezes at 0 °C and boils at 100 °C. The conversion of Fahrenheit to Celcius is

$$°C = \frac{(°F - 32)}{1.8}$$

Kelvin scale A temperature scale based on measurement of molecular motion. At zero Kelvin (absolute zero), all molecular motion theoretically ceases. Kelvin degrees are the same size as Celsius degrees, and the conversion is

$$K = °C + 273$$

absolute zero The temperature at which all molecular motion would theoretically stop, measured as −273.18 °C or 0 K.

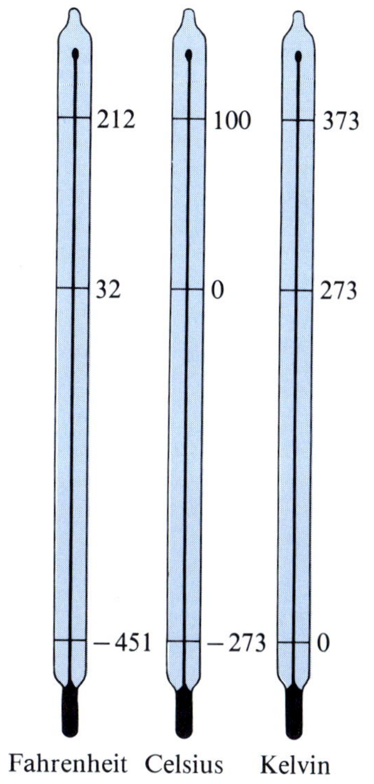
212
32
−451
100
0
−273
373
273
0
Fahrenheit
Celsius
Kelvin

FIGURE 1.17
The relationship between the three common temperature scales is shown in this figure.

TESTING YOURSELF

Celsius, Fahrenheit, and Kelvin Temperature Scales

1. Using the freezing and boiling points of water as reference points, a temperature change of 100 °C equals a change of how many Fahrenheit degrees?
2. Consider a temperature of 40 °C. This is 40 ° above the freezing point of water. How many Fahrenheit degrees would it be above the freezing point of water?
3. What is the Fahrenheit temperature of the sample discussed in question 2?
4. What Celsius temperature corresponds to −40 °F?

Answers **1.** 180 °F **2.** 72 °F **3.** 104 °F **4.** −40 °C

1.9 Calories and Specific Heat

The calorie is the amount of heat required to change 1 g of water 1 °C.

Specific heat is the amount of heat required to change 1 g of a material 1 °C. The specific heat of water is exactly 1 cal/g °C.

The **calorie (cal)** is defined as the amount of heat required to raise or lower the temperature of 1 g of water 1 °C. The heat capacity or **specific heat** of a substance is the amount of heat required to change 1 g of a substance 1 °C. By definition then, the specific heat of water is 1.00 cal per gram for each degree Celsius temperature change. The specific heat of aluminum is 0.214 cal/g °C, and the specific heat of silver is 0.0558 cal/g °C. Few elements or compounds have a greater specific heat than water. The unit of heat energy in the SI is the **joule (J)**, which is equal to 4.18 cal. The calorie, however, is still commonly used in chemistry.

To compute the calories of heat gained or lost during a change of temperature, one multiplies the mass of the material (in grams) times the number of calories required to change 1 g of the material 1 °C (the specific heat) times the temperature change in degrees Celsius:

Heat gain or loss = (Mass)(Specific heat)(Temperature change)

$$\text{cal} = (\text{g})\left(\frac{\text{cal}}{\text{g}\ {}^\circ\text{C}}\right)({}^\circ\text{C})$$

EXAMPLE 1.26

Consider a 250-g sample (about 1 cup) of water at room temperature (20.0 °C) that has been placed in a microwave oven. After 1 min the temperature of the water is 80.0 °C. How much heat has been added to the water? The temperature of the water has changed by +60.0 °C. The total amount of heat added to the sample is then:

(Mass)(Specific heat)(Temperature change) = Heat gain

$$(250\ \text{g})\left(\frac{1\ \text{cal}}{\text{g}\ {}^\circ\text{C}}\right)(60.0\ {}^\circ\text{C}) = 15{,}000\ \text{cal}$$

$$= 1.50 \times 10^4\ \text{cal}$$

The **kilocalorie (kcal)** is an SI unit commonly used in the field of nutrition. It is equal to 1000 "small" calories and is sometimes called a large Calorie (abbreviated Cal, with a capital C). Dieters who refer to the "calories" in their food actually mean kilocalories. The larger size of this unit makes it more convenient for use in discussions of heat transfer in the body.

EXAMPLE 1.27

Suppose that you made a cup of tea out of the hot water discussed in the last example and drank it. The temperature of the water would ultimately fall from 80 °C to a body temperature of 37 °C. In kilocalories, how much heat would be transferred to your body?

The resulting temperature change would be 80.0 − 37.0 = 43.0 °C. Using the equation for heat gain, we have

$$(250\ \text{g})\left(\frac{1\ \text{cal}}{1\ \text{g}\ {}^\circ\text{C}}\right)(43.0\ {}^\circ\text{C}) = 10{,}750\ \text{cal}$$

$$= 10.75\ \text{kcal}$$

$$\approx 10.8\ \text{kcal} \quad \text{(Rounded result)}$$

Specific Heat Calculations

The specific heat of water is higher than practically all other substances. One gram of silver, for example, requires only 0.056 cal to change its temperature 1.0 °C (written 0.056 cal/g °C). One gram of iron requires 0.114 cal for the same temperature change. Other materials are similar: aluminum (0.214), air (0.240), wood (0.42), and glass (0.198) (Table 1.6).

Specific heat has much to do with local climate. Coastal regions near large amounts of water always have moderate climates, while inland areas removed from water usually exhibit large temperature extremes. The effect of specific heat is apparent to those of us who go swimming on a sunny spring day. The sand feels hot to bare feet, the air feels warm, and the water is cold. Since sand, air, and water are all exposed to the same sunlight, why do their temperatures differ?

The specific heat of water is much higher than that of rock (or sand) and air, so many more calories must be added or removed to change the water temperature than to change the temperature of rock or air. If equal heat energy from the sun is applied to rock, air, and water, the temperature of the water will change the least.

TABLE 1.6
Specific Heats of Some Common Materials

Material	Specific Heat (cal/g °C)
Aluminum	0.214
Copper	0.092
Gold	0.031
Iron	0.114
Mercury	0.033
Lead	0.030
Uranium	0.027
Silver	0.056
Tin	0.054
Water	1.00
Air	0.240
Sugar	0.274
Wood	0.42
Quartz	0.188
Paraffin	0.694
Ethyl alcohol	0.581
Glass	0.198
Granite	0.192

EXAMPLE 1.28

Consider the effect of 1 cal of heat energy from the sun striking the earth. What is the increase in temperature for three substances it will warm: rock (granite), air, and water?

(a) Rock (granite)

Specific heat of granite is 0.192 cal/g °C

$$\text{Temperature change} = \frac{\text{(Heat gain)}}{\text{(Mass)(Specific heat)}}$$

$$= \frac{(1.00\ \cancel{\text{cal}})}{(1.00\ \cancel{\text{g}})\left(\dfrac{0.192\ \cancel{\text{cal}}}{\cancel{\text{g}}\ {}^\circ\text{C}}\right)} = 5.21\ {}^\circ\text{C}$$

(b) Air

Specific heat of air is 0.240 cal/g °C

$$\text{Temperature change} = \frac{(1.00\ \cancel{\text{cal}})}{(1.00\ \cancel{\text{g}})\left(\dfrac{0.240\ \cancel{\text{cal}}}{\cancel{\text{g}}\ {}^\circ\text{C}}\right)} = 4.17\ {}^\circ\text{C}$$

(c) Water

Specific heat of water is 1.00 cal/g °C

$$\text{Temperature change} = \frac{(1.0\ \cancel{\text{cal}})}{(1.0\ \cancel{\text{g}})\left(\dfrac{1.00\ \cancel{\text{cal}}}{\cancel{\text{g}}\ {}^\circ\text{C}}\right)} = 1.00\ {}^\circ\text{C}$$

The results of Example 1.28 are summarized in this line graph:

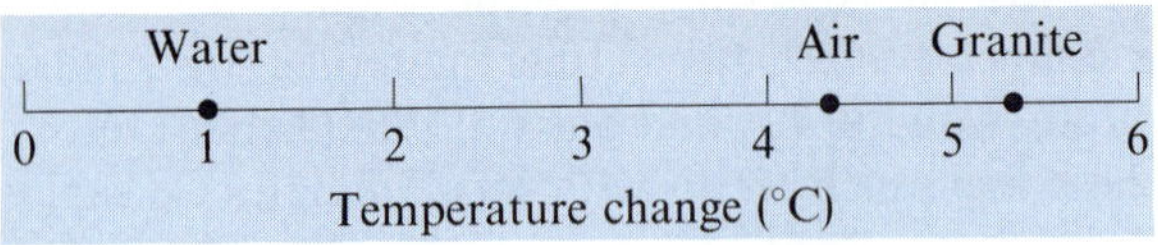
Water
Air
Granite
0
1
2
3
4
5
6
Temperature change (°C)

The temperature increase of granite is about five times that observed for water; the temperature increase of the air is about four times that of water. Since water is so hard to heat or cool, the presence of a large amount of it will moderate rapid changes in air temperature caused by sunshine or storms. San Francisco's location near the ocean gives it a much more moderate climate than St. Louis, although both are at about the same latitude.

NEW TERMS

calorie (cal) A unit of heat measurement equal to the amount of heat required to change the temperature of 1 g of water 1 °C.

Specific heat The number of calories required to change the temperature of 1 g of a material 1 °C.

joule (J) The SI unit for measurement of heat energy, equal to 4.18 cal.

kilocalorie (kcal) A unit of heat measurement equal to 1000 cal. In nutrition, the large Calorie is used instead of kilocalorie.

TESTING YOURSELF

Calories and Specific Heat

1. How many calories of heat will be required to heat 946 mL (1 quart) of water from room temperature (20 °C) to its sea level boiling point (100 °C)?
2. Suppose that you spilled 10.0 g of hot water at 77 °C on your hand. How many calories of heat would be given up as the water cooled to your body temperature of 37 °C?

Answers **1.** 75,680 cal = 75.7 kcal **2.** 400 cal

SUMMARY

Measurement systems—a part of our culture from the time mankind learned to count—have undergone two types of evolution. First, large units and small units of measure have been developed for convenience, to match the unit to the quantity being measured, and second, the standard for measurement systems has changed from regional and biological standards to international standards based on physical quantities reproducible worldwide.

The *Système Internationale* is the current international standard for measurement. Its units and their physical references are shown in Table 1.7. The addition of latin prefixes, which multiply and divide the base unit by powers of ten, provide an easy-to-use and consistent set of units for measurement of both small and large quantities.

Conversion between units of measure is easily accomplished by use of dimensional analysis; the measured quantity is *always* multiplied by a conversion factor. This conversion factor relates two equivalent units in the different units of measure.

$$(\text{Measured unit})\left(\frac{\text{Desired unit}}{\text{Measured unit}}\right) = \text{Desired unit}$$

In the next chapter we will begin to look at some of the behaviors of matter, and we will see some evidence that a far-reaching order may exist in the design of atoms and molecules.

TABLE 1.7
Common Units of the Système Internationale

Measurement	Unit	Symbol	Standard
Length	meter	m	Distance traveled by light in 1/299,792,458 sec
Volume	liter	L	1000 cm^3
Mass	kilogram	kg	Standard mass in Bureau of Standards
Temperature	degrees Celsius	°C	Boiling and freezing points of water
	degrees Kelvin	K	Vanishing point of an ideal gas
Heat	calorie	cal	Heat to change 1 g of water 1 °C
Time	second	sec	Time required for a certain number of vibrations of a specific type of cesium atom.

TERMS

Système Internationale d'Unite (SI)
meter (m)
metric system
kilometer (km)
centimeter (cm)
millimeter (mm)
dimensional analysis
conversion factor
liter (L)
cubic centimeter (cm^3)
milliliter (mL)
weight
mass
gram (g)
kilogram (kg)
density
specific gravity
accuracy
precision
significant figures
scientific notation
rounding
energy
heat
temperature
Fahrenheit scale
Celsius scale
Kelvin scale
calorie (cal)
specific heat
joule (J)
kilocalorie (kcal)

CHAPTER REVIEW

QUESTIONS

1. Which of the following is a unit of distance in the SI?
   a. calorie b. microgram c. milliliter d. millimeter e. liter
2. The metric prefix *milli* means
   a. 1/10 b. 1/100 c. 1/1000 d. 100 e. 1000
3. A British or **Imperial gallon** is defined as that volume containing 10 lb of water at 62 °F. An Imperial gallon is equal to 1.20 U.S. gallons. A U.S. gallon is 4 quarts, and a U.S. quart is equal to 946 mL. Which of the following conversion factors would you use to compute the number of U.S. gallons equal to 3.5 Imperial gallons?

   a. $\frac{1.00 \text{ Imperial gallon}}{1.20 \text{ U.S. gallons}}$ c. $\frac{1.20 \text{ U.S. gallons}}{1.00 \text{ Imperial gallon}}$

   b. $\frac{1.00 \text{ Imperial gallon}}{946 \text{ mL}}$ d. $\frac{1.00 \text{ U.S. gallon}}{946 \text{ mL}}$

   Remember:

$$(\text{Measured unit})\left(\frac{\text{Desired unit}}{\text{Measured unit}}\right) = \text{Desired unit}$$

4. Compute the answer to question 3. How many U.S. gallons in 3.5 Imperial gallons?
5. How many milliliters are equal to 3.5 Imperial gallons?
6. How many milliliters are equal to 41.2 L?
7. A block of white pine has dimensions of 5.00 cm × 3.00 cm × 10.00 cm. Its mass is 55.500 g. What is the density of this wood?
8. How many significant figures is one justified in reporting in question 7?
9. How many feet are in 1 km?
10. Consider a puddle with 5.0 L of water in it. The water temperature is 15 °C. How many calories of heat energy must be delivered by sunlight to raise the temperature of the water in the puddle to 20 °C?

Answers **1.** d **2.** c **3.** c **4.** 4.2 U.S. gallons **5.** 15,893 mL **6.** 41,200 mL **7.** 0.370 g/mL or cc **8.** 3 **9.** 3281 ft **10.** 25,000 cal

DIAGNOSTIC CHART

Blacken in all of the circles under the number of each question that you missed in the preceding chapter review questions. The diagnostic chart will help you identify concept areas that need more study.

	Questions									
Concepts	1	2	3	4	5	6	7	8	9	10
Length	○								○	
Metric prefixes		○		○	○	○			○	
Dimensional analysis			○	○	○	○			○	
Volume			○	○	○	○				
Mass and weight						○	○			
Density							○			
Significant figures							○	○		
Scientific notation									○	
Temperature scales										○
Calories and heat										○

EXERCISES

Length

1. To what physical standard was the original metric unit of length referenced?
2. Which is larger, the metric meter or the English yard?

Metric Prefixes

3. A 50-m Olympic swimming pool is how many centimeters long?
4. 1.00 in. is equal to 2.54 cm. How many millimeters is this?
5. 1.00 lb is equal to 453.6 g. How many milligrams is this?
6. 1.00 L of water has a mass of 1.00 kg. How many milligrams is this?

Dimensional Analysis

7. A fathom is 6.0 ft. Given that 1.0 ft = 12 in., how many inches is 1.0 fathom?
8. Given that 1.000 m = 39.37 in., how many meters is 1.0 fathom?
9. Given that 1.00 quart = 946 mL, how many liters of gasoline will a 21.0-gallon gasoline tank holds?
10. An aspirin tablet contains 5 grains of aspirin. If 15 grains equal 1 g, how many aspirin tablets can be made from 25 g of aspirin?
11. Perform the following conversions:
    a. 36 yards to in.
    b. 6.0 miles to ft
    c. 184 in. to ft
    d. 50 m to cm
    e. 60 km to m
    f. 3540 cm to m
    g. 420 cm to in.
    h. 4 ft to in.
    i. 30 cm to mm
    j. 12 m to ft

Volume

12. The Imperial gallon is the volume of 10.0 lb of water. To what measurement is the metric liter referenced?
13. Consider a stock-watering tank that is 0.5 m high, 1.0 m wide, and 2.0 m in length. When full of water, how many liters will this contain?
14. The cylinder of a hypodermic syringe has an inside radius of 0.5 cm and a length of 5.0 cm. How many milliliters of solution will it hold?
15. The label on a soda can indicates a volume of 10 fluid oz. How many milliliters does the soda can contain?

    Useful conversion factors: 1.0 quart = 946 mL
    1.0 quart = 32 fluid oz

Mass and Weight

16. To what substance was the SI unit of mass originally referenced?
17. Given that an Imperial gallon is 1.2 U.S. gallons and that an Imperial gallon is the volume of 10 lb of water, how much will a U.S. gallon of water weigh?

18. At sea level, 1.000 lb has a mass of 453.6 g. Using this conversion factor, what is the mass of a 110-lb person?
19. A person weighs 150 lb, and the correct dosage of a drug is given as 1.5 mg per kilogram of body weight. How many milligrams of the drug should be given?

Density

20. A liquid sample is known to be one of the materials listed below. 25.0 mL of this material has a mass of 19.80 g. What is its probable identity?

Substance	Density (g/mL)
Alcohol (a solvent)	0.78
Ethylene glycol (antifreeze)	1.11
Acetone (a solvent)	0.79
Chloroform (a solvent)	1.50
Octane (from gasoline)	0.70

21. The density of gasoline is about 0.70 g/mL. An automobile's gasoline tank will hold 20.0 gallons. What is the weight, in pounds, of 20 gallons of gasoline?

    Useful conversion factors: 1 quart = 946 mL
    1 lb = 453.6 g

22. Consider a bar of aluminum 2.00 cm × 5.00 cm × 1.00 m in length. Using data from Table 1.6, determine the mass of this aluminum bar as expressed in kilograms.
23. Question 20 lists several organic or carbon-based liquids. Two of these—chloroform and octane—will not mix with water. Which of these will float on top of the water?
24. Using data from Table 1.4, determine the mass of the following:
    a. 25mL of isopropanol
    b. 1 L of lead
    c. 35 mL of chloroform
    d. 120 cc of gold

Significant Figures

25. What is the weight in ounces of 1.00 quart of water?

    Useful conversion factors: 1.000 quart = 946.36 mL
    1.0000 lb = 453.59 g
    16.000 oz = 1.0000 lb

    Based on the accuracy of the data, how many significant figures are appropriate for your answer?
26. Look again at question 22. How many significant figures is one justified in reporting in this result? What is the factor limiting the accuracy of the result?
27. The density of aluminum is 2.702 g/mL. Consider again the problem outlined in question 22. How many significant figures is one now justified in reporting for the mass of the block?

Scientific Notation

28. Express the following numbers in scientific notation:

21	431	45,000	0.1
210	431.1	4,500,000	0.0045
2100	5905	4,500,000,000	0.000002

29. Convert the following numbers to regular decimal notation:

4.5×10^4	3.541×10^3	7.59×10^2
4.5×10^4	3.541×10^{-1}	7.59×10^{-4}
4.5×10^{-1}	3.451×10^6	7.59×10^2

30. A water molecule has a mass of about 2.99×10^{-23} g. How many water molecules are present in 1.000 L (1000 g) of water?
31. 6.02×10^{23} calcium atoms have a mass of 40.1 g. What is the mass of one calcium atom?
32. Using significant figures and proper rounding techniques, compute answers to the following problems:
    a. (34.2)(3) = ?
    b. (87)/(31.3) = ?
    c. 42.86 + 6.1 = ?
    d. 143.227 + 18.33 = ?

Temperature Scales

33. The temperature of the earth increases about 30 °C per kilometer of depth. If the surface temperature is 20 °C, what would you expect the temperature to be at a depth of 3 km? Express this value in degrees Fahrenheit.
34. At a temperature of 10 km above the surface of the earth, the temperature is about 230 °K.
    a. What is the temperature in degrees Celsius?
    b. What is the temperature in degrees Fahrenheit?
35. Reduced atmospheric pressure causes water to boil at lower temperatures at high elevations. At an elevation of 5000 ft, water boils at about 95 °C. To what Fahrenheit temperature does this correspond?
36. Hypothermia sets in when one's body temperature falls below about 90 °F. To what Celsius temperature does this correspond?
37. Perform the following temperature conversions:
    a. 38 °C to °F
    b. 32 °F to °C
    c. 63 °C to K
    d. 27 °F to K
    e. 300 K to °C
    f. 84 °C to °F

Calories and Heat

38. How many calories of heat will be required to heat 1 quart (946 mL) of water from room temperature (20 °C) to its sea level boiling point?

39. Suppose that the body temperature of a 110-lb (50-kg) person fell to 85 °F (29.5 °C). How many calories of heat must be added to bring the body temperature back to normal (37 °C)?

40. Suppose that you spilled 0.1 g of hot water at 100 °C on your hand. How many calories of heat would be given up by the water as it cooled to body temperature (37 °C)?

41. Suppose that a 0.1-g aluminum screw at 100 °C dropped into your hand. How many calories of heat would be given off as it cooled to body temperature?

CHAPTER

2

Some Basic Principles of Chemistry

Atoms and molecules of all types have much in common. Similarities in their chemical and physical behavior indicate that a far-reaching order exists in their design.

OBJECTIVES

After completing this chapter, you should be able to

- Distinguish among the solid, liquid, and gaseous states of matter and define heat of fusion and heat of vaporization.
- Differentiate between chemical and physical change.
- Differentiate among mixtures, compounds, and elements.
- Define the terms *atomic weight* and *atomic mass*.
- Identify chemical families within the periodic table of the elements.

2.1 Physical Properties of Matter

Our understanding of chemistry is based on two types of information—things we can see and measure and things we infer because of these observations. No one has ever seen an atom or a molecule, yet we believe in their existence because we can use these ideas to explain the behavior of the materials that surround us.

Matter is a term used to describe the materials that make up the universe. Matter is anything that occupies space and has mass. The air we breathe, the food and water we ingest, the materials that clothe and house us, and the minerals that form the earth are all matter.

Matter is anything that occupies space and has mass.

This chapter begins by exploring some of the easily observable physical properties of matter. Physical properties are those characteristics that can be observed and measured without changing the chemical composition of the material. These physical properties are different from those made as the chemical composition of a material is altered. Based on the physical and chemical properties of matter, a natural classification into three groups can be made: mixtures, compounds, and elements. The chapter concludes by considering some trends in physical and chemical behavior that are apparent when elements are arranged in order of relative atomic weight. When arranged in this manner, the chemical and physical properties of elements appear to vary in a predictable and periodic manner.

States of Matter

Matter exists in three states: solid, liquid, and gaseous. A **solid** has a definite volume and a definite shape. A **liquid** has a definite volume, but its shape is determined by the container in which it is placed. A **gas** has no definite shape or volume and will spread out to fill any container in which it is placed (Figure 2.1).

Matter exists in three states: solid, liquid, and gas.

Most materials exist in all three of these physical states, although the temperatures required for conversion from one state to another are in some cases extreme. Oxygen, for example, is found as a gas at normal temperature and pressure. It will liquify, however, at −183 °C and will freeze into a solid

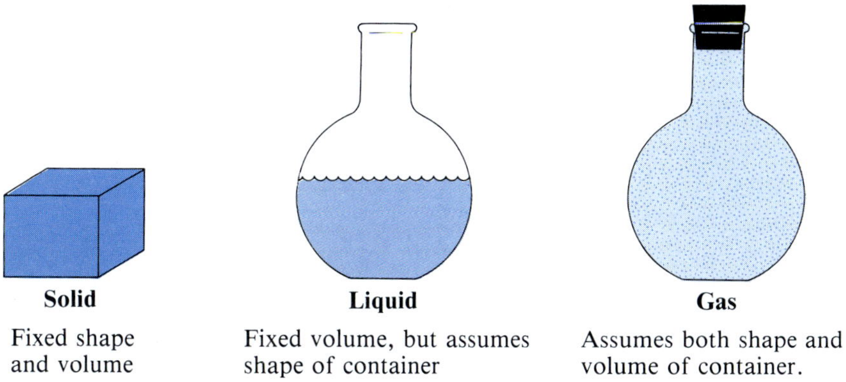
Solid
Fixed shape and volume
Liquid
Fixed volume, but assumes shape of container
Gas
Assumes both shape and volume of container.

FIGURE 2.1
The three states of matter differ in the way in which shape and volume are determined by the container the material is placed in.

TABLE 2.1
Melting and Boiling Points of Some Common Materials

Material	Melting Point (°C)	Boiling Point (°C)
Water	0	100
Acetone	−95	56.5
Ethyl alcohol	−117	78.5
Ethylene glycol (antifreeze)	−17.4	197.2
Aluminum	659.7	2057
Copper	1083	2336
Iron	1535	3000

Changes of state occur at the melting and boiling points.

at −218 °C. Iron is a solid at normal temperatures, but will melt at 1535 °C and will vaporize at 3000 °C (Table 2.1). Water is unique in that the temperatures at which it will change from solid to liquid (its **melting point**) and liquid to gas (its **boiling point**) are produced by normal weather conditions and sunlight. Few other substances will change physical state under normal temperatures found on the earth.

Physical Change: Heat of Fusion

Figure 2.2 shows ice in a glass of lemonade and a steam iron. What do these two objects have in common? Both use the fairly substantial amounts of energy involved in change of physical state to serve a useful purpose.

The energy required to melt ice is 80 cal of heat per gram

The energy required to change from the solid to the liquid state is called *heat of fusion.*

Ice is placed in the lemonade to cool it and to maintain the cooler temperature. Why does it do such a good job? The change from liquid to solid water takes place at 0 °C and requires removal of 80 cal for each gram of liquid water converted to ice. In the same manner, addition of 80 cal of heat will change 1 g of ice at 0 °C to 1 g of water at 0 °C. The energy required to change between the solid and liquid states is called the **heat of fusion**, which has units of calories per gram. Water is unique in that its heat of fusion is higher than most other substances. Because of this, ice does not melt rapidly.

FIGURE 2.2
A steam iron and the ice cubes in a glass of lemonade have a good deal in common. They both use the energy changes involved in change of physical state to serve a useful purpose.

Snow often remains far into the summer at high elevations. Why is this so?

The amount of heat that must be added to melt a sample (or removed to freeze it) can be calculated by multiplying the mass of the sample by the heat of fusion:

$$\text{Heat of solid-liquid or liquid-solid conversion} = (\text{Mass})(\text{Heat of fusion})$$

$$\text{cal} = (\text{g})\left(\frac{\text{cal}}{\text{g}}\right)$$

EXAMPLE 2.1

An ice pack contains 250 g of ice (about 1 cup) at 0 °C. As this ice pack melts, how many calories of heat will be removed from the person upon which the ice pack is placed?

(a) Melting the ice

$$\begin{aligned}\text{Heat of conversion} &= (\text{Mass})(\text{Heat of fusion})\\ &= (250.0\ \text{g})\left(\frac{80\ \text{cal}}{1.0\ \text{g}}\right)\\ &= 20{,}000\ \text{cal} = 20\ \text{kcal}\end{aligned}$$

Thus, 20 kcal of heat will be removed from the person holding the ice pack as it melts. Raising the resulting 250.0 g of ice water to body temperature will then require more heat:

(b) Warming the water

$$\begin{aligned}\text{Heat} &= (\text{Mass})(\text{Specific heat})(\text{Temperature change})\\ &= (250.0\ \text{g})\left(\frac{1\ \text{cal}}{1\ \text{g}\ {}^\circ\text{C}}\right)(37.0\ {}^\circ\text{C})\\ &= 9250\ \text{cal} = 9.25\ \text{kcal}\end{aligned}$$

(a) Change of state		(b) Change of temperature	
Ice	Melt ice (20 kcal) → Liquid	Warm water (9.5 kcal) →	Liquid
0 °C	0 °C	← 37 °C →	37 °C

$$20 \text{ kcal} + 9.25 \text{ kcal} = 29.25 \text{ kcal}$$
$$\approx 29.3 \text{ kcal}$$

One can easily see why an ice pack would cool a sprain or other injury better than a cloth soaked in cold water. One cup of ice melted and brought up to body temperature will remove about 29.3 kcal; one cup of ice water will only remove about 1/3 as much heat as it warms to body temperature.

Physical Change: Heat of Vaporization

The energy required to change from the liquid to the gaseous state is called *heat of vaporization.*

Consider now the steam iron in Figure 2.2. Conversion from the liquid to gaseous state requires much more energy than the solid to liquid conversion—about 540 cal (0.54 kcal) per gram of water. Those of us who have been burned by steam from a tea kettle or steam iron are painfully aware of the **heat of vaporization**. The heat of vaporization is multiplied times the mass to obtain the amount of heat used in a liquid to gaseous conversion:

$$\text{Heat of liquid-gas or gas-liquid conversion} = (\text{Mass})(\text{Heat of vaporization})$$

$$\text{cal} = (\text{g})\left(\frac{\text{cal}}{\text{g}}\right)$$

What happens when we come into contact with steam? Two processes are involved. First, the steam (at 100 °C) changes into liquid at 100 °C. Second, the hot liquid cools to body temperature of 37 °C.

EXAMPLE 2.2

Compute the amount of heat that your body must absorb if 5 g of steam (about the same mass as a nickel) were to strike your arm, condense, and cool to body temperature.

(a) Condensing the steam

$$\text{Heat of conversion} = (\text{Mass})(\text{Heat of vaporization})$$
$$= (5.00 \text{ g})\left(\frac{540 \text{ cal}}{1.0 \text{ g}}\right)$$
$$= 2700 \text{ cal} = 2.70 \text{ kcal}$$

ENERGY AND WEATHER

Energy is transferred through our weather system by the energy changes involved in the change of physical state of water. Energy from the sun evaporates water in oceans and lakes. Each gram of water evaporated accepts 540 calories from the sun as it changes from the liquid to the gaseous state. Winds carry the gaseous water across the continents where, as the vapor rises and cools, condensation to the liquid state occurs. As each gram of water vapor condenses into liquid water, 540 calories of heat are released. This heat causes thunderheads to rise and warms the earth below. Tropical monsoon areas are warmer than can be accounted for simply by sunlight because of this added heat input from the condensation of huge amounts of water vapor.

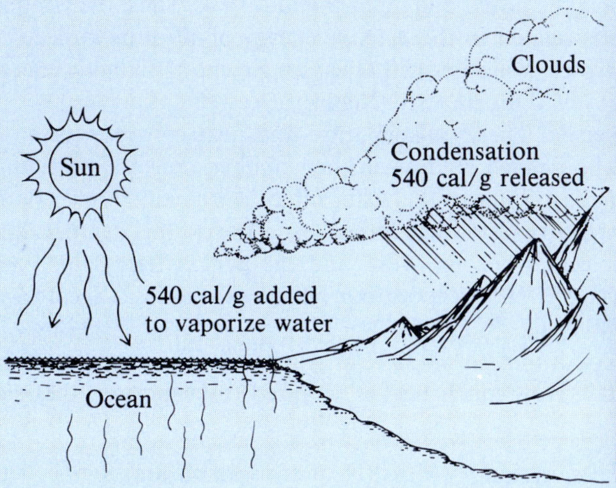
Clouds
Sun
Condensation
540 cal/g released
540 cal/g added
to vaporize water
Ocean

(b) Cooling the water

$$\text{Heat} = (\text{Mass})(\text{Specific heat})(\text{Temperature change})$$
$$= (5.00\ \text{g})\left(\frac{1\ \text{cal}}{1\ \text{g}\ {}^\circ\text{C}}\right)(63\ {}^\circ\text{C})$$
$$= 315\ \text{cal} = 0.315\ \text{kcal}$$

(a) Change of state — (b) Change of temperature

Steam —— Condense steam (2.70 kcal) ——→ Liquid —— Cool water (0.315 kcal) ——→ Liquid

100 °C | 100 °C ←—— 63 °C ——→ 37 °C

$$2.70\ \text{kcal} + 0.315\ \text{kcal} = 3.015\ \text{kcal}$$
$$\approx 3.02\ \text{kcal}$$

Steam irons are effective because the condensing steam transfers much more heat to the fabric than does a hot, dry iron. In the case of a burn caused by steam, about 90% of the energy used to cause the burn comes from the condensing steam. Some of us have tried (once) to use a steam iron to touch up a wrinkled pair of trousers or a skirt while we were wearing it. Few are willing to try it twice!

Temperature is an "Average" Quantity

Have you ever wondered how the sun can evaporate sea water to form clouds or how a wet sidewalk dries so rapidly after a rain? Evaporation does not require that the water boil, but only that individual molecules receive enough energy to leave the liquid and enter the air above. Since the temperature of a substance is related to the *average* energy of *all* of its molecules, on the whole, a few water molecules will always have enough kinetic energy to evaporate, even at low temperatures (Figure 2.3).

The idea of molecular energy distribution and heat of vaporization is familiar to anyone who has felt cold when they climbed out of a swimming pool, even on a warm day. Some of the water molecules covering one's body will have enough energy to evaporate. As they leave, the average temperature of the remaining molecules becomes less, and the person feels cold. Each gram of the thin layer of water evaporating off the skin takes 540 cal from the person (only 0.540 kcal).

Evaporation of water requires 540 cal per gram.

Under normal conditions about 1 L of water is evaporated from the body surface each day. How much heat is required to evaporate 1000 g of water?

$$\text{Heat of conversion} = (\text{Mass})(\text{Heat of vaporization})$$
$$= (1000\ \text{g})\left(\frac{540\ \text{cal}}{1\ \text{g}}\right)$$
$$= 540{,}000\ \text{cal} = 540\ \text{kcal}$$

If one's daily dietary intake is about 2500 kcal (Cal), more than 1/5 of this food energy is used simply to evaporate water.

About 20% of our food energy is used to evaporate water.

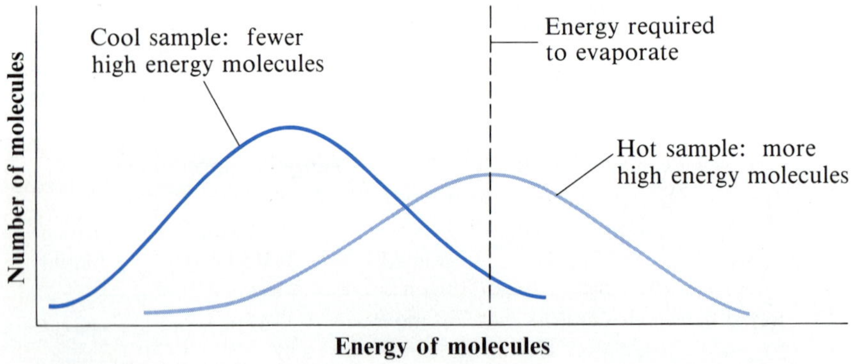
Cool sample: fewer high energy molecules
Energy required to evaporate
Hot sample: more high energy molecules
Number of molecules
Energy of molecules

FIGURE 2.3
Molecules in any sample come with a range of energies. Some molecules are barely moving; some have quite high energy. The temperature of the sample is related to the average energy of the molecules. This principle will be used in Chapter 11 to show how chemical reactions occur.

Evaporation continued over a moderate period of time will reduce the temperature of the body core. This is the reason that damp or wet clothing can so rapidly produce hypothermia (lowered body temperature), even in moderately warm weather.

NEW TERMS

matter Anything that has mass and occupies space.

solid A state of matter in which the material has a definite shape and volume.

liquid A state of matter in which the material has a definite volume, but assumes the shape of its container.

gas A state of matter in which the material assumes both the volume and shape of it container.

melting point The temperature at which a substance in the solid state is converted to the liquid state. The freezing point is the same temperature, with the change in the opposite direction.

boiling point The temperature at which a substance in the liquid state is converted to the gaseous state. The gaseous form of the substance will condense into a liquid at the same temperature.

heat of fusion The amount of heat that must be added to change 1 g of a material from its solid to liquid state or must be removed to change the material from liquid to solid. This value is 80 cal for water.

heat of vaporization The amount of heat that must be added to change 1 g of a material from its liquid to its gaseous state. This value is 540 cal for water.

TESTING YOURSELF

Physical Properties of Matter

1. Consider a 100-g popsicle. How many calories would be required to make this popsicle melt from ice at 0 °C to water at 0 °C?
2. Consider again the 100-g popsicle, now melted in your stomach. How many calories are required to bring this liquid from 0 °C to body temperature at 37 °C?
3. How many kilocalories (large Calories) did you use to eat this popsicle?
4. Suppose that you accidentally squirted 0.1 g of steam at 100 °C onto your hand. How many calories of heat will be given up to your hand as the steam condenses to water at 100 °C?
5. What is the total amount of heat absorbed by your hand as the 0.1 g of steam condenses and is cooled to body temperature at 37 °C?
6. Which hurts more: contact with 0.1 g of boiling water or 0.1 g of steam?
7. Nurses disinfect an area of skin before an injection with a swab soaked in alcohol. Why does it feel cold? (The boiling point of ethyl alcohol is 78.5 °C, and its heat of vaporization is 204 cal/g.)
8. Canteens used for hiking are often covered with canvas. If the canvas is kept damp, the water inside remains cool. Why does evaporation of the water from the canvas cool the water in the canteen?

Answers **1.** 8000 cal **2.** 3700 cal **3.** 11.7 kcal **4.** 54 cal **5.** 60 cal **6.** steam
7. Because alcohol evaporates very quickly, and your skin provides the heat to vaporize it.
8. The heat to evaporate the water in the canvas comes in part from the water in the canteen.

FIGURE 2.4
Life depends on the ability of water to exist in all three of its physical states at temperatures normally found at the earth's surface. A large fraction of the freshwater present at any time is stored as ice and snow, to be released slowly during the summer by solar energy. The hydrologic cycle depends on transport of water in the gaseous state from oceans to land and mountains.

2.2 Physical and Chemical Change and Conservation of Matter

Physical Change

Physical change involves a change in state, but not in chemical composition, and is usually easily reversible.

Physical change involves a change in *state*, but does not involve a change in chemical composition. For example, water is H_2O whether it is found as ice, liquid, or steam. Because of the moderate energy changes involved, physical changes are usually readily reversible. Water is evaporated from the oceans by heat from the sun, crosses the continents in the gaseous state, perhaps falls as snow, and ultimately melts to return to the ocean as water (Figure 2.4). As water moves through its three physical states in the *hydrologic cycle*, it retains the same chemical composition and remains H_2O. It has undergone physical change only. The metal in a welding rod temporarily changes from solid to liquid as the weld is made and then returns to the solid state. This is a physical change. Dry ice (carbon dioxide, CO_2) changes from a solid to a gas as it warms, but it still is CO_2. This also is a physical change.

Chemical Change and the Law of Conservation of Matter

Chemical change involves a change in the chemical composition of the material and is often not easily reversible.

Chemical change involves a change in the *composition* of the material. In general, chemical change is not easily reversible, and it usually involves a greater change in energy than does physical change.

Before
250 g H_2O at 10 °C
Lamp and fuel = 250.00 g
After
250 g H_2O at 40 °C
Lamp and fuel = 248.88 g

FIGURE 2.5
Alcohol in the stove combines with oxygen from the air to form carbon dioxide, water vapor, and heat. Energy is released by this chemical reaction. This energy heats the water.

Consider a simple alcohol stove used to heat some water (Figure 2.5). As the alcohol burns, molecules of alcohol react with molecules of oxygen from the air to form molecules of carbon dioxide and water vapor and to release energy that heats the water. (Molecules are groups of atoms connected or bonded together; they are discussed in Section 2.3.) This *chemical reaction* can be described by a simple word equation:

$$\text{Ethyl alcohol} + \text{Oxygen} \xrightarrow[\text{to produce}]{\text{Reacts}} \underset{\text{dioxide}}{\text{Carbon}} + \text{Water} + \text{Heat}$$

or, as we will see in Chapter 7, by a symbolic equation:

$$C_2H_6O + 3\ O_2 \longrightarrow 2\ CO_2 + 3\ H_2O + \text{Heat}$$

The numbers that precede the molecular formulas tell us that one molecule of alcohol reacts with three molecules of oxygen gas to form two molecules of carbon dioxide and three molecules of water. Note that the same number of each type of atom is present both before and after the reaction. Careful inspection of the reaction will reveal that both sides have two carbon atoms, six hydrogen atoms, and seven oxygen atoms (see Chapter 7).

The **law of conservation of matter** states that matter (atoms) cannot be created nor destroyed by ordinary means (see Chapter 7). The last three words provide an exception for nuclear reactions, which convert matter to energy. (Details of nuclear reactions are described in Chapter 13.) For this reason, chemical reactions *always* have as products the same number of each kind of atom that was present as a reactant. Chemical reactions simply rearrange the component atoms, often releasing or absorbing energy during the process (see Chapter 7).

The law of conservation of matter states that atoms cannot be created nor destroyed by ordinary means.

There are three important implications of this example. Most important is that a change of chemical composition has occurred as a result of the combustion process. Alcohol and oxygen have been converted to carbon dioxide and water vapor, thus a chemical change has taken place. The second implication has to do with reversibility. Physical changes are easily reversible. It is easy to melt ice and then freeze the water to obtain ice again. Carbon dioxide and water vapor, however, will not easily return to alcohol and oxygen if left alone. Chemical changes are in general not as easily reversible as physical changes. The third implication concerns energy change. In general, the energy changes involved in chemical change are 10 to 100 times greater than those observed for physical change of state alone.

NEW TERMS

physical change A change that usually requires only moderate amounts of energy, is reversible, and results in no change in the chemical identity of the substance.

chemical change A change in the chemical composition of a substance, that is usually not easily reversible and involves large changes in energy. Chemical change occurs as the result of a chemical reaction.

law of conservation of matter Matter (atoms) cannot be created nor destroyed by ordinary means.

TESTING YOURSELF

Physical and Chemical Change

1. Easy reversibility is a characteristic of what type of change?
2. Compute the amount of heat energy given up as 4.00 L (about 1 gallon) of rain condenses from the gaseous phase.
3. When 1 g of octane (a component of gasoline) is burned, carbon dioxide and water vapor are formed and 28,900 cal (28.9 kcal) of heat are produced. Compare this with the amount of heat required to vaporize 1.0 g of water (540 cal), and consider the reversibility of this reaction. Is combustion of gasoline a physical or chemical change?

Answers **1.** physical change **2.** 2160 kcal **3.** chemical change

2.3 Mixtures, Compounds, and Elements

Mixtures are composed of two or more components in *any* proportion.

Materials can be classified as mixtures, compounds, or elements. A large number of the materials we come in contact with in our everyday activities are **mixtures**. Mixtures can consist of any proportion of the materials that compose them. Concrete, for example, is a mixture of sand, gravel, cement, and water. The composition of concrete is variable, depending upon the desired strength and texture of the resulting material. More cement makes a stronger product, more sand a smoother texture. It is also possible to separate a mixture into its components. Gravel in concrete mix, for example, can be easily separated from cement and sand by sifting through a screen.

Compounds contain several elements in *fixed* proportions.

Salt, on the other hand, is a **compound**. It can only be separated into its component parts (sodium and chlorine) by chemical means. When salt is decomposed and the component materials collected, a clear-cut difference between mixtures and compounds emerges: the components are always exactly in the same proportions. We always observe 39.3% by weight sodium and 60.7% by weight chlorine in salt. Another example is water, which is always 11.1% by weight hydrogen and 88.9% oxygen. When sugar (sucrose) is decomposed, we always find 42.1% carbon, 6.4% hydrogen, and 51.5% oxygen. The simplest subdivision beyond which a compound cannot be easily broken is called an **element**, akin to an atomic building block. Sodium, chlorine, hydrogen, carbon, and oxygen are examples of elements.

An element is the simplest subdivision of matter.

This regularity of composition of compounds led John Dalton, an English school teacher, chemist, and meteorologist, to formulate the **law of definite proportions**. This law states that compounds are always composed of fixed proportions of elements.

The smallest particle into which an element can be subdivided and still retain its chemical properties is an **atom**. Atoms can be further subdivided into their component parts: neutrons, protons, and electrons (see Chapter 3). These atomic particles can be further subdivided into smaller particles, but not by ordinary means. For our every day needs, the definition of an atom as the smallest subdivision of matter still stands. **Molecules** are groups of atoms attached or bonded together in fixed proportions (see Chapter 5). For example, a group of 12 carbon atoms, 22 hydrogen atoms, and 11 oxygen atoms, properly arranged, form a molecule of the compound sucrose, a common sugar.

Molecules are groups of atoms in fixed proportions.

NEW TERMS

mixture A material that has no set proportion of its component substances and can be separated into its components.

compound A pure substance containing elements in definite and constant proportion. For example, sugar (sucrose) is always 42.1% carbon, 6.4% hydrogen, and 51.5% oxygen by weight.

element Basic substances that cannot be broken down by chemical means to simpler substances. Each element is unique.

law of definite proportions A compound always contains the same elements combined in the same proportions by weight.

atom The smallest particle into which an element can be subdivided and still retain its physical and chemical properties. It is a basic building block of matter. Atoms of each element are alike, but differ from those of other elements in size, mass, and chemical reactivity.

molecule A group of atoms bonded together in fixed proportions which comprise a compound. For example, two hydrogen atoms and one oxygen atom make a water molecule.

TESTING YOURSELF

Mixtures, Compounds, and Elements

Classify the following materials as mixtures, compounds, or elements:

1. Coffee, ready to drink
2. Sugar to put in the coffee ($C_{12}H_{22}O_{11}$)
3. The cast-iron frying pan in which you cook your eggs
4. The salt you put on your eggs
5. Orange juice
6. Gold plating on jewelry (Au)
7. Epsom Salts ($MgSO_4$)
8. Sea water

Answers **1.** mixture **2.** compound **3.** element **4.** compound **5.** mixture **6.** element **7.** compound **8.** mixture

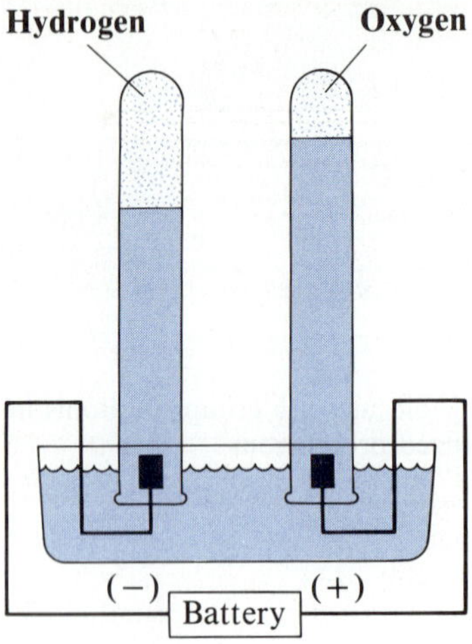
Hydrogen
Oxygen
(−)
(+)
Battery

FIGURE 2.6
Water can be decomposed into its component gases—hydrogen and oxygen—by the action of an electric current. Exactly twice as much hydrogen as oxygen by volume is produced.

2.4 Before It Was a Science: The Discovery of Order

Today there are 105 known elements; 91 of these are natural and 14 have been synthesized. Chemistry would be a difficult science if we had to memorize the properties of all of these elements and the way in which each combines with its fellow atoms to form compounds. Luck is with us, though. Similarities in chemical and physical behavior indicate that a far-reaching order exists in the design of atoms. The first inkling of this order came from classification of the elements according to their relative atomic weights.

Elemental Composition of Compounds

Our insight into the elemental composition of compounds awaited the Italian physicist Alessandro Volta's invention of the storage battery in 1800. Within about six weeks of Volta's discovery, two English chemists demonstrated that water could be separated into two gases by the action of an electric current. This separation technique, called **electrolysis**, is illustrated in Figure 2.6. Every time a sample of water is broken down into its component parts, exactly twice as much hydrogen gas as oxygen gas is produced. These two gases can be recombined (with the aid of a spark, for instance) to again form water. Exactly two volumes of hydrogen always combine with exactly one volume of oxygen. It was observed, in fact, that whenever gases combine to form compounds, they combine in small whole number ratios. One volume of ammonia gas, for example, is formed from three volumes of hydrogen gas and one volume of nitrogen gas.

When water is separated into its component parts, the volume of hydrogen produced is twice as great as the volume of oxygen.

These discoveries set the stage for determination of the relative weight of atoms. Consider a sample of water broken down into its components, hydrogen and oxygen:

Sample: Water

	Hydrogen	Oxygen
Composition by volume	2 units	1 unit
Composition by weight	11.11%	88.89%

A mass of 100 g of water contains 11.11 g of hydrogen and 88.89 g of oxygen. To compute the *relative* weight (or, more accurately, mass) of hydrogen and oxygen atoms, we need to know how many atoms are in each sample. Does the hydrogen sample have twice as many atoms as the oxygen sample because it occupies twice as much volume? Or are there more oxygen atoms because the mass of oxygen is about eight times as great as that of hydrogen?

Avogadro's Hypothesis

In 1811, an Italian named Amedo Avogadro hypothesized that the atoms or molecules that comprise a gas are very small compared to the space they occupy and are in constant motion. As these atoms or molecules move, they exert pressure by bumping against each other and against the walls of their container (Figure 2.7). He proposed that if a group of gas samples are maintained at the same temperature and pressure, *the volumes they occupy will*

Avogadro proposed that the number of atoms in a gas sample is directly proportional to its volume.

be in direct proportion to the number of gas atoms or molecules doing the bumping. If a sample of gas, such as the hydrogen in the experiment just discussed, had twice the volume of a sample of oxygen held under equal conditions of temperature and pressure, there would be twice as many hydrogen atoms as oxygen atoms.

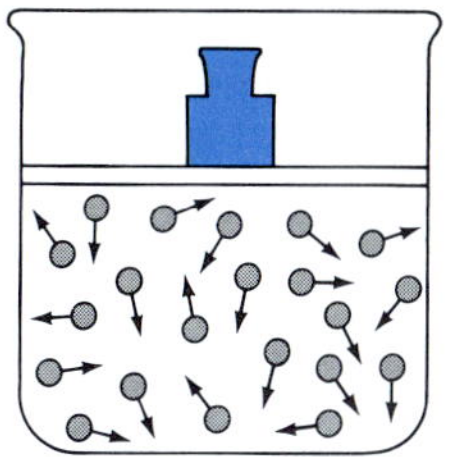

FIGURE 2.7
Avogadro proposed that the volume occupied by a gas is determined by the number of atoms or molecules bumping against the walls of its container. If temperature and pressure are the same, identical volumes of gas will contain identical numbers of atoms or molecules.

Relative Atomic Weight

Avogadro's hypothesis provided a way to determine the relative weights of hydrogen and oxygen atoms. If a sample of water does indeed decompose into twice as many hydrogen atoms as oxygen atoms, then we can compute how much heavier oxygen is than hydrogen.

EXAMPLE 2.3

If two volumes of hydrogen atoms comprise 11.11% of the total weight (or mass) of water, one volume of hydrogen atoms will comprise 5.555% of the sample weight, and an equal number of oxygen atoms will comprise 88.89% of the sample weight. In 100 g of water, there would be 88.89 g of oxygen (O) and 5.555 g of hydrogen (H).

$$\frac{88.89 \text{ g O}}{5.555 \text{ g H}} = 16.00$$

Thus, an oxygen atom is 16 times heavier than a hydrogen atom.

A similar experiment using ammonia gas indicates that nitrogen atoms are heavier than hydrogen atoms but lighter than oxygen atoms.

EXAMPLE 2.4

Sample: Ammonia Gas

	Hydrogen	Nitrogen
Composition by volume	3 units	1 unit
Composition by weight (or mass)	17.6%	82.4%

Since three volumes of hydrogen equal 17.6% of the weight of the sample, one volume of hydrogen should comprise 17.6/3 = 5.88% of the total sample weight. One volume of nitrogen was observed to comprise 82.4% of the total sample weight. In 100 g of ammonia, one volume of nitrogen (N) atoms would have a mass of 82.4 g, and each volume of hydrogen (H) atoms would have a mass of 5.88 g.

$$\frac{82.4 \text{ g N}}{5.88 \text{ g H}} = 14.0$$

Thus, we can predict that a nitrogen atom is 14 times heavier than a hydrogen atom.

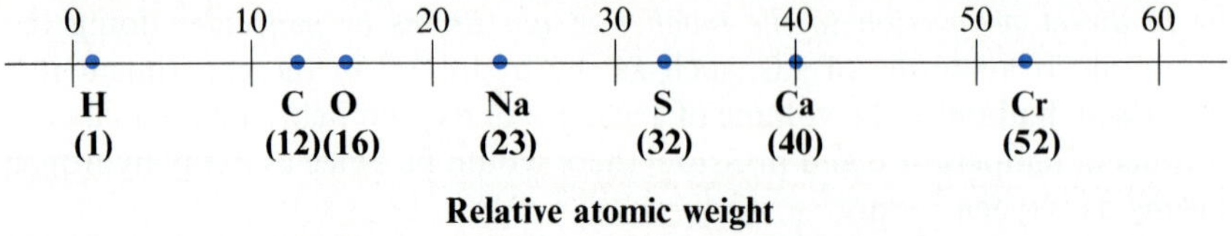
0
10
20
30
40
50
60
H
(1)
C O
(12)(16)
Na
(23)
S
(32)
Ca
(40)
Cr
(52)
Relative atomic weight

FIGURE 2.8
Relative atomic weights of a few elements.

Although the term *atomic weight* is actually incorrect, it is traditionally used instead of the more accurate term *atomic mass.*

Similar experiments with other compounds showed that it was possible to order elements in terms of their **relative atomic weight**. (*Atomic mass* is a more accurate term than *atomic weight*, but because of its widespread usage, *atomic weight* is generally accepted.) The best support for this idea came from its internal consistency. Once a scale of relative atomic weights had been developed for a number of elements, known elements could be measured in different compounds with the same result (Figure 2.8). For example, the relative atomic weight of carbon is 12, regardless of whether it is measured in carbon dioxide (CO_2), methane (CH_4), or sugar ($C_{12}H_{22}O_{11}$).

The atomic weight scale is now referenced to carbon-12 as 12.000 atomic mass units.

All elements were at first compared to the lightest element hydrogen (assigned atomic weight 1.00), as we have done in these sample calculations. However, it became increasingly more convenient to reference to oxygen as 16.000. This simple change caused hydrogen to have a relative atomic weight of 1.008. Even more recently, scientists have agreed to reference the atomic weight scale to an atom of carbon assigned a mass of exactly 12. As we will see in the next chapter, all elements consist of similar atoms having slightly different masses, called isotopes. Carbon-12 (written ^{12}C) is one such type of carbon atom. The basic unit used for atomic weight, the atomic mass unit (amu), is based on the mass of the ^{12}C atom and equals exactly 1/12 of its mass (see Chapter 3).

NEW TERMS

electrolysis The use of electrical current to break molecules into their component elements.

relative atomic weight The average mass of a sample of atoms of an element, as compared to an accepted standard. The original standard was the lightest atom, hydrogen, assigned an arbitrary atomic mass of 1.000. The reference was changed recently to the most common type of carbon atom, which was assigned a mass of exactly 12 atomic mass units. (The term *atomic mass* is used interchangeably and is actually more accurate.)

TESTING YOURSELF

Atomic Mass

1. A molecule of methane (CH_4), commonly called natural gas, consists of four hydrogen atoms and one carbon atom. It is 75% by mass carbon and 25% by mass hydrogen. What is the relative atomic mass of carbon as compared to hydrogen?
2. A molecule of hydrogen sulfide, which is responsible for rotten egg smell, consists of two hydrogen atoms and one sulfur atom. It is 5.92% by mass hydrogen and 94.08% sulfur. What is the relative atomic mass of sulfur as compared to hydrogen?

Answers **1.** 12 **2.** 32

2.5 Chemical Periodicity and Classification of the Elements

The process of science is composed of two tasks: (1) finding the information and (2) making sense out of it. In general, finding information is much easier than making sense out of it.

Consider the alphabetical listing of 24 common elements shown in Table 2.2. A considerable amount of information concerning the chemical and physical behavior of these elements is presented. In the next few pages we will use the idea of relative atomic weight to try to make sense out of this information. The generalization developed here, called *chemical periodicity*, permits easy prediction of chemical and phyical properties of elements and is one of the most powerful generalizations used in chemistry.

Chemical periodicity is the periodic repetition of physical and chemical properties with increasing atomic weight.

Chemical Periodicity

The number of hydrogen atoms with which an element will combine is well known. This information is displayed in column four of Table 2.2 in such a way that we can easily see the elements that behave in a similar manner.

TABLE 2.2
Alphabetical Arrangement of Selected Elements Showing Physical and Chemical Properties

			Number of H Atoms It Combines With					Conducts Electricity?		Density (g/mL)	
Element	Symbol	Atomic Weight	0	1	2	3	4	Yes	No	High	Low
Aluminum	Al	26.98				3		Yes		2.70	
Argon	Ar	39.94	0						No		0.00178
Arsenic	As	74.92				3			No	5.73	
Beryllium	Be	9.01			2			Yes		1.8	
Boron	B	10.81				3			No	3.33	
Bromine	Br	79.90		1					No	3.12	
Calcium	Ca	40.08			2			Yes		1.55	
Carbon	C	12.01					4		No	2.25	
Chlorine	Cl	35.45		1					No		0.00321
Fluorine	F	18.99		1					No		0.00169
Gallium	Ga	69.72				3		Yes		5.91	
Helium	He	4.00	0						No		0.00018
Krypton	K	83.30	0						No		0.00307
Lithium	Li	6.93		1				Yes		0.53	
Magnesium	Mg	24.31			2			Yes		1.74	
Neon	Ne	20.18	0						No		0.00089
Nitrogen	N	14.00				3			No		0.00125
Oxygen	O	16.00			2				No		0.00142
Phosphorus	P	30.97				3			No	2.20	
Potassium	K	39.10		1				Yes		0.87	
Selenium	Se	78.96			2				No	4.8	
Silicon	Si	28.09					4		No	2.33	
Sodium	Na	22.98		1				Yes		0.97	
Sulfur	S	32.06			2				No	2.07	

Aluminum combines with hydrogen to make the compound aluminum hydride, AlH_3, thus aluminum is listed as combining with three hydrogens. Oxygen combines with hydrogen to form water, H_2O, so oxygen is in the "2" column.

Table 2.2 also lists the ability of the elements to conduct electricity. These data are set into separate "yes" and "no" columns for easy identification. Density values are also set in two columns according to their value so we can identify groups of elements of high and low density.

Is there any regularity in the behavior of these elements? Can they be divided into families with similar chemical and physical properties? At this point, the data almost appear to be random.

A few generalizations are possible. The elements divide into two logical groups: those that conduct electricity and those that do not. Observation tells us that those elements that are good electrical conductors in general are also malleable (capable of being hammered into thin sheets without breaking). We call these elements **metals**. The remaining elements are called **nonmetals**, which do not conduct electricity and break when struck in their solid forms.

Metals are materials that conduct electricity and are malleable.

Let's use this new information to rearrange the elements to see if other generalizations become visible. In Table 2.3 the elements are divided into two groups: the metals, which conduct electricity, and the nonmetals, which do not conduct electricity.

Two additional generalizations are visible in this arrangement of the data. Metals appear to always combine with only one, two, or three hydrogens. This does not tell us too much, however, because several of the nonmetals also combine with one, two, or three hydrogens. But notice that only among the nonmetals are elements that combine with 0 or 4 hydrogens. Also notice that the densities of the metals are more similar to one another than those of the nonmetals. The densities of the metals range from a low of 0.534 g/mL for lithium to a high of 5.91 g/mL for gallium, a factor of about 11. In contrast, those of the nonmetals range from 0.00018 g/mL for helium to 5.73 for arsenic, a factor of about 32,000.

Perhaps there are other arrangements of the elements that will yield additional information. Table 2.4 provides another look at the same data, this time with the elements arranged in order of increasing atomic weight.

Suddenly several trends are obvious. The ability to combine with hydrogen appears to follow a regular cycle as atomic mass increases (see column four). There is something very basic about atomic weight which, in a repeatable or *periodic* way, determines the behavior of the elements.

Atomic weight (or mass) is the unifying theme in the periodic table.

Note that there is an inconsistency in the atomic weights of potassium and argon. To make them follow the same trends in chemical and physical behavior apparent for all other elements, their positions in the table have been reversed (argon, with a higher atomic weight, has been put first). For the time being let's assume that there is something about argon that makes it act a little lighter than it really is. We will consider this problem again in Chapter 5.

The elements now appear to follow a cycle comprised of eight elements. Table 2.5 is a reproduction of Table 2.4 showing gaps inserted at eight element intervals, with the initial break made right below helium. The elements in each group divide into one subgroup that conducts electricity and one

TABLE 2.3
Arrangement of Selected Elements into Metals and Nonmetals Showing Physical and Chemical Properties

			Number of H Atoms It Combines With					Conducts Electricity?		Density (g/mL)	
Element	Symbol	Atomic Weight	0	1	2	3	4	Yes	No	High	Low
Metals											
Aluminum	Al	26.98				3		Yes		2.70	
Beryllium	Be	9.01			2			Yes		1.85	
Calcium	Ca	40.08			2			Yes		1.55	
Gallium	Ga	69.72				3		Yes		5.91	
Lithium	Li	6.93		1				Yes		0.53	
Magnesium	Mg	24.31			2			Yes		1.74	
Potassium	K	39.10		1				Yes		0.87	
Sodium	Na	22.98		1				Yes		0.97	
Nonmetals											
Argon	Ar	39.94	0						No		0.00178
Arsenic	As	74.92				3			No	5.73	
Boron	B	10.81				3			No	3.33	
Bromine	Br	79.90		1					No	3.12	
Carbon	C	12.01					4		No	2.25	
Chlorine	Cl	35.45		1					No		0.00321
Fluorine	F	18.99		1					No		0.00169
Helium	He	4.00	0						No		0.00018
Krypton	K	83.30	0						No		0.00307
Neon	Ne	20.18	0						No		0.00089
Nitrogen	N	14.00				3			No		0.00125
Oxygen	O	16.00			2				No		0.00142
Phosphorus	P	30.97				3			No	2.20	
Selenium	Se	78.96			2				No	4.8	
Silicon	Si	28.09					4		No	2.33	
Sulfur	S	32.06			2				No	2.07	

subgroup that does not. The groups have been numbered, for reasons we will see later in this chapter.

We can now define **chemical periodicity** more precisely: it is the periodic recurrence of similar chemical properties with increasing atomic mass. Notice that the density of the elements in each group cycles from slightly low to high and then back to very low as atomic weight increases. Note that the third element in each group is the most dense. As atomic weight increases, a regular change is observed in all three properties: the number of hydrogens with which the element will combine, the electrical conductivity, and the density. This must be more than coincidental (see Chapter 4).

There is a minor irregularity near the bottom of the table, in the fourth group of elements. It appears that something is missing. This is the only group that does not have an element that will combine with four hydrogens.

TABLE 2.4
Selected Elements and Their Physical and Chemical Properties Arranged by Increasing Atomic Weight

			Number of H Atoms It Combines With					Conducts Electricity?		Density (g/mL)	
Element	**Symbol**	**Atomic Weight**	**0**	**1**	**2**	**3**	**4**	**Yes**	**No**	**High**	**Low**
Helium	He	4.00	0						No		0.00018
Lithium	Li	6.93		1				Yes		0.53	
Beryllium	Be	9.01			2			Yes		1.85	
Boron	B	10.81				3			No	3.33	
Carbon	C	12.01					4		No	2.25	
Nitrogen	N	14.00				3			No		0.00125
Oxygen	O	16.00			2				No		0.00142
Fluorine	F	18.99		1					No		0.00169
Neon	Ne	20.18	0						No		0.00089
Sodium	Na	22.98		1				Yes		0.97	
Magnesium	Mg	24.31			2			Yes		1.74	
Aluminum	Al	26.98				3		Yes		2.70	
Silicon	Si	28.09					4		No	2.42	
Phosphorus	P	30.97				3			No	2.20	
Sulfur	S	32.06			2				No	2.07	
Chlorine	Cl	35.45		1					No		0.00321
Argon	Ar	39.94	0						No		0.00178
Potassium	K	39.10		1				Yes		0.87	
Calcium	Ca	40.08			2			Yes		1.55	
Gallium	Ga	69.72				3		Yes		5.91	
Arsenic	As	74.92				3			No	5.73	
Selenium	Se	78.96			2				No	4.8	
Bromine	Br	79.90		1					No	3.12	
Krypton	K	83.30	0						No		0.00307

The real value of the discovery of order in nature is the power it gives us to make predictions. Consider the problem of the missing element. Several predictions can be made from Table 2.5:

1. If in fact the missing element exists, it should combine with four hydrogens.
2. Its atomic weight should be between that of gallium (69.72) and arsenic (74.92); an estimate of 72.3 is half way between.
3. Look at the unusual order in the electrical conductivity column. The first two elements in Group 2 conduct electricity. The first three elements in Group 3 conduct electricity. If this trend is followed, Group 4 should have four elements that conduct electricity. The missing element would conduct electricity and would therefore be a metal.
4. A regular downward trend in density is apparent from gallium to bromine. A density of about 5.8 g/mL would put the missing element between gallium and arsenic.

Before we look for this element, let's first consider the origins of the periodic table.

TABLE 2.5
Selected Elements Arranged by Increasing Atomic Weight and Divided into Groups Comprising Complete Cycles of Chemical Reactivity

			Number of H Atoms It Will Combine With					Conducts Electricity?		Density (g/mL)	
Element	Symbol	Atomic Weight	0	1	2	3	4	Yes	No	High	Low
Group 1											
Helium	He	4.00	0						No		0.00018
Group 2											
Lithium	Li	6.93		1				Yes		0.534	
Beryllium	Be	9.01			2			Yes		1.8	
Boron	B	10.81				3			No	3.33	
Carbon	C	12.01					4		No	2.25	
Nitrogen	N	14.00				3			No		0.00125
Oxygen	O	16.00			2				No		0.00142
Fluorine	F	18.99		1					No		0.00169
Neon	Ne	20.18	0						No		0.00089
Group 3											
Sodium	Na	22.98		1				Yes		0.971	
Magnesium	Mg	24.31			2			Yes		1.74	
Aluminum	Al	26.98				3		Yes		2.70	
Silicon	Si	28.09					4		No	2.42	
Phosphorus	P	30.97				3			No	2.20	
Sulfur	S	32.06			2				No	2.07	
Chlorine	Cl	35.45		1					No		0.00321
Argon	Ar	39.94	0						No		0.00178
Group 4											
Potassium	K	39.10		1				Yes		0.87	
Calcium	Ca	40.08			2			Yes		1.55	
Gallium	Ga	69.72				3		Yes		5.91	
Missing element		*(—)*					*(4)*	*(—)*		*(—)*	
Arsenic	As	74.92				3			No	5.73	
Selenium	Se	78.96			2				No	4.8	
Bromine	Br	79.90		1					No	3.12	
Krypton	K	83.30	0						No		0.00307

Periodic Table of the Elements

A Russian chemist named Dmitri Mendeleev spent much of his life attempting to find order in the behavior of the elements. Cards carrying the names and properties of the 63 elements known at that time were pinned to the walls of his laboratory, to be sorted and arranged over and over again until they finally were grouped in families of similar chemical behavior.

Conduct electricity			Do not conduct electricity					
Number of hydrogens	1	2	3	4	3	2	1	0
Group 1	——— Increasing atomic mass ———→							4 **He**
Group 2	6.9 **Li**	9.0 **Be**	10.8 **B**	12.0 **C**	14.0 **N**	16.0 **O**	19.0 **F**	20.2 **Ne**
Group 3	22.98 **Na**	24.31 **Mg**	26.98 **Al**	28.09 **Si**	30.97 **P**	32.06 **S**	35.45 **Cl**	39.94 **Ar**
Group 4	39.1 **K**	40.08 **Ca**	69.72 **Ga**	?	74.92 **As**	78.96 **Se**	79.90 **Br**	83.3 **Kr**

FIGURE 2.9
This table arranges the elements in horizontal groups or periods, each of which comprises a complete cycle of chemical behavior. Note the missing element below silicon. Will it be a metal or nonmetal? What is its approximate atomic mass?

Although Mendeleev dealt with 63 elements and we have listed properties of only 24 here, we can trace some of his logic by constructing a simple table that will group our elements in families of similar behavior. There are three properties with which we can easily work: (1) electrical conductivity, (2) number of hydrogens with which the element will combine, and (3) atomic weight (or mass).

Metals are located on the left-hand side of the periodic table.

The table presented in Figure 2.9 reflects the properties listed above. First, the table is broken horizontally into two parts: those elements that conduct electricity and those that do not (the stepped blue line divides them). Second, it is arranged so that one complete group of elements fits on a horizontal row in order of the number of hydrogens with which it will react (from one to four and back down to zero). Third, with the exception of argon and potassium, the elements are arranged in order of increasing atomic weight from upper left to lower right.

This **periodic table of the elements** is an extremely powerful tool. It organizes the elements into *periods*, or horizontal groups comprising one cycle of chemical behavior. Not only does it order the elements in terms of a predictable change in their properties, each vertical column forms a "chemical family" of similar chemical behavior. For example, all of the elements of the first vertical row are reactive metals that behave in a similar manner. Each reacts with one chlorine atom to form products analogous to ordinary table salt, sodium chloride (NaCl). These compounds are called "salts" and include LiCl, NaCl, and KCl. Potassium chloride (KCl) is often used as a salt substitute by people who must restrict their sodium intake. The inert gases are all grouped vertically in a family in the right-hand column of the table. Elements in the oxygen family—O, S, and Se—react in a similar fashion with hydrogen, forming the compounds H_2O, H_2S, and H_2Se. The table permits us to make some intelligent predictions about the behavior of otherwise unknown elements, given some knowledge of the behavior of one member of each vertical family of elements.

Elements of similar chemical behavior are found in vertical columns in the periodic table.

By now you have noticed the missing element. So did Dmitri Mendeleev. From his periodic table of the elements, Mendeleev in 1869 predicted the existence of a yet unknown element he called "eka-silicon" (Es). He predicted that this unknown element would have an atomic weight of about 72, would be a gray metal with a density of about 5.5, and would react with chlorine to form a compound of the form $EsCl_4$. In 1886, the German chemist Winkler

TABLE 2.6
Comparison of Mendeleev's Predicted "Eka-Silicon" and Winkler's Observed Germanium

	Mendeleev's Eka-Silicon (Predicted)	Winkler's Germanium (Observed)
Metallic character	Metal	Metal
Color	Gray	Gray-white
Density	5.5 g/cm^3	5.36 g/cm^3
Electrical conductivity	Yes	Yes
Atomic weight	72.3	72.6
Compound with Cl	$EsCl_4$	$GeCl_4$

isolated a new element that he called Germanium (Ge). The observed properties of this element, together with Mendeleev's predictions, are shown in Table 2.6. They are obviously one and the same.

To the chemist, the beauty of Mendeleev's table lies in the fact that it has a place for every element, both known and undiscovered. To the student, it gives the ability to predict the behavior of any element, given only knowledge of the physical or chemical properties of nearby elements. Most important is the similarity in chemical behavior within vertical "families" of elements.

Several elements were discovered after their existence was predicted by the periodic table.

The Transition Elements, Lanthanides, and Actinides

A problem that bothered Mendeleev was that, as the atomic weight increased, the horizontal rows or periods held progressively more elements. Mendeleev attempted to handle the problem by folding back some of the rows to place more than one element in a given block. The problem has been solved by simply widening the periodic table to allow insertion of the additional elements. Figure 2.10 presents the modern version of the periodic table. Note the differences and similarities between this table and the table in Figure 2.9. Not only are all the elements now presented, but three new groups have been added. First, three horizontal rows or periods have been widened to include 10 additional elements immediately following the beryllium-magnesium-calcium-strontium-barium-radium family. These 10 elements are called the *transition metals* because they provide a transition between two columns of metallic elements on the left and the remaining six columns of elements (many of which are nonmetals) on the right. As we will see later, the transition metals share an arrangement of their outer electrons not found with the elements in the two left-hand or six right-hand columns.

The transition elements, which are all metals, fit between two columns of metallic elements on the left and six columns of principally nonmetallic elements on the right.

A similar increase in period length occurs after lanthanum (atomic mass 138.9) with the insertion of a 14-member *lanthanide* or *rare earth* group, and after actinium (atomic mass 227) with the insertion of a 14-member *actinide* group. The lanthanides and actinides share an organization of their outer electrons in which up to 14 of their electrons are held in a special set of orbits. All of the actinide elements are radioactive; only three of them occur naturally. The remaining 11 have been artificially synthesized in nuclear reactors, cyclotrons, and nuclear explosions.

Notice that, as in Figure 2.9, the elements in Figure 2.10 are listed generally in order of increasing atomic weight and in vertical groups of similar

Transition Metals

1 H 1.00797																1 H 1.00797	2 He 4.0026
3 Li 6.939	4 Be 9.0122											5 B 10.811	6 C 12.01115	7 N 14.0867	8 O 15.9994	9 F 18.9984	10 Ne 20.183
11 Na 22.9890	12 Mg 24.312											13 Al 26.9815	14 Si 28.086	15 P 30.9738	16 S 32.064	17 Cl 35.453	18 Ar 39.948
19 K 39.102	20 Ca 40.08	21 Sc 44.956	22 Ti 47.90	23 V 50.942	24 Cr 51.996	25 Mn 54.9380	26 Fe 55.847	27 Co 58.9332	28 Ni 58.71	29 Cu 63.54	30 Zn 65.37	31 Ga 69.72	32 Ge 72.59	33 As 74.9216	34 Se 78.96	35 Br 79.909	36 Kr 83.80
37 Rb 85.47	38 Sr 87.62	39 Y 88.905	40 Zr 91.22	41 Nb 92.906	42 Mo 95.94	43 Tc (99)	44 Ru 101.07	45 Rh 102.905	46 Pd 106.4	47 Ag 107.870	48 Cd 112.40	49 In 114.82	50 Sn 118.69	51 Sb 121.75	52 Te 127.60	53 I 126.9044	54 Xe 131.30
55 Cs 132.905	56 Ba 137.34	57 *La 138.91	72 Hf 178.49	73 Ta 180.948	74 W 183.85	75 Re 186.2	76 Os 190.2	77 Ir 192.2	78 Pt 195.09	79 Au 196.967	80 Hg 200.59	81 Tl 204.37	82 Pb 207.19	83 Bi 208.900	84 Po (210)	85 At (210)	86 Rn (222)
87 Fr (223)	88 Ra (226)	89 **Ac (227)															

* Lanthanum Series

58 Ce 140.12	59 Pr 140.907	60 Nd 144.24	61 Pm (147)	62 Sm 150.35	63 Eu 151.96	64 Gd 157.25	65 Tb 158.924	66 Dy 162.50	67 Ho 164.930	68 Er 167.26	69 Tm 168.934	70 Yb 173.04	71 Lu 174.97

** Actinium Series

90 Th 232.038	91 Pa (231)	92 U 238.03	93 Np (237)	94 Pu (242)	95 Am (243)	96 Cm (247)	97 Bk (247)	98 Cf (249)	99 Es (254)	100 Fm (253)	101 Md (256)	102 No (256)	103 Lr (257)

FIGURE 2.10
A simplified version of the modern periodic table of the elements. This table provides space for the additional elements that lengthen the later periods. The atomic number is given above each element symbol, and below each symbol is the atomic weight.

chemical behavior. Notice also that the elements in this periodic table have been numbered for easy reference. This numbering system begins with the lightest element, hydrogen, as number 1 and progresses in unit steps to lawrencium, the heaviest element listed, as number 103. This reference number, called the *atomic number*, gained an important significance with the discovery of the proton, which is discussed in the next chapter.

The atomic number shows an element's location in the periodic table.

NEW TERMS

metals A group of elements that are good electrical conductors and are malleable.

nonmetals All elements that do not conduct electricity.

chemical periodicity A cyclic or periodic repeating of chemical and physical properties with increasing atomic weight caused by the arrangement of electrons around atoms.

periodic table of the elements A table in which elements are arranged generally in order of increasing atomic mass. The elements of each horizontal row comprise one period or cycle of chemical and physical properties, and the vertical columns comprise elemental families of like chemical behavior.

TESTING YOURSELF

Chemical Periodicity

1. By reference to the periodic table presented in Figure 2.10, which element do you think would react in a manner most similar to oxygen: magnesium, sulfur, phosphorus, or argon?
2. Sodium combines with chlorine to form salt, NaCl. What would you expect the formula of a compound of potassium and chlorine to be?
3. Helium is chemically nonreactive—it will not form compounds with any other compound. Would you expect neon (Ne) to be chemically reactive?

Answers **1.** sulfur, S **2.** KCl **3.** no

SUMMARY

This chapter dealt with some basic principles of chemistry, and with some of the experimental indications that there indeed must exist a far-reaching order in the structure of atoms.

Matter is a term used to describe the materials that make up the universe. Anything that has mass and occupies space is matter. Matter may be characterized by describing its physical properties—those properties that can be measured without changing the chemical composition of the material. Matter exists in three possible states—*solid*, *liquid*, and *gas*. The order of the component atoms and molecules in matter is highest in solids and lowest in gases. Energy must be added to move from solid to liquid and from liquid to gas. The energy required to change one gram of a material from solid to liquid is called the *heat of fusion*, and the energy required to change one gram of a material from liquid to gas is called the *heat of vaporization*. Physical change involves a change between these states, while chemical change involves a change in the atomic composition of the material.

Based on its composition, matter can be classified as *mixtures*, whose components are found in any ratio; as *compounds*, whose components are found in fixed ratios; or as *elements*, in which all atoms are of the same type.

When elements are grouped according to common physical properties and chemical behavior, "families of elements" appear. These family characteristics seem to repeat in an orderly fashion as the atomic weight (or mass) of the elements increases. The *periodic table of the elements* illustrates these family similarities and provides the unifying theme for all of chemistry. However, the idea of *chemical periodicity* introduced at the end of this chapter is based on observation. In subsequent chapters we will develop a rationale for this behavior: an atom with a predictable arrangement of electrons that determines its chemical properties.

TERMS

matter
solid
liquid
gas
melting point
boiling point
heat of fusion
heat of vaporization
physical change
chemical change
law of conservation of matter
mixture
compound
element
law of definite proportions
atom
molecule
electrolysis
relative atomic weight
metals
nonmetals
chemical periodicity
periodic table of the elements

CHAPTER REVIEW

QUESTIONS

1. A mass of 0.5 g of steam at 100 °C deposited in your hand hurts much more than the same mass of water at 100 °C. This is related to which of the following?
   a. heat of fusion
   b. specific heat of water
   c. heat of vaporization
   d. specific heat of steam
2. In the northern United States snow often remains in the mountains until July, even when daytime temperatures are well above freezing (see photo on p. 45). Which of the following properties of water is related to this occurrence?
   a. heat of fusion
   b. specific heat of water
   c. heat of vaporization
   d. specific heat of steam
3. A small snowman has a mass of 200 kg. Compute the number of calories of heat required to melt the snowman, assuming the temperature of the snow to be 0 °C.
4. Which type of change usually involves the most energy?
   a. chemical change b. physical change
5. Ice cream would be considered a
   a. mixture b. compound c. element
6. Natural gas is principally the compound methane, CH_4. When burned in a furnace or hot water heater, one methane molecule will react with two oxygen molecules to produce one molecule of carbon dioxide, one molecule of water (steam), and about 13,180 cal/g of methane reacting. This reaction is described by the chemical equation

   $$CH_4 + 2\,O_2 \longrightarrow CO_2 + 2\,H_2O + \text{Heat}$$

   How many liters of water could be heated from 5 °C to 50 °C by combustion to 10 g of natural gas (assuming 100% efficiency of the water heater)?

7. Hydrogen sulfide, H_2S, is a gas produced as eggs spoil. It has been found to consist of twice as many hydrogen atoms as sulfur atoms. If it is 5.88% hydrogen and 94.12% sulfur by weight, what is the relative atomic weight of sulfur?
8. Hydrogen reacts with oxygen to form the compound water, H_2O, and with sulfur to form the compound H_2S. By reference to the periodic table in Figure 2.10, how many hydrogen atoms would you expect to react with an atom of selenium, Se?
9. The physical properties of each family of elements generally change smoothly as one moves up or down in the periodic table. Given the following data, what would be a reasonable estimate of the melting point of the element potassium (K)?

Element	Melting point (°C)
Lithium (Li)	180.5
Sodium (Na)	97.8
Rubidium (Rb)	38.9
Cesium (Cs)	28.7

(HINT: Locate these elements in the periodic table.)

10. Iodine (I) is essential for proper operation of the thyroid gland. Sodium iodide is often added to table salt to ensure sufficient iodine in the diet. By reference to the periodic table, how many iodine atoms probably combine with one sodium atom to form sodium iodide?

Answers **1.** c **2.** a **3.** 1.6×10^7 **4.** a **5.** a **6.** 2900 g or 2.9 L **7.** 32.01 **8.** 2 **9.** 65 °C **10.** 1

DIAGNOSTIC CHART

Blacken in all of the circles under the number of each question you missed in the Chapter Review questions. The diagnostic chart will help you identify concept areas that need more study.

	Questions									
Concepts	1	2	3	4	5	6	7	8	9	10
Change of state	○	○	○			○				
Chemical and physical change				○		○				
Mixtures, compounds, and elements					○					
Atomic weight							○			
Chemical periodicity								○	○	○

EXERCISES

Change of State

1. You just filled an ice cube tray with water and put it in the freezer. Before ice cubes can form, the temperature of the water must drop from room temperature to ice cold. Compute the number of calories of heat that must be removed from 1 L of water to change its temperature from 20 °C to 0 °C.
2. For those same ice cubes, now compute the number of calories of heat that must be removed from 1 L of water at 0 °C to convert it to ice at 0 °C.

3. How many calories of heat will be required to melt 946 g of ice?

4. Consider 10.0 g of snow at 0 °C. How many calories of heat energy must the sun provide to melt this snow and warm it to 20 °C? Use these data:

$$\text{Heat of fusion of water} = 80 \text{ cal/g}$$
$$\text{Specific heat of water} = 1.00 \text{ cal/g °C.}$$
$$\text{Heat of vaporization of water} = 540 \text{ cal/g}$$

5. How many calories of heat energy must be provided to melt the snow in exercise 4 and vaporize it?

6. Suppose that 0.1 g of steam at 100 °C struck your hand. How many calories of heat would be given off as it condensed into water and cooled to body temperature?

7. Consider 10.0 g of snow at 0 °C. How much energy must the sun supply to melt and evaporate this material? (HINT: there are three processes it must go through:

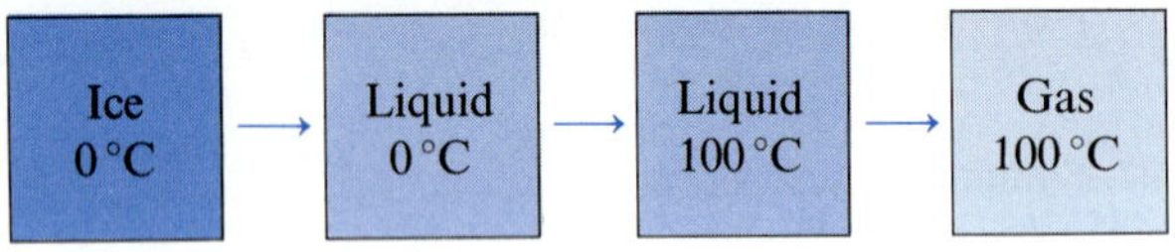
Ice
0 °C
Liquid
0 °C
Liquid
100 °C
Gas
100 °C

8. Explain why it is common in the spring to see snow on the ground for a few days while the sun is out and the air temperature is warm.

Chemical and Physical Change

9. Classify the following common phenomena as physical or chemical change.
   a. Burning leaves
   b. Melting chocolate
   c. Drying dishes in a dishwasher
   d. Dissolving sugar in water
   e. Starting a gas lawn mower
   f. Sunshine melting snow
   g. Eating a candy bar
   h. Burning a candle
   i. Making toast

10. When burned, propane combines with oxygen according to this chemical equation:

$$C_3H_8 + 5\,O_2 \longrightarrow 3\,CO_2 + 4\,H_2O + \text{Heat}$$

As a result, 16,430 cal of heat are produced for each gram of propane burned. How many grams of ice will 0.5 g of propane melt at 0 °C?

11. Two reactions actually occur in exercise 3 about the snowman:

$$\text{Ice} + \text{Heat} \longrightarrow \text{Liquid}$$

and

$$C_3H_8 + 5\,O_2 \longrightarrow 3\,CO_2 + 4\,H_2O + \text{Heat}$$

Which of these reactions is easily reversible?

12. Which type of reaction involves the greatest change in energy per gram of reactant involved?
   a. chemical change
   b. physical change

Mixtures, Compounds, and Elements

13. Which of the following has a *fixed* proportion of different types of atoms?
   a. mixture b. compound c. element

14. A material is found always to be 52% carbon, 13% hydrogen, and 35% oxygen. It is a
   a. mixture b. compound c. element

15. Classify the following materials as mixtures, compounds, or elements.
   a. Coffee with sugar in it
   b. Dried cocoa mix
   c. Salt
   d. Sugar
   e. Milk
   f. Gold ring
   g. Aluminum cookware
   h. Lemonade
   i. Silver dish
   j. Baking soda ($NaHCO_3$)

Atomic Weight

16. Bromine (Br) combines with hydrogen (H) to form a compound that has an equal number of bromine and hydrogen atoms. This compound is 98.76% Br and 1.24% H by weight. What is the relative atomic weight of bromine as compared to hydrogen?

17. Rubbing alcohol has the molecular composition C_3H_7OH. What is the weight of this molecule expressed in terms of the weight of hydrogen atoms?

18. Sugar (sucrose) has the molecular composition $C_{12}H_{22}O_{11}$. What is the weight of this molecule expressed in terms of the weight of hydrogen atoms?

19. Chlorine (Cl) combines with hydrogen to form a compound that has an equal number of chlorine and hydrogen atoms. This compound is 2.76% H and 97.23% Cl by weight. What is the relative atomic weight of chlorine as compared to hydrogen? Does the chlorine atom appear to be built of a whole number of hydrogen-sized building blocks?

Chemical Periodicity

The following problems may require reference to the periodic table in Figure 2.10.

20. Which of the following elements would behave chemically in a manner most similar to chlorine (Cl)?
   a. F b. Li c. N d. O e. Mg

21. Hydrogen forms compounds with elements of the seventh vertical column of the periodic table:

    HF, HCl, HBr, HAt.

    How many hydrogens do you predict would combine with the element iodine (I)?

22. Given the data below concerning boiling points of the fluorine (F) family, what is a reasonable estimate of the boiling point of bromine (Br).

Element	Boiling Point (°C)
F	−188.2
Cl	−34.7
Br	?
I	183

a. −200 °C b. −100 °C c. 0 °C d. 50 °C e. 150 °C

23. Given that one atom of lithium will combine with one atom of bromine, but that helium and neon will not combine to form compounds with any other atoms, with how many atoms of bromine would you expect krypton (Kr) to combine?

24. Given that magnesium (Mg) combines with oxygen to form the compound MgO, that calcium (Ca) combines with oxygen to form the compound CaO, and that strontium (Sr) combines with oxygen to form the compound SrO, predict the formula of a compound of barium (Ba) and oxygen.

25. One atom of nitrogen combines with three hydrogen atoms to form a molecule of ammonia, NH_3. Given this information, predict the formula of a compound of phosphorus (P) and hydrogen?

26. With how many magnesium atoms would you expect one atom of argon (Ar) to combine?

CHAPTER

3

The Structure of Atoms

Scientific understanding is gained from many well-designed experiments and from a few lucky accidents. This chapter describes the discovery of the structure of atoms and tells several stories that illustrate how important ideas are often deduced from relatively simple clues.

OBJECTIVES

After completing this chapter, you should be able to

- Describe the physical properties of an electron.
- Describe the "plum pudding" model of the atom.
- Differentiate between X-rays, alpha particles, beta particles, and gamma radiation.
- Describe the alpha particle scattering experiment of Rutherford, Geiger, and Marsden, and explain the idea of a nuclear atom that resulted from this experiment.
- Describe the physical properties of protons and neutrons.
- Sketch a reasonable model of the atom, and identify the various parts and the distances involved.
- Define isotope, and given both the atomic number and the atomic weight of an atom, compute the number of protons and neutrons in its nucleus.

3.1 Discovery of the Electron

The discovery of the nuclear atom is a story with two plots. It begins with some electrical discharge experiments performed in low pressure gases during the mid 1800s. From these experiments came the discovery of the electron and J. J. Thomson's "plum pudding" model of the atom. The second part of the story is about the discovery of natural radioactivity and the alpha particle scattering experiments that lead to the discovery of the atomic nucleus.

The conclusion, as is often the case in science, resulted in part from several lucky accidents. Scientific research still depends on people's ability to recognize the unexpected in everyday occurrences. As this chapter develops the nuclear model of the atom, it will show how important concepts are often deduced from relatively simple clues.

Experiments with the Crookes Tube

Until the turn of the last century, scientists believed in atoms but had little idea of how they were constructed. They had learned to determine the mass of atoms, and as we saw in the last chapter, this provided a unifying theme from which came the idea of chemical periodicity. Chemical reactions could be performed, and in some cases the products could be predicted with reasonable success. Chemistry was an *empirical* science, meaning that it was based on a large number of experimental observations. It still had no unifying principles that explained the behavior of all atoms. As we will see in this chapter, this situation changed drastically between about 1885 and 1925.

By the 1800s, the concept of **electrical charge** was well known. Apparently particles with positive charge and with negative charge both existed. For example, if one rubbed a glass rod with cat's fur or walked across a carpet on a dry day, an excess of one type of charged particle was accumulated. The object was therefore said to be "charged." Opposite electrical charges had been shown to attract each other. This was evidenced by the hair on a person's arm standing on end when a charged comb was brought near it or by a spark jumping from someone's hand to a door knob.

Electrical charges are of two types—positive and negative. Unlike electrical charges attract each other, like electrical charges repel.

Our first insight into the structure of atoms awaited a way of getting a few atoms isolated so that they could be observed. The development of a good vacuum pump in about 1885 set the stage for the observation of electrical charges moving through low pressure gases (Figure 3.1). Most of the air in a long glass tube was pumped out, forming a vacuum, and a generator that produced a high electrical charge was connected to metal electrodes

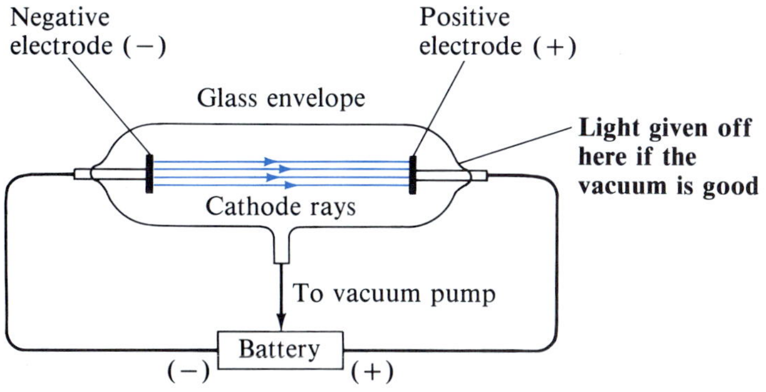
Negative
electrode (−)
Positive
electrode (+)
Glass envelope
Light given off
here if the
vacuum is good
Cathode rays
To vacuum pump
Battery
(−)
(+)

FIGURE 3.1
A gas discharge apparatus. When electrical charge is applied across the electrodes, something appears to move through the tube causing a glow to occur.

located inside the tube. When a negative charge was applied to one electrode and a positive charge to the other, a glow appeared in the tube.

What one observed was dependent upon how good the vacuum was inside the tube. If a small amount of gas were present, a faint gaseous glow was observed throughout the tube. However, if the vacuum was very good, the glow would fade out and the glass directly behind the positive electrode would give off a faint beam of green light. This light turned out to be both the key to discovery of the electron and an incorrect clue that lead to the discovery of X-rays and accidentally to the discovery of natural radioactivity.

Electrons were discovered by observing electrical discharges in a vacuum.

William Crookes devised a modification of the apparatus in Figure 3.1 that consisted of a tube with a cross-shaped metal target attached to the positive battery terminal. One form of Crookes's tube is illustrated in Figure 3.2. When the air was removed from the tube, a negative charge was applied to the right-hand electrode and a positive charge to the cross-shaped metal target. The glass behind the positive target would then give off a faint green light with a shadow of the target superimposed on it. It appeared that whatever rays had cast this shadow were moving from the negative electrode (called the *cathode*) toward the positive electrode (called the *anode*).

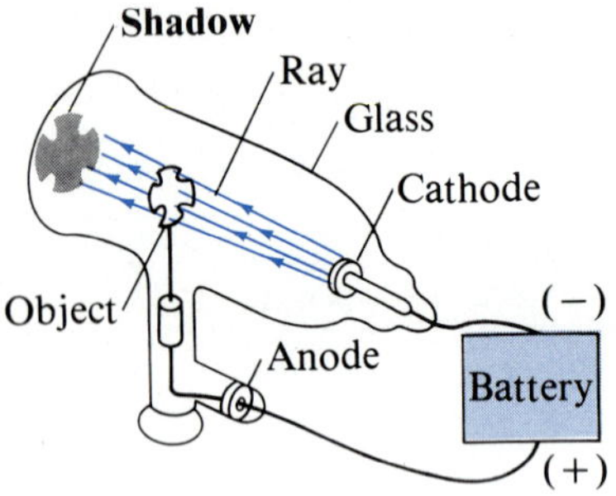
Shadow
Ray
Glass
Cathode
Object
(−)
Anode
Battery
(+)

FIGURE 3.2
In Crookes's experiment, a shadow of the target electrode was cast on the end of the tube by rays apparently emanating from the negative electrode (cathode). Since the rays appear to move from negative to positive, they must carry a negative charge.

What conclusions could be drawn from this experiment?

1. Something was moving through the tube to strike the glass to make it glow.
2. Whatever was moving appeared to move from the cathode (negative electrode) toward the anode (positive electrode) since the glow occured only behind the positive electrode.
3. Since unlike electrical charges attract each other, the moving rays must have been carrying a negative charge.
4. The negative particles were very tiny, since the presence of only a few air molecules impeded their flow through the tube. The glass glowed only when practically all of the air was removed from the tube.
5. The negative particles appeared to originate from the metal of the cathode. They may have come from the atoms that made up this metal.

Cathode rays move from the negative to the positive electrode. They must carry negative charge.

Since the rays originated at the negative electrode of the tube, or the cathode, and traveled in a straight line, they were called **cathode rays.** This term is still with us: TV tubes and computer displays are often called "CRTs" or cathode ray tubes.

Discovery of Negative Particles

About 22 years later, J. J. Thomson, working at Cambridge University in England, constructed a modified Crookes tube with two metal plates placed above and below the beam of cathode rays. Thomson's apparatus, somewhat similar in operation to a modern TV tube, is shown in Figure 3.3. When Thomson applied an electrical charge to the two plates, he was able to deflect the beam of cathode rays. A positive charge applied to the upper plate moved the beam up, a negative charge moved it down. Since unlike electrical charges attract each other, this experiment confirmed the earlier belief that cathode rays were negative.

By conducting similar experiments with different metal electrodes and other gases present in the tube, Thomson was able to show that the properties of the cathode ray were *independent of the nature of the material* used for the cathode in the Crookes tube. Cathode rays appeared to be fundamental chunks of negative matter that were possessed by all different kinds of atoms.

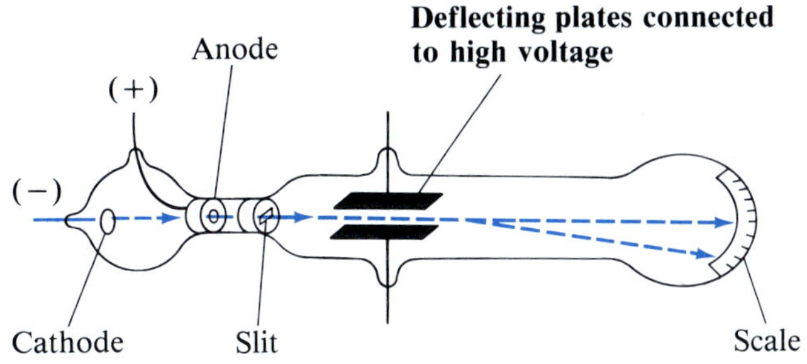
Deflecting plates connected to high voltage
Anode
(+)
(−)
Cathode
Slit
Scale

FIGURE 3.3
Using a modified Crookes tube similar to that shown here, Thomson was able to demonstrate that application of an electrical charge to the two deflecting plates would move the point of impact of the cathode ray beam up or down at the end of the tube. A positive charge applied to the upper plate moved the beam up, while a negative charge moved it down.

These fundamental chunks of negative matter were given the name **electrons** and were assigned a charge of −1, equal in magnitude to a positively charged hydrogen atom, only negative. Then in 1909 the American physicist Robert Millikan determined that electrons were actually negative particles with a mass of about 1/1837 that of a hydrogen atom, the lightest atom. Even on an atomic scale, these particles were tiny indeed.

Cathode rays are actually a stream of tiny negative particles called electrons. Electrons are 1/1837 as massive as a hydrogen atom.

NEW TERMS

electrical charge The presence of excess positive or negative particles in or on an object. Objects with opposite electrical charge attract each other, while objects of the same electrical charge repel each other.

cathode ray Particles of negative electricity traveling through a vacuum tube, from cathode to anode. We know cathode rays today as the beam that sweeps across the face of a television tube to produce the picture.

electron A light, negative particle that can be relatively easily removed from an atom. The mass of an electron is 1/1837 that of a hydrogen atom and its charge is −1.

TESTING YOURSELF

Cathode Rays and the Discovery of Electrons

Choose the correct answer in each set of parentheses.

1. A positive electrical charge will (attract, repel) a like electrical charge.
2. Cathode rays leave the (positive, negative) terminal in a cathode ray tube.
3. Since cathode rays move from negative to positive, they must carry a (positive, negative) electrical charge.
4. Since the movement of cathode rays is hindered by the presence of air molecules, cathode rays are probably (lighter, heavier) than air molecules.

Answers **1.** repel **2.** negative **3.** negative **4.** lighter

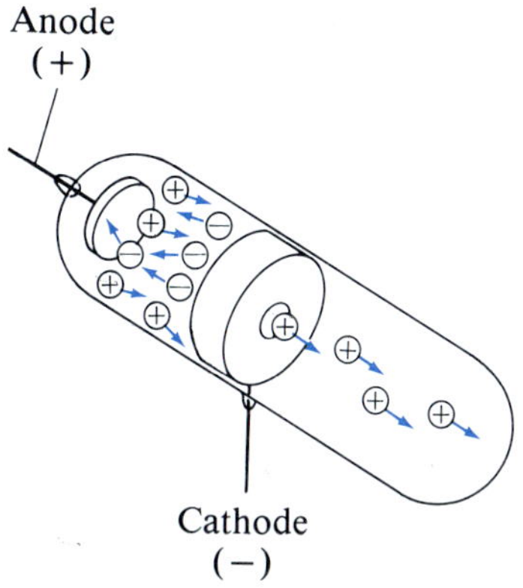
Anode
(+)
Cathode
(−)

FIGURE 3.4
Canal rays are observed only when a small amount of gas is present in the discharge tube, and they move in a direction opposite to that of the cathode rays (electrons). Canal rays appear to consist of positively charged particles left over after electrons are removed from gas molecules.

3.2 A Plum Pudding Model of the Atom

In 1886 the German physicist Goldstein discovered that if a small hole or "canal" was drilled into the cathode of a Crookes tube (Figure 3.4), another type of ray could be observed. A glowing stream would move through the canal in the *opposite* direction to the motion of the cathode rays. Goldstein found that these **canal rays** were observed only when some gas was still in the tube. When the gas pressure was further reduced, they simply disappeared.

He thus concluded that canal rays must be formed from the atoms of gas. In 1898 it was established that canal rays carried a charge opposite that of cathode rays and were relatively heavy. Canal rays were positive particles. Unlike cathode rays, the mass of the canal ray depended on the identity of the gas in the discharge tube. Measurements were made with hydrogen, helium, oxygen, and other gases. In each experiment, the mass of the particles comprising the canal ray was in agreement with the known value for that element.

Canal rays are the positive part of the atom that is left when electrons are removed.

This evidence suggested that atoms, which had been thought to be indivisible, must be composed of small negative particles (electrons) and a larger part that carried a positive charge (+1) and most of the mass of the atom (see Section 3.5).

Thomson then proposed that atoms must be composed of very small negative particles floating in a heavy sea of positive charge. He described the atom as similar to an English plum pudding—a filling of heavy positive charge with small "plums" of negative charge located here and there throughout the atom. The positive charge of the filling is balanced by the negative charge of the electrons to give an overall neutral charge to the atom. Today we might liken it to chocolate chip ice cream—a filling of vanilla ice cream with chocolate chip "electrons" distributed throughout. For 13 years, until the discovery of an atomic bullet with which to probe the atom, the **plum pudding model** illustrated in Figure 3.5 was science's best view of the atom.

J. J. Thomson proposed that atoms are like plum pudding—a thick, uniform, and heavy positive filling with tiny electron "plums" scattered throughout.

NEW TERMS

canal ray A stream of positive particles; neutral atoms from which one or more electrons have been removed.

plum pudding model A model of the atom representing it as a positive filling in which negative electrons ("plums") are floating. The design of the atomic model was similar to that of plum pudding popular in England, hence the name.

TESTING YOURSELF

Plum Pudding Model

1. When an electron is removed from a neutral atom, does the remaining atom carry a negative or positive electrical charge?
2. Based on his observations of moving positively charged atoms (canal rays), Thomson proposed a plum pudding model in which the plums were likened to what negative particles?
3. The "plums" were floating in a sea of which charge, positive or negative?

Answers **1.** positive **2.** electrons **3.** positive

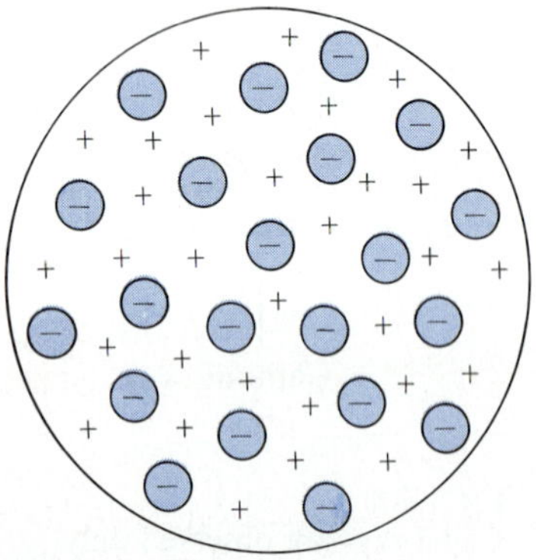

FIGURE 3.5
Thomson's plum pudding model of the atom involved a sea of heavy positive charge with small negative "plums" (electrons) floating in it.

3.3 Discovery of X-Rays

Natural radioactivity ultimately provided the atom-sized probe to explore the plum pudding atom. The discovery of natural radioactivity first began with the interest of a German physicist, Wilhelm Roentgen, in the ability of cathode rays to cause fluorescence in glass. **Fluorescence** is a soft glow of light given off when a substance is exposed to certain kinds of radiation. For example, white cloth and teeth fluoresce with a blue-white light when struck by ultraviolet light. As we have seen, the glass in cathode ray tubes fluoresces with a soft green glow when struck by cathode rays.

To observe the faint fluorescence that is produced by the cathode rays, Roentgen had darkened his laboratory. He was working with such a tube when a flash of light that did not come from the tube caught his eye. Quite a distance from the discharge tube was a glowing sheet of paper treated with a fluorescent material. When the discharge tube was turned off, the glow ceased. When Roentgen brought the paper closer to the tube, the rays that caused the paper to glow appeared to originate from the fluorescent spot in the glass. Roentgen concluded that the fluorescence created a form of radiation that could pass through the glass of the tube and strike materials outside. In fact, he found that if he took the coated paper into the next room, it still glowed whenever the cathode rays were in action. Apparently the radiation given off by the cathode ray tube would even penetrate walls. Roentgen's experiment is illustrated in Figure 3.6.

X-rays will penetrate most materials, and they cause some materials to fluoresce. They were used for medical purposes soon after their discovery.

Because he did not know what it was, Roentgen called this radiation **X-rays**, a name still used today. He found that X-rays would expose photographic plates and that objects placed between the cathode ray tube and a photographic plate would cast shadows of their internal structure on the plate. This occurred because some materials are "transparent" to X-rays (such as plaster and skin) and some are not (such as steel and bones). Roentgen soon toured Europe, with his wife serving as a subject to show the use of X-rays to take pictures of bones. The discovery of X-rays probably had a more immediate effect on medicine than any other discovery in the history of mankind. It also provided the spark that ultimately led to the discovery of natural radioactivity. Unfortunately, the health hazards encountered by those who were exposed to this radiation were not yet known.

We now know that X-rays are formed when high speed electrons are suddenly stopped. Medical and industrial X-ray machines use a high voltage to accelerate electrons, which then strike a solid metallic target, causing emission of X-rays. These X-rays can be focused to a degree by the shape of the target, and the X-ray beam can be limited in direction by placing a lead shield around the X-ray tube. The field of view of a modern X-ray machine is carefully limited to prevent X-ray exposure to patients and technicians beyond the desired area.

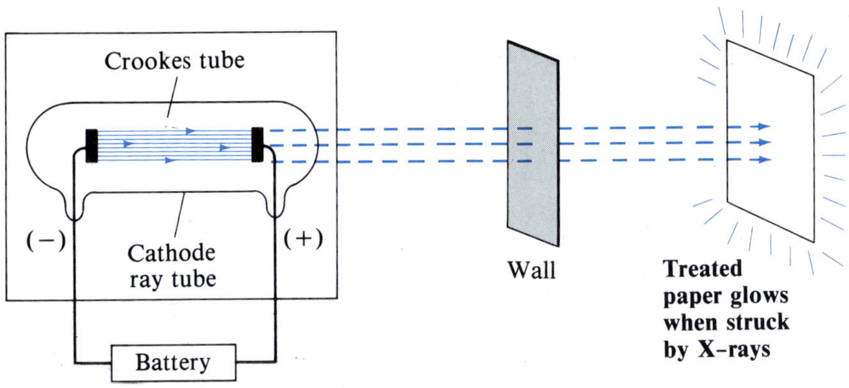
Crookes tube
(−)
Cathode
ray tube
(+)
Battery
Wall
Treated
paper glows
when struck
by X-rays

FIGURE 3.6
Roentgen discovered X-rays somewhat by accident when the high-voltage Crookes tube he was working with caused a paper treated with fluorescent material to glow. X-rays were even able to penetrate the wall of his laboratory.

NEW TERMS

fluorescence The emission of light when an object is struck by another form of light, such as ultraviolet or "black" light.

X-ray A highly penetrating ray produced as electrons slow up when they strike a material. Medical X-rays are produced today in a manner similar to that used by Roentgen.

3.4 Discovery of Natural Radioactivity

Henri Becquerel, a Professor of Physics at the Museum of Natural History in Paris, was interested by the report that X-rays appeared to orginate from the fluorescent spot on the wall of the cathode ray tube. Becquerel had investigated the fluorescence produced by light acting on chemical compounds. He reasoned that, if the fluorescence caused by cathode rays striking glass caused X-rays, then X-rays might also be produced by other forms of fluorescent glow.

Natural Radioactivity: A Lucky Accident

Ultraviolet light will also cause some materials to fluoresce.

To test this hypothesis, Becquerel chose as his fluorescent material some crystals of potassium uranyl sulfate, a complicated compound of potassium, uranium, sulfur, and oxygen. He knew from previous experiments that these crystals would fluoresce with a soft green glow when exposed to ultraviolet light. To detect the penetrating X-rays he hoped would be emitted, he prepared a photographic plate and wrapped it in heavy black paper to protect it from ordinary light. For a source of ultraviolet radiation, he chose sunlight, and he set the photographic plate outside his window with the crystals lying on the black paper wrapping. After several hours, he developed the photographic plate and was pleased to see grayish smudges appear on the plates wherever a crystal had lain (Figure 3.7). It appeared that sunlight was able to stimulate this crystal to fluoresce and, as it fluoresced, to give off penetrating radiation.

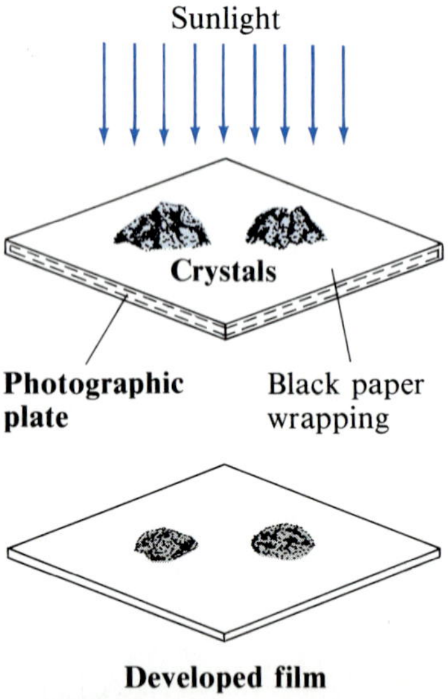
Sunlight
Crystals
Photographic plate
Black paper wrapping
Developed film

FIGURE 3.7
Becquerel attempted to stimulate the release of X-rays from the fluorescent material potassium uranyl sulfate by exposing it to the ultraviolet radiation of the sun. Radiation released from the crystal penetrated the black paper and exposed the photographic plate.

Becquerel continued his experiments, placing coins and other objects between the crystals and the photographic plate, and found that these objects cast shadows on the plate just as bones cast shadows using X-rays. He then placed a thin sheet of glass between the crystals and the photographic plate to make sure that the sun's heat had not driven some vapor from the crystal and through the paper to react with the photographic plate. Again the plate was darkened, as if the glass were not there. Becquerel concluded that the penetrating rays were produced by fluorescence, and he reported his discovery at a meeting of the French Academy of Sciences.

In the three days following Becquerel's report, the weather changed. When the photographic plates for Wednesday's experiments were ready, clouds came over the sun. Becquerel placed photographic plates, black paper, and crystals in a drawer to wait for sunny weather. There they lay for four days in the dark, where he was sure nothing would happen to them, since potassium uranyl sulfate would only glow when struck by ultraviolet radiation.

On Sunday Becquerel came to his laboratory, and because photographic emulsion used then did not last very long, he pulled the unused plates from his drawer and developed them to clean the plate for reuse. To his amazement, what he saw were blacker patches than he had ever observed before. Even without a source of ultraviolet radiation, the crystals seemed able to send out their rays.

Becquerel was perplexed at this observation. He thought that he had proved that the ultraviolet light from the sun caused the crystals to fluoresce and release X-rays, but the X-rays seemed to be released by the potassium uranyl sulfate even without exposure to light. Could the crystals store up the light, giving off the radiation sometime later? He stored some crystals away in darkness to see how long it might take for their penetrating rays to fade. Whenever he tested them, even weeks later, the penetrating rays poured out in undiminished intensity. He tried other fluorescent compounds. Whenever the material contained uranium, the penetrating rays were present, but fluorescent compounds composed of other elements would not emit the penetrating radiation. He tried uranium compounds that were not fluorescent. Curiously enough, these compounds also emitted the penetrating rays. One thing gradually became apparent: whenever uranium was present in the compound tested, the penetrating radiation was observed. Clearly, uranium (U) was the source of the radiation.

Some materials (uranium, for example) apparently produce X-rays without fluorescing. These materials are naturally radioactive.

Becquerel discovered that a uranium ore called pitchblende gave a photographic effect much stronger than the amount of uranium present in the ore would indicate. He suspected that this ore contained another element of much greater radiant power than uranium. The Austrian government donated a ton of pitchblende, and for two years Marie and Pierre Curie helped Becquerel separate this material. In the summer of 1898 a new element was finally isolated. This material, hundreds of times more active than uranium, was named *polonium* (Po) in honor of Marie Curie's native country, Poland. By the end of 1898 another even stronger material had been isolated. This material, hundreds of thousands of times more active than uranium, was called *radium* (Ra). Uranium, polonium, and radium ore have **natural radioactivity**; they spontaneously emit penetrating radiation and at the same time are converted to different elements. Radiation from radium sterilized seeds, healed surface cancers, and killed microbes. For the discovery and characterization of radiation, the Curies shared in 1904 Nobel Prize in physics with Becquerel, the person who had started them on this search. In 1911 Marie Curie received a second Nobel Prize, this time in chemistry, for the discovery of radium. Marie Curie was the first woman to win a Nobel Prize, and one of the few people in history to win two.

Bullets for an Atomic Probe

Although the public interest in radium was principally related to its curious ability to glow in the dark and its tremendous curative power for cancer, this discovery ultimately provided the atomic probe needed to investigate Thomson's "plum pudding" atom. In 1899, Ernest Rutherford, a native of New Zealand who had worked with Thomson in England and was now at McGill University in Canada, discovered that radiation from polonium had both weakly and strongly penetrating components. The weakly penetrating particle he called an **alpha (α) particle** and the strongly penetrating particle a **beta (β) particle.**

Naturally radioactive materials emit three types of radiation: alpha, beta, and gamma.

By 1900 the Curies and Becquerel had shown that beta particles were negative and of the same mass as electrons. Alpha particles were subsequently

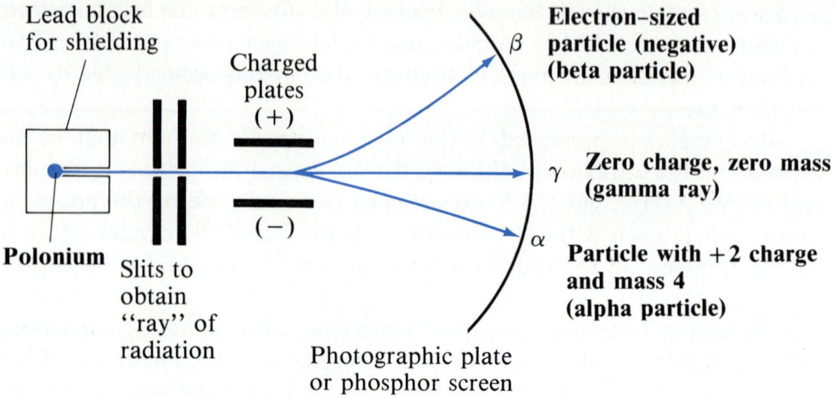
Lead block
for shielding
Charged
plates
(+)
(−)
Polonium
Slits to
obtain
"ray" of
radiation
Photographic plate
or phosphor screen
β
Electron-sized
particle (negative)
(beta particle)
γ
Zero charge, zero mass
(gamma ray)
α
Particle with +2 charge
and mass 4
(alpha particle)

FIGURE 3.8
Positively and negatively charged plates have an effect on alpha, beta, and gamma radiation. Alpha (α) particles (charge of +2) are deflected away from the positive electrode and beta (β) particles toward it. The uncharged gamma radiation is not deflected.

identified as helium atoms stripped of their electrons, having a charge of +2 and a mass four times that of hydrogen atoms (Figure 3.8). A third type of radiation was discovered. **Gamma (γ) radiation** appeared to carry neither electrical charge nor mass.

Positive alpha particles provide an "atomic gun" with which to probe the plum pudding atom.

The stage was now set to test Thomson's plum pudding model. At hand was a source of high speed, positively charged particles not much bigger than hydrogen atoms. As we shall see in the next section, perhaps these particles could be used to measure the distribution of positive charge in the plum pudding atom.

NEW TERMS

natural radioactivity The spontaneous emission of nuclear particles and penetrating radiation by naturally occuring unstable atoms. The radioactive atoms are changed to new elements during the process.

alpha (α) particles Relatively heavy particles having the mass of four hydrogen atoms that carry two positive charges and are fired out of some radioactive atoms.

beta (β) particles Negative, electron-sized particles fired out of a radioactive atom

gamma (γ) radiation Highly penetrating radiation made of particles with zero charge and zero mass that often accompany the emission of alpha and beta radiation by radioactive materials.

TESTING YOURSELF

Natural Radioactivity

1. The radiation Roentgen called X-rays differed from that observed by Becquerel in that it was produced by what?
2. Is this statement true or false: Becquerel found that penetrating radiation was produced by the fluoresence of certain materials.
3. Becquerel found that penetrating radiation was produced only by minerals that contained which element?
4. The Curies found what two elements in uranium ore that were very radioactive?
5. Three types of radiation were released by radioactive materials. One type was four times as heavy as a hydrogen atom and carried twice the charge of an electron, but positive. What was this particle?

6. A second type of radiation was electron-sized in mass and carried a single negative charge. What was this particle?
7. Gamma radiation carries what mass and what charge?

Answers **1.** cathode ray tube **2.** false **3.** uranium **4.** polonium and radium **5.** alpha **6.** beta **7.** zero, zero

3.5 Discovery of the Atomic Nucleus

Probing the Plum Pudding Atom

In 1907 Rutherford moved from McGill University in Canada to the University of Manchester in England. Here he met a young German, Hans Geiger, who had just received his Ph.D. from the University of Erlangen in Germany. Rutherford and Geiger turned their attention to a test of the plum pudding atom. The newly discovered alpha particles offered a chance to explore the uniform distribution of positive charge predicted by the plum pudding model because the alpha particle's path should be deflected as it passed a positive charge. Gold was chosen as the sample material because its atoms were large and carried a large positive charge (+79 hydrogen units). Also, gold, being a malleable metal, could be pounded into a thin sheet only a few thousand atoms thick.

Rutherford had found that alpha particles could be detected by observing the small flashes of light or **scintillations** they caused when they struck a small screen of zinc sulfide (ZnS). (The glowing material in a modern luminous dial wrist-watch is a tiny bit of radioactive substance mixed with zinc sulfide and painted onto the hands and numerals of the watch.) Geiger prepared a tube about 2 m long that could be pumped down to a very good vacuum (Figure 3.9). At one end he placed a small, cone-shaped glass tube with an alpha-emitting radioactive material in it, such as radium. In the middle of the tube was a slit to define the path of the alpha particles, and at the other end was a zinc sulfide screen that he could observe with a microscope. A thin sheet of gold foil could be tipped into the alpha particle path.

A scintillation is a flash of light caused by impact of an alpha particle on a fluorescent screen.

Geiger's microscope focused on a tiny part of the zinc sulfide scintillation screen and was mounted so that it could be run up and down along a millimeter scale. By looking through his microscope, Geiger could count the number of alpha particles striking at each position on the screen.

Geiger set up his apparatus and created a vacuum in the tube so that the alpha particles would have an unhindered path down the tube. A clear,

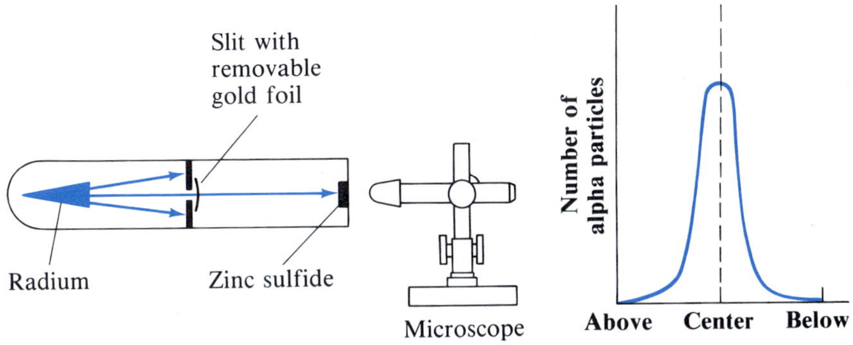
Slit with removable gold foil
Radium
Zinc sulfide
Microscope
Number of alpha particles
Above
Center
Below

FIGURE 3.9
Geiger's alpha particle apparatus. Data from the alpha particle scattering experiment is presented in the graph on the right. Very few alpha particles were deflected more than 1/2 degree as they passed through the gold foil.

sharply defined bar of light appeared on his screen. If he placed a thin sheet of gold foil across the slit, the bar blurred. Some of the alpha particles were slightly deflected as they passed through the gold atoms in the foil. Geiger plotted the rate at which alpha particles struck the screen in each position; the results of this experiment are plotted on the graph in Figure 3.9. He found that with nothing in the alpha particle beam, the beam was well defined and practically all of the alpha particles struck within an area defined by the slit. When a thin gold foil was placed across the slit, about two-thirds of the particles were deflected very slightly. Very few alpha particles, however, were deflected more than 1/2 degree.

Little deflection of alpha particles should occur if the atom's positive charge is evenly distributed. The plum pudding model seems reasonable.

These observations were consistent with what might be expected from Thomson's plum pudding atom. If the positive charge were evenly spread throughout the atom, little deflection should occur. It was somewhat like shooting a rifle bullet through a plastic bag full of jello. One would be quite surprised to see any significant deflection of the bullet. Thus, Rutherford and Geiger were not surprised with this result.

The Geiger-Marsden Experiment

Some months later Rutherford asked Geiger to work with an undergraduate student named Ernest Marsden and give him a chance at a little apprentice research. The job selected was to see if alpha particles were reflected from the front of a gold foil. It was unlikely that there would be any, since Geiger had just a few months earlier shown that the alpha particles went right through a thin sheet of gold foil with only very minor changes in direction. One would not expect a rifle bullet to bounce off jello. This, they thought, was a guarantee that Marsden's job would be a short one, but things turned out a little differently than expected.

A few alpha particles are reflected by thin gold foil. Rutherford said it was like having a cannon shell bounce off tissue paper.

Marsden aimed an alpha particle source at a piece of gold foil and set up a scintillation screen to check for reflected alpha particles (Figure 3.10). Between the source and the screen he placed a lead block to make sure that any scintillations observed would be due to alpha particles reflecting from the gold foil and not coming directly from the source. When Geiger and Marsden sat down in a dark room to observe through the microscope, there,

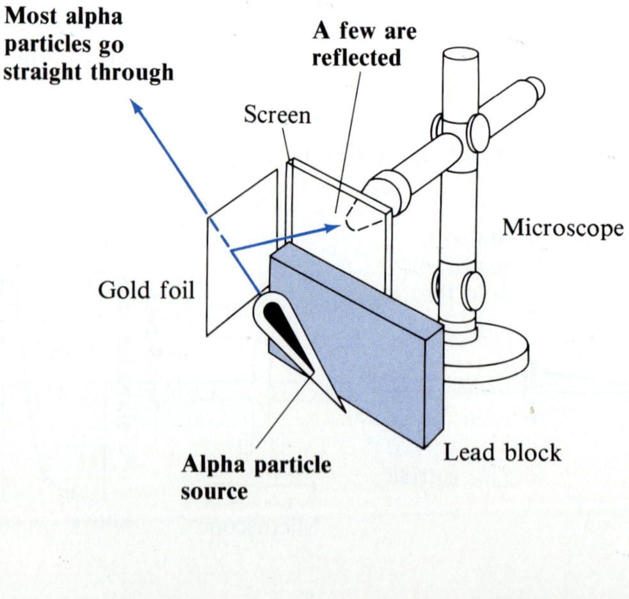
Most alpha particles go straight through
A few are reflected
Screen
Microscope
Gold foil
Lead block
Alpha particle source

FIGURE 3.10
The Geiger-Marsden experiment. This was an attempt to observe the reflection of alpha particles from the atoms in a gold foil. They did not expect to see any alpha particles reflected, but in fact, about 1 out of 8000 were reflected.

incredibly, were reflected alpha particles. Rutherford was astounded. He compared it to firing a 15-in. naval cannon at a piece of tissue paper and having the shell come back at you.

The number of alpha particles reflected back was very small compared to the number that passed through the gold foil. Only about 1 out of every 8000 particles striking the foil was reflected backward. Geiger reasoned that as each alpha particle made its way through the foil, it would pass close to (or through) many atoms. The repulsion of the positive alpha particles by the positive charge of the atom would cause the alpha particle to swerve a little as he had previously shown. But the probability of a greater than 90° deflection (as occurred in their experiment) was practically zero.

The Atomic Nucleus

Very few alpha particles are reflected (only about 1 in 8000); the rest pass through the gold with little or no deflection. Whatever reflects alpha particles must be both very massive and very small.

Rutherford finally decided that the only possible explanation for this abrupt reflection was an encounter between the positively charged alpha particle and some sort of tiny but dense "scattering center" in the gold foil. To turn an alpha particle as abruptly as seemed to be occurring would take a tremendous electrical force, and this probably meant that the alpha particle was approaching very close to something both relatively very heavy and very positive.

Knowing the charge and speed of the alpha particle and estimating the charge of the scattering center in the gold foil to be about 100 electron charges (but positive), Rutherford was able to calculate that the alpha particle would have to come within about 3×10^{-12} cm of the positively charged scattering center before it would be deflected back (Figure 3.11). Particles that did not come that close would be deflected but not greater than 90°.

Since the radius of an atom was generally accepted to be about 10^{-8} cm, this calculation was very surprising. The alpha particle apparently did not turn until it was about 3000 times closer to the center of the atom than were the edges of the same atom! Therefore, the heavy positively charged part of the gold atom must be only about 1/3000 of the total size of the atom rather than being evenly spread as predicted by Thomson's plum pudding model. This small kernel of mass and positive charge was called the **nucleus**. The rest of the space must be occupied by electrons which, because of their small mass. would not be able to deflect an alpha particle very much.

All of the positive charge and practically all of the mass of the atom reside in a tiny *nucleus* in the center of the atom. Electrons occupy the outer space.

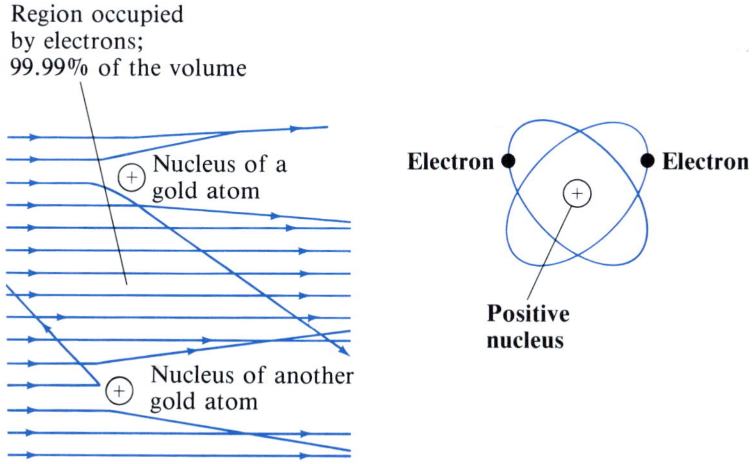
Region occupied by electrons; 99.99% of the volume
Nucleus of a gold atom
Nucleus of another gold atom
Electron
Electron
Positive nucleus

FIGURE 3.11
This cross section of a beam of alpha particles shows the reflection of alpha particles away from atomic nuclei. According to Rutherford, all of the positive charge and practically all of the mass of the atom must be concentrated in a small nucleus about 1/3000 the diameter of the atom.

TABLE 3.1
Some Important Subatomic Particles

Name	Symbol	Relative Charge	Atomic Mass (amu)
Electron	e^-	-1	1/1837
Proton	p	$+1$	1
Neutron	n	0	1
Alpha particle	α	$+2$	4
Beta particle	β	-1	1/1837

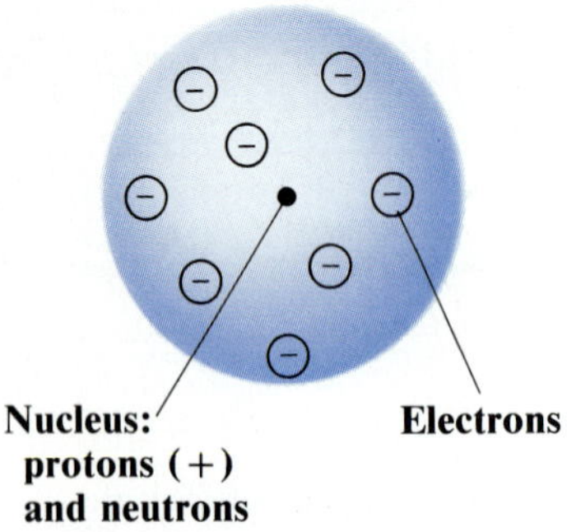
Nucleus:
protons (+)
and neutrons
Electrons

FIGURE 3.12
An atom's protons and neutrons are packed into a tiny nucleus about 1/3000 the diameter of the atom. The remaining space surrounding the nucleus is sparsely occupied by electrons.

Components of the Nuclear Atom

The discovery of the nuclear atom was followed by identification of the positive charge carrier in the nucleus as a **proton** with a charge of +1, and the subsequent discovery of the **neutron**, a zero-charged particle of about the same mass as a proton, which also resides in most nuclei (Table 3.1). Protons and neutrons are each about 1 amu. Recall from Chapter 2 that the **atomic mass unit (amu)** is equal to the mass of the carbon-12 atom, which has six protons and six neutrons.

Today we believe that atoms consist of a sparsely populated cloud of electrons surrounding an extremely tiny but dense nucleus (Figure 3.12). In this nucleus reside the positively charged protons and zero charged neutrons. The number of positive charges or protons in the nucleus is known as the **atomic number** (see Chapter 2). In a neutral atom, it is equal to the number of negative electrons surrounding the nucleus.

The atomic number of an element tells us the number of protons in its nucleus.

Isotopes

Isotopes are atoms of the same element with the same number of protons but a different number of neutrons

Isotopes are atoms of a particular element that contain the same number of protons and electrons, but a different number of neutrons. Thus, while their atomic numbers are identical, their mass numbers are different. The existence of isotopes explains why some elements appear to have fractional atomic masses. Chlorine, for example, has an atomic mass of 35.457. This is because about 75% of naturally occurring chlorine atoms have 17 protons and 18 neutrons (mass number 35) and about 24% of the atoms have 17 protons and 20 neutrons (mass number 37). The **atomic mass** listed for each element is the *average* mass of all of the naturally occurring isotopes of the element.

Atomic mass is the average mass of all of the isotopes of an element.

A simple symbolism is used to indicate the atomic number, atomic mass, and identity of an element or isotope. A fractional atomic mass indicates that the average atomic mass for the element is reported; whole number atomic masses refer to specific isotopes of the element. For example,

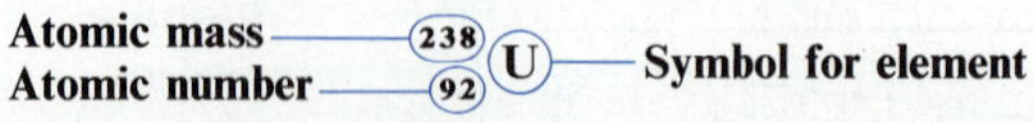
Atomic mass
238
Atomic number
92
U
Symbol for element

This represents an isotope of uranium with a mass of 238 amu and a nucleus containing 92 protons. The number of neutrons in the nucleus must then be 238 − 92 = 146 neutrons. A neutral atom of this element will have 92 electrons to balance the 92 positive charges in the nucleus.

EXAMPLE 3.1

Aluminum (Al) is the fourth most common element in the crust of the earth. An isotope of aluminum with an atomic mass of 27 is symbolized this way:

$$^{27}_{13}\text{Al}$$

a. How many protons are found in the nucleus of an atom of aluminum? From the above, we know that the atomic number is 13. There are thus 13 protons in the nucleus of an aluminum atom.

b. How many neutrons are found in the nucleus of an atom of aluminum?

Atomic mass is protons + neutrons	27
Subtract the number of protons	−13
The number of neutrons	14

c. How many electrons will be found in a *neutral* atom of aluminum? Since a neutral atom must have the same number of positive and negative charges, it must have an equal number of protons and electrons. Thus, it will have 13 electrons.

Recent research has indicated that neutrons and protons are not basic building blocks of matter, but in turn consist of even smaller subatomic particles. This interesting problem is the concern of nuclear physicists, however, and is not likely to affect our understanding of chemical reactions between atoms and molecules.

NEW TERMS

scintillation A flash of light given off when an atom is struck by a nuclear particle or gamma ray.

nucleus The tiny core of an atom, which contains neutrons, protons, and practically all of the mass of the atom. It is surrounded by a cloud of electrons.

proton A basic building block of nuclei, with a mass of 1 atomic mass unit (amu) and an electrical charge of +1.

atomic mass unit (amu) The basic reference for atomic mass measurements equal to 1/12 the mass of the carbon-12 atom. 1 amu = 1.66606×10^{-24} g.

neutron Another basic building block of nuclei, with a mass of 1 amu and an electrical charge of zero.

atomic number The number of protons (positive charges) in the nucleus of an atom. The atomic number is also the reference number used to specify the location of an element in the periodic table, and the number of electrons surrounding the nucleus of a neutral atom.

isotope A term used to designate an atom of a specific mass number. Atoms of a given element all have the same number of protons in their nuclei (atomic number), but may have different numbers of neutrons.

atomic mass The weighted average of the mass of all of the isotopes of a given element.

TESTING YOURSELF

Discovery of the Atomic Nucleus

1. Geiger checked the distribution of positive charge predicted by Thomson's "plum pudding" model by firing alpha particles through thin sheets of gold. Did he find significant deflection of the alpha particles?
2. What did Geiger then conclude about the distribution of the positive charge throughout the atom?
3. Marsden found that some alpha particles are reflected backward from gold foil. To turn an alpha particle this much in direction requires that it strike something relatively very heavy and probably positive. Are these heavy concentrations of charge very predominant in atoms? About what fraction of the alpha particles are reflected backward?
4. Can one explain the results of Geiger's and Marsden's experiments if all the positive charge and practically all of the mass of the atom is concentrated into a tiny chunk in the center of the atom?
5. What did Rutherford call this tiny chunk of positive material?
6. The nucleus is composed of protons and neutrons. Which carries a positive charge?
7. Which is more massive, a proton or a neutron?
8. An atom of aluminum (atomic mass 27, atomic number 13) has how many neutrons in its nucleus?
9. A neutral atom of aluminum has how many electrons surrounding its nucleus?
10. Since protons are all positive, and since like electrical charges repel each other, it seems reasonable that there would be a fairly strong repulsive force in nuclei with large numbers of protons. Neutrons (0 charge) appear to provide some of the "nuclear glue" necessary to hold these nuclei together. Look at the periodic chart in Figure 2.10. In the nuclei of light elements, there are about as many neutrons as protons. In heavy elements, is the fraction of neutrons much higher or lower?

Answers **1.** no, only very little **2.** distributed evenly **3.** no, 1 in 8000 **4.** yes **5.** nucleus **6.** proton **7.** They are the same. **8.** 14 **9.** 13 **10.** higher

SUMMARY

This chapter had two major goals. The first was to develop an understanding of our current view of the nuclear atom—an atom in which all of the positive charge and practically all of the mass is concentrated in a tiny nucleus about 1/3000 the diameter of the atom, while the remaining space is occupied by rapidly moving electrons. The atomic particles that comprise the nucleus are protons, which have a mass of one atomic mass unit (amu) and a charge of $+1$, and neutrons, which also have a mass of 1 amu but carry no electrical charge. The atomic number of an element is equal to the number of protons in the nucleus of an atom of the element. The atomic mass of a given atom is the sum of the masses of its neutrons and protons; the atomic mass of an element is the weighted average of the masses of all of its isotopes. Isotopes of an element are atoms with an equal number of protons but a different number of neutrons, and thus a different atomic mass. Electrons have a mass of 1/1837 amu and carry a -1 electrical charge. In a neutral atom, the number of protons in the nucleus equal the number of electrons in the cloud surrounding the nucleus.

The second goal of this chapter was to illustrate, by telling the stories of several of the important discoveries, the process by which scientific progress takes place. Scientific discovery depends upon keeping one's eyes open for unexpected trends in expected data.

Chemistry is the making and understanding of molecules. To understand how atoms connect together to make molecules, we must consider what does the connecting. In Chapter 4 we will examine how electrons are arranged around an atom's nucleus, which will lead us into an understanding of how atoms stick together or *bond* (Chapter 5).

TERMS

electrical charge
cathode
cathode ray
electron
canal ray
plum pudding model
fluorescence
X-ray
natural radioactivity
alpha (α) particles
beta (β) particles
gamma (γ) radiation
scintillation
nucleus
proton
atomic mass unit (amu)
neutron
atomic number
isotope
atomic mass

CHAPTER REVIEW

QUESTIONS

1. In a Crookes discharge tube, which electrode *attracted* the beam of cathode rays?
   a. positive
   b. negative
2. What electrical charge was carried by cathode rays?
   a. negative
   b. positive
3. Electrons have a mass that is
   a. equal to that of a proton.
   b. about 1/2000 that of a proton.
   c. about 2000 times that of a proton.
4. Which of the following statements about electrons is *not* true?
   a. Electrons carry a negative charge.
   b. Electrons are about as heavy as protons and neutrons.
   c. Electrons are thought to be in rapid motion.
   d. Some electrons are easily removed from atoms.
   e. Electrons occupy the outer part of atoms.
5. The "plum pudding atom" had
   a. its positive charge spread equally throughout the atom.
   b. its positive charge concentrated in the center of the atom.
   c. a positive charge bound to each electron.
6. Naturally radioactive materials emit three kinds of radiation called alpha, beta, and gamma. Which of these is heaviest and carries a positive electrical charge?
   a. alpha
   b. beta
   c. gamma
7. An alpha particle passing the nucleus of an atom would probably experience which of the following deflections?

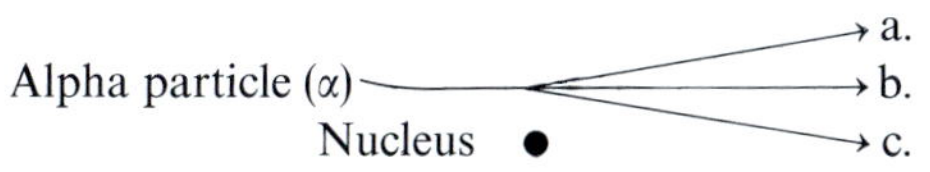
Alpha particle (α)
Nucleus
a.
b.
c.

8. Two isotopes of an element differ only in their
   a. symbol
   b. atomic number
   c. atomic mass
   d. number of protons
   e. number of electrons

Answers **1.** a **2.** a **3.** b **4.** b **5.** a **6.** a **7.** a **8.** c

DIAGNOSTIC CHART

Blacken in all of the circles under the number of each question you missed in the Chapter Review questions. The diagnostic chart will help you identify concept areas that need more study.

	Questions							
Concepts	1	2	3	4	5	6	7	8
Cathode rays	○	○	○	○				
Electrons					○	○		
Plum pudding atom							○	
Natural radioactivity								○

EXERCISES

Cathode Rays

1. A negatively charged metal plate placed near the path of a cathode ray beam will
   a. repel the beam.
   b. attract the beam.
2. Cathode rays are used today in
   a. computer monitors.
   b. TV sets.
   c. hospital electrocardiograph machines.
   d. none of the above
   e. all of the above

Electrons

3. Electrons carry a charge
   a. equal in magnitude but opposite in sign to a proton.
   b. equal to that carried by a proton.
   c. 1/2000 that carried by a proton.
4. Electrons are
   a. sometimes easily removed from atoms.
   b. never removed from atoms.

Plum Pudding Atom

5. Which experiment led to the idea that atoms have light negative parts and heavy positive parts?
   a. Crookes tube experiment
   b. Cathode ray charge-to-mass ratio experiment
   c. Canal ray experiment
   d. Geiger-Marsden alpha scattering experiment
6. Which of the following statements concerning canal rays is false?
   a. Canal rays appear to carry positive charge.
   b. Canal rays are about as massive as ordinary atoms.
   c. Canal rays are probably atoms stripped of one or more electrons.
   d. Canal rays are less dense than ordinary atoms.
   e. Canal rays are not observed when the tube is pumped down to a good vacuum.

Natural Radioactivity

7. Which of the following is the lightest particle?
   a. alpha
   b. beta
   c. gamma
8. Which of the following carries no electrical charge?
   a. alpha
   b. beta
   c. gamma
9. Which of the following is the most penetrating type of radiation?
   a. alpha
   b. beta
   c. gamma

The Nuclear Atom

10. Hans Geiger and Ernest Marsden performed an experiment (Figure 3.10) in which they attempted to reflect alpha particles from a thin gold foil. They found that about 1 in 8000 alpha particles were reflected back. What did they learn from this experiment?
    a. Gold foil is mostly empty space to alpha particles.

b. If an alpha particle passes enough atoms, its path will reverse.
c. The positive charge of an atom must be evenly spread through it.
d. The mass and positive charge of an atom must be concentrated in a tiny chunk in the center of the atom.

11. Sketched here is an experiment in which an object of unknown size and shape is hidden under a board. On the left is a tray of marbles that are released and then roll under the board and out the other side. From the directions in which they roll when they emerge, one can make predictions about the size, shape, and orientation of the object (or objects) hidden under the board. This experiment is analogous to the deflection caused by an electric field on subatomic particles, as shown in Figure 3.8. For parts (a) through (f), sketch what you think the hidden object(s) would look like.

12. Consider a neutral atom of the element phosphorus:

$$^{31}_{15}P$$

a. Atoms of this element have how many protons in their nucleus?
b. How many electrons does a neutral atom of phosphorus have?
c. How many neutrons does an atom of phosphorus have in its nucleus?

13. Copper has two naturally occurring isotopes in these proportions:

$^{63}_{29}Cu$	69.09%
$^{65}_{29}Cu$	30.91%

a. How many protons does each have in its nucleus?
b. How many neutrons does each have in its nucleus?
c. Based on this information, what is the *average* atomic weight of copper?

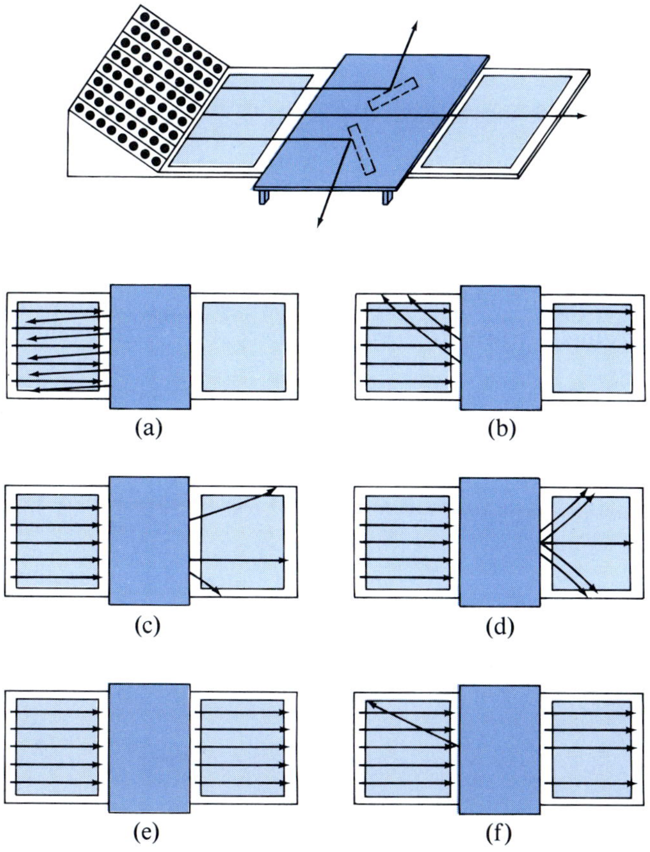
(a)
(b)
(c)
(d)
(e)
(f)

CHAPTER

4

Electrons and Chemical Periodicity

Nature speaks to us in curious ways. The colors of light given off by heated hydrogen gas led Niels Bohr to propose that atoms are like miniature solar systems, with electrons playing the role of planets and the nucleus acting as a sun. Although ultimately proven inaccurate, Bohr's model was the first step in the discovery of an extremely powerful tool called the electron filling series.

OBJECTIVES

After completing this chapter, you should be able to

- Define wavelength and frequency.
- Classify the regions of the electromagnetic spectrum in terms of wavelength and energy.
- Describe how Bohr's planetary electron idea explained why only certain colors of light were emitted by hot hydrogen gas.
- Sketch shapes for *s* and *p* orbitals.
- Use the electron filling series to predict the electron arrangement for neutral atoms up to atomic number 20.
- Use orbital notation to describe an atom's electron distribution.
- State Hund's rule and its significance for predicting the arrangement of electrons in an atom's orbitals.
- Sketch electron dot models for neutral atoms.

4.1 The Behavior of Electrons in Atoms

An atom-sized observer might tell us that, from his point of view, atoms appear to be mostly empty space and electrons, with a tiny nucleus containing protons and neutrons in the center. Suppose that a carbon atom could be enlarged to be a sphere as large as an ordinary gymnasium. In this analogy, the carbon nucleus with its six protons and six neutrons would be a tiny kernel in the center, about the size of a grape. In the remaining space—mostly empty—would be six gnat-sized electrons buzzing rapidly around.

There are two ways the gnat-sized electrons could use this space. They might fly randomly through it. Or they might be more territorial, each gnat or group of gnats staying pretty much in its own part of the gym. Our atom-sized observer reports that the latter is the case: the gnat-sized electrons are usually most stable when in groups of two, and there are a number of well-defined territories in which each pair of gnats spends most of their time. Gnats are seldom found in the "no man's land" between the territories.

Such is the case with electrons. Each pair of electrons is most likely to be found in its own territory or "cloud" in space around the nucleus. This territory is called an *orbital* (not to be confused with "orbit"). Orbitals never hold more than two electrons, and electrons are seldom found far outside their orbital. Orbitals can be classified in terms of energy. Those in which the electrons have the lowest energy are of the simplest shape, are closest to the nucleus, and are filled first as electrons are added to an atom.

Early in this century Niels Bohr, a Danish physicist, proposed that light is emitted as electrons fall from high energy paths far from the nucleus to lower energy paths closer to the nucleus. At the time, Bohr believed that electrons behaved quite a bit like planets, moving in several well-defined orbits around a tiny nuclear sun. To understand Bohr's proposal, we must know a little about light, which we look at in the first part of the chapter.

The second part of the chapter concerns Bohr's idea that electrons move in planetary orbits and his observation that only certain electron orbits appear to be available. Bohr presented some data that strongly supported this idea, but he was unable to explain why the number of possible orbits is limited. It is similar to the problem of the territorial gnats. One can identify their favored air space, but how can one explain why they behave in this manner? The reason became clear when it was discovered that particles as tiny as electrons also have some properties of waves.

When the wave properties of the electron were recognized, it quickly became possible to compute the shapes and sizes of orbitals or regions around the nucleus occupied by an electron of a given energy. A simple tool called the *electron filling series* enables both students and chemists to easily predict the electron arrangement for any atom, given only its atomic number. This principle also provides a clear explanation of chemical periodicity. It will be used as we consider the process of chemical bonding, which is presented in Chapter 5.

4.2 Light, Color, and Energy

Before we propose that light may behave like a wave, we must know something about waves and how they behave. Waves are a common part of our everyday experience. Figure 4.1 illustrates an experiment most of us have

FIGURE 4.1
This photograph illustrates several properties of waves: they are a repetitive phenomena, and one wave is usually followed by another just like it. Note the unusual pattern that occurs when waves cross. When crests cross, they add, but when a crest crosses a trough, they cancel and disappear. This phenomenon helps explain why electrons are found only in certain regions around a nucleus.

performed. Several pebbles were thrown in a lake. As each pebble struck the surface of the water, it created a wave that moved out in a circular pattern through the water. Two principles of interest to us are illustrated in this photograph:

1. Wave motion is repetitive. Each ring consists of a number of concentric waves, each evenly spaced from its neighbors. The distance between corresponding points on adjacent waves is the space occupied by one wave and is called the **wavelength**.
2. Wavelength is related to the frequency of wave generation. Waves may be generated by moving an object in the water. If the object is moved slowly, each wave has time to move a reasonable distance before the next wave is created. The distance from wave crest to wave crest (the wavelength) is long. On the other hand, if waves are created rapidly, the distance between waves is much shorter, and the wavelength is shorter.

Thus, **frequency**, or the number of waves created each second, is an important property of waves. It is measured in units of waves per second, or *hertz* (*Hz*).

For a given **wave velocity** (v), wavelength (λ) is inversely proportional to the frequency (f). Mathematically stated,

$$\text{Wavelength} = \frac{\text{Velocity}}{\text{Frequency}}$$

or, using mathematical symbols,

$$\lambda = \frac{v}{f}$$

For waves moving at a constant velocity, the wavelength becomes shorter as the frequency becomes higher (Figure 4.2). Note that the velocity of any form of wave motion is determined only by the material through which it moves and not by its frequency. For example, the low frequency sound waves from a tuba and the high frequency sound waves from a flute travel at the same speed through air, despite their difference in frequency. Consider how unusual a band concert would sound if the flute note arrived first.

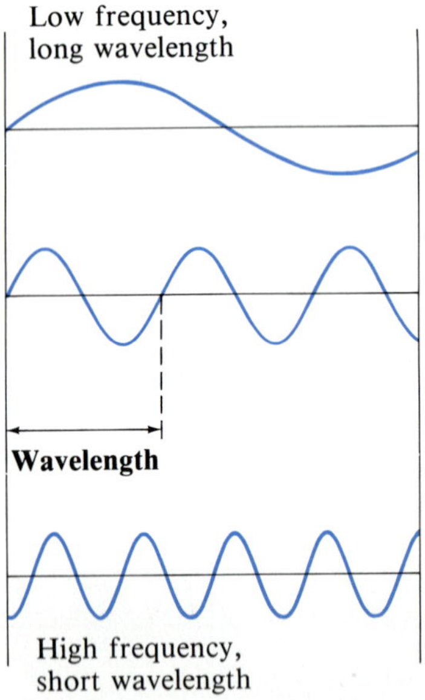
Low frequency,
long wavelength
Wavelength
High frequency,
short wavelength

FIGURE 4.2
Wavelength is defined as the distance between corresponding parts of adjacent waves. Wavelength is mathematically equal to the velocity of the wave divided by its frequency. Higher frequency waves have shorter wavelengths.

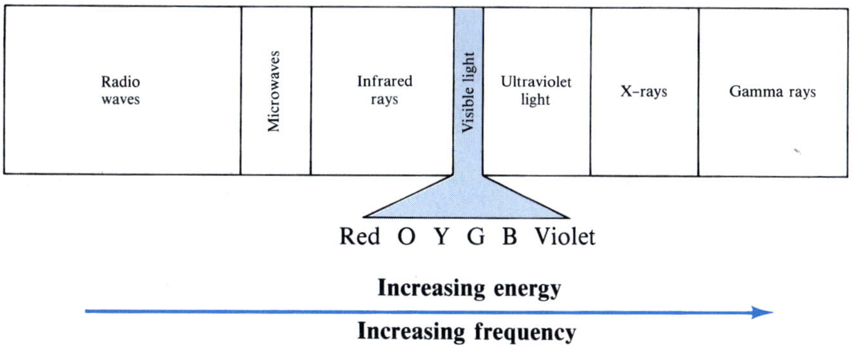
Radio waves
Microwaves
Infrared rays
Visible light
Ultraviolet light
X-rays
Gamma rays
Red O Y G B Violet
Increasing energy
Increasing frequency

FIGURE 4.3
The electromagnetic spectrum. The energy carried by electromagnetic radiation is directly related to the frequency of the radiation. Higher frequency, shorter wavelength radiation, such as X-rays and gamma rays, carries more energy. Radio waves are relatively low energy.

Now that we know a little about waves, let's look at how light behaves as a wave.

Light and Color

Light is a form of wave motion that has many properties in common with water waves and sound waves. Light, however, is **electromagnetic radiation**, a form of wave motion comprised of alternating electrical and magnetic fields. Light travels at an extremely high speed (3×10^8 m per sec), and because of its high frequency, it has very short waves. Red light has a wavelength of about 6.5×10^{-7} m, or 650 nanometers (nm). (A *nanometer* is 1×10^{-9} m.) Blue light has a wavelength of about 4.5×10^{-7} m, or 450 nm. The **spectrum** of visible light, arranged from long to short wavelength, comprises the colors red, orange, yellow, green, blue, and violet. Plate 1 in the color insert shows a beam of white light being broken into a spectrum of visible light by a prism. Some forms of electromagnetic radiation have longer waves, such as infrared rays, microwaves, and radio waves, while others have shorter wavelengths, such as ultraviolet rays, X-rays, and gamma rays (Figure 4.3).

Light and Energy

The energy carried by a light wave is directly proportional to the frequency of the wave. High frequency waves carry more energy than do lower frequency waves. This relationship is illustrated for the entire electromagnetic spectrum in Figure 4.3. Violet light, because of its shorter wavelength and higher frequency, carries more energy than red light. Thus, energy and frequency are directly proportional. This principle was proposed in the early 1900s by Max Planck, a German physicist, and is commonly known as **Planck's law**. It can be stated mathematically as a simple equation:

Light of shorter wavelength and higher frequency carries greater energy.

$$\text{Energy} = \text{A constant} \times \text{Frequency of electromagnetic radiation}$$

$$E = (h)(\nu)$$

Planck's law: $E = h\nu$

Planck's constant (h) is a very small number, but since the frequency of light waves is so large, visible light waves carry enough energy to start many chemical reactions. Planck believed that light waves are not continuous, but come in little packets of waves he called **photons**.

Photons are tiny packets of waves.

There are many examples of the relationship between energy and color or frequency of light in nature. Ultraviolet light, which because of its short wavelength and high frequency carries high energy, causes sunburn by breaking chemical bonds in skin molecules. Special lights used in tanning salons use ultraviolet light to produce a "controlled sunburn," or tan. Because broken chemical bonds in cellular molecules increase the chance of mutation of the cell, excessive exposure to ultraviolet light increases the probability of skin cancer.

NEW TERMS

wavelength The amount of space occupied by one wave; mathematically equal to the wave velocity divided by the frequency.

frequency The number of waves that crosses a certain point each second, measured in waves per second or hertz.

wave velocity The speed at which a wave moves through a material; it depends on the material and the type of wave, not the frequency of the wave.

electromagnetic radiation A form of wave motion comprised of alternating electrical and magnetic fields that travels at an extremely high speed (3×10^8 m/sec).

spectrum The colors of light given off by an object.

Planck's law A mathematical statement relating energy and frequency of electromagnetic radiation: the greater the frequency (and the bluer the color of light), the greater the energy.

photon A packet of light waves, sometimes called a "quantum."

TESTING YOURSELF

Waves and Light

1. Suppose that you are making waves in a tank of water by moving your hand up and down. If you move your hand more rapidly (higher frequency), will the resulting waves be shorter or longer?
2. When the frequency of the wave is increased, will it move faster or slower?
3. List the colors of the visible spectrum in order of decreasing wavelength.
4. As the wavelength of light decreases, does the energy carried by the light increase or decrease?
5. Which of these types of electromagnetic radiation carries the least energy: radio, television, blue light, X-rays, or microwaves?

Answers **1.** shorter **2.** no change in speed **3.** red, orange, yellow, green, blue, and violet **4.** increases **5.** radio

4.3 The Bohr Atom: Electrons May Behave Like Planets

In the spring of 1912, a few months after Rutherford's discovery of the nuclear atom, Niels Bohr visited Rutherford's laboratory. Bohr had recently completed a Ph.D. in physics at the University of Copenhagen and was spending a year of study abroad. During the four months Bohr spent at the University of Manchester, he became convinced that electrons must exist in

SPECTROPHOTOMETRY: CHEMICAL ANALYSIS WITH LIGHT

What do astronomers, metallurgists, and medical technologists have in common? They all use light to determine the composition of chemical samples.

Some of the most convenient and powerful qualitative (what it is) and quantitative (how much) analysis tools use light as their atomic or molecular probe. When atoms are heated, their electrons move to outer, higher energy orbitals. When these electrons fall back to lower energy orbitals, light of a color determined by the electron energy change is emitted. This *emission spectrum* or set of spectral lines is unique for each element and can be used to identify the element(s) present in the sample (see Figure 4.4). The source of light for *emission spectrophotometry* can be as close as a heated sample within the spectrophotometer or as far away as the sun or a distant star.

Because a specific energy is associated with the gap between the energy of two orbitals, it is possible to use light of this specific energy to promote an electron from a lower energy orbital to a higher energy orbital. The color of this light must be *exactly* right—its energy must be exactly that required to promote the electron. If white light is passed through a gaseous group of atoms, the atoms will absorb certain very narrow lines of color that exactly match their electron energy gaps, thus giving notice of their presence and identity. Gaseous elements surrounding the sun remove small spectral lines from the white light produced by the sun; these *Fraunhofer lines* are clues to the composition of the sun.

Because atoms and molecules in the solid or liquid state vibrate, rotate, and interact with each other, their electron energy states are somewhat broadened, and atomic absorption lines that would appear extremely narrow in a gaseous absorption spectrum become broader. As we learned earlier in this chapter, the light we call "white" is actually all of the colors of the visible spectrum present at one time. If some of these colors are missing, then we see the remaining spectrum as colored. This situation is used to great advantage in medical laboratories.

Absorption spectrophotometry uses the measurement of absorbed light to determine the number of light-absorbing molecules in the beam of colored light. Suppose, for example, that a blood sample were treated so that the iron in the hemoglobin formed a red-colored solution. Red solutions are red because they absorb blue and green light. An absorption spectrophotometer uses a white light source, a prism or diffraction grating element to disperse the white light into a spectrum, and a slit to select the desired color of light. A phototube detects the light; the phototube signal is amplified and placed on a meter or digital readout device (see figure). The measurement is made by passing the selected color of light first through a "blank" of distilled water and recording an arbitrary 100% transmission, then replacing the blank with the colored sample. The amount of blue-green light absorbed by the sample is related to the concentration of blue-green absorbing molecules in the sample and may be converted to a concentration value by reference to a calibration chart. Spectrophotometric measurements are quick, accurate, and inexpensive. They are the principal analytical chemistry analysis technique used in medicine today.

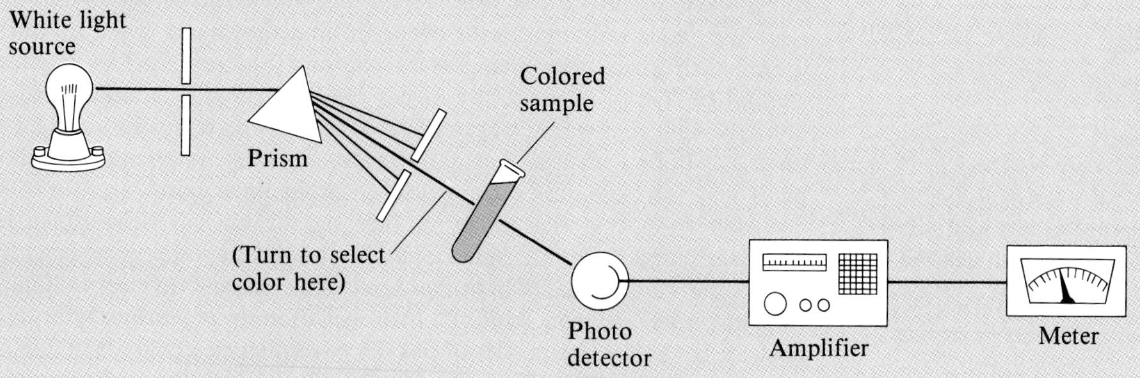
White light source
Prism
Colored sample
(Turn to select color here)
Photo detector
Amplifier
Meter

An absorption spectrophotometer

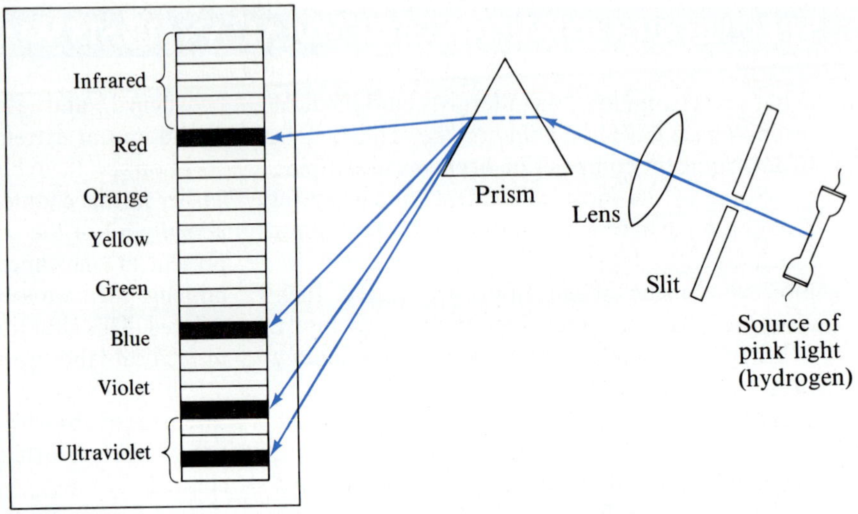

Infrared
Red
Orange
Yellow
Green
Blue
Violet
Ultraviolet
Prism
Lens
Slit
Source of
pink light
(hydrogen)

FIGURE 4.4
When excited by an electrical discharge, gaseous hydrogen emits a series of visible spectral lines or colors beginning with a bright red line and a blue-green line and converging to a group of violet lines. Similar converging series are observed in the infrared and ultraviolet regions. These are caused by electron jumps ending on the third and first electron orbits.

some sort of regular distribution around the newly discovered nucleus of the atom. Bohr returned to Denmark and spent the winter of 1912–1913 trying to fit experimental observations into some sort of useful structural model of the atom.

Heated hydrogen gas gives off only certain colors of light.

One of Bohr's principal items of experimental evidence was the light given off by heated hydrogen atoms. A hydrogen atom has only one electron and was therefore a simple atom with which to begin experimentation. When a sample of hydrogen gas was heated in an electrical discharge (Figure 4.4), a pink light was produced. When this light was passed through a prism and separated into its components, several discrete colors of light or **spectral lines** were observed. One of these lines was red, one blue-green, and several violet. This observation was in considerable contrast to the spectrum observed when a solid was heated to incandescence. Hot gas samples emit only certain colors of light, a **line spectrum**, while all hot solids emit a **continuous spectrum**, or band spectrum, giving off all colors of light.

Larger orbital jumps produce more energy and thus bluer light.

Bohr thought that movement of the hydrogen atom's electron might be responsible for the emission of light. He proposed that electrons might move around the nucleus much like planets orbit the sun; this was **Bohr's planetary electron model**. He further proposed that electrons may be free to jump from one orbit to another. Electrons might jump out to higher energy orbits when the gas is heated and then return to inner, lower energy orbits as the gas cools. Since at this time it was known that accelerated electrons produce electromagnetic radiation, Bohr proposed that the energy given up during these return jumps would appear as photons (Figure 4.5). The larger the amount of energy released, the higher the frequency of radiation and the bluer the light that would be produced. Since only certain colors of light are observed, Bohr reasoned that only certain electron jumps must be taking place, and therefore only certain electron orbits must exist.

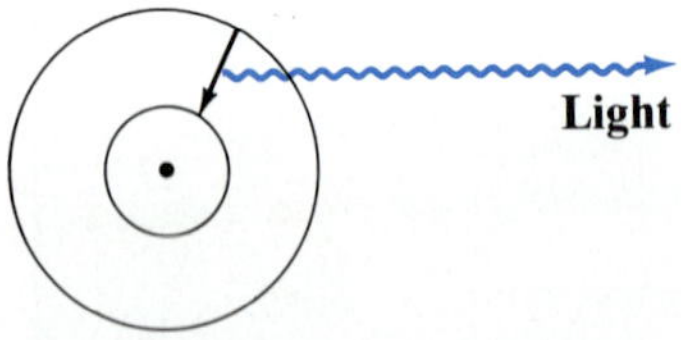

Light

FIGURE 4.5
Bohr's solar system model of the atom proposes that light is given off when an electron jumps from an outer high energy orbit to an inner, lower energy orbit. Since only certain colors or energies of light are observed, only certain electron orbits must be available.

After proposing that electrons orbiting the nucleus behave like tiny planets orbiting their sun, Bohr decided to test his hypothesis by using it to predict the wavelengths of light that *should* be emitted if an electron jumped from any outer electron orbit of a hydrogen atom to any orbit closer to the nucleus. He applied principles of physics governing electrical attraction and planetary motion, as well as the Planck relationship between energy and

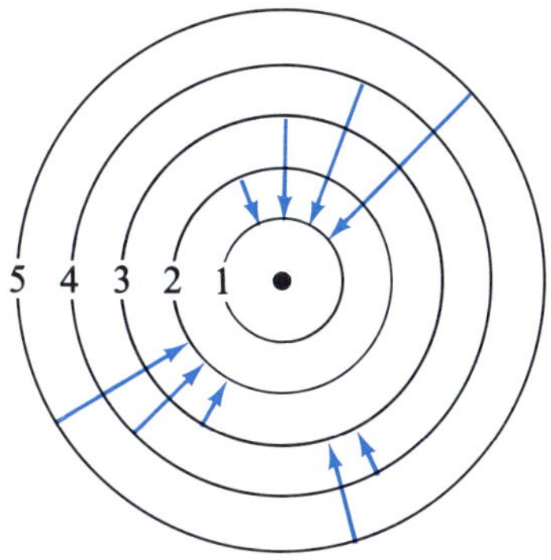
5 4 3 2 1

Bohr's orbits

Orbit Jump	Predicted Wavelength (Bohr's Model) (nm)	Observed Wavelength (nm)	Color of Light Observed
2→1	121.568	121.568	Ultraviolet
3→1	102.573	102.573	Ultraviolet
4→1	97.254	97.254	Ultraviolet
5→1	94.979	94.979	Ultraviolet
3→2	656.469	656.210	Red
4→2	486.273	486.070	Blue
5→2	434.172	434.010	Violet
4→3	1875.6	1875.1	Infrared
5→3	1282.1	1281.7	Infrared

FIGURE 4.6
Bohr's planetary electron theory was quite successful in predicting the wavelengths of light produced by hot hydrogen atoms.

frequency of light ($E = h\nu$), to develop an equation. The results of Bohr's calculations are shown in Figure 4.6. Note how closely the actual wavelengths are to those predicted by Bohr's equation.

Bohr developed a successful equation to predict the colors of light that would be produced by electron orbital jumps in heated hydrogen.

The excellent agreement caused many scientists to believe that electrons do behave just like tiny planets circling their "sun" in definite and well-specified orbits. This idea still lives in a number of general science texts. You may have encountered the statement that each orbit consists of a "shell" of electrons; two electrons are found in the first shell, eight in the second shell, eighteen in the third shell, and so on. As we will see shortly, this statement is only partially true.

The success of Bohr's planetary electron idea caused many to believe that electrons do orbit nuclei like planets around a sun.

Bohr's planetary electron theory had some problems. One was that it only worked for hydrogen atoms. More important, however, was that it worked only if one presumed that only certain orbits were possible. Bohr could not explain why only certain orbits appeared to exist. As we will see in the next section, Bohr received some help from a French graduate student, Louis de Broglie, a few years later.

NEW TERMS

spectral line A single color of light emitted by a hot gas atom.

line spectrum A set of distinct spectral lines emitted by a hot gas sample.

continuous spectrum A continuous band of all colors emitted by a hot solid.

Bohr's planetary electron model The model proposed by Niels Bohr suggesting that electrons rotate around a nucleus in a set of fixed orbits like planets around the sun. The model predicts that light is given off when electrons jump from outer orbits to inner orbits.

TESTING YOURSELF

The Bohr Atom

1. Which type of material when heated to a high temperature will give off only certain colors of light: solid, liquid, or gas?

2. Which is produced in question 1, a line or a continuous spectrum?
3. If light is produced when electrons jump from higher to lower orbits, an atom that gave off all colors of light must have
   a. only certain possible electron orbits.
   b. an infinite number of possible adjacent electron orbits.
4. Suppose that a hypothetical atom gives off a red, green, and violet line spectrum and that its orbits are arranged according to the energy chart shown below. Which jump would give off the red spectral line (least energy)?

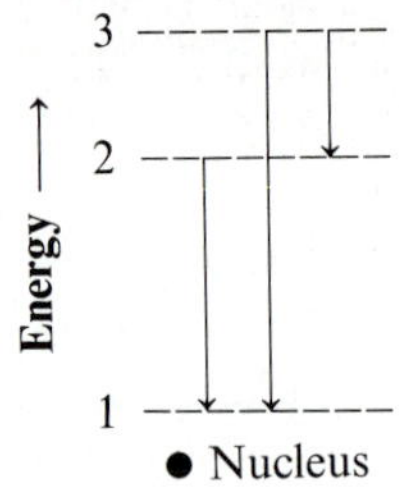
3
2
1
Energy ⟶
● Nucleus

5. Which jump in question 4 would give off the violet spectral line (greatest energy)?

Answers **1.** gas **2.** line **3.** b **4.** 3 to 2 **5.** 3 to 1

4.4 Orbitals

If electrons have wave properties, they should only be visible where the electron waves cross with their displacements in the same direction.

The middle 1920s were a fruitful time in atomic physics. Early in the decade, the French physicist Louis de Broglie proposed that objects as tiny as an electron should have observable wave properties and should behave in some ways like water waves (see Section 4.2). This was proven three years later by D. J. Davisson and L. H. Germer in the United States. This set the stage for Erwin Schrödinger, an Austrian physicist, to propose a new view of the distribution of electrons in atoms. Just as the water waves in Figure 4.1 added together when their crests crossed and canceled when their wave crests crossed the wave troughs, Schrödinger proposed that there are certain regions around the nucleus where electron waves add or reinforce and other areas where they cancel. He then went on to propose that the electron density would be high in those areas where the waves add and low where they cancel. He combined an equation from physics that described the three-dimensional movement of a normal wave with the electron wave idea and came up with an equation that related the energy of an electron to the probability of its being found in any given position around the atom.

The electron wave idea predicts regions around the nucleus called *orbitals* where an electron is most likely to be visible.

Schrödinger's equation, unlike Bohr's, could be applied to atoms with more than one electron. However, it gave results in terms of the *probability* of finding an electron instead of firmly locating the orbit of the electron as did Bohr (Figure 4.7). While Bohr had predicted well-defined "planetary" orbits for electrons, Schrödinger predicted "clouds" of various shapes surrounding the nucleus in which electrons of different energies could be found. Each enclosed a region in which the electron waves would reinforce each other. Some of these clouds were spherical, others were dumbbell shaped,

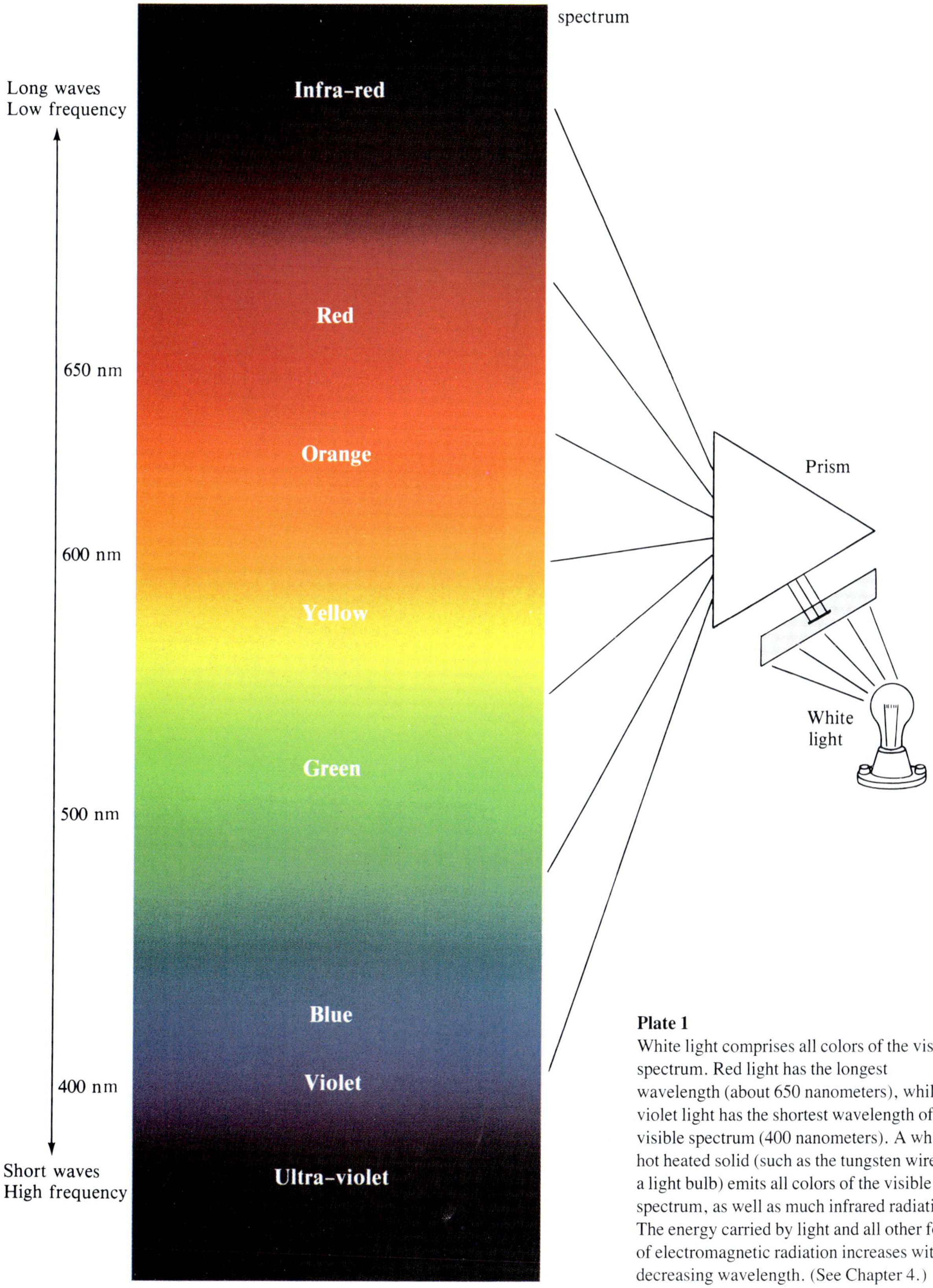
spectrum
Long waves
Low frequency
Infra-red
Red
650 nm
Orange
600 nm
Yellow
Green
500 nm
Blue
Violet
400 nm
Short waves
High frequency
Ultra-violet
Prism
White
light

Plate 1
White light comprises all colors of the visible spectrum. Red light has the longest wavelength (about 650 nanometers), while violet light has the shortest wavelength of the visible spectrum (400 nanometers). A white-hot heated solid (such as the tungsten wire in a light bulb) emits all colors of the visible spectrum, as well as much infrared radiation. The energy carried by light and all other forms of electromagnetic radiation increases with decreasing wavelength. (See Chapter 4.)

Plate 2
Medical X-rays were the first practical use of ionizing radiation. (See Chapter 3.) This X-ray shows a fractured tibia and fibula in the lower leg.

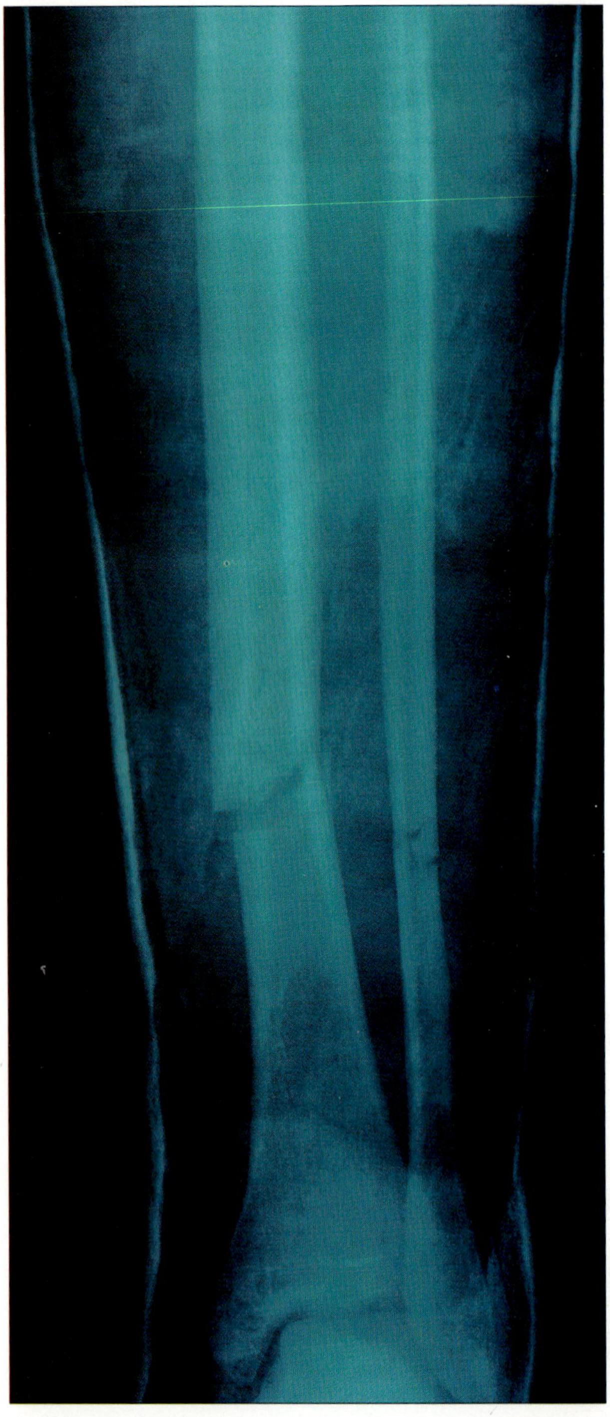

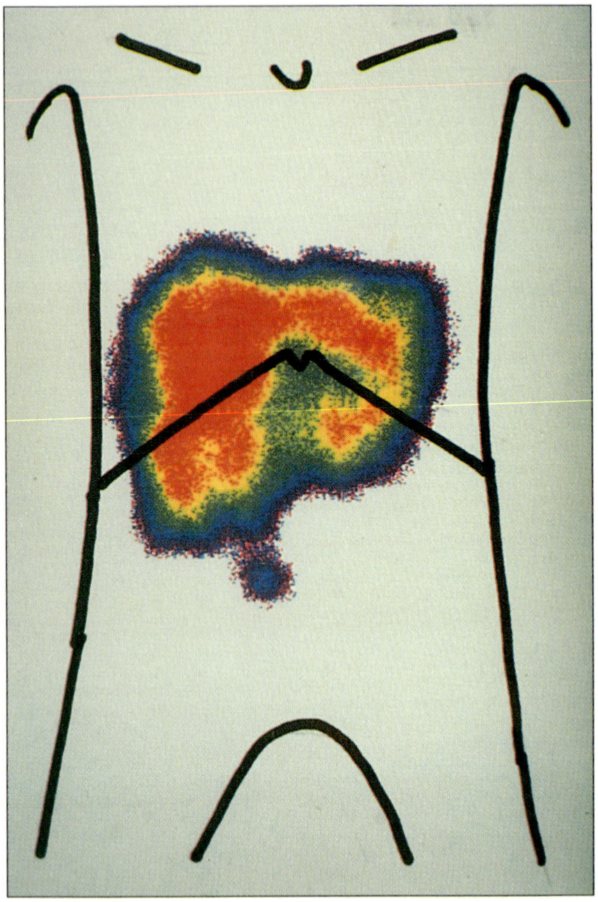

Plate 3
(Above) A liver scan conducted after injection of monoclonal antibodies labeled with radioactive iodine-131 shows the presence of hepatoma cancer. Radioisotope tracers provide a safe, nonintrusive, and highly effective diagnostic tool. (See Chapter 13.)

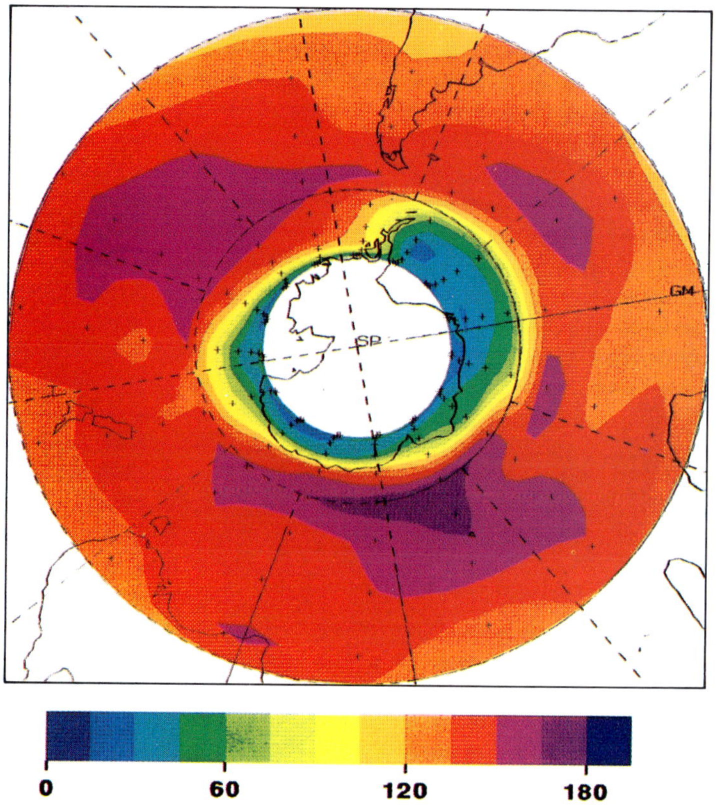
GM
SP
0
60
120
180

Plate 4

Depletion of ozone layer over Antarctica. (Top) Ozone values at an altitude of 50 millibars (about 19 km). (Bottom) Ozone values at 100 millibars altitude (about 15 km). These graphics from the NASA SAGE II satellite instrument show both the spatial and vertical depletion in ozone around the South Pole. The satellite images were obtained from September through October 1987. The color bars represent the pressure of ozone in nanobars. (Recall from the gas laws that $pv = nrt$; thus, pressure relates directly to volume.) (See Chapter 15.)

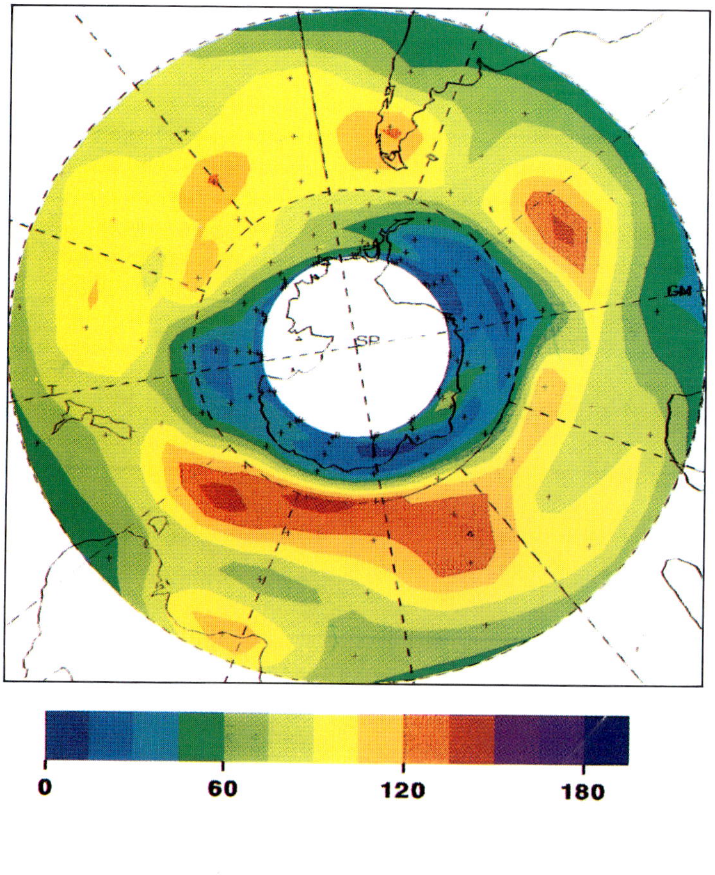
GM
SP
0
60
120
180

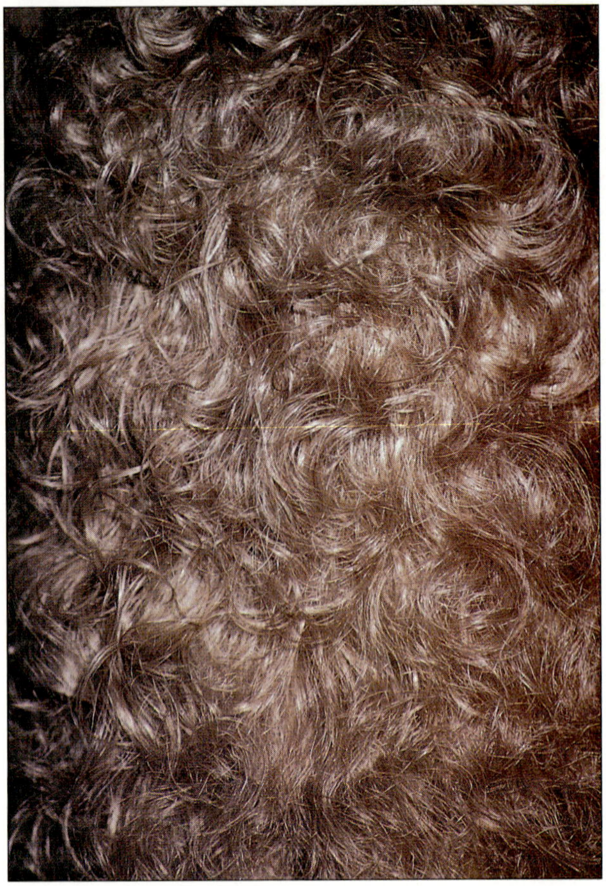

Plate 5
Hair is rich in disulfide bonds. Straight hair (left) can be subjected to a "perm," which reduces the disulfides to thiols. By rearranging the "reduced" hair on curlers, followed by an oxidation step, new disulfide bonds are created that hold the hair in the new curled shape (right). (See Chapter 18.)

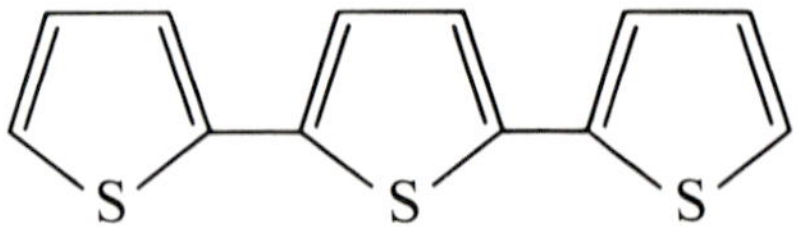

Marigold

Plate 6
Sulfur-containing molecules help to provide the unique fragrance of marigolds. (See Chapter 18.)

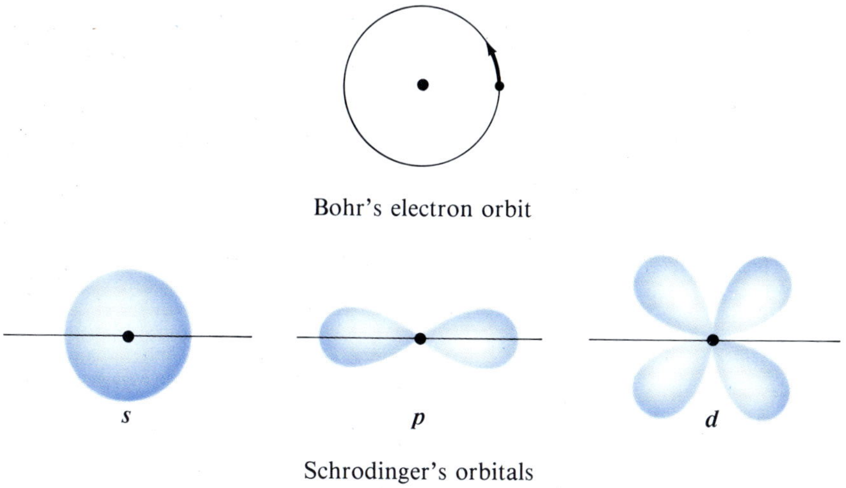
Bohr's electron orbit
s
p
d
Schrodinger's orbitals

FIGURE 4.7
In contrast to Bohr's planetary orbits (top), Schrödinger's electron wave model predicted a number of regions around the nucleus where electron waves would reinforce and where electrons would be found. These regions, or orbitals, have several different shapes: *s*, *p*, and *d*.

Schrödinger used *quantum numbers* to describe the possible electron orbitals, called *s*, *p*, and *d*. Lowest electron energies are associated with the simplest shapes.

and still others were shaped like four-leaf clovers. Each cloud could hold two electrons and was called an **orbital**.

Schrödinger's approach used three numbers called **quantum numbers** to describe each orbital surrounding an atom. The first number (n) described the average distance of each major group of electrons from the nucleus. These numbers corresponded to Bohr's orbit numbers. A second number (l) designated the shape of the orbital. A third number (m) designated the orbital's direction in space with respect to a hypothetical x,y,z three-dimensional axis centered on the nucleus. Schrödinger predicted several orbital shapes: a 0 shape number corresponding to a spherical, or ***s*** orbital; a shape number of 1 corresponding to a dumbbell-shaped orbital, designated ***p***; and a shape number of 2 predicting a four-leaf clover shaped orbital, designated ***d***.

Experimental support for Schrödinger's orbital model came from an ability to predict spectra and, as we will see in Chapter 6, from an ability to predict molecular shapes. The *s*, *p*, and *d* notation came from the terminology of those who studied spectra. Spectral lines were classified as "sharp" (*s*), bright or "principal" (*p*), or diffuse (*d*). Orbitals were named by the type of spectral line their electrons produced.

Schrödinger's quantum number theory made it possible to predict both the shape and direction of each orbital in each major electron group. Consider the first major group of electrons, those closest to the nucleus. For this group of electrons, only one type or shape of electron orbital is possible. This one has the simplest shape: a spherical orbital designated *s* (Figure 4.8).

For the second principal group of electrons, two orbital shapes are possible. The first shape is spherically symmetrical. It is called a 2*s* orbital (Figure 4.9). The next set of orbitals in the second electron group have a *p* or dumbbell shape. Three *p* orbitals are possible. These orbitals are shown in Figure 4.10. They are oriented on arbitrarily assigned *x*, *y*, and *z* axes and are called p_x, p_y, and p_z orbitals. Note that they do not overlap.

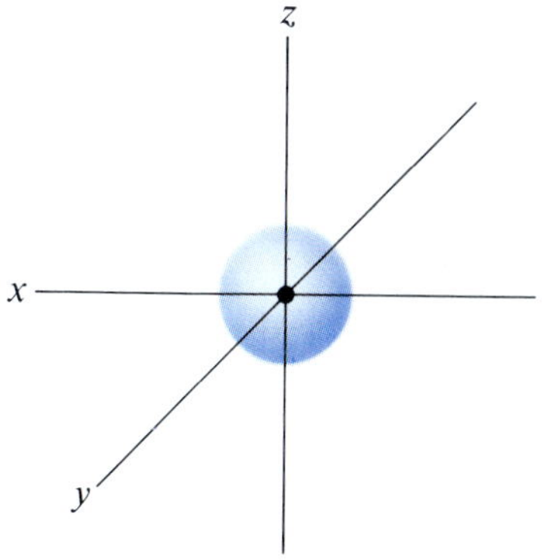
z
x
y

FIGURE 4.8
For the first principal electron group, only one orbital can exist. The computed shape for this orbital is spherically symmetrical, and it is called an *s* orbital.

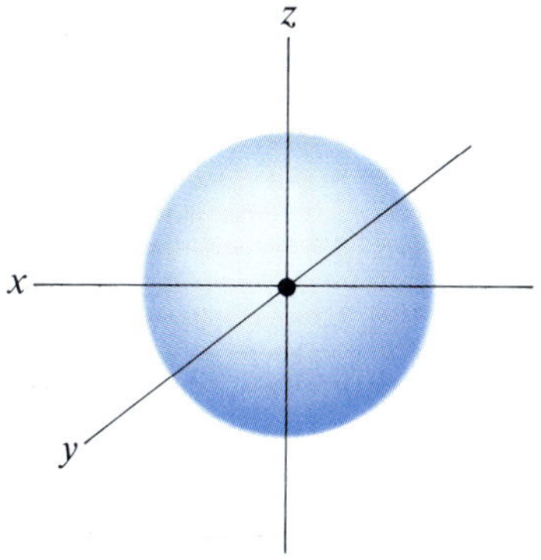
z
x
y

FIGURE 4.9
The first of the orbitals in the second principal group of electrons is spherically symmetrical, but larger than the *s* orbital of the first group.

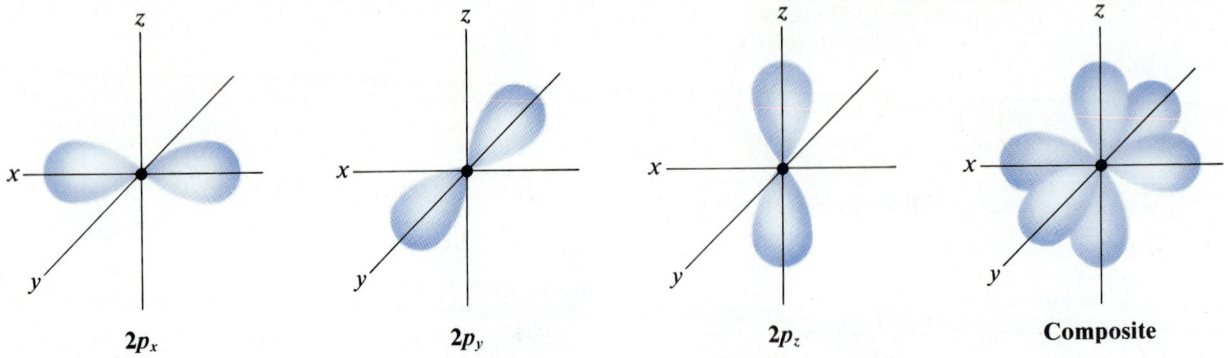
z
x
y
$2p_x$
$2p_y$
$2p_z$
Composite

FIGURE 4.10
Three p electron orbitals are possible. These orbitals are dumbbell-shaped and oriented at right angles to each other. The composite of all three is shown on the right.

The third major group of electrons has three different types of orbitals (Figure 4.11). It has one s orbital, three p orbitals, and five d orbitals, four of which have shapes resembling a four-leaf clover. No electrons are present in d orbitals in elements with atomic numbers less than 21. The transition elements introduced in Chapter 2 have d orbital electrons in their outer group (see the periodic table).

As a shorthand, each orbital can be designated by a number, called the **principal quantum number**, which indicates its major electron group or distance from the nucleus, and a letter (s, p, or d) describing its shape. Also, if necessary, a coordinate designation (x, y, or z) corresponding to an arbitrary coordinate system surrounding the nucleus can be added to describe its directional orientation (Figure 4.12). This coordinate designation is shown as a subscript on the shape indicator. For example, an electron may be described as being in a 1s orbital (first major electron group, spherical orbital); a $2p_x$ orbital (second major electron group, dumbbell-shaped orbital on the x axis); or a $4p_z$ orbital (fourth major electron group, dumbbell-shaped orbital, z axis).

NEW TERMS

orbital A region in the space of an atom occupied by as many as two electrons.

principal quantum number A set of numbers used to designate an electron's major electron group, its orbital shape, and its orbital direction.

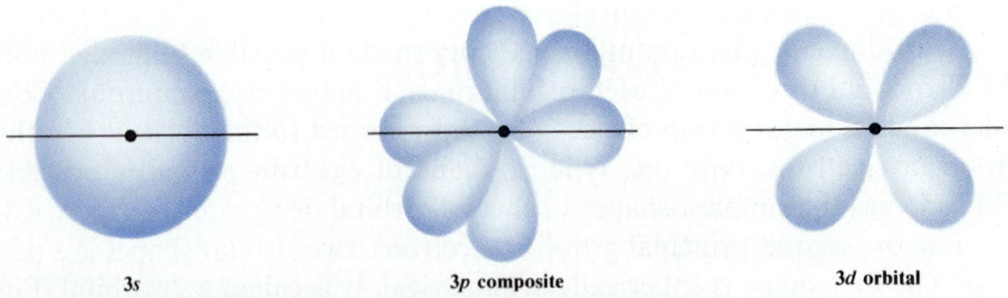
3s
3p composite
3d orbital

FIGURE 4.11
The third major electron group has three different types of orbitals—s, p, and d. Three p orbitals and five different d orbitals are possible.

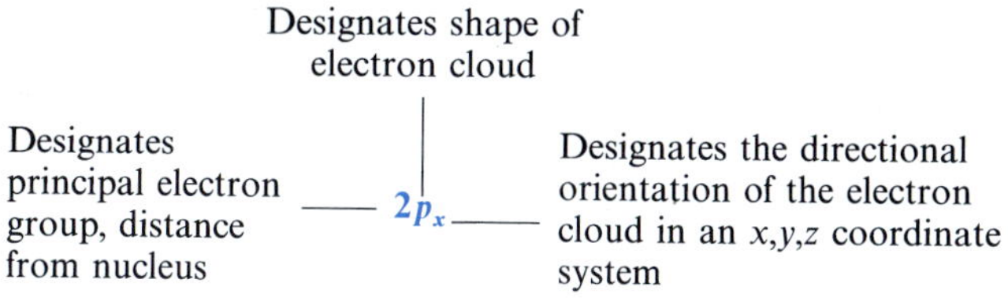
Designates shape of electron cloud
Designates principal electron group, distance from nucleus
$2p_x$
Designates the directional orientation of the electron cloud in an x,y,z coordinate system

FIGURE 4.12
Orbitals are represented by a convenient notation that designates the distance of the major electron group from the nucleus, the shape of the orbital, and the directional orientation of this orbital with respect to a coordinate system whose origin is at the nucleus.

4.5 Electron Filling Series

The **electron filling series**, or *Aufbau* rule, is an easy and useful way of remembering the sequence of orbital filling predicted by Schrödinger's quantum numbers. (*Aufbau* means "building up" in German.) This process of building up the electron structure of any atom assumes that electrons will first fill orbitals of lower energy. Each orbital will hold two electrons. These electrons differ slightly in energy, a difference caused by their direction of spin. Electrons always pair with another electron of opposite spin. The direction of electron spin is commonly indicated by a vertical arrow with the head up or down.

The electron filling series is an easy way to predict the orbital filling sequence.

A graphic way of presenting the electron filling series is shown in Figure 4.13. The electron arrangement for aluminum, atomic number 13, is illustrated as an example. The nucleus of an aluminum atom has 13 protons; hence a neutral atom of aluminum has 13 electrons. The small up and down arrows symbolize the opposite spins of the two electrons in each pair.

You can see from Figure 4.13 that the 4*s* orbital is actually lower in energy than the 3*d* orbital. Because of the more complex shape of the *d* orbitals, the simpler *s* orbital of the fourth electron group will fill before the *d* orbital of the lower group. A similar situation holds for the *d* orbital in the fifth, sixth, and seventh major electron groups.

A *d* orbital does not fill until the next higher group's *s* orbital is filled.

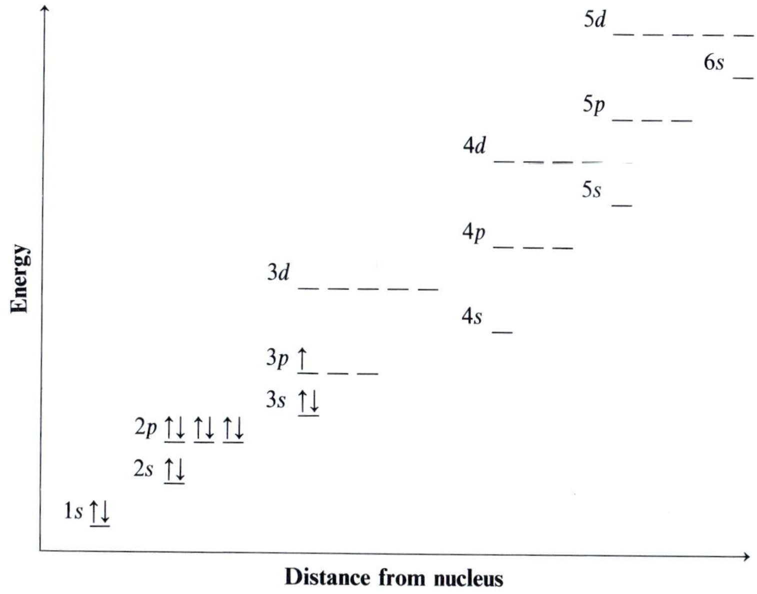
5d
6s
5p
4d
5s
4p
3d
4s
3p ↑
3s ↑↓
2p ↑↓ ↑↓ ↑↓
2s ↑↓
1s ↑↓
Energy
Distance from nucleus

FIGURE 4.13
The electron filling series is an easy way to remember the order of orbital filling. Orbitals each hold two electrons and fill from the lowest energy orbitals first. The example shown is for aluminum, atomic number 13. Aluminum's 13 electrons are shown as up and down arrows.

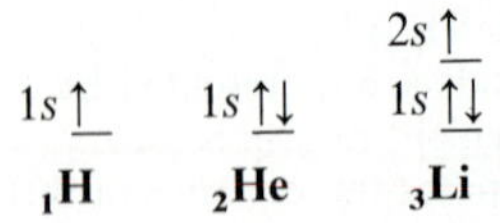
2s
1s 1s 1s
1H 2He 3Li

FIGURE 4.14
Electron energy diagrams for neutral atoms of hydrogen, helium, and lithum are illustrated in this figure. The number of electrons in a neutral atom is the same as the number of protons in the nucleus, as indicated by the atomic number. (It is normal practice to sketch only those orbitals actually occupied.)

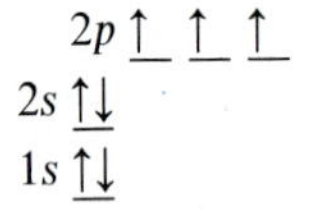
2p
2s
1s

FIGURE 4.15
An electron energy diagram for nitrogen. Note that the *p* orbital electrons are not paired.

The electron arrangement for neutral atoms of hydrogen, helium, and lithium is illustrated in Figure 4.14. The rules are simple: lowest energy orbitals fill first, with two electrons per orbital.

Figure 4.15 presents an electron filling series diagram for nitrogen, atomic number 7. Nitrogen, as a neutral atom, will have seven electrons. One begins by filling the 1*s* orbital with two electrons, then the 2*s* orbital with two more electrons. This process leaves only three electrons to place in the three 2*p* orbitals. Do these electrons go in separate orbitals, or will two of them pair together to occupy one orbital and the third remain by itself in a second orbital?

Hund's Rule

This question is answered by **Hund's rule**, which states that electrons will not pair in orbitals of the same energy until each orbital is singly occupied. This rule is easily understood in terms of electron–electron repulsion. We recall that the three *p* orbitals differ in directional orientation, but not in energy. Since electrons carry a negative charge, it seems reasonable that a lower energy situation would result if two electrons were in separate orbitals with different directional characteristics than if they occupied the same orbital. The same holds true for the *d* and *f* orbitals. Thus, the rule that electrons will not pair until each orbital of the same energy is occupied by one electron seems reasonable.

Orbital Notation

Orbital notation is a convenient way to show the electron arrangement in an atom.

In many cases it is convenient to express the electron configuration of an atom in a form that is more easily written than the electron filling series chart. In **orbital notation**, the orbital populations are simply listed; for example, nitrogen is written this way:

$$1s^2 2s^2 2p^3$$

The superscripts indicate the number of electrons in that orbital. Phosphorus, atomic number 15, should have 15 electrons in its neutral state. The Aufbau principle predicts that these electrons will be arranged as illustrated in Figure 4.16. Note that the electron configuration of the outermost principal electron group (the third group for phosphorus) has two *s* electrons and three *p* electrons. This is the same as that for nitrogen! Now locate the relative positions of nitrogen and phosphorus on the periodic table (see Figure 4.18). They are both in the same vertical column, or family of elements.

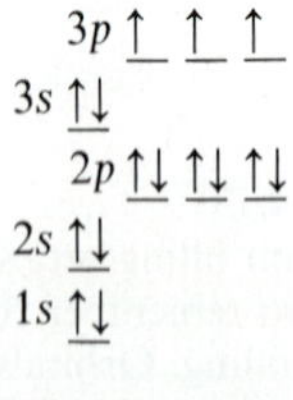
3p
3s
2p
2s
1s

FIGURE 4.16
An electron energy diagram for phosphorus. Note that the outer electron group arrangement for phosphorus is the same as that of nitrogen (without regard for the value of *n*): s^2p^3.

EXAMPLE 4.1

Sketch an electron energy diagram for a neutral atom of oxygen, $_8O$, and write its electron configuration in orbital notation.

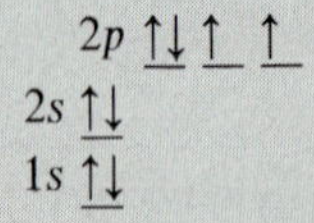
2p
2s
1s

$_8O$

$1s^2 2s^2 2p^4$

EXAMPLE 4.2

Sketch an electron energy diagram for a neutral atom of chlorine, $_{17}$Cl, and write its electron configuration in orbital notation.

3*p* ↑↓ ↑↓ ↑ $\qquad$ $_{17}$Cl
3*s* ↑↓
2*p* ↑↓ ↑↓ ↑↓
2*s* ↑↓
1*s* ↑↓ $\qquad$ $1s^2 2s^2 2p^6 3s^2 3p^5$

Electron Dots

Another useful method of indicating the arrangement of electrons around an atom is the **electron dot method**. This approach simply involves arranging dots representing the electrons in the outermost group symmetrically in four quadrants around the symbol for the element. The electron dot method is useful in predicting chemical bonds. Since bonds form only with outer group electrons of adjacent atoms, it has become common practice to indicate only the outer group electrons in electron dot diagrams. Note that electron dot diagrams do not distinguish between *s* and *p* electrons.

Electron dots are sometimes used to show the outer group electron arrangement around an atom. It is useful in predicting chemical bonds.

To determine the number of outer group electrons, you must first sketch an electron energy diagram, as shown in the following examples.

EXAMPLE 4.3

Construct an electron dot model for a neutral atom of the element sodium, $_{11}$Na. Remember that electron dot models only show outer group electrons.

3*p* _ _ _ } Outer
3*s* ↑ } group
2*p* ↑↓ ↑↓ ↑↓
2*s* ↑↓ $\qquad$ $_{11}$Na $\quad$ Ṅa
1*s* ↑↓

EXAMPLE 4.4

Construct an electron dot model for a neutral atom of the element sulfur, $_{16}$S.

3*p* ↑↓ ↑ ↑ } Outer
3*s* ↑↓ } group
2*p* ↑↓ ↑↓ ↑↓
2*s* ↑↓ $\qquad$ $_{16}$S $\quad$:S̤̈·
1*s* ↑↓

EXAMPLE 4.5

Construct an electron dot model for a neutral atom of the element oxygen, ${}_8O$.

$2p$ ↑↓ ↑ ↑ } Outer group

$2s$ ↑↓

$1s$ ↑↓

${}_8O$:Ö·

EXAMPLE 4.6

Construct an electron dot model for a neutral atom of the element argon, ${}_{18}Ar$.

$3p$ ↑↓ ↑↓ ↑↓ } Outer group

$3s$ ↑↓

$2p$ ↑↓ ↑↓ ↑↓

$2s$ ↑↓

$1s$ ↑↓

${}_{18}Ar$:Ar: (eight dots)

EXAMPLE 4.7

Construct an electron dot model for a neutral atom of the element neon, ${}_{10}Ne$.

$2p$ ↑↓ ↑↓ ↑↓ } Outer group

$2s$ ↑↓

$1s$ ↑↓

${}_{10}Ne$:Ne: (eight dots)

Note the similar arrangement of outer group electrons around oxygen and sulfur. Both oxygen and sulfur contain one filled *s* orbital, one filled *p* orbital, and two half-filled *p* orbitals. This similarity in electron arrangement results in similar chemical reactivity toward other atoms. From the point of view of an adjacent atom wishing to form a chemical bond, oxygen and sulfur look quite similar. Also note the similarity between argon and neon. These elements, members of the family called *inert* or *noble gases*, are both chemically nonreactive.

NEW TERMS

electron filling series A system by which orbitals are filled with electrons, beginning with orbitals of lowest energy.

Hund's rule Electrons will not pair (occupy the same orbital) until all orbitals of that energy have at least one electron. Electrons are negative and will stay as far apart as possible as long as possible.

orbital notation A shorthand for indicating the orbital "addresses" of electrons. An atom with two 1*s* electrons, two 2*s* electrons, and a 2*p* electron could have its orbital population expressed as $1s^2 2s^2 2p^1$.

electron dot method A graphic shorthand for writing the outer group electron complement of an atom. Used principally for showing electron sharing and electron transfer, but does not specify the orbitals involved.

TESTING YOURSELF

Orbitals and the Electron Filling Series

1. Which orbital looks like a dumbbell: *s*, *p*, or *d*?
2. According to Hund's rule, how many *p* orbitals would be occupied in a neutral atom of oxygen (atomic number 8)? How many would be filled?
3. What is the outermost electron in a neutral atom of sodium (Na)?
4. What is the electron configuration of a neutral atom of sulfur (S)?

Answers **1.** *p* **2.** 3, 1 **3.** 3*s* **4.** $1s^2 2s^2 2p^6 3s^2 3p^4$

4.6 Electrons and the Periodic Table

Since the early 1800s, attempts have been made to classify elements into families according to similarities in chemical reactivity and behavior. As we saw in Chapter 2, one of the most successful attempts was made by Dmitri Mendeleev; this has been passed down to us as a chart listing families of similar elements in vertical columns and was discussed in Chapter 2. Three rows of Mendeleev's chart are reproduced in Figure 4.17, along with the electron arrangements predicted for each of these elements by the electron filling series.

It is reasonable to expect that the chemical behavior of an atom would be related to how it appears to a neighboring atom, which of course is related to the arrangement of its outermost group of electrons. Mendeleev classified the elements in his chart according to similar chemical reactivity. As we view the electron arrangement predicted by the Aufbau principle, it is also apparent that elements in vertical columns have the same outermost *s* and *p* electron complement. Hydrogen, lithium, and sodium all have one outer level *s* electron. Oxygen and sulfur both have two outer level *s* electrons

FIGURE 4.17
Elements of similar chemical reactivity have identical electron arrangements in their *outermost* group of electrons.

H 1s ↑							**He** 1s ↑↓
Li 2p _ _ _ 2s ↑	**B** 2p _ _ _ 2s ↑↓	**Be** 2p ↑ _ _ 2s ↑↓	**C** 2p ↑ ↑ _ 2s ↑↓	**N** 2p ↑ ↑ ↑ 2s ↑↓	**O** 2p ↑↓ ↑ ↑ 2s ↑↓	**F** 2p ↑↓ ↑↓ ↑ 2s ↑↓	**Ne** 2p ↑↓ ↑↓ ↑↓ 2s ↑↓
Na 3p _ _ _ 3s ↑	**Mg** 3p _ _ _ 3s ↑↓	**Al** 3p ↑ _ _ 3s ↑↓	**Si** 3p ↑ ↑ _ 3s ↑↓	**P** 3p ↑ ↑ ↑ 3s ↑↓	**S** 3p ↑↓ ↑ ↑ 3s ↑↓	**Cl** 3p ↑↓ ↑↓ ↑ 3s ↑↓	**Ar** 3p ↑↓ ↑↓ ↑↓ ↑↓

PERIODS	GROUPS I	II											III	IV	V	VI	VII	VIII	
1	1 H $1s^1$																	2 He $1s^2$	1
2	3 Li $2s^1$	4 Be $2s^2$											5 B $2s^22p^1$	6 C $2s^22p^2$	7 N $2s^22p^3$	8 O $2s^22p^4$	9 F $2s^22p^5$	10 Ne $2s^22p^6$	2
3	11 Na $3s^1$	12 Mg $3s^2$											13 Al $3s^23p^1$	14 Si $3s^23p^2$	15 P $3s^23p^3$	16 S $3s^23p^4$	17 Cl $3s^23p^5$	18 Ar $3s^23p^6$	3
4	19 K $4s^1$	20 Ca $4s^2$	21 Sc $4s^23d^1$	22 Ti $4s^23d^2$	23 V $4s^23d^3$	24 Cr $4s^13d^5$	25 Mn $4s^23d^5$	26 Fe $4s^23d^6$	27 Co $4s^23d^7$	28 Ni $4s^23d^8$	29 Cu $4s^13d^{10}$	30 Zn $4s^23d^{10}$	31 Ga $4s^24p^1$	32 Ge $4s^24p^2$	33 As $4s^24p^3$	34 Se $4s^24p^4$	35 Br $4s^24p^5$	36 Kr $4s^24p^6$	4
5	37 Rb $5s^1$	38 Sr $5s^2$	39 Y $5s^24d^1$	40 Zr $5s^24d^2$	41 Nb $5s^14d^4$	42 Mo $5s^14d^5$	43 Tc $5s^24d^5$	44 Ru $5s^14d^7$	45 Rh $5s^14d^8$	46 Pd $4d^{10}$	47 Ag $5s^14d^{10}$	48 Cd $5s^24d^{10}$	49 In $5s^25p^1$	50 Sn $5s^25p^2$	51 Sb $5s^25p^3$	52 Te $5s^25p^2$	53 I $5s^25p^5$	54 Xe $5s^25p^6$	5
6	55 Cs $6s^1$	56 Ba $6s^2$	57 *La $6s^25d^1$	72 Hf $6s^25d^2$	73 Ta $6s^25d^3$	74 W $6s^25d^4$	75 Re $6s^25d^5$	76 Os $6s^25d^6$	77 Ir $6s^25d^7$	78 Pt $6s^15d^9$	79 Au $6s^15d^{10}$	80 Hg $6s^25d^{10}$	81 Tl $6s^26p^1$	82 Pb $6s^26p^2$	83 Bi $6s^26p^3$	84 Po $6s^26p^4$	85 At $6s^26p^5$	86 Rn $6s^26p^6$	6
7	87 Fr $7s^1$	88 Ra $7s^2$	89 **Ac $7s^26d^1$																

* Lanthanides

58 Ce $6s^24f^2$	59 Pr $6s^24f^3$	60 Nd $6s^24f^4$	61 Pm $6s^24f^5$	62 Sm $6s^24f^6$	63 Eu $6s^24f^7$	64 Gd $6s^25d^14f^7$	65 Tb $6s^24f^9$	66 Dy $6s^24f^{10}$	67 Ho $6s^24f^{11}$	68 Er $6s^24f^{12}$	69 Tm $6s^24f^{13}$	70 Yb $6s^24f^{14}$	71 Lu $6s^25d^14f^{14}$

** Actinides

90 Th $7s^26d^2$	91 Pa $7s^26d^15f^2$	92 U $7s^26d^15f^3$	93 Np $7s^26s^15f^4$	94 Pu $7s^25f^6$	95 Am $7s^25f^7$	96 Cm $7s^26d^15f^7$	97 Bk $7s^25f^9$	98 Cf $7s^25f^{10}$	99 Es $7s^25f^{11}$	100 Fm $7s^25f^{12}$	101 Md $7s^25f^{13}$	102 No $7s^25f^{14}$	103 Lw $7s^26d^15f^{14}$

FIGURE 4.18
The modern periodic table including the notation for electron configuration. Note that the electron configuration of the outermost electron group is similar throughout each chemical family (vertical column).

and four outer level p electrons. The inert gases (helium, neon, and argon in Figure 4.17) all have completely filled outermost electron groups. Outer group electron structures for all of the elements in the periodic table are shown in Figure 4.18. Note the identical outer group orbital configuration among members of each vertical family of elements.

Elements in vertical columns of the periodic table have similar outer group electron configurations. This is the reason for their similar chemical behavior.

Is it the number and arrangement of electrons in the outermost populated group of electrons that determines the chemical reactivity of the element? This is indeed the case, as we will see in the next chapter.

TESTING YOURSELF

Electrons and the Periodic Table

1. According to the periodic table (Figure 4.18), which two of the following elements would behave chemically in a similar manner: Li, C, Ne, Na, and Si?
2. What is the outer electron configuration of these two elements?
3. Using your knowledge of chemical periodicity, what is the outermost electron configuration of the element bromine (Br)?

Answers **1.** Li and Na **2.** s^1 **3.** s^2p^5

SUMMARY

An understanding of the arrangement of electrons in atoms is one of the most important concepts of chemistry. Because atoms bond together by transferring or sharing electrons, the chemical reactivity of an element is determined by the structure of its outermost group of electrons.

Because electrons generate light when they change their position suddenly, the color of light given off by an atom is a direct clue to the change in energy of its electrons. Niels Bohr suggested that the micro world of the atom might be organized in a manner similar to the solar system—its nucleus (sun) in the center surrounded by orbiting electrons (planets) in well-defined orbits. Bohr surmised that if electrons run in planetary orbits around their nucleus, heated atoms should only give off certain colors of light when their dislodged electrons drop back into their inner orbits. Bohr's model, however, only worked if one presupposed only certain electron orbits to be possible. Bohr could not explain why this was so. Louis de Broglie later suggested that objects as small as electrons would have observable wave properties. This idea was soon confirmed experimentally. This was the answer to Bohr's problem—if electrons have wave properties, they can only exist in orbits that are a whole number multiple of a wavelength in circumference. Otherwise, the electron wave would destructively interfere and the electron would not exist.

Application of the known principles of wave behavior to the planetary electron model yield a prediction that electrons would favor certain regions of space around the nuclus called *orbitals*. The shape of each region was dependent on the energy of the electron. Orbitals found in the common elements are of spherical shape (s orbitals), dumbbell shape (p orbitals), and four-leaf clover shape (d orbitals). The *electron filling series* shows graphically how the orbitals of an atom fill, beginning with the lowest energy (innermost) electron group and simplest orbital shape and working out toward higher energy electron groups and more complex orbital shapes. Because the atomic number of an element specifies the number of electrons in its neutral atom, one can use this information to predict the organization of electrons in any atom.

The electron filling series provides the rationale for the families of elements found in the periodic table. All elements in each vertical column or family have the same *outer group* electron configuration and thus similar chemical behavior. The periodic table is a unifying force in chemistry.

Now that you know quite a bit about the orbital arrangement of electrons in atoms, let's use this information in the next chapter to see how atoms bond together to form multiatom groups called molecules.

TERMS

wavelength
frequency
wave velocity
electromagnetic radiation
spectrum
Planck's law
photon
spectral lines
line spectrum
continuous spectrum
Bohr's planetary electron model
orbital
principal quantum number
electron filling series
Hund's rule
orbital notation
electron dot method

CHAPTER REVIEW

QUESTIONS

1. Which of these general statements relates frequency to wavelength for any type of wave motion?
   a. As frequency increases, wavelength increases.
   b. As frequency decreases, wavelength increases.
   c. As frequency increases, wavelength stays the same.
2. Which of the following colors of light has the shortest wavelength?
   a. blue b. red c. orange d. green e. yellow
3. Which of the following types of electromagnetic radiation carries the greatest energy?
   a. blue light d. microwaves
   b. radio waves e. heat (infrared) radiation
   c. Ultraviolet light
4. What is the process that occurs whenever two waves cross?
   a. interference c. ionization
   b. multiplication d. no interaction when waves cross
5. Some substances give off all colors of light when heated (a continuous spectrum), while others give off only certain colors (a line spectrum). The type of spectrum emitted is a clue to the physical state of the heated material. Which type of material emits a line spectrum?
   a. solid b. liquid c. gas
6. The emission of light by atoms appears to be related to the movement of which of the following components of the atom?
   a. protons b. neutrons c. electrons d. nuclei
7. Niels Bohr introduced one new and important idea to our understanding of the structure of atoms. Which of the following statements best describes Bohr's contribution?
   a. Bohr indicated that the nucleus of the atom is very small and surrounded by by empty space.
   b. Bohr's model permitted more accurate computation of spectral lines.
   c. Bohr's model tied the energy of the electrons to specific orbits around the nucleus.

d. Bohr's model predicted several different shapes for electron paths around the nucleus.

8. Schrödinger's model was able to predict the regions of space around a nucleus occupied by electrons of a given energy. His quantum numbers are keys to predicting the shapes occupied by electrons of different energy. In Schrödinger's third major group of electrons ($n = 3$), how many different types of orbitals are possible?
   a. 1 b. 2 c. 3 d. 4 e. 5

9. Which of the following shapes represents a "p" orbital?
   a.  b.  c. 

10. Using your knowledge of the electron filling series, predict the orbital in which the outermost electrons of a neutral atom of sulfur would be found.
    a. $2s$ b. $2p$ c. $3s$ d. $3p$ e. $3d$

11. The orbital notation for the electron distribution of a neutral atom is shown here

$$1s^2 2s^2 2p^6 3s^2 3p^2$$

    Which element is represented?
    a. O b. Si c. P d. Ca e. Cl

Answers **1.** b **2.** a **3.** c **4.** a **5.** c **6.** c **7.** c **8.** c **9.** b **10.** d **11.** b

DIAGNOSTIC CHART

Blacken in all of the circles under the number of each question you missed in the Chapter Review questions. The diagnostic chart will help you identify concept areas that need more study.

	Questions										
Concepts	1	2	3	4	5	6	7	8	9	10	11
Waves and light	○	○	○	○							
Continuous and line spectra					○						
Bohr atom						○	○				
Quantum numbers								○	○		
Electron filling series										○	○

EXERCISES

Waves and Light

1. In which of the following lists are the colors of the visible spectrum arranged in order of decreasing wavelength?
   a. Red, orange, yellow, green, blue, violet
   b. Violet, blue, green, yellow, orange, red
   c. Red, violet, orange, yellow, green, blue
   d. Violet, red, orange, yellow, green, blue

2. Which of the colors listed in question 1 is of greatest energy?

3. Why is it dangerous to look at an ultraviolet light?

Band and Line Spectra

4. The line spectra of a gaseous element is an excellent "fingerprint" of the element. The combination of spectral lines emitted by a gaseous element are unique for each element and can be used to identify the element. Geological samples, moon rocks, and metals are routinely analyzed in this way. The top four spectra in the figure shown here are emission spectra for several pure elements (Ca, Li, Sr, and Na). The spectrum at the bottom is a mixture of several of these elements. Which of the known elements are present in the mixture?

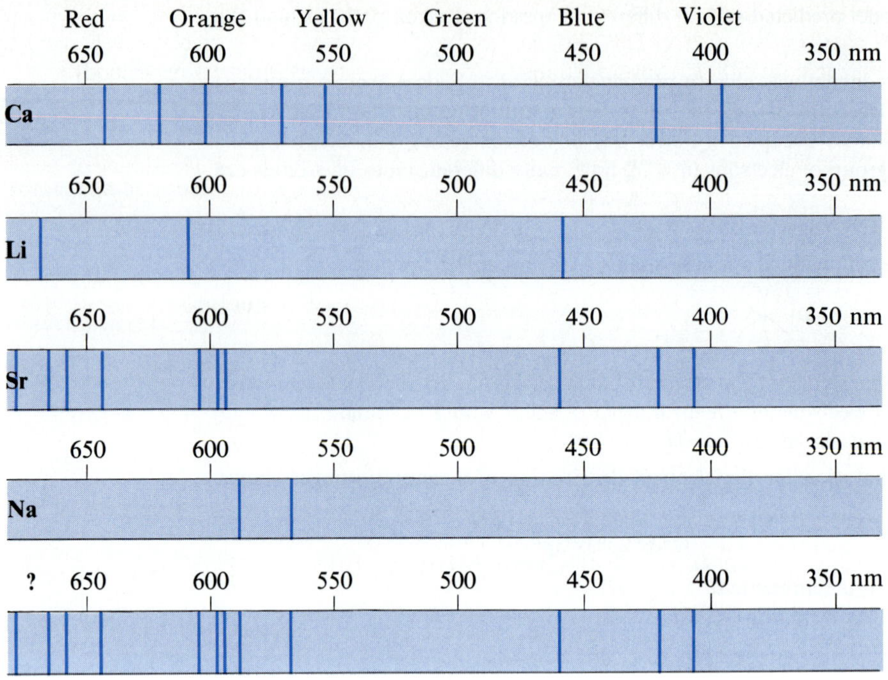

Red Orange Yellow Green Blue Violet
650 600 550 500 450 400 350 nm
Ca
Li
Sr
Na
?

5. Using the information from question 4, what color would you expect a flare containing calcium to be?
6. What color would you expect a sodium vapor street light to be?
7. What color would you expect fireworks containing strontium to be?
8. The color of fireworks is caused by a mixture of gunpowder and certain metals. Which metal will produce a yellow flame? Which will produce a red flame?
9. The figure below represents a Bohr model of an atom and a spectrum given off by this atom. Which of the spectral lines (a–f) would probably be caused by the orbit 4 to orbit 2 transition?

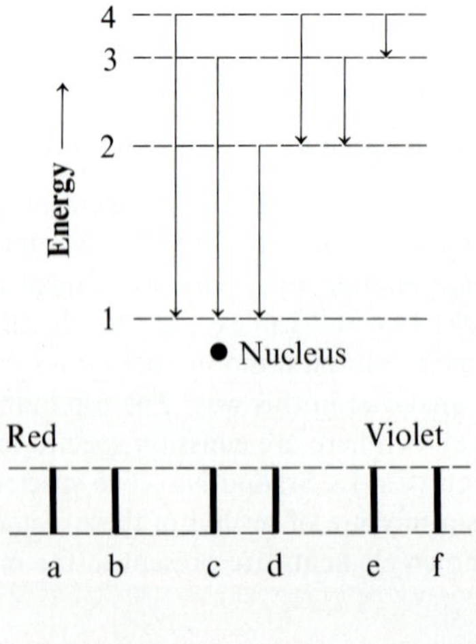

4
3
2
1
Energy ⟶
● Nucleus
Red
Violet
a b c d e f

Electron Filling Series

10. Using your knowledge of the electron filling series, which orbital would you expect the outermost electron of an atom of aluminum (atomic number 13) to occupy?
   a. 1*s* b. 2*s* c. 2*p* d. 3*s* e. 3*p*
11. The orbital notation for the electron distribution of a neutral atom is shown here.

$$1s^22s^22p^63s^23p^64s^2$$

   Which element is represented?
   a. O b. Si c. P d. Ca e. Cl
12. The outermost electron configuration of a neutral atom of oxygen would be
   a. s^2 b. s^2p^2 c. s^2p^4 d. s^2p^5 e. s^2p^6
13. Give the electron distribution for neutral atoms of the following elements:

Na	F
B	As
C	Ca
N	Be
S	Ne
Cl	

14. Is there any similarity among the electron distributions of Be, Mg, and Ca?
15. Using your knowledge of chemical periodicity, what is the outermost electron configuration of a neutral atom of the element argon (Ar).
   a. s^2 b. s^2p^2 c. s^2p^5 d. s^2p^6

16. One of the byproducts of nuclear fission is the radioactive isotope of strontium (^{90}Sr). This isotope, produced in reasonable quantities during atmospheric nuclear testing, has been spread widely over the earth by winds, washed from the atmosphere by rain, deposited on grass, ingested by livestock, and finally has found its way into the human food supply. Radioactive strontium, because of its similarity to an element of major physiological importance, has been easily incorporated into the human body and has caused a measurable increase in the incidence of leukemia worldwide, particularly in children. Using your knowledge of chemical periodicity, predict which of the following elements strontium replaces in the body.
    a. iron (plays an important part in oxygen transfer in the blood)
    b. iodine (important to the thyroid)
    c. calcium (an important component of bone)
    d. sodium and potassium (important to the nerve transmission mechanism)
    e. carbon (forms the basic skeleton of most of the molecules in our body)

CHAPTER 5

The Chemical Bond

There is a certain electron configuration with which atoms are most stable. They will transfer or share electrons to achieve this condition, and in the process, they become strongly attracted to one another. This forms the basis of the chemical bond.

OBJECTIVES

After completing this chapter, you should be able to

- State the configuration of the outer group of electrons that is common to all inert gas elements.
- Using the electron filling series and the stable electron configuration of the inert gases, predict the outer electron structure and ionic charge of stable ions up to atomic number 20.
- Use the electron filling series and the periodic table to predict balanced formulas for two-element ionic compounds.
- Explain the effect of ionic charge and ion size on the strength of an ionic bond, and explain why ionic compounds will conduct electricity when molten.
- Explain how nuclear charge and nucleus–electron distance affect the electron-attracting power of an atom.
- Define *ionization energy*, and explain how it varies across the periodic table.
- Describe the attractive and repulsive forces operating in a single covalent bond.
- Use electron filling series or electron dot models to predict the number of covalent bonds in a covalent compound.
- Define *electronegativity*, and use it to classify chemical bonds as covalent, polar covalent, or ionic.
- Explain why metals conduct electricity.

5.1 Electron Configuration of the Inert Gases

Seldom are isolated atoms found in nature. Sometimes they are bound to other atoms of the same element. For example, atoms of gold are bound to many other atoms of gold to form nuggets. Other elements exist as **diatomic molecules**, which are two atoms of the same element bound to each other in pairs. For example, oxygen and nitrogen exist in the atmosphere as O_2 and N_2. Most atoms, however, are bound to atoms of a different sort, such as carbon atoms bound to hydrogen atoms (natural gas, CH_4), oxygen atoms bound to hydrogen atoms (water, H_2O), and sodium atoms bound to chlorine atoms (table salt, NaCl). These bonds are quite strong, and significant amounts of energy are required to separate the atoms. Once separated, the component atoms quickly recombine with each other or with adjacent atoms capable of an independent existence. The component atoms are not chemically stable alone.

A few elements, however, exist independently in our environment, each atom by itself. These elements are called the **inert gases**, which we mentioned in Chapter 4. They are sometimes found in high concentrations, such as the helium deposits in the North Texas gas fields. The inert gas elements are unique. Each atom acts as an independent agent and is not dependent upon combination with other atoms to be chemically stable. Why do they behave differently than all other known elements?

As one might expect from their similar chemical behavior, the inert gas elements form a vertical column or "family" in the periodic table (Figure 5.1). All members of the inert gas family have one common characteristic: a complete complement of *s* and *p* electrons in their outermost group of electrons (Figure 5.2). This is called a **stable electron configuration**. (Helium, an apparent exception to this rule, technically qualifies since a full complement of *p* electrons in group 1 is, of course, zero.)

The inert gases are chemically stable as individual atoms and have a full complement of outer group *s* and *p* electrons.

The fact that the chemically stable inert gas elements all have a full complement of outer level *s* and *p* electrons suggests that this might also be

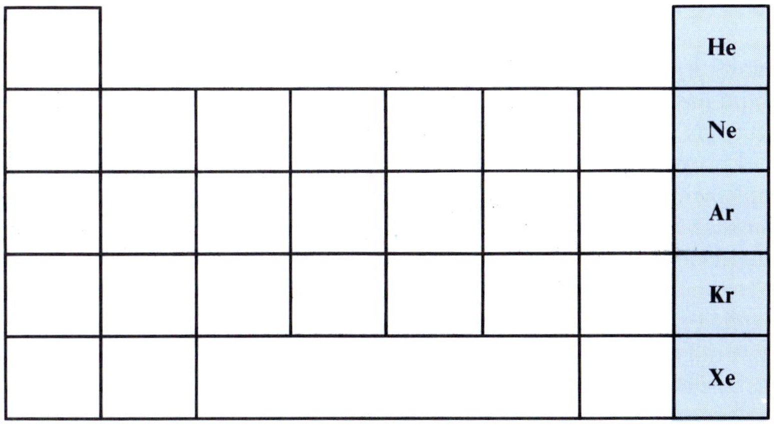
He
Ne
Ar
Kr
Xe

FIGURE 5.1
A family of elements known as the inert gases are unique in their ability to exist in a chemically stable form as individual atoms. The reason for this chemical stability is found in the similar configuration of their outermost group of electrons.

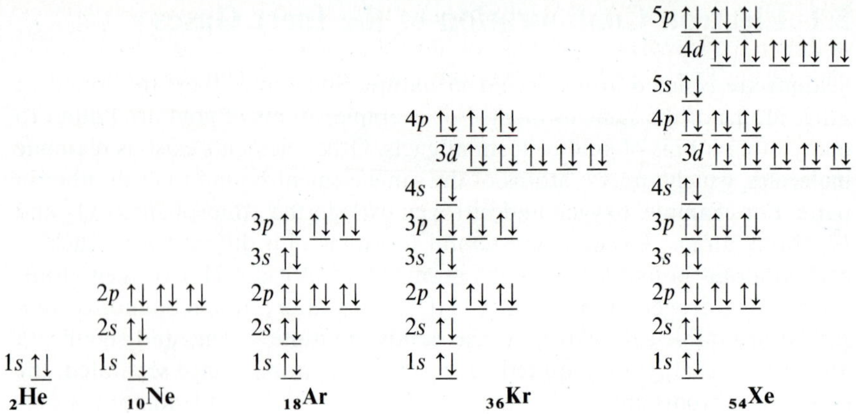
5p ↑↓ ↑↓ ↑↓
4d ↑↓ ↑↓ ↑↓ ↑↓ ↑↓
5s ↑↓
4p ↑↓ ↑↓ ↑↓
3d ↑↓ ↑↓ ↑↓ ↑↓ ↑↓
4s ↑↓
3p ↑↓ ↑↓ ↑↓
3s ↑↓
2p ↑↓ ↑↓ ↑↓
2s ↑↓
1s ↑↓
₂He
₁₀Ne
₁₈Ar
₃₆Kr
₅₄Xe

FIGURE 5.2
Each of the inert gas elements has a complete complement of *s* and *p* electrons in its outer electron group. Note that helium, with only two *s* electrons in its first electron group, qualifies for this rule since a full complement of *p* electrons for electron group 1 is zero.

a criterion for the chemical stability of other elements. Now we will examine how an element such as sodium, with only one outer group electron, or chlorine, with seven outer group electrons, can meet this criterion.

NEW TERMS

diatomic molecules Stable molecules comprised of two atoms of the same element. Oxygen (O_2), nitrogen (N_2), and chlorine (Cl_2) are examples of diatomic molecules.

inert gases A family of gases that are stable as individual atoms, all having a full complement of outer group *s* and *p* electrons. They are sometimes called the *noble gases.*

stable electron configuration A complete complement of outer-group *s* and *p* electrons. Often referred to as a noble gas electron complement.

5.2 Ionic Bonding: Electron Transfer as a Route to Chemical Stability

Figure 5.3 presents electron distribution diagrams for sodium and chlorine. Sodium has one *s* electron in its outermost electron group; chlorine has two *s* electrons and five *p* electrons in its outermost electron group. According to our proposed criterion for chemical stability outlined in Section 5.1 (a full complement of *s* and *p* electrons in the outermost electron group), neither atom should be chemically stable by itself. This fact is supported by experimental evidence.

However, suppose that it were possible to remove one electron from the sodium atom. This would leave it with no electrons in its third electron group and would make its second electron group (consisting of two *s* and six *p* electrons) the outermost set of electrons. Sodium would then qualify for chemical stability, based on our proposed rule. The electron configuration of sodium would be the same as the inert gas neon.

But what about chlorine? Chlorine has two *s* electrons and five *p* electrons in its outermost group of electrons. Suppose that the single *s* electron removed from the sodium atom were given to the chlorine atom. Chlorine would then have two *s* electrons and six *p* electrons in its outer electron

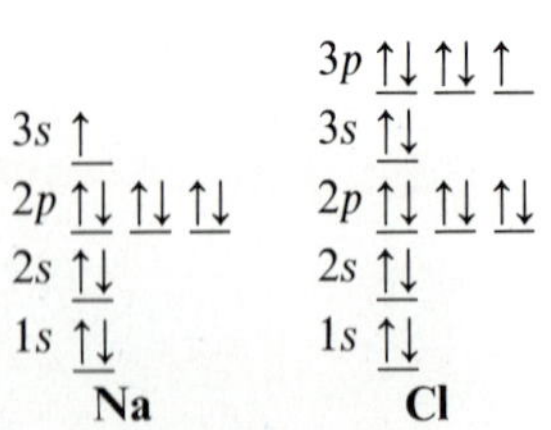
3p ↑↓ ↑↓ ↑
3s ↑
3s ↑↓
2p ↑↓ ↑↓ ↑↓
2s ↑↓
1s ↑↓
Na
Cl

FIGURE 5.3
The electron configuration of sodium (Na) and chlorine (Cl). Neither element has the full complement of *s* and *p* electrons in its outer electron group required for chemical stability.

group, which is the full complement of *s* and *p* electrons required for chemical stability. The chlorine atom would have the same electron configuration as the inert gas argon. **Electron transfer** from sodium to chlorine is illustrated in Example 5.1.

Atoms can transfer electrons from one to another to achieve a stable electron configuration.

EXAMPLE 5.1

Show the electron transfer of one electron from sodium (Na) to chlorine (Cl).

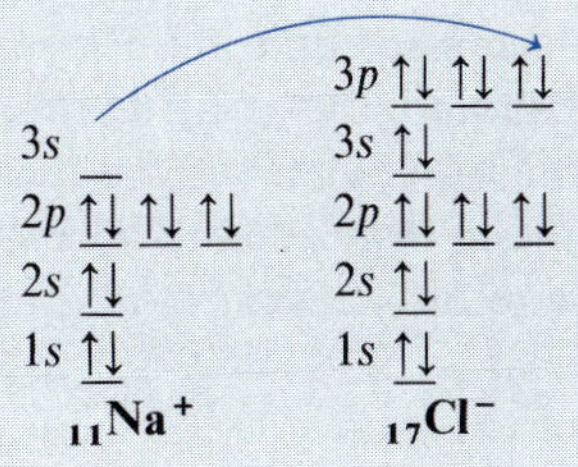
3p
3s
2p
2s
1s
$_{11}Na^+$
3p
3s
2p
2s
1s
$_{17}Cl^-$

What are the implications of such an electron transfer? Sodium, which has 11 protons (or 11 plus charges) in its nucleus, will now have only 10 electrons (or 10 minus charges) if one is removed, for an overall charge of +1. Chlorine, with 17 protons in its nucleus, will now have 18 electrons, for an overall charge of −1. The transfer of electrons results in an unbalanced electrical charge at both atoms. These charged atoms are called **ions**. A positively charged ion is called a **cation** (in this case, Na^+), and a negatively charged ion is called an **anion** (Cl^-).

An ion is a charged atom that results when an atom has more or less electrons than the number of protons in its nucleus.

The transfer of an electron from sodium to chlorine results in a full complement of outer group *s* and *p* electrons in each atom, but a +1 electrical charge will exist on the stable sodium ion and a −1 electrical charge on the stable chlorine ion (Example 5.2). These two ions will be held together by the electrical attraction between their unlike electrical charges. Since one sodium atom supplies the electron required by one chlorine atom and since the plus and minus charges add up to zero, the prediction of one sodium atom for each chlorine atom for table salt (NaCl) is reasonable. The **formula** NaCl indicates this one-to-one ratio of atoms. The fact that this is also the experimentally determined formula for the compound of sodium and chlorine lends credibility to our rule for chemical stability.

EXAMPLE 5.2

Using orbital notation, show electron transfer between sodium and chlorine.

Transfer

$$1s^22s^22p^63s^1 + 1s^22s^22p^63s^23p^5 \longrightarrow 1s^22s^22p^6 + 1s^22s^22p^63s^23p^6$$

$$\mathbf{Na} + \mathbf{Cl} \longrightarrow \mathbf{Na^+} + \mathbf{Cl^-}$$

Unstable as atoms — Stable as ions

Ions of unlike electrical charge are attracted to each other, forming ionic bonds.

This type of chemical bond, an attractive force between ions of unlike electrical charge, is called an **ionic bond** (or *electron transfer bond*). It results from transfer of electrons from one atom to another to produce a stable complement of *s* and *p* electrons for each atom involved.

Most atoms are chemically stable if they have an "octet" of electrons in their outer group.

Note that both the stable sodium ion and the stable chlorine ion have eight electrons in their outer group—an "octet." Before the discovery of orbitals it was observed that a set of eight outer electrons appeared to confer chemical stability, and the **octet rule** for chemical stability was developed. The rule that a full complement of outer group *s* and *p* electrons results in chemical stability is, however, more general in its application.

Aluminum flouride is a compound known to consist of one atom of aluminum for each three atoms of fluorine. Example 5.3 shows an electron transfer model for this compound. By giving up three electrons, the aluminum atom will achieve a stable complement of outer level *s* and *p* electrons. Thus, three fluorine atoms will be needed to accept these electrons. One predicts a compound consisting of one ion of Al^{3+} and three ions of F^-, with a compound formula of AlF_3. Note that the superscripts $^{3+}$ and $^{-}$ refer to the electrical charges of the ions.

EXAMPLE 5.3

Using orbital notation, show electron transfer between aluminum (Al) and fluorine (F).

Transfer

$$\begin{array}{ccccccc} & & 1s^22s^22p^5 & & & & 1s^22s^22p^6 \\ 1s^22s^22p^63s^23p^1 & + & 1s^22s^22p^5 & \longrightarrow & 1s^22s^22p^6 & + & 1s^22s^22p^6 \\ & & 1s^22s^22p^5 & & & & 1s^22s^22p^6 \\ \textbf{Al} & + & \textbf{F} & \longrightarrow & \textbf{Al}^{3+} & + & \textbf{F}^- \end{array}$$

Unstable as atoms — Stable as ions

Magnesium oxide, MgO, is a white compound that is a solid at normal temperatures and is known to consist of one atom of magnesium for each atom of oxygen. Example 5.4 presents an ionic bonding picture for this compound. An ion of Mg^{2+} and an ion of O^{2-} are predicted.

EXAMPLE 5.4

Using orbital notation, show electron transfer between magnesium (Mg) and oxygen (O).

Transfer

$$\begin{array}{ccccccc} 1s^22s^22p^63s^2 & + & 1s^22s^22p^4 & \longrightarrow & 1s^22s^22p^6 & + & 1s^22s^22p^6 \\ \textbf{Mg} & + & \textbf{O} & \longrightarrow & \textbf{Mg}^{2+} & + & \textbf{O}^{2-} \end{array}$$

Unstable as atoms — Stable as ions

At this point, the idea of ionic bonding appears to be quite successful. We can correctly predict the formula of a compound by considering each atom's complement of electrons, the electron filling series, and the full complement of outer group *s* and *p* electrons characteristic of the inert gases. Can other observable properties of common compounds be predicted by the ionic bonding idea?

NEW TERMS

electron transfer Transfer of an electron from one atom to another, generally to achieve a stable electron configuration.

ion A charged atom; an atom that has lost or gained electrons and no longer has the same number of electrons as the number of protons in its nucleus. Positive ions are called **cations**, and negative ions are called **anions**.

formula An abbreviated notation for a chemical compound that describes its composition in terms of the ratio of one ion to another. For example, the compound $MgCl_2$ has two chlorine ions for every magnesium ion.

ionic bond The attractive force between two oppositely charged ions.

octet rule Chemical stability is implied if an atom has a set of eight outer electrons (2*s*, 6*p* electrons). (Note that hydrogen does not obey this rule.)

TESTING YOURSELF

Predicting Formulas by Electron Transfer Bonding

1. What does a stable ion's outermost electron group have?
2. By application of the electron filling series and your knowledge of the outer electron configuration of stable ions and elements, predict the charges of stable ions of the following elements.

$_{3}Li$	$_{15}P$
$_{7}N$	$_{17}Cl$
$_{8}O$	$_{19}K$
$_{9}F$	$_{20}Ca$
$_{12}Mg$	

3. Organize the elements presented in question 2 into groups of the same ionic charge. Now locate the elements of each group in the periodic table. Can you state a generalization?
4. By application of your ability to predict ionic charge, predict balanced formulas for compounds of the following elements.

Lithium and fluorine	Aluminum and oxygen
Sodium and oxygen	Calcium and oxygen
Magnesium and sulfur	Potassium and fluorine
Aluminum and chlorine	

5. By use of the periodic table, predict balanced formulas for compounds of the following elements:

Potassium and iodine	Strontium and chlorine
Cesium and oxygen	Radium and bromine

Answers **1.** A full complement of *s* and *p* electrons **2.** Li, +1; N, −3; O, −2; F, −1; Mg, +2; P, −3; Cl, −1; K, +1; Ca, +2 **3.** Elements of similar ionic charge are in the same vertical columns or "chemical families" on the periodic table. **4.** LiF, Na_2O, MgS, $AlCl_3$, Al_2O_3, CaO, KF **5.** KI, Cs_2O, $SrCl_2$, $RaBr_2$

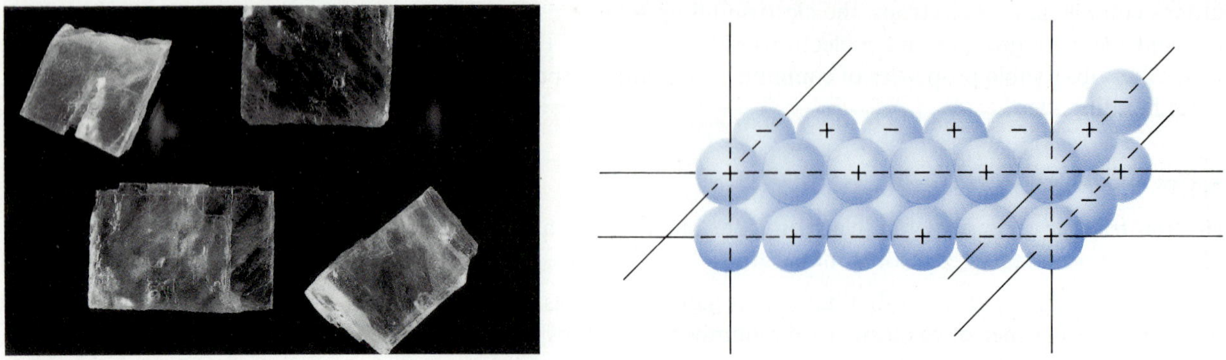

FIGURE 5.4
An ionic crystal. Since electrical forces act in all directions, the crystal is three-dimensional. Each ion is attracted to neighboring ions of opposite charge. This crystal is rock salt, NaCl.

5.3 Properties of Ionic Compounds

Crystalline Structure

The charged atoms or ions predicted by electron transfer form regularly arranged **ionic crystals** by the attraction of each ion to all neighboring ions of opposite charge. A photograph of some crystals and a schematic drawing of an ionic crystal's internal structure are shown in Figure 5.4. (In Chapter 14 we will examine carbon-based compounds which form crystals for a different reason.) Since electrical forces act in all directions, the crystal actually grows in three dimensions.

Bond Strength and Length

The strength of an ionic bond is related to bond length. Longer bonds are weaker.

The **bond strength** of an ionic bond is determined by two factors: (1) the magnitude of the charge difference between the two oppositely charged ions, and (2) the distance separating the two ions (bond length). A greater charge difference between the two ions results in greater attractive force, whereas a greater distance between the two ions results in less attractive force.

A good measure of the force holding ions together—the strength of the bond—is the temperature required to melt the compound. In the solid state, ions are held tightly in their position in the crystal lattice. When sufficient heat energy is added, vibration of the atoms overcomes the binding forces, and they are free to move independently.

Figure 5.5 presents melting point data for several compounds of sodium and the *halogens*: fluorine, chlorine, bromine, and iodine. Because of their larger atomic number and greater number of electrons, the halogen ions become larger as one moves down the periodic table from fluorine to iodine. (We will consider trends in ionic size in greater detail in Section 5.4.) Greater ionic diameter results in a greater distance between the center of the positive ion and the center of the negative ion, and thus a weaker attractive force or bond strength. A weaker bonded compound melts at a lower temperature, as is apparent from the graph in Figure 5.5.

Ionic Charge

The strength of an ionic bond is greater when the difference in charge of the two ions is greater.

The strength of an ionic bond is also related to the difference in charge between the two ions. The larger this charge differential, the greater the attraction between the two ions and the stronger the bond. Table 5.1 presents

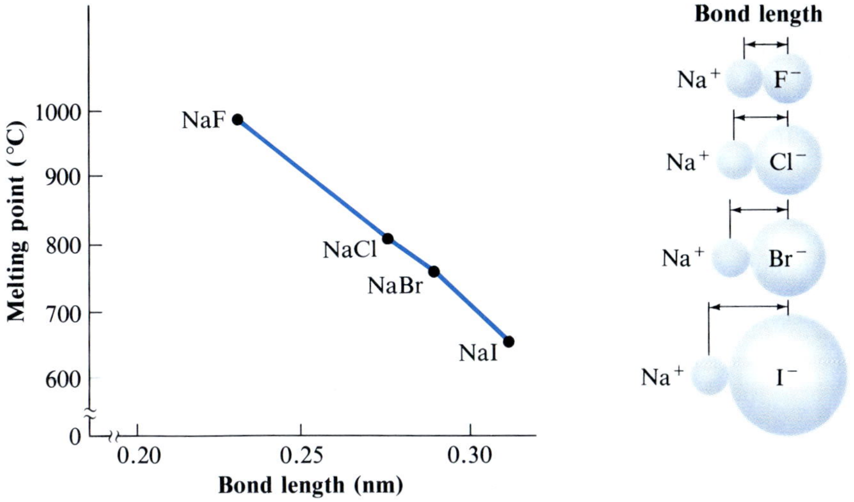
Melting point (°C)
1000
900
800
700
600
0
NaF
NaCl
NaBr
NaI
0.20
0.25
0.30
Bond length (nm)
Bond length
Na+
F−
Na+
Cl−
Na+
Br−
Na+
I−

FIGURE 5.5
A plot of melting point versus bond length for some halogen compounds. Bond strength decreases with increasing distance between the two attracting charges.

data for two groups of compounds. One of these groups is comprised of +2 and −2 ions, and the other of +1 and −1 ions. Note the significantly higher melting point, and therefore bond strength, of compounds formed of +2 and −2 ions than compounds of similar bond length composed of +1 and −1 ions.

Note again the effect of distance on bond strength exhibited in Table 5.1. Because the potassium ion (K^+) is slightly larger than the sodium ion (Na^+), the K—Cl bond length will be slightly greater than the Na—Cl bond length. Thus, we would expect KCl to have a weaker bond and to melt more easily. Such is the case. Likewise, a calcium ion (Ca^{2+}) is slightly larger than a magnesium ion (Mg^{2+}). Note that the apparent bond strength for CaO is less than for MgO.

Electrical Conductivity

Electrical conductivity is the ability of a material to transfer electrical charge and is a characteristic of molten ionic compounds. If ions are present in a compound and are free to move, electrical conductivity results. Electrical conductivity may be measured with a simple apparatus such as that shown

TABLE 5.1
Increased Melting Point (and thus Strength of Ionic Bond) with Increasing Charge Difference Between Positive and Negative Ions

Positive Ion	Radius (nm)	Negative Ion	Radius (nm)	Formula	Melting Point (°C)
+1 and −1 Ions					
Na^+	0.116	Cl^-	0.167	NaCl	801
K^+	0.227	Cl^-	0.167	KCl	776
+2 and −2 Ions					
Mg^{2+}	0.086	O^{2-}	0.126	MgO	2800
Ca^{2+}	0.114	O^{2-}	0.126	CaO	2580

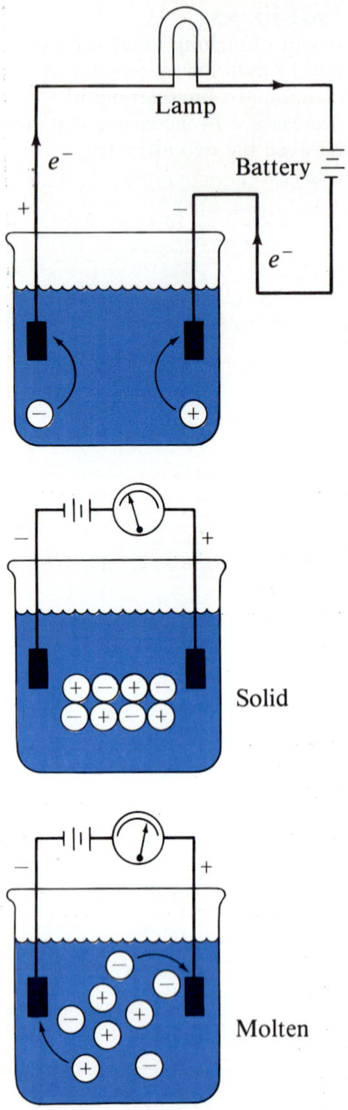

Lamp
e^-
Battery
+
−
e^-
−
+
Solid
−
+
Molten

FIGURE 5.6
A conductivity apparatus. Positive ions pick up electrons from the negative electrode (the cathode), while negative ions give up electrons to the positive electrode (the anode). The resulting motion of electrons in the external circuit causes the lamp to light (or a meter to move). Such an apparatus is used to check for the presence of ions in a sample.

in Figure 5.6. This apparatus uses a battery to force electrons through an external circuit composed of a light bulb and two electrodes. When the electrodes are pushed together, electrons flow through the circuit, causing the bulb to light.

If mobile ions are present in a substance placed between the two electrodes, the positive cations will be attracted toward the negative electrode, where they will pick up an electron. Conversely, the negative anions will be attracted toward the positive electrode, where they will give up an electron. This flow of electrons through the circuit will cause the lamp to light or a meter to move. Data gathered with this instrument are presented in Table 5.2. Note that these compounds appear to divide into two classes—those compounds that conduct electricity when melted and those that do not. Those that do not conduct electricity have much lower melting points and thus much weaker bonds than those that do conduct electricity. This probably indicates the existence of a different kind of bond between these atoms.

Ionic bonding theory will correctly predict the formulas of all of the compounds in Table 5.2 and the electrical conductivity of many of them. None of the compounds are expected to conduct electricity as solids, since the mobile charge carriers are tightly bound in the crystal and unable to move. When the compounds are melted, however, ions should be quite mobile and able to transfer electrical charge through the liquid.

Note that the last four compounds in Table 5.2 do not behave like the others, even though their formulas are correctly predicted by the electron transfer bonding theory. The melting points of these four compounds are much lower than the melting points of the other compounds. A bonding theory that explains the behavior of such compounds as sodium chloride (NaCl), magnesium oxide (MgO), and aluminum fluoride (AlF_3) does not apply well to compounds such as water (H_2O), carbon tetrachloride (CCl_4),

TABLE 5.2
Electrical Conductivity of Several Compounds

	Electrical Conductivity		
Compound	Solid	Molten	Melting Point (°C)
NaF	None	Good	980
NaCl	None	Good	801
NaBr	None	Good	755
NaI	None	Good	651
MgO	None	Good	2800
CaO	None	Good	2580
SrO	None	Good	2430
BaO	None	Good	1923
AlF_3	None	Good	1040
NH_3	None	None	−77
H_2O	None	None	0
CCl_4	None	None	−23
CH_4	None	None	−184

and methane (CH_4). In the next section, we will consider factors that should favor ionic bonding and that will allow us to differentiate between compounds that exhibit electron-transfer behavior and compounds that exhibit a different type of bonding.

NEW TERMS

ionic crystals A group of positive and negative ions stacked in a regular manner.

bond strength The ability of one atom to hold onto another. Stronger bonds hold atoms together more tightly, resulting in higher melting points for their compounds.

electrical conductivity The ability of a material to transfer electrical charge. A material that will conduct electricity has either mobile electrons or mobile anions and cations.

TESTING YOURSELF

Properties of Ionic Compounds

1. Consider crystals of two ionic compounds: KF and KI.
   a. In which compound is the distance between positive and negative ions greatest?
   b. Which compound would have the *highest* melting point?
2. Based on the differential in ionic charge, which of the two ionic compounds given below would have the strongest bond and therefore the highest melting point?

RbCl   CaS

Answers **1. a.** KI **b.** KF **2.** CaS

5.4 Factors Involved in Transfer of Electrons

The basic premise of ionic bonding is that electrons may be transferred from one atom to another, producing a stable inert gas electron configuration for both the resulting positive ion (that loses one or more electrons) and the negative ion (that gains one or more electrons). The ionic bond is caused by electrical attraction between the two opposite-charged ions. To understand how this can happen, we need to know a little about how atoms hold onto their electrons. Transfer of an electron from one atom to another is not a casual exchange.

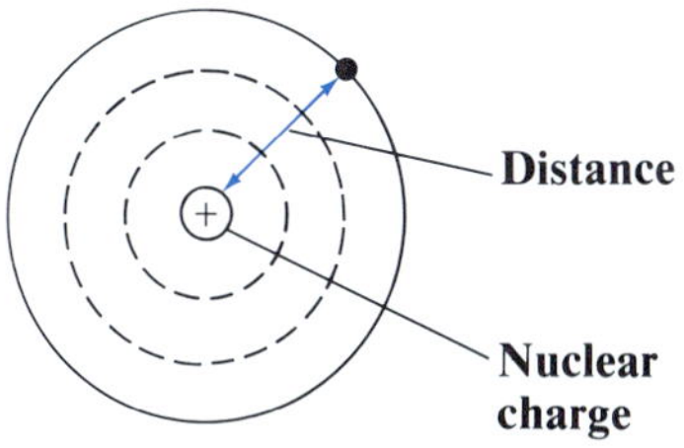
Distance
Nuclear charge

FIGURE 5.7
The attractive force by which an atom holds an electron is related to both the nuclear charge and the nucleus–electron distance. Moving the electron farther from the nucleus rapidly reduces the attractive force.

Effect of Nuclear Charge

All atoms hold their electrons by an electrical attraction between the positive charge of the nucleus and the negative charge of each electron. The actual force holding each electron is determined principally by two factors: (1) the number of positive charges held by the nucleus, called the **nuclear charge**, and (2) the distance between the electron and the nucleus (Figure 5.7).

The first ionization energy is the energy required to remove an electron from a neutral atom.

The energy required to remove the least tightly bound electron from a gaseous atom is called the **first ionization energy**. Ionization energy is often expressed in terms of **electron volts (eV)**. An ordinary flashlight battery supplies electrons with 1.5 eV of energy.

If all other factors remain constant, the attractive force exerted on the outermost electron of an atom is directly related to the positive charge of the nucleus. For example, a +10 nucleus will exert twice the force on its outermost electron that a +5 nucleus will exert. This effect is complicated by other factors, such as nucleus–electron distance, which do not remain exactly constant when nuclear charge is changed. It is, however, a reasonable generalization.

Effect of Nucleus–Electron Distance

Nucleus–electron distance is the most significant factor affecting an atom's electron-holding or attracting power. *The attractive force between two unlike electrical charges varies inversely with the square of the distance between them.* This can be written mathematically as

$$f = \frac{1}{d^2}$$

The electron-holding ability of an atom or ion is greatly reduced by increased nucleus–electron distance.

where f is the force and d is the distance. Doubling the distance results in a decrease in attractive force to 1/4 of the original value; tripling the distance results in a decrease to 1/9 of the original value.

Table 5.3 illustrates this effect by comparing the first ionization energy of neon with sodium and argon with potassium. In each case, the elements differ by only one unit of nuclear charge, but they differ by practically a factor of two in the nucleus to outer electron distance because the next electron is placed in a new major electron group. Note that as the distance from the nucleus to the outermost electron is approximately doubled, the energy required to remove the outer electron falls to approximately 1/4 of its original value.

Atomic Size and Position in the Periodic Table

The absolute value of the nuclear charge also affects the size of an atom. As the number of protons in the nucleus increases from left to right across each row of the periodic table, greater attractive force is exerted on each

TABLE 5.3
Effect of Nucleus–Electron Distance on Ionization Energy

Element	Nuclear Charge	Atomic Radius (nm)	Ionization Energy (eV)	Ratio
Ne	+10	0.071	21.47	0.24
Na	+11	0.154	5.12	
Ar	+18	0.098	15.68	0.28
K	+19	0.203	4.32	

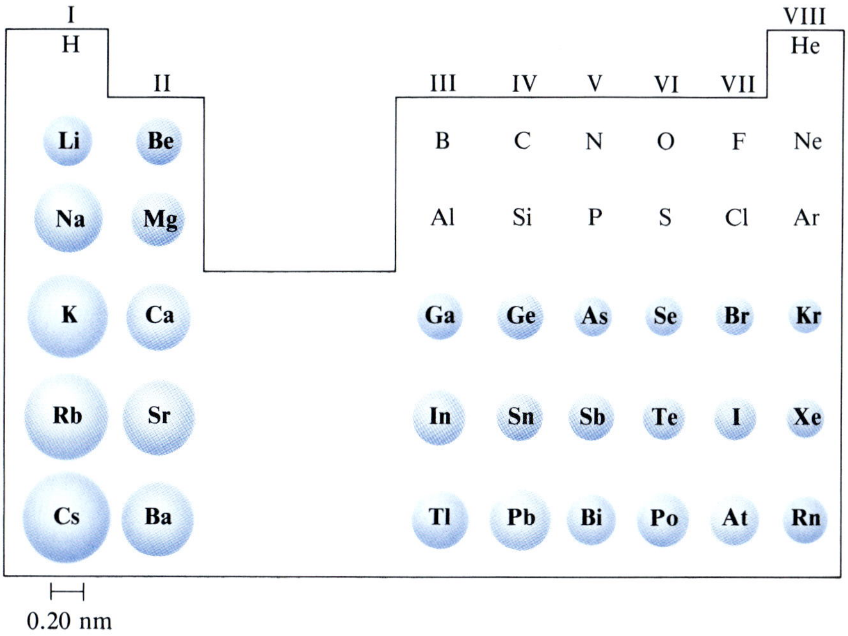
I
VIII
H
He
II
III
IV
V
VI
VII
Li
Be
B
C
N
O
F
Ne
Na
Mg
Al
Si
P
S
Cl
Ar
K
Ca
Ga
Ge
As
Se
Br
Kr
Rb
Sr
In
Sn
Sb
Te
I
Xe
Cs
Ba
Tl
Pb
Bi
Po
At
Rn
0.20 nm

FIGURE 5.8
Within a given row of the periodic table, the size of the electron cloud tends to decrease as the nuclear charge increases. Small atoms hold their electrons closer to the nucleus and, therefore, more tightly.

atom's electrons, and a shinkage of the electron cloud occurs. Figure 5.8 illustrates this effect for elements of the first three rows of the periodic table, up to atomic number 18. This decreased distance between the nucleus and the outer electrons results in a still greater attractive force. The overall result is a wide variation (a factor of two or three) in electron-holding ability from the left to the right in a given row of the periodic table.

For a given major electron group, the electron cloud shrinks with increasing nuclear charge.

Ions are quite different in size than their neutral parent atoms. Positive ions are positive because they contain more plus charges than minus charges. Each electron is attracted by more positive charge in the positive ion than in the neutral atom of the same element. The result is that a positive ion will hold its electrons closer than the corresponding neutral atom and will thus be smaller.

Positive ions are smaller than their corresponding neutral atoms while negative ions are larger.

Negative ions, on the other hand, have more minus charges than plus charges. Each electron is attracted by less positive charge than is the case for the neutral atom of the element, and the electron cloud expands. Negative ions are thus larger than their neutral atom counterparts.

Generalizations about the Periodic Table

Three factors—nuclear charge, nucleus–electron distance, and contraction of the electron cloud—allow us to make some useful generalizations about the relationship of an atom's position on the periodic table to its ability to hold onto its electrons.

1. Elements toward the right of a given row of the periodic table display both increasing nuclear charge and shrinking electron clouds and tend to hold their outer electrons tightly.
2. Elements near the top of the periodic table exhibit small nucleus–electron distance as compared to those near the bottom of the table. Since attractive force decreases with the square of the nucleus–electron distance

and increases linearly with increasing nuclear charge (atomic number), the energy necessary to remove an electron decreases rapidly as one moves down the periodic table.

3. The overall electron-attracting ability of an element increases as one moves up and to the right in the periodic table. This generalization is sketched in Figure 5.9.

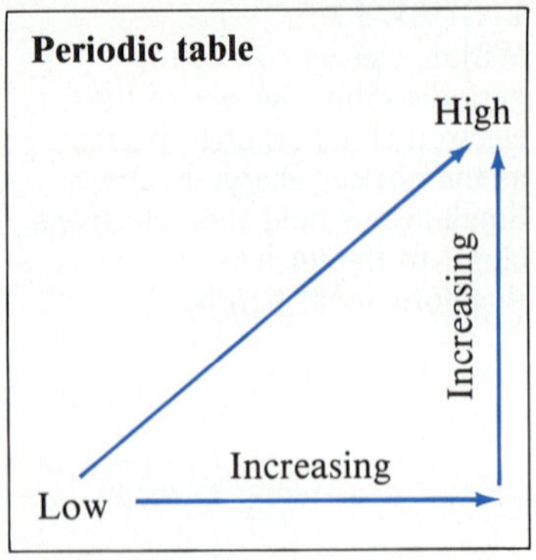
Periodic table
High
Increasing
Increasing
Low

FIGURE 5.9
The ability of an atom to hold onto its outer electrons is related to its position in the periodic table. Electron-attracting ability increases as one moves up and to the right in the periodic table.

Implications for Bonding

The type of chemical bond that forms between two atoms is determined by the *difference* in electron-holding ability of the two atoms involved. Table 5.4 lists the first ionization energy of several elements up to atomic number 36. In the third column are bars indicating the *difference* between the ionization energies of these elements involved in several compounds. The electrical conductivity of each of these compounds in the molten state is also indicated.

Compounds that exhibit electrical conductivity in the molten state (which confirms the presence of ions) have large differences among the ionization energy or electron-attracting ability of their component atoms. Those compounds with small differences among the electron-attracting ability of their component atoms do not give evidence of the presence of ions. They have melting points much lower than compounds of similar molecular weight that

Ionic compounds, which conduct electricity when molten, have large differences in ionization energy between their component elements.

TABLE 5.4
Difference in Ionization Energy and Electrical Conductivity of Some Compounds

Compound	First Ionization Energy (eV)	Difference Between Ionization Energies	Electrical Conductivity (Molten)
NaCl	Na = 5.14 Cl = 12.96	7.82 eV	Good
NaBr	Na = 5.14 Br = 11.8	6.66 eV	Good
MgO	Mg = 7.6 O = 13.56	5.96 eV	Good
CaO	Ca = 6.11 O = 13.56	7.45 eV	Good
AlF_3	Al = 5.96 F = 17.06	11.10 eV	Good
NH_3	N = 14.51 H = 13.50	1.01 eV	Poor
H_2O	H = 13.50 O = 13.56	0.06 eV	Poor
CCl_4	C = 11.23 Cl = 12.96	1.73 eV	Poor
CO_2	C = 11.23 O = 13.56	2.33 eV	Poor

conduct electricity when molten. They generally do not conform well to the behavior predicted by the electron transfer bonding theory, *even though their formulas can be correctly predicted by this theory.*

But should this be unexpected? Ionic bonding is based on the premise that one atom must remove electrons from another. This requires that the atom gaining the electron must have a much greater attraction for electrons than the atom losing an electron; a large difference must exist between the electron-attracting capability of the two atoms involved.

Two atoms must have a large difference in electron-attracting ability for electron transfer to occur.

The type of bond between two atoms is related to the difference in the electron-attracting ability or ionization energy of the atoms involved. When this difference is large, electron transfer will occur, and the compound will behave as an ionic compound. When this difference is small, electron transfer will not occur, and a different type of bond must be holding the atoms together. Another possibility for bonding will be considered in the next section.

NEW TERMS

nuclear charge The number of protons in the nucleus, expressed as the atomic number of the element.

first ionization energy The energy required to remove one electron from a neutral atom.

electron volt (eV) A unit describing the energy of an electron. A flashlight battery gives electrons 1.5 eV of energy.

TESTING YOURSELF

Factors Favoring Electron Transfer

1. Which factor most significantly affects an atom's ability to hold onto an electron: (a) nuclear charge or (b) nucleus–electron distance?
2. Which of the following atoms probably has the smallest radius: Rb, Na, Al, N, or F?
3. Which of the elements listed in question 2 would probably have the least hold on its outermost electron?
4. Of the elements listed in question 2, which would probably have the greatest attraction for its outermost electron?
5. Of the elements listed in question 2, which two would probably form the bond with greatest ionic character?
6. Which compound has the higher melting point: $Al_2{}^{3+}O_3{}^{2-}$ or K^+Cl^-?
7. Ionization energy increases as one moves from where to where in the periodic table?

Answers **1.** distance **2.** F **3.** Rb **4.** F **5.** Rb and F **6.** Al_2O_3 **7.** lower left to upper right

5.5 Covalent Bonding: Electron Sharing as a Route to Chemical Stability

Ionic bonding is favored by very large differences in the electron-attracting ability of the component atoms. But consider the opposite situation: where there is no difference in electron-attracting ability between the two atoms.

An excellent example of this is hydrogen gas, H_2. Hydrogen gas is a diatomic molecule composed of two atoms of the simplest element known. Hydrogen liquifies at −252 °C and freezes at −259 °C. It does not conduct electricity in the solid, liquid, or gaseous states. Its physical properties are much different from those compounds whose properties are predicted by an electron transfer bond.

1s ↑

FIGURE 5.10
The electron configuration of a hydrogen atom. It only has one electron.

The hydrogen atom does not have a stable electron configuration by itself, since it is short one electron of having a filled first electron group (Figure 5.10). Hydrogen needs one more electron to be chemically stable.

An electron transfer bonding model is not satisfactory for hydrogen gas. If one hydrogen atom were to remove an electron from another hydrogen atom, positive and negative hydrogen ions should result (Figure 5.11). Although this model would predict the correct formula for hydrogen gas, it provides a stable electron complement for only one of the two ions. Also, since there is no difference in the electron-attracting ability of the two hydrogen atoms, it is not reasonable to expect that one hydrogen should be able to remove an electron from the other.

Covalent bonding is caused by a common attraction of two positive nuclei for one or more shared pairs of electrons.

However, if the two hydrogen atoms could *share* each other's electrons, this sharing would result in a full complement of electrons for each atom, and according to our rule for chemical stability, a chemically stable situation would exist for both atoms. Figure 5.12 shows an electron probability distribution illustrating the overlap of the two 1*s* orbitals and a region of high electron density between the two hydrogen nuclei. The two electrons are shown as in a "flash" picture. The common attraction of both nuclei for the shared electrons creates a **covalent bond** between the two atoms. Two objects with no attraction for each other can be held together if both are

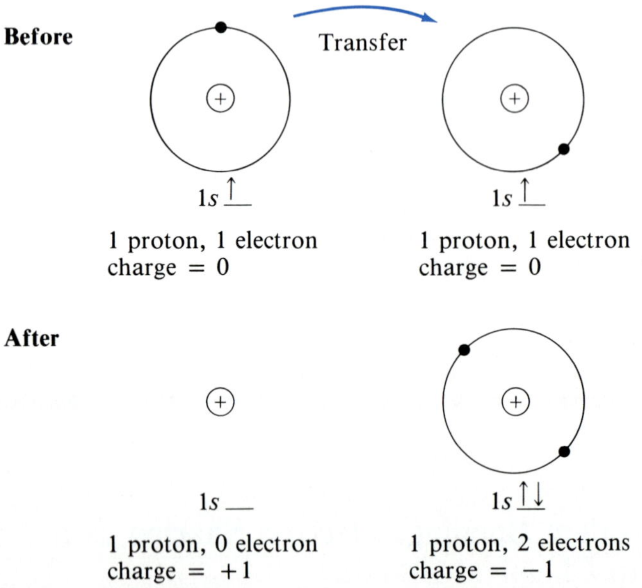
Before
Transfer
1s ↑
1s ↑
1 proton, 1 electron
charge = 0
1 proton, 1 electron
charge = 0
After
1s
1s ↑↓
1 proton, 0 electron
charge = +1
1 proton, 2 electrons
charge = −1

FIGURE 5.11
A hypothetical electron transfer bond for hydrogen gas is illustrated here. There is little reason to believe that one hydrogen atom could remove an electron from another hydrogen atom. This transfer will not take place.

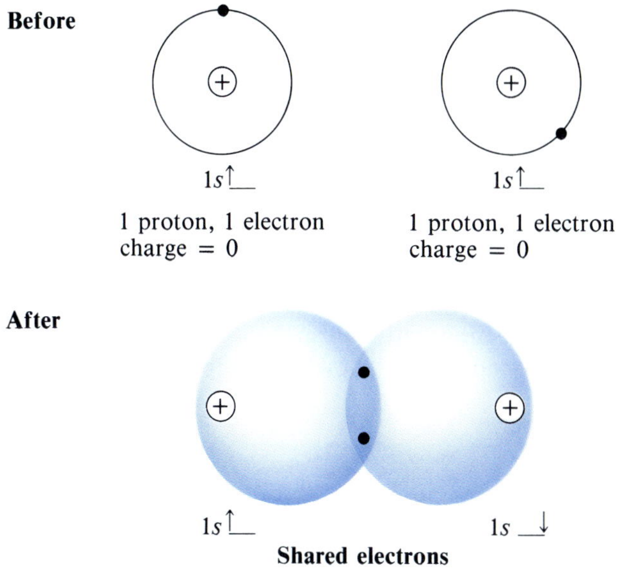
Before
1s↑
1 proton, 1 electron
charge = 0
1s↑
1 proton, 1 electron
charge = 0
After
1s↑
1s ↓
Shared electrons

FIGURE 5.12
An electron-sharing bond or covalent bond for hydrogen. By sharing electrons, both hydrogens achieve a stable electron configuration.

attracted to a shared third object, in this case, the shared electrons. Two youngsters sharing a lollipop are held together as long as the lollipop lasts.

Electron-sharing bonds of this type, in which the region of maximum electron density is located along the axis between the two atoms, are called **sigma bonds** and are designated by the Greek letter σ. Since only two electrons may occupy a given orbital or region in space, any additional pairs of electrons shared between the two atoms must occupy space above and below or beside the sigma bond. These bonds are called *pi* (π) *bonds*. Sigma and pi bonding will be discussed further in the organic chemistry part of this text (for example, see Chapter 16).

A sigma bond involves two electrons shared between adjacent nuclei. It is a linear bond.

NEW TERMS

covalent bond A chemical bond formed as two atoms share one or more pairs of electrons. The bond is due to the common attraction of each nuclei to the same pair(s) of electrons.

sigma (σ) bond Two electrons shared on the axis *between* two covalently bonded atoms.

5.6 Electronegativity: A Way to Predict Bond Type

Electronegativity and Bond Type

A convenient way to predict the ionic or covalent character of a chemical bond was suggested in the 1930s by Linus Pauling. Pauling's method assigned to each element an electron-attracting value he called **electronegativity**. These values ranged from a low of 0.7 for cesium to a high of 4.0 for

Bond type can be predicted by computing the difference in electronegativity across the bond.

fluorine. The ionic character of a chemical bond was predicted by calculating the difference in electronegativity of the two atoms involved. Large differences in electronegativity indicate a bond of high electron transfer character, whereas small differences indicate electron-sharing or covalent bonding. Table 5.5 presents information that correlates difference in electronegativity with the percentage of ionic or electron transfer character of a single chemical bond. Note that the trend in electronegativity is the same as that previously described for ionization energy: electronegativity values are lowest in the lower left of the periodic table and become greater as one moves up and to the right (Figure 5.13).

Difference in electronegativity can be used to predict the bond types present in a compound, as the following examples show.

TABLE 5.5
Percentage of Ionic Character in Relationship to Difference in Electronegativity

Difference in Electronegativity	Ionic Character (%)
0.1	0.5
0.2	1
0.3	2
0.4	4
0.5	6
0.6	9
0.7	12
0.8	15
0.9	19
1.0	22
1.1	26
1.2	30
1.3	34
1.4	39
1.5	43
1.6	47
1.7	51
1.8	55
1.9	59
2.0	63
2.1	67
2.2	70
2.3	74
2.4	76
2.5	79
2.6	82
2.7	84
2.8	86
2.9	88
3.0	89
3.1	91
3.2	92

EXAMPLE 5.5

Determine the type of bond in rubidium fluoride, RbF.

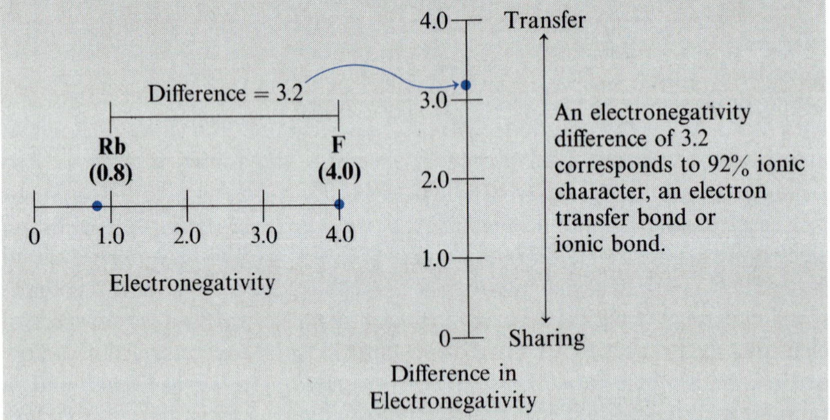
Difference = 3.2
Rb
(0.8)
F
(4.0)
0
1.0
2.0
3.0
4.0
Electronegativity
4.0
3.0
2.0
1.0
0
Transfer
Sharing
Difference in Electronegativity
An electronegativity difference of 3.2 corresponds to 92% ionic character, an electron transfer bond or ionic bond.

EXAMPLE 5.6

Determine the type of bond in methane, CH_4.

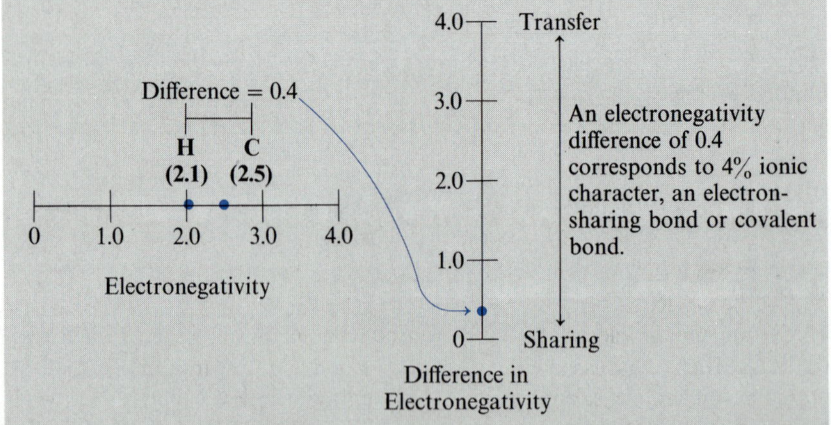
Difference = 0.4
H
(2.1)
C
(2.5)
0
1.0
2.0
3.0
4.0
Electronegativity
4.0
3.0
2.0
1.0
0
Transfer
Sharing
Difference in Electronegativity
An electronegativity difference of 0.4 corresponds to 4% ionic character, an electron-sharing bond or covalent bond.

Increasing electronegativity: Low → High

IA	IIA	IIIB	IVB	VB	VIB	VIIB	VIII	VIII	VIII	IB	IIB	IIIA	IVA	VA	VIA	VIIA	
H 2.1																	
Li 1.0	Be 1.5											B 2.0	C 2.5	N 3.0	O 3.5	F 4.0	
Na 0.9	Mg 1.2											Al 1.5	Si 1.8	P 2.1	S 2.5	Cl 3.0	
K 0.8	Ca 1.0	Sc 1.3	Ti 1.5	V 1.6	Cr 1.6	Mn 1.5	Fe 1.8	Co 1.8	Ni 1.8	Cu 1.9	Zn 1.6	Ga 1.6	Ge 1.8	As 2.0	Se 2.4	Br 2.8	
Rb 0.8	Sr 1.0	Y 1.3	Zr 1.4	Nb 1.6	Mo 1.8	Tc 1.9	Ru 2.2	Rh 2.2	Pd 2.2	Ag 1.9	Cd 1.7	In 1.7	Sn 1.8	Sb 1.9	Te 2.1	I 2.5	
Cs 0.7	Ba 0.9	La 1.1	Hf 1.3	Ta 1.5	W 1.7	Re 1.9	Os 2.2	Ir 2.2	Pt 2.2	Au 2.4	Hg 1.9	Tl 1.8	Pb 1.8	Bi 1.9	Po 2.0	At 2.2	

FIGURE 5.13 Electronegativities of the elements.

Some compounds, such as water (Example 5.7), involve bonds that are intermediate between the electron sharing or electron transfer extremes of Pauling's scale.

EXAMPLE 5.7

Determine the type of bond in water, H_2O.

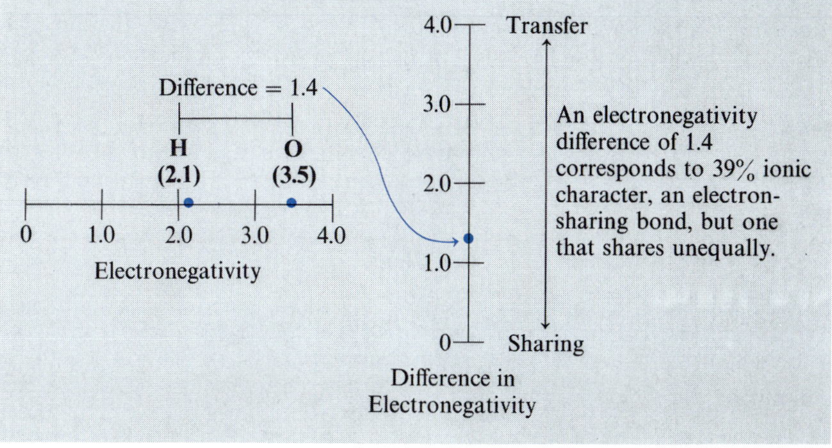
Difference = 1.4
H
(2.1)
O
(3.5)
0
1.0
2.0
3.0
4.0
Electronegativity
4.0
3.0
2.0
1.0
0
Transfer
Sharing
Difference in Electronegativity
An electronegativity difference of 1.4 corresponds to 39% ionic character, an electron-sharing bond, but one that shares unequally.

Such bonds are covalent, but with the electrons "shared" unequally. In the case of the hydrogen–oxygen bond found in water, the shared pair of electrons spend more time closer to the oxygen than to the hydrogen. This type of bond is called a **polar covalent bond**, or simply *polar bond*. It results in formation of an electrical **dipole** or region of unbalanced electrical charge. (Dipoles will be discussed further in Chapter 6.) Except for bonds between atoms of the same element (with no difference in electronegativity), all covalent bonds have some polar character.

Polar bonds are covalent bonds in which the electrons are unequally shared.

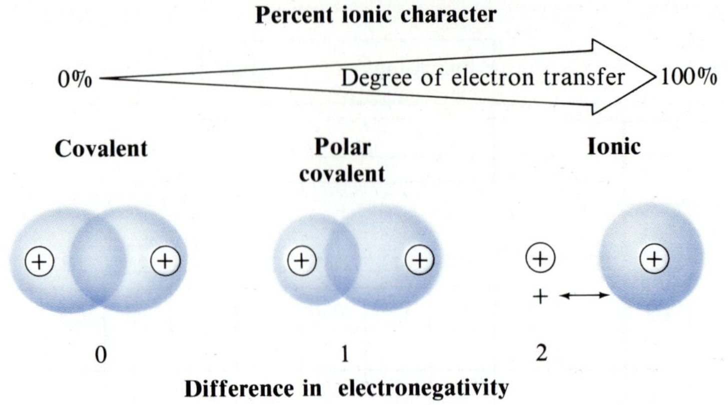
Percent ionic character
0%
Degree of electron transfer
100%
Covalent
Polar
covalent
Ionic
0
1
2
Difference in electronegativity

FIGURE 5.14
The relationship between electronegativity difference and bond type is a gradually changing scale, with no clear-cut division. Bonds with small differences in electronegativity are classified as electron sharing or covalent, while bonds with large differences in electronegativity are classified as electron transfer or ionic. Those bonds of intermediate electronegativity difference are classified as polar covalent.

Gradational Nature of Bond Types

Although chemical bonds span the entire spectrum from complete electron transfer (ionic bonding) through intermediate polar covalent bonds to complete electron sharing (covalent bonding), *there are no clear-cut dividing lines between the bond types.* An analogous situation exists in the growth of a person. A person can be classified as child, adolescent, and adult during his or her life, but it is difficult to draw clear-cut lines dividing these stages.

Just as people can be classified into three general categories by age, chemical bonds can be classified into three general categories depending on their degree of ionic character. In general, bonds with low electronegativity differences (up to 0.5) are classified as *covalent.* Bonds with large electronegativity differences (1.8 or more) are arbitrarily classified as *ionic*, and those of intermediate electronegativity difference as *polar covalent.* The dividing lines between the three bond types are, however, cloudy and subjective. The real situation, as illustrated in Figure 5.14, is one of gradually increasing ionic character of the bond as the electronegativity difference between the two elements becomes greater. The only pure covalent bonds are those with zero difference in electronegativity across the bond, those bonds between atoms of the same element.

NEW TERMS

electronegativity A measure of an atom's ability to attract electrons. One can predict the type of chemical bond that will form between two atoms by computing the difference in their electronegativities.

polar covalent bond A shared or covalent bond in which electrons are shared unevenly. For purposes of definition, bonds are considered to have polar properties when the electronegativity difference is 0.5 or greater.

dipole An electrical dipole is a region of unbalanced electrical charge. A dipole is a characteristic of all polar bonds. Dipoles also affect the solubility characteristic of a molecule, as we will see in the next chapter.

TESTING YOURSELF

Covalent Bonding and Classification of Bond Type

1. The attractive force in a covalent bond is between the nuclei of the two atoms and what?

2. What is the principal factor that determines whether a chemical bond will be ionic or covalent?
3. Use your knowledge of the electronegativity concept to classify the following compounds as ionic, polar covalent, or covalent.

   Methane, CH_4 (natural gas)
   Sulfur dioxide, SO_2
   Sodium iodide, NaI (iodized salt)
   Stannous fluoride, SnF_2 (toothpaste)

4. Predict bond type and formula for compounds of the following elements.

   Sodium and oxygen
   Hydrogen and sulfur
   Hydrogen and chlorine
   Carbon and fluorine

Answers **1.** electrons shared between the two nuclei **2.** difference in electronegativity between the two atoms in the bond **3.** methane is covalent, sulfur dioxide is polar covalent, sodium iodide is ionic; stannous fluoride is ionic **4.** Na_2O is ionic, H_2S is covalent, HCl is polar covalent, and CF_4 is polar covalent

5.7 Electron Dots: Used to Show Chemical Bonds

Covalent bonds are usually shown as lines between atoms or as adjacent dots representing the electron contributed to the bond by each participating atom. Unshared electrons are shown by placing dots representing the unshared or nonbonding electron pairs symmetrically around the "parent" atom. However, we live in a three-dimensional world, and as handy as electron dots are for text or blackboard, they usually are not a good representation of the shapes actually taken by the molecules, as we will see in Chapter 6.

Earlier in this chapter we encountered the compound ammonia, NH_3. The properties of this compound are quite different than those of known electron transfer compounds. The concept of electronegativity indicates that ammonia is, in fact, a covalent or electron-sharing molecule. This is shown in Example 5.8.

EXAMPLE 5.8

Show that ammonia (NH_3) has a polar covalent bond.

Element	Electronegativity
N	3.0
H	2.1
Difference =	0.9

An electronegativity difference of 0.9 corresponds to 19% ionic character, a polar covalent bond.

Now, using ammonia as an example, let's use the electron filling series to find the number of outer electrons each element in ammonia possesses, and then use electron dots to simply indicate the formation of bonds within the molecule (Figure 5.15).

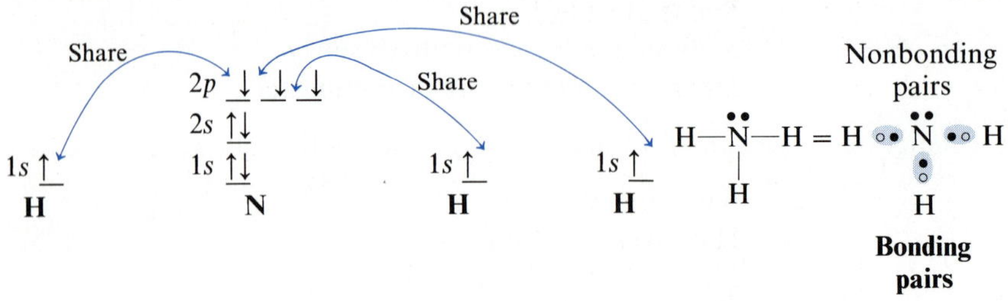
Share
Share
Share
2p
2s
1s
1s
1s
1s
H
N
H
H
Nonbonding pairs
Bonding pairs

FIGURE 5.15
Diagram showing electron transfer (on left) and a bonding model for ammonia (on right). Open circles indicate hydrogen electrons, closed circles, electrons brought in by the nitrogen atom. Shared electron pairs are shaded in blue. By sharing electrons, each atom has a full complement of *s* and *p* electrons in its outermost level.

Water is another compound that involves electron-sharing bonding (Example 5.9). The oxygen-hydrogen bond is actually polar covalent, with the electrons held more closely to the more electronegative oxygen (Figure 5.16).

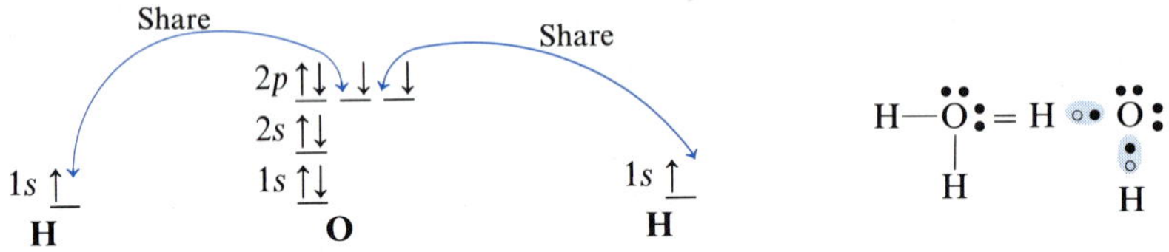
Share
Share
2p
2s
1s
1s
1s
H
O
H

FIGURE 5.16
Water is a polar covalent molecule. Two of its outer electrons form bonds with hydrogen atoms while two pairs of electrons remain unshared.

EXAMPLE 5.9

Show that water (H_2O) has a polar covalent bond.

Element	Electronegativity
O	3.5
H	2.1
Difference =	1.4

An electronegativity difference of 1.4 corresponds to 39% ionic character, thus, the O—H bond is polar covalent.

5.8 Coordinate Covalent Bonding

A coordinate covalent bond has its shared electrons both contributed by one atom.

It is not necessary that each atom participating in a covalent bond contribute one electron to the shared electron pair. When one atom contributes both electrons, the bond is called a **coordinate covalent bond**. Consider the compound sulfuric acid, H_2SO_4, illustrated in Example 5.10. The diagram at the bottom of this figure shows that each hydrogen shares its single electron with a half-filled oxygen orbital in a normal covalent bond. Two half-filled sulfur orbitals share electrons with two half-filled oxygen orbitals. The remaining two oxygen atoms form coordinate covalent bonds with sulfur; sulfur contributes both of the electrons to these bonds. By rearranging its outer electrons, oxygen presents three filled orbitals and a fourth empty orbital to the sulfur. In the next chapter, we will see that the *s* and *p* outer group orbitals are actually similar in energy and shape when only single bonds are involved.

EXAMPLE 5.10

The compound H_2SO_4 involves both covalent and coordinate covalent bonding. Only the outer group electrons are shown in this example. Note that each atom has a full complement of outer group *s* and *p* electrons. Oxygen and sulfur both have a stable "octet."

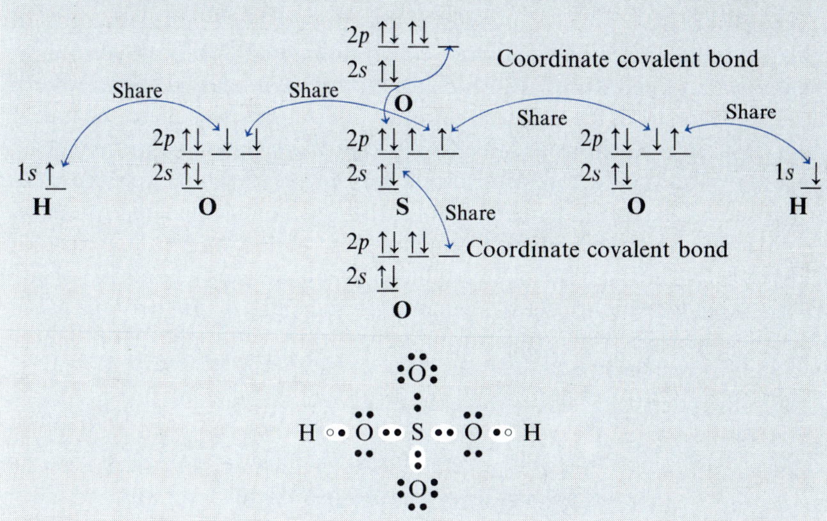
Coordinate covalent bond
Share
Share
Share
Share
Share
Coordinate covalent bond
1s
2s
2p
H
O
S
O
H
O
O

NEW TERM

coordinate covalent bond A covalent bond in which both electrons are contributed by one of the atoms.

TESTING YOURSELF

Electron Dots and Coordinate Covalent Bonding

1. Draw an electron dot model of hydrogen sulfide, H_2S.
2. Draw an electron dot model of methyl alcohol, CH_3OH.

H

H C O H

H

3. Draw an electron dot model of nitric acid, HNO_3. (HINT: this molecule involves coordinate covalent bonding.)

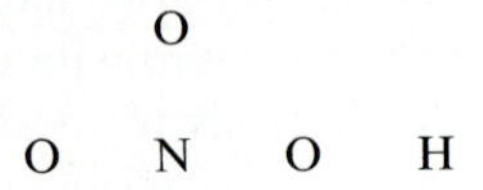
O
O N O H

Answers

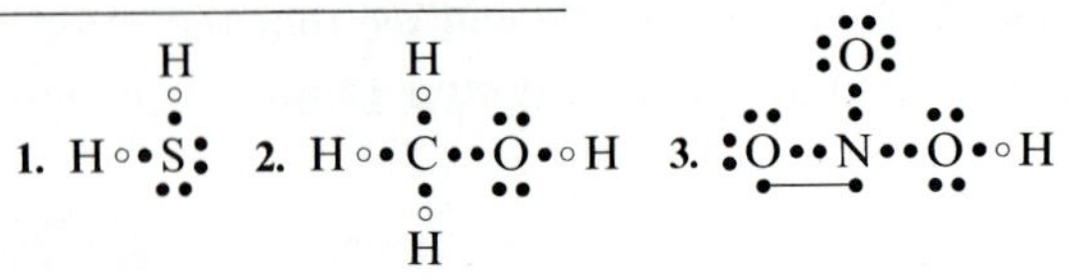
1.
2.
3.

5.9 Metallic Bonding

The metallic elements—elements that conduct electricity in the solid physical state—involve a different sort of bonding than has been discussed so far. To conduct electricity, mobile ions or electrons must be available. Bonding in metals is accomplished by each atom contributing one or more electrons to a common "sea of electrons" in which all atoms are immersed (Figure 5.17). Thus, each can share the required number of electrons to achieve chemical stability. Because electrons are free to move through the metal in this "sea," metals are good conductors of electricity. In molecules, however, the electrons are localized around specific atoms. So for **metallic bonding** to occur, the individual atom must have a low attraction for electrons—a low electronegativity. Notice that all of the metallic elements are on the left side of the periodic table (see Figure 5.13). Elements to the right of the periodic table will not let their electrons stray far enough to transfer to adjacent atoms.

Metallic bonding involves sharing of the same electrons by many atoms. Elements that participate in metallic bonding must have low electronegativity to permit this widespread sharing of electrons.

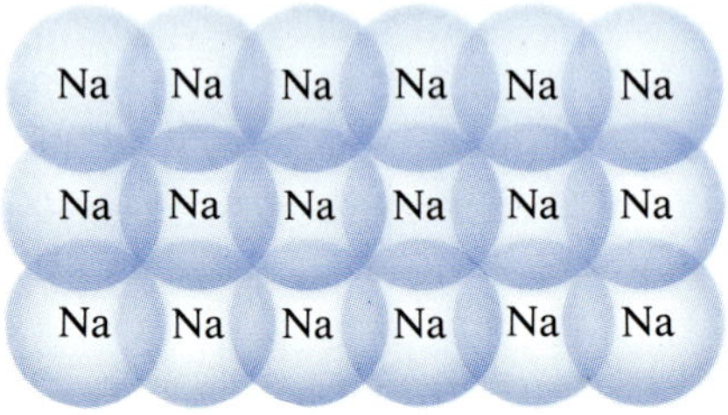
Na Na Na Na Na Na
Na Na Na Na Na Na
Na Na Na Na Na Na

FIGURE 5.17
Outer orbitals are shared and outer electrons can move from atom to atom in metallic bonding.

NEW TERM

metallic bonding A type of chemical bond in which atoms share electrons by each contributing to a "sea" of mobile electrons that surround all of the atoms. For this reason metals conduct electricity when in the solid state.

SUMMARY

Electrons are responsible for chemical bonding. Atoms can be held together by electrical attraction between two *ions*, which are atoms that have gained or lost one or more electrons to achieve a stable outer *electron configuration* and thus carry a net positive or negative charge. Atoms are also held together by common attraction toward one or more pairs of electrons that are "shared" between the two atoms involved. These two bonding theories—*ionic* and *covalent bonding*—are the extremes in the spectrum of bond types. Very few molecules, however, involve "pure" ionic or covalent bonding. Instead, the bonding electrons in most molecules are shared somewhat unequally to form *polar bonds.*

In the next chapter, we will see that unshared pairs of electrons occupy space around an atom and determine the directional orientation of the bonds and the symmetry of electrical charge distribution around the molecule. The unusual behavior of one of the most common molecules on earth—water—is determined by the distribution of electrical charge around its molecule. Life as we know it could not be sustained on our planet without this unusual behavior.

TERMS

diatomic
inert gases
stable electron configuration
electron transfer
ion
formula
ionic bond
octet rule
ionic crystals
bond strength
electrical conductivity
nuclear charge
first ionization energy
electron volt
covalent bond
sigma bond
electronegativity
polar covalent bond
dipole
coordinate covalent bond
metallic bonding

CHAPTER REVIEW

QUESTIONS

1. Which of the following statements best describes the electron configuration of an atom of any inert gas?
   a. The atom has a full complement of *s* electrons.
   b. The atom has a full complement of *p* electrons.
   c. Each available electron is matched with one of opposite spin.
   d. The outermost major group of electrons has a complete set of *s* and *p* electrons.
   e. Each orbital has at least one electron.
2. An ion is defined as
   a. a charged atom.
   b. a stable atom.
   c. a group of stable atoms.
   d. an atom with a complete set of *s* electrons.
   e. a group of stable atoms.
3. If phosphorus (atomic number 15) formed an ionic bond with another atom, the charge of the phosphorus ion would probably be
   a. -3 b. -2 c. -1 d. $+1$ e. $+2$
4. If magnesium (atomic number 12) formed an ionic bond with another atom, the charge of the magnesium ion would probably be
   a. -2 b. -1 c. $+1$ d. $+2$ e. $+3$
5. By application of your knowledge of the electron filling series and chemical periodicity, predict the probable charge of a stable ion of cesium, with atomic number 55.
   a. -2 b. -1 c. $+1$ d. $+2$ e. $+3$
6. A compound of calcium and fluorine would probably have the formula
   a. CaF b. Ca_2F c. CaF_2 d. CaF_4 e. Ca_3F_2

7. The type of chemical bond that exists between two atoms is determined principally by the difference in the atoms' ability to hold onto their outermost electrons. Several factors influence an atom's ability to hold an electron. Three of these are listed below. A tripling of which of these variables will cause the *greatest change* in the electron-holding capability of an atom?
   a. the charge of the nucleus
   b. the distance from nucleus to outermost electron
   c. the number of intermediate screening electrons
8. Which of the atoms listed below would probably have the *least* ability to hold onto its outer electrons?
   a. nitrogen d. rubidium, Rb
   b. hydrogen e. fluorine
   c. sulfur
9. Using your knowledge of the way the above-mentioned factors affect an atom's hold on its outermost electron, decide in which of the following compounds the atoms would be most strongly bound.
   a. LiBr b. KBr c. NaBr d. RbBr

If an electron transfer bond were to form between two of the elements listed below, which two elements would form the bond of greatest ionic character? List the positive ion of the compound as answer 10 and the negative ion as answer 11.

10. The more positive element would be
    a. N b. H c. S d. Rb e. F
11. The more negative element would be
    a. N b. H c. S d. Rb e. F
12. By reference to the data in Table 5.5 and the knowledge that an electronegativity difference greater than about 1.8 across a chemical bond results in observable ionic character, how many elements will form an ionic bond with the element carbon?
    a. 0 b. 1–5 c. 6–15 d. 15–25 e. more than 25
13. Carbon dioxide (CO_2) is a byproduct of human metabolism. What bond type(s) are probably involved in this molecule?
    a. covalent b. ionic c. both
14. The sketch below represents a molecule of hydrogen gas (H_2). The two hydrogen atoms are held together by a single electron-sharing bond. The electrons are in constant motion, and the region of high electron density is represented in the figure below by shading. The electrons and nuclei are shown as if a "flash" picture were taken. The arrows in the diagram represent attractive and repulsive forces. Which arrow represents a repulsive force?

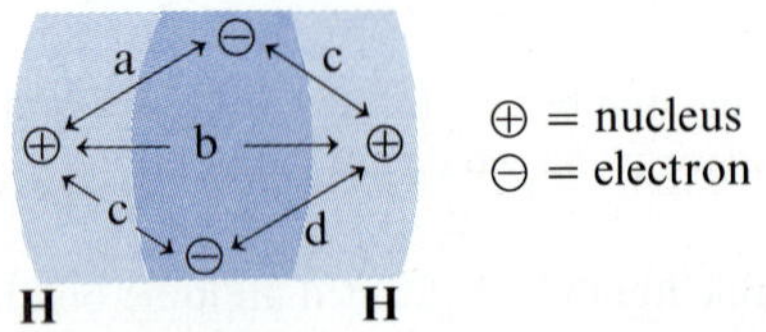
a
c
b
c
d
H
H
⊕ = nucleus
⊖ = electron

Answers **1.** d **2.** a **3.** a **4.** d **5.** c **6.** c **7.** b **8.** d **9.** a **10.** d **11.** e **12.** a **13.** a **14.** b

DIAGNOSTIC CHART

Blacken in all of the circles under the number of each question you missed in the Chapter Review questions. The diagnostic chart will help you identify concept areas that need more study.

Concepts	Questions 1	2	3	4	5	6	7	8	9	10	11	12	13	14
Periodicity					○									
Stability rules	○		○	○	○	○								
Ionic charge		○	○	○	○	○								
Formula prediction						○								
Ionic compounds								○						
Electron attraction							○	○	○	○	○			
Electron sharing										○	○			○
Electronegativity												○	○	

EXERCISES

Stability Rules

1. Atoms are chemically stable when their outer group of electrons has a full complement of *s* and *p* electrons. Why is a single atom of the inert gas helium (atomic number 2) chemically stable?

Prediction of Ionic Charge and Compound Formula

2. A neutral atom of aluminum (atomic number 13) has three electrons in its outer electron group. The neutral aluminum atom is not chemically stable.

   3*p* ↑ _ _
   3*s* ↑↓
   2*p* ↑↓ ↑↓ ↑↓
   2*s* ↑↓
   1*s* ↑↓

   a. Is it easier for aluminum to gain or to lose electrons to achieve chemical stability?
   b. What will be the charge of a stable ion of aluminum?

3. Predict the ionic charges of stable ions of the following elements.

Li	K
Ca	Mg
O	S
F	Cl

4. Use chemical periodicity to predict the ionic charges of stable ions of the following elements.

Cs	Se
Ba	Br

5. Predict balanced formulas for compounds of the following elements.

Lithium and fluorine	Aluminum and oxygen
Sodium and oxygen	Calcium and oxygen
Magnesium and sulfur	Potassium and fluorine
Aluminum and chlorine	Sodium and chlorine

6. Use your knowledge of chemical periodicity to predict balanced formulas for compounds of the following elements:

Radium and chlorine	Barium and iodine
Strontium and oxygen	Rubidium and sulfur

Properties of Ionic Compounds

7. Which is the most significant factor in determining the strength of an ionic bond?
   a. the relative ionic charges of the two atoms
   b. the distance separating the centers of the two atoms

8. Which of the following compounds would probably have the strongest bond?

   LiF NaCl KBr

9. Which of the following compounds would probably have the strongest bond?

   NaCl MgS

Electron Attraction

10. Which of the following ions might be expected to be smallest?

    Na^{+} Mg^{2+} Al^{3+}

11. What property of atoms that increases as one moves to the right in a given row of the periodic table explains question 10?
    a. nuclear charge
    b. number of electrons
    c. atomic weight

12. Which of the following neutral atoms might be expected to have the least hold on its outer electrons?

Li	Rb
Na	Cs
K	

13. What property of atoms that increases as one moves down in the periodic table explains question 12?
    a. nucleus–electron distance
    b. atomic number
    c. atomic weight
    d. nuclear charge

14. Given the following group of elements, which two would probably form the compound of greatest ionic character?

Li	K
C	Cl
O	Cs
F	

Electron Sharing and Covalent Bonding

15. When elements do not differ enough in electron-attracting ability for one atom to remove an electron from another, the atoms will share electrons to achieve a stable electron configuration. Common attraction of both nuclei for the shared electrons holds the atoms together in a covalent bond. Using your knowledge of the factors that determine an atom's electron-attracting ability, decide which two of the following atoms would form a bond of the greatest covalent character.

K	N
Al	O

16. Propane (C_3H_8) is a covalently bonded compound with this structural formula:

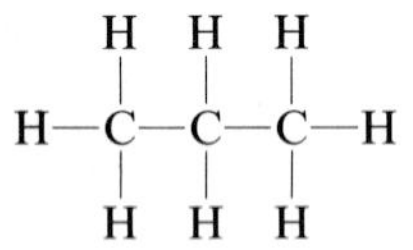

Sketch an electron dot model for this molecule.

17. Ammonia (NH_3) is also a covalently bonded molecule. Sketch an electron dot model for this molecule.

Electronegativity and Bond Type Prediction

18. Use your knowledge of the electron filling series and electronegativity to predict the formula and bond type for compounds of the following elements:

Aluminum and chlorine	Calcium and fluorine
Oxygen and hydrogen	Hydrogen and chlorine
Hydrogen and sulfur	Carbon and fluorine
Potassium and bromine	

19. Some compounds share more than one pair of electrons between adjacent atoms, forming a double or triple bond. Predict bond types and sketch electron dot models for the following:

Carbon dioxide (CO_2)	O C O
Acetylene (C_2H_2)	H C C H
Nitrogen gas (N_2)	N N
Oxygen gas (O_2)	O O
Chlorine gas (Cl_2)	Cl Cl
Formaldehyde (H_2CO)	H C H O

CHAPTER

6

Shapes and Polarities of Covalent Molecules

The behavior of a covalent molecule is determined to a large degree by the symmetry with which bonding and unshared pairs of electrons are distributed around it. This property is apparent in many natural phenomena: the sea is salty, ice doesn't sink, and we don't dissolve when we go out in the rain. A simple concept called molecular polarity helps us understand this kind of behavior.

OBJECTIVES

After completing this chapter, you should be able to

- Define the terms *dipole* and *molecular polarity.*
- By use of electron pair repulsion theory, predict the shape and molecular polarity of simple covalent molecules.
- Explain how ionic compounds are dissolved in polar solvents.
- Define the term *hydrogen bonding,* and explain how the strength of the hydrogen bond is related to the strength of the molecular dipole.
- Explain how the unusual behavior of water is caused by hydrogen bonding.

6.1 Shapes of Covalent Molecules

Have you ever read the label of an extract used as a flavoring in cooking? To do so is educational. Mint extract contains oil of spearmint, oil of peppermint, water, and 90% ethyl alcohol! It is 180 proof. Orange extract is only slightly less potent—80% ethyl alcohol or 160 proof. The alcohol, however, boils away during cooking, so intoxication cannot be the goal of the flavor manufacturers. So why is it included in such large proportions? We all have observed droplets of gasoline standing on water in a service station, and we all know that bacon grease cannot be removed from a pan without soap. What do these apparently unrelated phenomena have in common? They can all be explained by a property called *molecular polarity*, which is related to the shape of the molecule and the symmetry with which the electrical charge is distributed around it.

Valence Shell Electron Pair Repulsion

VSEPR: outer group electron pairs in molecules stay as far apart as possible.

Valence shell electron pair repulsion (VSEPR) theory is a large name for a relatively simple idea. The name consists of three parts. *Valence shell* means the outer group electrons, which we studied in Chapters 4 and 5. *Electron pair* refers to the fact that in a bonded molecule each orbital has two electrons, and thus all electrons are paired. Some pairs of electrons are located between the bonded atoms and thus provide a common negative charge toward which the two adjacent nuclei are attracted. These are called "bonding electrons." Other pairs of electrons do not participate in the bond, but do occupy space around the atom. These are called "nonbonding electrons." The last part of the term, *repulsion theory*, indicates that we expect repulsion between these negative pairs of electrons. Each pair will try to get as far away from the other pairs as possible, a fact that we will later see as extremely important in determining the three-dimensional shape of molecules.

Given these ideas, let us consider the shape of a water molecule. Figure 6.1 shows electron filling diagrams and electron dot models for water. Water possesses polar covalent bonds. Two electron pairs provide single bonds between the central oxygen and the two hydrogens. Two additional electron pairs are not shared, but also occupy space around the oxygen atom.

In Chapter 5 we considered atomic orbital models of atoms. Can this technique be applied to molecules? An atomic orbital picture of the water molecule is shown in Figure 6.2. There are some problems with this model, however. The model predicts O—H bonds between half-filled oxygen *p* orbitals and half-filled hydrogen *s* orbitals. Two oxygen orbitals (an *s* and a *p*) are filled with unshared electron pairs. The model predicts a 90° angle between the two hydrogen atoms. But interestingly, the actual angle between the two hydrogen atoms in a water molecule is 104.5°.

FIGURE 6.1
This electron filling diagram (left) and electron dot model (right) show that water is a polar covalent molecule with two pairs of unshared electrons.

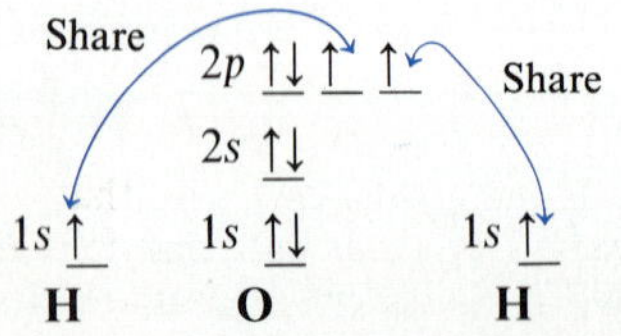
Share
2p ↑↓ ↑ ↑
Share
2s ↑↓
1s ↑
1s ↑↓
1s ↑
H
O
H

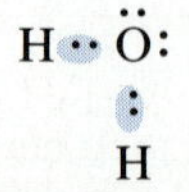
H
O
H

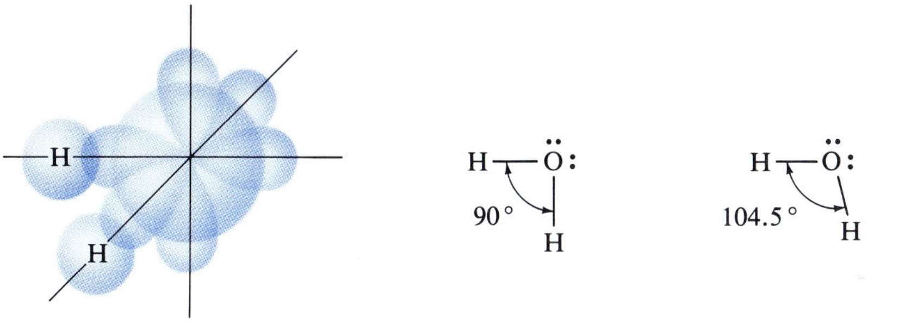
H
H
H—Ö:
90°
H
H—Ö:
104.5°
H
Atomic orbital model
Predicted bond angle
Actual bond angle

FIGURE 6.2
An atomic orbital model (left) for the water molecule. The model predicts a 90° bond angle between the two hydrogens. This model is not correct because the actual bond angle is 104.5°.

It is evident from inspection of Figure 6.2 that the atomic orbital space utilization around the oxygen atom is not particularly effective. Some regions are crowded and some are not. Since pairs of negative electrons will attempt to get as far away from each other as possible, it seems likely that the atomic orbitals will distort to use this space.

The water molecule has four pairs of electrons. Two pairs are involved in oxygen-hydrogen bonds, and the two remaining pairs belong to the oxygen. What geometric shape will allow four pairs of electrons to get as far apart from one another as possible? We can investigate this problem by tying four small balloons together. Although it is possible to force the balloons into a flat rectangular arrangement, if you allow them to move freely, the balloons always rearrange themselves to their "lowest energy" or farthest apart shape. This shape forms a four-sided pyramid or **tetrahedron** (Figure 6.3).

The VSEPR theory, or more simply, electron pair repulsion theory, predicts that each pair of electrons will be located at one corner of the tetrahedron. This situation is sketched for a water molecule in Figure 6.4.

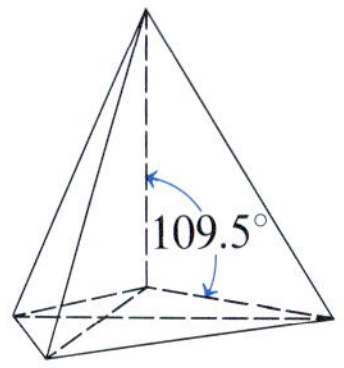
109.5°

FIGURE 6.4
When an atom forms a covalent bond, its pairs of outer electrons will assume the geometric shape that permits them to be as far apart as possible. For four pairs of electrons, this shape is a tetrahedron.

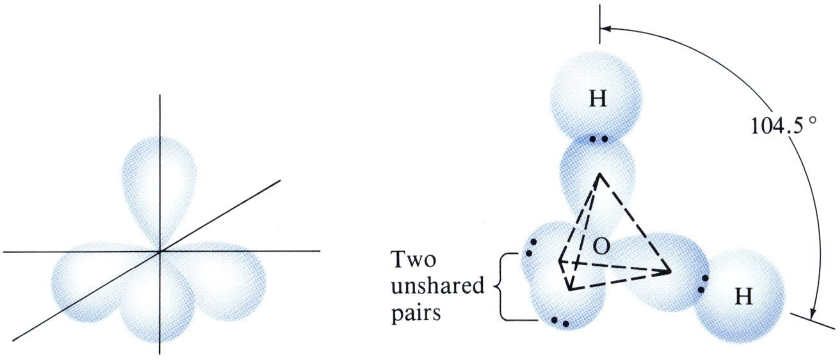
H
104.5°
O
Two unshared pairs
H
Balloon model
Electron arrangement

FIGURE 6.3
A four-sided pyramid or tetrahedron is the geometric shape that permits the farthest separation between four objects. Thus, four balloons (left) and four electron pairs in a water molecule (right) would both assume the same shape.

Note that the angle between the electron pairs and the center of the tetrahedron is 109.5°, quite close to the actually measured 104.5° for the water molecule. The difference results from mutual repulsion of the two unshared electron pairs which slightly distorts the molecule's bond angles and forces the hydrogen atoms slightly closer together.

Orbital Hybridization

Symmetrical atomic orbitals are distorted as they combine during bonding, a process called hybridization.

When a covalent bond forms between two atoms, the symmetrical orbital shapes of each unbonded atom are distorted as the atom shares one or more of its electrons with the adjacent atom. The resulting orbital shapes are called *hybridized orbitals.* This change in orbital shape that occurs during bonding is called **hybridization**.

Hybridized orbitals are named according to the set of orbitals from which they are constructed. In the case of water, the oxygen's set of four hybridized orbitals are made from one atomic *s* and three atomic *p* orbitals. They are thus called "sp^3 hybrid" orbitals (Figure 6.5).

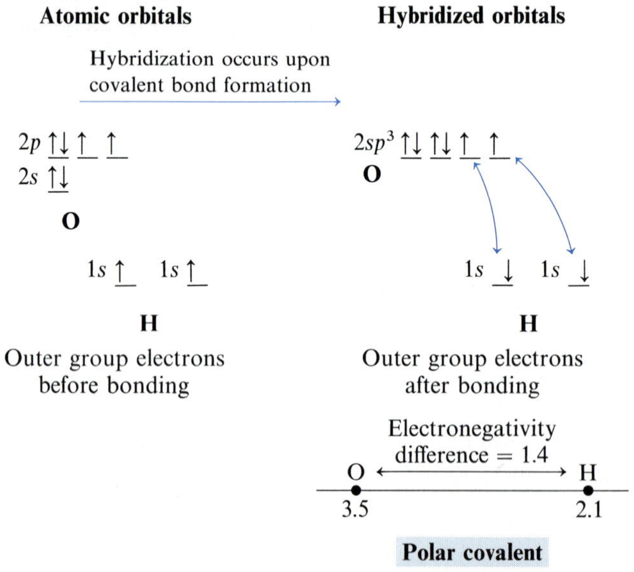
Atomic orbitals
Hybridized orbitals
Hybridization occurs upon covalent bond formation
2p
2s
O
1s 1s
H
Outer group electrons before bonding
2sp³
O
1s 1s
H
Outer group electrons after bonding
Electronegativity difference = 1.4
O
H
3.5
2.1
Polar covalent

FIGURE 6.5
Hybridization is a rearrangement of the atomic orbitals of an atom during bonding. The set of hybridized orbitals is named by counting the atomic orbitals that were used to make them—one *s* and three *p* orbitals in this case.

NEW TERMS

valence shell electron pair repulsion (VSEPR) theory The idea that pairs of outer group electrons will move to be as far apart as possible; it is used to predict the shape of an atom.

tetrahedron A symmetrical geometric shape with four vertices.

hybridization The change in orbital shape that occurs during covalent bonding.

TESTING YOURSELF

Covalent Molecules

Consider the compound carbon tetrafluoride, CF_4.

1. How many *bonding* pairs of electrons are found in the outer group of carbon's electrons?
2. How many *nonbonding* pairs of electrons are found in the outer electron group of carbon?
3. What is the hybridization of the carbon atom?
4. How many *bonding* pairs of electrons are found in the outer group of each fluorine?
5. How many *nonbonding* pairs of electrons are found in the outer electron group of each fluorine?
6. What is the hybridization of each fluorine atom?
7. What is the shape of the molecule?

Answers **1.** four **2.** none **3.** sp^3 **4.** one **5.** three **6.** sp^3 **7.** tetrahedral

6.2 Molecular Dipoles: A Consequence of Molecular Shape and Electronegativity

The three-dimensional sketch shown in Figure 6.3 illustrates an important property of water. The presence of hydrogen nuclei on one side of the molecule and unshared pairs of electrons on the other side results in an uneven distribution of electrical charge around the molecule. This is made more uneven by the greater electronegativity of oxygen.

The overall result of this molecular shape is to make the water molecule somewhat positive in the region of the hydrogen nuclei and somewhat negative in the region of the unshared electron pairs on the other side of the oxygen atom. This unbalanced distribution of charge is called an *electrical dipole* (Figure 6.6), which is represented by an arrow pointing across the

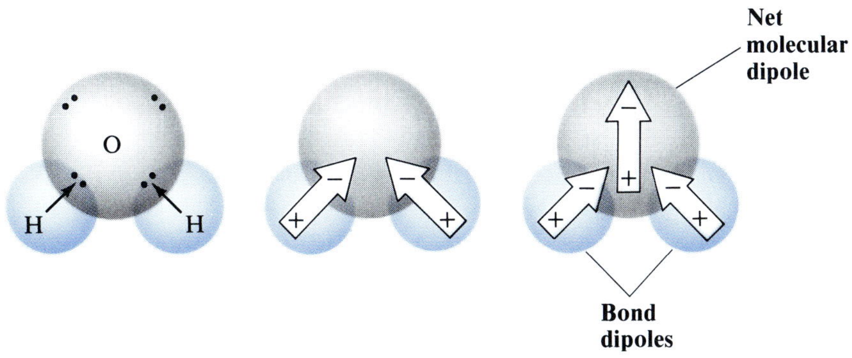
Net molecular dipole
O
H
H
Bond dipoles

FIGURE 6.6
The electrical charge in a water molecule is unevenly distributed. The molecule is slightly negative in the region of the unshared electron pairs and slightly positive near the hydrogen nuclei. The net overall charge of the molecule, however, is zero.

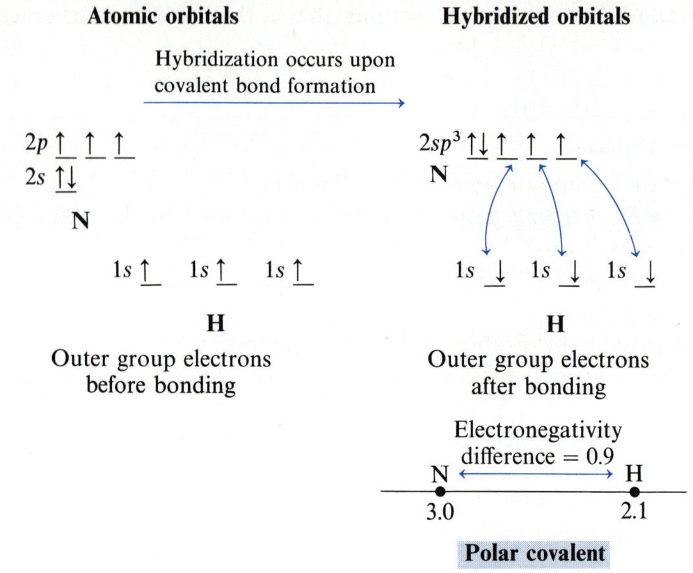
Atomic orbitals
Hybridized orbitals
Hybridization occurs upon
covalent bond formation
2p ↑ ↑ ↑
2s ↑↓
N
2sp³ ↑↓ ↑ ↑ ↑
N
1s ↑ 1s ↑ 1s ↑
1s ↓ 1s ↓ 1s ↓
H
H
Outer group electrons
before bonding
Outer group electrons
after bonding
Electronegativity
difference = 0.9
N
H
3.0
2.1
Polar covalent

FIGURE 6.7
Ammonia is a polar covalent molecule with one pair of unshared electrons. Because nitrogen has four pairs of electrons (3 shared pairs, 1 unshared pair), it will have a tetrahedral shape.

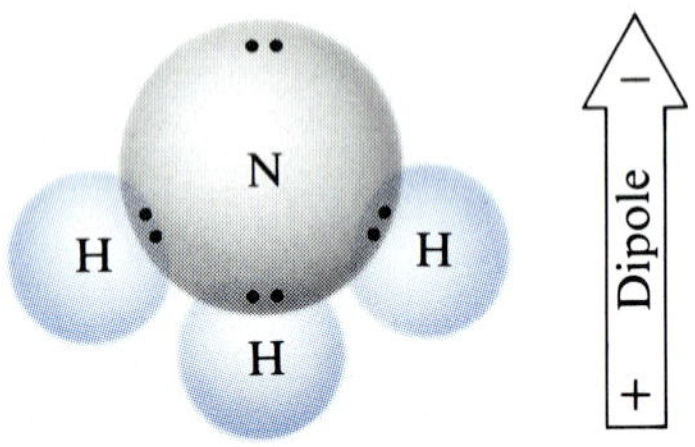
N
H
H
H
Dipole
−
+

FIGURE 6.8
Ammonia's unshared pair of electrons and its slightly polar N—H bond give it a non-symmetrical charge distribution and make it a polar molecule. The net molecular dipole is the sum of the individual bond dipoles.

molecule from the region of extra positive charge and toward the region of extra negative charge. This unbalanced charge is said to form a **molecular dipole**, indicated by an arrow which points toward the region of extra negative charge. Since the water molecule has a strong molecular dipole, it is considered to be a **polar molecule**.

The VSEPR and hybridization concepts can be used to predict the polar properties of any covalently bonded molecule. For example, ammonia (NH_3) is composed of nitrogen and hydrogen. The electronegativity difference across

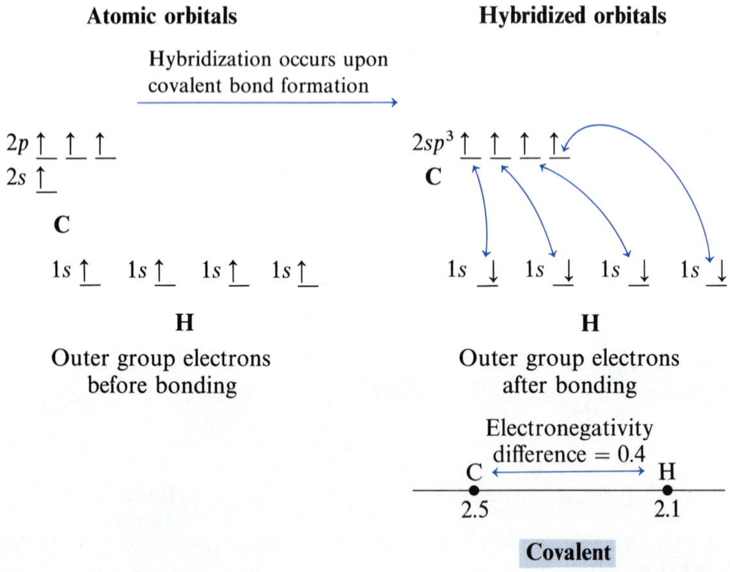
Atomic orbitals
Hybridized orbitals
Hybridization occurs upon
covalent bond formation
2p ↑ ↑ ↑
2s ↑
C
2sp³ ↑ ↑ ↑ ↑
C
1s ↑ 1s ↑ 1s ↑ 1s ↑
1s ↓ 1s ↓ 1s ↓ 1s ↓
H
H
Outer group electrons
before bonding
Outer group electrons
after bonding
Electronegativity
difference = 0.4
C
H
2.5
2.1
Covalent

FIGURE 6.9
Methane is a covalent molecule with no unshared pairs of electrons. When carbon participates in a covalent bond, one of the 2*s* electrons usually moves up to the vacant 2*p* orbital. This permits the carbon to form four bonds.

the N—H bond is 0.9, corresponding to about 19% ionic character. This bond is somewhat polar, although not as polar as the O—H bond. Electron filling diagrams and electron dot models of the ammonia molecule are shown in Figure 6.7.

Electron pair repulsion theory predicts that ammonia's four pairs of electrons will form a tetrahedron as water does. The ammonia molecule's unshared pair of electrons gives it a nonsymmetric charge distribution because there is no proton (hydrogen nucleus) to balance the negative charge of the unshared electron pair. The slightly polar N—H bond also contributes to this unbalance of electrical charge. The dipole in the ammonia molecule, which is the sum of the individual bond dipoles, points toward the pair of unshared electrons (Figure 6.8).

Methane (natural gas, CH_4) is composed of carbon and hydrogen. The electronegativity difference across the C—H bond is 0.4, which corresponds to about 4% ionic character. This is practically a pure covalent bond. The electron filling series for methane is shown in Figure 6.9.

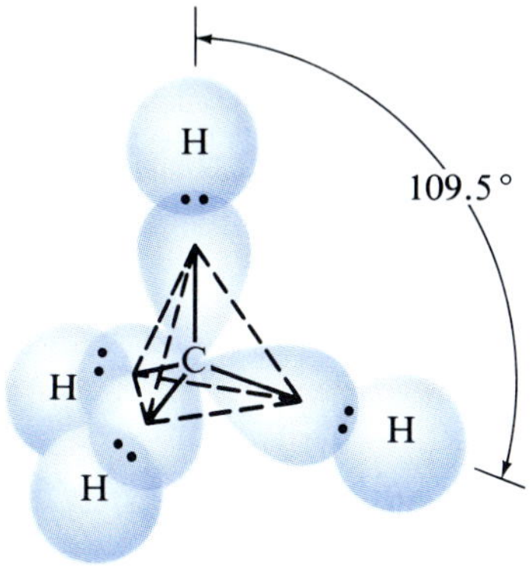
H
109.5°
C
H
H
H

FIGURE 6.10
The symmetrical distribution of electrical charge in methane makes it a nonpolar molecule. The individual bond dipoles are in opposite directions and cancel each others' effects.

Methane's four pairs of electrons form a tetrahedral structure, with a pair of electrons and a hydrogen nucleus at each corner (Figure 6.10). Although very small dipoles exist across each of the H—O bonds, the symmetry of the structure causes their contributions to cancel. The symmetry of the electron distribution results in a balanced electrical charge. Thus, methane is considered to be a **nonpolar molecule**.

Whenever a covalently bonded atom has a *total of four pairs* of electrons involved in single bonding and/or held as "lone pairs" (as in the case of oxygen), the four electron pairs will move as far apart as possible and the atom will exhibit sp^3 hybridization. This rule does not apply if more than one bond exists between two adjacent atoms. (Multiple covalent bonds will be discussed in Chapter 14.) Figure 6.11 shows structures for several situations in which the central atom will always have sp^3 hybridization.

sp^3 hybridization results when an atom has a total of four electron pairs involved in single bonding and/or nonbonded pairs.

The C—H structure illustrated by methane can be extended by adding carbons and hydrogens to enlarge the molecule. Figure 6.12 shows structural

sp^3 hybridization occurs upon covalent bond formation

F $2s^2 2p^2 2p^2 2p^1 \longrightarrow (sp^3)^2(sp^3)^2(sp^3)^1(sp^3)^1$

Example: Hydrogen fluoride H—F: (with lone pairs)

O $2s^2 2p^2 2p^1 2p^1 \longrightarrow (sp^3)^2(sp^3)^2(sp^3)^1(sp^3)^1$

Example: Water H—Ö: —H

N $2s^2 2p^1 1p^1 2p^1 \longrightarrow (sp^3)^2(sp^3)^1(sp^3)^1(sp^3)^1$

Example: Ammonia H—N̈—H, with H below N

C $2s^2 2p^1 2p^1 2p^0 \longrightarrow (sp^3)^1(sp^3)^1(sp^3)^1(sp^3)^1$

Example: Methane H—C—H, with H above and below C

FIGURE 6.11
Any atom having four pairs of outer electrons and participating only in single covalent bonds will have sp^3 hybridization, as shown here for fluorine, oxygen, nitrogen, and chloride.

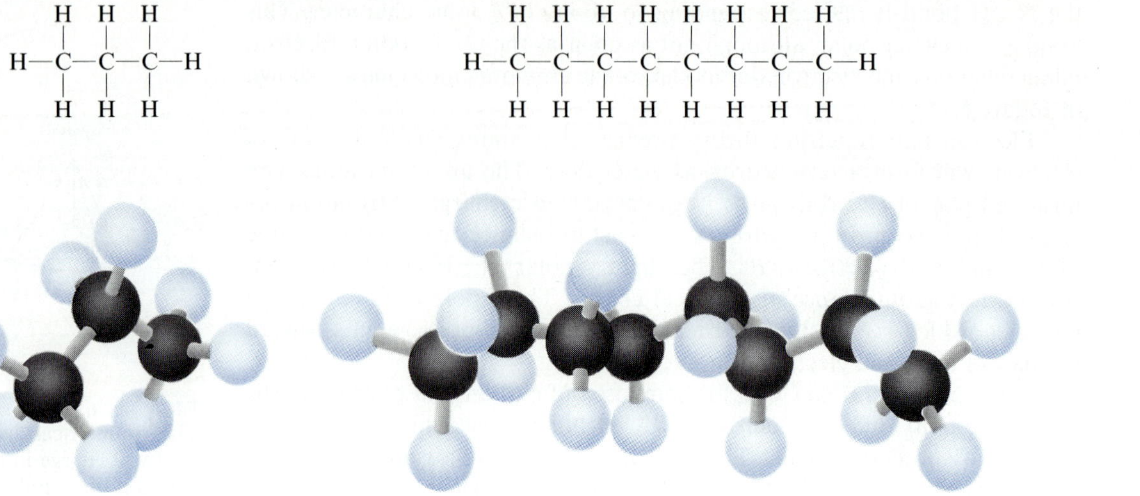

Propane

Octane

FIGURE 6.12
As these examples of propane and octane show, all hydrocarbons have nonpolar bonds and symmetrical electron distribution and are nonpolar.

diagrams for propane, a three-carbon hydrocarbon, and octane, an eight-carbon hydrocarbon. Propane is a common fuel; octane is better known as a component of gasoline. Since the bonds are practically nonpolar and the electrical charge is symmetrically distributed, these molecules are considered to be nonpolar.

NEW TERMS

molecular dipole An unbalanced distribution of electrical charge across a molecule, giving one side a more negative charge and one side a more positive charge.

polar molecule A molecule with a molecular dipole.

nonpolar molecule A symmetrical molecule with no unbalanced electrical charge.

TESTING YOURSELF

Predicting Molecular Polarity

1. Butane (C_4H_{10}) is a common fuel. Structural and ball-and-stick models of this molecule are shown below. Is this a polar or nonpolar molecule?

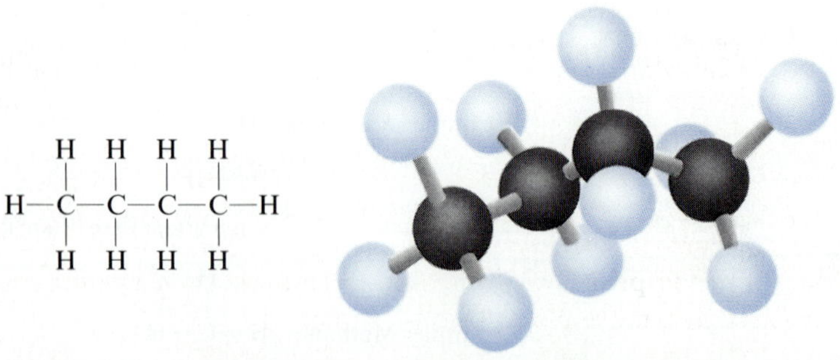

Butane

2. Ethylene glycol, the major component in antifreeze, has the molecular structure shown in structural and ball-and-stick models below. The dipoles in ethylene glycol point toward which of the lowercase letters shown here?

b c d
a
e
g f

Ethylene glycol

3. Carbon skeleton compounds sometimes bend completely around to form closed loops. The drawing below shows a compound of carbon, hydrogen, and oxygen that is important to the pine bark beetle. This molecule is a sex attractant used by the female pine bark beetle to lure the male beetle into her wings. It is used by forestry biologists to lure the male pine bark beetle into a trap, thus saving the pine trees from beetle damage. A structural drawing and a ball-and-stick model of this molecule are shown below. Does this molecule have any polar properties? If so, in which direction are they oriented?

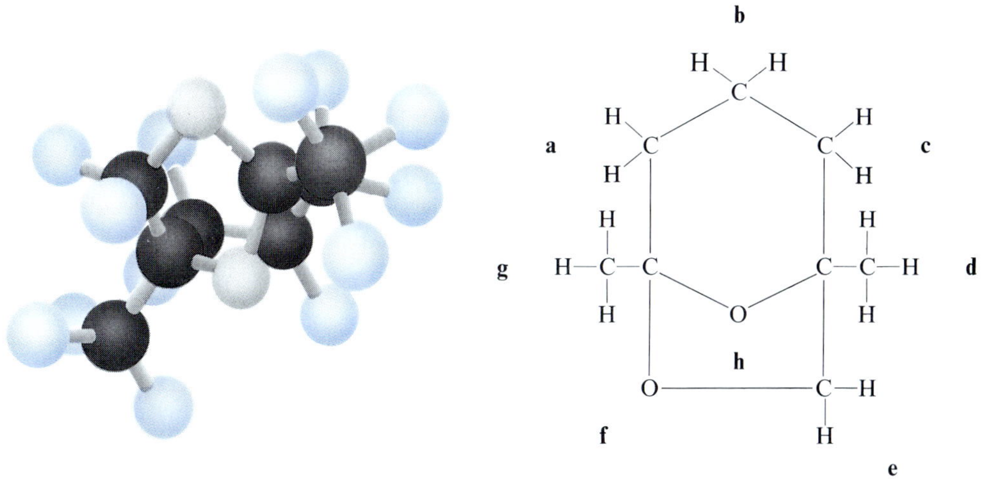
b
a
c
g
d
h
f
e

Since a dipole is caused by the presence of a pair of electrons unbalanced by a bonding nucleus, we can expect that any atom with one or more unshared pairs of electrons will contribute toward a molecular dipole. Notice that the presence of an oxygen atom is almost always a dead giveaway that a molecular dipole is present.

Answers **1.** nonpolar **2.** c and g **3.** f and h

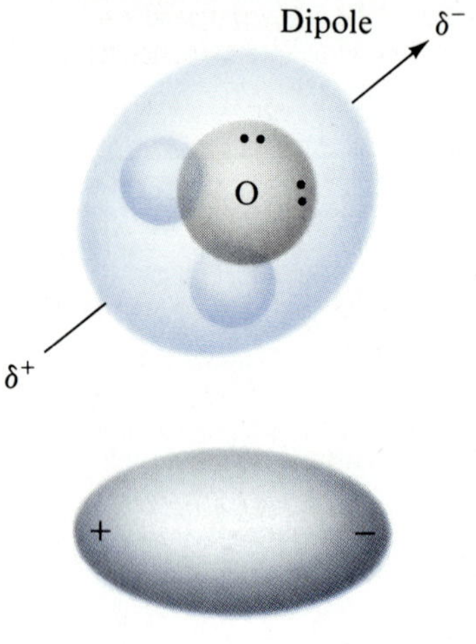
Dipole
δ⁻
O
δ⁺
+
−

FIGURE 6.13
A model of the polar water molecule. Polar molecules can be shown in a simplified manner as an oval or egg shape with positive and negative ends indicated by plus (+) and minus (−) charges. The Greek delta (δ) symbol indicates a partial charge, as compared to the unit charge of an ion.

6.3 Behavior of Polar Molecules

The symmetry of the electron distribution around a molecule and the resulting molecular dipole (or lack thereof) strongly influence the physical and chemical behavior of the molecule. Consider again the water molecule (Figure 6.13). The dipole in water is due to two factors: the presence of two pairs of unshared electrons and the highly polar O—H bond whose electrons reside considerably closer to the more electronegative oxygen than to the hydrogen.

As one might expect, because of the unlike charges at opposite ends, polar molecules are strongly attracted to one another. Figure 6.14 illustrates the sort of attraction that might exist between polar molecules. If the dipole is strong enough, the molecules actually form chains five to ten or more molecules in length, giving rise to significant changes in physical behavior.

Solubility and Molecular Polarity

The ability of one type of molecule to dissolve in another is related to the polarity of the molecules. The chain illustrated in Figure 6.14 is not selective as to its membership. Formation of this chain requires only that each molecule have a partially negative end and a partially positive end, that is, they must be polar. For example, ammonia and water are both very polar molecules. Ammonia mixes very well with water—89.9 g of ammonia will dissolve in 100 mL of water. We say it is very soluble in water.

Methane, on the other hand, is a nonpolar molecule. Since methane has no dipole, it cannot find a place in the chain. It is almost as if the polar molecules have formed an exclusive club. To join in their dance, one must have an electrical dipole. Methane, lacking this dipole, is squeezed out and excluded. Consequently, methane is very sparingly soluble in water—only 0.0064 g per 100 mL of water (Figure 6.15).

Nonpolar molecules, however, mix very well with other nonpolar molecules. Since no molecular dipoles are present, there is very little attraction or repulsion between their molecules. They move among each other attracted only by weak forces caused by induced charges between the molecules. These forces are called *van der Waals* or *London forces.*

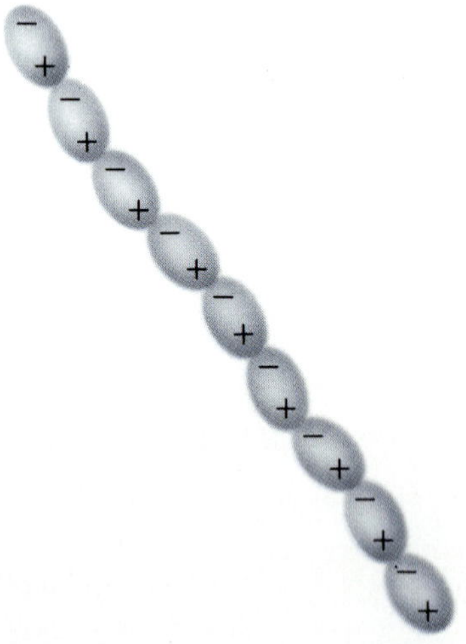

FIGURE 6.14
Polar molecules are attracted to each other and often form loosely linked chains several molecules long.

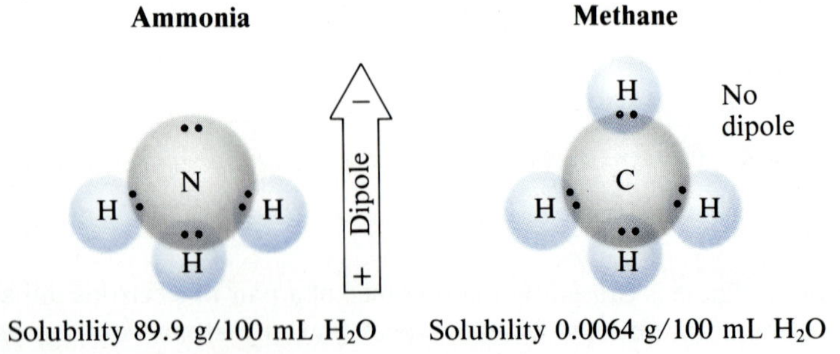
Ammonia
Methane
N
H
H
H
Dipole
−
+
C
H
H
H
H
No dipole
Solubility 89.9 g/100 mL H₂O
Solubility 0.0064 g/100 mL H₂O

FIGURE 6.15
Polar molecules, such as ammonia, are soluble in water, which is also a polar material. Nonpolar compounds, such as methane, are not very soluble in water. The solubility of a compound is determined by the similarity of the symmetry of its electrical charge distribution to that of its solvent.

Molecules with Polar and Nonpolar Parts

Some molecules—alcohols, for example—have both polar and nonpolar sections. A common compound used as a solvent and antiseptic in medicine is ethyl alcohol, C_2H_5OH (Figure 6.16).

Ethyl alcohol involves both covalent and polar covalent bonds. The electronegativity difference across a C—C bond is, of course, zero, making this bond 100% covalent. C—H bonds have only 4% ionic character and therefore are essentially nonpolar; the C—O bond has only 22% ionic character and is polar; and the O—H bond, the most polar bond in the molecule, has only 39% ionic character. A model of this molecule is shown in Figure 6.17. The polar O—H bond and, in particular, the two unshared pairs of electrons make the molecule very polar at the O—H end. However, from the C—O bond on down the carbon-hydrogen chain, the molecule is very symmetrical and nonpolar. This molecule has both polar and nonpolar properties and, therefore, is very soluble in both polar and nonpolar substances. (One end is soluble in polar materials, the other end in nonpolar materials.) For this reason, ethyl alcohol is a good solvent—a compound that can dissolve two mutually immiscible substances.

This explains why ethyl alcohol serves as a solvent in many flavor extracts. Flavors, such as oil of spearmint and oil of orange, have large nonpolar molecules. By using the nonpolar end of a relatively large number of alcohol molecules (80 to 90% by volume), the nonpolar flavor molecules can be dissolved. The polar ends of the alcohol molecules will then dissolve in water and thus mix the flavor into whatever is being cooked.

Soaps are another example of compounds that have both polar and nonpolar properties. Soaps and detergents consist of long nonpolar carbon-hydrogen chains with very polar groups attached to one end (Figure 6.18). The nonpolar group (called *lipophilic*) mixes easily with nonpolar grease, while the polar end (called *hydrophilic*) remains dissolved in water. The effect is to tie the grease molecules to the water molecules, thus enabling us to rinse the grease away in the water. Without soap there would be no interaction between water molecules and grease molecules, and no cleaning action would be possible. (For more on soap, see Chapter 20, specifically, Figure 20.11.)

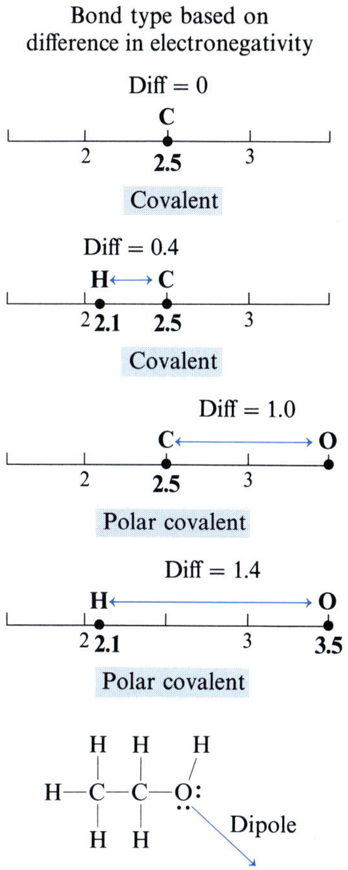
Bond type based on
difference in electronegativity
Diff = 0
C
2
2.5
3
Covalent
Diff = 0.4
H
C
2
2.1
2.5
3
Covalent
Diff = 1.0
C
O
2
2.5
3
Polar covalent
Diff = 1.4
H
O
2
2.1
3
3.5
Polar covalent
Dipole

FIGURE 6.16
Bond type and structure of ethyl alcohol. Note the direction of the dipole.

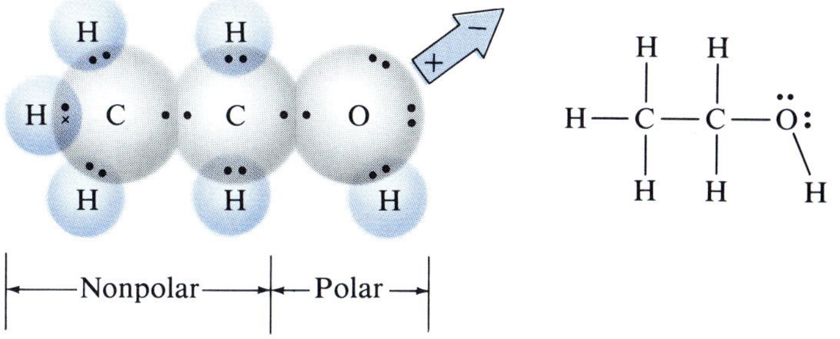
Nonpolar
Polar

FIGURE 6.17
The ethyl alcohol molecule has both polar and nonpolar parts. Alcohols all have the general structure R-OH, where R is a nonpolar hydrocarbon group. (Naming of common hydrocarbon groups is discussed in Chapters 14 and 15.)

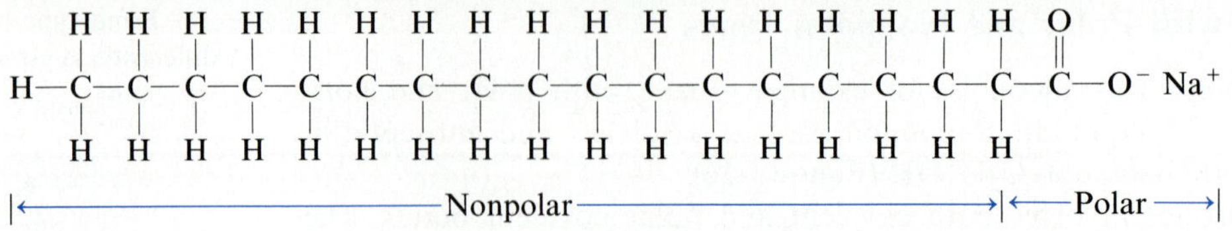
Nonpolar
Polar

FIGURE 6.18
The chemical structure of sodium stearate, a typical soap. This molecule has both polar (ionic, in this case) and nonpolar characteristics, and thus it will dissolve in both grease and water.

Dissolving Ionic Compounds

The process of dissolving ionic compounds in water is also related to the principle of molecular polarity. It is remarkable that a compound such as sodium chloride (NaCl)—in which the ionic bonds are so strong that a temperature of 801 °C is required to melt the solid crystal—may be easily dissolved in water at room temperature. Sodium chloride, however, will not dissolve in nonpolar substances such as carbon tetrachloride.

Ionic compounds will dissolve in polar solvents.

A sodium chloride crystal is held together by electrical attraction between the positive sodium ions and the negative chlorine ions. This attractive force, as indicated by the high melting point, is quite strong. If a crystal of sodium chloride is placed in water, however, the polar water molecules will surround the positive and negative ions, thus reducing the ionic attractive forces and permitting them to "float away" (Figure 6.19). An ion thus surrounded by water molecules is called a **hydrated ion**. The resulting mobility of hydrated ions is demonstrated by the excellent electrical conductivity of water solutions of ionic compounds. Solutions that conduct electricity are called **electrolytes**.

The effect of molecular polarity on the solubility of ionic compounds is illustrated by the data in Figure 6.20. The degree of solubility of an ionic compound is reflected by the number of mobile ions and, thus, by the electrical conductivity of the solution. This experiment involves measurement of the electrical conductivity of saturated solutions of sodium chloride in each of four solvents—water, methyl alcohol, ethyl alcohol, and normal-propyl alcohol. These molecules are similar in that they each contain a polar O—H group. However, the degree of molecular polarity is decreased by the addition of progressively larger nonpolar C—H units in place of the second hydrogen found in water. This set of molecules provides a sequence of gradually decreasing polar character. As one might predict, the electrical conductivity due to dissolved ions decreases rapidly as the solvent molecule

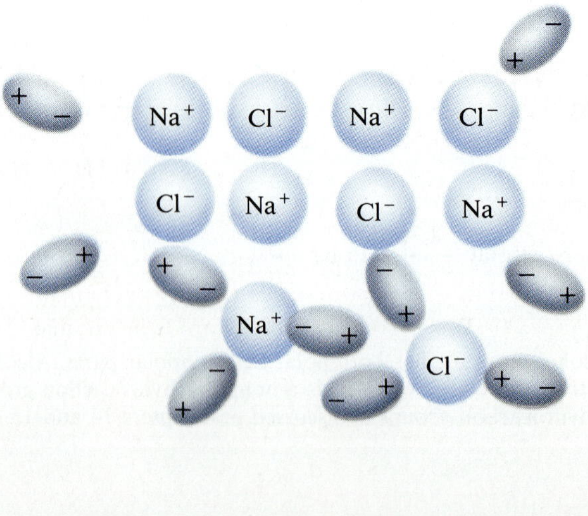
Na+
Cl−
Na+
Cl−
Cl−
Na+
Cl−
Na+
Na+
Cl−

FIGURE 6.19
Ionic compounds will usually dissolve in polar solvents such as water. Polar molecules surround the charged ions and cause them to float away.

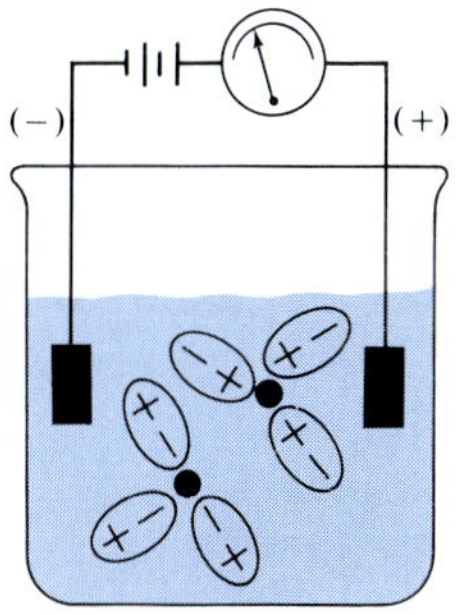
(−)
(+)

Solvent	Conductivity (arbitrary units) Solvent alone	Solvent saturated with NaCl
Water	120	440,000
Methyl alcohol	46	24,000
Ethyl alcohol	5.2	490
n-Propyl alcohol	2.5	40

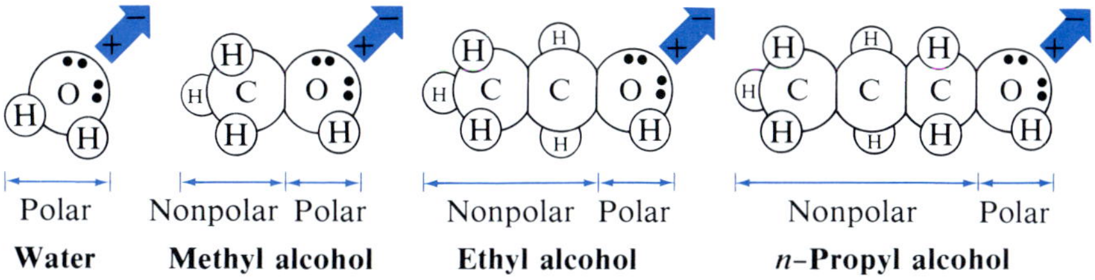
Polar
Water
Nonpolar Polar
Methyl alcohol
Nonpolar Polar
Ethyl alcohol
Nonpolar Polar
n-Propyl alcohol

FIGURE 6.20
The solubility of an ionic compound is determined by the polarity of the solvent molecule. Highly polar solvents do the best job of dissolving ionic compounds. The apparatus illustrated in the upper left corner of this figure uses a battery to force electrons onto one electrode (−) and to draw them in from the other electrode (+). If ions are present in the solution, negative ions will give up electrons and positive electrons will accept electrons, and electron flow will be noted in the external circuit. The conductivity of the solution is a measure of its ionic content.

becomes less polar and less able to neutralize the attractive forces between charged ions.

NEW TERMS

hydrated ion An ion surrounded by water molecules.

electrolyte A solution that contains ions and conducts electricity.

TESTING YOURSELF

Molecular Polarity and Solubility

1. Of the three molecules shown below, which will exhibit the most polar character?

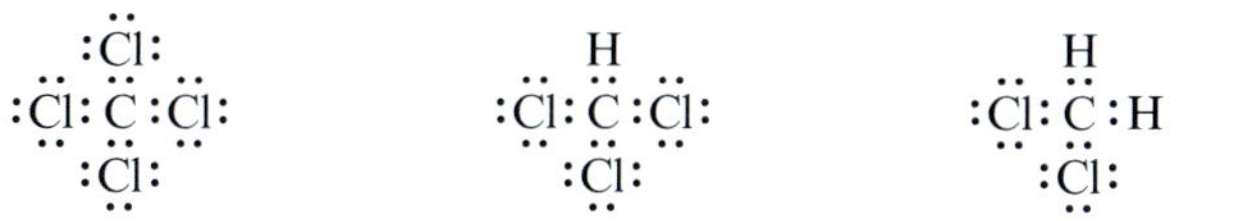

Carbon tetrachloride (tetrachloromethane) **Chloroform** (trichloromethane) **Methylene chloride** (dichloromethane)

2. Which of the following compounds do you think would be the best at dissolving table salt (sodium chloride)?

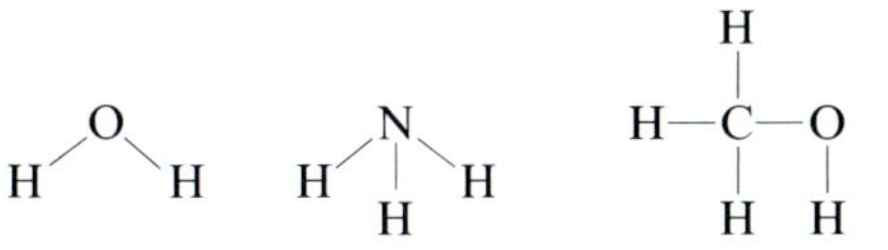

Water **Ammonia** **Methyl alcohol**

Answers **1.** methylene chloride **2.** water

6.4 Hydrogen Bonding

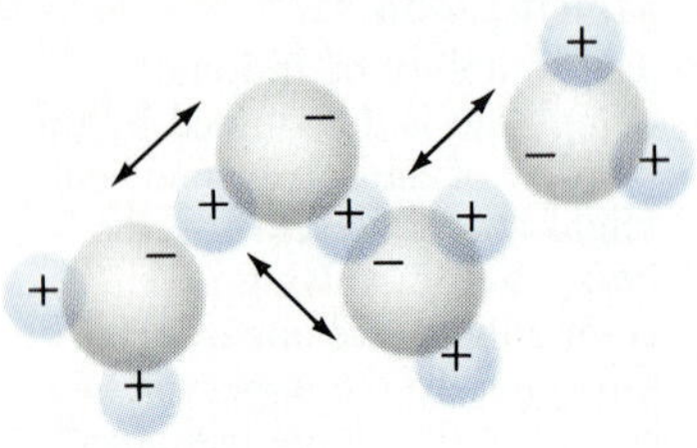

FIGURE 6.21
Hydrogen bonding causes water molecules to form clusters of 9 to 12 molecules. The resulting higher effective molecular weight causes a large increase in boiling point as compared to a nonpolar molecule of similar molecular weight.

In this and the previous chapter, we have studied covalent bonds. In these bonds there is an attraction from the positive nuclei of the bonding atoms toward one or more pairs of shared electrons. Another kind of bond exists that is considerably weaker and less substantial in that it does not involve shared electrons. It applies to only a few kinds of atoms, but in the instances that it is operating, it is an essential feature of life on our planet.

This bond is the **hydrogen bond**. Hydrogen bonding is a weak ionic bond that forms between oppositely charged parts of adjacent polar molecules. It is only noticeable if the dipoles of the participating molecules are strong. It could be more accurately described as a permanent dipole–dipole attraction. The hydrogen bond is best exemplified by water, in which two pairs of unshared electrons cause a significant negative area on one side of the oxygen. On the other side, electrons shared between the hydrogens and the oxygen lie much closer to the oxygen because of the difference in electronegativity (H = 2.1, O = 3.5; difference = 1.4, for a 39% ionic character). As a result, the hydrogens appear as positive "bumps" on the side of the oxygen opposite its unshared electron pairs. The tendency of water's hydrogen atoms to be attracted toward unshared electron pairs on adjacent water molecules is illustrated in Figure 6.21.

Hydrogen bonding has several important effects on the behavior of water. First, it significantly increases its boiling point because of the attraction of the water molecules to one another. A nonpolar molecule of methane (molecular weight 16) will boil at about −160 °C. Polar water (molecular weight 18), however, boils at 100 °C, a 260 °C difference.

Second, hydrogen bonding affects the freezing point of water. As water cools down, molecular motion becomes slower, and finally the water molecules align themselves with positive and negative ends together in much the same manner as one would expect for an ionic crystal. This regular arrangement is apparent if you take a close look at frost patterns (Figure 6.22). However, a regular arrangement of water molecules in ice takes more space than the random arrangement in the liquid state, and ice is a little less dense than water. For this reason, unlike almost all other substances, the solid form of water is less dense than its liquid form. Ice floats on water, whereas most other liquids will sink as they freeze.

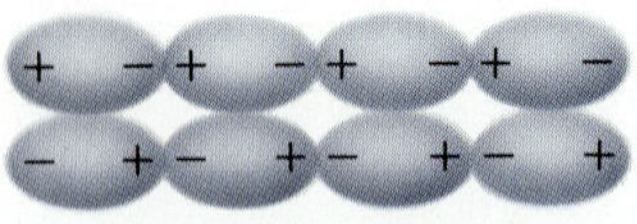
Regular alignment
of water molecules in ice

FIGURE 6.22
The crystalline form of ice is due to the alignment of water dipoles. This is responsible for the strength of ice and the shape of frost patterns. It also makes ice float.

NEW TERM

hydrogen bond A weak ionic bond between the negative region of a polar molecule and a positive hydrogen on an adjacent molecule.

6.5 Water: A Very Special Molecule

Water is an unusual molecule. It is present on earth in greater quantity than any other compound except silicon oxide. Its properties are well matched for the location of our planet with respect to the sun. Consider some of the ways the unique properties of water affect our lives.

Water as a Solvent

Because of its strong electrical dipole, water is unmatched in its ability to dissolve ionic compounds. Ionic compounds necessary to support human and animal life are transported dissolved in water in our blood. Sodium and potassium ions are necessary for nerve conduction. Sodium bicarbonate and carbonic acid regulate the acidity of blood. Many other ionic compounds necessary for life are dissolved in the blood.

On a larger scale, ionic compounds that inhibit plant life—sodium, magnesium, and calcium chlorides, for example—are washed from the soil by rainfall and ultimately accumulate in the ocean. Saline seep is a problem caused by the same effect that is taking agricultural land out of production in the West (Figure 6.23). Land that had been used for grazing or dry land farming and is now being irrigated experiences dissolving of soluble salts and a high concentration of these salts in drainage areas. The resulting high salt concentration renders the land useless for agriculture, and run-off water useless for irrigation.

Motion of water in the hydrologic cycle has moved most of the earth's ionic materials to the oceans.

Change of State

Without hydrogen bonding, a molecule of water's molecular weight (18) would probably boil at about −140 °C. Water's strong hydrogen bond, however, causes its molecules to stick together and raises its boiling point to 100 °C. If it were not for the hydrogen bond, water would always be a gas on our planet. Water will both freeze and evaporate at the temperatures

No other planet in our solar system exists at a temperature that will allow water to remain a liquid. Life on this planet depends on the hydrogen bond to keep water from boiling.

FIGURE 6.23
Saline seep is a serious problem in agricultural areas in the western United States. Irrigation of range or farmland in arid areas is in some cases dissolving and concentrating ionic salts that are toxic to plant life. Soil with a high salt concentration is no longer productive.

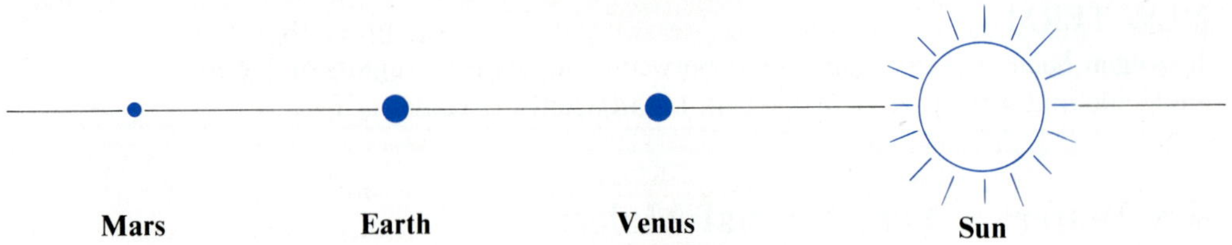
Mars
Earth
Venus
Sun

	Mars	Earth	Venus
Average surface temperature	−40 °C	+15 °C	+430 °C
Distance from sun	228×10^6 km	149×10^6 km	108×10^6 km

FIGURE 6.24
The earth is unique among planets in that water can exist in the solid, liquid, and gaseous states at temperatures normally found on the surface of the planet. Average surface temperatures and distances from the sun for the earth and its two neighboring planets are shown in this figure.

the sun places on our planet. Since there is only a 100 °C difference between the freezing and boiling points of water, if the earth were closer to the sun and thus hotter, all water would be in the gaseous form. If it were further from the sun and cooler, all water would be ice. Water on Venus is always in the gaseous state, and it is always frozen on Mars (Figure 6.24).

Density

Most substances contract as they cool and contract even more as they change from liquid to solid. This means that most substances are more dense as solids than as liquids. Water contracts as it cools, but as it freezes, the hydrogen bonds arrange water molecules into a precisely arranged crystal structure, which causes the water to expand slightly and become less dense. The fact that ice is less dense than liquid water is important to our planet. For example, what would happen if ice sank? In colder latitudes, the lakes and streams would begin to freeze on top where they come in contact with cold air, then the ice would sink, and more ice would form. Ultimately, the lake would be solid ice from the bottom to the top and would thus be unable to melt completely in the summer. Instead, what actually happens is that the less dense floating ice insulates the remaining water below from cold surface temperatures, and aquatic life is less affected by bitter winters because of the protective ice.

If the hydrogen bond did not cause water to expand and float as it freezes, lakes would freeze from the bottom to the top, and aquatic life would die during the winter.

Heat Capacity

Recall from Chapter 1 that water has one of the highest specific heats of any known material. One calorie of heat is required to change the temperature of 1 g of water 1 °C. In contrast, granite requires about 0.19 cal to change 1 g 1 °C, air requires 0.25 cal to change 1 g 1 °C, and copper requires 0.09 cal to change 1 g 1 °C. This special property of water affects us in two important ways.

The first involves maintenance of body temperature. About 75% of our body tissue is water, which requires 1 cal to change its temperature 1 °C. Bone has a much lower specific heat, on the order of 1/4 that of water. When you are outside on a cold day, which part of your body gets cold first? It's your fingers and toes, which have a much higher percentage bone content than the trunk. Removal of 100 cal from a 100-g section of finger, for example, would change its temperature 3 to 4 °C depending on the amount of fat and tissue surrounding the bone. On the other hand, removal of 100 cal from a 100-g section of torso would probably lower the temperature only about 1 °C. Surface area also works against us. Fingers and toes have more surface area for the volume they enclose and thus lose heat faster than the torso. This double effect caused by lower specific heat and larger surface area is why it is so easy for the fingers and toes to freeze (become frostbitten) while the body core remains warm.

This is also why hypothermia (a condition in which the temperature of the human body is below the normal 98.6 °F or 37.0 °C temperature) is so serious a condition. Once the temperature of the body core is lowered substantially (the chemical reactions that support life cannot continue if the temperature drops much below 80 °F, or 26.6 °C), it is difficult to add enough heat to bring the temperature up rapidly. (Hypothermia is discussed further in Chapter 10.)

Induced hypothermia is becoming a common practice for certain types of brain and open heart surgeries. When the body is sufficiently cooled, sensitive organs can go without blood and oxygen for longer periods—sometimes up to one hour—thus allowing time for the operation without causing damage to the organ.

Climate

Areas in close proximity to large bodies of water also profit from the high specific heat of water in the moderation of their climate. The earth receives an average of about 2 cal per square centimeter per minute in solar radiation from the sun. This amount of heat applied to 1 g of granite would raise its temperature about 10 °C, but applied to 1 g of water would only raise it 2 °C. Therefore, it takes much longer to change the temperature of a body of water than it does to change the temperature of the land. Large bodies of water absorb heat when the sun shines and give it back to the surrounding land and air when it is cloudy or dark. The result is a significant moderation in temperature extremes. For examples, consider Puget Sound in the state of Washington. It rarely freezes in Seattle, and rarely does the temperature exceed 30 °C. Due east 120 km in Wenatchee, Washington, at an elevation of about 200 m, the temperature extremes regularly range from −20 °C to more than 40 °C. It is separated from the coast by a mountain range and unprotected by a large body of water (Figure 6.25). A look at a map will convince you that any location of temperate climate in the world is close to a large body of water.

The high heat capacity of water moderates the climate in coastal regions.

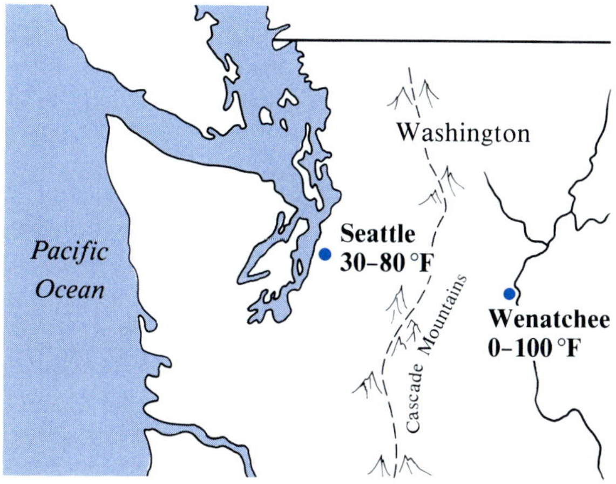
Washington
Pacific Ocean
Seattle
30–80 °F
Cascade Mountains
Wenatchee
0–100 °F

FIGURE 6.25
The presence of large bodies of water has a moderating effect on climate because of the large amount of heat gain or loss necessary to change the temperature of the water significantly. This map compares the normal temperature extremes at Seattle and Wenatchee in Washington state. Both cities are about the same latitude, and Wenatchee is only slightly higher (about 200 m). The presence of the large body of water in Puget Sound moderates the Seattle climate considerably.

SUMMARY

We have introduced much material in this chapter that can be summed up in three new major concepts. First, in covalently bonded molecules, electron pairs arrange themselves to be as far apart as possible. This is called *electron pair repulsion theory*. When atomic orbitals overlap during covalent bonding, the resulting orbitals are said to be *hybridized*. Second, if this arrangement results in an overall nonsymmetrical distribution of electrical charge across the molecule, an *electrical dipole* will result and the molecule will be *polar*. Third, interaction between molecules is determined by their polar properties. Polar molecules will mix freely with other polar molecules. Nonpolar molecules will mix freely with other nonpolar molecules. However, polar and nonpolar molecules will not dissolve in each other, unless a third type of molecule having both polar and nonpolar parts is present to bring the two together.

Molecular polarity determines much of the physical and chemical behavior of a molecule. We will find it of considerable importance as we proceed with our study of organic and biological chemistry.

TERMS

valence shell electron pair repulsion (VSEPR) theory
tetrahedron
hybridization
molecular dipole
polar molecule
nonpolar molecule
hydrated ion
electrolyte
hydrogen bond

CHAPTER REVIEW

QUESTIONS

1. Consider an atom that has three pairs of electrons. How can these electron pairs arrange themselves to stay as far apart as possible?
   a. In the form of a square
   b. In the form of a four-sided pyramid
   c. In the form of a flat triangle
   d. None of the above
2. Consider the compound ammonia, NH_3. The electronegativity of hydrogen is 2.1, the electronegativity of nitrogen is 3.0. What type of chemical bond will form between these two elements?
   a. covalent b. polar covalent c. ionic

   An electron filling series diagram for both nitrogen and hydrogen is shown below. Answers to several of the following questions can be drawn from this information.

Atomic orbitals ⟶ **Hybridized molecular orbitals**

$2p$ ↑ ↑ ↑		↑↓ ↑ ↑ ↑
$2s$ ↑↓		
	1s ↑ 1s ↑ 1s ↑	1s ↓ 1s ↓ 1s ↓
N	**H**	

3. How many *electron pairs* (from both nitrogen and the hydrogens) surround the nitrogen when bonding is complete?
   a. 1 b. 2 c. 3 d. 4 e. 5
4. Which atomic orbitals of nitrogen hold these electron pairs?
   a. $2s$ b. $2s$ and one $2p$ c. $2s$ and two $2p$'s d. $2s$ and three $2p$'s
5. When the nitrogen's atomic orbitals hybridize, what term is used to describe the resulting hybrid orbitals?
   a. s b. p c. sp d. sp^2 e. sp^3

6. A model of the ammonia molecule is shown on the right. It is
   a. polar with a dipole toward **a**
   b. polar with a dipole toward **b**
   c. nonpolar

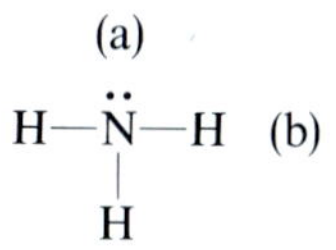

(a)
H—N—H (b)
H

7. Based on the polar properties of ammonia, would you expect it to be soluble in water?
   a. yes   b. no
8. Water molecules tend to stick together in clumps or chains because of weak bonds between the molecules. These bonds are shown by dashed lines in the sketch below. These bonds are called
   a. polar bonds
   b. ionic bonds
   c. covalent bonds
   d. hydrogen bonds

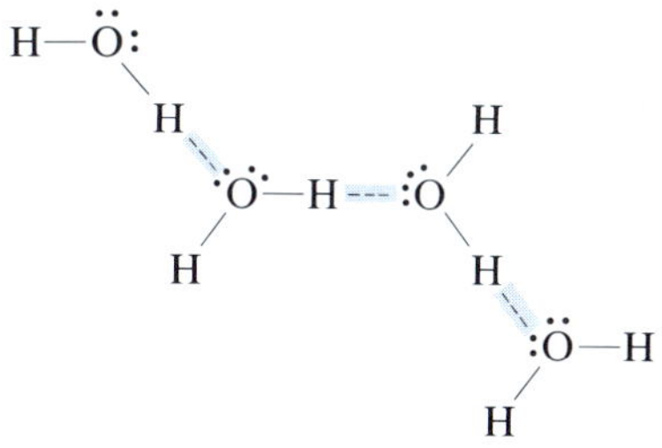

Answers **1.** c **2.** b **3.** d **4.** d **5.** e **6.** a **7.** a **8.** d

DIAGNOSTIC CHART

Blacken in all of the circles under the number of each question that you missed in the Chapter Review questions. The diagnostic chart will help you identify concept areas that need more study.

	Questions							
Concepts	**1**	**2**	**3**	**4**	**5**	**6**	**7**	**8**
Bond types		○						
Electron pair repulsion	○		○	○				
Hybridization					○	○	○	
Molecular polarity						○	○	
Solubility							○	
Hydrogen bonding								○

EXERCISES

Bond Types

1. Which of the elements listed below is central in the electronegativity scale and thus would be expected to form covalent bonds with practically any other element?
   a. oxygen   d. aluminum
   b. silicon   e. carbon
   c. hydrogen

2. What type of bond would form between carbon and hydrogen?
   a. ionic
   b. polar covalent
   c. covalent

3. What type of bond would form between oxygen and hydrogen?
   a. ionic
   b. polar covalent
   c. covalent

Electron Pair Repulsion and Hybridization

4. Consider the compound ethylene glycol, commonly known as antifreeze. Its molecular structure is sketched below.

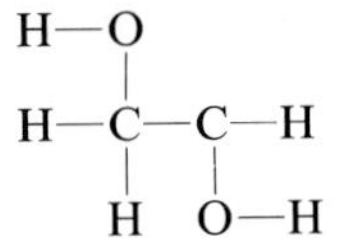

   a. What types of bonds exist in this molecule?
   b. Which atoms have *unshared* pairs of electons?
      a. hydrogen
      b. oxygen
      c. carbon
   c. What is the hybridization of each carbon atom?
      a. *s* c. sp^2
      b. *sp* d. sp^3
   d. How many dipoles are in this molecule?
      a. one
      b. none
      c. more than one
   e. Draw an arrow across the molecule to indicate the position of the dipole(s).
   f. Will this molecule be water soluble?

Molecular Polarity and Solubility

6. Electron dot models of several molecules are sketched below. Classify these molecules as polar or nonpolar. If they are polar, show the location of their dipole.

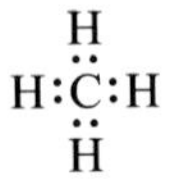

Methane

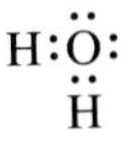

Water

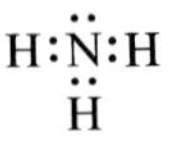

Ammonia

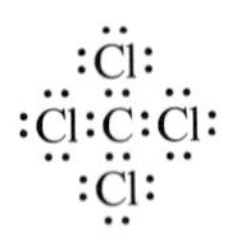

Carbon tetrachloride

7. Classify the following molecules as polar or nonpolar. Some molecules have both properties. In these cases, indicate the polar and nonpolar regions.

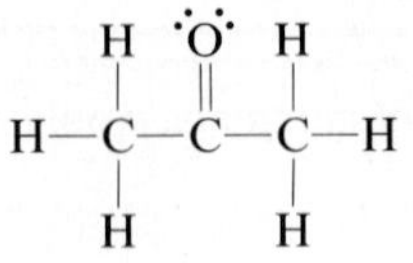

Acetone

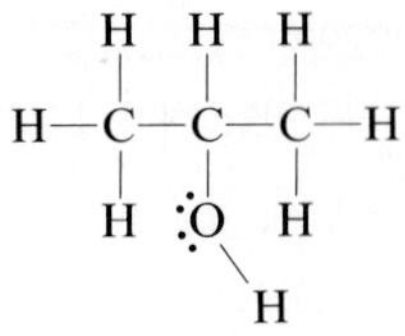

Isopropyl alcohol

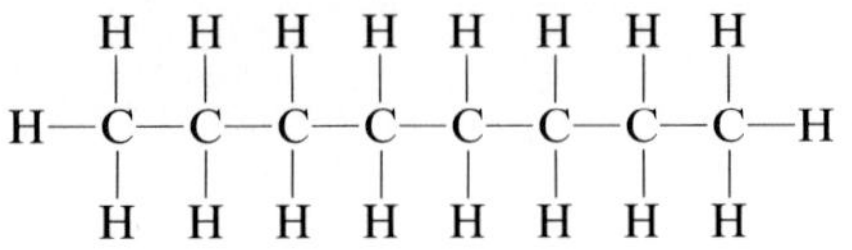

Octane

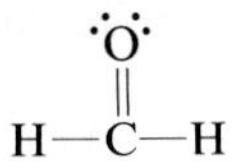

Formaldehyde

8. Why are ionic compounds often soluble in polar solvents but not in nonpolar materials?
9. Which of the molecules listed above in question 7 would be the poorest solvent for salt?

Hydrogen Bonding

10. What is the usual effect of hydrogen bonding on the boiling point of a compound?
    a. raises it
    b. lowers it
    c. no change
11. Which of the following molecules would probably have the greatest tendency to participate in hydrogen bonding? Recall that the electronegativity across a bond is a good measure of the polar character of the bond. (Note: watch for unshared electron pairs.)

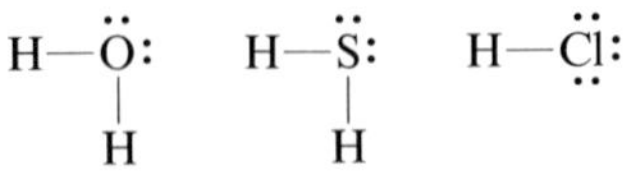

12. Why does water expand as it freezes? What might be some environmental consequences if this did not occur and ice were more dense than water? Most solids are more dense than their liquid counterparts.

CHAPTER

7

Chemical Formulas and Equations

This chapter begins by introducing some simple methods of naming ionic compounds and some conventions used in writing balanced chemical formulas. It shows how chemical periodicity may be used to supplement one's memory in formula prediction and how to write and balance equations that describe simple ionic reactions. A discussion of electron transfer reactions concludes the chapter.

OBJECTIVES

After completing this chapter, you should be able to

- Use the electron filling series and the electron transfer bonding model to predict charges of common ions.
- Define the term *oxidation number*, and give formulas and oxidation numbers for common and polyatomic ions.
- Use the periodic table to predict oxidation numbers of elements above atomic number 20.
- Balance charge to write correct formulas for compounds composed of common ions and polyatomic ions.
- Name compounds consisting of common ions and polyatomic ions.
- Write and balance chemical equations, given reactant and product identities.
- Define the terms *oxidation* and *reduction*, and use the electrochemical series to predict the direction of electron transfer during an electrochemical reaction.
- Write half-reactions for the oxidized and reduced species in electron transfer reactions.

7.1 Naming Chemical Compounds

The problem of naming chemical compounds was first faced by ancient scholars when names were given to the common elements and materials. Gold, for example, was known by the Latin name *aurum*, which meant "shining dawn." Mercury was known as "quicksilver" or "liquid silver."

Many compounds are commonly referred to today by names that are unrelated to their chemical composition. Water, for example, carries in its name no hint of its chemical composition, although to those who have had any contact with chemistry, it signifies the compound H_2O. Muriatic acid, commonly used to clean cement from bricks, is actually a water solution of hydrogen chloride, HCl. The lead ore galena is actually lead sulfide, PbS. Common or trivial names help us talk about a material, but do not give any hint of its composition. For anyone using chemistry, a systematic naming system that provides more information about the material is of considerable value. Note in Table 7.1 how closely related the chemical formulas are to the systematic names.

The first systematic system for naming chemical compounds was developed in the late 1700s.

The first serious attempt to systematize chemical nomenclature occurred in 1787 when four French scientists (Lavoisier, Morveau, Berthollet, and Fourcroy) published a book titled *Methode de Nomenclature Chimique*. This book suggested that compounds should be named according to their composition rather than after the person who discovered them, the town in which they were discovered, or how they looked. They further suggested that Latin and Greek element names be used for some chemical symbols, since they are more readily acceptable across political boundries (Table 7.2). Thus, the symbol for gold (*aurum*) became Au. Silver (*argentum*) became

TABLE 7.1
Common and Systematic Names for Selected Compounds

Common Name	Systematic Name	Formula
Blue vitriol	Cupric sulfate	$CuSO_4$
Galena	Lead sulfide	PbS
Chalcocite	Cupric sulfide	CuS
Hematite	Ferric oxide	Fe_2O_3

TABLE 7.2
Symbols for Some Elements Derived from Their Latin or Greek Names

Element	Latin Name	Symbol
Gold	**Au**rum	Au
Lead	**Plumb**um	Pb
Mercury	**H**ydra**g**yrum	Hg
Iron	**Fe**rrum	Fe
Silver	**Arg**entum	Ag
Potassium	**K**alium	K
Sodium	**Na**trium	Na
Tin	**Stan**num	Sn

Ag. Much of the nomenclature suggested in 1787 is still in use today and will be encountered later in this chapter.

For naming purposes, all chemical compounds are divided into two general classifications: *organic*, which includes most compounds based on a carbon skeleton, and *inorganic*, the remaining compounds. The rules presented in this chapter apply only to simple inorganic compounds—compounds that are principally ionic. The nomenclature of organic compounds will be considered in Chapter 14.

Organic compounds have a carbon skeleton.

7.2 Balanced Formulas

The electron transfer approach to bonding introduced in Chapter 5 is very successful in predicting the formulas of most **binary compounds**, which are those made of two elements, such as NaCl or H_2O. Electron transfer bonding involves transfer of electrons from one atom to another until all atoms involved have an inert gas electron configuration (a full complement of *s* and *p* electrons) in their outermost group of electrons. The charges of the resulting stable ions are then easily predicted by counting protons and electrons for each atom, and a compound formula may be written by balancing the number of atoms of each element until the overall formula charge is zero. This method of predicting compound formula is illustrated in Figure 7.1. The compound in this example is composed of magnesium and fluorine.

Binary compounds are comprised of two elements.

Figure 7.1 predicts formation of a magnesium +2 ion and two fluorine −1 ions. To obtain an overall balanced charge, one must have one magnesium ion and two fluoride ions; the correct formula of a compound of magnesium and fluorine is thus written MgF_2. Note that **subscript numbers** set below the line of type are used to indicate the proper number of each ion required to produce a balanced electrical charge. (The "1" is implied and therefore not written.)

Subscript numbers indicate the number of each type of atom needed to balance the electrical charge in a formula.

There are three conventions used in writing formulas for ionic compounds. The first involves knowing which way to transfer the electrons. This means checking the electronegativities for both elements. *The element with the lesser electronegativity will give up the electron and become the positive*

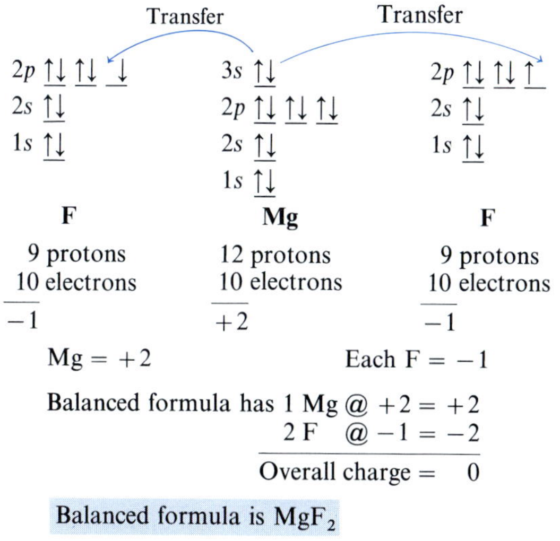
Transfer
Transfer
2p ↑↓ ↑↓ ↓
2s ↑↓
1s ↑↓
3s ↑↓
2p ↑↓ ↑↓ ↑↓
2s ↑↓
1s ↑↓
2p ↑↓ ↑↓ ↑
2s ↑↓
1s ↑↓
F
Mg
F
9 protons
10 electrons
−1
12 protons
10 electrons
+2
9 protons
10 electrons
−1
Mg = +2
Each F = −1
Balanced formula has 1 Mg @ +2 = +2
2 F @ −1 = −2
Overall charge = 0
Balanced formula is MgF2

FIGURE 7.1
This figure illustrates the use of the electron filling series to predict the formula of a compound. Since the electronegativities of magnesium and fluorine differ greatly (2.8 electronegativity units), electron transfer will occur. Two electrons are transferred from magnesium to give it a stable inert gas outer electron structure; one of these electrons is used to stabilize each of the two fluorine atoms.

ion. The second convention simply states that the more electropositive element is written first. Thus, it is correct to write the formula MgF_2, but the formula F_2Mg, although conveying the same chemical information, is not generally accepted. The third convention states that *subscript numbers following an element are used to balance charge within the formula.* Thus, MgF_2 is correct, while Mg2F is not. Large numbers set on the line of type are used to designate numbers of formula units in a balanced chemical equation, as we will see later in this chapter.

Oxidation Numbers

Oxidation number is the charge predicted for an ion by the electron transfer bonding theory. The sum of oxidation numbers in a balanced formula always equals zero.

Although we have been speaking of the positive or negative charge placed on an atom by loss or gain of electrons, a more general term correctly applicable even to covalent compounds is **oxidation number**. We would say that in the compound of magnesium and fluorine, the oxidation number of magnesium is +2 and the oxidation number of fluorine is −1. *The sum of the oxidation numbers in a balanced formula must always equal zero.*

Oxidation number is simply the charge predicted for an ion by the electron transfer bonding theory. In the case of ionic compounds, this number has a very real meaning. However, in the case of covalent compounds where electrons are shared rather than transferred, the concept of charge has little significance. Even with covalent compounds, however, a correct formula can be predicted by the electron transfer bonding theory. In this case, the oxidation number is useful as a bookkeeping device and indicates the number of shared electrons. For example, as we have seen in previous chapters, the electronegativity difference between oxygen and hydrogen is 1.4. Water is thus a polar covalent compound (Figure 7.2). Each hydrogen shares one electron with the more electronegative oxygen. The oxidation number of hydrogen is therefore +1; 1 because it shares one electron, and + because it is the more electropositive (less electronegative) element.

Note that if electron transfer occurred in this compound (it does not), the oxygen ion would have a charge of −2 and each hydrogen ion a charge of +1. The definition of oxidation number as the charge predicted for an atom by the electron transfer bonding theory is valid. By general agreement among chemists, the sign of an oxidation number is written *after* the number to avoid confusion with exponents in mathematical expressions. The oxidation number of hydrogen is written 1+ and that of oxygen 2−.

Regardless of bond type, the sum of the oxidation numbers in a balanced

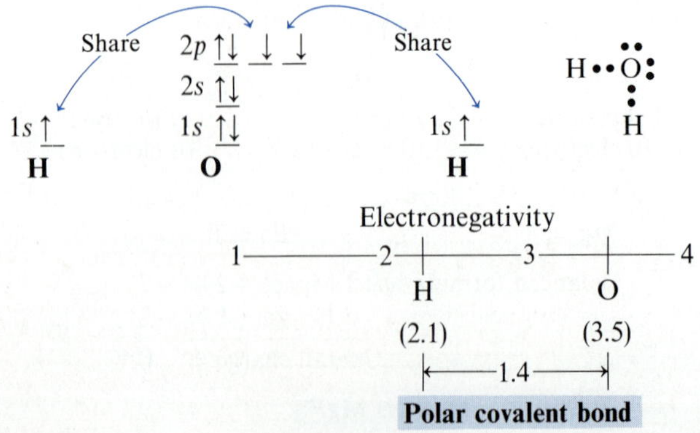
Share
Share
2p ↑↓ ↓ ↓
2s ↑↓
1s ↑
1s ↑↓
1s ↑
H
O
H
H••O:
H
Electronegativity
1
2
3
4
H
O
(2.1)
(3.5)
1.4
Polar covalent bond

FIGURE 7.2
The difference in electronegativity between oxygen and hydrogen is 1.4, thus water is a polar covalent compound. Because each hydrogen shares one electron with the more electronegative oxygen, the oxidation number of hydrogen is 1+. Oxygen shares two electrons so its oxidation number is 2−.

TABLE 7.3
Oxidation Numbers are Used to Balance Charge in Chemical Formulas, Regardless of Bond Type

Compound	Bond Type	Element	Oxidation Number	Sum of Oxidation Numbers
NaCl	Ionic	Na	1+	
		Cl	1−	0
HCl	Polar covalent	H	1+	
		Cl	1−	0
H_2O	Polar covalent	H	1+	
		O	2−	0
NH_3	Covalent	H	1+	
		N	3−	0
CO_2	Covalent	C	4+	
		O	2−	0
CCl_4	Covalent	C	4+	
		Cl	1−	0

formula is always zero. Table 7.3 shows how oxidation numbers are used to balance charge in chemical formulas, regardless of bond type. The numbers of each atom in the formula are adjusted with subscript numbers until the oxidation numbers total zero. In some cases, elements have more than one possible oxidation number.

To write a formula for a binary compound, one must (1) know the oxidation numbers of the component atoms and, (2) by use of subscript numbers, balance the number of positive and negative atoms to give an overall zero charge to the formula. When the numbers are small, as in the case of magnesium fluoride, the process is easy. When the numbers are larger, it helps to compute the least common multiple of the two oxidation numbers.

EXAMPLE 7.1

Write a balanced formula for a compound of aluminum and oxygen.

Oxidation numbers:

$$\text{Al} = 3+$$
$$\text{O} = 2-$$

Summation of oxidation numbers:

The least common multiple of 3 and 2 is 6. To gain an oxidation number of 6+ or 6−, we need

2 Al @ 3+ = 6+
3 O @ 2− = 6−
Sum = 0

Predicted formula:

Al_2O_3

For simple binary ionic compounds, a "criss-cross" rule helps to predict charge balance:

Al^{3+} O^{2-}

Al_2 O_3

Distinction Between Molecules and Compounds

Covalent compounds form well-defined groups of atoms called *molecules*. The formula tells the number of each kind of atom in the molecule.

Ionic compounds form crystals, the smallest repeating unit of which is described by the formula.

Perhaps by now you have noticed that we have been careful in our use of the words *molecule* and *compound*. Although water is both a compound and a molecule, sodium chloride, for example, is a compound but not a molecule. What is the difference? Water is a molecule with polar covalent bonds; it is a discrete unit composed of two hydrogen atoms and one oxygen atom. If extra hydrogen and oxygen atoms are present, more *molecules* of water are formed. The **molecular formula** of water (H_2O) tells the identity and number of each atom in the molecule. On the other hand, sodium chloride is an ionic compound. The basic unit of this compound is one sodium ion and one chlorine ion. If extra sodium and chlorine atoms are present, a larger three-dimensional crystal of sodium chloride is formed. The **compound formula** of sodium chloride (NaCl) tells the identity and elemental ratio of the *smallest repeating unit* of the compound: one sodium (Na) and one chlorine (Cl). This distinction becomes quite important when we begin to compute molecular or formula masses. The mass of a molecule of water is

$$2\text{ H} @ 1\text{ amu} + 1\text{ O} @ 16\text{ amu} = 18\text{ amu}$$

The formula weight of NaCl is

$$1\text{ Na} @ 22.99\text{ amu} + \text{Cl} @ 35.45\text{ amu} = 58.44\text{ amu}$$

even though the actual crystal of NaCl may be many repeating units in size.

EXAMPLE 7.2

Write a balanced formula for a *molecule* of carbon and sulfur.

Oxidation numbers:

$$C = 4+$$
$$S = 2-$$

Summation of oxidation numbers:

$$1\text{ C} @ 4+ = 4+$$
$$2\text{ S} @ 2- = \underline{4-}$$
$$\text{Sum} = 0$$

Predicted formula: CS_2

EXAMPLE 7.3

Write a balanced formula for a *compound* of potassium and oxygen.

Oxidation numbers:

$$K = 1+$$
$$O = 2-$$

Use "criss-cross" rule to help predict charge balance:

K^{1+} O^{2-}

K_2 O_1

Summation of oxidation numbers:

$$2\text{ K} @ 1+ = 2+$$
$$1\text{ O} @ 2- = \underline{2-}$$
$$\text{Sum} = 0$$

Predicted formula:

K_2O — The smallest repeating unit of the compound is 2 K and 1 O.

Elements with Multiple Oxidation Numbers

Some elements may have more than one oxidation number.

Some elements have more than one oxidation number depending on what compound they are in. Manganese, for example, has an oxidation number of 4+ in the compound manganese oxide (MnO_2), but an oxidation number of 7+ in potassium permanganate ($KMnO_4$). In most cases, elements with possible multiple oxidation numbers are found combined with elements having only one oxidation number, so it is a relatively simple task to compute the unknown oxidation number. Because of its high electronegativity, oxygen is always 2− except when bonded to itself in oxygen gas, in peroxides, or in oxygen-fluorine compounds. Elements in periodic group I (the lithium, sodium, and potassium family) are always 1+, and those in group II (beryllium, magnesium, and calcium family) are always 2+. (See Section 7.5 for more discussion of multiple oxidation numbers.)

EXAMPLE 7.4

(a) Determine the oxidation number of Mn in the compound manganese oxide, MnO_2.

$$\begin{aligned} 1\ \text{Mn} &= ? \\ 2\ \text{O} @ 2- &= 4- \\ \hline \text{Sum must} &= 0 \end{aligned}$$

$$\text{If} \quad \text{Mn} + 2\ \text{O} = 0$$
$$? + 4- = 0$$

$$\text{Thus,} \quad \text{Mn must} = \boxed{4+}$$

(b) Determine the oxidation number of Mn in potassium permanganate, $KMnO_4$.

$$\begin{aligned} 1\ \text{K} &= 1+ \\ 1\ \text{Mn} &= ? \\ 4\ \text{O} @ 2- &= 8- \\ \hline \text{Sum must} &= 0 \end{aligned}$$

$$\text{If} \quad \text{K} + \text{Mn} + \text{O} = 0$$
$$1+ + \ ? \ + 4(2-) = 0$$

$$\text{Thus,} \quad \text{Mn must} = \boxed{7+}$$

Elements with an Oxidation Number of Zero

All elements carry an oxidation number of zero when in their pure or elemental state. Thus, copper in a copper wire is Cu^0, iron in a railroad rail is Fe^0, and gold in a ring is Au^0. As discussed in Chapter 5, seven elements

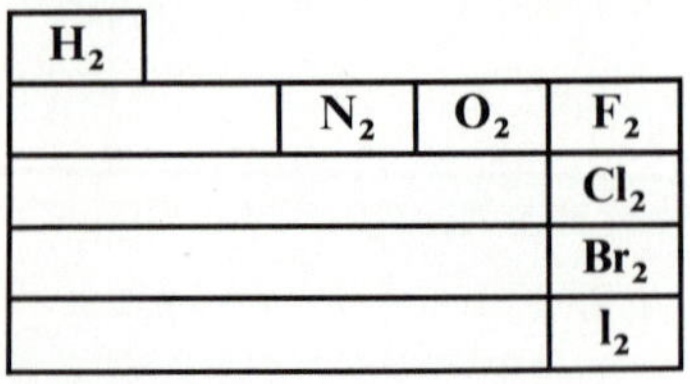
H_2
N_2
O_2
F_2
Cl_2
Br_2
I_2

FIGURE 7.3
Seven elements are found as diatomic molecules when not combined with another element.

occur naturally as *diatomic molecules* when not part of a compound. Oxygen, for example, is found as O_2. However, since oxygen is not combined with any other element, its oxidation number is still zero. The elements found in nature as diatomic molecules are identified in Figure 7.3.

Chemical Periodicity

The identical outermost electron configuration of elements in vertical columns of the periodic table makes possible prediction of formulas of compounds involving elements with large numbers of electrons. Consider, for example, a compound consisting of barium and iodine. Although you could work through the electron filling series for both of these elements, barium has 56 electrons and iodine has 53, making this a time-consuming job. However, barium is in the same vertical column of the periodic table as magnesium, and iodine is in the same vertical column as fluorine (Figure 7.4). The outer electron configuration of barium is thus the same as that of magnesium, and that of iodine is the same as fluorine.

The periodic table can be used to predict oxidation numbers, given knowledge of the oxidation number of one element in the vertical column.

Earlier in Figure 7.1 we predicted the oxidation number of magnesium to be 2+ and the oxidation number of fluorine to be 1−. The correct formula of a compound of these two elements is MgF_2. By applying the principles of chemical periodicity, we predict the oxidation number of barium to be 2+, and the oxidation number of iodine to be 1−. Thus, the correct formula of a compound of barium and iodine would be BaI_2. Using the periodic table in this manner makes possible the prediction of the formulas and properties of a large number of compounds by reference only to the electron configuration of the smallest element in the family.

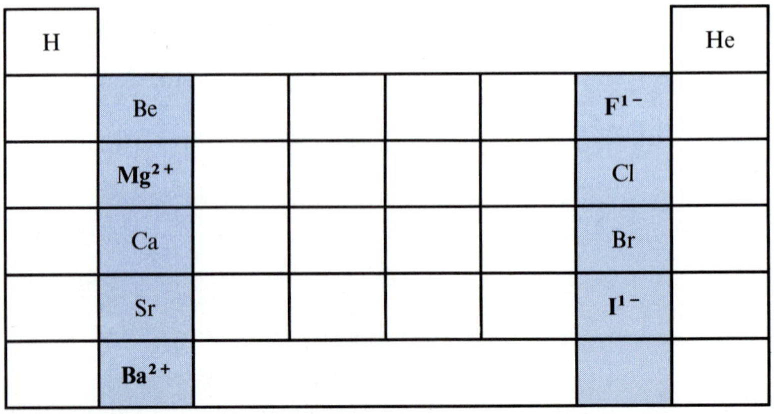
H
He
Be
F^{1-}
Mg^{2+}
Cl
Ca
Br
Sr
I^{1-}
Ba^{2+}

FIGURE 7.4
The periodic table can be used to predict oxidation numbers of most elements. Elements in the same chemical family (vertical column) have the same outermost electron configuration and almost always have the same oxidation number. Thus, if you know the oxidation number of one member of the family, you have an excellent chance of predicting the oxidation number of other elements in this family.

NEW TERMS

binary compounds A compound consisting of only two elements.

subscript numbers A number placed below and after the symbol for an element in a chemical formula indicating the number of that element in the balanced formula for the compound. For example, the subscript "2" in H_2O indicates two hydrogens. If a subscript number is not given, it is presumed to be "1".

oxidation number The charge predicted for an atom by the electron transfer bonding theory, even if the molecule is covalent; or the number of electrons shared by the atom in a covalent or polar covalent molecule.

molecular formula The formula of a covalent compound which gives the identity and number of each atom in the molecule.

compound formula The formula of an ionic compound, which gives the identity and number of each atom in the smallest repeating unit of the compound.

TESTING YOURSELF

Ionic Charge and Oxidation Numbers

1. Using the electron filling series and the rules for chemical stability developed in Chapter 4, predict the charge that should exist on stable ions formed from the following elements:

Li	Cl
Na	O
Mg	S
Ca	Al
F	

2. Predict the oxidation number of sulfur in the following compounds. (Recall that the sum of the oxidation numbers in a balanced formula is always zero.)

Hydrogen sulfide	H_2S
Sulfur dioxide	SO_2
Sulfuric acid	H_2SO_4
Magnesium sulfide	MgS

3. Predict the oxidation number of iron in the following compounds. (Note that the suffix following the stem name *ferr* for *ferrum* or iron changes from *ic* when iron is +3 to *ous* when iron is +2. This use of a variable suffix on an element name to indicate magnitude of charge is discussed in Section 7.5.)

Ferric chloride	$FeCl_3$
Ferric oxide	Fe_2O_3
Ferrous sulfide	FeS

4. Predict the oxidation number of nitrogen in the following compounds:

Nitric acid	HNO_3
Nitric oxide	NO
Nitrogen dioxide	NO_2
Sodium nitrate	$NaNO_3$

5. By use of the periodic table, predict the oxidation numbers of the following elements:

Radium	Ra
Iodine	I
Selenium	Se

Answers **1.** Li 1+, Na 1+, Mg 2+, Ca 2+, F 1−, Cl 1−, O 2−, S 2−, Al 3+ **2.** H_2S 2−, SO_2 4+, H_2SO_4 6+, MgS 2− **3.** $FeCl_3$ 3+, Fe_2O_3 3+, FeS 2+ **4.** HNO_3 5+, NO 2+, NO_2 4+, $NaNO_3$ 5+ **5.** Ra 2+, I 1−, Se 2−

7.3 Simple Binary Compounds

Names of all binary compounds end in *-ide*

Naming ionic binary or two-element compounds involves only two simple rules: (1) the name of the element with the most positive oxidation number is placed first, and (2) the **stem name** of the second element is followed by the suffix *-ide*. For example, a compound of sodium (oxidation number 1+) and chlorine (oxidation number 1−) would have the formula NaCl and would be called sodium chlor*ide*. In the same way, the compound of barium and iodine discussed at the end of the preceding section would have the formula BaI_2 and would be called barium iod*ide*. (Although BaI_2 contains three atoms, it contains only two elements and thus qualifies as a binary compound.) Table 7.4 illustrates the application of this principle to several common binary compounds.

TABLE 7.4
Formulas and Names of Some Common Compounds

First Element	Oxidation Number	Second Element	Oxidation Number	Balanced Formula	Compound Name
Na	1+	Cl	1−	NaCl	Sodium chloride
Al	3+	F	1−	AlF_3	Aluminum fluoride
Ca	2+	O	2−	CaO	Calcium oxide
H	1+	S	2−	H_2S	Hydrogen sulfide
K	1+	Br	1−	KBr	Potassium bromide
Al	3+	O	2−	Al_2O_3	Aluminum oxide
Mg	2+	I	1−	MgI_2	Magnesium iodide

To help in your studying, it is best to learn the system for naming compounds rather than trying to remember a large number of specific examples. It is sometimes helpful to place the general rule of each naming system on an index card to use for review.

NAMING COMPOUNDS

Binary Compounds

	Name of metallic element (+)	+	Stem name of second element (−)	+	*-ide* ending	=	Compound name
NaCl:	Sodium	+	chlor	+	ide	=	Sodium chloride
Al_2O_3:	Aluminum	+	ox	+	ide	=	Aluminum oxide

NEW TERM

stem name The first part of the element name that provides enough information to identify the element in a compound name. The stem name of sulfur, for example, is *sulf-*. Thus, the binary compound of hydrogen and sulfur has the name hydrogen *sulf*ide.

TESTING YOURSELF

Formulas and Names for Simple Binary Compounds

1. Write balanced formulas for the following binary compounds:

Potassium fluoride	Aluminum oxide
Magnesium chloride	Hydrogen fluoride
Sodium oxide	

2. Name the following simple binary compounds:

Li_2O	NaBr
Na_2S	$RaCl_2$
PbS	

Answers **1.** KF, $MgCl_2$, Na_2O, Al_2O_3, HF **2.** lithium oxide, sodium sulfide, lead sulfide, sodium bromide, radium chloride

7.4 Compounds Involving More Than Two Elements

So far we have considered only binary or two-element compounds. A large number of compounds, however, contain more than two elements, and some compounds contain groups of atoms, covalently bound together, that act as a single ion. These groups are called **polyatomic ions**, and although they are covalently bonded, the atoms that make up the group carry an overall excess or deficiency of electrons. The group itself acts as a single ion and bears a characteristic charge.

Polyatomic ions are covalently bonded groups that carry an electrical charge. The group acts as an ion.

Consider, for example, the hydroxide group (OH^-). This group, comprised of one oxygen atom and one hydrogen atom, carries a 1− overall charge and acts as an ion in its reactions with other elements and ions. The electron arrangement of the hydroxide group is shown in Figure 7.5. Note that the charge of a polyatomic ion is equal to the sum of the oxidation numbers of its components. Oxidation numbers refer only to individual atoms or ions.

A number of polyatomic ions commonly encountered in chemistry are listed in Table 7.5. Since it is time consuming to predict charges for all of these groups, we suggest that you simply memorize them. We will not ask you to memorize many things, but this is one situation where it is easier to memorize a few often-used facts. Listed with this group are several elements having oxidation numbers that are difficult to determine by examination of the electron filling series. Also listed are a number of common elements with very obvious oxidation numbers. Knowing these charges and oxidation numbers by memory will make the process of equation balancing (introduced later in Section 7.6) much easier.

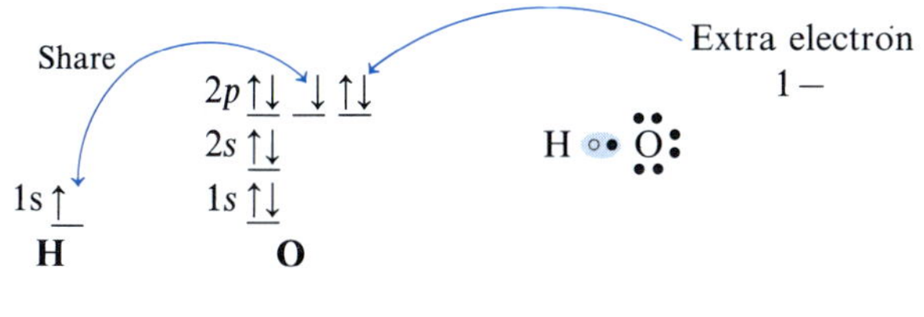

Share
Extra electron
1−
2p
2s
1s
1s
H
O
H
O

FIGURE 7.5
The polyatomic ion known as the hydroxide group (OH^-) carries a −1 charge due to the extra electron added to stabilize it. Although the O—H bond is polar covalent, the group itself acts as an ion.

TABLE 7.5
Oxidation Numbers of Some Common Ions and Polyatomic Ions

Oxidation Number	Ion Name	Formula	Oxidation Number	Ion Name	Formula
1+	Ammonium	NH_4^+	1−	Acetate	$C_2H_3O_2^-$
	Hydrogen*	H^+		Bromide*	Br^-
	Lithium*	Li^+		Bicarbonate	HCO_3^-
	Potassium*	K^+		Chlorate	ClO_3^-
	Silver	Ag^+		Chloride*	Cl^-
	Sodium*	Na^+		Cyanide	CN^-
	Cuprous	Cu^+		Fluoride*	F^-
				Hydroxide	OH^-
				Iodide*	I^-
				Nitrate	NO_3^-
				Thiocyanate	SCN^-
				Permanganate	MnO_4^-
2+	Beryllium*	Be^{2+}	2−	Carbonate	CO_3^{2-}
	Calcium*	Ca^{2+}		Chromate	CrO_4^{2-}
	Cobalt	Co^{2+}		Dichromate	$Cr_2O_7^{2-}$
	Cupric	Cu^{2+}		Oxygen* (oxide)	O^{2-}
	Ferrous	Fe^{2+}		Sulfate	SO_4^{2-}
	Lead	Pb^{2+}		Sulfite	SO_3^{2-}
	Magnesium*	Mg^{2+}		Sulfur* (sulfide)	S^{2-}
	Nickel	Ni^{2+}			
	Strontium	Sr^{2+}			
	Zinc	Zn^{2+}			
3+	Aluminum*	Al^{3+}	3−	Ferricyanide [Hexacyanoferrate(III)]	$Fe(CN)_6^{3-}$
	Ferric [Iron(III)]	Fe^{3+}		Phosphate	PO_4^{3-}
			4−	Ferrocyanide [Hexacyanoferrate(II)]	$Fe(CN)_6^{4-}$
				Silicate	SiO_4^{4-}

* The oxidation numbers of these ions can be easily determined from the periodic table.

Formulas involving polyatomic ions are written and their charges balanced in the same manner as simple binary compounds. For example, a compound consisting of the ferric ion (oxidation number 3+) and the nitrate polyatomic ion (charge −1) would be written $Fe(NO_3)_3$. Note the use of parentheses to indicate that three of the nitrate NO_3^- groups are present for one ferric ion. The use of parentheses and brackets together with subscript numbers to write balanced formulas for complex compounds is illustrated in Figure 7.6. When charge cannot be easily balanced by simple visual examination, one can compute the lowest common multiple of the positive and negative charges or oxidation numbers. In the example of Figure 7.6, the

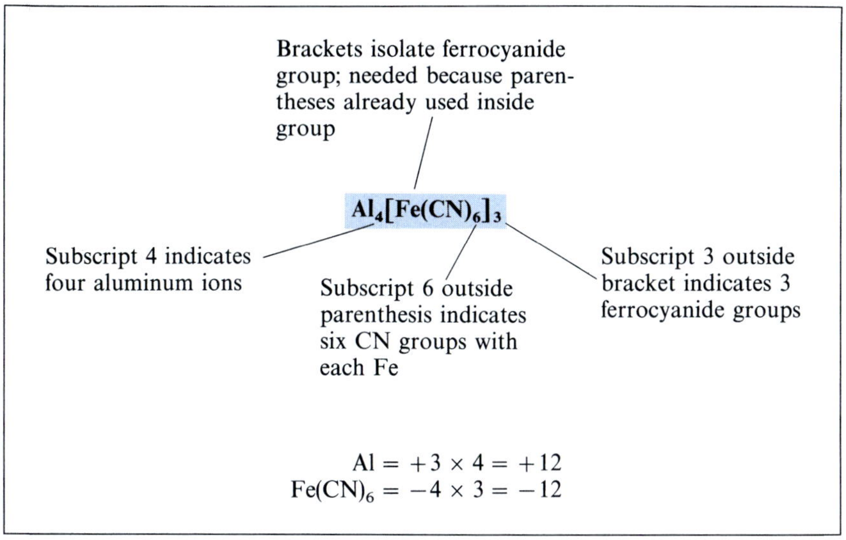
Brackets isolate ferrocyanide group; needed because parentheses already used inside group
Al4[Fe(CN)6]3
Subscript 4 indicates four aluminum ions
Subscript 6 outside parenthesis indicates six CN groups with each Fe
Subscript 3 outside bracket indicates 3 ferrocyanide groups
Al = +3 × 4 = +12
Fe(CN)6 = −4 × 3 = −12

FIGURE 7.6
Parentheses and brackets are used with subscript numbers to write balanced charges for compounds containing polyatomic ions.

lowest common multiple of the aluminum 3+ and the ferrocyanide 4− is 12. Thus, four aluminum ions (12+) and three ferrocyanide groups (12−) are required.

Generation of balanced charges for several compounds involving positive and negative polyatomic ions is illustrated in Table 7.6. In each case the principle is the same—the sum of the oxidation numbers of the simple ions and polyatomic ions involved must equal zero.

When the polyatomic ion is negative, as is usually the case, the compound is named by combining the positive ion name with the negative group name (Example 7.5).

EXAMPLE 7.5

Name these compounds.

	Name of positive ion	+	Name of negative group	=	Compound name
K_2CrO_4:	K^+ Potassium	+	CrO_4^{2-} Chromate	=	Potassium chromate
1LiOH:	Li^+ Lithium	+	OH^- Hydroxide	=	Lithium hydroxide
$NaHCO_3$:	Na^+ Sodium	+	HCO_3^- Bicarbonate	=	Sodium bicarbonate

When the polyatomic ion is positive (as is the case with the ammonium ion) and the negative ion is a simple one, the compound is named as if it were a normal binary compound. The ammonium ion counts as a single element.

TABLE 7.6
Some Examples of Compounds Involving Polyatomic Ions

Positive Group	Oxidation Number	Negative Group	Oxidation Number	Formula	Name
Na^+	1+	OH^-	1−	$NaOH$	Sodium hydroxide
NH_4^+	1+	Cl^-	1−	NH_4Cl	Ammonium chloride
K^+	1+	CrO_4^{2-}	2−	K_2CrO_4	Potassium chromate
Al^{3+}	3+	PO_4^{3-}	3−	$AlPO_4$	Aluminum phosphate
Li^+	1+	SO_4^{2-}	2−	Li_2SO_4	Lithium sulfate

EXAMPLE 7.6

Name this compound.

	Name of positive group	+	Name of negative ion	=	Compound name
NH_4Cl:	NH_4^+ Ammonium	+	Cl^- Chloride	=	Ammonium chloride

When both positive and negative ions are polyatomic groups, the compound name is a combination of the names of the two groups.

EXAMPLE 7.7

Name these compounds.

	Name of positive group	+	Name of negative group	=	Compound name
NH_4NO_3:	NH_4^+ Ammonium	+	NO_3^- Nitrate	=	Ammonium nitrate
$(NH_4)_2SO_4$:	NH_4^+ Ammonium	+	SO_4^{2-} Sulfate	=	Ammonium sulfate

NAMING COMPOUNDS

Inorganic Compounds with Polyatomic Groups

	Name of positive group or ion	+	Name of negative group or ion	=	Compound name
NH_4OH:	Ammonium	+	Hydroxide	=	Ammonium hydroxide
$CaCO_3$:	Calcium	+	Carbonate	=	Calcium carbonate

Naming of compounds containing polyatomic ions is generally not difficult. It does, however, require knowledge of the formula, oxidation number, and name of the polyatomic group. The best approach for a beginning chemistry student is simply to memorize these facts for the most common groups, as shown in the preceding "notecard."

NEW TERM

polyatomic ion A charged, covalently bound group of atoms that together act as an ion. Examples include the hydroxide group (OH^-) and the nitrate group (NO_3^-).

TESTING YOURSELF

Names and Formulas of Compounds with Polyatomic Ions

1. Write balanced formulas for the following compounds that involve polyatomic ions.

Potassium carbonate	Calcium phosphate
Sodium bicarbonate	Ammonium hydroxide
Lead nitrate	

2. Give names for the following compounds.

NH_4Cl	$PbCrO_4$
$Mg(OH)_2$	$Al(OH)_3$
$AgNO_3$	

Answers **1.** K_2CO_3, $NaHCO_3$, $Pb(NO_3)_2$, $Ca_3(PO_4)_2$, NH_4OH **2.** ammonium chloride, magnesium hydroxide, silver nitrate, lead chromate, aluminum hydroxide

7.5 Additional Systems of Naming Compounds

Compounds Containing Elements with Multiple Oxidation Numbers

As mentioned earlier, several elements have more than one possible oxidation number: for example, iron (Fe^{2+}, Fe^{3+}), carbon (C^{2+}, C^{4+}), and copper (Cu^{1+}, Cu^{2+}). There are three ways of handling these problems. We may, for example, simply use Latin and Greek prefixes to designate the number of each atom represented in the formula (Table 7.7). This method is used when the simplest form of the compound is a molecule.

TABLE 7.7
Common Prefixes Used to Designate Numbers of Atoms in Compound Names

mono-
di-
tri-
tetra-
penta-
hexa-
hepta-
octa-
nona-
deca-

EXAMPLE 7.8

Name these compounds.

Compound	Oxidation Number of Positive Element	Name
CO_2	C^{4+}	Carbon *di*oxide
CO	C^{2+}	Carbon *mon*oxide
CCl_4	C^{4+}	Carbon *tetra*chloride
N_2O_5	N^{5+}	*Di*nitrogen *pent*oxide

ADDING *-OUS* OR *-IC* SUFFIXES Since most multiple oxidation number elements have only two oxidation states, an outdated but still often used method involves application of the suffixes *-ous* or *-ic* to the Latin stem name of the element of multiple oxidation number. The **-ous** suffix indicates the lower, and the **-ic** the upper, of the two oxidation states. Neither, however, tells exactly the value of the oxidation number.

EXAMPLE 7.9

Name these compounds.

Compound	Oxidation Number of Positive Element	Name
FeO	2^+	Ferr*ous* oxide
Fe_2O_3	3^+	Ferr*ic* oxide
CuCl	1^+	Cupr*ous* chloride
$CuCl_2$	2^+	Cupr*ic* chloride

The *Stock method*, preferred today, indicates oxidation state with Roman numerals: Mn(II).

THE STOCK METHOD Another approach, presently preferred and recently adopted by the International Union of Pure and Applied Chemistry, was suggested by Albert Stock. It is known as the **Stock method**. This involves using the English name for the element with a multiple oxidation number and, immediately following this, parentheses containing a Roman numeral indicating the oxidation state of that element. The negative ion stem name and the *-ide* ending complete the name of a binary compound, while the regular rules apply for compounds of more than three elements.

EXAMPLE 7.10

Name these compounds.

Compound	Oxidation Number of Positive Element	Name
FeO	2^+	Iron(II) oxide
Fe_2O_3	3^+	Iron(III) oxide
CuCl	1^+	Copper(I) chloride
$CuCl_2$	2^+	Copper(II) chloride

Binary Compounds Containing Hydrogen

Binary compounds composed of hydrogen and a nonmetallic element are named according to the same system as other binary compounds. However, when dissolved in water, some of these hydrogen-containing compounds form acids and exhibit considerably different behavior than the pure gaseous or liquid compounds. Consider, for example, gaseous hydrogen chloride dissolved in water:

$$HCl_{(gas)} + H_2O \longrightarrow HCl_{(aqueous)}$$

The aqueous (water) solution of HCl exhibits properties of an acid (which is discussed in Chapter 11). It is named by combining the prefex *hydro-* (for hydrogen) with the stem name of the second element and the suffix *-ic*, followed by the word acid. This is another rule that is best memorized.

Compounds containing ionizable hydrogens are called *acids.*

NAMING COMPOUNDS

Binary Compounds Containing Hydrogen

	Prefix for hydrogen	+	Stem name of second element	+ ic +	The word acid	=	Acid name
HCl:	hydro	+	chlor	+ ic +	acid	=	Hydrochloric acid
HF:	hydro	+	fluor	+ ic +	acid	=	Hydrofluoric acid

NEW TERMS

-ous A suffix indicating the lower of two possible oxidation states. For example, in ferr*ous* sulfate the iron is 2+.

-ic A suffix indicating the higher of two possible oxidation states. For example, in ferric nitrate the iron is 3+.

Stock method Oxidation states are indicated by Roman numerals immediately following the name of the element. For example, Fe^{3+} is iron(III).

TESTING YOURSELF

More Compounds to Name

1. Using the -ous/-ic method, name the following compounds.

$FeCl_2$ $FeBr_3$
CuO $FeSO_4$

2. Using the Stock method, name the following compounds.

$CuCl$ $FeCrO_4$
$Fe(NO_3)_3$ $CuCO_3$

3. Name the following acids.

HCl HBr

Answers **1.** ferrous chloride, cupric oxide, ferric bromide, ferrous sulfate **2.** copper(I) chloride, iron(III) nitrate, iron(II) chromate, copper(II) carbonate **3.** hydrochloric acid, hydrobromic acid

7.6 Writing and Balancing Chemical Equations

Chemical nomenclature consists of two parts. One part, which we have just considered, is the common language that is used to speak and write about compounds. The second part, which comprises the remainder of this chapter, is the method of describing the relative numbers of reactant and product atoms or molecules involved in chemical reactions.

When molecules collide, chemical bonds are sometimes broken and new bonds are formed. When this occurs, the molecules reassemble in a different form. This rearrangement usually results in release of energy as well as the formation of new molecules.

The *law of conservation of matter* states that the same number of atoms must be present before and after a reaction.

As mentioned in Chapter 2, one of the basic foundations of modern science is the *law of conservation of matter*—matter can not be created or destroyed by chemical reactions. This law has important implications for the study of chemical reactions, since it implies that the amount of material present before a chemical reaction must equal the amount of material present after the reaction, although it may be in a different form. There is no change in the *number* of each type of atom present before and after a chemical reaction, although the atoms may be combined with different partners after the reaction. When describing a chemical reaction, it is important not only to correctly describe the composition of each reactant and product, but also to correctly describe the quantities of each type of atom or molecule involved in the reaction. Writing a **balanced chemical equation** for a chemical reaction involves writing correct formulas for both reactant and product and indicating the quantity of each atom or compound present before and after the reaction.

Balanced chemical equations indicate the quantity of each type of atom or molecule present before and after the reaction.

On the following pages, we will discuss a straightforward way of writing chemical equations that accurately describes the materials present before and after a chemical reaction and the relative amounts of each compound used or produced by the reaction. However, there is a subtle trap in this procedure. A number of factors are involved in making a chemical reaction take place: (1) introducing enough energy to break chemical bonds in the reactant molecules, (2) having the correct bonds break, and (3) recombining the compo-

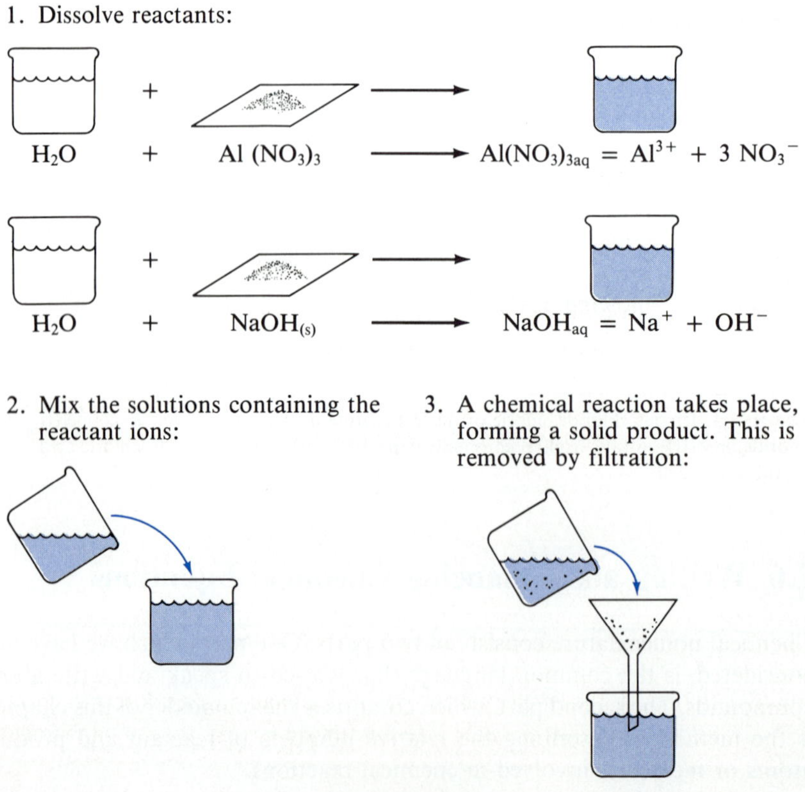
1. Dissolve reactants:
+
H_2O + $Al(NO_3)_3$ → $Al(NO_3)_{3aq}$ = Al^{3+} + 3 NO_3^-
+
H_2O + $NaOH_{(s)}$ → $NaOH_{aq}$ = Na^+ + OH^-
2. Mix the solutions containing the reactant ions:
3. A chemical reaction takes place, forming a solid product. This is removed by filtration:

FIGURE 7.7
Solutions of aluminum nitrate and sodium hydroxide are prepared by dissolving solid aluminum nitrate and solid sodium hydroxide in water. Conductivity measurements indicate that ions are present. The subscript $_{(aq)}$ means that the material is dissolved in water—an aqueous solution. The reaction is initiated by mixing the solutions containing the reactant ions. A white solid forms, which can be separated by filtering the mixture. New materials are formed; a chemical reaction has taken place.

nent atoms into the desired product molecules. These complicating factors make it possible to write equations that describe chemical reactions but that will not actually take place as written. In Chapter 10 we will discuss some of the factors that are involved in making a chemical reaction proceed. In this chapter we will be concerned only with the relatively basic procedures involved in describing the reaction.

Let's begin by considering one way to make aluminum hydroxide, $Al(OH)_3$. Aluminum hydroxide is a principal component of antacid tablets, and it can be synthesized by reacting a solution containing dissolved aluminum nitrate with a solution containing dissolved sodium hydroxide. Both solutions exhibit good electrical conductivity, indicating that ions are present. The aluminum nitrate solution contains Al^{3+} ions and NO_3^{1-} ions, and the sodium hydroxide solution contains Na^{1+} ions and OH^{1-} ions. The steps involved in conducting this reaction are illustrated in Figure 7.7.

The process of writing a balanced equation that will describe a chemical reaction can be divided into five parts. The first step involves clearly stating the identity of the compounds we intend to react and writing formulas with balanced charge for the reactants. This is usually accomplished by working from the names of the reactants.

A five-step process is involved in balancing chemical equations.

STEP 1. WRITE FORMULAS WITH BALANCED CHARGE FOR THE REACTANTS

a. Write the names of the reactants.

Aluminum nitrate + Sodium hydroxide

b. Write the symbols and oxidation numbers or charges for the ions involved.

$$Al^{3+}\ NO_3^{1-} + Na^{1+}\ OH^{1-}$$

c. Using parentheses as required and subscript numbers, write balanced formulas for each reactant.

$$Al(NO_3)_3 + NaOH$$

The second step involves predicting the identity of the reaction products. In an ionic reaction, it is possible for any positive ion to combine with any negative ion. It would not, of course, be possible for two positive ions or two negative ions to combine.

STEP 2. PREDICT THE IDENTITY OF THE REACTION PRODUCTS

a. Using a box as shown below, list all positive and all negative ions contributed by the reactant compounds.

Positive Ions	Negative Ions
Al^{3+}	NO_3^{1-}
Na^{1+}	OH^{1-}

Note that only the *identity* and *oxidation number of* each ion is shown. We have not yet determined the number of each that is required to complete any possible reaction.

b. Predict the identity of the reaction products. If a reaction takes place, we can assume that the products will be different from the reactants. In this case we predict that the aluminum ion will combine with the hydroxide ion and that the sodium ion will combine with the nitrate ion. Note that only ions of unlike charge can combine.

Positive Ions	Negative Ions
Al^{3+}	NO_3^{1-}
Na^{1+}	OH^{1-}

The third step involves writing formulas with balanced charges for the predicted reaction products. This procedure is similar to step 1 and involves *only* subscript numbers and parentheses.

STEP 3. WRITE FORMULAS WITH BALANCED CHARGE FOR THE PRODUCTS

a. Write the symbols and oxidation numbers for the ions involved.

$$Al^{3+}OH^{1-} + Na^{1+}NO_3^{1-}$$

b. Using parentheses and subscript numbers as required, write correct formulas for each product.

$$Al(OH)_3 + NaNO_3$$

Note that the ( )$_3$ around the OH group indicates that three of these groups are associated with the aluminum. In contrast, the subscript 3 following the oxygen in the nitrate group indicates only that there are three oxygens in sodium nitrate. If we wrote $AlOH_3$, this would indicate a compound consisting of one aluminum, one oxygen, and three hydrogens.

c. Name each of the products.

$$Al(OH)_3 \quad + \quad NaNO_3$$

Aluminum hydroxide + Sodium nitrate

We are now in a position to write a **balanced chemical equation** describing the predicted reaction. At this point we must consider the law of conservation of matter and balance the equation so that the same amount of material is shown entering as is shown leaving the reaction.

STEP 4. WRITE AND BALANCE THE CHEMICAL EQUATION

a. Write the chemical equation in both words and symbols. The arrow (→) indicates the predicted direction of the reaction.

Aluminum nitrate + Sodium hydroxide ⟶ Aluminum hydroxide + Sodium nitrate

$$Al(NO_3)_3 + NaOH \longrightarrow Al(OH)_3 + NaNO_3$$

b. At this point make sure that all formulas are correctly balanced.

c. Using numbers called *coefficients* preceding the compound, adjust the quantity of each compound to balance the equation. This is easiest if you start by considering the most complex product.

(1) One $Al(OH)_3$ requires three OH^{1-} groups. This means three NaOH must be present as reactants. Indicate this with a coefficient of 3 before the NaOH.

$$Al(NO_3)_3 + 3\ NaOH \longrightarrow Al(OH)_3 + NaNO_3$$

(2) At this point sufficient material is present on the reactant (left) side to provide for the $Al(OH)_3$ on the product (right) side. We have three $NO_3{}^{1-}$ groups and three Na^{1+} ions left over. This is exactly what is needed to make three $NaNO_3$ groups. Indicate this with a coefficient 3 before the $NaNO_3$.

$$Al(NO_3)_3 + 3\ NaOH \longrightarrow Al(OH)_3 + 3\ NaNO_3$$

STEP 5. CHECK THE EQUATION TO MAKE SURE THERE ARE EQUAL NUMBERS OF ATOMS ON THE REACTANT AND PRODUCT SIDES.

Element	Reactant	Product
Al	1	1
N	3	3
Na	3	3
H	3	3
O	12	12

If the charges have been correctly balanced, the element balance will come out even.

To summarize, there are three basic rules that must be followed carefully for success in balancing chemical equations:

There are three basic rules for balancing chemical equations.

1. Use subscript numbers *only* when balancing oxidation numbers in formulas. These numbers are placed *following* the group or ion to which they refer and below the line of type. If you use a subscript with a group of atoms (a polyatomic ion), put parentheses around the group to which the subscript refers.
2. Use coefficients *only* when balancing numbers of reactant and product molecules. These numbers are placed *preceding* the formula to which they refer and apply to all of the atoms in the formula.
3. Always check to make sure that the number of each type of atom on the reactant side is the same as on the product side of the reaction.

Several symbols are used to indicate the physical states of reactants and products and the addition of heat or removal of a product during a chemical reaction. These symbols are presented in Table 7.8. We can use these symbols to describe our example of the reaction between aluminum nitrate and sodium hydroxide. When solutions of these reactants were mixed, a white precipitate formed and settled to the bottom of the beaker. This precipitate,

TABLE 7.8
Common Symbols Used to Describe Reactants and Chemical Reactions

Symbol	Example	Meaning
(s)	$NaCl_{(s)}$	The material (NaCl) is in the solid state.
(l)	$H_2O_{(l)}$	The material (H_2O) is in the liquid state.
(g)	$O_{2(g)}$	The material (O_2) is in the gaseous state.
(aq)	$HCl_{(aq)}$	The material (HCl) is dissolved in water.
Δ	$\xrightarrow{\Delta}$	Heat is added to help the reaction proceed.
$\uparrow$	$H_{2(g)}\uparrow$	Hydrogen leaves the reaction as a gas.
$\downarrow$	$AgCl_{(s)}\downarrow$	Silver chloride leaves the reaction as a solid; it "precipitates out."

found by analysis to be aluminum hydroxide, is insoluble in water. The other product, sodium nitrate, remains dissolved as Na^+ and NO_3^- ions in solution. The formal equation describing this reaction is written:

$$Al(NO_3)_{3(aq)} + 3\ NaOH_{(aq)} \longrightarrow Al(OH)_{3(s)} + 3\ NaNO_{3(aq)}$$

Net ionic reactions show only the ions or molecules that react to form an insoluble product.

Sometimes equations describing reactions between aqueous ions are written as **net ionic equations**, which show only the reactions that take place to produce products that remove themselves from the ionic mixture by precipitating, by becoming gases, or by becoming covalent molecules. In this example, because only one insoluble product is produced and because sodium and nitrate ions remain in solution, the reaction of aluminum hydroxide and sodium hydroxide could be written:

$$Al^{3+} + 3\ NO_3^- + 3\ Na^+ + 3\ OH^- \longrightarrow Al(OH)_{3(s)}\downarrow + 3\ Na^+ + 3\ NO_3^-$$

It is apparent that the only reaction that took place was between the aluminum and the hydroxide. The net ionic equation is then

$$Al^{3+} + 3\ OH^- \longrightarrow Al(OH)_{3(s)}\downarrow$$

To write net ionic equations, one must know which products will remove themselves from the reaction. Table 7.9 gives some information that will help you make this prediction. The use of solubility tables to predict ionic reactions is illustrated in Example 7.11.

EXAMPLE 7.11

Consider this possible reaction between lead nitrate and sodium chloride:

$$Pb(NO_3)_2 + 2\ NaCl \longrightarrow PbCl_2 + 2\ NaNO_3$$

a. Will the reactant compounds be soluble in water?

Yes. All common nitrates are water soluble, as are all common compounds containing sodium.

b. Will either of the product compounds be insoluble, thus removing itself from the reaction mixture?

TABLE 7.9
Solubilities of Common Ionic Compounds in Water

Ion	Solubility Properties of Compounds Containing This Ion
NH_4^+	All common salts of the ammonium ion are soluble in water.
Na^+, K^+	All common salts of sodium and potassium are soluble in water.
NO_3^-	All common nitrates are water soluble.
Cl^-, Br^-, I^-	All chlorides, bromides, and iodides are soluble in water, except those of Ag, Hg_2^{2+}, and Pb. Lead chloride and lead bromide are slightly soluble in hot water.
SO_4^{2-}	All sulfates are soluble except $CaSO_4$, $BaSO_4$, $PbSO_4$, and Ag_2SO_4.
$C_2H_3O_2^-$	All acetates are soluble in water except iron(III) acetate.
S^{2-}	Only ammonium sulfide and sulfides of the lithium, sodium, and potassium family are water soluble.
PO_4^{3-}, CO_3^{2-}	Solubility rules for phosphates and carbonates are the same as for the sulfides.
OH^-	Same as for sulfides, except that Ca, Ba, and Sr hydroxides are slightly soluble.

Yes. All chlorides are soluble except those of lead, silver, and mercury. Lead chloride will be insoluble in water. Since all nitrates are water soluble, the sodium nitrate will remain in solution.

c. What is the net ionic reaction?

$$Pb^{2+} + 2\,Cl^- \longrightarrow PbCl_2\downarrow$$

EXAMPLE 7.12

Consider this possible reaction between ammonium sulfate and iron(II) chloride:

$$(NH_4)_2SO_4 + FeCl_2 \longrightarrow FeSO_4 + 2\,NH_4Cl$$

a. Will the reactant compounds be soluble in water?

Yes. All common ammonium compounds are water soluble, as are all common chloride compounds, except those of lead, silver, and mercury.

b. Will either of the product compounds be insoluble, thus removing itself from the reaction mixture?

No. All sulfates are soluble except those of calcium, barium, lead, and silver. Iron(II) sulfate will be water soluble. Since all ammonium compounds are water soluble, ammonium chloride will also be soluble.

c. What is the net ionic reaction?

No net reaction occurs.

NEW TERMS

balanced chemical equation A chemical equation that indicates the balanced formulas and the relative quantity of each element or compound present before and after the reaction.

net ionic equation A chemical equation describing an ionic reaction that is written to identify those ions that *leave* the reaction via precipitation, formation of a gas, or formation of nonreactive covalent substance.

TESTING YOURSELF

Balancing Chemical Equations

Complete and balance the following chemical equations:

1. $KOH + AgNO_3 \longrightarrow ? + ?$
2. $AgNO_3 + K_2CrO_4 \longrightarrow ? + ?$
3. Magnesium hydroxide + hydrogen chloride $\longrightarrow ? + ?$
4. What is the net ionic reaction for the combination of ammonium sulfate and copper(II) nitrate?
5. What is the net ionic reaction for the combination of sodium iodide and lead nitrate?

Answers **1.** $KOH + AgNO_3 \longrightarrow AgOH + KNO_3$ **2.** $2\ AgNO_3 + K_2CrO_4 \longrightarrow Ag_2CrO_4 + 2\ KNO_3$ **3.** $Mg(OH)_2 + 2\ HCl \longrightarrow MgCl_2 + 2\ HOH$ **4.** No net ionic reaction; copper(II) sulfate is water soluble. **5.** $Pb^{2+} + 2I^- \longrightarrow PbI_2$

7.7 Classification of Chemical Reactions

Chemical reactions can be classified into two general types. The first is reactions that involve transfer of electrons. Electron transfer reactions are further subdivided into three types: (a) synthesis, (b) decomposition, and (c) single replacement. The second general reaction type involves trade of ion partners. This simpler reaction is called *double replacement*. For example, the reaction that forms aluminum hydroxide is a double replacement reaction since the groups accompanying both the aluminum and sodium ions are replaced during the reaction. As will be seen on the following pages, the steps outlined in the previous section can be used to balance most electron transfer reactions.

Electron Transfer Reactions: Synthesis or Combination Reactions ($A + B \rightarrow AB$)

Synthesis reactions involve combination of two reactants to form a single product.

Synthesis or combination *reactions* involve a combination of two reactants to form a single product. An example of this type of reaction is an ordinary flashbulb, in which magnesium ribbon is ignited in an atmosphere of oxygen, producing magnesium oxide and a good deal of heat and light.

STEP 1. WRITE FORMULAS WITH BALANCED CHARGE FOR THE REACTANTS

(Remember that oxygen is a diatomic gas.)

$$\text{Magnesium} + \text{Oxygen}$$
$$Mg_{(s)} + O_{2(g)}$$

STEP 2. PREDICT THE IDENTITY OF THE REACTION PRODUCTS

Determination of the product of a synthesis reaction depends on analysis of the product. In this case, the product is magnesium oxide.

$$\text{Magnesium} + \text{Oxygen} \longrightarrow \text{Magnesium oxide}$$
$$Mg_{(s)} + O_{2(g)} \longrightarrow$$

STEP 3. WRITE FORMULAS WITH BALANCED CHARGE FOR THE PRODUCTS

Since the oxidation number of magnesium is 2+ and the oxidation number of oxygen (as oxide) is 2−, the balanced formula of magnesium oxide is MgO.

$$\text{Magnesium} + \text{Oxygen} \longrightarrow \text{Magnesium oxide}$$
$$Mg_{(s)} + O_{2(g)} \longrightarrow MgO_{(s)}$$

STEP 4. WRITE AND BALANCE THE CHEMICAL EQUATION

Since oxygen gas is a diatomic molecule (O_2), two magnesium atoms are required to react with one oxygen gas molecule, and two molecules of magnesium oxide will be produced.

$$\text{Magnesium} + \text{Oxygen} \longrightarrow \text{Magnesium oxide}$$
$$2\ Mg_{(s)} + O_{2(g)} \longrightarrow 2\ MgO_{(s)}$$

STEP 5. CHECK THE EQUATION TO MAKE SURE THERE ARE EQUAL NUMBERS OF ATOMS ON THE REACTANT AND PRODUCT SIDES

Element	Reactant	Product
Mg	2	2
O	2	2

Reactions that involve transfer of electrons from one atom to another are called **redox reactions** for *red*uction-*ox*idation. This electron transfer is reflected by a change in the oxidation number of the atoms. An atom that gives up electrons is said to be **oxidized**; an atom that gains electrons is said to be **reduced**. Since electrons must be both given and received for a reaction to take place, this type of reaction is called *oxidation-reduction* or, in a shorter and reversed form, a redox reaction. Redox reactions are of considerable importance in biochemistry and deserve introduction here.

Oxidation is the loss of electrons, while *reduction* is the gain of electrons.

Redox reactions involve transfer of electrons from one atom to another.

Consider the oxidation numbers of the example atoms before and after the reaction:

$$2\ Mg_{(s)} + O_{2(g)} \longrightarrow 2\ MgO_{(s)}$$

Element	Oxidation Number Before	Oxidation Number After	Reaction	Electron Transfer	Ratio gain:lost
Mg	0	2+	Oxidation	2 lost	2:2 = 1
O	0	2−	Reduction	2 gained	

Magnesium is oxidized in this reaction while oxygen is reduced. The electron accounting involved in such a reaction can be a useful aid in balancing the chemical equation. In this reaction, each magnesium atom lost two electrons while each oxygen atom gained two electrons. Thus, in the balanced reaction, the ratio of magnesium atoms to oxygen atoms should be one. Since oxygen gas occurs as O_2, each oxygen gas molecule must react with two magnesium atoms. Later in this chapter (Section 7.8) we will show how such electron transfer can be predicted. It is an important factor in both protection of building materials and in the production of energy in biochemical reactions.

Electron Transfer Reactions: Decomposition Reactions ($AB \rightarrow A + B$)

Decomposition reactions occur when one compound is broken apart to form two or more elements or compounds.

Decomposition reactions occur when one compound is taken apart to form two elements or compounds. An example of this is the decomposition of potassium chlorate to yield oxygen gas and potassium chloride. The same steps shown previously will produce a balanced chemical equation formula for decomposition reactions.

STEP 1. WRITE FORMULAS WITH BALANCED CHARGE FOR THE REACTANTS

Potassium chlorate is formed from potassium ions and chlorate groups. Since the oxidation number of potassium is 1+ and that of the chlorate group 1−, the balanced formula of potassium chlorate is $KClO_3$.

Potassium chlorate

$$K^{1+}\,ClO_3^{\,1-}$$

$$KClO_3$$

STEP 2. PREDICT THE IDENTITY OF THE REACTION PRODUCTS

Products of decomposition reactions are often difficult to predict directly and are usually determined by observation or analysis following the reaction. In this example, the products have been determined to be potassium chloride and oxygen gas.

$$\text{Potassium chlorate} \longrightarrow \text{Potassium chloride} + \text{Oxygen}$$

STEP 3. WRITE FORMULAS WITH BALANCED CHARGE FOR THE PRODUCTS

Remember that oxygen gas is a diatomic molecule.

Potassium chloride	Oxygen gas
$K^{1+}\,Cl^{1-}$	
KCl	O_2

STEP 4. WRITE AND BALANCE THE CHEMICAL EQUATION

Since one $KClO_3$ will decompose to give $1\frac{1}{2}$ O_2 molecules, we need to double all the quantities to maintain whole numbers of molecules.

$$2\,KClO_{3(s)} \longrightarrow 2\,KCl_{(s)} + 3\,O_{2(g)}$$

STEP 5. CHECK THE EQUATION TO MAKE SURE THERE ARE EQUAL NUMBERS OF ATOMS ON THE REACTANT AND PRODUCT SIDES

Element	Reactant	Product
K	2	2
Cl	2	2
O	6	6

STEP 6. CHECK FOR ELECTRON TRANSFER

Consider the oxidation states of the atoms before and after the reaction:

$$2\,KClO_{3(s)} \longrightarrow 2\,KCl_{(s)} + 3\,O_{2(g)}$$

	Oxidation Number				
Element	Before	After	Reaction	Electron Transfer	Ratio gain:lost
K	1+	1+	No change	0	
Cl	5+	1−	Reduction	6 gained	3:1
O	2−	0	Oxidation	2 lost	

In this reaction, the oxidation state of chlorine is reduced from 5+ to 1−, a change of six electrons. The oxidation number of the oxygen is increased (oxidation) from 2− to 0, a change of two electrons. Three oxygen atoms must oxidize to provide the electrons to reduce one chlorine atom. Thus, the ratio of oxygen to chlorine in the balanced equation must be three oxygens to one chlorine. Inspection of the balanced equation shows that this is in fact the case.

Electron Transfer Reactions: Single Replacement Reactions (A + BC → AB + C)

A *single replacement reaction* involves replacement of one compound by another. They occur when an active metal, such as sodium, is placed in water.

Single replacement reactions involve replacement of one element in a compound by another.

STEP 1. WRITE FORMULAS WITH BALANCED CHARGE FOR THE REACTANTS

$$\text{Sodium} + \text{Water}$$
$$Na + H_2O$$

STEP 2. PREDICT THE IDENTITY OF THE REACTION PRODUCTS

Prediction of the identity of reaction products in a single-replacement reaction requires some knowledge of the chemistry of the reacting species.

In this example, the product of the reaction between sodium and water is known to be sodium hydroxide and hydrogen gas.

$$\text{Sodium} + \text{Water} \longrightarrow \text{Sodium hydroxide} + \text{Hydrogen gas}$$

STEP 3. WRITE FORMULAS WITH BALANCED CHARGE FOR THE PRODUCTS

Remember that hydrogen gas is a diatomic molecule.

$$\begin{array}{ccc} \text{Sodium hydroxide} & + & \text{Hydrogen gas} \\ Na^{1+}\,OH^{1-} & + & H_2 \end{array}$$

STEP 4. WRITE AND BALANCE THE CHEMICAL EQUATION

Since reacting one sodium atom with one water molecule will only produce enough hydrogen for $\frac{1}{2}\,H_2$ molecule, all quantities are doubled to come out with whole numbers of molecules.

$$\begin{array}{ccccccc} \text{Sodium} & + & \text{Water} & \longrightarrow & \text{Sodium hydroxide} & + & \text{Hydrogen gas} \\ 2\,Na_{(s)} & + & 2\,H_2O_{(aq)} & \longrightarrow & 2\,NaOH_{(aq)} & + & H_{2(g)} \end{array}$$

STEP 5. CHECK THE EQUATION TO MAKE SURE THERE ARE EQUAL NUMBERS OF ATOMS ON THE REACTANT AND PRODUCT SIDES

Element	Reactant	Product
Na	2	2
H	4	4
O	2	2

STEP 6. CHECK FOR ELECTRON TRANSFER

Consider the oxidation states of the atoms before and after the reaction:

$$2\,Na_{(s)} + 2\,H_2O_{(aq)} \longrightarrow 2\,NaOH_{(aq)} + H_{2(g)}$$

	Oxidation Number				
Element	Before	After	Reaction	Electron Transfer	Ratio gain:loss
Na	0	1+	Oxidation	1 loss	1:1
H	1+	1+	No change		
H	1+	0	Reduction	1 gain	
O	2−	2−	No change		

In this reaction, sodium is oxidized and hydrogen is reduced. Note, however, that not all of the hydrogen is reduced. One hydrogen atom from each water molecule stays with the oxygen, forming the hydroxide group. Neither the oxygen nor its companion hydrogen participate in the electron transfer. Since one sodium atom will donate the electron required to reduce one hydrogen atom, the ratio of sodium atoms to hydrogen atoms participating in electron transfer should be 1:1. Since hydrogen gas exists only as the diatomic molecule H_2, two sodium atoms will react with two water molecules to produce two sodium hydroxides and one H_2 molecule.

Double Replacement Reactions (AB + CD → AD + CB)

Double replacement reactions are commonly encountered when aqueous solutions of compounds react with one another. They are probably the most common type of reaction in inorganic chemistry. An example is aqueous hydrogen chloride (hydrochloric acid) mixed with aqueous sodium hydroxide, which forms sodium chloride and water.

Double replacement reactions involve trading of ions; oxidation and reduction does not occur.

STEP 1. WRITE FORMULAS WITH BALANCED CHARGE FOR THE REACTANTS

Hydrogen chloride + Sodium hydroxide

$$H^{1+}Cl^{1-} \quad + \quad Na^{1+}OH^{1-}$$

$$HCl \quad + \quad NaOH$$

STEP 2. PREDICT THE IDENTITY OF THE REACTION PRODUCTS

+ Ions	− Ions
H^{1+}	Cl^{1-}
Na^{1+}	OH^{1-}

STEP 3. WRITE FORMULAS WITH BALANCED CHARGE FOR THE PRODUCTS

Hydrogen hydroxide (water)	Sodium chloride
$H^{1+}\ OH^{1-}$	$Na^{1+}\ Cl^{1-}$
H_2O	$NaCl$

STEP 4. WRITE AND BALANCE THE CHEMICAL EQUATION

$$\text{Hydrogen chloride} + \text{Sodium hydroxide} \longrightarrow \text{Sodium chloride} + \text{Water}$$

$$HCl_{(aq)} + NaOH_{(aq)} \longrightarrow NaCl_{(aq)} + H_2O_{(l)}$$

In this case, all coefficients are 1.

STEP 5. CHECK THE EQUATION TO MAKE SURE THERE ARE EQUAL NUMBERS OF ATOMS ON THE REACTANT AND PRODUCT SIDES.

Element	Reactant	Product
H	2	2
Cl	1	1
Na	1	1
O	1	1

STEP 6. CHECK FOR ELECTRON TRANSFER

Consider the oxidation states of the atoms before and after the reaction:

$$HCl_{(aq)} + NaOH_{(aq)} \longrightarrow NaCl_{(aq)} + H_2O_{(l)}$$

Element	Oxidation Number Before	Oxidation Number After	Reaction	Electron Transfer	Ratio
H	1+	1+	No change	0	
Cl	1−	1−	No change	0	
Na	1+	1+	No change	0	None
O	2−	2−	No change	0	
H	1+	1+	No change	0	

Not all reactions involve electron transfer. In a double replacement reaction, ions are exchanged between reacting compounds, but no electrons are transferred.

This reaction is somewhat unusual in that water is a product. Since water is a stable covalent compound, the ionic equation can be written:

$$H^+{}_{(aq)} + Cl^-{}_{(aq)} + Na^+{}_{(aq)} + OH^-{}_{(aq)} \longrightarrow Na^+{}_{(aq)} + Cl^-{}_{(aq)} + H_2O$$

and the net ionic equation will be

$$H^+{}_{(aq)} + OH^-{}_{(aq)} \longrightarrow H_2O$$

We will encounter this type of equation again in Chapter 11; compounds that donate hydrogen ion are called *acids* and compounds that accept hydrogen ion are called *bases.* If the base contains hydroxide, one product will always be water and the other is called a *salt*. The reaction of acid and base is called *neutralization.*

NEW TERMS

oxidation The loss of electrons in a chemical reaction. When iron rusts, it changes its oxidation state from 0 to +3 by giving up three electrons in a two step process, thus it is *oxidized.*

reduction Gain of electrons in a chemical reaction. When heated with carbon (coke), the iron Fe^{3+} in hematite (Fe_2O_3) gains electrons and is *reduced* to metallic iron, Fe^0

redox reaction A reaction that involves both oxidation and reduction.

TESTING YOURSELF

Chemical Reactions and Equations

1. Complete and balance the chemical equation describing the reaction of ammonium phosphate and ferric nitrate.
2. In the combustion of natural gas, which element is oxidized and which is reduced?

$$CH_4 + 2O_2 \longrightarrow CO_2 + 2\,H_2O$$

3. In the reaction of potassium with water, which element is oxidized and which is reduced?

$$2\ K + 2\ H_2O \longrightarrow 2\ KOH + H_2$$

Answers **1.** $(NH_4)_3PO_4 + Fe(NO_3)_3 \longrightarrow FePO_4 + 3\ NH_4NO_3$ **2.** $C^{4-} \longrightarrow C^{4+}$ = Oxidation, $O^0 \longrightarrow O^{2-}$ = Reduction **3.** $K^0 \longrightarrow K^{1+}$ = Oxidation, $2\ H^{1+} \longrightarrow H_2{}^0$ = Reduction, $O^{2-} \longrightarrow O^{2-}$ = No change

7.8 Electron Transfer Reactions

Electrons are important currency in chemical reactions. Electron transfer reactions can occur in one of two ways: the transfer of electrons is caused by a force outside the reaction or the transfer of electrons is spontaneous. An idea called the *electrochemical series* helps us predict which reactions will be spontaneous and which must be forced.

Forced Electrochemical Reactions

Electrochemical reactions may be forced by using a battery to move electrons.

Forced electrochemical reactions are important to many industries. Steel automobile bumpers are covered with a thin coat of protective chromium through an electrochemical process. Gold-plated wrist-watches and jewelry are covered with a thin layer of gold through a similar process.

How does this process of electroplating occur? Figure 7.8 shows a simple electroplating apparatus. It consists of a strip of copper metal, a solution containing copper ion (Cu^{2+}), a flashlight battery, and an object to be plated (in this case, a key). The battery acts simply as an electron pump to attract electrons into its positive pole and pumps them out of its negative pole.

Two reactions take place in this system. At the copper electrode, electrons are removed from copper atoms and pulled into the battery, causing copper ions to form:

$$Cu^0 - 2\ \text{electrons} \longrightarrow Cu^{2+}$$

This loss of electrons is called oxidation, as we learned in the previous section.

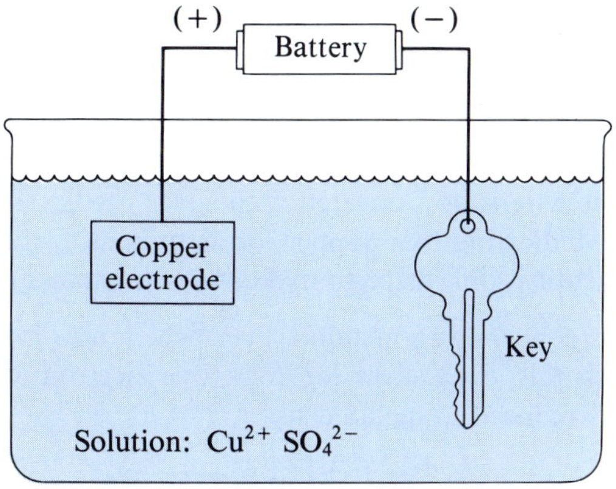
(+)
Battery
(−)
Copper electrode
Key
Solution: Cu^{2+} $SO_4{}^{2-}$

FIGURE 7.8
The process of electroplating involves a forced electrochemical reaction. Electrons forced onto the key attract and reduce metal ions from the solution; these reduced metal atoms provide a protective coating for the key.

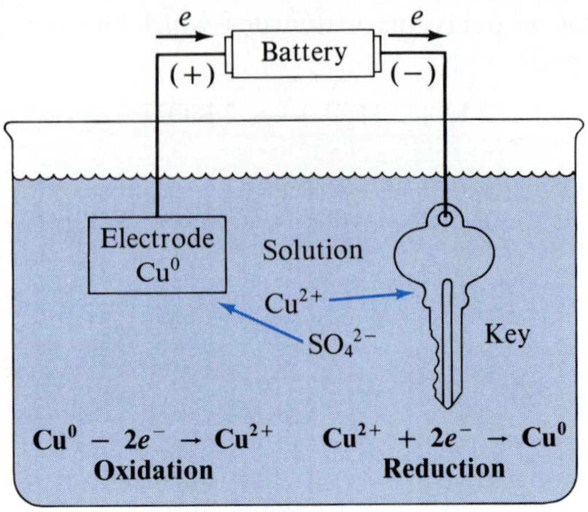
e
e
Battery
(+)
(−)
Electrode
Cu^0
Solution
Cu^{2+}
SO_4^{2-}
Key
$Cu^0 - 2e^- \rightarrow Cu^{2+}$
Oxidation
$Cu^{2+} + 2e^- \rightarrow Cu^0$
Reduction

FIGURE 7.9
The electroplating process involves production of metal ions at one electrode (oxidation) and reduction of metal ions at the other electrode. Deposited metal ions are removed from the solution, and an equal number are replaced by the oxidation process at the left-hand electrode. Since equal numbers of electrons enter and leave the electron pump (battery), equal numbers of ions are produced and deposited, and the charge balance of the solution remains neutral.

FIGURE 7.10
This outboard motor uses the electrochemical behavior of zinc for corrosion protection.

The battery forces electrons onto the key (Figure 7.9), which then attracts positive copper ions from the solution. As these positive ions strike the key, they each pick up two electrons (reduction) to become metallic copper.

$$Cu^{2+} + 2 \text{ electrons} \longrightarrow Cu^0$$

If the copper atoms are deposited slowly, a smooth, bright coating will result. Chrome plating of automobile bumpers and gold plating of jewelry involves similar processes, but with different metals.

Spontaneous Electron Transfer Reactions

A garbage can and an outboard motor have something in common. The garbage can is steel that has been galvanized, which means it has been dipped in molten zinc. The outboard motor has iron components. If you look carefully at the plate above the propeller in Figure 7.10, you will see a small zinc block. Why is zinc used in both of these situations? Both use principles of electrochemistry to protect something from oxidation.

Some reactions transfer electrons spontaneously.

Let's first consider a simpler problem. Figure 7.11 shows a "Christmas tree" suspended in a beaker of liquid. A few hours before this picture was taken, a tree-shaped piece of copper wire was suspended in a solution of silver nitrate. The wire is now covered with silver needles, and the solution has turned blue, indicating that copper ion is present in the solution. Two spontaneous reactions took place to make this Christmas tree:

1. Silver ion was converted to metallic silver, which now forms the needles. To accomplish this, each silver ion took one electron from some other atom and was reduced to metallic silver.

$$Ag^+ + 1 \text{ electron} \longrightarrow Ag^0$$

Through this gain of electrons, the silver was reduced. Where did the silver ions find these electrons?

2. Metallic copper atoms in the wire were converted to copper ion, which is presently in solution. To accomplish this, each copper atom had to give up two electrons.

$$Cu^0 - 2 \text{ electrons} \longrightarrow Cu^{2+}$$

Through this loss of electrons, the copper atoms were oxidized.

How did this happen? If copper atoms lost electrons and silver ions gained electrons, it makes sense that electrons were transferred from the copper atoms to the silver ions. To explain this, let's write chemical equations for addition of electrons to both silver and copper ions. Note that two reaction arrows have been drawn, indicating that electrons can either be added or removed. Also note that, by general agreement among chemists, these reactions are written so that the reduced species is on the right.

$$Ag^+ + 1 \text{ electron} \rightleftharpoons Ag^0$$

$$Cu^{2+} + 2 \text{ electrons} \rightleftharpoons Cu^0$$

Silver *ions* (Ag^+) are reduced by adding an electron (→), whereas silver *atoms* (Ag^0) are oxidized by removing an electron (←). However, silver *ions* have a greater attraction for electrons than do neutral atoms of copper metal. Silver ions striking copper atoms will actually remove electrons from the copper, causing the silver ions to reduce to metallic silver and the copper atoms to oxidize to copper ion, which floats out into the solution causing a blue color. The silver ion reaction proceeds as a *reduction* reaction, and the copper atom reaction as an *oxidation* reaction.

$$Ag^+ + 1 \text{ electron} \longrightarrow Ag^0 \text{ (Reduction)}$$

$$Cu^{2+} + 2 \text{ electrons} \longleftarrow Cu^0 \text{ (Oxidation)}$$

Note that, to conserve an electron balance, two silver ions must be reduced to accept the two electrons produced by the oxidation of one copper atom.

$$2\,Ag^+ + 2 \text{ electrons} \longrightarrow 2\,Ag^0$$

$$\uparrow$$

$$Cu^{2+} + 2 \text{ electrons} \longleftarrow Cu^0$$

The direction of electron transfer can be predicted by a simple table that classifies these reactions in terms of their ability to attract electrons (Table 7.10). Silver is above copper in this table; electrons are actually pulled from the copper reaction to the silver reaction with an electrical force of 0.45 volts.

FIGURE 7.11
A copper "Christmas tree" suspended in a solution containing silver ion (Ag^+) will quickly grow metallic silver needles. As this occurs, the clear silver solution turns a characteristic copper(II) ion blue. It appears that silver ions are taking electrons from metal Cu^0 atoms.

Half-Reactions

Neither of the two reactions shown in Table 7.10 can take place by itself. One reaction frees electrons (the oxidation reaction), while the other reaction (the reduction reaction) captures them. It is analogous to a person who has a lot of money and wishes to become a philanthropist. Having the money

Table 7.10
Electron Transfer Reactions Arranged in Order of Their Tendency to Gain Electrons

	Reaction	Force of Reaction (volts)
Increasing tendency to gain electrons ↑	$Ag^+ + 1\text{ electron} \longrightarrow Ag^0$	0.45
	$Cu^{2+} + 2\text{ electrons} \longleftarrow Cu^0$	

does not make him or her a philanthropist; giving it to good causes does. The key is giving. To give money away, someone must be available to accept it. Likewise, a chemical reaction cannot give electrons away (oxidation) until another reaction is available to accept the electrons (reduction). As we discussed in Section 7.7, that is why reduction and oxidation always occur together in redox reactions.

A *half-reaction* describes either the oxidation or reduction component of a redox reaction.

Each of the two reactions shown in Table 7.10 is *half* of a redox pair. Thus, they are each called **half-reactions**. The complete reaction is described by writing each of the half reactions in the forward direction and then adding them together:

$$2\ Ag^+ + 2\text{ electrons} \longrightarrow 2\ Ag^0$$

$$Cu^0 \longrightarrow Cu^{2+} + 2\text{ electrons}$$

$$2\ Ag^+ + 2\text{ electrons} + Cu^0 \longrightarrow 2\ Ag^0 + Cu^{2+} + 2\text{ electrons}$$

Note that equal numbers of electrons appear on each side of the reaction; these can be removed to show the net reaction.

$$2\ Ag^+ + Cu^0 \longrightarrow 2\ Ag^0 + Cu^{2+}$$

Electrochemical Series

On the basis of a series of experiments similar to the silver-copper experiment described above, many elements and ions have been arranged in order of their tendency to gain electrons. Data for a number of common elements and ions are presented in Table 7.11. This table is called the **electrochemical series**. Again by general agreement among chemists, the reduction of hydrogen ion to hydrogen gas

The *electrochemical series is used* to predict the direction of electron transfer.

$$2\ H^+ + 2\text{ electrons} \rightleftharpoons H_2{}^0$$

has been assigned a zero reference position in the table. Because the half-reactions in this table have been arranged in order of increasing tendency to gain electrons, *any half-reaction will take electrons from any half-reaction below it in the table.* The force by which electrons will be transferred is calculated by computing the algebraic difference between the voltages shown for each of the two half-reactions.

Now we are ready to consider again the problem of the garbage can and the trim tab on the boat drive. Why is zinc involved in both of these? The garbage can is made of steel, and the stern drive and boat motor have steel components. Iron (which combines with carbon to form steel) will rust

TABLE 7.11
Part of the Electrochemical Series

	Reaction	Tendency to Gain Electrons as Compared to Hydrogen (volts)
Increasing tendency to gain electrons ↑	Au^{3+} + 3 electrons ⇌ Au^0	1.50
	$Cl_2{}^0$ + 2 electrons ⇌ 2 Cl^-	1.36
	Pt^{2+} + 2 electrons ⇌ Pt^0	1.20
	Ag^+ + 1 electron ⇌ Ag^0	0.79
	Fe^{3+} + 1 electron ⇌ Fe^{2+}	0.77
	O_2 + 2 H_2O + 4 electrons ⇌ 4 OH^-	0.40
	Cu^{2+} + 2 electrons ⇌ Cu^0	0.34
	2 H^+ + 2 electrons ⇌ $H_2{}^0$	0.00
	Pb^{2+} + 2 electrons ⇌ Pb^0	−0.13
	Fe^{2+} + 2 electrons ⇌ Fe^0	−0.44
	Zn^{2+} + 2 electrons ⇌ Zn^0	−0.76
	Al^{3+} + 3 electrons ⇌ Al^0	−1.66
	Mg^{2+} + 2 electrons ⇌ Mg^0	−2.37

in the presence of oxygen and water. The process is clear from the electrochemical series.

	Reactions	Force of Reaction (volts)
Increasing tendency to gain electrons ↑	O_2 + 2 H_2O + 4 electrons ⟶ 4 OH^-	0.40
	Fe^{2+} + 2 electrons ⟵ Fe^0	−0.44
		↕ 0.84

The electrical force moving the electrons (0.84 volts) is substantial. The reaction will proceed at a relatively rapid rate. The net reaction is shown below. Note that two iron atoms must oxidize to reduce one oxygen gas molecule (O_2).

$$O_2 + 2\,H_2O + 4\text{ electrons} \longrightarrow 4\,OH^-$$

$$2\,Fe^0 \longrightarrow 2\,Fe^{2+} + 4\text{ electrons}$$

$$O_2 + 2\,H_2O + 4\text{ electrons} + 2\,Fe^0 \longrightarrow 4\,OH^- + 2\,Fe^{2+} + 4\text{ electrons}$$

Since four electrons appear on each side of the equation, these may be removed for simplification:

$$O_2 + 2\,H_2O + 2\,Fe^0 \longrightarrow 4\,OH^- + 2\,Fe^{2+}$$

This is the first step in the process of rusting. Metallic iron (Fe^0) is converted into Fe^{2+}, which is subsequently oxidized to Fe^{3+} forming iron(III) oxide, or Fe_2O_3 (rust).

There are several ways to protect iron from rusting. Since the reaction requires both oxygen and water, rusting can be prevented by keeping oxygen and water away from the iron. Painting is the easiest way to do this. Note that automobile fenders only rust where the paint has chipped and the iron is exposed. Keeping the iron dry is another way. This is why the Air Force stores old aircraft in the Arizona desert.

It is difficult to keep garbage cans and boat motors away from water and oxygen, and they are subjected to enough scratching that one cannot depend on paint for protection. This is where the zinc comes in. Zinc is considerably more prone to oxidation than is iron, as illustrated by its position in the electrochemical series.

	Reaction	Force of Reaction (volts)
Increasing ↑	$O_2 + 2\,H_2O + 4$ electrons $\rightleftharpoons$ $4\,OH^-$	0.40
tendency to	$Fe^{2+} + 2$ electrons $\rightleftharpoons$ Fe^0	−0.44
gain electrons	$Zn^{2+} + 2$ electrons $\rightleftharpoons$ Zn^0	−0.76

Zinc's great tendency to give up electrons means that it will preferentially oxidize as compared to iron. If any iron atoms are oxidized, they will immediately remove electrons from zinc atoms to return the iron to its metallic or elemental state.

	Reaction	Force of Reaction (volts)
Increasing ↑ tendency to	$Fe^{2+} + 2$ electrons $\longrightarrow$ Fe^0	−0.44
gain electrons	$Zn^{2+} + 2$ electrons $\longleftarrow$ Zn^0	−0.76

Thus, metallic zinc, which is not necessary for structural strength, is sacrificed to protect the iron. If you rub your hand across a garbage can, a black material will come off onto your skin. This is zinc oxide (ZnO) that has been produced as the zinc coating oxidizes. Likewise, the zinc block will preferentially oxidize to protect the steel parts of the outboard motor. When the zinc block begins to be eaten away, it is simply replaced. The zinc is called a "sacrificial anode."

Elements high on the scale of ability to attract electrons are often found in their pure state.

The electrochemical series also explains why some metals are used for jewelry and some are not. Note the position of gold (Au) at the top of Table 7.11. Gold ions can remove electrons from any element on the list—and no ion on the list can remove electrons from gold atoms. Thus, gold will not oxidize, and it preserves its bright finish even after prolonged contact with air, water, and moisture and ions from the body. Likewise, platinum (Pt) is extremely resistant to oxidation. Silver (Ag) is slightly more likely to oxidize, and those who have had experience with silver-plated tableware know that it must be polished from time to time. In contrast, consider copper jewelry. This is protected by a thin coat of varnish, which, when worn through, permits oxidation. Copper jewelry oxidizes very rapidly and must be polished every time it is used.

Spontaneous electron transfer reactions also provide part of the energy to sustain our bodies. Assimilation of oxygen into the bloodstream involves reaction of oxygen gas (O_2) with iron-containing hemoglobin in the blood to transport the oxygen from the lungs to cells throughout the body. Adenosine triphosphate (ATP) plays an important part in many electron transfer reactions involved in metabolism. We will return to this topic during our study of biochemistry in the latter part of this book.

NEW TERMS

half-reaction The oxidation or reduction component of a redox reaction, stated as a separate reaction.

electrochemical series A listing of half-reactions in order of increasing tendency to gain electrons used to predict whether a given electron transfer reaction will proceed. Any half-reaction can remove electrons from any half-reaction below it in the series.

TESTING YOURSELF

Electron Transfer Chemistry

1. As aluminum oxidizes, a thin layer of aluminum oxide (Al_2O_3) forms on the metal surface. This layer is reasonably strong, is air-tight and water-tight, and protects the surface of the metal. Because of its relatively low cost, light weight, and resistance to further oxidation once the protective coating is formed, aluminum is a commonly used building material. Suppose that aluminum metal (Al^0) is exposed to oxygen and water. Will it be oxidized to aluminum ion (Al^{3+})? Use the electrochemical series presented in Table 7.11 to decide.
2. Printed circuits used in electronic instruments, radios, and TVs have a phenolic or fiberglass base that is coated with copper metal. To form the wires on the printed circuit, a protective coating is photographically placed over the copper where a wire is desired, and the remaining copper is etched away. One common etchant is iron(III) chloride. By reference to the electrochemical series, do the following:
   a. Determine which two reactions take place
   b. Determine which element is oxidized and which ion is reduced
   c. Calculate how much force is placed on the electrons as they are transferred from metal atom to ion
   d. Determine how many iron(III) ions are required to oxidize one copper atom from metallic copper to Cu^{2+}

Answers **1.** yes **2. a.** Fe^{3+} + 1 electron $\longrightarrow$ Fe^{2+}, Cu^{2+} + 2 electrons $\longleftarrow$ Cu^0 **b.** iron(III) is reduced, copper oxidized **c.** 0.43 volts **d.** two

SUMMARY

This chapter's principal goal was development of some chemical communication skills. These fall into four categories: (1) writing correctly balanced formulas for simple compounds, (2) naming these compounds, (3) writing balanced chemical equations that describe chemical reactions, and (4) predicting the transfer of electrons in redox reactions.

Chemical reactions are balanced by a five step process:

1. Write formulas with balanced charge for the reactants.
2. Predict the identity of the reaction products.
3. Write formulas with balanced charge for the products.
4. Write and balance the chemical equation.
5. Check the equation to make sure there are equal numbers of atoms on the reaction and product sides.

The *electrochemical series* lists *half-reactions* for common elements and ions in order of their tendency to gain electrons. This series is used to predict the direction of electron transfer in *redox reactions* and to identify the *oxidation* and *reduction* half-reactions.

In the next chapter, the formula and equation skills developed here will be combined with some simple arithmetic to predict the amounts of reactants and products involved in chemical reactions.

TERMS

binary compounds
subscript number
oxidation number
molecular formula
compound formula
stem name
-ide suffix
polyatomic ion
-ous suffix
-ic suffix
Stock method
balanced chemical equation
net ionic equation
oxidation
reduction
redox reaction
half-reaction
electrochemical series

CHAPTER REVIEW

QUESTIONS

1. The element potassium is represented by which symbol?
   a. P b. Na c. K d. Po e. Km
2. Consider the compound iron(III) sulfide. The correct formula for this compound could be written

$$Fe_xS_y$$

   in which x and y are subscript numbers.
   a. In the balanced formula for ferric sulfide, what will subscript x be?
   b. In the balanced formula for ferric sulfide, what will subscript y be?
3. Predict the oxidation number of nitrogen in sodium nitrate, $NaNO_3$.
   a. $+5$ b. $+3$ c. 0 d. -3 e. -5
4. Using the Stock method, name the compound $FeSO_4$.
5. Consider the reaction between silver nitrate and sodium chloride:

$$AgNO_3 + NaCl \longrightarrow ? + ?$$

   In the balanced equation describing this reaction, what will be the numerical coefficient preceding the silver nitrate?
   a. 1 b. 2 c. 3 d. 4 e. 5
6. What is the reaction in question 5?
   a. synthesis reaction
   b. decomposition reaction
   c. single replacement reaction
   d. double replacement reaction.

7. Consider the reaction between silver nitrate and ammonium chromate:

$$AgNO_3 + (NH_4)_2CrO_4 \longrightarrow ? + ?$$

In the balanced equation describing this reaction, what will be the numerical coefficient preceding the silver nitrate?
a. 1 b. 2 c. 3 d. 4 e. 5

8. What will be the identity of the nitrogen-containing product of the reaction shown in question 7?
a. NH_4 b. NH_4NO_3 c. $(NH_4)_2NO_3$ d. $AgNH_4$ e. NO_4

9. Consider the reaction between lead nitrate and sodium sulfate:

$$Pb(NO_3)_2 + Na_2SO_4 \longrightarrow PbSO_4 + 2\ NaNO_3$$

One of the compounds listed in this equation is insoluble in water. By reference to the solubility data in Table 7.9, which compound is insoluble in water?

10. By reference to the electrochemical series presented in Table 7.11, explain why prospectors sometimes find nuggets of metallic gold, but never nuggets of metallic aluminum.

11. The following equation describes an attempt to dissolve metallic iron by placing it in a solution of hydrochloric acid. (Remember that it is possible to write equations for reactions that will not occur.) Whether or not the reaction will proceed as written depends on whether electrons can be transferred in the required direction.

$$Fe^0 + 2\ HCl \longrightarrow FeCl_2 + H_2$$

a. Which element should be oxidized in this reaction?
b. Which element should be reduced in this reaction?
c. Based on the data in the electrochemical series presented in Table 7.11, will this reaction proceed as written?

Answers 1. c **2. a.** 2 **b.** 3 **3.** a **4.** iron(II) sulfate **5.** a **6.** d **7.** b **8.** b **9.** $PbSO_4$ **10.** No ions are able to remove electrons from gold. Therefore it will remain as a metal and will be found as nuggets. Practically all ions will remove electrons from aluminum, so it will always be found as a compound with other ions. **11. a.** Fe **b.** H^+ **c.** The reaction will proceed as written. Hydrogen ion will remove electrons from metallic iron atoms.

DIAGNOSTIC CHART

Blacken in all of the circles under the number of each question you missed in the Chapter Review questions. The diagnostic chart will help you identify concept areas that need more study.

	Questions										
Concepts	1	2	3	4	5	6	7	8	9	10	11
Naming simple compounds	○				○						
Balancing formulas		○	○	○	○	○		○			
Polyatomic ions				○	○			○			
Oxidation numbers		○	○	○	○			○			
Balancing equations							○	○			
Solubility prediction									○		
Reaction classification						○					
Electron transfer											○

EXERCISES

Naming Compounds, Balancing Formulas, and Predicting Solubility

The chart below lists a number of negative ions on its horizontal axis and a number of positive ions on its vertical axis. Solution of the problems in this section require that you work on a separate piece of paper. There are 180 individual problems in this chart. For each block in the chart, do the following.

1. Write a balanced formula for the compound formed by the positive and negative ion represented by the block.
2. Name the compound formed by this positive and negative ion.
3. Use the data presented in Table 7.7 to predict the water solubility of this compound.

Balancing Equations, Classifying Reactions, and Predicting Solubility

The chart on the opposite page lists a number of compounds on its horizontal axis and another group of compounds on its vertical axis. Each box in the chart represents a reaction between the compound on the horizontal axis and the compound on the vertical axis. Solution of the problems in this section require that you work on a separate piece of paper. There are 53 individual problems in this chart. For each block in the chart, do the following.

4. Decide if both reactant compounds are water soluble. (For reactions to take place in water solution, both reactant compounds must be water soluble.)
5. If both reactant compounds are water soluble, write a

Negative Ion

Positive Ion	Br^-	HCO_3^-	ClO_3^-	F^-	OH^-	NO_3^-	CO_3^{2-}	CrO_4^{2-}	S^{2-}	SO_4^{2-}	MnO_4^-	PO_4^{3-}
H^+												
NH_4^+												
Li^+												
K^+												
Ag^+												
Na^+												
Ca^+												
Co^+												
Cu^{2+}												
Fe^{2+}												
Pb^{2+}												
Mg^{2+}												
Sr^{2+}												
Al^{3+}												
Fe^{3+}												

balanced equation describing the reaction that might occur between the two compounds.

6. Decide which product compounds are *not* water soluble. For reactions producing nonsoluble products, write an equation describing the net ionic reaction that takes place.

Compound A \ Compound B	HCl	NH_4OH	$(NH_4)_2S$	K_2SO_4	CaS	Na_3PO_4
$AgNO_3$						
$Pb(NO)_2$						
$NaNO_3$						
$Cu(NO_3)_2$						
$Ni(NO_3)_2$						
$LiNO_3$						
$Fe(C_2H_3O_2)_3$						
$Ca(NO_3)_2$						
$Al(NO_3)_3$						

Electron Transfer Reactions

7. Several reactions are shown below. For each reaction, determine the following.
   a. The type of reaction: synthesis, decomposition, single replacement, or double replacement.
   b. Which element is oxidized and which is reduced.
   c. Whether the reaction will proceed spontaneously as written.

$$FeCl_2 + Sn^0 \longrightarrow SnCl_2 + Fe^0$$

$$FeCl_2 + Zn^0 \longrightarrow ZnCl_2 + Fe^0$$

$$Pb^0 + 2\ HCl \longrightarrow PbCl_2 + H_2{}^0$$

$$2\ Ag^0 + 2\ HCl \longrightarrow 2\ AgCl + H_2{}^0$$

8. Iron is converted from its ore in the form of iron (III) oxide to metallic iron by a process called *smelting* in which the iron(III) oxide is reacted with carbon monoxide. The carbon monoxide is formed by blowing oxygen from air over coke (carbon) in a high temperature blast furnace. The metallic iron melts after being formed and runs out the bottom of the furnace.

   Formation of carbon monoxide:

$$O_2 + 2\ C \longrightarrow 2\ CO$$

   Reduction of iron ore:

$$Fe_2O_3 + 3\ CO \longrightarrow 2\ Fe^0 + 3\ CO_2$$

   a. Does oxidation or reduction take place in the first reaction? If so, which element is oxidized and which is reduced? Write half-reactions for the oxidation and reduction reactions.
   b. Does oxidation or reduction take place in the second reaction? If so, which element is oxidized and which is reduced? Write half-reactions for the oxidation and reduction reactions.

CHAPTER 8

Chemical Arithmetic: Weight Relationships in Chemical Reactions

OBJECTIVES

After completing this chapter, you should be able to

- Compute the percentage composition of a compound given balanced formula and atomic mass data.
- Define the term *mole*.
- Use the mole concept to compute empirical formulas from percentage composition data.
- Use the mole method to compute amounts of reactant and product involved in a chemical reaction.

Knowledge of atomic mass (or weight) and compound formula permits us to compute the mass of molecules. This chapter introduces a method of computing the percentage composition of a molecule, a "mole" method for working with large groups of molecules, and a method of computing the relative amounts of reactants and products involved in chemical reactions.

8.1 An Introduction to Chemical Arithmetic: Percentage Composition

This chapter deals with the "how much" aspect of chemistry—the application of simple mathematics to help us predict the composition of known compounds, the formulas of unknown compounds, and the yield of chemical reactions. One of the simplest quantitative relationships in chemistry is that of **percentage composition**, the percent by mass represented by each element in the compound. The elemental percentage composition of a compound can be calculated if its formula and the atomic weights of its constituent atoms are known. (Recall our discussion of atomic weight in Chapter 2.) This chapter begins with consideration of percentage composition problems. Given elemental percentage composition, the problem can be worked in reverse to determine the experimental or *empirical* formula of a compound. The *mole* is a conversion factor between atom-sized atomic mass units and easily measured grams. The mole concept helps us with empirical formula calculations and provides a practically foolproof way of computing amounts of reactants and products involved in chemical reactions.

Consider the water molecule, which contains one oxygen atom and two hydrogen atoms. The **molecular weight** of the water molecule is the sum of the atomic weights (expressed in atomic mass units, or amu) of its component atoms.

The term *molecular weight* describes the mass of a molecule.

2 hydrogens	@ 1.00 amu	=	2.00 amu
1 oxygen	@ 16.00 amu	=	16.00 amu
Molecular weight		=	18.00 amu

The percentage by weight of each element present in the compound can then be computed by finding what fraction of the total molecular weight is contributed by each element.

$$\%\ \text{hydrogen} = \frac{2.00\ \text{amu hydrogen}}{18.00\ \text{amu total}} \times 100 = 11.1\%$$

$$\%\ \text{oxygen} = \frac{16.00\ \text{amu oxygen}}{18.00\ \text{amu total}} \times 100 = 88.9\%$$

(Recall from Chapter 2 that the term *atomic weight* is used even though *atomic mass* is a more accurate term. The same goes for *molecular weight* and *formula weight*, which are used instead of their more accurate counterparts, *molecular mass* and *formula mass*.)

Sodium chloride (NaCl), commonly known as table salt, is quite different from water in that it is an ionically bound three-dimensional lattice of sodium ions and chloride ions. Although we can correctly say that a water molecule is composed of one oxygen atom and two hydrogen atoms, the limits to the sodium chloride "molecule" are really the limits of each individual crystal of salt. Thus, it is more appropriate to think of sodium chloride as having a formula weight. The formula weight expresses the composition of the ionic compound just as the molecular weight of water expressed its composition.

The term *formula weight* describes the mass of the smallest repeating unit of an ionic compound.

1 sodium	@ 22.99 amu	=	22.99 amu
1 chlorine	@ 35.45 amu	=	35.45 amu
Formula weight		=	58.44 amu

The percentage by weight of each element in this compound can then be computed in the same manner as water:

$$\% \text{ sodium} = \frac{22.99 \text{ amu sodium}}{58.44 \text{ amu total}} \times 100 = 39.34\% \qquad \% \text{ chlorine} = \frac{35.45 \text{ amu chlorine}}{58.44 \text{ amu total}} \times 100 = 60.66\%$$

Because nitrogen is an important element for plant nutrition, it is a common component of most fertilizers. Consider three commonly used fertilizers: ammonium sulfate, ammonium nitrate, and anhydrous ammonia. Since transportation cost is a significant factor in the cost of a fertilizer, all other factors being equal, the fertilizer with the highest percentage of nitrogen would be chosen over the others as the most cost effective. Which of these three compounds has the highest percentage of nitrogen?

EXAMPLE 8.1

Calculate the percentage of nitrogen in ammonium sulfate, $(NH_4)_2SO_4$.

2 nitrogens	@ 14.00 amu	=	28.00 amu
8 hydrogens	@ 1.00 amu	=	8.00 amu
1 sulfur	@ 32.06 amu	=	32.06 amu
4 oxygens	@ 16.00 amu	=	64.00 amu
Formula weight		=	132.06 amu

$$\% \text{ nitrogen} = \frac{28.00 \text{ amu nitrogen}}{132.06 \text{ amu total}} \times 100 = 21.20\%$$

EXAMPLE 8.2

Calculate the percentage of nitrogen in ammonium nitrate, NH_4NO_3.

2 nitrogens	@ 14.00 amu	=	28.00 amu
4 hydrogens	@ 1.00 amu	=	4.00 amu
3 oxygens	@ 16.00 amu	=	48.00 amu
Formula weight		=	80.00 amu

$$\% \text{ nitrogen} = \frac{28.00 \text{ amu nitrogen}}{80.00 \text{ amu total}} \times 100 = 35.00\%$$

EXAMPLE 8.3

Calculate the percentage of nitrogen in anhydrous ammonia, NH_3. (The term *anhydrous* means "no water.")

1 nitrogen	@ 14.00 amu	=	14.00 amu
3 hydrogens	@ 1.00 amu	=	3.00 amu
Formula weight		=	17.00 amu

$$\% \text{ nitrogen} = \frac{14.00 \text{ amu nitrogen}}{17.00 \text{ amu total}} \times 100 = 82.35\%$$

AgriBasics

FIGURE 8.1
Anhydrous ammonia is transported in pressurized tanks. Its high percentage of nitrogen makes it a cost-effective fertilizer to transport.

Anhydrous ammonia (with about 82% nitrogen by weight) is clearly the most efficient of these three compounds in terms of pounds (or grams) of nitrogen added to the soil for each pound (or gram) of fertilizer transported. Although ammonia is gaseous at normal temperatures and must be stored, transported, and injected into the soil from pressurized tanks, the fact that 82% of the load transported is usable nitrogen has made this a common method of fertilization. Anhydrous ammonia tanks are a common sight in farming communities (Figure 8.1).

NEW TERMS

percentage composition The percentage by weight of each element in a compound, computed from formula or molecular weights. It is equal to the number of atomic mass units of the element in question present in the formula, divided by the total formula or molecular weight.

molecular weight The sum of the atomic weights of the elements comprising one molecule of a covalent compound.

formula weight The sum of the atomic weights of the elements comprising the simplest formula of an ionic compound.

TESTING YOURSELF

Molecular and Formula Weight and Percentage Composition

1. Compute the molecular or formula weight of the following compounds:

   Water, H_2O
   Sugar (sucrose) $C_{12}H_{22}O_{11}$
   Epsom salts, $MgSO_4$
   Carbon tetrachloride, CCl_4

2. By application of your knowledge of bonding, which of the values given for question 1 are formula weights and which are molecular weights?

3. Compute the percentage, by weight, of the indicated element in each of the following compounds:

% Ca in calcium carbonate, $CaCO_3$
% O in water, H_2O
% Mg in magnesium hydroxide, $Mg(OH)_2$
% H in propane, C_3H_8

Answers **1.** Water, 18; Sugar, 342; Epsom salts, 120.37; carbon tetrachloride, 153.8
2. molecular weights: water, sugar, and carbon tetrachloride; formula weights; Epsom salts
3. 40.0%, 88.9%, 41.7%, 18.2%

8.2 The Mole: A "Chemist's Dozen"

Although the atomic mass unit (amu) is convenient for describing the mass of atoms and for computation of percentage composition, it is an inconvenient unit to use because of its small size. In fact, it is so small that it takes 6.02×10^{23} (602,000,000,000,000,000,000,000) atomic mass units to make 1 g. The best balances have a sensitivity of only about 1×10^{-8} g, many multiples of ten larger than the atomic mass unit, and cannot detect individual atoms or molecules. If we had a larger unit available, we would be able to weigh pure substances and determine the number of atoms or molecules present.

To solve this problem, we can assemble a large number of atoms or molecules together so that there is enough mass to measure, even on an insensitive balance. If 6.02×10^{23} amu have a mass of 1.000 g, then 6.02×10^{23} molecules of ammonia (at 17 amu per molecule) will have a mass of 17.00 g. Similarly, 6.02×10^{23} hydrogen atoms will have a mass of 1.00 g, and the same number of water molecules (H_2O, molecular weight = 18 amu) will have a mass of 18.00 g.

Avogadro's number is the number of atomic mass units that equal 1 g, which is 6.02×10^{23}.

One mole of atomic mass units has a mass of 1 g.

Recall from Chapter 2 that this number, 6.02×10^{23}, is called **Avogadro's number** after the Italian scientist Amedeo Avogadro. It is sort of a "chemist's dozen." A **mole** (abbreviated *mol*) is 6.02×10^{23} atoms or molecules. Although the number of items in a mole is large and awkward, the usefulness of the mole lies in the fact that the mass (in grams) of a mole of atoms is numerically equal to the atomic weight as expressed in atomic mass units (Table 8.1).

We can appreciate the huge number of atoms or molecules represented by a mole, and the minute size of atoms and molecules, by computing the volume that would be occupied by a mole of ordinary marbles. One mole

TABLE 8.1
Numerical Equivalence of Atomic Weight to Mass of One Mole

Substance	Atomic Weight (Mass) of One Atom or Molecule	Mass of 1 Mol (6.02×10^{23}) of Atoms or Molecules
Hydrogen (H)	1 amu	1 g
Helium (He)	4 amu	4 g
Ammonia (NH_3)	17 amu	17 g
Water (H_2O)	18 amu	18 g

of marbles would fill a cube 84,437,000 marbles in length, width, and height. Since the volume of a rectangular solid is equal to its length × width × height, such a cube would contain $(84{,}437{,}000)^3 = 602{,}002{,}623{,}600{,}000{,}000{,}000{,}000$ marbles, or approximately 1 mol of marbles. Since a standard marble is about 0.5 inches in diameter, the length of each side of the cube would be:

$$(84{,}437{,}000 \text{ marbles})(0.5 \text{ in. per marble}) = 42{,}218{,}000 \text{ in.}$$

$$(42{,}218{,}000 \text{ in.})\left(\frac{1.0 \text{ ft}}{12 \text{ in.}}\right) = 3{,}518{,}000 \text{ ft}$$

$$(3{,}518{,}000 \text{ ft})\left(\frac{1.0 \text{ mile}}{5280 \text{ ft}}\right) = 666.3 \text{ miles}$$

If a cube of these dimensions were placed over the central United States (Figure 8.2), it would occupy an area roughly bounded by Billings, Montana; Minneapolis, Minnesota; Little Rock, Arkansas; and Santa Fe, New Mexico. It would be about 666 miles high.

A mole is a lot of marbles. It is also a lot of atoms or molecules. But atoms are so small that a mole of copper atoms, for example, would only occupy a cube 1.92 cm on a side. The important thing about the mole is not the number of atoms or molecules involved, but that this number of atoms or molecules is always the same and always has the same mass (expressed in grams) as the atomic or molecular weight of the atom or molecule concerned (expressed in atomic mass units).

A mole of atoms or molecules has the same mass, in grams, as the atomic or molecular mass of that atom or molecule.

Although the mole is principally used to calculate weight relationships in chemical reactions, it can also be used to compute the number of atoms or molecules of a material present in a sample of known mass, as shown in the following examples.

EXAMPLE 8.4

Compute the weight of one mole of carbon dioxide, CO_2.

$$\begin{aligned} 1 \text{ C} @ 12.0 \text{ g/mol} &= 12.0 \text{ g} \\ 2 \text{ O} @ 16.0 \text{ g/mol} &= 32.0 \text{ g} \\ \hline 1 \text{ mol } CO_2 &= 44.0 \text{ g} \end{aligned}$$

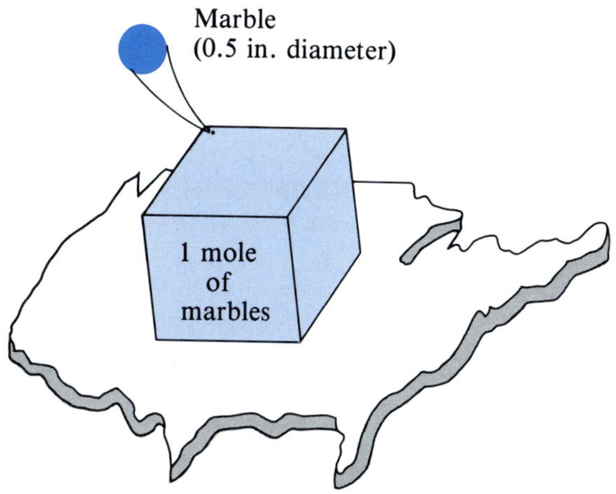
Marble
(0.5 in. diameter)
1 mole of marbles

FIGURE 8.2
A mole is a very large number. One side of a cube containing one mole of marbles would cover a large fraction of the area of the United States. In contrast, a mole of copper atoms would fill less than one cubic inch. The mole concept allows us to deal with groups of atoms large enough to conveniently weigh.

EXAMPLE 8.5

Compute the number of moles represented by 27 g of water.

$$\begin{aligned} 2\text{ H} @ \ \ 1.0\text{ g/mol} &= \ \ 2.0\text{ g} \\ 1\text{ O} @ \ 16.0\text{ g/mol} &= 16.0\text{ g} \\ 1\text{ mol H}_2\text{O} &= 18.0\text{ g} \end{aligned}$$

$$\text{Conversion factor: } \frac{1.00\text{ mol}}{18.0\text{ g}} = 1$$

$$(27.0\text{ g H}_2\text{O})\left(\frac{1.00\text{ mol}}{18.0\text{ g}}\right) = 1.50\text{ mol H}_2\text{O}$$

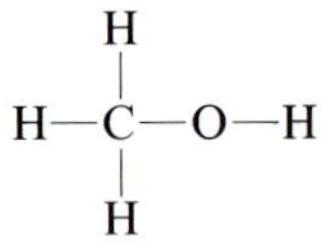

Methyl alcohol

EXAMPLE 8.6

Compute the number of grams equal to 0.60 mol of methyl alcohol, CH_4O.

$$\begin{aligned} 4\text{ H} @ \ \ 1.0\text{ g/mol} &= \ \ 4.0\text{ g} \\ 1\text{ C} @ \ 12.0\text{ g/mol} &= 12.0\text{ g} \\ 1\text{ O} @ \ 16.0\text{ g/mol} &= 16.0\text{ g} \\ 1\text{ mol} &= 32.0\text{ g} \end{aligned}$$

$$\text{Conversion factor: } \frac{32.0\text{ g}}{1.00\text{ mol}} = 1$$

$$(0.60\text{ mol})\left(\frac{32.0\text{ g}}{1.00\text{ mol}}\right) = 19.2\text{ g CH}_4\text{O}$$

Although the mole concept is usually used to compute grams or moles of a material present, it can also be used to compute the number of atoms or molecules present in a sample of known mass.

EXAMPLE 8.7

Determine the number of copper atoms in a 3.10-g sample of metallic copper, Cu.

Since 1 atom of Cu = 63.54 amu, then 1 mol of Cu atoms = 63.54 g.

$$\text{Conversion factor: } \frac{1\text{ mol Cu atoms}}{63.54\text{ g}}$$

And we know that 1 mole = 6.02×10^{23} atoms.

$$\text{Conversion factor: } \frac{6.02 \times 10^{23}\text{ Cu atoms}}{1\text{ mol}}$$

a. Find moles of Cu atoms present in 3.10-g sample.

$$(3.10\text{ g})\left(\frac{1\text{ mol Cu atoms}}{63.54\text{ g}}\right) = 0.0488\text{ mol Cu atoms}$$

b. Find atoms of Cu present in 0.0488 mol.

$$(0.0488 \text{ ~~mol~~})\left(\frac{6.02 \times 10^{23} \text{ Cu atoms}}{1 \text{ ~~mol~~}}\right) = 2.94 \times 10^{22} \text{ Cu atoms}$$

NEW TERMS

Avogadro's number The number of atomic mass units that equal 1.00 g, which is 6.02×10^{23}.

mole (mol) 6.02×10^{23} atoms or molecules. One mole of atoms or molecules has the same mass in grams as the atomic mass of an individual atom or molecule in atomic mass units.

TESTING YOURSELF

Moles

1. Compute the mass (in grams) of 1 mol of each of these compounds:

   $NaHCO_3$ $Pb(OH)_2$
   $CaCO_3$ $PbCrO_4$

2. Compute the number of moles represented by each of the following:

   16.0 g sulfur, S
   67.45 g aluminum, Al
   22.4 g calcium oxide, CaO
   365.4 g nickel nitrate, $Ni(NO_3)_2$

3. Compute the number of grams represented by each of the following:

   2.50 moles cobalt, Co
   12.0 moles carbon, C
   0.41 moles calcium carbonate, $CaCO_3$
   2.50 moles sucrose, $C_{12}H_{22}O_{11}$

4. How many atoms are contained in each of the following:

   20.0 g of calcium, Ca
   35.6 g of tin, Sn

Answers **1.** 84.02 g, 100.1 g, 241.2 g, 323.2 g **2.** 0.500 mol, 2.50 mol, 0.399 mol, 2.000 mol **3.** 147 g, 144 g, 41 g, 855 g, **4.** 3.01×10^{23}, 1.81×10^{23}

8.3 Empirical Formulas

Empirical formulas are based on experimentally determined percentage composition.

A simple application of the mole concept is involved in the computation of **empirical formulas**. An empirical formula is simply a statement of the simplest ratio of atoms found in the compound. In contrast, a *molecular formula* is a statement of the exact number of each kind of atom in a molecule. The molecular formula of glucose, for example, is $C_6H_{12}O_6$, while its empirical formula is CH_2O. Because one is a multiple of the other, both contain the same percentage by weight of C, H, and O.

Empirical formulas are computed from percentage composition data for the compound in question. Calculations of this type are relatively common in identification of unknown compounds in organic chemistry. The percentage carbon, hydrogen, nitrogen, and oxygen are experimentally determined; the empirical formula is computed; and then a molecular arrangement of the component atoms is devised that fits both the atomic ratio and the rules for bonding. As an example, let's calculate the empirical formula of a compound that has been analyzed and found to contain 17.65% hydrogen and 82.35% nitrogen.

a. First, suppose that we had exactly 100.00 g of this material. This sample would contain

$$(0.1765)(100.00\ \text{g}) = 17.65\ \text{g H}$$

$$(0.8235)(100.00\ \text{g}) = 82.35\ \text{g N}$$

Note that, by arbitrary choice of a 100.00-g sample, the computation is very simple.

b. We can find the relative ratio of hydrogen atoms to nitrogen atoms by first finding the number of moles of each element present in the 100.00-g sample. For hydrogen, 1.00 g = 1.00 mol. The conversion factor from grams to moles will be

$$\frac{1.00\ \text{mol H}}{1.00\ \text{g H}} = 1$$

Now we can compute the number of moles of hydrogen present in 17.65 g.

$$(\text{g})\left(\frac{\text{mol}}{\text{g}}\right) = \text{mol}$$

$$(17.65\ \text{g})\left(\frac{1.00\ \text{mol H}}{1.00\ \text{g H}}\right) = 17.65\ \text{mol H}$$

Likewise, we can compute the number of moles of nitrogen present in 82.35 g. Since 1.00 mol N = 14.00 g N, the conversion factor is

$$\frac{1.00\ \text{mol N}}{14.00\ \text{g N}} = 1$$

$$(82.35\ \text{g})\left(\frac{1.00\ \text{mol N}}{14.00\ \text{g N}}\right) = 5.88\ \text{mol N}$$

c. Then find the atomic ratio by dividing the smallest value into all of the other values:

$$\frac{17.65\ \text{mol H}}{5.88\ \text{mol}} = 3\ \text{H}$$

$$\frac{5.88\ \text{mol N}}{5.88\ \text{mol}} = 1\ \text{N}$$

d. The simplest atomic ratio or empirical formula is thus one nitrogen for each three hydrogens, or NH_3. The compound is ammonia.

EXAMPLE 8.8

Find the empirical formula of a compound found to be 41.69% magnesium, 54.88% oxygen, and 3.43% hydrogen.

a. If we had exactly 100.00 g of this material, the sample would contain

$$(0.4169)(100.00 \text{ g}) = 41.69 \text{ g Mg}$$

$$(0.5488)(100.00 \text{ g}) = 54.88 \text{ g O}$$

$$(0.0343)(100.00 \text{ g}) = 3.43 \text{ g H}$$

b. Find the number of moles of each element present in the 100-g sample.

$$\frac{1.00 \text{ mol Mg}}{24.31 \text{ g Mg}} = 1$$

$$(41.69 \text{ g Mg})\left(\frac{1.00 \text{ mol Mg}}{24.31 \text{ g Mg}}\right) = 1.715 \text{ mol Mg}$$

$$\frac{1.00 \text{ mol O}}{16.00 \text{ g O}} = 1$$

$$(54.88 \text{ g O})\left(\frac{1.00 \text{ mol O}}{16.00 \text{ g O}}\right) = 3.43 \text{ mol O}$$

$$\frac{1.00 \text{ mol H}}{1.00 \text{ g H}} = 1$$

$$(3.43 \text{ g H})\left(\frac{1.00 \text{ mol H}}{1.00 \text{ g H}}\right) = 3.43 \text{ mol H}$$

c. Find the ratio by dividing the smallest mole value (1.715) into all other values:

$$\frac{1.715 \text{ mol Mg}}{1.715 \text{ mol}} = 1 \text{ Mg}$$

$$\frac{3.43 \text{ mol O}}{1.715 \text{ mol}} = 2 \text{ O}$$

$$\frac{3.43 \text{ mol H}}{1.715 \text{ mol}} = 2 \text{ H}$$

d. The simplest atomic ratio is MgO_2H_2, but this formula is more commonly written $Mg(OH)_2$.

EXAMPLE 8.9

Find the empirical formula of a compound found to be 69.94% iron and 30.06% oxygen.

a. If we had exactly 100 g of this material, the sample would contain

$$(0.6994)(100.00 \text{ g}) = 69.94 \text{ g Fe}$$

$$(0.3006)(100.00 \text{ g}) = 30.06 \text{ g O}$$

b. Find the number of moles of each element present in the 100.00-g sample.

$$\frac{1.00 \text{ mol Fe}}{55.85 \text{ g Fe}} = 1$$

$$(69.96 \text{ g Fe})\left(\frac{1.00 \text{ mol Fe}}{55.85 \text{ g Fe}}\right) = 1.252 \text{ mol Fe}$$

$$\frac{1.00 \text{ mol O}}{16.00 \text{ g O}} = 1$$

$$(30.06 \text{ g O})\left(\frac{1.00 \text{ mol O}}{16.00 \text{ g O}}\right) = 1.879 \text{ mol O}$$

c. Find the ratio by dividing the smallest mole value (1.252) into all other values:

$$\frac{1.252 \text{ mol Fe}}{1.252 \text{ mol}} = 1 \text{ Fe}$$

$$\frac{1.879 \text{ mol O}}{1.252 \text{ mol}} = 1.5 \text{ O}$$

d. The simplest formula is $FeO_{1.5}$. The compound contains 1.5 times as many oxygen atoms as iron atoms. Since compound formulas are commonly written with whole-number subscripts, all subscripts are simply multiplied by two.

Element	Simplest Ratio	Ratio × 2
Fe	1	2
O	1.5	3

Thus, the empirical formula for the compound is Fe_2O_3. The compound is iron(III) oxide, commonly known as "rust."

NEW TERM

empirical formula The simplest ratio of atoms in a compound, usually derived from experimental percentage composition data.

TESTING YOURSELF

Empirical Formulas

1. Compute the empirical formula of a compound composed of 22.26% iron and 77.74% oxygen. Give the name for this compound, using the Stock system to indicate the oxidation state of the iron.
2. Compute the empirical formula of a compound with the following elemental composition. What is the name of this compound?

$$Na = 27.38\%$$
$$H = 1.20\%$$
$$C = 14.28\%$$
$$O = 57.14\%$$

Answers **1.** FeO, iron(II) oxide **2.** $NaHCO_3$, sodium bicarbonate

8.4 Weight Relationships in Chemical Reactions

A balanced chemical equation conveys much information about the reaction. Ammonia is manufactured industrially by reacting nitrogen from the air with hydrogen gas. This is called the *Haber process.*

$$N_2 + 3\,H_2 \longrightarrow 2\,NH_3$$

This chemical equation states that one molecule of nitrogen gas (N_2) will react with three molecules of hydrogen gas (H_2) to form two molecules of ammonia (NH_3). The weight relationships involved in this reaction can be determined by computing the molecular weight of each component of the reaction:

N_2	+	$3\,H_2$	$\longrightarrow$	$2\,NH_3$
2 N @ 14.00 amu		6 H @ 1.00 amu		2 N @ 14.00 amu = 28 amu
				6 H @ 1.00 amu = 6 amu
28 amu	+	6 amu	=	34 amu

In this example, 28 amu of nitrogen gas reacting with 6 amu of hydrogen gas will yield 34 amu of ammonia. When manufacturing ammonia, the reactants are used exactly in the ratio of 28 amu of nitrogen to 6 amu of hydrogen. Recall, however, that it is impossible to weigh 28 amu of nitrogen gas or 6 amu of hydrogen gas because balances are not sensitive enough to such small amounts. By scaling up the reaction from atomic mass units to grams using moles, the masses become sufficient to measure with conventional laboratory balances.

Consider this reaction again, but this time start with 28 grams (or 1 mol) of nitrogen gas.

N_2	+	$3\,H_2$	$\longrightarrow$	$2\,NH_3$
1 mole = 28 g	+	3 moles = 6 g	=	2 moles = 34 g

According to the balanced equation, one mole of nitrogen gas will react with three moles of hydrogen gas to yield two moles of ammonia. We can use this balanced equation in a four-step procedure called the **mole method** to compute the amounts of any reactant or any product in any chemical reaction. Although there are several methods available for computing amounts of product and reactant involved in a chemical reaction, the mole method is able to handle any problem at any level of chemistry. It is a little like cutting down trees. Sometimes an axe will work. Sometimes a saw is satisfactory. But you can always count on a bulldozer to do the job. There are quicker ways to solve some weight relationship problems in chemistry, but the mole

The *mole method* will solve any type of weight relationship problem in chemistry.

method always works in all types of problems. Because of its versatility, we will concentrate our attention on this method of chemical calculation.

Example of the Mole Method: Ammonia

To demonstrate the mole method, let's suppose we had 7.0 g of nitrogen. How much hydrogen would be required to react completely with it, and how much ammonia would be produced?

The mole method involves four basic steps.
Step 1. Write a balanced equation describing the chemical reaction.

STEP 1. WRITE A BALANCED EQUATION DESCRIBING THE CHEMICAL REACTION

$$\text{Nitrogen} + \text{Hydrogen} \longrightarrow \text{Ammonia}$$

$$N_2 \quad + \quad 3\,H_2 \quad \longrightarrow \quad 2\,NH_3$$

In the remaining steps, we will work from this equation to determine first the number of moles and then the number of grams of each unknown reactant and product.

$$N_2 \quad + \quad 3\,H_2 \longrightarrow 2\,NH_3$$

N_2	$3\,H_2$	$2\,NH_3$
7.0 g	? g	? g
↓	↑	↑
? mol ⟶	? mol ⟶	? mol

Step 2. Compute the number of moles of known compound present.

STEP 2. COMPUTE THE NUMBER OF MOLES OF KNOWN COMPOUND PRESENT

We started with 7.0 g of nitrogen gas (N_2). The molecular weight of N_2 is 28 amu or 28 g/mol.

$$\text{Conversion factor: } \frac{1\text{ mol } N_2}{28\text{ g } N_2} = 1$$

Find the number of moles of nitrogen gas present as reactant:

$$(7.0\text{ g } N_2)\left(\frac{1\text{ mol } N_2}{28\text{ g } N_2}\right) = 0.25\text{ mol } N_2$$

N_2	$3\,H_2$	$2\,NH_3$
7.0 g	? g	? g
↓	↑	↑
0.25 mol ⟶	? moles ⟶	? moles

Step 3. Predict the number of moles of unknown compound required.

STEP 3A. PREDICT THE NUMBER OF MOLES OF THE UNKNOWN COMPOUND REQUIRED

This prediction is based on the balanced equation and the number of moles of known compound present. The unknown compound may be any other reactant or product involved in the reaction. After the number of moles of one compound is determined, the number of moles of all other compounds will be fixed. H_2 will be our first unknown compound in this example.

From the balanced equation, we know that 1 mol N_2 will react with 3 mol H_2.

$$\text{Conversion factor: } \frac{3 \text{ mol } H_2}{1 \text{ mol } N_2} = 1$$

$$(0.25 \cancel{\text{mol}}\ N_2)\left(\frac{3 \text{ mol } H_2}{1 \cancel{\text{mol}}\ N_2}\right) = 0.75 \text{ mol } H_2$$

$$\begin{array}{ccccc} N_2 & + & 3\,H_2 & \longrightarrow & 2\,NH_3 \\ 7.0\text{ g} & & ?\text{ g} & & ?\text{ g} \\ \downarrow & & \uparrow & & \uparrow \\ 0.25\text{ moles} & \longrightarrow & 0.75\text{ moles} & \longrightarrow & ?\text{ moles} \end{array}$$

STEP 4A. COMPUTE THE NUMBER OF GRAMS OF UNKNOWN COMPOUND

Step 4. Compute the number of grams of the unknown compound.

First compute the mole weight of the compound.

$$\text{Conversion factor: } \frac{2.00 \text{ g } H_2}{1.0 \text{ mol } H_2} = 1$$

Then compute the number of grams of unknown compound required.

$$(0.75 \cancel{\text{mol}}\ H_2)\left(\frac{2.00 \text{ g } H_2}{1.0 \cancel{\text{mol}}\ H_2}\right) = 1.5 \text{ g } H_2$$

$$\begin{array}{ccccc} N_2 & + & 3\,H_2 & \longrightarrow & 2\,NH_3 \\ 7.0\text{ g} & & 1.5\text{ g} & & ?\text{ g} \\ \downarrow & & \uparrow & & \uparrow \\ 0.25\text{ mol} & \longrightarrow & 0.75\text{ mol} & \longrightarrow & ?\text{ mol} \end{array}$$

We can compute the amount of ammonia produced by this reaction by simply repeating the last two steps of the sequence. Note that you should always refer back to the original known compound. If you do this, any error in the calculations will not be carried throughout the entire problem, and you can catch the mistake in the last step when you check for conservation of mass.

STEP 3B. PREDICT THE NUMBER OF MOLES OF THE NEXT UNKNOWN COMPOUND PRODUCED

From the balanced equation, we know that 1 mol N_2 will react to produce 2 mol NH_3.

$$\text{Conversion factor: } \frac{2 \text{ mol } NH_3}{1 \text{ mol } N_2} = 1$$

$$(0.25 \cancel{\text{mol}}\ N_2)\left(\frac{2 \text{ mol } NH_3}{1 \cancel{\text{mol}}\ N_2}\right) = 0.50 \text{ mol } NH_3$$

$$\begin{array}{ccccc} N_2 & + & 3\,H_2 & \longrightarrow & 2\,NH_3 \\ 7.0\text{ g} & & 1.5\text{ g} & & ?\text{ g} \\ \downarrow & & \uparrow & & \uparrow \\ 0.25\text{ mol} & \longrightarrow & 0.75\text{ mol} & \longrightarrow & 0.50\text{ mol} \end{array}$$

STEP 4B. COMPUTE THE NUMBER OF GRAMS OF UNKNOWN COMPOUND

a. Compute the mole weight of the compound NH_3.

$$\begin{aligned} 1 \text{ mol N} &= 14.00 \text{ g} \\ 3 \text{ mol H} &= \underline{\ 3.00 \text{ g}} \\ 1 \text{ mol } NH_3 &= 17.00 \text{ g} \end{aligned}$$

$$\text{Conversion factor: } \frac{17.00 \text{ g } NH_3}{1.00 \text{ mol } NH_3} = 1$$

b. Compute the grams of unknown compound produced.

$$(0.50 \text{ mol } NH_3)\left(\frac{17.00 \text{ g } NH_3}{1.00 \text{ mol } NH_3}\right) = 8.5 \text{ g } NH_3$$

$$\begin{array}{ccccc} N_2 & + & 3\,H_2 & \longrightarrow & 2\,NH_3 \\ 7.0 \text{ g} & & 1.5 \text{ g} & & 8.5 \text{ g} \\ \downarrow & & \uparrow & & \uparrow \\ 0.25 \text{ mol} & \longrightarrow & 0.75 \text{ mol} & \longrightarrow & 0.50 \text{ moles} \end{array}$$

According to the law of conservation of matter, the number of grams of reactant should equal the number of grams of product. This provides a good check on our calculations.

$$\begin{array}{ccc} N_2 + 3\,H_2 & \longrightarrow & 2\,NH_3 \\ 7.0 \text{ g} + 1.5 \text{ g} & = & 8.5 \text{ g} \end{array}$$

Application of the Mole Method to a Double Replacement Reaction

The mole method can handle the weight relationships in *any* chemical reaction. Let's now look at an example that looks more complex, but is really just larger. Mole problems can have more parts, but making them larger does not make them more difficult.

Silicon carbide (SiC) is chemically an extremely inactive compound, and its crystals are almost as hard as diamond. It is usually called by the trade name "carborundum," and it is used widely as an abrasive in grinding stone and sandpaper. Silicon carbide is manufactured by reacting sand (silicon oxide) with coke (carbon) under conditions of extremely high temperature.

$$\text{Silicon oxide} + \text{Carbon} \longrightarrow \text{Silicon carbide} + \text{Carbon monoxide}$$

$$SiO_{2(s)} + 3\,C_{(s)} \longrightarrow SiC_{(s)} + 2\,CO_{(s)}$$

Suppose that we had 100.0 g of silicon oxide. How much carbon would be required to react completely with this material, and how much silicon carbide and carbon monoxide would be produced?

STEP 1. WRITE A BALANCED EQUATION DESCRIBING THE CHEMICAL REACTION

$$\text{Silicon oxide} + \text{Carbon} \longrightarrow \text{Silicon carbide} \longrightarrow \text{Carbon monoxide}$$

$$SiO_2 + 3\,C \longrightarrow SiC \longrightarrow 2\,CO$$

STEP 2. COMPUTE THE NUMBER OF MOLES OF KNOWN COMPOUND PRESENT

We started with 100.0 g of silicon oxide, SiO_2.

$$\begin{array}{ccccccc}
SiO_2 & + & 3\,C & \longrightarrow & SiC & + & 2\,CO \\
100.0\text{ g} & & ?\text{ g} & & ?\text{ g} & & ?\text{ g} \\
\downarrow & & \uparrow & & \uparrow & & \uparrow \\
?\text{ mol} & \longrightarrow & ?\text{ mol} & \longrightarrow & ?\text{ mol} & \longrightarrow & ?\text{ mol}
\end{array}$$

First compute the mole weight of this compound:

$$\begin{aligned}
1\text{ mol Si} &= 28.09\text{ g} \\
2\text{ mol O} &= 32.00\text{ g} \\
1\text{ mol } SiO_2 &= 60.09\text{ g}
\end{aligned}$$

$$\text{Conversion factor: } \frac{1\text{ mol } SiO_2}{60.09\text{ g } SiO_2} = 1$$

Then compute the number of moles represented by the sample:

$$(100.0\text{ g SiO})\left(\frac{1\text{ mol } SiO_2}{60.09\text{ g } SiO_2}\right) = 1.664\text{ mol } SiO_2$$

$$\begin{array}{ccccccc}
SiO_2 & + & 3\,C & \longrightarrow & SiC & + & 2\,CO \\
100.0\text{ g} & & ?\text{ g} & & ?\text{ g} & & ?\text{ g} \\
\downarrow & & \uparrow & & \uparrow & & \uparrow \\
1.164\text{ mol} & \longrightarrow & ?\text{ mol} & \longrightarrow & ?\text{ mol} & \longrightarrow & ?\text{ mol}
\end{array}$$

STEP 3A. PREDICT THE NUMBER OF MOLES OF UNKNOWN COMPOUND REQUIRED

From the balanced equation, we know that 1 mol SiO_2 will react with 3 mol C.

$$\text{Conversion factor: } \frac{3\text{ mol C}}{1\text{ mol } SiO_2} = 1$$

$$(1.664\text{ mol } SiO_2)\left(\frac{3\text{ mol C}}{1\text{ mol } SiO_2}\right) = 4.992\text{ mol C}$$

$$\begin{array}{ccccccc}
SiO_2 & + & 3\,C & \longrightarrow & SiC & + & 2\,CO \\
100.0\text{ g} & & ?\text{ g} & & ?\text{ g} & & ?\text{ g} \\
\downarrow & & \uparrow & & \uparrow & & \uparrow \\
1.164\text{ mol} & \longrightarrow & 4.992\text{ mol} & \longrightarrow & ?\text{ mol} & \longrightarrow & ?\text{ mol}
\end{array}$$

STEP 4A. COMPUTE THE NUMBER OF GRAMS OF UNKNOWN COMPOUND

First compute the mole weight of the compound.

$$1.00\text{ mol C} = 12.01\text{ g}$$

$$\text{Conversion factor: } \frac{12.01\text{ g C}}{1.0\text{ mole C}} = 1$$

Then compute the number of grams of unknown compound required.

$$(4.992\ \text{mol C})\left(\frac{12.01\ \text{g C}}{1.0\ \text{mol C}}\right) = 59.95\ \text{g C}$$

$$\begin{array}{ccccccc}
SiO_2 & + & 3\,C & \longrightarrow & SiC & + & 2\,CO \\
100.0\ \text{g} & & 59.95\ \text{g} & & ?\ \text{g} & & ?\ \text{g} \\
\downarrow & & \uparrow & & \uparrow & & \uparrow \\
1.164\ \text{mol} & \longrightarrow & 4.992\ \text{mol} & \longrightarrow & ?\ \text{mol} & \longrightarrow & ?\ \text{mol}
\end{array}$$

At this point we have determined that 59.95 g of carbon are required to react completely with 100.0 g of silicon oxide. Let us now compute the amounts of silicon carbide and carbon monoxide that will be produced by this reaction. Since these are additional unknowns in the same reaction, we can simply return to steps 3 and 4 of the procedure.

STEP 3B. PREDICT THE NUMBER OF MOLES OF UNKNOWN COMPOUND PRODUCED

From the balanced equation, we know that 1 mol SiO_2 will react to produce 1 mol SiC.

$$\text{Conversion factor: } \frac{1\ \text{mol SiC}}{1\ \text{mol } SiO_2} = 1$$

$$(1.664\ \text{mol } SiO_2)\left(\frac{1\ \text{mol SiC}}{1\ \text{mol } SiO_2}\right) = 1.664\ \text{mol SiC}$$

$$\begin{array}{ccccccc}
SiO_2 & + & 3\,C & \longrightarrow & SiC & + & 2\,CO \\
100.0\ \text{g} & & 59.95\ \text{g} & & ?\ \text{g} & & ?\ \text{g} \\
\downarrow & & \uparrow & & \uparrow & & \uparrow \\
1.164\ \text{mol} & \longrightarrow & 4.992\ \text{mol} & \longrightarrow & 1.664\ \text{mol} & \longrightarrow & ?\ \text{mol}
\end{array}$$

STEP 4B. COMPUTE THE NUMBER OF GRAMS OF UNKNOWN COMPOUND

First compute the mole weight of the compound:

$$\begin{aligned}
1\ \text{mol Si} &= 28.09\ \text{g} \\
1\ \text{mol C} &= 12.01\ \text{g} \\
\hline
1\ \text{mol SiC} &= 40.10\ \text{g}
\end{aligned}$$

$$\text{Conversion factor: } \frac{40.01\ \text{g SiC}}{1.00\ \text{mol SiC}} = 1$$

Then compute the grams of unknown compound produced:

$$(1.664\ \text{mol SiC})\left(\frac{40.10\ \text{g SiC}}{1.00\ \text{mol SiC}}\right) = 66.73\ \text{g SiC}$$

$$\begin{array}{ccccccc}
SiO_2 & + & 3\,C & \longrightarrow & SiC & + & 2\,CO \\
100.0\ \text{g} & & 59.95\ \text{g} & & 66.73\ \text{g} & & ?\ \text{g} \\
\downarrow & & \uparrow & & \uparrow & & \uparrow \\
1.164\ \text{mol} & \longrightarrow & 4.992\ \text{mol} & \longrightarrow & 1.664\ \text{mol} & \longrightarrow & ?\ \text{mol}
\end{array}$$

Consider now the other product, carbon monoxide (CO).

STEP 3C. PREDICT THE NUMBER OF MOLES OF UNKNOWN COMPOUND PRODUCED

From the balanced equation, we know that 1 mol SiO_2 will react to produce 2 mol CO.

$$\text{Conversion factor: } \frac{2\text{ mol CO}}{1\text{ mol SiO}_2} = 1$$

$$(1.664\ \cancel{\text{mol}}\ \text{SiO}_2)\left(\frac{2\text{ mol CO}}{1\ \cancel{\text{mol}}\ \text{SiO}_2}\right) = 3.328\text{ mol CO}$$

$$\begin{array}{ccccccc} \text{SiO}_2 & + & 3\text{ C} & \longrightarrow & \text{SiC} & + & 2\text{ CO} \\ 100.0\text{ g} & & 59.95\text{ g} & & 66.73\text{ g} & & ?\text{ g} \\ \downarrow & & \uparrow & & \uparrow & & \uparrow \\ 1.164\text{ mol} & \longrightarrow & 4.992\text{ mol} & \longrightarrow & 1.664\text{ mol} & \longrightarrow & 3.328\text{ mol} \end{array}$$

STEP 4C. COMPUTE THE NUMBER OF GRAMS OF UNKNOWN COMPOUND

First compute the mole weight of the compound:

$$\begin{aligned} 1\text{ mol C} &= 12.01\text{ g} \\ 1\text{ mol O} &= 16.00\text{ g} \\ 1\text{ mol CO} &= 28.00\text{ g} \end{aligned}$$

$$\text{Conversion factor: } \frac{28.00\text{ g CO}}{1.00\text{ mol CO}} = 1$$

Then compute the grams of unknown compound produced:

$$(3.328\ \cancel{\text{mol}}\ \text{CO})\left(\frac{28.00\text{ g CO}}{1.00\ \cancel{\text{mol}}\ \text{CO}}\right) = 93.22\text{ g CO}$$

$$\begin{array}{ccccccc} \text{SiO}_2 & + & 3\text{ C} & \longrightarrow & \text{SiC} & + & 2\text{ CO} \\ 100.0\text{ g} & & 59.95\text{ g} & & 66.73\text{ g} & & 93.22\text{ g} \\ \downarrow & & \uparrow & & \uparrow & & \uparrow \\ 1.164\text{ mol} & \longrightarrow & 4.992\text{ mol} & \longrightarrow & 1.664\text{ mol} & \longrightarrow & 3.328\text{ mol} \end{array}$$

According to the law of conservation of matter, the number of grams of reactant should equal the number of grams of product.

$$\begin{array}{rcl} \text{SiO}_2 + 3\text{ C} & \longrightarrow & \text{SiC} + 2\text{ CO} \\ 100.0\text{ g} + 59.95\text{ g} & = & 66.73\text{ g} + 93.22\text{ g} \\ 159.95\text{ g} & = & 159.95\text{ g} \\ \textbf{Mass reactants} & = & \textbf{Mass products} \end{array}$$

NEW TERM

mole method Use of the mole concept to compute masses of reactants and products involved in chemical reactions.

TESTING YOURSELF

Mole Problems

1. When propane is burned in air, the following reaction takes place:

$$C_3H_8 + 5\ O_2 \longrightarrow 3\ CO_2 + 4\ H_2O + \text{heat}$$

   a. Suppose that we burned 4.40 g of propane. How many moles of propane were burned?
   b. How many moles of oxygen gas would be required to react with this amount of propane? How many grams?
   c. How many moles of carbon dioxide gas would be produced by this reaction? How many grams?
   d. How many moles of water would be produced?
   e. How many grams of water would be produced?
   f. Compare the total mass of reactants (propane and oxygen) with the total mass of products (carbon dioxide and water).

2. The compound urea is a common fertilizer and is manufactured by reacting ammonia and carbon dioxide:

$$2\ NH_3 + CO_2 \longrightarrow (NH_2)_2CO + H_2O$$

   a. Suppose that we had 340 g of ammonia. How many grams of carbon dioxide would be required to react with this amount of ammonia?
   b. How many grams of urea would be produced?
   c. How many grams of water would be produced as a byproduct?

3. Aspirin is manufactured by reacting salicylic acid with acetic anhydride:

$$\text{Salicylic acid} + \text{Acetic anhydride} \longrightarrow \text{Aspirin} + \text{Acetic acid}$$

$$C_7H_6O_3 + C_4H_6O_3 \longrightarrow C_9H_8O_4 + HC_2H_3O_2$$

   a. Suppose that we had 100.0 g of salicylic acid. How many grams of acetic anhydride would be required to react completely with this material?
   b. How many grams of aspirin would be produced, assuming 100% yield?
   c. How many grams of acetic acid would be produced as a byproduct?

Answers **1. a.** 0.100 mol **b.** 0.500 mol, 16.0 g **c.** 0.300 mol, 13.2 g **d.** 0.400 mol **e.** 7.20 g **f.** 20.4 g reactants, 20.4 g products **2. a.** 440 g **b.** 600 g **c.** 180 g **3. a.** 73.91 g **b.** 130.4 g **c.** 43.47 g

SUMMARY

Two major concepts were introduced in this chapter. The first dealt with the use of atomic weight and compound formula to predict the percentage composition of compounds. *Percentage composition* is determined by simply computing the fraction of the atomic weight of the compound due to each component. *Empirical formulas* are determined by computing the *ratio* of the number of moles of each atomic component in a sample of known mass. The second major concept concerned the *mole method*—a method of predicting the amounts of reactant and product involved in a chemical reaction. The mole method allows calculation of weight relationships through four steps:

Step 1. Write a balanced equation describing the chemical reaction.

Step 2. Compute the number of moles of the known compound present.

Step 3. By reference to the balanced equation and the number of moles of known compound present, predict the number of moles of the unknown compound required.

Step 4. Compute the number of grams of the unknown compound.

In the next chapter we will see how the mole method is used to express the concentration of solutions.

TERMS

percentage composition
molecular weight
formula weight
Avogadro's number
mole
empirical formula
mole method

CHAPTER REVIEW

QUESTIONS

1. What is the formula weight (in atomic mass units) of the compound sodium bicarbonate, $NaHCO_3$?
2. Chalk is calcium carbonate, $CaCO_3$. Compute the following for this compound.
   a. The percentage by weight calcium
   b. The percentage by weight carbon
   c. The percentage by weight oxygen
3. How many moles of sulfur are present in 48.00 g of this element?
4. A cup of water has a volume of about 250 mL and a mass of about 250 g. How many moles of water are present in this amount?
5. Remember that Avogadro's number is 6.02×10^{23}. How many molecules of water are present in the cup described in question 4?
6. Consider a compound with the following composition:

Na	32.38%
S	22.57%
O	45.06

   What is the empirical formula of this compound?
7. Methane, commonly known as natural gas, combines with oxygen when it burns to form carbon dioxide and water vapor. The reaction is quite efficient, and natural gas furnaces and water heaters produce little if any colored smoke (due to the presence of unburned carbon particles). On cold days, one can see condensed water vapor rising from chimneys.

$$CH_4 + 2\,O_2 \longrightarrow CO_2 + 2\,H_2O$$

   a. Consider a sample of 96.0 g of methane. How many moles of methane are present?
   b. How many moles of oxygen will be required to react with this methane?
   c. How many grams of oxygen will be required to react with this amount of methane?
   d. Consider the amount of oxygen required. Why is it important to provide a good air supply to a furnace?
8. Consider the problem of combustion of natural gas discussed in question 7.
   a. With combustion of 96 g of methane, how many moles of carbon dioxide will be produced?
   b. How many grams of carbon dioxide will be produced?

9. When 96.0 g of methane are burned,
   a. How many moles of water vapor will be produced?
   b. How many grams of water vapor will be produced?
10. Calculate the total amount of material entering and leaving the reaction of question 7.

CH_4 +	$2\,O_2$	$\longrightarrow$	CO_2		$+ 2\,H_2O$	
96.0 g +	__g	=	__g	+	__g	
	? g	=	? g			
	Reactants	=	Products			

Answers **1.** 84 amu **2. a.** 40%, **b.** 12%, **c.** 48% **3.** 1.5 mol **4.** 13.9 mol H_2O **5.** 8.36×10^{24} molecules **6.** Na_2SO_4 **7. a.** 6 mol CH_4, **b.** 12 mol O_2, **c.** 384 g O_2 **8. a.** 6 mol CO_2, **b.** 264 g CO_2 **9. a.** 12 mol H_2O, **b.** 216 g H_2O **10.** Reactants, 480 g; products, 480 g

DIAGNOSTIC CHART

Blacken in all of the circles under the number of each question that you missed in the Chapter Review questions. The diagnostic chart will help you identify concept areas that need more study.

	Questions									
Concepts	1	2	3	4	5	6	7	8	9	10
Formula weight	○	○		○	○		○	○	○	○
Percentage composition		○								
Moles			○	○	○	○	○	○	○	○
Avogardo's number					○					
Empirical formulas						○				
Weight relationships							○	○	○	○

EXERCISES

Formula and Molecular Weight

Compute the mass, (in grams) of 1 mol of each of the following compounds.

1. Sodium chloride, $NaCl$
2. Potassium hydroxide, KOH
3. Lead hydroxide, $Pb(OH)_2$
4. Hydrogen sulfide, H_2S
5. Silver nitrate, $AgNO_3$
6. Lead chromate, $PbCrO_4$
7. Zinc sulfate, $ZnSO_4$
8. Aluminum oxide, Al_2O_3

Percentage Composition

Find the elemental percentage composition of the following compounds.

9. Natural gas, CH_4
10. Octane, C_8H_{18}
11. Sucrose, $C_{12}H_{22}O_{11}$
12. Rust, Fe_2O_3
13. Lye, $NaOH$
14. Ethyl alcohol, C_2H_5OH

Moles

Compute the number of moles of material represented by each of the following samples.

15. 16.0 g of sulfur
16. 11.17 g of iron
17. 67.45 g of aluminum
18. 95.31 g of copper
19. 53.55 g of aluminum chloride, $AlCl_3$
20. 22.40 g of calcium oxide, CaO
21. 366.5 g of nickel nitrate, $Ni(NO_3)_2$
22. 48.95 g of ammonium bromide, NH_4Br

Compute the number of grams represented by each of the following.

23. 2.5 mol of cobalt, Co
24. 0.30 mol of silver, Ag
25. 0.50 mol of mercury, Hg
26. 0.75 mol of uranium, U
27. 12.0 mol of carbon, C
28. 2.0 mol of potassium hydroxide, KOH
29. 0.41 mol of calcium carbonate, $CaCO_3$
30. 2.5 mol of sucrose, $C_{12}H_{22}O_{11}$
31. 0.15 mol of aluminum sulfate, $Al_2(SO_4)_3$

Avogadro's Number

How many atoms are contained in each of the following?

32. 20.0 g of calcium, Ca
33. 12.7 g of copper, Cu
34. 35.6 g of tin, Sn
35. 50.76 g of iodine, I

Empirical Formulas

Compute empirical formulas for each of the following compounds, and name each compound.

36. Cu = 79.88%
    O = 20.12%
37. Cu = 33.88%
    N = 14.94%
    O = 51.18%
38. Li = 28.98%
    O = 66.80%
    H = 4.21%
39. Na = 28.36%
    C = 14.81%
    S = 39.55%
    N = 17.28%
40. K = 24.74%
    Mn = 34.76%
    O = 40.50%

Weight Relationships

41. Zinc chromate is a component of some types of primer paint. This material can be synthesized by reacting zinc nitrate with sodium chromate:

Zinc nitrate	+	Sodium chromate	⟶	Zinc chromate	+	Sodium nitrate
$Zn(NO_3)_2$	+	Na_2CrO_4	⟶	$ZnCrO_4$	+	$2NaNO_3$

Suppose that we had 37.87 g of zinc chromate (0.200 mol), and sufficient sodium chromate to react completely with the zinc nitrate.

a. How many moles of zinc chromate would be produced?
b. How many grams of zinc chromate would be produced?
c. How many moles of sodium nitrate byproduct would be produced?
d. How many grams of sodium nitrate byproduct would be produced?

42. Zinc is commonly found as the ore sphalerite, ZnS. Tennessee, New York, Idaho, Arizona, Utah, and Colorado are leading producers of zinc in the United States. The refining process involves first heating the ore with air to form zinc oxide. This process is called "roasting."

Zinc sulfide	+	Oxygen	⟶	Zinc oxide	+	Sulfur dioxide
2 ZnS	+	3 O_2	⟶	2 ZnO	+	2 SO_2

a. Suppose that we were to follow 1000 g of zinc sulfide through the refining process. How many moles is 1000 g of ZnS?
b. How many moles of oxygen would be required to react with this amount of zinc sulfide?
c. How many grams of oxygen would be required to react with this amount of zinc sulfide?
d. How many grams of zinc oxide would be produced as a result of roasting 1000 g of ZnS?
e. How many grams of sulfur dioxide (SO_2) would be produced by this reaction? This last step is a serious problem for the zinc industry. SO_2 is highly toxic. Plant life within a few miles of this refining operation has been seriously affected or killed. Recent air quality restrictions have required installation of "scrubbers" to remove the toxic SO_2 before it reaches the atmosphere.

43. Zinc oxide resulting from the oxidation of zinc ore (discussed in exercise 42) can be converted to metallic zinc by heating the zinc oxide with powdered coal (carbon) at temperatures of 1200–1300 °C.

$$ZnO + C \longrightarrow Zn + CO$$

Suppose that one began with 1000 g of ZnO.

a. How many grams of carbon (coal) would be required to react completely with this material?
b. How many grams of metallic zinc would be produced?
c. How many grams of byproduct carbon monoxide (CO) would be produced?

44. A major component of automobile gasoline is octane, C_8H_{18}. When octane burns in the presence of adequate oxygen, it combines with oxygen to produce carbon dioxide and water vapor.

$$2\ C_8H_{18} + 25\ O_2 \longrightarrow 16\ CO_2 + 18\ H_2O$$

Consider the combustion of 1 gal (3.78 L) of octane. The density of octane is 0.70 g/mL; 1 gal has a mass of about 2650 g.

a. How many moles of octane are present in 2650 g of the material?
b. How many moles of oxygen are required to completely burn this amount of octane?
c. How many grams of oxygen are required to completely burn this amount of octane?
d. How many grams of carbon dioxide will be produced?
e. How many grams of water will be produced?
f. Can you give a reason why automobile exhaust systems rust out relatively rapidly, particularly if the car is driven short distances in the city and never has a chance to heat up well?

CHAPTER 9

Solutions, Colloids, and Suspensions

Practically all chemical reactions take place in solutions. This chapter is concerned with methods of describing solution concentration and ways the behavior of solutions differs from that of pure substances.

OBJECTIVES

After completing this chapter, you should be able to

- Define the terms *solute* and *solvent*.
- Differentiate among solutions, colloids, and suspensions.
- Compute solution concentrations in moles per liter.
- Describe the effect a dissolved material has on the freezing and boiling points of water.
- Describe the effect of solute concentration on osmosis.

9.1 Characteristics of Solutions, Colloids, and Suspensions

A *solution* is a mixture in which ion or molecule-sized components are evenly distributed.

Solutions are mixtures in which one substance is dissolved in another. They are very important to chemists because most chemical reactions take place in solutions.

Solutions have some unique properties. Because of its salt content, seawater freezes at a lower temperature than freshwater. Addition of automobile antifreeze (ethylene glycol) to an automobile's coolant increases the boiling point as well as lowers the freezing point. The tendency of solutions separated only by a thin membrane to transfer molecules in an attempt to equalize their concentrations is called *osmosis*; this is the force that moves water up tall trees.

Knowledge of the amount of dissolved material is important to those who work with solutions. This chapter deals with a number of ways of expressing solution concentration and some of the unique physical properties of solutions.

Solutions

A *solute* is the substance to dissolve, while the solvent is the substance in which the solute is being dissolved.

Solutions are homogeneous mixtures in that any ratio of the components is often possible. To make a solution, one pure substance called the **solute** is mixed with and dissolved into another substance called the **solvent**. The resulting solution will be clear and uniform throughout (Figure 9.1). The molecules or ions of solute are distributed evenly among the molecules of solvent. This requires that solute and solvent be similar enough in their molecular polarity to interact well. Solutions are clear, will not separate on standing, and will not scatter light.

We usually think of solutions as materials dissolved in the most common solvent, water. For example, sugar dissolved in water is a solution. Salt, when dissolved in water, is separated into its component sodium and chloride ions, which are distributed uniformly through the solvent. However, solutions need not be liquid at room temperature. Brass, for example, is a solution of zinc (the solute) dissolved in copper (the solvent). When solid, the mixture of zinc and copper has strength and resistance to chemical attack not possible with either of its parent metals.

Colloids

Some materials appear to be solutions, but closer examination shows they have different characteristics. Smoke, for example, is a mixture of carbon particles and air, in which the carbon particles are more or less uniformly

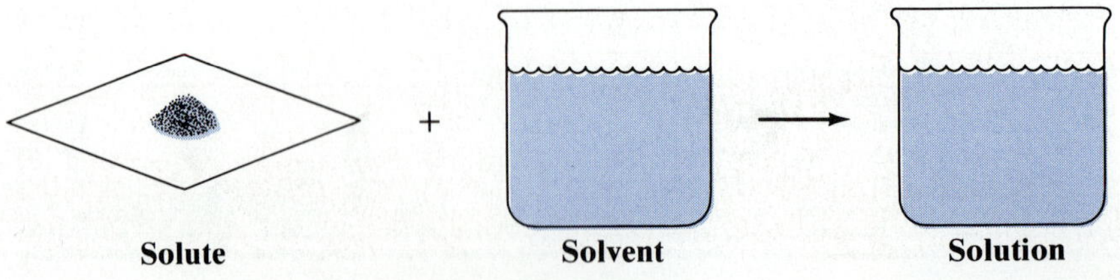
Solute
+
Solvent
Solution

FIGURE 9.1
Solutions are mixtures in which one substance (the solute) is dissolved into another substance (the solvent). In true solutions, the solute molecules are dispersed evenly throughout the solvent. A solution is clear, will not separate on standing, and will not scatter light.

distributed. However, the carbon particles are large enough to reflect or scatter light; they are not truly *dissolved*. The mixture is not clear.

Milk is another example of this class of material. When normal pasteurized milk is allowed to stand, it will separate into a layer with high butterfat content (cream) and a layer with high water content. Homogenization is a process during which the mixture is vibrated rapidly by ultrasonic waves, thus breaking the nonwater soluble fats into tiny clumps that will disperse evenly throughout the mixture. The particle size is still large enough to scatter light and the resulting mixture is cloudy. Both milk and smoke are examples of **colloids**, mixtures in which the dispersed material is evenly distributed but broken into larger than ion- or molecule-sized chunks. Even when colloidal mixtures are extremely dilute and appear clear, the tiny particles will scatter light. This is called the *Tyndall effect* and is a common test for the presence of colloidal particles (Figure 9.2).

A *colloid* is a mixture in which components that are larger than ions or molecules are evenly distributed.

Colloidal mixtures are distinguished from solutions by particle size. In a solution, the dissolved particles are molecule size—less than 10^{-7} cm in diameter. In a colloid, the particles are 10 to 1000 times larger—as large as 10^{-4} cm (100 μm). These particles are small enough that normal molecular motion within the gas or liquid will overcome the gravitational force and will keep the particles evenly distributed. A searchlight beam is visible in clear night air because it reflects off colloidal dust particles in the atmosphere. The random movement of these particles is called **Brownian motion**.

Colloidal particles are too small to be affected by gravity.

Suspensions

Suspensions are mixtures in which the solute particles are sufficiently larger than colloidal particles that gravitational force will ultimately cause them to separate. The size of particles in suspensions are generally on the order of 0.001 to 0.01 mm. The particles may be uniformly distributed when mixed, but will eventually separate. For example, many of the components of lemonade are held in suspension, which will settle if left standing too long. That's why you need to stir it before pouring a glass.

Suspensions are mixtures in which the suspended particles are large enough to be separated by gravity.

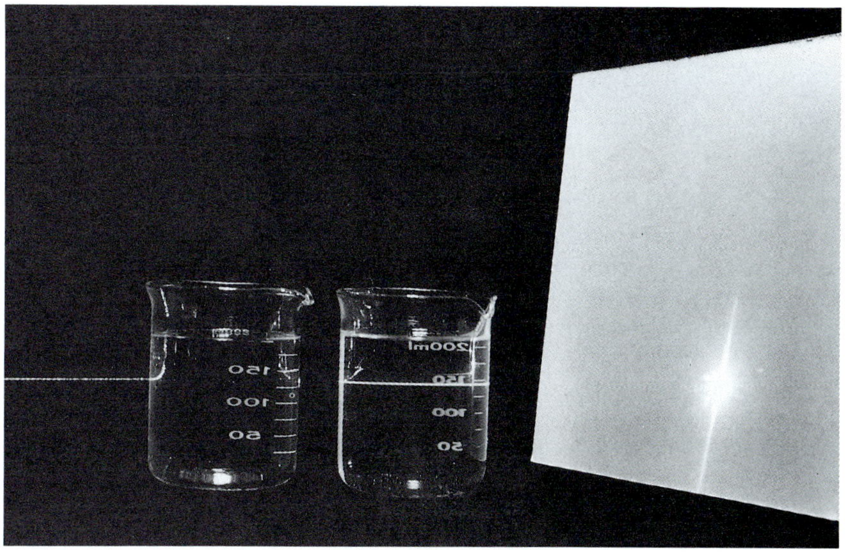

FIGURE 9.2
Light from a laser (entering from the left) passes through two beakers and strikes a white card. Particles in a solution (left) are so small that it appears clear even when light passes through it. In contrast, a colloidal mixture (right) consists of solute particles large enough to scatter a light beam and make it visible when it passes through the container. This scattering of light by colloids is called the *Tyndall effect*.

Separating Suspensions, Solutions, and Colloids

Suspensions can be separated by a process called **filtration** (Figure 9.3). In the laboratory, filtration is usually accomplished by pouring the mixture through a funnel lined by a paper with small pores. The suspended material remains on the filter paper, and the solvent passes through. Drip coffee makers use filtration to separate the suspended coffee grounds from the solution (coffee). Suspended particles in industrial processes, such as the combustion products from a coal-fired electric generating plant, are often separated by filtration in a "baghouse" filter. This filter, as the name implies, is a building full of large bags through which the smoke passes. A smaller scale application of this principle is used in home vacuum cleaners.

FIGURE 9.3
Insoluble particles suspended in a mixture can be separated by a process called *filtration*. The mixture is passed through a porous filter paper; the suspended solid is trapped on the paper and the solution passes through.

In contrast, solutions cannot be separated into their component parts through filtration. They require a special process called **distillation** (Figure 9.4). In this process, the material having the lower boiling point is selectively vaporized and removed from the mixture, then reclaimed by condensation. Distillation is commonly used to separate crude oil (petroleum) into components of differing molecular weight—kerosene, gasoline, lubricating oils, and heavy hydrocarbons used for asphalt.

The process of distillation is keyed to the fact that, as a material is boiled, its temperature remains the same until the change of state from liquid to gas

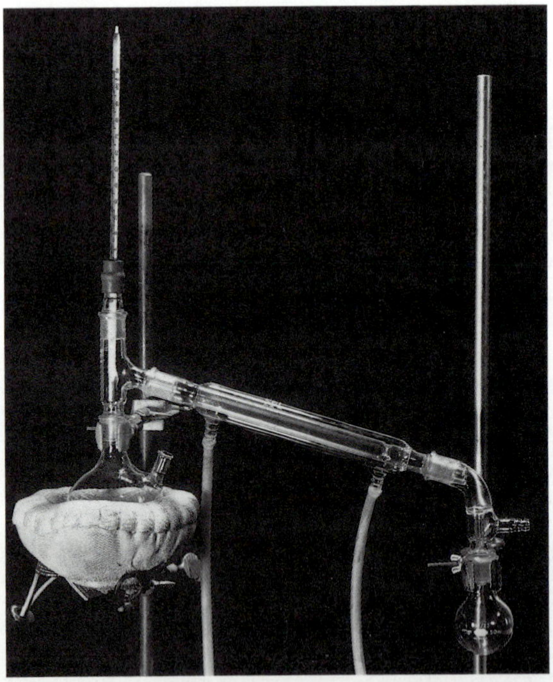

FIGURE 9.4
Components of a mixture can be separated by distillation if their boiling points are different. This figure shows a laboratory distillation apparatus. The heating mantle below the distillation flask heats the mixture. Rising vapor travels through the water-cooled condenser, condenses from gas to liquid, and runs into the collecting flask. Note the thermometer, which monitors the temperature of the escaping gas. Mixtures are separated by changing the collecting flask each time the boiling point rises and a new component leaves the solution.

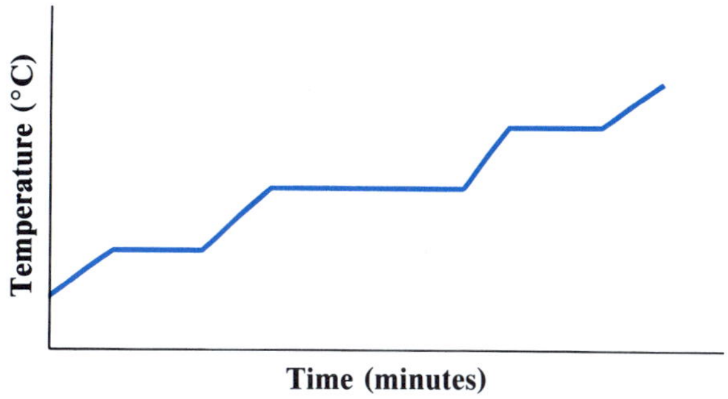
Temperature (°C)
Time (minutes)

FIGURE 9.5
The process of distillation is based on the fact that the temperature of a material that is changing its physical state will not change until the change of state is complete. This graph of the distillation of benzene shows these "plateaus" in the temperature rise with time.

is completed. For example, the boiling point of benzene is 80.2 °C, and its heat of vaporization is 94.3 cal per gram. If a mixture containing benzene and other materials with higher boiling points is heated, the solution temperature will not rise above 80.2 °C until enough energy has been added to evaporate all of the benzene. The vaporized material is reclaimed by condensation; by changing collecting containers each time the solution temperature rises, one can separate the volatile components of a mixture (Figure 9.5).

Colloids can be separated by distillation if the colloidal particles are nonvolatile, but they lend themselves well to other methods of purification as well. As colloidal particles bump against each other, they develop static electrical charges in the same manner as a person scuffing his shoes across a carpet on a dry day. These electrical charges tend to force the colloidal particles apart, preventing "clumping" or condensation that would make them large enough to settle out of the mixture. Because they carry an electrical charge, a device called an *electrostatic precipitator* is an extremely effective method of removing colloidal particles from air (Figure 9.6). Air containing colloidal particles is passed through a strong electrical field formed between plates of the precipitator. The charged colloidal particles are attracted to the plates, their electrical charges are neutralized, and they clump together and fall to a collecting bin at the bottom of the precipitator.

Colloids can be separated from the solvent by neutralizing their electrical charge.

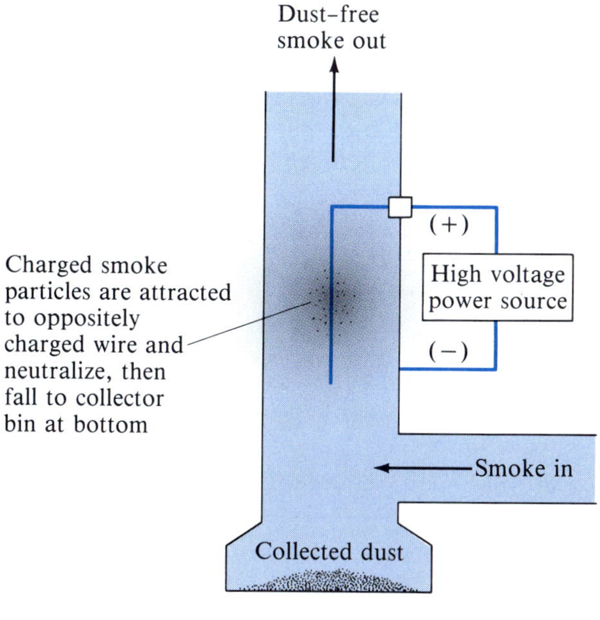
Dust-free smoke out
(+)
High voltage power source
(−)
Charged smoke particles are attracted to oppositely charged wire and neutralize, then fall to collector bin at bottom
Smoke in
Collected dust

FIGURE 9.6
An electrostatic precipitator separates colloidal particles from a gas stream by neutralizing the static electrical charges that hold the particles together. The neutral particles then fall into a collecting bin below the precipitator. Electrostatic precipitators are also used in some of the more expensive home air purifiers.

FIGURE 9.7
Ions in salt water neutralize the electrical charge of tiny colloidal soil particles carried by freshwater as the river reaches the sea. Uncharged, the soil particles no longer can remain suspended in the water, and settle to the bottom. River deltas are formed in this manner, as the river immediately deposits soil carried thousands of miles in freshwater.

Neutralization of colloidal particles is also partially responsible for the formation of river deltas (Figure 9.7). Mud consists of tiny colloidal particles of soil that are held suspended in the river partly by the electrical charges on the surfaces of the particles. When the river reaches the ocean, sodium, magnesium, and chloride ions in the salt water neutralize the charges on the colloidal mud particles, causing the mud to settle out. The Nile River delta in the Mediterranean Sea (above) and the Mississippi River delta near New Orleans are two examples of this effect.

Colloids can also be purified or separated from ionic species and small molecules by a process called **dialysis**. This process is illustrated in Figure 9.8. The colloidal mixture, along with smaller impurity molecules, is placed in a bag with very small pores. Cellophane would be an example of this type of material. This bag is placed in a container through which water or some other pure solvent passes. The ions and small impurity molecules diffuse

Dialysis is separation of colloidal particles by diffusion of smaller solvent molecules through a porous filter.

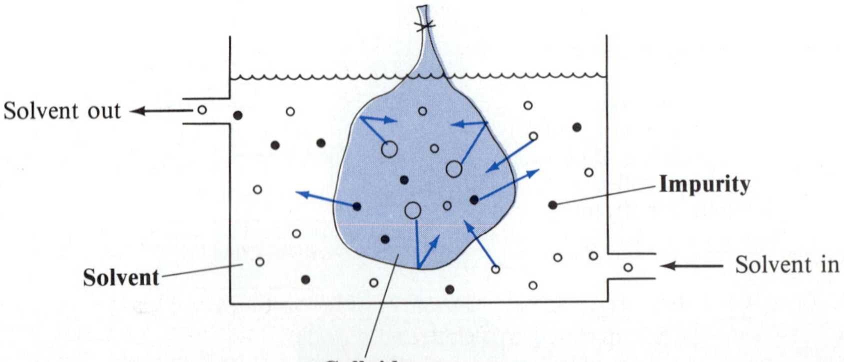
Solvent out
Impurity
Solvent
Solvent in
Colloid

FIGURE 9.8
Dialysis is a process in which colloidal particles are held within a permeable membrane, while ions and small molecules are free to diffuse through the pores of the membrane. By continually supplying a clean solution outside the membrane, the diffusion of unwanted molecules is always from the inside to the outside of the membrane. A kidney dialysis machine works by this principle.

through the pores in the bag into the external solution, but the colloidal particles are too large and remain in the bag. Eventually, the colloidal material is all that is left in the bag with solvent molecules. A kidney dialysis machine works in much the same way. Blood is passed through a tube that is immersed in a solution of the same ionic content as blood. Impurities in the blood diffuse through the dialysis membrane into the external solution, which is continually being replaced. Since the external solution has the same ionic content as blood, any diffusion of this solution into the blood causes no problem. (For more on dialysis, see Chapter 30.)

NEW TERMS

solution A mixture of two or more substances (solvent and solute) that is clear and uniform throughout. Molecules or ions of solute are dispersed evenly throughout the mixture.

solute The material that is dissolved in the solvent.

solvent A pure substance in which another material (the solute) is dissolved.

colloid A mixture in which the solute is evenly distributed, but present in particles of sufficient size to reflect or scatter light.

Brownian motion Constant, random motion characteristic of colloidally suspended particles.

suspension Mixtures in which the solute particles are large enough that gravitational force will eventually cause them to separate.

filtration The process of separating a suspended material from a mixture by passing it through a filter material.

distillation A process in which the solution is heated until the material of lower boiling point boils, leaves the solution, and is trapped and condensed.

dialysis Separation of colloidal materials from smaller ionic and molecular impurities by diffusion of impurity ions or molecules through a membrane that will not pass the colloidal particles.

TESTING YOURSELF

Solutions, Colloids, and Suspensions

1. What characteristic of the dissolved particles distinguishes solutions from colloids?
2. Place the terms *solution*, *colloid*, and *suspension* according to their relative particle size on the scale below.

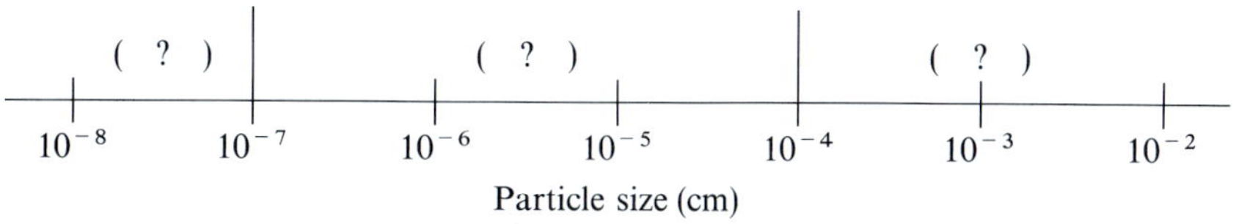
(?)
(?)
(?)
10^{-8}
10^{-7}
10^{-6}
10^{-5}
10^{-4}
10^{-3}
10^{-2}
Particle size (cm)

3. Even when they are extremely dilute, colloidal particles will scatter _____.
4. Can colloids be separated by filtration?
5. Can suspensions be separated by filtration?

6. Electrostatic precipitation is useful with gaseous
   a. suspensions b. colloids c. solutions
7. Does an ordinary furnace filter separate colloidal or suspended material in the air?
8. The process of dialysis works because of the difference in what characteristic of colloid versus solvent molecules?

Answers **1.** size **2.** solution, colloid, particle **3.** light **4.** no **5.** yes **6.** colloids **7.** suspended **8.** size

9.2 Solution Concentration

Solutions are a mixture of solute and solvent. The **concentration** of a solution is an expression of the ratio of solute to solvent. Solution concentration can be expressed in one of four ways: (1) weight (or mass) of solute per unit volume of solution; (2) percentage of the solution that is solute, by weight or by volume; (3) the number of moles of solute per liter of solution; or (4) the number of moles of a specific ion per liter of solution.

Concentration by Weight per Unit Volume

Concentration by weight per unit volume is a simple system to use, but it tells little about the number of molecules present.

The earth and environmental sciences often express solution concentration in terms of the weight of solute per unit weight or volume of solution. For example, the Pacific Ocean contains 35.9 g of dissolved salt in every 1000 g of seawater. Air pollutants are often measured in terms of milligrams or micrograms of pollutant per liter or cubic meter of air.

Concentration by Percentage

The medical and biological sciences often express concentration of solute as a percentage by weight or volume of the total solution. If a solution consists of a liquid solvent and a solid solute, the concentration can be expressed in *weight percent*, as seen in the following three examples.

EXAMPLE 9.1

A solution of sodium bicarbonate ($NaHCO_3$) was prepared by dissolving 5.00 g in 100 g of water. What is the concentration of sodium bicarbonate in weight percent?

$$\text{Weight percent} = \frac{\text{Grams of solute}}{\text{Grams of solute} + \text{Grams of solvent}} \times 100$$

$$= \frac{5.00\text{ g NaHCO}_3}{5.00\text{ g NaHCO}_3 + 100\text{ g H}_2\text{O}} \times 100$$

$$= \boxed{4.76\%}$$

EXAMPLE 9.2

A potassium chloride (KCl) solution is 5% by weight potassium chloride. What is the composition of 50 g of this solution?

$$\text{Weight of KCl} = (0.05)(50\text{ g}) = \boxed{2.5\text{ g}}$$

$$\text{Weight of } H_2O = 50\text{ g} - 2.5\text{ g} = \boxed{47.5\text{ g}}$$

EXAMPLE 9.3

An *isotonic* saline solution is 0.9% sodium chloride (NaCl) by weight. Such solutions are commonly used in medicine as a medium for intravenous feeding of nutrients or drugs. As we will see later in this chapter, the ionic concentration of this solution is the same as that of blood. To introduce solutions of greater or lesser ionic concentration causes the blood cells to either rupture or dehydrate. If 1000 g of isotonic saline solution were prepared, what would its composition be?

$$\text{Weight of NaCl} = (0.009)(1000\text{ g}) = \boxed{9.0\text{ g}}$$

$$\text{Weight of } H_2O = 1000\text{ g} - 9.0\text{ g} = \boxed{991\text{ g}}$$

If both solvent and solute are liquids, the concentration is often expressed as *volume percent*. For example, rubbing alcohol is a solution of water and isopropyl alcohol. It normally is about 70% alcohol by volume. To make 1000 mL of rubbing alcohol, one would mix 700 mL of isopropyl alcohol and 300 mL of water. Volume percent is a convenient measure, since only calibrated glassware is required for preparation of solutions. For this reason it is a common unit in hospitals.

If both components of a mixture are liquids, percent by volume is a simple concentration unit to use.

EXAMPLE 9.4

Suppose that you had 500 mL of isopropyl alcohol and needed to prepare a solution of 70% alcohol to use as rubbing alcohol. How much water would you add?

Since the volume of alcohol equals 0.7 times the total volume of the solution, then

$$\frac{\text{Volume of alcohol}}{0.7} = \text{Total volume}$$

$$\frac{500\text{ mL}}{0.7} = 714\text{ mL}$$

And since the volume of the alcohol plus the water equals the total volume, then

$$\text{Total volume} - \text{Volume of alcohol} = \text{Volume of } H_2O$$

$$714\text{ mL} - 500\text{ mL} = \boxed{214\text{ mL}}$$

When two different liquids are mixed, the resulting volume is sometimes slightly less than the sum of the component volumes. It is a little like pouring marbles and golf balls together. The marbles will fill in the spaces between the golf balls. Small molecules will often pack in among the larger molecules.

Weight/volume percent is another common percentage concentration unit, but it is misnamed because mass and volume units do not cancel to give a true percentage. This unit is used when one unit is a solid and one a liquid. It is commonly encountered in medicine because of its ease of use, as shown in the following example.

EXAMPLE 9.5

A nurse is asked to prepare a 2% boric acid solution that will be used as a mild disinfectant. She uses 4.0 g of boric acid. What volume of water should she use to dissolve this material to make a solution that has a weight/volume percent of 2%?

$$\text{Weight of boric acid} = 0.02 \times \text{Volume of } H_2O$$

$$\frac{4.0 \text{ g boric acid}}{0.02} = 200 \text{ mL } H_2O$$

Concentration by percentage is a tricky unit, since one must be sure to specify whether percent by weight, percent by volume, or percent by weight/volume is intended. Unfortunately, too often the concentration is simply expressed as "percent."

Concentration by Moles

Although expressing concentration by weight per unit volume or by percentage is extremely convenient, from a chemist's standpoint both are poor systems because they do not tell anything about the number of moles of solute present. To calculate the number of moles, an additional mathematical step is necessary. For example, a solution that is 10% by weight sodium chloride would contain 100 g NaCl per 1000 g of solution. This corresponds to

$$(100 \text{ g NaCl})\left(\frac{1.0 \text{ mol NaCl}}{58.45 \text{ g NaCl}}\right) = 1.71 \text{ mol NaCl}$$

Likewise, a solution that was 10% by weight sucrose would contain 100 g $C_{12}H_{22}O_{11}$ per 1000 g of solution. This corresponds to

$$(100 \text{ g sucrose})\left(\frac{1.0 \text{ mol sucrose}}{342 \text{ g sucrose}}\right) = 0.292 \text{ mol sucrose}$$

Although the two solutions have equal concentrations in terms of weight percent, they have greatly different numbers of reactable ions or molecules present.

Concentration by moles per unit volume, or molarity, gives the number of molecules present. It is the preferred unit in chemistry.

The unit of concentration most commonly used in chemistry is moles of solute per liter of total solution. This unit is called **molarity**, appreviated *M*. Preparation of a solution of known molarity involves four steps:

1. The number of grams of material corresponding to the desired number of moles is computed.
2. The solute is weighed.
3. The weighed solute is placed in a 1000-mL volumetric flask, and enough water is added to dissolve the solute.
4. Additional water is added to bring the volume to exactly 1000 mL.

This procedure is illustrated in Figure 9.9. Calculations of molar concentration are shown in the following three examples.

EXAMPLE 9.6

Suppose that you wished to prepare 1 L of a solution of sodium chloride, with a concentration of 0.300 mol/L. How much sodium chloride would be required?

$$\begin{aligned} 1\ \text{Na} \ @\ 22.99\ \text{g/mol} &= 22.99\ \text{g/mol} \\ 1\ \text{Cl} \ @\ 35.45\ \text{g/mol} &= \underline{35.45\ \text{g/mol}} \\ \text{NaCl} &= 58.44\ \text{g/mol} \end{aligned}$$

$$\text{Conversion factor} = \frac{58.44\ \text{g NaCl}}{1.00\ \text{mol NaCl}} = 1$$

$$(0.300\ \cancel{\text{mol}}\ \text{NaCl})\left(\frac{58.44\ \text{g NaCl}}{1.00\ \cancel{\text{mol}}\ \text{NaCl}}\right) = 17.5\ \text{g NaCl}$$

The sodium chloride would then be dissolved in water and diluted to exactly 1 L.

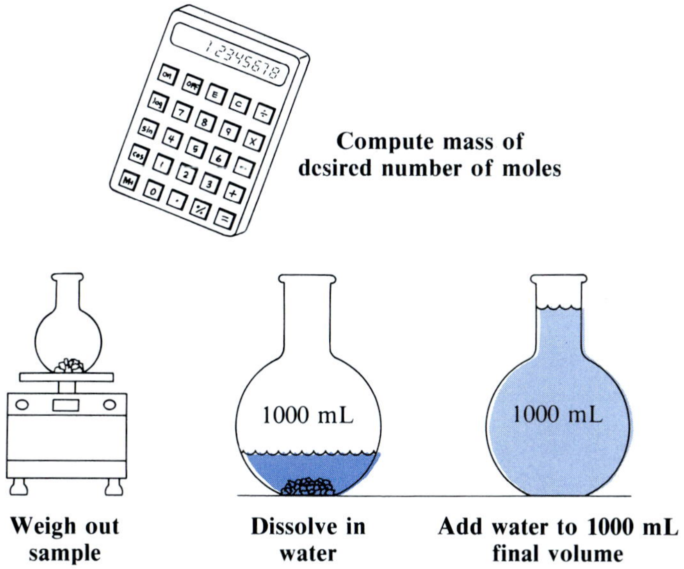
Compute mass of
desired number of moles
1000 mL
1000 mL
Weigh out sample
Dissolve in water
Add water to 1000 mL final volume

FIGURE 9.9
Steps in preparation of a solution of known molarity.

EXAMPLE 9.7

Find the molar concentration of a solution prepared by dissolving 50 g of sodium bicarbonate ($NaHCO_3$) or baking soda in water and diluting to a final volume of 1000 mL.

a. Find the number of moles of solute present.

$$\begin{aligned} 1\ \text{Na} @ 22.99\ \text{g/mol} &= 22.99\ \text{g/mol} \\ 1\ \text{H} @ 1.00\ \text{g/mol} &= 1.00\ \text{g/mol} \\ 1\ \text{C} @ 12.01\ \text{g/mol} &= 12.01\ \text{g/mol} \\ 3\ \text{O} @ 15.99\ \text{g/mol} &= \underline{47.97\ \text{g/mol}} \\ NaHCO_3 &= 83.97\ \text{g/mol} \end{aligned}$$

$$\text{Conversion factor: } \frac{1.00\ \text{mol}\ NaHCO_3}{83.97\ \text{g}\ NaHCO_3} = 1$$

$$(50.0\ \cancel{\text{g}}\ NaHCO_3)\left(\frac{1.00\ \text{mol}\ NaHCO_3}{83.97\ \cancel{\text{g}}\ NaHCO_3}\right) = 0.595\ \text{mol}\ NaHCO_3$$

b. Find the number of moles per liter of solution.

$$0.595\ \text{mol}/1000\ \text{mL} = 0.595\ M\ \text{concentration}$$

EXAMPLE 9.8

Find the molar concentration of a solution prepared by dissolving 30.0 g of acetic acid ($HC_2H_3O_2$) in water and diluting to a final volume of 1000 mL.

a. Find the number of moles of solute present.

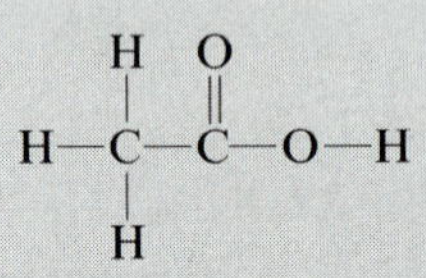
H O
H—C—C—O—H
H

$$\begin{aligned} 4\ \text{H} @ 1.00\ \text{g/mol} &= 4.00\ \text{g/mol} \\ 2\ \text{C} @ 12.01\ \text{g/mol} &= 24.02\ \text{g/mol} \\ 2\ \text{O} @ 15.99\ \text{g/mol} &= \underline{31.98\ \text{g/mol}} \\ HC_2H_3O_2 &= 60.00\ \text{g/mol} \end{aligned}$$

$$\text{Conversion factor: } \frac{1.00\ \text{mol}\ HC_2H_3O_2}{60.0\ \text{g}\ HC_2H_3O_2} = 1$$

$$(30.0\ \cancel{\text{g}}\ HC_2H_3O_2)\left(\frac{1.00\ \text{mol}\ HC_2H_3O_2}{60.0\ \cancel{\text{g}}\ HC_2H_3O_2}\right) = 0.500\ \text{mol}\ HC_2H_3O_2$$

b. Find the number of moles per liter solution.

$$0.500\ \text{mol}/1000\ \text{mL} = 0.500\ M\ \text{concentration}$$

NEW TERMS

concentration A measure of the amount of solute per unit of solution, expressed as mass per unit volume, weight percent, volume percent, or moles per liter.

molarity The concentration of a solution in moles of solute per liter of solution. The preferred unit of concentration in chemistry.

TESTING YOURSELF

Solution Concentration

1. How many grams of sucrose would be needed to prepare 500 g of a solution that was 10% by weight sucrose? How many grams of water would you use to prepare this solution?
2. Suppose that you wished to prepare 1000 g of a 0.9% solution by weight of sodium chloride in water. How much salt and how much water would you use?
3. Compute the number of grams of solute required to prepare 1 L of each of the following materials:
   a. 1.0 *M* NaOH solution
   b. 0.5000 *M* KCl solution
   c. 0.40 *M* acetic acid, $HC_2H_3O_2$
4. What would be the concentration in moles per liter of the following solutions?
   a. 1.00 L of a solution containing 4.0 g NaOH
   b. 1.00 L of a solution containing 33.8 g silver nitrate, $AgNO_3$
   c. 500 mL of a solution containing 63 g HNO_3
   d. 200 mL of a solution containing 9.8 g phosphoric acid, H_3PO_4

Answers **1.** 50 g sucrose, 450 g water **2.** 9.0 g salt, 991 g water **3. a.** 40 g **b.** 37.28 g **c.** 24 g **4. a.** 0.10 *M* **b.** 0.199 *M* **c.** 2.0 *M* **d.** 0.50 *M*

9.3 Colligative Properties of Solutions

Colligative properties depend only on the number of solute particles dissolved in a solution, not upon their identity.

Colligative properties are those properties of a solution that depend only on the concentration of solute particles, not on their chemical identity. The change in boiling and freezing points of a solution caused by the presence of nonvolatile solutes (those that will not evaporate from the solution) are two examples of these properties. Osmotic pressure is also an example of a colligative property.

Change in Boiling Point

Boiling and freezing points of a pure solvent are changed by the presence of an impurity.

When a solute is added to a pure solvent, the solute particles interfere with the ability of the solvent molecules to escape from the liquid. The amount of energy required to push the solvent molecules out of the solution is increased, and the boiling point increases. This increase in boiling point is directly related to the *concentration* of solute particles, not to their chemical identity.

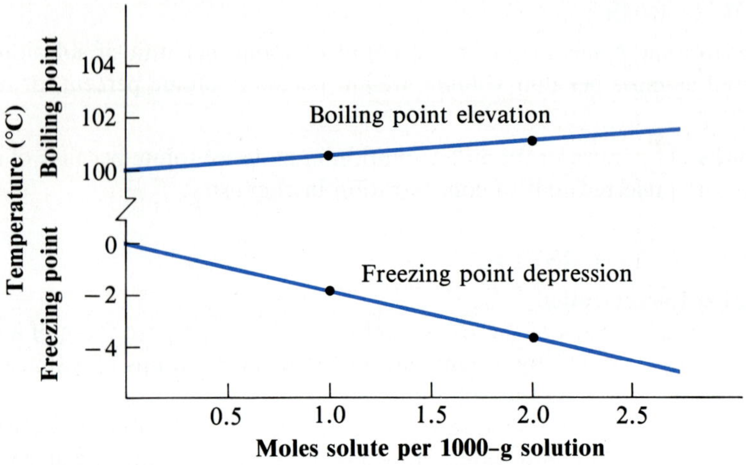
Temperature (°C)
Boiling point
Freezing point
104
102
100
0
−2
−4
Boiling point elevation
Freezing point depression
0.5
1.0
1.5
2.0
2.5
Moles solute per 1000-g solution

FIGURE 9.10
Both the freezing and boiling points of water are changed by the presence of solute particles. The presence of antifreeze (ethylene glycol) mixed with water in an automobile radiator both raises the boiling point (a plus in summer when driving in hot weather) and lowers the freezing point to protect the automobile cooling system in cold weather.

Change in Freezing Point

Likewise, the presence of solute particles within a solution interferes with the tendency of solvent molecules to line up in an organized manner, which is necessary for the solvent to freeze. Thus, a lower temperature is necessary before the solvent molecules will form a solid. Again, this effect is dependent on the concentration of solute particles, not on their chemical identity. The effect of solute concentration on the freezing and boiling points of water is illustrated in Figure 9.10.

Although the effect of solute particles on the freezing and boiling points of water is not related to the chemical identity of the solute, the presence of the solute may change the chemical reactivity of the resulting solution. Ethylene glycol is used as an automobile antifreeze for several reasons. Its boiling point of 197 °C ensures that it will not distill out of the radiator if a vapor leak occurs. Most important, its covalent structure ensures that it will not react with the iron of the motor block at normal operating temperatures.

Ethylene glycol is a relatively heavy molecule (molecular weight = 62) and must be manufactured. Another compound used as antifreeze is ordinary salt, sodium chloride. Sodium chloride has two advantages. First, it is inexpensive, and second, although its formula weight (58.44) is about the same as ethylene glycol, it ionizes when placed in water. Thus, you get twice as many particles from a mole of sodium chloride than from a mole of ethylene glycol, and the depression of freezing point is twice as great for the same number of moles of sodium chloride as for ethylene glycol (Figure 9.11). For these reasons, sodium chloride is the preferred antifreeze to spread on icy roads. As we saw in Chapter 7, however, sodium chloride does not have the nonreactive nature of ethylene glycol. Those who live in the northeast, midwest, and mountain areas of the country where salt is used on roads are painfully aware of the corrosive action of sodium chloride on metal. This is why it is not used in automobile radiators.

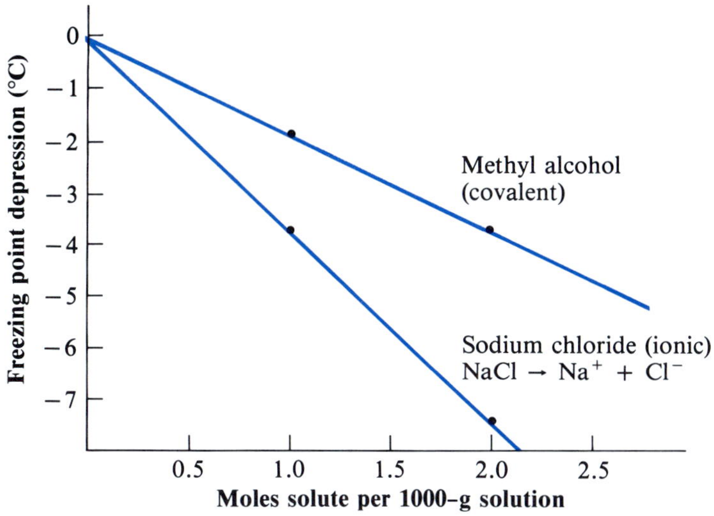
0
−1
−2
−3
−4
−5
−6
−7
Freezing point depression (°C)
Methyl alcohol
(covalent)
Sodium chloride (ionic)
NaCl → Na⁺ + Cl⁻
0.5
1.0
1.5
2.0
2.5
Moles solute per 1000-g solution

FIGURE 9.11
Since the reduction of freezing point caused by a solute is determined only by the number of particles present, a compound that will ionize is far more effective at freezing point reduction than a covalent compound. This is one of the principal factors responsible for the fact that coastal waterways do not freeze as rapidly as do inland (freshwater) lakes and streams.

Osmotic Pressure

Osmosis is the process by which solutions pass small molecules through a membrane.

Osmosis is a process by which two solutions separated by a semipermeable membrane pass small molecules through the membrane to equalize the pressure their solvent exerts on each side of the membrane. Osmosis is extremely important to biological processes, since it is thc way by which water and some nutrients are passed into cells. (Many nutrients are passed through the membrane by another process called *active transport*, in which other molecules assist their passage through the membrane. Active transport is further discussed in Chapter 24.)

The process of osmosis can be understood most easily by considering the greatly magnified view of a semipermeable membrane such as a rubber balloon, a plant cell wall, or cellophane (Figure 9.12). Small holes or pores in the membrane permit small solvent molecules to move back and forth, but do not permit the large solute molecules to pass.

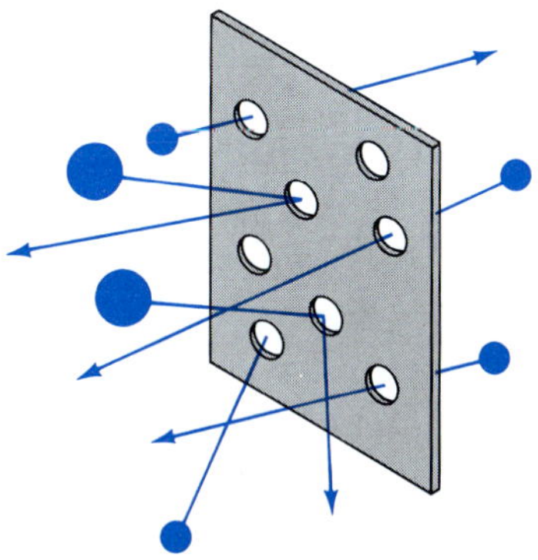

FIGURE 9.12
A semipermeable membrane permits small solvent molecules to pass relatively freely from side to side, but will not allow larger solute molecules to pass.

Consider now the U-shaped tubes illustrated in Figure 9.13. In part (a), more water has been placed on one side of the tube than the other. Since the pressure exerted on the membrane is greater on the right (high pressure) side than on the left side, more water molecules move from right to left than from left to right. Gradually, water molecules accumulate on the left until the water column heights are equal, and the pressure exerted on both sides of the membrane is the same, as in part (b). Water molecules continue to pass back and forth through the membrane, but at the same rate to the left and to the right. This situation of equal but opposing motion is called *equilibrium* (more about this in Chapter 10).

In part (c), large solute molecules have been added to the right hand column of the U-shaped tube. Although the water column heights are equal at this point, and there should be an equal number of molecule–barrier collisions from each side, most of the collisions on the left (solvent only) side

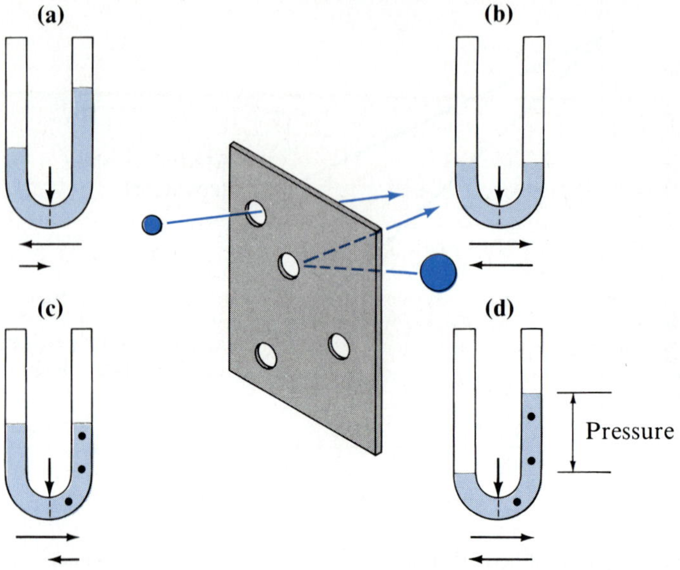
(a)
(b)
(c)
(d)
Pressure

FIGURE 9.13
A demonstration of osmotic pressure. A semipermeable membrane is located at the bottom of the U. (a) More water is in the right tube than in the left. The higher pressure on the right side will force more water molecules to move from right to left than from left to right. (b) Soon, enough water molecules move across the barrier to produce equilibrium. (c) Large solute molecules have been added to the right tube. Now more water molecules move from left to right. (d) The pressure head causes equal motion in the reverse situation.

result in passage of a solvent molecule through the barrier. However, some of the collisions on the right side are due to solvent molecules and some are due to large solute molecules. Only those collisions of solvent molecules result in transfer through the barrier. The rate of solvent molecule motion is greater to the right—from pure solvent to solution—than from solution to pure solvent. As water molecules accumulate on the right, the column gradually rises, and the pressure increases on the water molecules attempting right-to-left movement. Eventually enough additional water molecules accumulate on the solute side to equalize the pressure on the two sides of the barrier. The water molecules then pass at equal rates in both directions, and an equilibrium situation results.

Osmotic pressure is related to the difference in concentration of solute molecules on opposite sides of a permeable membrane.

The difference in column height or pressure necessary to equalize the flow of solvent molecules in both directions is called the **osmotic pressure**. This pressure is related *only* to the concentration of solute molecules present, not to the chemical identity of the solute. Thus, it is a colligative property.

The movement of water up the stem of a plant is a good example of osmosis. Some of the force that moves water from root to leaf originates from the cells of the plant's root system. These cells have a high concentration of nonvolatile solute molecules. Water thus moves preferentially through the root membrane from the soil to the root system. The increased pressure within the root system forces water up the stem of the plant or trunk. Transpiration at the leaf also plays an important part in water movement. As water transpires from leaf cells, the solution concentration increases within the cell. This causes water to be pulled from the lower part of the plant through a system of small tubes. The osmotic pressure of a plant system is high—

OSMOSIS: A BACTERIA KILLER

Osmosis is important to human and microbiological systems in some ways you may not expect. For example, aftershave lotion is more than just something to make your face smell good. Shaving often causes minor cuts, which can become infected by the bacteria on the surface of the skin. Aftershave lotions contain alcohol. The presence of the alcohol outside of the bacterial cells results in movement of water from inside the cells toward the outside to equalize the solute concentration on each side of the cell wall. The resulting removal of water dehydrates the bacteria and causes it to die. The same principle applies in the use of salt as a preservative for meat, which was used long before refrigeration was invented. The high salt concentration outside of the bacterial cells on the meat likewise causes migration of water from inside the cells toward the outside, again causing dehydration and death of the bacteria.

enough to lift water from the ground to the top of a tall tree. The rate of transport is also rapid—often as much as 40 m per hour. This is about the same speed as the tip of a sweep second hand on a classroom clock. The importance of osmotic transport across the root membrane becomes clear when you consider that a four-month-old rye plant can have as much as 10,000 km of roots.

Osmotic pressure is also a factor in the behavior of blood cells. Blood cells contain a fairly high concentration of glucose and ionic salts—sodium and potassium ions, bicarbonate ions, chloride ions, and others. If a blood cell is placed in water, water molecules will attempt to migrate in through the cell wall until the concentration of solute particles inside the cell is equal to that on the outside of the cell. This massive infusion of water molecules into a cell causes it to rupture because it cannot stretch to contain the additional water. This process of rupturing is called **hemolysis** (Figure 9.14).

The process by which cells burst when too many water molecules infuse through the cell wall is called *hemolysis.*

In contrast, if blood cells are placed in a solution of greater salt content than their internal environment, migration of water molecules will occur from inside the cell to the more highly concentrated solution outside the cell. The cell will become dehydrated and shrink, a condition known as **crenation**.

If water molecules migrate outside of a cell to a region of higher salt concentration, the resulting dehydration of the cell is called *crenation.*

For this reason, it is important that solutions injected intravenously have the same ionic concentration as blood plasma. An antibiotic or glucose solution used intravenously must also contain salts to bring the ionic concentration up to the point where it is equal to or **isotonic** with blood. As we saw in Example 9.3, an isotonic saline solution is 0.9% by weight NaCl. A **hypotonic** solution has a lower solute (salt) concentration than blood, while a **hypertonic** solution has a higher salt concentration than blood.

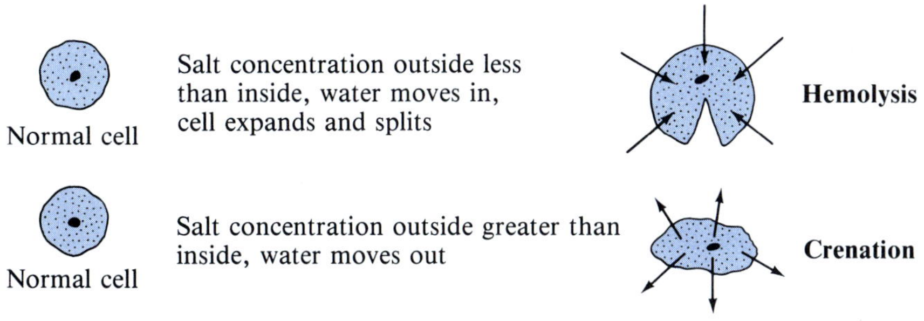

Normal cell
Salt concentration outside less than inside, water moves in, cell expands and splits
Hemolysis
Normal cell
Salt concentration outside greater than inside, water moves out
Crenation

FIGURE 9.14
It is important that blood cells remain in a solution in which the concentration of solute particles is approximately the same as the environment inside the cell. If the cell fluid is more concentrated than the external solution, water will enter the cell faster than it leaves, ultimately causing the cell to stretch and rupture (top). If the surrounding solution is more concentrated than the cell fluid, water will leave the cell, causing it to become dehydrated or crenated (bottom).

NEW TERMS

colligative properties Properties of solutions that are affected only by the concentration of the solute, not by its chemical identity.

osmosis A process in which two solutions separated by a semipermeable membrane attempt to equalize their concentrations by passing small molecules through the membrane. Biological systems use osmosis to pass water and some nutrients between cells.

osmotic pressure A measure of the pressure driving the transfer of solvent molecules across a membrane from a solution of low solute concentration to a solution of high solute concentration. Osmotic pressure is the pressure that would have to be applied to the solution on the more concentrated side of the membrane to equalize the flow of solvent molecules in both directions.

hemolysis The swelling and rupturing of a cell that occurs when blood cells are placed in a solution of less than isotonic concentration, and water moves in through the cell membrane.

crenation The shrinkage and dehydration of a cell that occurs when blood cells are placed in a solution of more than isotonic concentration, and water moves out through the cell membrane.

isotonic As applied to blood, a solution that has an ionic concentration equal to that of blood.

hypotonic As applied to blood, a solution that has an ionic concentration less than that of blood.

hypertonic As applied to blood, a solution that has an ionic concentration greater than that of blood.

TESTING YOURSELF

Colligative Properties of Solutions

1. Does the presence of dissolved salts cause seawater to boil at a higher or lower temperature than freshwater?
2. A 100-g bag of which of the following salts would do the best job of melting ice on a sidewalk?

$$\text{NaCl:}\quad \text{Formula weight} = 58.45\ \text{g/mol}$$

$$\text{KCl:}\quad \text{Formula weight} = 74.55\ \text{g/mol}$$

3. Ethylene glycol has about the same molecular weight (62) as sodium chloride. If you had 100 g of each, which would lower the freezing point of 100 g of water the most?
4. For osmosis to occur, must the solute molecules be larger or smaller than the solvent molecules?

Answers **1.** higher **2.** NaCl, more particles/gram **3.** NaCl, because it ionizes to give twice as many particles **4.** larger

SUMMARY

Because of water's polar properties and because it is a liquid at the common temperatures found on earth, water is the solvent of preference for all biological systems.

Solutions are mixtures in which one substance (the *solute*) is dissolved in another (the *solvent*). True solutions are homogeneous mixtures, the molecules or ions of solute being distributed evenly through the solvent. Solutes that clump into particles larger than molecules but that are kept in suspension by thermal motion of the solvent molecules are said to be in the *colloidal state.* Larger clumps of solute that settle on standing form *suspensions.*

The presence of solute particles in a solution is quantified by *concentration,* which tells us the relative amounts of solute and solvent in a solution. There are several ways of expressing concentration. Concentration by weight (or mass) per unit volume is a common unit used in the biological and environmental sciences. Air pollutants are measured in terms of the number of milligrams or micrograms of pollutant per liter or cubic meter of air. The medical and biological sciences often express concentration by percent—weight percent or weight/volume percent. These concentration units are easy to use in a hospital or biological laboratory because the measurements refer only to mass or mass and volume and not to the molecular weight of the material. Because they are not compound specific, they are of little use to chemists. Concentration in moles per liter of solution, or *molarity,* is the concentration unit of preference to chemists who must know the relative amounts of each reactant present in a reaction.

Some of the behaviors of solutions are determined only by the concentration of solute particles present—not by their chemical identity. These properties, which include freezing point depression, boiling point elevation, and *osmotic pressure,* are called *colligative properties.*

The next chapter will use the concept of solution concentration to help us understand some of the factors that affect the rate of chemical reactions.

TERMS

solution
solute
solvent
colloid
Brownian motion
suspension
filtration
distillation
dialysis
concentration
molarity
colligative property
osmosis
osmotic pressure
hemolysis
crenation
isotonic
hypotonic
hypertonic

CHAPTER REVIEW

QUESTIONS

1. When a small amount of sugar is stirred into water, the resulting liquid is colorless and clear. A light beam passed through the container is not visible. What is this liquid?
   a. solution b. colloid c. suspension
2. In the situation described in question 1, the sugar is a
   a. solute b. solvent c. solution
3. Milk of Magnesia is a commonly used antacid and laxative. It often comes as a white liquid, which must be shaken well before pouring. This liquid is a
   a. solution b. colloid c. suspension
4. Milk of Magnesia is made by mixing water and magnesium hydroxide, $Mg(OH)_2$, which is a white solid at room temperature. It is converted to magnesium oxide when its temperature is elevated above 350 °C. It is practically insoluble in both cold and hot water. One could best separate $Mg(OH)_2$ from water by
   a. filtration b. distillation c. dialysis

5. A solution of sodium bicarbonate ($NaHCO_3$) is needed for use as an eye wash. Its concentration is to be 3% by weight. How much $NaHCO_3$ must be used to make up 2 L of this solution?
6. A volume of 250 mL of isopropyl alcohol is mixed with 750 mL of water. What is the percent by volume alcohol of this solution?
7. A solution was prepared by dissolving 100 g of sodium chloride in water, and diluting it to a final volume of 1000 mL. What is the concentration of this solution in moles per liter NaCl?
8. Two mugs are filled with crushed ice. One is then filled with water, the other with root beer. As the ice melts, both cool. Which will be coldest?
9. The freezing point of water is decreased 1.86 °C for each mole of solute added to 1000 g of water. Which of the following materials would lower the freezing point the most when added to 1000 g water?
   a. 10 g of methyl alcohol, CH_3OH
   b. 10 g of ethyl alcohol, CH_3CH_2OH
   c. 10 g of isopropyl alcohol, $CH_3CHOHCH_3$
   d. 10 g of sugar, $C_{12}H_{22}O_{11}$
10. An amount of 10 g of sodium chloride will lower the freezing point of 1000 g of water more than any of the materials listed in question 9. This is related to
    a. formula weight b. bond type c. density
11. Suppose that coastal flooding causes seawater to saturate the ground where plants are growing. The presence of salt water around the root system will
    a. increase the plant's ability to move water up its stalk.
    b. decrease the plant's ability to move water up its stalk.
    c. cause no change in the plant's growth.

Answers **1.** a **2.** a **3.** c **4.** a **5.** 6 g **6.** 25% **7.** 1.71 mol **8.** root beer, because of dissolved solutes **9.** methyl alcohol, because of lowest molecular weight **10.** bond type **11.** b

DIAGNOSTIC CHART

Blacken in all of the circles under the number of each question missed in the Chapter Review questions. The diagnostic chart will help you identify concept areas that need more study.

	Questions										
Concepts	**1**	**2**	**3**	**4**	**5**	**6**	**7**	**8**	**9**	**10**	**11**
Solutions, colloids, and suspensions	○	○	○								
Separation techniques				○							
Solution concentration					○	○	○				
Colligative properties								○	○	○	
Osmosis											○

EXERCISES

Solutions, Colloids, and Suspensions

1. By reference to the solubility table in Chapter 7 (Table 7.9), determine which compounds formed from the following ions would form true solutions when a small amount is mixed with water and which could be separated from the water by filtration.

		Negative Ion				
		Cl^-	NO_3^-	SO_4^{2-}	S^{2-}	OH^-
Positive Ion	Ag^+					
	Pb^{2+}					
	NH_4^+					
	Na^+					
	K^+					
	Fe^{3+}					

2. A liquid has a light purple color but appears clear. Describe a test you might use to determine if it is a true solution or a colloid.

Separation Techniques

3. Acetone is a colorless liquid that boils at about 56 °C and is soluble in water. Describe a technique by which the acetone could be separated from the water.
4. Sodium chloride is a white crystalline material that melts at 801 °C and boils at 1413 °C. It is very soluble in water. Describe a technique by which the salt could be separated from the water.
5. Concentrated vanilla flavoring contains about 35% ethyl alcohol. Vanilla is not directly soluble in water, but it is soluble in alcohol. Alcohol is soluble in water so it brings the vanilla flavor along when mixed in water. Ethyl alcohol boils at about 78.5 °C. A sample of this material is placed in a beaker, which is heated at a constant rate by a hot plate. The temperature of the material is recorded every 20 sec; these data are shown in the following table. Divide the data into natural stages and briefly explain what is happening at each stage. (HINT: You may wish to make a graph of the temperature versus time.)

Time	Solution Temperature (°C)
0 sec	20
20 sec	25
40 sec	30
1 min	35
20 sec	40
40 sec	45
2 min	50
20 sec	55
40 sec	60
3 min	65
20 sec	70
40 sec	75
4 min	78.5
20 sec	78.5
40 sec	78.5
5 min	86
20 sec	92
40 sec	98
6 min	100
20 sec	100
40 sec	100
7 min	100

Solution Concentration

6. The table below lists a number of substances. Please compute the number of grams of solute required to prepare each of the indicated solutions.
7. Vinegar is 5% by weight acetic acid, CH_3COOH. What is the acetic acid concentration in vinegar, given in moles per liter?

		Solute		
		Common Salt (NaCl)	Sugar ($C_{12}H_{22}O_{11}$)	Rubbing Alcohol ($CH_3CHOHCH_3$)
Solution	1000 g, 5% by weight			
	1000 mL, 0.1 *M*			
	500 mL, 0.5 *M*			
	1500 mL, 0.2 *M*			

8. Wine is 12% ethyl alcohol by volume.
   a. How many milliliters of alcohol are in 250 mL of wine (approximately 1 cup)?
   b. Ethyl alcohol has a density of 0.789 g/mL. How many grams of alcohol are in 1.0 L of wine?
   c. What is the concentration of alcohol in wine in moles per liter?

Colligative properties

9. Automobile radiators usually carry a sticker indicating that they must contain an antifreeze solution for proper operation, winter or summer. Why should one not use ordinary water during the summer?
10. Why can coastal waterways remain open during winter weather while nearby freshwater streams are frozen over?
11. The freezing point of water is decreased by 1.86 °C for each mole of solute dissolved in 1000 g of water. Please compute the freezing point of solutions prepared by dissolving the following amounts of solute in 1000 g of water:
12. What is the freezing point of a solution containing 50 g of sodium chloride? NaCl ionizes when placed in water: $NaCl \rightarrow Na^+ + Cl^-$.

Osmosis

13. Consider a cell in a plant leaf. This cell contains water, sugars, and large biochemical molecules that cannot easily move through the cell membrane. As heat from the sun causes water molecules to evaporate from the leaf, will the concentration of the solution within the cell increase or decrease?
14. Plant leaves remain crisp because their cells are filled with water. If a plant is not watered, and the cellular water content decreases from evaporation, the plant will wilt. Lettuce placed in water remains crisp. However, when vinegar salad dressing is placed on the lettuce, it rapidly wilts. How does solution concentration and osmosis explain this occurrence?

		Solute		
		Methyl Alcohol (CH_3OH)	Sugar ($C_{12}H_{22}O_{11}$)	Ethylene Glycol (CH_2OHCH_2OH)
Solution	50 g solute + 1000 g water			
	100 g solute + 1000 g water			
	500 g solute + 500 g water		(will not dissolve)	

15. Consider a patient who has been given an intravenous feeding of a glucose solution that is less concentrated in salts than blood. The fact that the IV solution is not isotonic with blood will harm the patient. Will blood cells tend to absorb water and split because of this condition, or will they dehydrate as water moves out of the cells and into the blood stream?
16. In the last century sailing ships often stored their meat supplies by packing them in salt and water. Thus packed, the meat would not spoil by bacterial action. The bacteria died as a result of the presence of the saltwater. Did the bacteria absorb water and explode, or did their internal cellular water move outside into the brine causing the bacteria to dehydrate?
17. Although the bacteria in milk will cause it to sour rapidly if stored at room temperature, fudge, which is made of milk and sugar, will not sour easily if stored at room temperature. The difference is the presence of a large amount of sugar. Likewise, canned fruit stored in a sugar syrup will not spoil. What does the presence of a high concentration of sugar do to the motion of water into or out of a bacterial cell?

CHAPTER

10

Energy: The Driving Force in Chemical Reactions

For molecules to react, they must first collide with enough energy to break chemical bonds. The rate of reaction is determined by how often the reacting molecules collide, the energy with which they collide, and the chance that a given collision will break the proper chemical bonds for reaction to occur. A relatively simple idea called molecular collision theory helps us predict these effects.

OBJECTIVES

After completing this chapter, you should be able to

- Define the term *activation energy.*
- From energy diagrams, identify the activation energy and useful energy produced by a chemical reaction.
- Differentiate between exothermic and endothermic reactions in terms of the energy changes involved.
- Apply the molecular collision model to predict the effect of changes in temperature and concentration on reaction rate.
- Describe the effect of increasing molecular complexity on reaction rate.
- Explain the action of a catalyst on the energy relationships involved in a chemical reaction.
- Define the term *enzyme.*
- By reference to an energy diagram of the reaction, predict the relative abundances of reactant and product molecules at equilibrium.
- By reference to the equilibrium constant, predict the predominant direction in which a reaction will proceed.

10.1 The Chemical Reaction: It Starts with a Collision

Figure 10.1 illustrates a scene familiar to campers—the lighting of a gasoline lantern. The lantern pumps a fine spray of gasoline slowly into a fabric bag, where it mixes with oxygen from the air, reacts to form carbon dioxide and water vapor, and gives off sufficient heat to excite the electrons around the metal atoms that coat the fabric. As these electrons fall back to lower energy states, they give off light—many colors of light, since heavy atoms with many electrons are used—and a bright, white light results.

The example of the lantern points to an obvious question. Why do we have to light the lantern? Why won't gasoline and air combine spontaneously to give off heat and light?

Molecular Collision Theory

Molecular collision theory is based on the idea that, to react, molecules must collide and that they must collide with enough energy to break chemical bonds.

An idea that has contributed greatly to our understanding of chemical reactions is called **molecular collision theory**. This theory is based on two simple assumptions: For a chemical reaction to take place, the reacting molecules must (1) collide and (2) collide with sufficient energy to break some chemical bonds so that new bonds can be formed (Figure 10.2). As simple as this idea seems, it helps us make many useful predictions about chemical reactions.

Gasoline molecules and oxygen molecules do not react when they mix at normal temperatures because few molecules collide with sufficient energy to break chemical bonds. However, when the mixture of gasoline and oxygen is heated with a match, a few molecules in the vicinity of the match pick

Coleman
THE Powerhouse

FIGURE 10.1
A gasoline lantern needs a little help to begin producing light.

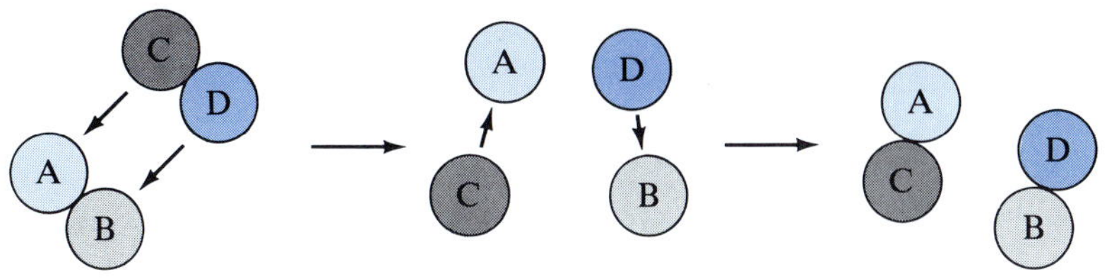

C
D
A
B
A
D
C
B
A
D
C
B

FIGURE 10.2
Two molecules must collide with sufficient energy to break chemical bonds for a chemical reaction to take place.

up sufficient energy to break chemical bonds as they collide. These molecules give off enough energy while reacting to cause bonds in adjacent molecules to break, and the reaction spreads quickly across the lantern mantle. The reaction will be self-sustaining as long as reactant molecules (gasoline and oxygen) are available.

Energy Relationships and Diagrams

Energy relationships in a chemical reaction are easy to visualize if we construct an energy diagram that correlates the total system energy with time. Such a diagram is presented in Figure 10.3. This chart shows that before this particular reaction began, the total potential energy of the system was 60 arbitrary energy units. When 15 units of energy were added to increase the molecular velocity to the point that the molecules would break chemical bonds upon collision, the total system energy was increased to 75 units. This energy added to start the reaction is called the **activation energy**. The products of the reaction have less energy than do the reactants; energy is given off as reactant molecules react together and change to product molecules. In this example, the energy retained by the product molecules was 30 energy units; 45 units of energy were given off during the reaction. If 15 of these units were used to break bonds in an adjacent set of molecules to start them reacting, 30 units would be available to do useful work.

Activation energy is the additional energy required to break chemical bonds.

Since it gave off energy, this reaction is called **exothermic**. Once exothermic reactions are started, they will sustain themselves if the energy released exceeds the activation energy necessary to start and if the heat is

Exothermic reactions release energy.

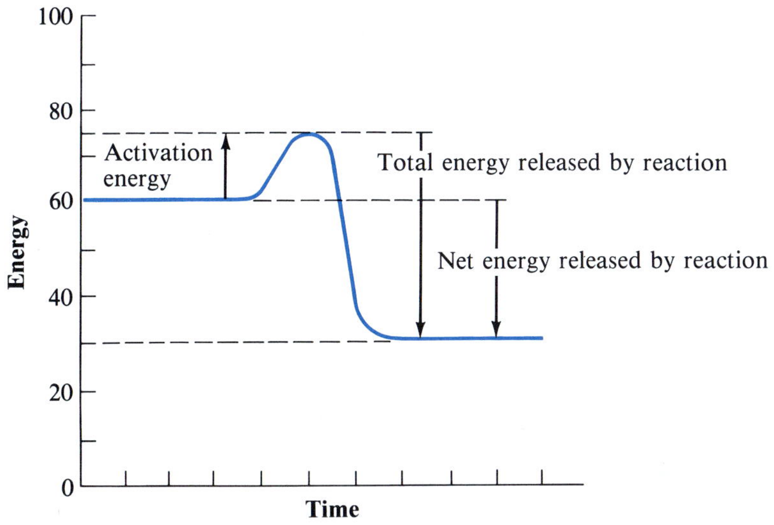

100
80
60
40
20
0
Energy
Time
Activation energy
Total energy released by reaction
Net energy released by reaction

FIGURE 10.3
The energy relationships in a chemical reaction can be conveniently presented on an energy graph. This example shows an exothermic reaction, which gives off more energy than is required to start it.

ROCKETS

Rockets accelerate by firing light weight, high speed molecules out of their motors. The criteria used in choosing rocket fuels are based on common sense. First, a large difference must exist between the energy of the reactant fuel molecules and the product exhaust molecules. The larger this difference, the more exothermic the reaction and the more velocity will be imparted to the departing product molecules. Second, the reactants must be reasonably convenient to handle, and third, it must be possible to shut off the supply of reactants to control the rocket motor.

The reaction between oxygen and hydrogen to produce water has an extremely large difference between the energy of the reactants and the energy of the product water molecules, and it produces gaseous water molecules as a product. The reaction looks like it might be a good candidate for a rocket motor.

$$2\,H_2 + 2\,O_2 \longrightarrow 2\,H_2O$$

Hydrogen and oxygen, however, have very low boiling points, and must be stored at extremely low temperatures to keep them in the liquid state (oxygen boils at $-180\,°C$, hydrogen at $-253\,°C$). Large insulated tanks are required to store these fuels. For this reason, liquid hydrogen and liquid oxygen are used only with large rockets where space is available and large amounts of energy are required. The space shuttle main engines use liquid hydrogen and liquid oxygen.

When smaller amounts of controlled thrust are needed, less energy is demanded from the reaction, and convenience factors become important. The attitude control thrusters on the space shuttle, for example, use methyl hydrazine as a fuel and nitrogen dioxide as an oxidizer. The products of this reaction are hydrogen, nitrogen, carbon dioxide, and ammonia:

$$\underset{\text{Liquids}}{N_2O_4 + 4\,N_2H_4} \longrightarrow \underset{\text{Gases}}{2\,NH_3 + 4\,N_2 + H_2 + 4\,H_2O}$$

Both reactants are liquids at normal temperatures and can be pumped into the combustion chamber in controlled amounts. The two reactants spontaneously react when mixed, so the combustion chamber does not require an igniter. The system is very simple and reliable and can be turned on and off as required to maintain the spacecraft's position.

Solid fuel rockets use solids for fuel and oxidizer, which are mixed and fill the inside of the rocket engine. Once ignited, a solid fuel rocket will burn until its reactants are gone. There is no control once the combustion has begun. On the other hand, these fuels are easily stored for extended periods of time. Solid fuels are used when long term storage is important and when it is not necessary to change the rocket motor's reaction rate once it has been ignited. The booster rockets for the space shuttle used solid fuel rockets, as do military ICBM rockets and air-to-air missiles. Propellant is checked only once every several years in military solid fuel rockets.

contained in the reaction area. You can put out a match by blowing the hot molecules away from the fuel. The rate of a reaction is related to the amount of energy produced during the reaction. If much energy is produced, a large number of adjacent molecules will have their bonds broken and will react, and the reaction will spread rapidly. On the other hand, if each set of molecules only releases enough energy to start one additional set of molecules reacting, the reaction will proceed slowly.

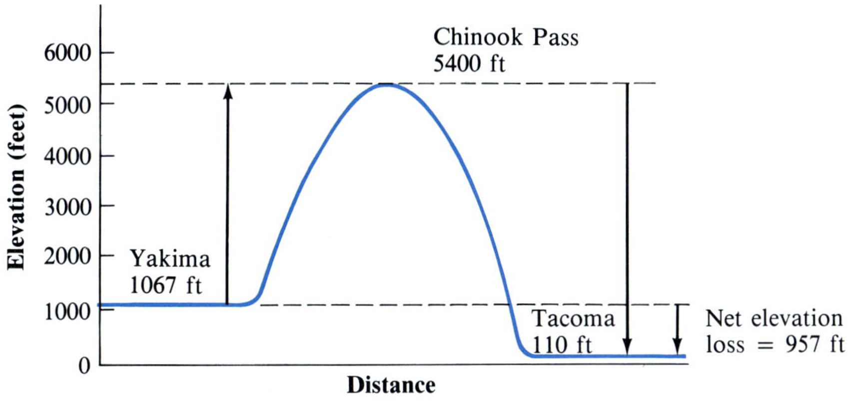

Chinook Pass
5400 ft
6000
5000
4000
3000
2000
1000
0
Elevation (feet)
Yakima
1067 ft
Tacoma
110 ft
Net elevation
loss = 957 ft
Distance

FIGURE 10.4
Highway routes are similar to chemical reactions in their energy relationships. If one could eliminate friction, an automobile would ultimately give off energy on a trip from Yakima to Tacoma, Washington, but energy must first be added to get over Chinook pass. Note that more energy is given off on the downhill side than is required on the uphill side. This might be considered an "exothermic" drive.

Some reactions do not release enough energy to break the bonds of adjacent reactant molecules. These reactions—for example, frying an egg—require continual input of energy to continue reacting. Such reactions are called **endothermic**.

Endothermic reactions require continual input of energy to maintain reaction.

To better understand the reaction energy diagram, let's look at an analogous situation in transportation. Consider, for example, the route between Yakima and Tacoma, Washington. The elevation of Yakima is 1067 ft above sea level and that of Tacoma, 110 ft. This appears to be a simple downhill trip, in which energy would be released. However, the highway crosses the Cascade Mountains, a major mountain range, at Chinook pass near Mount Rainier, having an elevation of 5400 ft. A great deal of energy (analogous to activation energy) must be added to reach the pass before energy can be reclaimed on the downhill path (Figure 10.4).

Another route through the mountains goes north from Yakima, crossing the Cascades at Snoqualmie pass, elevation 3010 ft. Although longer, this route has some advantages. Would you expect it to have any effect on the *net* elevation change experienced by an automobile making the trip? No, the overall elevation change is the same: 1067 ft to 110 ft, a change of 957 ft. Is the activation energy affected? Which route would probably be easier to drive? We will consider these questions again when we discuss catalyzed reactions later in the chapter.

Although activation energy is a property of all chemical reactions, the activation energy required by some reactions is so low that the simple movement of molecules at room temperature is enough to overcome the activation energy barrier. For example, the reaction by which milk sours has a low enough activation energy to proceed at a reasonably rapid rate at room temperature. Refrigerating the milk slows up the molecular motion to the point that the number of molecules having sufficient energy to break chemical bonds as they collide is very small, and the milk can be stored much longer before spoiling.

For a chemical reaction to occur, the reactant molecules must collide. This collision must be of sufficient energy to provide the activation energy required to break chemical bonds. Anything that will increase the number of effective collisions per unit time will cause a chemical reaction to speed up. Conversely, anything that will decrease the number of effective collisions per unit time will cause the reaction to slow down. In Sections 10.2 and 10.3, we will consider four factors that affect the rate of a chemical reaction: (1) reactant concentration, (2) temperature, (3) the presence of a catalyst, and (4) the effect of molecular complexity.

NEW TERMS

molecular collision theory The idea that for a chemical reaction to occur, molecules must (a) collide and (b) collide with sufficient energy to break chemical bonds. Changes in reaction rate will occur if a change is made in any factor that determines collision rate or collision energy.

activation energy The energy added to molecules so that they will collide with sufficient energy to break chemical bonds.

exothermic A chemical reaction that gives off energy.

endothermic A chemical reaction that requires a continual input of energy to keep going.

TESTING YOURSELF

Energy Relationships

1. In the following energy diagram, identify the three arrows as:
   a. energy released by the reaction
   b. activation energy
   c. useful energy available from the reaction

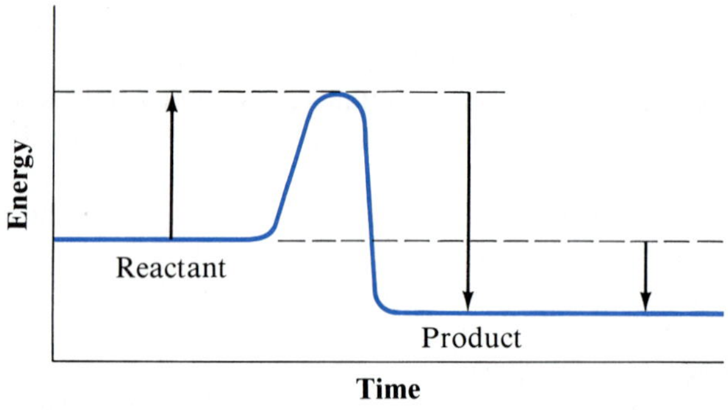
Energy
Reactant
Product
Time

2. Would the reaction sketched in question 1 be endothermic or exothermic?

Answers **1.** from left to right, b, a, c **2.** exothermic

10.2 Effect of Concentration on Reaction Rate

Change in concentration is a widely used tool to control the rate of a chemical reaction. For example, the fuel to air mixture in an automobile engine is compressed by a factor of about 8.5 before ignition to increase the chance of a gasoline–oxygen collision, thereby increasing the rate at which combustion occurs. An oxyacetylene welding torch achieves a temperature of about 3000 °F by reacting pure oxygen with acetylene. In contrast, when an

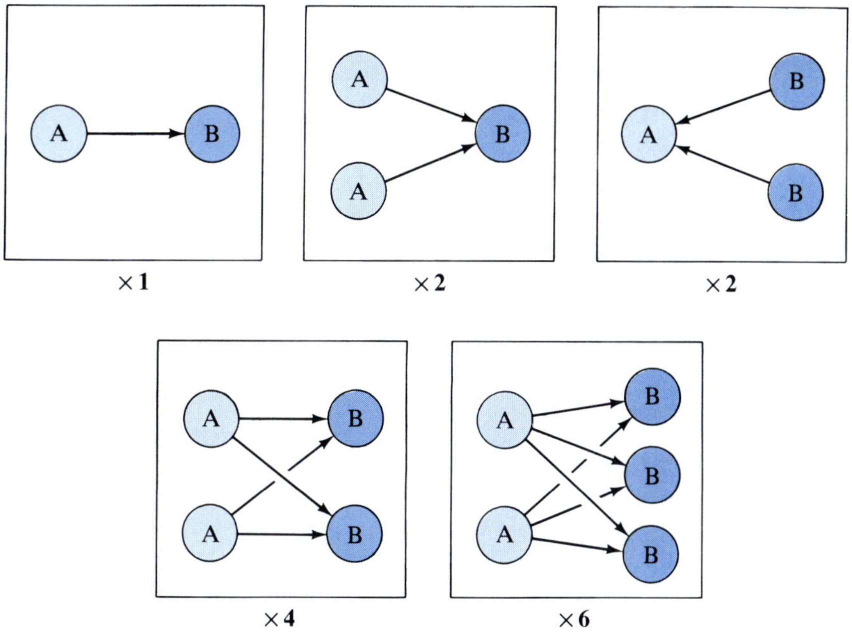
A
B
×1
A
A
B
×2
B
A
B
×2
A
B
A
B
×4
A
B
B
A
B
×6

FIGURE 10.5
When the concentration of either reactant A or reactant B is increased, the probability of an A–B collision increases, as shown by the numbers below each box.

acetylene flame burns in air (which is 20% oxygen), it has a relatively cool, smokey flame.

The probability of collision between reactant molecules is related to the amount of space between them in the reaction container, other factors remaining constant. If the number of reactant molecules in a unit volume is increased, the probability of a collision will increase and the rate of the reaction will increase. In a simple single-step reaction, doubling the concentration of either reactant will double the collision probability and will thus double the rate of the reaction (Figure 10.5). Doubling the concentration of both reactants will increase the rate of the reaction by a factor of four (2 × 2), and tripling it, a factor of 9 (3 × 3).

Increasing reactant concentration increases the probability of molecular collision.

10.3 Effect of Temperature on Reaction Rate

The terms *temperature* and *energy of molecular motion* are practically interchangable. When heat energy is added to a substance, the random jostling motion of the molecules increases. More reactant molecules collide per unit time, and most of these collisions are more energetic.

However, not all molecules in a sample at a given temperature move at the same rate. Molecules are a little like people in this sense. In any group of people, we could calculate an average weight. Although this average would be very useful for making calculations based on the whole group (for example, how much they might eat and how many could fit on a boat), the computed average weight could not be expected to fit too many individuals. The situation is similar with molecules. The temperature of a sample is determined by the *average* rate of molecular motion. Some molecules are moving faster, some slower (Figure 10.6). The significance of this is related to activation energy. Even when the temperature is low, a few molecules will have

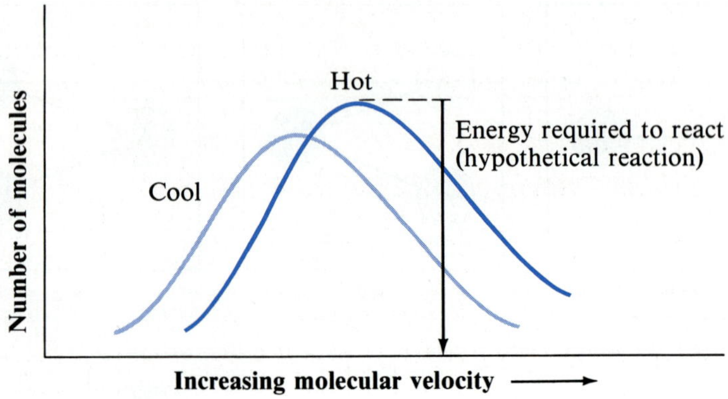
Hot
Energy required to react
(hypothetical reaction)
Cool
Number of molecules
Increasing molecular velocity

FIGURE 10.6
At any given temperature, some molecules are moving faster than the average and some slower. This graph illustrates the distribution of molecular energy at two different temperatures. Even at low temperatures, a few molecules will have enough energy to break chemical bonds when they collide.

sufficient energy to react. Milk will slowly sour, even in a refrigerator (Figure 10.7). Likewise, although water must be heated to 100 °C to boil, a few molecules will have sufficient energy to evaporate even at low temperatures. This is why wet pavement dries shortly after a rainstorm.

Increased temperature increases the rate of molecular motion, which increases both the collision rate and the chance that a collision will result in broken bonds.

Increasing the temperature has two effects on the rate of a chemical reaction. First, the number of collisions per unit time increases, causing the reaction rate to increase. Second, the increased molecular energy results in a higher percentage of collisions of sufficient energy to break chemical bonds and initiate a reaction. The multiplying effect of these two factors on the overall reaction rate makes it very dependent on temperature. A 10 °C change in temperature can change the rate of a reaction by as much as a factor of two. That is why the relatively small decreases in temperature encountered in a refrigerator greatly extends the storage life of perishable food. Our bodies also use this effect to enable blood cells to fight infection more effectively.

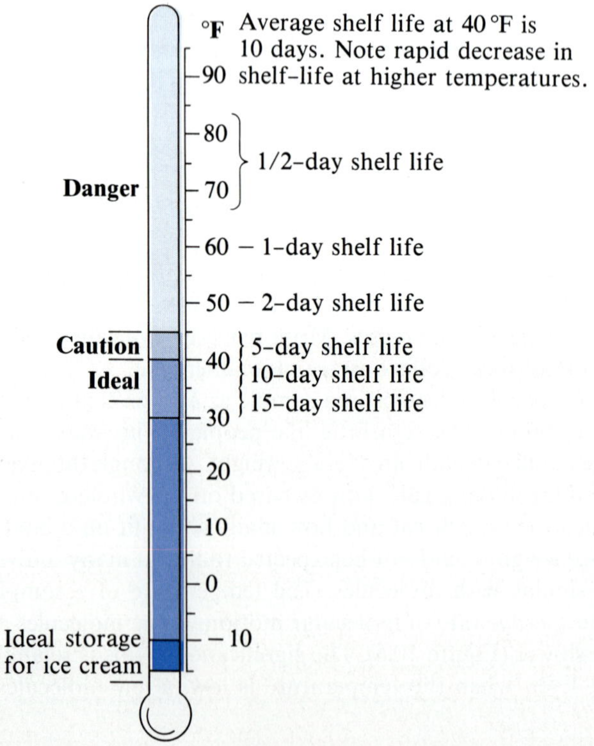
°F
Average shelf life at 40 °F is
10 days. Note rapid decrease in
shelf-life at higher temperatures.
90
80
1/2-day shelf life
Danger
70
60 – 1-day shelf life
50 – 2-day shelf life
Caution
40
5-day shelf life
10-day shelf life
Ideal
15-day shelf life
30
20
10
0
Ideal storage
for ice cream
– 10

FIGURE 10.7
The rate of reaction slows rapidly as temperature is lowered. A decrease in temperature from 80 °F to 40 °F results in an increase of about 30 times in the storage life of milk.

The few degrees of increase in body temperature that occur with a fever cause the reaction by which blood cells attack bacteria or virus cells to proceed at a much faster rate.

An increase in temperature of 10 °C will approximately double the rate of a reaction.

TESTING YOURSELF

Concentration and Temperature Effects

1. Epoxy cement comes in two tubes. To use them, you measure out equal volumes of the two materials, mix them, and wait for them to harden.
   a. The instructions indicate that the cement will set up in about 1 hour at room temperature. They also indicate that the cement should not be used at low temperatures. Suggest a reason for this.
   b. What would probably happen to the setup time if you diluted one or both materials with a nonreactive solvent? Can this situation be explained by molecular collision theory?
2. A four-cycle automobile engine has four distinct piston movements. In the first movement, the piston moves down, drawing air and fuel into the cylinder. In its second movement, it moves up toward the top of the cylinder, compressing the air and fuel vapor. As it reaches the top, the spark plug fires. The resulting explosion drives the piston down, its third movement. The fourth movement occurs as the piston again moves up, driving out the reaction products through an open exhaust valve. The sequence then repeats.
   a. What does compression of the air and fuel vapor do to the concentration of these reactants? How does it affect the rate of reaction when the reactants are ignited?
   b. How would a high compression engine that compressed 11 volume units to 1 volume unit (11:1) differ in reaction temperature from a lower compression engine that only compressed the air to fuel mixture to one-eighth its original volume? Air is about 80% nitrogen and 20% oxygen. At high temperatures, these two gases react together to form toxic gaseous nitrogen oxides, which are in large part responsible for smog pollution. Which engine would probably produce the most nitrogen oxide—the 11:1 or the 8:1 compression ratio motor?
   c. Which of the three types of energy described in Figure 10.3 is provided by the spark plug?

Answers **1. a.** The epoxy will not set up unless the temperature is high enough for the reacting molecules to collide with sufficient energy to break chemical bonds. **b.** Changing the concentration of one of the reacting molecules will change the rate of the reaction. If the reactant is diluted, molecular collision will occur between reacting molecules less often, and the reaction will slow down. **2. a.** Compression of the air and fuel mixture increases the reaction rate when the mixture is ignited. **b.** The higher compression engine will operate at a faster reaction rate and higher combustion temperature and will therefore produce more nitrogen oxides. **c.** activation energy

10.4 Hypothermia: A Reaction Rate Problem

Hypothermia, or lowering of the temperature of the body core, is a vivid demonstration of the effect of temperature on reaction rate. It is the most frequent killer of outdoor sportsmen. A relatively small decrease in the temperature of the body core will result in reduced rates for the reactions that

Hypothermia is the slowing of the body's chemical reactions by decreased body temperature.

sustain the heart and brain. Normal activity can usually be maintained as long as the temperature of the body core is maintained above 90 °F. Between 90 °F and 85 °F (32 °C and 29 °C), uncontrolled shivering, slurred speech, and incoherence occur, and below 85 °F the individual must have an external source of heat or death may result. Below 75 °F (24 °C) death will almost certainly result.

Hypothermia is often called the "killer of the unwary" because it is most often encountered in air temperatures from 30 °F to 50 °F (−1 °C to 10 °C). Hypothermia is particularly dangerous because it is often ignored by companions of the individuals affected. Despite the symptoms and because of the lowered reaction rate in the brain, victims will often steadfastly claim that they are "fine."

Two conditions often leading to hypothermia are wet or damp clothing and immersion in cold water. A person with wet clothing who allows body heat to evaporate the water has not considered the heat of vaporization of water—540 cal/g. Evaporation of 1 pint of water (about 500 g) from clothing will require

$$(500\ \text{g})(540\ \text{cal/g}) = 270{,}000\ \text{cal of heat}$$

This is only 270 kcal, an amount that should be easily replaced by normal metabolism. However, if it is removed more rapidly than metabolism can replace it, the temperature of the body core and the rate of metabolism will both be lowered. A person who weighs 70 kg (70,000 g, or about 150 lb) would have each gram of his or her body lose

$$\frac{(270{,}000\ \text{cal})}{(70{,}000\ \text{g})} = 3.86\ \text{cal/g}$$

If you assume that the specific heat of body tissue averages about 0.9 cal/g °C, each gram of body tissue would have its temperature lowered about

$$\begin{aligned}(3.86\ \text{cal/g})/(0.9\ \text{cal/g}\ ^\circ\text{C}) &\approx 4.3\ ^\circ\text{C}\\ &\approx 7.7\ ^\circ\text{F}\end{aligned}$$

This alone would put an individual near the 90 °F (32 °C) danger line, and the additional cooling effect of wind could easily cool the person down below this point.

Even more dangerous is immersion in cold water. In this case, the individual is not fighting the problem of evaporation, but instead the body's attempt to transfer heat to the surrounding water. Table 10.1 shows survival times expected for various immersion times over a common range of water temperatures.

Although the initial reaction of anyone involved in a boating accident is to swim ashore, the motion of cold water past the limbs increases the heat loss and can decrease survival time. Swimming is not recommended unless you can reach a nearby craft, a fellow survivor, or a floating object on which you can lean or climb. The swimming range of an individual wearing a personal flotation device (PFD) in 55 °F water is only about 0.85 miles. If shore is close by, swim for it quickly. If not, try to get back into the boat or on top of it.

TABLE 10.1
Hypothermia Chart (U.S. Coast Guard)

Water Temperature (°F)	Time to Exhaustion or Unconsciousness	Expected Time of Survival
32.5	Under 15 min	Under 15–45 min
32.5–40	15–30 min	30–90 min
40–50	30–60 min	1–3 hr
50–60	1–2 hr	1–6 hr
60–70	2–7 hr	2–40 hr
70–80	3–12 hr	3 hr–indefinite
Above 80	Indefinite	Indefinite

Cooling of the body core is sometimes deliberately employed in medicine to prepare a person for surgery that involves temporarily stopping the heart. By slowing the reactions that sustain life, sufficient oxygen can be stored in the cells to keep the patient alive for a few minutes until the heart repair is complete and the circulation restarted. Body temperature is then returned to normal by circulating the blood through a warming solution.

Hypothermia is sometimes deliberately induced to slow a person's metabolism for surgery.

NEW TERM

hypothermia A condition of lowered body temperature that causes a decrease in the chemical reactions that support body functions and life. The chemical reactions in the human body operate properly only in a narrow range of temperatures centered on 98.6 °F (37 °C).

TESTING YOURSELF

1. Consider a person who falls into a stream at a temperature of 5 °C (about 40 °F). After climbing out and shaking off, 1 quart (946 g) of water remains in his clothes. He then walks briskly to warm up and dry the water from his clothes. Presume that this person has a mass of 70 kg and that the specific heat of his body is 1.0.
   a. How many calories of heat will be required to heat the quart of water in his clothes from 5 °C to the point of evaporation?
   b. How many calories of heat will be required to evaporate 946 g of water?
   c. How many calories total must be supplied by this person's body heat to warm and evaporate the water?
   d. If this evaporation takes place faster than the body's metabolism can replace the heat, the heat will be given up by the body and the body temperature will be lowered. Assuming a body mass of 70 kg and an average specific heat of 1.0, how much would the person's body temperature be lowered by this heat loss?
   e. Assuming an initial body temperature of 37 °C, what would the person's new body temperature be?

Answers **1. a.** 89,870 cal **b.** 510,840 cal **c.** 600,710 cal **d.** 8.58 °C **e.** 28.42 °C

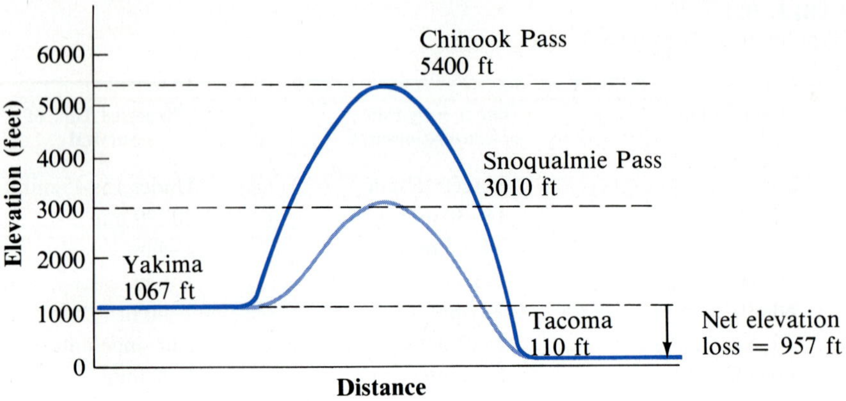
Chinook Pass
5400 ft
Snoqualmie Pass
3010 ft
Yakima
1067 ft
Tacoma
110 ft
Net elevation
loss = 957 ft
Elevation (feet)
6000
5000
4000
3000
2000
1000
0
Distance

FIGURE 10.8
The route with the lowest pass in this case is the easiest to travel, since it requires less "activation energy" (see Figure 10.4). In an analogous way, a catalyst provides an alternate route for a chemical reaction.

10.5 Effect of Catalysts on Reaction Rate

Catalysts speed up a chemical reaction by providing an alternate path of lower activation energy.

Catalysts are compounds that speed up chemical reactions. Catalysts are usually specific for a given reaction; that is, a given catalyst only works for one kind of reaction.

The action of a catalyst is best understood by applying the activation energy concept. Earlier in this chapter (Section 10.1) we spoke of the "activation energy" necessary to move a car over the crest of the Cascade Mountains on the route from Yakima to Tacoma, Washington (see Figure 10.4). At the conclusion of that discussion, we mentioned that an alternate, although slightly longer, route was possible and that this highway crossed the crest of the mountains at a lower elevation. The implication, of course, was that it is easier to travel the road with the lower pass, because it is the route with the lowest activation energy (Figure 10.8).

Before continuing our discussion of catalysts, an important point about our presentation should be made. Through the rest of this chapter we will be discussing some factors that apply to *all* chemical reactions. Rather than use specific reactions as examples, and risk having the reader believe that the principles apply only to these reactions, we will use capital letters A, B, C, etc. as generalized names for reactant and product molecules. For example, gasoline and oxygen (reactants) react to produce carbon dioxide and water vapor (products):

$$\text{Gasoline} + \text{Oxygen} \longrightarrow \text{Carbon dioxide} + \text{Water}$$

$$\text{A} + \text{B} \longrightarrow \text{C} + \text{D}$$

In this example, increasing the concentration of oxygen will increase the chance of a gasoline–oxygen collision and will thus increase the reaction rate. In general terms, increasing the concentration of reactant B will increase the probability of a collision between reactant A and reactant B and will thus increase the rate of the reaction. The use of capital letters to represent compounds will also help us write some very simple mathematical relationships later in the chapter.

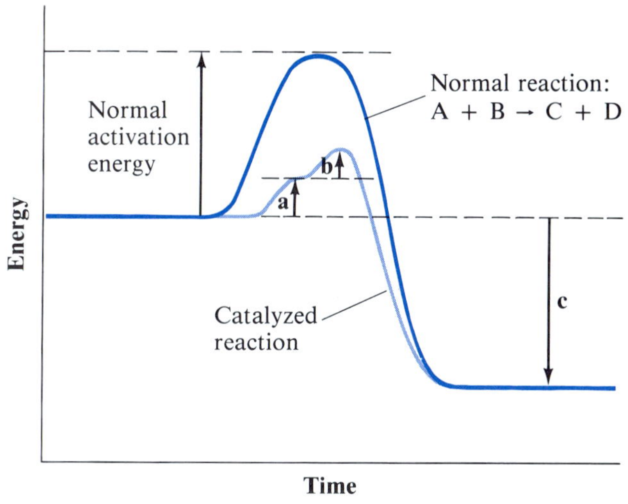

Normal activation energy
Normal reaction:
A + B → C + D
a
b
c
Catalyzed reaction
Energy
Time

FIGURE 10.9
A catalyzed reaction proceeds faster because less activation energy is required. (a) Reactant A combines with the catalyst to form the intermediate compound A-catalyst. A smaller activation energy is required. (b) Intermediate A-catalyst reacts easily with reactant B, forming product AB and freeing the catalyst. (c) The useful energy liberated by the reaction is the same regardless of the route, but less activation energy is required by the catalyzed route.

A catalyst simply provides an alternate, lower energy route for a chemical reaction (Figure 10.9). Consider a generalized reaction in which reactants A and B react to form product AB:

$$A + B \longrightarrow AB$$

Introduction of an appropriate catalyst that will react with component A results in the formation of an intermediate compound, A-catalyst, sometimes called an *activated complex*.

$$A + \text{Catalyst} \longrightarrow \text{A-catalyst}$$

This reaction requires very little activation energy and proceeds rapidly. The compound A-catalyst, however, must be carefully chosen so that it will react with compound B using very little activation energy. If this is the case, the reaction with B will also proceed rapidly.

$$B + \text{A-catalyst} \longrightarrow AB + \text{Catalyst}$$

At the conclusion of this reaction, the product AB has been formed and the catalyst has been freed to react again with another molecule of compound A.

Notice, however, that far less activation energy has been required, and more molecules at any given temperature will have sufficient energy to break the required bonds as they collide (Figure 10.10). The reaction will proceed much more rapidly on the catalyzed route than when operating without a catalyst.

It is apparent that a catalyst must be designed specifically for each reaction that must be catalyzed. Such is the case in our body biochemistry. Most biochemical reactions that make life possible are reactions between complex molecules that proceed slowly. Without biochemical catalyst molecules

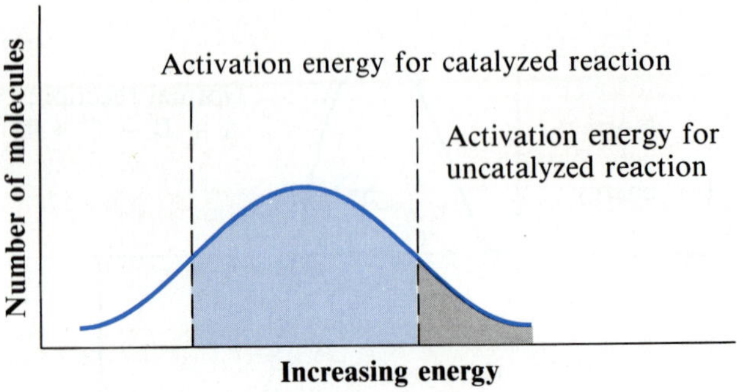
Activation energy for catalyzed reaction
Activation energy for uncatalyzed reaction
Number of molecules
Increasing energy

FIGURE 10.10
A catalyzed reaction has a lower activation energy requirement than an uncatalyzed reaction. At any given temperature, more molecules can meet the activation energy requirement for the catalyzed reaction, and the reaction will go faster.

known as **enzymes**, life as we know it could not exist because the reactions would be too slow. Enzymes are highly complicated molecules that react to provide low energy pathways for biochemical reactions. Thus, a reaction that would proceed too slowly at 98 °F to keep a person alive will proceed at an adequate rate in the presence of the proper enzymes. About 1000 different enzymes have been discovered, and most of these catalyze specific biochemical reactions. For example, the presence of certain kinds of cancer cells in the body can be detected by identification of the specific enzyme necessary for the growth of these cells in the bloodstream. If this enzyme is not present, the cancer cells cannot grow. A number of diseases are caused by absence of specific enzymes. Some children, for example, cannot drink milk because they lack a specific enzyme necessary for its digestion. A soybean substitute is prescribed for these children.

NEW TERMS

catalyst A compound that participates in a chemical reaction to provide an alternate reaction pathway of lower activation energy, thus increasing the rate of the reaction. The catalyst forms an intermediate compound with one reactant and is released following the second reaction.

enzyme A complex molecule that acts as a catalyst in biochemical reactions.

10.6 Effect of Molecular Complexity on Reaction Rate

Complex molecules are less likely to break properly at the first collision; they react more slowly.

The rate at which chemical reactions proceed is also related to the complexity of the molecules involved. The reactions by which our growth processes occur, which involve extremely complex molecules, proceed very slowly. In contrast, the reaction between sodium chloride and silver nitrate that you may have observed in the laboratory is practically instantaneous. These reactions differ in two ways: in the complexity of the molecules involved and in the type of bonds involved. The sodium chloride–silver nitrate reaction is a reaction between ions. Since the electrical attraction in an ionic bond is nondirectional, the direction from which the reactants approach each other is not important and any collision is likely to result in reaction (Figure 10.11).

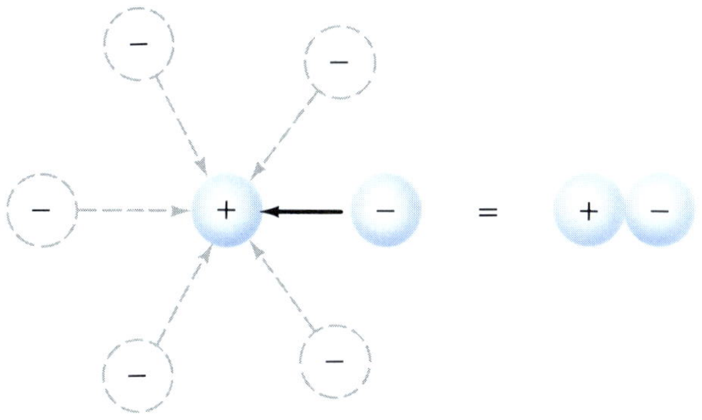
−
−
−
+
−
=
+ −
−
−

FIGURE 10.11
Reactions between ions are usually very rapid because the electrical attraction between the positive and negative ion allows one ion to approach the other from any direction.

This can be envisioned by considering two velcro-covered balls. No matter which way they hit one another, they will stick together.

On the other hand, covalently bonded molecules—particularly complex covalently bonded molecules—must collide with the proper orientation so that specific bonds are broken and specific atoms are placed adjacent to one another to form new bonds. As a result, only a very small fraction of the molecular collisions that occur result in reaction, and the reaction rate is much slower than an ionic reaction conducted under similar conditions of temperature and concentration. The more complex the reactant molecules, the more carefully they must be oriented for a reaction to occur, and the less likely it is that a given molecular collision will result in reaction.

Figure 10.12 illustrates the reaction between carbon dioxide and water to form carbonic acid, a reaction that takes place when carbon dioxide is dissolved in water to make soda pop. The carbonic acid formed gives soda pop its sharp taste, while the unreacted carbon dioxide gives it its "fizz." Since this reaction does not proceed too rapidly, and particularly since it will also proceed in the reverse direction, the concentration of carbonic acid is never very high in soda pop.

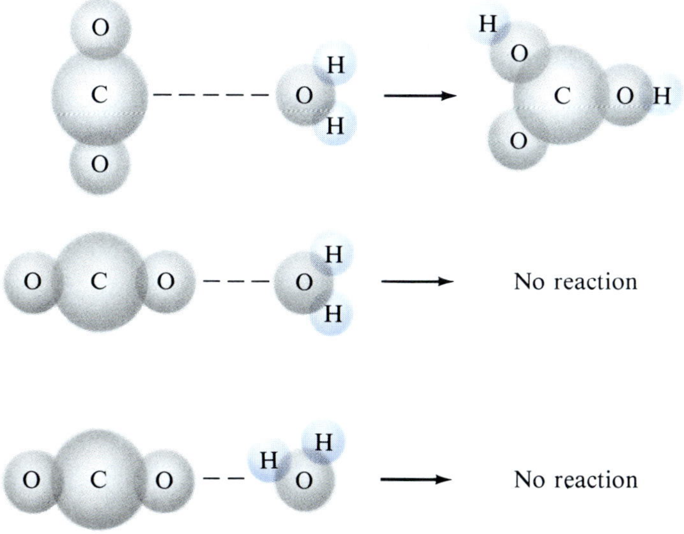
O
C
O
H
O
H
H
O
C
O H
O
O C O
H
O
H
No reaction
O C O
H
H
O
No reaction

FIGURE 10.12
The reaction between carbon dioxide (CO_2) and water (H_2O) to form carbonic acid (H_2CO_3), which is used to make soda pop. The molecules will only react (top) if oriented in the correct manner.

TESTING YOURSELF

Catalysts and Molecular Complexity

1. A catalytic converter unit, placed in the exhaust system of an automobile, uses a platinum catalyst to help convert unburned hydrocarbons to carbon dioxide and water vapor. The converter will last a long time if unleaded gasoline is used.
   a. Why doesn't the converter run out of platinum catalyst?
   b. Why can't leaded fuel be used in a car with a catalytic converter?
   c. It is dangerous to operate small trucks with catalytic converters in grain fields, because the converter runs hot enough to start a fire. Why does it run hot?
2. Pudding is made by boiling the ingredients (reactants) and stirring it to keep the mixture at a uniform temperature, until enough molecules are linked together to make the resulting mixture semisolid. Contrast this to the rapid reaction between vinegar and baking soda. Which system do you think involves the most complex molecules?

Answers **1. a.** The converter does not run out of platinum catalyst because the catalyst is regenerated by the catalyzed reaction. **b.** Leaded fuel will "poison" the catalyst—it will irreversibly bind the platinum catalyst, which will no longer be available to catalyze subsequent reactions **c.** Catalytic converters get hot because the heat of the catalyzed reaction is deposited in the metal container holding the catalyst. **2.** The components of pudding react more slowly than the vinegar and baking soda because the molecules in the pudding are more complex and have a lower probability of colliding with the proper orientation to react. Many collisions may be necessary before the reaction occurs.

10.7 Equilibrium: Reactions That Go Both Directions

Most chemical reactions can proceed both forward and backward, although one direction is usually favored. Consider again our highway example (Figure 10.13). It is possible to drive from Tacoma to Yakima but it is much easier, in terms of energy expended, to drive from Yakima to Tacoma.

Chemical reactions are similar. Although they proceed predominantly in one direction (usually the direction in which they release energy), at any

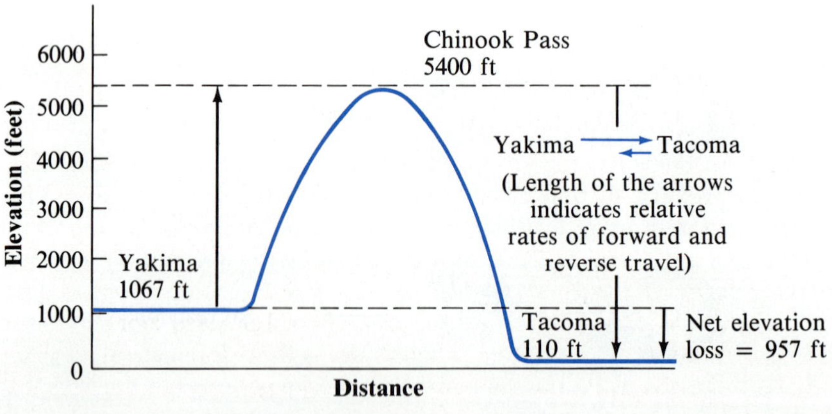
Chinook Pass
5400 ft
Elevation (feet)
6000
5000
4000
3000
2000
1000
0
Yakima
1067 ft
Yakima ⇄ Tacoma
(Length of the arrows indicates relative rates of forward and reverse travel)
Tacoma
110 ft
Net elevation loss = 957 ft
Distance

FIGURE 10.13
Although it is possible to drive from Tacoma to Yakima, it is easier, in terms of energy expended, to drive from Yakima to Tacoma. Similarly, in chemical reactions, a few molecules have sufficient energy to overcome the reverse activation energy barrier. The number of molecules reacting in the reverse direction per unit time is related to the height of the reverse activation energy barrier and to the concentration of product molecules.

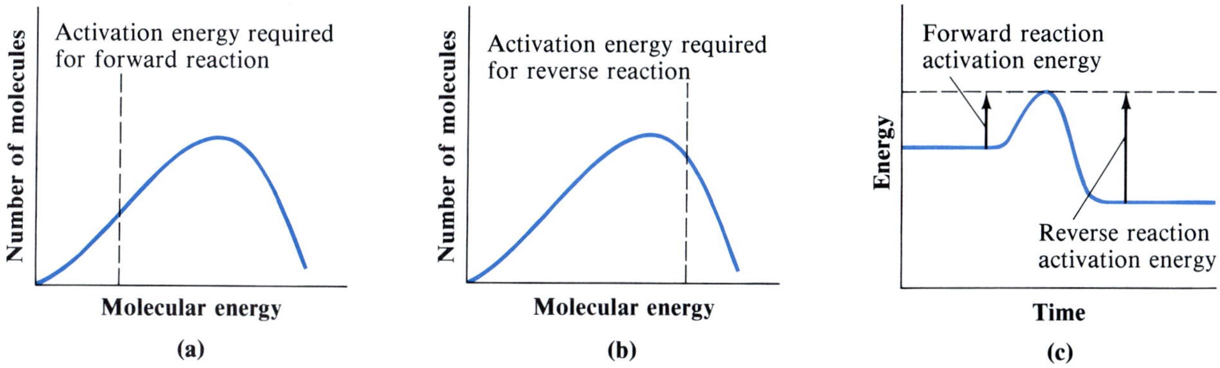
Activation energy required for forward reaction
Number of molecules
Molecular energy
(a)
Activation energy required for reverse reaction
Number of molecules
Molecular energy
(b)
Forward reaction activation energy
Energy
Reverse reaction activation energy
Time
(c)

FIGURE 10.14
The temperature of a group of molecules is related to the *average* rate of molecular motion. Some molecules move with more energy, some with less. (a) In this graph the temperature is high enough that a large number of molecules have sufficient energy to overcome the forward activation energy barrier. (b) Note that only a few molecules have sufficient energy to overcome this higher reverse activation energy barrier. (c) The reaction will move rapidly in the forward direction, slowly in the reverse direction.

given time, a few molecules of product will have sufficient energy to overcome the reverse reaction activation energy barrier. They will react to form reactant molecules. This is illustrated by reaction arrows pointing in both the forward and reverse directions; the relative length of the arrows indicates the relative rates of forward and reverse directions.

$$A + B \rightleftarrows C + D$$

The same factors that govern the rate of the forward reaction govern the rate of the reverse reaction, as illustrated in Figure 10.14. Since the temperature at which both forward and reverse reactions are conducted is the same, the most important factor is reactant concentration.

When a reaction begins, the concentrations of reactants A and B are large, and the concentrations of products C and D are essentially zero. Since the C and D concentrations are extremely low, the rate of reverse reaction is also practically zero. However, as the reaction proceeds, with reactants A and B reacting to form C and D, the C plus D concentration will increase and the probability of a C–D collision will increase. Similarly, the concentrations of A and B decrease as they react to form C and D, and the rate of the forward reaction begins to decrease because of the decreased chance of an A–B collision.

Eventually a point will be reached when the concentrations of C and D are high enough to cause a reverse reaction rate equal to the forward reaction rate maintained by the now practically depleted reactants A and B. This condition, when the forward rate equals the reverse rate, is called **equilibrium**.

The concentration of reactants (A and B) and products (C and D) at equilibrium depends completely on the relative heights of the forward and reverse activation energy barriers. If the forward activation energy is less than the reverse activation energy, the reaction will proceed more easily in the forward direction, and product molecules will predominate at equilibrium. Conversely, if the reverse activation energy is less than the forward

Equilibrium is the situation in which the rate of a chemical reaction in the forward direction equals that in the reverse direction.

The relative concentration of reactants and products at equilibrium depends on the relative heights of the forward and reverse activation energy barriers.

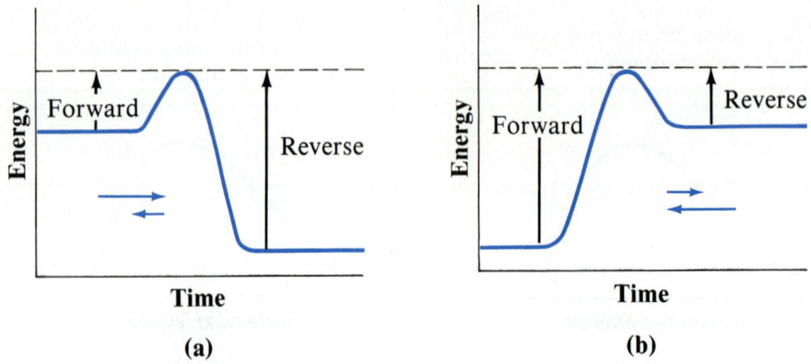
Energy
Forward
Reverse
Time
(a)
Energy
Forward
Reverse
Time
(b)

FIGURE 10.15
The predominant direction of a chemical reaction is determined by comparison of its forward and reverse activation energies.
(a) Reactions can proceed more easily in the forward direction, and product molecules predominate in its reaction mixture at equilibrium.
(b) Reactions can also proceed more easily in the reverse direction, and reactant molecules predominate in its reaction mixture at equilibrium.

activation energy, the reaction will proceed more easily in the reverse direction, and reactant molecules will predominate at equilibrium (Figure 10.15).

The reaction of weak acids with water, which maintains the proper acidity of our blood and many other solutions, is an example of a reaction that proceeds faster in its reverse direction. We will discuss weak acids in more detail in Chapter 11.

NEW TERM

equilibrium The condition in which the rate of a chemical reaction is the same in the forward and reverse directions. Reactant and product concentrations are seldom equal at equilibrium.

10.8 The Equilibrium Constant: A Useful Chemical Tool

The *equilibrium equation* helps us predict the relative concentrations of reactant and product when equilibrium is reached.

The relative concentration of reactant and product molecules in any reversible reaction can be predicted easily by use of a relationship called the **equilibrium equation**. This equation is simply a mathematical statement of the molecular collision theory. The equilibrium equation is based on three assumptions:

1. The reaction is of the general form

$$A + B \rightleftharpoons C + D$$

2. The temperature remains constant so that changes in concentration are the only variables.
3. The reaction rate is directly proportional to the number of collisions per unit time between reacting molecules.

Consider first the forward reaction—the reaction between molecules A and B. The rate of the forward reaction is directly proportional to the chance of a collision between molecule A and molecule B and therefore directly proportional to the concentrations of these two reactants. If the concentration of B is held constant and the concentration of A is doubled, the rate of the reaction will double. If the concentration of A is tripled, the rate of the reaction will triple. Stated in mathematical terms for reactant A, the forward rate, abbreviated R_f, is proportional to ($\sim$) the concentration of reactant A, abbreviated [A]. The square brackets [ ] indicate the concentration of the reactant, expressed in moles per liter.

$$R_f \sim [A]$$

By the same logic, the rate forward is proportional to the concentration of reactant B:

$$R_f \sim [B]$$

Since the forward rate of the reaction is directly proportional to the concentration of each reactant, it will be proportional to the product of the two. (Remember, while tripling either reactant concentration increases the reaction rate by a factor of three, tripling *both* increases the rate by a factor of 3 × 3, or 9.)

$$R_f \sim [A][B]$$

To make this into an equality, we simply multiply the right side of the equation by a proportionally constant (k_f) that converts R_f into a real number indicating the number of moles reacting each second in 1 L of solution.

$$R_f = k_f[A][B]$$

This constant, k_f, is called the *forward rate constant*. The mathematical statement simply says that the rate at which the reaction proceeds forward is equal to a constant times the concentration of reactant A times the concentration of reactant B. The constant is really a measure of the molecular orientation factor affecting reaction rate as well as a measure of the activation energy. If very accurate molecular orientation is required for a collision to result in reaction, very few collisions will result in reactions, and the forward rate constant will be small. But if the collision requirements are noncritical, a large percentage of collisions will result in reaction, and the forward rate constant will be large. By the same token, a large forward activation energy will result in a small forward rate constant, and a small forward activation energy in a larger forward rate constant.

The forward rate constant is an indication of how easily the reaction goes forward.

Using similar logic, one can develop an expression for the reverse reaction rate:

$$R_r = k_r[C][D]$$

The definition of equilibrium is that situation in which the rate of the forward reaction equals the rate of the reverse reaction:

$$R_f = R_r$$

Now, if one substitutes the expressions for R_f and R_r given above, equilibrium can be expressed by the following equation:

$$R_f = k_f[A][B] \quad \text{and} \quad R_r = k_r[C][D]$$

$$k_f[A][B] = k_r[C][D]$$

Dividing both sides of the equation by k_r, we have

$$\frac{k_f[A][B]}{k_r} = \frac{\cancel{k_r}[C][D]}{\cancel{k_r}}$$

$$\frac{k_f[A][B]}{k_r} = [C][D]$$

Dividing both sides of the equation by [A][B], we have

$$\frac{k_f[A][B]}{k_r[A][B]} = \frac{[C][D]}{[A][B]}$$

$$\frac{k_f}{k_r} = \frac{[C][D]}{[A][B]}$$

At this point we have produced an algebraic equality that has constants on one side and reactant and product concentrations on the other side. Let's first consider the side with the constants.

The equilibrium constant is the ratio of the forward and reverse rate constants.

The quotient of two constants is itself a new constant, the **equilibrium constant**, so we can write

$$K_{\text{equilibrium}} = \frac{k_f}{k_r}$$

An equilibrium constant greater than one indicates that the products predominate at equilibrium.

An equilibrium constant less than one indicates that the reactants predominate at equilibrium.

Note that this constant, abbreviated K_{eq}, is determined by the values of the forward rate constant and the reverse rate constant. If the forward rate constant is larger than the reverse rate constant, indicating that the reaction will go forward more easily than backward, the fraction k_f/k_r will be greater than 1, and the equilibrium constant, K_{eq}, will thus be greater than 1. Conversely, if the reverse rate constant is larger than the forward rate constant, indicating that the reaction will go more easily backward than forward, the fraction k_f/k_r will be less than 1, and the equilibrium constant will be less than 1. To summarize,

If $K_{eq} > 1$	If $K_{eq} < 1$
Forward reaction predominates	Reverse reaction predominates

(The > and < signs mean greater than and less than, respectively.) Note that on the concentration side, product concentrations are in the numerator while reactant concentrations are in the denominator.

$$K_{eq} = \frac{k_f}{k_r} = \frac{[C][D]}{[A][B]} = \frac{\text{Products}}{\text{Reactants}}$$

The meaning of this is now clear. If K_{eq} is greater than 1, the reaction goes mainly forward, and product molecules will predominate in the equilibrium mixture. If K_{eq} is less than 1, reactant molecules will predominate in the equilibrium mixture. This equation is the *equilibrium equation.* It is a useful tool for predicting which direction reactions will proceed and, ultimately, for calculating concentrations of the various species at equilibrium. Equilibrium constants for most common reactions have been determined and are readily available in such references as the *Handbook of Chemistry and Physics.*

The equilibrium constant is also applicable to chemical reactions involving coefficients other than 1. Consider, for example, the reaction

$$A + 2B \rightleftharpoons C + 3D$$

Since the molecules B and D occur more than once, the equilibrium constant equation would be written:

$$K_{eq} = \frac{[C][D][D][D]}{[A][B][B]} = \frac{[C][D]^3}{[A][B]^2}$$

In general, for any chemical reaction such as

$$aA + bB \rightleftharpoons cC + dD$$

where the lowercase letters represent multipliers, the equilibrium constant can be written

$$K_{eq} = \frac{[C]^c[D]^d}{[A]^a[B]^b}$$

We will use the equilibrium constant in Chapter 11 to estimate the pH or hydrogen ion concentration of weak acids and buffer solutions.

NEW TERMS

equilibrium equation A mathematical expression relating concentrations of reactants and products at equilibrium for a given reaction (*see* equilibrium constant).

equilibrium constant A mathematical constant, K_{eq}, that enables one to predict the predominant direction of a reaction.

$$K_{eq} = \frac{[A][B]}{[C][D]}$$

TESTING YOURSELF

The Equilibrium Constant

Compounds known as *weak acids* ionize when they react with water, donating a hydrogen ion to the water molecule. This results in formation of the hydronium ion, H_3O^+. The general reaction is

$$\text{Weak acid} + H_2O \rightleftharpoons H_3O^+ + \text{Negative acid ion}$$
$$A \quad + \quad B \rightleftharpoons C \quad + \quad D$$

This table shows the equilibrium constants for four common weak acids.

Compound	Found In	Equilibrium Constant
Acetic acid	Vinegar	0.000977
Ascorbic acid	Vitamin C	0.0044
Carbonic acid	Soda pop	0.0000238
Citric acid	Lemon juice	0.0466

The general equilibrium equation for weak acids is

$$K_{eq} = \frac{(H_3O^+ \text{ concentration})(\text{Negative acid ion concentration})}{(\text{Acid concentration})(\text{Concentration of } H_2O)}$$

1. Which of these compounds has the greatest tendency to ionize?
2. Which is the "weakest" acid—the compound with the least tendency to ionize?

3. If you placed 1 mol of each of these compounds in a separate liter of water, in which solution would the concentration of hydronium ion be the greatest?

Answers **1.** citric acid **2.** carbonic acid **3.** citric acid

SUMMARY

The *molecular collision model* is a powerful tool for predicting the effect changes in concentration, temperature, and molecular complexity have on the rate of a chemical reaction. In general, anything that increases the rate or energy of molecular collision, or enhances the probability that a given collision will be of the proper orientation to break the appropriate chemical bonds, also increases the rate of the reaction. Most chemical reactions will proceed in both a forward and reverse direction; the predominant direction is determined by the relative activation energies required by the forward and reverse directions. Almost all reactions ultimately reach a point when the forward and reverse directions proceed at the same rate; this is called *equilibrium.* The relative abundance of product and reactant molecules at equilibrium is determined by the relative rates of the forward and reverse reactions. The *equilibrium constant* (K_{eq}) is a mathematical statement of the relationship between product and reactant molecule concentrations at equilibrium. In the next chapter we will use the equilibrium constant to predict some of the behaviors of weak acids encountered in biological systems.

TERMS

molecular collision theory
activation energy
exothermic
endothermic
hypothermia
catalyst
enzyme
equilibrium
equilibrium equation
equilibrium constant

CHAPTER REVIEW

QUESTIONS

1. Is a campfire an endothermic or exothermic reaction?
2. When an egg is fried, small molecules are joined to form large molecules. This process is called *polymerization.* The stove provides energy to this reaction.
   a. Is the reaction endothermic or exothermic?
   b. What is the energy supplied by the stove called?
3. Milk sours rapidly when left out on a kitchen counter. However, this same milk will last several days if refrigerated. Which factors are at work here?
   a. activation energy
   b. concentration
   c. molecular collision rate
   d. (a) and (c)
   e. (a) and (b)
4. Gunpowder burns slowly if a few grams are placed on a table and lighted with a match. However, this same gunpowder burns very rapidly if enclosed in a gun barrel where the heat of combustion cannot escape. Enclosing the combustion area in a very small chamber increases which of the following factors affecting reaction rate?
   a. activation energy
   b. concentration
   c. molecular collision rate
   d. (a) and (c)
   e. (b) and (c)

5. Our molecular collision model for chemical reactions was based on the premise that three conditions must be satisfied for a chemical reaction to take place:
   a. The molecules must collide.
   b. They must collide with sufficient energy to break chemical bonds.
   c. They must collide with the proper orientation to break the appropriate chemical bonds.

   An oxygen-acetylene welding torch produces a high temperature through a reaction of acetylene and oxygen. If acetylene is burned in air, a smokey yellow flame results. If mixed with pure oxygen, an intense blue flame results. Which of the above factors explains the difference in reaction rate when pure oxygen is used?
6. Ionic reactions proceed more rapidly in general than reactions involving covalent molecules. Which of the factors listed in question 5 best explains this?

The graph below is an energy diagram that represents the changes that occur during a chemical reaction. Questions 7–11 refer to this graph.

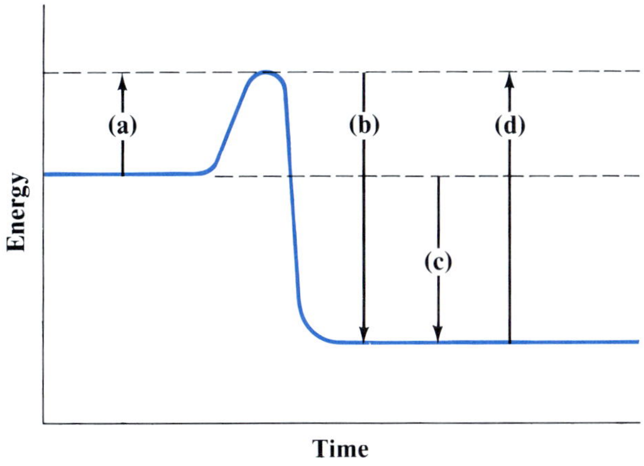
Energy
(a)
(b)
(d)
(c)
Time

7. Which arrow on the graph represents the useful energy that can be withdrawn from the reaction?
8. Which arrow on the graph represents the activation energy of the reverse reaction?
9. Would this reaction be considered endothermic or exothermic as it would proceed in the forward direction?
10. Consider a catalyzed reaction that proceeds in the forward direction. The presence of the catalyst changes which of the variables in the graph, thus increasing the rate of the reaction?
11. Vinegar is a 5% solution by weight of acetic acid in water. Acetic acid reacts with water according to this equation:

$$\underset{\text{Acetic acid}}{HC_2H_3O_2} + \underset{\text{Water}}{H_2O} \rightleftharpoons \underset{\text{Hydronium ion}}{H_3O^+} + \underset{\text{Acetate ion}}{C_2H_3O_2^-}$$

    The equilibrium constant for this reaction is about 1×10^{-3}. Which compound will predominate in the equilibrium mixture—acetic acid or hydronium ion?

Answers **1. a.** exothermic **b.** activation energy **2.** endothermic **3.** d. **4.** b **5.** a **6.** c **7.** c **8.** d **9.** exothermic **10.** a **11.** acetic acid

DIAGNOSTIC CHART

Blacken in all of the circles under the number of each question you missed in the Chapter Review questions. The diagnostic chart will help you identify concept areas that need more study.

	Questions										
Concepts	1	2	3	4	5	6	7	8	9	10	11
Energy relationships in chemical reactions	○	○	○		○		○	○	○	○	
Factors affecting reaction rate			○	○	○	○	○	○	○	○	
Equilibrium											○

EXERCISES

Chemical Reactions and Reaction Rates

1. By use of the concepts of activation energy and molecular collision theory, explain why a campfire will usually start if one uses small, closely spaced sticks as fuel, but will not start if a match is applied to a few large logs.
2. Why can you put out a fire by pouring water on it? Earlier we identified three factors that must occur for a reaction to take place. Which of these is removed when you put water on a fire?
   a. The molecules must collide.
   b. They must collide with sufficient energy to break chemical bonds.
   c. They must collide with the proper orientation to break the appropriate chemical bonds.
3. Fires are also extinguished by flooding the area with carbon dioxide or another noncombustible gas such as halon. In years past, carbon tetrachloride (CCl_4) was used for this purpose. However, CCl_4 fumes are heavy and remain in closed areas where they have been used. They have caused a number of fatalities when people enter the area following the fire, because oxygen cannot return to the area until the carbon tetrachloride fumes are removed. For this reason, CCl_4 extinguishers have been outlawed for use in boats and other closed areas. Which of the factors listed in exercise 2 explains the fire-extinguishing properties of these noncombustible gases?
4. Latex house paint bears the warning on its label that it is not to be applied if the outside temperature is below 50 °F. If one ignores this warning and paints when the temperature is colder, the paint will not set up. Which of the factors listed in exercise 2 explains why this problem occurs?
5. Milk will sour rapidly if left at room temperature for several hours, but it lasts for several weeks if properly refrigerated. Lowering the storage temperature affects which of the factors necessary to support a chemical reaction listed in exercise 2?
6. Which of the factors listed in exercise 2 explain why an elevated body temperature (a fever) helps cells in the blood fight infection more effectively?
7. Which of the factors listed in exercise 2 explains why the chemical reactions that support thought processes slow down when a person becomes hypothermic?
8. Suppose that you had a solution of reactant A with a concentration of 0.1 mol/L and a solution of reactant B with a concentration of 0.2 mol/L. Suppose 90-mL samples of these two solutions would react completely in 100 sec.
   a. If the concentration of reactant A were tripled, to 0.3 mol/L, how long would it take for the reaction to be completed?
   b. If the concentration of reactant A were changed to 0.5 mol/L, and the concentration of reactant B were changed to 0.4 mol/L, how many seconds should the reaction require?
9. What role does a catalyst play in a catalyzed chemical reaction?

Equilibrium

10. The equilibrium constant permits us to predict both the predominant direction of a reaction and the relative concentrations of reactant and product molecules at equilibrium. Equilibrium constants describing the ionization of several weak acids are shown in the following table.
    a. Which of these compounds has the greatest tendency to ionize?
    b. If you placed 1 mol of each of these compounds in a separate liter of water, in which solution would the concentration of hydronium ion (H_3O^+) be the greatest?

Compound	Found In	Equilibrium Constant
Acetic acid	Vinegar	0.000997
Ascorbic acid	Vitamin C	0.0044
Carbonic acid	Soda pop	0.0000238
Citric acid	Lemon juice	0.0466

The general equilibrium equation for weak acids is

$$K_{eq} = \frac{(H_3O^+ \text{ conc.})(\text{Negative ion conc.})}{(\text{Acid conc.})(\text{Conc. of } H_2O)}$$

11. Consider a sample of vinegar, which is 5% by weight acetic acid. The acetic acid reacts with water, according to this general reaction:

$$\underset{\text{Weak acid}}{HC_2H_3O_2} + H_2O \rightleftharpoons H_3O^+ + \underset{\text{Negative acid ion}}{C_2H_3O_2^-}$$

   a. What will be the predominant direction of the reaction?
   b. Sodium acetate strongly ionizes when placed in water, forming sodium ions and acetate ions:

$$NaC_2H_3O_2 \longrightarrow Na^+ + C_2H_3O_2^-$$

12. Suppose that you added sodium acetate to vinegar, thereby increasing the concentration of acetate ion. Which way would the reaction be encouraged to go? (HINT: recall molecular collision theory.) Which molecular collisions would be more likely to happen with increased acetate ion concentration? What would happen to the concentration of H_3O^+? Would it increase or decrease?

CHAPTER

11

Acids and Bases

Acids are molecules containing ionizable hydrogen atoms. When such a molecule (an acid) collides with another molecule having a greater attraction for hydrogen (a base), the ionizable hydrogen is captured by the base. It is essential to living things for the concentration of the ions resulting from these reactions to be carefully controlled.

OBJECTIVES

After completing this chapter, you should be able to

- Define the terms *acid*, *base*, and *salt*.
- Describe the hydronium ion.
- Differentiate between strong and weak acids in terms of the parent molecule's tendency to ionize.
- Define the term *pH*, and use this term to describe hydrogen ion concentrations in acidic, neutral, and basic solutions.
- Describe the function of a chemical indicator, and explain the use of several indicators used to determine the pH of a solution.
- Solve titration problems using the mole method.
- Describe the function of a chemical buffer.
- By use of the equilibrium equation, estimate the pH of buffer solutions.

11.1 Acids: Hydrogen Ion Donors

Acids are common compounds that all of us encounter every day. But what characterizes a molecule as an acid? What do all acids have in common? Table 11.1 lists a number of compounds that carry the name *acid* and several of their properties. All conduct electricity to some degree (some better than others), indicating that all ionize to some extent. All have a sour taste (although strongly ionizing acids are very dangerous to taste except in extremely dilute solutions), indicating that they undergo similar reactions with our sensory molecules.

Since all of these compounds behave in a reasonably similar manner, they must have something in common. The element common to all members of the group is hydrogen.

To better understand this common group of compounds, let's look first at the history of its discovery. The fact that water solutions of table salt (NaCl) and many other compounds conduct electricity was well known by the middle 1800s, although early experimentation was crude. The great English scientist Cavendish investigated the conductivity of compounds dissolved in rainwater by passing an electrical charge through the solution into his body and estimating the shock he received. Cavendish found that dilute solutions of compounds similar to sodium chloride would transfer electrical charge, while solutions of many other compounds would not. The mechanism of this charge transfer was unknown.

Svante Arrhenius: The Discovery of Ions

Svante Arrhenius was a young graduate student at the University of Upsala in Sweden in the early 1880s. His doctoral research dealt with the behavior of dilute solutions of salts. To explain the electrical conductivity of these solutions, Arrhenius proposed that the molecules came apart into positively and negatively charged parts he called *ions*. He proposed that when positively and negatively charged electrodes are placed in a solution, the positively charged particles move toward the negative electrode, picking

Svante Arrhenius believed that molecules containing hydrogen ionized in water to give up hydrogen ions. He called these molecules *acids*. This idea was not well accepted at the time.

TABLE 11.1
Common Acids

Compound	Formula	Found in	Electrical Conductivity	Taste
Citric acid	$HC_6H_7O_7$	Lemon juice	Fair	Sour
Phosphoric acid	H_3PO_4	Gatorade	Fair	Sour
Acetic acid	$HC_2H_3O_2$	Vinegar	Fair	Sour
Boric acid	H_3BO_3	Eyewash	Fair	Sour
Oxalic acid	$H_2C_2O_2$	Rhubarb	Fair	Sour
Carbonic acid	H_2CO_3	Soda pop	Fair	Sharp
Hydrochloric acid	HCl	Stomach acid	Good	Sour in dilute solutions; dangerous to taste
Sulfuric acid	H_2SO_4	Industrial uses	Good	
Nitric acid	HNO_3	Industrial uses	Good	

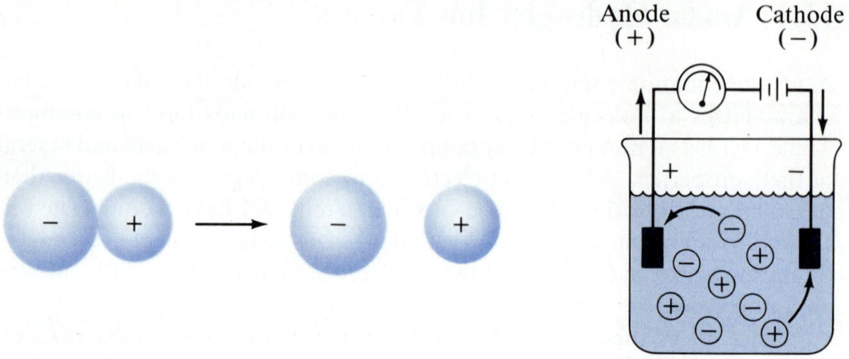
Anode
(+)
Cathode
(−)
−
+
−
+
+
−

FIGURE 11.1
Svante Arrhenius proposed that molecules could come apart into positively and negatively charged parts. These parts (called ions) could move through a solution, thus carrying electrical charge from one electrode to another. This "revolutionary" idea was not accepted at the time.

up negative charges. Likewise, the negatively charged particles move toward the positive electrode, giving up negative charges (see Chapter 3). Electrical charge is thus transferred through the solution (Figure 11.1).

Unfortunately, Arrhenius's idea was considered outlandish by his professors. After a 4-hour thesis examination, he was given a "fourth class" rating for his dissertation project, but still won his Ph.D. principally because his grades were good otherwise.

As we know now, Arrhenius was right. But we should be careful in our judgment of his professors. It was 23 years until the electron was discovered, and Arrhenius had challenged the accepted dogma of the indivisibility of molecules. Acceptance of Arrhenius's work took almost 20 years, but to the credit of the scientific community, he was awarded the Nobel prize in 1904.

Arrhenius proposed that **acids** are molecules that ionized—to a strong or weak extent—to give up hydrogen ions (Figure 11.2). Thus, acids are hydrogen ion donors. The **Arrhenius model for an acid** fits the evidence quite well. Since the element present in all acids is hydrogen, and since all obviously ionize because their solutions conduct electricity, it seems logical that it is the hydrogen atom that is ionizing. Figure 11.3 shows the ionization reactions that could theoretically be predicted for several common acids based on Arrhenius's theory.

A free hydrogen ion (H^+) does not have the stable complement of outer group *s* and *p* electrons.

Arrhenius, however, was unaware that a hydrogen ion, a bare proton, could not exist by itself in a water solution. We now know that to be chemically stable, an atom or ion must have a full complement of *s* and *p* electrons in its outermost electron group. Arrhenius did not know this. His predictions

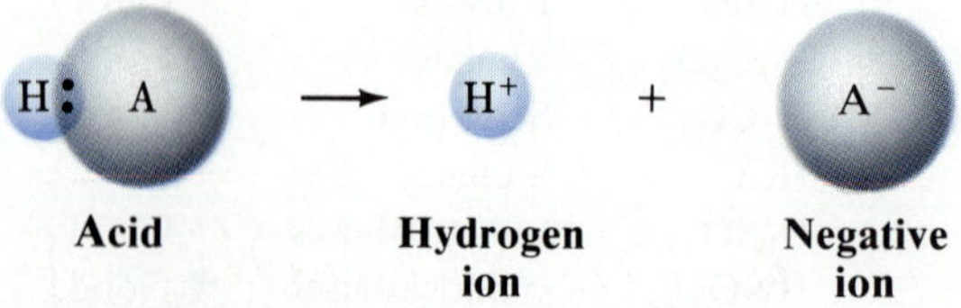
H: A
H+
+
A−
Acid
Hydrogen ion
Negative ion

FIGURE 11.2
The Arrhenius model is the idea that acids are molecules that ionize to give up hydrogen ions.

Phosphoric acid:	H_3PO_4	$\longrightarrow$	$H^+ + H_2PO_4^-$
Acetic acid:	$HC_2H_3O_2$	$\longrightarrow$	$H^+ + C_2H_3O_2^-$
Hydrochloric acid:	HCl	$\longrightarrow$	$H^+ + Cl^-$
Sulfuric acid:	H_2SO_4	$\longrightarrow$	$H^+ + HSO_4^-$
Nitric acid:	HNO_3	$\longrightarrow$	$H^+ + NO_3^-$

FIGURE 11.3
Arrhenius predicted that acid molecules ionize (indicated by arrows) to liberate hydrogen ions (H^+). Some acids ionize more easily than others; solutions of these conduct electricity better because more ions are present.

shown in Figure 11.3 could not possibly be correct. Hydrogen cannot exist as a stable positive ion (H^+) because it would be stripped of *all* of its electrons. The behavior of acids—hydrogen ion donors—must be a special case in the behavior of ions. Arrhenius was on the right track, but what really happens to the hydrogen ion?

Brønsted and Lowry: There Are No Free Protons

In 1923 a Danish chemist, Johannes Brønsted, and an English chemist, Thomas Lowry, independently proposed a solution to this problem. The **Brønsted–Lowry model for an acid** proposed that there are no free protons; as hydrogen ions are donated by one molecule, they are accepted by another. They do not float around in a free state, but are handed-off from one molecule to another as they come in contact, much in the same manner as the hand-off of a football from quarterback to fullback. Brønsted and Lowry proposed that acid molecules donate hydrogens to other molecules, which themselves then became ions.

The *Brønsted–Lowry model* proposed that hydrogen ions do not float free, but are handed from molecule to molecule like a football.

The most available receptor molecule in this exchange is water. Recall that water is a very polar molecule, with two hydrogens sharing electrons on one side, and two unshared pairs of electrons on the other side. If one of these unshared pairs of electrons on a water molecule bumps up against a loosely held hydrogen on an acid molecule, the hydrogen is essentially "captured" by the electrons of the water molecule, forming a stable positively charged molecule, H_3O^+. This is called the **hydronium ion** (see Chapter 10). Ionized hydrogen in a water solution is not a free H^+ ion, but is in fact always found as part of the hydronium ion (Figure 11.4). The covalent bond between the oxygen and the new or donated hydrogen is called a *coordinate covalent bond* (see Chapter 6). It is a covalent bond just as are the other two O—H bonds in the hydronium molecule, but both electrons are provided by the oxygen. As soon as the third hydrogen ion has arrived, all three O—H bonds are identical. The positive hydronium ion and the remaining anion from the acid molecule provide the positive and negative ions necessary to conduct electricity through the solution. This explains Arrhenius's original supposition that ions are responsible for electrical conduction. His only incorrect prediction was that the hydrogen ion would exist as a single charged H^+ ion.

A *hydronium ion* (H_3O^+) is a water molecule that has accepted a hydrogen ion.

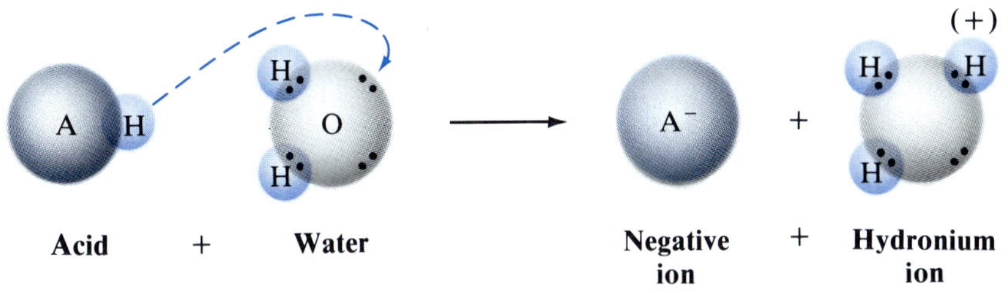
(+)
A
H
H
O
H
A⁻
+
H
H
H
Acid
+
Water
Negative ion
+
Hydronium ion

FIGURE 11.4
In water solutions, acid molecules donate hydrogen ions to water molecules as the water and acid molecules collide. The curved arrow shows the final location of the hydrogen ion. The resulting hydronium ion behaves as an ordinary positive ion.

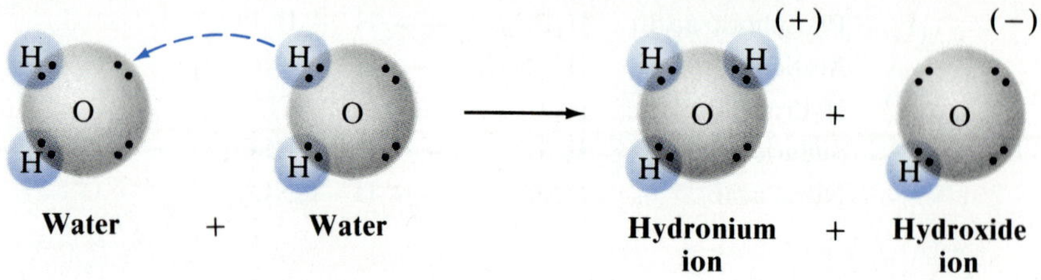
(+)
(−)
H
O
H
H
O
H
H
H
O
H
+
O
H
Water
+
Water
Hydronium ion
+
Hydroxide ion

FIGURE 11.5
If two water molecules collide with the proper orientation and energy, a proton may be transferred from one to the other, forming hydronium and hydroxide ions. This does not happen very often.

The hydronium ion is even present to a small extent in pure water. If two water molecules collide with the correct orientation and energy, a proton will transfer from one water molecule to the other, causing formation of a hydronium ion and a hydroxide ion (Figure 11.5). This reaction between water molecules requires breaking a polar covalent O—H bond of only 1.4 electronegativity difference, and it does not occur easily. At 25 °C, one hydronium and hydroxide ion pair will exist for every 550 million normal water molecules. This is a small fraction, but it is extremely important in providing a controlled amount of hydronium ion needed by the chemical reactions that support life. Too much hydronium ion will attack and destroy many molecules that support life; too little will not allow them to conduct their normal reactions of metabolism and growth. It is a fine line on which we live, as will be seen later in this chapter.

NEW TERMS

acid A hydrogen ion donor; a molecule or ion that will ionize to give up one or more hydrogen ions.

Arrhenius model for an acid The idea proposed by Svante Arrhenius that acids are molecules which, when placed in water, ionize to produce hydrogen ions.

Brønsted–Lowry model for an acid The idea independently recognized by Brønsted and Lowry that an acid cannot donate a hydrogen ion unless there is another molecule—a hydrogen ion acceptor—present to accept the proton. The hydrogen ion acceptor is called a *base.*

hydronium ion The ion H_3O^+ that exists in water solutions because the hydrogen ion (H^+) is not stable by itself.

TESTING YOURSELF

Acids

1. Which of the following compounds could not possibly be an acid?

$$H_2O \quad CuSO_4 \quad NaHCO_3 \quad CH_3COOH$$

2. Why is your selection in question 1 correct?
3. In the following reaction, which compound is the acid?

$$HNO_3 + NH_4OH \longrightarrow NH_4NO_3 + H_2O$$

Answers **1.** $CuSO_4$ **2.** An acid must have a donable hydrogen **3.** HNO_3

11.2 Bases: Hydrogen Ion Acceptors

An acid is a hydrogen ion donor—a molecule or ion that will ionize to give up one or more hydrogen ions. However, this is only half of the picture. For an acid to donate a hydrogen ion, another molecule must be present to receive the hydrogen ion. Brønsted and Lowry's picture of an acid required two parts. They called the hydrogen ion donor an acid and the hydrogen ion acceptor a **base**.

A hydrogen ion acceptor is called a *base*.

Acid: A molecule or ion that donates one or more hydrogen ions.

Base: An atom, molecule, or ion that receives hydrogen ions.

In the reaction of an acid with water, the water molecule serves as the hydrogen ion acceptor or base.

$$\underset{\text{Acid}}{HA} + \underset{\text{Base}}{H_2O} \longrightarrow \underset{\text{Hydronium ion}}{H_3O^+} + \underset{\text{Negative ion}}{A^-}$$

The negative ion, an anion, is represented by the general symbol A^-.

Conjugate Acids and Bases

Defining a molecule as an acid or base is not a simple task, since molecules and ions differ in their ability to donate or attract hydrogen ions. For example, if nitric acid is placed in water, water molecules will take protons from the nitric acid molecules, forming hydronium ions and nitrate ions. Water molecules have a stronger attraction to hydrogen ions than nitrate ions do; the water molecule is the hydrogen ion acceptor and therefore the base. HNO_3 is the acid, since it is the hydrogen ion donor.

$$\underset{\text{Acid}}{HNO_3} + \underset{\text{Base}}{H_2O} \longrightarrow \underset{\text{Hydronium ion}}{H_3O^+} + \underset{\text{Nitrate ion}}{NO_3^-}$$

Because many acid–base reactions can proceed both in the forward and reverse directions (but more easily in one direction), the products of an acid–base reaction themselves can serve as acids and bases for the reverse reaction. The above reaction of nitric acid and water proceeds in both the forward and reverse directions. In the reverse reaction, the hydronium ion is the acid and the nitrate ion the base.

$$HNO_3 + H_2O \longleftarrow \underset{\text{Acid}}{H_3O^+} + \underset{\text{Base}}{NO_3^-}$$

Since hydronium ion and nitrate ion are products of the initial acid–base reaction, they are called the **conjugate acid** and the **conjugate base**.

$$\underset{\text{Acid}}{HNO_3} + \underset{\text{Base}}{H_2O} \rightleftharpoons \underset{\text{Conjugate acid}}{H_3O^+} + \underset{\text{Conjugate base}}{NO_3^-}$$

When an acid loses a proton, the remaining ion is called a conjugate base because it will act as a base if a proton donor becomes available. When a

When an acid loses a hydrogen ion, the remaining negative ion is called a *conjugate base*.

base accepts a proton, the resulting compound is called a conjugate acid because it will act as an acid if a proton acceptor becomes available.

A base that has accepted a hydrogen ion becomes a *conjugate acid.*

As an example, let's look at ammonia, a common household cleaner. It has a strong attraction for hydrogen ions, so if placed in water, some protons are transferred from the water molecules to the ammonia molecules. In this case, the ammonia molecule is the hydrogen ion acceptor and therefore the base. Water, the hydrogen ion donor, is the acid.

H_2O	+	NH_3	$\leftrightarrows$	NH_4^+	+	OH^-
Acid		Base		Ammonium ion		Hydroxide ion
Hydrogen ion donor		**Hydrogen ion acceptor**				

The reaction proceeds more easily to the left, thus, equilibrium favors the species on the left side of the reaction. More ammonia is present dissolved in water as NH_3 than as NH_4^+.

Classification of Molecules as Acids or Bases

It is possible to classify molecules and ions in terms of their ability to capture a hydrogen ion.

Classification of a molecule or ion as an acid or base is a question similar to that of determining bond type by the electronegativity difference. It depends on who the partner is. Whichever molecule or ion gives up a hydrogen in any given reaction is considered to be the acid, and whichever molecule or ion gains a hydrogen ion is considered the base. Table 11.2 lists a number of common ions and molecules arranged upward in order of increasing ability to capture a hydrogen ion and downward in order of increasing tendency to give up a hydrogen ion. All reactions are written in the forward direction as hydrogen ion capturing reactions.

The logic of this table is the same as that used in Chapter 5 when electronegativity differences were presented as a means of predicting the direction

TABLE 11.2
Common Molecules and Ions Listed in Order of Their Tendency to Capture or Give Up a Hydrogen Ion

Molecule or Ion		Molecule or Ion
Strongest Base		
Hydroxide ion	$OH^- + H^+ \rightleftarrows H_2O$	Water
Carbonate ion	$CO_3^{2-} + H^+ \rightleftarrows HCO_3^-$	Bicarbonate ion
Ammonia	$NH_3 + H^+ \rightleftarrows NH_4^+$	Ammonium ion
Bicarbonate ion	$HCO_3^- + H^+ \rightleftarrows H_2CO_3$	Carbonic acid
Acetate ion	$C_2H_3O_2^- + H^+ \rightleftarrows HC_2O_3O_2$	Acetic acid
Hydrogen phosphate ion	$H_2PO_4^- + H^+ \rightleftarrows H_3PO_4$	Phosphoric acid
Sulfate ion	$SO_4^{2-} + H^+ \rightleftarrows HSO_4^-$	Bisulfate ion
Water	$H_2O + H^+ \rightleftarrows H_3O^+$	Hydronium ion
Nitrate ion	$NO_3^- + H^+ \rightleftarrows HNO_3$	Nitric acid
Chloride ion	$Cl^- + H^+ \rightleftarrows HCl$	Hydrochloric acid
Bisulfate ion	$HSO_4^- + H^+ \rightleftarrows H_2SO_4$	Sulfuric acid
Perchlorate ion	$ClO_4^- + H^+ \rightleftarrows HClO_4$	Perchloric acid
		Strongest Acid

↑ Increasing Tendency to Capture Hydrogen Ions

↓ Increasing Tendency to Give Up Hydrogen Ions

of electron transfer in ionic bonding (see Table 5.4). It is also similar to the reasoning used in Chapter 7 when the electrochemical series was used to predict the direction of electron transfer in oxidation-reduction reactions (see Table 7.11).

Hydrogen ions are transferred from species of lower capturing ability to species of higher capturing ability.

Remember that for a molecule or ion to act as an acid—to donate a hydrogen ion—another molecule or ion (a base) must be present to accept it. To use this table, you must choose two reactions, one below the other. The lower reaction will proceed to the left, giving up a hydrogen ion. The upper reaction will move to the right, gaining a hydrogen ion. The compound at the lower right will be the acid or the hydrogen ion donor; the compound at the upper left will be the base or hydrogen ion acceptor.

The information presented in Table 11.2 can help us predict whether a given molecule or ion will capture or donate hydrogen ions when placed with another molecule or ion.

EXAMPLE 11.1

Suppose that hydrogen chloride is placed in water. Will hydrogen chloride give up a hydrogen ion to water, forming hydronium ion and chloride ion?

Predicted reaction:

$$HCl + H_2O \longrightarrow H_3O^+ + OH^-$$

The applicable parts of Table 11.2 are shown here to allow comparison of hydrogen ion attracting ability:

↑ Increasing Tendency to Capture Hydrogen Ions	**Base**		
	Water	$H_2O + H^+ \rightleftarrows H_3O^+$	Hydronium ion
	Chloride ion	$Cl^- + H^+ \rightleftarrows HCl$	Hydrochloric acid
			Acid ↓

Since the ability of a molecule to attract hydrogen ions increases as one moves *up* this scale, water should be able to remove a hydrogen ion from hydrogen chloride. The uppermost reaction in the table will proceed in the forward direction, and the lower reaction in the reverse direction. The molecule or ion higher in the table will gain the hydrogen ion. The hydrogen ion moves up the table. The reaction will proceed as predicted.

$$HCl + H_2O \longrightarrow H_3O^+ + Cl^-$$

EXAMPLE 11.2

If acetic acid is placed in water, will the acetic acid molecule donate a hydrogen ion to water to make the hydronium ion?

Predicted reaction:

$$HC2H_3O_2 + H_2O \longrightarrow H_3O^+ + C_2H_3O_2{}^-$$

Use Table 11.2 to compare the hydrogen ion attracting ability:

↑ Increasing Tendency to Capture Hydrogen Ions	**Base**			
	Acetate ion	$C_2H_3O_2^- + H^+$	$HC_2C_3O_2$	Acetic acid
		✕ **No!**		
	Water	$H_2O + H^+$	H_3O^+	Hydronium ion
				Acid ↓

For the predicted reaction to occur, acetic acid would have to transfer a hydrogen ion *down* the table to water. The reaction, however, prefers to go in the other direction. Any acetate ion present in water will take a hydrogen from any available hydronium ion to make water and the molecular form of acetic acid. Any acetate ion present in water is simply waiting for an opportunity to collide with a hydronium ion and form acetic acid.

$$HC_2H_3O_2 + H_2O \longleftarrow H_3O^+ + C_2H_3O_2^-$$

There is a subtle point to be made here. Note that we did *not* say that acetic acid will not react with water to form hydronium ion and acetate ion. We said that any free acetate ion will, upon encountering a hydronium ion, immediately form acetic acid and water. There will always be a small but measurable amount of hydronium ion and acetate ion present in a water solution of acetic acid, even though the equilibrium favors the left side of the equation.

$$HC_2H_3O_2 + H_2O \rightleftarrows H_3O^+ + C_2H_3O_2^-$$

Acetic acid is a *weak acid*, that is, it is a poor donor of hydrogen ions. (More about this in Section 11.3.)

EXAMPLE 11.3

Suppose that phosphoric acid were placed in a solution containing hydroxide ion. Can hydroxide ion take a hydrogen ion away from phosphoric acid?

Predicted reaction:

$$H_3PO_4 + OH^- \longrightarrow H_2O + H_2PO_4^-$$

Comparison of hydrogen ion attracting ability:

↑ Increasing Tendency to Capture Hydrogen Ions	**Base**				
	Hydroxide ion	$OH^- + H^+$	$\rightleftarrows$	HOH	Water
	Hydrogen phosphate ion	$H_2PO_4^- + H^+$	$\rightleftarrows$	H_3PO_4	Phosphoric acid
					Acid ↓

$$\begin{array}{ccccc} HA & + & H_2O \rightleftharpoons & H_3O^+ & + A^- \\ 0.83\ M - [x] & & & [x] & [x] \end{array}$$

Now, by substituting these values into the equation for the acid equilibrium constant, we get

$$K_a = K_{eq}(55.5\ M) = \frac{[H_3O^+][A^-]}{[HA]}$$

$$1.7 \times 10^{-5} = \frac{[x][x]}{0.83 - [x]}$$

At this point, to simplify the math considerably, we can assume the $[x]$ value in the denominator to be equal to zero. We can do this because the ionized amount $[x]$ is very small with respect to the concentration of the un-ionized acid HA. If this assumption is made, and the very small error it creates is accepted, the problem becomes

$$1.7 \times 10^{-5} = \frac{[x][x]}{0.83}$$

$$(1.7 \times 10^{-5})(0.83) = [x]^2$$

$$1.41 \times 10^{-5} = [x]^2$$

$$3.75 \times 10^{-3} = [x]$$

The hydronium ion concentration is 3.75×10^{-3} mol/L, and the pH is slightly less than 3. Methods of exactly computing pH will be discussed later in Section 11.6 of this chapter.

Note that the amount of hydronium ion formed, and therefore the amount of acetic acid ionized, is

$$3.75 \times 10^{-3}\ M = 0.00375\ M,$$

When compared to the initial amount of un-ionized acetic acid present,

$$\frac{0.00375\ M}{0.83\ M} = 0.0045 = 0.45\%$$

which means that only 0.45% of the original acid was ionized. Less than half of 1% of the acetic acid was in the ionized state at any one time. This is why it is regarded as a weak acid, and why one can make the simplifying assumption that, for practical purposes, the original acid concentration is not changed when it is dissolved in water.

In contrast, the K_a for a strong acid is much larger than 1.0, and for all practical purposes, almost complete ionization takes place. If one begins with 0.1 M HCl, the resulting solution will contain 0.1 M H_3O^+ and 0.1 M Cl^-.

$$\begin{array}{lcccc} & HCl + H_2O & \rightleftharpoons & H_3O^+ & + \ Cl^- \\ \text{Start:} & 0.1\ M & & 0 & 0 \\ \text{Finish:} & 0 & & 0.1\ M & 0.1\ M \end{array}$$

The equilibrium is so far to the right that for practical purposes, one can assume that there is no un-ionized acid. The pH of this solution is 1.0.

The Water Ionization Constant

The ionization of pure water is a special case that produces a useful ionization constant that can be used to compute hydroxide ion concentration if hydronium concentration is known, or **pOH** if pH is known.

$$H_2O + H_2O \rightleftharpoons H_3O^+ + OH^-$$

The equilibrium equation for the ionization of water is

$$K_{eq} = \frac{[H_3O^+][OH^-]}{[H_2O][H_2O]}$$

We know that the concentration of water is about 55.5 mol/L and that in each liter only about 0.0000001 mol of hydronium ion and hydroxide ion are present. The equilibrium is very much to the left. For practical purposes, *no* water is ionized. Armed with this assumption, one can further assume (as was done earlier) that the water concentration is constant, and that

$$K_{eq}[H_2O][H_2O] = [H_3O^+][OH^-]$$
$$K_{eq}(\text{Constant}) = [H_3O^+][OH^-]$$

Since both H_3O^+ and OH^- in water are $1 \times 10^{-7}\ M$, this reduces to

$$K_{eq}(\text{Constant}) = (1 \times 10^{-7})(1 \times 10^{-7}) = 1 \times 10^{-14}$$

The *water ionization constant*, K_w, is used to compute hydronium or hydroxide ion concentrations in water solutions.

Since the product of any two constants is a third constant, a new constant can be created. This constant is called the **water ionization constant, K_w**. The product of the H_3O^+ concentration times the OH^- concentration in any water solution will always be 1×10^{-14}. This is the value of the water ionization constant:

$$K_w = [H_3O^+][OH^-] = 1 \times 10^{-14}$$

Consider now our acetic acid sample that had a hydronium ion concentration of $3.75 \times 10^{-3}\ M$. What is the OH^- concentration in this solution?

$$K_w = 1 \times 10^{-14} = [H_3O^+][OH^-]$$
$$= (3.75 \times 10^{-3})[OH^-]$$

$$[OH^-] = \frac{1 \times 10^{-14}}{3.75 \times 10^{-3}} = 2.67 \times 10^{-12}\ M$$

Since addition of logarithms is equivalent to multiplication, a similar generalization can be stated for computation of pOH when pH is known:

$$pH + pOH = 14$$

If the pH is 4.0, the pOH will be

$$pOH = 14 - pH = 10.$$

Acidic solution	**Basic solution**
$HIn + H_2O \leftrightarrows H_3O^+ + In^-$	$HIn + H_2O \rightleftarrows H_3O^+ + In^-$
High hydronium ion concentration HIn predominates	Low hydronium ion concentration In^- predominates

FIGURE 11.9
The form in which a chemical indicator is found depends on the hydronium ion concentration in the solution. When hydronium ion is predominant (acid solution), indicator ions (In^-) will have a good chance of colliding with H_3O^+ and picking up a hydrogen. The molecular form of the indicator, HIn, will predominate. On the other hand, if the presence of base makes H_3O^+ scarce, the chance of a hydronium to In^- collision is small, and most of the indicator will be in the In^- form.

Chemical Indicators and Equilibrium

A **chemical indicator** is a compound having a color that is dependent upon the pH of the solution in which it resides. Indicators are relatively common in nature as plant pigments and are found in grape juice, red cabbage, and tea. You have probably noticed the change in the color of tea when lemon juice (an acid) is added. Grape juice changes color when it is diluted with water, raising its pH. As you have perhaps guessed, these indicators are themselves acids or bases that have different colors in their molecular or ionized form.

The formula of an indicator may be expressed in the general form HIn, where H stands for an ionizable hydrogen and In for the remainder of the indicator molecule. When a hydrogen ion is attached to the indicator molecule, one type of bond structure exists and the indicator is a certain color. When the hydrogen is removed from the molecule, a different bond structure results and the colors of light absorbed by the indicator change.

Chemical indicators behave like weak acids. The color of an indicator changes as the molecule ionizes.

The color of an indicator solution is thus dependent on which form of the indicator is present—the form with the hydrogen or the form in which the hydrogen has been removed. The indicator methyl orange, for instance, is red in acid solution. The ionizable hydrogen is attached to the molecule, which is in the form HIn. The indicator is yellow in a basic solution, when the ionizable hydrogen has been removed by a base. Methyl orange changes color at about pH 3.5 (Figure 11.9).

Since different indicators have different tendencies to ionize, the hydrogen ion concentration at which the indicator will change differs from indicator to indicator. It is possible to use several chemical indicators to estimate the hydronium ion concentration of unknown solutions (see Figures 11.10 and 11.11).

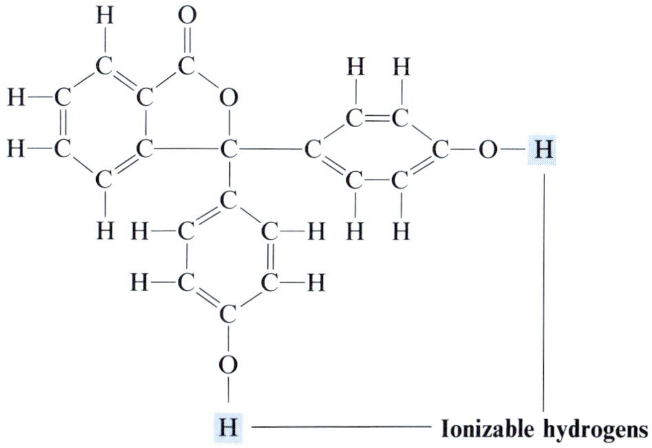
Ionizable hydrogens

FIGURE 11.10
The molecular structure of the indicator phenolphthalein is shown in this figure. When one of the two indicated hydrogens are removed by a base, the indicator turns from colorless to red.

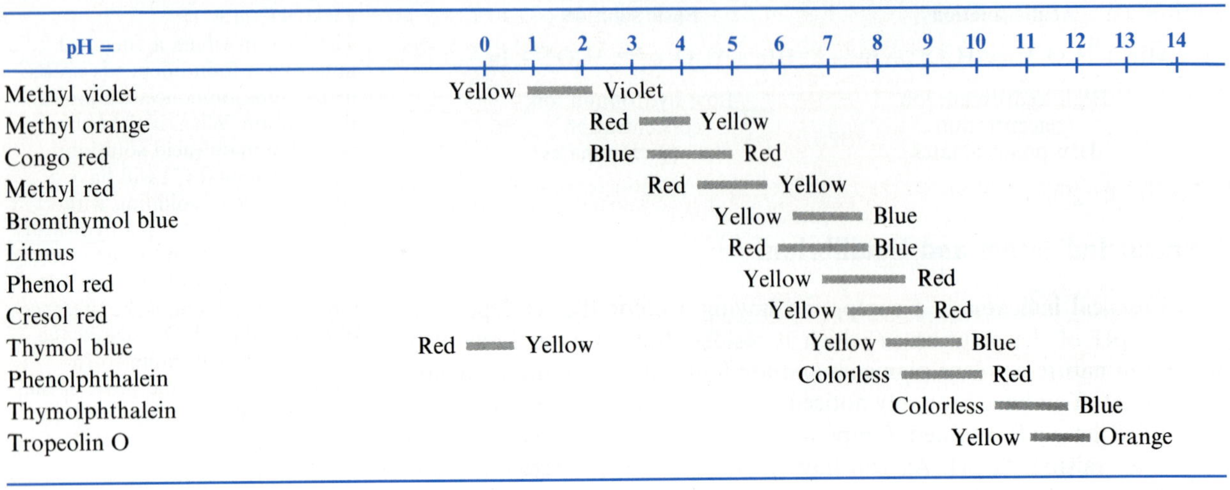
pH =
0 1 2 3 4 5 6 7 8 9 10 11 12 13 14
Methyl violet
Methyl orange
Congo red
Methyl red
Bromthymol blue
Litmus
Phenol red
Cresol red
Thymol blue
Phenolphthalein
Thymolphthalein
Tropeolin O
Yellow — Violet
Red — Yellow
Blue — Red
Red — Yellow
Yellow — Blue
Red — Blue
Yellow — Red
Yellow — Red
Red — Yellow
Yellow — Blue
Colorless — Red
Colorless — Blue
Yellow — Orange

FIGURE 11.11
The hydronium ion concentration at which a number of common chemical indicators change color is shown on this chart. This information can be used to estimate the hydronium ion concentration of unknown solutions.

NEW TERMS

pH A measure of hydronium ion concentration which equals $-\log[H_3O^+]$

acid equilibrium constant (K_a) The equilibrium constant for a weak acid, which is a combination of the water concentration (considered to be constant) and the equilibrium constant

pOH A measure of hydroxide ion concentration which equals $-\log[OH^-]$.

water ionization constant (K_w) The product of hydronium ion and hydroxide ion concentration in water solutions, which is always 1×10^{-14}.

chemical indicator A compound whose color is dependent on the hydronium ion concentration in its solution. Most indicators have just two colors and will indicate whether the hydronium ion concentration is above or below a certain $[H_3O^+]$ value.

Material	pH
Apples	2.9–3.3
Beans	5.0–6.0
Butter	6.1–6.4
Gooseberries	2.8–3.0
Lemons	2.2–2.4
Limes	1.8–2.0
Oranges	3.0–4.0
Milk	6.3–6.6

TESTING YOURSELF

pH, Equilibrium, and Indicators

1. Which of the materials listed in the margin is most basic?
2. Which of the materials listed in the margin is most acidic?
3. Consider a liquid that gives the results listed below when tested with a number of indicators. What is its approximate pH?

Indicator	Color
Congo red	Red
Methyl red	Yellow
Bromthymol blue	Blue
Phenol red	Red
Cresol red	Orange
Thymol blue	Yellow
Phenolphthalein	Colorless

4. Formic acid, found in bee stings, has an acid equilibrium constant of approximately 1×10^{-4}. What is a reasonable estimate for the hydronium ion concentration in a 1 *M* solution of formic acid?

$$HCOOH + H_2O \rightleftarrows H_3O^+ + HCOO^-$$

5. The hydronium ion concentration in the stomach is about 1×10^{-2} *M*, or about pH 2. This concentration is maintained by production of HCl by the parietal cells in the stomach lining. What is the OH^- concentration of this stomach acid?

Answers **1.** milk **2.** limes **3.** pH 8 **4.** 1×10^{-2} *M* or pH 2 **5.** 1×10^{-12} *M*

11.5 Acid–Base Reactions and Volumetric Analysis

The Relationship of Acids and Bases to Salts

Since acids and many hydroxide-containing bases ionize to provide the components necessary to make water, acid–base reactions (neutralization reactions) are considered to be a special case among chemical reactions. Consider the reaction between hydrochloric acid and sodium hydroxide:

$$HCl + NaOH \longrightarrow NaCl + H_2O$$

HCl	NaOH	NaCl	H_2O
Acid	Base	Sodium chloride (a salt)	Water

The net ionic reaction is

$$H^+ + OH^- \longrightarrow HOH$$

The reaction between nitric acid and potassium hydroxide is similar:

$$HNO_3 + KOH \longrightarrow KNO_3 + H_2O$$

HNO_3	KOH	KNO_3	H_2O
Acid	Base	Potassium nitrate (a salt)	Water

Again, the net ionic reaction is

$$H^+ + OH^- \longrightarrow HOH$$

Reactions between acids and hydroxide-containing bases result in the production of water and a compound known as a **salt**. In the above reactions, sodium chloride and potassium nitrate are both salts. If the water is evaporated from the solution containing the metal ions and chloride or nitrate anions, the salt will form an ionic crystal. Many important compounds are salts. Table 11.3 lists several common ones along with their uses and the acids and bases from which they are formed.

Salts are the metal-containing product of an acid–base reaction.

TABLE 11.3
Common Salts

Salt	Name	Use	Formed from Acid	Formed from Base
$NaCl$	Sodium chloride	Table salt	HCl	$NaOH$
KCl	Potassium chloride	Salt substitute	HCl	KOH
NH_4NO_3	Ammonium nitrate	Fertilizer	HNO_3	NH_4OH
Na_2CO_3	Sodium carbonate	Washing soda	H_2CO_3	$NaOH$
$AlCl_3$	Aluminum chloride	Deodorant	HCl	$Al(OH)_3$
$CaCl_2$	Calcium chloride	Desiccant	HCl	$Ca(OH)_2$
$Sr(NO_3)_2$	Strontium nitrate	Red color in fireworks	HNO_3	$Sr(OH)_2$
$MgSO_4$	Magnesium sulfate	Epsom salts	H_2SO_4	$Mg(OH)_2$

Net Ionic Equations

Since acid–base reactions are really reactions between compounds that ionize in water, the reactions are more accurately written with the reactants and ionizable products expressed as ionic species.

$$H^+ + NO_3^- + K^+ + OH^- \longrightarrow K^+ + NO_3^- + H_2O$$

(Note that the transfer of H^+ from the nitric acid to the hydroxide ion is actually accomplished by first forming a hydronium ion with water, and then reacting the hydronium ion with the hydroxide ion. Acid–base reactions are often simplified to show just the transfer of the hydrogen ion.) Since all reactants except the hydrogen and hydroxide remain in their ionic state, the net ionic reaction is simply

$$H^+ + OH^- \longrightarrow H_2O$$

Since water has only a very slight tendency to ionize, the chance for the reaction to proceed in the reverse direction is very slight. As a result, acid–base reactions tend to go forward until all hydrogen and hydroxide reactants have been converted to water.

Although an ionic equation is a more accurate representation of what actually occurs during an acid–base reaction, equations are usually written to show the molecular species.

Volumetric Analysis

Volumetric mole problems are similar to weight-relationship mole problems.

Acid–base reactions are important in the digestive system of the human body. The digestive process involves hydrochloric acid (HCl), which is produced in the parietal cells of the stomach. Gastric juice contains a fair amount of hydrochloric acid. This acid can be analyzed by volumetric analysis, as will be shown later in this section. *Volumetric analysis* is an analysis technique that depends on measurement of volumes rather than masses.

Tension, among other factors, can cause overproduction of HCl, resulting in a high acid concentration in the stomach and some discomfort. Antacid tablets are a common remedy to this problem, although not a valid long term solution. Most antacid tablets contain aluminum hydroxide, $Al(OH)_3$, or magnesium hydroxide, $Mg(OH)_2$. Aluminum hydroxide is usually favored to neutralize stomach acid since it contains a larger fraction of hydroxide and since magnesium hydroxide also acts as a laxative.

$3\,HCl$	$+\ Al(OH)_3$	$\longrightarrow$	$AlCl_3$	$+\ 3\,H_2O$
Stomach acid	Antacid (base)		Salt	Water

Prolonged exposure of stomach and small intestine lining to excessive acid concentrations results in ulceration of the lining. One of the diagnostic tests usually ordered by physicians when they suspect gastric ulcers is analysis of the gastric juice for acid concentration. To accomplish this, a sample of gastric juice is pumped from the patient's stomach. A measured amount of this is reacted with a solution of a base (for example, NaOH), and the concentration of acid is calculated using standard mole problem procedures.

The progress of the reaction can be followed by placing a few drops of a chemical indicator in the solution (Figure 11.12). Bromthymol blue, for example, turns from yellow to blue as the solution changes from acid to basic, with a green color existing at pH 7. The pH changes rapidly as the equivalence point is reached, which is the point at which the amount of hydronium ion present is exactly equalled by the amount of base. Thus, indicators that change color near pH 7, such as phenolphthalein at pH 8, give almost equal results. Phenolphthalein is a commonly used indicator since its change from colorless to red is very distinctive.

The process of reacting a measured volume of one reactant with another is called **titration**. If the concentration of one reactant is known, the amount of the other material present can be computed, as illustrated in the following examples.

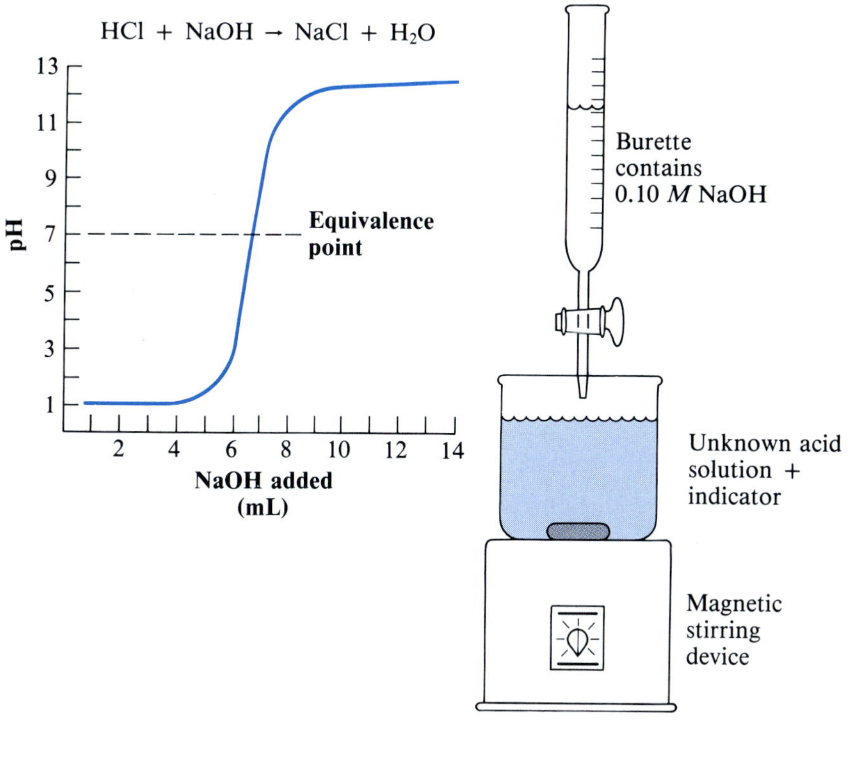
HCl + NaOH → NaCl + H2O
pH
13
11
9
7
5
3
1
Equivalence point
2 4 6 8 10 12 14
NaOH added (mL)
Burette contains 0.10 M NaOH
Unknown acid solution + indicator
Magnetic stirring device

FIGURE 11.12
Concentration of an unknown acid solution can be determined by reacting a known quantity of the acid with a measured amount of base. This process, called *titration*, can be followed either with a pH meter or by a chemical indicator. A chart produced by a recording pH meter is shown in the figure. At the point where the number of moles of NaOH added equals the number of moles of HCl present in the sample, the solution should be composed of NaCl and H_2O, and the pH should be about 7.

EXAMPLE 11.4

A 5.00-mL sample of gastric juice is diluted with about 100 mL of water to provide a reasonable volume for the reaction to take place. Several drops of bromthymol blue indicator are added. NaOH solution at a concentration of 0.100 mol/L is then slowly added from a burette until the indicator turns from yellow to green. At this point 12.1 mL of 0.100 *M* NaOH solution has been added. How many grams of HCl are in the gastric juice sample?

a. Write a balanced equation for the reaction.

$$\text{HCl} + \text{NaOH} \longrightarrow \text{NaCl} + \text{H}_2\text{O}$$

b. Find moles of known compound (NaOH).

1000 mL = 0.1000 mol NaOH
12.1 mL were used

$$(12.1\ \cancel{\text{mL}})\left(\frac{0.1000\ \text{mol NaOH}}{1000\ \cancel{\text{mL}}}\right) = 0.00121\ \text{mol NaOH}$$

$$\text{HCl} + \text{NaOH} \longrightarrow \text{NaCl} + \text{H}_2\text{O}$$

? g HCl ↑ ? mol HCl ⟵ **0.00121 mol NaOH** ↓ 12.1 mL 0.100 *M* NaOH

HCl	+	NaOH	⟶ NaCl + H_2O
? g HCl		12.1 mL 0.100 *M* NaOH	
↑		↓	
? mol HCl	⟵	**0.00121 mol NaOH**	

c. Predict moles unknown compound (HCl).

$$\text{HCl} + \text{NaOH} \longrightarrow \text{NaCl} + \text{H}_2\text{O}$$

From the balanced equation:

1 mol HCl requires 1 mol NaOH

$$(0.00121\ \cancel{\text{mol}}\ \text{NaOH})\left(\frac{1\ \text{mol HCl}}{1\ \cancel{\text{mol}}\ \text{NaOH}}\right) = 0.00121\ \text{mol HCl}$$

HCl	+	NaOH	⟶ NaCl + H_2O
? g HCl		12.1 mL 0.100 *M* NaOH	
↑		↓	
0.00121 mol HCl	⟵	0.00121 mol NaOH	

d. Find grams of HCl:

1 H	@	1.00 g
1 Cl	@	35.45 g
HCl:		36.45 g = 1 mol

$$(0.00121\ \cancel{\text{mol}}\ \text{HCl})\left(\frac{36.45\ \text{g}}{1.00\ \cancel{\text{mol}}}\right) = 0.044\ \text{g HCl}$$

HCl	+	NaOH	⟶ NaCl + H_2O
0.044 g HCl		12.1 mL 0.100 *M* NaOH	
↑		↓	
0.00121 mol HCl	⟵	0.00121 mol NaOH	

EXAMPLE 11.5

Vinegar is normally about 5% by weight acetic acid. Suppose that 10.0 mL (10.0 g) of vinegar is diluted with about 100 mL of water to provide an appropriate volume for the reaction. Several drops of bromthymol blue indicator is added. The indicator is yellow, indicating an acidic solution. NaOH solution, 0.1000 mol/L, is then added from a burette until the indicator turns green. Total volume NaOH: 49.0 mL. What is the percent by weight acetic acid?

a. Write a balanced equation for the reaction.

$$HC_2H_3O_2 + NaOH \longrightarrow NaC_2H_3O_2 + H_2O$$

b. Find moles of known compound (NaOH).

1000 mL = 0.1000 mol NaOH
49.0 mL were used

$$(49.0\ \cancel{\text{mL}})\left(\frac{0.1000\ \text{mol NaOH}}{1000\ \cancel{\text{mL}}}\right) = 0.00490\ \text{mol NaOH}$$

$$HC_2H_3O_2 \quad + \quad NaOH \longrightarrow NaC_2H_3O_2 + H_2O$$

? g acid ↑ ⟵ ? mol acid ⟵ 0.00490 mol NaOH ⟵ ↓ 49.0 mL 0.100 *M* NaOH

c. Predict moles of unknown compound (acetic acid) present.

$$HC_2H_3O_2 + NaOH \longrightarrow NaC_2H_3O_2 + H_2O$$

From the balanced equation:

$$1\ \text{mol}\ HC_2H_3O_2\ \text{requires}\ 1\ \text{mol NaOH}$$

$$(0.00121\ \cancel{\text{mol}}\ \text{NaOH})\left(\frac{1\ \text{mol}\ HC_2H_3O_2}{1\ \cancel{\text{mol}}\ \text{NaOH}}\right) = 0.0049\ \text{mol}\ HC_2H_3O_2$$

$$HC_2H_3O_2 \quad + \quad NaOH \longrightarrow NaC_2H_3O_2 + H_2O$$

? g acid ↑ ⟵ 0.0049 mol acid ⟵ 0.0049 mol NaOH ⟵ ↓ 49.0 mL 0.100 *M* NaOH

d. Find grams of acetic acid.

1 H @	1.00 =	1.00	
2 C @	12.00 =	24.00	
3 H @	1.00 =	3.00	
2 O @	16.00 =	32.00	
		60.00 g	= 1.00 mol

$$(0.0049\ \cancel{\text{mol}})\left(\frac{60.00\ \text{g}}{1.00\ \cancel{\text{mol}}}\right) = 0.294\ \text{g acetic acid}$$

$HC_2H_3O_2$ +	$NaOH \longrightarrow NaC_2H_3O_2 + H_2O$
0.294 g acid	49.0 mL 0.100 *M* NaOH
↑	↓
0.0049 mol acid ⟵	0.0049 mol NaOH

e. Compute percent by weight acetic acid in a 10.0 g vinegar sample.

$$\left(\frac{0.294 \text{ g acid}}{10.0 \text{ g sample}}\right) \times 100 = 2.94\% \text{ acetic acid}$$

This vinegar is weak.

Normality in Volumetric Analysis

Normality is the molarity of a solution expressed in terms of reactable hydrogen ions.

Volumetric analysis is one of the most commonly used techniques in analytical chemistry. To make the process simpler, a shortcut unit called **normality** has been developed. Normality is simply the molarity of a solution expressed in reactable hydrogen ions or hydroxide ions. A 1.0 *M* solution of HCl has one mole per liter reactable hydrogen, thus it is 1.0 normal, or 1.0 *N*. On the other hand, a 1.0 *M* solution of sulfuric acid, H_2SO_4, has two moles of reactable hydrogen and is 2.0 *N*.

The value of normality is based on the net ionic equation for an acid–base reaction:

$$H^+ + OH^- \longrightarrow H_2O$$

One mole of hydrogen ion reacts with one mole of hydroxide ion to form one mole of water. Thus, at the equivalence point of the titration,

$$\text{Number of moles of } H^+ = \text{Number of moles of } OH^-$$

The number of moles of hydrogen ion placed in the solution equals

$$\left(\frac{\text{Moles of reactable H}}{\cancel{\text{Liters}} \text{ of reactable H}}\right)(\cancel{\text{Liters}} \text{ added}) = \text{Moles of reactable hydrogen}$$

Normality is the *number of hydrogen equivalents* per liter of solution.

$$(\text{Normality of acid})(\text{Liters of acid}) = \text{Equivalents of hydrogen}$$

Since, at the equivalence point,

$$\text{Equivalents of hydrogen} = \text{Equivalents of hydroxide}$$

a simple equality can be set up relating normality and volume of both acid and base solutions:

$$\text{Equivalents of hydrogen} = \text{Equivalents of hydroxide}$$

$$(\text{Normality of acid})(\text{Liters of acid}) = (\text{Normality of base})(\text{Liters of base})$$

The next two examples show the use of this equation for volumetric analysis.

EXAMPLE 11.6

A volume of 50.0 mL of an unknown base solution is titrated to neutralization with 35.0 mL of 0.100 *N* acid solution. What is the normality of the basic solution?

$$(\text{Normality of acid})(\text{Liters of acid}) = (\text{Normality of base})(\text{Liters of base})$$

$$(0.1\ N)(0.0350\ \text{L}) = (\text{Normality of base})(0.0500\ \text{L})$$

$$\text{Normality of base} = \frac{(0.1\ N)(0.0350\ \cancel{\text{L}})}{(0.0500\ \cancel{\text{L}})} = 0.07\ N$$

Note that the identity of the acid and base are no longer important—just the number of reactable hydrogens and hydroxide groups.

Often milliliters and milliequivalents are used instead, since these units are more convenient in the laboratory, as shown in the next example.

EXAMPLE 11.7

A volume of 14.0 mL of 0.25 *N* acid is used to neutralize 45.0 mL of unknown base. What is the normality of the basic solution?

$$\begin{pmatrix}\text{Normality}\\ \text{of acid}\end{pmatrix}\begin{pmatrix}\text{Milliliters}\\ \text{of acid}\end{pmatrix} = \begin{pmatrix}\text{Normality}\\ \text{of base}\end{pmatrix}\begin{pmatrix}\text{Milliliters}\\ \text{of base}\end{pmatrix}$$

$$(0.25\ N)(14.0\ \text{mL}) = (\text{Normality of base})(45.0\ \text{mL})$$

$$\text{Normality of base} = \frac{(0.25\ N)(14.0\ \cancel{\text{mL}})}{(45.0\ \cancel{\text{mL}})} = 0.0778\ N$$

The value of the normality system is its ease of use. Its disadvantage is that one can lose sight of the chemistry involved.

NEW TERMS

salt The metal-containing product of an acid–base reaction.

titration A process in which a measured volume of a known solution is reacted with a solution of unknown concentration to determine its concentration. The point at which the unknown is completely consumed is usually identified by a chemical indicator that changes color at the "end point."

normality The concentration of a solution in moles per liter of reactable hydrogen ion. For example, a 1.0 *M* solution of HCl is 1.0 *N*, and a 1.0 *M* solution of the strong acid H_2SO_4 is 2.0 *N*.

TESTING YOURSELF

Acid–Base Reactions

1. Predict the products of each of the following acid–base reactions, identify and name the salt produced, and balance the equation describing the reaction.
   a. Nitric acid + Magnesium hydroxide $\longrightarrow$
   b. Citric acid + Calcium hydroxide $\longrightarrow$
2. Suppose that 25.0 mL of an unknown hydrochloric acid solution was completely neutralized by addition of 40 mL of 0.1 *M* sodium hydroxide solution:

$$HCl + NaOH \longrightarrow NaCl + H_2O$$

   a. How many moles of NaOH were used to complete the reaction?
   b. How many moles of HCl were present in the sample?
   c. If 25.0 mL of the unknown sample contained 0.004 mol HCl, how many moles would be contained in 1000 mL of this material?
3. A common antacid tablet contains aluminum hydroxide, $Al(OH)_3$. Suppose that a 2.0-g tablet of this material is completely neutralized by 64.0 mL of 0.2 *M* HCl:

$$Al(OH)_3 + 3\ HCl \longrightarrow AlCl_3 + 3\ HOH$$

   a. How many moles of HCl were used to complete the reaction?
   b. How many moles of aluminum hydroxide were present in the tablet?
   c. How many grams of aluminum hydroxide were present in the tablet?
   d. What is the percent by weight of aluminum hydroxide in the 2.0-g tablet?
4. A 20.0-mL sample of soda pop containing phosphoric acid, H_3PO_4, is titrated with 16.5 mL 0.100 *N* base. What is the normality of the soda pop sample?
5. What is the normality of a solution of acetic acid that is 0.25 *M*?

$$HC_2H_3O_2 + H_2O \longrightarrow H_3O^+ + C_2H_3O_2^-$$

Answers **1. a.** $2\ HNO_3 + Mg(OH)_2 \longrightarrow Mg(NO_3)_2 + 2\ HOH$

Acid, Base, Salt (Magnesium nitrate), Water

b. $2\ HC_6H_7O_7 + Ca(OH)_2 \longrightarrow Ca(C_6H_7O_7)_2 + 2\ HOH$

Acid, Base, Salt (Calcium citrate), Water

2. a. 0.004 mol **b.** 0.004 mol **c.** 0.16 mol **3. a.** 0.0128 mol **b.** 0.00427 mol **c.** 0.333 g **d.** 16.6% **4.** 0.0825 *N* **5.** 0.25 *N*

11.6 Buffers and Equilibrium: The Hydrogen Ion in Biological Systems

Buffers are compounds that maintain solution pH reasonably constant with addition of moderate amounts of acid or base. They are mixtures of a weak acid and the salt of this acid.

Buffers: A Way to Stabilize pH

Buffers are chemical systems that maintain the pH of a solution at a relatively constant level with addition of moderate amounts of either acid or base. A buffer is composed of approximately equal amounts of a weak acid

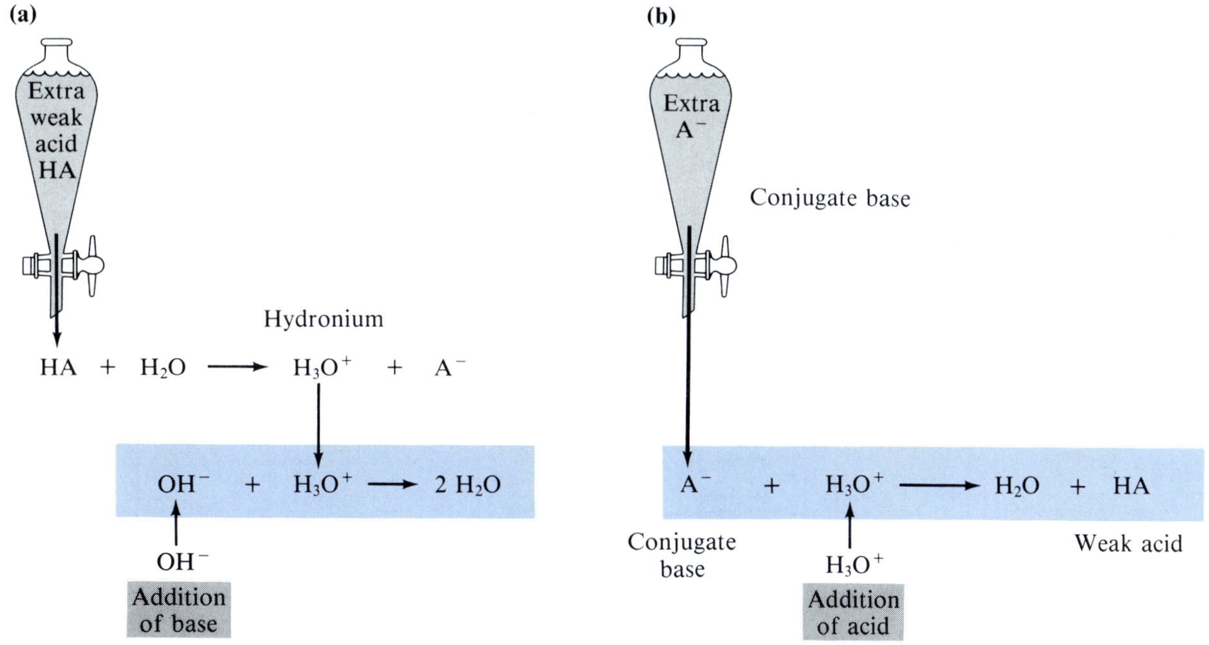

(a)
Extra weak acid HA
Hydronium
HA + H₂O → H₃O⁺ + A⁻
OH⁻ + H₃O⁺ → 2 H₂O
OH⁻
Addition of base
(b)
Extra A⁻
Conjugate base
A⁻ + H₃O⁺ → H₂O + HA
Conjugate base
H₃O⁺
Addition of acid
Weak acid

FIGURE 11.13
A buffer consists of reserve amounts of both a weak acid and the salt (conjugate base) of this acid. (a) As long as the reserve of weak acid (HA) holds out, incoming base will be neutralized by hydronium ion, and the pH will not rise. (b) Addition of acid results in formation of water and weak acid, HA. As long as the reserve of the conjugate base A^- holds out, incoming acid will be neutralized, and the pH will not rise.

and the salt of this acid. The salt is formed by the reaction of the weak acid with a strong base. For example, the potassium salt of the weak acid acetic acid is formed by the reaction

$$HC_2H_3O_2 + KOH \longrightarrow K^+C_2H_3O_2^- + H_2O$$

Weak acids have a slight tendency to ionize, while the negative ions from the salts of these acids have a strong tendency to capture hydronium ions when they collide with them, forming the molecular form of the acid; A^- thus serves as an ion trap (forming HA) for H^+:

$$\underset{\text{Weak acid}}{HA} + H_2O \rightleftarrows H_3O^+ + \underset{\text{Conjugate base}}{A^-}$$

Consider a solution containing relatively large amounts of both the weak acid (HA) and the salt of this acid (which contains the conjugate base A^-). The weak acid does not contribute many hydronium ions to the solution, and the conjugate base is standing ready to sweep up any stray hydronium ions that come its way. However, suppose that a moderate amount of additional strong acid or strong base were added to the solution. How would these components react (Figure 11.13)?

Note that buffers can regenerate themselves. The product of the reaction of the weak acid that neutralizes the incoming base is the conjugate base A^- ion, which will react with hydronium ion. As the conjugate base neutralizes incoming acid, the original weak acid is produced (Figure 11.14).

The fact that buffers normally have about equal reserves of base absorber (HA) and acid absorber (A^-) makes possible a simple calculation with the acid equilibrium constant:

$$HA + H_2O \rightleftarrows H_3O^+ + A^-$$

$$K_a = K_{eq}[H_2O] = \frac{[H_3O^+][A^-]}{[HA]}$$

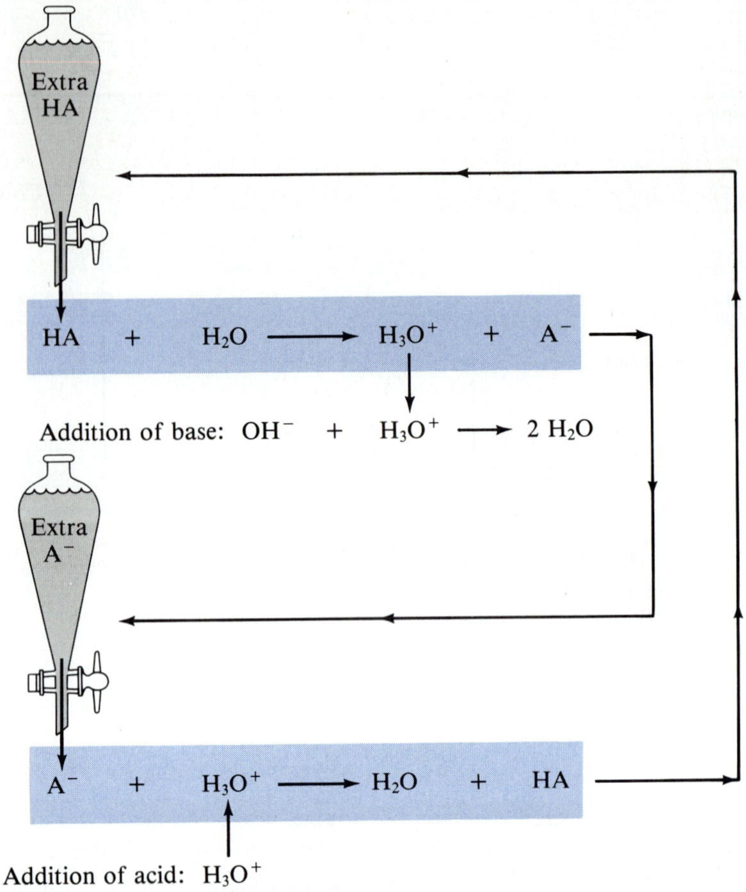
Extra
HA
HA + H2O → H3O+ + A−
Addition of base: OH− + H3O+ → 2 H2O
Extra
A−
A− + H3O+ → H2O + HA
Addition of acid: H3O+

FIGURE 11.14
Buffers tend to regenerate themselves if they are alternately subjected to excess acid and excess base. A buffer will hold the pH relatively constant until one of its reserve components runs out.

In the special case in which the weak acid concentration [HA] and the concentration of salt $[A^-]$ are equal, these values cancel in the equilibrium expression:

$$K_a = \frac{[H_3O^+]\cancel{[A^-]}}{\cancel{[HA]}}$$

We are left with

$$K_a = [H_3O^+]$$

The hydronium ion concentration of the buffer mixture is equal numerically to the value of the acid equilibrium constant. Buffers of a desired pH may be prepared by selection of the proper weak acid–salt combination.

EXAMPLE 11.8

What is the pH of a buffer solution made with 0.10 *M* lactic acid and 0.10 *M* sodium lactate (the salt of lactic acid)?

$$K_a = \frac{[H_3O^+]\cancel{(0.10\ M\ A^-)}}{\cancel{0.10\ M\ HA}} = 1.4 \times 10^{-4}\ \text{mol/L}$$

$$[H_3O^+] = 1.4 \times 10^{-4}\ \text{mol/L}$$

$$\text{pH} = 3.9$$

TABLE 11.4
Acid Equilibrium Constants for Some Common Weak Acids

Compound	Found in	Ionization Reaction	K_a	pK_a
Acetic acid	Vinegar	$HC_2H_3O_2 + H_2O \longrightarrow H_3O^+ + C_2H_3O_2^-$	1.7×10^{-5}	4.76
Ascorbic acid	Vitamin C	$HC_6H_7O_6 + H_2O \longrightarrow H_3O^+ + C_6H_7O_6^-$	7.9×10^{-5}	4.10
Butyric acid	Rancid butter	$HC_4H_7O_2 + H_2O \longrightarrow H_3O^+ + C_4H_7O_2^-$	1.5×10^{-5}	4.82
Carbonic acid	Soft drinks	$H_2CO_3 + H_2O \longrightarrow H_3O^+ + HCO_3^-$	7.9×10^{-7}	6.10
Citric acid	Citrus fruit	$HC_6H_7O_7 + H_2O \longrightarrow H_3O^+ + C_6H_7O_7^-$	8.4×10^{-4}	3.07
Formic acid	Bee sting	$HCHO_2 + H_2O \longrightarrow H_3O^+ + CHO_2^-$	1.8×10^{-4}	3.74
Lactic acid	Milk	$HC_3H_5O_3 + H_2O \longrightarrow H_3O^+ + C_3H_5O_3^-$	1.0×10^{-4}	3.85
Oxalic acid	Rhubarb	$H_2C_2O_4 + H_2O \longrightarrow H_3O^+ + HC_2O_4^-$	6.5×10^{-2}	1.19
Tartaric acid	Wine	$H_2C_4H_4O_6 + H_2O \longrightarrow H_3O^+ + HC_4H_4O_6$	1.1×10^{-3}	2.96

Table 11.4 lists the ionization reactions and K_a values for a number of weak acids. It also lists a new value called **pK_a**. Just as pH is the negative log of the hydronium ion concentration, pK_a is the negative log of the acid ionization constant, K_a. The pK_a unit is useful because it is the solution pH at which the acid will transfer a hydrogen ion. It is also useful as a rating scale for hydrogen ion acceptors—the higher the number, the higher the attraction for hydrogen ions. The term pK_a is commonly used in organic and biochemistry.

Computing pH with Pocket Calculators

Pocket calculators equipped with a log function provide a very easy way to compute pH. In water solutions, pH is defined as

$$pH = -\log[H_3O^+]$$

Note that the formal definition of pH is also written as $-\log[H^+]$, but the bare hydrogen ion will not exist by itself in water solutions.

In Example 11.8, we saw that the computed hydronium ion concentration was

$$[H_3O^+] = 1.4 \times 10^{-4} \text{ mol/L}$$

To compute the pH with a pencil and logarithm table requires four steps:

1. Separate the concentration into number and exponent terms. Note that addition of logarithms is equivalent to multiplication of the numbers.

$$\log[H_3O^+] = \log 1.4 + \log 10^{-4} \text{ mol/L}$$

2. Determine the logarithm of the number and exponent terms by use of a logarithm table. The logarithm of 1.4 is 0.146, and the logarithm of 10^{-4} is -4.

$$\log[H_3O^+] = 0.146 + (-4) \text{ mol/L}$$

3. Add the two numbers.

$$\log[H_3O^+] = -3.854 \text{ mol/L}$$

4. Multiply the resulting number by -1.

$$-\log[H_3O^+] = (-1)(-3.854) \text{ mol/L}$$

$$pH = 3.85$$

Computation of pH with a calculator is much easier. Enter 1.4×10^{-4} into the calculator and press the "log" button. The display will present the number -3.85387196. Multiply by -1 and reduce to the proper number of significant figures to get pH = 3.85, as above. (When using logarithms, the number of significant figures are counted to the right of the decimal point.)

EXAMPLE 11.9

Compute the pH of a buffer solution of 1.0 *M* acetic acid and 0.50 *M* sodium acetate (the salt and conjugate base). The equilibrium constant for the ionization of acetic acid is 1.7×10^{-5}.

$$K_a = 1.7 \times 10^{-5} = \frac{[H_3O^+](0.50\ M\ A^-)}{1.0\ M\ HA}$$

$$[H_3O^+] = \frac{(1.7 \times 10^{-5})(1.0\ M)}{0.50\ M} = 3.4 \times 10^{-5}$$

Since the hydronium ion concentration is slightly greater than 10^{-5} mol/L, we would estimate the pH to be between 4 and 5. Pressing the log key on our calculator gives a result of -4.468; expressed to the proper number of significant figures, the result is pH 4.47.

Buffers in Living Systems

The concentration of hydronium ion is important to all living organisms. Many plants are quite pH sensitive and will grow only in soils of suitable pH range. Peas and beans, for example, grow best when the soil pH is in the range 6–8.

The range of pH is important to humans, too. The pH of blood must be maintained within a few hundredths of the value 7.40. Numerous stresses are put on our acid–base balance by the simple process of living. Exercise, for example, produces fair quantities of lactic acid and carbon dioxide, both of which tend to lower blood pH.

Carbon dioxide is produced as a byproduct of cell metabolism. This is dissolved in the bloodstream for the trip to the lungs where it is exhaled and oxygen picked up for the return trip. The dissolved carbon dioxide reacts with water in the bloodstream to form carbonic acid:

$$CO_2 + H_2O \rightleftarrows H_2CO_3$$

At any moment about 1000 times more CO_2 is carried simply dissolved in the blood than is converted to carbonic acid. There is an important implication in this—if something reacts with the carbonic acid in the blood to reduce the acid concentration and raise the blood pH, a large reserve of carbon dioxide is immediately available to form additional carbonic acid.

The carbonic acid–bicarbonate ion buffer is one of the systems that helps maintain blood pH at 7.40.

Blood also contains a fair amount of bicarbonate ion (for example, as sodium bicarbonate). This ion is the salt of the weak acid carbonic acid, and

ACID RAIN

Coal and other fossil fuels that we burn to heat our homes and to provide other energy needs contain small amounts of sulfur, often as much as 6 to 8%. When sulfur is burned with oxygen, the colorless, suffocating gas sulfur dioxide (SO_2) is produced. A total of 95% of our sulfur emissions into the atmosphere are in the form of sulfur dioxide.

$$S + O_2 \longrightarrow SO_2$$

The effects of sulfur dioxide on human health can be illustrated by the 4000 human deaths that occurred in London, England, in 1952 during a five-day period of polluted, stagnant air. The maximum concentration of sulfur dioxide at that time was recorded at an astonishing 1.34 parts per million (ppm). For comparison, the U.S. standard for maximum sulfur dioxide concentration is 0.14 ppm for a 24-hour period and 0.03 ppm annually. Concentrations as high as 0.12 ppm for a 24-hour period have been observed in Los Angeles, California.

The environmental effects of sulfur dioxide, however, are even more far-reaching. Sulfur dioxide reacts with oxygen in the atmosphere to form sulfur trioxide:

$$SO_2 + O \longrightarrow SO_3$$

Sulfur trioxide will react with water vapor to form sulfuric acid:

$$SO_3 + H_2O \longrightarrow H_2SO_4$$

Sulfuric acid can remain suspended in the atmosphere, forming an aerosol haze, or it will wash down with the next rain.

The reaction of gaseous sulfur dioxide with building materials, particularly limestone and marble (both forms of calcium carbonate), has irreparably damaged many buildings and monuments since the turn of the century.

$$H_2SO_4 + CaCO_3 \longrightarrow H_2O + CO_2 + CaSO_4$$

Rain water is normally slightly acidic because of the presence of dissolved carbon dioxide, which reacts with the water to form small amounts of the weak acid carbonic acid (H_2CO_3). The normal pH of rainwater is about 5.7. The presence of sulfuric acid in the atmosphere, however, pushes the pH of rain water down to the range of 3 to 5. This ultimately results in a lowering of the pH of freshwater streams and lakes and killing aquatic life.

Combustion of coal is one of the principal sources of heat for electric power plants. The mining of only low sulfur coal and scrubbing of sulfur compounds from power plant effluent can be a significant part of the expense of electricity production today.

---

it is present at about 20 times the concentration of carbonic acid. Together, these two compounds form a highly effective buffer system in our blood (Figure 11.15).

Blood normally contains bicarbonate ion (HCO_3^-) at a concentration of about 0.026 mol/L and carbonic acid (H_2CO_3) at about 0.0013 mol/L. Referring to the equilibrium constant for carbonic acid,

$$K_a = \frac{[H_3O^+][HCO_3^-]}{[H_2CO_3]} = 7.9 \times 10^{-7}$$

$$K_a = \frac{[H_3O^+](0.026\ M)}{0.0013\ M} = 7.9 \times 10^{-7}$$

$$[H_3O^+] = \frac{(7.9 \times 10^{-7})(0.0013\ M)}{0.026\ M} = 3.95 \times 10^{-8}\ M$$

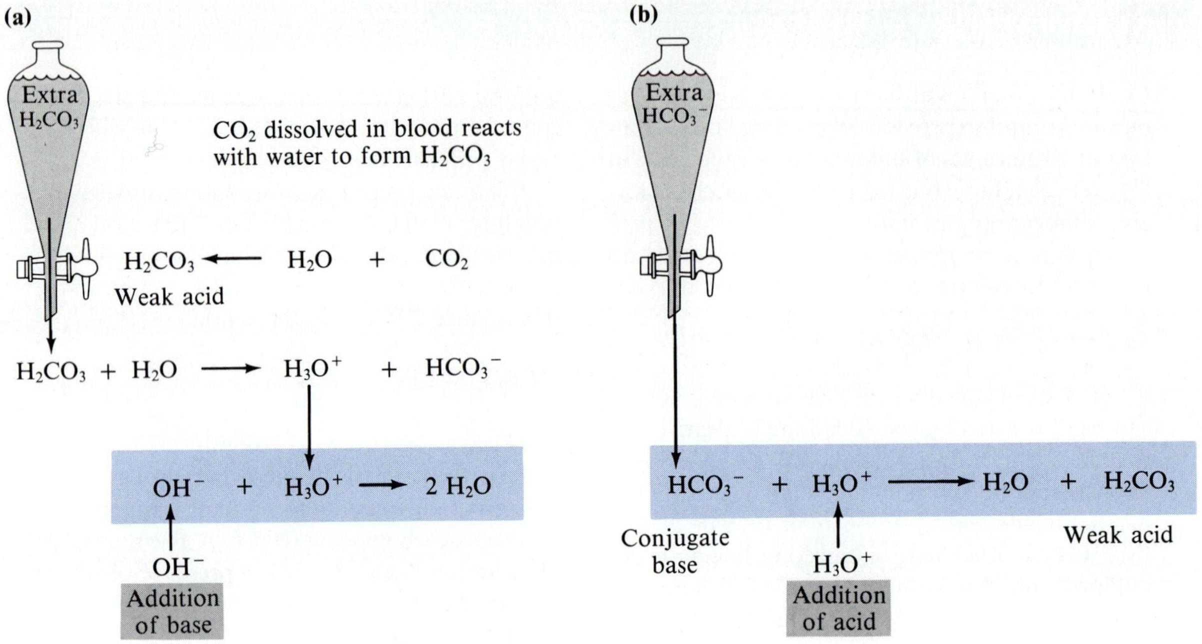

(a)
Extra H_2CO_3
CO_2 dissolved in blood reacts with water to form H_2CO_3
$H_2CO_3 \leftarrow H_2O + CO_2$
Weak acid
$H_2CO_3 + H_2O \rightarrow H_3O^+ + HCO_3^-$
$OH^- + H_3O^+ \rightarrow 2\ H_2O$
OH^-
Addition of base
(b)
Extra HCO_3^-
$HCO_3^- + H_3O^+ \rightarrow H_2O + H_2CO_3$
Conjugate base
H_3O^+
Addition of acid
Weak acid

FIGURE 11.15
The carbonic acid/bicarbonate buffer system holds the pH of the blood relatively constant with addition of moderate amounts of either acid or base. (a) Addition of base results in formation of water and additional HCO_3^-; as long as the reserve of H_2CO_3 holds out, incoming base will be neutralized and the pH will not rise. Dissolved CO_2 provides a reserve to generate additional H_2CO_3 as needed. (b) Addition of acid results in formation of water and weak acid, H_2CO_3. As long as the reserve of HCO_3^- holds out, incoming acid will be neutralized, and the pH will not rise.

The use of a calculator gives us a pH value of 7.4 which is maintained by our blood's buffer system.

The carbonic acid–bicarbonate buffer is a typical weak acid–salt buffer system. The large reserve of bicarbonate ion and dissolved carbon dioxide from which carbonic acid can be formed gives this buffer a large capability for reaction with either acid or base.

The respiratory system responds to changes in blood pH. A high concentration of CO_2 causes increased carbonic acid concentration and lower blood pH, triggering an increased respiration rate.

The respiratory system is partly controlled by the concentration of CO_2 in the blood. Strenuous exercise increases the concentration of CO_2, thus increasing the concentration of H_2CO_3 and lowering blood pH. The decrease in blood pH results in increased respiratory rate and more rapid removal of CO_2 through the lungs. This reduces the concentration of carbonic acid and thereby raises the blood pH to an acceptable level. If you overbreathe, or hyperventilate, CO_2 will be exhaled and its level will fall below normal. The blood pH will increase, and you may feel light-headed. Increase in blood pH above normal levels is called *alkalosis*, whereas a decrease in blood pH below normal levels is called *acidosis*. Alkalosis often produces fainting, which lowers the respiratory rate and allows the blood pH to return to a lower, more normal level.

The bicarbonate–carbonic acid buffer system is only one of the hydrogen ion control systems present in the body. Blood pH is also maintained by a phosphate ion buffer system and by binding of hydronium ions to proteins in the blood plasma. The kidneys perform a similar function in that they are able to vary the pH of the urine from about 4.5 to 8, thus either discarding or holding hydronium ion to return the blood to pH 7.40.

NEW TERMS

buffer A mixture of roughly equal parts weak acid and the salt of the acid (its conjugate base) that maintains the pH of a solution reasonably constant with addition of moderate amounts of either acid or base.

pK_a The logarithm of the acid ionization constant. The pK_a is the solution pH at which the acid will transfer a hydrogen ion.

TESTING YOURSELF

Buffers

1. A buffer system is formed with tartaric acid ($H_2C_4H_4O_6$) and potassium tartrate ($K^+HC_4H_4O_6^-$) during fermentation of wine. Suppose that the concentrations of the acid and its salt were approximately equal. What is a reasonable estimate of the pH of the wine at this point in its production?
2. A buffer system is formed by mixing equal concentrations of citric acid and sodium citrate, which ionizes to provide the citrate ion. What is a reasonable estimate of the pH of this solution?

Answers **1.** pH = 2.96 **2.** pH = 3.07

11.7 Lewis Acids and Bases

Both the Arrhenius and Brønsted–Lowry theories viewed acids as hydrogen ion donors and bases as hydrogen ion acceptors. G. N. Lewis, an American chemist who proposed the electron-pair theory of covalent bonding, suggested a more generalized view of acid–base reactions. While both the Arrhenius and Brønsted–Lowry theories focused on transfer of the hydrogen ion, Lewis focused on the electron pair that accepted the proton.

Consider the reaction of water with ammonia, NH_3:

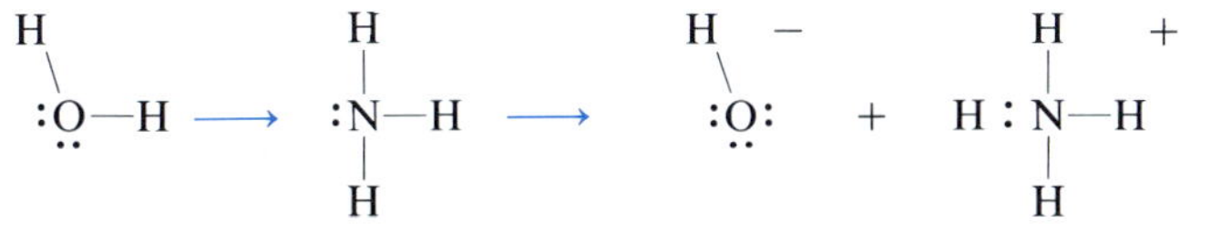

Water (acid) | Ammonia (base) | Hydroxide ion | Ammonium ion

When one of a water molecule's hydrogens collides with the unshared electron pairs of an ammonia molecule, transfer of the hydrogen ion sometimes occurs. In this case, the water molecule is the hydrogen ion donor or acid, while the ammonia molecule is the hydrogen ion acceptor, or base. A covalent bond is formed between hydrogen and nitrogen, using the nitrogen's unshared pair of electrons. The focus in this explanation of the reaction is upon transfer of the hydrogen ion.

Now consider this reaction from another point of view. The unshared electron pair on ammonia's nitrogen could be viewed as a "covalent bond waiting to happen." If this pair of electrons were to encounter an atom of appropriate electronegativity with an empty orbital, a covalent bond could form. When the unshared electron pairs of the ammonia bump against one of the water molecule's hydrogens, temporarily dislodging it, the resulting hydrogen ion has an empty orbital and can then accept nitrogen's pair of electrons to form a covalent bond:

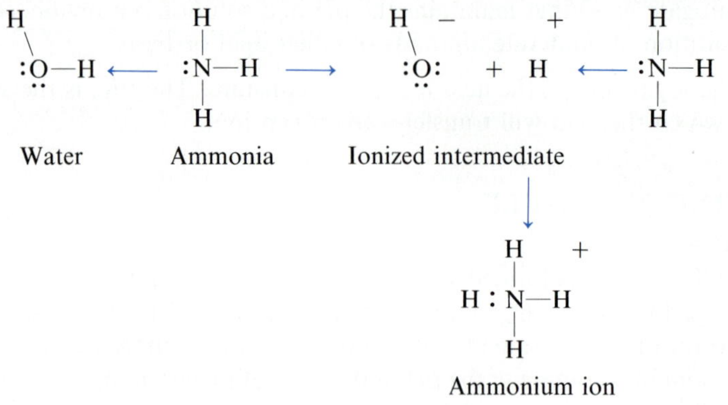
Water
Ammonia
Ionized intermediate
Ammonium ion

A *Lewis acid* is an electron pair acceptor, while a *Lewis base* is an electron pair donor.

In this explanation for the reaction, the electrons for the covalent bond are donated by the nitrogen of the ammonia molecule. The ammonia molecule is an *electron pair donor*, or a **Lewis base**. The hydrogen ion is an *electron pair acceptor*, or a **Lewis acid**.

A Lewis base need not use its unshared electron pair to form a covalent bond only with hydrogen. Any molecule with an empty orbital capable of accepting the electron pair to form a covalent bond can play the part of a Lewis acid. Many reactions in organic and biochemistry are easily explained by tracing the path of an unshared electron pair as it seeks to form a covalent bond. We will encounter these reactions beginning in Chapter 14.

NEW TERMS

Lewis base A molecule or ion that can form a covalent bond with another species by donating a pair of electrons.

Lewis acid A molecule or ion that can form a covalent bond with another species by accepting a pair of electrons.

SUMMARY

Water is the second most prevalent molecule on earth. Water ionizes very slightly, producing hydrogen ions and hydronium ions. Since the hydrogen ion (a bare proton) is not stable by itself, it actually exists as *hydronium ion* (H_3O^+) in aqueous solutions.

This chapter introduced two definitions of an *acid*—the early hydrogen ion donor idea of *Arrhenius* and the more accurate hydronium ion concept developed independently by *Brønsted* and *Lowry*. One must remember that an acid is only half of the picture; for a hydrogen ion to be donated, an acceptor ion or molecule must be present. These acceptor molecules or ions are called *bases*. *Strong acids* donate hydrogen ions more easily than does water; *weak acids* hold their hydrogen ions more tightly than does water. The *pH scale* is a convenient way of expressing hydrogen ion (actually, hydronium ion) concentration in solutions. A pH of 7 is neutral (the concentrations of hydronium ion and hydroxide ion are equal). Values of pH less than 7 are classified as acidic, while values from 7 to 14 are classified as basic. *Chemical indicators* change colors at particular pH values and are used to estimate solution pH.

Acid–base reactions are of great value in analytical chemistry. The *volumetric analysis* technique introduced in this chapter is probably the most common wet chemistry technique in medical and biological laboratories.

Control of hydronium ion concentration with moderate additions of acid or base is accomplished by a *buffer system*, which is a mixture of a weak acid and the *salt* of this acid. The concept of the equilibrium constant introduced in Chapter 10 can be used to predict the behavior of buffer systems.

TERMS

acid	pOH
Arrhenius model of an acid	water ionization constant (K_w)
Brønsted–Lowry model of an acid	chemical indicator
hydronium ion	salt
conjugate acid	titration
conjugate base	normality
strong acid	buffer
weak acid	pK_a
pH	Lewis base
acid equilibrium constant (K_a)	Lewis acid

CHAPTER REVIEW

QUESTIONS

1. Which of the following compounds could never be considered an acid?
   a. H_3PO_4 b. H_2O c. KCl d. $H_2C_2O_2$

Refer to these two reactions to answer questions 2–5.

a. $HCl \longrightarrow H^+ + Cl^-$

b. $HCl + H_2O \longrightarrow H_3O^+ + Cl^-$

2. Which of these reactions represents the Arrhenius view of an acid?
3. Which of these reactions represents the Brønsted–Lowry view of an acid?
4. Which of these reactions involves hydronium ion?
5. One compound is acting as a base in one of the reactions listed above. What is the formula of this compound?
6. What is the basic difference between a strong acid and a weak acid, if equal quantities of each are placed in identical volumes of water?
7. A solution of boric acid has a hydronium ion concentration of 0.000001 mol/L.
   a. What is the pH of this solution?
   b. Do you have enough information to tell if boric acid is a weak acid or a strong acid?
   c. What is the OH^- concentration in this solution, in moles per liter?
   d. What is the pOH of this solution?
8. A solution of unknown pH was tested with a number of chemical indicators. The results of these tests are indicated in the table to the right. By reference to Figure 11.11, what is a reasonable estimate of the pH of the solution?

Indicator	Color
Methyl violet	Violet
Methyl orange	Yellow
Congo red	Blue
Methyl red	Yellow
Bromthymol blue	Yellow
Litmus	Red
Phenolphthalein	Colorless

9. A volume of 10.0 mL of stomach acid containing HCl was titrated with 35.0 mL of 0.100 *M* NaOH, resulting in exact neutralization.

a. Write the chemical equation describing this reaction.
b. Identify the salt resulting from this reaction.
c. Determine the number of moles of base added.
d. Determine the number of moles of HCl present.
e. Determine the concentration of the acid sample in moles per liter.

10. Consider a buffer compound made of equal amounts of carbonic acid (H_2CO_3) and the bicarbonate ion (HCO_3^-). By reference to Table 11.4, what is a reasonable estimate of the pH of this buffered solution?

Carbonic acid: H_2CO_3

$\downarrow$

Buffer system equilibrium: $H_2CO_3 + H_2O \rightleftharpoons H_3O^+ + HCO_3^-$

$\uparrow$

Sodium bicarbonate: $NaHCO_3 \longrightarrow Na^+ + HCO_3^-$

Answers **1.** c **2.** a **3.** b **4.** b **5.** H_2O **6.** Strong acid produces more hydronium ion than weak acid **7. a.** pH 6 **b.** no **c.** $[OH^-] = 1 \times 10^{-8}\ M$ **d.** pOH = 8 **8.** pH 6 **9. a.** $HCl + NaOH \rightarrow NaCl + H_2O$ **b.** NaCl is salt **c.** 0.0035 mol base added **d.** 0.0035 mol acid **e.** acid concentration = 0.35 *M* **10.** pH = about 6; actually 6.1

DIAGNOSTIC CHART

Blacken in all of the circles under the number of each question you missed in the Chapter Review questions. The diagnostic chart will help you identify concept areas that need more study.

	Questions									
Concepts	1	2	3	4	5	6	7	8	9	10
Definitions	○	○	○	○						
Acids and bases				○	○	○				
$[H_3O^+]$ and pH						○	○			
Chemical indicators								○		
Acid–base reactions									○	
Weak acids and buffers										○

EXERCISES

Basic Concepts

1. Describe the Arrhenius model for an acid.
2. How does the Brønsted–Lowry model for an acid differ from Arrhenius's view?
3. Both Arrhenius and Brønsted–Lowry viewed bases as playing a similar role. What is the function of a base?

Acids and Bases

4. Strong and weak acids are differentiated by their tendency to donate hydrogen ions. By reference to the information in Table 11.2, would the addition of sulfuric acid (H_2SO_4) to water be expected to produce hydronium ions?

$$H_2SO_4 + H_2O \overset{?}{\rightleftharpoons} H_3O^+ + SO_4^{2-}$$

After deciding which direction this reaction will proceed, identify the compound acting as a base in the reaction.

5. Ammonia gas is very soluble in water. When anhydrous ammonia is injected into the soil to provide nitrogen fertilizer, the gaseous ammonia dissolves rapidly in the soil moisture. When water and ammonia molecules are present together, the following reaction is possible:

$$H_2O + NH_3 \overset{?}{\rightleftharpoons} NH_4^+ + OH^-$$

a. By reference to the information in Table 11.2, which direction will be predominant in this reaction?

b. After deciding which direction this reaction will proceed, identify the compounds acting as acid and base.

6. Nitric acid, HNO_3, is a strong acid.
   a. Write the reaction expected when nitric acid is dissolved in water.
   b. Suppose that 0.2 mol of nitric acid are placed in 1.0 L of water. What will be the hydronium ion concentration of the resulting solution?

7. Carbonic acid, H_2CO_3, is formed when carbon dioxide is dissolved in water:

$$H_2O + CO_2 \longrightarrow H_2CO_3$$

The resulting carbonic acid will then react with surrounding water molecules:

$$H_2CO_3 + H_2O \overset{?}{\rightleftharpoons} H_3O^+ + HCO_3^-$$

   a. By reference to the information in Table 11.2, which direction will be predominant in this reaction?
   b. After deciding which direction this reaction will proceed, identify the compounds acting as acid and base.
   c. Is carbonic acid a strong acid or a weak acid?

Hydronium Ion Concentration $[H_3O^+]$ and pH

8. The term pH is used to describe the concentration of hydronium ion in a solution. Classify solutions with the following pH values as acidic, neutral, or basic.

pH 1.5	pH 7.5
pH 3.5	pH 9.0
pH 5.5	pH 11.0
pH 7.0	pH 13.8

9. Which of the pH values listed in question 8 represents the largest concentration of hydronium ion?

10. Consider a solution of pH 9.0.
    a. What is the hydronium ion concentration of this solution in moles per liter?
    b. What is the hydroxide ion concentration of this solution in moles per liter?
    c. What is the pOH of this solution?
    d. Is the solution acidic, neutral, or basic?

11. Rounded to one significant figure, gastric juice has a pH of 1. By what factor must one reduce the hydronium ion concentration to raise the pH to 2?

12. Rounded to one significant figure, orange juice has a pH of 4, black coffee a pH of 5, and household ammonia used for cleaning windows, a pH of 11.
    a. How many times more concentrated is the hydronium ion in orange juice than in black coffee?
    b. How many times more concentrated is the hydronium ion in orange juice than in household ammonia?

Chemical Indicators

13. Explain the function of a chemical indicator in an acid–base reaction.

14. What is the relative concentration of the molecular and ionized forms of a chemical indicator as it changes color?

15. By reference to Figure 11.11, estimate the K_a for the indicator phenolphthalein.

Acid–Base Reactions

16. Compute the number of grams of solute required to prepare 1 L of each of the following solutions:
    a. 0.10 *M* NaOH solution
    b. 0.50 *M* KCl solution
    c. 0.40 *M* acetic acid solution ($HC_2H_3O_2$)
    d. 0.15 *M* $AgNO_3$ solution
    e. 0.60 *M* NaCl solution

17. Compute the number of grams of solute required to prepare each of the following solutions:
    a. 500 mL 0.10 *M* KOH
    b. 200 mL 0.20 *M* NaF
    c. 100 mL 1.0 *M* LiCl

18. What would be the concentration, in moles per liter, of the following solutions?
    a. 1.00 L of solution containing 4 g NaOH
    b. 1.00 L of solution containing 33.8 g $AgNO_3$
    c. 500 mL of solution containing 63 g HNO_3
    d. 200 mL of solution containing 9.8 g H_3PO_4

19. Predict the products of each of the following acid–base reactions, name the salt produced, and balance the equation describing the reaction. Some acids will donate all of their hydrogens to a base; some only one or a few. The ionizable hydrogens are highlighted in the reactions shown below.
    a. Hydrochloric acid + Sodium hydroxide ⟶
       $HCl + NaOH \longrightarrow$
    b. Nitric acid + Magnesium hydroxide ⟶
       $HNO_3 + Mg(OH)_2 \longrightarrow$
    c. Boric acid + Ammonium hydroxide ⟶
       $H_3BO_3 + NH_4OH \longrightarrow$
    d. Sulfuric acid + Calcium hydroxide ⟶
       $H_2SO_4 + Ca(OH)_2 \longrightarrow$
    e. Citric acid + Sodium hydroxide ⟶
       $HC_6H_7O_7 + NaOH \longrightarrow$

20. Suppose that 25.0 mL of a hydrochloric acid solution of unknown concentration was completely neutralized by 40.0 mL of 0.10 *M* sodium hydroxide solution.

$$HCl + NaOH \longrightarrow NaCl + H_2O$$

    a. How many moles of NaOH were used to complete the reaction?

b. How many moles of HCl were present in the 25.0-mL sample?
c. How many moles of HCl would 1.0 L of this solution contain?

21. A common antacid tablet contains aluminum hydroxide, $Al(OH)_3$. Suppose that a 2.0-g tablet of this material is found to be completely neutralized by 64.0 mL of 0.20 *M* HCl.
    a. Write the balanced equation for this reaction.
    b. How many moles of HCl were used to complete the reaction?
    c. How many moles of aluminum hydroxide were present in the table?
    d. How many grams of aluminum hydroxide were present in the tablet?
    e. What was the percentage by weight of aluminum hydroxide in the 2.0-g tablet?

22. A number of popular soft drinks contain phosphoric acid, H_3PO_4. Suppose that the phosphoric acid present in a 10.0-mL sample of a soft drink could be neutralized by addition of 40.0 mL of 0.15 *M* NaOH solution. What is the concentration of phosphoric acid in the sample in moles per liter?

$$H_3PO_4 + 3\ NaOH \longrightarrow Na_3PO_4 + 3\ H_2O$$

23. Suppose that a 25.0-mL sample of sulfuric acid could be neutralized by addition of 80.0 mL 0.25 *M* KOH. What is the concentration of the sulfuric acid sample in moles per liter?

24. A common drain cleaner contains potassium hydroxide, KOH. One tablet of this cleaner weighs 3.0 g. If it is dissolved in water and titrated with 0.10 *M* hydrochloric acid, 428 mL of the HCl solution are required to neutralize the KOH. What is the percentage by weight of KOH in the drain cleaner?

Weak Acids and Buffers

25. What would be the hydronium ion concentration in a solution of citric acid of concentration 1.072 *M*? Assume that a negligible amount of the acid ionizes. (You may wish to refer to Table 11.4 for K_a values.)

26. What would be the hydronium ion concentration in a solution of carbonic acid of concentration 0.114 *M*? Again assume that a negligible amount of acid ionizes.

27. What would be the hydronium ion concentration in a solution of oxalic acid of concentration 0.154 *M*? Again assume that a negligible amount of acid ionizes.

28. What is the role of a buffer in an acid–base system?

29. What are the two components of an acid–base buffer system?

30. Which part of the buffer system is active when acid is added to the buffered solution? What is the product of this reaction? What happens when base is added to the buffered solution?

31. What is meant by the "capacity" of a buffer system? How is this related to the concentrations of the two components of the buffer?

32. Consider the buffer system formed by carbonic acid and sodium bicarbonate. Suppose that carbonic acid and sodium bicarbonate were both present at a concentration of 0.1 *M*.

Carbonic acid: H_2CO_3 [0.1 *M*]

Buffer system equilibrium: $H_2CO_3 + H_2O \rightleftharpoons H_3O^+ + HCO^{3-}$

Sodium bicarbonate: $NaHCO_3 \longrightarrow Na^+ + HCO_3^-$ (0.1 *M*)

a. By reference to the K_a value for carbonic acid shown in Table 11.4, compute the hydronium ion concentration of this buffer solution.
b. By use of a pocket calculator or logarithm table, compute the pH of this solution.

33. Consider the buffer system formed by citric acid and sodium citrate. Suppose that the concentration of citric acid is 0.20 *M* and the concentration of sodium citrate is 1.68 *M*. The K_a value for citric acid is 8.4×10^{-4}.

Citric acid: $HC_6H_7O_7$ (0.2 *M*)

Buffer system equilibrium: $HC_6H_7O_7 + H_2O \rightleftharpoons H_3O^+ + C_6H_7O_7^-$

Sodium citrate: $NaC_6H_7O_7 \longrightarrow Na^+ + C_6H_7O_7^-$ (1.68 *M*)

a. Compute the hydronium ion concentration of this buffer solution.
b. By use of a pocket calculator or logarithm table, compute the pH of this solution.

CHAPTER

12

The Behavior of Gases

As we have seen, gases expand to fill the space they occupy. This unique characteristic of gases makes it possible for them to be constantly available to sustain life on earth—oxygen for the animal kingdom and carbon dioxide for plants. There are some simple relationships between the volume of a gas and its temperature and pressure that help us to predict its behavior.

OBJECTIVES

After completing this chapter, you should be able to

- Correct gas volume data for changes in temperature.
- Correct gas volume data for changes in pressure.
- Use the combined gas law to correct gas volume data for changes in temperature and pressure.
- Given a number of moles of gas, use the ideal gas law to compute the volume this sample would occupy at specified temperature and pressure.
- Define the term *partial pressure*, and compute percentage composition of a gas from partial pressure data.

Gases are quite unlike solids or liquids. The volume of a gas is governed by three factors: (1) how many molecules (or moles) of the gas are present, (2) the temperature of the gas, and (3) the pressure exerted on the gas. Solids have a fixed volume and a fixed shape, and liquids have a fixed volume but assume the shape of their container. Gases, however, compress or expand to fill any container of any shape. We put these special properties of gases to good use. Large numbers of moles of oxygen and acetylene gas are stored in pressurized cylinders for use in welding. Hospitals store large amounts of pure oxygen under high pressure for medical use. Our bodies use changes in volume and pressure to move carbon dioxide out of the lungs and to pull fresh air containing oxygen into our bodies where it can be transported to sites of metabolism in the cells.

In this chapter we consider the problem of predicting changes in gas volume when temperature and pressure are changed and look at ways of computing the number of moles of gas present in either a pure or mixed sample. But first, let's examine how molecules behave as gases.

12.1 Kinetic Molecular Theory

The behavior of gases can be understood if we assume that gases differ from solids and liquids in the relative freedom of their molecules. In solids, atoms or molecules are tightly bound to adjacent atoms or molecules throughout the material, forming a structurally solid material. In liquids, molecules are relatively free to slide across one another, but enough intermolecular attraction exists to hold the molecules together in a group. In gases, however, each molecule is essentially free to wander by itself, and there is a great deal of space between molecules. The rate of molecular wandering—the molecular velocity—is dependent only on the temperature of the gas sample and the mass of the molecule. Higher temperatures result in higher molecular velocities and higher kinetic energy of the molecules.

The speed at which molecules move is dependent on temperature.

The **kinetic molecular theory** is another way of saying "moving molecule theory." *Kinetic energy* is the energy of a moving object, and it depends on both mass and velocity. In this theory, gases are viewed as consisting of tiny independent molecules wandering randomly through a container at a speed related to their temperature. As gas molecules strike the wall of their container and rebound, they exert pressure on the container wall. This pressure is related both to the number of collisions per each unit of wall area per unit time and to the average energy of these collisions. The pressure exerted on the container walls by a gas sample therefore involves two factors: (1) the number of gas molecules in the container per unit volume and (2) their energy of motion (Figure 12.1).

The pressure exerted by gas molecules is dependent on the energy with which they strike the side of their container.

NEW TERMS

kinetic molecular theory The idea that molecules in gases are in constant motion.

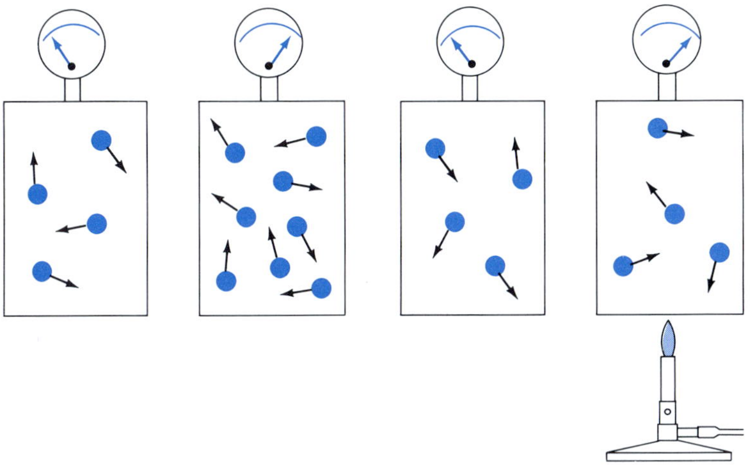

FIGURE 12.1
The pressure exerted by a gas depends on the number of collisions over each unit of wall area per unit of time and on the average energy of these collisions. Doubling either the number of molecules or their average energy of motion (by increasing the temperature) will double the pressure (indicated by gauges along the top).

12.2 Relationship of Volume to Temperature and Pressure

Volume and Temperature

Gases expand when heated. In Chapter 1 we introduced the relationship between the volume and temperature of a gas sample. Recall that the volume of a gas increases in a linear fashion as the temperature increases (see Figures 1.15 and 1.16). Data from this experiment are reproduced in Figure 12.2. From these data, one could predict that the gas volume would become smaller and smaller as the temperature is reduced, until eventually the gas would "vanish." This experiment predicted a "vanishing point" of −273 °C, a temperature known as *absolute zero*. This behavior is consistent with a model that attributes pressure to the movement of molecules against the side of the container. Theoretically, at absolute zero molecular motion ceases and the gas sample is simply compressed to zero volume by outside pressures. Recall from Chapter 1 that the absolute temperature scale is called the *Kelvin* scale. Any Celsius temperature can be converted to Kelvin by simply adding

Absolute zero is the vanishing point of an ideal gas.

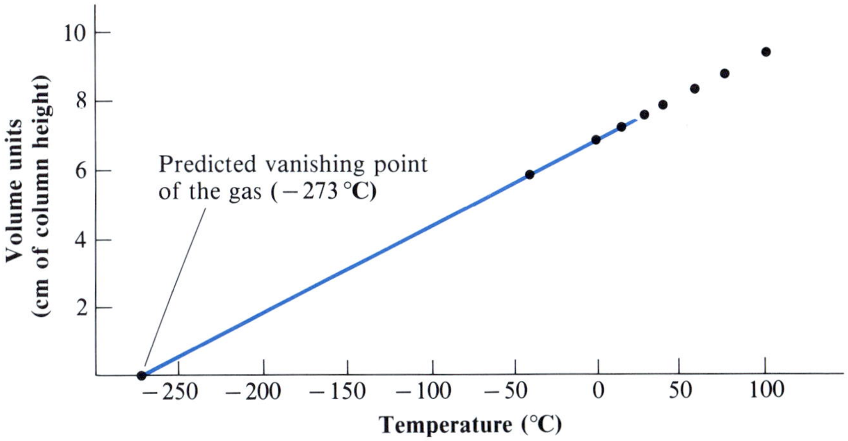

Volume units (cm of column height)
10
8
6
4
2
Predicted vanishing point of the gas (−273 °C)
−250
−200
−150
−100
−50
0
50
100
Temperature (°C)

FIGURE 12.2
The volume of a gas sample increases in direct proportion to its temperature. This graph predicts that the gas would vanish if its temperature fell to −273 °C. This temperature is known as absolute zero, or zero Kelvin.

273. Kelvin degrees are the same size as Celsius degrees, only the zero point is changed (see Figure 1.17).

In reality, when molecules are pushed this close together, intermolecular forces take over and the gas becomes a liquid before it vanishes. This "real" behavior, however, does not affect the behavior of the gas as long as it remains a gas, and the mathematical relationship between absolute temperature and volume is valid and useful.

The volume of a gas varies in direct proportion to its absolute temperature. This is known as *Charles' law.*

It is apparent from the data in Figure 12.1 that gas volume is directly proportional to absolute temperature. We can use this proportionality to predict gas sample volumes when the temperature is changed. This relationship is known as **Charles' law**. The "law" states that at constant pressure, the volume of a given mass of gas is directly proportional to its absolute temperature. In other words, if a gas is heated, it will expand; if it is cooled, it will contract, as long as the pressure remains constant. Expressed mathematically, this is

$$\frac{\text{Initial volume}}{\text{Initial temperature}} = \frac{\text{Final volume}}{\text{Final temperature}}$$

$$\frac{v_1}{t_1} = \frac{v_2}{t_2}$$

Solving for v_2,

$$v_2 = v_1\left(\frac{t_2}{t_1}\right)$$

Therefore, the new volume equals the old volume times the fractional change in temperature (t_2/t_1).

EXAMPLE 12.1

A balloon had a volume of 125 mL at room temperature (20 °C). It was cooled to 0 °C, with the pressure remaining constant. What is its new volume at 0 °C?

a. Convert to absolute temperature.

$$20\,^\circ\text{C} = 20 + 273 = 293\ \text{K}$$

$$0\,^\circ\text{C} = 273\ \text{K}$$

b. What is the fractional change in temperature?

$$\frac{t_2}{t_1} = \frac{273\ \not{\text{K}}}{293\ \not{\text{K}}} = 0.932$$

c. Compute the new volume.

$$v_2 = v_1\left(\frac{t_2}{t_1}\right)$$

$$v_2 = 125\ \text{mL} \times 0.932 = 116.5\ \text{mL}$$

EXAMPLE 12.2

A gas sample had a volume of 50.0 mL at 20 °C. It was then heated to 50 °C, pressure remaining constant. What is the new volume at 50 °C?

a. Convert to absolute temperature.

$$20\,^\circ\text{C} = 20 + 273 = 293\ \text{K}$$

$$50\,^\circ\text{C} = 50 + 273 = 323\ \text{K}$$

b. What is the fractional change in temperature?

$$\frac{t_2}{t_1} = \frac{323\ \text{K}}{293\ \text{K}} = 1.10$$

c. Compute the new volume.

$$v_2 = v_1\left(\frac{t_2}{t_1}\right)$$

$$v_2 = 50\ \text{mL} \times 1.10 = 55\ \text{mL}$$

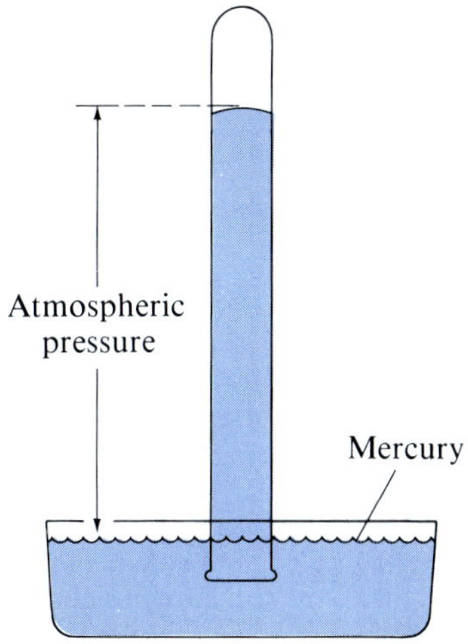
Atmospheric pressure
Mercury

FIGURE 12.3
A mercury barometer can be constructed by inverting a standard glass tube full of mercury into a mercury pool. The atmospheric pressure can be described in terms of the length of mercury column it can support.

Volume and Pressure

There are several ways to measure pressure, but the most common is to use the liquid metal mercury as a standard. Mercury is very dense for a liquid (13.6 g/cm^3). Figure 12.3 shows an experiment in which a standard glass tube about 800 mm long is filled with mercury and then inverted into a beaker containing mercury. The mercury does not run out of the tube; atmospheric pressure caused by the presence of several miles of air molecules pressing down from above the earth's surface pushes on the surface of the mercury in the beaker and holds the mercury up in the tube. Atmospheric pressure is usually indicated in terms of the height of a standard column of mercury that can be supported. At sea level, this is normally about 760 mm. Water can be used in place of mercury, but the column must be 13.6 times longer, or about 33 ft. Some early barometers were constructed as water-filled glass tubes running from the basement to the third floor of a home. One simply went upstairs to read the barometric pressure.

The idea of atmospheric pressure and this method of measuring it was discovered by the Italian Torricelli. In recognition of his work, the unit millimeters of mercury, or mm Hg, is now known as the **Torr**. Normal atmospheric pressure is 760 Torr; this is sometimes referred to as 1.0 **atmosphere**, abbreviated **atm**. Although we will use Torr as a pressure unit in this chapter, many chemists prefer instead to work with the atmosphere as their pressure unit.

The Torr is equal to the pressure exerted by a column of mercury 1 mm high. Normal atmospheric pressure is 760 Torr.

The relationship between gas volume and pressure is straightforward. If gas pressure is proportional to the number of gas molecule collisions with a given area of container wall per unit time, then compressing a number of gas molecules into half the space should double the number of collisions per unit time. Doubling the number of collisions should double the pressure

The product of pressure and volume for a gas sample is always the same: $p \times v =$ a constant. This is called *Boyle's law.*

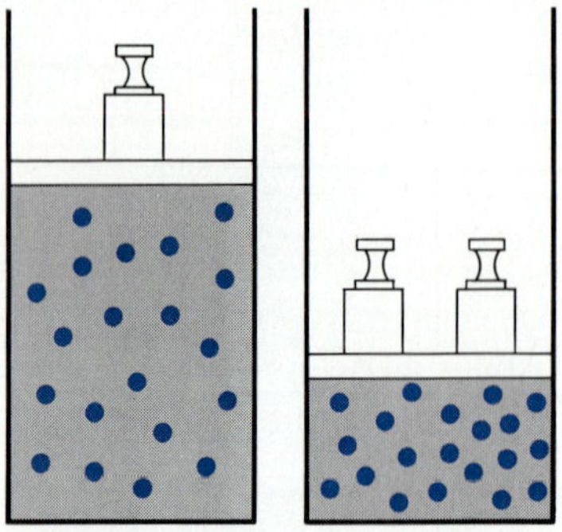

FIGURE 12.4
When a gas is compressed into one-half its original volume, the number of molecular collisions within a unit area of the container wall will double. The pressure will be twice as great. Likewise, if the gas is compressed into one-fourth its original volume, the pressure will be four times as great.

TABLE 12.1
Pressure-Volume Data for a Given Gas Sample

Pressure (Torr)	Volume (mL)	Pressure × Volume
700	100	70,000
1400	50	70,000
2800	25	70,000

(Figure 12.4). Data from the experiment of Figure 12.4 are presented in Table 12.1. It is interesting to note that for any set of pressure-volume data, the *product* of the pressure and volume values is always a constant, as seen in the third column of the table.

This relationship between gas volume and pressure was discovered by the English scientist Robert Boyle in the seventeenth century, and it is known today as **Boyle's law**. Since, for a given sample of a gas, the pressure-volume product is always a constant, Boyle's law permits us to easily compute the new volume when the gas is subjected to a different pressure.

$$p_1v_1 = p_2v_2 = \text{A constant}$$

Solving for final volume,

$$\text{Final volume} = \text{Initial volume}\left(\frac{\text{Initial pressure}}{\text{Final pressure}}\right)$$

$$v_2 = v_1\left(\frac{p_1}{p_2}\right)$$

EXAMPLE 12.3

A balloon contains 1000 mL of gas at 760 Torr. The weather changes, with a resulting change of barometric pressure to 740 Torr. What is the new volume of the balloon?

$$v_2 = v_1\left(\frac{p_1}{p_2}\right)$$

$$v_2 = (1000\ \text{mL})\left(\frac{760\ \cancel{\text{Torr}}}{740\ \cancel{\text{Torr}}}\right) = 1027\ \text{mL}$$

EXAMPLE 12.4

An air pump contains 1000 mL of air at 760 Torr. What pressure must be exerted to compress the air to a volume of 500 mL?

$$p_1v_1 = p_2v_2$$

$$p_2 = (p_1)\left(\frac{v_1}{v_2}\right)$$

$$= (760\ \text{Torr})\left(\frac{1000\ \cancel{\text{mL}}}{500\ \cancel{\text{mL}}}\right) = 1520\ \text{Torr}$$

Standard Temperature and Pressure

Standard temperature and pressure (STP) is 273 K and 760 Torr.

Since the volume of a gas sample depends on both temperature and pressure, and since it is unlikely that two laboratories in the world operate at exactly the same temperature and pressure, it is common practice to convert all gas volume data to **standard temperature and pressure**. This is commonly known as **STP**, or "standard conditions." Standard pressure has been agreed upon as 1.0 atm or 760 Torr, and standard temperature as 0 °C or 273 K.

NEW TERMS

Charles' law A relationship stating that when the temperature of a gas sample is changed, its volume will change an amount proportional to the change in absolute temperature.

Torr A unit of pressure equal to the pressure exerted by a standard column of mercury 1.00 mm in height.

atmosphere (atm) Normal atmospheric pressure, where 1.00 atm equals 760 Torr.

Boyle's law A relationship stating that the product of pressure times volume for any gas sample is always a constant.

standard temperature and pressure (STP) Conditions of 0 °C (273 K) and 760 Torr standardized worldwide.

TESTING YOURSELF

Charles' and Boyle's Laws

1. A temperature of −40 °C corresponds to what Kelvin temperature?
2. A child's inflatable toy was filled with air to atmospheric pressure, 760 Torr, at a temperature of 10 °C. Its volume was 3.50 L. The toy was left outside, and the temperature the next morning was −15 °C (+5 °F). The atmospheric pressure was the same. What would be the new volume of the gas in the toy?
3. Suppose that a bicycle pump will hold 1500 mL of air when the plunger is up and the atmospheric pressure is 700 Torr. What pressure must be exerted on the plunger to compress the air to a volume of 500 mL?

Answers **1.** 233 K **2.** 3.19 L **3.** 2100 Torr

12.3 Combined Gas Laws

Charles' law, which relates gas volume to temperature, and Boyle's law, which relates gas volume to pressure, can be combined to provide a means of correcting for changes in both temperature and pressure simultaneously.

$$\text{Final volume} = \text{Initial volume} \times \text{Correction factor}$$

To correct for changes in temperature:

$$v_2 = v_1 \times \left(\frac{t_2}{t_1}\right)$$

To correct for changes in pressure:

$$v_2 = v_1 \times \left(\frac{p_1}{p_2}\right)$$

To correct for changes in temperature and pressure simultaneously:

$$v_2 = v_1 \times \left(\frac{t_2}{t_1}\right)\left(\frac{p_1}{p_2}\right)$$

Rearranging to general form:

$$\frac{v_1 p_1}{t_1} = \frac{v_2 p_2}{t_2}$$

This new equation, which shows the relationship among pressure, volume, and temperature, is called the **combined gas law**. The equation can be solved for any of the six variables included, as long as the others are known.

The *combined gas law* permits correction for changes in both temperature and pressure.

EXAMPLE 12.5

A helium-filled high altitude weather balloon containing 3000 L of gas is launched at sea level ($p = 760$ Torr and $t = 20\,°C$). At an altitude of 10 km the balloon encounters a pressure of 210 Torr and a temperature of $-53\,°C$. What will be the volume of the balloon at this altitude?

a. Convert to Kelvin temperature:

$$20\,°C = 20 + 273 = 293\text{ K}$$

$$-53\,°C = -53 + 273 = 220\text{ K}$$

b. Compute the new volume:

$$\frac{v_1 p_1}{t_1} = \frac{v_2 p_2}{t_2}$$

Solving for final volume,

$$v_2 = \frac{v_1 p_1 t_2}{t_1 p_2}$$

$$v_2 = \frac{(3000\text{ L})(760\text{ Torr})(220\text{ K})}{(210\text{ Torr})(293\text{ K})}$$

$$v_2 = 8152\text{ L}$$

Meterological balloons are only partially filled at launch. At an altitude of 10 km the gas volume is already almost three times the launch volume. The balloon will eventually reach an altitude where it will burst; the weather reporting equipment is then lowered by parachute.

EXAMPLE 12.6

A steel-belted automobile tire is filled with air at a pressure of 2000 Torr (24 psi above normal atmospheric pressure) at a temperature of 15.5 °C (60 °F). It is then driven for several hours in hot weather and reaches a temperature of 49 °C (120 °F). What will be the new gas pressure within the tire? (Discount the slight expansion of the tire.)

a. Convert to absolute temperature scale:

$$15.5\,°C = 15.5 + 273 = 288.5\text{ K}$$

$$49\,°C = 49 + 273 = 322\text{ K}$$

b. Compute the new pressure:

$$\frac{v_1 p_1}{t_1} = \frac{v_2 p_2}{t_2}$$

Solving for final pressure,

$$p_2 = \frac{v_1 p_1 t_2}{v_2 t_1}$$

Because volume remains constant, the volumes cancel.

$$p_2 = \frac{(2000\text{ Torr})(322\text{ K})}{(288.5\text{ K})} = 2232\text{ Torr}$$

This is an increase of

$$\frac{2232 \text{ Torr}}{2000 \text{ Torr}} \times 100 = 111.6\%$$

$$\approx 112\%$$

This corresponds to about 28.5 psi above normal atmospheric pressure. The tire pressure increased by about 4.5 psi.

NEW TERM

combined gas law A combination of Charles' law (compensating for temperature change) and Boyle's law (compensating for pressure change).

TESTING YOURSELF

Combined Gas Law

1. Consider a sample of gas of volume 200 mL at 20 °C and 760 Torr. What volume will this gas occupy at 0 °C and 700 Torr?
2. An inflatable beach ball has a volume of 14.0 L at a temperature of 25 °C and a pressure of 760 Torr. This ball is taken up into the mountains where the atmospheric pressure is 640 Torr and the temperature is 10 °C. Will the ball expand or contract? What will be the new volume of the ball?

Answers **1.** 202 mL **2.** expand; 15.8 L

12.4 The Ideal Gas Law

The general form of the combined gas law implies that the quantity

$$\frac{(\text{Pressure})(\text{Volume})}{(\text{Temperature})}$$

is always a constant.

$$\frac{v_1 p_1}{t_1} = \frac{v_2 p_2}{t_2} = \text{A constant}$$

The value of this constant is directly proportional to the number of moles of gas present. For example, if a given sample of gas at any constant temperature and pressure occupies a volume of 2.0 L, twice that many molecules or moles of the gas would occupy a volume of 4.0 L under the same conditions of temperature and pressure. Thus,

$$\frac{(\text{Pressure})(\text{Volume})}{(\text{Temperature})} = n \times \text{A constant}$$

where n is the number of moles of gas.

Since the constant relates gas volume, temperature, pressure, and the number of moles of gas present, it has been given a more descriptive name and its own symbol. It is called the **universal gas constant**, ***R***, and it relates gas volume to temperature, pressure, and number of moles of gas present.

TABLE 12.2
Values for R for Four Different Sets of Volume and Pressure Units

R	Volume Unit	Pressure Unit
$0.0821 \frac{\text{L atm}}{\text{mol K}}$	Liters	Atmospheres
$82.1 \frac{\text{mL atm}}{\text{mol K}}$	Milliliters	Atmospheres
$62.4 \frac{\text{L Torr}}{\text{mol K}}$	Liters	Torr
$0.0624 \frac{\text{mL Torr}}{\text{mol K}}$	Milliliters	Torr

The value of R varies depending on which units are used (Table 12.2). The most commonly used values are

$$R = 62.4 \frac{\text{L Torr}}{\text{mol K}} \quad \text{and} \quad R = 0.0821 \frac{\text{L atm}}{\text{mol K}}$$

The *ideal gas law* permits calculation of volume, temperature, pressure, or moles of gas when three of the four variables are known.

Using R, we can now state the **ideal gas law** mathematically as

$$\frac{pv}{t} = nR$$

or in its more common form,

$$pv = nRt$$

The ideal gas law is principally useful for computing volumes, temperatures, pressures, and moles of gas present when three of the four variables are known. It applies only to ideal gases. An **ideal gas** has infinitely small molecules and behaves exactly as predicted by kinetic molecular theory. In reality, gas molecules are not infinitely small and will ultimately liquify when cooled. However, when in the gaseous state, practically all materials behave as ideal gases and obey the gas laws.

EXAMPLE 12.7

What volume will be occupied by 0.200 mol of gas at 27 °C and a pressure of 650 Torr (approximately the atmospheric pressure in Denver)?

a. Convert to absolute temperature scale:

$$27\,°\text{C} = 27 + 273 = 300\text{ K}$$

b. Apply the ideal gas law:

$$pv = nRt$$

$$v = \frac{nRt}{p}$$

$$v = \frac{(0.200 \cancel{\text{mol}})\left(62.4 \dfrac{\text{L} \cancel{\text{Torr}}}{\cancel{\text{mol}}\,\cancel{\text{K}}}\right)(300 \cancel{\text{K}})}{(650 \cancel{\text{Torr}})}$$

$$v = 5.76 \text{ L}$$

Note how all the units cancel except for liters.

An important application of the ideal gas law is that it can be used to calculate the volume of gas at standard pressure and temperature.

EXAMPLE 12.8

Use the ideal gas law to find how many liters 1.0 mol of gas occupies at STP.

$$pv = nRt$$

$$v = \frac{nRt}{p}$$

$$v = \frac{(1.0 \cancel{\text{mol}})\left(62.4 \dfrac{\text{L} \cancel{\text{Torr}}}{\cancel{\text{mol}}\,\cancel{\text{K}}}\right)(273 \cancel{\text{K}})}{760 \cancel{\text{Torr}}}$$

$$v = 22.4 \text{ L}$$

EXAMPLE 12.9

Zinc will react with hydrochloric acid to produce zinc chloride and hydrogen gas:

$$\text{Zn} + 2\ \text{HCl} \longrightarrow \text{ZnCl}_2 + \text{H}_2$$

According to this reaction, 1 mol of zinc reacts to produce 1 mol of hydrogen gas. Suppose that 6.53 g of zinc (0.100 mol) is reacted with excess HCl, producing hydrogen gas. What volume of hydrogen gas, at 20 °C and 750 Torr, would be produced by this reaction?

a. Convert to absolute temperature:

$$20\ ^\circ\text{C} = 20 + 273 = 293 \text{ K}$$

b. Apply the ideal gas law:

$$pv = nRt$$

$$v = \frac{nRt}{p}$$

$$v = \frac{(0.10 \cancel{\text{mol}})\left(62.4 \dfrac{\text{L} \cancel{\text{Torr}}}{\cancel{\text{mol}}\,\cancel{\text{K}}}\right)(293 \cancel{\text{K}})}{750 \cancel{\text{Torr}}}$$

$$v = 2.44 \text{ L of hydrogen gas}$$

NEW TERMS

universal gas constant (*R*) A constant in the ideal gas law that relates pressure, volume, temperature, and number of moles of gas present.

ideal gas law An equation relating gas volume to number of moles of gas present, temperature, and pressure:

$$pv = nRt$$

ideal gas A theoretical gas composed of infinitely small molecules that behaves exactly as predicted by kinetic molecular theory.

TESTING YOURSELF

Ideal Gas Law

1. What volume would be occupied by 0.5 mol of natural gas (CH_4) at 20 °C and 760 Torr?
2. A 2.00-L sample of a gas is measured at 20 °C and a pressure of 740 Torr. Assuming that the gas sample is pure oxygen, how many moles of oxygen gas are present in the sample?
3. Suppose that the sample container used for question 2 had a mass of 250.0 g when evacuated and a mass of 252.59 g when containing the oxygen sample at the conditions outlined above. Can you compute the molecular weight of oxygen from this data? What is the experimentally determined molecular weight of oxygen from this experiment?

Answers **1.** 12.0 L **2.** 0.0809 mol **3.** 32.0 g/mol

12.5 Partial Pressure

The last concept to be introduced in this chapter is *partial pressure*. At the beginning of this chapter, we developed the idea that gas pressure is due to collisions of moving gas molecules with the container walls. The pressure observed is the total pressure, the sum of all of the collisions of all of the molecules in the sample (Figure 12.5).

The total pressure exerted by a gas is the sum of the *partial pressures* exerted by each component of the gas mixture.

Partial pressure is simply the pressure of each individual component of the gas mixture. Consider a sample of air, at normal atmospheric pressure of 760 Torr. Air is not a pure substance but a mixture with the composition shown in Table 12.3. The total pressure is the sum of the partial pressures exerted by each of these components.

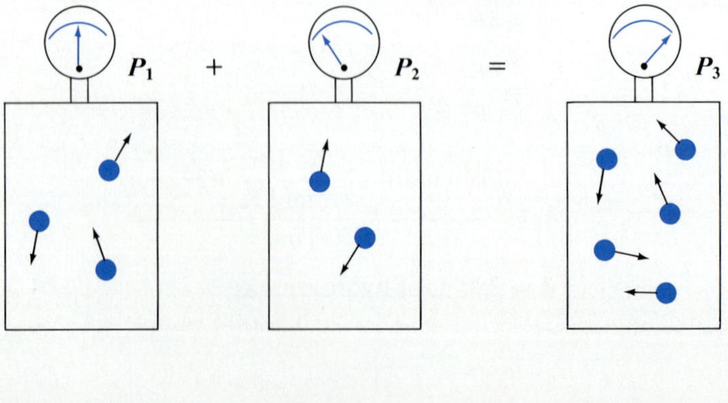
P_1
+
P_2
=
P_3

FIGURE 12.5
Each gas in a separate container exerts its own pressure called *partial pressure* that depends on the amount of gas present. When all of the gases are placed in the same container, each contributes to the total pressure.

TABLE 12.3
Composition of Air and Partial Pressure Exerted by Each Component at Standard Atmospheric Pressure

Component	Amount (%)	Partial Pressure at STP (Torr)
Nitrogen	78.03	593.0
Oxygen	20.99	159.5
Argon	0.94	7.14
Carbon dioxide	0.03	0.23
Other gases	Trace	0.13
	Total pressure	760

A major value of the partial pressure concept is that it permits application of the ideal gas law to mixed gases. Consider, for example, the combustion of propane in a camp stove.

$$C_3H_8 + 5\ O_2 \longrightarrow 3\ CO_2 + 4\ H_2O$$

A camp stove is not fed with pure oxygen, but with air that is about 21% oxygen. The partial pressure of oxygen (p_{O_2}) in air at standard pressure is about 159 Torr. Suppose that you wished to burn 4.4 g (0.1 mol) of propane. (For comparison, a nickel has a mass of about 5 g.) From inspection of the balanced equation, five times as many moles (0.5 mol) of oxygen will be required. The volume of air required is shown in the following example.

EXAMPLE 12.10

Using the preceding data, calculate the air volume required at 20 °C.

Given: Air at 760 Torr; partial pressure of oxygen is 159 Torr; temperature is 20 °C.

Find: Volume of air required to give 0.5 mol oxygen gas.

a. Convert to absolute temperature:

$$20\ °C = 20 + 273 = 293\ K$$

b. Apply the ideal gas law:

$$v = \frac{nRt}{p}$$

$$v = \frac{(0.50\ \cancel{mol})\left(62.4\ \dfrac{L\ \cancel{Torr}}{\cancel{mol}\ \cancel{K}}\right)(293\ \cancel{K})}{159\ \cancel{Torr}}$$

$$v = 57.5\ \text{L of air}$$

This is a considerable amount of air. To completely burn 4.4 g of propane requires an amount of air that would fill an ordinary automobile gas tank. If air is restricted, hydrocarbons such as propane will burn to form poisonous carbon monoxide instead of carbon dioxide. It is important that an adequate air supply be provided.

The concept of partial pressure helps explain the important transfer of oxygen to cells in our bodies and the subsequent elimination of carbon dioxide. The general reaction by which cells produce energy involves the oxidation of glucose:

$$C_6H_{12}O_6 + 6\ O_2 \longrightarrow 6\ CO_2 + 6\ H_2O + \text{Energy}$$

During respiration, oxygen is consumed and carbon dioxide is produced. Both gases are transported by the bloodstream, where they are dissolved in the blood. Although both venous and arterial blood is moved through the body in large veins and arteries, at both the delivery sites (the cells) and in the lungs, the blood vessels split into a large number of very fine capillaries. The walls of these capillaries provide a large amount of surface area and are formed by a thin membrane through which gas molecules can pass.

In our respiratory system, oxygen and carbon dioxide move from areas of higher partial pressure to lower partial pressure.

In this system, gases move across the membranes toward the side of least concentration, or least partial pressure. Figure 12.6 illustrates the movement of gases through the human respiratory and circulatory systems. Atmospheric air has a partial pressure of oxygen (p_{O_2}) of about 159 Torr and a partial pressure of carbon dioxide (p_{CO_2}) of about 0.3 Torr. Because breathing does not cause a complete exchange of air within the alveoli of the lungs, the p_{O_2} in the alveoli is about 100 Torr and the p_{CO_2} is about 40 Torr. However, venous blood returning to the lung has a p_{O_2} of about 40 Torr and a p_{CO_2} of about 46 Torr. The carbon dioxide moves toward the area of least concentration (the atmosphere of the lungs), while the oxygen from the lungs moves toward the region of lowest oxygen concentration (the blood). When the oxygenated (arterial) blood reaches the cells, another capillary system provides for transfer between the blood system and the cells. The p_{CO_2} in the cells is about 46 Torr, and it moves toward the bloodstream. The p_{O_2} in the cells, however, is about 40 Torr, and so oxygen moves from the blood toward the cells.

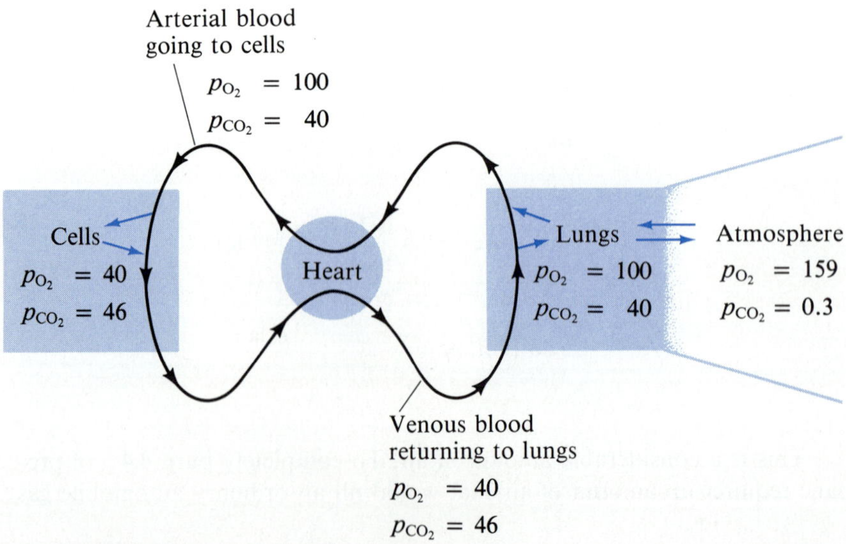
Arterial blood
going to cells
p_{O_2} = 100
p_{CO_2} = 40
Cells
p_{O_2} = 40
p_{CO_2} = 46
Heart
Lungs
p_{O_2} = 100
p_{CO_2} = 40
Atmosphere
p_{O_2} = 159
p_{CO_2} = 0.3
Venous blood
returning to lungs
p_{O_2} = 40
p_{CO_2} = 46

FIGURE 12.6
The movement of oxygen and carbon dioxide in the human circulatory and respiratory systems. The transfer of gases dissolved in the blood is toward the region of least concentration. The exchange across the capillary barrier is so effective that the blood comes to complete equilibrium with the transfer site as it passes. (Pressure units are in Torr.)

NEW TERM

partial pressure The pressure, in Torr or other pressure units, exerted by each component of a mixture of gases. The total of the partial pressures of all gases will equal the total pressure of the gas.

TESTING YOURSELF

Universal Gas Law and Partial Pressure

1. Suppose that you breath in 1.0 L of air at sea level (760 Torr) and 20 °C. If 20.99% of the total pressure is due to oxygen gas, for a p_{O_2} of 159.5 Torr, how many moles of oxygen would be present in the 1.0-L sample of air?
2. Suppose that you climbed to about 3000 m elevation in the mountains and that the total pressure there is 400 Torr. The p_{O_2} will be 20.99% of this, or 83.96 Torr. How many moles of oxygen will you receive in a 1.0-L breath? Can you now explain why you pant and tire easily when climbing high in the mountains?
3. One of the authors likes to fish in a mountain lake near Bozeman, MT. Its elevation is about 7000 ft, and the total atmospheric pressure is about 560 Torr. The outboard motor's pistons each take in about 90 mL of air on each intake stroke. If air is 20.99% oxygen and the air temperature is 20 °C, answer these questions.
   a. What is the partial pressure of oxygen at this altitude?
   b. How many moles of oxygen are in 90 mL of air at this altitude?
   c. The motor seems much more powerful at sea level. How many moles of oxygen are in 90 mL of air at sea level?
   d. What is the ratio of moles of oxygen in a 90-mL intake stroke at 7000 ft to the same value at sea level? Can you explain why the motor seems less powerful at high altitude?

Answers **1.** 0.0087 mol **2.** 0.0045 mol **3. a.** 117.5 Torr **b.** 0.000578 mol **c.** 0.000785 mol **d.** 0.736 The motor is only about $\frac{3}{4}$ as powerful as at sea level.

SUMMARY

Gas is a state of matter in which atoms or molecules are very small and far apart with respect to the volume they occupy. These atoms or molecules are in constant motion, their rate of motion determined by their temperature, an idea called *kinetic molecular theory*. The volume occupied by these molecules is determined by the absolute temperature, the number of moles of molecules present, and the pressure exerted on them. *Charles' law* relates the volume of gas to its temperature, and *Boyle's* law relates its volume to its pressure. The *ideal gas law* is a mathematical expression that relates all four of these variables and permits calculation of any of the four, given data for the other three. *Partial pressure* is that component of the total gas pressure exerted by one component of the gas mixture. It is important to the respiration and metabolism of the human body.

TERMS

kinetic molecular theory
Charles' law
Torr
atmosphere (atm)
Boyle's law
standard temperature and pressure (STP)
combined gas law
universal gas constant
ideal gas law
ideal gas
partial pressure

CHAPTER REVIEW

QUESTIONS

1. Does the kinetic molecular theory predict that gas molecules are attracted to each other, or does it predict that they move independently?
2. What Kelvin temperature corresponds to −10 °C?
3. A gas sample occupies 2.0 L at a temperature of 0 °C. What volume will it occupy if it warms up to 20 °C? (Assume constant pressure.)
4. A gas sample occupies 2.0 L at a temperature of 20 °C and a pressure of 720 Torr. What volume will it occupy if the pressure goes up to 770 Torr?
5. A sample of gas at 25 °C and 700 Torr occupies a volume of 1200 mL. What volume will this occupy at standard temperature and pressure?
6. The amount of 0.1 mol of propane gas (C_3H_8) is compressed in a 1.0-L storage bottle. If the temperature is 20 °C, what is the pressure in the storage container, expressed in Torr?
7. Sometimes patients who need large amounts of oxygen are placed in pressure chambers. Suppose a pressure chamber were operated at 20 °C and 1.5 times atmospheric pressure (1050 Torr), and the chamber were filled with pure oxygen. What would be the partial pressure of oxygen in this chamber?
8. For question 7, how many times greater will the oxygen availability be than under normal conditions at 20 °C?
9. An aircraft experienced a power reduction attempting to take off when the outside temperature was 115 °F (45 °C). How much less oxygen is drawn in on a given intake stroke, as compared to a temperature of 15 °C? (Assume constant pressure.)

Answers **1.** Molecules act independently **2.** 263 K **3.** 2.15 L **4.** 1.87 L **5.** 1012 mL **6.** 1828 Torr **7.** 1050 Torr **8.** 6.58 times as much oxygen **9.** 0.9 times as much oxygen as at 15 °C.

DIAGNOSTIC CHART

Blacken in all of the circles under the number of each question you missed in the Chapter Review questions. The diagnostic chart will help you identify concept areas that need more study.

	Questions								
Concepts	**1**	**2**	**3**	**4**	**5**	**6**	**7**	**8**	**9**
Kinetic molecular theory	○								
Kelvin temperature		○							
Volume–temperature relationships			○						○
Volume–pressure relationships				○					
Combined gas law					○				
Ideal gas law						○			
Partial pressure							○	○	○

EXERCISES

Celsius and Kelvin Temperature

1. Convert the following Celsius temperatures to Kelvin.

   0 °C   20 °C   50 °C   127 °C

2. Convert the following Kelvin temperatures to Celsius.

   310 K   233 K   373 K   33 K

Volume–Temperature Relationships

3. Consider a gas sample of volume 250 mL and temperature 37 °C. What would be the volume of this sample if the temperature is increased to 100 °C and the pressure remains constant?
4. Suppose that you inhaled 1.0 L of air at −30 °C (−22 °F) and held this air long enough to warm it to body temperature, 37 °C. What would be the new volume of air in the lungs, assuming constant pressure?

Volume–Pressure Relationships

5. Express the following pressures in Torr.

   1.5 atm   0.30 atm   0.90 atm

6. Express the following pressures in atmospheres (atm).

   750 Torr   640 Torr   9.5 Torr

7. A gas sample had a volume of 550 mL at a pressure of 690 Torr. What would be the volume of this sample at a pressure of 1 atm, assuming constant temperature?
8. A tire had a pressure of 2000 Torr when filled at 20 °C. One week later the outside temperature was −10 °C. The volume of the tire changed very little. What is its new pressure?

Combined Gas Law

9. Consider a weather balloon filled with 500 L of helium at 26 °C and 750 Torr. At an altitude of 10 km the atmospheric pressure is about 210 Torr and the temperature is about −43 °C. What will the volume of the balloon be at this altitude?
10. A 5.0-L gas sample was measured at 27 °C and 700 Torr. What will its volume be at STP?

Ideal Gas Law

11. What volume would be occupied by 0.4 mol of natural gas (CH_4) at 20 °C and 760 Torr?
12. Consider the natural gas sample of question 11.
   a. What would be the mass of this gas sample?
   b. What would be the density (in grams per liter) of natural gas at this temperature and pressure?
   c. The density of air at room temperature (20 °C) and 1 atm is about 1.2 g/L. Is natural gas lighter or heavier than air?
13. Natural gas reacts with oxygen from the air to form carbon dioxide, water vapor, and heat:

$$CH_4 + 2\,O_2 \longrightarrow CO_2 + 2\,H_2O + \text{Heat}$$

   Suppose that you had a 8.0-L container filled with pressurized methane (CH_4) to a pressure of 8.0 atm. The temperature is 20 °C.
   a. How many moles of methane are in the container?
   b. How many moles of oxygen gas will be required to react completely with this amount of methane?
   c. What volume of oxygen gas, at 20 °C and 1 atm, will be required to react with the methane in this container?
   d. Since air is only 21% oxygen, what volume of *air* will be required to react with the methane in this container?
14. Suppose that 9.0 g of water are vaporized, and the resulting steam is heated to 200 °C. If this steam is held in a 3.0-L vessel that contains no other gas, what will be the pressure in the container?

Partial Pressure

15. When one inhales dry air, water vapor is added as the air moves through the nose and throat. If the air is too dry, water is removed from the tissues faster than it can be replaced. The table below shows the partial pressure of oxygen, carbon dioxide, water vapor, and nitrogen in the atmosphere, as the air enters and leaves the lungs.

Gas	Atmosphere (Torr)	Inhaled Air (Torr)	Exhaled Air (Torr)
Oxygen	159	149	116
CO_2	0.3	0.3	28
H_2O	6	47	47
Nitrogen	595	564	569
Totals	760	760	760

   a. Why is the partial pressure of oxygen less as it is drawn into the lung than in the atmosphere?
   b. Why is the partial pressure of nitrogen higher than all of the other components?
   c. Carbon dioxide is produced in cell metabolism and is carried dissolved in the blood to the lung, where it diffuses into the air and is exhaled. What percentage of the molecules leaving the lung are CO_2?
16. Methane (CH_4) or natural gas is burned as fuel in furnaces and hot water heaters in this reaction:

$$CH_4 + 2\,O_2 \longrightarrow CO_2 + 2\,H_2O$$

   a. Suppose that one began with 4.0 g of methane. How many moles are represented? Use the ideal gas law to find the volume of this gas at STP.
   b. How many moles of oxygen are required for complete combustion of this amount of methane?
   c. The methane is mixed with air before it enters the burner. The p_{O_2} at 760 Torr (atmospheric pressure) is about 159 Torr. What volume of oxygen at this partial pressure will be required to completely burn 4.0 g of methane?
   d. The burner has inlets for methane and air. Which must be bigger?

CHAPTER

13

Nuclear Chemistry

Alternately condemned and praised, nuclear reactions have the potential to provide energy to sustain civilization or to ensure its destruction. Since atoms are without conscience, the decision is ours. The best decisions are made with a clear understanding of the basic principles involved.

OBJECTIVES

After completing this chapter, you should be able to

- Describe the properties of alpha, beta, and gamma radiation.
- Predict the product of a nuclear decay that emits alpha or beta radiation.
- Define the term *half-life*.
- Describe the interaction of alpha, beta, and gamma radiation with matter.
- Describe the operation of gas ionization and scintillation detectors of nuclear radiation.
- Describe two ways nuclear radiation interacts with human tissue and the probability of each occurring.
- Identify the sources of natural background radiation.
- Describe the process of nuclear fission.

The science of nuclear chemistry has two "birthdays." The first was the discovery of natural radioactivity in 1896 by Henri Becquerel and the subsequent discovery of radium and polonium by Marie and Pierre Curie (see Chapter 3). From these discoveries came the probe used to discover the nuclear atom and a new and effective treatment for cancer. The second birthday occurred in 1938 with the discovery that when uranium atoms of mass 235 are bombarded with neutrons, some atoms split, releasing a tremendous amount of energy. From this discovery came synthetic radioisotopes, nuclear medicine, power reactors, and ultimately, nuclear weapons.

Although you may have heard much about radiation, you may be surprised to realize that only one-third of the radiation we are exposed to comes from manmade sources. The rest comes from naturally radioactive materials in the earth and air and from cosmic radiation. It is part of living on the earth. This chapter will present some basic principles of nuclear chemistry that explain the behavior of both naturally occurring and synthetic radioactive materials. It will examine the interaction of this radiation with living things and will illustrate its use as a diagnostic and therapeutic tool in medicine and as a source of energy.

13.1 Nuclear Decay: Transmutation and Half-Life

Neutron–Proton Ratio: "Nuclear Glue"

All radioactive materials are composed of elements having unstable nuclei. At some time in their existence, they attempt to become more stable by expelling small chunks of matter, principally *beta particles* (high speed electrons), *positrons* (electron-sized particles of positive charge), *alpha particles* (helium nuclei), and *gamma radiation* (a high energy electromagnetic wave, or photon, with a shorter wavelength than visible light) (see Chapter 3). This process of emitting radioactive particles is called **radioactive decay**.

Stable nuclei have a favored ratio of protons to neutrons. Unstable or radioactive nuclei will emit mass and positive or negative charge to move toward the stable ratio.

Every element has one or more favored combinations of neutrons and protons that make up its nucleus. Any departure from these combinations results in an unstable nucleus and ultimately in radioactive decay. All nuclei are made up of protons and neutrons. Hydrogen is the only exception to this rule; its most common isotope has only one proton and no neutrons. But consider a helium nucleus. It contains two protons and two neutrons. The two positive protons, which have the same electrical charge, should greatly repel each other and should try to be much farther apart than the diameter of the nucleus would permit. Something must neutralize this repulsive force and hold the protons together in the nucleus. Neutrons are part of the "nuclear glue" that holds the unit together. Except for hydrogen, most light elements have about one neutron for every proton. Carbon (mass 12, atomic number 6), for example, has six protons and six neutrons. Oxygen has eight protons and eight neutrons. Calcium has 20 protons and 20 neutrons, for a mass number of 40. As one moves to higher atomic numbers and more protons in the nucleus, the relative number of neutrons increases (Table 13.1). Cobalt has 27 protons and 32 neutrons, for a mass number of 59. Gold has 79 protons and 118 neutrons, for a mass number of 197. Atoms with many protons have a higher percentage of neutrons to balance the repulsive forces

TABLE 13.1
Neutron–Proton Ratio for Some Common Isotopes

Element	Protons	Neutrons	Neutrons per Proton	
C	6	6	1	
O	8	8	1	
P	15	16	1.07	
S	16	16	1	
Ag	47	60	1.28	
I	53	74	1.40	
Hg	80	122	1.53	
Pb	82	126	1.54	
Th	90	142	1.58	Nuclei with more than 83 protons are always unstable.
U	92	146	1.59	

between protons. The ratio of neutrons to protons in stable nuclei ranges from 1.0 (equal quantities) in light elements to about 1.54 in lead and bismuth. There are no stable elements with more protons than bismuth, which has an atomic number of 83.

Nuclei need not be heavy to be unstable. They must only lack the proper combination of neutrons and protons. The process of radioactive decay bears some similarity in its energy relationships to ordinary chemical reactions. The radioactive nucleus is stressed by an unfavorable neutron to proton ratio and, as a result, exists in a high energy state. Sooner or later the nuclear particles interact to produce an internal energy greater than the "surface tension" that holds the outer skin of the nucleus together. At this point the unstable nucleus fires out some of its material and reverts to a lower energy state. The process of nuclear decay is called **transmutation**, the transformation of one element to another. The reactant and product nuclei involved in a radioactive decay are called **parent** and **daughter nuclei**. The daughter nucleus is what remains after the nuclear particle is released (Figure 13.1).

Negative Beta Decay

Beta(−) decay occurs when there is a deficiency of protons in the nucleus, and it results in conversion of a neutron to a proton.

Negative beta decay occurs in unstable nuclei that are neutron rich compared to the stable isotope of the element. Negative beta emission results from the conversion of one of the parent nuclei's neutrons to a proton (Figure 13.2).

Negative beta, or beta(−), decay results in the creation of a daughter nucleus with an atomic number that is *one more* than the parent (because

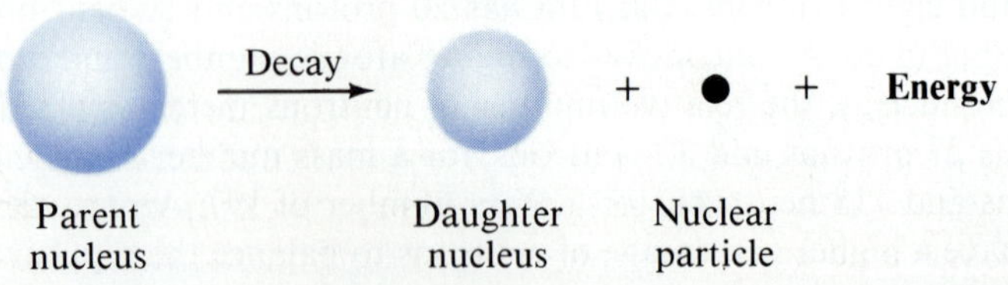
Decay
+
+
Energy
Parent nucleus
Daughter nucleus
Nuclear particle

FIGURE 13.1
The energy relationships involved in nuclear decay are similar to those of ordinary chemical reactions. The product nucleus has less energy than the reactant nucleus. The change in energy shows up in a nuclear reaction as the mass and energy of the expelled particle.

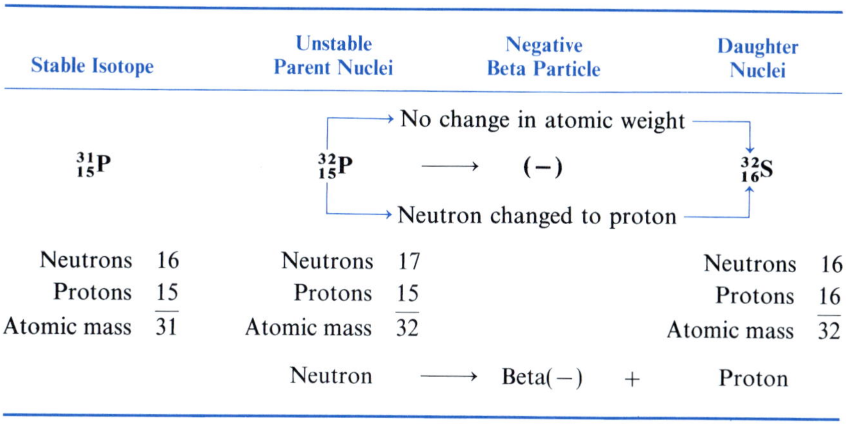
Stable Isotope
Unstable Parent Nuclei
Negative Beta Particle
Daughter Nuclei
No change in atomic weight
$^{31}_{15}P$
$^{32}_{15}P$
(−)
$^{32}_{16}S$
Neutron changed to proton
Neutrons 16
Protons 15
Atomic mass 31
Neutrons 17
Protons 15
Atomic mass 32
Neutrons 16
Protons 16
Atomic mass 32
Neutron ⟶ Beta(−) + Proton

FIGURE 13.2
Chart showing negative beta decay of phosphorus-31 (^{31}P).

of the creation of an additional proton), and an atomic mass that remains essentially unchanged. Since a proton weighs about the same as a neutron, the small mass of the ejected beta particle (1/1837 amu) does not significantly affect the atomic mass.

Beta particle emission is accompanied by the emission of a *neutrino*, an extremely small, uncharged particle that is ejected in a direction opposite to that of the beta particle. The energy released in the decay is split (although not necessarily equally) between the beta particle and the neutrino, so beta particles exhibit a spectrum of different energies for a given decay.

EXAMPLE 13.1

Show the negative beta decay of iodine-131 (^{131}I).

$$^{131}_{53}I \longrightarrow {}^{0}_{-1}\text{Beta} + {}^{131}_{54}\text{Xe} + \text{Neutrino}$$

One of iodine's neutrons has been converted to a proton and a beta particle. Note that the atomic number (number of protons) of the daughter nucleus indicates one more proton than the parent.

EXAMPLE 13.2

Show the negative beta decay of cobalt-60 (^{60}Co).

$$^{60}_{27}\text{Co} \longrightarrow {}^{0}_{-1}\text{Beta} + {}^{60}_{28}\text{Ni} + \text{Neutrino}$$

One of cobalt's neutrons has been converted to a proton and a beta particle. The atomic number of the daughter nucleus is one greater than the parent. Note, however, that the atomic mass remains the same. The total number of protons and neutrons has not changed.

Positron Decay

Positive beta particles, called *positrons*, are emitted from unstable nuclei that are neutron poor compared to the stable isotope of the element. **Positive beta**, or beta(+), **decay**, called **positron emission**, results from the conversion of a proton to a neutron in the parent nucleus (Figure 13.3).

Positron emission occurs when there is a deficiency of neutrons in the nucleus, as compared to the stable isotope of the element. It results in conversion of a proton to a neutron.

FIGURE 13.3
Positron decay occurs in the radioactive isotope sodium-23 (^{23}Na).

Stable Isotope	Unstable Parent Nuclei	Positive Beta Particle	Daughter Nuclei
$^{23}_{11}Na$	$^{22}_{11}Na$ ⟶	(+)	$^{22}_{10}Ne$

No change in atomic weight ($^{22}_{11}Na$ → $^{22}_{10}Ne$)

Proton changed to neutron ($^{22}_{11}Na$ → $^{22}_{10}Ne$)

	Stable Isotope	Unstable Parent Nuclei	Daughter Nuclei
Neutrons	12	11	12
Protons	11	11	10
Atomic mass	23	22	22

$$\text{Proton} \longrightarrow \text{Beta}(+) + \text{Neutron}$$

Positron decay results in the creation of a daughter nucleus with an atomic number that is *one less* than the parent, and an atomic mass that remains unchanged. Since a neutron has about the same mass as a proton, the small mass of the ejected positron (1/1837 amu) does not significantly affect the mass of the nucleus. Positrons, like negative beta particles, are essentially electrons, but carry a +1 charge instead of a −1 charge. Positron emission, like beta emission, is accompanied by neutrino emission; the decay energy is split between the positron and the neutrino as in the case of negative beta decay. Positron-emitting isotopes are rare in nature, since naturally occurring nuclei tend instead to be stable or neutron rich. Very useful positron-emitting isotopes, however, are routinely created in nuclear reactors for medical and research use.

EXAMPLE 13.3

Show the positron decay of iron-52 (^{52}Fe).

$$^{52}_{26}\text{Fe} \longrightarrow {}^{0}_{+1}\text{Positron} + {}^{52}_{25}\text{Mn} + \text{Neutrino}$$

One of iron's protons has been converted to a neutron. Note that the number of protons in the daughter isotope is one less than in the parent, but the atomic mass is the same.

EXAMPLE 13.4

Show the positron decay of phosphorus-30 (^{30}P).

$$^{30}_{15}\text{P} \longrightarrow {}^{0}_{+1}\text{Positron} + {}^{30}_{14}\text{Si} + \text{Neutrino}$$

Again, one of the parent isotope's protons has been converted to a neutron. The atomic number of the daughter isotope is one less than the parent, but the atomic mass is unchanged.

Alpha Decay

Alpha decay is most common in heavy nuclei and results in loss of two protons and two neutrons.

Nuclei that are heavy and have high nuclear charge often decay by emission of an alpha particle (two protons and two neutrons). This emission, called **alpha decay**, results in the creation of a daughter nucleus with an

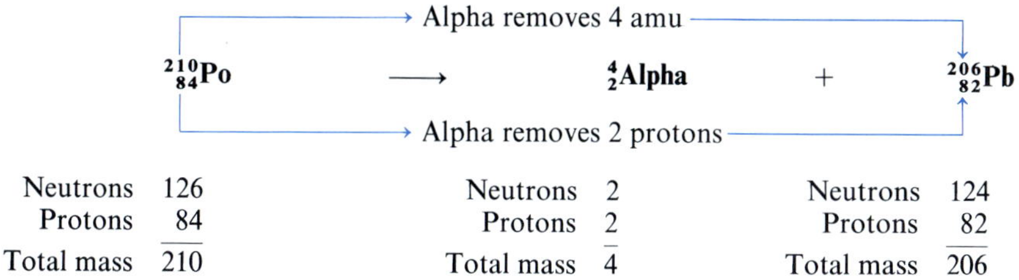
Alpha removes 4 amu
$^{210}_{84}$Po ⟶ $^{4}_{2}$Alpha + $^{206}_{82}$Pb
Alpha removes 2 protons
Neutrons 126
Protons 84
Total mass 210
Neutrons 2
Protons 2
Total mass 4
Neutrons 124
Protons 82
Total mass 206

FIGURE 13.4
This diagram shows the unstable polonium-210 (^{210}Po) nuclide undergoing alpha decay.

atomic number that is two less than the parent nucleus and an atomic mass that is four less than the parent (Figure 13.4).

EXAMPLE 13.5

Show the alpha decay of uranium-238 (^{238}U).

$$^{238}_{92}\text{U} \longrightarrow {}^{4}_{2}\text{Alpha} + {}^{234}_{90}\text{Th}$$

Loss of an alpha particle removes two charge units and four atomic mass units from the uranium nucleus. The daughter product has an atomic number of 90 and an atomic mass of 234.

EXAMPLE 13.6

Show the alpha decay of radium-226 (^{226}Ra).

$$^{226}_{88}\text{Ra} \longrightarrow {}^{4}_{2}\text{Alpha} + {}^{222}_{86}\text{Rn}$$

The radium nucleus also loses two protons and two neutrons during an alpha decay. The daughter product has two less charge units and four less mass units.

Nuclear Decay: A Generalization

The changes that occur during beta(−), positron, and alpha decay are generalized in Figure 13.5. Beta(−) decay produces a daughter nuclei with one more proton and one less neutron than the parent nuclei, for a change of +1 in atomic number and no change in atomic weight. Positron decay produces a daughter nuclei with one more neutron and one less proton than

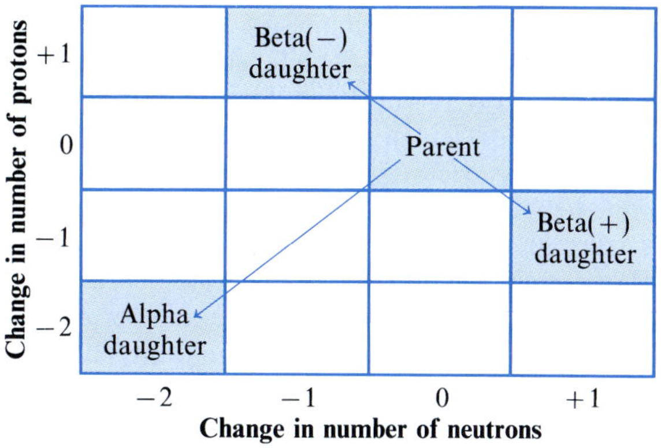
Beta(−)
daughter
Parent
Beta(+)
daughter
Alpha
daughter
+1
0
−1
−2
Change in number of protons
−2
−1
0
+1
Change in number of neutrons

FIGURE 13.5
A memory aid for predicting the product of alpha, beta (−), and beta (+) or positron decay.

the parent nuclei, for a change of −1 in atomic number and no change in atomic weight. Alpha decay produces a daughter nuclei with an atomic number two less than the parent, and an atomic weight four less than the parent.

Gamma Radiation

Gamma radiation is electromagnetic radiation like light, but of much greater energy and is produced by energy changes within the nucleus.

Gamma radiation resembles light, except that its energy is much higher and its wavelength much shorter (see Chapter 3). Gamma radiation originates as nuclear particles change from higher to lower energy levels within the nucleus, much as visible light is given off when electrons fall from higher to lower electron energy levels around the nucleus. Energy changes within the nucleus, however, are much larger than the usual changes in electron energy. While visible light is produced by changes in electron energy on the order of one or two electron volts (eV), gamma radiation results from changes within the nucleus that are on the order of 10^5 to 10^7 eV. In contrast, the chemical reactions in a flashlight battery produce an energy of 1.5 eV.

Gamma radiation is usually given off by a daughter nucleus that has just released an alpha or beta particle as it rearranges itself to a more stable configuration within its nucleus. This rearrangement is usually fast enough that gamma emission is simply said to accompany the original alpha or beta decay. While almost all alpha decay is accompanied by gamma radiation, some beta decay (which causes less change in the parent nucleus) produces stable daughter nuclei that do not emit gamma radiation. These beta decays, however, are in the minority, and most beta and positron decays are accompanied by gamma radiation.

Because the energy of gamma radiation is directly related to the energy level spacing of the daughter nucleus, this radiation provides an energy spectrum that is characteristic for a particular nucleus. This creates a "fingerprint" for that nucleus just as the visible spectrum of an excited gas provides a "fingerprint" for that particular element.

Half-Life

The decay of unstable nuclei is a random process, in many ways analogous to the death of humans. We cannot usually predict when a given person will die, but we can accurately predict the average number of deaths per year in a large population. Life insurance companies are thus able to adjust policies to meet expected payments. When we deal with atoms, we are usually working with a very large number—often many times more than the entire population of the earth. Thus, average predictions made for the behavior of unstable atoms are very accurate estimates of what actually occurs.

The period required for half of a sample of radioactive nuclei to decay is called the *half-life.*

Suppose that an unstable nucleus is under such stress that the probability is 20% that within one year it will decay. Suppose further that we have a sample of 1000 of these nuclei. In the first year, 20% or 200 of them will decay. In the second year, each unstable atom again has a 20% chance of decaying, so of the remaining 800, 160 will decay. In the third year the same situation holds true: 20% of the nuclei present at the beginning of the year will decay. This on-going situation can be easily summarized by the concept of **half-life**, which is the period necessary for the number of nuclei decaying during a unit time to fall to half of its original amount. Half-life is a general concept that can be applied any time during the nuclei's life (Table 13.2).

TABLE 13.2
Decay of a Radioactive Element with a Half-Life of 3 Years

Time (years)	Nuclei Present at Beginning of Year	Nuclei Decaying During Year
0	1000	200
1	800	160
2	640	128
3	512	102
4	410	82
5	328	66
6	262	52
7	210	42

(Half-life brackets: 200 to 102; 128 to 66; 82 to 42.)

Note in Table 13.2 that between year 0 and year 3, the population falls from 1000 to 512 and the decay rate falls from 200 to 102, a reduction to about one-half. Between year 2 and year 5 (a three-year period), the population falls from 640 to 328 and the decay rate falls from 128 to 66, a reduction to about one-half. The half-life of this isotope is thus about 3 years.

A graph of these data is presented in Figure 13.6. A smooth curve is formed, with approximately one-half of the unstable nuclei decaying in the first 3-year period. In the second 3-year period, the number of nuclei decaying again falls approximately by one-half. In the third 3-year period, another reduction to one-half occurs. Half-life is a characteristic property of radioactive nuclei that, like the type of decay and the energies of radiation emitted, can be used to identify the radioactive isotope. It can also be used to date ancient objects and to determine the length of storage time required for nuclear wastes.

Nuclear Dating

Radioactive carbon-14 (^{14}C), with 6 protons and 8 neutrons in its nucleus, has a half-life of 5760 years. It is formed by the collision of cosmic radiation with nitrogen in the atmosphere. Some of this radioactive carbon is eventually combined with oxygen to form carbon dioxide. Plants "breathe" and metabolize this radioactive carbon dioxide in the same manner as normal carbon

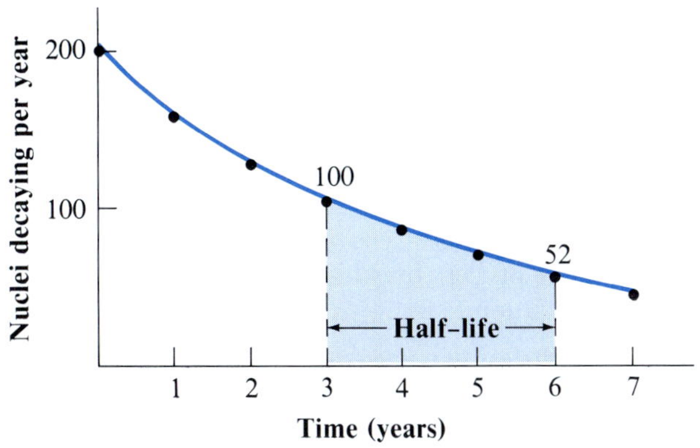
Nuclei decaying per year
200
100
100
52
Half-life
1
2
3
4
5
6
7
Time (years)

FIGURE 13.6
The half-life of a radioisotope is the time required for half of it to decay. If you select any point on the graph, 3 years later the activity is about one-half as much. Thus, the half-life of this example is 3 years.

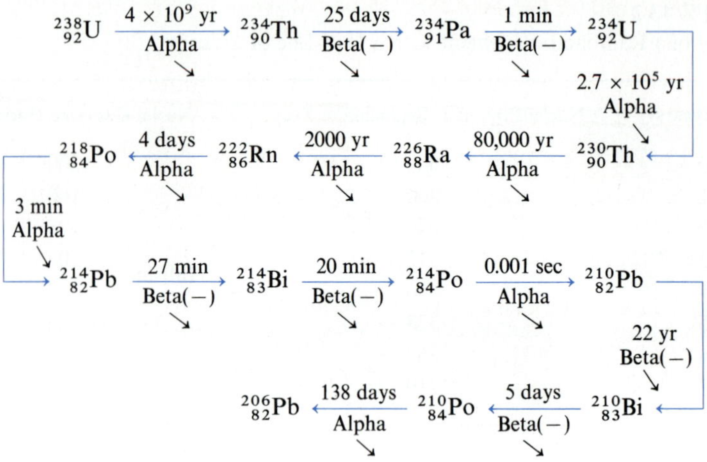

$^{238}_{92}U$
4 × 10⁹ yr Alpha
$^{234}_{90}Th$
25 days Beta(−)
$^{234}_{91}Pa$
1 min Beta(−)
$^{234}_{92}U$
2.7 × 10⁵ yr Alpha
$^{230}_{90}Th$
80,000 yr Alpha
$^{226}_{88}Ra$
2000 yr Alpha
$^{222}_{86}Rn$
4 days Alpha
$^{218}_{84}Po$
3 min Alpha
$^{214}_{82}Pb$
27 min Beta(−)
$^{214}_{83}Bi$
20 min Beta(−)
$^{214}_{84}Po$
0.001 sec Alpha
$^{210}_{82}Pb$
22 yr Beta(−)
$^{210}_{83}Bi$
5 days Beta(−)
$^{210}_{84}Po$
138 days Alpha
$^{206}_{82}Pb$

FIGURE 13.7
The decay series for uranium-238 (^{238}U). All of the isotopes shown here are present in a natural sample of uranium ore. The age of the ore may be determined by measuring the ratio of ^{238}U (the parent) to ^{206}Pb (the final daughter in the series).

dioxide, and eventually bind some of this carbon in their molecules. The percentage of radioactive carbon in a plant will be relatively constant as long as the plant is alive. However, when the plant dies and stops absorbing new radioactive carbon, the amount of ^{14}C present will decrease at a rate consistent with the 5760-year half-life.

Naturally radioactive carbon (^{14}C) present in the atmosphere is taken up by living plants. Comparison of the amount of radioactive ^{14}C in old and new tissue gives the age of the material, called *radiocarbon dating*.

Suppose, for example, that we discovered an ancient wooden tool. A small sample of the wood from this tool could be counted with a radiation detector to determine the amount of ^{14}C present. Then a similar sample of wood from a recently cut tree could be counted. If one-half of the negative beta emissions were measured for the ancient tool per gram of wood compared to the new sample, we could conclude that its age was about one half-life or 5700 years old. This process is called **radiocarbon dating**. Radiocarbon dating has been used to estimate the age of materials from the Egyptian civilization, the Dead Sea scrolls, and other ancient artifacts.

Dating of ancient objects can also be done using radioactive materials other than carbon-14. Nuclear dating has been used to estimate the age of meteorites, rocks forming the crust of the earth, and lunar surface rocks. Both uranium and thorium decay through a series of intermediate unstable nuclides until stable lead is finally produced (Figure 13.7). This is called a **decay series**. We can estimate the age of a rock containing uranium by measuring the ratio of uranium-238 (^{238}U) to stable lead (^{206}Pb) and considering the half-lives of ^{238}U and the other members of the decay series in our calculation. Estimates of the age of the earth based on this evidence indicate that about 4.5 billion years have elapsed since the uranium-containing rocks were formed. Estimates of the age of the moon by analysis of lunar rock samples and the age of the solar system by analysis of meteorites yield a similar value.

NEW TERMS

radioactive decay The process by which an unstable nucleus emits alpha or beta and gamma radiation.

transmutation The process of conversion of one element to another by radioactive decay.

parent nuclei (or isotope) The unstable atom prior to radioactive decay.

daughter nuclei (or isotope) The product of the radioactive decay. The nucleus of a specific isotope is sometimes called a *nuclide.*

negative beta decay Radioactive decay involving emission of a negative beta particle. The daughter product of a negative beta decay will have one less neutron and one more proton than the parent isotope.

positive beta decay (positron emission) Radioactive decay involving emission of a positron. The daughter product of positron decay will have one less proton and one more neutron than the parent isotope.

alpha decay Radioactive decay involving emission of an alpha particle. The daughter product of alpha decay will have two less protons and two less neutrons than the parent isotope.

half-life The time required for half of the radioactive nuclei present to undergo radioactive decay.

radiocarbon dating Determination of the age of an object by measuring the amount of radioactive carbon present as compared to the amount in a similar living sample. Nuclear dating can also be based on other isotopes, such as thorium or uranium and lead.

decay series A group of successively decaying isotopes from one parent isotope. Several such decays are often required before the product isotope is stable (usually lead).

TESTING YOURSELF

Radioactive Decay and Half-Life

1. Thorium-232 ($^{232}_{90}Th$) is an abundant naturally occurring radioisotope on the earth. Approximately 1.2 g of every 100 kg of the earth's crust is ^{232}Th. This radioisotope has a half-life of 14 billion years and decays by emission of an alpha particle. What is the product of the decay of ^{232}Th?
2. Potassium-40 ($^{40}_{19}K$) is also a naturally occurring radioisotope. Approximately 0.01% of all potassium on the earth is radioactive. ^{40}K has a half-life of about 1.4 billion years and emits a beta(−) particle. What is the product of the decay of ^{40}K?
3. Zinc-65 (^{65}Zn) is a positron emitter that is produced in nuclear reactors. What is the product of the decay of ^{65}Zn?
4. The following table presents time and count rate data describing the decay of a radioisotope.

Time (days)	Disintegrations per Minute
0	7000
1	5550
2	4409
4	2275
5	2204
7	1388

Plot this data and estimate the half-life of the radioisotope.

Answers **1.** ^{228}Ra **2.** ^{40}Ca **3.** ^{65}Cu **4.** 3 days

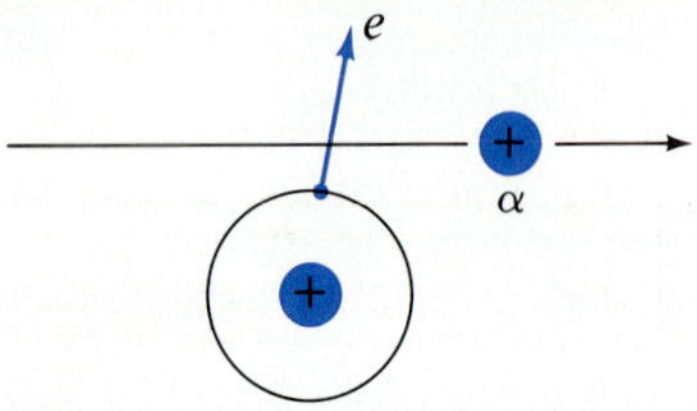

e
α

FIGURE 13.8
Alpha particles lose their energy by knocking electrons from the atoms they pass. Since an alpha particle's relatively large charge, large mass, and low velocity create a high probability of interaction, alpha particles lose their energy rapidly as they pass through matter.

13.2 Interaction of Radiation with Matter

The way in which nuclear radiation interacts with matter is determined by the charge, mass, and velocity of the radiation. The greater the charge and mass of a nuclear particle and the more slowly it passes an atom, the greater the chance of interaction.

Alpha particles, with a +2 charge and mass of 4 amu, interact with both nuclei and electrons. Because of the small size of nuclei and the large amount of space between them, collision with a nucleus is not common. (Recall from Chapter 3 the problems Geiger, Marsden, and Rutherford had trying to bounce alpha particles from gold foil.) However, the alpha particle's high charge and relatively low velocity make it a good candidate for an interaction with an electron. Although a fair amount of energy (10 to 20 eV) is required to remove an electron from an atom, an alpha particle starting out with an energy of 5 million eV can knock electrons from about 0.25 to 0.5 million atoms before expending all its energy. Electrons are common in all matter, and the chance of interaction with a large, slow, and highly charged projectile like an alpha particle is great. For example, the electrons in a few centimeters of air or a single layer of plastic wrap will stop most alpha particles (Figure 13.8).

Beta particles also dislodge electrons, but not as effectively as do alpha particles.

A beta particle, in contrast, is much smaller than an alpha particle. It has only half the charge (but negative) and about 1/7000 the mass. Since the energy of motion (kinetic energy) of an object is equal to one-half the mass times its velocity squared, to have the same energy as an alpha particle, a beta particle must move about 83 times faster. As a result, the beta particle's interaction with the nuclei and electrons it passes is much less than that of an alpha particle. Its range of penetration, therefore, is much greater. Beta particles will penetrate several millimeters of aluminum or several feet of air.

Gamma radiation has no mass and no charge and travels at the speed of light, so it is by far the most penetrating of the three general types of nuclear radiation. Gamma radiation will penetrate several feet of concrete or lead. However, the very factors that make gamma radiation so penetrating result in less damage (ionization) along its path (Figure 13.9).

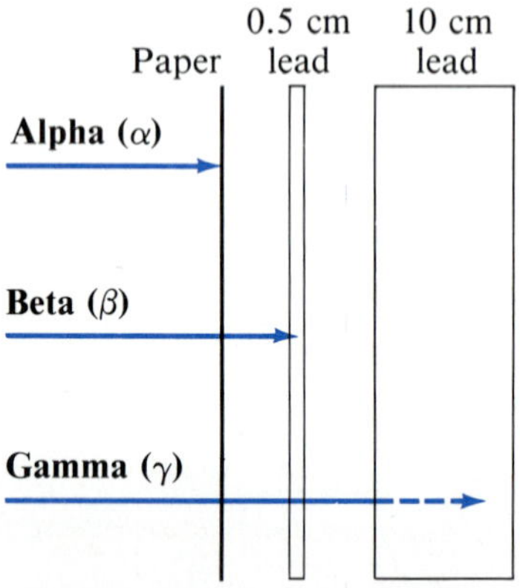

Paper
0.5 cm lead
10 cm lead
Alpha (α)
Beta (β)
Gamma (γ)

FIGURE 13.9
The relative penetrating ability of the three major types of nuclear radiation is illustrated in this figure. Alpha particles interact the most and are most easily stopped. Beta particles and positrons interact less and penetrate much farther. Gamma radiation has the greatest penetrating ability but causes the least ionization.

13.3 Detection of Nuclear Radiation

Detection of nuclear radiation is closely tied to its interaction with matter. There are three general types of radiation detectors. One is based on photographic film. The second involves collection of free electrons formed as radiation passes through a solid or a gas. The third measures light produced as electrons disturbed by the passage of radiation fall back to lower energies in their parent atoms.

Photographic Detection of Radiation

Radiation interacts with photographic film by causing ionization of atoms in the silver-containing compound that coats the film. During development of the film, silver atoms near these ionized sites are reduced to metallic silver, leaving a dark mark on the film. Photographic film was used by Becquerel when he discovered natural radioactivity. Today it is routinely

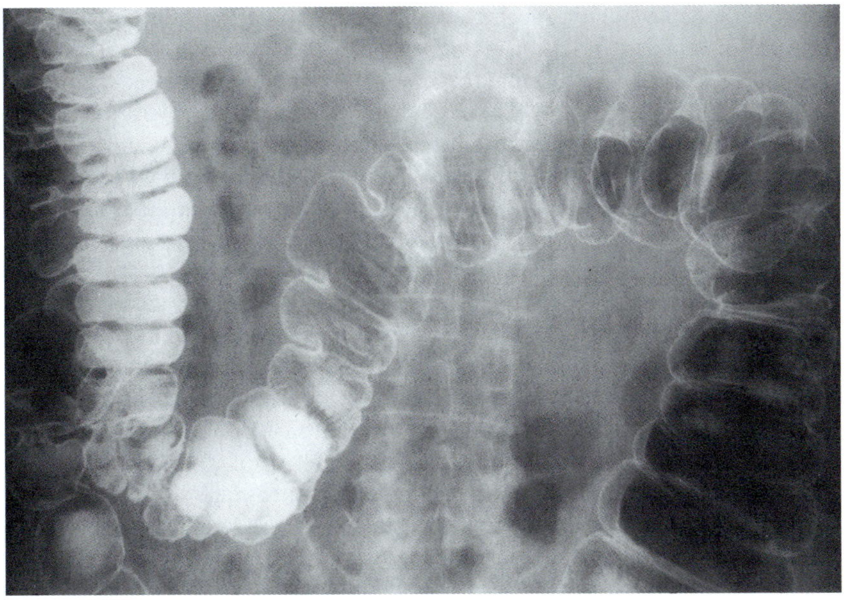

FIGURE 13.10
Film is still one of the most commonly used detectors of radiation. X-rays are absorbed by barium compounds; the colon is visible in this photograph because the subject has been given an enema containing a solution of barium sulfate.

used for medical and dental X-rays (Figure 13.10). Samples of film are often worn in small plastic badges by X-ray technologists to measure the wearer's exposure to gamma and beta radiation. Once every few weeks the film is replaced, and the used film is returned to a central processing facility, where it is developed and the radiation dose measured by the amount of darkening of the film.

Gas Ionization Detectors

Gas ionization detectors operate by collecting the electrons and positive ions formed as ionizing radiation passes through a gas. Probably the best known gas ionization detector is the **Geiger–Müller counter** or Geiger counter (Figure 13.11). This detector consists of a gas-filled tube with a conductive wall and a metallic center conductor. When ionizing radiation passes through the tube, ion pairs (electrons and positive ions) are formed. The electrons are attracted toward the positively charged central electrode while the positive

Geiger–Müller counters will detect the passage of a single beta particle, but are less efficient for detection of gamma radiation because of the low density of its detecting gas.

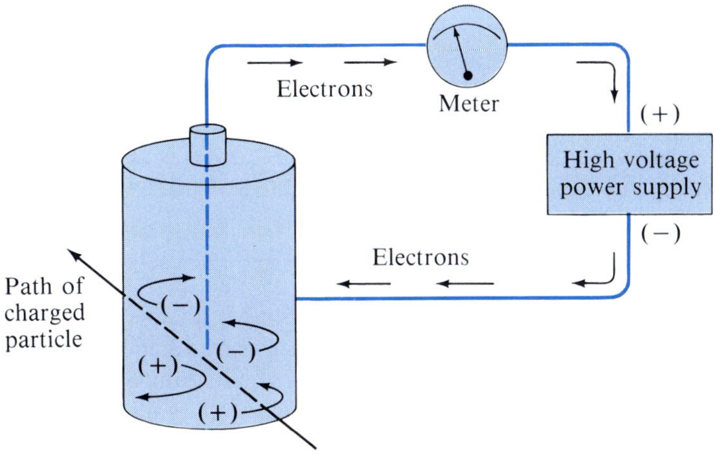
Electrons
Meter
(+)
High voltage power supply
(−)
Electrons
Path of charged particle
(−)
(−)
(+)
(+)

FIGURE 13.11
The Geiger–Müller tube is a gas ionization detector. Ions and electrons formed as radiation passes through the tube are collected and counted to determine the radiation level.

ions are collected at the negatively charged wall of the tube. Each time a particle of ionizing radiation passes through the tube, a group of electrons are collected, passed through the counting circuit and battery, and then moved to the outer case of the tube where they recombine with the positive ions. A high positive charge on the collecting electrode attracts the free electrons so rapidly toward the central wire that they strike and ionize other gas molecules on the way in. This results in collection of a large number of electrons (on the order of 10 billion) each time a nuclear particle passes through the tube. This *gas multiplication* means that a Geiger–Müller (G-M) counter requires little subsequent amplification to observe its signal and is thus both very sensitive and quite inexpensive (Figure 13.12).

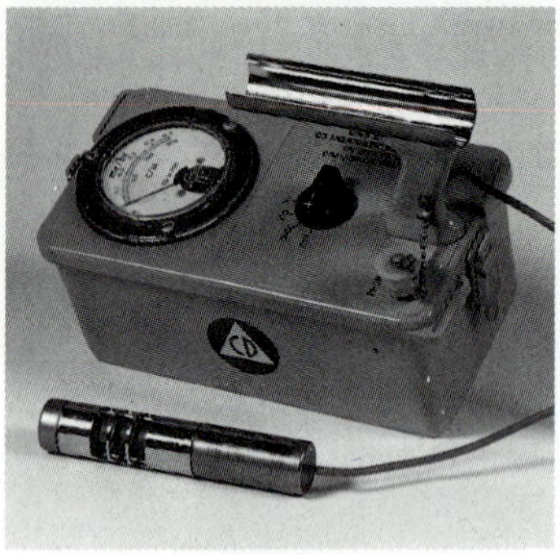

FIGURE 13.12
Geiger–Müller detectors are reasonably inexpensive and rugged. They are widely used as detectors for survey instruments.

A short period of time (about 200 microseconds) is required to sweep the positive ions from the tube after the passage of each particle, and during this time, the G-M tube is "dead." As a result, G-M detectors do not work well in areas of high radiation. Ion chambers, similar in design to G-M detectors but operated at about 25 V in contrast to the 900 to 1200 V used with a G-M detector, collect only "primary" ion pairs. They are thus less sensitive and able to handle larger radiation levels. High level survey instruments usually make use of ionization chambers.

Since G-M detectors depend on gas ionization to operate, they are most sensitive to those types of radiation that cause the greatest degree of ionization. As a result, the G-M detector is an excellent alpha particle detector if it is equipped with a thin enough window to allow the alpha particles to enter. It is also very effective for beta particles; beta detectors have a more rugged window since the beta particles are more penetrating than alpha particles. The G-M detector, however, is inefficient for gamma radiation. The low density gas in its detecting chamber interacts poorly with an uncharged gamma ray.

Scintillation Detectors

A third type of radiation detector, the **scintillation detector**, is effective for all three of the principal types of radiation, but it is most commonly used with gamma radiation. Scintillation detectors are considerably more expensive than gas ionization detectors, hence their principal applications lie in detection of gamma radiation, where the gas ionization detector performs poorly.

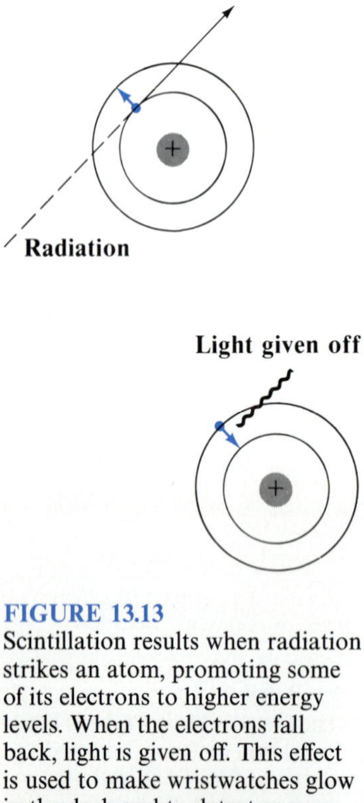
Radiation
Light given off

FIGURE 13.13
Scintillation results when radiation strikes an atom, promoting some of its electrons to higher energy levels. When the electrons fall back, light is given off. This effect is used to make wristwatches glow in the dark and to detect gamma radiation in medical and nuclear laboratories.

When nuclear radiation interacts with an atom, electrons are ionized or promoted to higher energy levels. When these electrons fall back to their normal lower energy states, energy is released as visible or ultraviolet light. This flash of light is called a *scintillation*, and its color depends on the energy level spacing of the atom involved. Although most atoms scintillate in the ultraviolet region of the spectrum, a few exhibit this effect in the visible region and are useful radiation detectors (Figure 13.13). Wristwatches that glow in the dark use a mixture of a radioactive material and a scintillator; radiation from the radioactive material strikes atoms and electrons in the scintillator, causing them to give off light. A number of years ago a mixture of alpha and high energy beta-emitting nuclides were used for this purpose; more recently, low energy beta-emitting nuclides such as tritium (^3_1H) have been used. Thus, an older luminous watch will give a good count on a G-M detector (beta particles will go through the watch crystal), while in new luminous watches, the crystal absorbs the weaker particles.

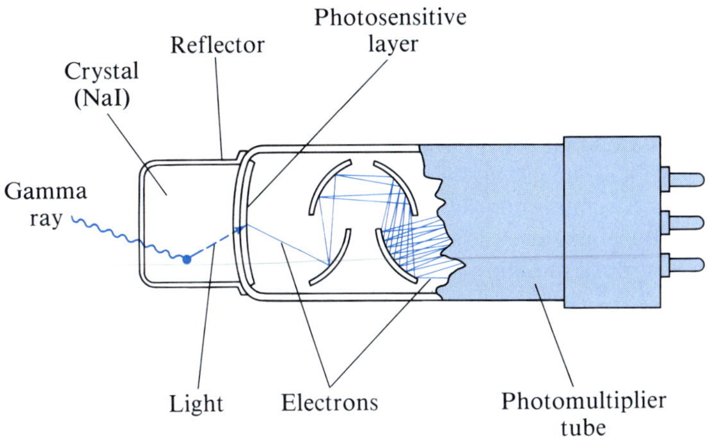
Photosensitive layer
Reflector
Crystal (NaI)
Gamma ray
Light
Electrons
Photomultiplier tube

FIGURE 13.14
The dense detector crystal in a scintillation detector provides a high probability of gamma interaction, and thus good detection efficiency for gammas. If the crystal is large enough to stop the gamma ray, the brightness of the flash of light produced will be proportional to the gamma ray energy. Thus, the scintillation detector not only detects the presence of a gamma ray, but also measures its energy. Different radioisotopes in a given sample can be differentiated by this procedure.

Scintillation detectors are good detectors of gamma radiation because they can be made using transparent solids rather than gases. The greater density of a solid compared to a gas increases the probability of a gamma ray interaction. Most scintillation detectors use a clear, dense crystal of sodium iodide (NaI) as the detector, which is viewed with a very sensitive light detector called a photomultiplier (Figure 13.14). Flashes of light caused by gamma interactions in the crystal are detected by the photomultiplier, amplified, and counted. This detector has the added advantage of being energy sensitive. The brightness of the flash of light depends on the number of interactions the gamma ray has while being absorbed in the crystal and therefore depends on its energy. The gamma ray spectrum of a radioisotope can be used to "fingerprint" and identify the element in just the same way the emission spectrum of a gas provides a fingerprint and positive identification of that element.

Gamma scintillation detectors are commonly used in medicine, since the gamma radiation from a "tracer" isotope injected into a patient is absorbed only slightly by tissue and can be readily detected outside the patient's body. This application of radiation is discussed in more detail in the next section.

NEW TERMS

Geiger–Müller counter A gas ionization detector operated at high enough voltage that considerable multiplication occurs as electrons are collected within the Geiger tube. G-M counters are very sensitive to beta radiation and to alpha radiation if properly equipped. Also called "Geiger detector."

scintillation detector A detector used principally for gamma rays. Photons striking the detector knock electrons from atoms; as these electrons fall back to lower energy positions in their atoms, light is given off. This light is observed by a photomultiplier tube, and the passage of the particle is recorded.

TESTING YOURSELF

Detection of Nuclear Radiation

1. A very high voltage (900–1000 V) is used to collect electrons produced by the passage of radiation through a Geiger–Müller detector. Why is such a high voltage used when only a few volts will collect most electrons produced by the radiation?

2. Why are G-M detectors "dead" for a short period of time following the passage of nuclear radiation?
3. A G-M survey meter has a window about as thick as a piece of heavy aluminum foil. This is strong enough to protect the detector from damage if it is treated carefully. Which type of radiation will not be detected by such an instrument: alpha, beta, or gamma?
4. Why is a scintillation detector more efficient than a G-M detector for detection of gamma radiation?
5. Which type of radiation penetrates the farthest into matter: alpha, beta, or gamma?

Answers **1.** The high voltage accelerates the electrons and causes gas multiplication. **2.** The tube is "dead" until the positive ions are swept out of the detector. **3.** alpha **4.** Dense crystal interacts better with zero charged gamma. **5.** gamma

13.4 Radiation Biology and Nuclear Medicine

The most probable interaction of radiation with living matter involves ionization of water molecules.

Far less probable, but far more serious, is the possibility of radiation damage to cellular or reproductive DNA. Genetic modification of a cell is called *mutation.*

Nuclear radiation interacts with plant and human tissue in two ways, both involving use of the radiation to break chemical bonds or to remove electrons from atoms or molecules. The first is through interaction with water molecules. Since a high percentage of all tissue is water, the probability of this interaction is high. It results in formation of hydrogen ions and hydroxide ions, as well as other electron-deficient groups, which react with the adjacent molecules in the cell. If enough ions are formed by the passage of radiation, the result is usually death of the cell. The second interaction is far less probable, but potentially more serious. It involves collision with the cells' genetic messenger, the DNA molecule. Since the amount of DNA or genetic material in the cell is hundreds of thousands of times less than the amount of water, this interaction is rare. But when it does occur, the effect on the cell is serious since it may appear in subsequent generations. This modification of genetic information, called **mutation**, usually results in deformation or death of the cell. Some forms of cancer are believed to be caused by mutation of normal cells. Humans and animals exposed to moderate amounts of radiation show a measurably greater incidence of cancer.

Effects of Radiation on Humans

Large radiation doses affect rapidly growing tissue the most. The most sensitive system in the human body is the bone marrow, which produces red blood cells.

The effect of radiation on humans involves the interaction with both water and genetic material. Large radiation doses affect rapidly growing tissue the most. One such tissue, bone marrow, produces red blood cells, which have a "half-life" of about one month and must be continually replaced. A radiation dose of about 450 rad (see definition of a rad on p. 338) will damage the bone marrow beyond repair in about 50% of the cases, producing death within about a month. This dosage is called the $\mathbf{LD_{50}}$, which stands for lethal dose for 50% of the population. Data for estimates of the LD_{50} have been gathered by experimentation with sheep and other animals, from accidents in the nuclear weapons program and nuclear explosions in Japan at the close of World War II, and from accidental exposure of a native population during atmospheric nuclear testing in the South Pacific in the 1950s.

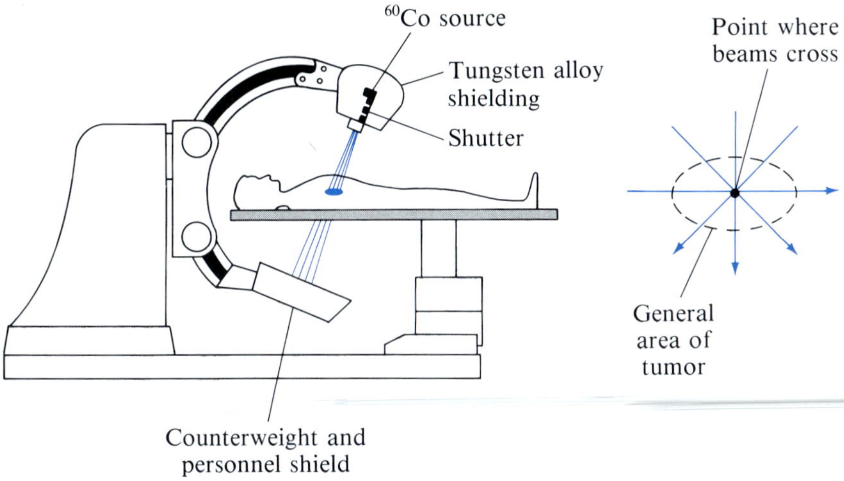

^{60}Co source
Tungsten alloy shielding
Shutter
Counterweight and personnel shield
Point where beams cross
General area of tumor

FIGURE 13.15
Gamma radiation from ^{60}Co is routinely used for treatment of tumors. By rotating the gamma source around the patient, the radiation dose to any segment of tissue will be small, except for the tissue at the center of rotation where the beams always cross.

Gastric and intestinal tissue and the nervous system are roughly two and four times, respectively, more radiation resistant than bone marrow.

Large, localized radiation doses are sometimes used to selectively destroy sections of diseased tissue, such as tumors. The highly penetrating gamma radiation from cobalt-60 (^{60}Co) is routinely used today for this type of treatment. Radioactive cobalt (half-life 5.3 years) is produced in a nuclear reactor and installed within a large shield. The gamma radiation is emitted from this shield in a carefully focused beam. By rotating the source around the patient, a superficial dose is received except at the center of rotation where the beams always cross. By carefully positioning the patient, it is possible to selectively destroy a small tumor deep within the body without subjecting the patient to surgery (Figure 13.15).

Large, focused radiation doses are used to destroy cancerous tissue.

Biological Tracers

Radioactive **tracers** are "tagged atoms" that can be traced through a system by observation of the radiation they emit. One of the very early uses of radioactive tracers was made by a scientist who suspected the left-over food in his boarding house was showing up the next day in a different dish. One day he sprinkled a small amount of a radioactive isotope on some scraps and then brought a Geiger counter to dinner the next day. The hash was radioactive, and he changed boarding houses. One would not get away with this use of tracers today. Tracers are, however, widely used in medicine to evaluate the function of thyroid and kidneys and to locate tumors.

Some elements are concentrated by certain parts of the body. Iodine, for example, is rapidly concentrated in the thyroid gland. The thyroid uses iodine to produce thyroxin, a compound that is involved in the control of the metabolic rate. To test the level of thyroid activity, a patient drinks a solution of sodium iodide containing radioactive iodine (^{131}I). The thyroid area is later scanned by a scintillation detector, and the amount of radioactive iodine that has been taken up is measured. This amount is normally between 15% and 45% of the ingested dose. A scan of the area can be taken by coupling the scintillation detector to a computer that records the count. By placing a focusing shield around the detector, only a very small area of tissue is viewed at a time. By slowly moving the detector across the patient,

Radioactive *tracers* are used in diagnostic medicine.

a picture of the distribution of the radioactive iodine can be drawn. If the patient is rotated 90° and the procedure repeated, a three-dimensional view of the thyroid gland results. The scan clearly defines the position of the thyroid should surgery be necessary and identifies areas of high or low iodine uptake. (Also see Plate 3 in color insert.)

Other radioisotopes provide access to different systems within the body. Technetium-99m is a short-lived (6.5-hour half-life), gamma-emitting isotope that tends to concentrate in the brain. Since fast-growing tumors absorb technetium to a different degree than normal brain cells, this provides a means of determining the size and location of brain tumors.

Units of Radiation

Several units are used to define an amount of radiation. Of the four units listed below, one is based on the rate of decay of the radioisotope, while the other three tell us the amount of energy deposited in air or living tissue as the radiation passes through.

1. The **curie (Ci)** was originally defined as the radioactivity of 1 g of radium, and it amounts to 3.7×10^{10} disintegrations per second. This decay rate represents a large amount of radioactive material. More practical units are the *millicurie* (*mCi*), 3.7×10^7 disintegrations per second, and the *microcurie* (μCi), 3.7×10^4 disintegrations per second.
2. The **roentgen (R)** was the earliest unit used to describe the amount of energy deposited in matter by the passage of radiation. The roentgen is the amount of X-radiation or gamma radiation that will produce 1.61×10^{12} ion pairs in 1 g of air (a volume of about 773 mL). As a comparison, a normal chest X-ray exposes you to approximately 0.1 R.
3. The **rad (radiation absorbed dose)** references the radiation dose to tissue rather than air, as is the case with the roentgen. For practical purposes, the rad and roentgen are very similar in size. A whole body dose of 450 rad would be fatal to approximately 50% of the population. This dose is called the LD_{50}.
4. The **rem (roentgen equivalent man)** is a radiation dose unit that permits comparison of radiation exposure from alpha, beta, and gamma radiation. Because of the different penetrating and ionizing abilities of these types of radiation, it is difficult to compare directly the effect of 1 R of gamma radiation with, for example, a 1-R dose of alpha radiation. Exposure to a small number of alpha particles could produce the same amount of damage (and radiation dose in rem) as exposure to a much larger number of X-rays or gamma rays. The number of rem's is additive regardless of the type of radiation involved. We are exposed to about 10 millirems per month from cosmic and natural background radiation. Federal law permits workers in the nuclear industry ten times this exposure, or 300 millirems in a given three-month period.

Radiation-Induced Mutations

The effect of ionizing radiation on the genetic process was demonstrated in the late 1920s by exposure of the fruit fly drosophila to X-radiation. This particular strain of fly reproduces rapidly, and it is possible to follow many generations in a short period of time. Flies exposed to

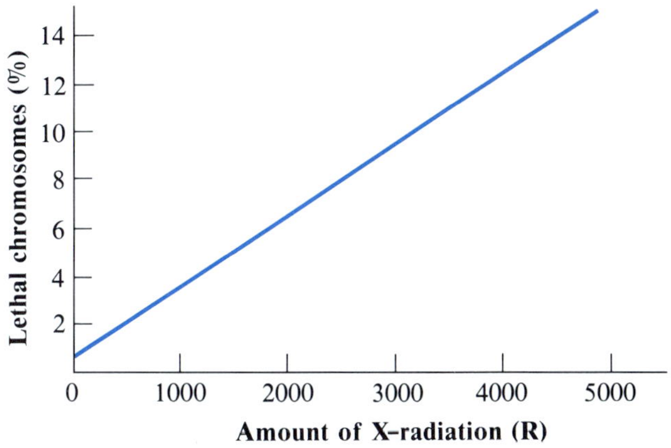
Lethal chromosomes (%)
14
12
10
8
6
4
2
0
1000
2000
3000
4000
5000
Amount of X-radiation (R)

FIGURE 13.16
Experiments with fruit flies have shown that the rate of mutation is directly related to the amount of exposure to radiation. However, since the genetic cells are such a small fraction of the material in an organism, very large radiation doses must be used to make this effect visible.

large doses of X-radiation experienced many more lethal mutations than did similar flies not exposed to radiation. The number of mutations was approximately proportional to the radiation exposure. Thus, doubling the radiation dose approximately doubled the number of mutations. This relationship is illustrated in Figure 13.16.

Two important points are evident from this figure. One is the size of the radiation dose—up to 5000 R. Since a dose of 450 R has a 50% chance of killing a person, these are large doses indeed. Because the amount of genetic material is small, the chance of interaction of radiation with a DNA molecule is small. The second point is even more important: even when manmade radiation is reduced to zero, mutations still occur.

There are two reasons that the fruit flies would show an observable mutation rate when manmade radiation is reduced to zero. One is that artificial radiation accounts for only part of the radiation that the flies were exposed to. Natural background radiation due to cosmic radiation and radioactive elements in the atmosphere and the earth's crust contribute a fixed background level of radiation to which everything is exposed. Natural background radiation, however, accounts for less than 1% of the spontaneous mutations that take place. Most spontaneous mutations result from damage to genetic materials by random collision with other molecules, from illness of the mother, and other factors.

Even low radiation doses have a finite possibility of producing mutation. However, less than 1% of spontaneous mutations are due to natural background radiation.

Background Radiation

About 40% of the natural **background radiation** we receive is cosmic in origin—radiation emitted by the sun and other stars that continually bombards the earth from space. Most of the remaining natural background radiation comes from radioactive elements in the earth's crust. The most prevalent of these is potassium-40 (^{40}K), a beta emitter with a 1.3-billion-year half-life. About 0.01% of the potassium found on the earth is ^{40}K; it is present to the extent of about 3 g per 10 kg of soil. Since potassium is an important element in body fluids, we all carry a small amount of radioactive potassium in us at all times.

Natural background radiation is caused by cosmic rays, and by naturally occurring radioactive materials in air and earth.

Potassium is not the only naturally occurring radioisotope, although it is the most predominant. Thorium-232 (^{232}Th), an alpha emitter with a half-life of 14-billion-years, is present to the extent of about 12 g per 1000 kg of soil. Uranium-238 (^{238}U), an alpha emitter with a half-life of 4.5 billion years, is present at about 4 g per 1000 kg of soil. Radium-226 (^{226}Ra) is common

TABLE 13.3
Radiation Exposure for One Year from Natural and Man-Made Sources

	Amount	
	Millirem/year	**Percentage**
Natural sources		
Cosmic radiation	50.0	25.8
The earth	47.0	24.2
Building materials	3.0	1.5
Inhaled in air	5.0	2.6
Elements found naturally in human tissues	21.0	10.8
Subtotal	126.0	64.9
Medical sources		
Diagnostic X-rays.	50.0	25.7
Radiotherapy	10.0	5.1
Internal diagnosis	1.0	0.5
Subtotal	61.0	31.3
Other manmade sources		
Nuclear power industry	0.85	0.4
Luminous watch dials, TV tubes, industrial wastes	2.0	1.0
Fallout from nuclear testing	4.0	2.0
Subtotal	6.85	3.4
Total	193.85	

enough that about 2 g will be found in the top foot of soil in each square mile of the earth's surface.

Elevation is an important factor in cosmic ray intensity. Since cosmic rays are partially absorbed by the atmosphere, a higher cosmic ray background exists at higher altitudes. Denver, for example, has about 1/3 greater cosmic ray background than a city at sea level.

About one-third of our normal radiation exposure is from medical sources. Two-thirds is from natural sources, and about 0.4% from the nuclear power industry.

We are all exposed to radiation from a number of sources (Table 13.3). About 65% of our annual radiation exposure comes from cosmic radiation and from radiation emitted by naturally occurring radioisotopes in the soil and air. About 17% of our annual radiation exposure comes from medical procedures, principally chest and dental X-rays, and another 17% comes from luminous watch dials, television sets, radioactive fallout from nuclear weapons testing, and the nuclear power industry. Since the atmospheric nuclear test ban treaty in the early 1960s, the amount of radioactive fallout has been decreasing.

During the period from World War II to the early 1960s, testing of nuclear weapons in the atmosphere contributed a measurable amount to the natural background radiation. A small (1000-ton TNT equivalent) nuclear weapon produces about 2 oz of fission products. Shortly after an explosion, this material has a gamma ray activity comparable to about 30,000 tons of radium. Although many fission products have short half-lives, cesium-137 (33-year half-life) and strontium-90 (25-year half-life) remain in the evironment for a long time. For comparison purposes, the nuclear weapons dropped on Japan near the close of World War II were equivalent to about 20,000 tons of TNT and yielded the radiation equivalent of about 600,000 tons of radium.

Living with Radiation

Radiation has always been a part of our environment, and the human body has learned to live with it. Cellular damage due to radiation is handled in much the same manner as a mild sunburn—the damaged cell is simply replaced. As long as the body's ability to replace the cells exceeds the rate at which they are damaged, no serious harm results. One might then assume that a "safe" level of radiation exposure is a level at which the body's repair mechanism can easily keep up with the necessary cell replacement.

Except for possible genetic damage, the "safe" radiation exposure is one at which the body's repair mechanism can keep up with the necessary cell replacement.

Genetic damage, however, is a different situation. Although the possibility of genetic damage by radiation is many thousands of times less than the possibility of cell damage through water ionization, it has more serious implications. Suppose that a person had one million reproductive cells and that one of these was damaged. The probability of this particular cell being the one involved in reproduction might be only one in a million, but if it is the one used, it will not matter that there were 999,999 perfectly good reproductive cells that might have been used. So we could say that the only "safe" radiation exposure is no radiation at all. However, it is impossible to escape from natural background radiation, and our bodies have adapted to its presence. Medical applications of radiation provide additional exposure. We must weigh the medical value of detection of tuberculosis, a decayed tooth, or a stomach ulcer against the additional radiation exposure involved. Except for pregnant women, there is little question that the medical value of an X-ray far exceeds any small risk that might result. On the other hand, there is no point in exposing ourselves needlessly to radiation. For this reason, radiologists and X-ray technicians wear lead aprons when working, and the dentist leaves the room when the dental X-ray machine is turned on. These people work with radiation everyday, while the patient may have only one X-ray every few years.

A similar trade-off may be facing us in the future with respect to the nuclear power industry. Doubling the size of this industry would increase the background radiation by about 0.85 millirem—from 193.85 millirem to 194.7 millirem per year, an increase of about 0.4%. We must decide if the increased radiation exposure would be balanced by the social, economic, and health values of the availability of additional electrical power. These decisions must be made on the basis of sound technical information as well as social and political factors.

NEW TERMS

mutation Damage to the genetic mechanism of a cell, causing it to reproduce in a different form.

LD_{50} A radiation dose that will be lethal to 50% of the population being tested. The term LD_{50} is also used for toxic chemicals and drugs.

tracer A radioactive isotope used to follow the passage of the element or molecule in question through a biological or physical system.

curie (Ci) An amount of radioactivity equal to 3.7×10^{10} disintegrations per second.

roentgen (R) A unit used to describe the amount of energy deposited in 1 g of air by X-rays or gamma radiation.

rad (radiation absorbed dose) A unit that describes the amount of energy deposited in tissue by X-rays or gamma radiation. A whole body dose of 450 rad will be fatal to 50% of the population.

rem (roentgen equivalent man) A unit that equates radiation damage caused by alpha, beta, and gamma radiation. We are exposed to about 10 millirem per month from natural sources of radiation.

background radiation Radiation received from naturally radioactive elements in the atmosphere and the earth's surface and from cosmic radiation. About 65% of our annual radiation dose comes from background radiation.

TESTING YOURSELF

Interaction of Radiation with Matter

1. What are the two principal ways radiation interacts with living tissue?
2. Which of your two answers to question 1 is the most probable type of interaction?
3. What is the principal difference between the two radiation units, curie and rem?
4. Radiation causes the greatest damage to fast-growing cell systems. Which of the following biological systems is most sensitive to radiation: bone marrow, gastrointestinal tract, or central nervous system?
5. About 65% of the background radiation we receive comes from natural sources. The remainder is manmade.
   a. What are the two principal sources of natural background radiation?
   b. Of the manmade radiation sources, which is the principal contributor to our background radiation dose?

Answers **1.** water ionization and DNA damage **2.** water ionization **3.** The curie tells disintegration rate of the radioisotope; the rem tells biological damage done by the passage of radiation. **4.** bone marrow **5. a.** cosmic rays and natural radioactive material in the soil and air **b.** medical X-rays

13.5 Nuclear Power

The Importance of Energy

Until fairly recently in human history, the earth only had one source of energy: the sun. Solar energy striking the planet each year amounts to about 5.1×10^{21} British thermal units or Btu (a Btu is the amount of heat required to heat 1 lb of water 1 °F). This is about 23,000 times the entire world's use of energy in 1980. About 30% of this energy, however, is immediately lost by reflection. About 47% is directly converted to heat, which warms our atmosphere, land, and oceans. About 23% is used to evaporate water from oceans and lakes, thus driving the hydrologic cycle, the weather, and the distribution of water over the globe. Less than 1% is used by plants to support photosynthesis (Figure 13.17).

Except for nuclear energy, the sun is directly or indirectly the source of all energy on the earth.

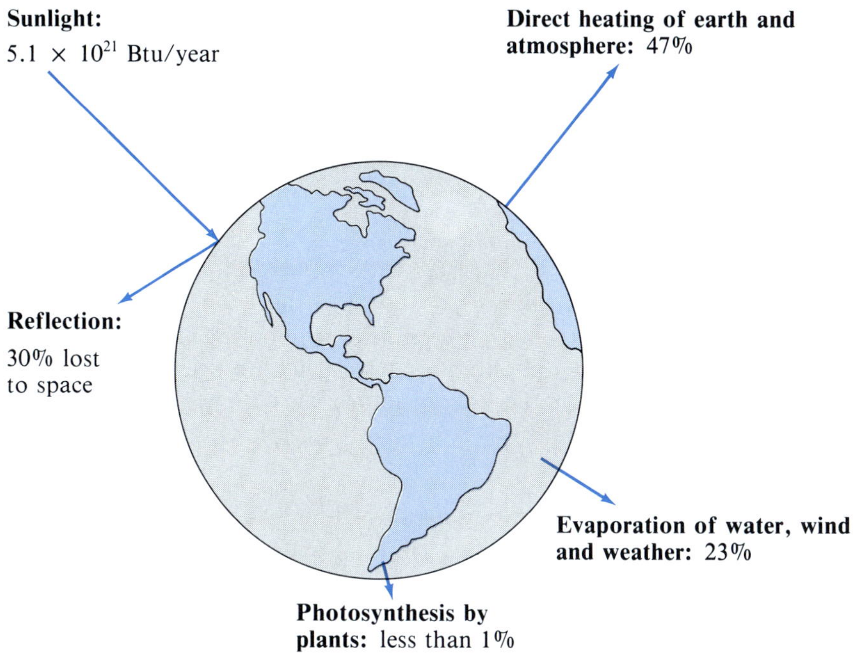
Sunlight:
5.1 × 10[21] Btu/year
Direct heating of earth and atmosphere: 47%
Reflection:
30% lost to space
Evaporation of water, wind and weather: 23%
Photosynthesis by plants: less than 1%

FIGURE 13.17
The sun is our planet's principal source of energy, although less than 1% is used to support photosynthesis.

In regions outside of the equatorial zone, keeping warm has always been a major concern. By the Middle Ages, the growth of cities in Europe had been limited by the distance one could carry firewood. The problems were so severe that some areas were completely deforested and no trees could grow, so the people were forced to move on.

By about A.D. 1500, civilization had learned to supplement the direct products of photosynthesis with water power and coal. Mankind had found the key to nature's savings account—fossil fuels.

The chart illustrated in Figure 13.18 shows where the United States gets its energy from today. The "savings account" of fossil fuels provides an important part of our income.

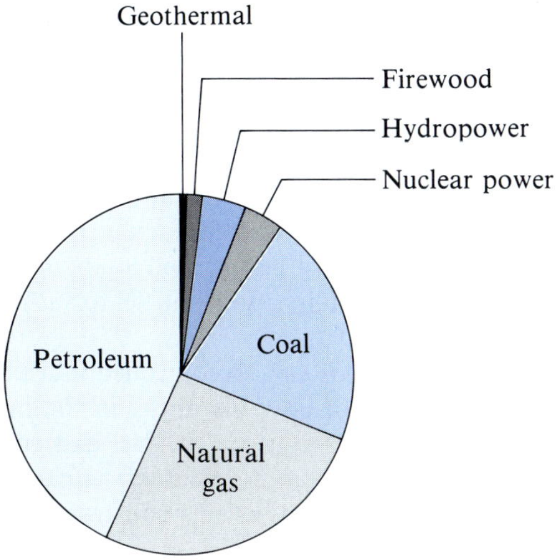
Geothermal
Firewood
Hydropower
Nuclear power
Coal
Petroleum
Natural gas

Energy Source	Percentage
Petroleum	43.1%
Natural gas	25.9
Coal	21.6
Hydropower	4.1
Nuclear power	4.0
Firewood	1.1
Geothermal	0.02

FIGURE 13.18
Sources of energy for the United States in 1980. Fossil fuels are the principal source of energy.

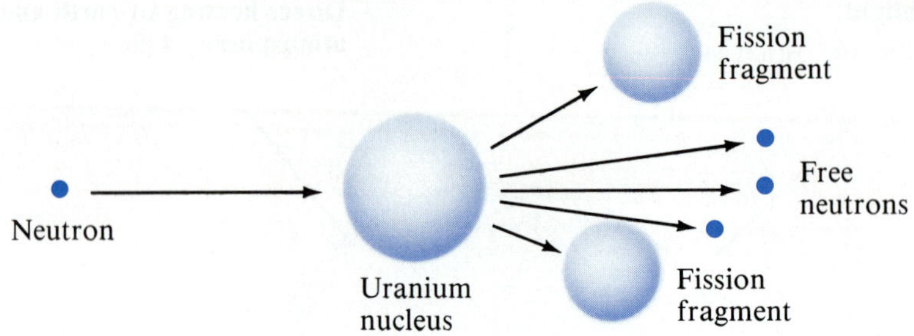
Fission
fragment
Free
neutrons
Neutron
Uranium
nucleus
Fission
fragment

FIGURE 13.19
Fission of a nucleus is illustrated in this figure. Uranium-235, the only naturally occurring nuclide that will spontaneously fission, splits when struck by a neutron to form two smaller nuclei (fission products) and several high speed neutrons. A considerable amount of energy is also released during the fission process. The fission products are highly radioactive.

Coal has been used for several centuries to supplement the direct products of photosynthesis, and oil and natural gas have been important during this century. Now another energy source has become available: nuclear fission.

Nuclear Fission

Nuclear power produces about two million times the energy per gram of fuel burned as compared to carbon.

Only about 0.7% of naturally occurring uranium will fission. To conduct a fission reaction in a power reactor requires enrichment of the fuel to about 4% fissionable uranium. Nuclear weapons use about 90% enriched fuel.

The amounts of energy produced by most chemical reactions are relatively small. For example, the reaction of a carbon atom and two oxygen atoms to form carbon dioxide produces less than 2 eV, an amount of energy similar to that produced by a flashlight battery. Thus, many carbon atoms must be burned to produce useful heat. In contrast, some unstable radioactive heavy nuclei will split when struck by a slow neutron, producing 100 million times more energy than burning carbon. This splitting of the nuclei is called **nuclear fission**. These nuclei, principally uranium-235, are relatively rare and isolated enough from each other that the fission reaction does not occur spontaneously in nature on a useful scale. Only about 0.7% of all naturally occurring uranium is fissionable ^{235}U. However, the uranium resources available to the United States are substantial and have created great interest in this form of power.

Nuclear fission involves the spontaneous splitting of a uranium nucleus when struck by a slow neutron. The fission process results in release of several high speed neutrons, a considerable amount of energy, and formation of two smaller atoms called *fission* fragments (Figure 13.19). Each fission reaction produces a number of high speed neutrons. Each of these neutrons can trigger another fission event. If the fissionable nuclei are close enough together, a neutron from one fission event will strike a nearby fissionable atom, causing another fission. If enough fissionable atoms are close enough together, the process will continue in a **chain reaction** (Figure 13.20). The amount of fissionable material needed to sustain a chain reaction is called the *critical mass*. Naturally occurring uranium contains only about 0.7% fissionable ^{235}U; the remainder has uranium mass 238. Naturally occurring uranium will not sustain a chain reaction. Reactor fuel must be enriched to about 4% ^{235}U to provide a high enough concentration of fissionable nuclei to sustain a chain reaction.

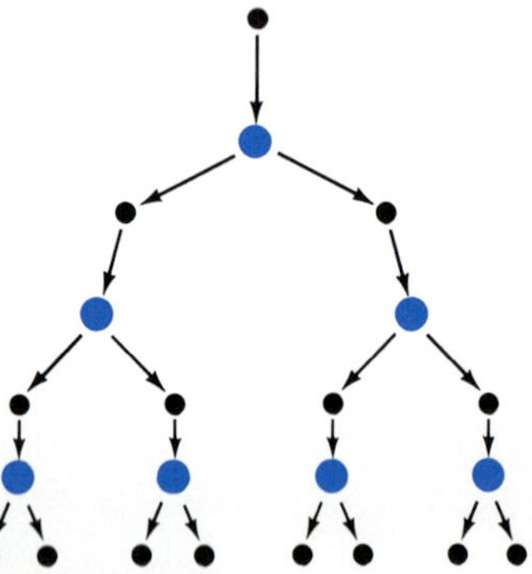

FIGURE 13.20
A chain reaction will occur if neutrons generated during the fission of a nucleus strike another fissionable nucleus. The chance that a chain reaction will be self-sustaining depends on the presence of additional fissionable nuclei.

The rate at which a chain reaction proceeds can be controlled in a nuclear reactor by adjusting the concentration of unstable nuclei in the fuel and by placing neutron-absorbing control rods between the fuel elements. The chance of neutron capture by a ^{235}U nucleus is considerably enhanced if the neutron is moving slowly. Nuclear reactors are usually constructed by

NUCLEAR WASTE

The amount of high level nuclear waste stored in the United States has increased by about one-third from 1979 to 1987 (see table). The amount of commerically generated low level waste from hospitals and industry increased by a factor of about 2.4 during this period, while the amount of defense-related low level waste increased by a factor of 1.6. By the year 2000, radioactive waste from fission reactors in nuclear power plants in this country will amount to about 500,000 cubic feet, enough to cover a city block to a depth of about 10 feet.

However, the basic problem with **nuclear waste** is not its volume, which is relatively small (see table). The real danger lies in the long half-lives of isotopes produced during the fission process, which are highly radioactive and must be very carefully transported and securely stored for several thousand years (see figure). Some progress has been made toward solution of this problem. Radioactive wastes are currently being combined with ceramic materials to form an insoluble ceramic solid that will have very little possibility of dissolving in groundwater should its container break after burial. Certain stable geological rock formations are also being considered for long-term torage of radioactive wastes. Most commercial high level waste is used in fuel elements from power reactors, which are being stored in pools in the reactor buildings until the site of a national underground storage facility is agreed upon by Congress.

Volume of Waste in Thousands of Cubic Meters

	Commercial		Defense	
	1979	1987	1979	1987
High level waste	4.3	6.8	283	382
Low level waste	515	1261	1470	2380

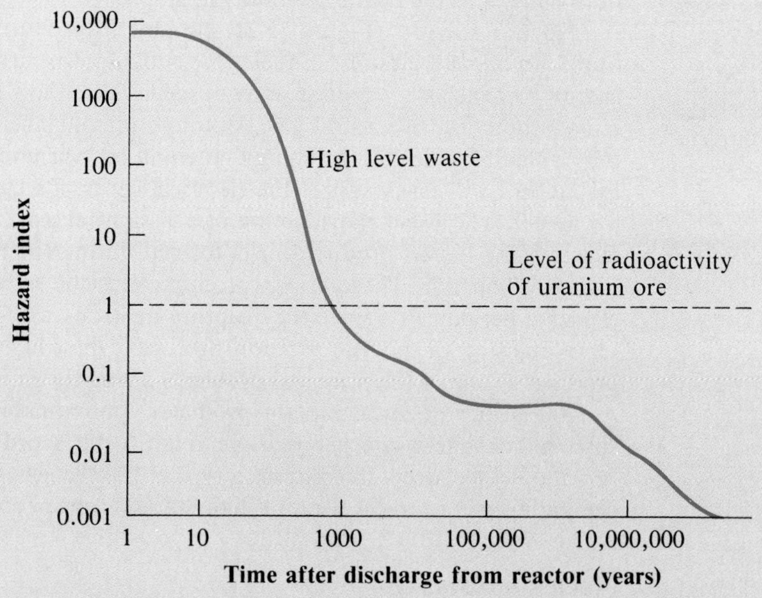
10,000
1000
100
10
1
0.1
0.01
0.001
Hazard index
High level waste
Level of radioactivity of uranium ore
1
10
1000
100,000
10,000,000
Time after discharge from reactor (years)

The most radioactive nuclear waste decays relatively rapidly. After storage for about 1000 years, the level of radioactivity is about the same as that of the uranium ore when it was originally mined.

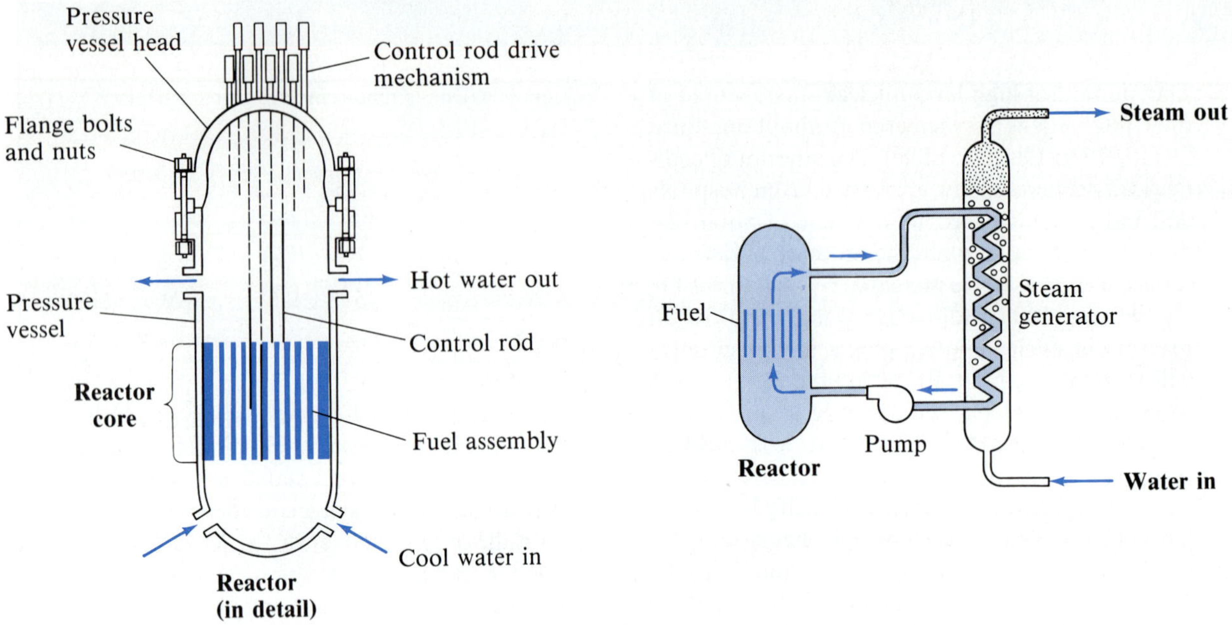
Pressure vessel head
Control rod drive mechanism
Flange bolts and nuts
Hot water out
Pressure vessel
Control rod
Reactor core
Fuel assembly
Cool water in
Reactor (in detail)
Steam out
Fuel
Steam generator
Pump
Reactor
Water in

FIGURE 13.21
A reactor core contains a number of fuel elements that hold the fissionable fuel. These are placed close enough together that neutrons emitted by the fission of a nucleus have a good chance of striking another fissionable atom. The neutrons are slowed by water, which also acts as a coolant as it is pumped through the reactor core. Steam produced as the coolant circulates through a secondary boiler powers a turbine, which generates electricity.

placing the fuel elements containing the fissionable atoms in a grid or lattice with water in between. The water acts as a **moderator**, slowing the neutrons and increasing the change that they will be captured by another ^{235}U nucleus. It also provides a safety factor: if the chain reaction proceeds too rapidly, generating too much energy and heat, the water will boil out. With the reactor's moderator thus removed, the chance of neutron capture decreases drastically, and the chain reaction will stop.

Nuclear reactors (Figure 13.21) are able to produce a large amount of heat with a small amount of fuel. One cubic foot of uranium has the same energy content as 1.7 million tons of coal, 7.2 million barrels of oil, or 32 billion cubic feet of natural gas. Although present nuclear reactors extract only a small fraction of the energy potential of their nuclear fuel, one truckload of fuel will easily supply the electrical needs of a city of 200,000 people for a year. The basic shortcoming of a fission reactor, however, lies in the fact that the fission product nuclei formed during the nuclear reaction are highly radioactive. These fission products must be separated from the fuel elements periodically, since they capture neutrons without fissioning themselves and slow the nuclear reaction. Storage of these highly radioactive waste products is one of the principal problems facing the nuclear industry today. A 1000-megawatt nuclear plant produces approximately 100 cubic feet of solid radioactive waste per year—enough to fill a ordinary pick-up truck bed about 3 feet deep. In contrast, a typical 1000-megawatt coal-fired electric generating plant produces more than 900,000 tons of solid waste each year.

Nuclear Fusion

While nuclear *fission* involves release of energy by splitting of heavy, unstable nuclei, energy may also be released by *fusing* light-weight nuclei. This process is called **nuclear fusion**. When the hydrogen isotope deuterium (one proton and one neutron) is subjected to a high enough temperature, it will fuse with another deuterium to form a helium nucleus. In the process, it will release much more energy than the fission of a uranium nucleus.

$$^{2}_{1}H + ^{2}_{1}H \xrightarrow{\text{Fusion}} ^{4}_{2}He + \text{Energy}$$

The temperature required to sustain this reaction is as much as 100,000,000 °C or more. Obviously, this makes it difficult to sustain in a controlled manner. The fusion reaction of deuterium to produce helium is the source of the sun's enormous energy. It is also the energy source in thermonuclear weapons. If fusion can be harnessed, its deuterium fuel could be inexpensively obtained from seawater. About 1/6000 of all naturally occurring hydrogen is deuterium, and although this fraction seems small, water in the oceans contains about 35×10^{12} tons of deuterium. If only 1% of the deuterium could be withdrawn for fusion fuel, the energy released would amount to about 500,000 times the energy of the world's initial supply of fossil fuels. It is, for practical purposes, an inexhaustible source of fuel.

The fusion reaction may eventually burn deuterium from sea water. If 1% of the available deuterium were used in fusion reactors, the energy released would be 500,000 times that of our initial supply of fossil fuels.

Another advantage of fusion power is that, unlike fission reactions, the by-products are stable isotopes. There is no problem of storage of radioactive wastes. Promising research is underway concerning electromagnetic containment of fusion reactions. It is hoped that safe containment of this reaction will be a reality before our fossil reserves are exhausted.

Nuclear Weapons

Nuclear power was born in wartime and was first used as an explosive. At the close of World War II, two nuclear weapons were dropped on Japan. One of these involved fission of uranium-235. The uranium in the weapon was enriched to about 90% ^{235}U by a gaseous diffusion process to provide a high enough concentration of fissionable nuclei to cause the chain reaction to grow exponentially and explode. The other weapon used fission of plutonium-239, an artificial element created by bombardment of uranium-238 with neutrons in a nuclear reactor. Although different in their overall design, the two weapons were conceptually similar. To sustain a chain reaction, one unstable nucleus must be close enough to catch a neutron from each fission event. If about 4% of the uranium atoms are ^{235}U, the critical mass has been reached and a chain reaction can be sustained. If the concentration of unstable atoms is increased to 90%, a run-away chain reaction and a tremendous explosion will result. By splitting the fuel into two or more parts that are well separated, a critical mass is not achieved and no explosion occurs. The weapon is exploded by driving the fuel components together with an explosive charge to form a critical mass and a run-away chain reaction (Figures 13.22 and 13.23).

Fuel used in nuclear reactors is not enriched sufficiently to serve as fissionable material for nuclear weapons.

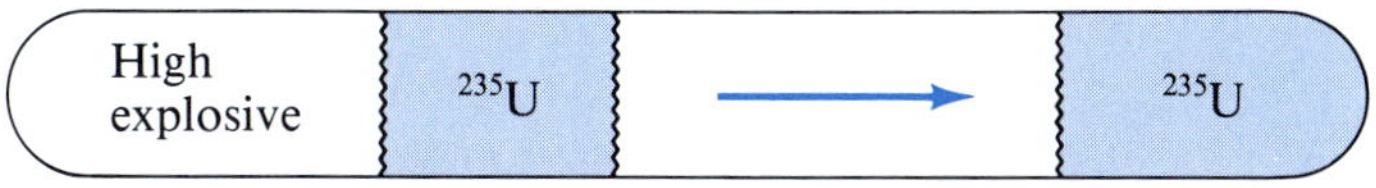

High explosive
235U
235U

FIGURE 13.22
Nuclear weapons are made to explode by pushing two smaller than critical mass chunks of fissionable material together very rapidly with explosives. The arrow shows the motion of one part of the fissionable material.

FIGURE 13.23
This is a thermonuclear bomb that was dropped accidentally during an aircraft accident over Spain in 1966. This weapon uses a small fission bomb to trigger a much larger fusion explosion. The parachute is to slow its descent to allow the aircraft time to leave the area. The nose of the weapon, crushed when it landed in an irrigation ditch, is visible in the photograph. (Photo courtesy Department of Energy.)

The atmospheric test ban treaty has reduced the amount of worldwide radioactive fallout.

Testing of nuclear weapons is a difficult problem. Fission products produced during a chain reaction in a nuclear reactor are normally trapped in the fuel elements for later separation and storage. But in nuclear weapons, these radioactive products are blown into the atmosphere by the explosion and are spread around the earth by prevailing winds. The amount of radioactivity produced by a nuclear explosion is so large that it is difficult to conceive. Although much of the radioactivity is short lived, a "small" nuclear explosion such as occurred in Japan produced radioactive material equivalent to about 30,000 tons of radium. Radioactive fallout from nuclear weapons testing in the Pacific Ocean following World War II caused a significant increase in the incidence of leukemia worldwide (Figure 13.24). The atmospheric

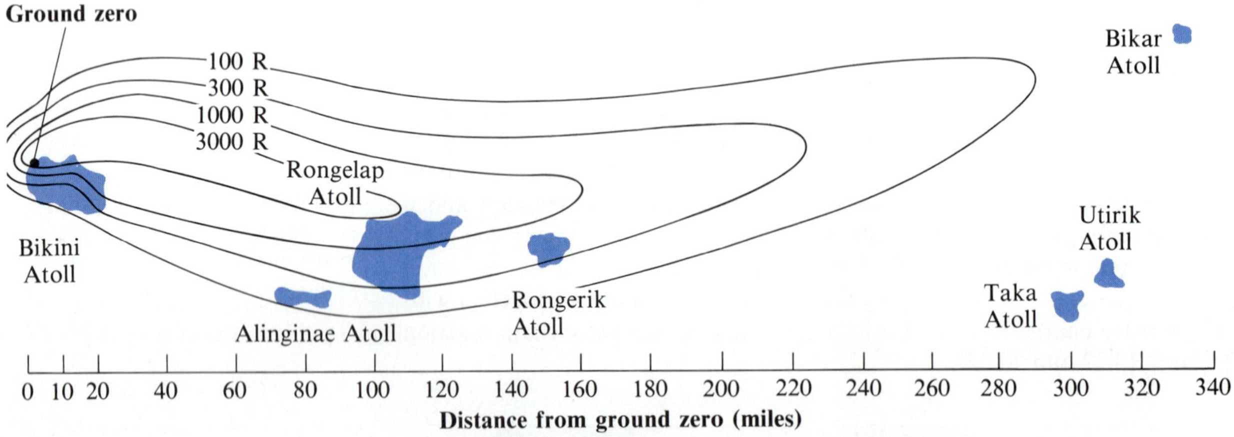
Ground zero
100 R
300 R
1000 R
3000 R
Rongelap Atoll
Bikar Atoll
Bikini Atoll
Alinginae Atoll
Rongerik Atoll
Utirik Atoll
Taka Atoll
0 10 20 40 60 80 100 120 140 160 180 200 220 240 260 280 300 320 340
Distance from ground zero (miles)

FIGURE 13.24
Nuclear explosions produce substantial amounts of radioactive material that is spread by prevailing winds. This map shows the distribution of fallout from a thermonuclear explosion conducted in the South Pacific in 1950. The contour lines show radiation dose in roentgens per hour 96 hours after the explosion. One end of the populated atoll of Rongelap, 100 miles away, was in the path of the major fallout pattern.

test ban treaty developed during the Kennedy administration is observed by most countries today; nuclear testing is now carried out entirely underground.

NEW TERMS

nuclear fission Splitting of a heavy nucleus to produce two lighter "fission fragment" nuclei, a number of free neutrons, and a large amount of energy. Fission is triggered when an unstable nucleus absorbs a neutron.

chain reaction A series of very rapid reactions that will occur among fissionable atoms if they are close enough together.

moderator A material that slows neutrons. Slow neutrons are more readily captured by fissionable nuclei than are fast neutrons.

nuclear fusion Fusing of two light nuclei together to form a heavier reaction. When very light nuclei are fused together, the product nucleus has less energy than the two "reactant" nuclei; considerable energy is released in the fusion reaction.

nuclear waste Fission fragment nuclei are highly radioactive and extremely dangerous. They are referred to as "nuclear waste," or in the case of a weapons explosion, as "fallout."

TESTING YOURSELF

Nuclear Power

1. What percentage of the sunlight striking the earth is used to heat soil, atmosphere, and water?
2. Fission of a uranium nucleus produces about how many times more energy than the reaction of a carbon atom and two oxygen atoms to form carbon dioxide?
3. About what fraction of naturally occurring uranium is fissionable?
4. A moderator provides what function in a nuclear reactor?
5. About what fraction of a reactor's fuel must be fissionable uranium to support a chain reaction?
6. About what fraction of a nuclear fuel must be fissionable to support a run-away chain reaction useful in an atomic weapon?
7. What is the best long-term source of deuterium fuel for fission reactors?
8. About how long must nuclear waste from a reactor be stored for its radioactivity to fall to that of the ore when it was initially mined?

Answers **1.** about 47% **2.** 100,000,000 **3.** about 0.7% **4.** The moderator slows up neutrons, increasing their chance of capture by fissionable nuclei. **5.** about 4% **6.** about 90% **7.** seawater **8.** about 1000 years

SUMMARY

Nuclear chemistry is likely to play an important part in two aspects of our lives—in diagnostic and therapeutic medicine and in energy production. The use of diagnostic tracers and radiation treatment for cancer is well established and has saved far more lives than were taken by the use of nuclear weapons at Hiroshima and Nagasaki at the end of World War II. The use of nuclear chain reactions to generate energy is more controversial. Although the fuel for fission reactors is in reasonable supply, containment and long-term storage of the radioactive waste products produced by these reactions are the cause of much national concern. The possibility of safe fusion power, using readily available deuterium as a fuel and producing stable and safe by-products, lies somewhere in the future.

To live with radiation and to vote in an informed manner on nuclear-related issues requires an understanding of basic principles, which are not controversial. Natural radioactivity, or the spontaneous decay of unstable nuclei, has been present on the earth since its beginning. The age of our part of the universe has been estimated by nuclear dating techniques which presume an initially pure source of uranium and measure the amount of stable lead subsequently created by nuclear decay. *Radioactive decay* occurs by three means—by emission of a *negative beta particle*, by emission of a *positive beta particle* (a positron), and by emission of an *alpha particle*. Many radioactive decays are accompanied by emission of gamma radiation, which is a form of high energy electromagnetic radiation. Gamma radiation is produced by changes in the internal energy of the nucleus in much the same manner as light is produced by motion of electrons around an atom.

Alpha particles, because of their large mass and +2 charge, interact very strongly with matter and penetrate only a very short distance (measured in thousands of atoms) into materials. Beta particles, because of their smaller size and lesser charge, penetrate as much as a centimeter into materials such as wood and tissue. Gamma radiation, carrying no charge and no mass, penetrates well into matter, and is thus of greatest biological danger.

Radiation interacts with living tissue in two ways. The most common type of interaction involves ionization of a water molecule. If enough water molecules are ionized in a cell, the cell will die. A more serious but much less likely kind of interaction involves interaction of the radiation with genetic material in the cell. This interaction creates broken or damaged genes and *mutation* of the cell. Most mutations are lethal, that is, the cell simply dies. Nonlethal mutations can produce uncontrolled cell growth called cancer. Because radiation has the possibility of causing cancer, unnecessary exposure to radioactive materials or X-rays should be avoided. Realize, however, that two-thirds of the radiation we receive comes from natural sources—naturally radioactive materials in the soil and air and in cosmic rays.

For most of us, the most important chemistry is that involved in supporting life—the reactions through which green plants convert sunlight, water, and carbon dioxide to sugars and foods and the reactions that support our lives. These reactions share two common threads: (1) the energy they use is ultimately traceable to the arrival of sunlight from the sun, and (2) the molecules involved in these reactions almost all have a carbon atom skeleton. The next part of this book will look at the common molecules and reactions involved in the support of life processes.

TERMS

radioactive decay
transmutation
parent nuclei (or isotope)
daughter nuclei (or isotope)
negative beta decay
positive beta decay (positron emission)
alpha decay
half-life
radiocarbon dating
decay series
Geiger–Müller counter
scintillation detector
mutation
LD_{50}
tracer
curie (Ci)
roentgen (R)
rad (radiation absorbed dose)
rem (roentgen equivalent man)
background radiation
nuclear fission
chain reaction
moderator
nuclear fusion
nuclear waste

QUESTIONS

CHAPTER REVIEW

1. Radon gas, produced by the decay of natural radioisotopes, is a health hazard when it is trapped in basements or living areas. Radon-220 ($^{220}_{86}Rn$) decays by alpha emission. What is the daughter product of this alpha decay?

$$^{220}_{86}\text{Rn} \longrightarrow \alpha + ?$$

2. Phosphorus-32 is commonly used as a tracer isotope in agricultural experiments. This isotope decays by emission of a negative beta particle. What is the daughter product of this decay?

$$^{32}_{15}\text{P} \longrightarrow \beta(-) + ?$$

3. The following table presents time and count rate data describing the decay of a radioisotope. By examination of the data, estimate the half-life of this isotope.

Time (days)	Decay Rate (counts/min)
0	7000
1	5550
2	4409
4	2775
5	2204
7	1388

4. Nuclear radiation can be classified according to charge, mass, and penetrating ability. Of the three predominant types of radiation—alpha, beta, and gamma—which is greatest in two of these categories? What are the two categories?

5. Why is a very high collecting voltage (on the order of 900 V) used to collect electrons in a Geiger–Müller tube when only a few volts will collect the electrons formed by the passage of ionizing radiation?

6. If a G-M detector was equipped with a *very* thin window, to which type of radiation would the detector be most sensitive and why?

7. Which type of radiation is able to penetrate most deeply into animal tissue?

8. About two-thirds of the radiation exposure we receive each year is from natural sources, while the remainder is from medical uses of radiation. What are the two major sources of natural background radiation?

9. Deuterium, a potential fuel for fusion reactors, is found in what?

10. A small chunk of fissionable ^{235}U will not explode, but a large chunk will. In fact, chemical plants that process used reactor fuel limit the size and spacing of reaction vessels to prevent large amounts of ^{235}U from coming together. What principle of fission is involved here?

Answers **1.** $^{228}_{88}Ra$ **2.** $^{40}_{20}Ca$ **3.** 3 days **4.** Alpha particles have the highest charge and greatest mass. **5.** The G-M detector uses high voltage to cause gas multiplication. **6.** A G-M detector is most sensitive to alpha radiation if the radiation can penetrate the detector window. **7.** gamma radiation **8.** cosmic rays and naturally radioactive material in the soil **9.** water **10.** A critical mass of fissionable nuclei is required to support a chain reaction.

DIAGNOSTIC CHART

Blacken in all of the circles under the number of each question you missed in the Chapter Review questions. The diagnostic chart will help you identify concept areas that need more study.

	Questions									
Concepts	1	2	3	4	5	6	7	8	9	10
Nuclear decay	○	○	○							
Interaction of radiation and matter				○	○					
Detection of nuclear radiation					○	○				
Radiation biology							○	○		
Nuclear power									○	○

EXERCISES

Nuclear Decay

1. Predict the daughter products of the following nuclear decay reactions.
   a. $^{32}_{15}P \longrightarrow$ Beta(−) + ?
   b. $^{45}_{20}Ca \longrightarrow$ Beta(−) + ?
   c. $^{90}_{38}Sr \longrightarrow$ Beta(−) + ?
   d. $^{65}_{30}Zn \longrightarrow$ Beta(+) + ?
   e. $^{235}_{92}U \longrightarrow$ Alpha + ?
   f. $^{234}_{94}Pu \longrightarrow$ Alpha + ?
   g. $^{212}_{83}Bi \longrightarrow$ Alpha + ?
   h. $^{22}_{11}Na \longrightarrow$ Beta(+) + ?
2. The table below presents time and count rate data describing the decay of a radioisotope. Plot these data and estimate the half-life of the isotope.

Time (hours)	Decay Rate (counts/min)
0	10,000
1	8,000
4	4,000
5	3,200
7	2,000
8	1,600

3. Phosphorus-32 (^{32}P) is a beta-emitting isotope sometimes used as an agricultural tracer. This isotope has a half-life of 14 days. Thus, 14 days after its delivery, only 1/2 as much is on hand, and 28 days after delivery, only 1/4 as much is left. Suppose that when the sample was shipped, it had an activity of 100,000 disintegrations per minute. By plotting the decay information on a graph, estimate the activity of the sample 3 weeks (or 21 days) after shipping.

Interaction of Radiation and Matter

4. Describe the properties of alpha, beta, and gamma radiation in terms of these three characteristics:
   a. electrical charge
   b. mass
   c. penetrating ability
5. Why does gamma radiation penetrate matter much better than alpha radiation?
6. Because of their relatively large charge and mass, alpha particles interact with practically every atom they pass. Suppose that an alpha particle has an energy of 1 million eV. Suppose further that 20 eV of energy are required to remove an electron from each atom the alpha particle passes.
   a. How many atoms can the alpha ionize before its energy is gone?
   b. The diameter of an atom is about 1×10^{-8} cm. If the alpha particle were to travel in a straight line, how far would it penetrate into a group of closely packed atoms before its energy were gone?

Detection of Nuclear Radiation

7. Why is a scintillation detector more sensitive to gamma radiation than a Geiger–Müller detector?
8. Geiger counters stop counting when exposed to high levels of radioactivity. This "dead time" is due to the formation of large numbers of positive ions in the detector following the passage of a nuclear particle. What must occur for the detector to recover and be ready to detect another particle?
9. What limits the sensitivity of photographic film badges to alpha radiation?

Radiation Biology

10. Radiation damage in living cells is caused principally by the interaction of radiation with water and with the genetic material in the cell. Which type of interaction is most likely to happen?
11. Which type of radiation causes the most damage to a cell: alpha, beta, or gamma?
12. Which type of radiation has the least interaction with biological material: alpha, beta, or gamma?
13. Which type of radiation is most likely to cause damage to internal organs within a person or animal: alpha, beta, or gamma?
14. The radiation from a chest X-ray is about 0.1R and is focused on the chest area. Suppose that a person were exposed to a whole-body radiation dose of this level. How many such whole-body doses would be required to produce the LD_{50} dose to the bone marrow?
15. The curie and the rem are both radiation units. How do they differ?

Nuclear Power

16. What is the role of a "moderator," and why is it required to sustain a chain reaction in a nuclear reactor?
17. Why must fission product atoms be separated from reactor fuel before all of the uranium has had time to fission?
18. How does nuclear *fusion* differ from nuclear *fission*?
19. Power reactors and nuclear weapons differ in the concentration of fissionable material in their fuel. Power reactor fuel requires about how much (by percent) fissionable material, while a run-away chain reaction in a weapon requires about how much fissionable material?
20. Does the earth have more available fuel for fission or for fusion?

PART

II

Organic and Biological Chemistry

(*Opposite page*)
This photo of a woman climbing up through Silver Grotto in the Grand Canyon illustrates a stark contrast between two areas of chemistry: living and nonliving; organic and inorganic. The chemical structure of the sandstone is based on a framework of silicon atoms; it is inanimate and without life. In contrast, the chemical structure of the woman is based principally on molecules with a framework of carbon atoms. Carbon chemistry, the topic of the second part of this book, is the chemistry of life.

In the first part of this book, you examined some of the rules governing chemistry and you surveyed the elements that make up our earth. The emphasis for the remainder of the book will be on the molecules of life. The composition of living organisms and the earth's crust are strikingly different. For comparison, Table 1 lists the most abundant elements found in the earth's crust and in the human body. Living systems selectively take up matter from their external environment rather than simply accumulating matter representative of the surroundings. For example, carbon is a relatively scarce element in the crust but an important and abundant one in organisms.

A century and a half ago, scientists believed the living (biotic) world and the nonliving (abiotic) world were mutually exclusive in a chemical sense. They thought that the substances of the nonliving world could not be made into the substances of living organisms and that the molecules of living organisms possessed a unique and vital force that was lacking in the compounds of the abiotic world. Scientists of that era made a clear distinction between the chemistry of the inorganic nonliving substances and the chemistry of organic materials from organisms. That distinction, reflected in the *vital force theory*, served as the first major division in chemistry, a division into inorganic and organic chemistry. This original organic chemistry was the chemistry of compounds from living systems.

TABLE 1
Comparison of the Abundance of the Elements in the Earth's Crust and the Human Body

Earth's Crust		Human Body	
Element	**Mole %**	**Element**	**Mole %**
O	47	H	63
Si	28	O	25.5
Al	7.9	C	9.5
Fe	4.5	N	1.4
Ca	3.5	Ca	0.31
Na	2.5	P	0.22
K	2.5	S	0.08
Mg	2.2	Na	0.07

Adapted from Lehninger's *Principles of Biochemistry*, 1982, and *Nutrition*, by Williams and Caliendo, 1984.

Several experiments that look place during the 1820s to 1840s established that the vital force theory was incorrect. In 1828, the German chemist Friedrich Wöhler attempted a synthesis of ammonium cyanate (an inorganic compound) from ammonia and cyanic acid, which were other inorganic compounds. The product of this particular synthesis was urea, a component of urine, an obvious organic compound. Wöhler's results were inconsistent with the vital force theory, but several additional experiments of this type were performed before the vital force theory was fully abandoned. Chemists began to look at all matter as though a single set of principles applied to it. The door to modern organic chemistry was opened. Today, *organic chemistry* is considered to be the chemistry of carbon and all the compounds that contain it.

Since antiquity, humans have used natural products that surround them. The early experiments of Wöhler and others showed that the synthesis of organic compounds was possible, and scientists enthusiastically began making them. Today many thousands of *new* compounds are made annually. A major goal of modern organic chemistry is the development of techniques to provide laboratory synthesis of useful molecules, including synthetics and mimics of natural compounds.

Organic chemistry directly impacts each of us in countless ways. From the clothes we wear, medicines we take, preservatives we use, herbicides and pesticides that protect our food supplies, and plastics that are used for containers and building, we are indebted to organic chemistry for food, fiber, and shelter. We often find organic chemists making novel compounds that do "better" than nature. Wonder fibers, such as those found in Goretex (Figure 1a) and Hollofil, provide winter clothing that protect us against cold and water better than goose down. Goretex is also used to make NASA space suits (Figure 1b). There are now substitutes for aspirin (Figure 2). Chemists are developing ways to make hydrocarbon fuels in the laboratory to reduce reliance on ever scarcer foreign sources. The use of chemicals from the reproductive system to control fertility has made a dramatic influence on the lives of many people (Figure 3).

(a)

(b)

FIGURE 1
Goretex is used in (a) all-weather gear and (b) in high-tech NASA space suits.

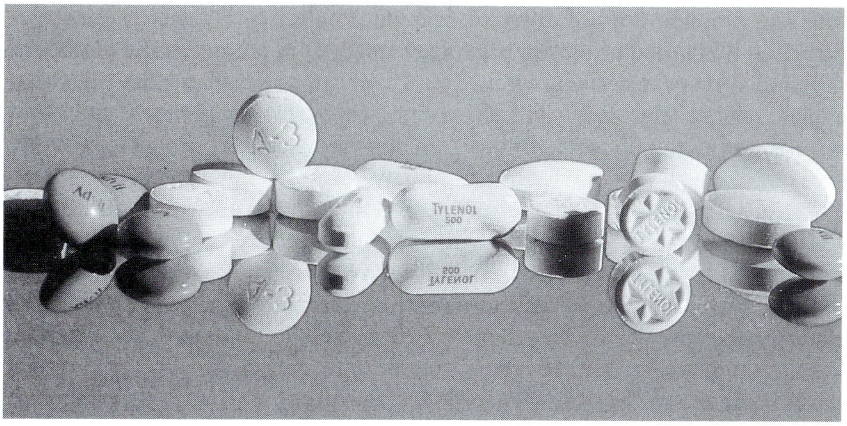
A-3
TYLENOL
500

FIGURE 2
Many varieties of aspirin substitutes are now available.

At the same time we are reminded of the mistakes that have been caused by chemistry: DDT and PCBs in the environment, depletion of the ozone layer by use of freons, and medicines such as thalidomide and DES with horrible side effects. How molecules influence life, and indeed the molecular basis of life itself, is the new frontier of chemistry. In the past, distinctions have been made between the chemistry of inorganic materials and organic materials. Today there are also distinctions between organic chemistry and biochemistry.

Biochemistry is the chemistry of living systems. Organisms and their chemical basis are fascinating topics, but they do not differ from the nonliving part of the world in any basic or fundamental way. All of the physical and chemical rules you have learned so far apply to biochemistry. Atomic struc-

FIGURE 3
Birth control pills have brought enormous changes to our society, allowing couples to plan when and if to have children.

ture and bonding are not different in biomolecules. The basic carbon structures and functional groups of organic chemistry are present in the molecules of living systems. Energy is vital to biochemical systems, and the rules that explain energy relationships in physics and chemistry are the same rules that govern energy relationships in biochemistry. Biochemistry differs from other areas of chemistry primarily in complexity.

Biochemistry has become an integral part of biology and the health sciences. Many of the terms you will learn here are used daily in health care. But your study of biochemistry will do more than help prepare you for the present, it will also give you a glimpse of the future. Biotechnology is removing some of the limitations that health care professionals now face. Through genetic engineering, human genes can be introduced into bacteria. When these bacteria are grown, they produce the product of that human gene. Already human insulin and other human proteins have been produced by this technique. More will quickly follow. Eventually the technology will allow the introduction of new or altered genes into humans. Biotechnology provides the potential for a brave new world.

CHAPTER 14

An Introduction to Organic Chemistry

The unique position of carbon for bonding is a reflection of its electronegativity. In this chapter the three most common ways for carbon to participate in bonding will be explored. As a result of these bonding arrangements, the shapes and acid–base character of compounds can be predicted. Novel incorporation of other atoms in the carbon framework will be introduced by the concept of functional groups.

OBJECTIVES

After completing this chapter, you should be able to

- Identify common elements that will form covalent bonds with carbon.
- Define the term *alkyl group.*
- Define the term *functional group.*
- Classify organic compounds into families by identification of their functional groups—the hydrocarbons, alcohols, ethers, ketones, aldehydes, carboxylic acids, organohalogens, and amines.
- Determine the structure and shape of the common functional groups.

14.1 Carbon's Electronegativity Makes It Unique

Carbon is essential to life. It is the major structural atom of the compounds found in living systems.

All living things have one basic characteristic in common. Whether from plant or animal, most molecules of living organisms are based on carbon. Consider the compounds shown in Figure 14.1. Glucose, a carbohydrate, is used by plants and animals as a source of quick energy. Acetic acid is an intermediate compound used in human metabolism and is the "sour" ingredient in vinegar. The amino acid lysine is synthesized by plants and is used by plants and animals to make protein molecules. It is an "essential" amino acid for humans; we cannot synthesize it but we require it for protein synthesis. Notice that all three of these compounds are based on carbon.

Although carbon makes up less than 0.03% of the earth's crust, its importance to living systems makes it a very important element. The central position of carbon on the electronegativity scale (Figure 14.2) permits it to form covalent bonds with practically all elements. Because carbon normally has four bonds, there is an unlimited number of three-dimensional, covalently bonded molecular skeletons based on this atom. The number of known molecules based on carbon exceeds all other known molecules of all other elements. Living organisms are composed principally of molecules made of carbon, hydrogen, nitrogen, phosphorus, oxygen, sulfur, and (in the case of marine organisms) chlorine.

In the following sections we will preview the special arrangements of atoms that are found in organic chemistry. These special arrangements, which occur over and over again, are called **functional groups**. A functional group is a group of one or more atoms that may substitute for an alkyl group's missing hydrogen. Because of the electron distribution and electronegativity of its component atoms, it often dramatically changes both the

Glucose
Blood sugar

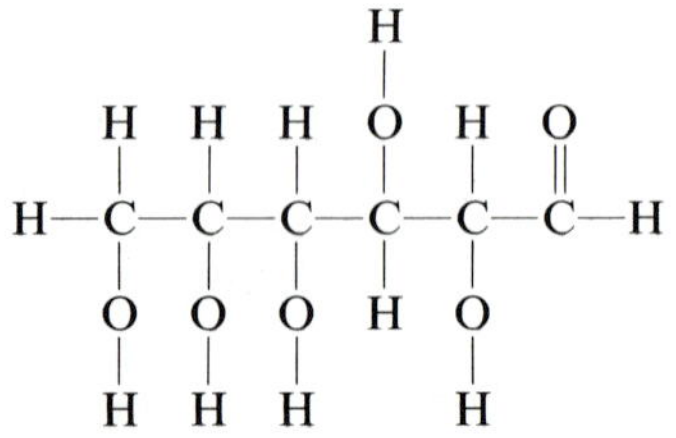

Acetic acid
Found in vinegar

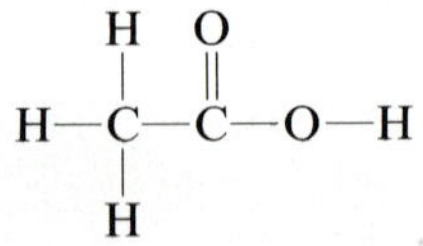

Lysine
A building block for proteins

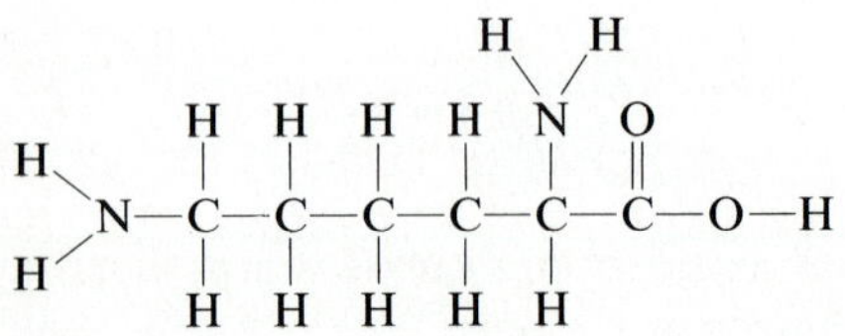

FIGURE 14.1
Three common compounds found in living things, all having something in common: they are built on a carbon skeleton.

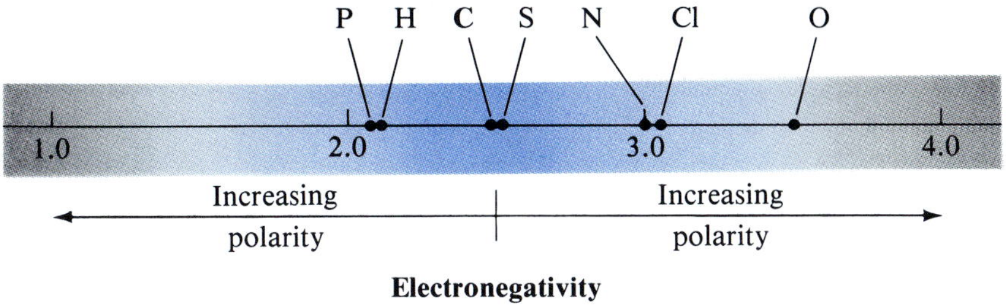
P H C S N Cl O
1.0
2.0
3.0
4.0
Increasing polarity
Increasing polarity
Electronegativity

FIGURE 14.2
Because of its central location in the electronegativity scale, carbon will form covalent bonds with hydrogen, sulfur, nitrogen, chlorine, and oxygen, all of which are elements common to living things and abundant on the earth. The difference in electronegativity between two atoms determines the degree of polarity of the bond and, to a large extent, the physical and chemical behavior of the bond.

physical properties and chemical reactivity of the parent hydrocarbon. Much of the chemical behavior of organic compounds can be directly related to functional groups.

NEW TERM

functional group A particular combination and arrangement of atoms; when attached to a hydrocarbon, unique physical and chemical properties are given to the molecule.

14.2 Bonding and Structure

Electron Dot Structures Revisited

In this section we discuss how the shapes of organic molecules (and indeed, most molecules) are readily predicted by application of the electron dot structures and VSEPR (valence shell electron pair repulsion) theory previously introduced in Chapter 6.

Electron dot structures are readily constructed using the valence electrons for each atom and applying some simple bonding rules.

Application of electron dot structures to predict three-dimensional features of molecules is best approached by some very simple rules, which we will call the ABCs of bonding.

A = The sum of the valence (outer shell) electrons of the atoms in the molecule.

B = Sum of electrons needed so that each atom has an inert gas electron configuration. This number is 8 for all atoms except hydrogen, which is 2.

C = **B** − **A** = Number of *bonding electrons* (these electrons must appear between atoms).

D = **A** − **C** = Number of electrons "left over" as nonbonding electrons.

For most considerations, hydrogen supports one bond, oxygen supports two bonds, nitrogen supports three bonds, and carbon supports four bonds. Halogens, which include fluorine, chlorine, bromine, and iodine, are single bonded.

EXAMPLE 14.1

Determine the arrangement of electrons for each atom and draw the electron dot structure for dichloromethane, CH_2Cl_2.

	Cl	Cl	H	H	C		Outer shell electrons
A =	7	+ 7	+ 1	+ 1	+ 4	=	20
B =	8	+ 8	+ 2	+ 2	+ 8	=	28
C =	28 − 20					=	8
D =	20 − 8					=	12

If two electrons are assigned per bond, four bonds are produced.

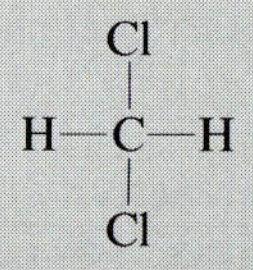

The carbon atom now has eight electrons around it and has an inert gas electronic structure. Each hydrogen atom has two electrons and is satisfied. But what about each Cl? By sharing, each has two electrons, but each needs six more. These are the nonbonding electrons, and each Cl atom receives them from the pool of nonbonding electrons. The electron dot structure for dichloromethane is now completed as follows:

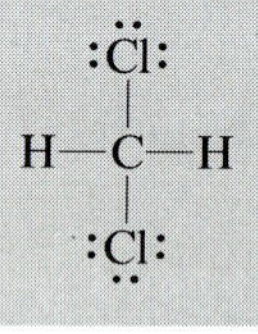

EXAMPLE 14.2

Determine the arrangement of electrons for each atom and draw the electron dot structure for the carbonate ion, CO_3^{2-}.

	O	O	O	C	Charge		Outer shell electrons
A =	6	+ 6	+ 6	+ 4	= 22 + 2*e*	=	24
B =	8	+ 8	+ 8	+ 8		=	32
C =	32 − 24					=	8
D =	24 − 8					=	16

Bonds must exist between C and O, and because there are eight bonding electrons, we must consider three bonds:

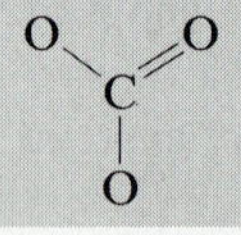

We could also divide the eight electrons equally over the three C—O bonds:

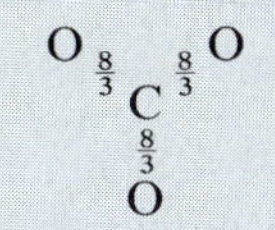

The carbon atom now has eight electrons around it; no more nonbonding electrons are needed. One oxygen has four electrons around it and needs four more, while the other two oxygen atoms have only two and need six more. The total number of nonbonding electrons is $4 + 6 + 6 = 16$ (D), and the structure for CO_3^{2-} must be as follows:

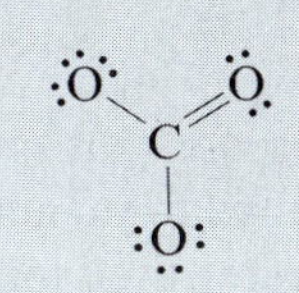

Note that we could have written the structure for CO_3^{2-} in Example 14.2 as any of the following:

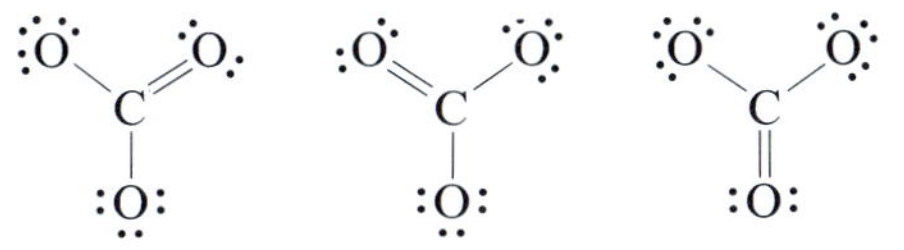

When two or more equivalent structures can be drawn for a molecule, these are called **resonance structures**. Resonance structures demonstrate that electrons are *delocalized*, that is, they are not totally localized between bonds. Notice that the *double bond* electrons have the ability to be delocalized. Single bonds will always have definite positions in the molecule. Compounds that can have resonance structures are always more stable than any single contributor could be.

EXAMPLE 14.3

Determine the arrangement of electrons for each atom and draw the electron dot structure for acetylene, C_2H_2.

	H	C	C	H		Outer shell electron
A =	1	+ 4	+ 4	+ 1	=	10
B =	2	+ 8	+ 8	+ 2	=	20
C =	20 – 10				=	10
D =	10 – 10				=	0

There will be five two-electron bonds and no nonbonding electrons in this molecule.

$$H—C{\equiv}C—H$$

Shapes From Electron Dot Structures and VSEPR Theory

If lone pair electrons are treated as "bonds", the shape of a molecule is readily predicted using the ideal orientation of bonds in space so that they can be as far from one another as possible.

Some simple rules allow us to easily convert an electron dot structure into a three-dimensional structure. These rules are summarized in the following list.

1. Recall from Chapter 6 that the arrangement of electrons will be tetrahedral if the following electron dot structures are present around a central atom, A. For example,

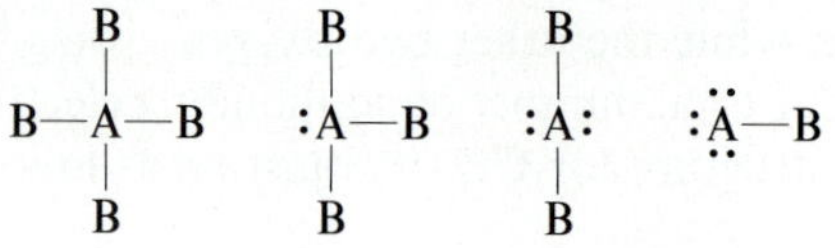

The hybridization of an atom can be predicted from Lewis electron dot structures.

Here the B stands for some atom that is bonding with A. Bond angles will be around 109°. In these arrangements, repulsions between lone pairs of electrons (two dots) are greater than lone pair–bond repulsions, which in turn are greater than bonding pair–bonding pair repulsions. Thus, one might expect the following examples:

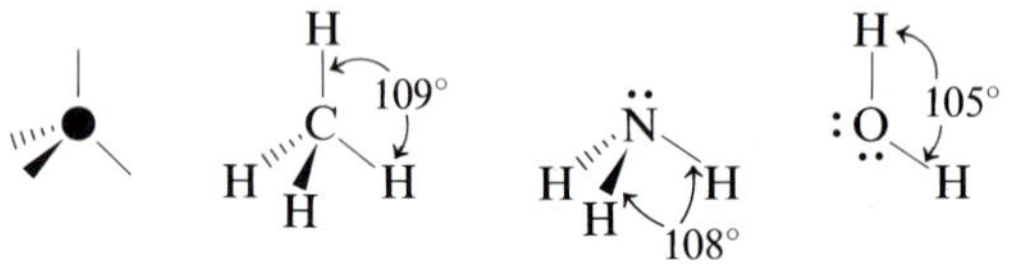
109°
108°
105°

2. Molecules will be planar about the central atom if the following electron dot structures are present.

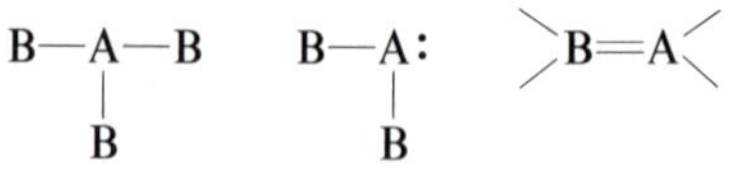

In general the bond angles will be about 120° as illustrated by these molecules. (See Chapter 15 for an explanation of the *p*-orbitals in the first molecule.)

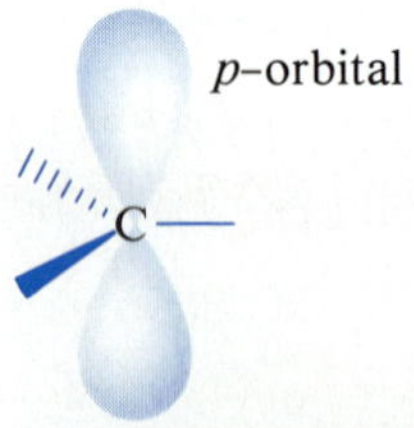
p-orbital

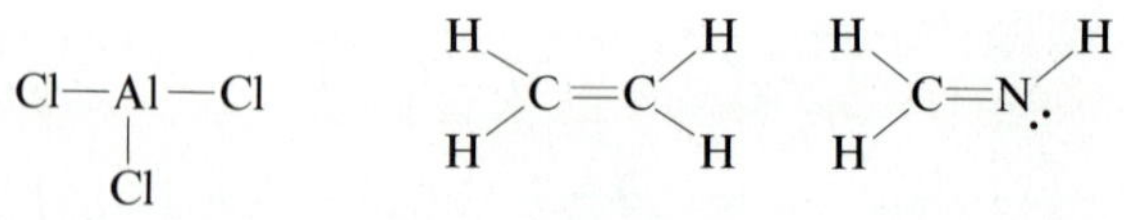

3. Molecules will be linear about the central atom if the electron dot structure has one of these general forms:

B—A—B B=A=B B≡A—B B≡A:

Bond angles for this type will be 180°. Some examples of linear molecules are shown here. (See Chapter 15 for an explanation of the *p*-orbitals in the first molecule.)

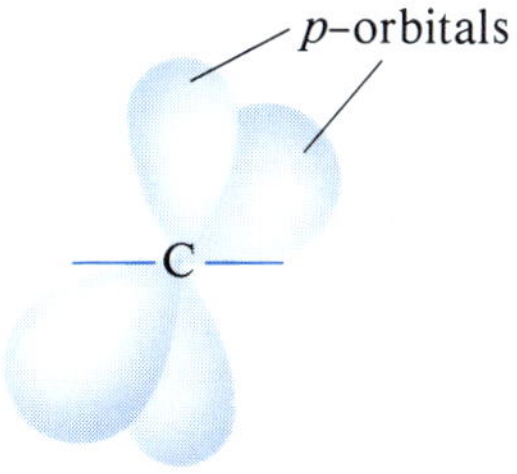
p-orbitals
C

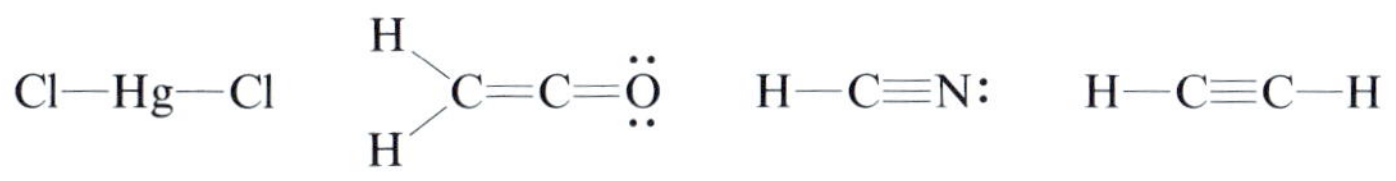
Cl—Hg—Cl
H
C=C=O
H
H—C≡N:
H—C≡C—H

Lewis Acid and Base Character from Electron Dot Structures

Chapter 11 introduced several models for explaining the behavior of acids and bases. The Brønsted–Lowry model defined acids as molecules or ions that give up a hydrogen ion to an acceptor molecule, which is called a base. But at the end of Chapter 11, a more general acid–base model was introduced that is extremely useful in organic and biochemistry: the **Lewis theory**.

Any atom that has lone pairs of electrons associated with it can function as a *Lewis base*, an electron donor. This atom can be an individual atom, anion, or an atom in a molecule. Molecules containing double or triple bonds also function as Lewis bases because the electrons associated with multiple bonds are not tightly attached to particular atoms and are available for reaction. In a similar definition, atoms (either alone, as cations, or in molecules) that accept electrons are classified as *Lewis acids*. By this definition we can include the Brønsted–Lowry acid, H^+, as a Lewis acid since it will seek electrons; the Brønsted–Lowry base, OH^-, is an electron donor (Lewis base). We can also include anions such as Cl^- and O^{2-} as bases and cations such as Ag^+ and K^+ as acids. By this definition, water and ammonia will function as Lewis bases because they have lone pairs of electrons. The Lewis theory of acids and bases will be seen in the chapters to follow as fundamental to much of the chemical behavior of organic and biological molecules. Examples of Lewis acid–base chemistry are seen in the following examples:

Electron pair ↘   Needs electrons ↙

$$HO^- + H^+ \longrightarrow H_2O$$
$$Cl^- + Ag^+ \longrightarrow AgCl$$

NEW TERMS

resonance structure Representations of the electronic arrangement of a molecule for which more than one Lewis electron dot structure can be drawn. The "real" electron distribution for the molecule will be a hybrid of each of the separate resonance structures.

Lewis theory A theory of electronic arrangement in a molecule to accommodate eight electrons around most atoms (hydrogen will have two electrons).

TESTING YOURSELF

Electron Dot Structures

1. Draw the electron dot structures and give the molecular shape around the highlighted atom in each of the following molecules.

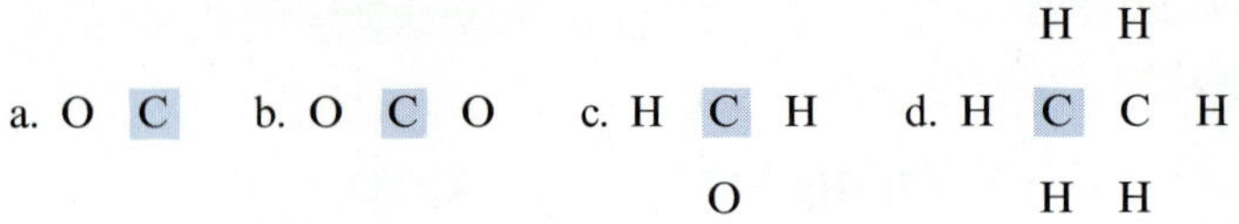

2. Give the shape (tetrahedral, planar, or linear) for the highlighted atom in each of the following structures.

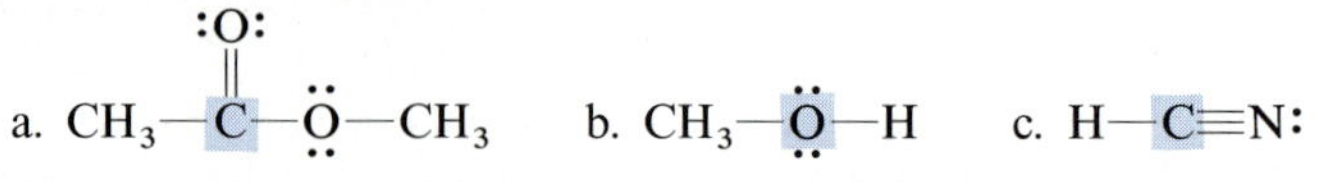

Answers **1. a.** :O≡C: (linear) **b.** $\ddot{\underset{..}{O}}$=C=$\ddot{\underset{..}{O}}$ (linear)

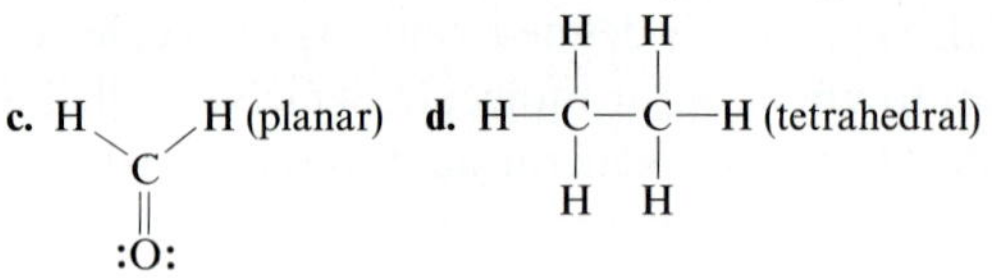

2. a. planar **b.** tetrahedral **c.** linear

14.3 Introduction To Functional Groups

Hydrocarbons

Hydrocarbons are simple molecules that are constructed of only carbon and hydrogen. The C—H bond is very nonpolar.

Hydrocarbons are the simplest carbon-based compounds—they are made only of carbon and hydrogen. Four-bonded carbon atoms form the structural skeleton of the molecule, and all bonds not connected to other carbon atoms are connected to hydrogen atoms. The electronegativity difference across a carbon–hydrogen bond is only 0.4 electronegativity units; the bond is practically purely covalent in nature. Thus, hydrocarbons are generally very nonpolar, and as a consequence, they are insoluble in polar solvents such as water. This feature will be of immense importance in our later study of biochemistry and will be discussed in greater detail in later chapters (for example, see Chapter 20).

Hydrocarbons take many forms, from the simplest methane (Figure 14.3) to the very large, complex molecules found in crude oil and tars. In methane, a single carbon atom is bonded to four hydrogen atoms.

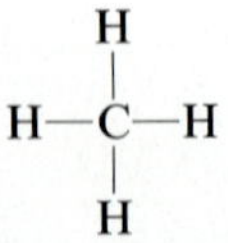

FIGURE 14.3
Methane (CH_4), commonly known as natural gas, is the simplest hydrocarbon.

By adding other carbon atoms to the skeleton, larger hydrocarbon molecules can be created. Figure 14.4 shows several simple hydrocarbons. Note that all of the hydrocarbons discussed so far have single bonds between the carbon atoms and names with the suffix *-ane*. The prefixes of each name reflect the number of carbons in the chain. The family of single-bonded hydrocarbons is called the **alkane** family. They are also known as **saturated hydrocarbons** because the carbon atoms are unable to accept additional hydrogens.

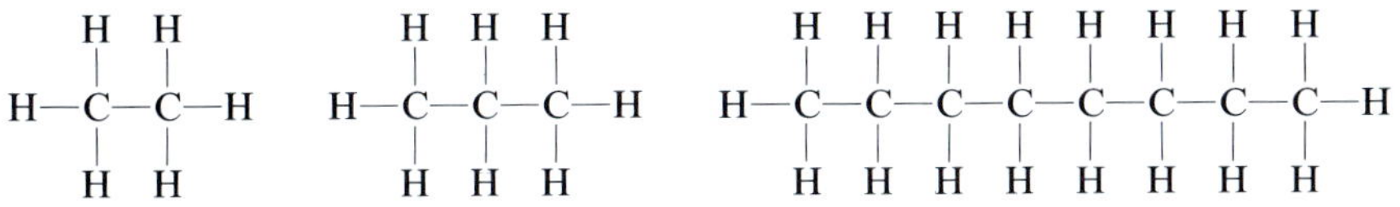

Ethane

Propane
A fuel

Octane
Component of gasoline

FIGURE 14.4
Some common hydrocarbons.

Some hydrocarbons have double or triple bonds between the structural carbon's atoms. These are called **unsaturated hydrocarbons** because they are able to accept additional hydrogen or other atoms. One can remember this by noting that saturated carbon systems have "full" bonds—that is, each of the four carbon bonds is attached to a different atom. Hydrocarbons with a double bond between adjacent carbons are called **alkenes**; those with a triple bond are called **alkynes** (Figure 14.5).

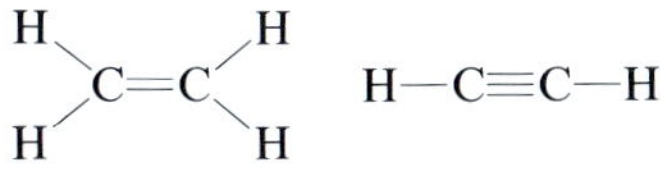

Ethene
A simple alkene

Ethyne (acetylene)
A simple alkyne

FIGURE 14.5
Hydrocarbons with double and triple bonds between structural carbons are called alkenes and alkynes, respectively. They are more reactive than their single-bonded counterparts.

Functional Groups Having Other Atoms

Hydrocarbons provide the basic structural unit for all molecules found in organic chemistry and biochemistry. The chemical and physical behavior of hydrocarbon molecules can be changed significantly by replacing one or more hydrogen atoms with functional groups made of other abundant elements—oxygen, hydrogen, chlorine, and nitrogen.

When a hydrogen atom is removed from a hydrocarbon, the resulting fragment is called an **alkyl group** (Figure 14.6). Alkyl groups are represented by the letter R in chemical structures. Alkyl groups have no stable existence, but they are a useful way of keeping some consistency in a system of organizing and naming compounds that could otherwise become quite complex.

Earlier in this chapter we identified several elements found in living organisms. These elements are listed in Table 14.1. They are all reasonably abundant on the earth, and all have electronegativity values close enough to carbon that they easily form covalent bonds with carbon.

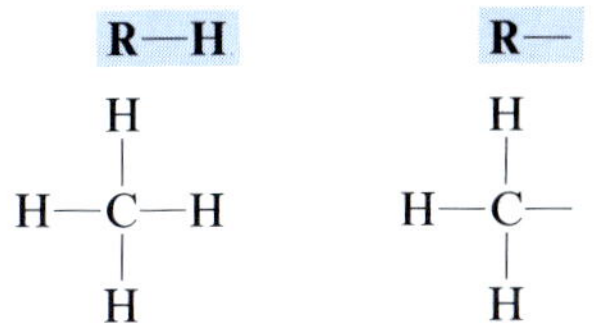

Methane
(CH_4)
A hydrocarbon

Methyl group
(CH_3—)
An alkyl group

FIGURE 14.6
An alkyl group is a hydrocarbon with one hydrogen removed. The "open bond" is often filled with an atom or group other than hydrogen. Alkyl groups are often referred to by the symbol R—, where R represents any hydrocarbon group and the dash represents the open bond that awaits a new partner. Any hydrocarbon can thus be represented by the general form R—H.

TABLE 14.1
Common Elements Found in Living Organisms.

Element	Electronegativity: Value	Electronegativity: Difference from Carbon	Number of Bonds
Oxygen	3.5	1.0	2
Hydrogen	2.1	0.4	1
Carbon	2.5	0.0	4
Sulfur	2.5	0.0	2
Chlorine	3.0	0.5	1
Nitrogen	3.0	0.5	3
Phosphorus	2.1	0.0	3, 5

TABLE 14.2
Some Common Functional Groups and the Families of Compounds They Create

Functional Group	General Formula	Family Name
—H	R—H	Hydrocarbon
—O—H	R—O—H	Alcohol
—S—H	R—S—H	Thioalcohol (thiol)
—O—	R—O—R′*	Ether
—C(=O)—	R—C(=O)—R′	Ketone
—C(=O)—H	R—C(=O)—H	Aldehyde
—C(=O)—O—H	R—C(=O)—O—H	Carboxylic acid
—C(=O)—O—R	R—C(=O)—O—R′	Ester
—Cl:	R—Cl:	Organohalogen
—NH₂ (—N(H)H)	R—N(H)H	Amine
—C(=O)—N<	R—C(=O)—N<	Amide

* When more than one alkyl group is involved in a compound, the second and third groups are designated R′ and R″.

The functional group concept allows for easy classification of compounds according to structure. This turns out to be also closely related to chemical behavior.

From the set of elements in Table 14.1, one can construct functional groups that substitute easily for one of the hydrogen atoms in a hydrocarbon molecule, and in doing so significantly change the behavior of the molecule. This change in behavior is so pronounced that the newly formed compounds have been given new family names. Some common functional groups involving oxygen, chlorine, nitrogen, and sulfur are shown in Table 14.2. The remainder of the chapter surveys the common organic functional group families.

Alcohols

Alcohols are recognized by the general structure:

R—OH

Substitution of the —OH functional group for a hydrogen atom changes a hydrocarbon into an **alcohol**; thus, the generalized form for an alcohol is R—O—H. Figure 14.7 shows the substitution of an alcohol group for a hydrogen in two simple hydrocarbons, methane and propane.

Sulfur is in the same periodic family as oxygen, so one might reasonably expect that a sulfur-containing compound similar to an alcohol might be

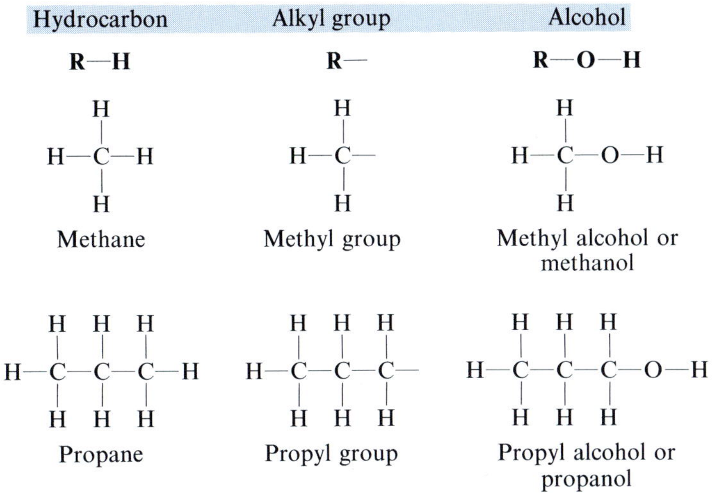
Hydrocarbon
Alkyl group
Alcohol
R—H
R—
R—O—H
Methane
Methyl group
Methyl alcohol or methanol
Propane
Propyl group
Propyl alcohol or propanol

FIGURE 14.7
The formation of two simple alcohols by substitution of the —OH functional group. Notice how the alkyl group and the functional group combine to form the new class of molecules.

found. As Figure 14.8 shows, this is indeed the case. The sulfur analogs of alcohols are called *thiols* and have the functional group —SH. The chemical and physical properties of alcohols and thiols will be discussed further in Chapter 18.

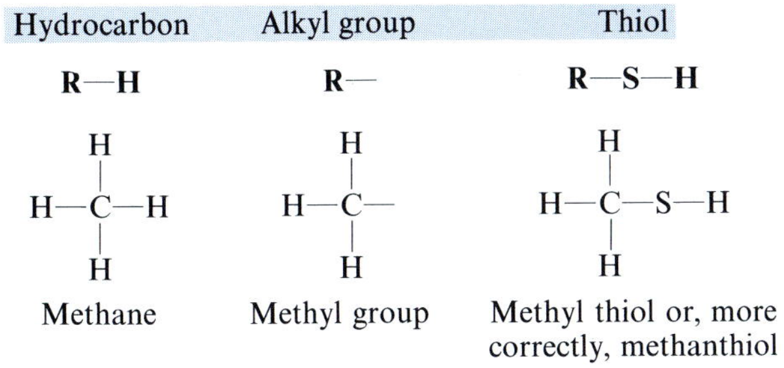
Hydrocarbon
Alkyl group
Thiol
R—H
R—
R—S—H
Methane
Methyl group
Methyl thiol or, more correctly, methanthiol

FIGURE 14.8
The creation of a simple thiol by the substitution of the —SH functional group.

Ethers

An oxygen atom, which has two bonds, can form a bridge between two hydrocarbon groups. The resulting compound is called an **ether**. The ether functional group is shown in Figure 14.9. The ethers will be discussed further in Chapter 18.

Ethers are recognized by the general structure:

R—O—R

As you can see from the electron dot structures in Figure 14.9, the oxygen atom of ethers can be classified by the following generalized structure:

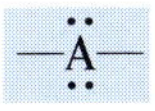

The oxygen's electrons are in a tetrahedral arrangement in an ether.

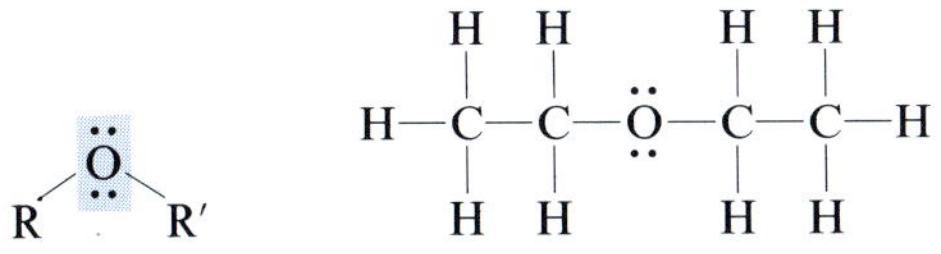

General formula of an ether **Diethyl ether**

FIGURE 14.9
An ether is composed of two alkyl groups attached to an oxygen atom.

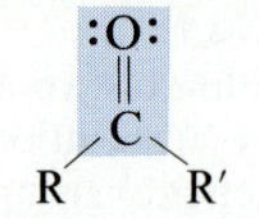

General formula of a ketone

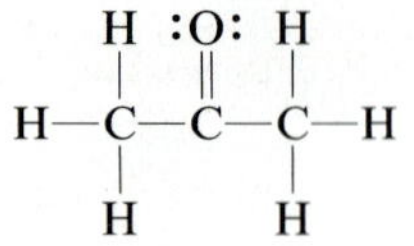

Dimethyl ketone (acetone)

FIGURE 14.10
The ketone family is characterized by two alkyl groups connected to a carbonyl functional group.

Aldehydes and Ketones

Ketones are characterized by the **carbonyl** group, which is an oxygen atom double bonded to a carbon. The remaining two bonds of the carbon are available to connect to other fragments of the molecule. When each connects to an alkyl group, the resulting compound is called a **ketone** (Figure 14.10).

From application of the electron dot structures in Figure 14.10, we find that the carbonyl group can be generalized by the following structure:

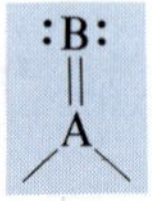

Aldehydes are close relatives of the ketone family. Aldehydes, however, have an alkyl group and a hydrogen atom connected to the carbonyl group instead of two alkyl groups (Figure 14.11). The details of this chemistry of these functional groups will be discussed in Chapter 19.

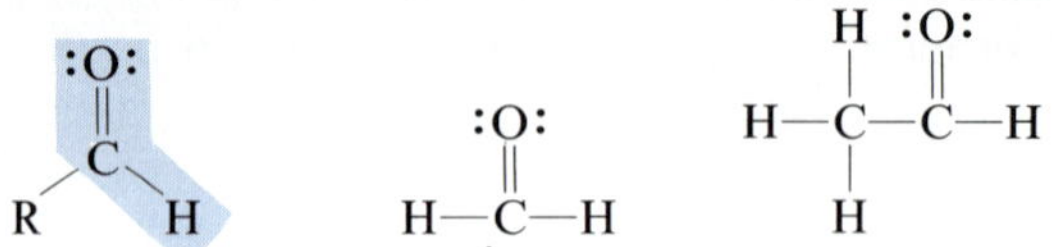

General formula of an aldehyde **Formaldehyde** **Acetaldehyde**

FIGURE 14.11
Aldehydes are characterized by an alkyl group and a hydrogen attached to a carbonyl functional group.

Carboxylic Acid

Carboxylic acids are similar to the ketone-aldehyde families except that an —OH group is connected to one side of the carbonyl carbon atom (Figure 14.12). A **carboxyl group**, however, is *not* a carbonyl and a hydroxide; the combination of one carbon atom, two oxygen atoms, and one hydrogen atom forms a unique functional group having unique properties. The physical and chemical behavior of carboxylic acids will be discussed in Chapter 20.

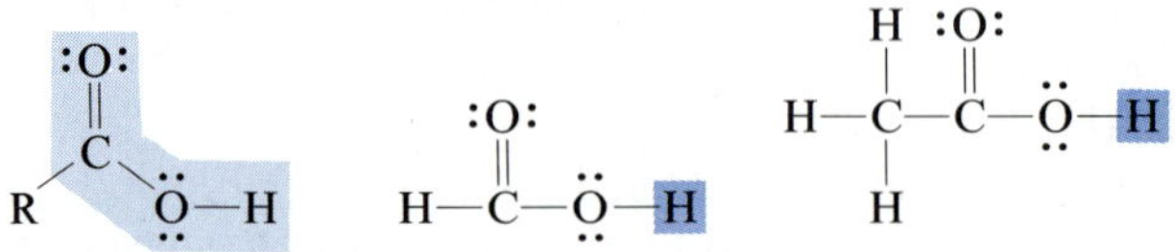

General formula of carboxylic acid **Formic acid** **Acetic acid**

FIGURE 14.12
Carboxylic acids are characterized by the presence of the carboxyl group. An alkyl group or hydrogen atom is connected to the carboxyl group. The highlighted hydrogen atom is slightly ionizable and is responsible for the acidity of this class of compounds.

Organohalogens

Twelfth in elemental abundance in the crust of the earth is chlorine, which forms one covalent bond with carbon. Chlorine is a member of the family of elements called *halogens*. Chlorine and the other members of the halogen family—fluorine, bromine, and iodine—can substitute for hydrogen atoms in a hydrocarbon to form a family of compounds known as **organohalogens** (Figure 14.13). The differences in electronegativity between carbon and the halogens are small enough that the bonds are still covalent. Note

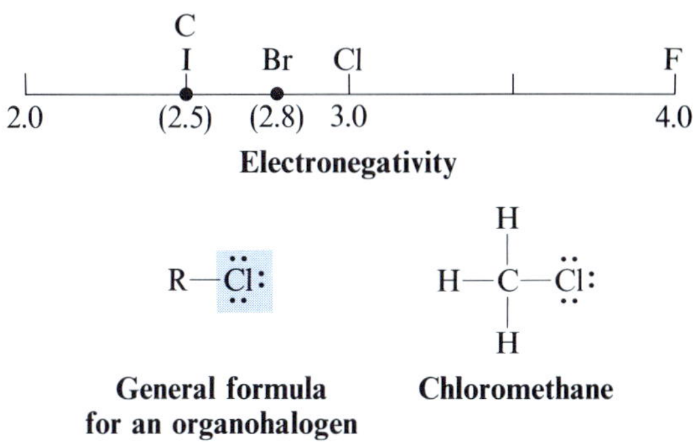
C
I
Br
Cl
F
2.0
(2.5)
(2.8)
3.0
4.0
Electronegativity
General formula for an organohalogen
Chloromethane

FIGURE 14.13
Chlorine is a member of the family of atoms (Group VII) known as the halogens. Although the atoms differ in electronegativity, they are all one electron short of a stable octet of electrons. The halogens can covalently bond to carbon. The organohalogens are compounds in which an alkyl group bonds to a halogen.

that more than one halogen can be connected to a carbon atom (Table 14.3). Thus, chlorinated hydrocarbons often contain more than one chlorine. If more than one carbon atom is present, these chlorines need not connect to the same carbon.

Amines

The *amino group* has nitrogen as its central atom. Nitrogen has a pair of unshared electrons that significantly affect the behavior of the group. The ammonia molecule is the parent compound of the amines. **Amines** are compounds formed by substitution of one or more alkyl groups for hydrogens in the ammonia (Table 14.4). An amine is the organic analog of ammonia, just as an alcohol can be considered as the organic analog of water.

The unshared electron pair on the nitrogen atom can serve as a hydrogen ion acceptor in the same manner that an unshared pair of electrons on the

Amines are recognized as the organic equivalents of ammonia and have one of these general structures:

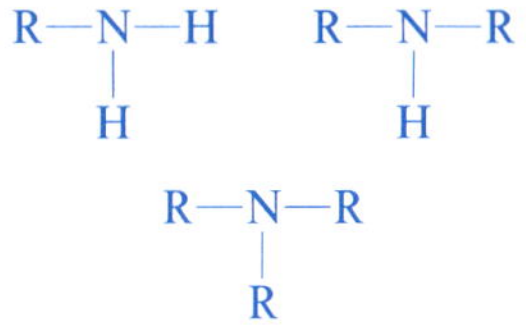

TABLE 14.3
Chlorinated Hydrocarbons

Compound Name	Chemical Structure
Chloromethane	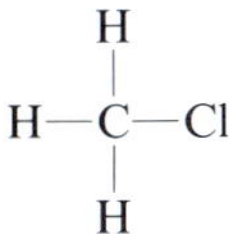
Dichloromethane	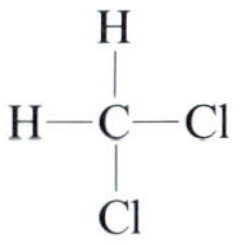
Trichloromethane (chloroform)	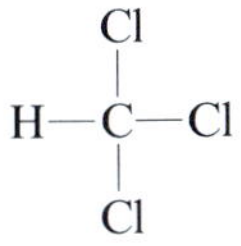
Tetrachloromethane (carbon tetrachloride)	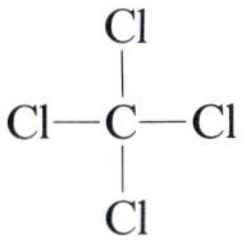

TABLE 14.4
Amines

Compound Name	Chemical Structure
Ammonia	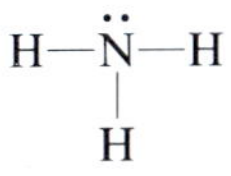
Methylamine	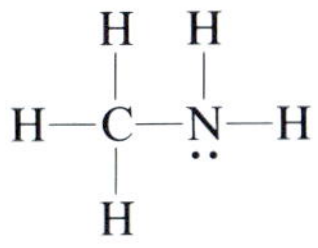
Dimethylamine	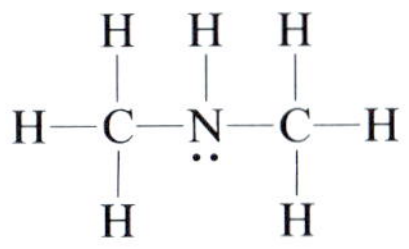
Trimethylamine	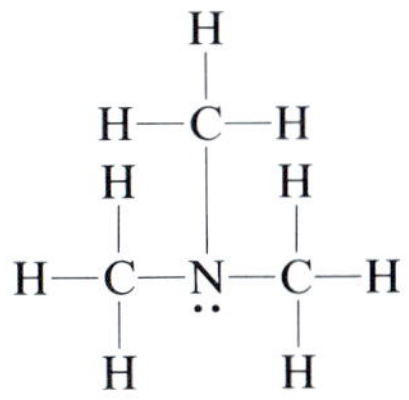

oxygen atom in water serves as a hydrogen ion acceptor to form hydronium ion. An amine is an "organic base" in the sense of a Lewis base—an amine is able to donate a pair of electrons.

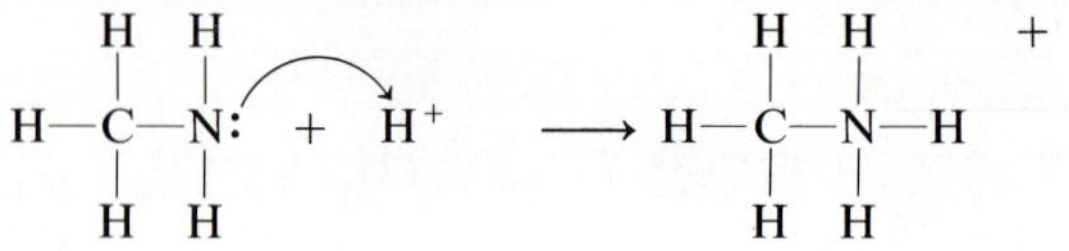

Lewis base **Lewis acid**

Amines will be discussed in further detail in Chapter 21.

Esters and Amides

The only two groups in Table 14.2 that we have not yet discussed are the esters and amides. Although very important to both organic chemistry and biochemistry, they are both simply modifications of the carboxylic acid functional group. Details of the chemistry of the esters and amides will be discussed under appropriate headings in Chapter 20. However, at this point it is appropriate to show the origins of these two functional groups:

Ester

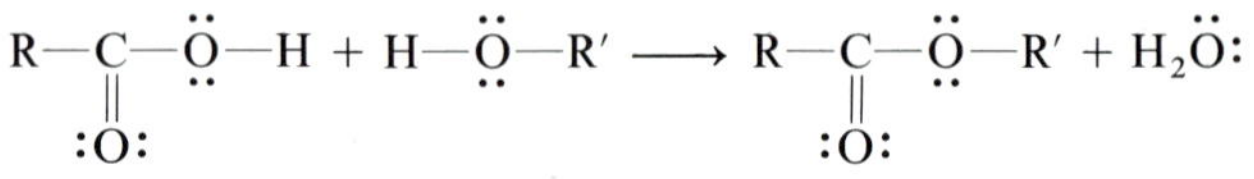

Acid Alcohol Ester Water

Amide

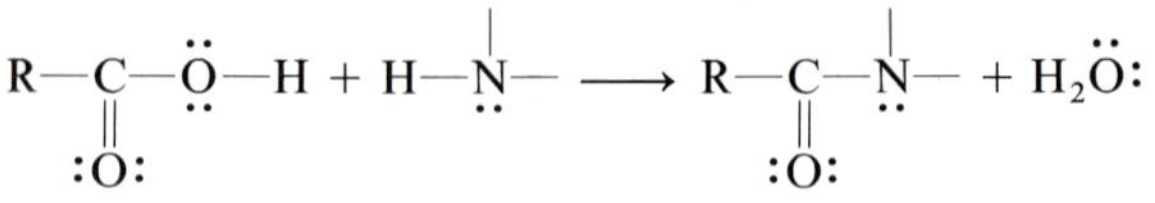

Acid Amine Amide Water

NEW TERMS

hydrocarbon A compound consisting only of the elements hydrogen and carbon. Hydrocarbons are almost always nonpolar.

alkane A family of hydrocarbon compounds having only single bonds between carbon atoms in the molecular skeleton.

saturated hydrocarbon A hydrocarbon containing only single carbon to carbon bonds.

unsaturated hydrocarbon A hydrocarbon containing either double or triple carbon to carbon bonds.

alkene A family of hydrocarbon compounds having at least one double bond between adjacent carbon atoms in the molecular skeleton.

alkyne A family of hydrocarbon compounds having at least one triple bond between adjacent carbon atoms in the molecular skeleton.

alkyl group A hydrocarbon with one hydrogen removed prior to bonding with another atom or group of atoms. Alkyl groups exist only on paper as a tool for naming and for explaining organic reactions. Alkyl groups are often represented by the symbol R.

alcohol An organic molecule of the form R—OH, where R is an alkyl group and the functional group is the —OH.

ether A family of organic compounds formed when an oxygen atom serves as a bridge between two alkyl groups.

carbonyl A family of organic compounds composed of a carbon double-bonded to an oxygen. The two remaining carbon bonds may be connected to other atoms or alkyl groups.

ketone A family of organic compounds formed when an alkyl group is connected to each of the two remaining carbon bonds of the carbonyl group.

aldehyde A family of organic compounds formed when an alkyl group is placed on one of the carbon bonds of a carbonyl group and a hydrogen is placed on the other.

carboxylic acid An organic compound containing one or more carboxyl groups.

carboxyl group A group of compounds composed of a carbonyl group with an —OH connected to one of the free carbon bonds. It is the "trademark" of organic acids.

organohalogen An organic compound in which a halogen (fluorine, chlorine, bromine, or iodine) has replaced one or more hydrocarbon hydrogens.

amine An organic compound in which a nitrogen is the central member of the functional group. Since nitrogen atoms have three bonds, only one of which connects to the parent hydrocarbon, it is possible to substitute additional alkyl groups on this nitrogen.

TESTING YOURSELF

Bonding and Functional Groups

1. Which of these elements makes the least polar bond with carbon: oxygen, hydrogen, nitrogen, or chlorine?
2. Which of the elements listed in question 1 will make an ionic bond with carbon?
3. Structural diagrams for several compounds are shown here. Identify the family of organic compounds to which each belongs.

a. 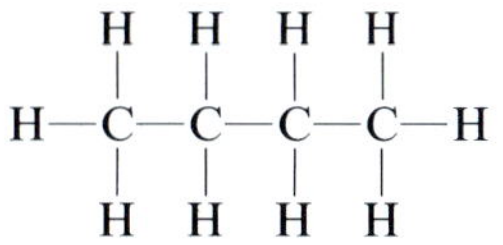

b. 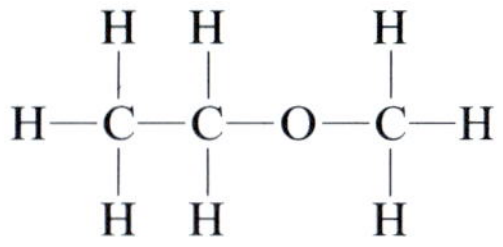

c. 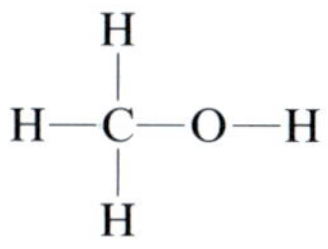

d. 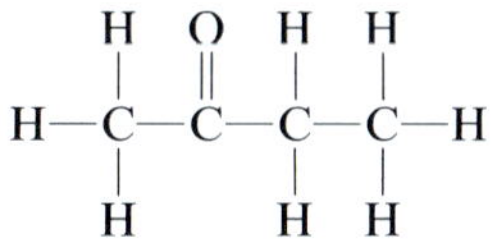

e. 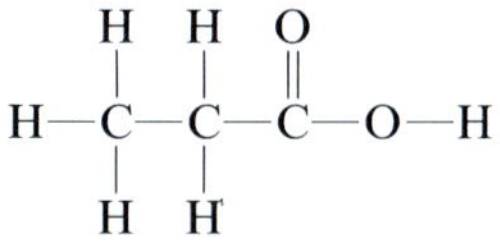

f. 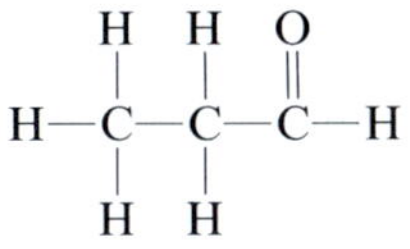

g. 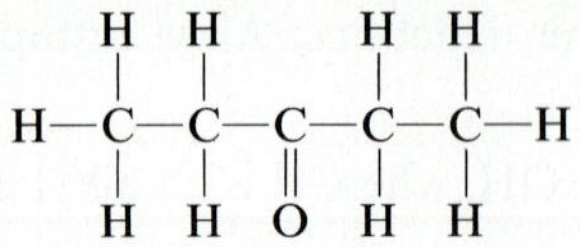

i. 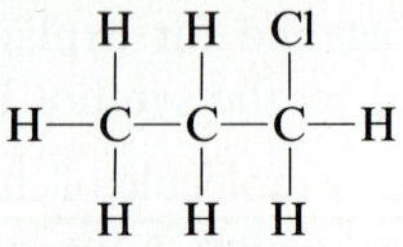

h. 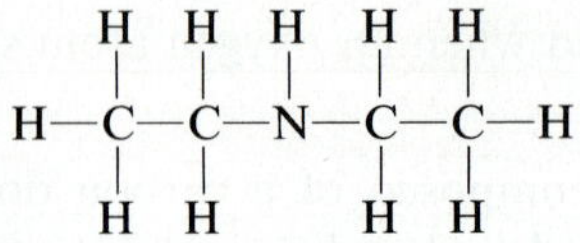

Answers **1.** hydrogen **2.** none **3. a.** hydrocarbon **b.** ether **c.** alcohol **d.** ketone **e.** carboxylic acid **f.** aldehyde **g.** ketone **h.** amine **i.** organohalogen

SUMMARY

The chemical and physical behavior of a hydrocarbon molecule is drastically changed if one or more of its hydrogen atoms is replaced by special atoms or groups of atoms called *functional groups*. Carbon-based compounds can be classified into families of compounds of similar structure and reactivity based on the presence of these functional groups. The specific families of compounds surveyed in this chapter and to be discussed in detail in the chapters that follow include the hydrocarbons (alkanes, alkenes, alkynes, and aromatics), organohalogen compounds, alcohols, phenols, ethers, thiols, aldehydes, ketones, acids, esters, amines, and amides.

The shapes of molecules and functional groups can be predicted from the Lewis electron dot structures. The numbers of bonds and electron lone pairs on an atom will determine the best way to situate them about the atom for minimal interference with one another. This simple approach gives good agreement with observed shapes. From the electron dot structures, predictions about the Lewis acid–base character of the molecule can also be made. According to this theory, Lewis acids will be electron acceptors while Lewis bases will be electron donors.

In the next chapter the simplest hydrocarbons, the alkanes, will be discussed. Highlights of this chapter will include the origins of nomenclature that will carry through the rest of the chapters, the concept of the alkyl group, and an introduction to the concept of molecular conformation—the preferred shapes of molecules.

TERMS

functional group	alkene	aldehyde
resonance structure	alkyne	carboxylic acid
Lewis theory	alkyl group	carboxyl group
hydrocarbon	alcohol	organohalogen
alkane	ether	amine
saturated hydrocarbon	carbonyl	
unsaturated hydrocarbon	ketone	

CHAPTER REVIEW

QUESTIONS

1. 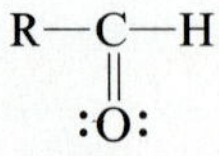 is a shorthand expression for a(n)

   a. acid b. ketone c. aldehyde d. alcohol

2. The carbon atom in the general structure shown in question 1 will have what shape?
   a. tetrahedral b. planar c. linear

3. An ether is best represented by which generalized structure?
   a. R—Ö—R (with lone pairs) b. R—Ö—H c. R—C(=O)—ÖH d. R—C(=O)—R

4. An amine is uniquely characterized by the presence of
   a. hydrogen b. carbon c. oxygen d. nitrogen

5. An alkyne has what type of bond?
   a. single b. double c. triple d. quadruple

6. An alcohol is represented by which generalized structure?
   a. R—X: b. R—OH c. R—O—R d. R—NH_2

7. A ketone is most similar to an
   a. alcohol b. aldehyde c. acid d. amine

8. A ketone will have how many R groups attached to the carbonyl group?
   a. 0 b. 1 c. 2 d. 3

9. 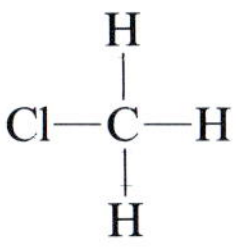 is classified as a(n)
   a. organohalogen b. amine c. hydrocarbon d. alkene

10. In the structure below, which numbered carbon atom(s) will have a linear geometry?

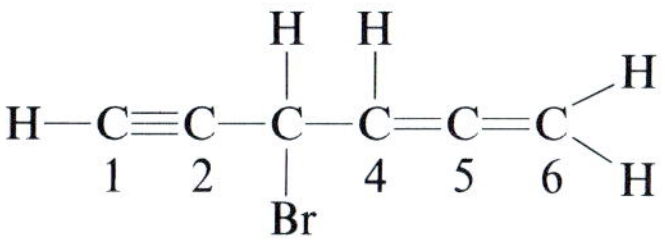
H H
H
H—C≡C—C—C=C=C
1 2 4 5 6
H
Br

   a. 1 b. 1 and 2 c. 1, 2, and 5 d. 1, 2, 4, and 6

Answers **1.** c **2.** b **3.** a **4.** d **5.** c **6.** b **7.** b **8.** c **9.** a **10.** c

DIAGNOSTIC CHART

Blacken in all of the circles under the number of each question that you missed in the Chapter Review questions. The diagnostic chart will help you identify concept areas that need more study.

	Questions									
Concepts	1	2	3	4	5	6	7	8	9	10
Structure	○	○	○	○	○	○	○	○		○
Alkyl groups								○		
Functional groups	○	○	○	○	○		○		○	

EXERCISES

1. Draw structures for the first ten straight chain hydrocarbons.
2. How does an alkene differ from an alkyne?
3. Give general structures for each of the following.
   a. alcohol d. ether
   b. aldehyde e. acid
   c. ketone f. amine
4. Draw the electron dot structures and determine the geometry of each highlighted atom in the following examples. Classify each compound according to its functional group.

```
     H H H              H
a. H C C C H      d. H C C H
     H H H              H O

     H
b. H C O H        e. Br C N
     H
                        H
     H O          f. H C Cl
c. H C C O H            H

     H     H              H H
g. H C  O  C H      h. H C N H
     H     H              H
```

(Highlighted atoms: a. the third C; b. C; c. both C; d. the first C; e. C; f. C; g. O; h. C)

5. Draw all of the arrangements of atoms that you can think of for $C_4H_{10}O$. What will be the functional group classification? (HINT: Each carbon needs four bonds, each hydrogen needs one bond, and each oxygen needs two bonds.)
6. The R group that is so convenient in organic chemistry can also be incorporated into electron dot structures. Assume that an R group can be treated like a hydrogen (it has one electron and needs two). Use the electron dot methods to obtain formulas for each of the following:

```
a. R X (where X is any halogen)      e. R N R
b. R O H                                  H
c. R C H                             f. R C R
     O                                    O
d. R C O H                           g. R O R
     O
```

CHAPTER

15

Alkanes and Cycloalkanes: Single-Bonded Hydrocarbons

The simplest carbon-based compounds involve only one other element—hydrogen. This family group—the hydrocarbons—is divided into three subgroups by the presence of single, double, or triple bonds between the carbon atoms. This chapter will consider the single-bonded hydrocarbons, the alkanes and cycloalkanes.

OBJECTIVES

After completing this chapter, you should be able to

- Identify members of the alkane and cycloalkane families.
- Name representative members of each family.
- Identify the factors that limit the reactivity of the alkanes and cycloalkanes.
- Predict the products of some representative reactions of single bonded hydrocarbons.

15.1 Structure and Physical Properties

Tetrahedral Shape

From the electron dot structures discussed in previous chapters, it is simple to determine that a carbon atom of an alkane will be tetrahedral. In Chapter 6 we developed the concept of hybridization for the structure of water. Based on the Lewis electron dot structures and the rules given in the last chapter, we can now state that any atom having a tetrahedral arrangement of atoms or electrons about it will be called sp^3 hybridized.

As for atomic orbitals, the hybrid orbitals need two electrons per orbital. Since each of the hybrid orbitals have one electron, they share their electrons. In the case of methane, this means that one electron is accepted from each hydrogen atom.

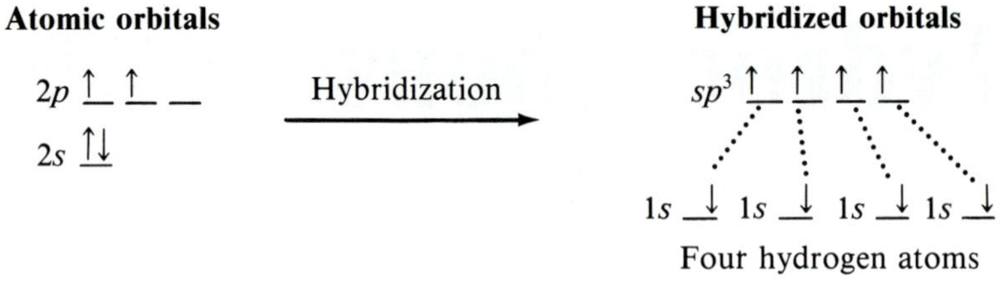
Atomic orbitals
Hybridized orbitals
2p
2s
Hybridization
sp³
1s
1s
1s
1s
Four hydrogen atoms

Methane

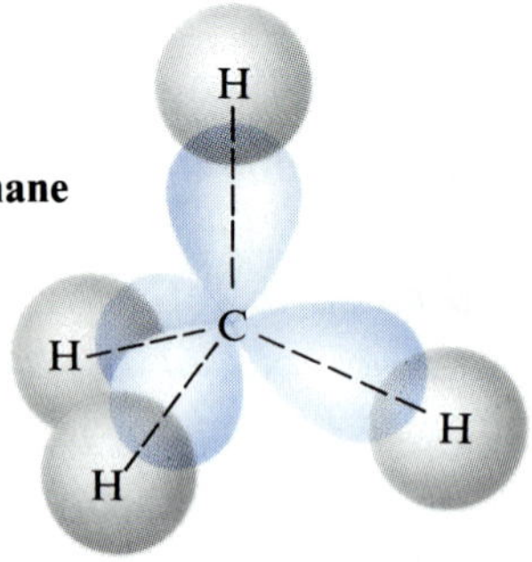
H
H
C
H
H

The Carbon–Hydrogen Bond

Carbon and hydrogen are very close in electronegativity. Bonds between these elements are therefore almost purely covalent, and there is no strong dipole effect due to large differences in electronegativity. The molecules are attracted to each other by *London forces*, attractive forces caused by *induced dipoles* (Figure 15.1). In these attractions, the positively charged atomic nucleus of one molecule attracts the electrons of another molecule. This induced dipole is only temporary, and the two molecules can then change roles. Since the London attractive forces are dependent on inducing a dipole in a molecule, the larger the molecule (the more electrons and protons present), the more important is the attractive force. Thus, the attraction between two molecules of pentane is greater than the attraction between two molecules of methane.

Because of the nonpolar nature of the C—H bond, there is little reason to expect that hydrocarbons will be soluble in water.

Boiling Point

In general, within a particular functional group, the boiling points of compounds will increase with increasing carbon number.

A consequence of this increase in attractive force with molecular weight, as well as the simple increase in weight itself, is an increase in boiling point

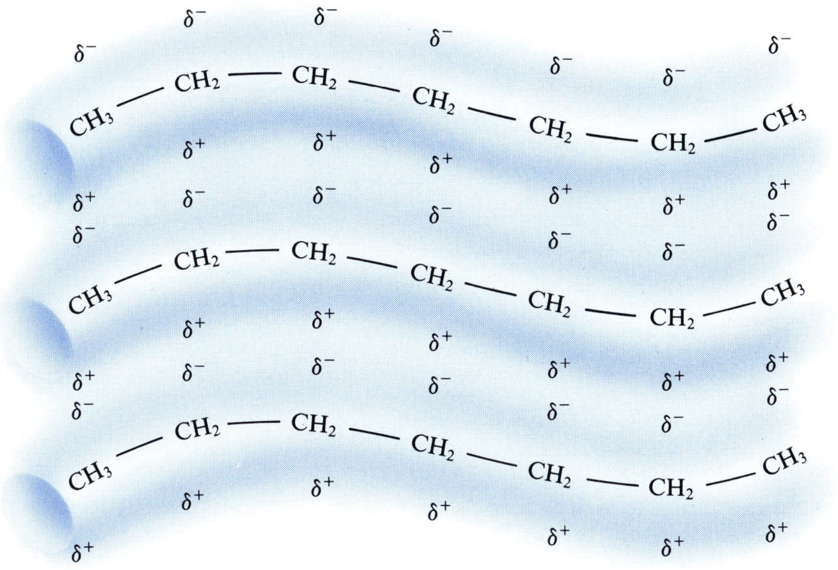

δ⁻
CH₃ — CH₂ — CH₂ — CH₂ — CH₂ — CH₂ — CH₃
δ⁺

FIGURE 15.1
Schematic representation of a hydrocarbon. Because there is little difference in electronegativity between carbon and hydrogen, attractive forces between molecules containing only carbon and hydrogen have a unique origin. Induced dipoles attract these molecules to one another. The electron cloud associated with one molecule is attracted toward the nuclear charge of another. This "instantaneously" changes so that the two molecules change roles; there are no permanent dipoles.

with increasing number of carbon atoms (Figure 15.2). Alkanes with fewer than five carbon atoms are gases at room temperature.

Solubility in Water

In Chapter 6 the importance of hydrogen bonding to the structure and properties of water was discussed. Since hydrocarbon molecules are attracted to one another by London forces and since breaking a hydrogen bond is an energy-losing process, it is not difficult to see why hydrocarbons do not dissolve in water. It is simply not an energetically useful process. For a hydrocarbon molecule to dissolve, it would have to break hydrogen bonds

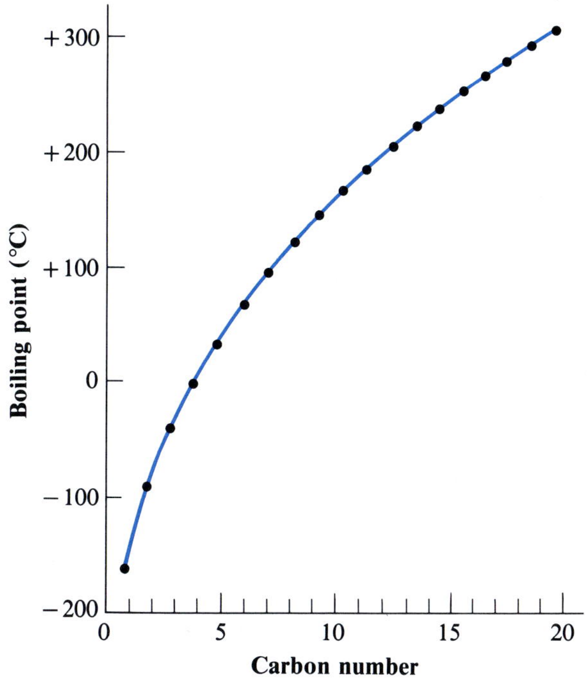

+300
+200
+100
0
−100
−200
Boiling point (°C)
0
5
10
15
20
Carbon number

FIGURE 15.2
The boiling point of hydrocarbon compounds increases as the number of carbon atoms increases. This is because of the mutual attraction of the electrons of one molecule for the nuclei of the other molecule.

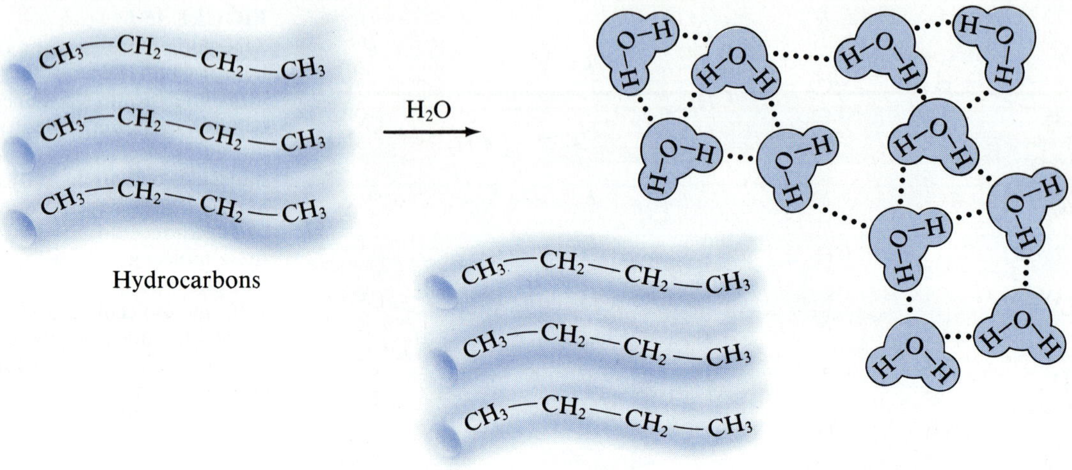
CH3—CH2—CH2—CH3
CH3—CH2—CH2—CH3
CH3—CH2—CH2—CH3
Hydrocarbons
H2O
CH3—CH2—CH2—CH3
CH3—CH2—CH2—CH3
CH3—CH2—CH2—CH3

FIGURE 15.3
Simple energy considerations explain the failure of hydrocarbons to dissolve in water. To dissolve, two things would have to happen: (a) many of the hydrogen bonds holding the water molecules together would have to be broken, and (b) the London forces holding the nonpolar hydrocarbon molecules together would also have to be broken. Since no new bond comparable to hydrogen bonds would form, this would be a net loss of bonding.

in a water lattice as well as lose its own association with other hydrocarbon molecules (Figure 15.3).

The effect of these ideas of solubility will be seen over and over again in both organic chemistry and biochemistry. In Chapter 6 we presented the notion that polar molecules dissolve in polar solvents and that nonpolar molecules dissolve in nonpolar solvents. A generalization that evolves from this can be expressed as "like dissolves like." Thus, nonpolar hydrocarbon compounds dissolve better in other hydrocarbon material, and polar compounds dissolve better in polar materials, such as water. This is why, for example, you can use warm water to clean up water-based acrylic paints, but turpentine (a hydrocarbon) is used to clean up oil paints.

TESTING YOURSELF

Physical Properties

1. Discuss the physical states of Cl_2, Br_2, and I_2 (gas, liquid, and solid) in terms of the London forces attracting the molecules to one another.
2. Imagine that you are stacking logs. You have two kinds of logs to stack: straight logs (1 ft in diameter and 4 ft long) and T-shaped logs (1 ft in diameter, 3 ft long, with a 1-ft branch piece in the center). Which stack better: the piles of straight logs or the piles of T-shaped logs? Apply this same reasoning to molecules.

Answers **1.** The order of size, numbers of electrons and numbers of protons follow the trend: $I_2 > Br_2 > Cl_2$. Thus, the London attractive forces follow the same order. **2.** You would be able to stack the straight logs better. There is more uniform surface contact between logs. The same holds for molecules; "odd shaped" molecules cannot maintain good contact with one another.

15.2 Alkanes and Their Nomenclature

The hydrocarbon family can be divided into four groups: (1) compounds that have single bonds between all carbons, (2) compounds in which single-bonded carbon atoms form a ring, (3) compounds that have one or more double bonds between adjacent carbons, and (4) compounds that have one or more triple bonds between adjacent carbons. This chapter will consider

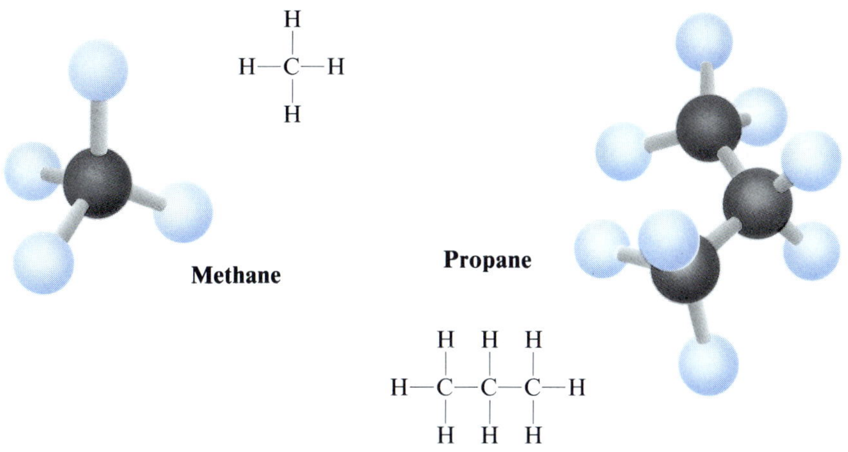
H
H—C—H
H
Methane
Propane
H H H
H—C—C—C—H
H H H

FIGURE 15.4
The alkanes are the simplest hydrocarbons. All carbons have single bonds to adjacent carbon or hydrogen atoms. Examples include methane and propane, both of which are used as heating fuels.

hydrocarbons with single bonds between carbons. Chapter 16 will introduce multiply-bonded carbon compounds.

The family called the **alkanes** contains the simplest hydrocarbons, which are characterized by the formula C_nH_{2n+2}. All carbon atoms have single bonds to other carbon atoms; all other carbon bonds are terminated with hydrogen atoms. Most of the important hydrocarbons have less than ten carbons. Methane and propane, two of the simplest hydrocarbons, are illustrated in Figure 15.4. Methane is the principal component of natural gas and occurs naturally in areas of decaying plants. It is a major component of marsh gas found around stagnant swamps. Microorganisms found in soil, water, and termites produce a significant amount of the methane found in our atmosphere. This could become a significant environmental problem. The only natural method to remove atmospheric methane is by *photooxidation*, a complex process of atmospheric oxidation using light energy.

Alkanes are hydrocarbons containing only single sigma bonds.

The carbon atom in single-bonded alkane hydrocarbons always assumes a structure in which the carbon is at the center of a tetrahedron and the four pairs of shared electrons are at the vertices. The bond angles are about 109°. All alkanes have sp^3 hybridized carbons atoms. Look again at the structures of the methane and propane molecules in Figure 15.4. Notice that each carbon atom in the two molecules is tetrahedral in shape.

The Homologous Series

Hydrocarbons are usually named according to the number of carbon atoms in their longest continuous carbon chain. The number of carbons in the chain provides the stem name, and the suffix for all of the alkanes is always *-ane*. For example, see how the name of this five-carbon alkane relates to its structure:

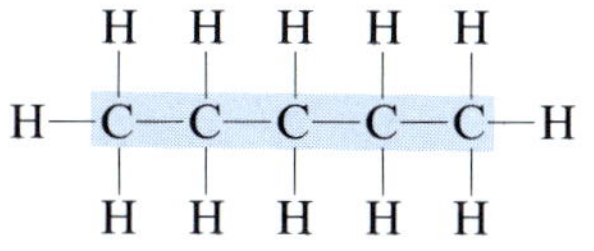
H H H H H
H—C—C—C—C—C—H
H H H H H

Stem name + *ane* = Pent *ane*

Names and formulas for the first ten alkanes are shown in Table 15.1. Most compounds we will study are derived from these ten basic hydrocarbons.

TABLE 15.1
Names and Structural Formulas for Simple Hydrocarbons Having up to Ten Carbons

Number of Carbons	Stem Name	Hydrocarbon Name	Structure
1	Meth	Methane	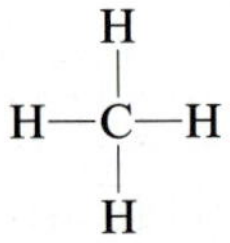
2	Eth	Ethane	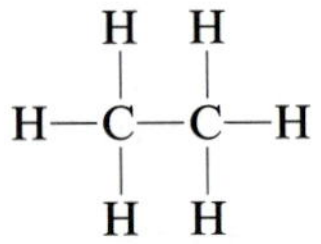
3	Prop	Propane	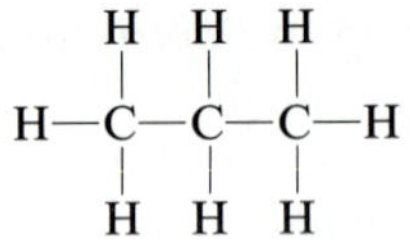
4	But	Butane	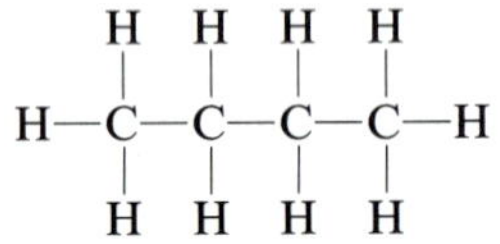
5	Pent	Pentane	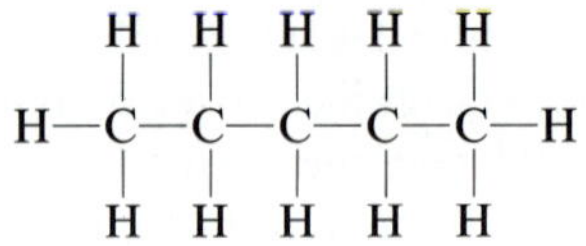
6	Hex	Hexane	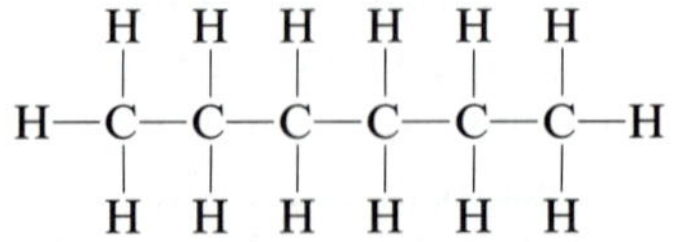
7	Hept	Heptane	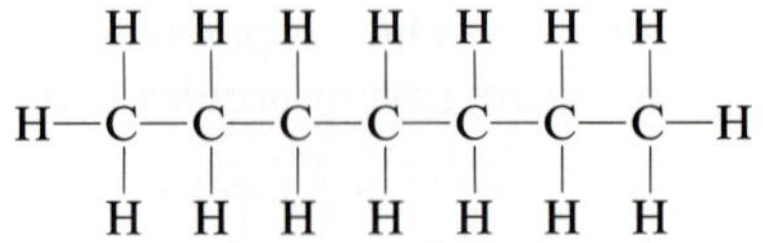
8	Oct	Octane	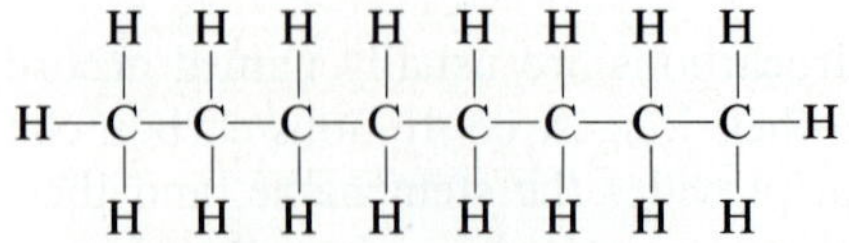
9	Non	Nonane	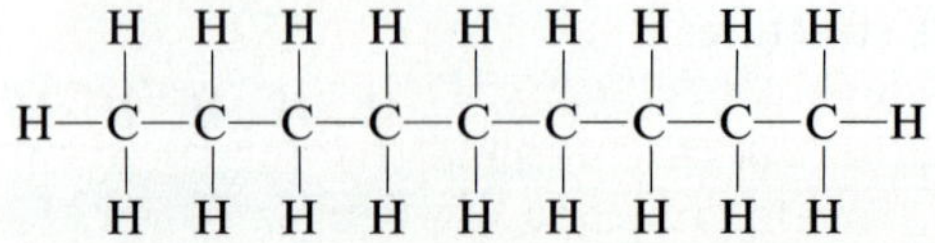
10	Dec	Decane	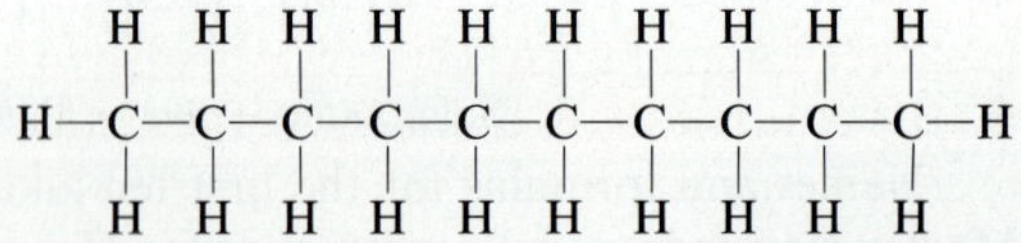

You have seen a "ball-and-stick" model of methane in Figure 15.4. This is a very useful representation of a chemical structure because it shows the shape of the molecule, whereas a **structural formula** shows only the connections of atoms. How do chemists, teachers, and students convey information about bonding in chemical formulas? A number of methods are available, depending on the information needed. Unfortunately, the ball-and-stick models, which are the most accurate, are also the least used because they require artistic drawings or photographs to do justice to the representation. What other ways can we show molecules?

In Figure 15.5 several representations are shown. A **perspective formula** is an attempt to couple the simplicity of a structural formula with the shape of a ball-and-stick model. A **condensed formula** is a way to simplify the structural formula. It is the easiest to write or type, but gives up the three-dimensional aspect. Finally, a **line formula** further simplifies the drawing of chemical structures, but it requires more "mental bookkeeping" because bonds to hydrogen are not shown. Each carbon has four bonds, and if the bond is to hydrogen, it is not shown. Line drawings are used extensively for ring compounds (discussed in a later section of this chapter).

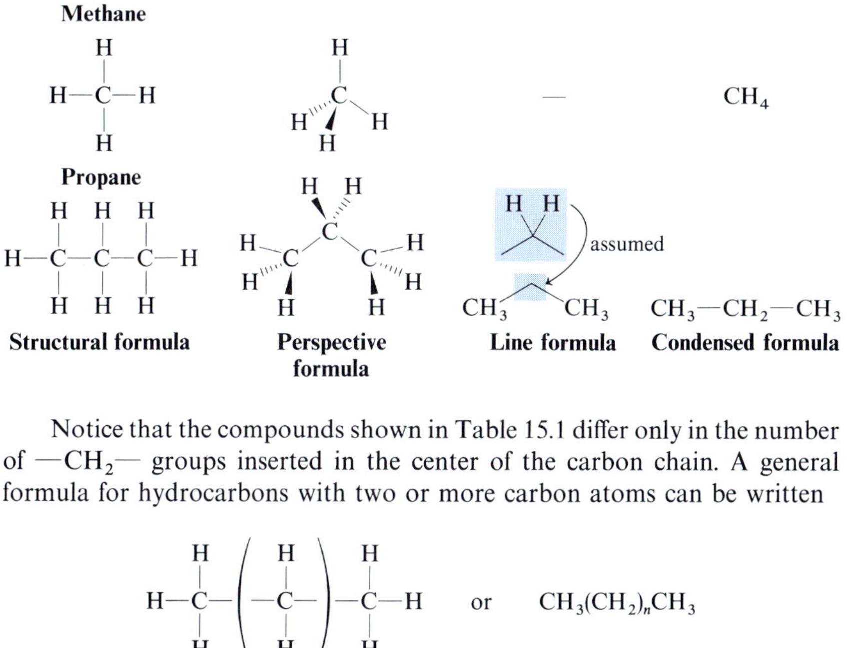
Methane
—
CH_4
Propane
assumed
CH_3 CH_3
$CH_3—CH_2—CH_3$
Structural formula
Perspective formula
Line formula
Condensed formula

FIGURE 15.5
Several methods of representing chemical structures, using two alkanes as examples.

Notice that the compounds shown in Table 15.1 differ only in the number of $—CH_2—$ groups inserted in the center of the carbon chain. A general formula for hydrocarbons with two or more carbon atoms can be written

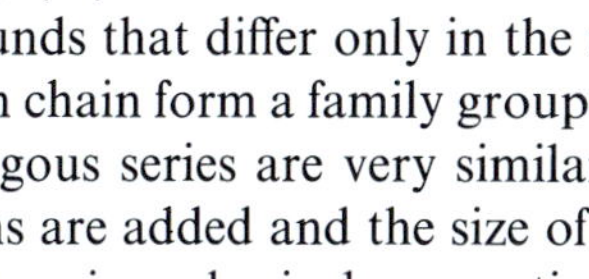

or $CH_3(CH_2)_nCH_3$

where $n = 1, 2, 3, \ldots$. A more general formula for an alkane is

$$C_nH_{2n+2}$$

where $n = 1, 2, 3, \ldots$.

Compounds that differ only in the number of $—CH_2—$ groups inserted in the carbon chain form a family group called a **homologous series**. Members of a homologous series are very similar in their chemical reactivity. But as more carbons are added and the size of the molecules increases, they exhibit gradually changing physical properties. Figure 15.2 showed that there is a regular relationship between the boiling point and the number of carbon atoms in the hydrocarbon chain.

A series of compounds differing only in the insertion of a $—CH_2—$ group is called a *homologous series.*

Isomers

Until now we have assumed that all of a hydrocarbon's carbon atoms are connected in a simple unbranched chain. This need not always be the case. It is possible to assemble several different structures having different bonding patterns with the same set of atoms. Consider the case of the four-carbon hydrocarbon, butane, which can be represented three different ways:

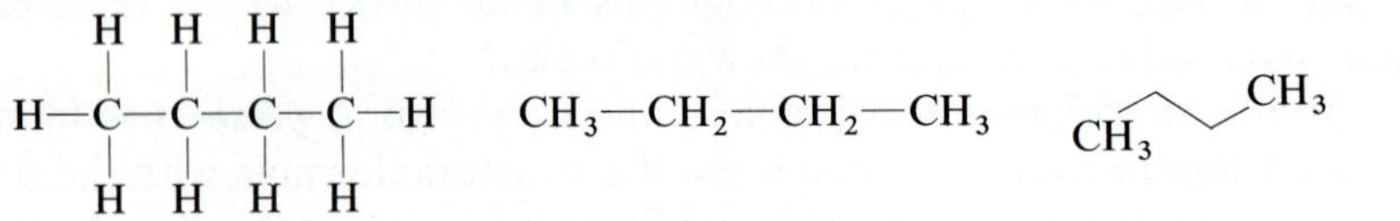

Structural **Condensed** **Line**

If butane contains four carbon atoms, then it must contain ten hydrogen atoms according to the formula

$$C_nH_{2n+2} \quad \text{where } n = 4$$

$$C_4H_{10}$$

Are there other arrangements of these atoms that can be made with this set of atoms? Consider the following four-carbon hydrocarbon:

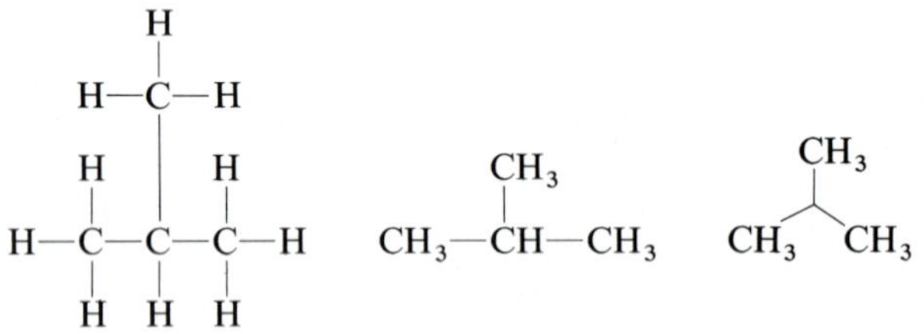

Structural **Condensed** **Line**

Isomers are compounds having the same empirical formula (ratio of atoms) but different molecular formulas. These compounds will differ in shape and properties, and are really different compounds.

The formula of this branched compound is also C_4H_{10}. Molecules formed by different arrangements of the same atoms are called **isomers**. The linear and branched butane four-carbon compounds are both isomers of butane. The linear molecule is called *normal butane.* One accepted name for the second molecule is *isobutane*. The opportunities for isomerization are limited for butane due to its small number of carbon atoms. An alternative structure formed by placing a CH_3 group on top of an end carbon simply yields the same normal butane molecule again.

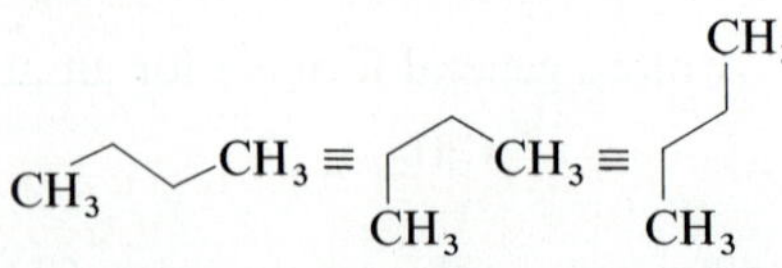

As the number of carbons in a hydrocarbon increases, the number of possible isomers increases very rapidly. In fact, isomerization is not only possible but common if the compound is larger than three carbons. For example, as shown in the figure at the top of page 385, the five-carbon compound pentane has three isomers (a, b, and c). When the number of carbon atoms reaches thirty, the number of isomers amounts to over four billion! Thus, the potential for many different organic compounds is enormous.

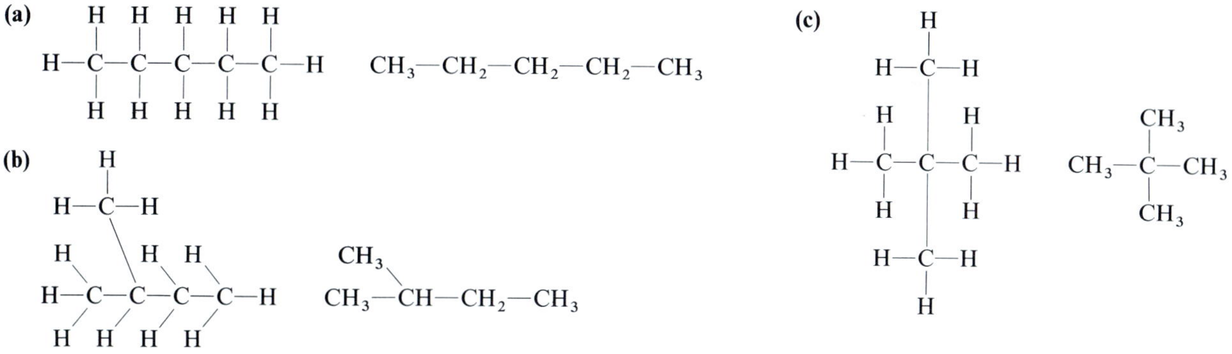
(a)
(b)
(c)

Three isomers of pentane

Because of the importance of London interactions to the boiling point of straight chain hydrocarbons, they will have higher boiling points than branched hydrocarbons (isomers). In the branched hydrocarbons there is less chance for continuous surface contact between molecules because the branched shapes are more "ball-like" and are less able to stack together. The effect of branching on boiling point can be seen in Figure 15.6.

The more branched a compound, the less it is able to interact with a neighboring compound. This will result in lower boiling points for liquids and lower melting points for solids.

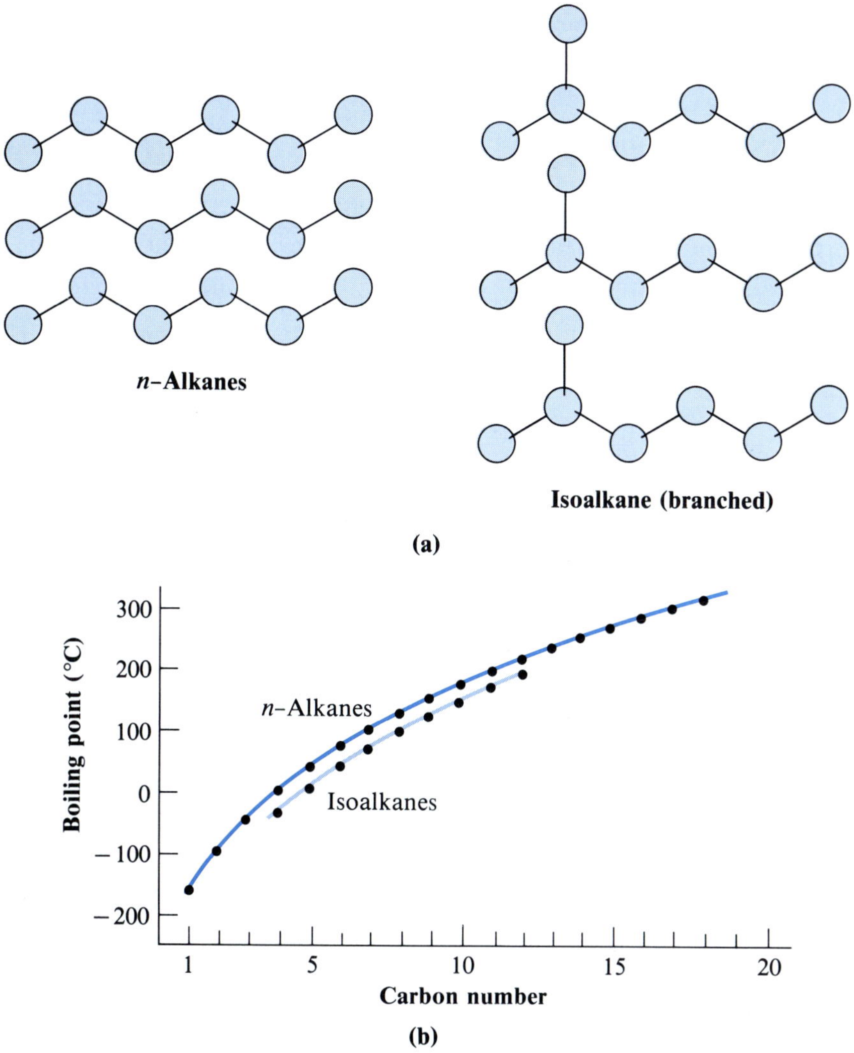
n-Alkanes
Isoalkane (branched)
(a)
Boiling point (°C)
300
200
100
0
– 100
– 200
n-Alkanes
Isoalkanes
1
5
10
15
20
Carbon number
(b)

FIGURE 15.6
(a) Because of their stackable shapes, linear hydrocarbon molecules (such as *n*-alkane) interact with one another better than branched hydrocarbons (such as isoalkane). (b) The branched hydrocarbons thus have decreased London forces between molecules and resulting lower boiling points.

NEW TERMS

alkane A family of hydrocarbons having only single carbon to carbon bonds characterized by the general formula C_nH_{2n+2}.

structural formula A representation of a structure that emphasizes the bond connection between atoms.

perspective formula A representation of a chemical structure that conveys the three dimensions of a ball-and-stick model and has some of the simplicity of the structural formula.

condensed formula A condensed representation of a chemical structure that leaves out the vertical bonds and shows the whole structure set on one line, such as $CH_3CH_2CH_3$.

line formula A simplified representation of a structural formula in which many of the C—H bonds are not shown.

homologous series A family of compounds differing only by the number of CH_2 groups in the formula. The series is represented by the general formula C_nH_{2n+2}.

isomer Variations of a particular compound having the same molecular formula but different arrangements of atoms and bonds.

TESTING YOURSELF

Chemical Structures

1. Draw the perspective formula for ethane.
2. Convert the structural formula of the following compound into a line structure.

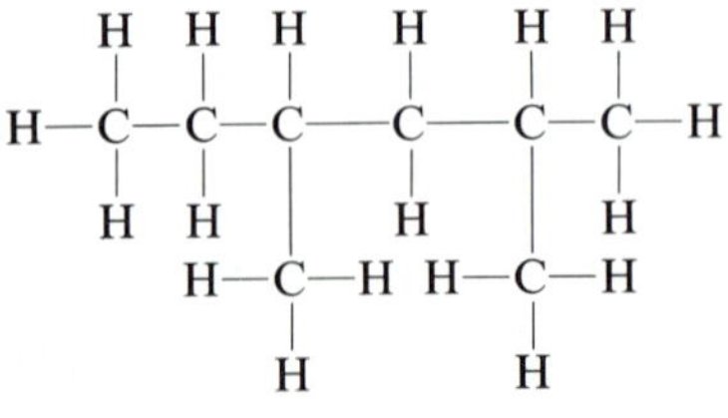

3. Convert this line structure to a condensed structural formula.

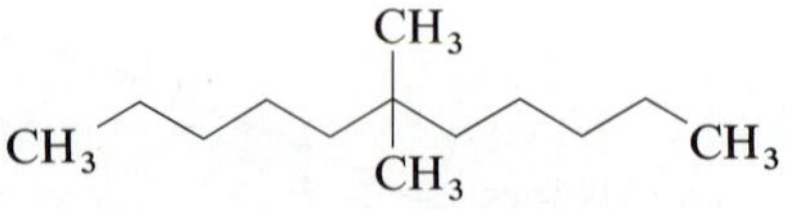

Answers **1.**

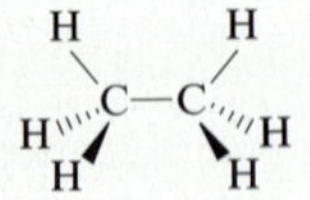

2.

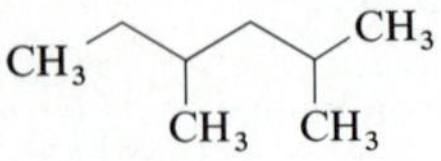

3.

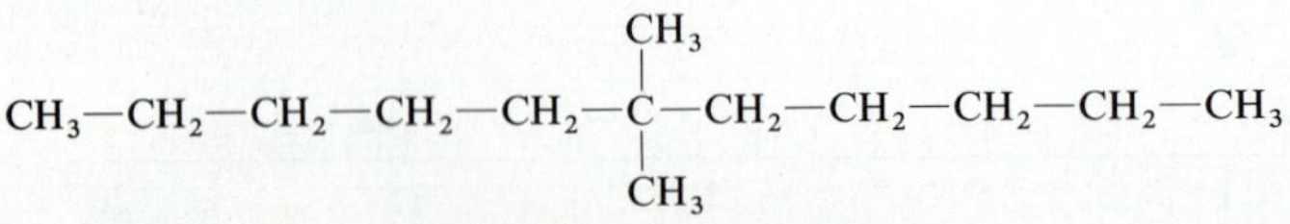

or

$CH_3CH_2CH_2CH_2CH_2C(CH_3)_2CH_2CH_2CH_2CH_2CH_3$

15.3 Alkyl Groups and Nomenclature

Simple Alkyl Groups

Because so many potential isomers exist, there must be a logical system for naming them. The solution to the problem of naming isomers lies in the use of the alkyl group. As we showed in the last chapter, an **alkyl group** is a hydrocarbon molecule minus a hydrogen. In place of the hydrogen is an "open bond" that can be attached to other carbon chains. Alkyl groups carry the stem name of their parent alkane, but end with the suffix *-yl* instead of *-ane*. We can see in Table 15.2 that the alkyl group is directly related to the parent hydrocarbon alkane. From the names of the compounds in the homologous series, we know the names of the corresponding alkyl groups.

Starting with propane, we find it possible to make isomeric alkyl groups. Depending on the hydrogen removed, we can make different alkyl groups.

TABLE 15.2
Names and Formulas for Some Common Alkyl Groups

Number of Carbons	Stem Name	Alkyl Group Name	Structure	Ball-and-Stick Model
1	Meth	Methyl	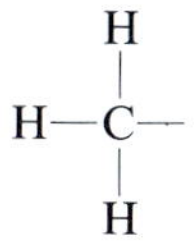	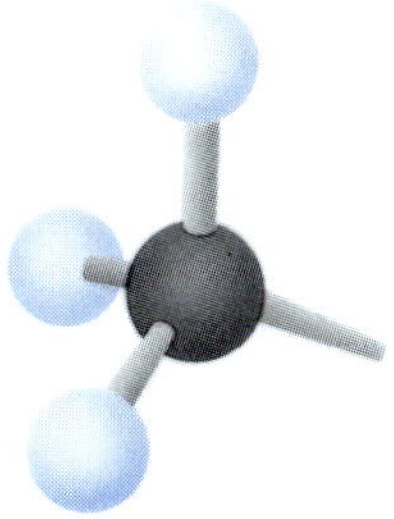
2	Eth	Ethyl	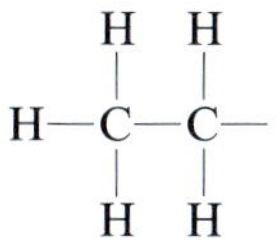	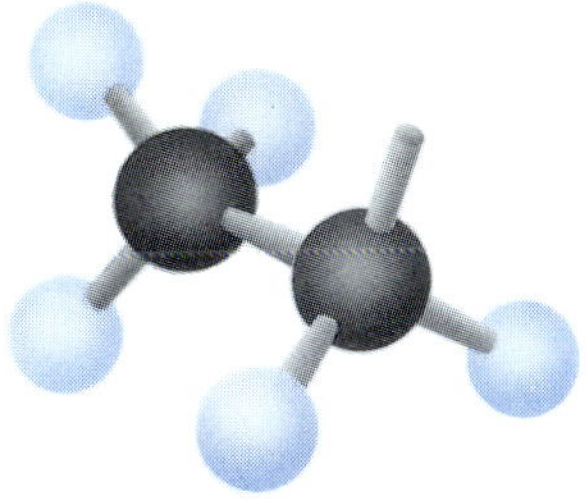
3	Prop	Propyl	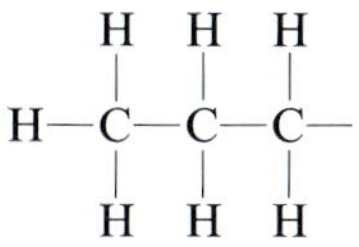	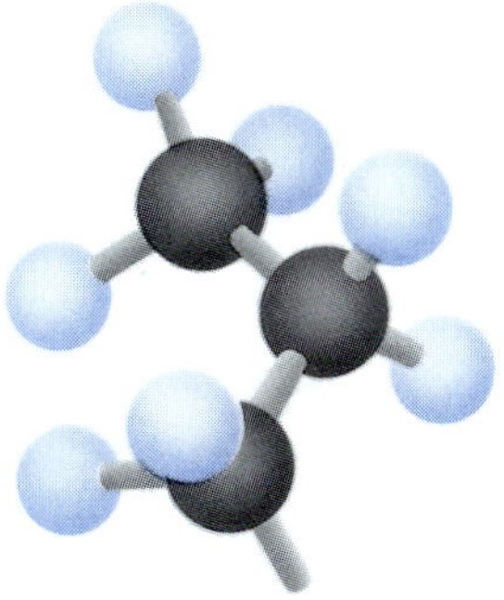

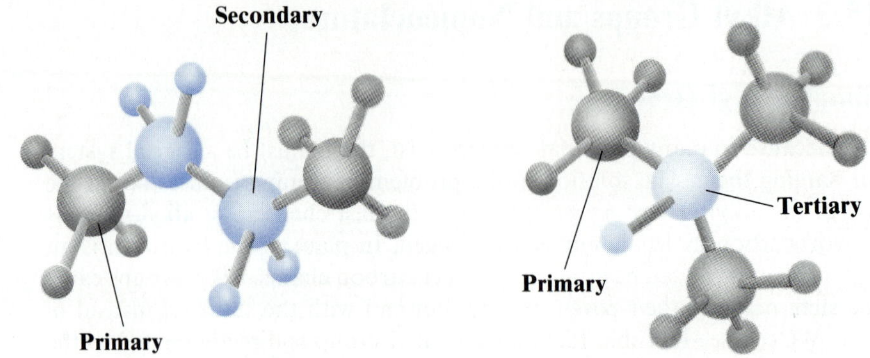
Secondary
Primary
Tertiary
Primary

FIGURE 15.7
Carbon atoms are classified as primary, secondary, or tertiary according to the number of other carbon atoms to which they are bonded, as shown in these two isomers of butane.

Primary, Secondary, and Tertiary Carbons

Carbons of an alkyl group are designated as **primary carbons** if they connect to only one additional carbon. Primary carbons are on the end of a chain. **Secondary carbons** bond to two adjacent carbons, and **tertiary carbons** bond to three adjacent carbons. The carbons in the two isomers of butane shown in Figure 15.7 are indicated as primary, secondary, or tertiary. These are abbreviated *p*, *s*, and *t*, respectively.

Hydrogens atoms are classified according to the type of carbon to which they attached and are given equivalent names. They are designated as *primary* if they connect to a primary or an end carbon in any hydrocarbon chain, as *secondary* if they connect to a secondary carbon, or as *tertiary* if they connect to a tertiary carbon. In the following propane structure, the highlighted hydrogens are primary and all other hydrogens are secondary.

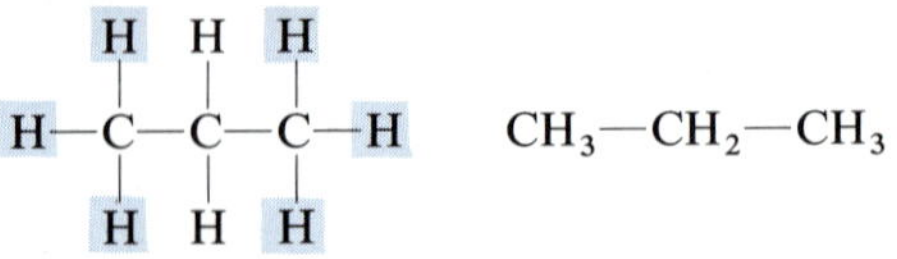

Propane

Suppose now that one of the primary hydrogens is removed to form the *normal propyl* group:

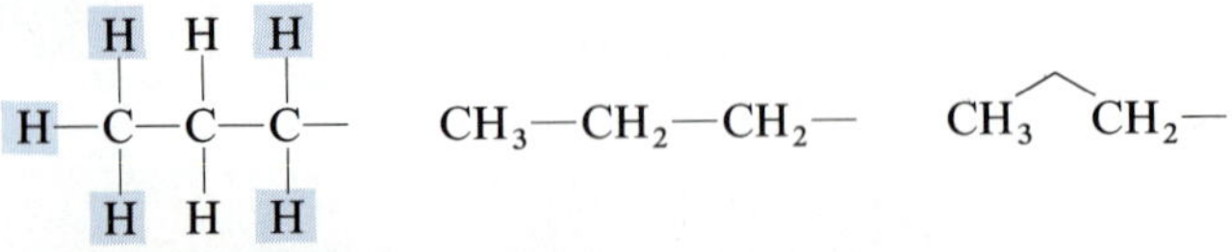

***n*-Propyl group**

The *n* in *n*-propyl stands for *normal.* If one of the secondary hydrogens is removed (the only other possibility for propane), the *isopropyl* group results.

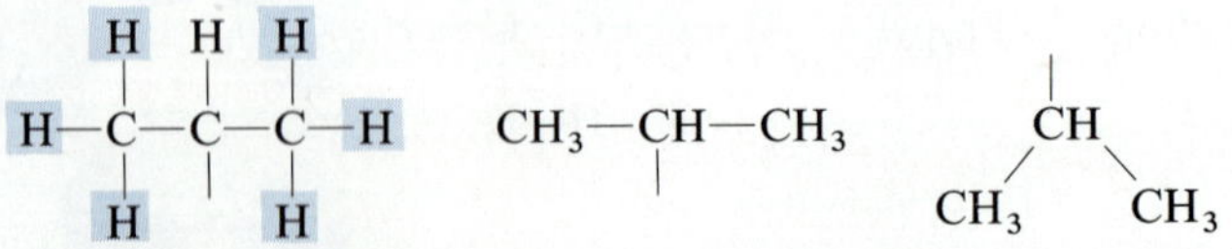

Isopropyl group

The situation is similar for butane. If one of the primary hydrogens is removed from butane, the *normal butyl* (or *n*-butyl) group results.

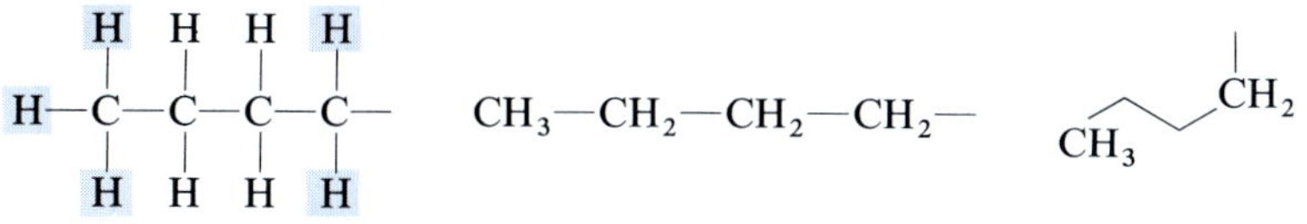

***n*-Butyl group**

Now recall that butane has two isomers, normal butane and isobutane. When one of the secondary hydrogens of *n*-butane is removed, the name isobutyl cannot be used because isobutane is a different compound. Instead, the resulting alkyl group is known as the *secondary butyl* (or *sec*-butyl) group.

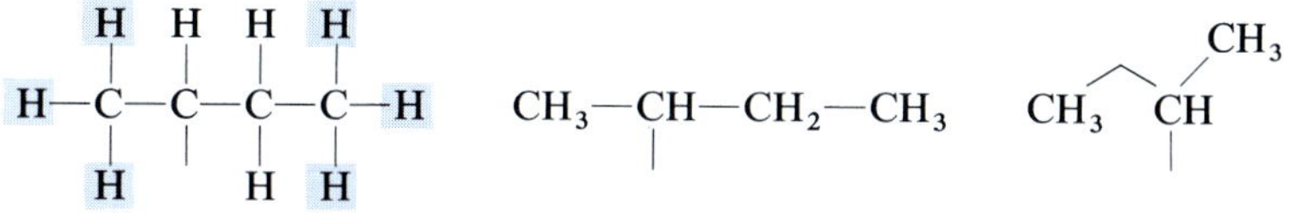

***sec*-Butyl group**

Consider now the case of isobutane:

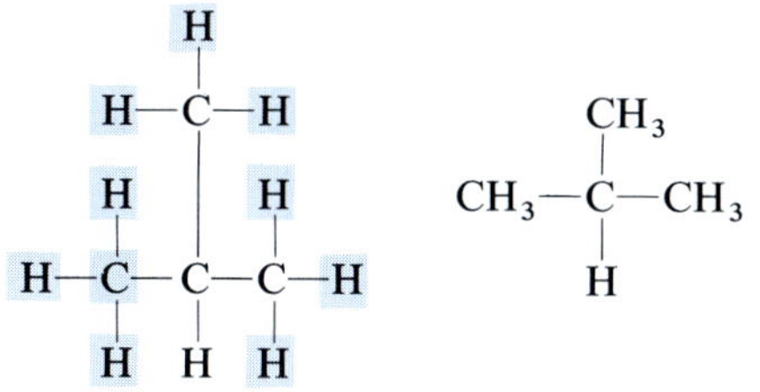

All of the hydrogen atoms in this compound are primary, except one, which is connected to a tertiary carbon. (It is the only hydrogen that is not highlighted.) If any primary hydrogen is removed, the *isobutyl* group results. This is the alkyl group (isobut*yl*) obtained from the alkane (isobut*ane*) by removal of a primary hydrogen.

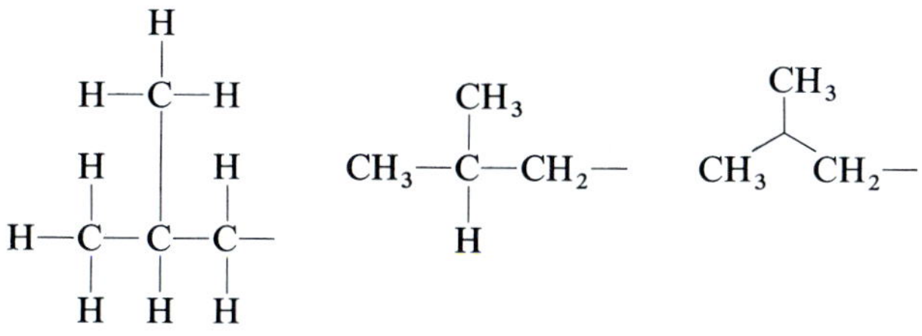

Isobutyl group

If, on the other hand, the sole tertiary hydrogen is removed, the *tertiary butyl* (or *tert*-butyl) group results.

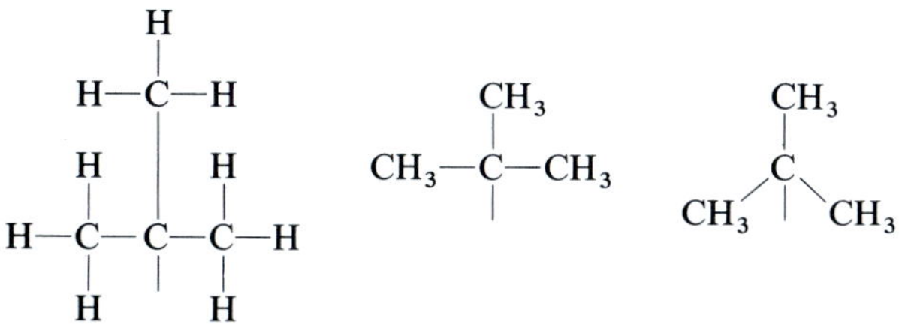

***tert*-Butyl group**

Alkyl groups that can be derived from the isomers of butane are summarized in Table 15.3.

TABLE 15.3
Isomers of Butane and Corresponding Alkyl Groups

Alkane	Alkyl Group	Molecular Structure
$CH_3-CH_2-CH_3$ Propane	$CH_3-CH_2-CH_2-$ *n*-Propyl group $CH_3-\underset{\vert}{C}H-CH_3$ Isopropyl group	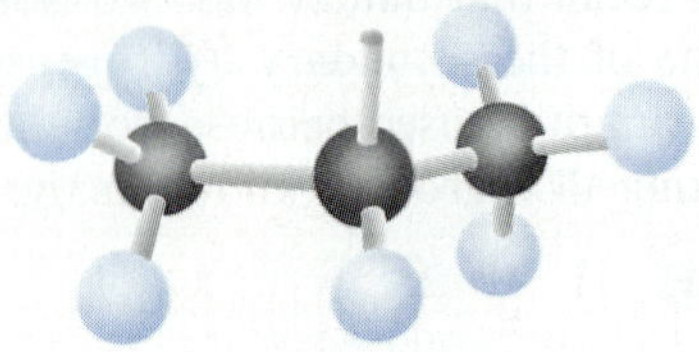
$CH_3-CH_2-CH_2-CH_3$ *n*-Butane	$CH_3-CH_2-CH_2-CH_2-$ *n*-Butyl group	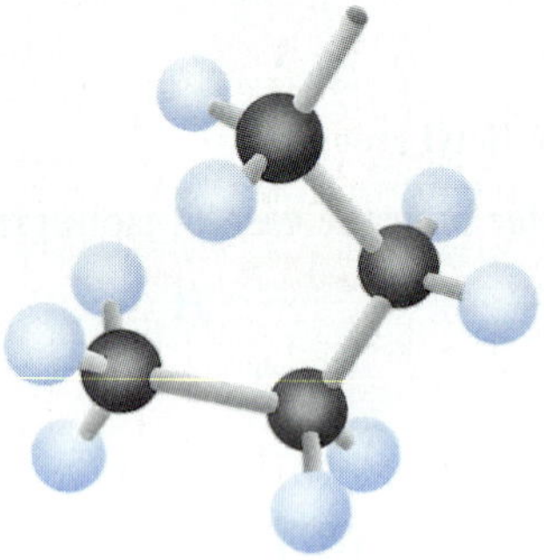
	$CH_3-\underset{\vert}{C}H-CH_2-CH_3$ *sec*-Butyl group	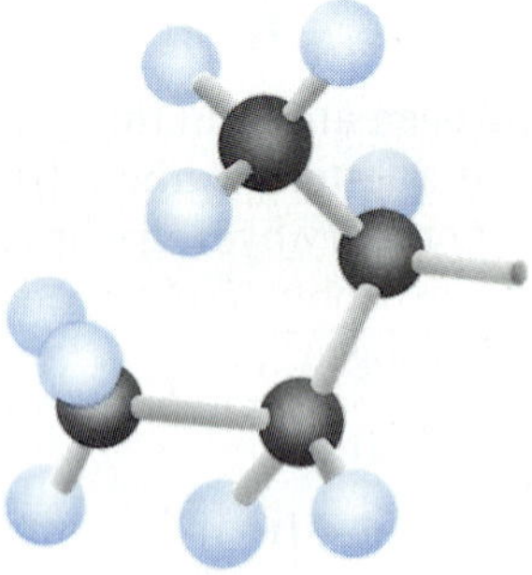
$CH_3-\overset{\overset{\displaystyle CH_3}{\vert}}{\underset{\underset{\displaystyle H}{\vert}}{C}}-CH_3$ Isobutane	$CH_3-\overset{\overset{\displaystyle CH_3}{\vert}}{\underset{\underset{\displaystyle H}{\vert}}{C}}-CH_2-$ Isobutyl group	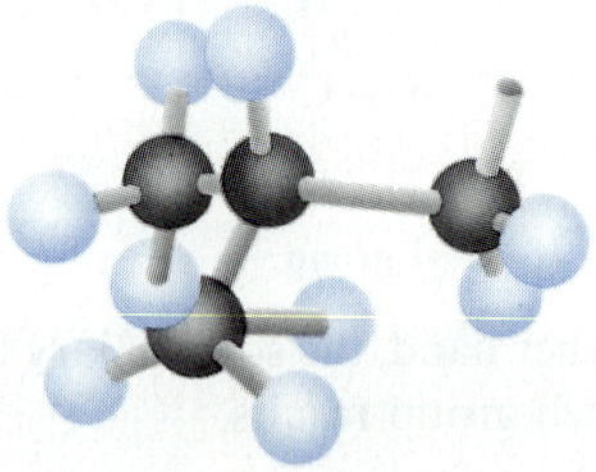
	$CH_3-\overset{\overset{\displaystyle CH_3}{\vert}}{\underset{\vert}{C}}-CH_3$ *tert*-Butyl group	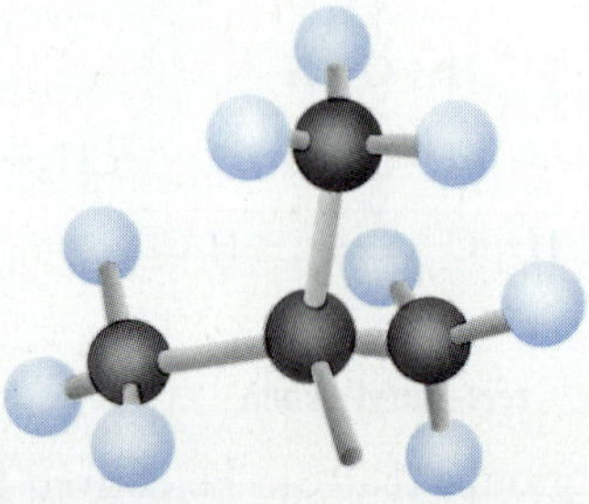

General Rules of Nomenclature

Branched carbon skeleton compounds have alkyl groups substituted for hydrogens on one or more of the carbons of the parent straight-chain compound. The resulting compounds are named by a system developed by the International Union of Pure and Applied Chemistry (IUPAC) in 1949. The IUPAC is a group through which chemists worldwide agree upon uniform ways of handling terms and communicating within their profession. The IUPAC system for naming hydrocarbons involves several simple steps. These steps will be illustrated by naming the two relatively simple carbon compounds shown below.

In the late 1800s the need for a unified system of nomenclature became apparent. By 1949 the international group IUPAC was formed and the basic rules were formulated, which are continuously updated.

EXAMPLE 15.1

Name this hydrocarbon using the IUPAC system.

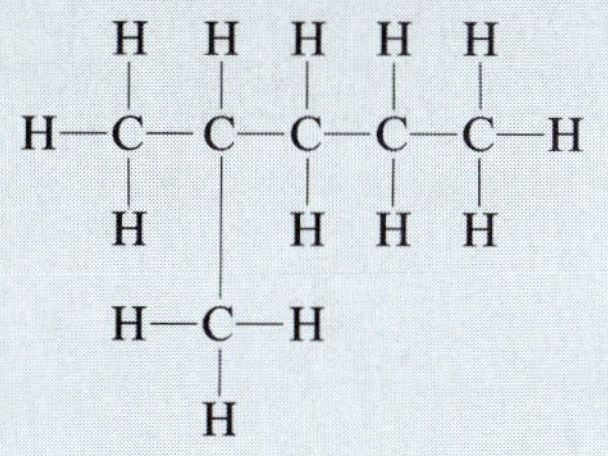

$$CH_3—\underset{\displaystyle CH_3}{\underset{|}{CH}}—CH_2—CH_2—CH_3$$

a. Find the longest *continuous* carbon chain. Number each carbon in the central chain, beginning at the end closest to the first substituent. The stem name is derived from the name of the parent alkane.

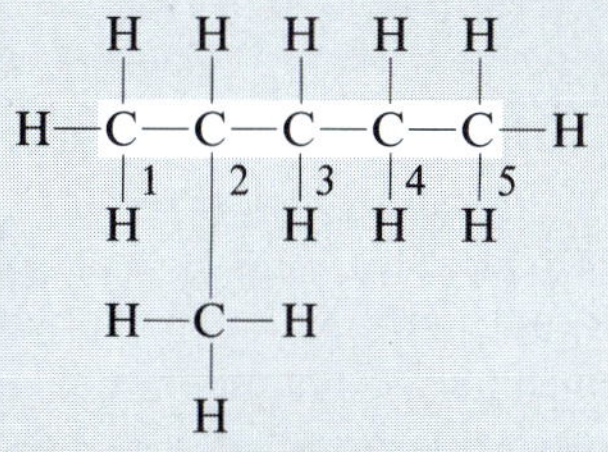

The parent compound is a 5-carbon *pentane*.

b. Find each alkyl group side chain. Assign it a name and indicate to which carbon it is attached.

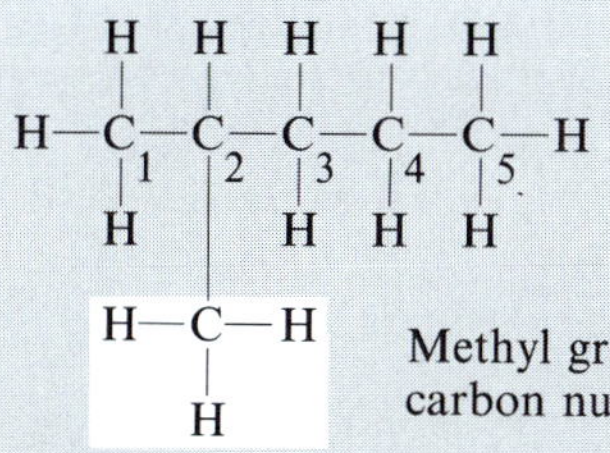

Methyl group, attached to carbon number 2.

c. Number and name the substituents. If the same substituent occurs more than once, the number of each carbon of the parent alkane to which it is attached is given, and the number of substituent groups involved is indicated by a prefix such as *di-*, *tri-*, *tetra-*, and so on.

The compound has a methyl group attached to carbon number 2. The substituent will be called 2-methyl.

d. Name the compound, beginning with the side chains in alphabetical order and ending with the name of the parent compound. Follow these rules:
   a. *Always* use commas (,) between numbers.
   b. *Always* use dashes (-) between numbers and words.
   c. Do not leave spaces in the name.
   d. When alphabetizing substituent groups, ignore the prefixes *di-*, *tri-*, and so on.
   e. Use the lowest possible numbers.

The compound is thus named 2-methylpentane.

EXAMPLE 15.2

Name this branched hydrocarbon using the IUPAC system.

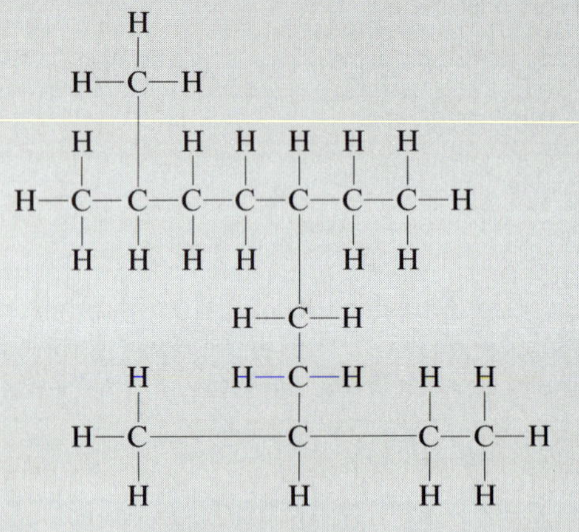

a. Find the longest *continuous* chain. Number each carbon in the central chain, beginning at the end closest to the first substituent. Name the parent compound as a normal alkane. Note that the longest chain need not be straight on the drawing. The stem name of the compound is derived from the name of the parent alkane.

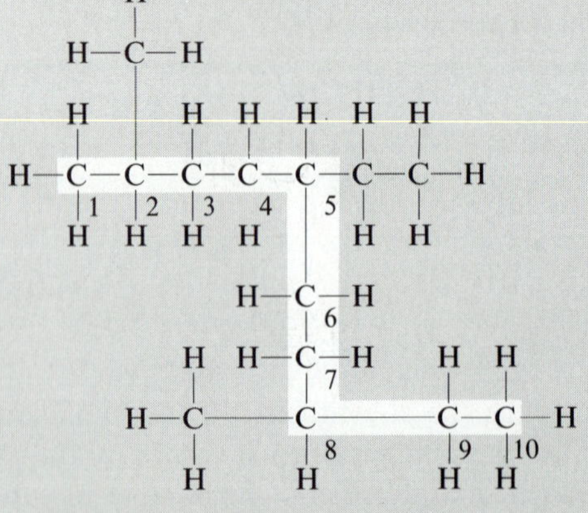

The longest continuous carbon chain in this compound is ten carbons in length; the parent compound is thus a *decane*. This chain is highlighted in the drawing. Identification of the longest chain requires care.

b. Find each alkyl group side chain. Assign it a name and indicate to which carbon it is attached.

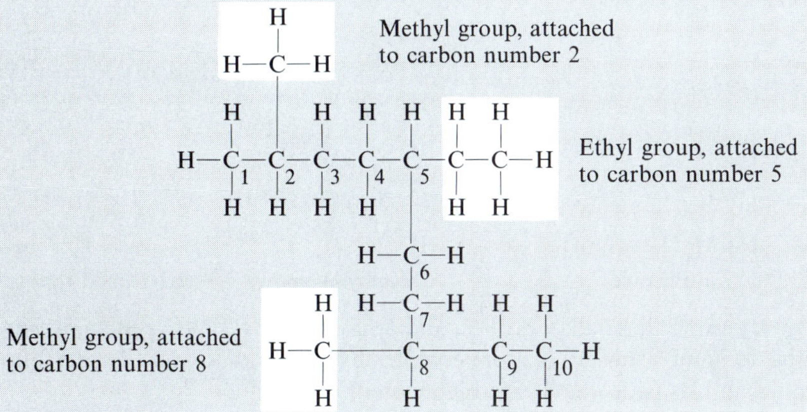
Methyl group, attached to carbon number 2
Ethyl group, attached to carbon number 5
Methyl group, attached to carbon number 8

c. Number and name the substituents.

This compound has methyl groups attached to carbon numbers 2 and 8, and an ethyl group attached to carbon number 5. The substituents will be called

2,8-dimethyl and 5-ethyl

d. Name the compound, beginning with the side chains in alphabetical order and ending with the name of the central chain.

Since *e* (ethyl) comes before *m* (methyl) in the alphabet, the compound will be named 5-ethyl-2,8-dimethyldecane (Figure 15.8).

Recall that prefixes are not used when alphabetizing substituent groups.

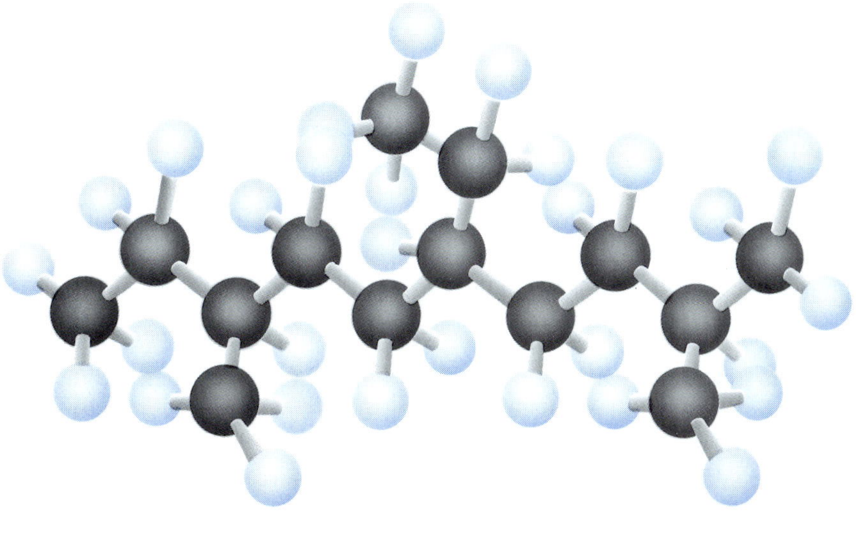

FIGURE 15.8
A ball-and-stick model of 5-ethyl-2,8-dimethyldecane.

The same process as shown in Examples 15.1 and 15.2 can be used to name the isomers of pentane. Note that each compound has the formula C_5H_{12}.

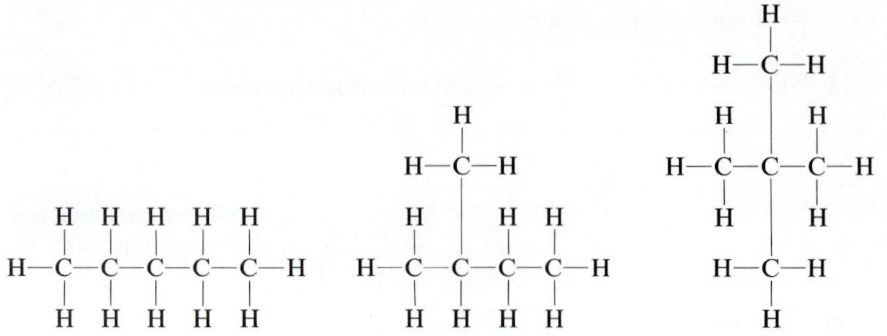

***n*-Pentane** **2-Methylbutane** **2,2-Dimethylpropane**

By using these alkyl groups, and applying the IUPAC system, it is possible to name a variety of more complex hydrocarbons. Consider the following examples (side chains are highlighted):

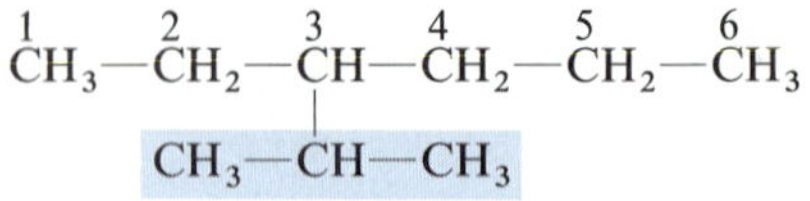

Long chain: hexane (six carbons)
Side chain: isopropyl group
Name: 3-isopropylhexane

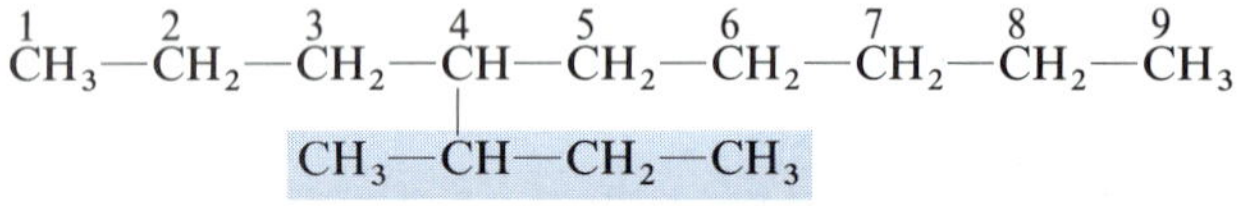

Long chain: nonane (nine carbons)
Side chain: *sec*-butyl group
Name: 4-*sec*-butylnonane

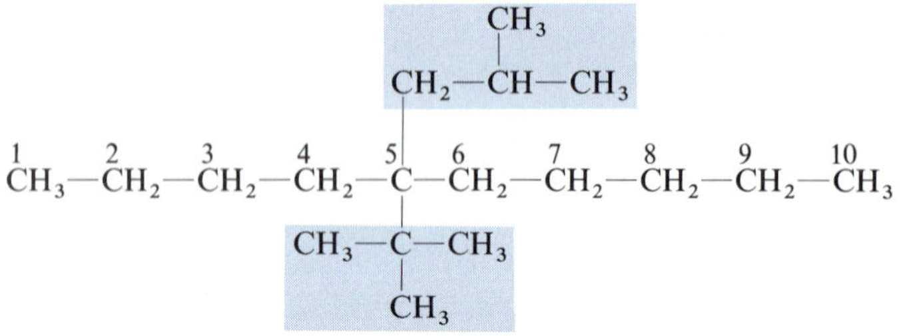

Long chain: decane (ten carbons)
Side chains: isobutyl group and *tert*-butyl group
Name: 5-isobutyl-5-*tert*-butyldecane

NEW TERMS

alkyl group A hydrocarbon group made up of a hydrocarbon minus one of its hydrogen atoms. This group is named from the alkane by replacing the *-ane* ending with *-yl*

primary carbon A carbon atom that is bonded to only one other carbon.

secondary carbon A carbon atom that is bonded to two other carbons.

tertiary carbon A carbon atom that is bonded to three other carbons.

TESTING YOURSELF

Naming Alkanes Using Alkyl Group Nomenclature

1. Give names for each of the following compounds.

a. 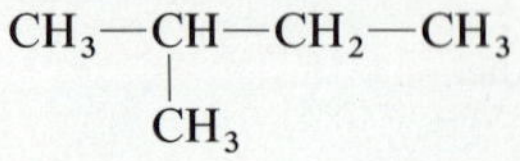

b. 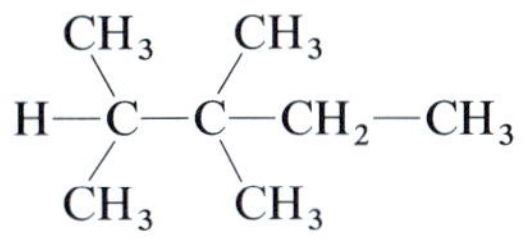

c. 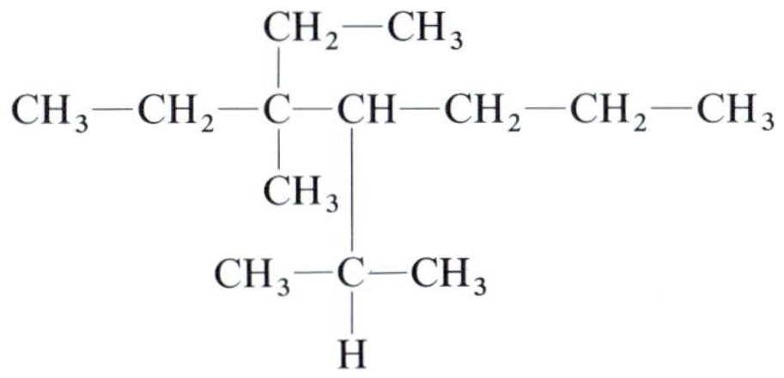

2. Draw structures for each of the following compounds.
   a. 4-ethyl-2-methylheptane
   b. 3,3,5-trimethyloctane
   c. 4-isopropyl-4-*tert*-butyloctane

Answers **1. a.** 2-methylbutane **b.** 2,3,3-trimethylpentane **c.** 3-ethyl-4-isopropyl-3-methylheptane

2. a. 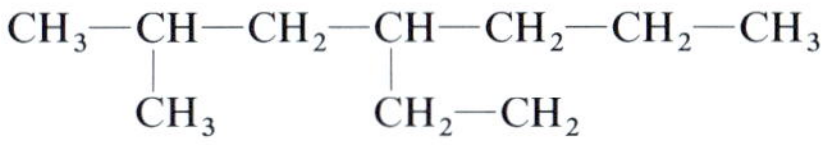

b. 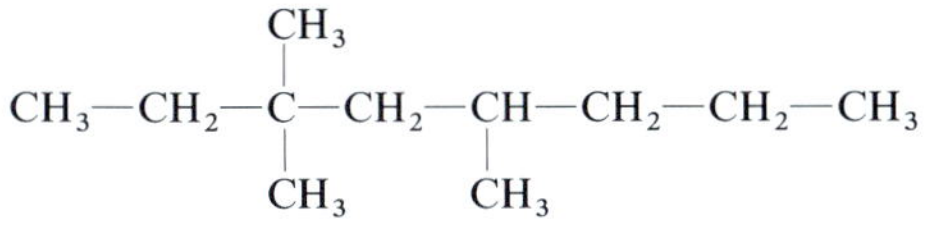

c. 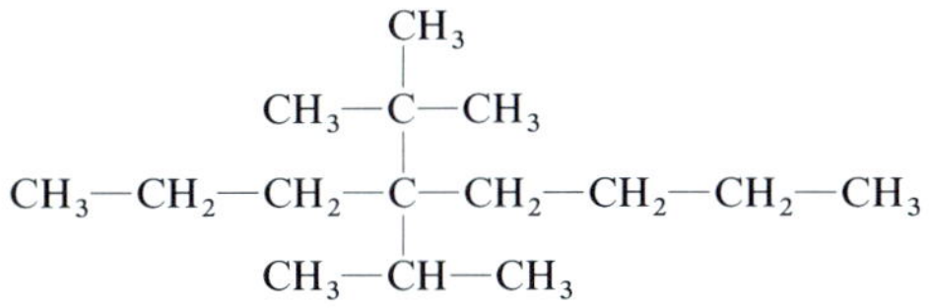

15.4 Cycloalkanes

So far in this chapter we have considered straight chain hydrocarbons as well as branched hydrocarbons having attached alkyl group side chains. Because of the tetrahedral shape of the single-bonded carbon atom, these are really "zigzag" in shape. A third possibility exists. Under proper conditions, alkane molecules can circle around to join head and tail, forming a ring. Compounds containing rings are quite common in nature.

Cycloalkanes are named according to the number of carbons in their ring in the same manner that normal alkanes are named according to the number of carbons in their longest continuous chain. Cyclic compounds have the prefix *cyclo-* preceding their name. A three-carbon cycloalkane, for example, is cyclopropane (Figure 15.9). A ten-carbon cycloalkane is cyclodecane.

Cycloalkanes can have as few as three members or can become very large.

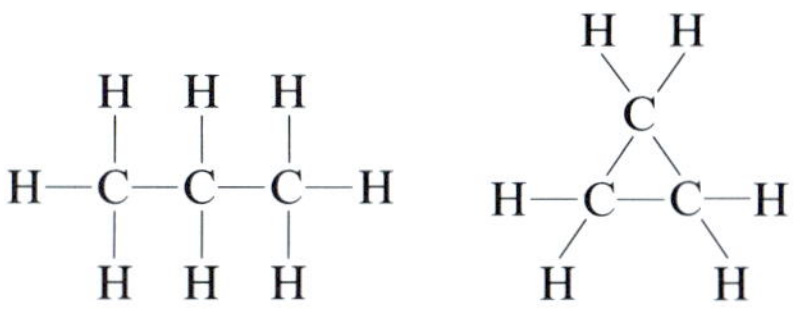

Propane **Cyclopropane**

FIGURE 15.9
In the cycloalkane family, the skeletal carbon framework forms a ring.

Structure

Early thoughts about ring compounds were based on the assumption that they were planar compounds. From this assumption one had only to consider the amount of deviation from the stable 109° tetrahedral shape to understand the strain built into a ring compound. Cycloalkanes have some special structural problems, since in the case of three- and four-membered rings, the 109° bond angle characteristic of the tetrahedral carbon cannot be achieved. In cyclopropane, for example, the three carbons form an equi-

TABLE 15.4
Some Common Cycloalkanes

Name	Required Bond Angle	Actual Bond Angle	Planar Structure	Ball-and-Stick Model
Cyclopropane	60°	60°		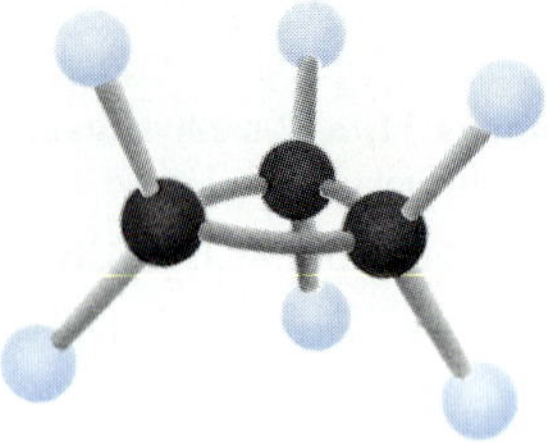
Cyclobutane	90°	88°		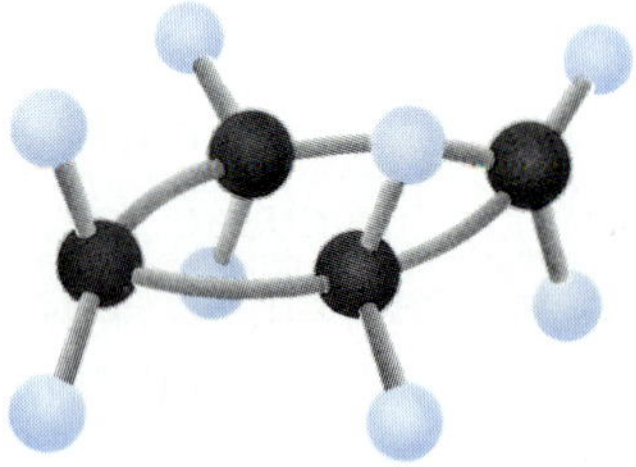
Cyclopentane	108°	105°	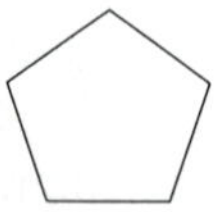	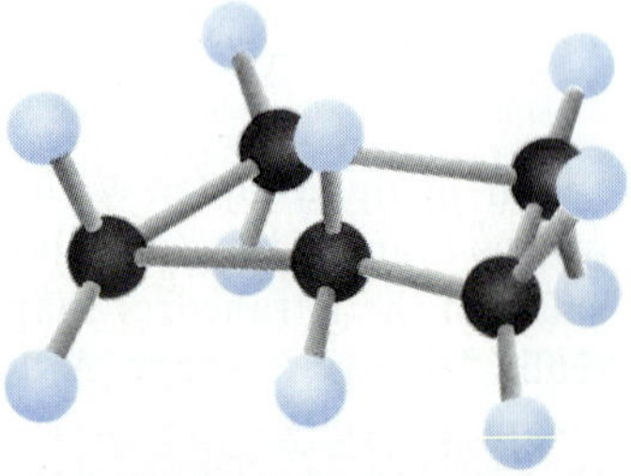
Cyclohexane	120°	109°	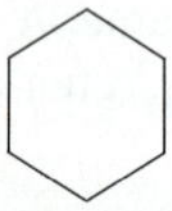	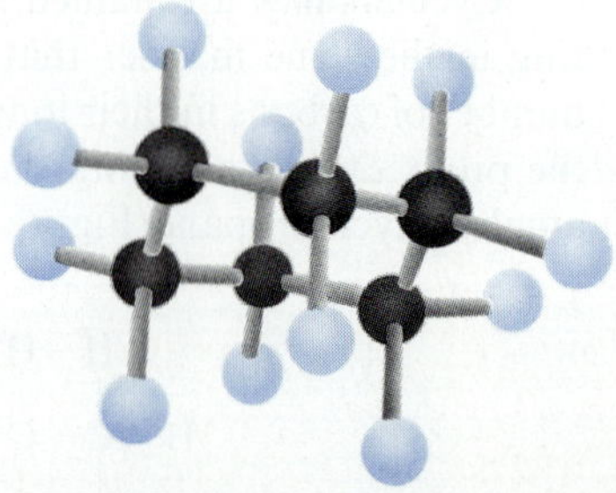

lateral triangle in which the bond angles are forced to be 60° (Table 15.4). As you might expect, the cyclopropane ring is highly strained. Larger cycloalkanes are less strained and are therefore more stable. Indeed, by applying this logic, you could predict that cyclopentane would be the most stable ring because its actual bond angle is very close to its required bond angle (Table 15.4). However, this is not observed.

The reason that this simple theory does not work is because, aside from cyclopropane, none of the rings are planar. Table 15.4 shows the planar structures of several of the common rings, as well as more realistic ball-and-stick models.

The cycloalkanes lend themselves well to a shorthand method of representation based on combining the geometric shape of the ring with a line drawing. This **geometric structure** can be shown by an example.

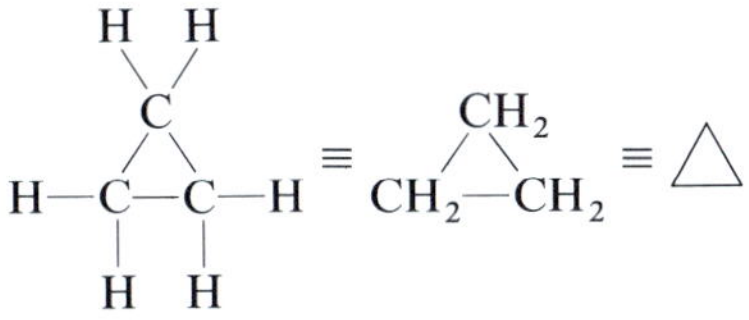
H H
C
H—C—C—H
H H
≡
CH₂
CH₂—CH₂
≡

When this shorthand method is used, it is understood that there is a four-bonded carbon atom at each vertex of the figure and that each carbon has two hydrogens as well as two bonds to adjacent carbon atoms.

Cycloalkanes can be represented by the general formula C_nH_{2n} where $n = 3, 4, 5, 6, \ldots$. Cyclopentane, for example, is C_5H_{10}.

Nomenclature

When atoms other than hydrogen or alkyl groups are substituted for hydrogens in a cycloalkane, they have a well-defined position relative to other groups on the ring. If two groups are on the same side of a ring, they are said to be **cis** to each other. If they are on opposite sides of the ring, they are said to be **trans**. Substituted cycloalkanes can thus exist as cis or as trans isomers of the same molecular formula.

Nomenclature of cycloalkanes includes cis and trans designations, where cis refers to groups on the same side and trans to groups on opposite sides.

Figure 15.10 shows two isomers of 1,3-dimethylcyclobutane. In Figure 15.10a, the two substituted methyl groups are on the same side of the molecule; this is the cis form of the molecule and is called *cis*-1,3-dimethylcyclobutane. In Figure 15.10b the methyl groups are on opposite sides of the molecule; this is the trans form and is called *trans*-1,3-dimethylcyclobutane.

A chemical shorthand is commonly used to indicate the position of substituents on a molecule. Bonds to groups "above" the ring are shown as solid wedges (◄), while bonds to groups "below" the ring are shown as dashed lines (⋯). This is shown in Figure 15.10.

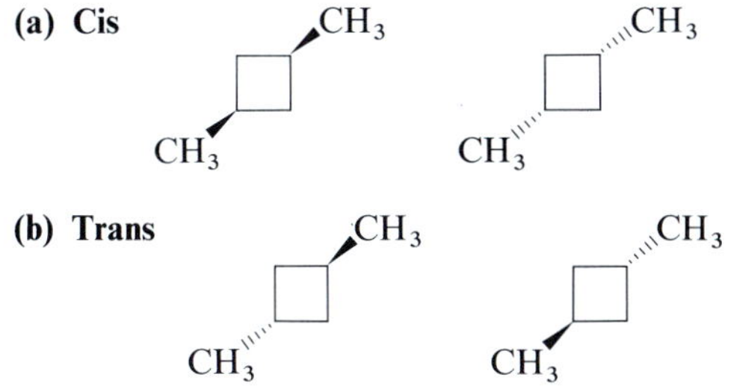
(a) Cis
CH₃
CH₃
CH₃
CH₃
(b) Trans
CH₃
CH₃
CH₃
CH₃

FIGURE 15.10
Two isomers of dimethylcyclobutane. (a) When groups are on the same side of the ring, the isomer is called a cis isomer. (b) When groups are on opposite sides of the ring, the isomer is called a trans isomer.

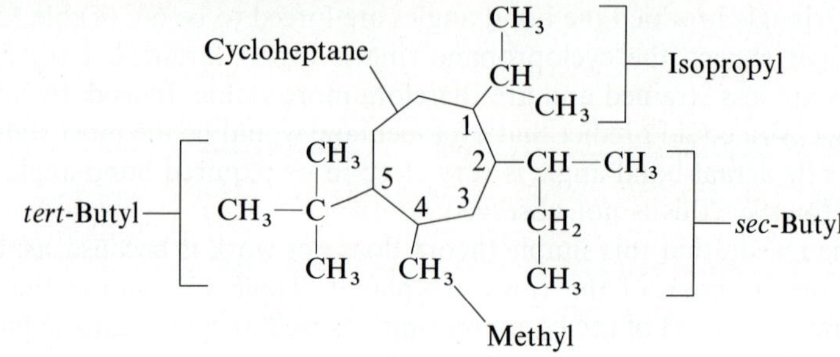
Cycloheptane
Isopropyl
tert-Butyl
sec-Butyl
Methyl

FIGURE 15.11
An example of nomenclature of a cycloalkane having a variety of alkyl substitutions.

The steps for naming substituted cycloalkanes are straightforward. When the cis and trans relationship of substituent groups is known, it is included in the name. Consider the following cyclobutanes:

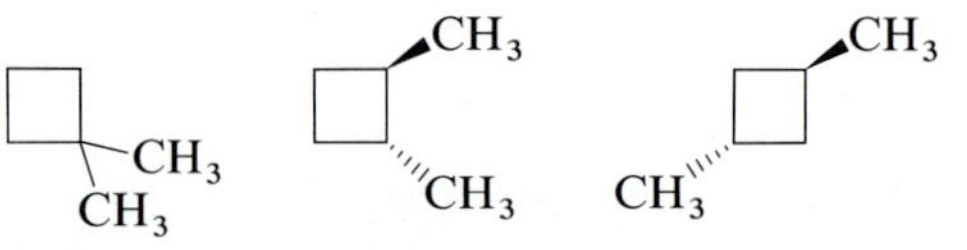

(a) **(b)** **(c)**

To name these cycloalkanes, first determine the parent hydrocarbon. Since all three of the compounds have four carbons in their ring, they are therefore cyclobutanes.

Next, you need to determine the alkyl groups and give each a name and number. Since there is no beginning or end of a ring, you start the numbering in such a way as to create the lowest possible numbers for the substituents. Compound (a) has two methyl groups on the first carbon. Thus, its name is 1,1-dimethylcyclobutane. Compound (b) has two methyl groups that are located on the first and second carbons. Note that they are trans to each other. This compound is thus called *trans*-1,2-dimethylcyclobutane. Compound (c) has two methyl groups that are located on the first and third carbons. They are also trans to each other. Its name is therefore *trans*-1,3-dimethylcyclobutane.

The nomenclature of cycloalkanes with different alkyl substitutions is shown in Figure 15.11. When compounds have more than one kind of alkyl group, these groups are listed in alphabetical order.

2-*sec*-butyl-5-*tert*-butyl-4-methyl-1-isopropylcycloheptane

If cis and trans information is not given in the example, it is not included in the name.

NEW TERMS

cycloalkane A hydrocarbon compound with single carbon–carbon bonds, in which the skeletal carbons form a ring.

geometric structure A geometric form representing a molecule; carbon atoms are assumed to be at each vertex and hydrogens are not shown.

cis A prefix used to designate two similar groups on the same side of a molecule.

trans A prefix used to designate two similar groups on opposite sides of a molecule.

TESTING YOURSELF

Naming Cycloalkanes

1. Draw structures for each of the following molecules.
   a. 1-methyl-3-*n*-propylcyclohexane
   b. 1-isopropyl-1-isobutylcyclopentane
   c. 2-*tert*-butyl-1,5-diethylcyclooctane
2. Name the following compounds.

a. 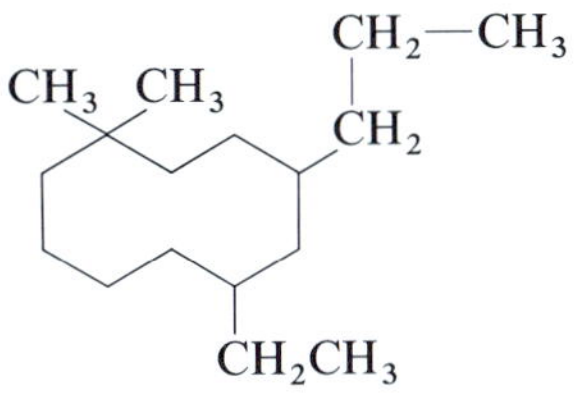
CH2—CH3
CH3 CH3
CH2
CH2CH3

c. 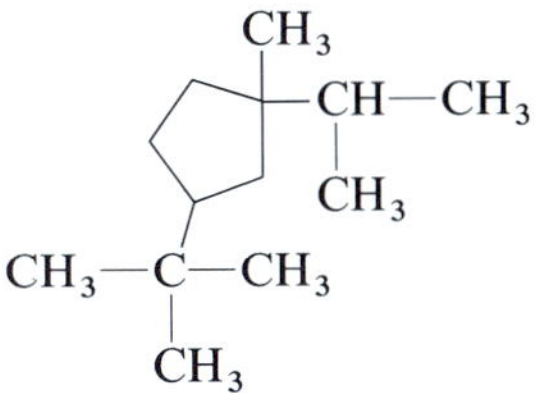
CH3
CH—CH3
CH3
CH3—C—CH3
CH3

b. 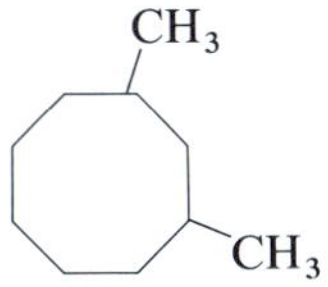
CH3
CH3

3. Identify the following substituents as cis or trans.

a. 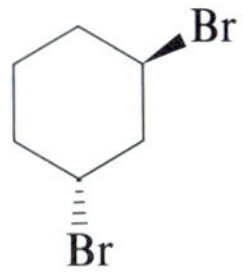
Br
Br

d. 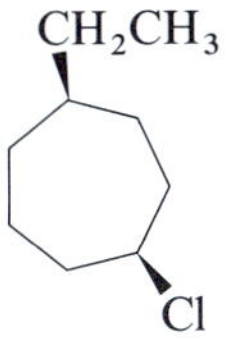
CH2CH3
Cl

b. 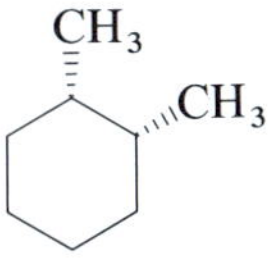
CH3
CH3

e. 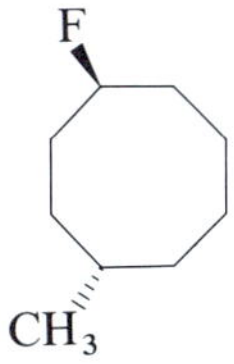
F
CH3

c. 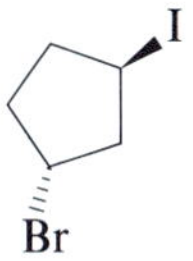
I
Br

Answers 1. a. 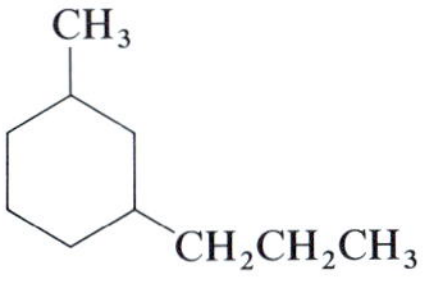
CH3
CH2CH2CH3

b. 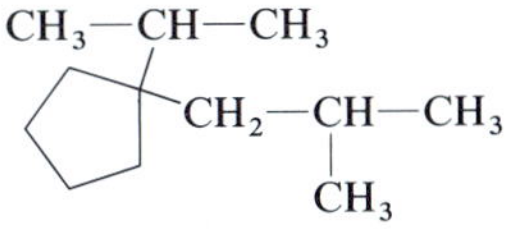
CH3—CH—CH3
CH2—CH—CH3
CH3

c. 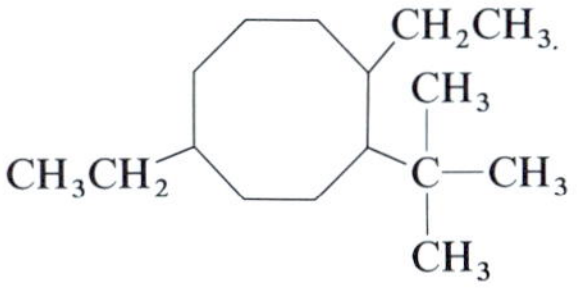
CH2CH3
CH3
CH3CH2
C—CH3
CH3

2. a. 6-ethyl-1,1-dimethyl-4-propylcyclodecane **b.** 1,3-dimethylcyclooctane **c.** 3-*tert*-butyl-1-isopropyl-1-methylcyclopentane **3. a.** trans **b.** cis **c.** trans **d.** cis **e.** trans

15.5 Conformations of Alkanes and Cycloalkanes

Alkanes

With free rotation about the C—C single bond, molecules can have their atoms in a number of different positions with respect to each other. These arrangements are called *conformations* and are *not* different compounds.

Because there is free rotation around a carbon–carbon single bond, atoms in a molecule can have different spatial orientations. These different orientations of the atoms are called **conformations**. Consider the molecule ethane. Several notably different conformations can be easily seen for ethane (Figure 15.12). When hydrogen atoms of ethane are directly behind one another, we call the conformation *eclipsed*. When the hydrogens are as far apart as possible, we call the conformation *staggered*.

You can see in Figure 15.13 that there are alternating staggered and eclipsed conformations every 60°. However, one hydrogen atom cannot be distinguished from another, so many conformations appear the same. This can be seen in an energy diagram where the "front carbon" is allowed to spin, and the interactions between hydrogen atoms are examined.

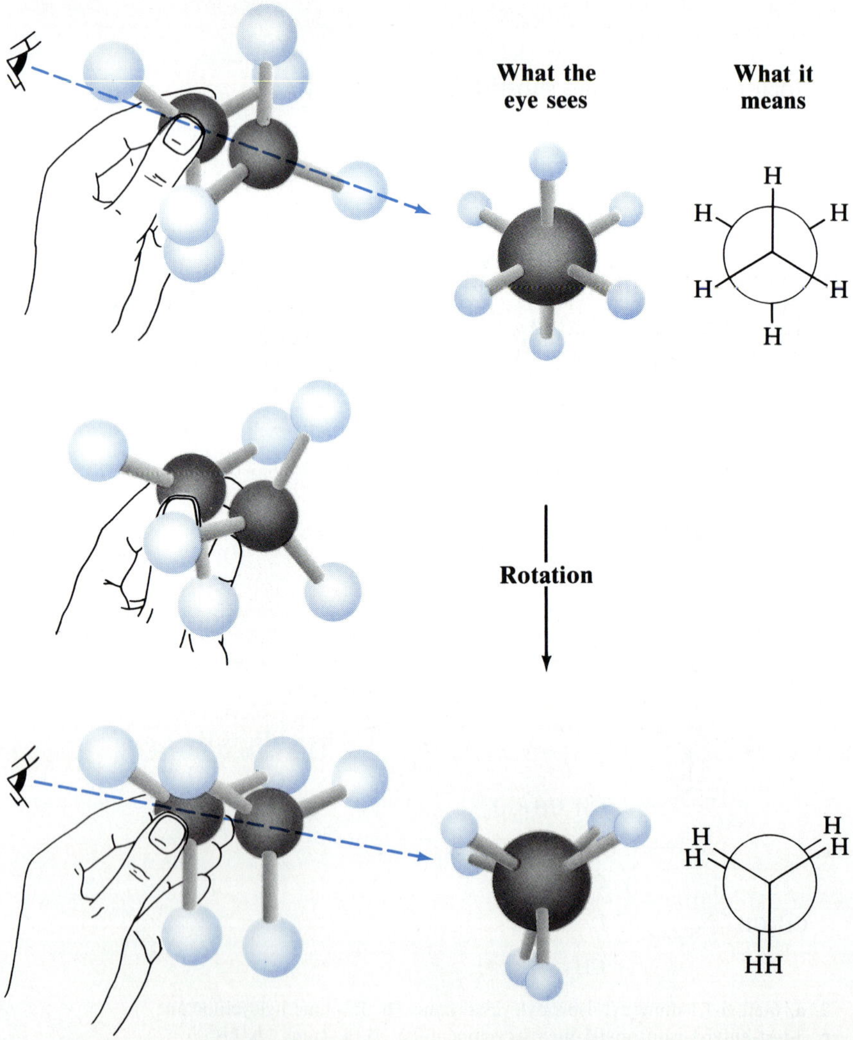
What the eye sees
What it means
H
H
H
H
H
H
Rotation
H
H
H
H
HH

FIGURE 15.12
Conformations of ethane. Notice how the relative positions of the atoms change by simply rotating around the carbon–carbon bond. Since no bonds are broken, each rotation is simply a different orientation of the *same* molecule.

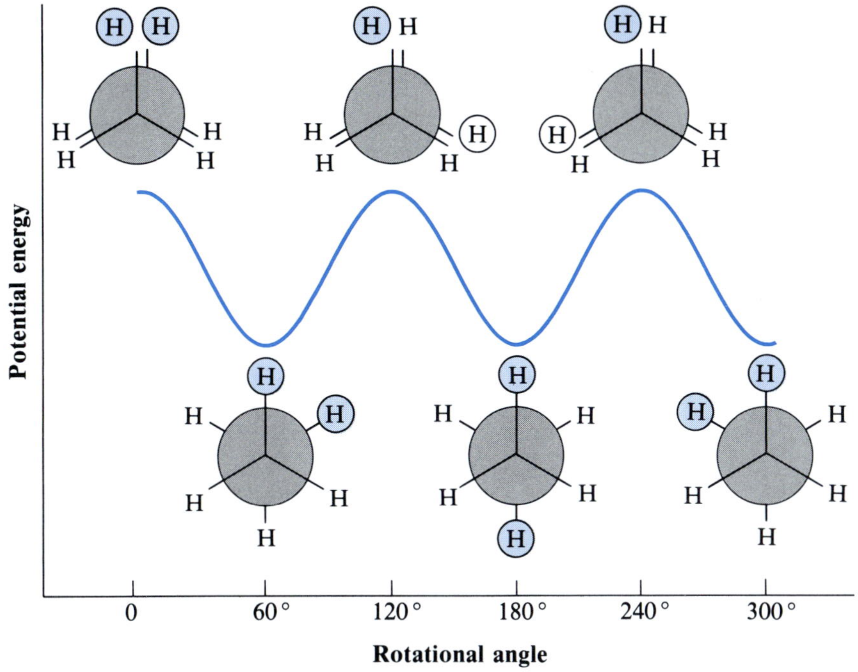
Potential energy
0
60°
120°
180°
240°
300°
Rotational angle

FIGURE 15.13
The relative energies of different conformations of ethane. When hydrogen atoms are directly behind one another, the conformation that has the highest energy is the least stable.

Now let's look at another molecule that shows larger differences in its conformations as it is rotated. Consider *n*-butane by "looking down" the bond between the second and third carbons (the C-2 to C-3 bond) (Figure 15.14). For butane we see that after 180° of rotation, the same conformation occurs. However, a greater variety of important conformations exist in space for butane than for ethane. If you accept the common sense notion that when groups are closer together there are more repulsions (higher energy and less stability), you could make the qualitative guess that some conformations are better than others (Figure 15.15). We see in the figure two kinds of eclipsed conformations and two kinds of staggered conformations.

The conformation that has the largest groups farthest apart is the most stable.

From this brief introduction to the concept of molecular shape, we can see that some forms of a molecule are more stable than other forms of the

FIGURE 15.14
Conformations of *n*-butane. By rotating about the central C—C bond, distinctly different stable conformations of butane can be made.

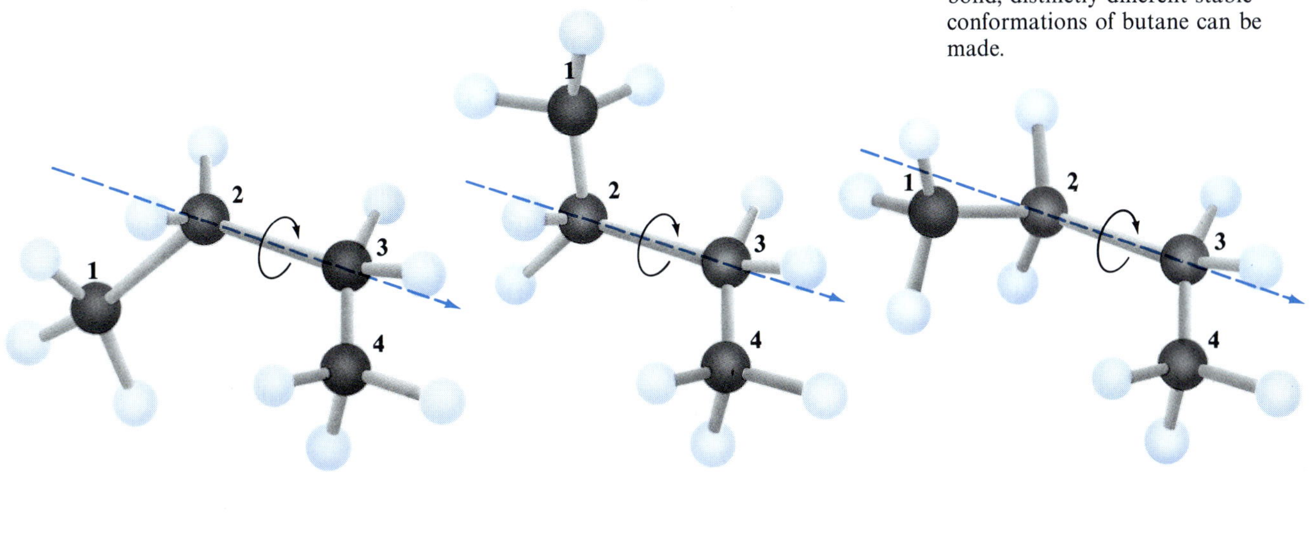

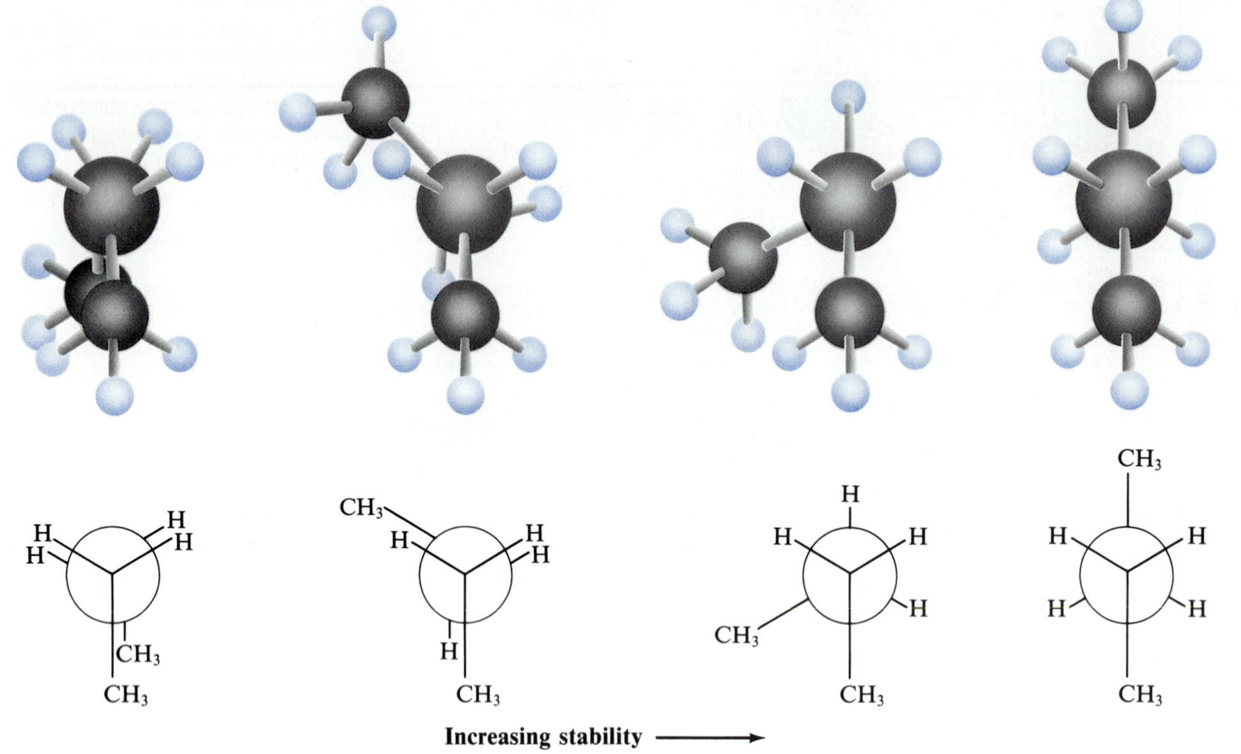
Increasing stability

FIGURE 15.15
Some of the conformations of butane, showing a qualitative order of stability. The conformation having the two methyl groups behind one another is least stable, followed by the conformation with a hydrogen and a methyl behind one another. The conformation having the two methyl groups farthest apart is the most stable. Of intermediate stability is the conformation in which the two methyl groups are close to but not directly in front or behind a hydrogen atom.

same molecule. In biochemistry we will see many examples of molecular shape playing a crucial role in the reactions of life processes.

Cycloalkanes

In the earlier section on cycloalkanes (Section 15.4), we saw that rings were not planar. Just as for noncyclic hydrocarbons, the cycloalkanes also have preferred shapes—orientations of the atoms that give the most stable conformation.

To better understand the effects of conformations on ring systems, let's look at the example of cyclohexane. For cyclohexane to maintain bond angles of 109°, it assumes what is called a *chair conformation* (Figure 15.16). A result of this shape is that all of the hydrogen atoms in cyclohexane assume a staggered arrangement. This results in two different orientations for the hydrogen atoms—*axial* and *equatorial.* These arrangements are shown in Figure 15.16. When only hydrogen atoms are involved on the carbons, there are six axial hydrogen orientations and six equatorial hydrogen orientations. If we replace a hydrogen with an alkyl group, this is changed. In one conformation there is only one axial group, five axial hydrogens, and six equatorial hydrogens. In the other conformation, there are six axial hydrogens, five equatorial hydrogens, and one equatorial alkyl group. The conformation with equatorial alkyl groups is more stable.

The cyclohexane molecule can undergo conformational changes such that groups attached to the ring will have the more stable equatorial orientation.

NEW TERM

conformation The three-dimensional shape of a molecule emphasizing the relative orientation of atoms in space.

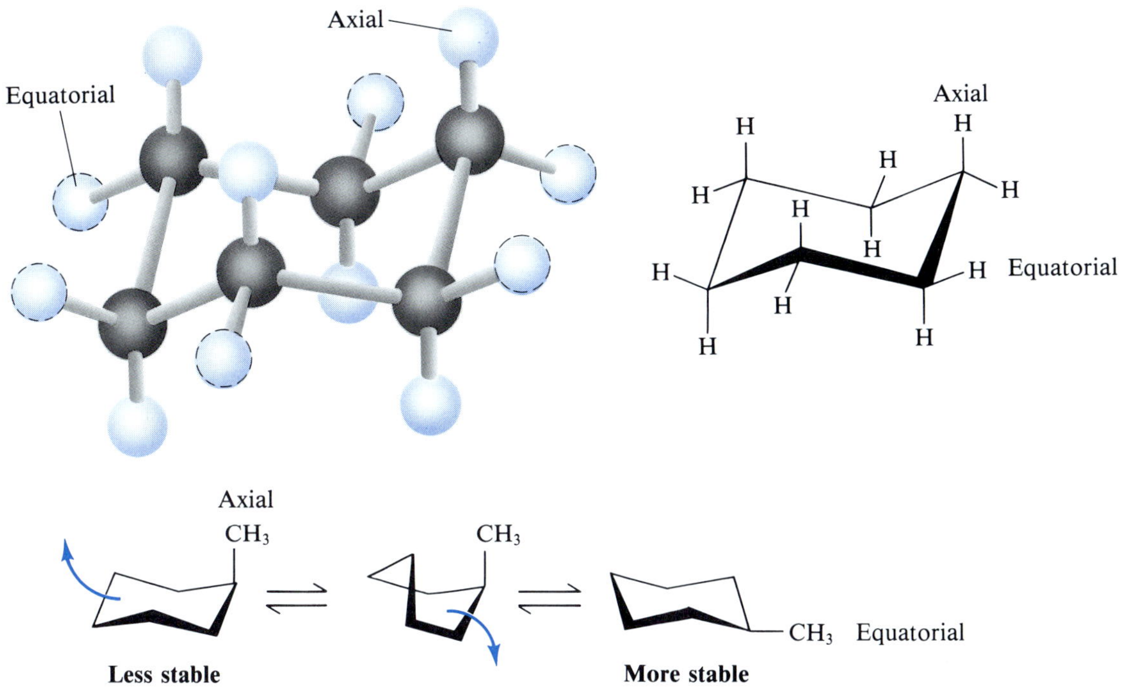
Axial
Equatorial
Axial
H
H
H
H
H
H
H
H
H Equatorial
H
H
Axial
CH3
CH3
CH3 Equatorial
Less stable
More stable

FIGURE 15.16
The chair conformation of cyclohexane. The chair form does not have any hydrogen atoms directly behind one another.

TESTING YOURSELF

Conformations of Molecules

1. Looking at the C1—C2 bond of propane, draw its major conformations and show which is *least* stable.
2. Which conformation for *cis*-1-ethyl-2-methylcyclohexane is more stable, and why?

a.

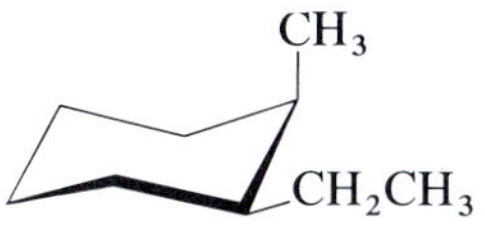
CH3
CH2CH3

b.

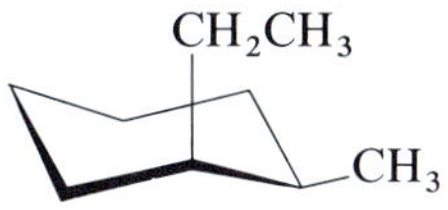
CH2CH3
CH3

Answers 1.

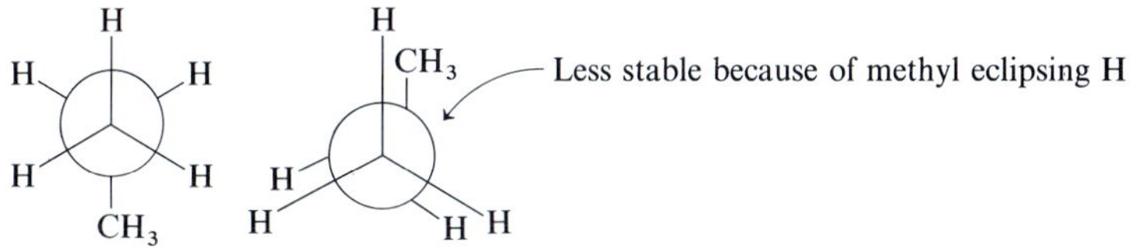
H
H
H
H
H
CH3
H
CH3
Less stable because of methyl eclipsing H
H
H
H H

2. (a) is more stable because the *larger* ethyl group is equatorial.

15.6 Chemical Reactivity of Alkanes and Cycloalkanes

Effects of Sigma Bonding

Alkanes and cycloalkanes have only single bonds and are thus relatively nonreactive.

The bonding of alkanes can be considered as very protected bonding. As we can see in Figure 15.17, the electrons involved in the bonding are shared directly between the atoms (recall that we call this a *sigma* bond). It is difficult for reagents to get to these electrons. There are also no "loose" electrons,

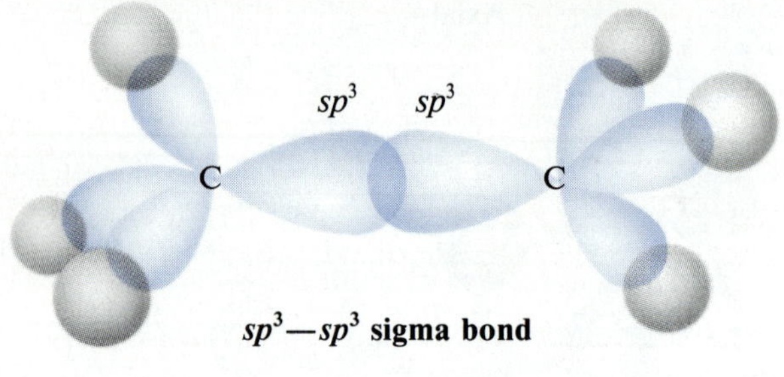

sp^3 sp^3
C
C
sp^3—sp^3 sigma bond

FIGURE 15.17
The chemical stability of the alkanes is due to the "protected" location of the shared bonding electrons between the bonded atoms. All electrons are involved in sigma bonding. There are no nonbonded or other available electrons to participate in reactions.

such as nonbonding electrons to participate in reactions. And without interaction with electrons, there is no reaction. Thus, we observe that alkanes are very resistant to normal chemical reactions.

Combustion of Alkanes

About 90% of all petroleum products and natural gas is used for fuel—to heat our homes, to generate electricity, and to power our automobiles and aircraft. Propane, a common fuel for camp stoves and soldering torches, reacts with oxygen according to the following equation:

$$\begin{array}{ccccccccc} & & H & & H & & H & & \\ & & | & & | & & | & & \\ H & - & C & - & C & - & C & - & H \\ & & | & & | & & | & & \\ & & H & & H & & H & & \end{array} + 5\,O_2 \longrightarrow 3\,CO_2 + 4\,H_2O + \text{Heat}\ (\Delta)$$

where Δ indicates heat that is released from the reaction.

In a similar way, all alkanes undergo combustion, forming carbon dioxide (CO_2) and water as products. Considerable heat is given off in the process because the sum of the energy of the bonds of the reactant molecules is greater than that of the product molecules. The excess bond energy is released as heat.

Since the combustion of every alkane follows the same general reaction, a generalized statement of the reaction that will apply in all cases can be useful. To help you in your studies, simply remember the generalized reaction and apply it to the specific example in question rather than trying to remember a large number of specific examples. General reactions are presented in notecard format throughout the rest of the book.

GENERAL REACTION

Combustion of Alkanes

$$R{-}H + O_2 \longrightarrow CO_2 + H_2O + \text{Heat}\ (\Delta)$$

Examples

$$\begin{array}{ccccccccccc} & & H & & H & & H & & H & & \\ & & | & & | & & | & & | & & \\ H & - & C & - & C & - & C & - & C & - & H \\ & & | & & | & & | & & | & & \\ & & H & & H & & H & & H & & \end{array} + 6.5\,O_2 \longrightarrow 4\,CO_2 + 5\,H_2O + \text{Heat}\ (\Delta)$$

$$\begin{array}{ccccc} CH_3 & - & CH & - & CH_3 \\ & & | & & \\ & & CH_2 & & \\ & & | & & \\ & & CH_3 & & \end{array} + 8\,O_2 \longrightarrow 5\,CO_2 + 6\,H_2O$$

THE GREENHOUSE EFFECT

Combustion of carbon fuel results in the formation of carbon dioxide. For every carbon atom burned, one molecule of carbon dioxide is produced. The natural abundance of carbon dioxide in our atmosphere has increased during this century as a result of the massive amounts of hydrocarbons burned as fuel for heating, power, and transportation. In the United States one-third of all carbon dioxide produced is from automobile emissions. The increased levels of carbon dioxide contribute to the *greenhouse effect*.

This name aptly describes the results of the effect. As in a greenhouse, which captures and keeps heat within its glass roof and walls, the carbon doxide (and other pollutant gases) in the atmosphere performs the same function. The short wavelength solar energy from outer space is able to pass through the gas molecules in the atmosphere. But energy radiated from the earth back to outerspace is of longer wavelength. Thus, these wavelengths are absorbed by the polluting gas molecules and do not leave our atmosphere. This trapped energy remains as heat.

Studies show that there has been an increase in worldwide temperature over the past century (see figure). Atmospheric levels of carbon dioxide have only been measured for short a period of time because the methods are relatively new. However, for the 30 years or so that data have been collected, an increase has been seen.

The concentration of methane, another "greenhouse gas," has been steadily increasing, too. Indeed, during the past decade, atmospheric levels have been increasing by 1.5% per year. Fossil methane, created at the same time as petroleum products from ancient plants, is deposited in pockets of the earth's mantle. These gases escape

(*continued*)

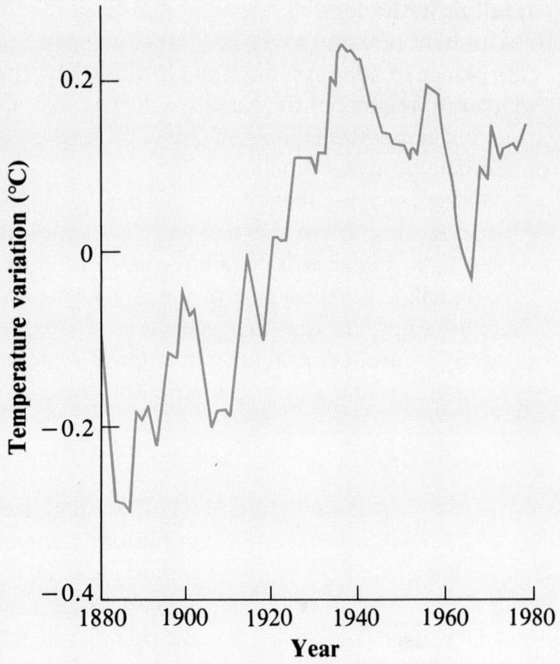

Temperature variation (°C)
0.2
0
−0.2
−0.4
1880
1900
1920
1940
1960
1980
Year

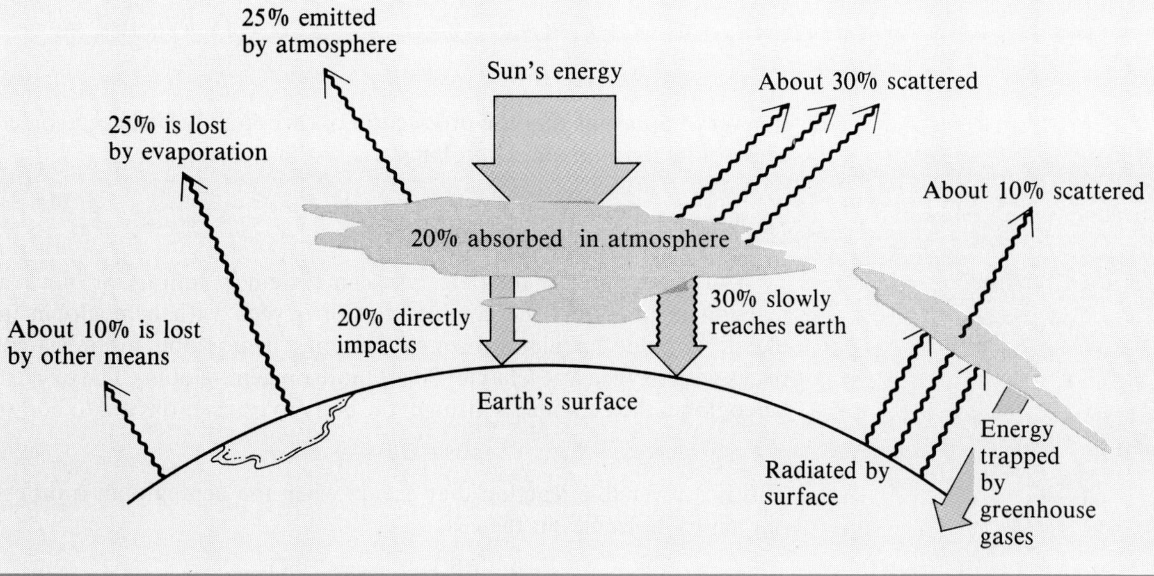

25% emitted by atmosphere
Sun's energy
About 30% scattered
25% is lost by evaporation
About 10% scattered
20% absorbed in atmosphere
30% slowly reaches earth
About 10% is lost by other means
20% directly impacts
Earth's surface
Radiated by surface
Energy trapped by greenhouse gases

during drilling and mining. By examining the ratios of carbon-14 to carbon-12 in the methane gases in our atmosphere, scientists believe about 15–20% comes from fossil sources. Modern methane comes from animal digestion processes, termites, and the vast wetlands and marshes of the world.

What are the long term consequences of the greenhouse effect? Some scientists suggest the most dire of consequences: that our farmland will become deserts and the polar ice caps will melt, resulting in the flooding of our coastal cities. These researchers point to the measurable warming of our planet to support this view. In this view, the temperate regions of the earth would be most affected. Temperatures in some parts of the world may increase by as much as 5 °C.

Others suggest that the earth's heating is localized and that there is in fact an overall cooling of the earth. Plants will use a large amount of the excess carbon dioxide we are producing. However, there are estimates that the earth loses an acre of forest every second, primarily in tropical regions. It would seem prudent to recognize that these forests are important resources for many immediate needs as well as long term storehouses for carbon. We need to ensure their survival. An interesting suggestion has been made that industrialized nations might consider renting rain forests from less developed countries!

Support for the notion that plants help reverse the heating trend is available. Palm Springs, California, is located in a desert area. As in many cities with large areas covered by concrete, the temperature of Palm Springs was regularly increasing. In the 1970s there was a trend toward construction of golf courses. Since that time, the temperature of that small oasis in the desert has dropped by 2–3 °F.

No matter what the source or makeup of the gas, nor whether the problem is immediate or long range, there is no question that we need to study the greenhouse problem. We cannot go on forever polluting our air, land, and water without expecting a consequence for these actions. We cannot continue to burn fossil fuels and to destroy the tropical forests. We have an obligation to those in our future to give them a future.

If alkanes are burned with insufficient oxygen, as sometimes happens in automobile engines or heating systems that have insufficient draft, carbon monoxide (CO) is produced instead of carbon dioxide. In a balanced equation we can see that less oxygen is needed to form carbon monoxide than carbon dioxide (3.5 instead of 5 O_2).

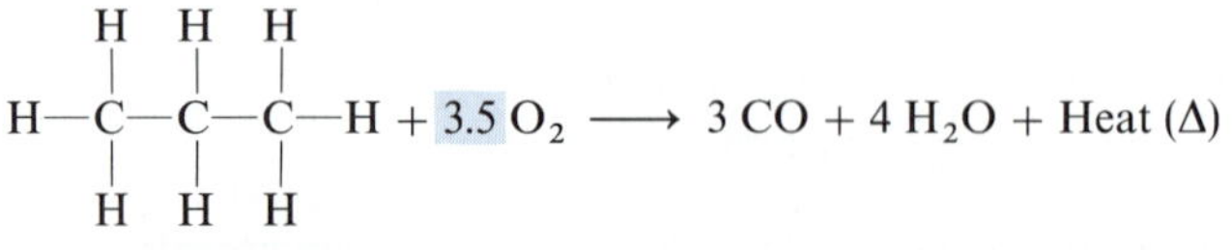

It is very important that the production of carbon monoxide be avoided because of its harmful effects on humans.

Carbon Monoxide Poisoning

Carbon monoxide interferes with an extremely important biochemical reaction in humans—the combination of oxygen with hemoglobin in the blood. As blood circulates through the lungs, hemoglobin (abbreviated Hb) reacts with oxygen (see Chapter 30 for more on hemoglobin). The oxygenated hemoglobin then circulates through the body to transfer oxygen to body cells.

$$Hb + O_2 \longrightarrow HbO_2$$

This is a reversible reaction that occurs when the hemoglobin transfers the oxygen to molecules in the cell.

$$Hb + O_2 \rightleftharpoons HbO_2$$

Carbon monoxide undergoes a similar reaction with hemoglobin:

$$Hb + CO \longrightarrow HbCO$$

This reaction is also reversible:

$$Hb + CO \rightleftharpoons HbCO$$

However, the equilibrium constant for the carbon monoxide reaction is about 200 times greater than that for the reaction of hemoglobin with oxygen. This means that the reaction of carbon monoxide with hemoglobin will produce a product that holds onto its oxygen about 200 times more tightly than the product of the reaction of oxygen with hemoglobin.

At equilibrium,

$$Hb + CO \xrightleftharpoons{200 \times \text{ more tightly held than } O_2} HbCO$$

$$Hb + O_2 \rightleftharpoons HbO_2$$

The result is that if hemoglobin molecules are tightly bound to carbon monoxide, they would no longer be available to transport oxygen.

Extremely small concentrations of carbon monoxide in the air can compete very effectively with oxygen for the hemoglobin molecule. Once bound, the carbon monoxide will not easily release the hemoglobin. The oxygen transport mechanism of the organism is severely restricted, often causing coma or death. One treatment for carbon monoxide poisoning is to greatly increase the concentration of oxygen in the blood by having the victim breathe pure oxygen or placing the victim in a high pressure chamber saturated with oxygen. The resulting high oxygen concentration makes it possible for the $Hb + O_2$ reaction to compete with the $Hb + CO$ reaction for available hemoglobin, even when the equilibrium constant is in favor of the lesser concentrated CO.

FIGURE 15.18
Animals that live in water-scarce areas, such as this (a) camel and (b) Australian plains mouse, generate their own water as a by-product of metabolism.

Biological Combustion of Alkanes

A large part of the food we eat contains fats and lipids (details of the structure and reactions of these molecules are discussed in Chapters 24 and 28). These have a large hydrocarbon component, as well as one or more functional groups as part of their structure. Since combustion of hydrocarbon fuels produces water as a by-product, it is not unreasonable to assume that the fuel we burn (or metabolize) in our body will give up a significant amount of water as a by-product of this metabolism. Many animals, including humans, dispose of excess water as perspiration and urine. However, many desert mammals use their metabolic water as a substitute for drinking water. The camel does not store water in its hump; rather, it stores fatty compounds rich in hydrocarbons that, when "burned" by the body metabolism, produce both energy and water. Cockroaches and desert mice cannot be killed by withholding water (Figure 15.18). They make their own!

Halogenation

In our discussion of the chemical reactivity of the alkanes (Section 15.6), we suggested that these hydrocarbons are considered chemically nonreactive. This is not completely true. Aside from combustion, one of the most important chemical reactions of alkanes is the **halogenation reaction**, a process by which

Organohalogen compounds (R—X) are very important to the chemical industry.

a carbon–hydrogen bond is replaced by a carbon–halogen bond. The products resulting from these reactions are called **organohalogen** compounds (sometimes called *alkylhalides*). The general reaction for halogenation can be stated as follows:

GENERAL REACTION

Halogenation of Hydrocarbons

$$R{-}H + X_2 \xrightarrow[\text{or light}]{\text{Heat}} R{-}X + H{-}X$$

Example

$$CH_4 + Cl_2 \xrightarrow{\text{Heat}} CH_3Cl + HCl$$

The halogenation reaction has a limited but very important use. The limitation is primarily related to the indiscriminate nature of the process—a halogen can replace *any* C—H bond. For example, four different products for the reaction of methane (CH_4) with chlorine (Cl_2) could occur, as shown here.

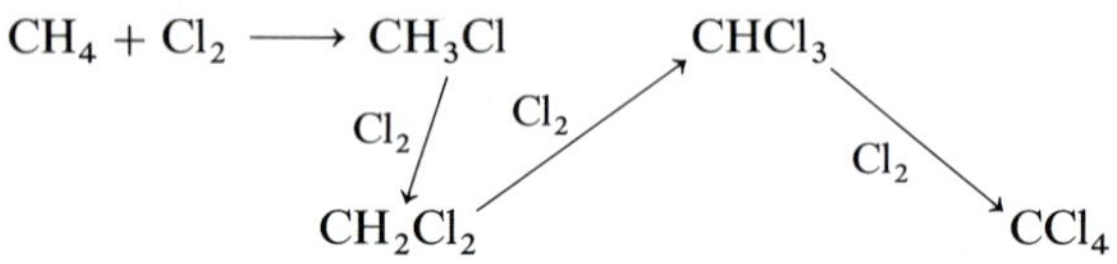

Excess chlorine continues to react with each new by-product. The Cl_2 near each arrow indicates that the reaction is occurring in the presence of chlorine.

A mixture of products is always formed from the halogenation of alkanes. As long as this mixture is easily separated into its component parts (by distillation, for example), it creates no special problems since most organohalogen compounds have some use. However, we can see the potential problems by considering the four products from the methane reaction. It takes little imagination to see the complexity that would result from the halogenation of a more complex hydrocarbon such as pentane.

Nomenclature of Organohalogens

The halogen imparts no special priority to the nomenclature process. A halogen is treated as an alkyl group.

Organohalogen compounds have no special nomenclature. Indeed, the rules already established for hydrocarbons are used here as well. Since the four halogens are chlorine, bromine, iodine, and fluorine, the prefixes *chloro-*, *bromo-*, *iodo-*, and *fluoro-* are used if a halogen is part of a compound. They are given no special priority in the naming scheme. This is best shown by examples.

	Name
$CH_3{-}\underset{\displaystyle Br}{\underset{\vert}{CH}}{-}CH_2{-}CH_2{-}CH_3$	2-*Bromo*pentane (not 4-bromopentane)
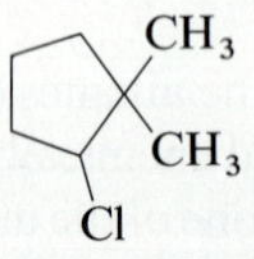	1,1-Dimethyl-2-*chloro*cyclopentane (*not* 1-chloro-2,2-dimethylcyclopentane)

As with all of the nomenclature developed to this point, we always strive for the lowest arrangement of numbers.

Artificial or manmade organohalogen compounds have widespread uses in our everyday life. They are used widely as pesticides and herbicides, as solvents and refrigerants, as medicines and anesthetics, and as plastic and rubber. Examples of some of the common organohalogen compounds and their uses are listed in Tables 15.5 through 15.8.

TABLE 15.5
Organohalogen Pesticides and Herbicides

Name	Description	Structure
DDT	A powerful insecticide. Outlawed because of toxic effects on humans and wildlife.	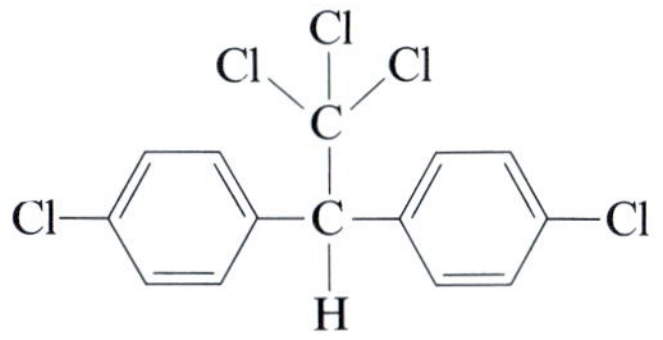
Lindane	Insecticide and soil poison. Now is used primarily for treatment of seeds. Found to cause cancer and birth defects.	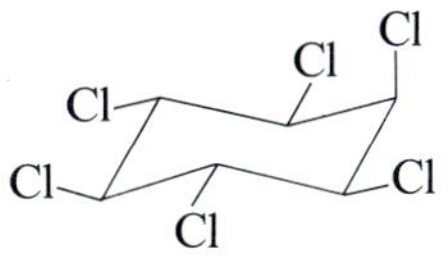
Dieldrin	Insecticide for garden pests. It is now used in place of DDT for control of DDT-resistant mosquitos. Much more toxic than DDT.	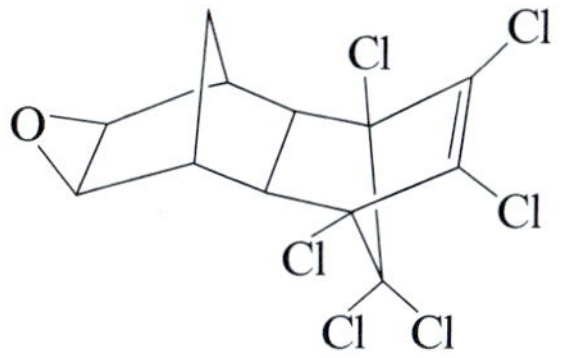
Endrin	The most toxic of all chlorinated insecticides.	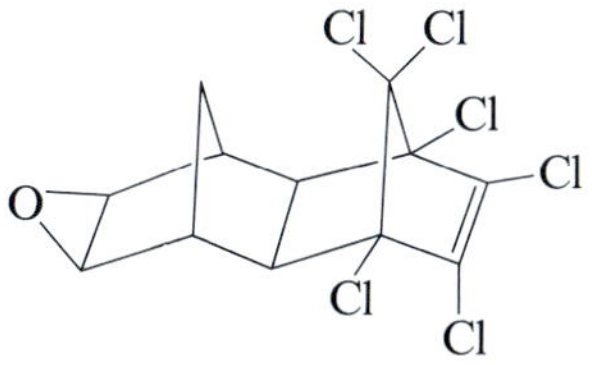
2,4-D	Ingredient in sprays to kill broadleaf plants such as dandelion. Makes the plant take up water too fast, causing a rapid growth of stem but no root. An ingredient of Agent Orange used as a defoliant in Vietnam.	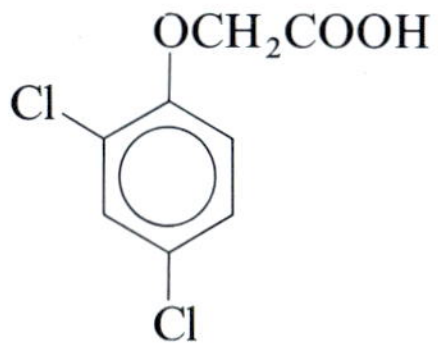

TABLE 15.6
Organohalogen Solvents

Name	Description	Structure
Carbon tetrachloride	Once used as a cleaner, it is now known to be toxic and a potential cancer-causing agent (carcinogen).	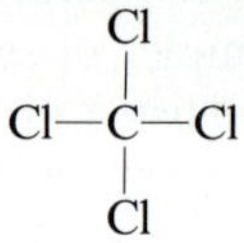
Chloroform	Once used as an inhalation anesthetic, this material is now known to be toxic and has been replaced by a variety of other compounds.	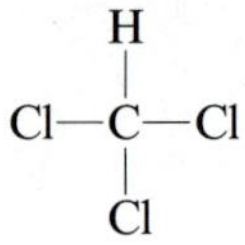
Freon-12	Refrigerant fluid. Once used in propellant sprays, but this use has been discontinued because of its potential harm to the atmospheric ozone layer.	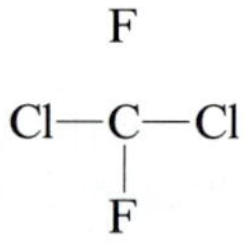
Dichloromethane	Commonly called methylene chloride. A very common solvent.	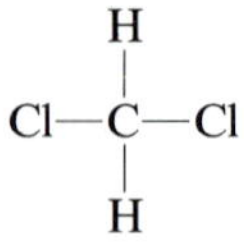
Tetrachloroethylene	One of the common solvents that has replaced carbon tetrachloride in the dry cleaning industry.	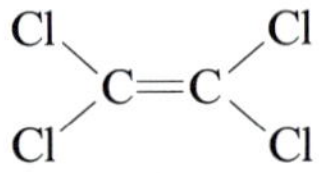
PCBs	The *poly*c*hlorinated* *b*iphenyls (or PCBs) were once commonly used as insulating, nonflammable components in transformers and other electrical devices. They have proven to be harmful, and their use has been discontinued.	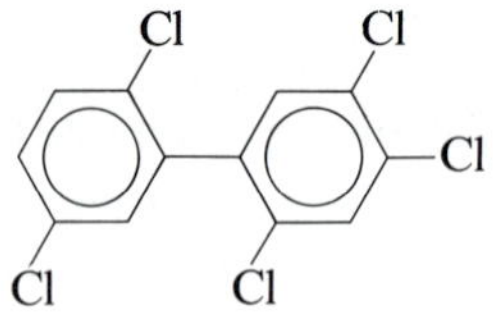

TABLE 15.7
Medical Uses of Organohalogen Compounds

Name	Description	Structure
Halothane	A widely used local anesthetic that has the added value of being nontoxic and nonflammable.	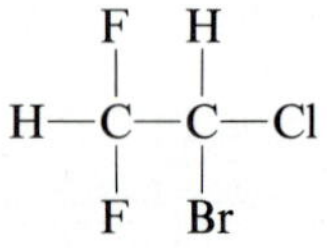
Tetrachloroethylene	An effective treatment for hookworm.	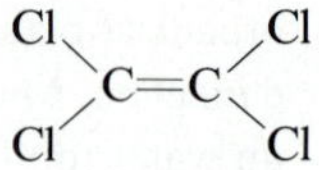
Mitotane	This compound is structurally similar to DDT and is used in the treatment of brain cancer.	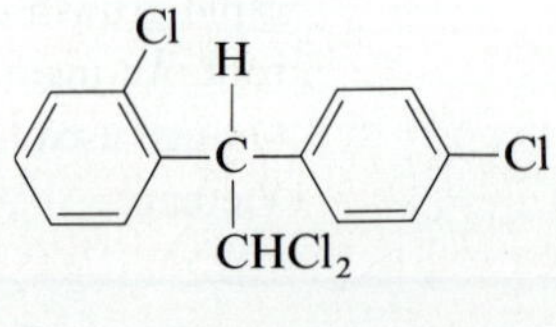

TABLE 15.8
Organohalogen Polymers

Name	Description	Monomer	Structure
PVC	Plastic used in making records and plastic pipe.	$CH_2{=}CHCl$	$-\!(CH_2-\overset{Cl}{\overset{\vert}{C}}H)_n\!-$
Teflon	Chemically resistant insulation for cooking utensils.	$F_2C{=}CF_2$	$-\!(CF_2-CF_2)_n\!-$
Neoprene	Synthetic rubber.	$CH_2{=}\overset{Cl}{\overset{\vert}{C}}-CH{=}CH_2$	$-\!(CH_2-\overset{Cl}{\overset{\vert}{C}}{=}CH-CH_2)_n\!-$

DEPLETION OF THE OZONE LAYER AND CFCs

The stratosphere is one of the middle layers of our atmosphere and is found at distances of 15 to 50 km above the earth's surface. Within the stratosphere is a layer of ozone (O_3) that surrounds the earth. The ozone plays an important role in absorbing harmful UV radiation. If this protection decreased, we would be bombarded with sufficient additional UV radiation that the incidence of skin cancer would dramatically increase. This is the immediate important role of the ozone to us.

One of the leading culprits in the destruction of the ozone layer is the widely used class of compounds called the *chlorofluorocarbons* or CFCs. These are chlorinated and fluorinated hydrocarbons that are very useful in refrigerants, aerosols, cleaning agents, fire extinguishers, and insulation foams. On the surface of ice crystals in the clouds of the stratosphere, the CFCs tend to come apart to liberate chlorine atoms. As you recall from earlier discussions on bonding, the chlorine atom has seven electrons, and to fulfill the inert gas electronic structure, it will react. In the atmosphere it reacts with ozone in the following way:

$$:\underset{\cdot\cdot}{\overset{\cdot\cdot}{Cl}}\cdot + O_3 \longrightarrow ClO\cdot + O_2$$

$$ClO\cdot + ClO\cdot \longrightarrow Cl_2O_2$$

$$Cl_2O_2 + \text{Light energy} \longrightarrow ClO_2 + :\underset{\cdot\cdot}{\overset{\cdot\cdot}{Cl}}\cdot$$

$:\underset{\cdot\cdot}{\overset{\cdot\cdot}{Cl}}\cdot$ repeats the process.

This is called a *chain reaction* because a reactant is liberated each time that can repeat the process. You can see that because of the millions of tons of CFCs used each year, it will not take too much for it to make a measurable impact on our atmosphere.

There has been much interest in the popular press about the ozone "hole" over Antarctica. This observation is consistent with the theory that the reactions taking place require cold surfaces. Computer-enhanced satellite images of Antarctica show the depletion graphically (see Plate 4 in color inserts). Much interest has been shown in the development of so-called safe CFCs; however, these cause the same effects, but at a rate of about one-fifth of the presently used CFCs. The development of materials that can serve the same roles is a high priority research area for leading chemical companies.

The United States and 31 other nations have signed an agreement known as the Montreal Protocol, which mandates a freeze and then regular reduction in use of CFCs until they are eliminated. Twelve western European countries that have been major producers of ozone destroying chemicals have recently taken an aggressive attitude to help curb their production. They have completely banned both production and use over the next ten years. If the United States follows their lead, over two-thirds of the total chemicals in this class will be removed by the year 2000. As optimistic as this appears, predictions are still not good for the atmosphere. It has been suggested that even if we stopped all use immediately, the ozone level would not reach 1985 levels until about 2050.

NEW TERMS

halogenation reaction A reaction of an alkane with halogen that is catalyzed by heat or light, in which a C—H bond is replaced by a C—X bond.

organohalogen A class of organic compounds characterized by carbon–halogen bonds.

TESTING YOURSELF

Chemical Reactivity and Organohalogens

1. Write a balanced equation for each of the following reactions:

   a. $CH_3—CH_2—CH_3 + O_2 \longrightarrow$
   (Propane)

   b. 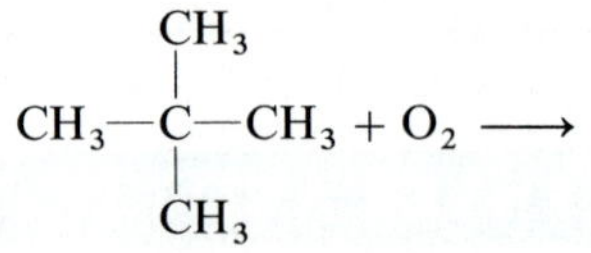

2. Draw and name all of the monobromination products from the bromination of pentane.

3. Draw and name all of the dibromination products from the bromination of pentane.

Answers **1. a.** $CH_3—CH_2—CH_3 + 5\,O_2 \longrightarrow 3\,CO_2 + 4\,H_2O$
b. $CH_3—C(CH_3)_2—CH_3 + 8\,O_2 \longrightarrow 5\,CO_2 + 6\,H_2O$ **2.** 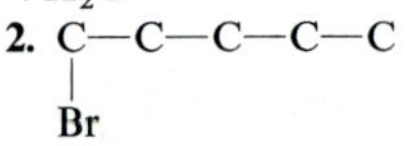

1-Bromopentane

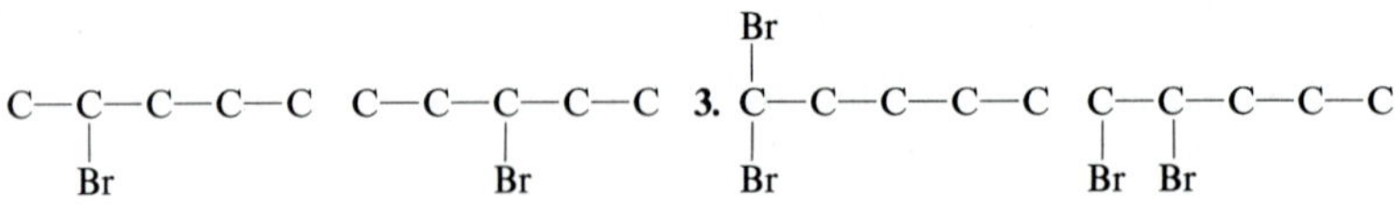

2-Bromopentane 3-Bromopentane 1,1-Dibromo- 1,2-Dibromo-

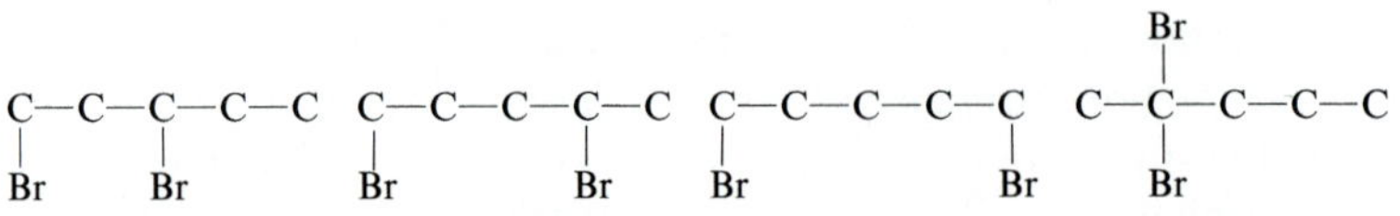

1,3-Dibromo- 1,4-Dibromo- 1,5-Dibromo- 2,2-Dibromo-

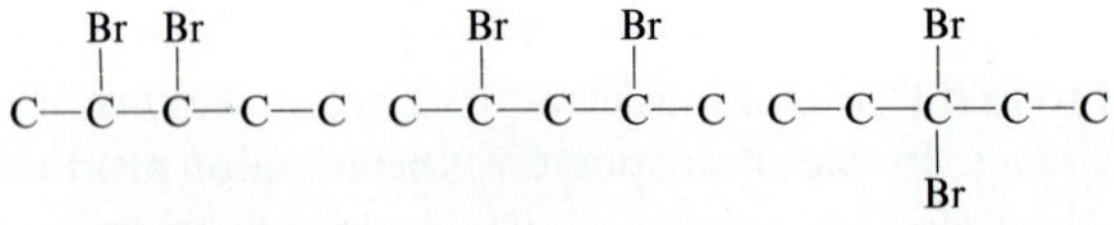

2,3-Dibromo- 2,4-Dibromo- 3,3-Dibromo-

15.7 Health-related Products Based on Hydrocarbon Structures

In this chapter we have surveyed the alkanes, cycloalkanes, and organohalogen compounds. All three of these groups are important in the medical and allied health fields. Thus, it is appropriate to conclude the chapter with a number of health-related applications.

Alkanes

The nonreactivity of hydrocarbons is evidenced by their scarcity in products for human use. However, it is precisely the nonreactivity and nonsolubility of high-boiling-point hydrocarbon fractions that make *petrolatum* (mineral oil) useful as a lubricant. Because it is an indigestible oil, there is only slight absorption from the intestinal tract. Yet the oil is able to soften stools in cases of minor constipation. Some fat-soluble vitamins (see Chapter 27) dissolve in mineral oil; thus, prolonged use might deprive the body of this type of vitamin.

Cycloalkanes

The most common health-related use of cycloalkanes can be found in the application of cyclopropane as an anesthetic. Cyclopropane is a fast-acting anesthetic having a wide margin of health safety associated with its use. Unfortunately, one danger is associated with its use—it is highly explosive. This has greatly restricted the use of cyclopropane as an anesthetic.

A cyclopropane derivative, *chrysanthemic acid*, is one of the basic structural units found in the *pyrethrins*. These are naturally occurring compounds found in the pyrethrum flower and have very pronounced insecticidal activity. Indeed, most of the common insect sprays for household use have synthetic pyrethrins (or analogs) as the active ingredient.

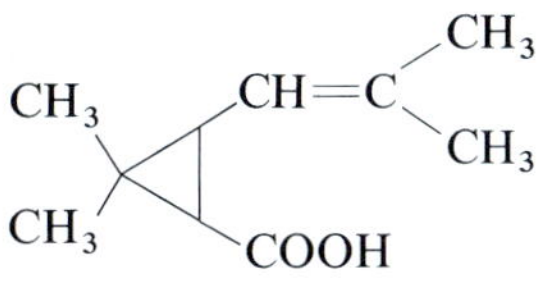
CH3
CH=C
CH3
CH3
CH3
COOH

Chrysanthemic acid

As one might expect, the highly strained bond angles in cyclopropane result in this being a molecule that readily breaks apart. Cyclopropane is the only cycloalkane that has ring rupturing as part of its normal chemical behavior. When it ruptures, it reacts with other molecules to form new compounds, as shown here.

$$\text{Cyclopropane} + H_2 + \text{Catalyst} \longrightarrow \text{Propane}$$

$$\text{Cyclopropane} + Br_2 \longrightarrow \text{1,3-Dibromopropane}$$

$$\text{Cyclopropane} + HCl \longrightarrow \text{1-Chloropropane}$$

The instability of the three-membered ring in cyclopropane will be seen in later chapters on the chemistry of *epoxides*, which are three-membered oxygen-containing analogs of cyclopropane.

Cyclopropane

O

Epoxide

As for the alkanes, there are numerous compounds having carbon ring systems as part of their structure. Examples of these compounds will be discussed in sections more appropriate to the functional groups involved in the structure.

ARTIFICIAL BLOOD

The great number of components in blood and its myriad functions make it highly improbable that a permanent substitute for blood could ever be found. However, we are often in need of short term supplies of blood, and even with all of the care taken to maintain blood, there are many problems associated with blood transfusions. One problem is keeping a supply of fresh blood available for emergency use, since "old" blood is not as effective as an oxygen carrier. Although blood does many things other than transport oxygen, this is an essential function, and a synthetic blood subsitute has been shown to be useful for short term applications.

A salt solution of *perfluorodecalin* and *perfluorotripropylamine* has been marketed in Japan as Fluosol-DA.

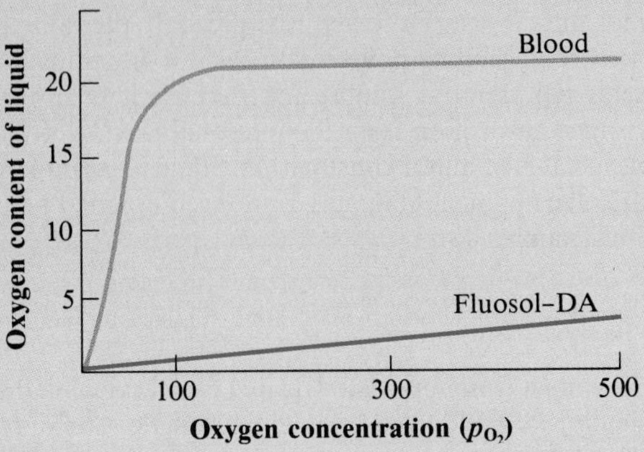
Blood
Fluosol-DA
Oxygen content of liquid
20
15
10
5
100
300
500
Oxygen concentration (p_{O_2})

Comparison of blood and Fluosol-DA for carrying oxygen.

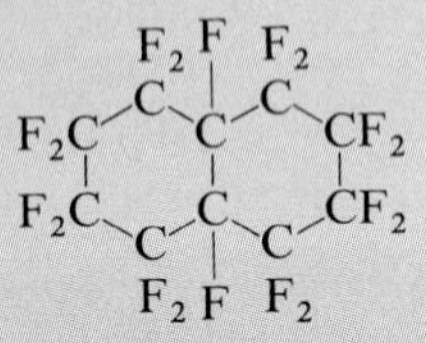

Perfluorodecalin

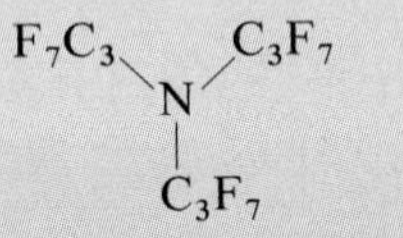

Perfluorotripropylamine

Although not as efficient as hemoglobin for transporting oxygen (see figure), the fluorocarbons are able to dissolve fairly substantial quantities of oxygen for transport.

Is this new product the answer to the problem of blood supply? It is a start. But problems remain and are now being studied. For example, what happens to the fluorocarbons in the body? Like teflon, the molecules are supposed to be very resistant to chemical activity; however, chemical evidence shows that this is not true. Also, what effect do the small impurities in the synthetic material have on the body over the long term? Finally, is this *really* an effective substitute for blood? There are mixed clinical data.

One potentially useful outcome of the studies may be the application of these materials to cancer treatment. Cancer and tumor cells have low oxygen content. This makes it difficult to treat the cells with chemotherapy or radiation therapy. There are encouraging data to suggest that the artificial oxygen carriers (which are much smaller than hemoglobin) can be targeted to the cancer cells.

Whatever the outcome of these studies, it remains a positive sign of the ingenuity of science that we can even approach these questions. By trying to mimic the natural processes in the human body, we find out more about how it works. We can then try to find natural ways to assist body processes to fight disease and injury.

Organohalogen Compounds

Compounds in this family have found wide use as anesthetics. In Table 15.7, we presented *halothane* (Fluothane) as an example of a fast-acting and widely used organohalogen compound. The fluorocarbons, an important group of organohalogens, are used extensively as refrigerant components and were once common as inert gases for spray containers. They are now being banned from use, and producers are phasing out their production. There is evidence that use of these compounds is directly related to the destruction of the ozone layer of our atmosphere.

One of the most exciting uses of highly fluorinated hydrocarbons (meaning all the C—H bonds are replaced by C—F bonds) is their ability to act as temporary substitutes for blood. Although the fluorocarbons cannot carry out the functions of blood, they are capable of dissolving large quantities of oxygen gas. Thus, the important function of carrying oxygen through the system can be maintained. A product called *Fluosol-DA* is being tested in the United States for possible use in emergency treatment of rapid blood loss. Of particular value might be its ability to serve as a temporary emergency blood supply without the need for matching blood types. Also, the storage capabilities would allow for this material to be available at all times in almost all places.

The ability for these compounds to dissolve large quantities of oxygen is dramatically seen in Figure 15.19, in which a rat shown totally submerged in liquid fluorocarbon is still breathing!

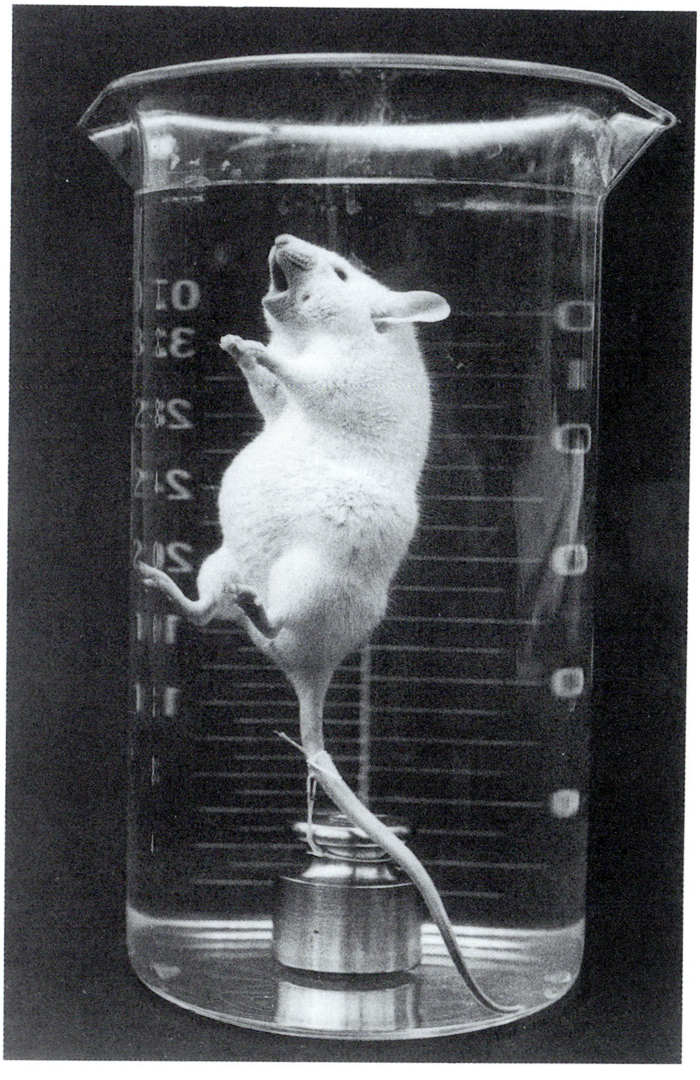

FIGURE 15.19
Fluorocarbons effectively dissolve large amounts of oxygen. There is enough oxygen available that an animal can "breathe" while completely submerged.

TESTING YOURSELF

Reactions of Cyclopropanes

1. Draw structures for the reactants and products of the following "word reactions."

a. 1,2,3-Trimethylcyclopropane + H_2 + Catalyst ⟶ 3-Methylpentane

b. Cyclopropane + Cl_2 ⟶ 1,3-Dichloropropane

c. Cyclopropane + HBr ⟶ 1-Bromopropane

Answers 1. a.

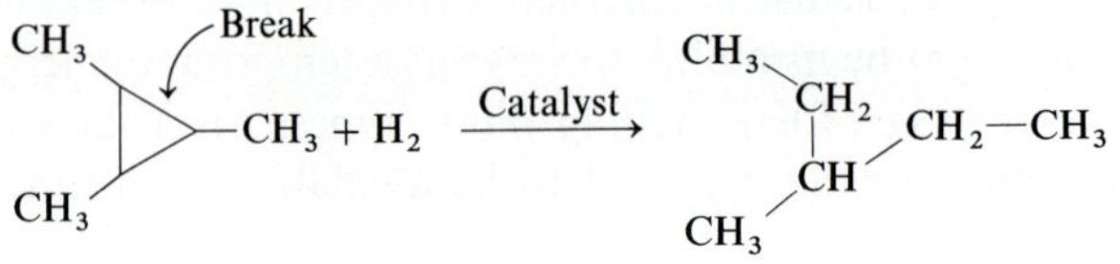
Break
CH_3
$CH_3 + H_2$
CH_3
Catalyst
CH_3
CH_2
$CH_2—CH_3$
CH
CH_3

b.

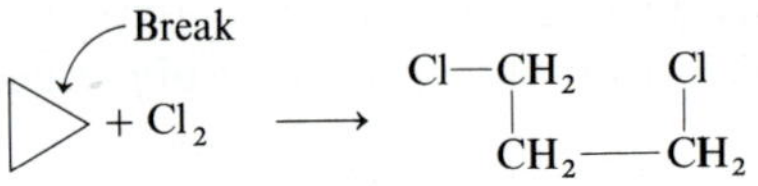
Break
+ Cl_2
Cl—CH_2
Cl
CH_2—CH_2

c.

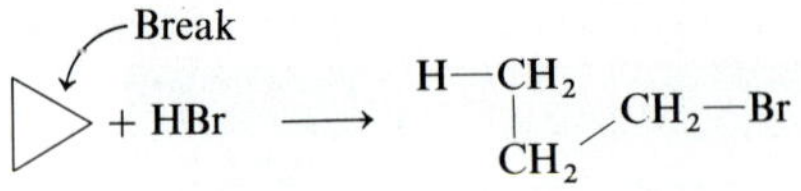
Break
+ HBr
H—CH_2
CH_2—Br
CH_2

SUMMARY

Alkanes and *cycloalkanes* are the simplest of the organic hydrocarbons, having only C—C and C—H single bonds. These compounds have physical properties largely due to the London forces attracting molecules to one another. The electrons forming the bonds between the carbon atoms and between the carbon and hydrogen atoms are sufficiently tightly held that these molecules are not very reactive.

The regularly increasing series of alkanes is called a *homologous series*, and names for this series provide the basis for nomenclature of organic compounds. *Alkyl groups*, derived from the hydrocarbons, provide a unique method of naming complex organic structures. *Isomers* are molecules having the same formula (in the case of hydrocarbons, the same number of hydrogens and carbons), but different structures. They are conveniently named by combining the names of members of the homologous series and alkyl groups.

The shape of a molecule is reflected in the *conformation*—the direction in space for the bonds between C—C and C—H. Some conformations are of lower energy (and thus are more favorable) than other conformations.

The simplest reaction of hydrocarbons is combustion (reaction with oxygen to form carbon dioxide and water) and *halogenation* (forming organohalogen compounds). The *organohalogen* compounds have extensive industrial use.

REACTION SUMMARY

Combustion of Alkanes with Adequate Oxygen

$$R—H + O_2 \longrightarrow CO_2 + H_2O + \text{Heat } (\Delta)$$

Halogenation of Hydrocarbons

$$R—H + X_2 \xrightarrow[\text{or light}]{\text{Heat}} R—X + HX$$

TERMS

alkane
structural formula
perspective formula
condensed formula
line formula
homologous series
isomer
alkyl group
primary carbon
secondary carbon
tertiary carbon
cycloalkane
geometric structure
cis
trans
conformation
halogenation reaction
organohalogen

CHAPTER REVIEW

QUESTIONS

1. Carbon atoms in alkanes and cycloalkanes (and in all carbon-containing compounds) always have how many bonds?
   a. 1 b. 2 c. 3 d. 4

2. The best name for 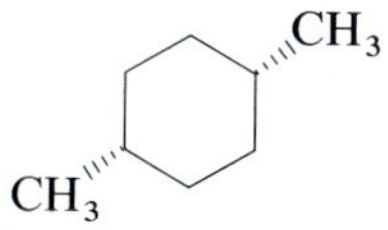 is
   a. *cis*-dimethylcyclohexane
   b. 1,4-dimethylcyclohexane
   c. *trans*-dimethylcyclohexane
   d. *cis*-1,4-dimethylcyclohexane

3. The best group name for 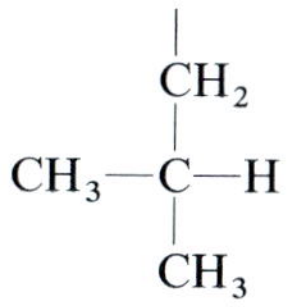 is
   a. isobutyl b. *tert*-butyl c. *sec*-butyl d. isopropyl

4. The best name for 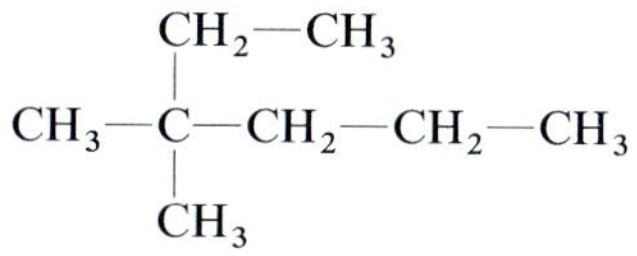 is
   a. 4-ethyl-4-methylpentane
   b. 2-methyl-2-propylbutane
   c. 3,3-dimethylhexane
   d. 4,4-dimethylhexane

5. Hydrocarbons are
   a. insoluble in water
   b. composed of carbon and hydrogen
   c. both (a) and (b)
   d. none of these

6. The series of hydrocarbons of the general formula C_nH_{2n+2} is called
   a. hydrocarbon series
   b. increasing series
   c. homologous
   d. alkane series

7. Bonding in an alkane is
   a. ionic
   b. covalent
   c. sigma
   d. hydrogen bonding
   e. both (b) and (c)

8. The hydrocarbons are ________ in water.
   a. insoluble b. soluble c. reactive d. none of these

9. The major carbon compound formed from the *incomplete combustion* of a hydrocarbon in air is
   a. carbon dioxide b. carbon monoxide c. alkyl chains d. water

10. The general formula for a ring hydrocarbon is
   a. C_nH_n b. C_nH_{2n} c. C_nH_{2n+2} d. C_nH_{2n-2}

Answers **1.** d **2.** d **3.** a **4.** c **5.** c **6.** c **7.** e **8.** a **9.** b **10.** b

DIAGNOSTIC CHART

Blacken in all of the circles under the number of each question you missed in the Chapter Review questions. The diagnostic chart will help you identify concept areas that need more study.

	Questions									
Concepts	**1**	**2**	**3**	**4**	**5**	**6**	**7**	**8**	**9**	**10**
Nomenclature		○		○						
Alkyl groups		○	○	○						
Physical properties					○			○		
Reactions									○	
Structure	○	○				○	○			○

EXERCISES

Structure

1. What is the hybridization of carbon atoms in the alkanes? What is the shape corresponding to this hybridization?

2. We often hear the phrase "straight-chain hydrocarbon." What is meant by this, and in what way is it an incorrect phrase?

3. Draw the two chair conformations of isopropylcyclohexane. Which one will have the lower energy (be more stable)?

Nomenclature

4. Give IUPAC names for each of the following hydrocarbon alkanes.

a.
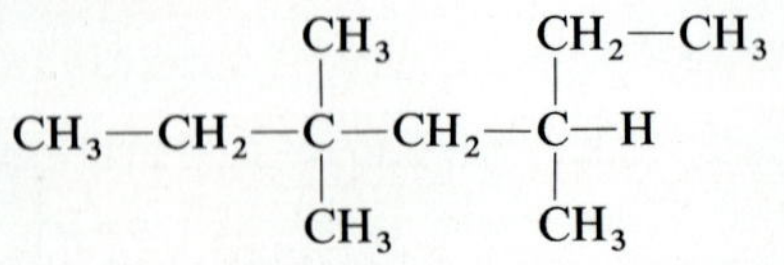

b.
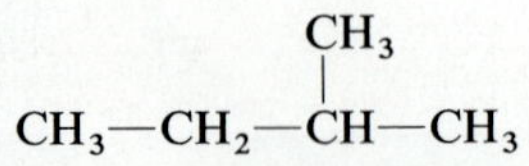

c.
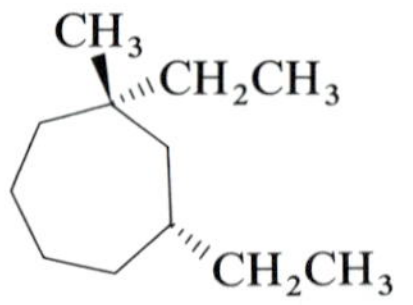

d.
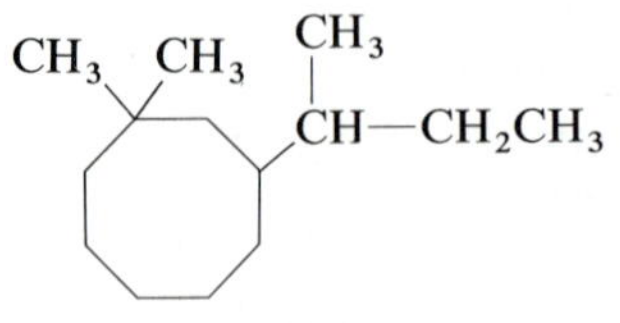

e.
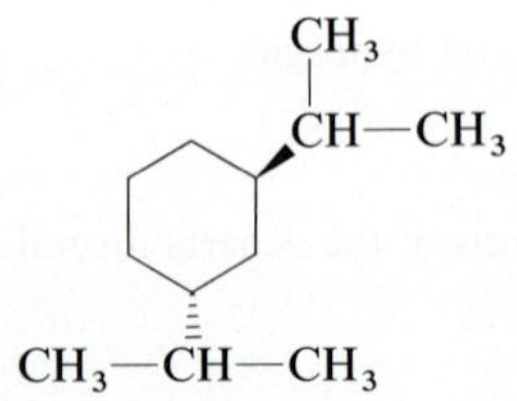

f.
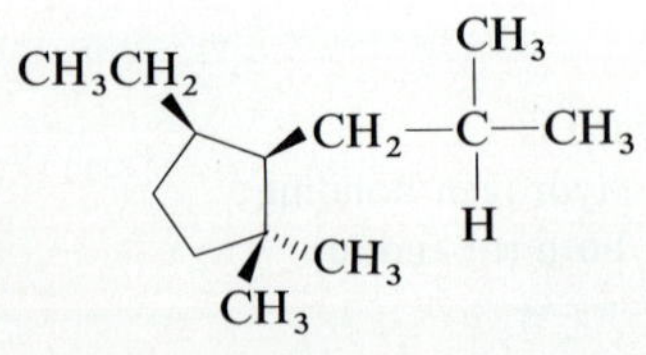

g. 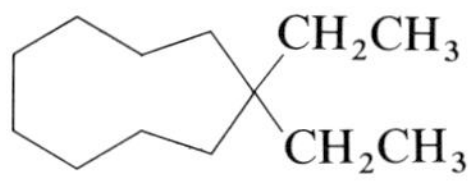

h. 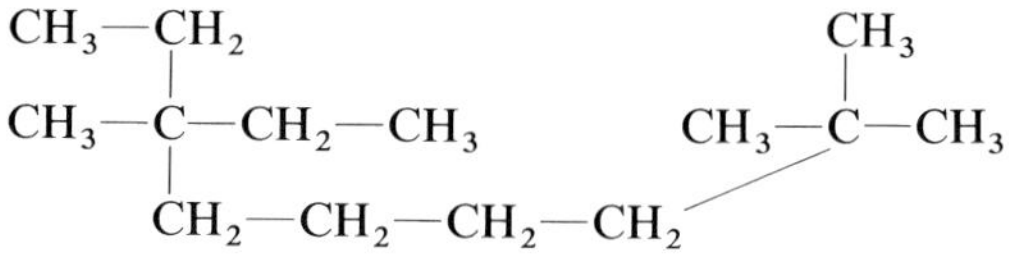

i. 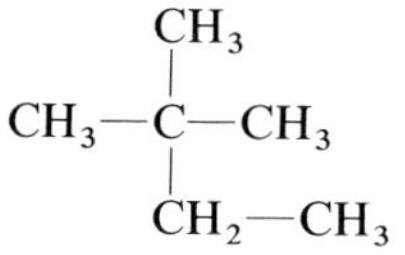

j. 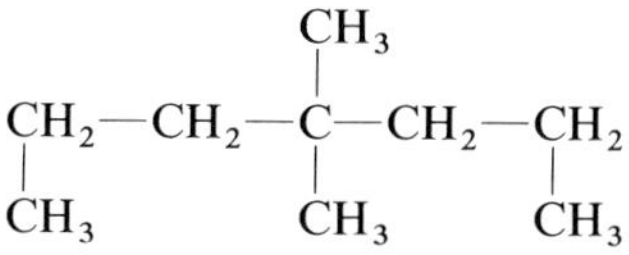

5. Give IUPAC names for each of the following simple hydrocarbons.

a. 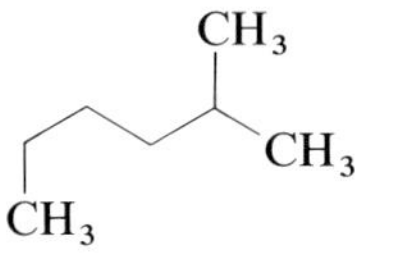

b. 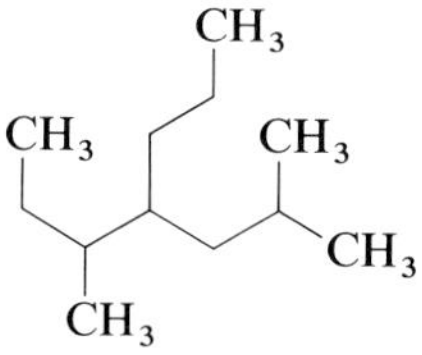

c. 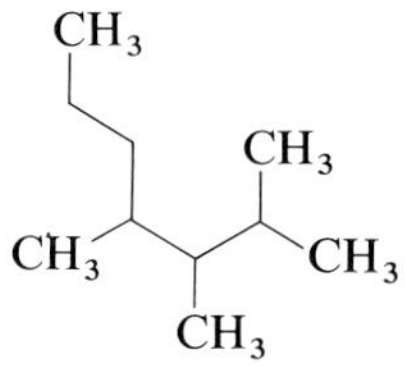

d. 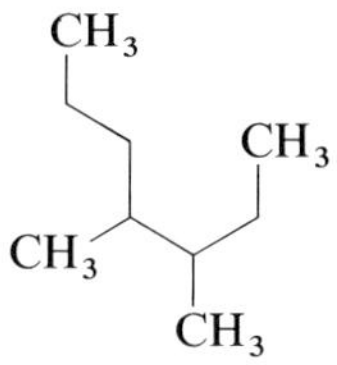

e. 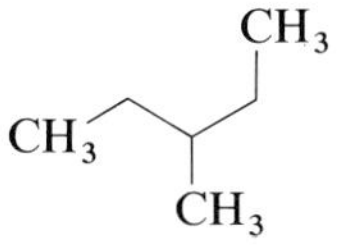

f. 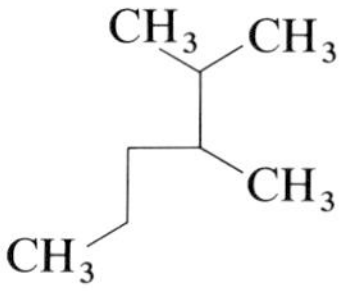

6. Give IUPAC names for each of the following simple hydrocarbons.

a. 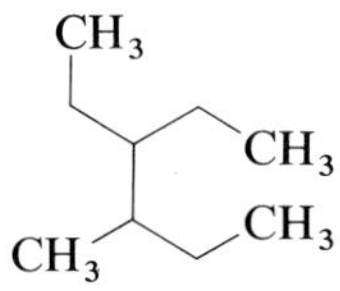

b. 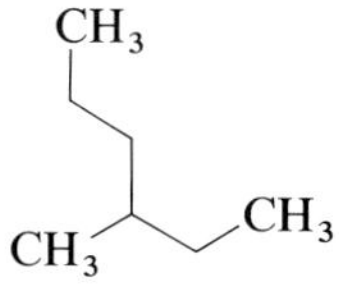

c. 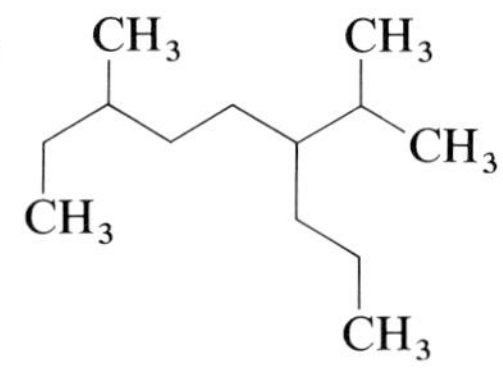

d. 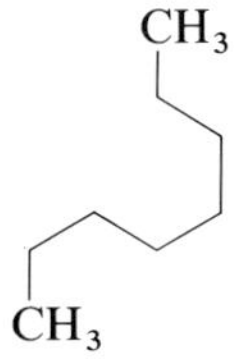

e. 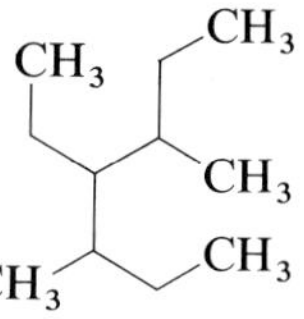

f. 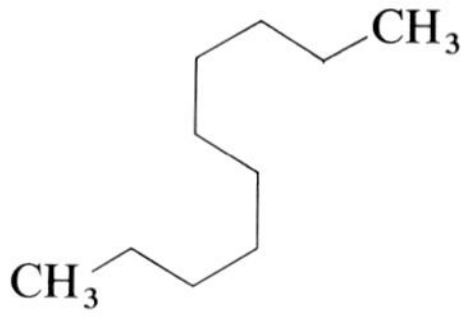

7. Give IUPAC names for each of the following cycloalkanes.

a. 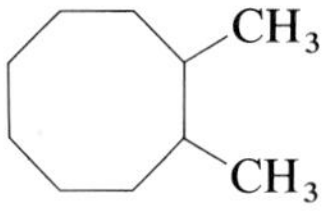

b. 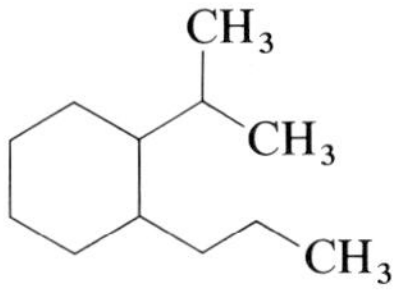

c. 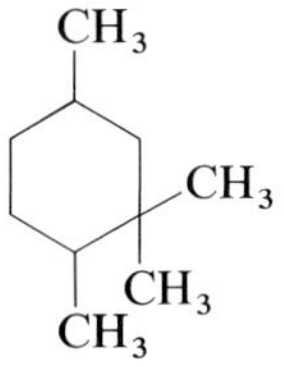

d. 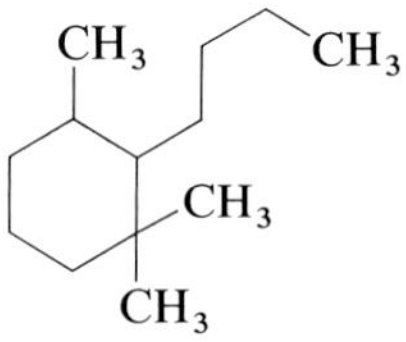

e. 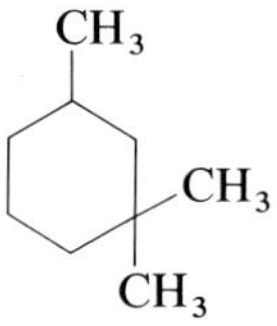

f. 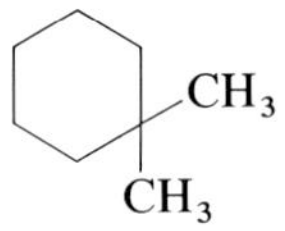

8. Give IUPAC names for each of the following halogenated hydrocarbons (organohalogens).

a. 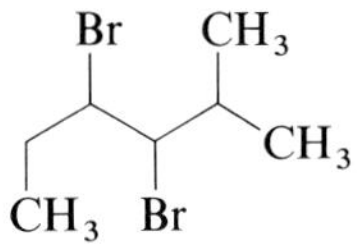

b. 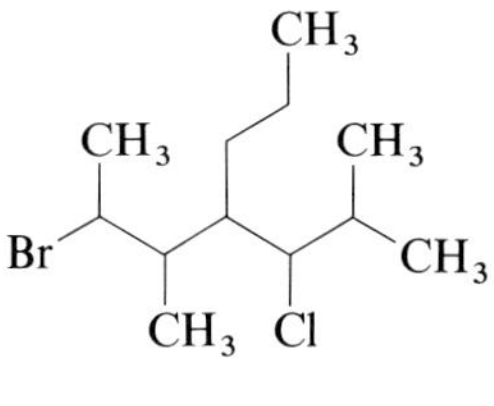

c. 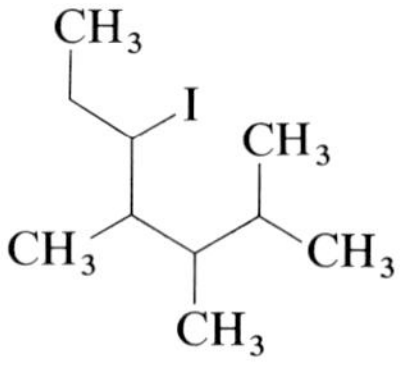

d. 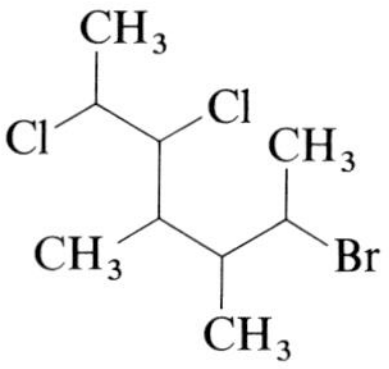

9. Give IUPAC names for each of the following halogenated hydrocarbons.

a. 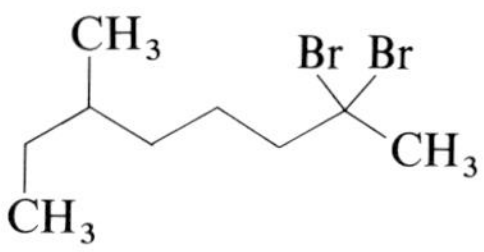

b. 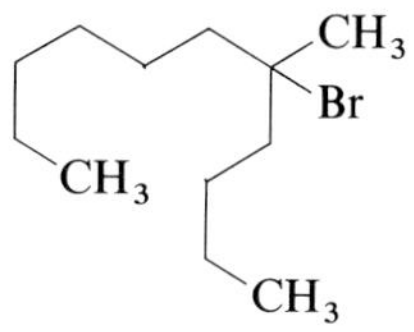

c. 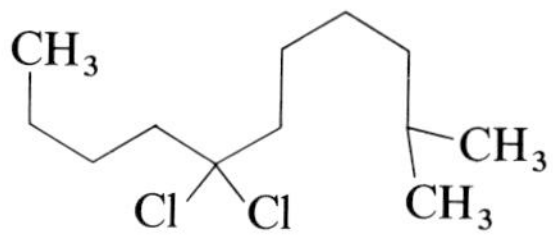

10. Give IUPAC names for each of the following halogenated cycloalkanes.

a. 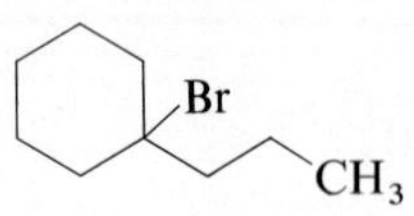

b. 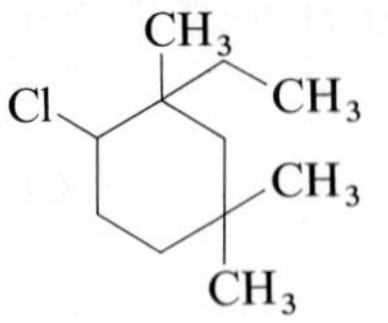

c. 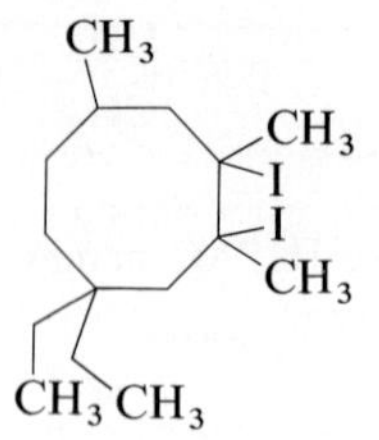

11. Give IUPAC names for each of the following.

a. 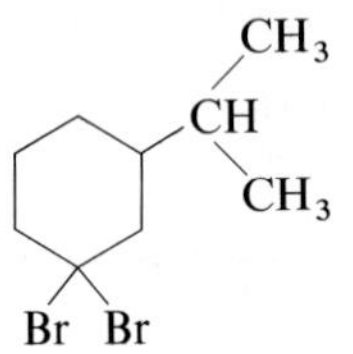

b. 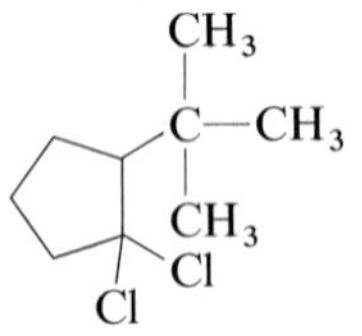

c. 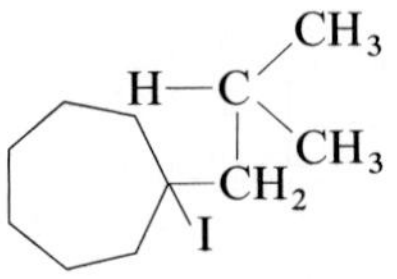

d. 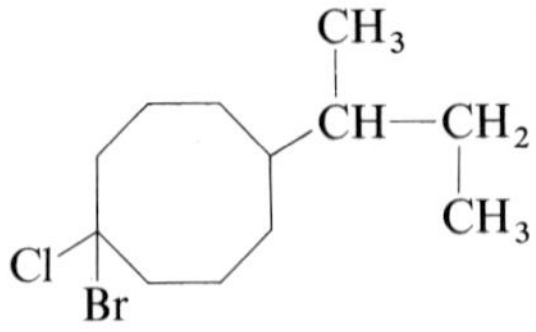

12. Draw structures corresponding to the following IUPAC names.
    a. 3,3-Dimethyl-4-isopropyloctane
    b. *cis*-1,2-Dimethylcyclohexane
    c. 2,2,3,3-Tetramethylhexane
    d. 3-*tert*-Butylheptane
    e. *trans*-2,3-Diisopropyl-1,1-dimethylcycloheptane
    f. 1-Methyl-1-*sec*-butylcyclohexane
    g. 1,1-Diethylcyclopentane
    h. *cis*-1,3-Diisopropylcyclohexane
    i. 4-*tert*-Butyloctane
    j. 1,1,3,3-Tetramethyl-4-isopropylcyclooctane
13. Draw and name the isomers of C_6H_{14}.
14. Draw and name all of the isomers of dimethylcyclohexane. (Do not forget cis and trans isomers.)
15. Draw a ball-and-stick model for 2,2-dimethylpropane.
16. Assign the status (primary, secondary, or tertiary) for each of the hydrogens in 2,2-dimethylbutane.
17. Draw and name all of the C_6H_{12} isomers (ring compounds).

Reactions

18. Why are hydrocarbons generally so unreactive?
19. Give products and balance the equations for the following combustion reactions.
    a. Propane + Sufficient oxygen $\longrightarrow$
    b. Isobutane + Sufficient oxygen $\longrightarrow$
    c. Hexane + Insufficient oxygen $\longrightarrow$
    d. Methylcyclohexane + Sufficient oxygen $\longrightarrow$
    e. Cyclopentane + Chlorine $\xrightarrow{\Delta}$
    f. Ethane + Bromine $\xrightarrow{\Delta}$
20. Give products for each of the following reactions.

a. 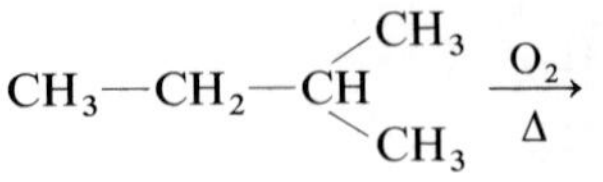

b. 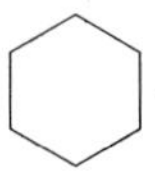 $+ Cl_2 \xrightarrow{\text{Light}}$

c. 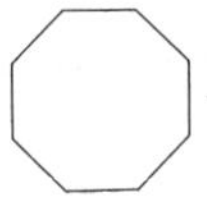 $+ O_2 \xrightarrow{\Delta}$

d. $CH_4 + Br_2 \xrightarrow{\Delta}$

21. What is wrong with the following names?
    a. 2, 4-Diethylpentane
    b. *cis*-3,4-Dimethylcylohexane
    c. 2-Isopropylbutane
    d. 1,5-Diethylcylopentane
    e. 2,2-Dibromo-4,4,5,5-tetramethylhexane
22. Are the following compounds isomers, or are they the same compound? (HINT: If in doubt, name each compound. Isomers will have different names.)

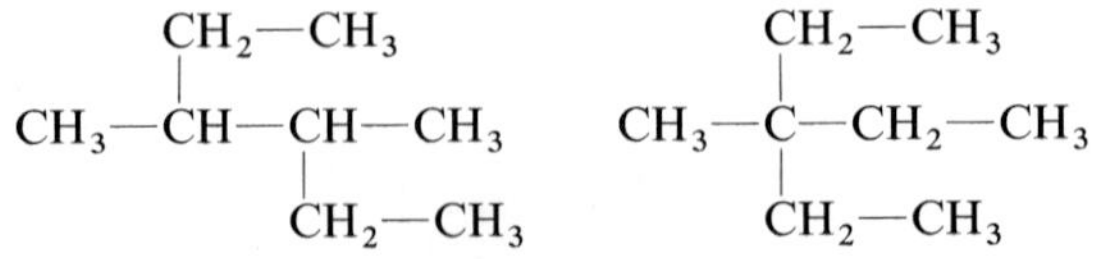

23. How many possible products could you produce from the dibromination of butane? Draw structures for each and supply an unambiguous name for each.
24. Explain why you might guess that 1-iodopropane would have a higher boiling point than 1-chloropropane.
25. Based on your ideas of electronegativity, which highlighted proton is more easily removed? Why?

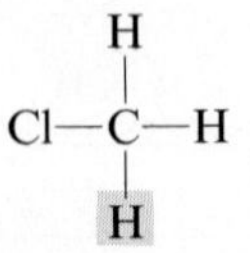
a.

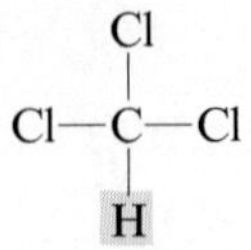
b.

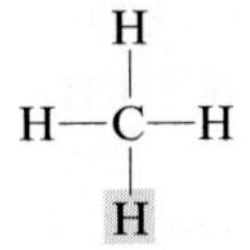
c.

CHAPTER 16

Unsaturated Hydrocarbons

Hydrocarbons with a double or triple bond between adjacent carbons are much more reactive than members of the single-bonded alkane family. The presence of multiple bonds also restricts rotation around the bond, giving these molecules well-defined shapes that are advantageous in biological systems.

OBJECTIVES

After completing this chapter, you should be able to

- Identify members of the alkene and alkyne families.
- Explain how sp^2 hybridization limits rotation within the molecule.
- Name representative members of the alkene and alkyne families.
- Explain why molecules with pi bonding are more subject to reaction than molecules with single-bonded carbons.
- Predict the products of common oxidation, addition, and polymerization reactions of the alkenes and alkynes.

16.1 Introduction to the Unsaturated Hydrocarbons

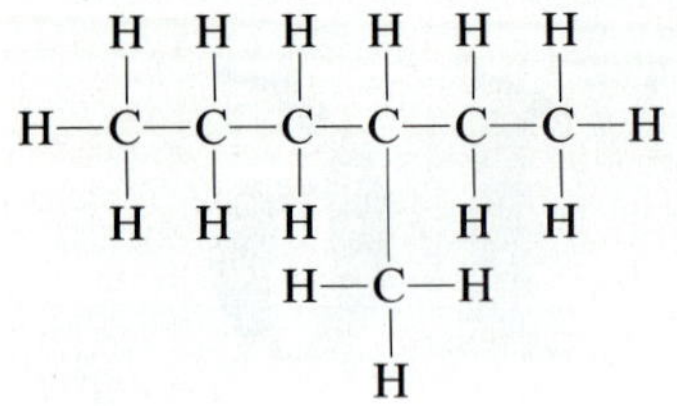

FIGURE 16.1
Saturated hydrocarbons are molecules that cannot accept any more atoms. Each carbon atom is bonded to four other atoms.

The hydrocarbons introduced in the last chapter—the alkanes and cycloalkanes—are known as *saturated* hydrocarbons because all carbon–carbon bonds are single bonds and every carbon bond is connected to a different atom. No more atoms can be placed on any carbon atom of the molecule; it is "saturated" (Figure 16.1).

Another class of hydrocarbon molecules exists as well. These molecules do not have a different atom connected to every bond; they are known as **unsaturated** hydrocarbons. Figure 16.2a shows structural diagrams and ball-and-stick models of two hydrocarbons. One of these molecules is saturated; you will recognize it as ethane. The second molecule is similar to ethane but has one important difference: it contains a double bond. It does not have enough hydrogen atoms to fill all four of the available carbon bonds. To achieve chemical stability, the unfilled orbitals share electrons with one another. This results in another bond between the carbon atoms. Organic compounds having a double bond between carbon atoms are called **alkenes.**

Alkenes and *alkynes* are classified as *unsaturated* hydrocarbons. They have double and triple bonds, respectively.

Hydrocarbons are not limited to double bonds between carbons. If two less hydrogen atoms are present, a triple bond will result between the carbon atoms. Hydrocarbons with one or more triple bonds are called **alkynes.** Examples of compounds containing triple bonds are shown in Figure 16.3.

Unsaturated hydrocarbons have characteristics that make them different

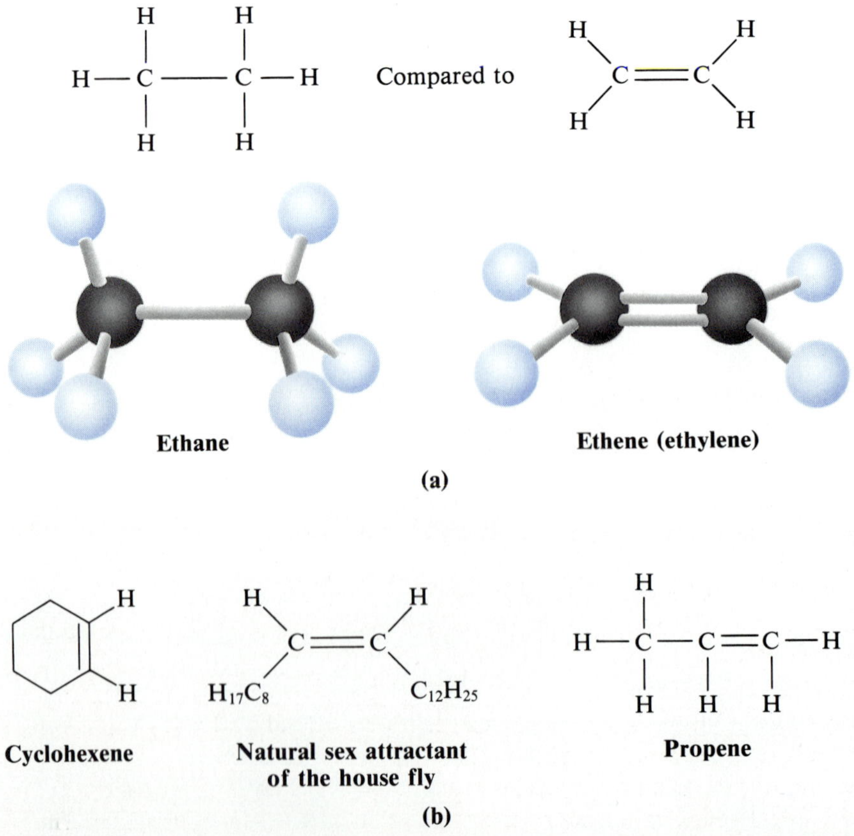
Compared to
Ethane
Ethene (ethylene)
(a)
Cyclohexene
Natural sex attractant of the house fly
Propene
(b)

FIGURE 16.2
Alkenes (a) Ethane (on the left) is saturated (an alkane), whereas ethene (or ethylene) (on the right) is unsaturated. It is the simplest member of the alkene class of hydrocarbons. Ethylene is extensively used as an industrial intermediate and to help ripen fruit. (b) Other examples of manmade and natural alkenes.

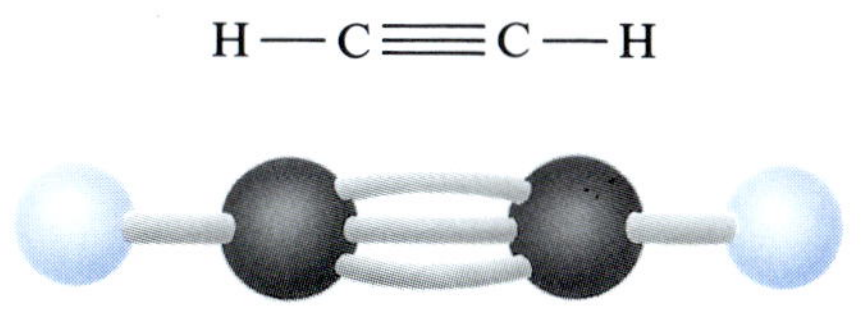
H—C≡C—H

Acetylene

(a)

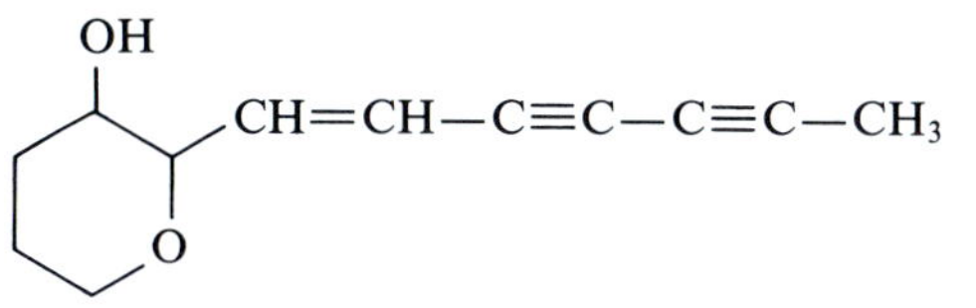
OH
CH=CH—C≡C—C≡C—CH3
O

Ichthyothereol
(natural product used by Indians of the Amazon as arrow-tip poison)

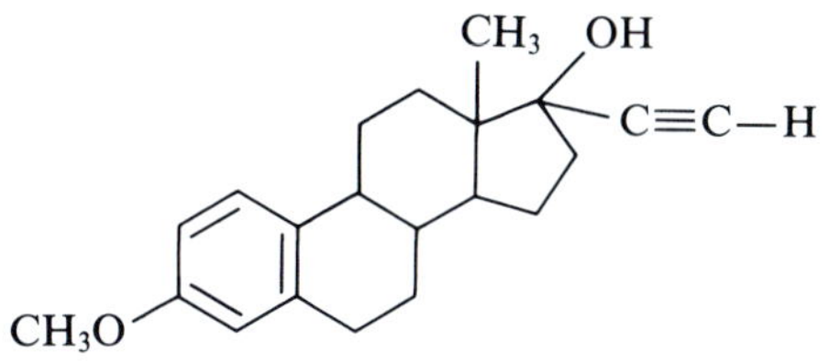
CH3
OH
C≡C—H
CH3O

Mestranol
(synthetic steroid used in birth control)

(b)

FIGURE 16.3
Alkynes (a) The simplest member of the alkyne series is acetylene. Note its triple bond. The alkyne functional group is not found to a large extent in naturally occurring molecules. (b) Examples of a natural and a manmade alkyne.

from saturated hydrocarbons. The extra one or two bonds between carbon atoms hold its electrons quite loosely, and they are unprotected. These electrons, acting as Lewis bases, are available for reaction with Lewis acids. Molecules with double and triple bonds are likely to react at the site of their multiple bond. Thus, much of the chemical behavior of the alkenes and alkynes can be understood as examples of Lewis acid–base chemistry.

NEW TERMS

unsaturated One or more double or triple bonds exist between carbon atoms in the molecule.

alkene A hydrocarbon having at least one double bond between carbon atoms.

alkyne A hydrocarbon having at least one triple bond between carbon atoms.

16.2 Structure And Physical Properties

Double-Bonded Molecules: The Alkenes

HYBRIDIZATION AND STRUCTURE In Chapter 14 we saw how structures for double-bonded compounds could be predicted from electron dot methods. It was shown that when atoms have a double bond, the atom holding the double bond has a planar geometry of its bonds to other atoms. Molecules of this type are classified as having sp^2 hybridization. This hybridization of atomic orbitals means that one s orbital of carbon and two p orbitals of carbon are "mixed" to give three new hybridized orbitals. The concept of orbital mixing was shown in Chapter 6 in the example about the tetrahedral shape of the water molecule, which has sp^3 hybridization. The

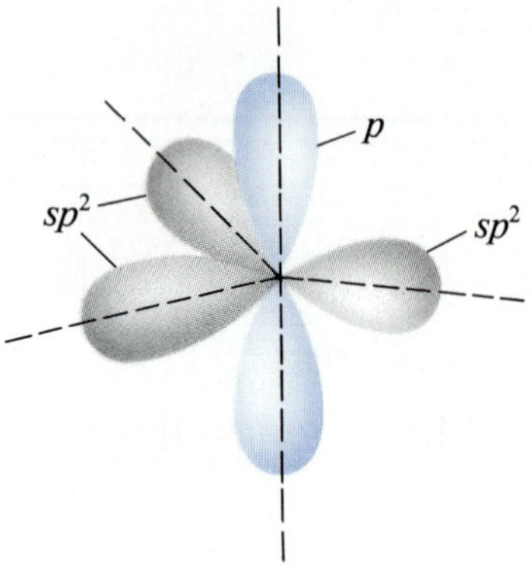
p
sp²
sp²

FIGURE 16.4
The sp^2 hybrid orbital. The three equivalent hybridized orbitals are in a plane with a bond angle of 120°. The p orbital is perpendicular to the plane formed by the hybridized orbitals.

sp^3 hybridization was also seen in Chapter 15 for the alkanes. The same approach is now used to construct the sp^2 hybridization.

$$2p \; \underline{\uparrow}\; \underline{\uparrow}\; \underline{\;\;} \xrightarrow{\text{Not hybridized}} 2p \; \underline{\uparrow}$$

$$2s \; \underline{\uparrow\downarrow} \xrightarrow{\text{Hybridized}} sp^2 \; \underline{\uparrow}\; \underline{\uparrow}\; \underline{\uparrow}$$

Atomic orbitals

We know from VSEPR theory that the three sp^2 orbitals arrange themselves to be as far apart as possible. This results in a flat, planar shape having bond angles between the orbitals of 120°. The remaining p orbital is perpendicular to the plane formed by the three hybrid orbitals (Figure 16.4).

The double bond of an alkene is made up of two different kinds of bonds. In the case of ethylene, for example, two of the hybrid orbitals on each carbon atom form single bonds with hydrogen atoms (see Figure 16.2). The third hybrid orbital forms a sigma bond with the other carbon atom. But what of the remaining p orbital electrons? The two adjacent p orbitals overlap to form another kind of bond called a **pi bond (π-bond)** (Figure 16.5). Since these electrons share a space between two atoms, they are attracted by both nuclei. This attractive force contributes to the increased strength of the double bond. However, since these electrons are not directly between the two nuclei, they do not contribute quite as much strength as do the sigma bonded electrons. A double bond is stronger than a single bond, but not twice as strong.

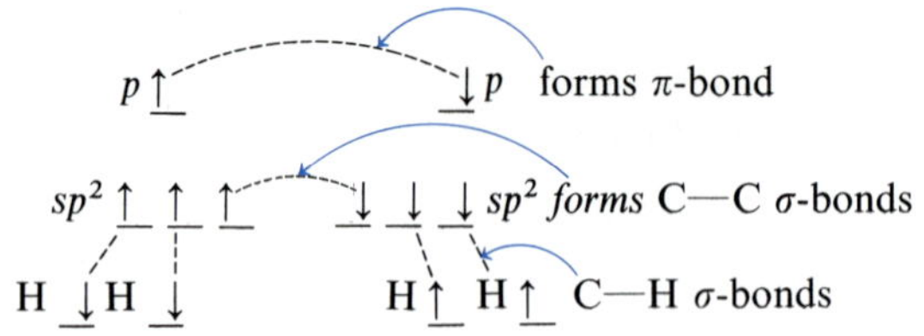
p ↑
↓ p
forms π-bond
sp² ↑ ↑ ↑
↓ ↓ ↓ sp²
forms C—C σ-bonds
H ↓ H ↓
H ↑ H ↑
C—H σ-bonds

FIGURE 16.5
The two bonds of the alkene double bond are not the same. The double bond is made up of one sigma bond and one pi bond. The pi bond, formed by overlap of adjacent p orbitals, has its electrons more loosely held in space.

There is *no* rotation about a C=C bond.

LACK OF ROTATION ABOUT THE DOUBLE BOND Another characteristic of the double bond is implied by the ball-and-stick model of ethylene shown in Figure 16.2. Single-bonded molecules can rotate about the sigma bonds holding the atoms together, but this rotation cannot occur in double-bonded molecules (Figure 16.6). The additional pi bond firmly fixes the carbon atoms with respect to one another; there can be no free rotation about the carbon–carbon double bond.

Boiling points of alkenes show a regular increase within a homologous series.

BOILING POINT AND CARBON NUMBER The boiling points of a series of alkenes are quite similar to those for the corresponding alkanes (Figure 16.7). This is because the only attractive forces between the molecules are the London forces. Because we are comparing the same number of carbon atoms and only two fewer hydrogens, there is little difference between the alkanes and the corresponding alkenes.

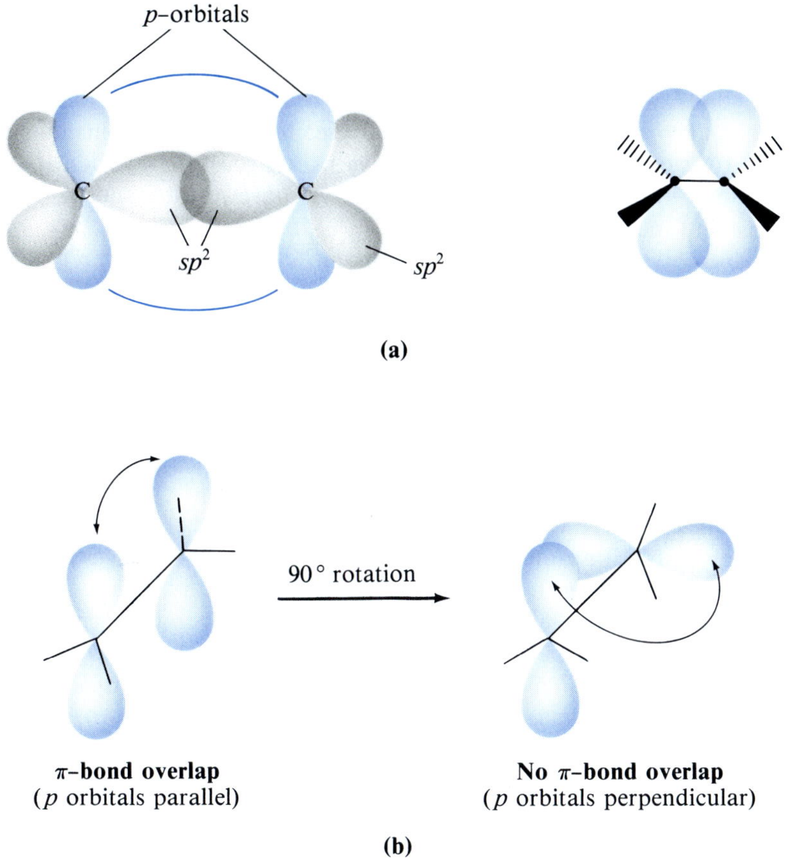

p-orbitals
C
C
sp^2
sp^2
(a)
90° rotation
π-bond overlap
(p orbitals parallel)
No π-bond overlap
(p orbitals perpendicular)
(b)

FIGURE 16.6
(a) The C—C double bond gives a special rigidity to this portion of the alkene because of the overlap of the adjacent *p* orbitals to form the pi bond. (b) If one rotated about the C—C double bond, the overlap of adjacent *p* orbitals would be lost.

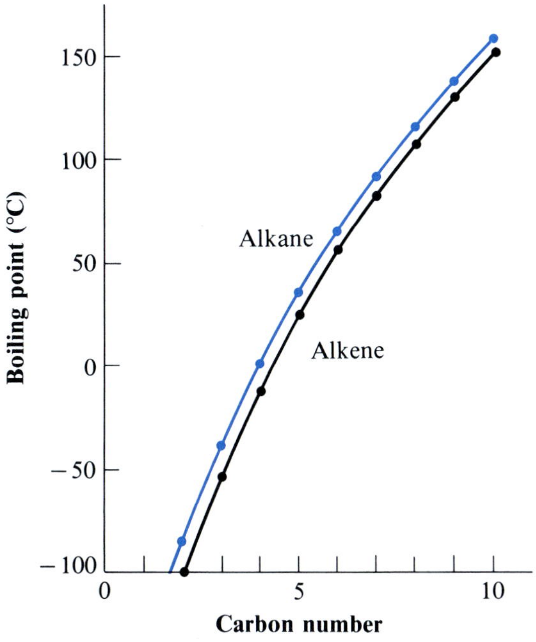

150
100
50
0
−50
−100
Boiling point (°C)
Alkane
Alkene
0
5
10
Carbon number

FIGURE 16.7
Boiling points of simple alkenes closely parallel those of the corresponding alkanes.

Triple-Bonded Molecules: The Alkynes

The carbon atoms involved in the pi bonds of the alkyne triple bond have *sp* hybridization.

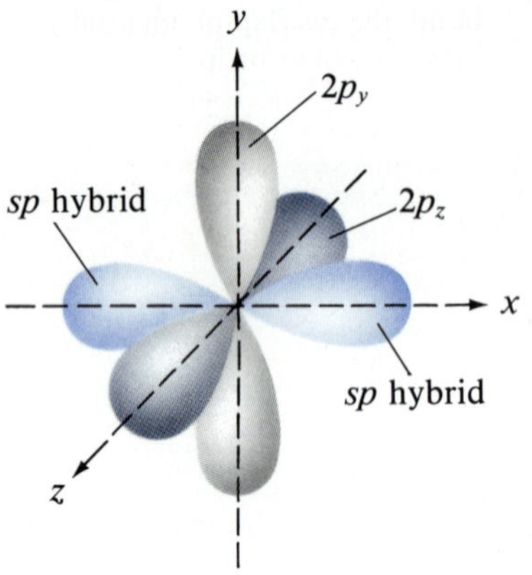

y
$2p_y$
sp hybrid
$2p_z$
x
sp hybrid
z

sp-hybridized carbon

HYBRIDIZATION AND STRUCTURE For carbon compounds containing triple bonds, we have already shown that the geometry about the carbon atoms tend to be in a linear arrangement (see Chapter 14). Atoms involved in this type of bonding are said to be *sp* hybridized. The hybridization is determined in exactly the same way that we have already shown for the alkanes and the alkenes.

Since only two sigma bonds are present, only one *s* and one *p* orbital must hybridize. The hybridization is thus *sp*.

$$2p\ \underline{\uparrow}\ \underline{\uparrow}\ \underline{\ \ } \xrightarrow{\text{Not hybridized}} 2p\ \underline{\uparrow}\ \underline{\uparrow}$$

$$2s\ \underline{\uparrow\downarrow} \xrightarrow{\text{Hybridized}} 2sp\ \underline{\uparrow}\ \underline{\uparrow}$$

Atomic orbitals

The geometric shape that will place two pairs of electrons as far apart as possible is a line. For example, acetylene (H—C≡C—H) is a *linear* molecule.

One of the carbon hybrid orbitals will bond to a hydrogen, the other to the adjacent carbon atom. The two *p* orbitals left over are perpendicular to each other. These each contain an electron that can participate in pi bonding as its orbital overlaps an adjacent *p* orbital. The triple bond is comprised of one sigma bond and two pi bonds, with the two pi bonds located at right angles to each other (Figure 16.8).

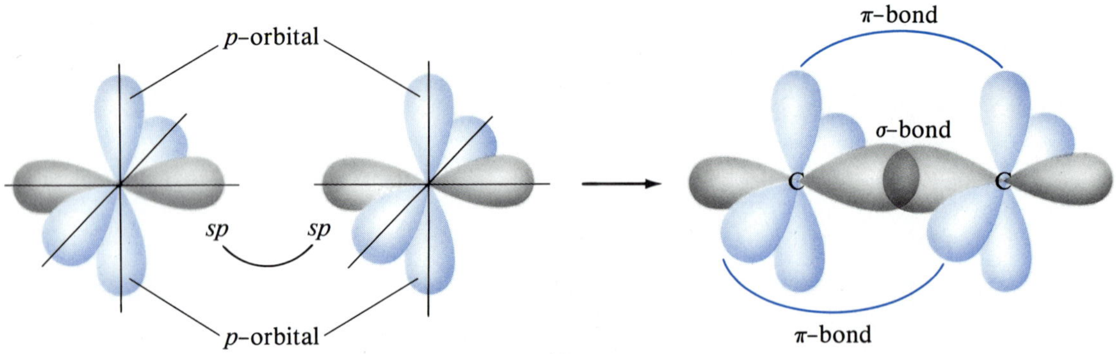

p-orbital
sp
sp
p-orbital
π-bond
σ-bond
C
C
π-bond

(a)

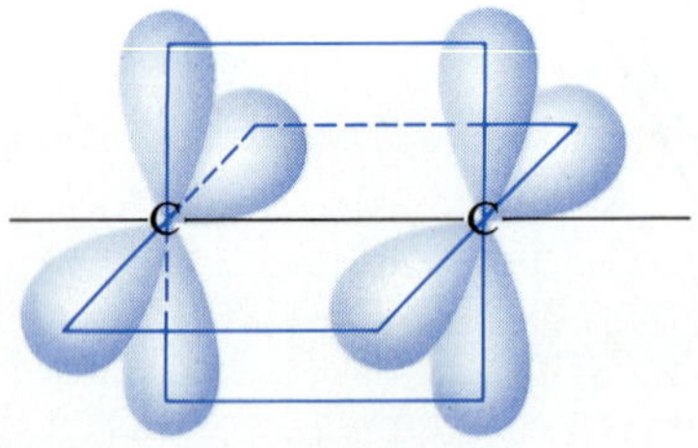

C
C

(b)

FIGURE 16.8
(a) Bonding in a triple-bonded molecule. The bonds consist of a single sigma bond and two pi bonds. (b) The lobes of the *p* orbitals overlap to form two pi bonds.

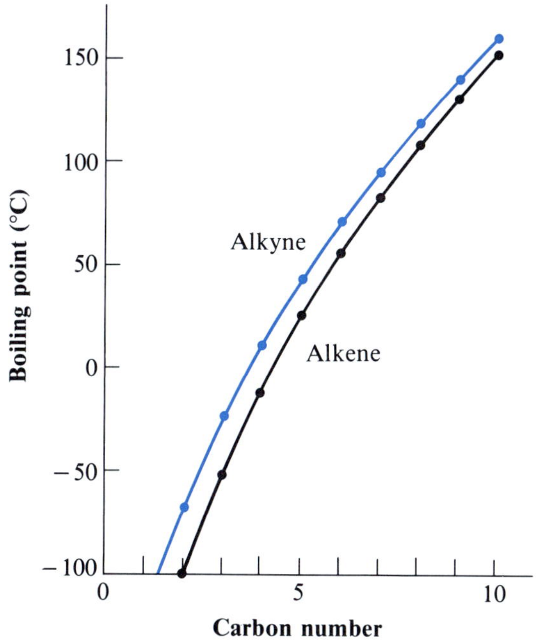
Boiling point (°C)
150
100
50
0
−50
−100
0
5
10
Carbon number
Alkyne
Alkene

FIGURE 16.9
The boiling points of alkynes are generally a little higher than those of the alkenes.

Triple-bonded molecules are much like double-bonded molecules, except that there are two pi bonds instead of one. The pi electrons are still on the "outside" of the molecule and are available for reaction. The pi bonds also restrict motion about the C—C sigma bond, although rotation of the entire linear molecule is still possible. Because of their linear geometry, triple-bonded molecules have no isomers. Only cyclic systems and alkenes have geometric isomers.

BOILING POINTS AND CARBON NUMBER The linear alkynes have physical properties similar to the alkenes; however, they generally have slightly higher boiling points (Figure 16.9). This observation is nicely in accord with our discussions about London forces. Because the alkyne is linear and the alkene has a bent shape (120°), the alkynes are easier to stack next to one another. This is similar to the discussion in Chapter 15 on the lower boiling points of the branched hydrocarbons.

The boiling points of alkynes are similar to those for the corresponding alkenes.

Isomers of Alkenes: Cis and Trans Relationships

One of the consequences of the geometry imposed by the bonding of alkenes is the existence of cis and trans isomers of alkenes. The same principles apply as for the cycloalkanes, for which totally free rotation around the C—C bond is also restricted. The terminology is the same, too. The cis isomer has similar groups on the *same side* of the pi bond, while the trans isomer has similar groups on *opposite sides* of the pi bond (Figure 16.10). The cis and trans isomers are essentially different compounds.

Cis alkenes have similar groups on the *same* side of the pi system:

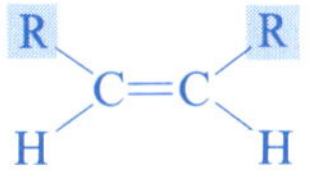
R
R
C=C
H
H

while trans isomers have similar groups on *opposite* sides:

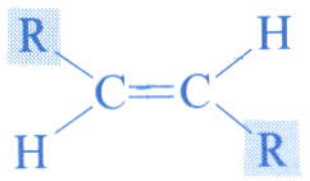
R
H
C=C
H
R

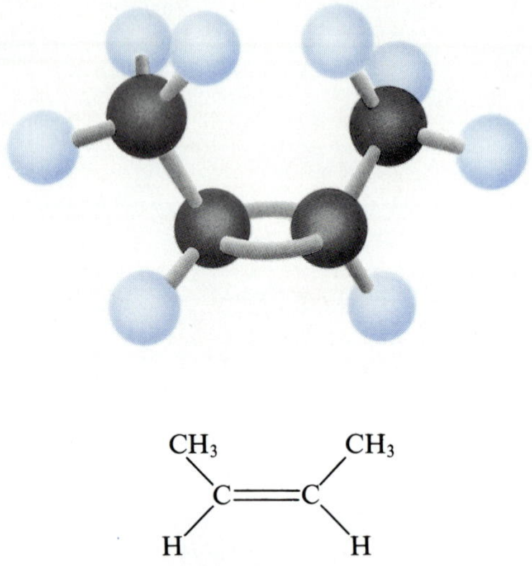

***cis*-2-Butene**
(boiling point 3.7 °C)

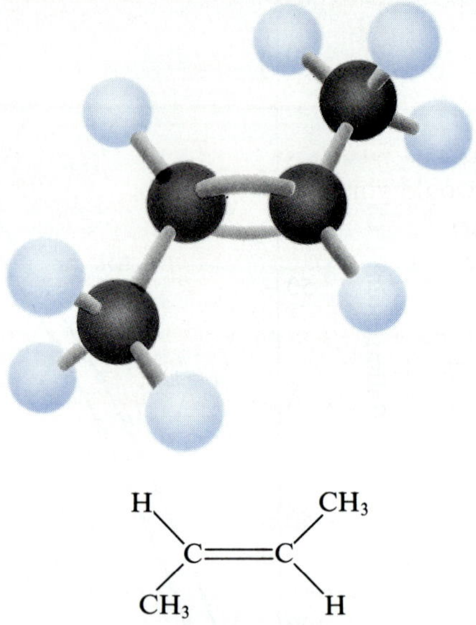

***trans*-2-Butene**
(boiling point 0.9 °C)

FIGURE 16.10
The geometric consequences of cis and trans isomers for alkenes. Since the pi bond prevents rotation about the C—C bond, the cis and trans isomers are truly different compounds and cannot be converted from one to the other by simple physical methods.

Fats and Oils

The presence of multiple bonds can often be seen in the physical behavior of some types of compounds. **Triglycerides**, or **triacylglycerols**, are complex esters that contain long carbon chains as part of their structure (Figure 16.11). If these hydrocarbon chains are saturated with hydrogen, the triacylglycerol is usually a solid at room temperature and is often classified as a *fat*. If the hydrocarbon chain contains double bonds (that is, it is unsaturated), the melting point is lowered and the material is often liquid at room temperature. This is an *oil*. Thus, the physical differences between fats and oils at room temperature are often due to the double bonds.

Unsaturated triacylglycerols are thought to be less likely to participate in the formation of fatty deposits in the blood vessels. Unsaturated cooking oils are therefore generally recommended for health reasons. (See Chapters 24 and 28 for more on fats and oils.)

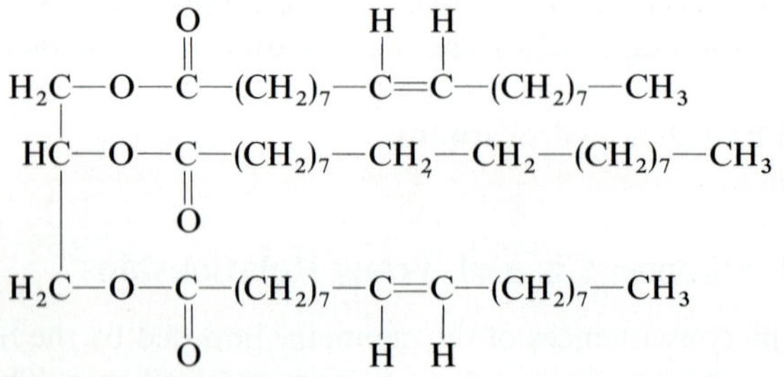

FIGURE 16.11
A general structure of a triacylglycerol. Many unsaturated fats and oils have double bonds in the hydrocarbon portion of the molecules.

NEW TERMS

pi bond (π-bond) A bond formed by overlap of unhybridized *p* orbitals of two adjacent atoms. No more than two pi bonds can exist between two adjacent atoms.

triglyceride (triacylglycerol) A complex ester having a long hydrocarbon chain.

TESTING YOURSELF

Properties and Structures of Alkenes and Alkynes

1. Why would you expect the water solubility of octane and 1-octene to be similar?
2. Why are there cis and trans isomers of 3-hexene but not of 3-hexyne?

Answers **1.** Both molecules are hydrocarbons whose attractions are limited to London forces. For both there is not sufficient ability for hydrogen bonding to break up this arrangement of water. **2.** The alkyne C—C triple bond is linear and thus cannot have isomers.

16.3 Nomenclature

Naming the Alkenes

Nomenclature for the alkenes is similar to that for the alkanes. Cis and trans isomers must be distinguished by including these terms in their names. As we have already seen for the cycloalkanes, the cis isomers have similar groups on the same side of the bond, while trans isomers have similar groups on opposite sides. In Figure 16.12 we show that in the nomenclature of alkenes, the cis or trans label is applied to the longest continuous carbon chain containing the double bond. Thus, if the longest continuous chain is on the same side of the pi bond, we call it a cis isomer; if it is on opposite sides, we call it a trans isomer.

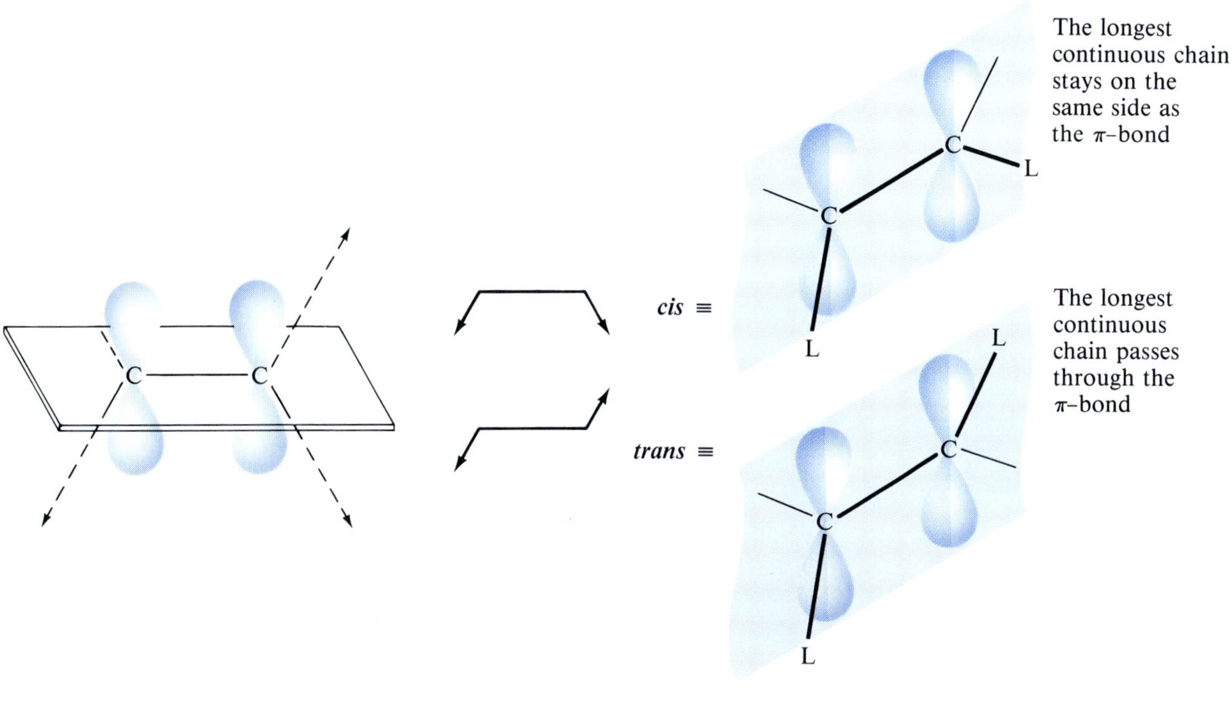

C
C
cis ≡
trans ≡
C
C
L
L
The longest continuous chain stays on the same side as the π-bond
C
C
L
L
The longest continuous chain passes through the π-bond

FIGURE 16.12
How alkene isomers are distinguished. The cis isomer has the longest continuous chain on the same side of the pi bond while the trans isomer has this chain passing through the pi bond.

STRANGE HYDROCARBONS

It may be hard to believe, but organic chemists have a sense of humor too! As a result of interest in the preparation and chemical behavior of strained hydrocarbon systems, a number of unique compounds have been prepared. Because many of these have invoked some inner artistic image, they have received unusual names that express their strange-looking shapes. See if you can find the reason for the names in the few examples presented here.

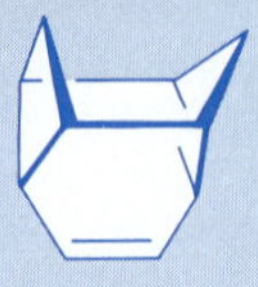

Felicene

Pterodactyladiene

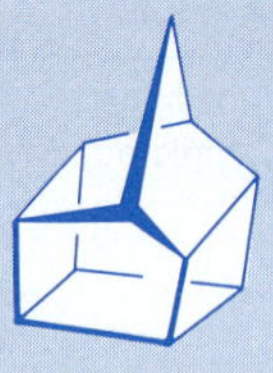

Churchane

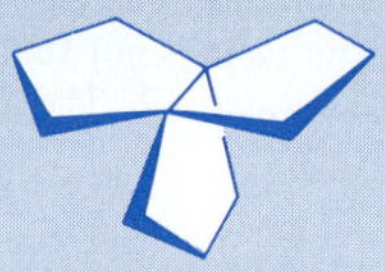

Propellane

Alkene nomenclature is similar to alkane nomenclature, except that there is an *-ene* ending, and the position of the double bond must be designated.

The rules for naming alkenes can be summarized as follows.

1. Find the longest carbon chain containing the double bond. This defines the parent compound.
2. Number the carbons from the end that places the double bond between the smallest possible numbers. Designate the location of the double bond with the lowest of these two numbers.
3. If enough information is available about the molecule to decide if it is cis or trans, make this designation.
4. Name substituted alkyl groups, as was done for the alkanes.
5. Replace the *-ane* ending of the parent alkane with *-ene*.

EXAMPLE 16.1

Name this alkene.

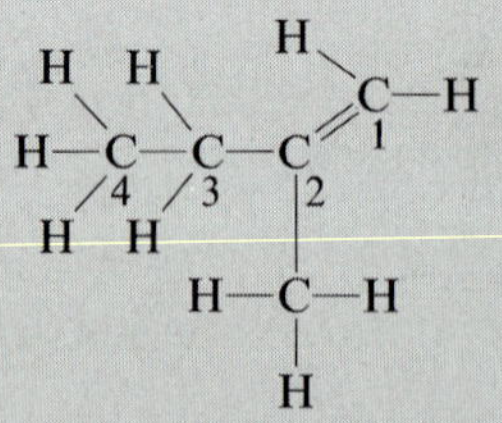

The longest chain that contains the double bond has four carbons. The parent compound is thus a butene. If we number from the left, the double bond is between carbons 3 and 4. If we number from the right, the double bond is between carbons 1 and 2. Numbering from the right gives the lowest number for the location of the double bond. The compound is a 1-butene. The methyl group is attached to carbon 2. The name of this compound is thus 2-methyl-1-butene.

Listed here are some more alkene structures with their correct names.

$CH_3—CH{=}CH—CH_2—CH_3$ **2-Pentene**

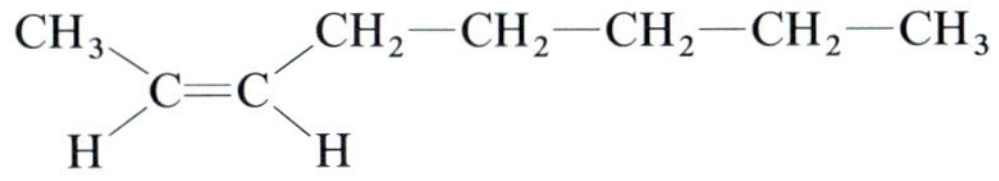

***cis*-2-Octene**

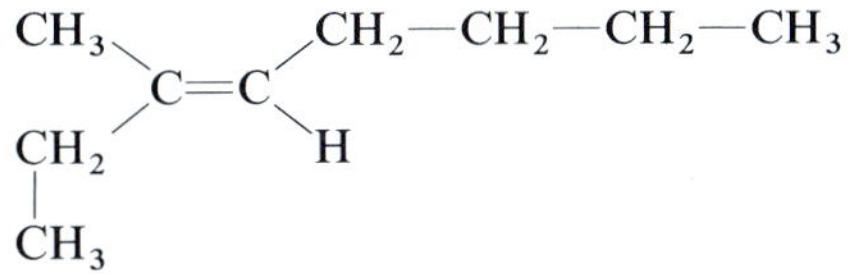

***trans*-3-Methyl-3-octene**

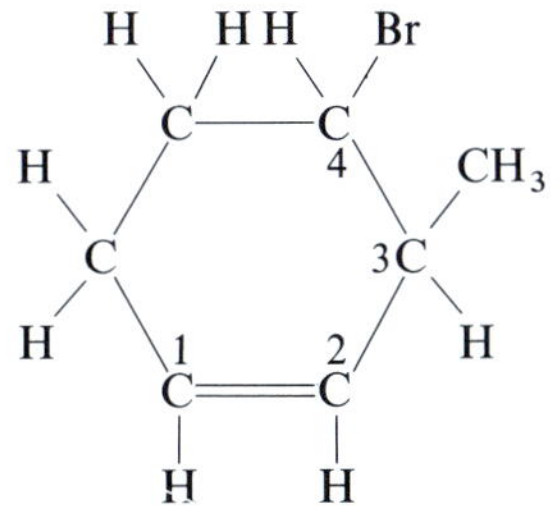

4-Bromo-3-methylcyclohexene

Polyunsaturated Alkenes

Often a carbon chain has more than one double bond. Multiple unsaturation is reflected in nomenclature as well as in unique chemical reactivity. A **polyunsaturated** compound has more than one double bond.

Polyunsaturated molecules have more than one double or triple bond.

The number of double bonds can be described by the prefixes *di-*, *tri-*, *tetra-*, and so on. The name also has to tell *where* the double bonds are located. Nomenclature rules are very similar to those already defined for simple alkenes and are most readily demonstrated by an example.

EXAMPLE 16.2

Name this polyunsaturated alkene.

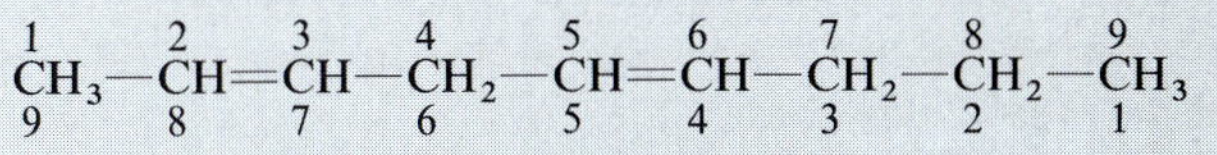

The longest chain containing the double bonds has nine carbons, so the parent compound is nonene. Since there are two double bonds, we add *di-* to the name, to get nonadiene. The lowest numbers for the locations of the double bonds are 2 and 5. Thus, the name for this compound is 2,5-nonadiene.

Listed here are some more polyunsaturated alkenes and their correct names.

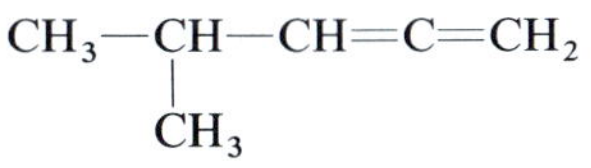

4-Methyl-1,2-pentadiene

$CH_3—CH{=}CH—CH{=}CH_2$ **1,3-Pentadiene**

(continued)

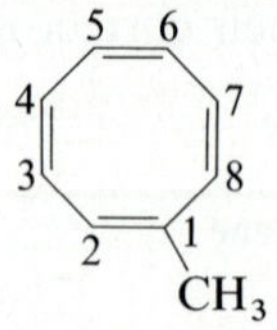
5
6
4
7
3
8
2
1
CH_3

1-Methyl-1,3,5,7-cyclooctatetraene

Naming the Alkynes

Alkyne nomenclature is similar to that for alkenes, except that there is an *-yne* ending.

The IUPAC rules for naming alkynes are similar to those for the alkanes and alkenes, except that the suffix *-yne* is used to indicate the presence of the triple bond. (See the rules listed for alkenes on p. 430.) Briefly summarized, alkyne names are constructed this way:

Name = Substituents and locations + Location of triple bond + Parent compound + −*yne*

Listed here are some alkyne structures with their correct names.

H—C≡C—H **Ethyne (acetylene)**

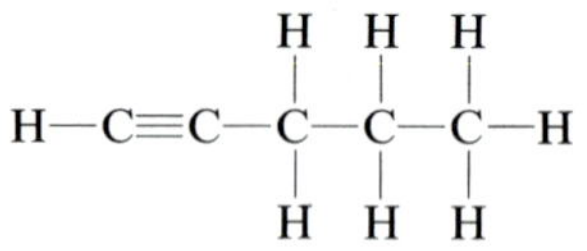

1-Pentyne

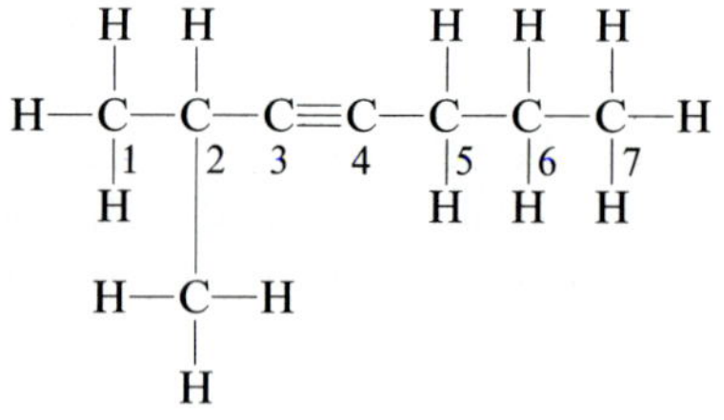
1
2
3
4
5
6
7

2-Methyl-3-heptyne

NEW TERM

polyunsaturated Molecules having more than one double or triple bond.

TESTING YOURSELF

Naming Alkenes and Alkynes

1. Name the following alkenes.

a. 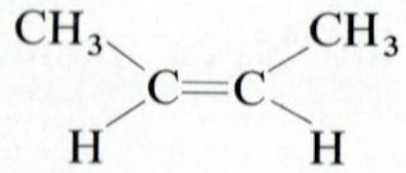

b. 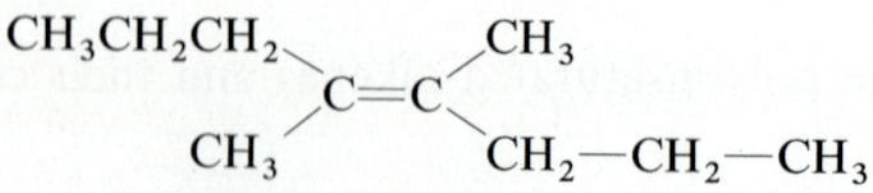

c. 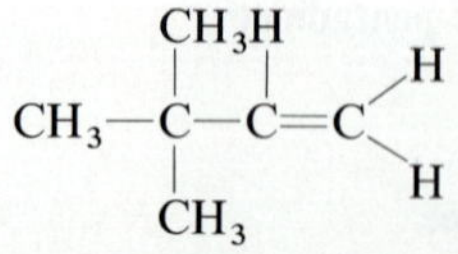

d.

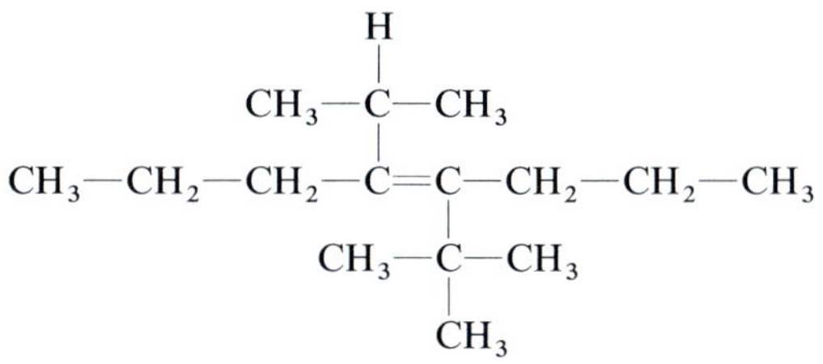

2. Name the following polyunsaturated alkenes.

a.

b.

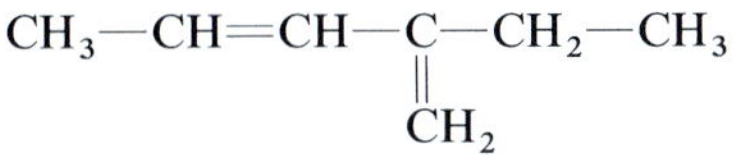

c. $CH_2{=}CH{-}CH{=}CH{-}CH{=}CH{-}CH_2{-}CH_2{-}CH_3$

d.

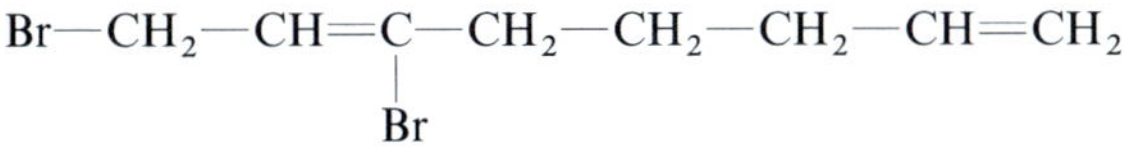

3. Name the following alkynes.

a.

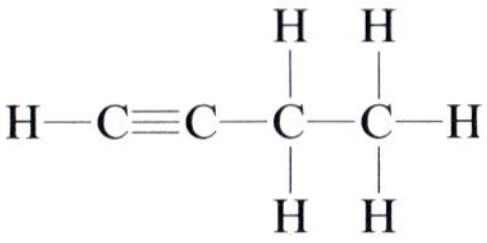

b.

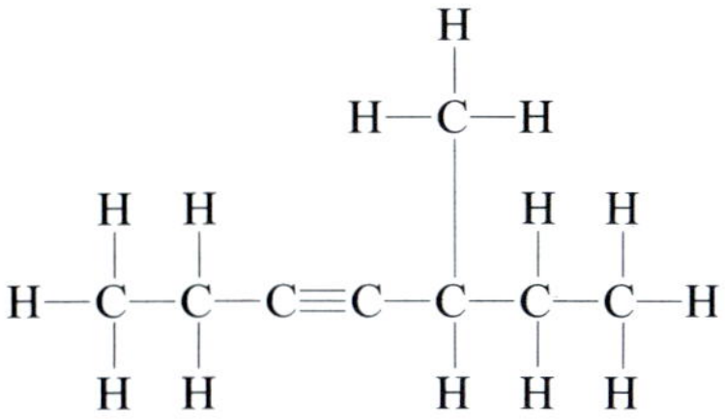

c.

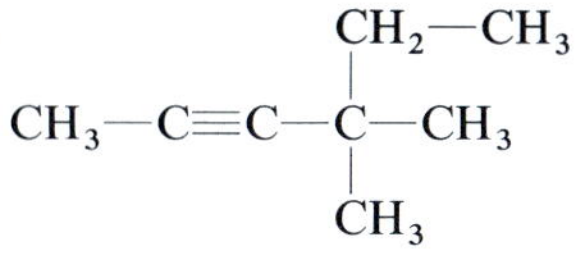

d.

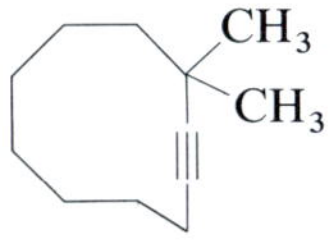

Answers **1. a.** *cis*-2-butene **b.** *trans*-4,5-dimethyl-4-octene **c.** 3,3-dimethyl-1-butene **d.** 4-isopropyl-5-*tert*-butyl-4-octene **2. a.** 1,3-cyclohexadiene **b.** 2-ethyl-1,3-pentadiene **c.** 1,3,5-nonatriene **d.** 6,8-dibromo-1,6-octadiene **3. a.** 1-butyne **b.** 5-methyl-3-heptyne **c.** 4,4-dimethyl-2-hexyne **d.** 3,3-dimethylcyclodecyne

16.4 Chemical Reactivity of Alkenes

The exposed pi bonding electrons of the alkenes give this family of compounds enhanced chemical reactivity. We will examine four types of common reactions in this section.

The pi electrons of the alkene and alkyne identify them as Lewis bases, and thus the unsaturated compounds are reactive to Lewis acids.

1. Combustion, with or without sufficient oxygen.
2. Addition reactions, with symmetric or nonsymmetric reactants.
3. Oxidation reactions, adding oxygen or hydroxide.

4. Polymerization reactions, in which the molecules form extremely long and often cross-linked molecules.

Combustion Reactions

We have already seen that one reaction characteristic of all hydrocarbons is combustion with oxygen. Since alkenes are hydrocarbons, it is no surprise to observe the same reaction. This reaction can be generalized as follows.

GENERAL REACTION

Combustion of an Alkene with Sufficient Oxygen

$$C_nH_{2n} + 1.5n\ O_2 \longrightarrow n\ CO_2 + n\ H_2O$$

Alkene, Oxygen → Carbon dioxide, Water

Example

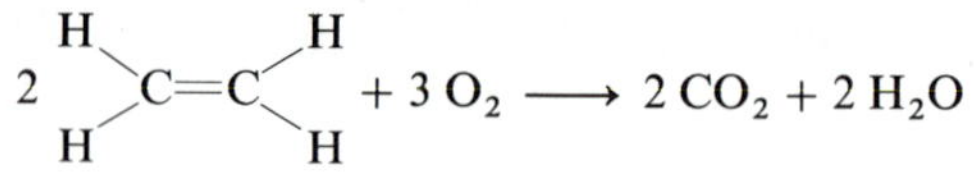

Combustion produces carbon dioxide and water vapor. If insufficient oxygen is present, alkenes will also burn to produce poisonous carbon monoxide.

Addition Reactions

In *addition* reactions, new atoms add to the double bond, converting it to a single bond.

More important than combustion, however, is *addition* of other elements or groups to the double bond. These reactions divide into two classes: (1) reactions in which the same substituent is added to each carbon of the double bond, and (2) reactions in which different substituents are added to each carbon of the double bond. Reactions observed for the alkene family are also characteristic of the alkyne family of compounds. However, in the case of the alkynes, two pi bonds are available for reaction.

Addition reactions are of the general form

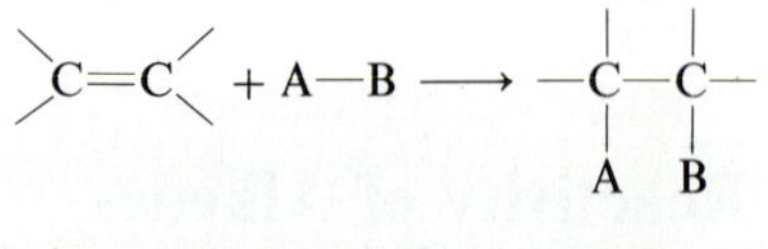

When an alkene undergoes an addition reaction, the double bond is lost. Reactions take place at the double bond.

If the reactant A—B is symmetrical (for example, H—H or Br—Br), the addition to the double bond is symmetrical. For example, hydrogen gas reacts with an alkene (an unsaturated hydrocarbon) in the presence of a metal catalyst such as platinum (Pt) to produce a saturated hydrocarbon.

A NOTE ABOUT WRITING EQUATIONS IN ORGANIC CHEMISTRY There are a number of ways to write reactions or equations in organic chemistry, but all should do the same thing: (1) show *what* reacted, (2) show special conditions, and (3) show products. Here are two general forms for the same reaction:

$$A + B \xrightarrow{\text{Conditions}} C + D$$

$$A \xrightarrow[\text{Conditions}]{B} C + D$$

Note that the second reactant (B) can be shown either before the reaction arrow or above it. You will see it done both ways throughout this and other textbooks.

ADDITION OF HYDROGEN (HYDROGENATION) The reaction of ethene with hydrogen in the presence of a catalyst can be shown in two equivalent ways (according to the preceding discussion).

$$\underset{H}{\overset{H}{>}}C=C\underset{H}{\overset{H}{<}} + H_2 \xrightarrow[\text{(catalyst)}]{Pt} H-\underset{\underset{H}{|}}{\overset{\overset{H}{|}}{C}}-\underset{\underset{H}{|}}{\overset{\overset{H}{|}}{C}}-H$$

$$\underset{H}{\overset{H}{>}}C=C\underset{H}{\overset{H}{<}} \xrightarrow[\text{Pt (catalyst)}]{H_2} H-\underset{\underset{H}{|}}{\overset{\overset{H}{|}}{C}}-\underset{\underset{H}{|}}{\overset{\overset{H}{|}}{C}}-H$$

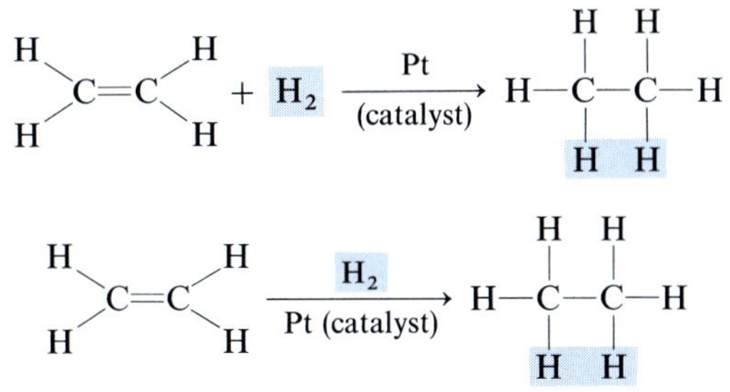

Ethene   **Ethane**

A reaction in which hydrogen is added to a double bond is called hydrogenation. **Hydrogenation** is used commercially to convert liquid oils to fats. As the number of double bonds is reduced, the melting point of the fat increases. Such a reaction, called *hardening*, is used to convert vegetable oils into margarine. Although oils and fats are complex esters (we encountered these earlier in this chapter), the functional group modification is the reduction of an alkene to an alkane.

The hydrogenation reaction can be generalized as follows.

The hydrogenation of an oil to a fat is called *hardening*, because a liquid is converted to a solid.

GENERAL REACTION

Reaction of an Alkene with Hydrogen

$$>C=C< + H_2 \xrightarrow[\text{(catalyst)}]{Pt} -\underset{\underset{H}{|}}{\overset{|}{C}}-\underset{\underset{H}{|}}{\overset{|}{C}}-$$

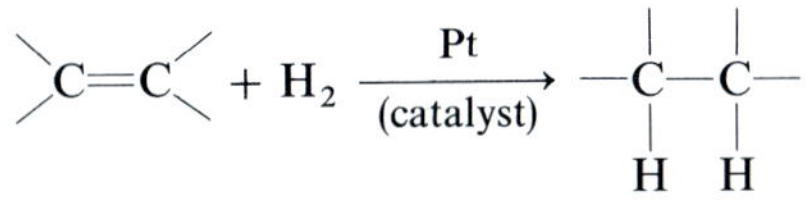

Example

$$CH_3CH=CHCH_3 + H_2 \xrightarrow[\text{(catalyst)}]{Pt} CH_3CH_2CH_2CH_3$$

EXAMPLE 16.3

Show the hydrogenation of vegetable oil.

$$\begin{array}{l}H_2C-O-\overset{O}{\underset{\|}{C}}-(CH_2)_7-\overset{H}{\underset{}{C}}=\overset{H}{\underset{}{C}}-(CH_2)_7-CH_3 \\ HC-O-\underset{\underset{O}{\|}}{C}-(CH_2)_7-CH_2-CH_2-(CH_2)_7-CH_3 \\ H_2C-O-\underset{\underset{O}{\|}}{C}-(CH_2)_7-\underset{\underset{H}{|}}{C}=\underset{\underset{H}{|}}{C}-(CH_2)_7-CH_3\end{array} + H_2 \xrightarrow[\text{(catalyst)}]{Pt}$$

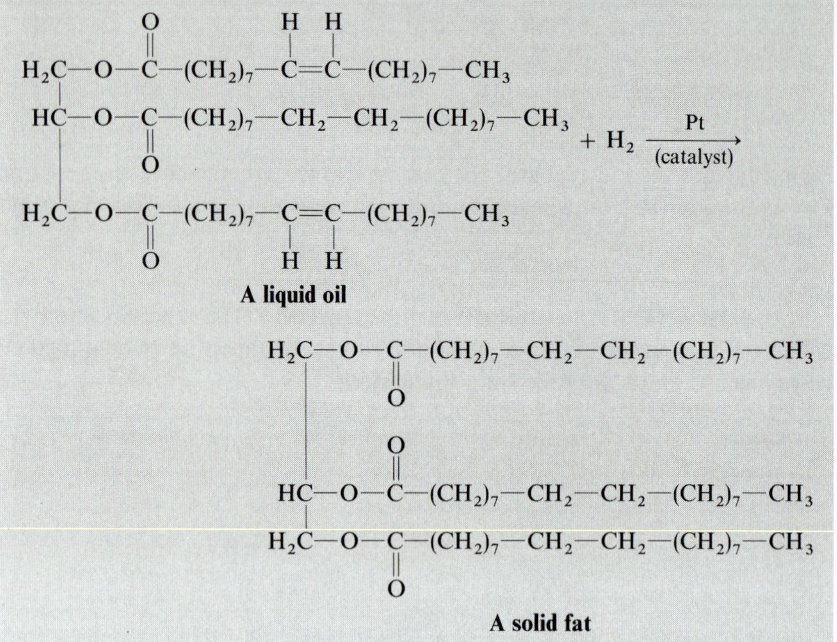

A liquid oil

$$\begin{array}{l}H_2C-O-\underset{\underset{O}{\|}}{C}-(CH_2)_7-CH_2-CH_2-(CH_2)_7-CH_3 \\ HC-O-\underset{\underset{O}{\|}}{C}-(CH_2)_7-CH_2-CH_2-(CH_2)_7-CH_3 \\ H_2C-O-\underset{\underset{O}{\|}}{C}-(CH_2)_7-CH_2-CH_2-(CH_2)_7-CH_3\end{array}$$

A solid fat

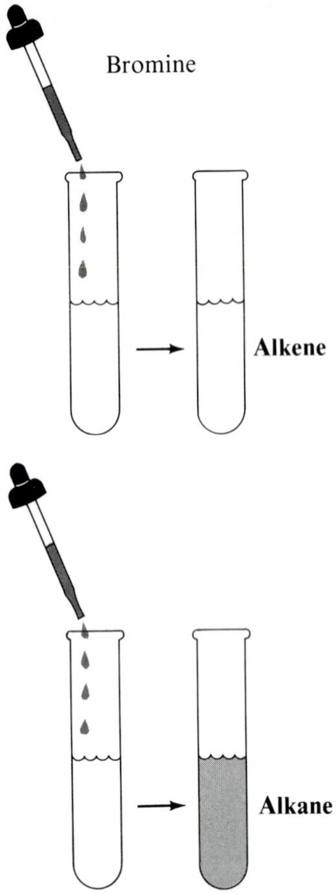

FIGURE 16.13
The addition of bromine to an alkene. The loss of the bromine color is an indication that the compound is an alkene; this observation is often used as a qualitative test for alkenes. If no double bond is present, such as in an alkane, the bromine color will not go away. The addition of bromine will also give a positive test for an alkyne.

ADDITION OF HALOGENS (HALOGENATION) The double bond of an alkene reacts rapidly with chlorine or bromine at room temperature without a catalyst. These are halogens so this reaction is called a **halogenation** reaction. Thus, for example, we would observe:

$$\underset{H}{\overset{H}{\diagdown}}C=C\underset{H}{\overset{H}{\diagup}} + Br_2 \longrightarrow H-\underset{\underset{Br}{|}}{\overset{\overset{H}{|}}{C}}-\underset{\underset{Br}{|}}{\overset{\overset{H}{|}}{C}}-H$$

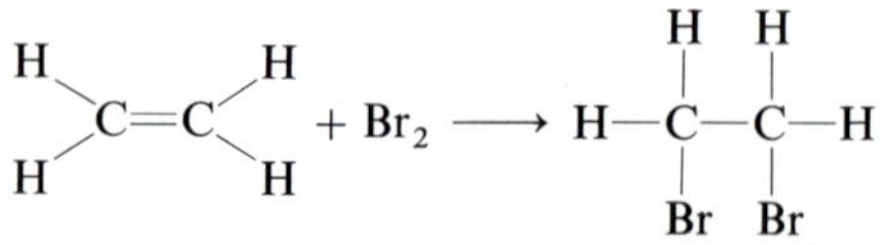

The relative reactivity of the halogens in this situation is

$$F > Cl > Br > I$$

Fluorine reacts explosively with alkenes, while iodine will not add to an alkene double bond under normal conditions. It makes sense, then, that chlorine (Cl_2) and bromine (Br_2) are the commonly used reactants.

Bromine, a dark brown liquid with dangerous fumes, is often used as a test reagent for identification of an alkene (Figure 16.13). It is usually prepared as a dilute solution in an inert solvent such as carbon tetrachloride. A small sample of the suspected alkene is placed in a test tube, and a few drops of the deeply colored bromine solution are added. If a double bond is present, the bromine will add to this bond and the brown bromine color will disappear. If double bonds are not present, the bromine will not react and the solution will remain brown.

The halogenation reaction can be generalized as follows.

16.4 CHEMICAL REACTIVITY OF ALKENES

> **GENERAL REACTION**
>
> Reaction of an Alkene with a Halogen

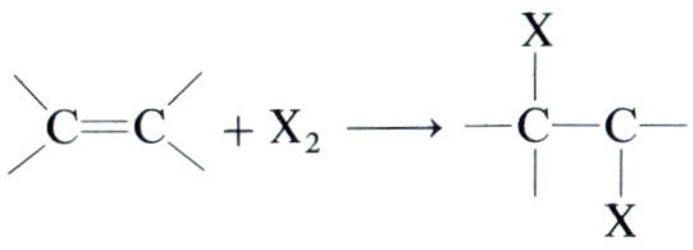

>
> $$\begin{array}{c}\\ \diagdown\diagup\\ C=C\\ \diagup\diagdown\end{array} + X_2 \longrightarrow \begin{array}{c}X\\ |\\ -C-C-\\ |\\ X\end{array}$$ where X is any halogen.
>
> Example
>
> $$CH_3CH=CHCH_3 + Br_2 \longrightarrow H-\underset{\underset{H}{|}}{\overset{\overset{H}{|}}{C}}-\underset{\underset{H}{|}}{\overset{\overset{Br}{|}}{C}}-\underset{\underset{Br}{|}}{\overset{\overset{H}{|}}{C}}-\underset{\underset{H}{|}}{\overset{\overset{H}{|}}{C}}-H$$

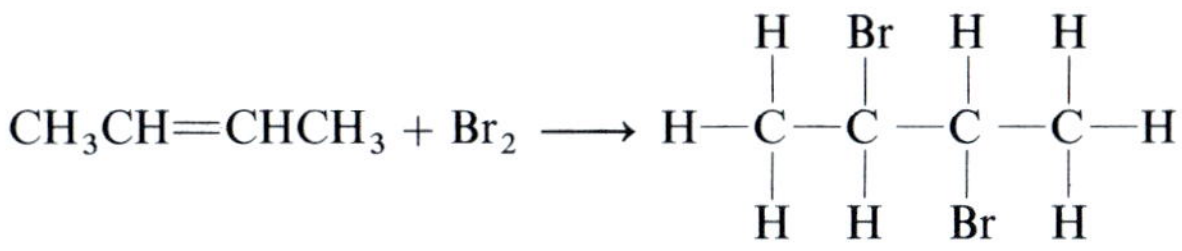

ADDITION OF HYDROCHLORIC ACID Nonsymmetric reactants, such as hydrochloric acid (HCl), react with double bonds in the same manner as the symmetric additions of H_2 and Br_2 discussed earlier. However, this reaction has two possible isomer products. Consider the example of the reaction of HCl with propene.

$$\underset{H}{\overset{H}{\diagdown}}C=C\underset{H}{\overset{CH_3}{\diagup}} + HCl \longrightarrow H-\underset{\underset{H}{|}}{\overset{\overset{H}{|}}{C}}-\underset{\underset{Cl}{|}}{\overset{\overset{H}{|}}{C}}-\underset{\underset{H}{|}}{\overset{\overset{H}{|}}{C}}-H$$

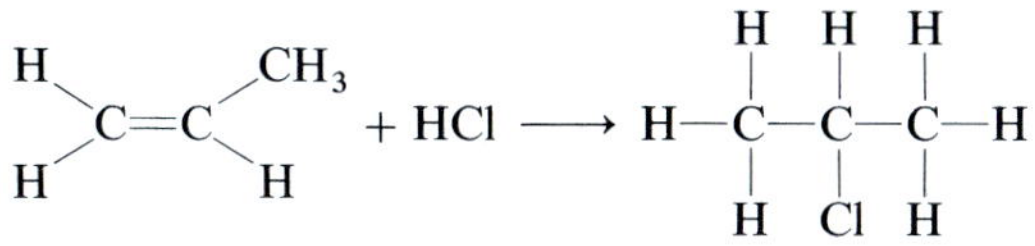

Propene   **2-Chloropropane** (principal product)

This reaction shows the chlorine connecting to the center carbon, while the hydrogen from the HCl bonds to the outer carbon. The other possibility would be for the chlorine to attach to the outer carbon and the hydrogen to the center carbon.

$$\underset{H}{\overset{H}{\diagdown}}C=C\underset{H}{\overset{CH_3}{\diagup}} + HCl \longrightarrow H-\underset{\underset{Cl}{|}}{\overset{\overset{H}{|}}{C}}-\underset{\underset{H}{|}}{\overset{\overset{H}{|}}{C}}-\underset{\underset{H}{|}}{\overset{\overset{H}{|}}{C}}-H$$

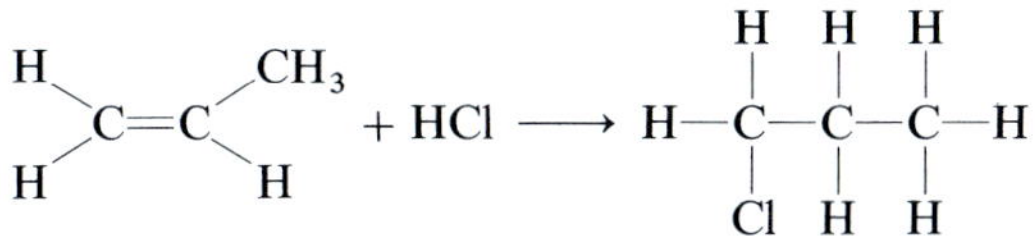

Propene   **1-Chloropropane** (minor product)

Both products shown above are possible, although the 2-chloropropane isomer predominates.

ADDITION OF WATER (HYDRATION) In the presence of acid, water will add to an alkene double bond:

$$\underset{H}{\overset{H}{\diagdown}}C=C\underset{H}{\overset{CH_3}{\diagup}} + H_2O \xrightarrow[\text{(catalyst)}]{H^+} H-\underset{\underset{H}{|}}{\overset{\overset{H}{|}}{C}}-\underset{\underset{\overset{O}{|}\atop H}{|}}{\overset{\overset{H}{|}}{C}}-\underset{\underset{H}{|}}{\overset{\overset{H}{|}}{C}}-H$$

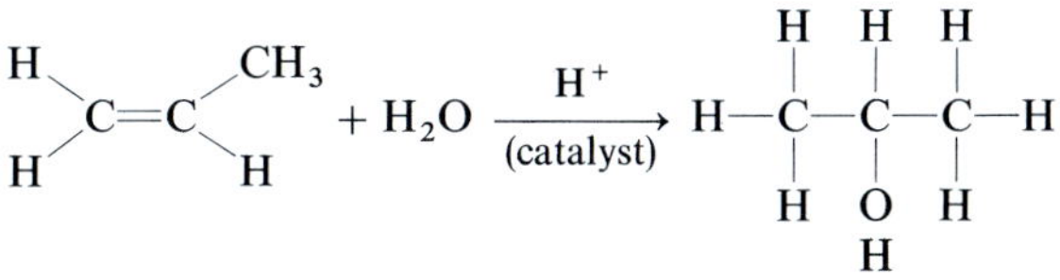

Propene   **2-Propanol** (or isopropyl alcohol)

This addition of water to the double bond is called **hydration** and results in the formation of an alcohol. The product of the hydration of propene is 2-propanol, or isopropyl alcohol. The compound is commonly known as rubbing alcohol. However, two different isomer products are possible here, also. The other product would be formed according to this reaction:

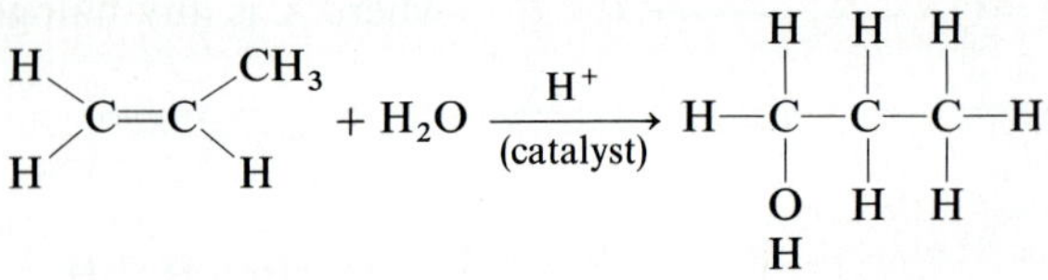

Propene **1-Propanol (or *n*-propyl alcohol)**

With hydration, the 2-propanol (isopropyl alcohol) predominates as the product. This type of reaction between ethene and water is commercially used to prepare millions of gallons of ethanol (ethyl alcohol) each year.

MARKOVNIKOV'S ADDITION Organic chemists are often accused of being a little careless in the balancing of the chemical equations that describe organic reactions. There is, however, a reason for this. Inorganic chemists predict with reasonable certainty the products of their reactions. In organic chemistry, a number of reactions are almost always possible, and one can only indicate the relative probability that a certain product will be synthesized. The Russian chemist Vladimer Markovnikov (1838–1904) stated a general rule that helps us predict the structure of the principal product when a nonsymmetric molecule containing hydrogen is reacted with a double bond. Markovnikov observed that, when a nonsymmetric molecule such as HCl or H_2O adds to a double bond, the double-bonded carbon with the most hydrogens will get the additional hydrogen. **Markovnikov's rule** was a statement of fact based on observation. Markovnikov was unable to explain why the rule worked. However, he should not be judged too harshly. The electron was discovered only about seven years before his death, and the ideas of orbitals and hybridization developed more than twenty years after he had died. Markovnikov's rule is, however, still useful today.

The H^+ in a reaction with a double bond will add to the carbon atom that has the most hydrogen atoms attached to it; this is called *Markovnikov's rule.*

GENERAL REACTION

Reaction of an Alkene with a Nonsymmetric Reactant (Markovnikov's Rule)

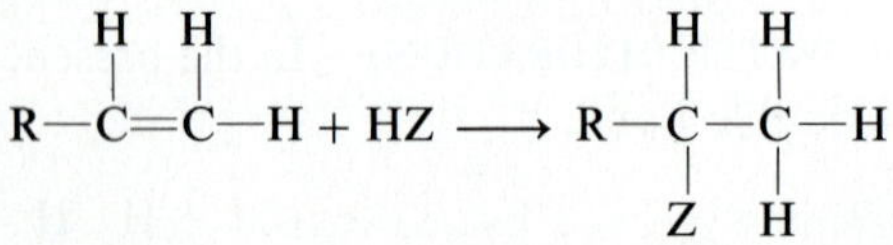

In this reaction, Z is Cl or OH (HZ is a "generic" reagent). Notice that the carbon that has the most hydrogens to begin with gets another.

We have discussed the notion that nonsymmetric reagents add to nonsymmetric alkenes in a particular way. We have given this concept the name

Markovnikov's addition. Let us now generalize these two reactions; the addition of HX (hydrohalogenation, or the addition of H$^+$ and X$^-$, where X = Cl, Br, I) and addition of water (hydration, or the addition of H$^+$ and OH$^-$).

GENERAL REACTION

Addition of an Alkene with a Nonsymmetric HX

$$\text{>C=C<} + HX \longrightarrow \begin{array}{c} H \\ | \\ -C-C- \\ | \\ X \end{array}$$

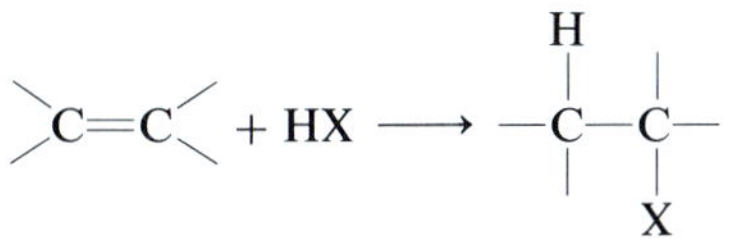

where X is any halogen. The H will add to the carbon bearing the most hydrogens.

Example

$$\begin{array}{c} H \\ \diagdown \\ H \end{array} C=C \begin{array}{c} CH_2-CH_3 \\ \diagup \\ H \end{array} + HX \longrightarrow \begin{array}{ccc} H & CH_2-CH_3 \\ | & | \\ H-C-C-H \\ | & | \\ H & X \end{array}$$

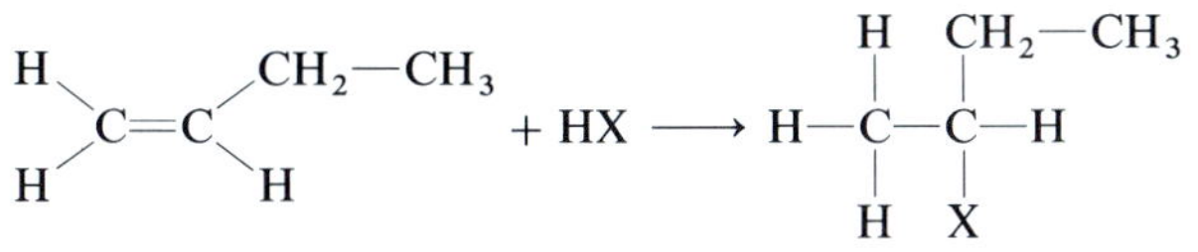

GENERAL REACTION

Addition of an Alkene with Water

$$\text{>C=C<} + H_2O \xrightarrow[\text{(catalyst)}]{H^+} \begin{array}{c} H \\ | \\ -C-C- \\ | \\ OH \end{array}$$

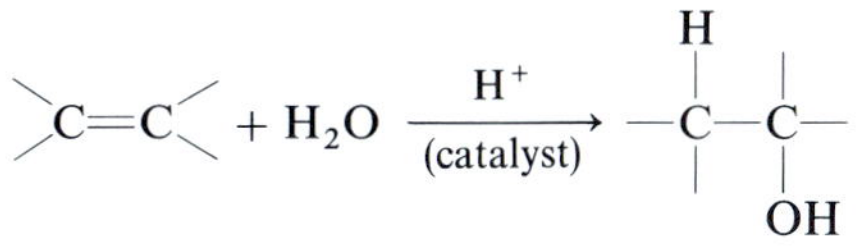

The H will add to the carbon bearing the most hydrogens.

Example

$$\begin{array}{c} H \\ \diagdown \\ H \end{array} C=C \begin{array}{c} CH_2-CH_2-CH_3 \\ \diagup \\ H \end{array} + H_2O \xrightarrow[\text{(catalyst)}]{H^+} \begin{array}{ccc} H & CH_2-CH_2-CH_3 \\ | & | \\ H-C-C-H \\ | & | \\ H & OH \end{array}$$

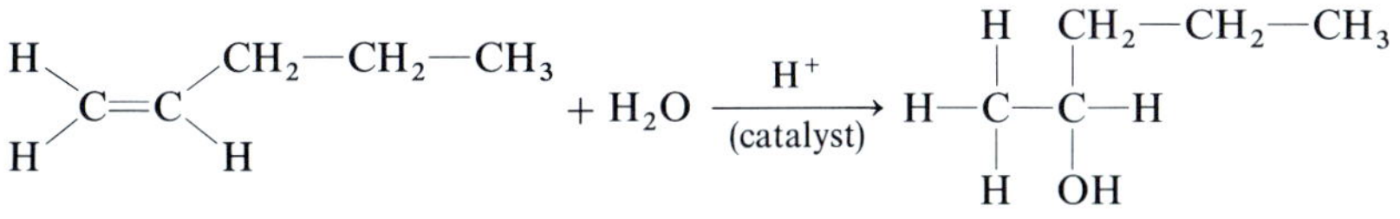

We have already stated that the major product from the reaction of propene with water and acid catalyst is 2-propanol. As an example of how the Markovnikov rule applies, let us reexamine this reaction.

$$\begin{array}{c} H \\ \diagdown \\ H \end{array} C=C \begin{array}{c} CH_3 \\ \diagup \\ H \end{array} + H_2O \xrightarrow[\text{(catalyst)}]{H^+} \begin{array}{ccc} H & H & H \\ | & | & | \\ H-C-C-C-H \\ | & | & | \\ H & O & H \\ & | \\ & H \end{array}$$

$$\quad\text{1}\quad\text{2}$$

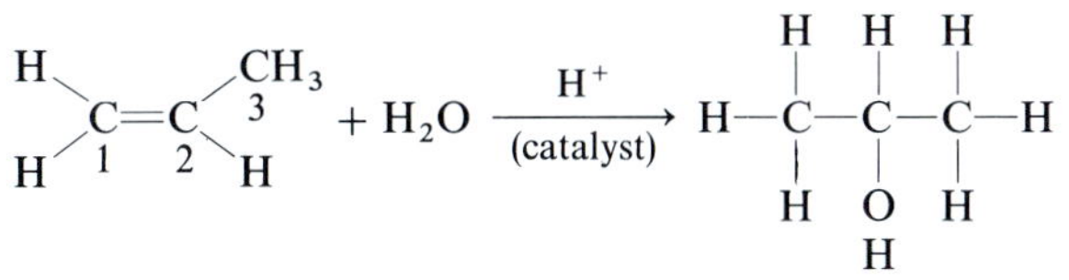

Propene **2-Propanol**

There are three carbons in the propene molecule; the double bond is between carbon 1 and carbon 2. Carbon 1 has two hydrogens attached to it; carbon 2 has one hydrogen and a methyl group attached to it. Markovnikov's rule predicts that the hydrogen from the reactant (H—OH) should go to the carbon that already has the most hydrogens. This is carbon 1. Carbon 2 must then get the —OH group.

We have seen that alkenes undergo addition reactions, both with symmetrical and unsymmetrical reagents. A few additional examples of the addition of unsymmetrical reagents will be useful to illustrate the scope of these reactions. Remember that an unsymmetrical reagent can add to any alkene; however, if it is an unsymmetrical alkene, we must apply the Markovnikov rule to ascertain the correct product.

EXAMPLE 16.4

Show the addition of water to fumaric acid to form malic acid

$$\text{HOOC—CH=CH—COOH} + H_2O \longrightarrow \text{HOOC—}\underset{}{\overset{\text{H}}{\overset{|}{\text{C}}}}\text{H—}\underset{\underset{\text{OH}}{|}}{\text{C}}\text{H—COOH}$$

Fumaric acid **Malic acid**

The addition of water to an alkene to form malic acid will be seen in more detail later in the biochemistry part of this book (see Chapter 28). This reaction is part of the *Krebs cycle* (or citric acid cycle or tricarboxylic acid cycle). This is one of the important biochemical sequences in the human body that converts carbon compounds to energy through biological combustion, giving off carbon dioxide as a by-product.

EXAMPLE 16.5

Predict the reaction of hydrochloric acid with the cycloalkene 1-ethylcycloheptene.

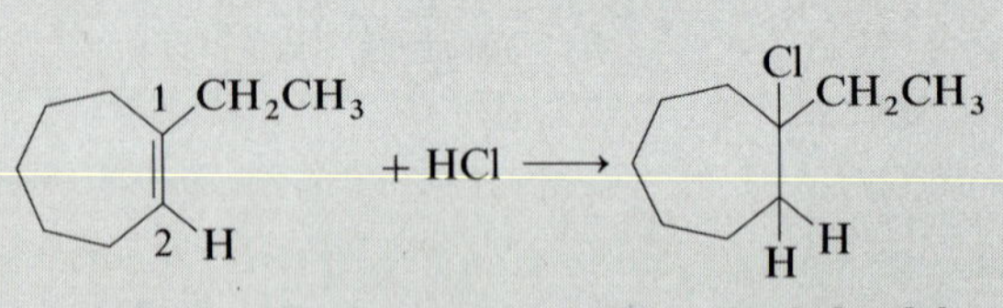
1
CH_2CH_3
2
H
+ HCl
Cl
CH_2CH_3
H
H

1-Ethylcycloheptene **1-Chloro-1-ethylcycloheptane**

In Example 16.5, note that because the pi bond was lost during the reaction and the product is a halogenated alkane, the naming system is based here only on the location of the substituents. As predicted, the hydrogen atom went to the double-bonded carbon that had the most hydrogens (carbon 2). Even though this was only one hydrogen, it was one more than carbon 1 had. The chlorine then went to the more substituted carbon.

EXAMPLE 16.6

Predict the product of the reaction of HBr with 2-methyl-2-pentene.

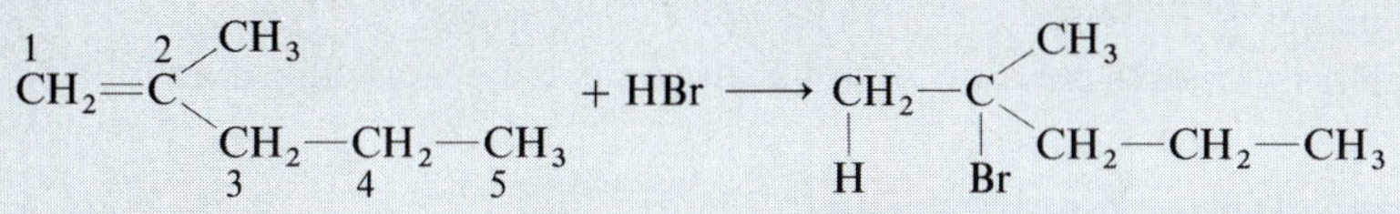

Notice that the alkene bond in Example 16.6 is between C-1 and C-2. The C-1 is directly attached to two hydrogen atoms, while C-2 is directly attached to two carbon atoms. Thus, C-2 is more substituted than C-1. From this we would predict that the H^+ adds to C-1 while the Br^- attaches to C-2.

Oxidation Reactions of Alkenes

A third type of chemical reactivity in alkenes is oxidation. Double bonds are highly susceptible to attack by oxidizing agents. These include oxygen, ozone, potassium permanganate, and potassium chromate. For example, rubber has a large number of carbon–carbon double bonds. The presence of sunlight will act as a catalyst to induce oxygen from the air to combine with these bonds. This is what causes deterioration of rubber tires. Tires should not be exposed to direct sunlight during extended periods of storage because of this breakdown process (Figure 16.14).

The type of oxidation that may occur depends on the strength of the oxidizing agent. For example, when a dilute water solution of potassium permanganate ($KMnO_4$) reacts with an alkene, —OH groups are added to the double bond.

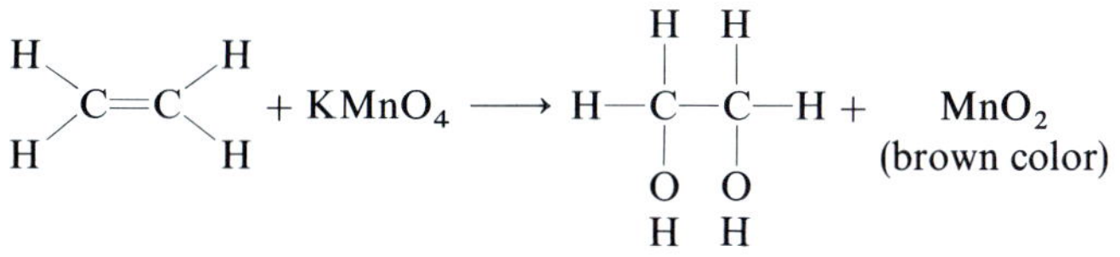

Ethene **Ethylene glycol**

FIGURE 16.14
The "checking" or deterioration of tires is caused by oxidation catalyzed by sunlight and atmospheric pollutants of the double bonds in the rubber.

The result is a dihydroxy alcohol, 1,2-ethandiol in this case. This compound is commonly called *ethylene glycol*, which is a well-known antifreeze. The visual formation of the brown precipitate is sometimes used as a test for alkenes.

An oxidation reaction with an alkene can be generalized as follows.

GENERAL REACTION

Oxidation Reaction of an Alkene

$$\text{C}=\text{C} + KMnO_4 \longrightarrow -\overset{|}{\underset{OH}{C}}-\overset{|}{\underset{OH}{C}}- + MnO_2$$

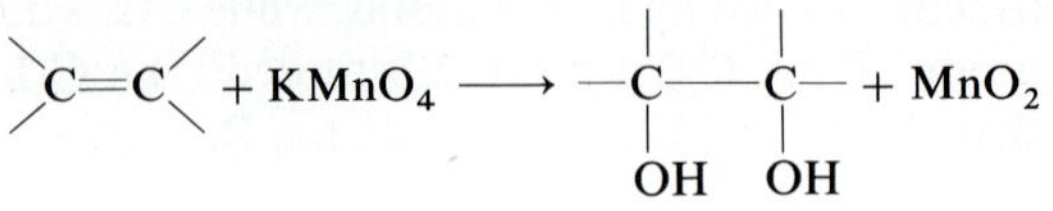

Example

$$\underset{H}{\overset{CH_3}{C}}=\underset{H}{\overset{CH_3}{C}} + KMnO_4 \longrightarrow CH_3-\underset{OH}{\overset{H}{C}}-\underset{OH}{\overset{H}{C}}-CH_3$$

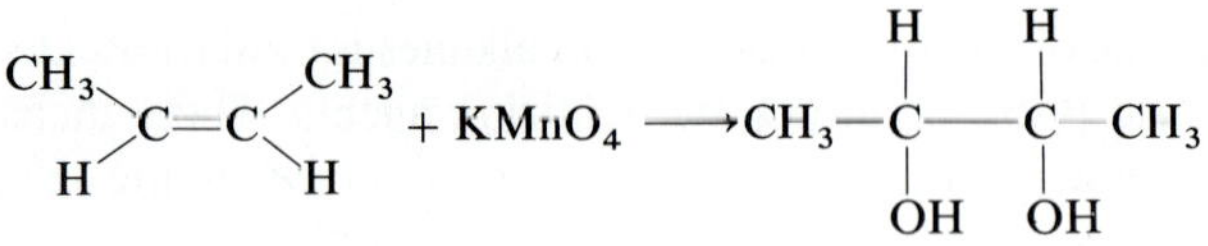

In Chapter 18 we will see another important oxidation reaction of alkenes, the epoxidation reaction.

Polymerization of Alkenes

Polymers are large, chained molecules that can be formed from alkenes.

Polymers are extremely large molecules formed of many thousands of small, repeating molecules called **monomers**. Polymers are common in both natural and synthetic materials (Table 16.1). Some examples include starch, Teflon in cookware, orlon in clothes, synthetic rubber in tires, polypropylene in containers, polyvinyl chloride (PVC) in pipes, bottles, and containers, and polystyrene in foam plastics and molded items. Alkenes are excellent building blocks for polymers because the pi bond provides two opportunities to bond to adjacent molecules.

Consider the polymerization of ethylene (ethene) to form polyethylene.

$$CH_2=CH_2 \xrightarrow{\text{(catalyst)}} CH_3-CH_2-(CH_2-CH_2)_n-CH_2-CH_3$$
Ethylene **Polyethylene**

The resulting polymer is a mixture of threads of polymer of varying length. It is used in making plastic film, pipe, and containers. Another example is rubber, which is a natural polymer of isoprene.

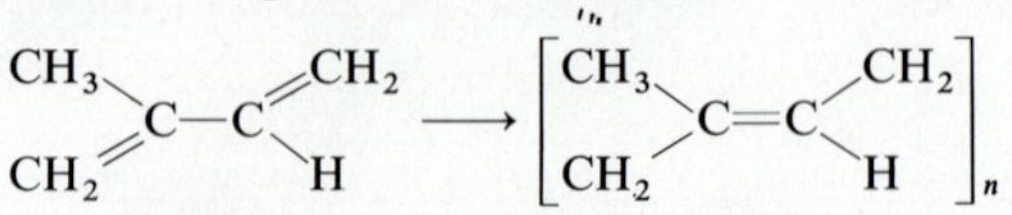

$$\underset{CH_2}{\overset{CH_3}{C}}=\underset{H}{\overset{CH_2}{C}} \longrightarrow \left[\underset{CH_2}{\overset{CH_3}{C}}=\underset{H}{\overset{CH_2}{C}} \right]_n$$
Isoprene **Rubber**

Note that the square bracket enclosing the polymer indicates repetition of this unit *n* times.

Some interesting observations about polymers can be made at this point. The next time you hold something made of polyethylene, such as a plastic cup or dish, take the time to feel its texture. You will notice that the material

TABLE 16.1
Some Common Polymers

Monomer	Polymer	Uses
Ethylene 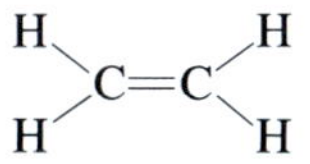	Polyethylene 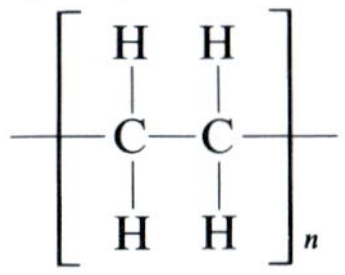	Plastic film, plastic pipe, plastic containers
Propylene 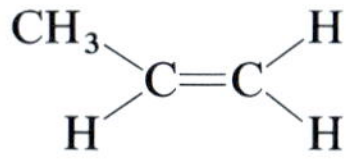	Polypropylene 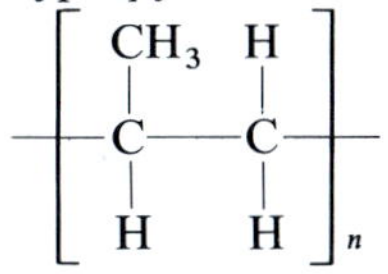	Molded plastic objects, rope, synthetic fiber
Vinyl chloride 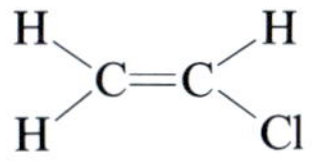	Polyvinyl chloride (PVC) 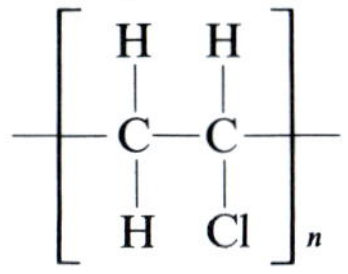	Plastic pipe, plastic containers, plastic rope
Dichloroethylene 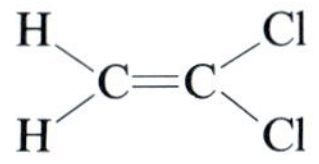	Polydichlorethylene 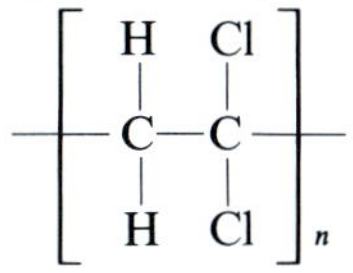	Plastic food wrap, tubing
Acrylonitrile 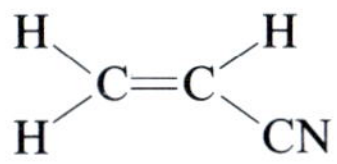	Polyacrylonitrile 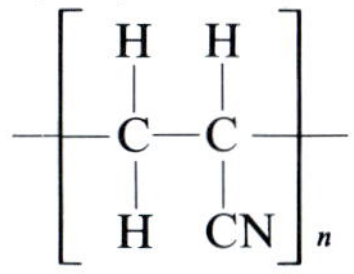	Orlon, fabrics
Tetrafluoroethylene 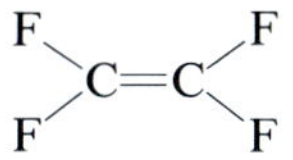	Polytetrafluoroethylene 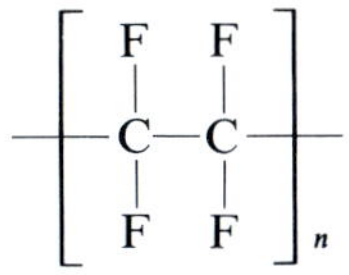	Teflon

feels "waxy." Wax is primarily a hydrocarbon. The reason that polyethylene holds water is readily understood by considering Figure 15.3 in the last chapter where we saw that hydrocarbons and water are not soluble in one another.

Polytetrafluoroethylene is a remarkable material. Most of us know it as the coating on "nonstick" cooking utensils marketed as Teflon,® among other names. Why is it so inert at high temperatures? In the last chapter we saw that simple hydrocarbons are relatively nonreactive because all of their bonds are single bonds. There are no "loose" electrons. The C—C single bond is quite strong. Because of the high electronegativity of fluorine, the C—F bond is very strong. In many respects the polytetrafluoroethylene molecule looks like polyethylene, except that it has F atoms in place of

H atoms. The F atoms form a protective cover over the internal backbone of carbon atoms. Thus, the normally strong C—C system is additionally protected.

NEW TERMS

hydrogenation The addition of hydrogen to an alkene or alkyne. An H is added to both carbon atoms of the double or triple bond, while a pi bond is lost

halogenation The addition of halogen to an alkene or alkyne. A halogen is added to both carbon atoms of the double or triple bond, while a pi bond is lost

hydration The addition of water to an alkene or alkyne. An H is added to one of the carbon atoms of the double or triple bond and an OH is added to the other carbon atom. A pi bond is lost.

Markovnikov's rule The generalization used to account for the way an unsymmetrical reagent adds to an unsymmetrical alkene. The positively charged reagent (often H^+) will add to the carbon atom directly attached to the greater number of hydrogens.

polymer A complex compound resulting from the end-to-end union of a large number of smaller units (monomers).

monomer The smallest repeating unit from which polymers are made.

TESTING YOURSELF

Reactivity of the Alkenes

1. Predict and name products for each of these hydrogenation reactions.

   a. $CH_3-\underset{\underset{H}{|}}{C}=\underset{\underset{H}{|}}{C}-CH_2-CH_3 + H_2 \xrightarrow[\text{(catalyst)}]{\text{Pt}}$

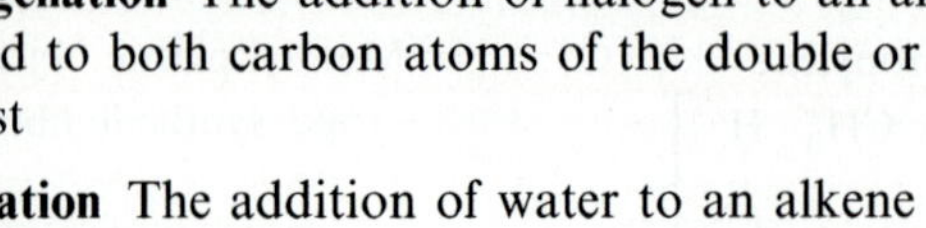

   b. $\underset{CH_3}{\overset{CH_3}{\diagdown}}C=C\underset{CH_3}{\overset{CH_3}{\diagup}} + H_2 \xrightarrow[\text{(catalyst)}]{\text{Pt}}$

   c. $CH_3-\underset{\underset{CH_3}{|}}{\overset{\overset{CH_3}{|}}{C}}-\overset{\overset{H}{|}}{C}=C\overset{\diagup H}{\diagdown H} + H_2 \xrightarrow[\text{(catalyst)}]{\text{Pt}}$

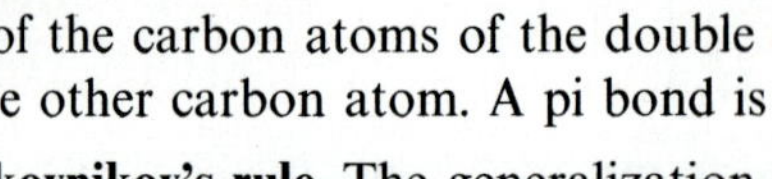

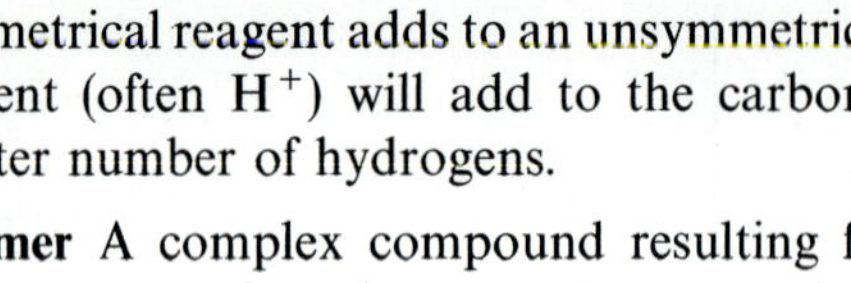

2. Predict and name the products of the following reactions of alkenes with halogens.

   a. $CH_3-CH_2-\underset{\underset{H}{|}}{C}=\underset{\underset{H}{|}}{C}-CH_3 + Br_2 \longrightarrow$

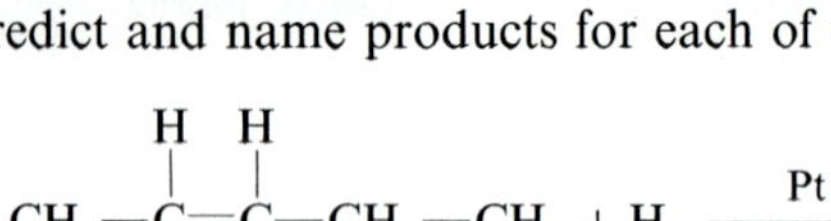

   b. $CH_3-CH_2-\underset{\underset{CH_3}{|}}{\overset{\overset{H}{|}}{C}}-\overset{\overset{H}{|}}{C}=C\overset{\diagup H}{\diagdown H} + Cl_2 \longrightarrow$

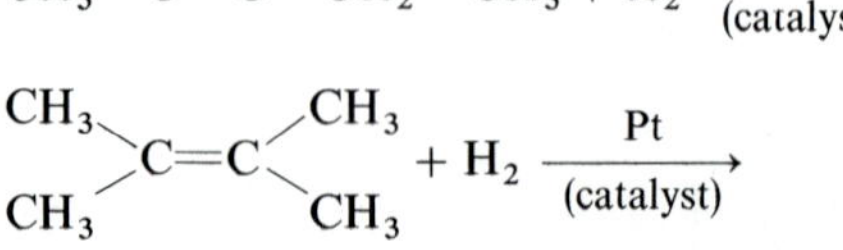

3. Predict products for the following reactions of alkenes with nonsymmetric reactants.

a. 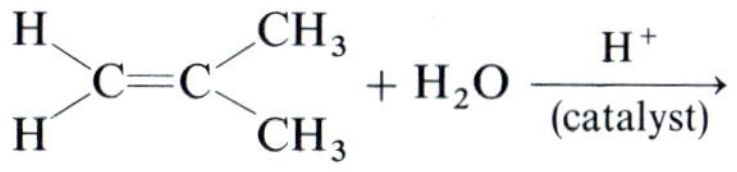

b. 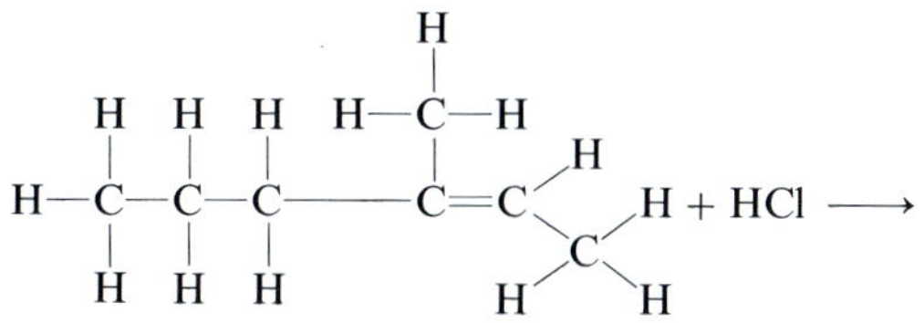

Answers **1. a.** pentane **b.** 2,3-dimethylbutane **c.** 2,2-dimethylbutane
2. a. 2,3-dibromopentane **b.** 1,2-dichloro-3-methylpentane

3. a. 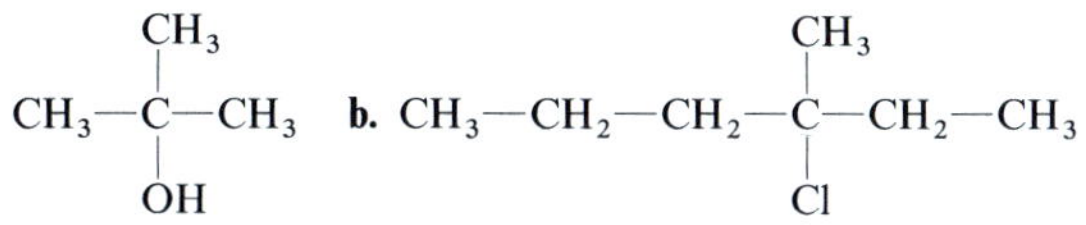

16.5 Chemical Reactivity of Alkynes

Alkynes undergo all of the reactions characteristic of pi-bonded hydrocarbons discussed in the alkene section, except that there are twice as many pi bonds to react. (Recall that the triple bond has one sigma bond and *two* pi bonds.) Markovnikov's rule predicts the outcome of reactions with nonsymmetric reactants just as it does for the alkenes.

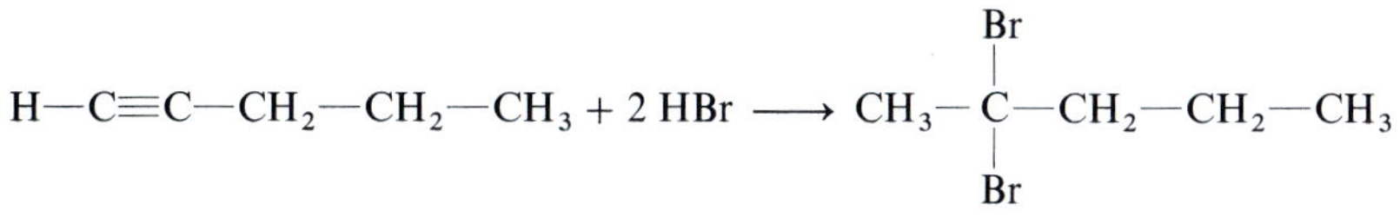

1-Pentyne **2,2-Dibromopentane**

For the case of hydrogen addition in alkynes, it is possible, by use of selective catalysts, to stop an addition to a triple bond when only one of the pi bonds has reacted. This produces a double-bonded material with definite cis and trans properties. This procedure has value in artificial synthesis of biological compounds such as insect sex attractants.

EXAMPLE 16.7

Predict the product of the reaction of this alkyne in the presence of hydrogen.

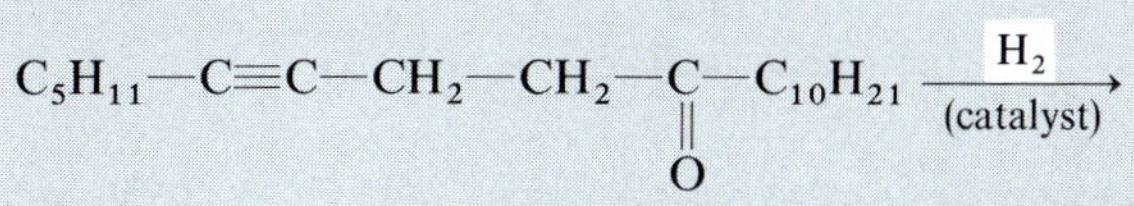

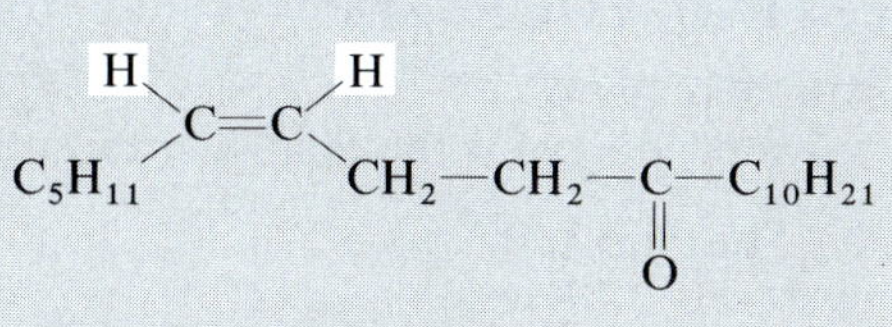

The product is the sex attractant for the douglas fir tussock moth.

The simplest alkyne, acetylene, is an extremely useful industrial chemical. Many other compounds are made using it as a starting material. Acetylene can be made by reacting carbon (coke) with lime (calcium oxide) in the presence of heat (Δ):

$$C + CaO \xrightarrow[(2500\,^{\circ}C)]{\Delta} \underset{\textbf{Calcium carbide}}{CaC_2} + \underset{\textbf{Carbon monoxide}}{CO}$$

When calcium carbide is placed in water, the following reaction takes place:

$$CaC_2 + H_2O \longrightarrow H{-}C{\equiv}C{-}H + CaO$$

The acetylene thus liberated can be burned or can be used as a raw material for synthesis of a more complicated compound.

$$H{-}C{\equiv}C{-}H + 3\,O_2 \longrightarrow 2\,CO_2 + 2\,H_2O$$

Early automobile headlamps and miners' helmet lamps used this source of acetylene for their fuel. Miners had to give up these kinds of lanterns because they sometimes caused explosions in mines that were naturally filled with methane.

The alkynes are not as widely distributed in nature as are the alkenes. However, several synthetic materials having the alkyne functional group have made a major impact on society. Synthetic derivatives of the natural female sex hormones estrogen and progesterone (see Chapter 25) are the major constituents of the widely used oral contraceptive known as the "pill." A common feature of these synthetic materials is the incorporation of an acetylene group.

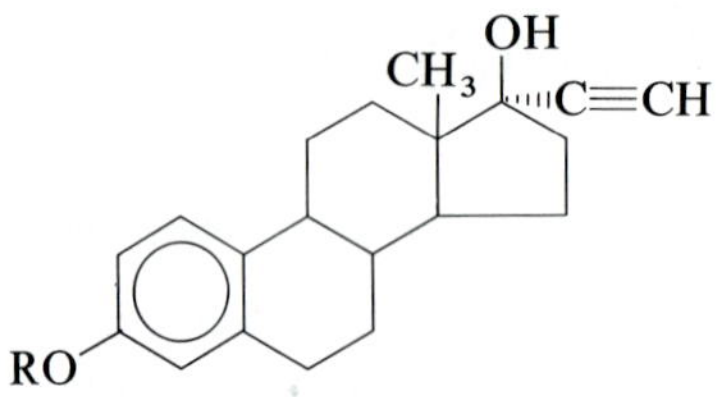
OH
CH3
C≡CH
RO

Ethinyl estradiol (R=H) or mestranol (R=CH$_3$)

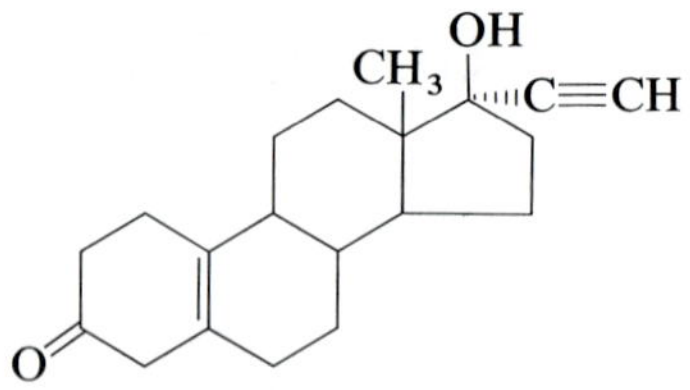
OH
CH3
C≡CH
O

Norethynodrel

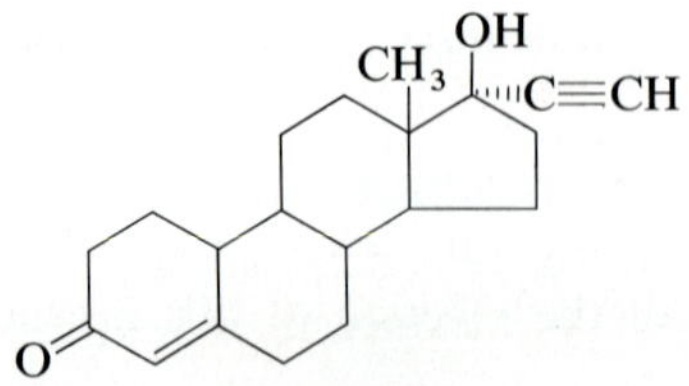
OH
CH3
C≡CH
O

Norethindrone (norethisterone)

TESTING YOURSELF

Alkyne Reactions

1. Give products for the following reactions

   a. $\text{2-hexyne} + \underset{\text{(excess)}}{Br_2} \longrightarrow$

   b. $\text{1-octyne} + H_2 \xrightarrow[\text{(catalyst)}]{}$

c. 1-octyne + HCl $\longrightarrow$
(excess)

Answers 1 **a.** 2,2,3,3-tetrabromohexane **b.** 1-octene will be the first product. Under controlled conditions, the reaction can stop here, but with excess hydrogen the alkene can be further reduced to octane. **c.** By Markovnikov's rule we would predict the H to add to C-1 and the Cl to C-2. The product from addition of two HCl molecules will be 2,2-dichlorooctane.

16.6 Polyunsaturated Alkenes

A hydrocarbon containing more than one double bond is a *polyunsaturated alkene*. For most polyunsaturated alkenes, the chemical reactions already discussed can be easily applied. The major difference is that more reagent can be added (note the addition of *two* bromine atoms in the example below).

EXAMPLE 16.8

Show the products of these reactions with polyunsaturated alkenes.

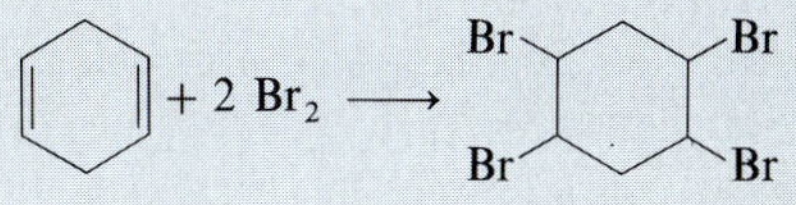
+ 2 Br2
Br
Br
Br
Br

$$CH_2{=}CH{-}CH_2{-}CH{=}CH_2 + 2\,KMnO_4 \longrightarrow$$

$$\underset{HO}{CH_2}{-}\underset{OH}{CH}{-}CH_2{-}\underset{HO}{CH}{-}\underset{OH}{CH_2}$$

However, a special class of polyunsaturated compounds exists called the **conjugated polyenes**. *Conjugation* means that there are alternating double and single bonds. Why are these special? This is most readily seen by examining the simplest conjugated polyene, 1,3-butadiene. In Figure 16.15 we can see the orbitals of this molecule and gain insight to the special nature of a conjugated system.

There is electron delocalization in polyunsaturated systems of alternating double and single bonds.

It is easy to see the overlap of the *p*-orbitals between C-1 and C-2 that give one of the double bonds and between C-3 and C-4 that give the second double bond. But why can't there also be overlap between C-2 and C-3? Actually, there is. The electrons in a conjugated system are *not* localized between specific atoms, but are rather spread out or *delocalized* over the

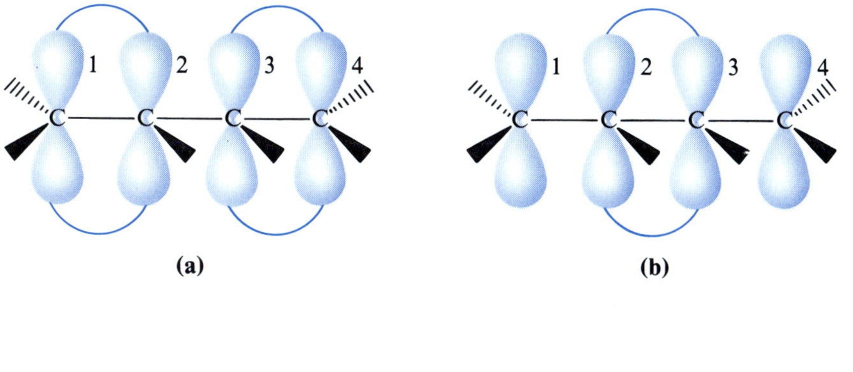
1
2
3
4
C
C
C
C
(a)
1
2
3
4
C
C
C
C
(b)

FIGURE 16.15
A conjugated polyene. (a) The orbitals of 1,3-butadiene. (b) A possibility for *p* orbital overlap exists between the central two *p* orbitals.

entire pi framework. This delocalization of electrons leads to greater stability—a concept that will be seen again the next chapter.

NEW TERM

conjugated polyene A polyene for which there are alternating double and single bonds.

16.7 Interesting Unsaturated Compounds

Many important natural products and biologically significant synthetic molecules have unsaturated compounds as part of their structure. A number of these molecules will be discussed in greater detail in other chapters. Presented here are a few molecules that represent some of the many unsaturated

TERPENES AND VAN GOGH

The van Gogh legacy is the great art that he left behind. But we know that Vincent van Gogh was a troubled man who sometimes exhibited bizarre behavior. Recent medical sleuthing has come up with a "chemical" explanation for much of his behavior—he was addicted to terpenes. As a heavy drinker of absinthe flavored with wormwood, he ingested large amounts of the terpene *thujone.*

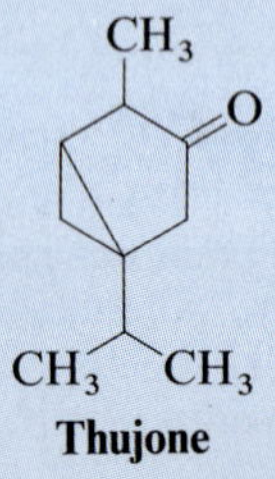

CH_3
O
CH_3
CH_3

Thujone

This compound is now thought to have caused many of the disorders known to have troubled van Gogh. He apparently also became dependent on sniffing the odors of terpenes that he encountered everyday in the turpentine so common to painters. This addiction finally reached the stage where he required camphor to be in his pillows and mattress in order to sleep. Camphor is also known to cause many of the same symptoms as thujone. As a result of his addiction to terpenes, he was even known to eat the paint and turpentine mixtures used in his art. Before he was finally committed to an asylum, it is reported that he had to be restrained from drinking a quart of turpentine.

Self-portrait of Vincent van Gogh with his bandaged ear which he cut off as part of his bizarre behavior, possibly brought on by his terpene addiction.

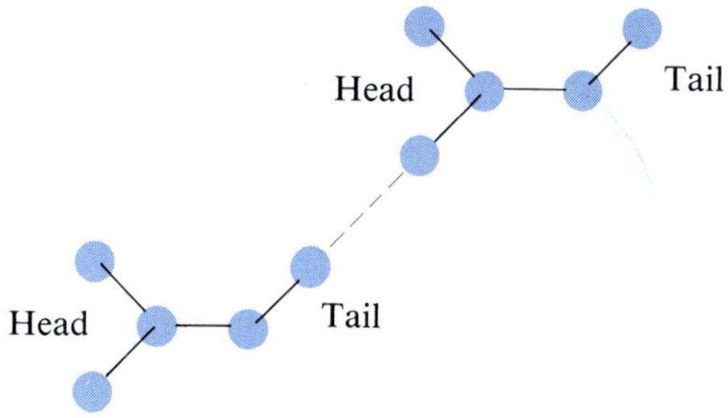
Head
Tail
Head
Tail

(a)

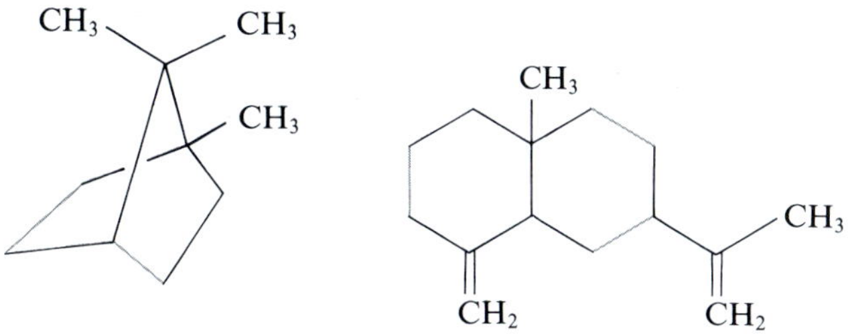
CH_3
CH_3
CH_3
CH_3
CH_3
CH_2
CH_2

Camphene **β-Selinene**

(b)

FIGURE 16.16
Structure of terpenes. (a) Isoprene units are often incorporated into terpenes in a head-to-tail fashion, as shown here. (b) Examples of two terpenes showing the head-to-tail arrangement of isoprene units.

compounds found in daily life, including compounds that make up turpentine, vitamin A, fats and oils in our diet, and important chemical communicators called pheromones.

Terpenes

The large class of naturally occurring compounds known as **terpenes** are well represented as flavorings and fragrances. The terpenes are characterized by their carbon framework, which appears to be made up of repeating isoprene units. We should note here that it is *not* the positioning of double bonds that is important, but rather the general arrangement of carbon atoms. These are often found in a head-to-tail fashion (Figure 16.16).

A number of terpenes are common to our experience, including pine oil, lemon oil (Figure 16.17), and the oils in celery and ginger. A synthetic compound structurally related to vitamin A (shown below) is now marketed under the name *acutane.* It is used for aggressive treatment of acne. Although the mechanism of action is not understood, the compound has been shown to be very effective in reducing skin oils responsible for acne. The product is a potent terategen (causes abnormalities in the fetus if used during pregnancy). However, it is not mutagenic.

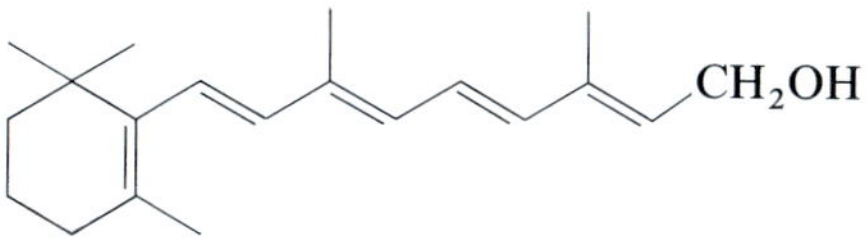
CH_2OH

Vitamin A

(a)

(b)

FIGURE 16.17
Two sources of common terpenes. (a) Pine oil. (b) Lemon oil.

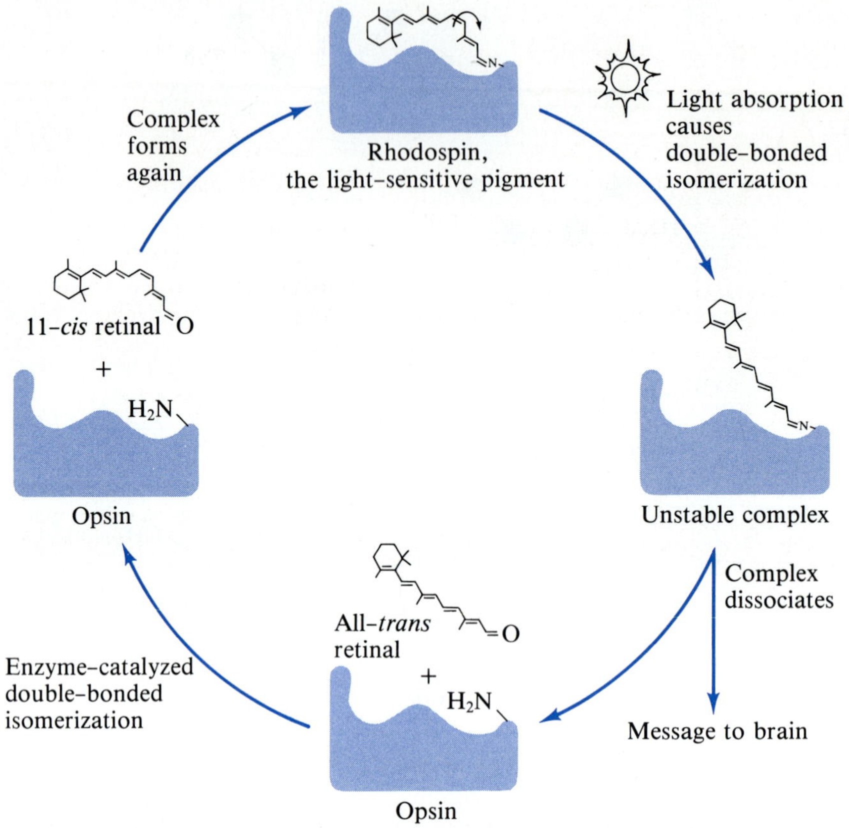

Complex forms again
Rhodospin, the light-sensitive pigment
Light absorption causes double-bonded isomerization
N
11-*cis* retinal
O
+
H_2N
Opsin
Unstable complex
N
Complex dissociates
Message to brain
All-*trans* retinal
O
+
H_2N
Opsin
Enzyme-catalyzed double-bonded isomerization

FIGURE 16.18
Vitamin A is converted to *retinal*, the key molecule in the chemistry of vision. Light energy converts a cis double bond to a trans double bond as part of the process involved in vision.

The chemistry of vision is also controlled by compounds similar to vitamin A. Light converts 11-*cis* retinal to all-*trans* retinal, and this conversion is then responsible for the sensation of vision (Figure 16.18).

Fatty Acids

Fatty acids are organic acids having a large hydrocarbon portion. Some of the fatty acids have no double bonds (saturated), some have one double bond (unsaturated), and some have more than one double bond (polyunsaturated). Since fatty acids are the major components of fats and oils, we can see where some of the terminology common to advertising comes from. There is a strong emphasis on reducing saturated fats from our diets and increasing the amounts of polyunsaturated oils. In an earlier section of this chapter, we showed that hydrogenation of an oil (containing unsaturation) results in a fat (the hardening process).

An essential fatty acid is one that we must have in our diets because we cannot produce our own. Arachidonic acid is a fatty acid necessary for the biosynthesis of prostaglandins and leukotrienes (see Chapter 24). Each plays an important role in the function of the human body (Figure 16.19).

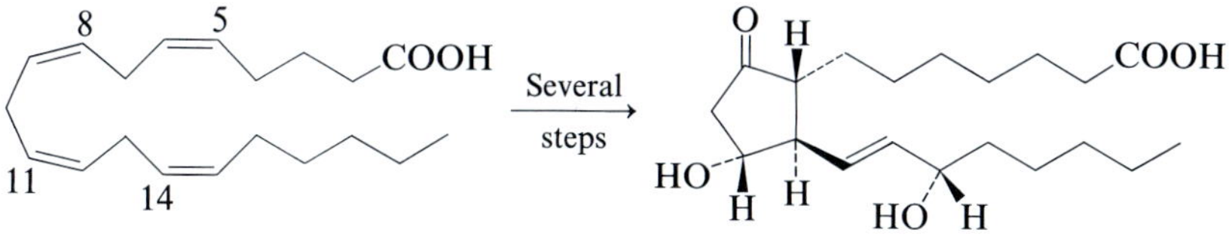
8
5
COOH
11
14
Several
steps
O
H
COOH
HO
H
H
HO
H

Arachidonic acid

Aspirin blocks the conversion of arachidonic acid to prostaglandins and leukotrienes

A prostaglandin

Active in inflammations, fever, regulates blood pressure, can be used as a decongestant

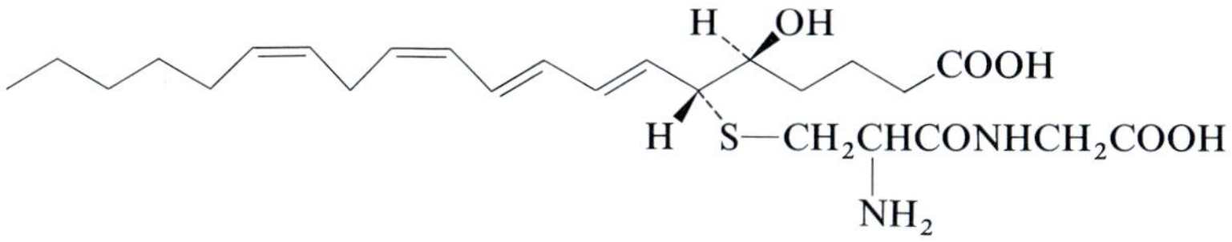
H
OH
COOH
H
S—CH₂CHCONHCH₂COOH
NH₂

A leukotriene

Responsible for asthma attacks

FIGURE 16.19
Arachidonic acid is an important unsaturated acid. It is the natural precursor to the prostaglandins and the leukotrienes. The prostaglandins have important physiological properties, including those associated with pain and fever. One of the roles of aspirin in controlling these conditions is that it hinders formation of the prostaglandins. The leukotrienes are directly responsible for many allergic responses.

Pheromones

Naturally occurring compounds used for communication by plants and animals are called **pheromones**. Some of these compounds signal for alarm, some for paths to food, and some for purposes of mating. Many of these pheromones are rich in unsaturation. An example of a pheromone is the sex attractant for the silkworm moth, shown here.

$$CH_3—CH_2—CH{=}CH—CH—(CH_2)_8—CH_2OH$$

Note that this compound has only one double bond; however, this makes all the difference between a compound that is biologically active and one that is not.

NEW TERMS

terpene A naturally occurring compound that contains isoprene units.

pheromone A naturally occurring compound produced by an organism for the purpose of chemical communication.

SUMMARY

An *unsaturated* molecule does not have enough atoms to provide four bonds for all of the carbon atoms. Carbon atoms in this situation form one or more multiple carbon–carbon bonds. There is only room enough between two carbon atoms for one sigma bond. When more than one bond forms between two adjacent atoms, the second and third bonds must use space above and below and on either side of the sigma bond. These bonds involve unhybridized *p* orbitals and are called *pi* (π) *bonds*. Since pi-bonding electrons are not shielded between the bonding atoms, but are highly visible from outside the molecule, they are readily susceptible to chemical attack by other molecules or ionic groups. Hydrocarbons with pi bonds are highly reactive in comparison to the sigma-bonded alkane family. Systems for naming double- and triple-bonded hydrocarbons (*alkenes* and *alkynes*, respectively) follow the same procedure as developed for the single-bonded alkane family, with the exception of procedures to designate and locate the multiple bond(s).

Polymerization is a characteristic of compounds that have one or more multiple bonds between adjacent carbon atoms. *Polymers* are an extremely important class of compounds, both from a biological and an industrial point of view. We will see more of these in later chapters.

Many natural products are unsaturated, and representative examples are found in terpenes, fatty acids, and pheromones.

REACTION SUMMARY*

Combustion of an Alkene with Sufficient Oxygen

$$C_nH_{2n} + 1.5n\, O_2 \longrightarrow n\, CO_2 + n\, H_2O$$
Alkene   Oxygen   Carbon dioxide   Water

Reaction of an Alkene with Hydrogen

$$\text{C=C} + H_2 \xrightarrow[\text{(catalyst)}]{Pt} -\underset{H}{\overset{}{C}}-\underset{H}{\overset{}{C}}-$$

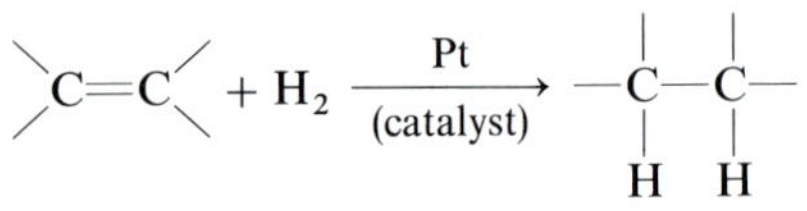

Reaction of an Alkene with a Hydrogen

$$\text{C=C} + X_2 \longrightarrow -\underset{X}{\overset{X}{C}}-\overset{}{C}-$$

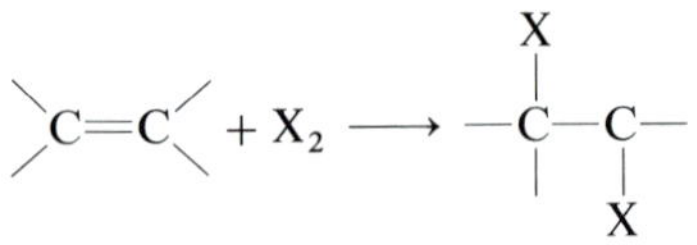

where X is any halogen.

Addition of an Alkene with a Nonsymmetric HX

$$\text{C=C} + HX \longrightarrow -\underset{X}{\overset{H}{C}}-\overset{}{C}-$$

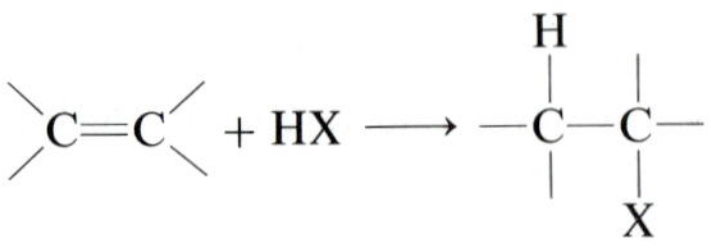

where X is any halogen. The H will add to the carbon with the most hydrogens.

Addition of an Alkene with Water

$$\text{C=C} + H_2O \xrightarrow[\text{(catalyst)}]{H^+} -\underset{OH}{\overset{H}{C}}-\overset{}{C}-$$

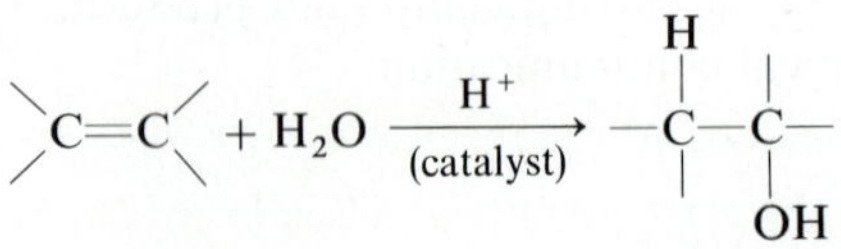

The H will add to the carbon with the most hydrogens.

---

* The reactions of the alkynes are the same as those of the alkenes, except that (1) two moles of reactant add to the triple bond and (2) the reaction of water does not apply to alkyne chemistry.

TERMS

unsaturated
alkene
alkyne
pi bond (π-bond)
triglyceride (triacylglycerol)
polyunsaturated
hydrogenation
halogenation
hydration
Markovnikov's rule
polymer
monomer
conjugated polyene
terpene
pheromone

CHAPTER REVIEW

QUESTIONS

1. The carbon atoms involved in the double bond of an alkene have which hybridization?
   a. *sp* b. sp^2 3. sp^3
2. The most reasonable name for 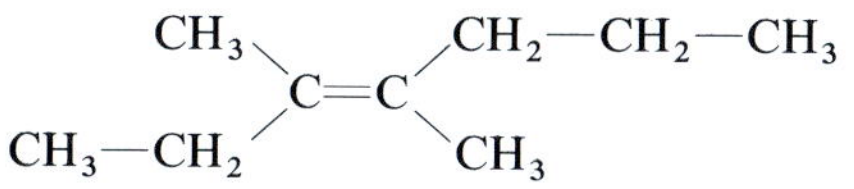 is
   a. *cis*-2-ethyl-3-methyl-2-hexene
   b. *trans*-2-ethyl-3-propyl-2-butene
   c. *trans*-1,4-dimethyl-3-heptene
   d. *trans*-3,4-dimethyl-3-heptene
3. The higher reactivity of an alkene or alkyne, as compared to an alkane, is due to
   a. sigma bonds b. pi bonds c. hydrogen bonds d. none of these
4. The bond angle associated with the hybrid orbitals of a carbon involved in a triple bond is
   a. 180° b. 120° c. 109° d. 45°
5. In the reaction of 1-butene with HCl, the H of the HCl will become attached to carbon number
   a. 1 b. 2 c. 3 d. 4
6. If you had a polymer of the general structure 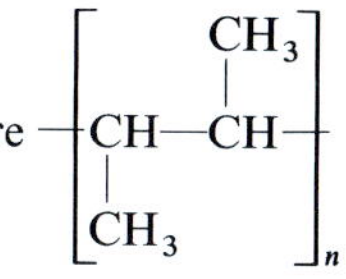 the monomer would most likely be:
   a. 1-propene b. 1-butene c. 2-butene d. 2-pentene
7. Combustion of an alkene with sufficient oxygen will produce
   a. carbon dioxide and water
   b. carbon monoxide and water
   c. only carbon dioxide
   d. only carbon monoxide
8. Reaction of cyclohexene with bromine would give
   a. 1,2-dibromocyclohexene
   b. 1,1-dibromocyclohexane
   c. 1,2-dibromocyclohexane
   d. 1,1,2,2-tetrabromocyclohexane
9. 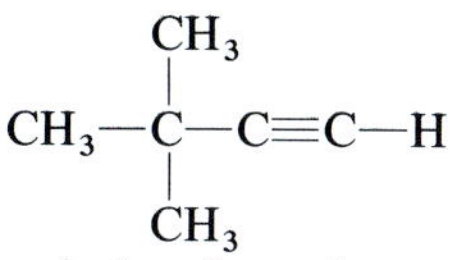 is best named by IUPAC as
   a. isobutyl acetylene b. 2,2-dimethyl-3-butyne c. 3,3-dimethyl-1-butyne

10. For 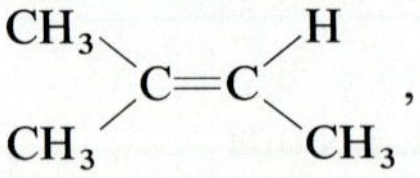 , what carbon atom will be attached to OH after action with H_2O and H^+?
a. C-4 b. C-3 c. C-2 d. C-1

Answers **1.** b **2.** d **3.** b **4.** a **5.** a **6.** c **7.** a **8.** c **9.** c **10.** c

DIAGNOSTIC CHART

Blacken in all of the circles under the number of each question that you missed in the Chapter Review questions. The diagnostic chart will help you identify concept areas that need more study.

	Questions									
Concepts	**1**	**2**	**3**	**4**	**5**	**6**	**7**	**8**	**9**	**10**
Structure	○	○	○	○		○	○			○
Nomenclature		○				○	○		○	
Reaction			○		○	○		○	○	
Hybridization	○		○	○						○

EXERCISES

Structure of Alkenes and Alkynes

1. The carbon atom participating in the double bond is ________ hybridized.
2. What are the two different bonds in a double bond?
3. Why do you think *trans*-cyclohexene would not be a stable compound?
4. The carbon atom participating in the triple bond is ________ hybridized.
5. What kinds of bonds are found in the triple bond?
6. Why could cyclopentyne not exist?
7. Compare the structural features of an alkene with an alkyne.

Nomenclature

8. Give IUPAC names for each of the following structures.

a. 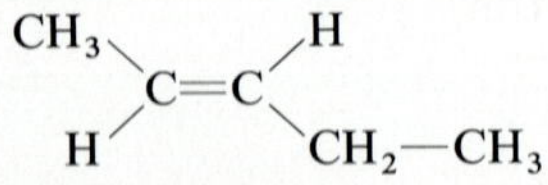

b. 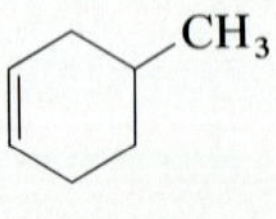

c. 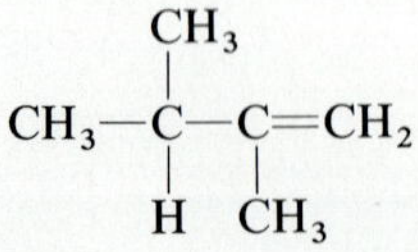

d. 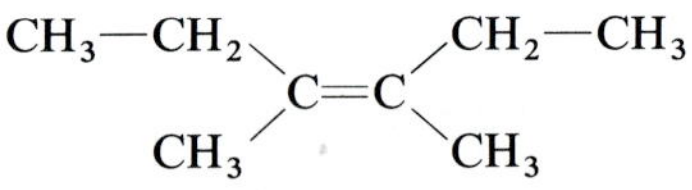

e. $CH_3—CH_2—C{\equiv}C—CH_3$

f. 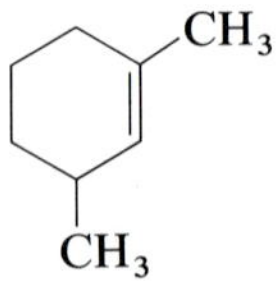

g. 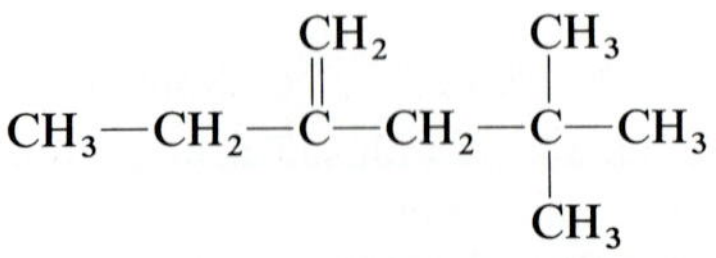

h. 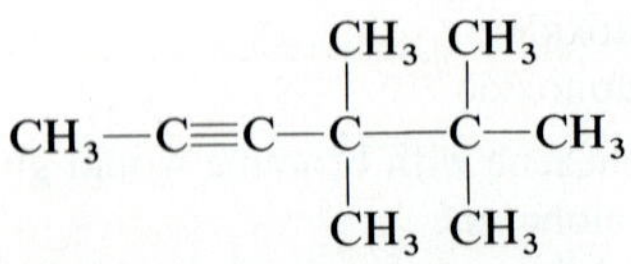

i. 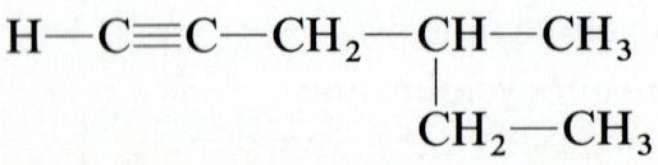

j. 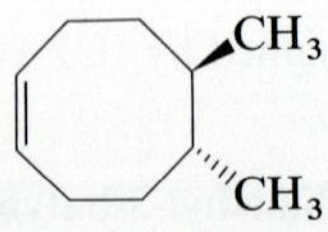

k.

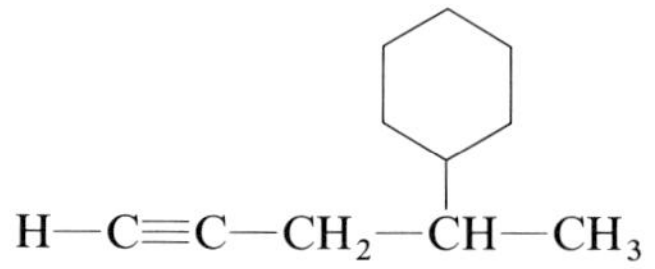

9. Draw structures for each of the following names.
   a. 1,2-dimethyl-*cis*-3-hexene.
   b. 5,5-dimethyloct-1-en-3-yne.
   c. 1,3,3-trimethyl-2-ethylcyclohexene.
   d. 3,3-dimethyl-1-hexyne
   e. 5,5-dibromo-2-hexyne
   f. 1,1,1-trichloro-*cis*-3-octene
   g. *cis*-3,4-dimethylcyclohexene
   h. 3,4-dimethyl-*cis*-3-octene
   i. *trans*-2,2-dimethyl-3-nonene
   j. 2,2-dimethyl-4-decyne
   k. 3,3-diethylcyclohexene
   l. 3,4-dimethyl-*trans*-3-hexene
   m. 1,3,4,4-tetramethylcycloheptene

Reactions

10. Why are alkenes and alkynes more reactive than alkanes?
11. What will be the products of an alkene that is burned with excess oxygen?
12. If an alkene is reacted with HBr, what is the electron acceptor?
13. Give products (with names) for each of the following reactions:
   a. 4-ethyl-4-decene + HBr
   b. 1,2-dimethylcyclopentene + Cl_2
   c. 2-butyne + Excess H_2 and catalyst
   d. Cyclohexene + Water and a trace of acid for catalyst
   e. Cyclopentene + Dilute permanganate
   f. 1,2-dimethylcyclohexene + Hydrogen gas and catalyst

g.

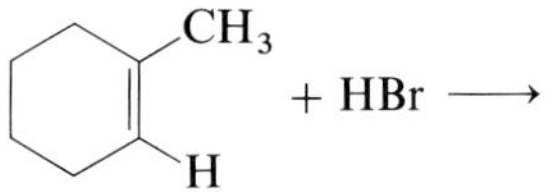

h.

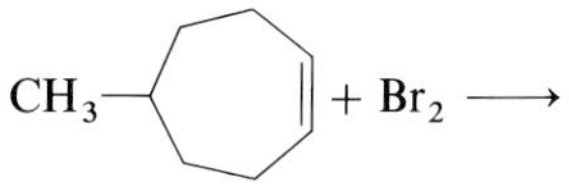

i.

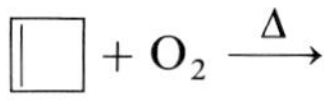

j.

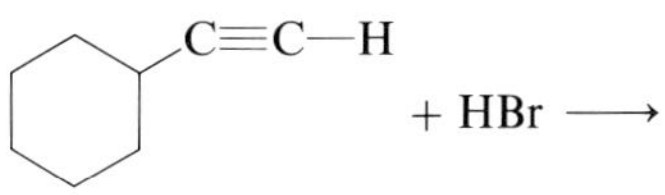

14. Starting with cyclooctene, suggest how you would prepare the following.

a.

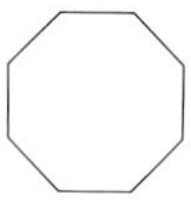

b.

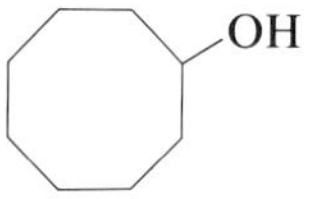

c.

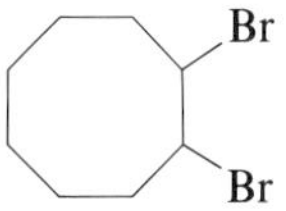

15. Considering electronegativity values, how would you predict that Br—Cl would react with 1-propene? (HINT: Would Br^+ or Cl^+ act as the electrophile? Why?)
16. Give products for each of the following reactions.

a.

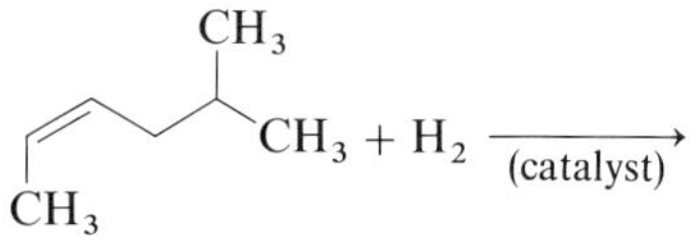

b.

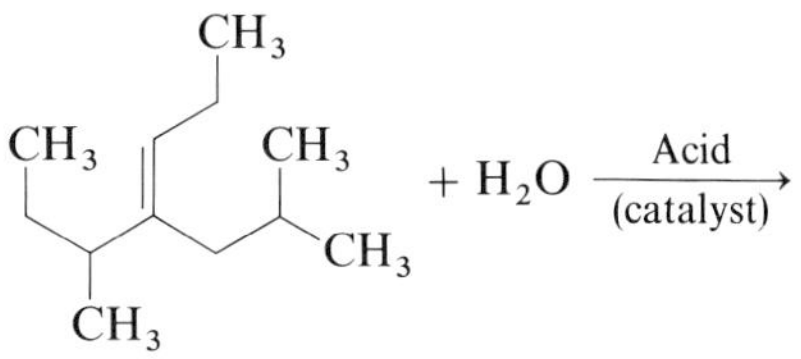

c.

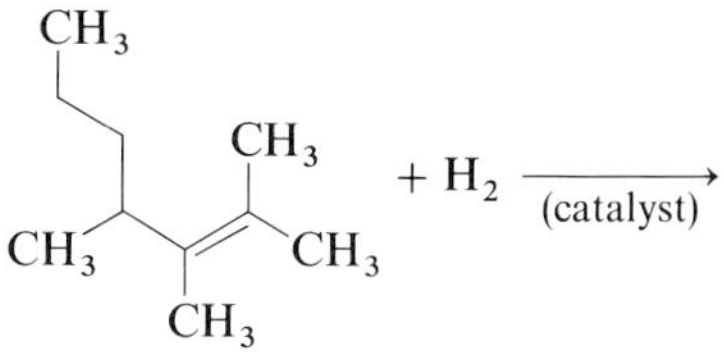

d.

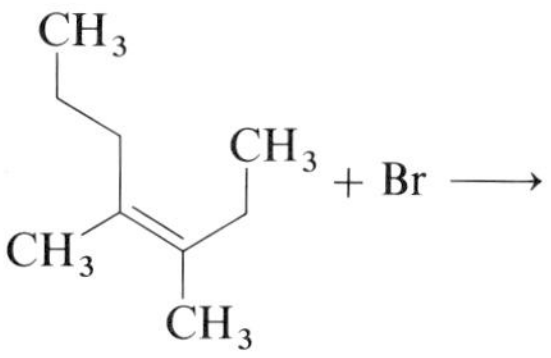

e.

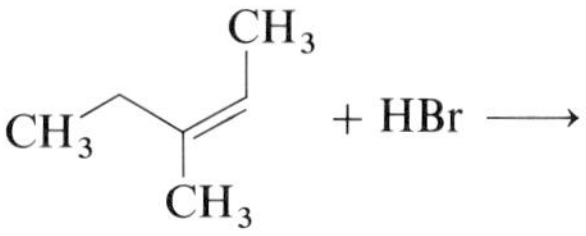

f.

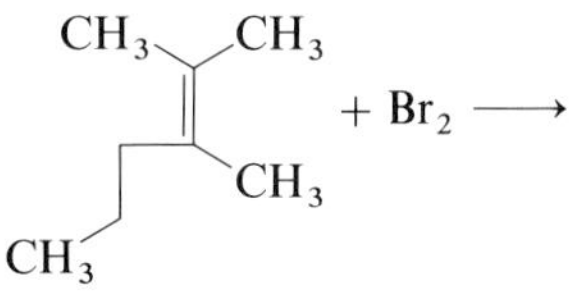

CHAPTER

17

Aromatic Hydrocarbons

OBJECTIVES

After completing this chapter, you should be able to

- Recognize conjugated double bonds, and explain why they differ in bond strength from normal double bonds.
- Recognize resonance structures, and explain what is meant by *delocalization* of pi electrons.
- Name some common aromatic compounds.
- Predict products of representative aromatic reactions.

In the not too distant past, a chemist named Kekule sat drowsily dreaming at his fireplace. As he dreamed, the flames turned into tiny carbon atoms, six of which joined hands to dance. As the chain wound through the flame, the end carbons suddenly joined hands to start a circle dance. Thus began our insight into the structure of aromatic compounds. Aromatics have unusual chemical stability because of the way their carbon atoms share electrons with their neighbors.

In the early years of investigating products from plants, fragrant materials such as clove, almond, and wintergreen were observed to produce interesting organic compounds. Because the structures of these distinctive smelling compounds showed sufficient similarities, they were classified as **aromatic**. Later, however, many other compounds that were *not* fragrant were shown to have similar structural and chemical properties, and these were also classified as aromatic. Soon, the term had less to do with aroma than with classification by structural and chemical features. This terminology remains. In the following sections we shall present details to help us understand the reasons for placing the compounds in the same class and the chemical basis for the classification.

17.1 Structure: Resonance and Electron Delocalization

Benzene is the "parent" chemical compound of the aromatic family; it has the formula C_6H_6. A number of possible structures have been suggested for benzene, including the following.

C_6H_6 is the formula for benzene.

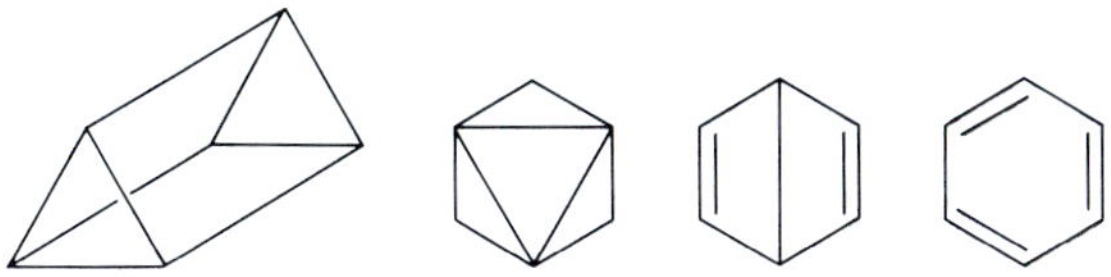

The structure on the far right, first proposed by Friedrich Kekule in 1865 after a dream, is closest to the structure as we now know it. Benzene represented in this way would really be 1,3,5-cyclohexatriene. This is nothing more than a cyclic triene, and we would expect it to behave as an alkene.

Between the time of Kekule and the present, there have been (and continue to be) many studies about benzene. As in many aspects of science, "right" or "wrong" answers are the products of many years of study and many people. A number of features have thus become known about benzene that have helped us understand its structure and properties. It is now known that benzene has two unusual features. One is that it is a *conjugated* molecule, that is, the double bonds alternate with single bonds in its structure (see Chapter 16). Benzene's double bonds are between alternating carbon atoms. The other feature is that benzene's structure can be written in two different ways, with double bonds between different carbons in each case.

Complete benzene structure

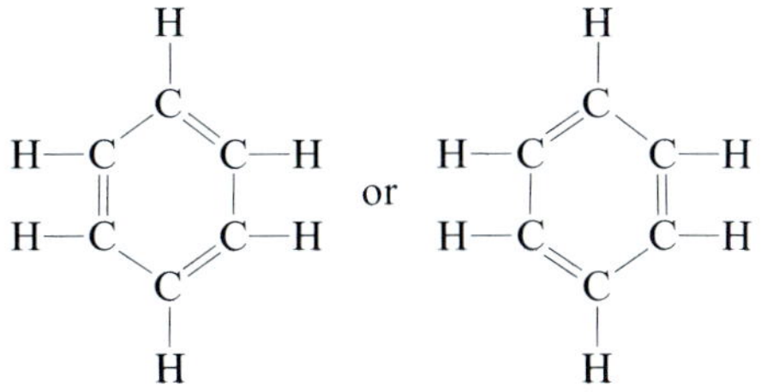
H
C
H—C
C—H
or
H—C
C—H
C
H

Simplified version

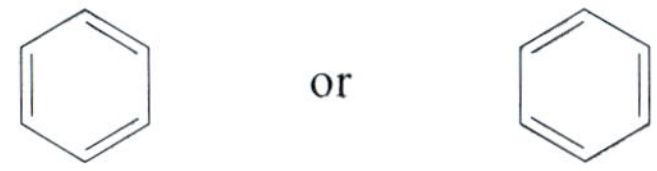
or

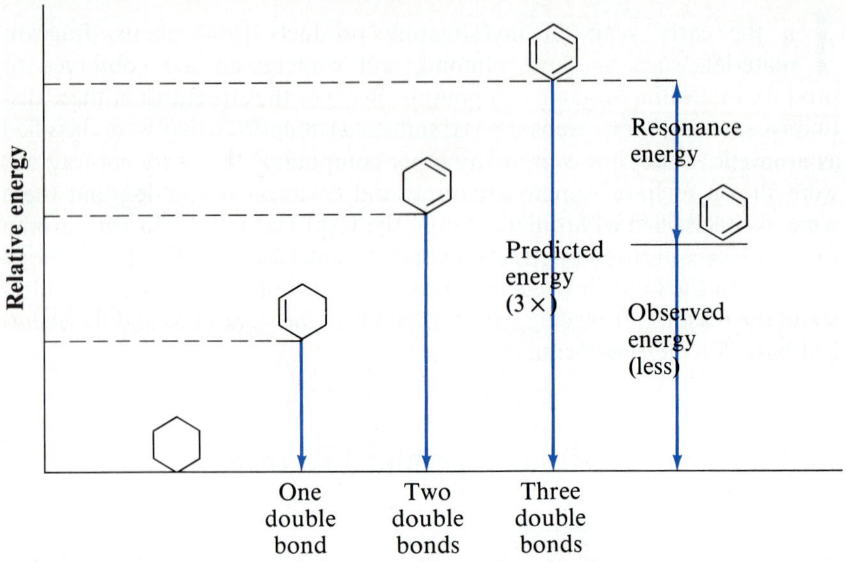
Relative energy
Resonance energy
Predicted energy (3 ×)
Observed energy (less)
One double bond
Two double bonds
Three double bonds

FIGURE 17.1
The resonance energy of benzene can be calculated from the predicted heat of hydrogenation. (See text for details.)

Resonance energy is the difference in the predicted energy and the observed energy of a molecule.

From our understanding of Chapter 16, we might correctly assume benzene to have some additional stability because of the delocalization of electrons associated with the conjugation. However, benzene is *much* more stable than one might predict from just considering it as a simple 1,3,5-cyclohexatriene. This extra stability is responsible for the lack of reactivity of benzene to reagents that normally react with alkenes. This extra stability is called **resonance energy** (see Chapter 14).

Figure 17.1 shows how the resonance energy is determined. There is a certain amount of heat liberated when cyclohexene is reduced to cyclohexane. You would expect three times as much heat to be liberated when 1,3,5-cyclohexatriene is reduced to cyclohexane (the predicted energy). But *less* than the predicted amount is actually observed. The difference between what is predicted and what is observed is the resonance energy. This energy is believed to be associated with the special stability achieved by the complete overlap of electrons in benzene's pi bond system.

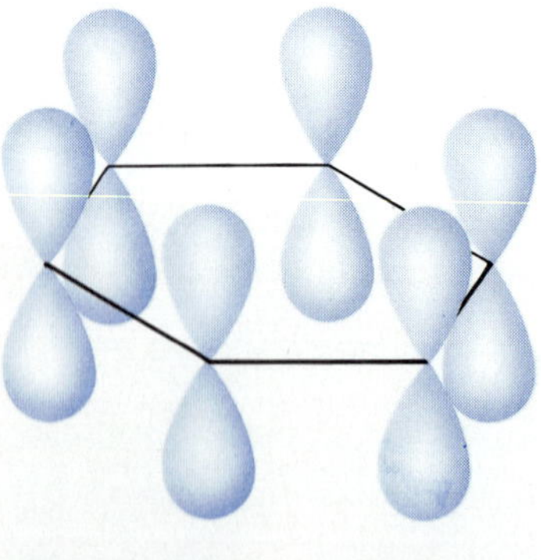

FIGURE 17.2
Alternating pi bonds in benzene produce a delocalized electron cloud of pi electrons.

Recall from Chapter 14 that when two different Lewis electron dot structures can be drawn for a compound, the two structures are called contributing or *resonance structures*, and it is understood that the real structure is something between the two proposed structures. Only compounds with alternating double bonds exhibit this property. When two or more resonance forms of a molecule can be drawn, the molecule is *always* more stable than would otherwise be predicted. Why is this so? In Chapter 16 we showed that conjugated double bonds result from adjacent pi orbitals. The adjacent pi electron clouds form a *delocalized* orbital involving all of the pi bonded carbon atoms. Such is the case with benzene. Benzene is much more resistant to reaction than one would expect because of the stability of this delocalized orbital (Figure 17.2). Thus, an aromatic compound must have a ring system with alternating double and single bonds.

A general rule (Huckel's rule) for aromatic character can be expressed by the simple equation:

$$N = 4n + 2$$

where N is the number of electrons in the pi bonds and n is a whole number (1, 2, 3, 4, . . .). Considering benzene, we see that there are six pi electrons in the double bonds. Thus,

$$N = 4n + 2$$
$$6 = 4n + 2$$
$$4 = 4n$$
$$1 = n$$

Since n here is a whole number, benzene fits the definition for an aromatic compound.

Benzene is often represented with the pi bonded electrons shown as a ring.

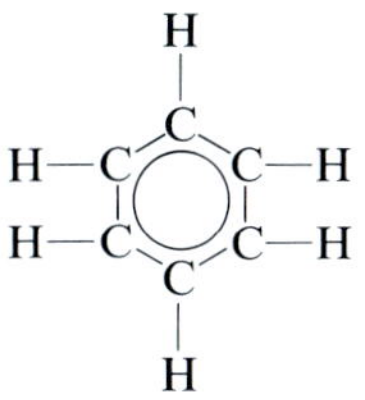

or

(shorthand)

This representation is chemically more correct because the structure shows the electron delocalization, but most texts continue to use the traditional alternating double bond resonance structure for convenience. It is instructive to examine benzene by the "ABC" rules of bonding presented in Chapter 14.

EXAMPLE 17.1

Give an electron dot structure for benzene.

C C C C C C H H H H H H

A = 4 + 4 + 4 + 4 + 4 + 4 + 1 + 1 + 1 + 1 + 1 + 1 = 30

B = 60

C = 30

D =

First place the electrons in the C—H bond (two each)

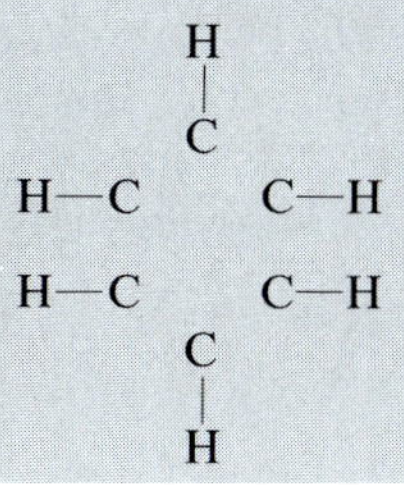

We now have 18 electrons left to be distributed among the 6 *equivalent* C—C bonds. Why not exactly share the electrons? That would require 3 electrons per bond.

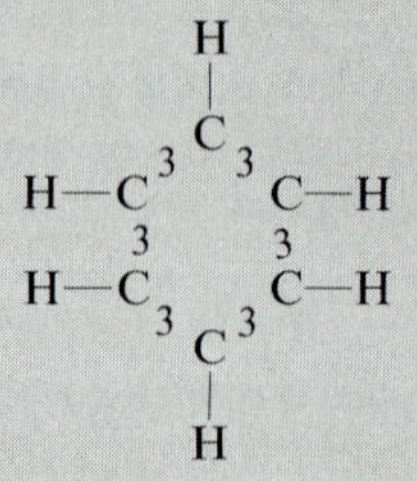

From this structure we see that it is the equivalent of the average of the two resonance structures!

NEW TERMS

aromatic A class of ring compounds that have alternating double bonds and subsequent pi electron delocalization. An aromatic compound will obey the $4n + 2$ rule.

benzene The simplest aromatic hydrocarbon.

resonance energy The energy due to delocalization of electrons.

TESTING YOURSELF

Resonance and Aromatic Character

1. Draw the resonance structures for the bicarbonate ion, 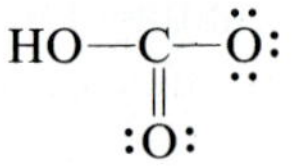

2. Show that cyclobutadiene, , is aromatic.

Answers 1. 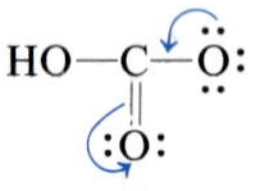 and 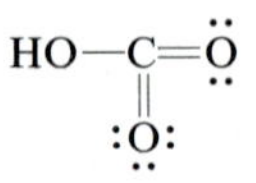

2. There are 4 π-electrons: $4 = 4n + 2$,
$4n = 2$, $n = 1/2$, not a whole number; thus cyclobutadiene is *not* aromatic.

17.2 Nomenclature

Monosubstituted Benzenes

Since the "double bonds" of benzene are not located between specific carbon atoms, a different approach to nomenclature is used for aromatic compounds. This nomenclature often involves using the alkyl group name followed by the word benzene. For example,

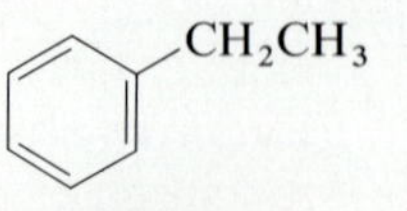

Ethylbenzene

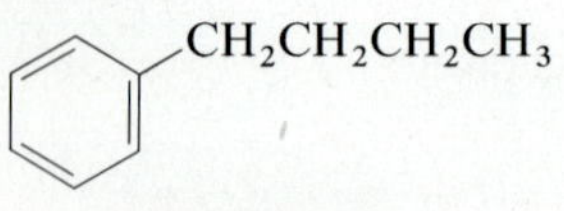

Butylbenzene

A *monosubstituted* benzene is one in which *one* group is attached to the benzene ring. Common names take priority for some of these simple substituted benzenes. Two common monosubstituted compounds are *toluene* (methylbenzene) and *cumene* (isopropylbenzene).

Common names are important to the nomenclature of aromatic compounds.

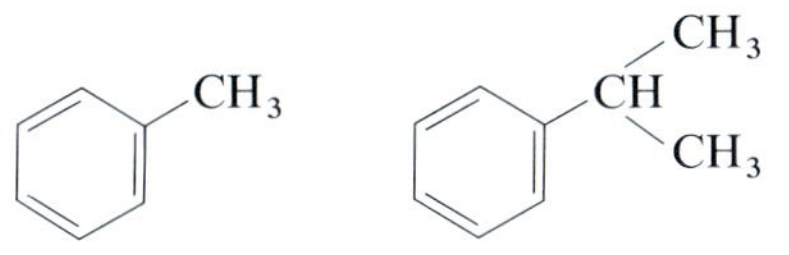
CH3
CH3
CH
CH3

Toluene (methylbenzene) **Cumene** (isopropylbenzene)

The nomenclature of a number of other simple benzene derivatives is based on the name of the group attached to the benzene ring. The halogens, when attached to a benzene ring, are called *bromo* (Br), *chloro* (Cl), *iodo* (I), and *fluoro* (F). The NO_2 group is called *nitro*. When the SO_3 group is attached to a benzene ring, as in SO_3H, the resulting compound has the common name *benzenesulfonic acid.*

EXAMPLE 17.2

Name these monosubstituted benzenes.

Alkyl group + Benzene   Halogen prefix + Benzene

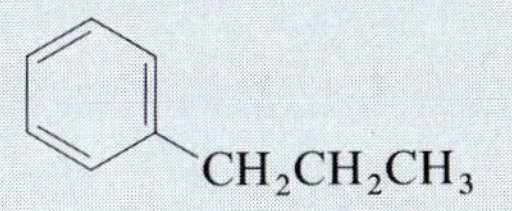
CH2CH2CH3

Propylbenzene

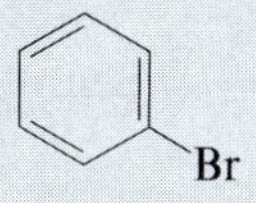
Br

Bromobenzene

Disubstituted Benzenes

In *disubstituted* benzenes two carbons have side groups attached. A unique system of nomenclature has evolved for these benzenes that requires one side group to be used to designate a *parent* aromatic compound. The second group is then assigned a position relative to the parent group. Two ways of assigning positions can be used. The first simply uses numbers, starting with the "parent" position as 1, and moving in the direction to give the second substituent the lowest possible number. For example,

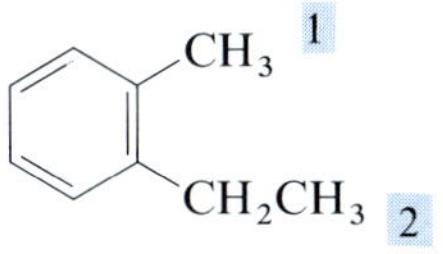
CH3
1
CH2CH3
2

is known as 2-ethyltoluene (not 5-ethyltoluene or 2-methylethylbenzene). The methyl group (the parent compound is toluene) is designated as position 1.

The second method uses the prefixes **ortho- (*o*-)**, **meta- (*m*-)**, and **para- (*p*-)** to designate positions with respect to the parent group. These are shown in Figure 17.3. To use this nomenclature, the parent becomes the base name, and the second group is simply given a position and name using the prefix *o*-, *m*-, or *p*-. If one group is part of a *common* name, it becomes the parent.

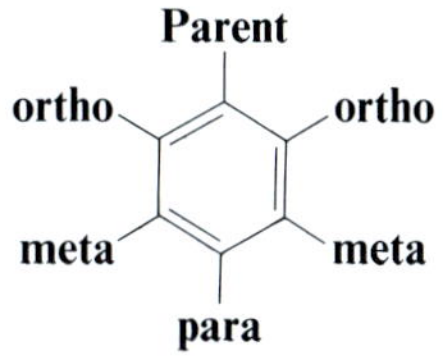
Parent
ortho
ortho
meta
meta
para

FIGURE 17.3
The prefixes *ortho-*, *meta-*, and *para-* can be used to designate positions on a benzene ring.

EXAMPLE 17.3

Name these disubstituted benzenes.

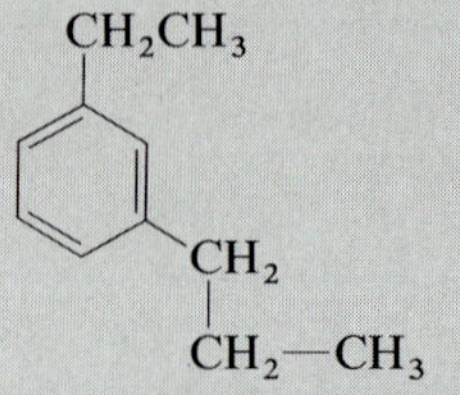
CH2CH3
CH2
CH2—CH3

The correct name is 3-ethylpropylbenzene or *m*-ethylpropylbenzene. Note that we used the propylbenzene as the parent compound. We could have used the ethylbenzene, and then the name would have been 3-propylethylbenzene or *m*-propylethylbenzene.

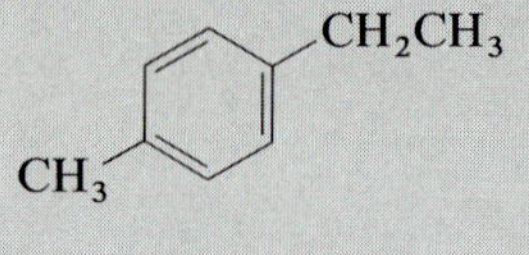
CH2CH3
CH3

4-Methylethylbenzene or *p*-methylethylbenzene would be an incorrect name because "methylbenzene" has the common name *toluene*. Thus, the correct names would be either 4-ethyltoluene or *p*-ethyltoluene.

Polysubstituted Benzenes

When more than two side groups are attached to the benzene ring, as in *polysubstituted* benzenes, numbers are used to show the position relative to the parent compound. As in previously discussed rules of nomenclature, one counts around the ring in a manner to get the lowest possible carbon numbers for the substituents.

EXAMPLE 17.4

Name these polysubstituted benzenes.

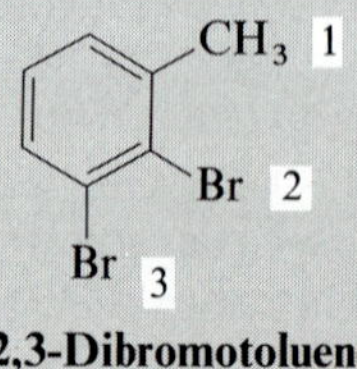
CH3 1
Br 2
Br 3

2,3-Dibromotoluene

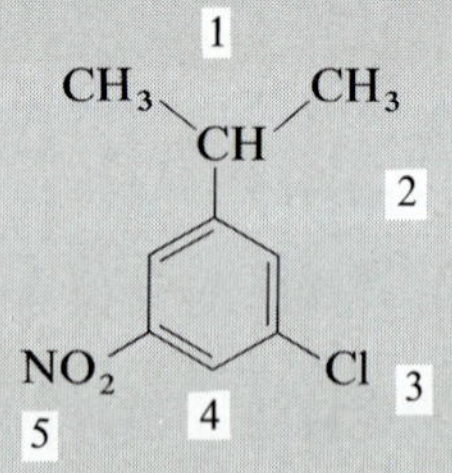
1
CH3 CH3
CH
2
NO2
Cl 3
4
5

3-Chloro-5-nitrocumene

Additional Nomenclature

Sometimes the benzene ring appears as a substituent group on a larger compound. When the benzene ring is used this way, it is called a **phenyl** group. Another common benzene derivative is the **benzyl** group, which is a phenyl group with a CH_2 attached.

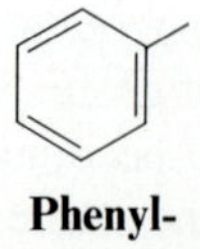

Phenyl-

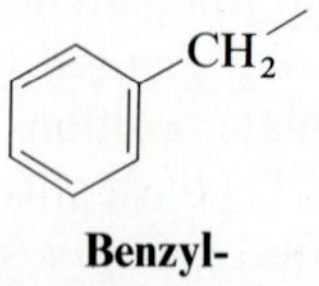
CH2

Benzyl-

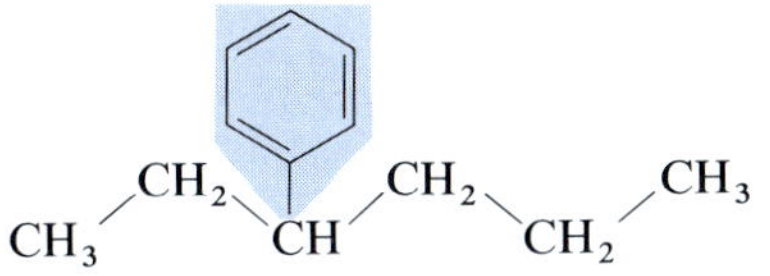
CH₂ CH₂ CH₃
CH₃ CH CH₂

3- Phenyl hexane

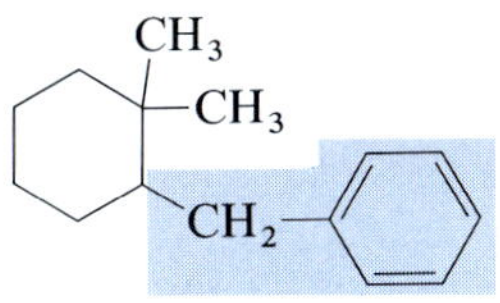
CH₃
CH₃
CH₂

1,1-Dimethyl-2- benzyl cyclohexane

A summary of aromatic nomenclature can be seen in the following examples.

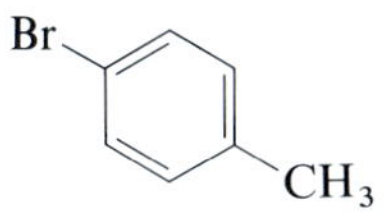
Br
CH₃

***p*-Bromotoluene**

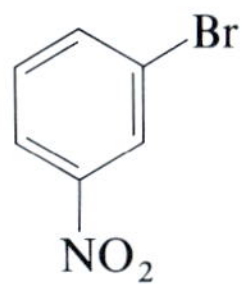
Br
NO₂

***m*-Nitrobromobenzene**
(or *m*-bromonitrobenzene)

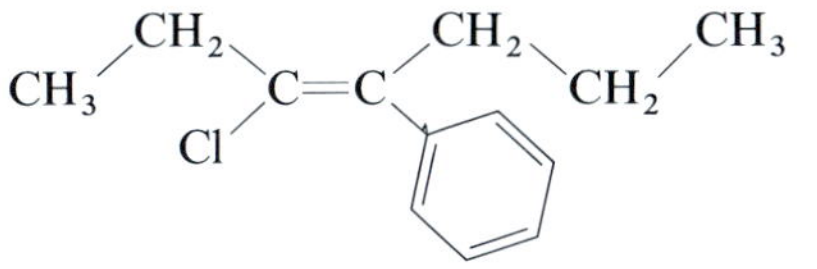
CH₂ CH₂ CH₃
CH₃ C=C CH₂
Cl

3-Chloro-4-phenyl-3-heptene

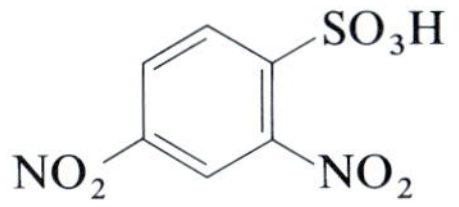
SO₃H
NO₂
NO₂

2,4-Dinitrobenzenesulfonic acid

Now that we know the rules of nomenclature for the aromatics, we can give the names and structures of the major components of the oils of clove, almond, and wintergreen that were mentioned at the beginning of this chapter. (Note that eugenol is a common name.)

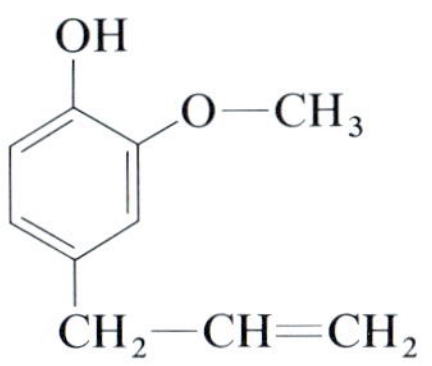
OH
O—CH₃
CH₂—CH=CH₂

Eugenol
(oil of clove)

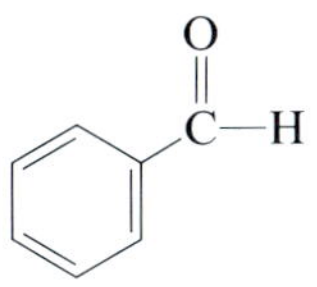
O
C—H

Benzaldehyde
(oil of almond)

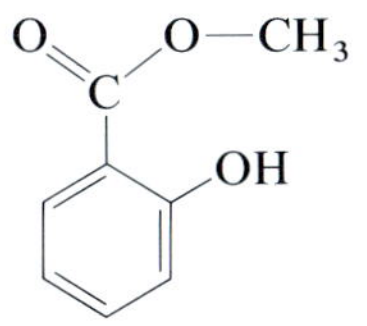
O O—CH₃
C
OH

Methyl salicylate
(oil of wintergreen)

NEW TERMS

***ortho-* (*o*-)** A prefix used to designate substituent position on a benzene ring. The ortho position is immediately adjacent to the parent substituent.

***meta-* (*m*-)** A prefix used to designate substituent position on a benzene ring. The meta position is second from the parent substituent.

***para-* (*p*-)** A prefix used to designate substituent position on a benzene ring. The para position is across from the parent substituent.

phenyl A benzene ring when used as an alkyl group attached to a larger molecule.

benzyl The alkyl group derived from toluene by loss of a hydrogen from the methyl group.

TESTING YOURSELF

Nomenclature

1. Name these substituted benzene compounds.

a. 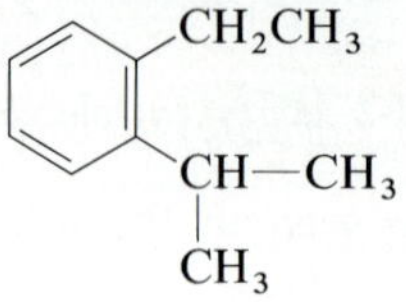

c. 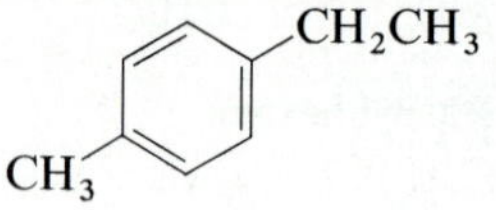

d. 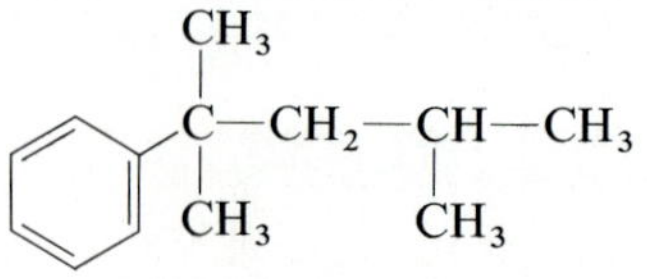

b. 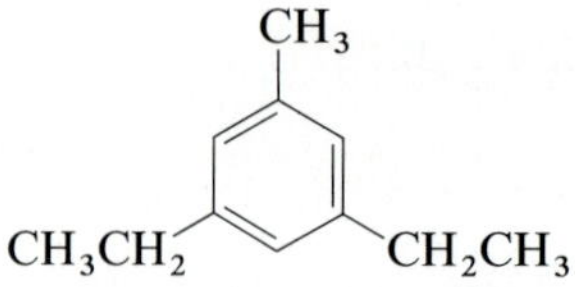

2. Draw structures for each of the following names.
   a. *o*-ethylbutylbenzene
   b. 2,3-dimethyltoluene
   c. 2-ethyl-4,5-dibutylcumene
3. Name these compounds, which have substitutions other than alkyl.

a. $CH_3—C{\equiv}C—CH_2—CH_2—NO_2$

c. 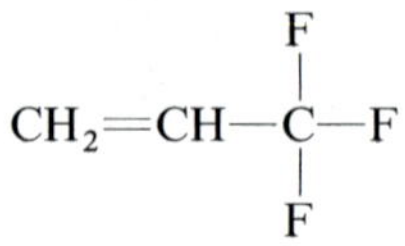

b. 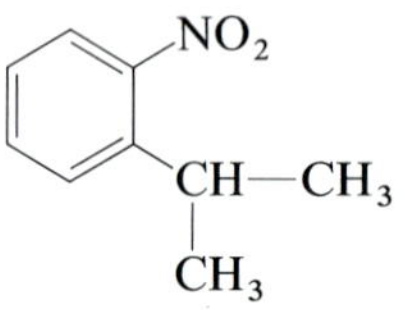

d. 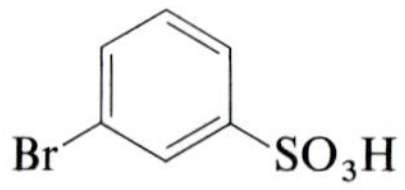

Answers **1. a.** *o*-ethylcumene or 2-ethylcumene **b.** 3,5-diethyltoluene **c.** *p*-ethyltoluene or 4-ethyltoluene **d.** 2-phenyl-2,4-dimethylpentane **2. a.** 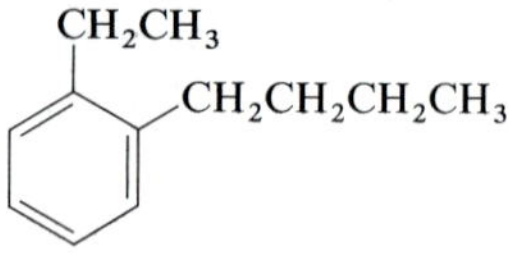

b. 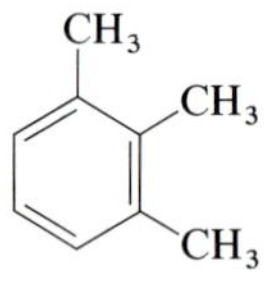 **c.** 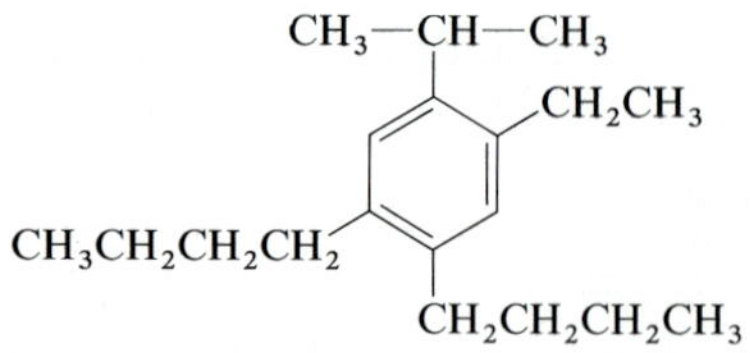

3. a. 5-nitro-2-pentyne **b.** *o*-nitrocumene or 2-nitrocumene **c.** 3,3,3-trifluoro-1-propene **d.** *m*-bromobenzenesulfonic acid

17.3 Aromatic Reactions: A Consequence of Overlapping Pi Bonds

Substitution Reactions

The delocalized nature of pi bonding electrons in benzene makes it unreactive to normal alkene reactions. Most of the alkene reactions discussed in the last chapter are *addition* reactions. The alkene double bond is lost as a result of the reaction. If benzene were to participate in an addition reaction, its double bond(s) would be lost, and pi electron delocalization would no longer be possible. The resulting compound would no longer be aromatic (Figure 17.4).

Lewis acids are *electrophiles* (abbreviated E^+), which means electron "lovers." They attack the pi electrons of the aromatic ring. However, rather than adding to the ring, the electrophile *substitutes* for a hydrogen atom. The replacement of H by E^+ is the reason that most aromatic reactions are called **electrophilic substitution reactions**.

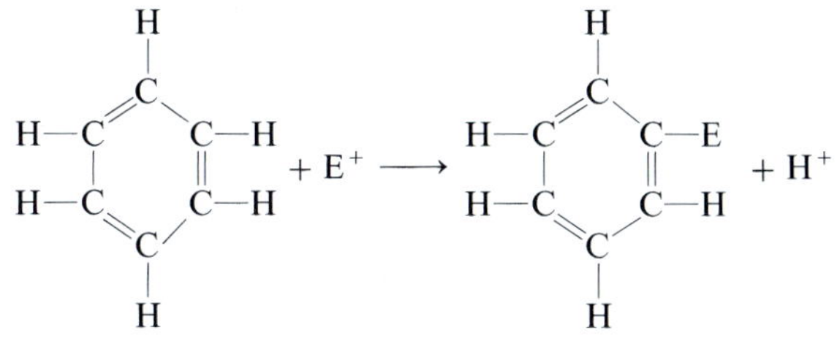

What kinds of electrophiles (E^+) will react with benzene? We know, for instance, that Br_2 *will not* react with benzene. Yet, Br_2 and $FeBr_3$ are an effective source of Br^+. There are special reaction conditions required for aromatic reactions. Table 17.1 lists some of the special reaction conditions needed to generate the electrophiles.

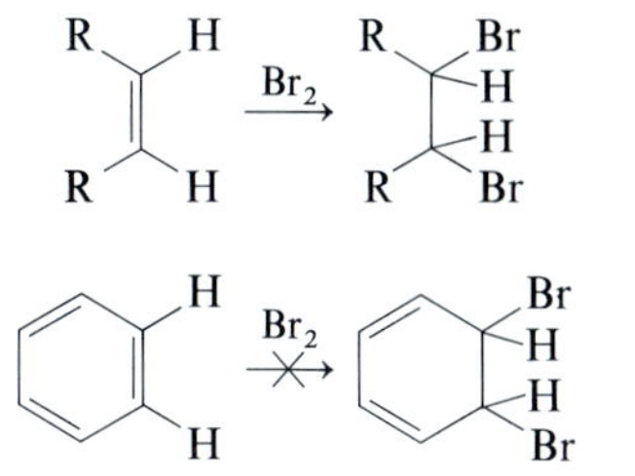

FIGURE 17.4
As we saw in Chapter 16, alkenes generally react by addition reactions. The double bond of an aromatic compound does not do this because if there was an addition across one of the double bonds, the benzene ring would lose its resonance stability.

GENERAL REACTION

Substitution Reaction of an Aromatic with an Electrophilic Group (E^+)

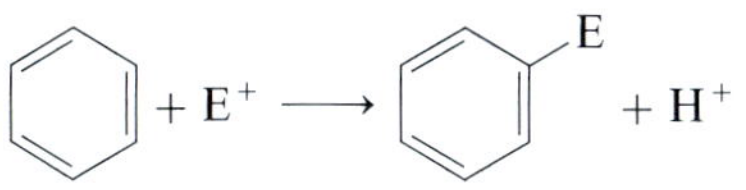

Examples

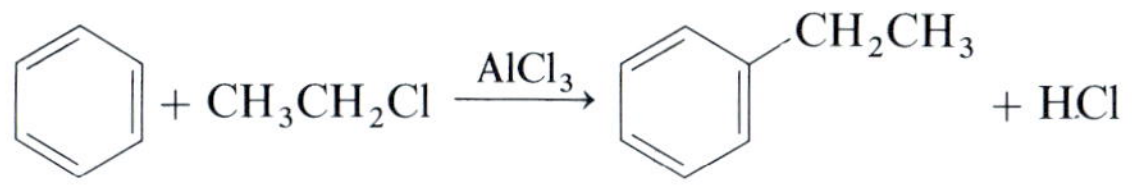

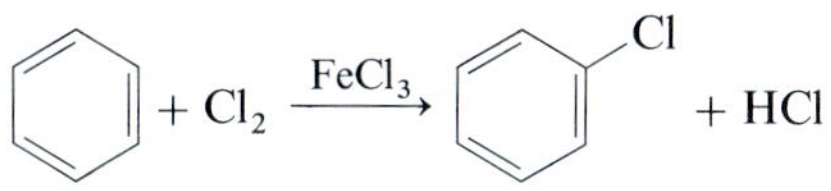

TABLE 17.1
Reaction Conditions and Electrophiles

Reaction Conditions	Identity of Electrophile
$X_2 \xrightarrow[\text{(catalyst)}]{FeX_3}$	X
$H_2SO_4 \xrightarrow{\Delta}$	SO_3H
$HNO_3 + H_2SO_4 \longrightarrow$	NO_2
$RX \xrightarrow[\text{(catalyst)}]{AlCl_3}$	R (only for methyl and ethyl)

Substitution of a Ring Already Having Substituents

Often a benzene ring already has one or more groups attached to it before it is reacted. For example, in a reaction of toluene with some E^+ reagent, where would the E^+ substitute?

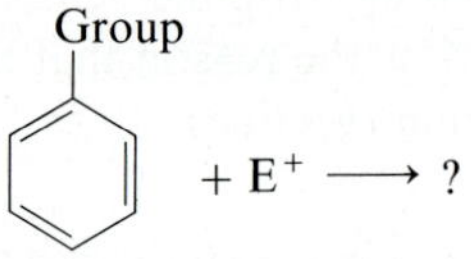
Group
+ E+ ⟶ ?

The group already present on a substituted benzene ring directs the position of attack by a new reagent.

After many years of observation and testing, chemists have determined that the group or groups already on the ring directs where the E^+ substitutes and that these groups can be classified into two categories. One category of groups always directs the incoming E^+ reagent to an ortho and/or para position—these are called *ortho–para directors.*

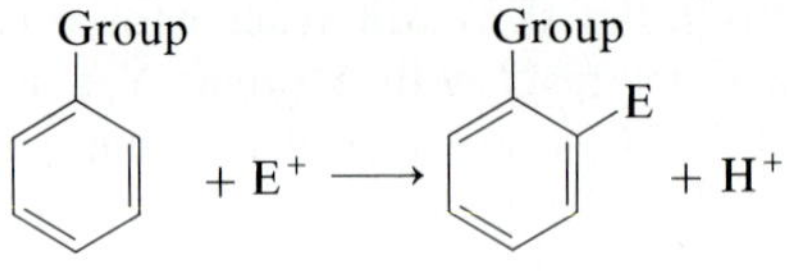
Group
+ E+ ⟶
Group
E
+ H+

ortho

or

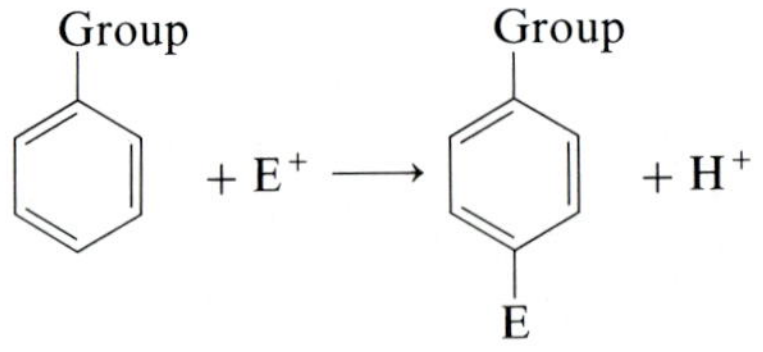
Group
+ E+ ⟶
Group
E
+ H+

para

TABLE 17.2
Directing Groups on a Benzene Ring

Symbol	Name or Prefix
Ortho-para directing groups	
—R	Alkyl
—OH	Hydroxy
—OR	Alkoxy
—X	Halogen
$—NH_2$	Amine
Meta directing groups	
$—NO_2$	Nitro
—C(=O)—OH	Carboxyl
—C(=O)—R	Ketone
—C(=O)—H	Aldehyde
$—SO_3H$	Sulfonic acid
—CN	Nitrile

Except for alkyl groups, the ortho–para directing groups have one or more pairs of unshared electrons associated with the atom directly attached to the aromatic ring.

The second type of group directs toward the meta position and is called *meta directors.*

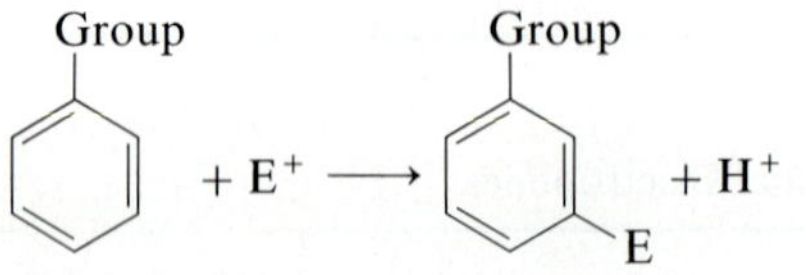
Group
+ E+ ⟶
Group
E
+ H+

meta

In the meta directing groups, a common characteristic is that the atom directly attached to the benzene ring is multiply bonded to an electronegative atom. Table 17.2 lists a number of these common **directing groups**.

When reactions are carried out on benzene rings that are already substituted, there are two steps we must think about: (1) what is the E^+ and (2) where will it substitute? The first question is answered in exactly the same way that was done in the previous section—the reagents that are used determine the identity of E^+. Once the identity of E^+ is known, we have only to decide what kind of substituent there is on the ring. This tells where the E^+ will attach itself.

EXAMPLE 17.5

Show products for the following reactions.

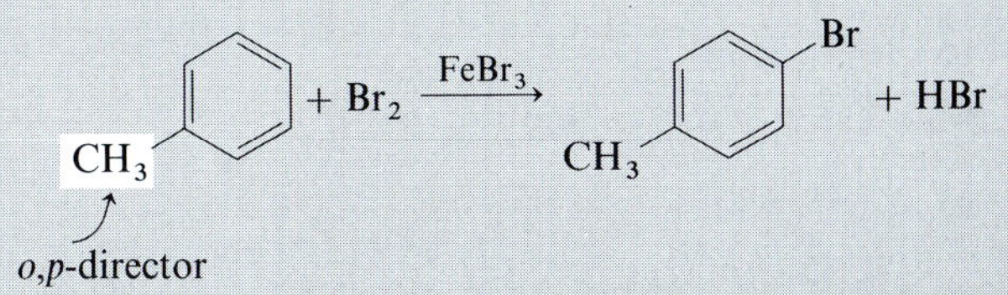

+ Br2
FeBr3
CH3
o,p-director
Br
CH3
+ HBr

You would also be correct in giving the product as 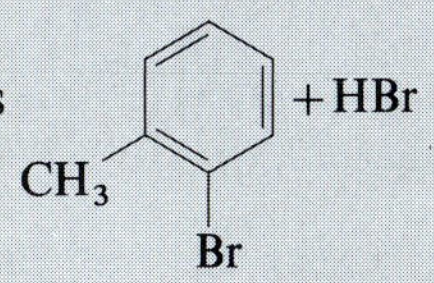

CH3
Br
+HBr

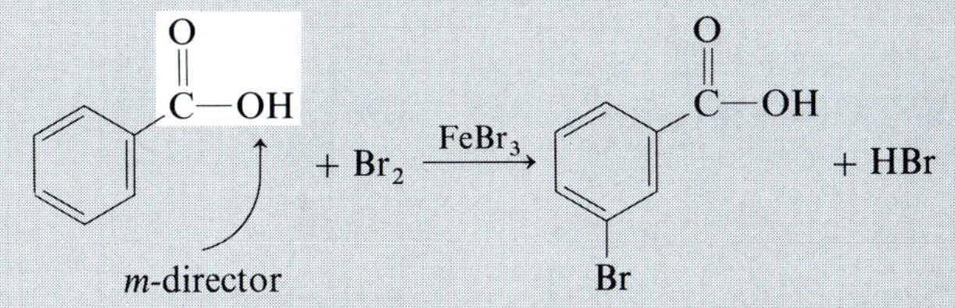

O
C—OH
m-director
+ Br2
FeBr3
O
C—OH
Br
+ HBr

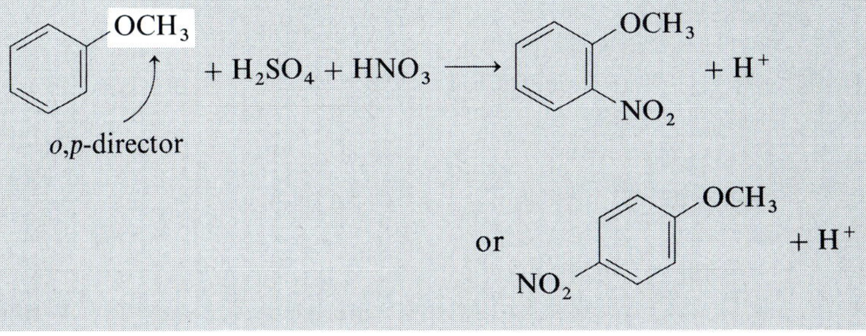

OCH3
o,p-director
+ H2SO4 + HNO3
OCH3
NO2
+ H+
or
OCH3
NO2
+ H+

Oxidation Reactions

The aromatic ring does not readily undergo oxidation. When ring oxidation does occur, particularly in biological oxidation reactions, the products are often *phenols* (see Chapter 18 for details). Phenols are benzene rings with hydroxyls attached and many are quite toxic.

Carbon side chains on an aromatic ring are readily oxidized.

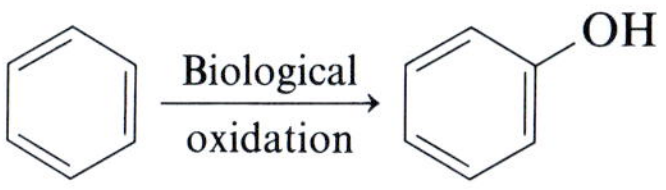

Biological
oxidation
OH

Phenol

Alkyl-substituted benzene rings undergo side-chain oxidation to benzoic acid and derivatives (see Chapter 20 for more details).

INTERESTING AROMATIC COMPOUNDS

A fundamental question for organic chemists has been the size of a ring that can maintain alternating double bonds and still provide special "aromatic" properties. Novel hydrocarbons have been constructed to test these questions, and in each case the application of the $4n + 2$ pi electron rule has been found to hold true.

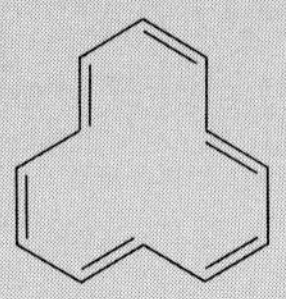

12 pi electrons

$4n + 2 = 12$

$4n = 10$

$n = 2.5$

(not a whole number)

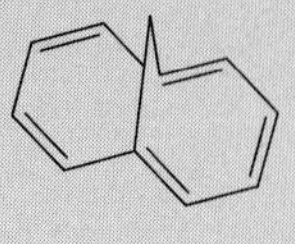

10 pi electrons

$4n + 2 = 10$

$4n = 8$

$n = 2$

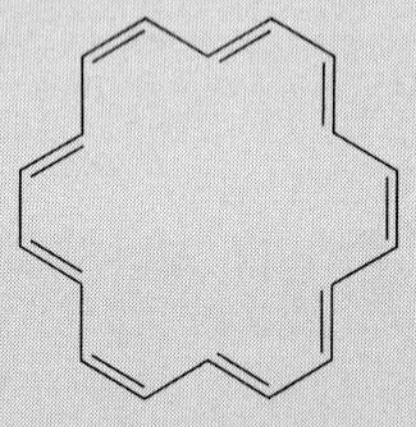

18 pi electrons

$4n + 2 = 18$

$4n = 16$

$n = 4$

There has been thought that the rule would fail for molecules having $n > 5$. In the past year the rule has been tested, and it is now known that it holds for larger values of n. The test molecules also happen to have some biological and potential medicinal significance.

Chlorophyll and heme are complex aromatic (porphyrin) systems that contain nitrogen and are able to attach to metals such as magnesium and iron. There has been interest in putting other metals in the cavities of these porphyrins, but the available space is too small for many metals of interest.

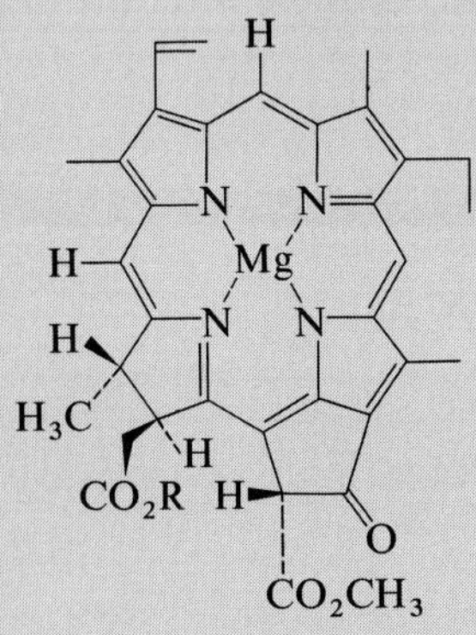
H
N
N
Mg
H
N
N
H
H_3C
H
CO_2R
H
O
CO_2CH_3

Chlorophyll

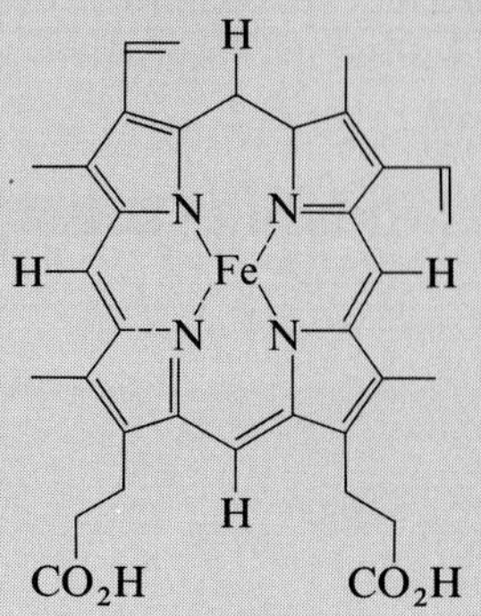
H
N
N
Fe
H
H
N
N
H
CO_2H
CO_2H

Heme

Recently additional space in the center of the molecule was provided when chemists synthesized *texaphyrin*, a molecule that can incorporate cadmium. (The name comes from Texas, the "Lone Star" State.)

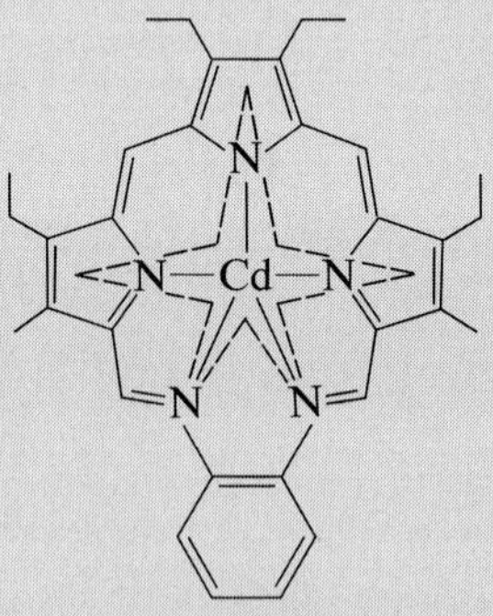
N
N
Cd
N
N
N

Texaphyrin

The cadmium system absorbs more light in the red region of the spectrum. This spectral region is very interesting because it can excite oxygen to a special, highly reactive state called *singlet oxygen*. Singlet oxygen has the ability to kill tumor cells, but it has been difficult to get this oxygen into the tumor cells. A possible future use of molecules such as texaphyrin might be to incorporate them into tumor cells, then excite them to help transfer energy to oxygen. This might provide a unique way to supply singlet oxygen to tumor cells.

Source: *New Scientist* (1989), Feb., p. 32.

GENERAL REACTION

Side Chain Oxidation

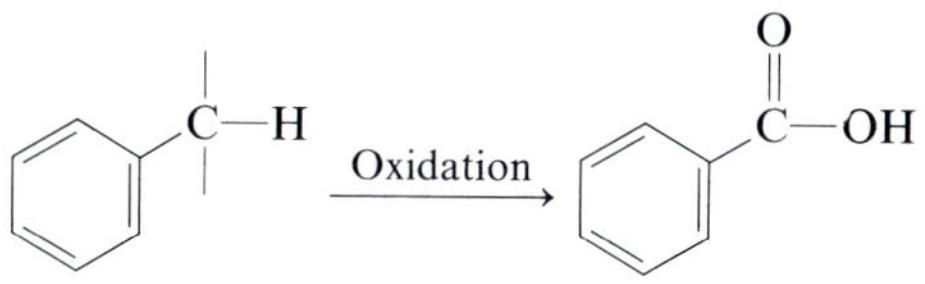

Example

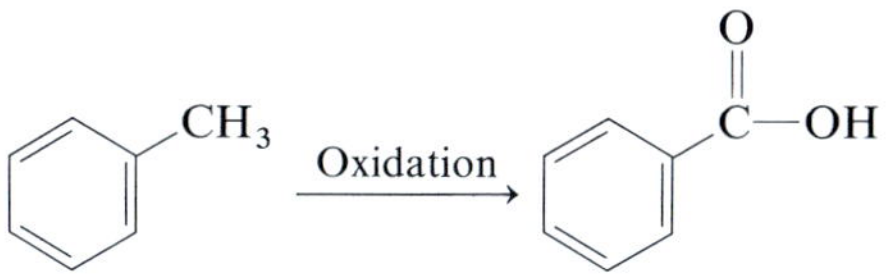

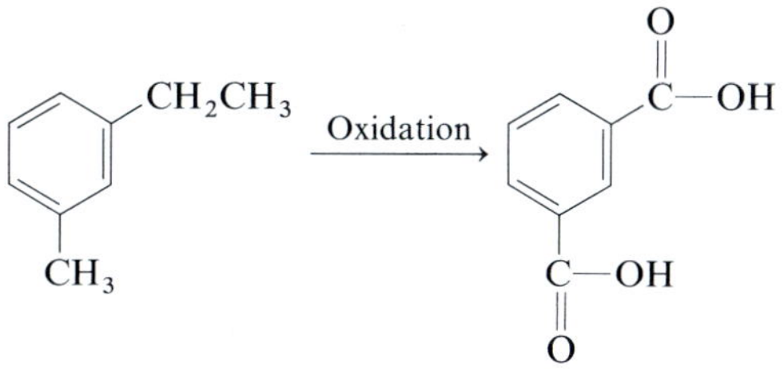

Benzoic acid derivatives are less toxic than phenol derivatives. Thus, we find that toluene is less harmful to us than benzene because the metabolic oxidation product of toluene is benzoic acid.

NEW TERMS

electrophilic substitution reaction Reaction in which an electrophile (E^+) substitutes for a H on an aromatic ring.

directing groups The groups already on a benzene ring that direct the position of attachment of electrophiles.

TESTING YOURSELF

Aromatic Reactions

1. Give products for each of the following electrophilic substitution reactions.

a.  $+ H_2SO_4 \xrightarrow{\Delta}$

b.  $+ CH_3Cl \xrightarrow{AlCl_3}$

c.  $+ Br_2 \xrightarrow{FeBr_3}$

d.  $+ HNO_3 + H_2SO_4 \xrightarrow{\Delta}$

2. Give the products expected from the following reactions with rings that already have substituents.

a. 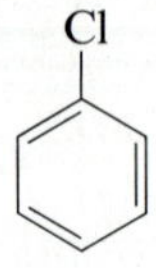
Cl
$+ HNO_3 + H_2SO_4 \xrightarrow{\Delta}$

b. 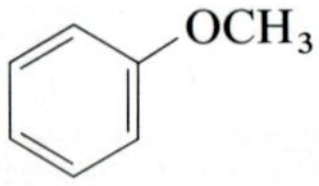
OCH₃
$+ H_2SO_4 \xrightarrow{\Delta}$

c. 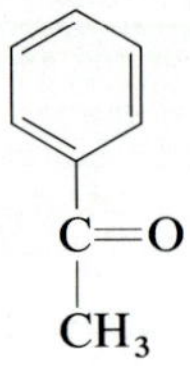
C=O
CH₃
$+ Br_2 \xrightarrow{FeBr_3}$

Answers **1. a.** 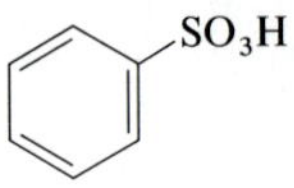
SO₃H
b. 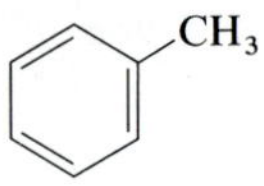
CH₃
c. 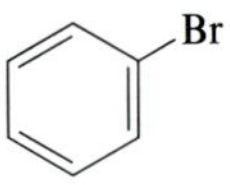
Br
d. 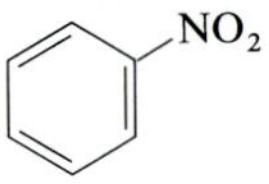
NO₂

2. a. 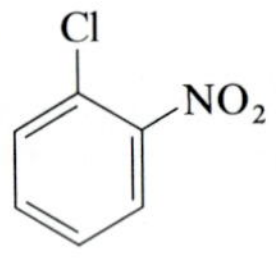
Cl
NO₂
+ para
b. 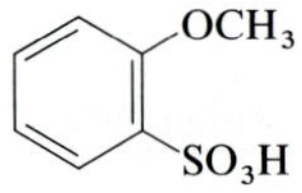
OCH₃
SO₃H
+ para
c. 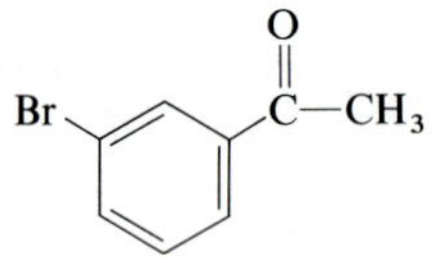
O
Br
C—CH₃

17.4 Fused-Ring Aromatic Systems

Aromatic character is due to the overlap of adjacent pi-bonding electrons in a ring compound. Although the six-carbon system is most common, linked rings and other ring-sized aromatics are also found in nature. As with mono rings, the major characteristic determining aromatic character (and thus substitution rather than addition reactivity) is the presence of $4n + 2$ pi electrons in the ring. For example, note the number of pi electrons in these fused-ring compounds.

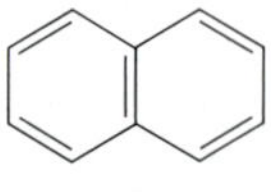

Naphthalene
(mothballs)

$n = 2$

$4n + 2 = 10$ pi elections

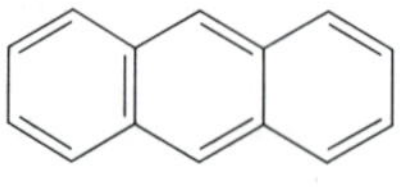

Anthracene

$n = 3$

$4n + 2 = 14$ pi electrons

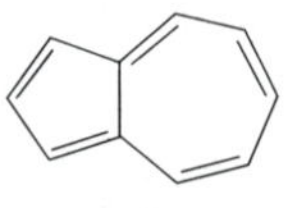

Azulene

$n = 2$

$4n + 2 = 10$ pi electrons

Some fused-ring aromatic hydrocarbons have been found to cause cancer in humans

A number of fused-ring aromatic compounds are known to be toxic or carcinogenic (cancer causing). Often these compounds are "bent" aromatics, that is, they are multiple rings fused together in such a way as to not be linear. For example,

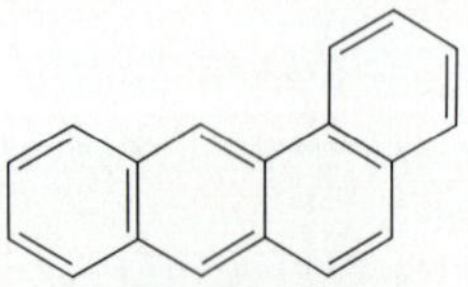

Benzanthracene

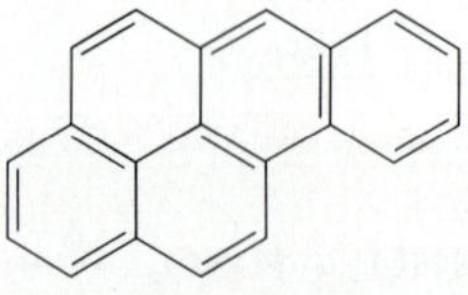

Benzpyrene

Benzpyrene is one of the major substances found in cigarette smoke and has been implicated as a potent carcinogen. It should be understood that benzpyrene itself is not the problem. However, it is easily converted by the body's oxidation processes to a compound that tightly binds to DNA and causes mutations. (Chapter 26 has more details of DNA and mutations.)

17.5 Important Aromatic Hydrocarbons

One of the most notable examples of a biological aromatic compound is *estradiol*, a major female sex hormone. Additional discussion of the role of this biological compound is in Chapter 24.

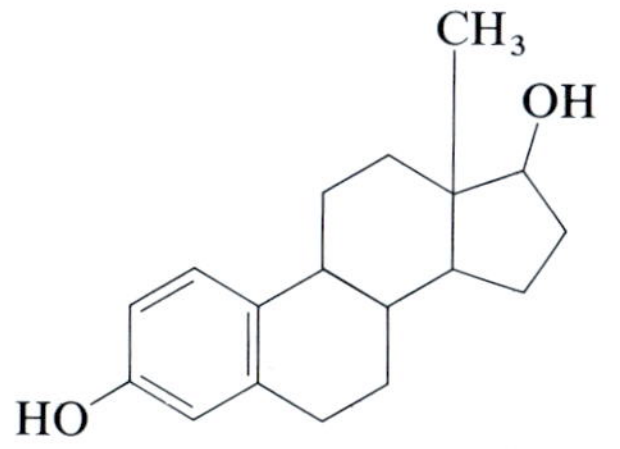

CH3
OH
HO

Estradiol

Flavorings are well represented by aromatic compounds. Three examples were presented at the introduction of this chapter (clove, almond, and wintergreen). Additional examples include the flavorings from anise, vanilla, and thyme.

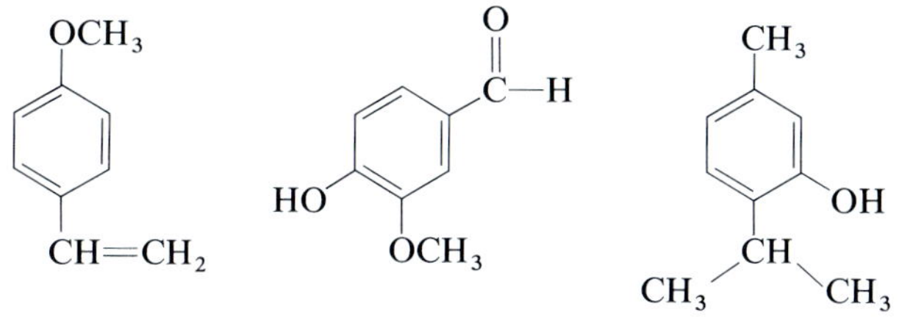

OCH3
CH=CH2
O
C—H
HO
OCH3
CH3
OH
CH
CH3
CH3

Anethole (anise) **Vanillin** (vanilla) **Thymol** (thyme)

Vitamin E is an aromatic compound with the chemical name α-tocopherol. It is an excellent antioxidant and has been shown to have antisterility properties in some animals.

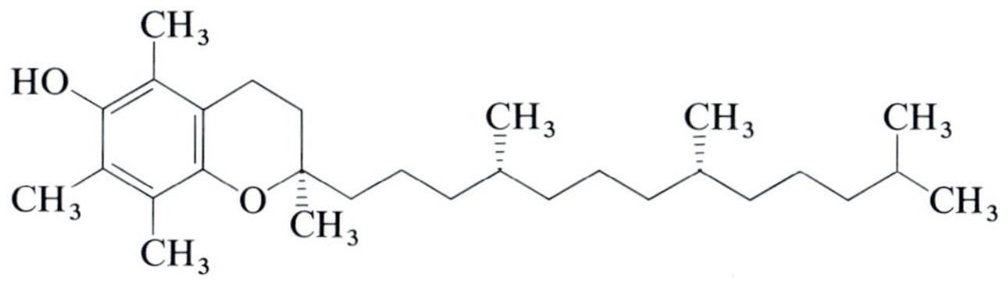

CH3
HO
CH3
O
CH3
CH3
CH3
CH3
CH3
CH3

α-Tocopherol

Warfarin is an aromatic compound that is a potent anticoagulant. One of the major uses of this compound and its derivatives is rat poison. When mice or rats ingest foods laced with this ingredient, they tend to hemorrhage and die.

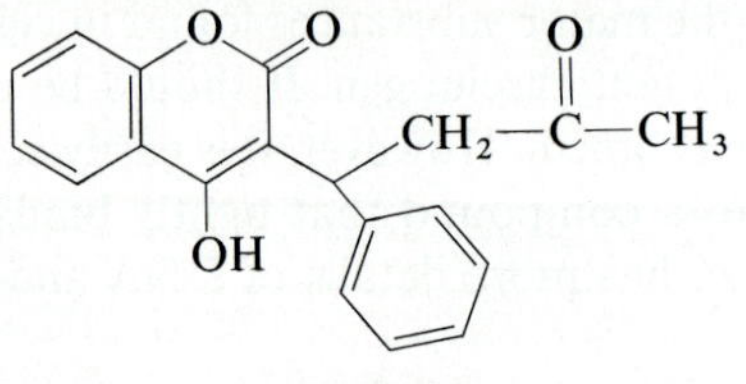

Warfarin

From these few examples it is readily apparent that —OH or —OR groups connected to an aromatic ring are quite common for biologically active compounds. In the next chapter we will be discussing in detail compounds of this type, and additional examples will be provided.

In Chapter 21 we will encounter a number of aromatic ring compounds that have elements other than carbon making up the ring. For example, an analog of benzene containing nitrogen is called pyridine.

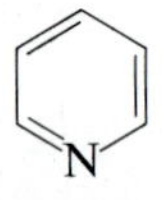

Pyridine

Ring compounds containing atoms other than carbon are called *heterocyclic*.

SUMMARY

Conjugated double bonds give a special stability, called *resonance energy*, to molecules that contain them. This is particularly demonstrated in *aromatic* molecules. These are ring systems with alternating double and single bonds that obey the Huckel rule. Because of the special stability of these molecules, they will not undergo simple addition reactions to the double bonds. Instead, the molecules undergo aromatic *substitution reactions*, where an electrophile (E^+) substitutes for a hydrogen. Another unique reaction of an aromatic molecule is *side chain oxidation*.

REACTION SUMMARY

Substitution Reaction of an Aromatic with an Electrophilic Group (E^+)

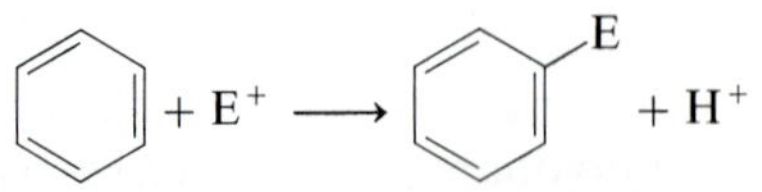

Side Chain Oxidation

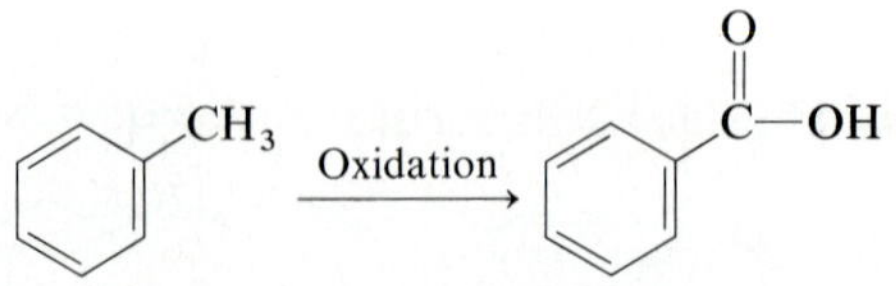

TERMS

aromatic
benzene
resonance energy
ortho-
meta-
para-
phenyl
benzyl
electrophilic substitution reaction
directing group

QUESTIONS

CHAPTER REVIEW

1. 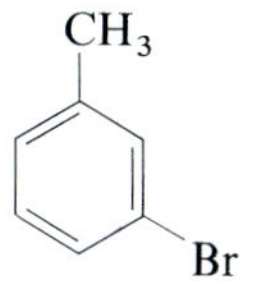 is best named

   a. 3-methylbromobenzene c. *p*-bromotoluene
   b. *o*-methyl-*p*-bromobenzene d. *m*-bromotoluene

2. Benzene derivatives undergo what type of reactions?
   a aromatic electrophilic substitution c. electrophilic addition
   b. aromatic nucleophilic d. nucleophilic addition

3. When a benzene ring is used as an alkyl group, it is called
   a. benzyl b. aromaticyl c. phenyl d. none of these

4. When the reactants $HNO_3 + H_2SO_4$ and heat are used with benzene, the resulting product is
   a. nitrobenzene c. sulfobenzene
   b. benzenesulfonic acid d. no reaction

5. For a reaction of ethylbenzene, the ethyl is considered what type of directing group?
   a. ortho b. ortho–para c. meta d. ortho–meta

6. 2,3-Dichlorobromobenzene is

   a. 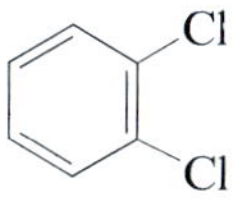
   c. 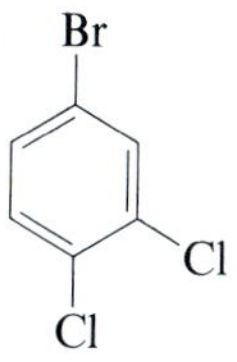
   d. 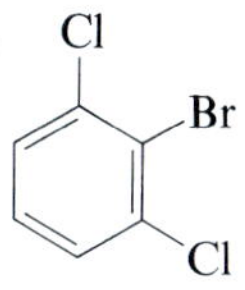
   b. 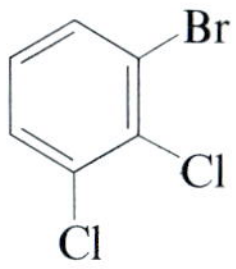

7. Which of the following is *not* aromatic?

   a. 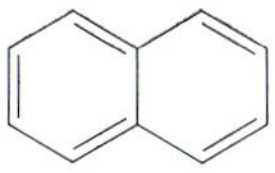
   c. 
   b. 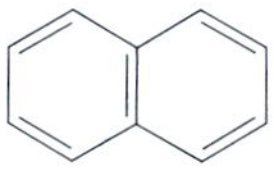
   d. 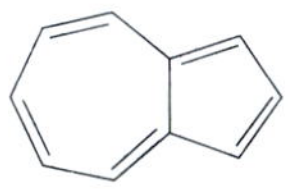

8 .The carbon atoms in a benzene ring are ________ hybridized.
   a. *sp* b. sp^2 c. sp^3

9. Which of the following resonance structures is most likely *not* to contribute to the stability of azulene?

   a. 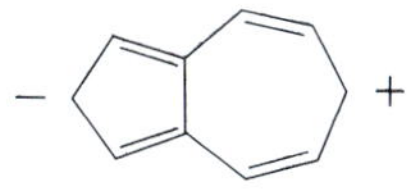
   c. 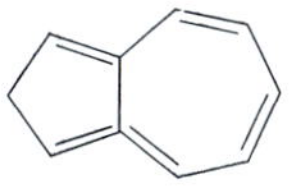
   b. 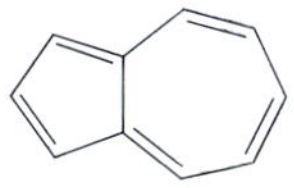
   d. 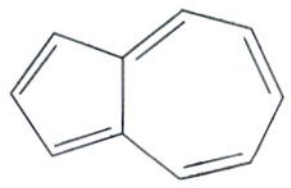

10. Reaction of nitrobenzene with Br_2 and $FeBr_3$ will give as the major product
    a. *o*-nitronitrobenzene c. *o*-nitrobromobenzene
    b. *m*-nitronitrobenzene d. *m*-nitrobromobenzene

Answers **1.** d **2.** a **3.** c **4.** a **5.** b **6.** b **7.** c **8.** b **9.** c **10.** d

DIAGNOSTIC CHART

Blacken in all of the circles under the number of each question you missed in the Chapter Review questions. The diagnostic chart will help you identify concept areas that need more study.

	Questions									
Concepts	1	2	3	4	5	6	7	8	9	10
Structure						○	○	○	○	
Nomenclature	○		○	○	○	○				○
Reactions		○		○	○					○
Hybridization								○		

EXERCISES

Nomenclature

1. Name the following compounds.

a. 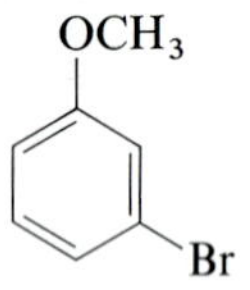

OCH_3
Br

g. 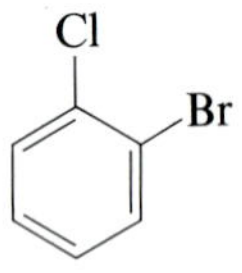

Cl
Br

b. 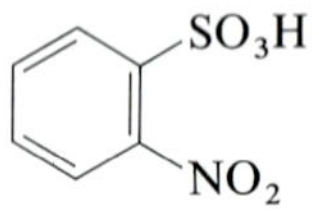

SO_3H
NO_2

h. 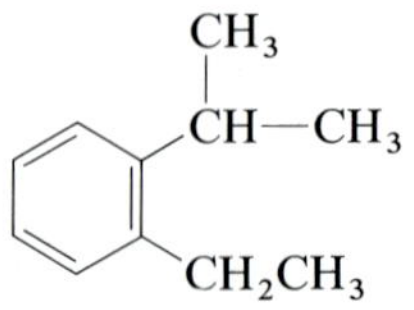

CH_3
$CH—CH_3$
CH_2CH_3

c. 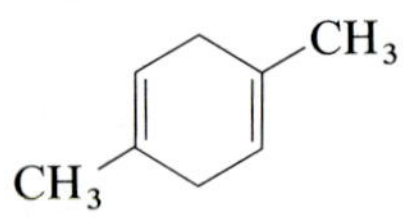

CH_3
CH_3

i. 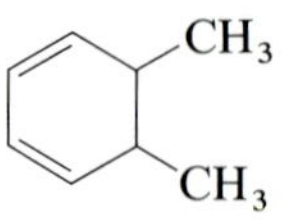

CH_3
CH_3

d. 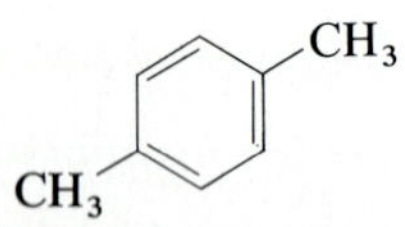

CH_3
CH_3

e. 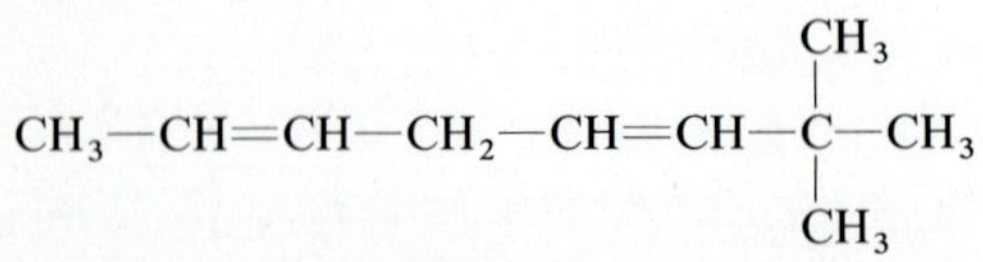

CH_3
$CH_3—CH=CH—CH_2—CH=CH—C—CH_3$
CH_3

f. 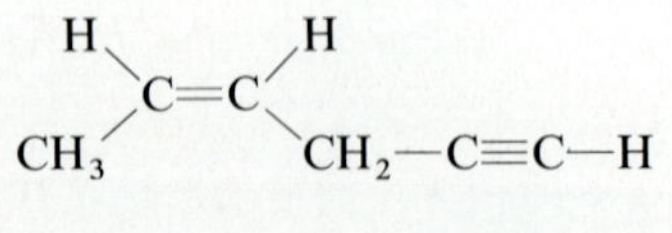

H
H
C=C
CH_3
$CH_2—C≡C—H$

j. 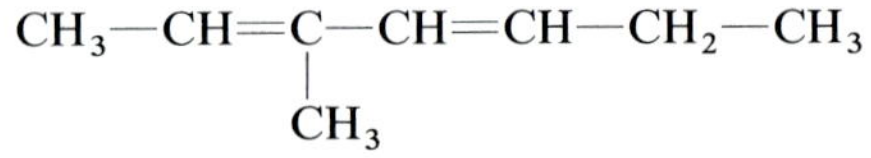

$CH_3—CH=C—CH=CH—CH_2—CH_3$
CH_3

2. Draw structures for the following compounds.
   a. *o*-bromocumene
   b. 2,4,6-dinitrotoluene
   c. *p*-chloronitrobenzene
   d. *p*-ethylbenzenesulfonic acid
   e. *o*-ethyltoluene
   f. 4-benzyl-1-octene
   g. 1,3-dibromobenzene
   h. *m*-cyanotoluene
   i. 2,3-dibromotoluene
   j. 3-phenylcyclohexene

3. Name the following compounds.

a. 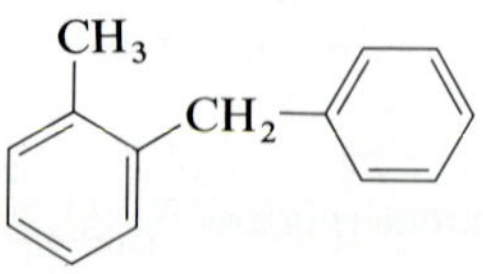

CH_3
CH_2

d. 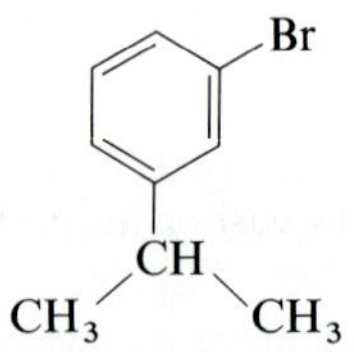

Br
CH
CH_3
CH_3

b. 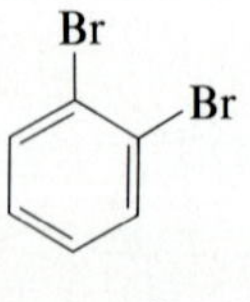

Br
Br

e. 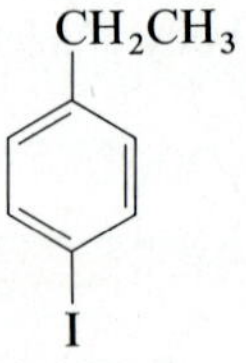

CH_2CH_3
I

c. 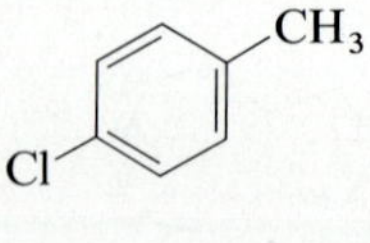

CH_3
Cl

f. 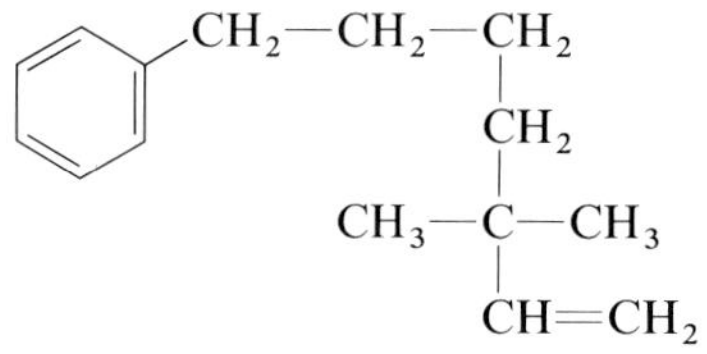

Resonance

4. Draw resonance structures for the nitrate ion and naphthalene.
5. Explain what is meant by "resonance structures."
6. What is meant by the term *resonance energy*?
7. From memory, list the electrophilic agent (E^+) that is formed from the following reagents.
   a. $Br_2 + FeBr_3$
   b. H_2SO_4 and heat (Δ)
   c. $H_2SO_4 + HNO_3$
   d. $CH_3CH_2Cl + AlCl_3$
8. From memory, label each of the following groups as a meta-director or as an ortho–para director.
   a. $CH_3—CH_2—$
   b. $CH_3—C(=O)—$
   c. $—C≡N$
   d. Br—
   e. HO—
   f. $—NO_2$
   g. $—O—CH_2—CH_3$
   h. $—SO_3H$
   i. $—C(=O)—H$
9. Give the structure of the major product from each of the following reactions. (NOTE: For reactions of ortho-para substituted benzene derivatives show either the ortho *or* the para product.)

a. 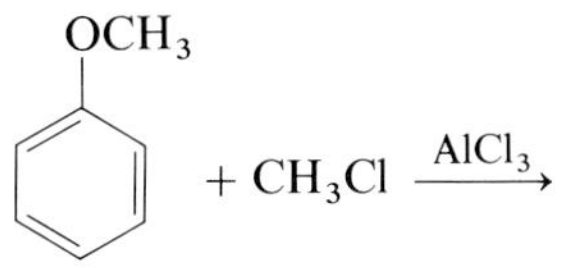

b. 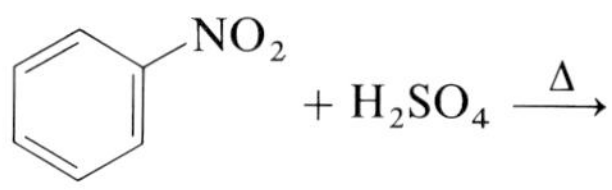

c. 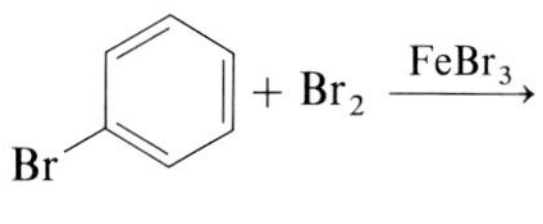

d. 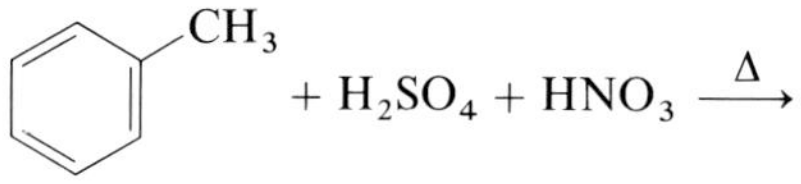

e. 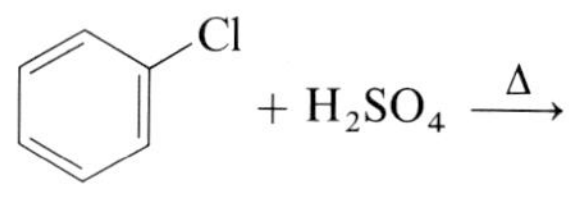

f. 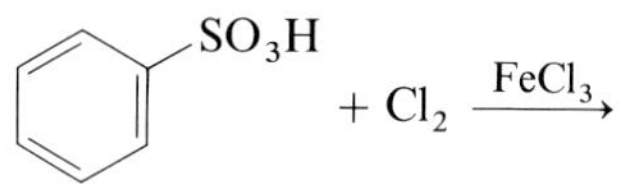

10. Give the structure of the major product from each of the following reactions.

a. 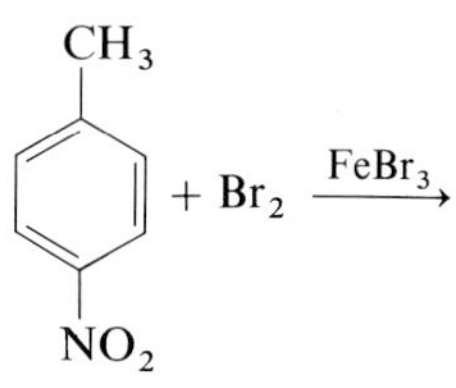

b. 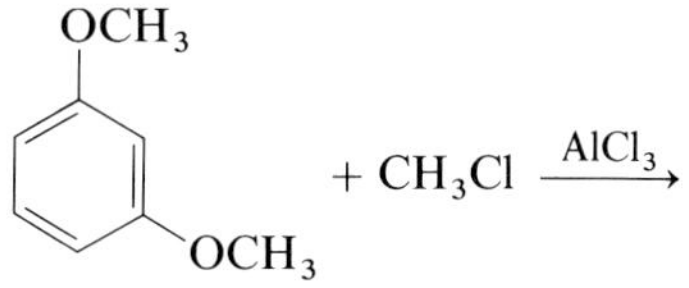

c. 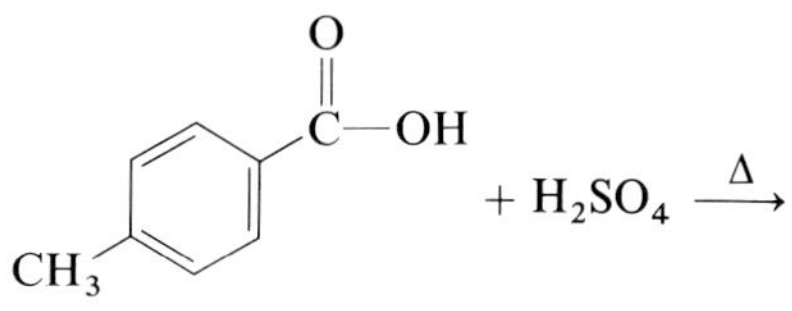

d. 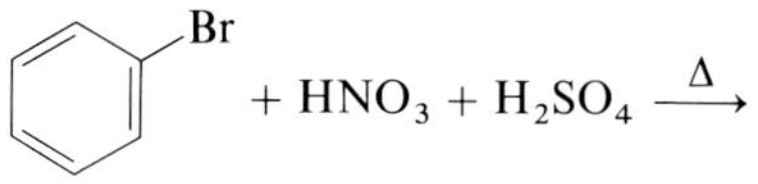

e. 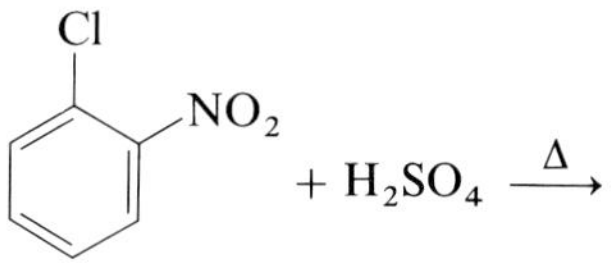

f. 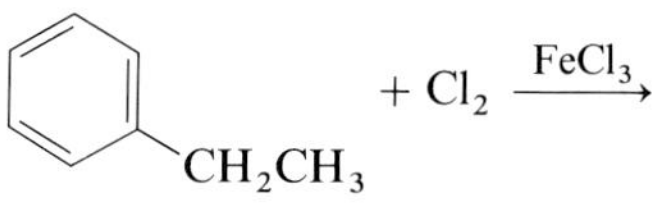

g. 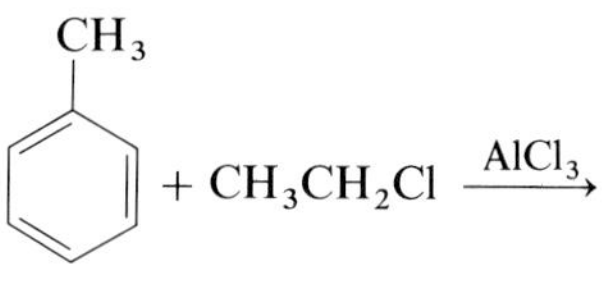

h. 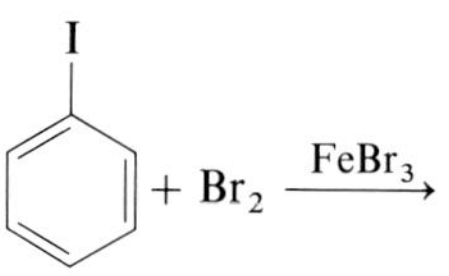

CHAPTER

18

Alcohols, Phenols, Ethers, and Thiols

OBJECTIVES

After completing this chapter, you should be able to

- Identify and name alcohols and phenols.
- Describe the solubility similarities between alcohols and water.
- Predict the products of common reactions of alcohols.
- Show how to prepare alcohols.
- Identify and name ethers.
- Predict the products of common reactions of ethers.
- Identify and name members of the thiol family.
- Predict the products of the common reactions of thiols.

Water is one of the most essential compounds for survival on this planet. As a consequence of hydrogen bonding and polarity, its physical properties are different than might be expected from its molecular weight. In this chapter we explore some of the organic analogs of water: alcohols, phenols, ethers, and sulfur-containing thiols. These molecules are the first of the organic compounds that start to exhibit physical and chemical properties different from hydrocarbons.

Oxygen is the most abundant element in the earth's crust, and water is one of the most abundant oxygen-containing molecules. In this chapter we introduce the oxygen-containing functional groups most related to water: those found in *alcohols*, *phenols*, and *ethers*. Alcohols and phenols are compounds in which one of the hydrogens of water is replaced with a carbon-containing group. Ethers are organic compounds that result from replacing both of the hydrogens with carbon-containing groups. Also included here are the *thiols*, which one can suggest as the sulfur equivalent of alcohols; they are based on hydrogen sulfide (the sulfur equivalent of water). The relationship between water and these functional groups can be summarized as follows:

$$\underset{\textbf{Water}}{\text{H—O—H}} \xrightarrow{\text{Replace an H with R}} \underset{\textbf{Alcohol}}{\text{R—O—H}} \xrightarrow{\text{Replace the second H}} \underset{\textbf{Ether}}{\text{R—O—R}'}$$

(If R is aromatic,
this is a phenol).
(If OH is replaced by
SH, this is a thiol).

There are structural similarities among water, alcohol, and ethers:

H—O—H	Water
H—O—R	Alcohol
R—O—R	Ether

18.1 Alcohols

Structure and Physical Properties

From application of the rules for Lewis electron dot structures, an **alcohol** can be characterized by this general structure:

$$\text{R—}\ddot{\underset{\cdot\cdot}{\text{O}}}\text{—H}$$

where R can be an alkyl or cycloalkyl (but not aromatic) group and —OH is the hydroxyl group. The oxygen atom of the alcohol has the same bonding characteristics that we have seen for water: sp^3 hybridization with two lone pairs of electrons.

Hydrogen Bonding and Boiling Point

For all of the functional groups discussed so far, a similar small increase of boiling point has occurred as the carbon number increases. We now begin to see a change in this relationship. The alcohol functional group (—OH) dramatically influences the physical properties of alcohols. First, notice the influence of the —OH group on boiling point (Figure 18.1). The curves for alcohol and alkane diverge greatly. To explain this, let's reexamine an analogous molecule—water.

Although having a molecular weight of only 18, water is a liquid at room temperature and must be heated to 100 °C at atmospheric pressure to be converted to a gas. In contrast, compare these properties to those of a similar molecular weight hydrocarbon. Methane, with a molecular weight of 16, is a gas at room temperature and doesn't liquify until it reaches the very low temperature of −162 °C. It is apparent that the —OH in the water plays a special role in its high boiling point. We would expect the similar structure of an alcohol to be reflected in similar unique properties for alcohols. This

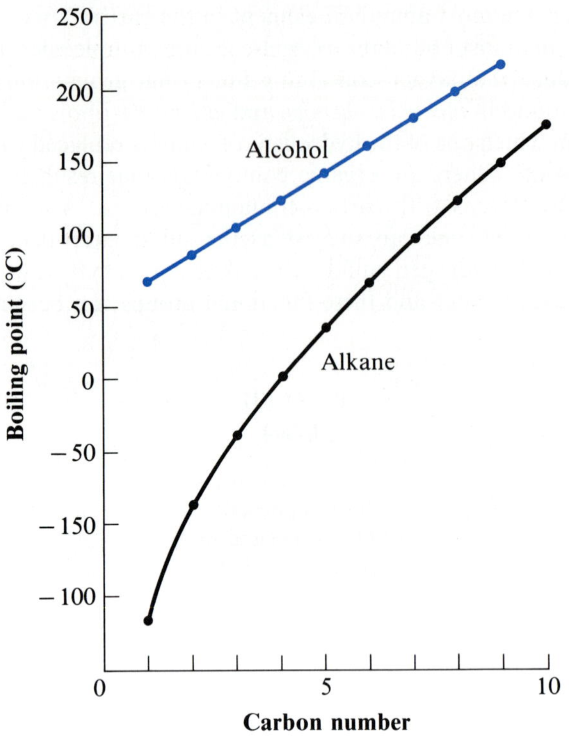
250
200
150
100
50
0
−50
−150
−100
Boiling point (°C)
Alcohol
Alkane
0
5
10
Carbon number

FIGURE 18.1
The relationship between carbon number and boiling point for the alcohols as compared with the alkanes. For the lower carbon number alcohols the —OH group plays an important role through its hydrogen bonding and dominates the physical properties. As the carbon number increases, an alcohol more resembles an alkane and the influence of the —OH group decreases. The physical properties of long carbon chain alcohols more closely resemble the parent hydrocarbon.

Hydrogen bonding in alcohols influences the boiling points of the lower carbon number alcohols.

is observed. The hydrogen bonding capabilities of water and alcohols are responsible for their similarities in physical properties (Figure 18.2).

Energy is needed for the alcohol molecules to free themselves from the weak hydrogen bonding forces that hold them together in the liquid state. Thus, alcohols have higher boiling points than we would expect based on the molecular weight.

It is apparent from Figure 18.1 that the hydroxyl group plays a more important role in determining the boiling points of the small carbon chain alcohols than for the longer chained alcohols. As the carbon chain becomes longer, the hydroxyl portion loses its *relative* importance and the molecule behaves more and more like a hydrocarbon. This ability of long hydrocarbon chains to mask functional group properties is repeated in a number of biologically important molecules. This often allows a polar portion of a molecule

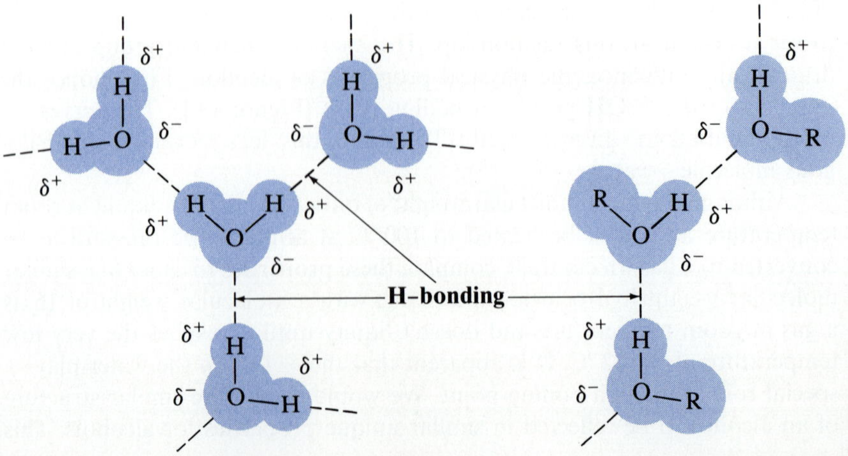
H-bonding

FIGURE 18.2
The structures of water and alcohols are very similar, as is their tendency to hydrogen bond. Because there is only one hydrogen–oxygen bond per alcohol molecule, there is less hydrogen bonding (dashed lines).

to function in a hydrocarbon-like environment. We will see a number of examples of this interplay between hydrocarbon (nonpolar) character and polar functional groups throughout our study of biochemistry.

Solubility: Like Dissolves Like

Since alcohol so closely resembles water in its ability to participate in hydrogen bonding, it is no surprise that water and alcohol have similar solubility properties. Figure 18.3 shows the influence of hydrogen bonding on the ability of alcohols to dissolve in water. The compounds methanol (CH_3—OH) and ethanol (CH_3CH_2—OH) have infinite solubility in water. Just as the hydroxyl group has its main effect on the boiling points of the smaller alcohols, it also influences the solubility of these lower carbon alcohols more. As the hydrocarbon portion of the alcohol increases in size, the alcohol starts to behave more like a hydrocarbon and thus becomes more insoluble (Figure 18.3).

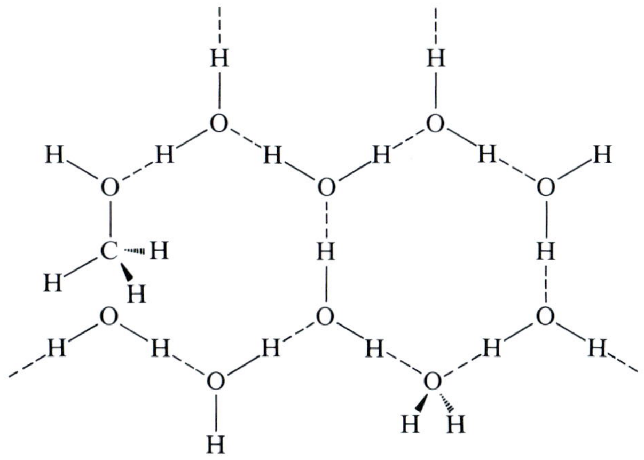

Methanol in water

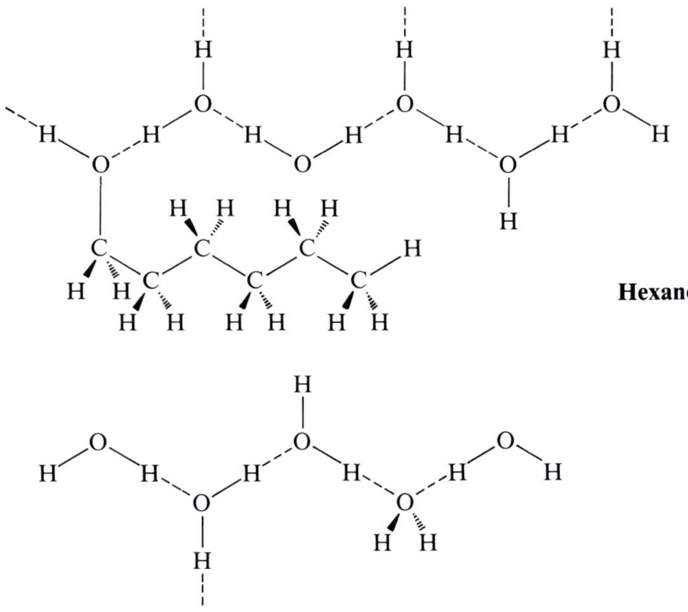

Hexanol in water

FIGURE 18.3
The polarity and the hydrogen bonding (dashed lines) of the —OH group helps an alcohol to dissolve in water. As the number of carbon atoms in the carbon chain increases, an alcohol starts to resemble a hydrocarbon and the solubility decreases until it becomes insoluble in water. (Compare this to Figure 15.3.)

Nomenclature of Alcohols: The Priority of the Hydroxyl Group

In the nomenclature of alcohols, the *-ol* ending is used and a number is given for the position of the —OH group. This group takes on a numbering priority.

In previous chapters, we have developed organized ideas and rules for naming simple hydrocarbons and organohalogen compounds. With these rules, unambiguous names can be given to complex molecules. The same ideas can be applied to naming alcohols as well, with only one major change: the —OH group takes *priority* over alkyl groups or halogens in nomenclature. This simply means that a way must be found to give the position of the —OH group special significance. Here are the basic nomenclature rules for alcohols:

1. When the —OH group is present in a hydrocarbon, the *-e* ending of alkan*e*, alken*e*, and alkyn*e* is replaced by *-ol*.
2. The numbering is assigned such that the —OH group gets the lowest possible number.
3. If an alkene and an alkyne also contain the —OH group, the name is expressed as:

________-en ________-ol
↑ Position of double bond; ↑ Position of —OH group

________-yn ________-ol
↑ Position of triple bond; ↑ Position of —OH group

EXAMPLE 18.1

Name the following alcohols.

$CH_3—OH$ Methanol

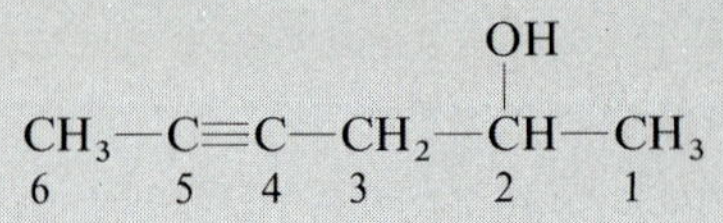
4-Hexyn-2-ol

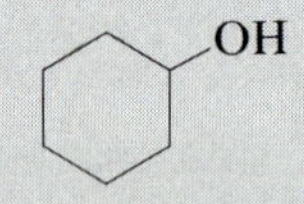
Cyclohexanol

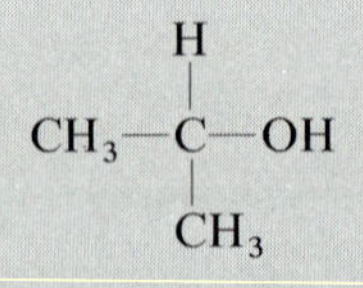
2-Propanol

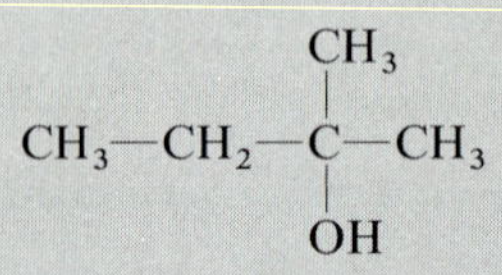
2-Methyl-2-butanol

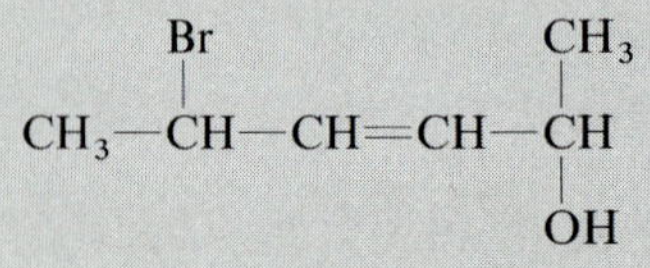
5-Bromo-3-hexen-2-ol

Some common manmade alcohols as well as a number of naturally occurring alcohols are shown in Table 18.1. Notice that some of them are known by a common name as well as an IUPAC name.

TABLE 18.1
Examples of Common and Naturally Occurring Alcohols

Name	Description	Structure
Common Alcohols		
Methanol (wood alcohol)	A solvent and an intermediate for other compounds. Several billion pounds are converted to formaldehyde each year.	$CH_3—OH$
Ethanol (grain alcohol)	The fermentation alcohol (ethyl alcohol) is used in alcoholic beverages and as a solvent.	$CH_3—CH_2—OH$
Butanol (isomer mixture)	A large quantity of butanol is made each year, and is used as a solvent and plastizer.	$C_4H_9—OH$
Natural Products		
Menthol	Found in peppermint oils. Used in throat sprays, cough drops, and inhalers.	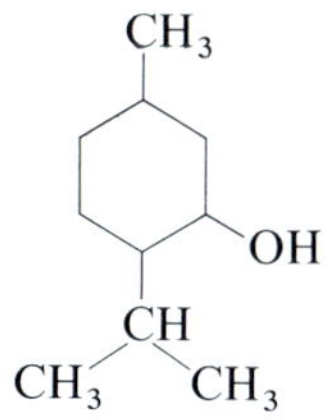
Citronellol	Rose oil	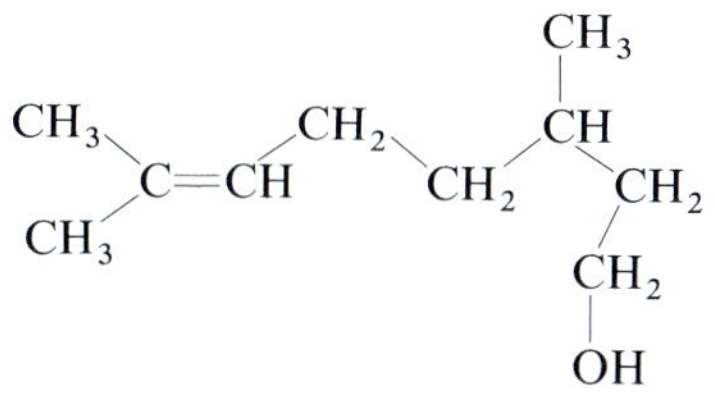
Cholesterol	Necessary for the preparation of steroids. Found in all body tissues, especially the brain. A main constituent of gallstones.	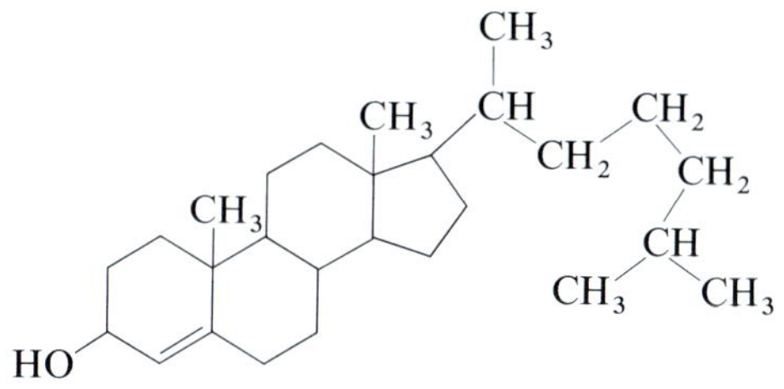
Grandisol	Boll weevil sex attractant	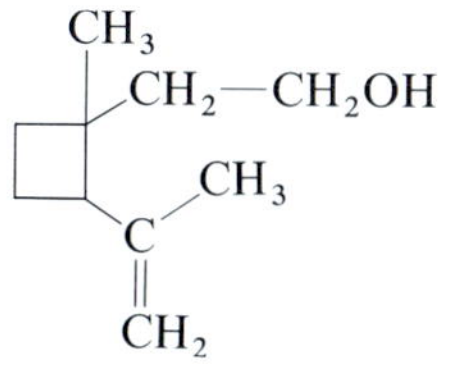

Reactions of Alcohols

As our study has progressed from the chemistry of alkanes to the alkenes, alkynes, and aromatics, the chemical reactivity was shown to be related to the availability of electrons or the *Lewis basicity*. This is a direct reflection of the Lewis base character of many of the functional groups. We will see this property repeated over and over again in our study of organic chemistry and biochemistry.

In the reactions of the alkenes, alkynes, and aromatics, the electrophiles (represented by E^+ in Chapter 17) attack the electrons in the pi bonds. This behavior initiates most of the chemical reactivity of the unsaturated hydrocarbons. If an alcohol is considered to be a Lewis base as a result of the presence of lone pairs of electrons on the oxygen, it follows that alcohols will react with Lewis acids (electrophiles).

At this point it is appropriate to point out the ability of an alcohol to also function as a Brønsted–Lowry acid. By giving up the hydrogen attached to the —OH group (much like water can give up H^+), an alcohol can function as an acid.

$$\text{R—O—H} \rightleftarrows \text{R—O}^- + \text{H}^+$$

As with water, the equilibrium favors the undissociated alcohol.

Alcohols are prepared by the hydration of alkenes.

DEHYDRATION: AN ELIMINATION REACTION When an alcohol reacts with a strong acid (often H_2SO_4), **dehydration**, or the elimination of water, occurs and an alkene is formed.

EXAMPLE 18.2

Show the reaction product of these alcohols with H_2SO_4.

(a)

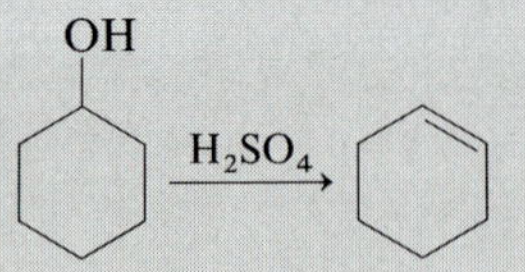
OH
H2SO4

(b)

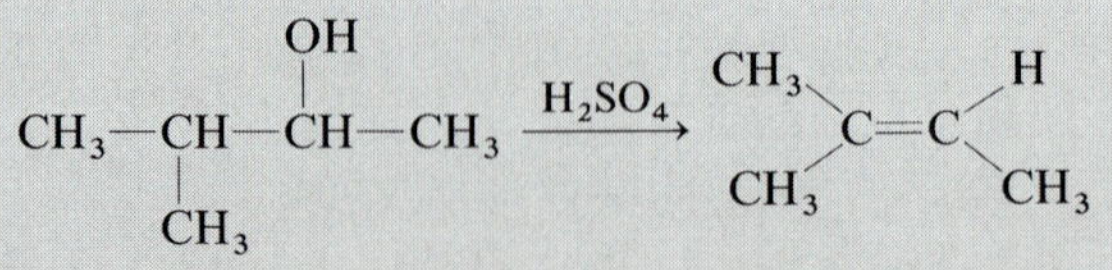
OH
CH3—CH—CH—CH3
CH3
H2SO4
CH3
H
C=C
CH3
CH3

Notice that another product seems possible for (b):

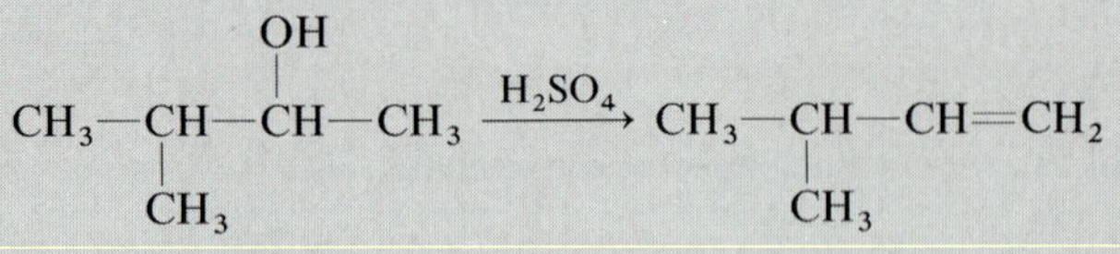
OH
CH3—CH—CH—CH3
CH3
H2SO4
CH3—CH—CH=CH2
CH3

Note that the second product for Example 18.2b is *not* found in the actual reaction. In general, it is observed that the *more substituted* alkene is favored in dehydration reactions. What do we mean by more substituted? Examine the atoms directly attached to the double bond carbon atoms—the fewer the hydrogens, the more substituted the alkene.

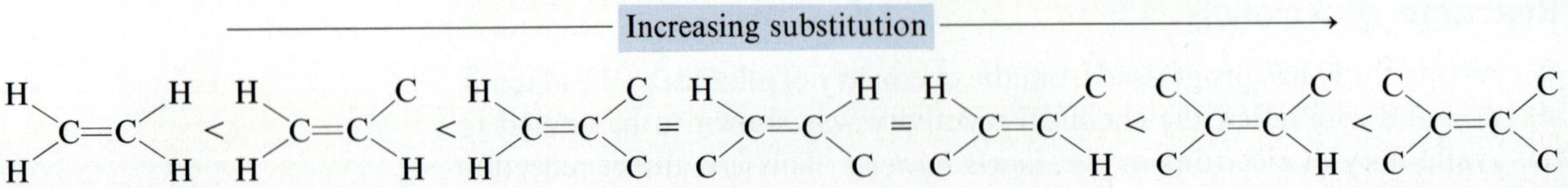
Increasing substitution

One of the important steps in the biological conversion of organic materials to carbon dioxide and water is the dehydration of citric acid. Essentially all organic compounds are converted to citric acid as part of human and animal metabolism. The details of this process will be discussed further in Chapter 28.

The dehydration of an alcohol is the reverse of the hydration of an alkene. Loss of water from an alcohol forms an alkene.

$$\text{HOOC—CH(H)—C(OH)(COOH)—CH}_2\text{—COOH} \xrightarrow{-H_2O} \text{HOOC(H)C=C(CH}_2\text{COOH)(COOH)}$$

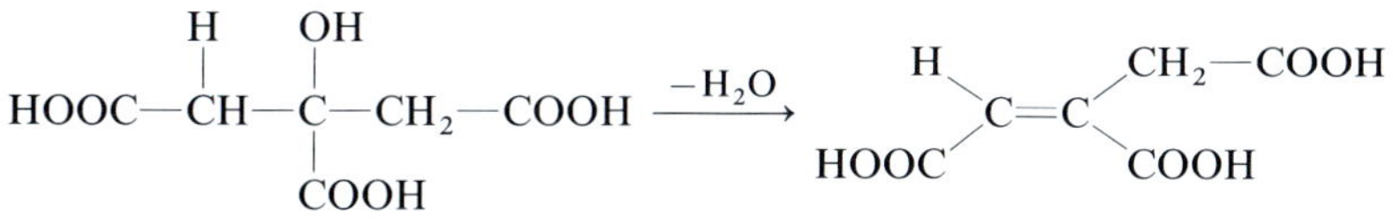

Citric acid → cis-Aconitic acid

GENERAL REACTION

Dehydration of an Alcohol to Form an Alkene

$$\underset{\underset{\text{OH}}{|}}{\overset{\overset{\text{H}}{|}}{R-C}}-\underset{\underset{R}{|}}{\overset{\overset{R}{|}}{C-R}} \xrightarrow{H^+} \underset{R}{\overset{R}{>}}C=C\underset{R}{\overset{R}{<}} + H_2O$$

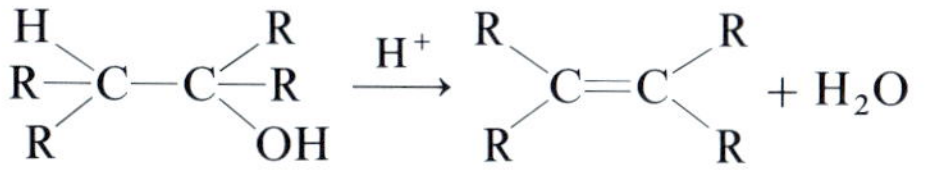

$$-\underset{\underset{H}{|}}{C}-\underset{\underset{OH}{|}}{C}-$$

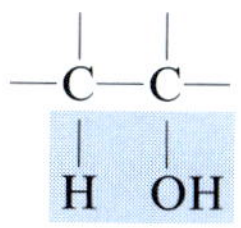

Note that the —OH and —H must be on adjacent carbon atoms.

Example

Cyclohexanol + $H_2SO_4 \longrightarrow$ Cyclohexene

1-Pentanol + $H_2SO_4 \longrightarrow$ 1-Pentene

We have already seen in Example 18.2 that when more than one product is possible from the dehydration reaction, the most substituted alkene is formed. A possible mechanism for this reaction is shown in Figure 18.4.

FORMATION OF ALKYL HALIDES Alcohols are often used as intermediates for preparing alkyl halides (organohalogen compounds):

$$R-OH \longrightarrow RX$$

This reaction can be accomplished by reacting the alcohol with HX. Hydrogen iodide and hydrogen bromide have been found experimentally to be the best hydrogen halides for this reaction. Phosphorus pentachloride (PCl_5) is also useful for this conversion.

An alcohol can be converted to an organohalogen.

EXAMPLE 18.3

Show the alkyl halide products of these reactions.

$$CH_3-CH_2-OH \xrightarrow{HI} CH_3-CH_2-I + H_2O$$

$$CH_3-\underset{\underset{CH_3}{|}}{CH}-OH \xrightarrow{PCl_5} CH_3-\underset{\underset{CH_3}{|}}{CH}-Cl$$

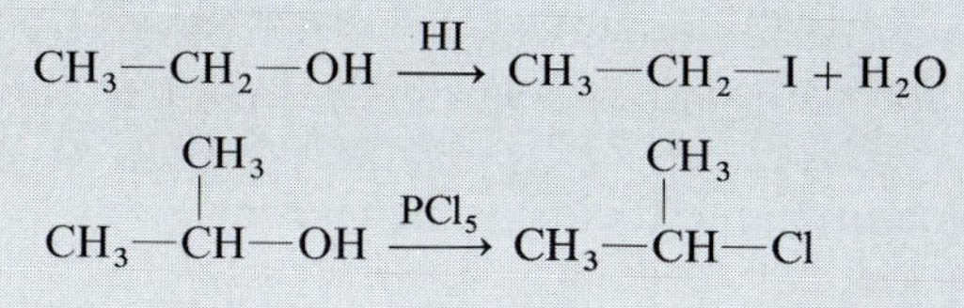

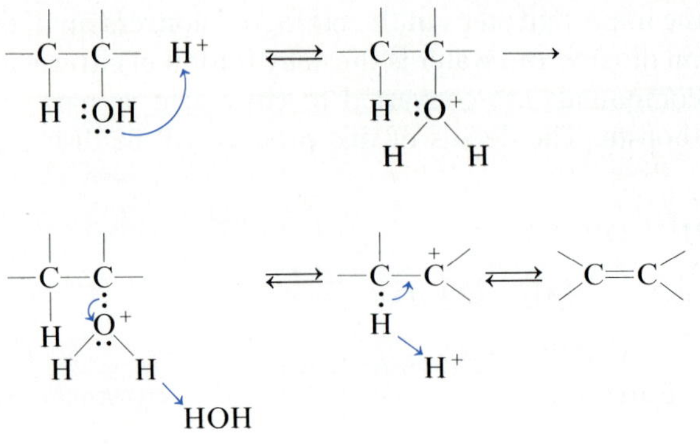

Loss of water
Loss of proton
Alkene

FIGURE 18.4 A mechanism for alcohol dehydration. The proton (H^+), a Lewis acid and electrophile, attacks the oxygen lone pair of electrons (Lewis base and nucleophile). The protonated —OH comes off as water, leaving a *carbocation*. This, in turn, loses an adjacent proton to form an alkene. (R groups are not shown.)

GENERAL REACTION

Conversion of an Alcohol to an Organohalogen

$$R{-}OH + H{-}X \longrightarrow R{-}X + H_2O$$
$$R{-}OH + PCl_5 \longrightarrow R{-}Cl + H_2O + POCl_3$$

Examples

$$(CH_3)_2CH{-}OH + HBr \longrightarrow \text{2-Bromopropane} + H_2O$$

$$\text{Cyclopentanol} + PCl_5 \longrightarrow \text{Chlorocyclopentane} + H_2O + POCl_3$$

The Lucas test differentiates small carbon chain alcohols by differences in reactivity.

TESTING FOR PRIMARY, SECONDARY, AND TERTIARY ALCOHOLS Because of the similarity between an alcohol and water, smaller alcohol molecules (about six carbon atoms maximum) tend to be water soluble. The replacement of the hydroxyl (—OH group) by chloride results in a water-insoluble alkyl chloride.

$$\underset{\text{Soluble in water}}{R{-}OH} \longrightarrow \underset{\text{Insoluble in water}}{R{-}Cl}$$

The **Lucas test** uses the differences in solubility between small molecular weight alcohols and the corresponding organohalogen compounds (alkyl halides) to test whether an alcohol is primary, secondary, or tertiary. (Notice that the alkyl group of the alcohol is the alkyl group found in the organohalogen compound.) A *primary alcohol* is one in which the carbon bearing the —OH group is attached to only hydrogens or one alkyl group. A *secondary alcohol* has the same carbon atom attached to two other alkyl groups, and a *tertiary alcohol* has the same carbon atom attached to three other alkyl groups. A mixture of HCl and $ZnCl_2$ is called the *Lucas reagent* and reacts with an alcohol according to the following reaction:

$$R{-}OH \xrightarrow[\text{(Lucas reagent)}]{HCl + ZnCl_2} R{-}Cl$$

A tertiary alcohol reacts almost instantly. A secondary alcohol will take 5 to 10 minutes to react, and a primary alcohol will not react at all within 10 minutes.

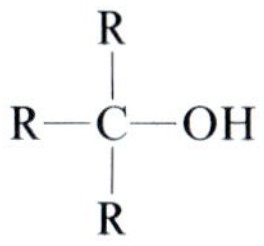

Tertiary alcohol

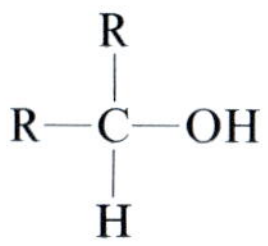

Secondary alcohol

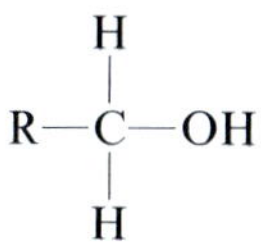

Primary alcohol

Fast ⟵⟶ Slow

Special Alcohols: Polyhydroxylic Compounds

There are many compounds that contain more than one—OH group per molecule. These kinds of compounds are named in the same way as ordinary alcohols except that each —OH is given a number to show its position, and the prefixes *di-*, *tri-*, and so on are used to designate the number of —OH groups. Common examples include ethane-1,2-diol (commonly called ethylene glycol) and propan-1,2,3-triol (commonly known as glycerine or glycerol).

Alcohols with more than one —OH group are called *polyhydroxylic.* Compounds with two —OH groups are often called *glycols.*

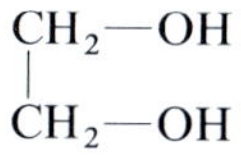

Ethylene glycol
(a major component of antifreeze)

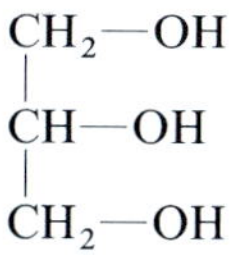

Glycerine (glycerol)
(helps give some soaps a soft feeling)

An important characteristic of these polyhydroxylic compounds is their water solubility. This property is expected since they closely resemble water, and the large ratio of —OH to hydrocarbon negates much of the nonpolar hydrocarbon character. The water solubility of ethylene glycol makes it possible to add it to water, in any proportion, for antifreeze. Also, its high boiling point (187 °C) prevents it from evaporating from the hot water in a car's radiator.

The common name **glycol** is used for alcohols having two —OH groups. The antifreeze properties of glycols are not unique to human use. Many butterflies produce a mixture of glycols, which circulate in their body fluids before cold weather arrives. This "butterfly antifreeze" allows them to survive the winters.

Carbohydrates are an important class of naturally occurring compounds and will be studied in Chapter 23. The carbohydrates, including the simple sugars glucose and fructose, are examples of polyhydroxylic alcohol compounds.

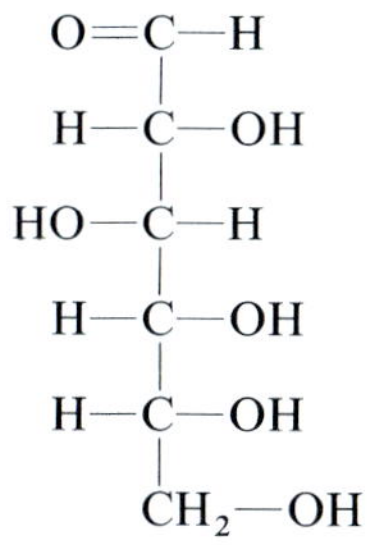

Glucose

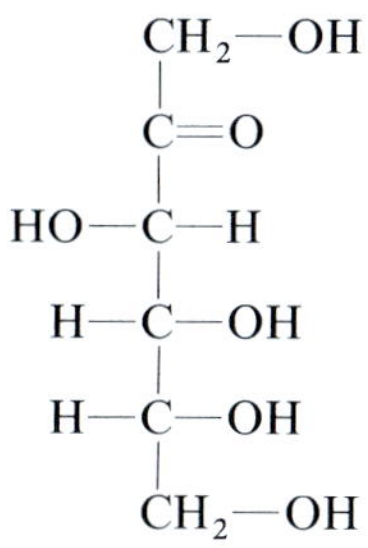

Fructose

ALCOHOLIC BEVERAGES

The fermentation of sugar generates ethanol, or ethyl alcohol. The first examples of alcoholic beverages were probably the result of natural yeasts acting on the sugar in honey to form mead. There is evidence that this drink was known as long as 10,000 years ago. It is reasonable to suspect that the early interest in alcohol had its origins in the observation that fermented foods kept better—they did not spoil. Even the ancients had problems with maintaining supplies of pure water, and fermented juices were found less likely to make a person sick. The ancient Egyptians wrote of the use of beer and wine as medicine. Later as humans were able to control this process, specialized methods were developed to make drinks.

Natural fermentation will generally not produce ethanol contents greater than 15%, as this is about the limit of tolerance for the yeast to alcohol. To obtain greater alcohol content, the enthanol must be distilled from the solution. Since ethanol has a lower boiling point than water, this provides greatly purified ethanol.

Most alcoholic beverages have unique flavors and qualities. These are generally due to substances other than ethanol, which in its pure form has no flavor. The unique tastes come from the materials fermented, the addition of flavorings, or the wooden containers that the beverage is aged in. Here are a few of the common groups of alcoholic beverages.

Table Wines These wines, sometimes called *still wines*, are widely produced throughout countries of the temperate zone, generally from various varieties of grapes. Many of these wines are named for the locations in which they were first made. Examples include Burgundy, Bordeaux, Claret, Tokay, Rhine, Zinfindel, Sauterne, Chablis, and Chianti. Wine can also be prepared from other fruits and berries. Most table wines have alcohol contents in the range of 10 to 14%.

Sparkling Wines Champagne (from the Champagne area of France) and other similar wines contain carbon dioxide.

Dessert Wines Sometimes classified as *fortified wines*, these wines have additional alcohol added to them, bringing the alcohol content to about 20%. Representatives of these wines include Sherry, Port, Muscatel, Madeira, and Marsala.

Aromatized Wines Addition of various herbs and spices give wines such as Vermouth and Dubonnet their unique flavors and odors.

Beer Beer is formed by the fermentation of barley with the addition of hops for flavoring. Lager beer generally has an alcohol content of 3 to 6%, while stout has a content of 4 to 8%.

Liquors These can contain up to 50% alcohol (100 Proof) as well as a variety of flavorings.

Brandy Distilled from wines, the generic name *brandy* includes cognac, absinth, and Benedictine.

Whiskey These include a number of beverages prepared from fermented corn, rye, and barley. Bourbon is a corn-fermented product, while Scotch is primarily fermented from barley.

Rum Prepared from fermented molasses and sugar cane.

Gin Essentially an alcohol solution flavored with juniper.

Vodka Prepared from fermented potatoes and is essentially an unflavored solution of ethanol.

Sake The Japanese drink from fermented rice (is also classified as a wine).

The process of fermenting sugars to produce ethanol is an important reaction of the carbohydrates. The details of the *fermentation* process (a reaction carried out by yeast) will be discussed in Chapter 28. Because of the potential use of ethanol as a fuel (such as in Gasohol, a mixture of one part ethanol and nine parts gasoline), as a disinfectant, and as a component in alcoholic beverages, it is important to recognize here that ethanol, or ethyl (grain) alcohol, is readily produced by fermentation of glucose and other carbohydrates.

$$\text{Glucose} \xrightarrow{\text{Fermentation}} \text{Ethanol}$$

NEW TERMS

alcohol A functional group characterized by the general formula R—O—H.

dehydration The formation of an alkene from an alcohol as a result of the loss of water.

Lucas test A test for identifying whether an alcohol is primary, secondary, or tertiary by its rate of conversion to a chloroalkane.

glycol A dihydroxylic alcohol, that is, one with two —OH groups.

TESTING YOURSELF

Alcohols

1. Name the following simple alcohols.

a. 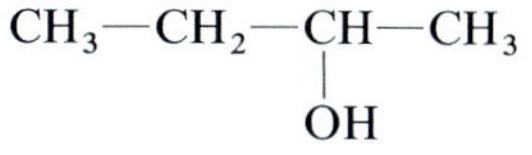

b. $CH_2{=}CH{-}CH_2{-}OH$

c. 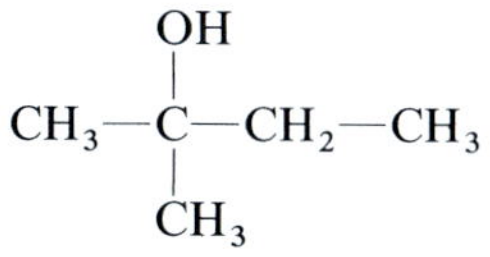

d. 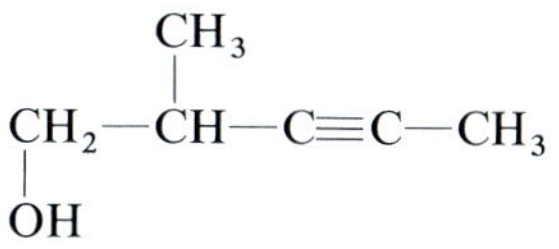

e. 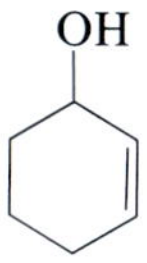

f. 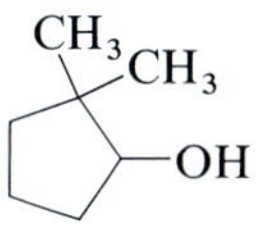

2. Draw structures for each of the following compounds.
   a. 5,5-dibromo-3-hepten-2-ol
   b. 3-hexyn-2-ol
   c. 4,4-dichlorocyclooctanol
3. Give the major product for each of the following dehydration reactions.

a. 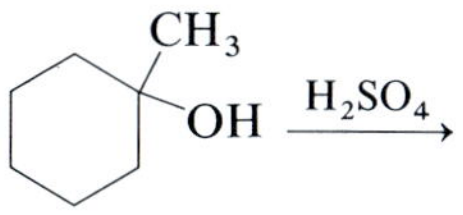

b. 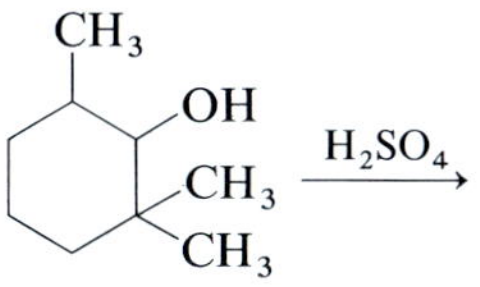

c. 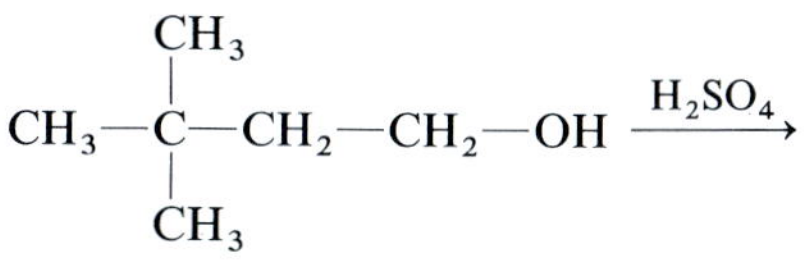

4. Give products for each of the following.

a. 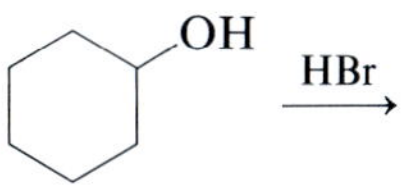

b. 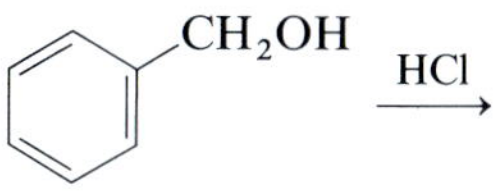

c. 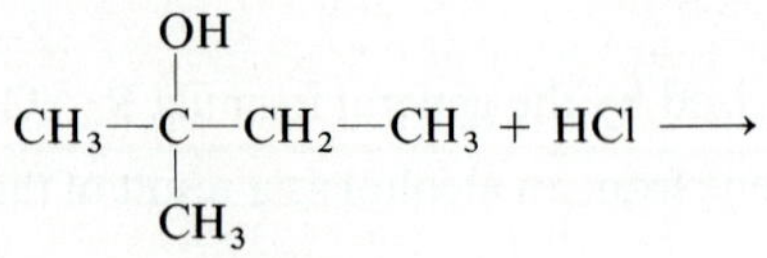

d. $CH_3-CH_2-OH \xrightarrow{\text{Lucas reagent}}$

e. 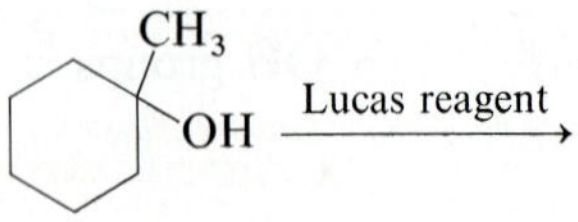

f. 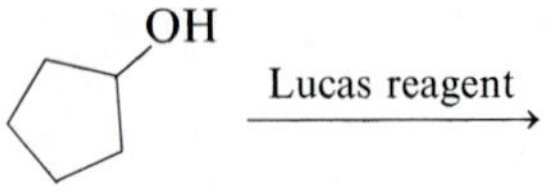

Answers **1. a.** 2-butanol **b.** 2-propen-1-ol **c.** 2-methyl-2-butanol **d.** 2-methyl-3-pentyn-1-ol **e.** 2-cyclohexenol **f.** 2,2-dimethylcyclopentanol

2. 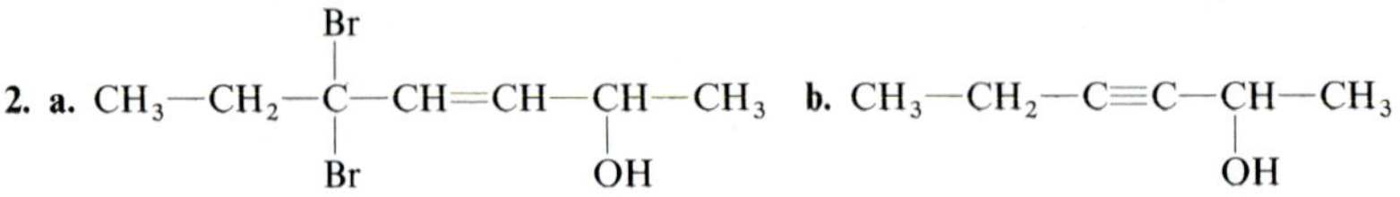

c. 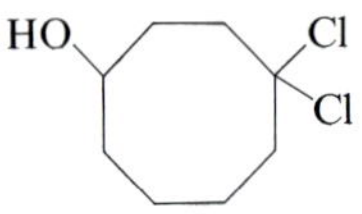

3. a. 1-methylcyclohexene **b.** 1,3,3-trimethylcyclohexene **c.** 3,3-dimethyl-1-butene

4. a. bromocyclohexane **b.** benzyl chloride **c.** 2-chloro-2-methylbutane **d.** No reaction within 10 minutes. **e.** Almost immediate conversion to the 1-chloro-1-methyl-cyclohexane and formation of two layers. **f.** The formation of two layers will take between 5 and 10 minutes.

18.2 Phenols

Structure

Phenols are characterized as a benzene ring with an —OH group attached.

The **phenols** appear similar to alcohols but are identified by the R group being an aromatic ring. The most common molecule of the phenol class of compounds is phenol itself.

Alcohol **Phenol**

Carbolic acid is an old name for phenol, and although it does not receive much modern usage, the term is still seen on occasion. It might be of interest to note that the IUPAC has suggested that phenol be named *benzenol*; however, this has not met with much acceptance among practicing chemists.

Nomenclature: Use Aromatic Rules

Nomenclature of phenols follows the rules of aromatic chemistry.

We have already seen the extensive IUPAC nomenclature for aromatic compounds. The same ideas as applied to toluene (methylbenzene) are used for the nomenclature of phenols. Phenol is used as the parent compound, and other compounds are named as derivatives of phenol. Thus, one can use either the ortho, para, and meta nomenclature or the numbering methods. In all cases, the positions of new groups are relative to the position of the —OH group of phenol.

EXAMPLE 18.4

Name the following phenols.

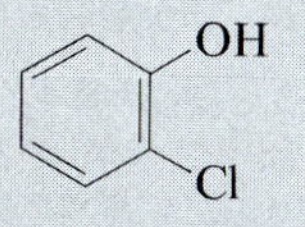
OH
Cl

o-Chlorophenol or 2-chlorophenol

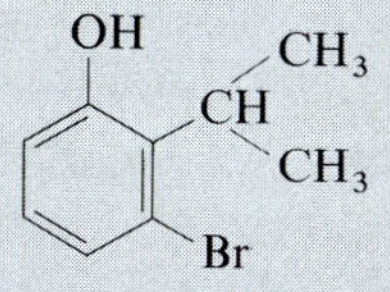
OH
CH3
CH
CH3
Br

3-Bromo-2-isopropylphenol

Reactions of Phenols

ACIDITY OF THE —OH BOND: A QUALITATIVE LOOK An important difference between alcohols and phenols can be found in the relative acidities of the —OH hydrogen. In fact, phenol is sometimes called carbolic acid to reflect its substantial acidity. This is qualitatively demonstrated by the comparisons shown in Table 18.2. Alcohols and phenols both react with metallic sodium to produce anions and hydrogen gas. Sodium hydroxide (NaOH) is not a strong enough base to remove the hydrogen from an alcohol, but it is strong enough to do so from phenols. Sodium bicarbonate ($NaHCO_3$), a very weak base, will not react with either an alcohol or a phenol.

Phenols are much more acidic than alcohols.

ACIDITY OF THE —OH BOND: A QUANTITATIVE LOOK A precise measure of the acidity or basicity of a substance is sometimes needed in chemical calculations. In Chapter 11 we learned to write equations to represent equilibria. The reaction of a phenol with water releases a proton (H^+) to the water

TABLE 18.2
Comparison of Alcohol and Phenol Acidity

Alcohol Reactions	Phenol Reactions
$2\ R{-}OH + 2\ Na \longrightarrow 2\ R{-}O^-\ Na^+ + H_2$*	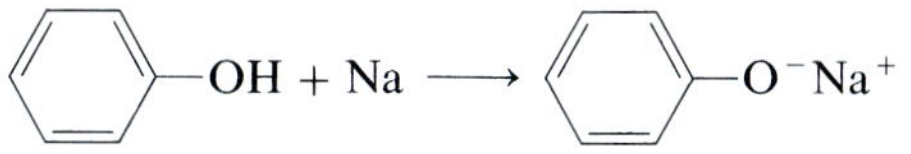OH + Na →
O⁻Na⁺
$R{-}OH + NaOH \longrightarrow$ No reaction	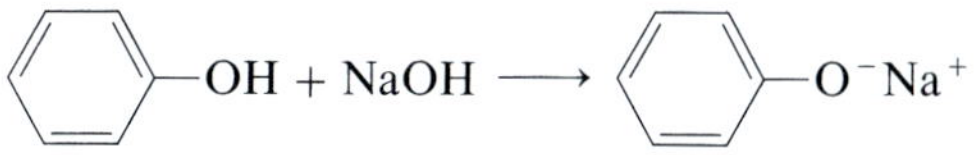OH + NaOH →
O⁻Na⁺
$R{-}OH + NaHCO_3 \longrightarrow$ No reaction	OH + NaHCO3 → No reaction

* This reaction is more correctly an oxidation–reduction reaction.

and forms a hydronium ion and a phenol anion. One can write an equilibrium reaction that shows this acidic nature of the phenol:

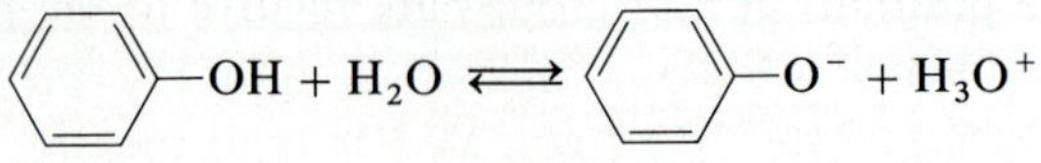

and thus

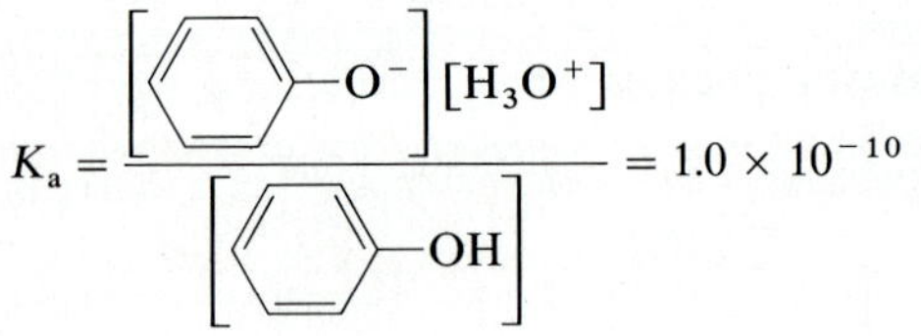

$$pK_a = -\log K_a = 10$$

TABLE 18.3
Relative pK_a Values for Common Functional Groups

Functional Group	pK_a Value
Alkane	50
Water	16
Alcohol	16–18
Phenol	10
HCN	9
HF	3
HCl	−2
H_2SO_4	−5

(This value is determined by experimentation.) Just as pH was determined to be $-\log[H_3O^+]$, $pK_a = -\log(K_a)$. Since $pK_a = -\log(K_a)$, the *weaker acids* have the *higher* pK_a values. In fact, every pK_a difference of one unit means a difference in acidity of 10 times! Some representative pK_a values are given in Table 18.3. This table shows that the pK_a values for alcohols are on the order of 16 to 18, while those for phenols are about 10. Thus, phenols are about a million times (10^6) stronger acids than alcohols are.

We can qualitatively explain the difference in acidities between alcohols and phenols by using the concept of resonance. Figure 18.5 shows the extensive delocalization of electrons into the ring for the anion of phenol. As a general rule, increased stability is associated with increased resonance or electron delocalization. Thus, the loss of a proton (H^+) from phenol leaves a more stabilized anion than the anion from an alcohol.

Because the —OH group in phenol is attached to a carbon that has sp^2 hybridization, phenols do *not* undergo dehydration or conversion to alkyl halides as observed for alcohols. Thus, at this point, it seems that only acidity distinguishes the —OH group of phenols from that of alcohols. However, as we will see in the following section, the presence of the aromatic ring on phenols will allow some interesting aromatic substitution reactions of phenols.

AROMATIC SUBSTITUTION REACTIONS Phenols react like other aromatic compounds except that they are strikingly more reactive. The phenolic

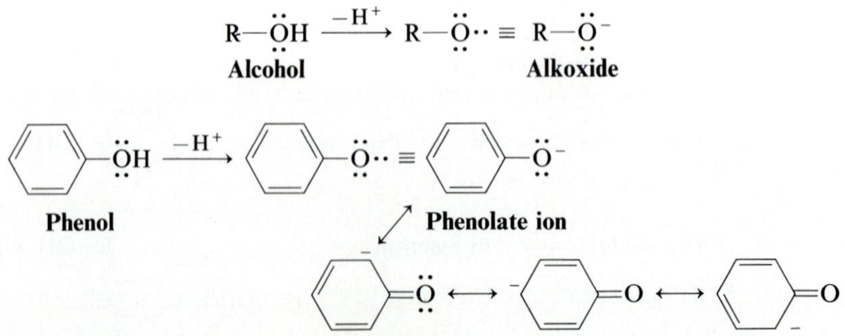
Alcohol
Alkoxide
Phenol
Phenolate ion

FIGURE 18.5
Phenols are more acidic than alcohols. The anions of phenol (phenolate ion) are more stable than the anions of alcohols (alkoxides).

—OH group is a powerful activator (ortho–para director) of the benzene ring for aromatic electrophilic substitution reactions, and aromatic substitution reactions of phenols tend to take place very easily. The general reaction for an aromatic substitution reaction of phenol is shown here, where E^+ can be any of the electrophiles discussed in Chapter 17.

Phenols undergo aromatic substitution reactions, with the —OH group acting as a powerful ortho–para director.

GENERAL REACTION

Aromatic Electrophilic Substitution of a Phenol

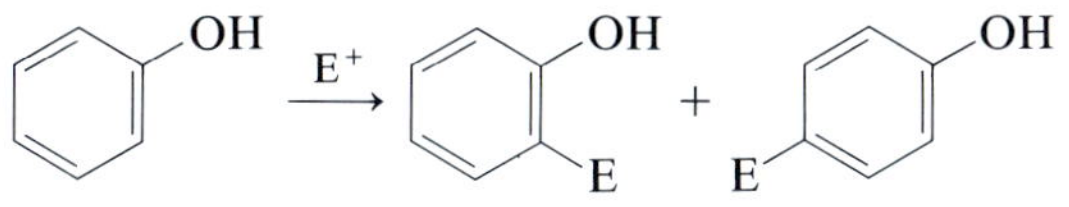

Examples

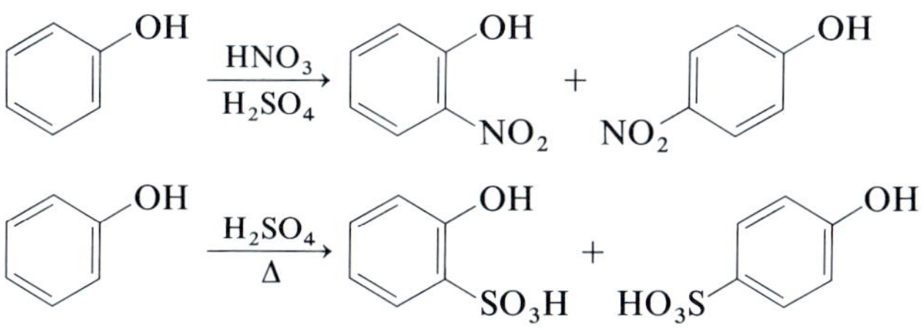

Special Phenols

Phenols are found in a variety of natural and synthetic products. Phenolic materials can be responsible for burning and blistering (as in poison ivy) and also can be used as highly effective antiseptics. One of the earliest uses of phenol as an antiseptic occurred in 1867 when Joseph Lister found that a solution of phenol killed bacteria. Examples of phenols are given in Table 18.4.

NEW TERMS

phenol Any aromatic ring with an —OH group is classified as a phenol. Phenol is also the name for the simplest member of this class.

TESTING YOURSELF

Phenols

1. Name the following phenols.

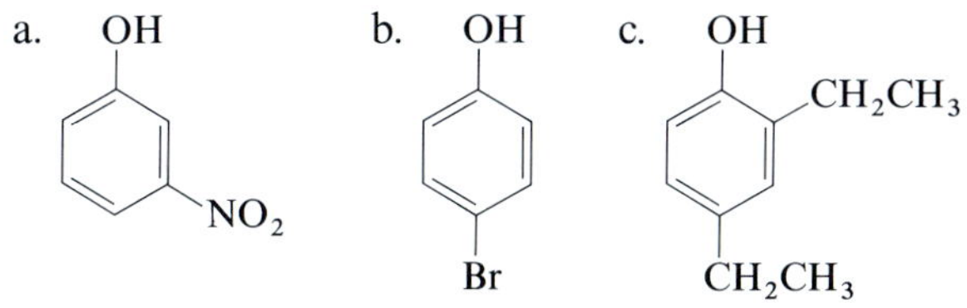

2. Which is more acidic, 2-propanol or *m*-nitrophenol? (Draw both.)

TABLE 18.4
Examples of Common Phenols

Name	Description	Structure
Methyl salicylate	Oil of wintergreen	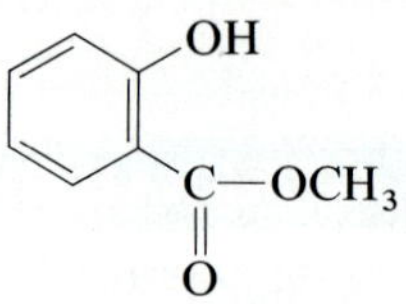
Urushiol	Constituent of poison ivy	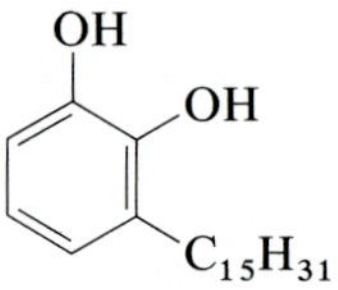
Guaiacol	Common expectorant	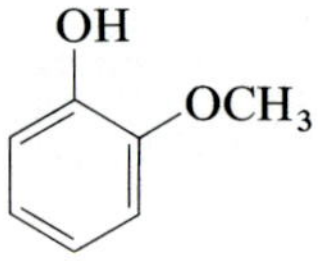
Vanillin	Component of vanilla	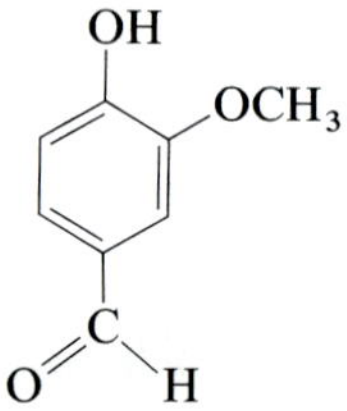
Eugenol	Oil of clove	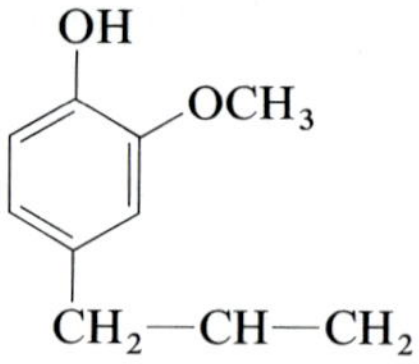
Thymol	Constituent of mint	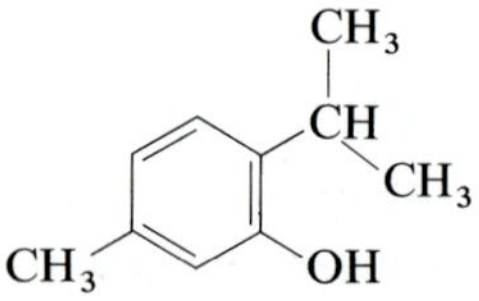
2-Phenylphenol	Constituent of Lysol	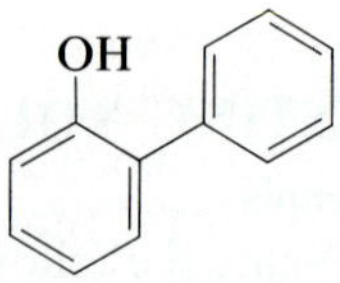

3. Give products for the following phenol reactions.

a. 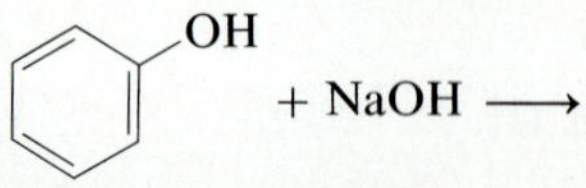

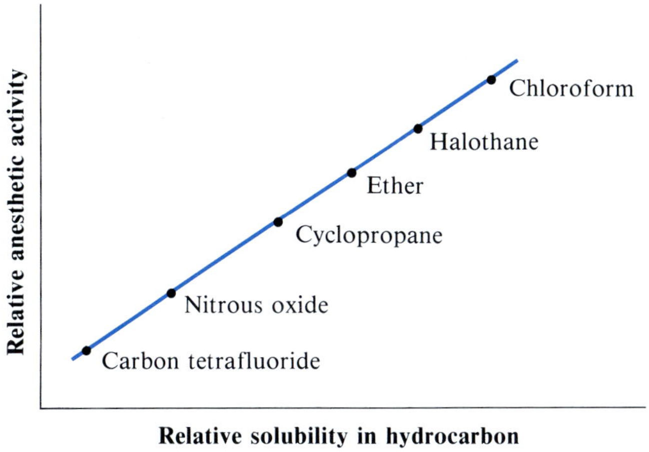
Relative anesthetic activity
Chloroform
Halothane
Ether
Cyclopropane
Nitrous oxide
Carbon tetrafluoride
Relative solubility in hydrocarbon

FIGURE 18.9
The relationship of hydrocarbon solubility and anesthetic activity for a variety of compounds. There is a direct correlation between the ability of a substance to act as an anesthetic and its solubility in hydrocarbon.

channels. When certain kinds of molecules (such as ether) are dissolved in the hydrocarbon layer (or fat layer) of the cell, the ions cannot pass through. There is a relationship between compounds' solubility in hydrocarbon and their effectiveness as anesthetics. Figure 18.9 shows that this relationship does exist.

NEW TERMS

ether An organic functional group characterized by the bonding

$$R-\ddot{\underset{\cdot\cdot}{O}}-R$$

alkoxide The basic and nucleophilic intermediates formed from an alcohol or phenol by removal of the proton of the O—H bond.

epoxide A three-membered, oxygen-containing ring.

carcinogenic Capable of inducing the formation of cancer cells.

TESTING YOURSELF

Ethers

1. Give the correct names for the ethers obtained from the following simple Williamson syntheses.

a.
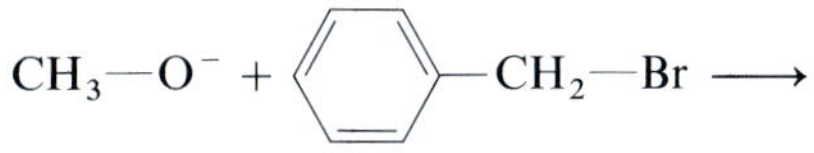

b.
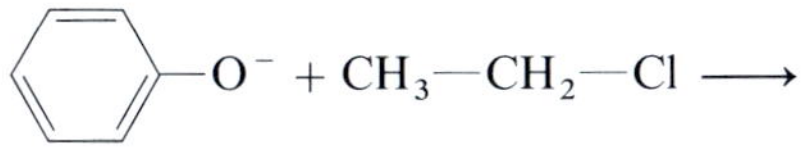

c. $Br-CH_2-CH_2-CH_3 + CH_3-\underset{\displaystyle O^-}{\underset{|}{CH}}-CH_3 \longrightarrow$

2. Give products for each of the following ether-forming reactions.

a.
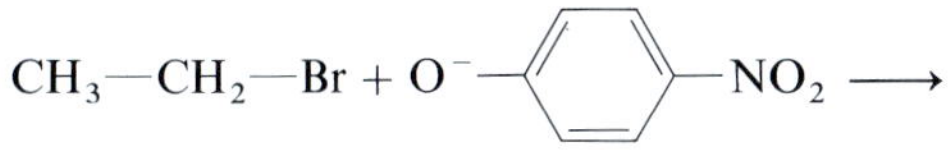

b. 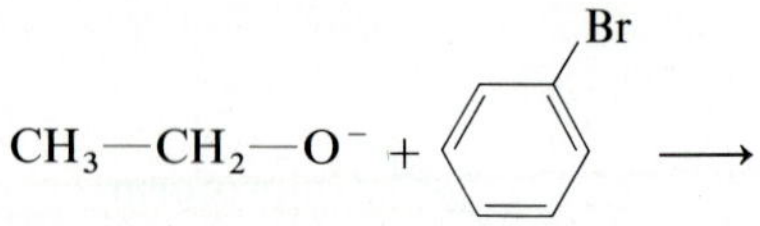

c. Cyclohexene + Peroxybenzoic acid ⟶

d. 2,3-Dimethyl-2-butene + Peroxyacetic acid ⟶

Answers **1. a.** benzyl methyl ether **b.** ethyl phenyl ether **c.** isopropyl propyl ether **2. a.** ethyl *p*-nitrophenyl ether **b.** no reaction **c.** 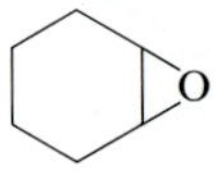 **d.** 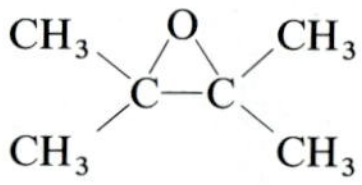

18.4 Thiols: Sulfur Equivalents of Alcohols

Structure

Thiols are the sulfur equivalent of alcohols:

R—S—H

Sulfur is less able to hydrogen bond than oxygen, so boiling points are lower.

Thiols are similar in structure to alcohols except that oxygen is replaced by sulfur. Figure 18.10 shows a comparison of the boiling points of the thiols with the alcohols. Their properties differ in that the electronegativity of sulfur is considerably less than that of oxygen (Figure 18.11). From an examination of the electronegativity table, you would not expect a large dipole in the C—S bond nor would you expect that hydrogen bonding would be important in the thiols. So even though sulfur is heavier than oxygen, the boiling points of the thiols are lower than those of the alcohols.

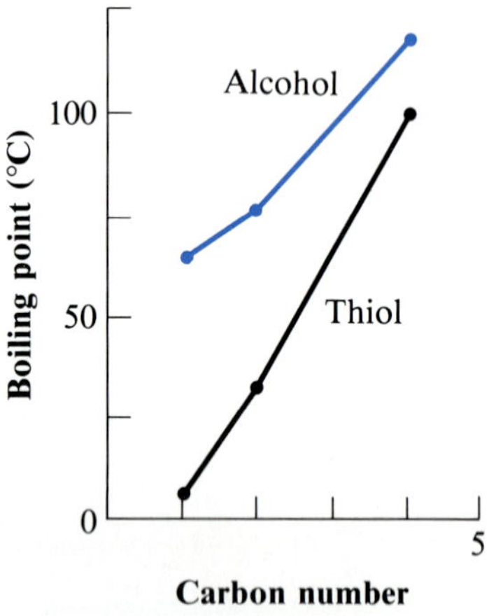
Alcohol
Thiol
100
50
0
5
Boiling point (°C)
Carbon number

FIGURE 18.10
Comparison of boiling points of thiols and alcohols. Thiols of equivalent carbon number boil at lower temperature than the corresponding alcohols. There is less hydrogen bonding in thiols than in alcohols.

Nomenclature

The rules for naming the sulfur analogs of alcohols are exactly like those for alcohols, except that the suffic *-ol* is replaced by *-thiol*. Several examples demonstrate this nomenclature.

Alcohol	Thiol
$CH_3—CH_2—OH$ **Ethanol**	$CH_3—CH_2—SH$ **Ethanthiol**
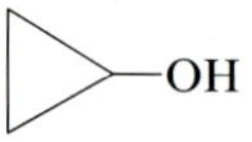 **Cyclopropanol**	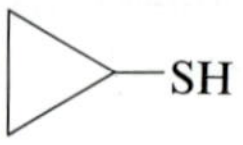 **Cyclopropanthiol**
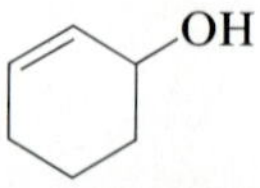 **2-Cyclohexenol** (or 2-cyclohexen-1-ol)	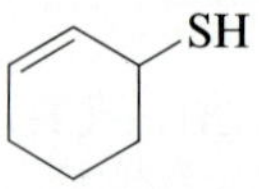 **2-Cyclohexenthiol**

(Notice in 2-cyclohexenol that 1- can be "understood" to be part of the nomenclature if not listed.)

An older nomenclature system describes thiols as *mercaptans*. Using this system of nomenclature, the compounds shown above would be named

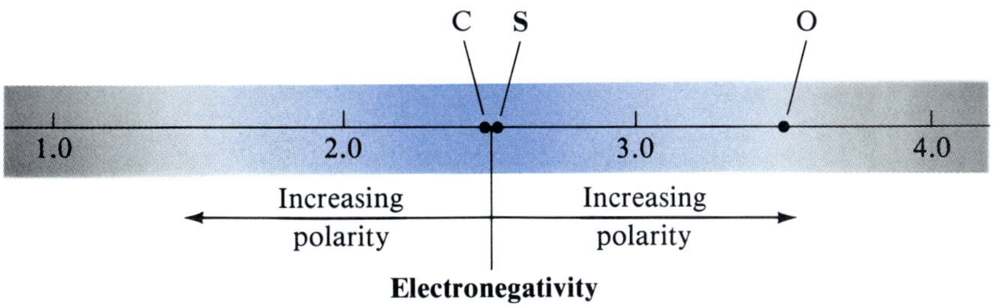
C S
O
1.0
2.0
3.0
4.0
Increasing polarity
Increasing polarity
Electronegativity

FIGURE 18.11
The electronegativity of sulfur (2.58) is very similar to that for carbon (2.55) and considerably less than that of oxygen (3.44). Thus, there is very little polarization of the C—S bond. The sulfur in a thiol also causes its hydrogen bonding to be greatly diminished.

ethylmercaptan, cyclopropylmercaptan, and 2-cyclohexenylmercaptan, respectively. In biochemistry, particularly in the chemistry of amino acids, the —SH grouping found in some amino acids is referred to as the *sulfhydryl group*.

Reactions

Oxidation of thiols produces disulfides.

OXIDATION Of particular interest and importance to our later understanding of biochemistry is the oxidation of a thiol to a **disulfide**, a compound with two sulfurs as its functional group.

GENERAL REACTION

Thiol Oxidation

Disulfide linkage

$$2\ R{-}S{-}H \xrightarrow{\text{Oxidation}} R{-}S{-}S{-}R + H_2O$$

Disulfide

Examples

$$CH_3{-}CH_2{-}SH \xrightarrow{\text{Oxidation}} CH_3{-}CH_2{-}S{-}S{-}CH_2CH_3$$

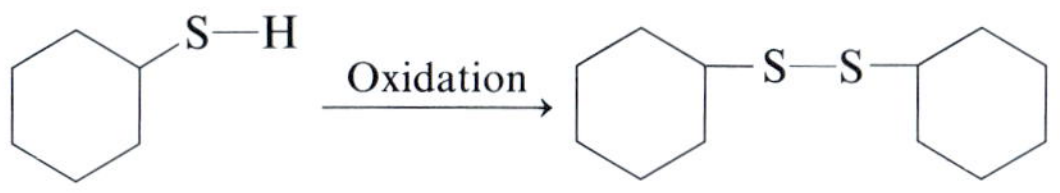
S—H
Oxidation
S—S

A thiol can be oxidized to a disulfide by any of a number of oxidizing agents, including oxygen. In a reverse reaction, it is also possible to reduce a disulfide to a thiol.

GENERAL REACTION

Disulfide Reduction

$$R{-}S{-}S{-}R \xrightarrow{\text{Reduction}} 2\,R{-}S{-}H$$

Examples

$$CH_3CH_2{-}S{-}S{-}CH_2{-}CH_3 \xrightarrow{\text{Reduction}} 2\,CH_3{-}CH_2{-}SH$$

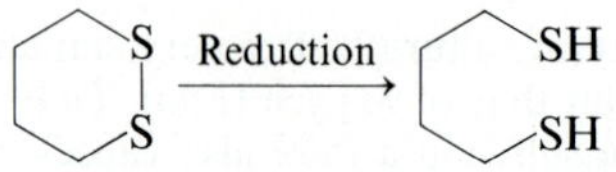
S
S
Reduction
SH
SH

Disulfide linkages are very important in the human body for maintaining the correct structures of proteins and enzymes. A protein found in hair has a large number of disulfide linkages, and by reducing these linkages, free thiol functional groups can be created. If the orientation of these free thiol groups is changed, then new disulfide linkages are created when hair is reoxidized. These processes are responsible for hair taking on a new shape as is done in permanent waving (Figure 18.12). (Also see Plate 5 in the color insert.)

Later in our study of biochemistry we shall encounter reactions involving a naturally occurring disulfide-containing compound called *lipoic acid.* During the biological transformations involving lipoic acid, an oxidation–reduction process occurs.

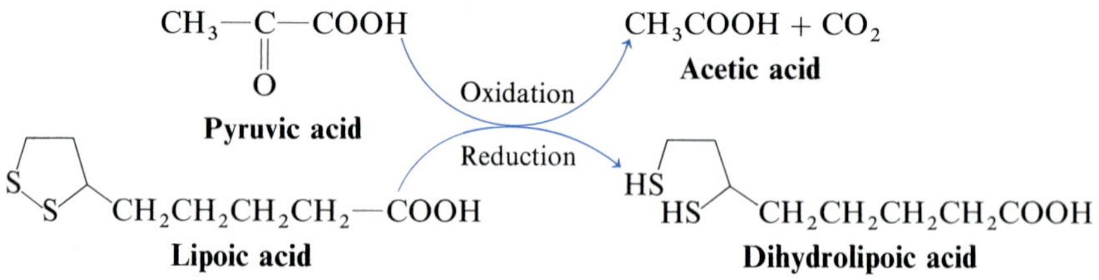
CH3—C—COOH
O
Pyruvic acid
Oxidation
CH3COOH + CO2
Acetic acid
Reduction
S
S
CH2CH2CH2CH2—COOH
Lipoic acid
HS
HS
CH2CH2CH2CH2COOH
Dihydrolipoic acid

Sulfur of thiols react with heavy metal ions such as Hg, As, and Pb.

THIOL REACTIONS WITH MERCURY One of the underlying reasons for the toxicity of mercury is its affinity for sulfur. Mercury salts react with thiols to form very stable compounds.

GENERAL REACTION

Reaction of Mercury with a Thiol

$$2\,R{-}S{-}H + Hg^{2+} \longrightarrow R{-}S{-}Hg{-}S{-}R + 2\,H^+$$

Examples

$$2\,CH_3{-}SH + Hg^{2+} \longrightarrow CH_3{-}S{-}Hg{-}S{-}CH_3 + 2\,H^+$$

$$\begin{matrix} CH_2{-}SH \\ | \\ CH_2{-}SH \end{matrix} + Hg^{2+} \longrightarrow \begin{matrix} CH_2{-}S \\ | \\ CH_2{-}S \end{matrix}\!\!\!>\! Hg + 2\,H^+$$

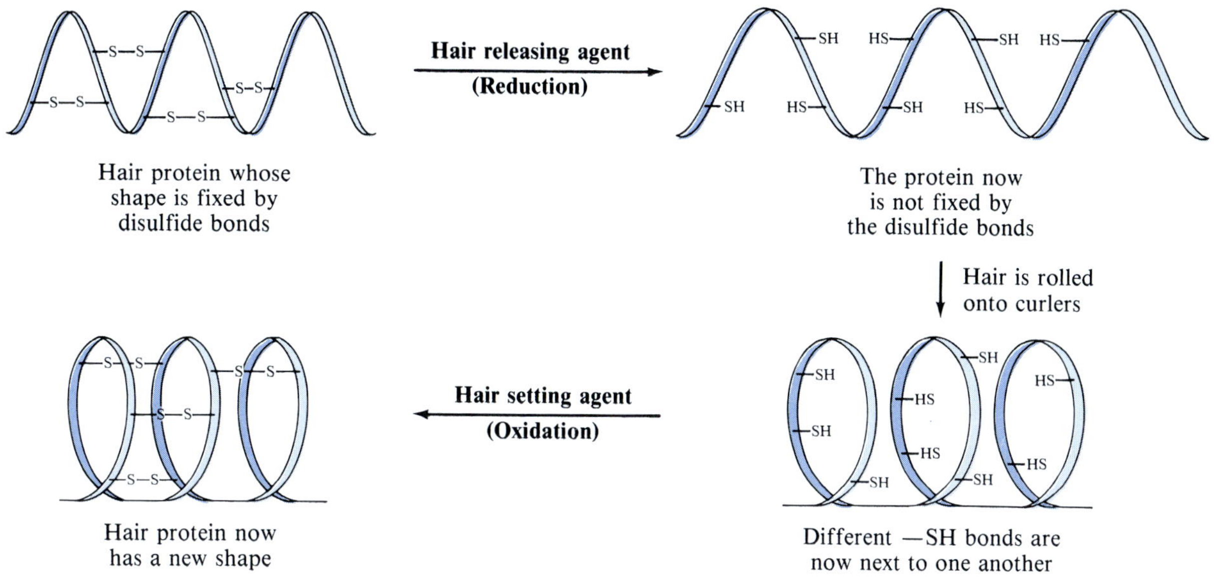
Hair releasing agent
(Reduction)
Hair protein whose shape is fixed by disulfide bonds
The protein now is not fixed by the disulfide bonds
Hair is rolled onto curlers
Hair setting agent
(Oxidation)
Hair protein now has a new shape
Different —SH bonds are now next to one another

FIGURE 18.12
The process by which hair is "permanent waved" is nothing more complicated than a reduction–oxidation process (See Plate 5.)

Enzymes are important protein molecules responsible for carrying out the extremely large and diverse number of reactions that allow our body to function properly (see Chapter 27). Many enzymes have thiol groups associated with them. If a thiol group on an important enzyme is attacked by mercury and removed from further chemical reaction, the ability of the enzyme to function properly is impaired. This can have very detrimental effects on our health. In later chapters we shall see why this is so damaging.

As a consequence of the ability of sulfur to coordinate with heavy metals such as mercury, lead, and arsenic (which are often toxic), a sulfur derivative of glycerine was developed by the British during World War I to neutralize the poison gas *lewisite*. This compound, abbreviated BAL for "British antilewisite" effectively ties up heavy metals with the sulfur atoms. This removes the metals from further biological reaction and effectively neutralizes their poisonous character. BAL is still used for treating severe cases of arsenic and mercury poisoning and is currently marketed as *dimercaprol*.

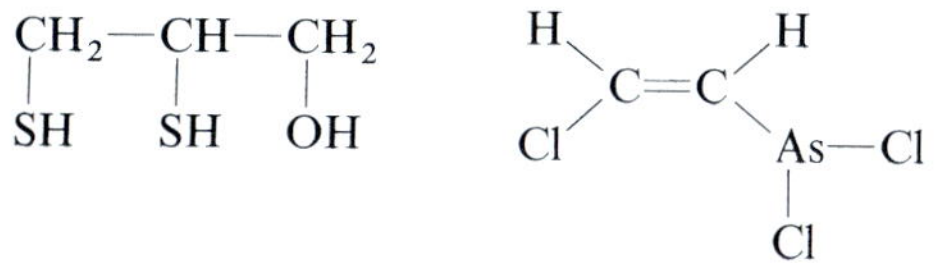

BAL **Lewisite**

Special Thiols

Sulfur-containing compounds are abundant in nature. There is often an association of sulfur with some disagreeable odors: the smell of hydrogen sulfide (H_2S) from rotten eggs, the essence of skunk, and the odor of garlic and onion. However, lest we always think of sulfur in a negative way, we should also be aware that the clean smell of the marigold and the aroma of coffee are both due to sulfur-containing compounds (Figure 18.13 and Plate 6 in color insert).

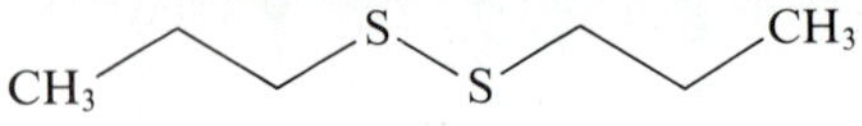
S
S
CH_3
CH_3
Onion

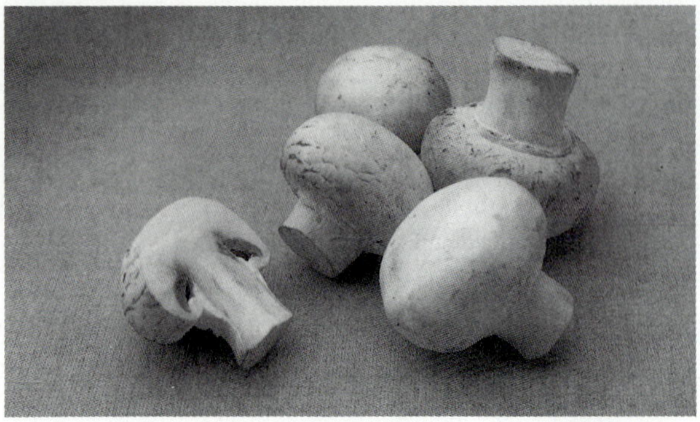

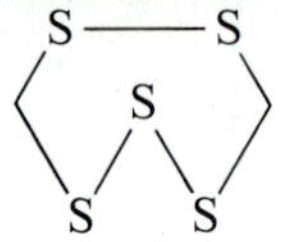
S
S
S
S
S
Mushroom

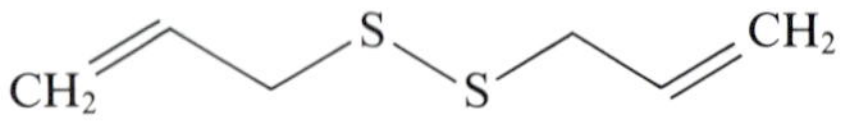
S
S
CH_2
CH_2
Garlic

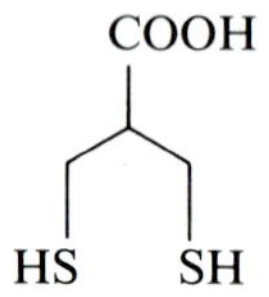
COOH
HS
SH
Asparagus

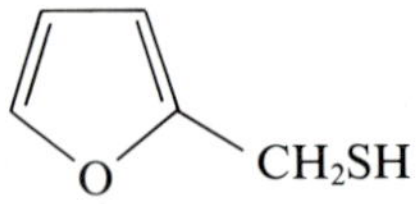
O
CH_2SH
Coffee

FIGURE 18.13
Some common natural substances that are rich in sulfur-containing compounds (Also see Plate 6 in color insert.)

GARLIC AND ONIONS

In 1609 it was written

> Garlic then have power to save
> from death
> Bear with it though it maketh
> unsavory breath,
> And scorn not garlic like some
> that think
> It only maketh men wink
> and drink and stink.*

Garlic and onion contain special thiols that have been suggested as having medicinal value. Many folk medicines have regularly employed these plants in their cures. There is a basis for this widespread use. Extracts of garlic and onions have been shown to have antifungal and antibacterial activity, as well as the ability to keep blood from clotting. In the writings of the ancients, garlic was prescribed as a therapeutic for ailments as diverse as headaches and tumors. No less an authority on antibiotics than Louis Pasteur studied these properties of garlic. Extracts of garlic juice can be diluted to less than one part in 100,000 and still actively inhibit growth of a number of bacteria.

* From Sir John Harrington, in the "Englishman's Doctor."

NEW TERMS

thiol The sulfur analog of an alcohol in which oxygen is replaced by sulfur.

disulfide The product of thiol oxidation, having the structure R—S—S—R.

TESTING YOURSELF

Thiols

1. Give an appropriate name for each of the following compounds:

a. 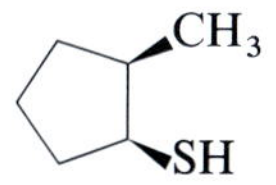

c. 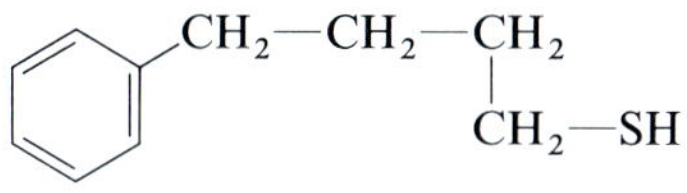

b. CH_3—CH—CH_2CH_3 (with SH on the CH)

2. Suggest products for each of the following reactions:

a. CH_3—CH_2—SH $\xrightarrow{\text{Oxidation}}$

b. CH_2—CH—CH_2OH + Hg^{2+} ⟶ (SH on CH_2 and SH on CH)

Answers **1. a.** *cis*-2-methylcyclopentanthiol **b.** 2-butanthiol **c.** 4-phenylbutanthiol **2. a.** CH_3—CH_2—S—S—CH_2—CH_3 **b.** 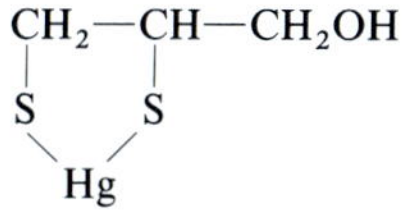

18.5 Antibiotics: Ethers and Alcohols Bind Metals

Many antibiotics are heavily oxygenated, either with the hydroxyl group or with ether oxygens. One of the mechanisms by which these biological molecules exhibit their activity is by their ability to bind metal ions. Compounds

A number of antibiotics contain ether and alcohol functional groups that allow them to specifically bind metals.

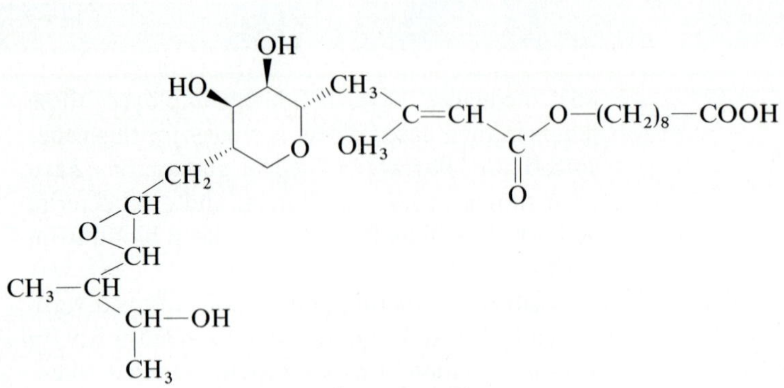

Pseudomonic acid

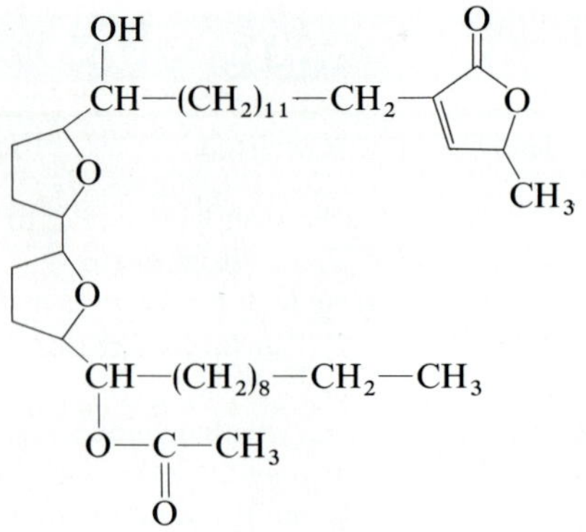

Uvaricin

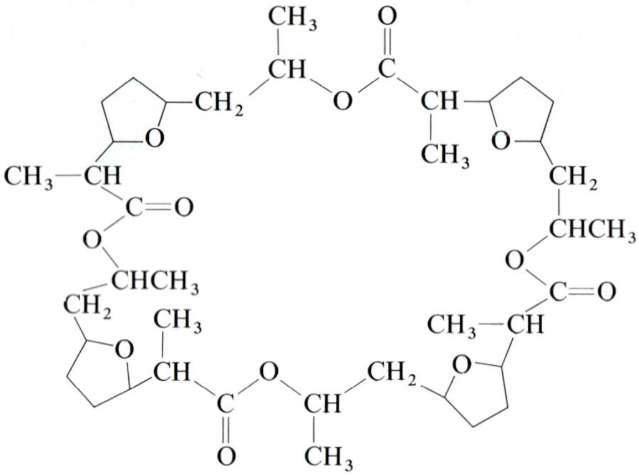

Nonactin

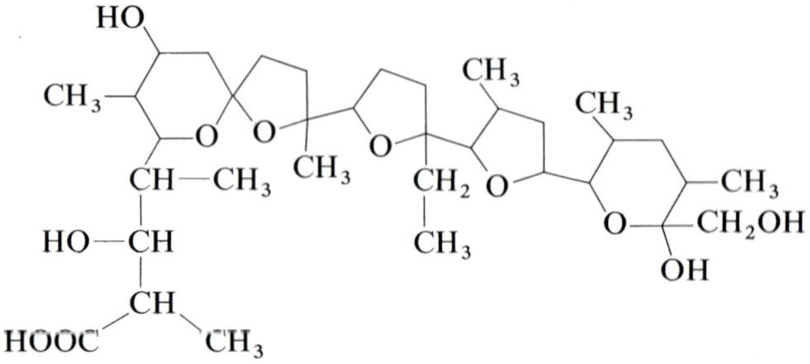

Monensin

FIGURE 18.14
Examples of ionophores, which are ethers and alcohols that bind metals.

that bind to metal ions are called *ionophores.* This ionophoric activity of antibiotics removes essential metals from the bacteria and therefore inhibits their growth, leading to eventual death of the invading bacterium. Examples of some of these interesting molecules are given in Figure 18.14.

A representative example of antibiotic activity can be found in the ionophoric compound *nonactin.* Nonactin efficiently transports sodium ions into bacterium. This results in a dramatic build-up of the Na^+ concentration. The only way for the bacterium to reduce this concentration is to dilute it by absorbing water (recall the discussion of osmosis in Chapter 8). This dilution process contributes to the death of the cell.

SUMMARY

Alcohols and *phenols* are characterized by the presence of the *hydroxyl group* as the important functional group. This group is responsible for the solubility of small- and medium-sized alcohols in water. Phenols are more acidic than alcohols. Alcohols can lose water through *dehydration* to yield alkenes and can be converted to organohalogen compounds. Phenols will not undergo either of these reactions. *Ethers* have the oxygen placed between two carbon atoms. The ether analog of the cyclopropane (*epoxide*) is an important functional group in biological reactions. *Thiols* are the sulfur analogs of alcohols.

REACTION SUMMARY

Dehydration of an Alcohol to Form an Alkene

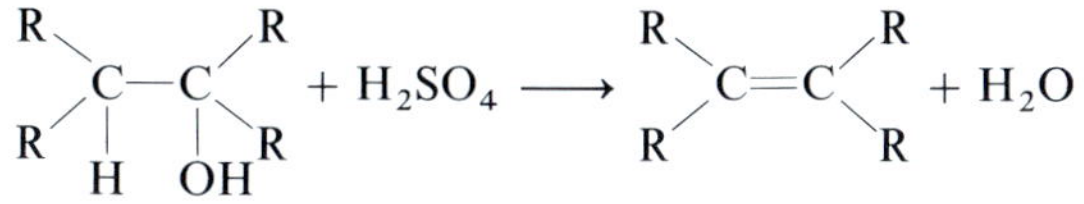

Conversion of an Alcohol to an Organohalogen

$$R{-}O{-}H + HX \longrightarrow RX + H_2O$$

$$R{-}O{-}H + PCl_5 \longrightarrow R{-}Cl + HCl + POCl_3$$

Aromatic Electrophilic Substitution of a Phenol

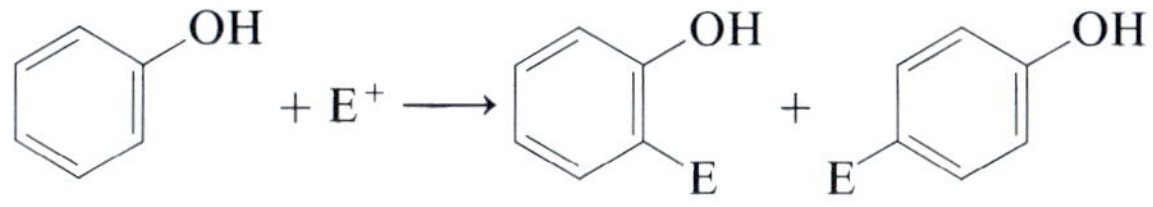

Williamson Ether Synthesis (Step 2)

$$R{-}O^- + R'X \longrightarrow R{-}O{-}R' + X^-$$

Peroxyacid Oxidation of an Alkene

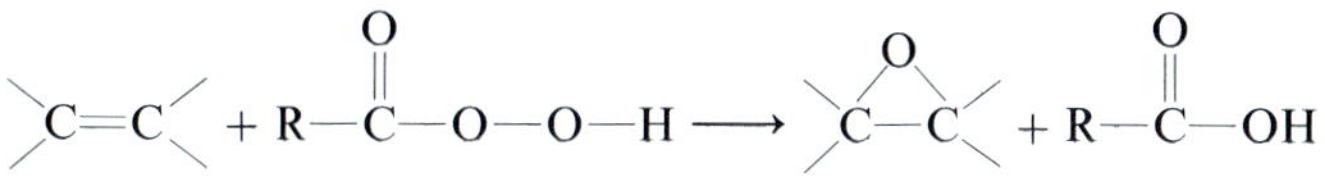

Thiol Oxidation

$$2\,R{-}S{-}H \xrightarrow{\text{Oxidation}} R{-}S{-}S{-}R + H_2O$$

Reaction of Mercury with a Thiol

$$2\,R{-}S{-}H + Hg^{2+} \longrightarrow R{-}S{-}Hg{-}S{-}R + 2\,H^+$$

TERMS

alcohol
dehydration
Lucas test
glycol
phenol
ether
alkoxide
epoxide
carcinogenic
thiol
disulfide

CHAPTER REVIEW

QUESTIONS

1. Which has the most acidic proton?
   a. thiol b. phenol c. alcohol d. ether
2. 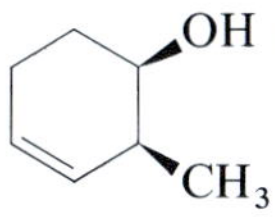 is best named
   a. *cis*-4-hydroxy-3-methylcyclohexene
   b. *trans*-4-hydroxy-3-methylcyclohexene
   c. *cis*-2-methyl-3-cyclohexenol
   d. *trans*-2-methyl-3-cyclohexenol

3. In the following dehydration reaction, what will the major product be?

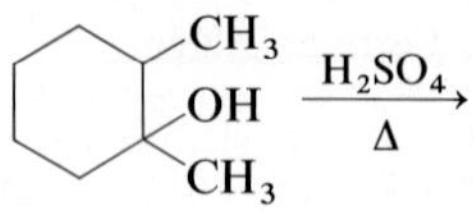

a. 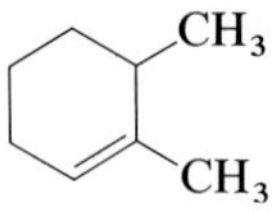 b. 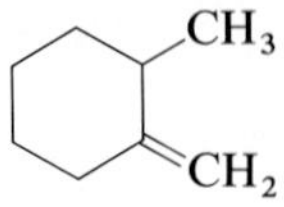 c. 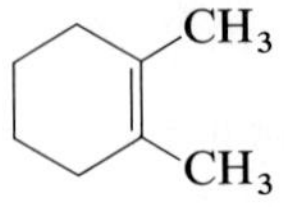 d. 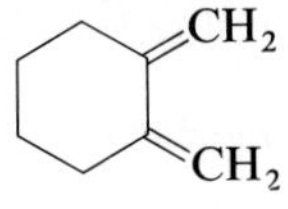

4. The best way to prepare 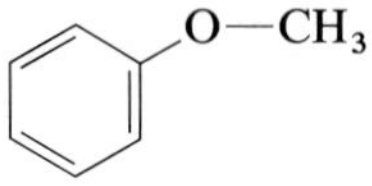 is

a. 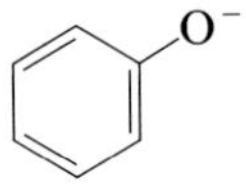 + CH_3Br b. CH_3O^- + 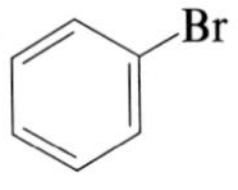

5. The most water soluble material will most likely be

a. 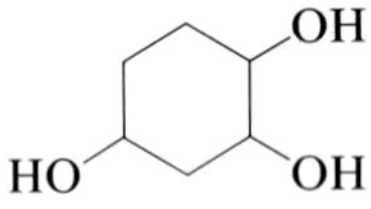 c. 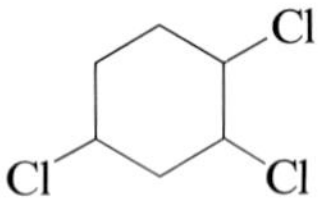

b. 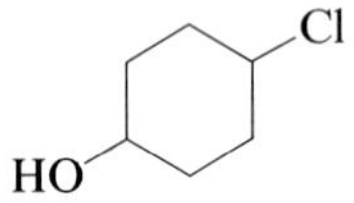 d. 

6. What is the most reasonable product from the following reaction of the phenol?

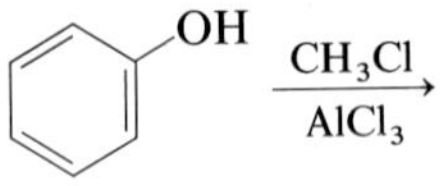

a. 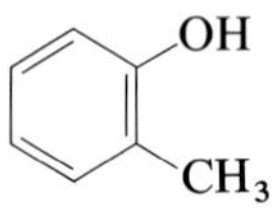 b. 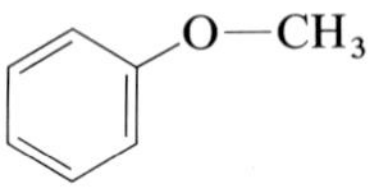 c. 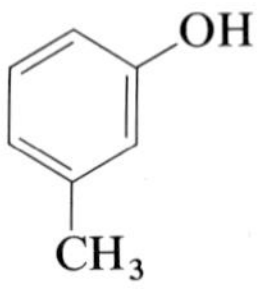 d. 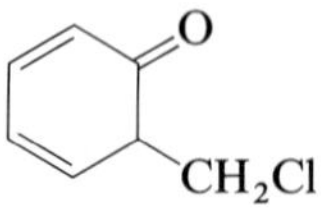

7. Assuming that each is water soluble, which of the following alcohols will react most quickly to the Lucas test?

a. 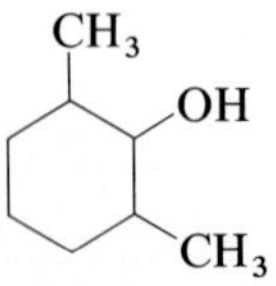 b. 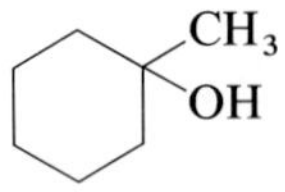 c. 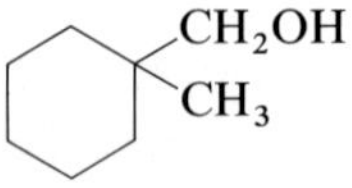 d. 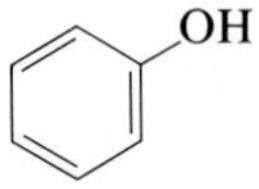

8. The high boiling points of alcohols, as compared to the corresponding alkane hydrocarbons, are due to
   a. hydrogen bonding c. water solubility
   b. heavy oxygen atoms d. none of these
9. Which of the following functional group series is ranked according to *increasing* boiling points?
   a. diethyl ether, ethane, ethanol, ethanthiol
   b. ethane, ethanol, diethyl ether, ethanthiol
   c. ethane, diethyl ether, ethanthiol, ethanol
   d. diethyl ether, ethane, ethanthiol, ethanol

10. The acidities of ________ can be understood by resonance concepts.
    a. alcohols b. phenols c. thiols d. ethers

Answers **1.** b **2.** b **3.** c **4.** a **5.** a **6.** a **7.** b **8.** a **9.** c **10.** b

DIAGNOSTIC CHART

Blacken in all of the circles under the number of each question that you missed in the preceding Chapter Review questions. The diagnostic chart will help you identify concept areas that need more study.

Concepts	Questions 1	2	3	4	5	6	7	8	9	10
Structure						○	○	○	○	○
Physical properties	○				○		○	○	○	
Nomenclature		○								
Reactions	○		○	○	○	○	○			○

EXERCISES

Functional group identification

1. For each of the following compounds, identify the phenolic, alcoholic, or ether groups that are present.
   a. Morphine
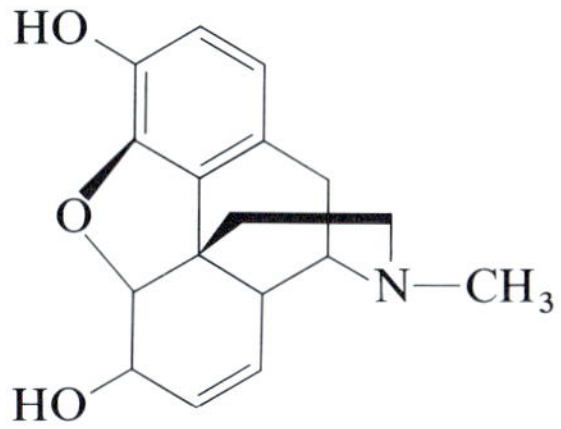
   b. Estradiol
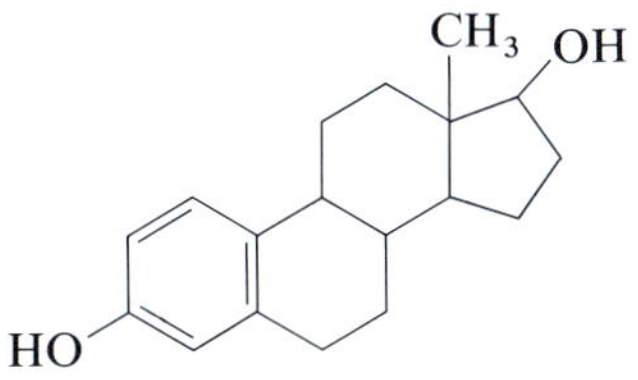
   c. Civet constituent
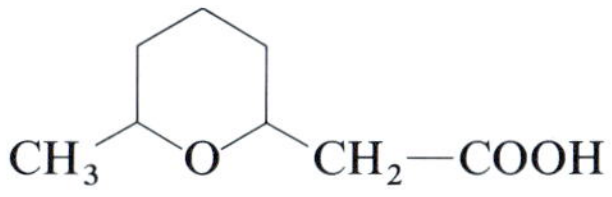
   d. Mescaline
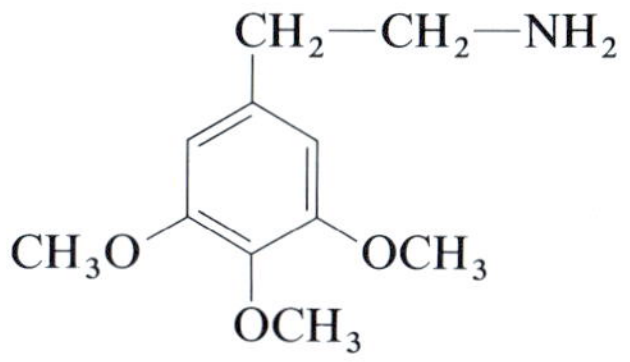

2. Identify each of the following primary, secondary, or tertiary alcohols.
   a.
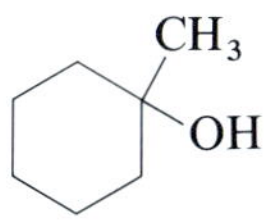
   b.
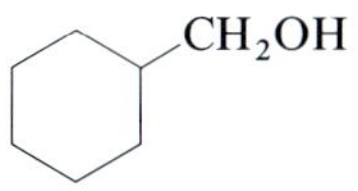
   c.
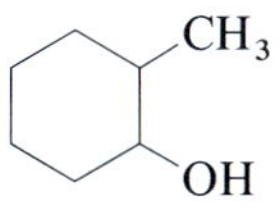
   d.
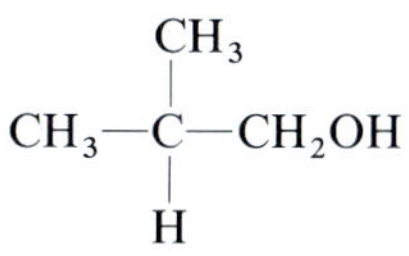
   e.
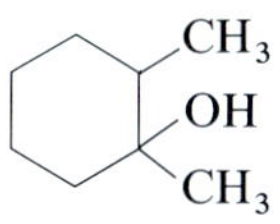
   f.
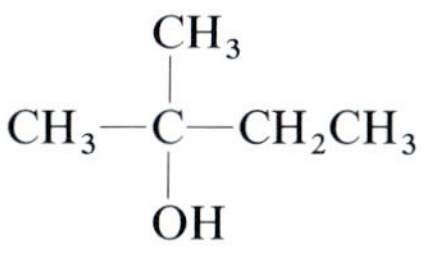

3. Draw a general structure for each of the following:
   a. phenol
   b. disulfide
   c. ether
   d. thiol
   e. epoxide
   f. tertiary alcohol
   g. primary alcohol
   h. secondary alcohol

4. Draw Lewis electron dot structures for the following.
   a.
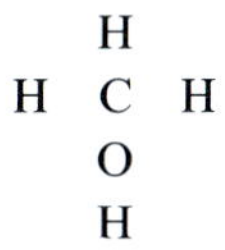
   c.
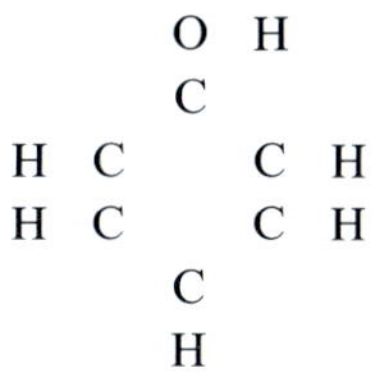
   b.
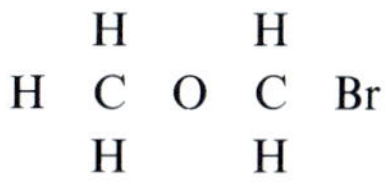

Physical Properties

5. Why are alcohols more water soluble than alkanes? Than ethers?
6. Even though a chlorine is "heavier" than the hydroxyl group, the hydroxy-substituted alkanes (alcohols) have higher boiling points than the chlorine-substituted alkanes (organohalogens). Why?
7. Given the formula C_2H_6O, we can draw an alcohol and an ether. Draw the structures. Why does the ether boil at a much lower temperature than the alcohol?
8. Butanthiol has an —S—H bond, while butanol has an —O—H bond. The thiol has a higher molecular weight, yet a lower boiling point. Explain.

Nomenclature

9. Give IUPAC names for each of the following compounds.

a. 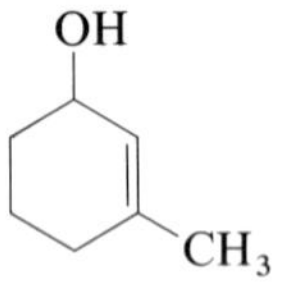

b. 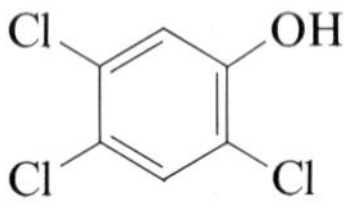

c. 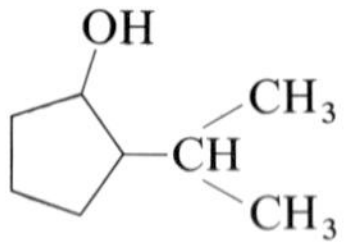

d. 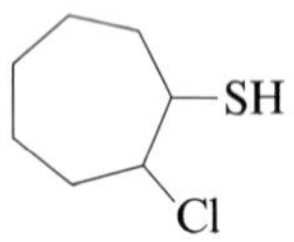

e. 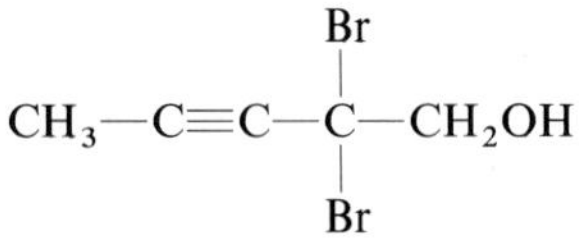

f. 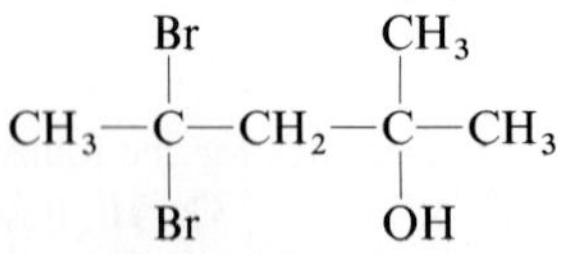

g. 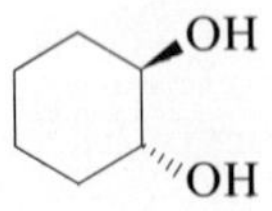

h. 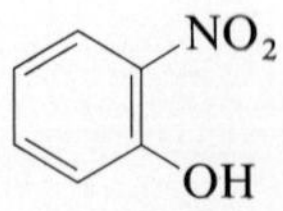

i. 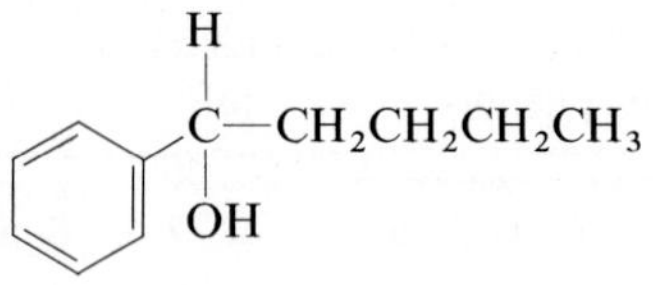

j. 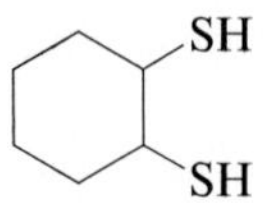

k. 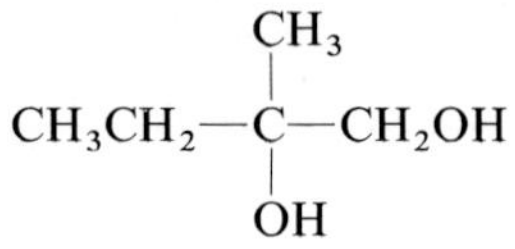

l. 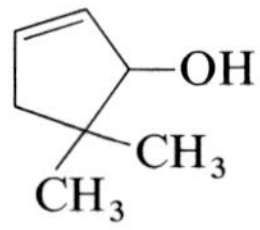

m. $H—C\equiv C—CH_2—CH_2—SH$

n. $CH_3—CH_2—O—CH_2CH_2CH_3$

o. 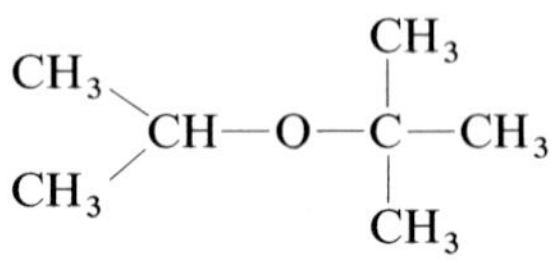

10. Draw structures for each of the following compounds.
   a. 2-chloro-3-hexanol
   b. *cis*-2-methylcyclohexanol
   c. *trans*-1,3-cyclohexandiol
   d. *p*-bromophenol
   e. isobutyl methyl ether
   f. 3-cyclohexenthiol
   g. 2,3-dimethyl-2-pentanol
   h. benzyl isopropyl ether
   i. 3,4-dimethylphenol
   j. 2-bromo-2-methyl-3-hexanthiol
   k. 3-bromo-2-heptanthiol
   l. *sec*-butyl isobutyl ether
   m. 4-methylcyclooctanol
   n. 3,3-dibromo-3-propan-1-ol
   o. 2,4,6-trinitrophenol
   p. 1-hexene-5-yne-3-ol
11. Give names for each of the following compounds.

a. 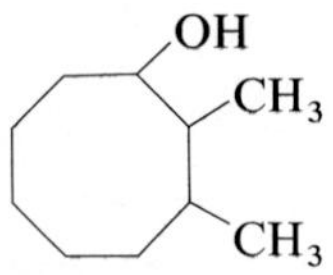

b. 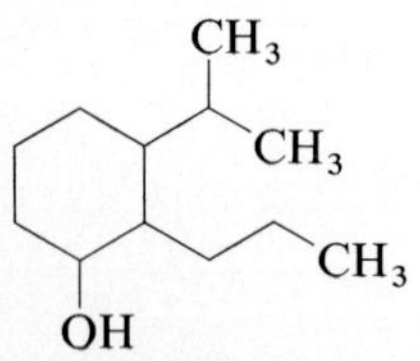

c. 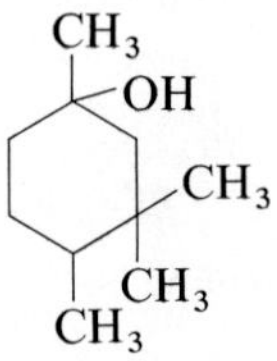

d. 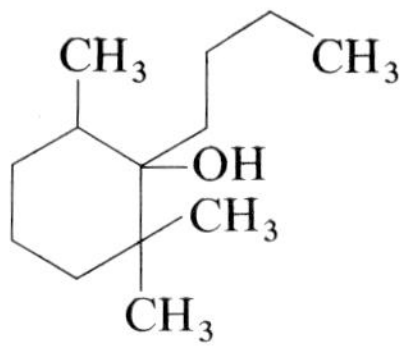
CH3
CH3
OH
CH3
CH3

e. 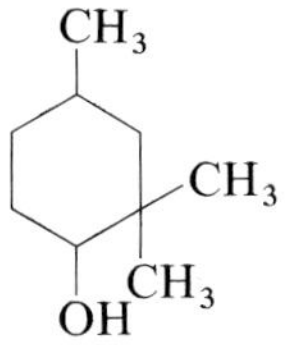
CH3
CH3
CH3
OH

f. 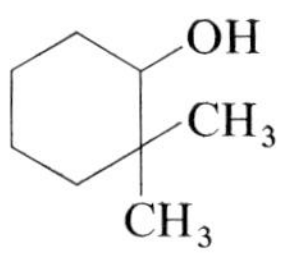
OH
CH3
CH3

g. 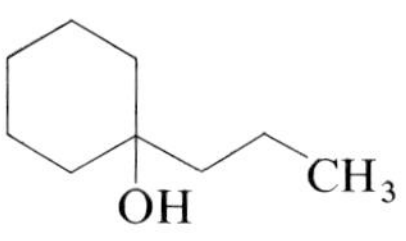
OH
CH3

h. 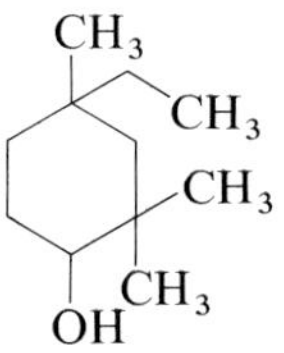
CH3
CH3
CH3
CH3
OH

i. 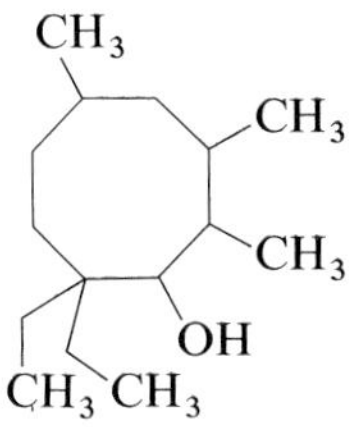
CH3
CH3
CH3
OH
CH3 CH3

Reactions

12. Give products for each of the following reactions:

a. 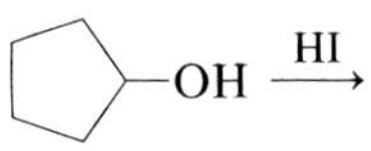
OH
HI

b. $CH_3CH_2{-}OH \xrightarrow{Na}$

c. 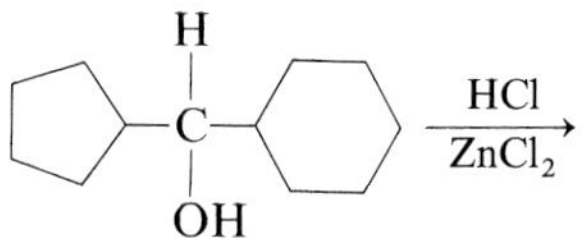
H
C
OH
HCl
ZnCl2

d. 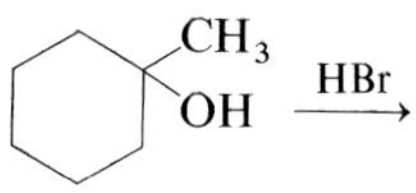
CH3
OH
HBr

e. $CH_3{-}CH_2{-}Br \xrightarrow{CH_3{-}O^-}$

f. 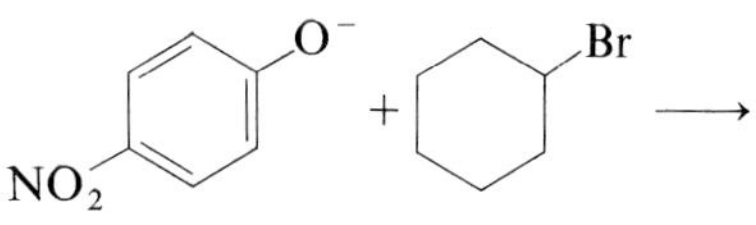
O−
NO2
+
Br

g. 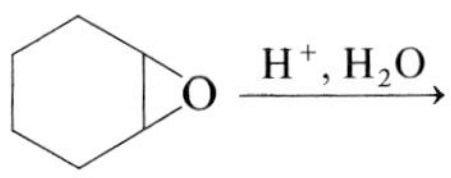
O
H+, H2O

h. 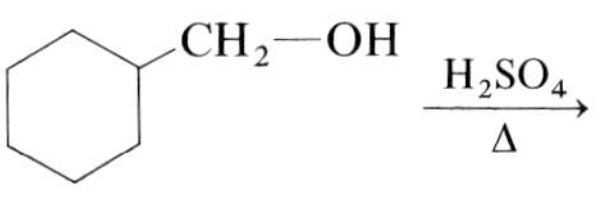
CH2—OH
H2SO4
Δ

i. 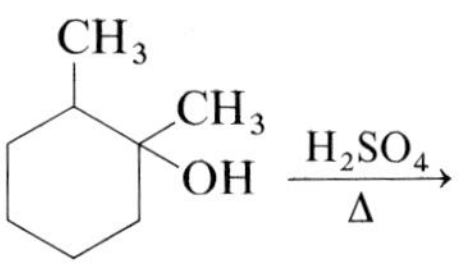
CH3
CH3
OH
H2SO4
Δ

j. $CH_3OH \xrightarrow[2.\ CH_3Br]{1.\ Na}$

k. 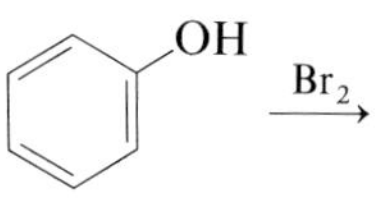
OH
Br2

l. 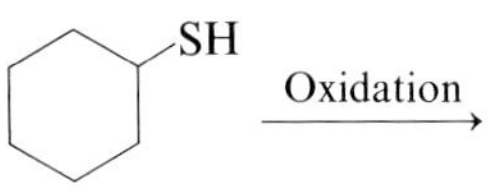
SH
Oxidation

m. 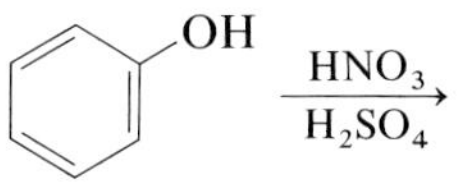
OH
HNO3
H2SO4

n. 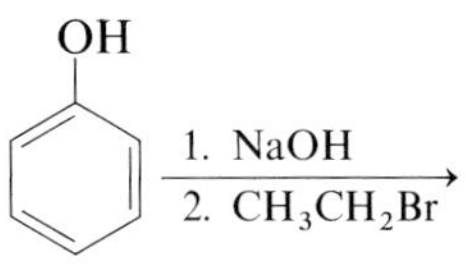
OH
1. NaOH
2. CH3CH2Br

13. Examine the alcohols in problem 2 and consider the dehydration of each; (a) determine the hydrogens adjacent to the carbon bearing the hydroxyl group that might be lost to dehydration and (b) give the major alkene product expected.

14. A compound has the formula $C_4H_{10}O$. It reacts with metallic sodium to give hydrogen gas. The compound reacts under conditions of the Lucas test to give an *immediate* reaction and formation of a second layer. Draw the structure of the compound.

15. Give the major product from each of the following alcohol reactions with sulfuric acid (dehydration reactions).

a. 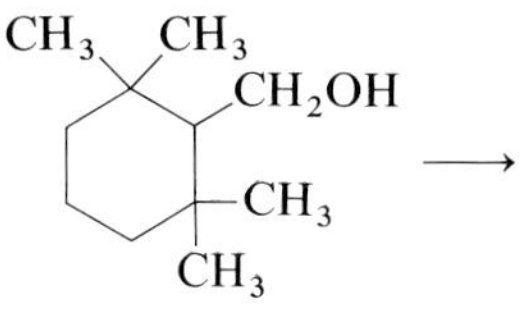
CH3
CH3
CH2OH
CH3
CH3

b. 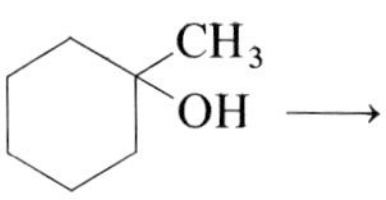
CH3
OH

c. 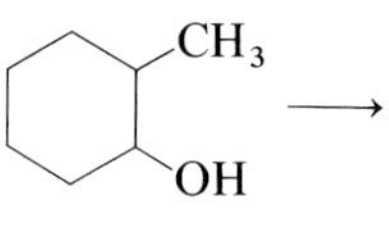
CH3
OH

d. 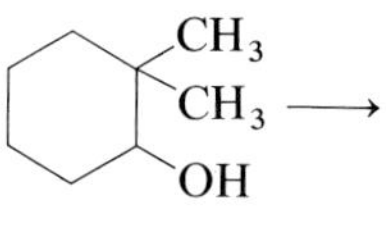
CH3
CH3
OH

e. $CH_3{-}\underset{|}{CH}{-}CH_2{-}CH_3 \longrightarrow$ (OH on the second carbon)

$\qquad\quad OH$

f. $(CH_3)(CH_3)CH{-}\underset{|}{CH}{-}CH_2CH_3 \longrightarrow$

$\qquad\qquad\qquad OH$

16. Give products from each of the following ether-forming reactions.

a.
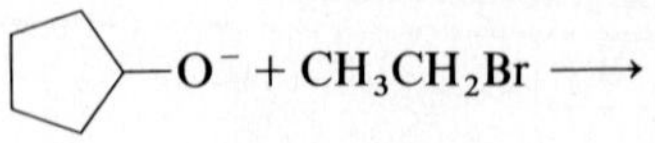

b.
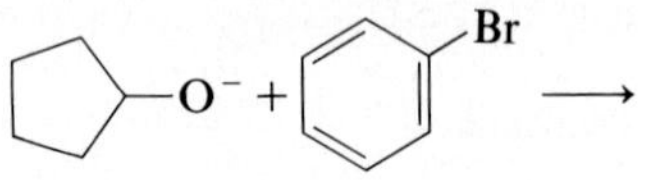

c.
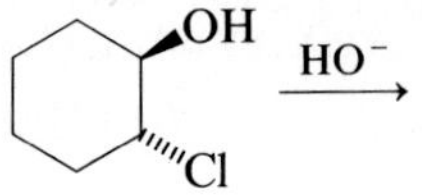

d.
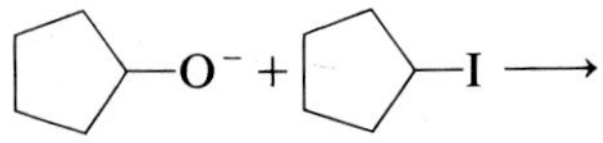

e.
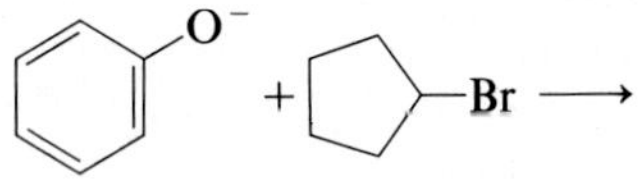

f.
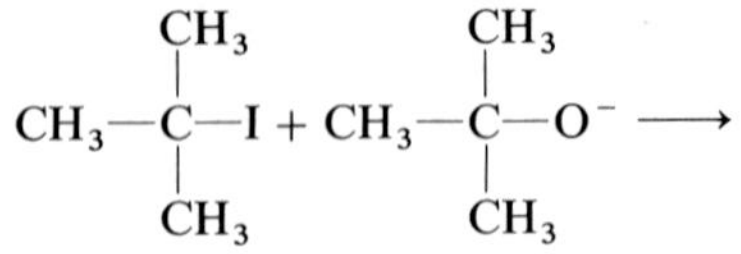

g.
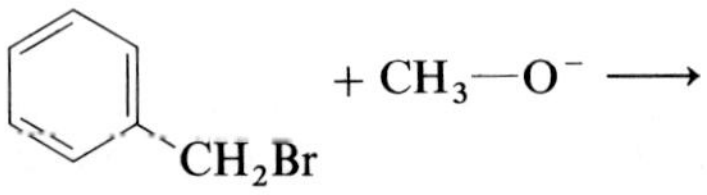

h. $CH_3-CH_2-CH_2-I + CH_3-CH_2-O^-$

17. Give products for each of the following phenol reactions.

a.
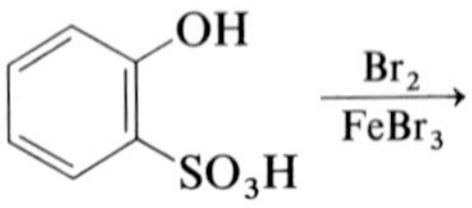

b
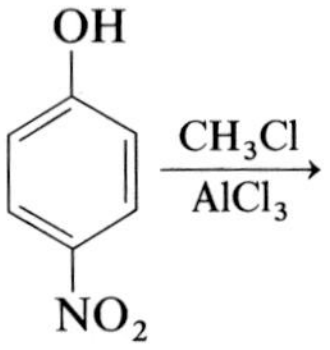

c.
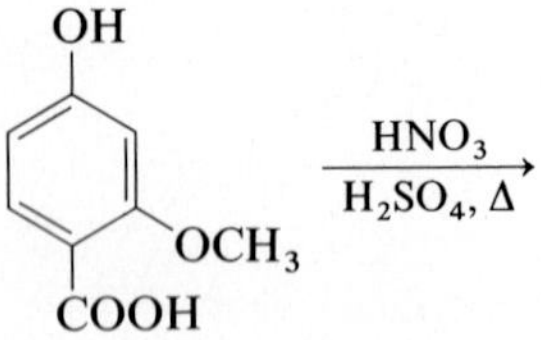

d.
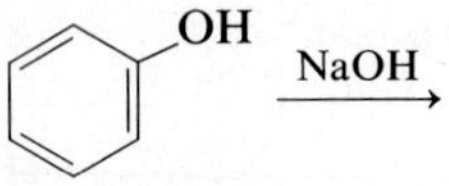

e.
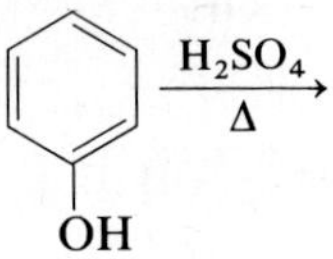

f.
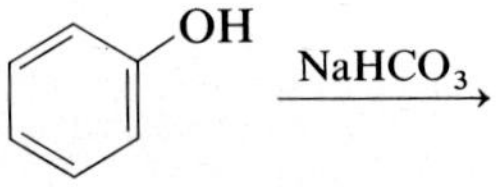

18. Give products for each of the following thiol reactions.

a. $CH_3-SH + HgCl_2 \longrightarrow$

b. $CH_3-CH_2-S-S-CH_2CH_3 \xrightarrow{\text{Reduction}}$

c.
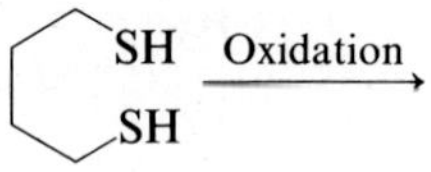

d.
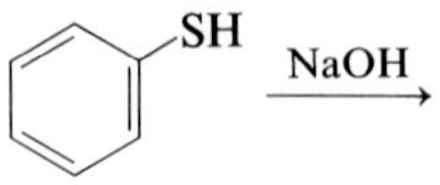

Miscellaneous

19. Name these compounds.

a.
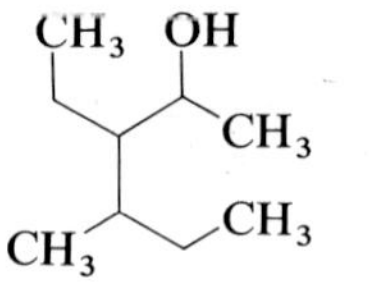

b.
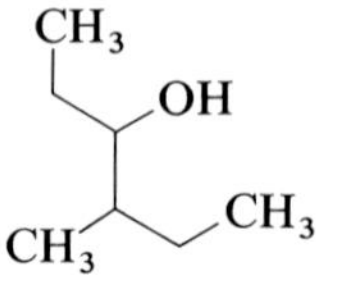

c.
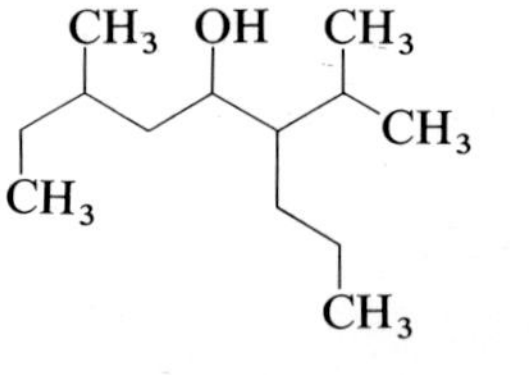

d.
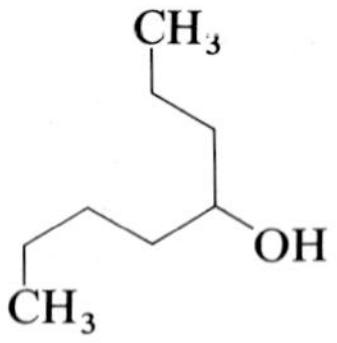

e.
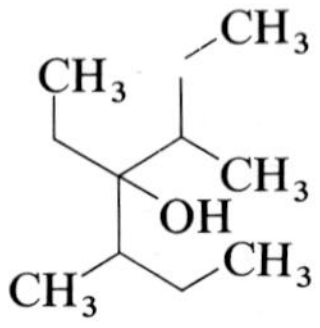

f.
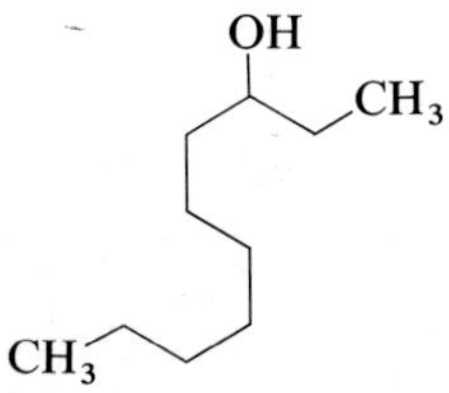

CHAPTER

19

Carbonyl Group and Its Compounds: Aldehydes and Ketones

In the last chapter we examined alcohols, phenols, and ethers—each having carbon–oxygen single bonds. In this chapter we examine a functional group, the carbonyl group, that has a carbon–oxygen double bond. The aldehydes and ketones are the family representatives of this functional group. The greater electronegativity of oxygen versus carbon gives the carbonyl group some unique chemical characteristics.

OBJECTIVES

After completing this chapter, you should be able to

- Explain how the reactivity of the carbonyl group is similar to the reactivity of an alkene and discuss how it is different.
- Give names to representative members of the aldehyde and ketone families.
- Predict products of common reactions of carbonyl compounds.
- Show how carbonyl compounds can be prepared from alcohols.
- Recognize the structures of acetyls and ketals and show where they come from.

19.1 Structure and Properties of the Carbonyl Group

Structure

The *carbonyl group* is structurally similar to an alkene.

The difference in electronegativity between an oxygen and a carbon gives the carbonyl a polarization.

In many respects, the **carbonyl group** is similar to the alkene functional group. While an alkene has two carbon atoms joined by a double bond, a carbonyl group has a carbon and an oxygen atom attached by a double bond (Figure 19.1). The difference between the chemical behavior of these two functional groups is due to the greater electronegativity of the oxygen atom. The double bond is polarized *toward* the oxygen atom; thus, the carbonyl group undergoes many reactions as a result of the carbon atom bearing a partial positive charge and the oxygen having a partial negative charge. This polarization can be represented by resonance structures (Figure 19.2).

The carbonyl group is common to both the **aldehydes** and **ketones**. The general structures of these compounds are given below, where R is used to designate an alkyl or cycloalkyl group. The symbol R′ simply suggests that the side groups may be (but are not necessarily) different. The shorthand notations Ar and Ar′ designate aromatic groups.

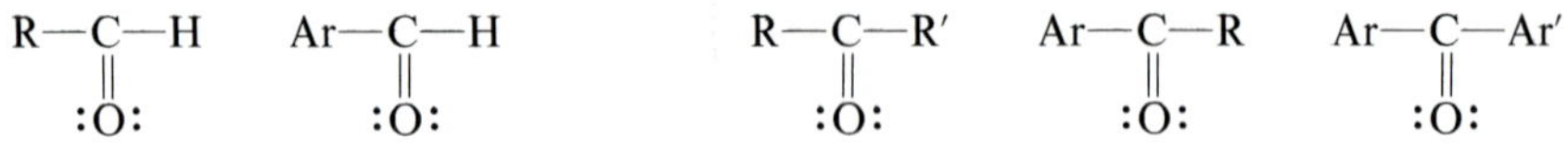

R—C—H
:O:
Ar—C—H
:O:
R—C—R′
:O:
Ar—C—R
:O:
Ar—C—Ar′
:O:

Aldehydes **Ketones**

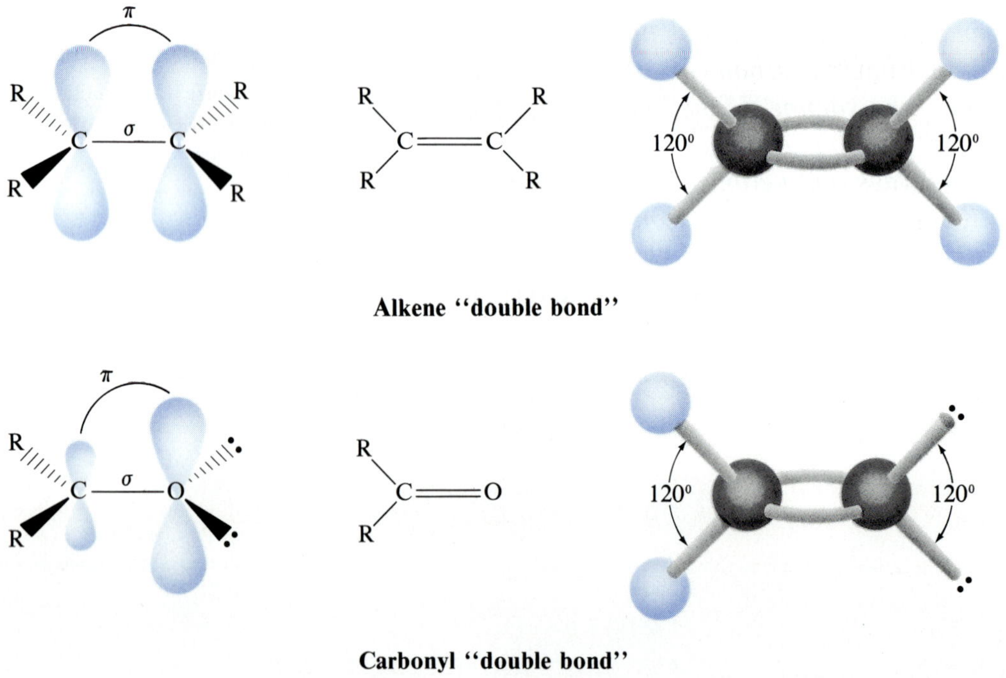

π
R
R
σ
C
C
R
R
R
R
C
C
R
R
120°
120°
Alkene "double bond"
π
R
σ
C
O
R
R
C
O
R
120°
120°
Carbonyl "double bond"

FIGURE 19.1
A comparison of an alkene and a carbonyl group. Although both functional groups have a double bond, the polarization of the carbonyl due to the higher electronegativity of oxygen results in a very different chemical behavior for the carbonyl group. The larger orbitals for the oxygen in the carbonyl group represent its greater electron density.

Physical Properties: Dipole–Dipole Interaction

In the previous chapter the dramatic influence of intermolecular hydrogen bonding of alcohols on their boiling points was observed. The carbonyl group, although polarized, cannot participate in the same kind of interaction because there are no O—H bonds. However, as a result of the polarity of the carbonyl group, a weak attraction occurs:

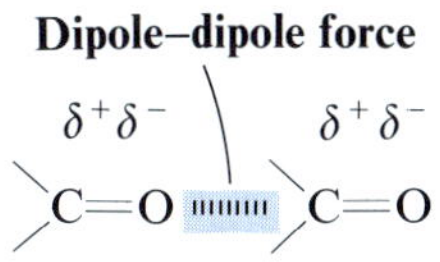
Dipole–dipole force
δ+ δ−
δ+ δ−
C=O
C=O

δ^+ and δ^- mean relative positive and negative charge

This weak attractive force, called a **dipole–dipole force**, helps to explain why the carbonyl compounds have higher boiling points than alkanes of similar molecular weight. However, the intermolecular attractive forces are much less than would result from hydrogen bonding. Figure 19.3 shows the boiling points of a homologous series of primary alcohols and their corresponding aldehydes. A similar relationship exists between secondary alcohols and ketones.

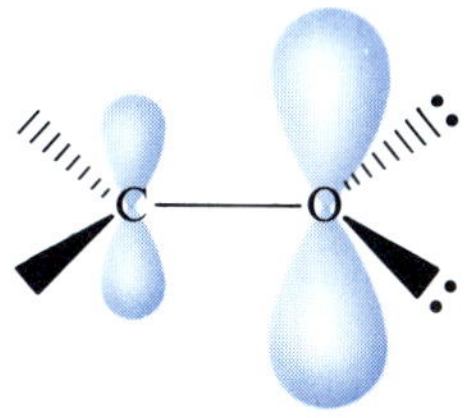

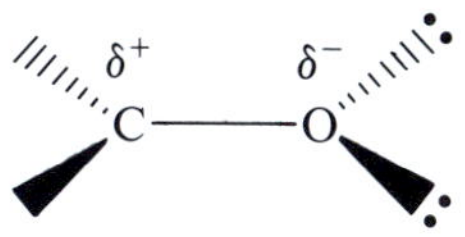

The π-electrons are polarized toward the more electronegative oxygen atom

FIGURE 19.2
Resonance structures of the carbonyl group. The polarization of the carbonyl group can be represented by a resonance structure that has a positive charge on the carbon atom and a negative charge on the oxygen atom.

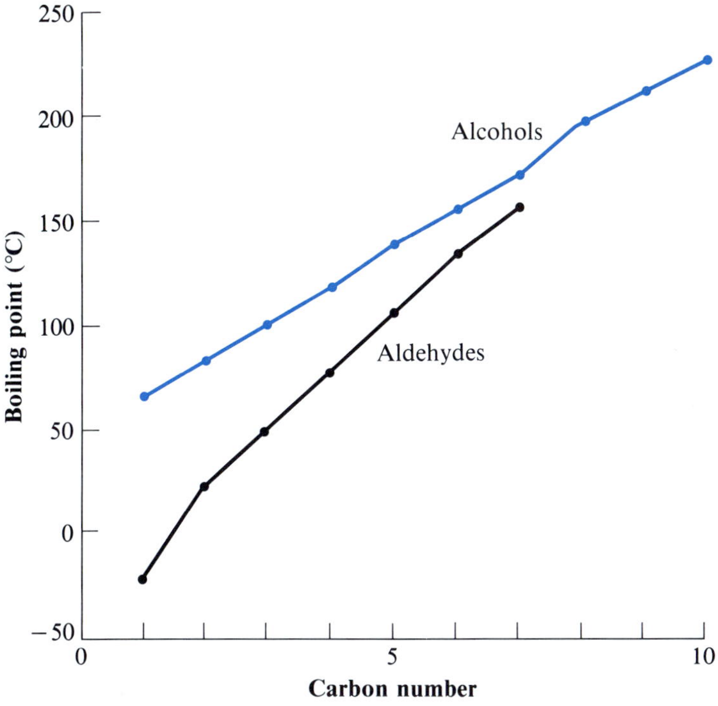
250
200
150
100
50
0
−50
0
5
10
Boiling point (°C)
Carbon number
Alcohols
Aldehydes

FIGURE 19.3
The boiling points of alcohols and aldehydes versus carbon number. The dipole–dipole interactions are not nearly as important as hydrogen bonding in determining boiling point. Again notice that as the carbon number increases, the importance of the functional group in dominating physical properties decreases.

NEW TERMS

carbonyl group The functional group characterized by a carbon–oxygen double bond:

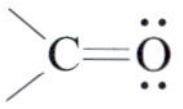

aldehyde A class of carbonyl-containing compounds of the general structure

$$\text{Ar}-\underset{\underset{:\ddot{O}:}{\|}}{C}-H \qquad R-\underset{\underset{:\ddot{O}:}{\|}}{C}-H$$

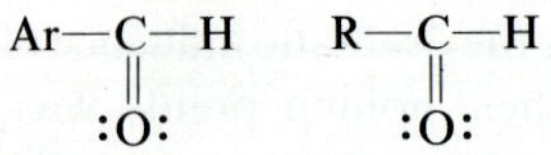

The carbonyl group is always on a terminal carbon atom.

ketone A class of carbonyl-containing compounds of the general structure

$$R-\underset{\underset{:\ddot{O}:}{\|}}{C}-R' \qquad \text{Ar}-\underset{\underset{:\ddot{O}:}{\|}}{C}-\text{Ar}' \qquad \text{Ar}-\underset{\underset{:\ddot{O}:}{\|}}{C}-R$$

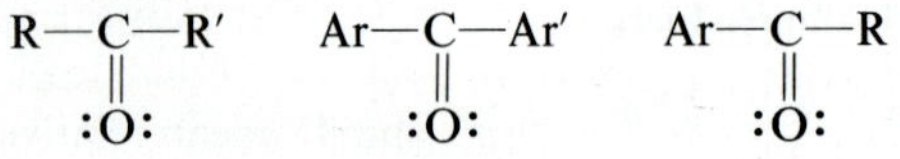

The carbonyl group is not on a terminal carbon atom.

dipole–dipole force An attractive force between two molecules as a result of the polarity of molecules.

TESTING YOURSELF

Structure and Properties
1. Show the Lewis electron dot structure for formaldehyde,

$$\underset{H}{\overset{H}{>}}C=O$$

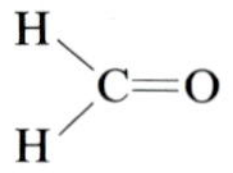

2. Which would have the stronger dipole–dipole interaction, H—Cl⋯⋯H—Cl or H—I⋯⋯H—I? Why?
3. Why does HI have a higher boiling point than HCl?

Answers 1. $\underset{H}{\overset{H}{>}}C=\ddot{\underset{..}{O}}:$

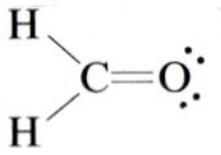

2. H—Cl because Cl is more electronegative and the H—Cl bond is more polarized than the H—I bond.
3. The molecular weight of HI is greater than that of HCl.

19.2 Nomenclature of Carbonyl Compounds

Aldehydes are named using the same general rules already described for alcohols, with the exception that the *-ol* ending of the alcohol is replaced by *-al*. Because of the structure of an aldehyde, the carbonyl group is always on a terminal carbon (carbon number 1), and all other substitutents are numbered from this position.

In nomenclature, aldehydes have an -al ending and ketones have an -one ending

EXAMPLE 19.1

Name this aldehyde.

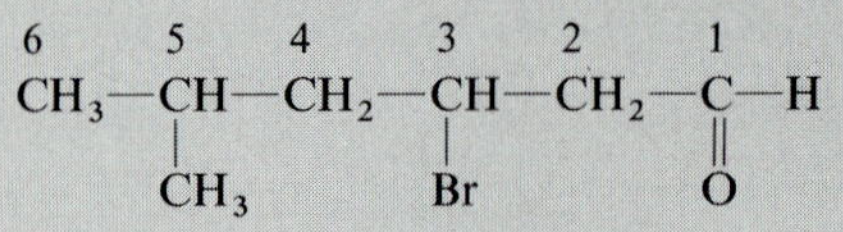

$$\overset{6}{C}H_3-\underset{\underset{CH_3}{|}}{\overset{5}{C}H}-\overset{4}{C}H_2-\underset{\underset{Br}{|}}{\overset{3}{C}H}-\overset{2}{C}H_2-\underset{\underset{O}{\|}}{\overset{1}{C}}-H \qquad \text{3-Bromo-5-methylhexan}\boxed{\text{al}}$$

Ketones also follow similar nomenclature, the *-ol* ending is replaced by *-one*. The carbonyl group takes priority in naming over all other functional groups studied so far. If there is also an alcohol group present as a substituent, the OH group is given as *-hydroxy*.

EXAMPLE 19.2

Name this ketone.

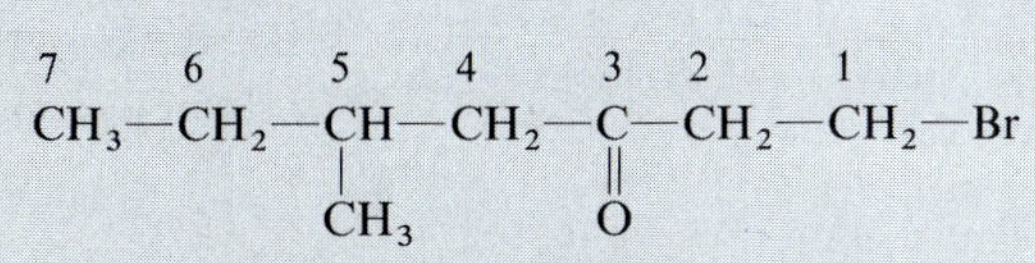

1-Bromo-5-methyl-3-heptanone

Listed here are a variety of examples of aldehyde and ketone structures along with their correct names. Make sure you understand how each was named before continuing on.

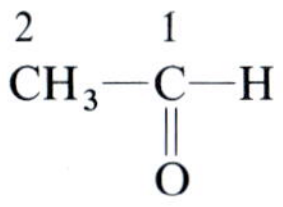

Ethanal

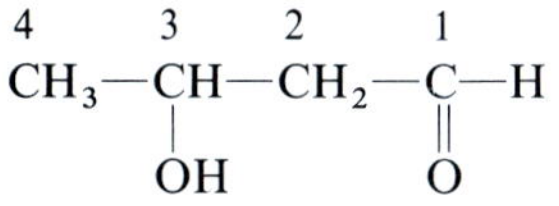

3-Hydroxybutanal

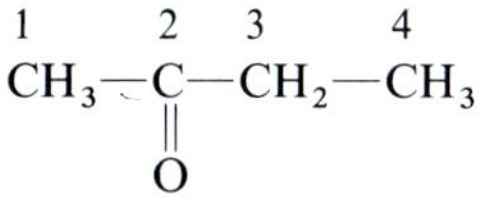

2-Butanone

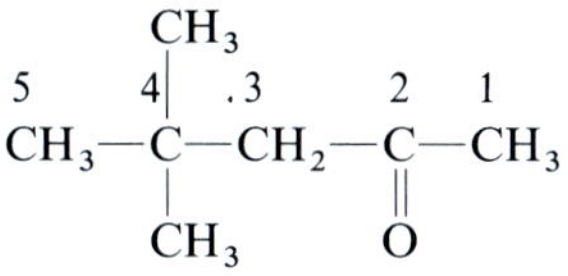

4,4-Dimethyl-2-pentanone
(not 2,2-dimethyl-4-pentanone)

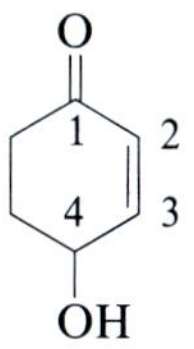

4-Hydroxy-2-cyclohexenone

Notice that the carbonyl group has a higher priority than the alcohol. Always start numbering the carbons of a ring compound at the attachment point of the highest priority functional group.

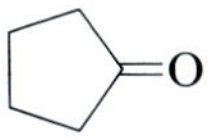

Cyclopentanone

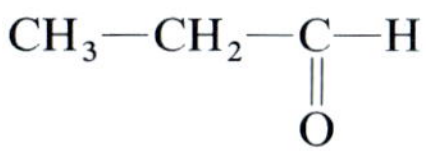

Propanal

There are also several carbonyl compounds whose common names have attained sufficiently wide acceptance that they should be learned as the "official" names (Table 19.1).

TABLE 19.1
Common Aldehydes and Ketones

Name	Description	Structure
Formaldehyde	The simplest aldehyde. A 40% water solution (formalin) is useful for preserving biological samples. It is a component of wood smoke, and the smoking of foods depends on formaldehyde to kill bacteria.	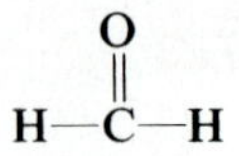 $$H-\underset{\underset{\|}{O}}{\overset{\|}{C}}-H$$
Acetaldehyde	A simple aldehyde formed from the oxidation of ethanol. It is responsible for the "hangover" from drinking. It is associated with the nutty flavor of sherry wine.	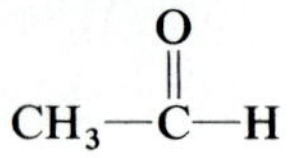 $$CH_3-\overset{O}{\underset{\|}{C}}-H$$
Benzaldehyde	The simplest aromatic aldehyde. It contributes to the aroma of almonds.	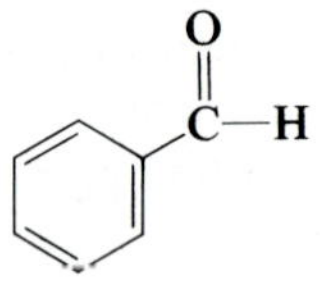 C_6H_5-CHO
Cinnemaldehyde	A fragrant compound of cinnamon	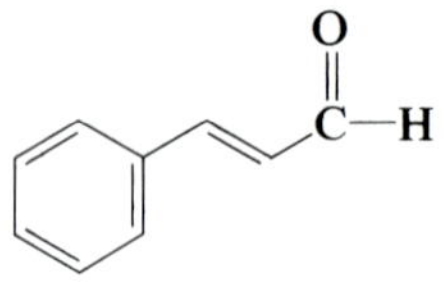 $C_6H_5-CH=CH-CHO$
Acetone	The simplest ketone. An important solvent. Acetone is found as a "ketone body" for diabetics.	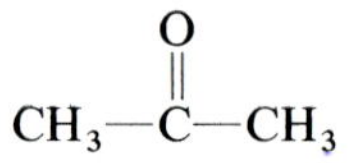 $$CH_3-\overset{O}{\underset{\|}{C}}-CH_3$$
Butanedione	The simplest diketone. It is one of the components in the odor of butter. It is associated with the unpleasant odor of sweat since it is produced by fermentation of perspiration.	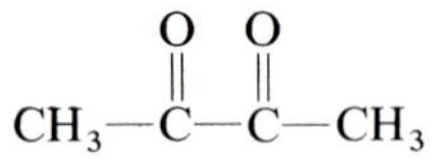 $$CH_3-\overset{O}{\underset{\|}{C}}-\overset{O}{\underset{\|}{C}}-CH_3$$
Glucose	A common sugar	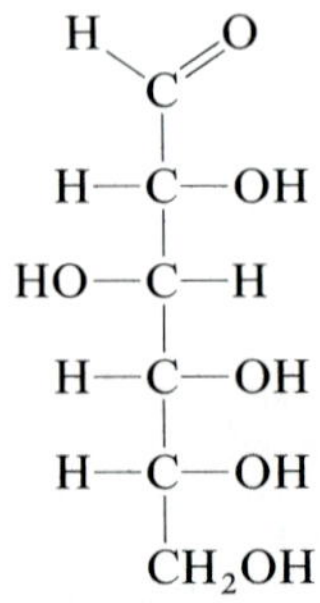 H−C(=O)−; H−C−OH; HO−C−H; H−C−OH; H−C−OH; CH_2OH
Glyceraldehyde	An intermediate in carbohydrate metabolism	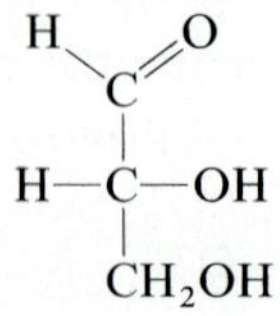 H−C(=O)−; H−C−OH; CH_2OH

(*continued*)

Name	Description	Structure
Dihydroxyacetone	**An intermediate in carbohydrate metabolism**	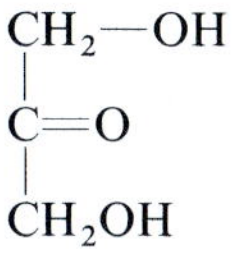 CH2—OH C=O CH2OH
Camphor	From the camphor tree. Used in medicine, although most is used to make film and celluloid.	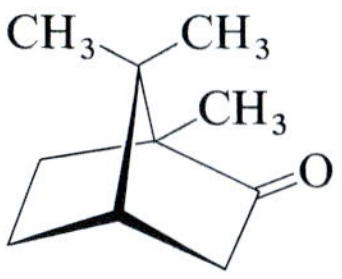 CH3 CH3 CH3 O
Civetone	A natural scent from the civet cat	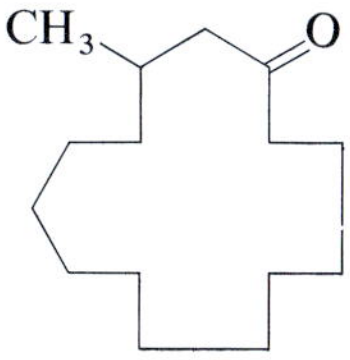 CH3 O
Jasmone	A compound obtained from the jasmine flower. The fragrance is used for perfumes.	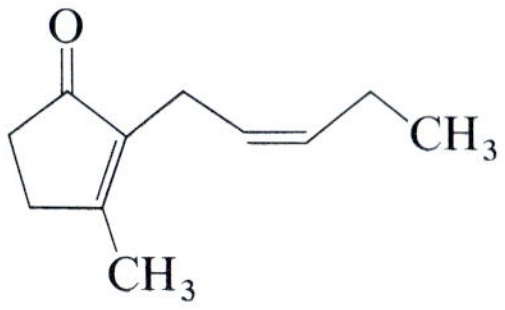 O CH3 CH3
Vanillin	From the vanilla bean (both a phenol and an aldehyde)	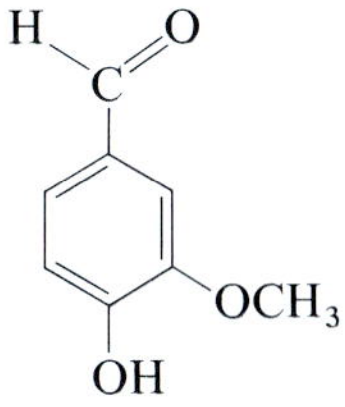 H C O OCH3 OH
Progesterone	A female sex hormone	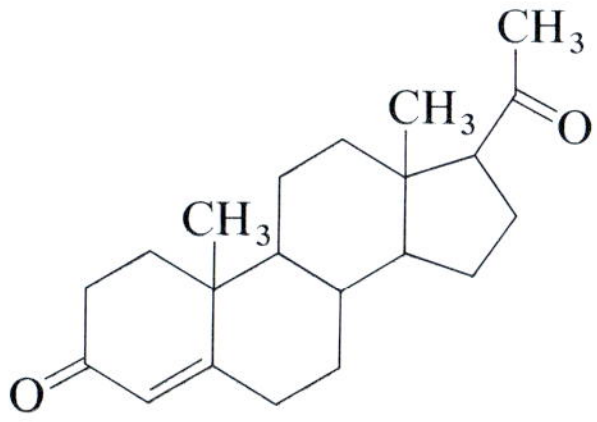 CH3 CH3 O CH3 O
Testosterone	A male sex hormone	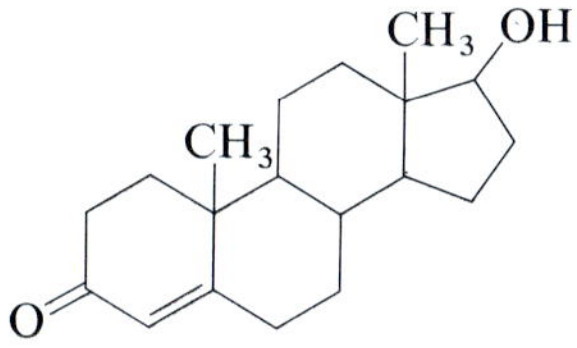 CH3 OH CH3 O

FORMALDEHYDE

Although formaldehyde is actually a gas at room temperature, most people who have worked with formaldehyde have used it in a diluted, liquid solution called *formalin*. In the past, formalin was commonly used in all biology labs as a preservation solution for biological specimens. Many foams, insulations, and bonding materials also contained formaldehyde. Much of its use has now declined with the recent finding that it is a hazardous carcinogen.

A common industrial use of formaldehyde is found in the polymerization reaction with phenol to form *Bakelite*. Bakelite is a special polymer used for coatings and plastics.

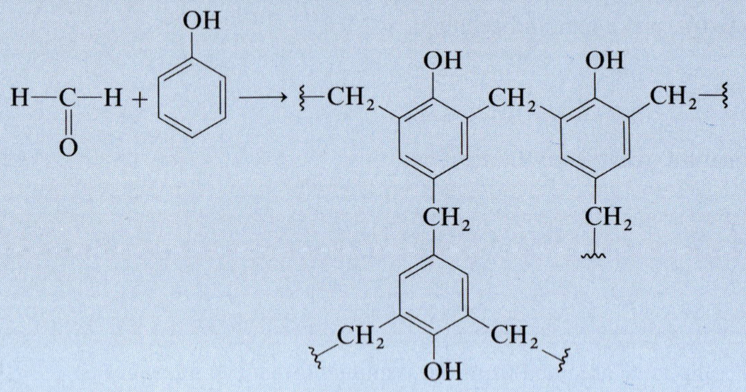

Bakelite

TESTING YOURSELF

Nomenclature

1. Give the correct name of each of the following.

a. 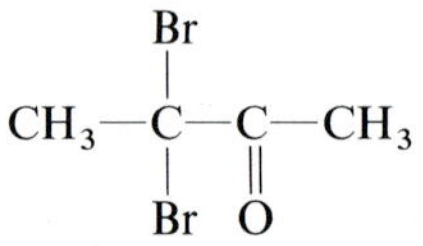

c. 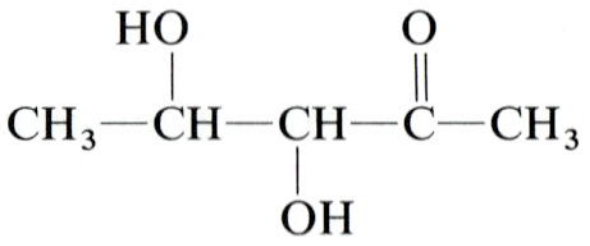

b. 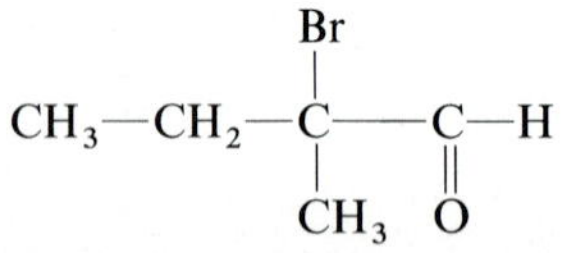

2. Draw structures for each of the following.
   a. *m*-nitrobenzaldehyde
   b. 2-bromocyclohexanone
   c. 2-cyclopentenone

Answers **1. a.** 3,3-dibromo-2-butanone **b.** 2-bromo-2-methylbutanal **c.** 3,4-dihydroxy-2-pentanone **2. a.**

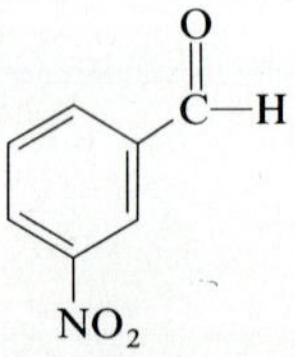

b. 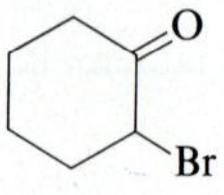

c. 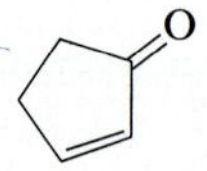

19.3 Reactions of Carbonyl Compounds

Additions to the Carbonyl: Double Bond Chemistry

As for all of the functional groups studied so far, the carbonyl group exhibits unique and important chemical behavior. We will now examine some of the important reactions of the carbonyl groups, along with examples to show their importance. Recall the addition of E^+ and Nu^- (an electrophile and a nucleophile) to an alkene. In this same addition to a carbonyl group, it is easy to predict where the E^+ and Nu^- will attach. Since the carbonyl group has a partial negative charge on the oxygen, the E^+ will preferentially add to the oxygen atom. In turn, the negatively charged nucleophile preferentially adds to the partially positively charged carbonyl carbon atom. This can be seen in the following general reaction.

The addition reactions of a carbonyl can be related to the reactions of an alkene.

GENERAL REACTION

Addition of an Electrophile and a Nucleophile to a Carbonyl

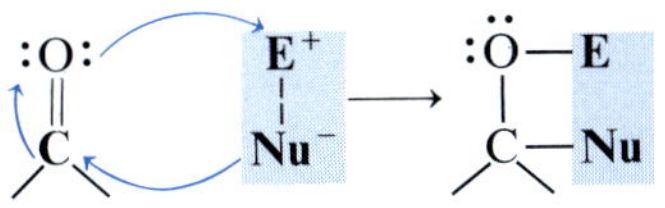

:O:
C
E+
Nu−
:Ö—E
C—Nu

Examples
(Details of these reactions will be discussed on the pages that follow.)

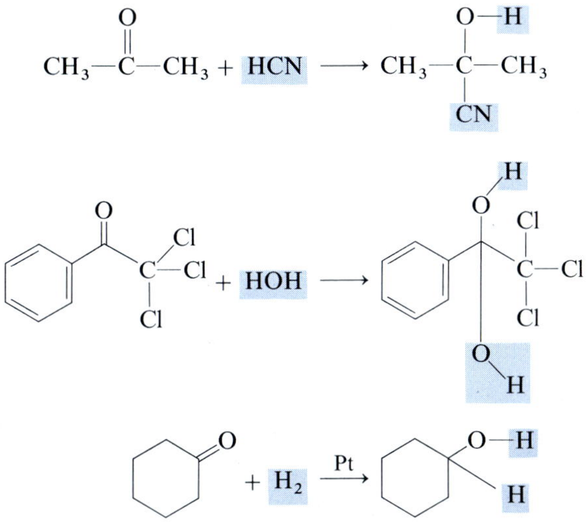

O
CH3—C—CH3 + HCN
O—H
CH3—C—CH3
CN
O
Cl
C—Cl
Cl
+ HOH
H
O
Cl
C—Cl
Cl
O
H
O
+ H2
Pt
O—H
H

GENERAL REACTION

Cyanohydrin Formation

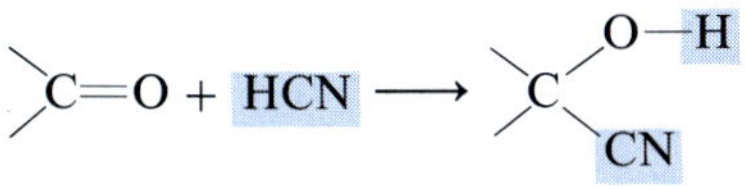

C=O + HCN
O—H
C
CN

cyanohydrin

(continued)

Addition of HCN to a carbonyl gives a *cyanohydrin.*

Examples

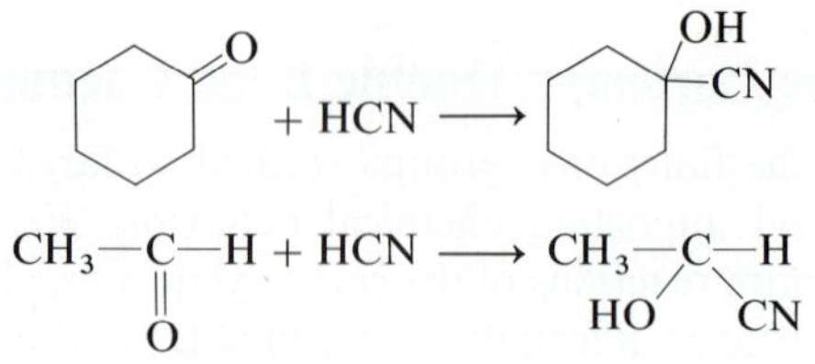
+ HCN ⟶
OH
CN
CH₃—C—H + HCN ⟶ CH₃—C—H
O
HO
CN

In this general reaction the stability of the aromatic ring can be seen. Alkenes and carbonyl groups will reduce much easier than a benzene ring.

CYANOHYDRIN AS AN INSECT DEFENSE MECHANISM Several insects make use of the reversible nature of the addition of HCN to the carbonyl group in their defense methods. For example, the millipede, *Apheloria corrigata*, uses the hydrolysis of *mandelonitrile*, a **cyanohydrin**, to ward off attackers (Figure 19.4). The insect uses an enzyme to catalyze the reaction that produces hydrogen cyanide. This deadly gas inhibits the ability of any organism to use oxygen, resulting in death. There are a number of organic compounds containing the cyanide group, —C≡N, these are often classified as *nitriles*.

CYANOHYDRIN IN LAETRILE The reputed anticancer drug, laetrile, has for its major constituent a complex compound called *amygdalin*. This com-

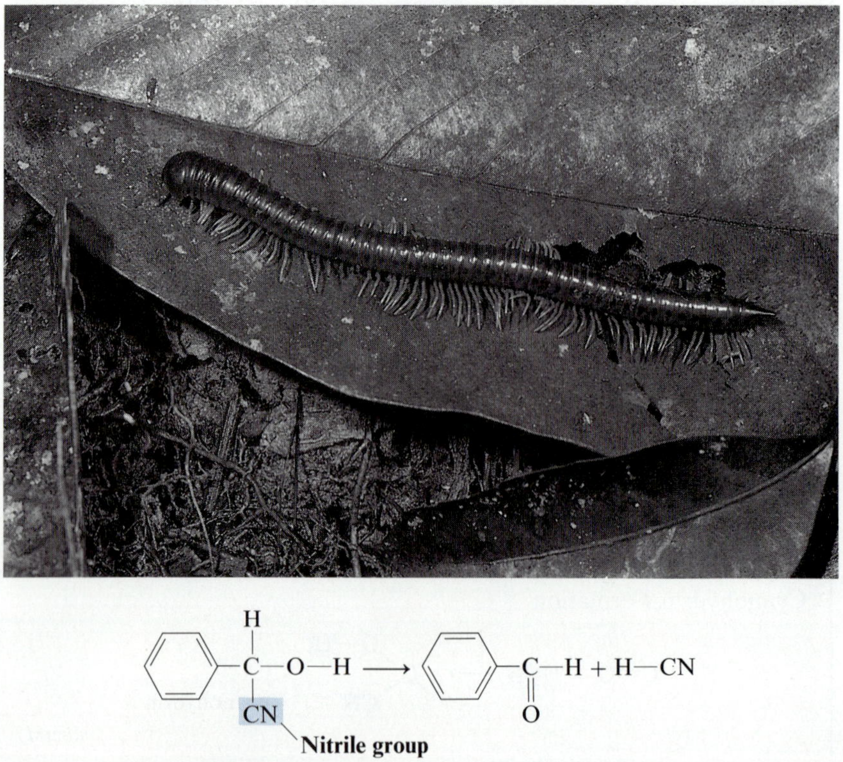
H
C—O—H ⟶
C—H + H—CN
O
CN
Nitrile group

FIGURE 19.4
A millipede produces a nitrile-containing compound that readily comes apart to form benzaldehyde and hydrogen cyanide (HCN). The hydrogen cyanide is a very toxic gas.

pound has been known for many years as a constituent of bitter almond. Hidden in this complex structure is the cyanohydrin of benzaldehyde. The compound has not been demonstrated to be effective as a cancer treatment, and its continued use can result in chronic poisoning by HCN.

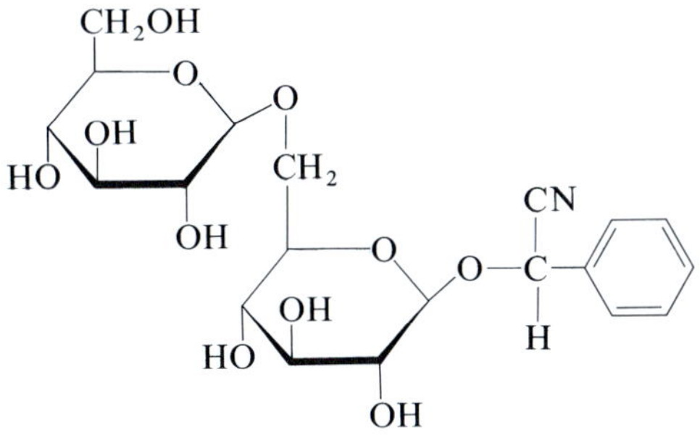
CH₂OH
O
O
OH
HO
CH₂
OH
CN
O
O—C
OH
H
HO
OH

Amygdalin

HYDRATE FORMATION A **hydrate** of a carbonyl group results from addition of water to the carbon–oxygen double bond, much like we saw in Chapter 16 for the addition of water to the carbon–carbon double bond. Because of the electronegativity difference between carbon and oxygen, the H will add to oxygen and the OH will add to carbon. It is generally difficult to obtain a hydrate since the equilibrium favors the reverse reaction, but certain aldehydes will give a stable product. The most well-known example is the reaction of *chloral* with water to form *chloral hydrate.*

Addition of water to a carbonyl gives a *hydrate.*

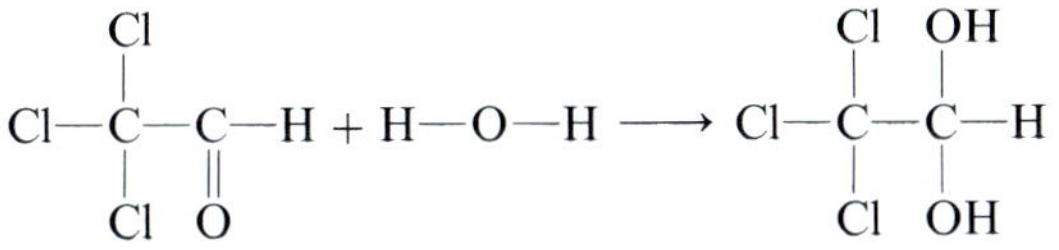

Chloral **Chloral hydrate**

Chloral hydrate, sometimes known as "knockout drops," is used as a hypnotic and in drug-withdrawal treatments.

GENERAL REACTION

Hydrate Formation

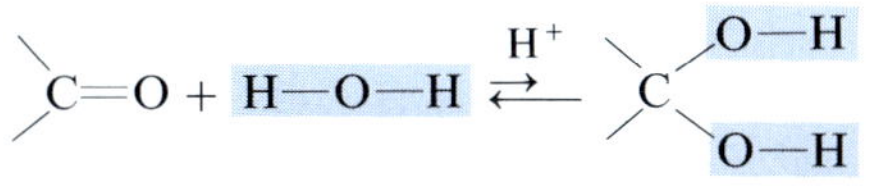

Examples

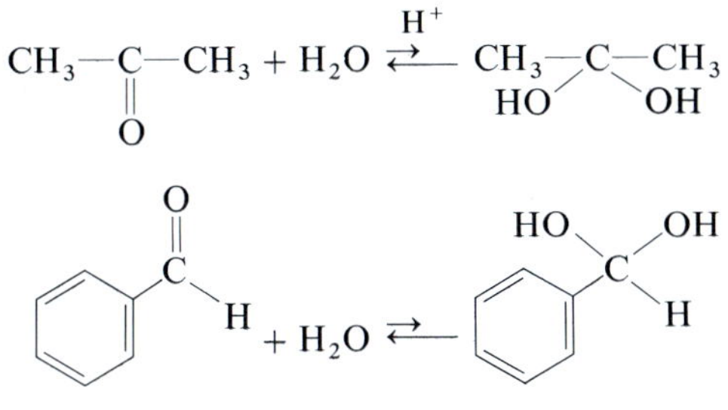

Reduction of a carbonyl gives an alcohol.

ALCOHOL FORMATION Reduction of an aldehyde gives a primary alcohol, while reduction of a ketone gives a secondary alcohol. Reducing agents include lithium aluminum hydride ($LiAlH_4$), sodium borohydride ($NaBH_4$) or dilute hydrogen and a catalyst.

GENERAL REACTION

Carbonyl Reduction

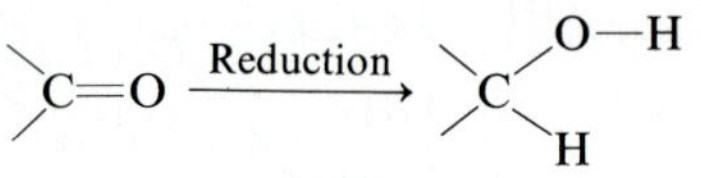

Reduction

Examples

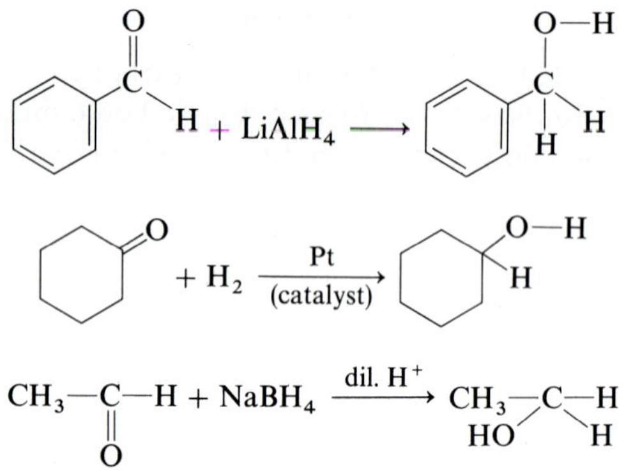

+ LiAlH4
+ H2
Pt
(catalyst)
+ NaBH4
dil. H+

Acetals and ketals are formed from the reaction of a carbonyl compound with an alcohol.

FORMATION OF ACETALS AND KETALS We have seen several general reaction types demonstrating that the carbonyl carbon is readily attacked by nucleophiles. Reactions of the carbonyl group are often enhanced by acid catalysis. This is due to the increased positive charge associated with the carbon atom after the oxygen has been protonated. The effect of acid catalysis can be shown by the mechanism described in Figure 19.5.

If a carbonyl compound is treated with an alcohol in the presence of a trace of acid, aldehydes will give a **hemiacetal** product (Figure 19.6), while ketones will give a **hemiketal** component. The first step takes place as already shown in Figure 19.5, where Nu: is H—O—R. In Figure 19.6 the terms *acetal* and *hemiacetal* are used, but the reaction can also be considered for

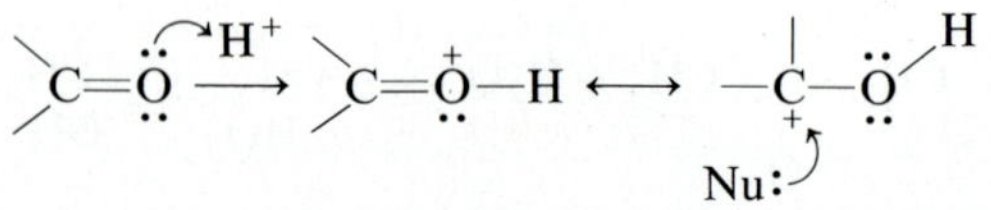

The electrons on the oxygen atom attack the proton.

The oxygen atom does not hold a positive charge well; thus, the electrons are removed from the carbon atom by resonance.

A nucleophile can now attack the positively charged carbon atom.

FIGURE 19.5
Acid catalysis of carbonyl reactions. Reaction of the negative end of a carbonyl group with H^+ results in a real positive charge on the carbon. This increases the ability of the carbon atom to react with nucleophiles.

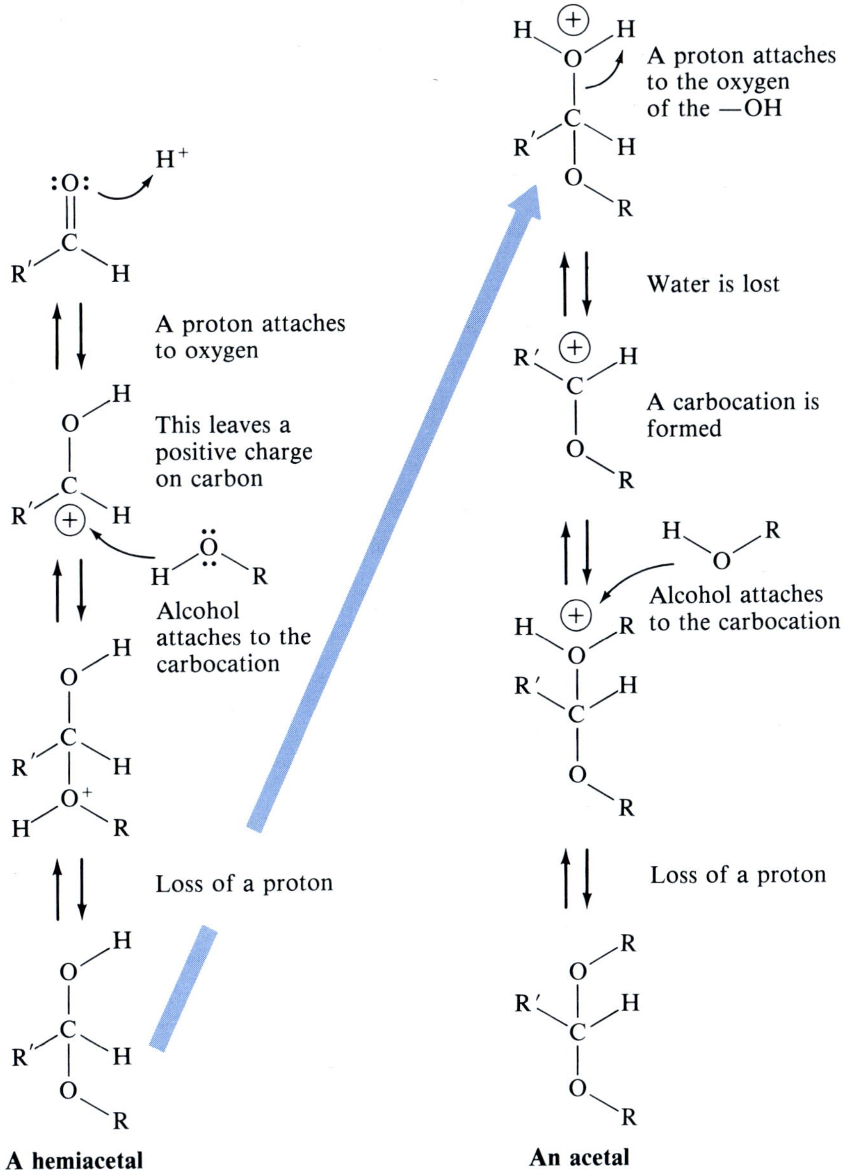
A proton attaches to oxygen
This leaves a positive charge on carbon
Alcohol attaches to the carbocation
Loss of a proton
A hemiacetal
A proton attaches to the oxygen of the —OH
Water is lost
A carbocation is formed
Alcohol attaches to the carbocation
Loss of a proton
An acetal

FIGURE 19.6
A mechanism for hemiacetal and acetal formation. The reaction requires a series of proton additions and loss of water. The same reaction with a ketone would provide a hemiketal and ketal.

ketal and hemiketal. The hemiacetals and hemiketals are susceptible to further reaction with additional alcohol and acid catalyst to give acetals and ketals, respectively. **Acetals** and **ketals** are compounds formed from alcohols reacting with aldehydes and ketones, respectively. They have the general structures:

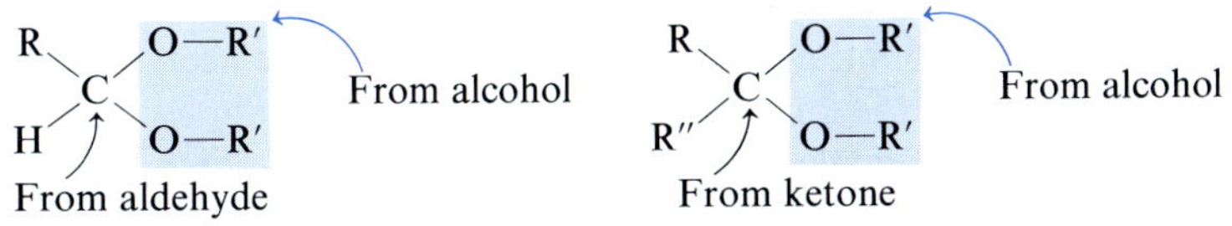
From alcohol
From aldehyde
From alcohol
From ketone

There has been a recent trend away from using the two different names, and correct nomenclature in the future will not include the name ketal. All molecules of this general type will be classified as acetals. Because there are some cases in which the older nomenclature is still used, we will employ it in this text.

A carbonyl compound in the presence of both excess alcohol and acid will react to produce the acetals and ketals.

GENERAL REACTION

Acetal and Ketal Formation

$$\begin{matrix} R \\ R' \end{matrix}\!\!>C{=}O + \underset{\text{excess}}{R''OH} \xrightarrow{H^+} \begin{matrix} R \\ R' \end{matrix}\!\!>C<\!\!\begin{matrix} O{-}R \\ O{-}R \end{matrix} + H_2O$$

Examples

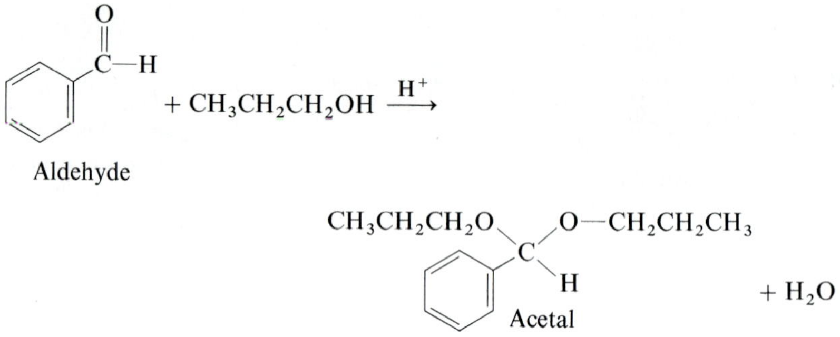
O
C—H
+ CH3CH2CH2OH
H+
Aldehyde
CH3CH2CH2O
O—CH2CH2CH3
C
H
+ H2O
Acetal

$$\underset{\text{Ketone}}{\begin{matrix} CH_3 \\ CH_3 \end{matrix}\!\!>C{=}O} + CH_3CH_2OH \xrightarrow{H^+} \underset{\text{Ketal}}{\begin{matrix} CH_3 \\ CH_3 \end{matrix}\!\!>C<\!\!\begin{matrix} O{-}CH_2CH_3 \\ O{-}CH_2CH_3 \end{matrix}} + H_2O$$

In the preceding general reaction, if R or R′ is H, the product is an acetal. If R and R′ are alkyl or aromatic, the product is a ketal. The R″ is the alkyl group belonging to the alcohol.

The formation of acetals and ketals from carbonyl compounds and the ability to reverse the reaction are excellent examples of the application of **Le Chatelier's principle**. To make an equilibrium reaction greatly favor the reaction in one direction or another, a way must be found to upset the equilibrium. Consider the equilibrium formation of an acetal or ketal:

$$>C{=}O + 2\,ROH \rightleftharpoons >C<\!\!\begin{matrix} OR \\ OR \end{matrix} + H_2O$$

To upset the equilibrium in favor of the formation of the acetal/ketal, water must be removed as it is formed. For the reaction to regain equilibrium, the reactants are forced to make more water (and, at the same time, acetal/ketal). Eventually all of the carbonyl is used up and is converted to the product. In the reverse reaction, an excess of water is added to the acetal/ketal. To establish an equilibrium, the reaction tries to get rid of some of the water, and in the process the carbonyl compound and alcohol are reformed. Some examples of these reactions are shown here.

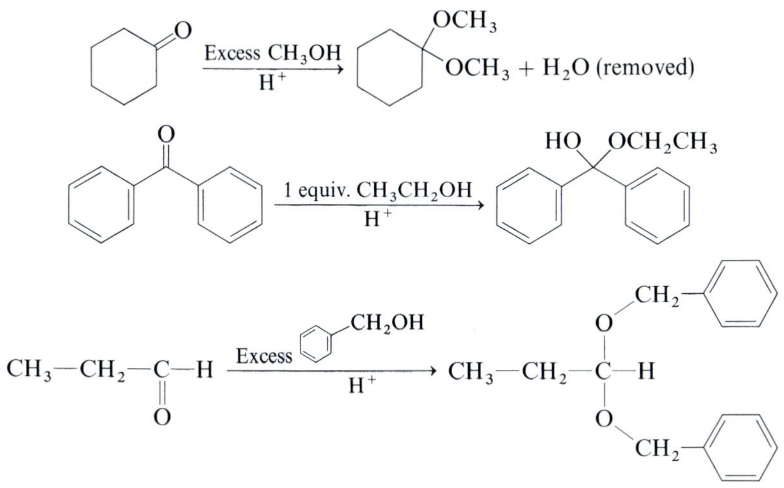
Excess CH3OH
H+
OCH3
OCH3 + H2O (removed)
1 equiv. CH3CH2OH
H+
HO OCH2CH3
Excess CH2OH
H+
CH3—CH2—C—H

NEW TERMS

cyanohydrin A carbonyl derivative made by the addition of H—C≡N to C=O.

hydrate A carbonyl derivative, usually unstable, made by the addition of water to a carbonyl.

hemiacetal The product formed from the acid-catalyzed reaction of an aldehyde with one equivalent (1 equiv.) of an alcohol.

hemiketal The product formed from the acid-catalyzed reaction of a ketone with one equivalent of an alcohol.

acetal The product formed from the acid-catalyzed reaction of an aldehyde with two equivalents of an alcohol.

ketal The product formed from the acid-catalyzed reaction of a ketone with two equivalents of an alcohol.

Le Chatelier's principle When a stress is applied to a system in equilibrium, the system will respond in a way to counteract the stress and try to reestablish equilibrium.

TESTING YOURSELF

Carbonyl Reactions

1. Give products for the following addition reactions.

a. 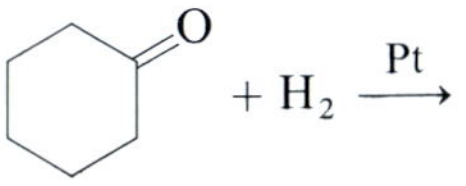
+ H2
Pt

c. 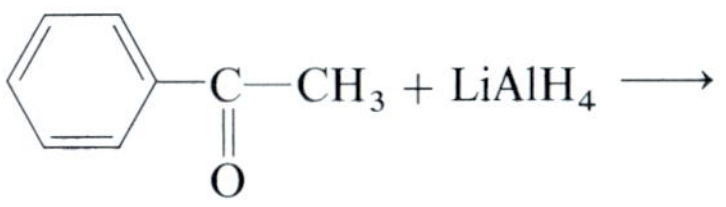
C—CH3 + LiAlH4

b. 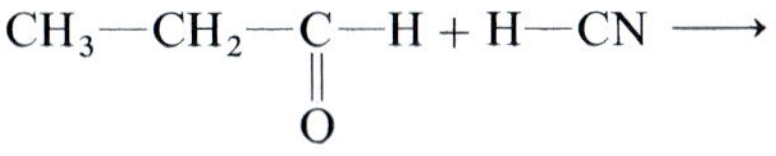
CH3—CH2—C—H + H—CN

d. 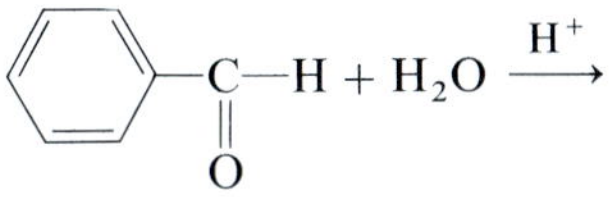
C—H + H2O
H+

2. Draw the structure for the major product expected from each of the following reactions.

a.

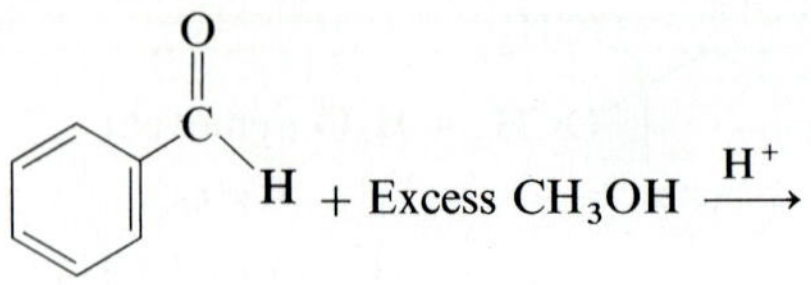
O
C
H
+ Excess CH3OH
H+

b.

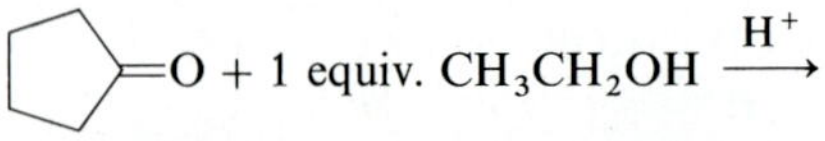
O + 1 equiv. CH3CH2OH
H+

c.

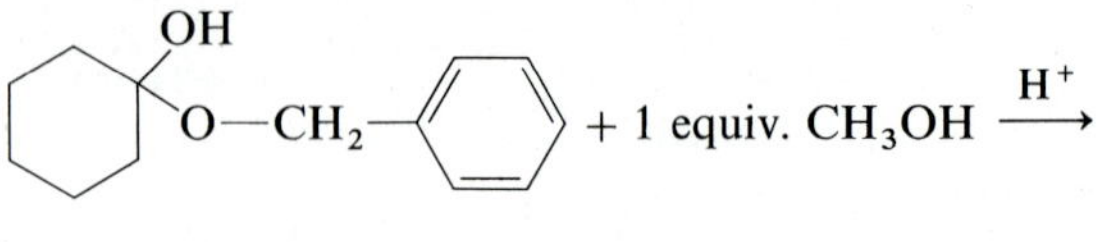
OH
O—CH2
+ 1 equiv. CH3OH
H+

d.

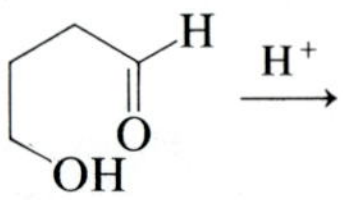
H
H+
O
OH

3. In the following general reaction what will happen to the equilibrium if a large excess of alcohol is used?

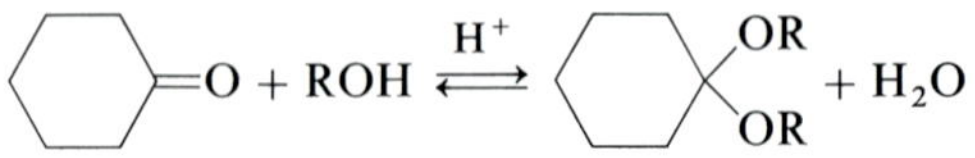
O + ROH
H+
OR
OR
+ H2O

Answers **1. a.**

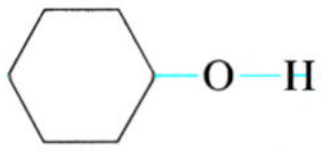
O—H

b.

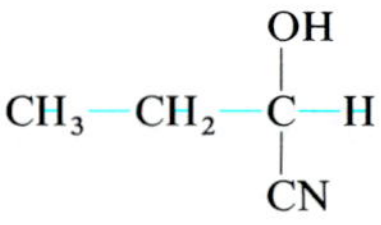
OH
CH3—CH2—C—H
CN

c.

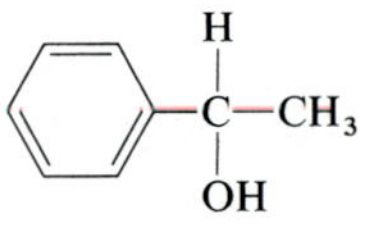
H
C—CH3
OH

d.

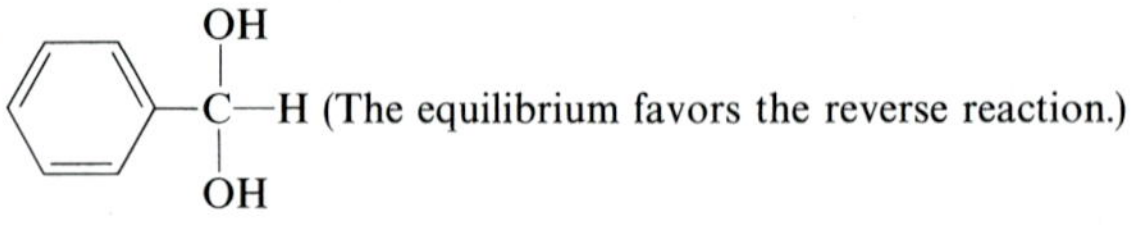
OH
C—H (The equilibrium favors the reverse reaction.)
OH

2. a.

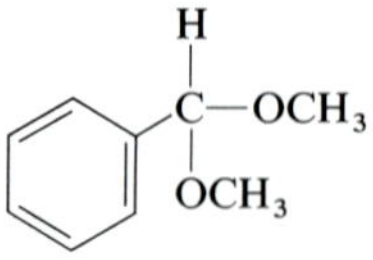
H
C—OCH3
OCH3

b.

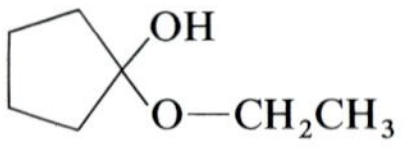
OH
O—CH2CH3

c.

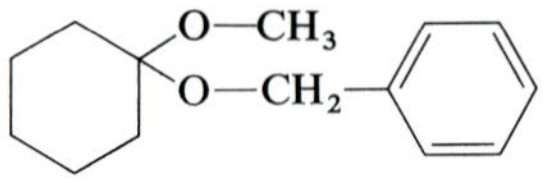
O—CH3
O—CH2

d.

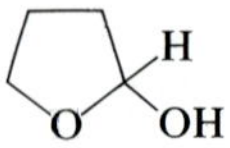
H
O
OH

Note that **2d** is an "internal" hemiacetal. This reaction would occur faster than two separate molecules coming together because of a very favorable proximity of reactive sites. This is similar to the easy reaction found for epoxide formation. **3.** The equilibrium will be shifted toward formation of the ketal.

19.4 Natural Examples of Acetals and Ketals

Carbohydrates

Carbohydrates are examples of complex polyhydroxylic carbonyl compounds.

The chemistry of acetals and ketals is very prominent in the **carbohydrates** (details of carbohydrate chemistry will be discussed in Chapter 23). Hemiacetals are formed by the reaction of an aldehyde with an alcohol. (Note that in question 2d of Testing Yourself at the end of the last section, the formation of an "internal" hemiacetal was shown to be a favorable process.) Sugars are examples of carbohydrates. These natural products are polyhydroxylic aldehydes or ketones (having more than one hydroxy group).

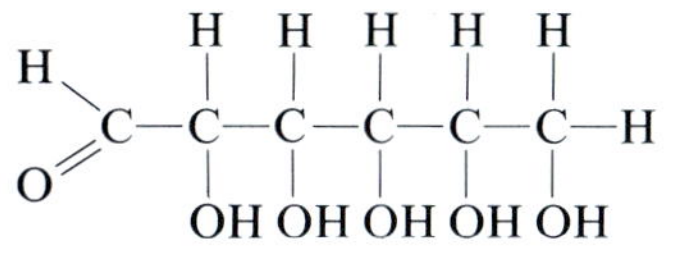

An aldehyde sugar

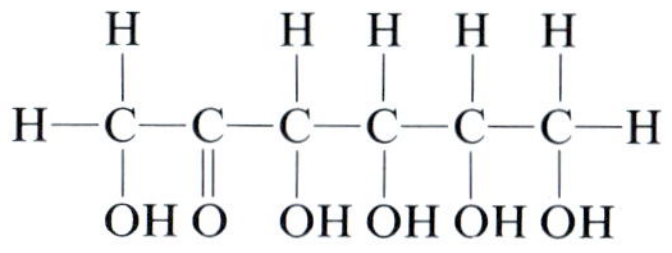

A ketone sugar

There are both carbonyl and hydroxyl groups within the same molecule; thus, it is not surprising that these undergo hemiacetal and hemiketal formation intramolecularly (that is, within the same molecule). Because of the energies involved in forming rings (including the probability that two ends can come together), five- and six-membered rings are formed more readily than most other ring systems. Thus, in the hemiacetal or hemiketal forms of sugars, only five- and six-membered rings are found. This ring-forming tendency is shown in Figure 19.7.

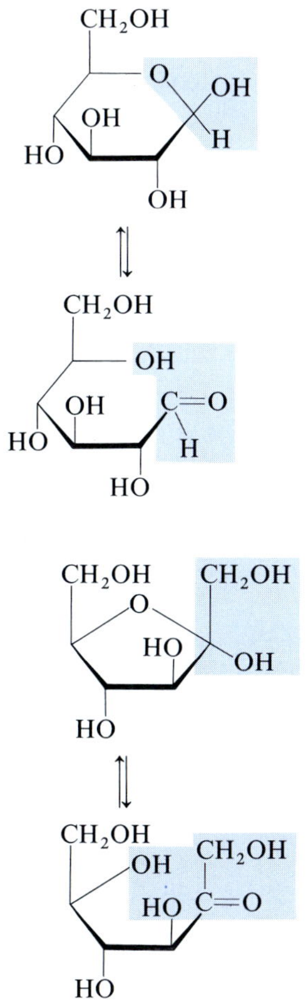

Hemiacetal formation with glucose

Hemiketal formation with fructose

FIGURE 19.7
The hemiacetal and hemiketal forms of carbohydrates.

Pheromones: Other Examples of Natural Ketals

As methods and techniques to isolate and identify small amounts of material have improved, chemists have become increasingly involved with **pheromones**—chemicals produced by an organism to communicate information to other members of the same species. This study of chemical communication has been most highly developed for insects; however, increasing attention is now being given to animals.

Some species of bark beetles have ketals as their *sex pheromones*, chemicals that aggregate and excite insects for mating. The female pine bark beetle *Dendroctonus brevicomis* releases a compound, given the trivial name of brevicomin, to attract male beetles to her "nuptial chamber" in the tree (Figure 19.8). The tree rarely survives these massive attacks of beetles, and many of the forests of the northwest have been devastated by the beetles. (See Plate 8 in color inserts for devastation of Yellowstone National Park.)

If the structure of brevicomin is dissected, one can see the origins of the ketal. This approach has been taken as part of its successful laboratory synthesis.

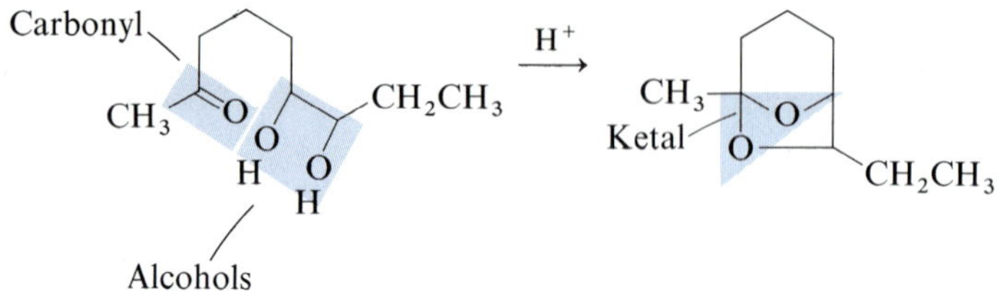
Carbonyl
CH_3
O
O
H
O
H
CH_2CH_3
Alcohols
H^+
CH_3
Ketal
O
O
CH_2CH_3

A pheromone that induces aggressive behavior in the house mouse (*Mus musculas*) has been recently characterized and is surprisingly similar to brevicomin. A bicyclic ketal is also found as the sex pheromone of the Dutch elm beetle, *Scolytus multistriatus* (Figure 19.9). This beetle has been largely re-

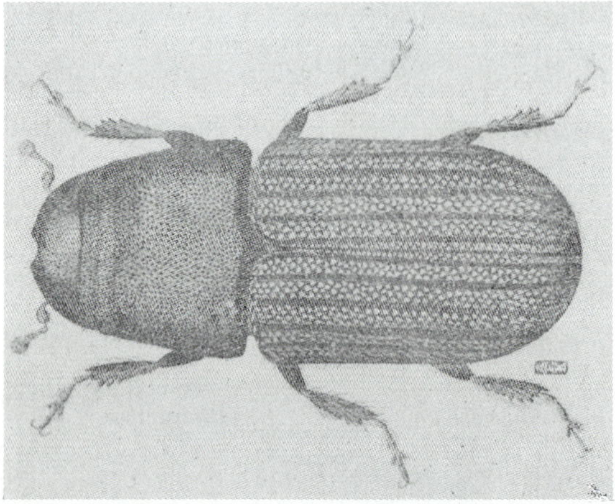

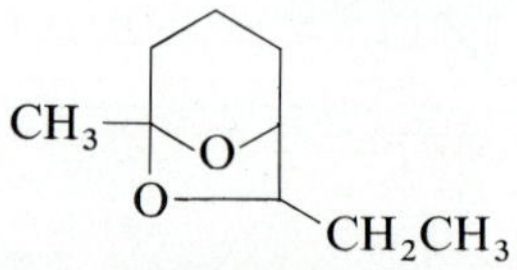
CH_3
O
O
CH_2CH_3

Brevicomin
(sex pheromone of the pine bark beetle)

FIGURE 19.8
The female pine bark beetle bores into a tree to make a nest. When in the tree, she gives off the pheromone brevicomin to attract males to the nest for mating.

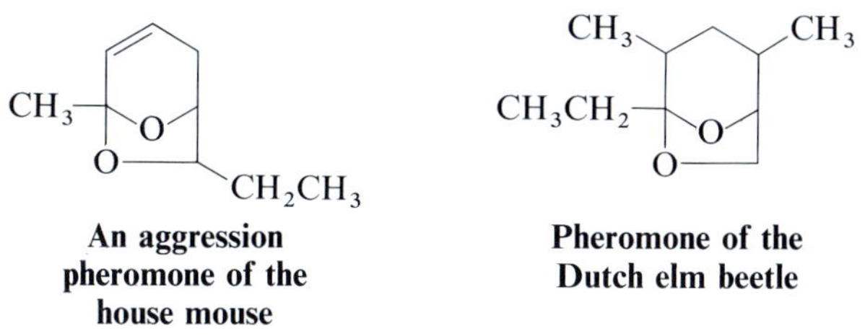
CH3
O
O
CH2CH3
An aggression pheromone of the house mouse
CH3
CH3
CH3CH2
O
O
Pheromone of the Dutch elm beetle

FIGURE 19.9
The common house mouse and the beetle responsible for spreading Dutch elm disease both have pheromones based on a ketal structure.

sponsible for the widespread destruction of the stately elm trees of the eastern part of the United States. (As for the pine bark beetle, the beetles themselves do not kill the elm trees, but carry a fungus that does.)

NEW TERMS

carbohydrate A naturally occurring carbonyl-containing polyhydroxylic compound.

pheromones Chemicals produced by organisms to communicate with other members of the same species.

19.5 Preparation Reactions of Carbonyl Compounds

Oxidation–Reduction Reactions

Aldehydes and ketones are most often formed by the **oxidation** of alcohols. Primary alcohols form aldehydes, while secondary alcohols form ketones. Tertiary alcohols are resistant to oxidation. These general oxidation patterns are shown in Figure 19.10. Typical oxidizing agents include a host of chromium reagents, including CrO_3 and H_2CrO_4. Sometimes just the oxygen in air can carry out an oxidation. In the remainder of this chapter,

Oxidation of primary alcohols gives aldehydes and of secondary alcohols gives ketones. Tertiary alcohols are resistant to oxidation.

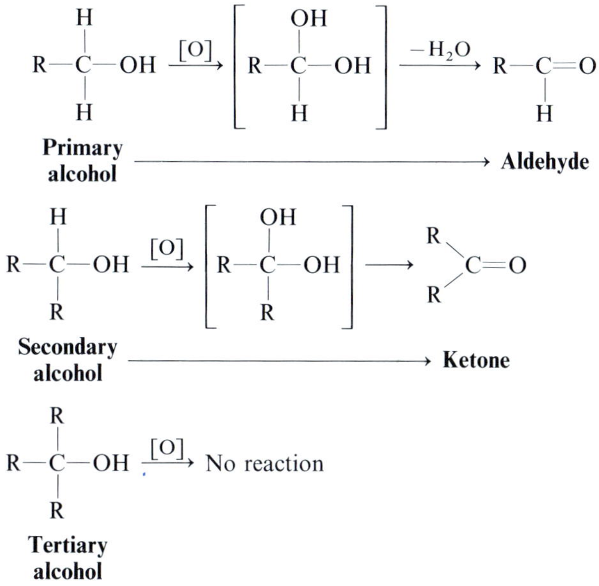
[O]
−H2O
Primary alcohol
Aldehyde
[O]
Secondary alcohol
Ketone
[O]
No reaction
Tertiary alcohol

FIGURE 19.10
A generalized oxidation scheme. This scheme is not intended to show *how* an oxidation takes place. Rather, the purpose is to clearly show the need for hydrogens attached to the carbon if oxidation is to take place.

a "generic" oxidizing agent in a reaction will be designated as [O] over the reaction arrow. The oxidation of ethanol forms the basis of the breath tests given by police to drivers suspected to be "driving while intoxicated." In the road test, a standard solution of orange-yellow dichromate ion ($Cr_2O_7^{2-}$) is the oxidizing agent and the alcohol is the reducing agent. The alcohol is oxidized while the chromium is reduced to a green-colored Cr^{3+} ion. The change in color is measured, and the level of alcohol in the driver is determined.

The road analysis of alcohol content relies on the fact that the ethanol ingested becomes rapidly equilibrated with water in the body. Ethanol easily diffuses through cell membranes, and the partial pressure of ethanol in the breath is a very good measure of the amount in the blood.

The reverse reaction, the **reduction** of carbonyl compounds to alcohols, is the method used for many alcohol preparations. For these reactions, typical reducing agents include sodium borohydride ($NaBH_4$), lithium aluminum hydride ($LiAlH_4$), and H_2/Pt. Note that the catalytic reduction of a carbonyl group is similar to the catalytic reduction of an alkene.

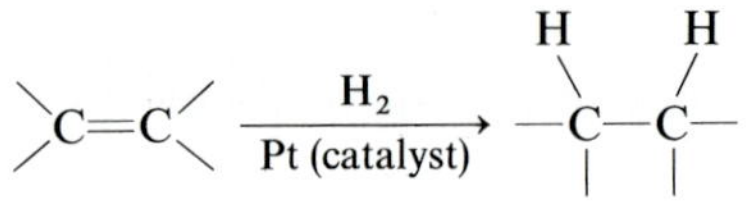

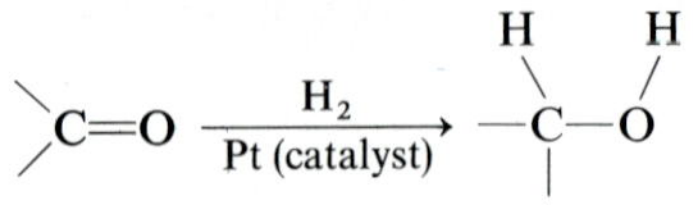

Although $LiAlH_4$ and $NaBH_4$ are both sources of hydride ion (H^-) and will reduce carbonyl compounds, *neither* will reduce an isolated alkene. The reason for this is that the hydride ion acts as a nucleophile and will attack the partially positively charged carbonyl carbon atom. The alkoxide ion formed in this reaction is very basic and will remove the proton from the water that is added after the reaction is over.

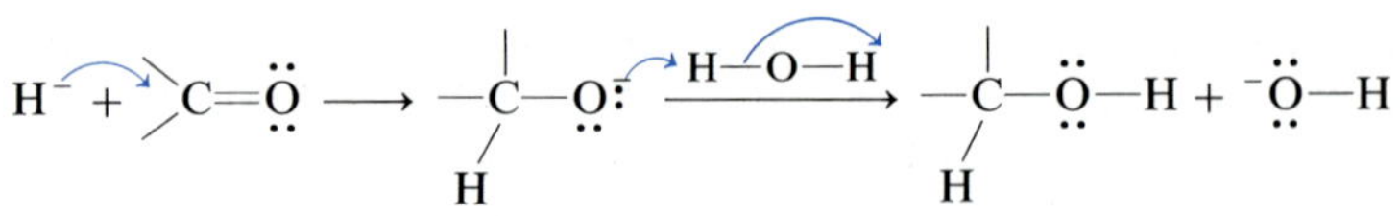

An alkoxide is a stronger base than ^-OH

Just as the general symbol [O] is used for an oxidizing agent, the general symbol [H] is used for a reducing agent. Aldehydes are reduced to primary alcohols, and ketones are reduced to secondary alcohols. This is the reverse of the oxidation process. Thus, the reversible nature of these oxidation–reduction reactions is apparent (Figure 19.11).

Two important points should be reviewed and reinforced at this time. First, every time an oxidation is carried out, there must also be a reduction. Second, in the complete sense, an oxidation is a loss of electrons while a reduction is a gain of electrons. This will become more apparent in other

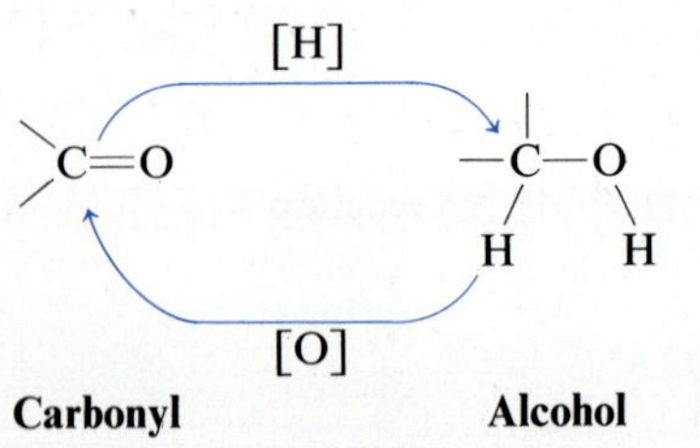
[H]
[O]
Carbonyl
Alcohol

FIGURE 19.11
The reversible nature of the oxidation–reduction process. A carbonyl compound can be reduced to an alcohol and an alcohol can be oxidized to a carbonyl compound.

specific oxidation reactions. However, the concept is very important for biological processes because electron movement in the living organism operates by the oxidation–reduction process. For example, in the important oxidation–reduction processes that keep the body operating, a carbonyl reduction takes place after strenuous exercise. When we become so winded from exertion that we feel we can't continue, we often feel muscle discomfort. This is due to the need for more oxygen to help in the conversion of NADH to NAD^+ (the acronym for the biological oxidizing agent). If more oxygen is not available, the body finds another way to make the conversion—it uses the NADH to reduce pyruvic acid:

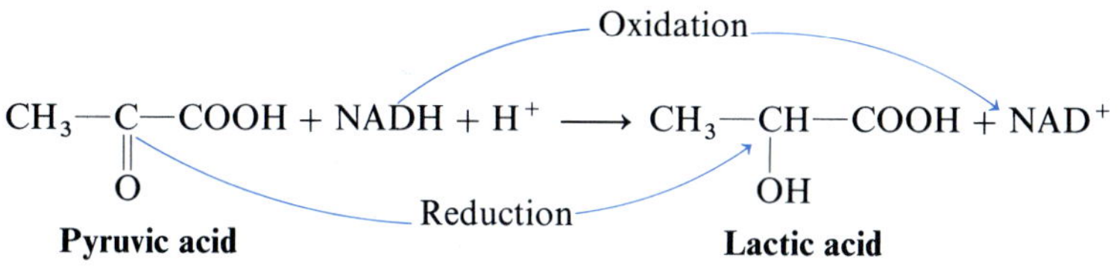
Oxidation
CH3—C—COOH + NADH + H+ ⟶ CH3—CH—COOH + NAD+
O
OH
Reduction

Pyruvic acid **Lactic acid**

The buildup of this *lactic acid* in the muscle tissue is responsible for the muscle pain we experience.

Tests for Aldehydes

Oxidation reactions are easy tests for aldehydes, since aldehydes are easily oxidized to acids.

Aldehydes are reactive toward further oxidation while ketones are resistant. This difference in reactivity towards oxidizing agents provides a test to determine whether a carbonyl compound is an aldehyde or a ketone. In general, the oxidation of an aldehyde can be shown as:

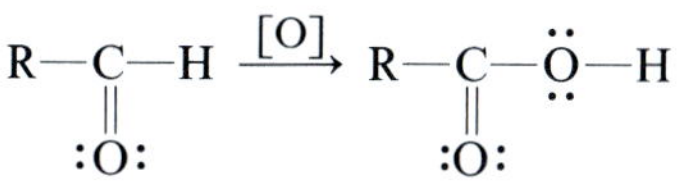
R—C—H [O]⟶ R—C—O—H
:O:
:O:

Aldehyde **Organic acid**

As already mentioned, for every oxidation that takes place, there must also be a reduction. This idea of *coupled reactions* is extremely important for biological processes.

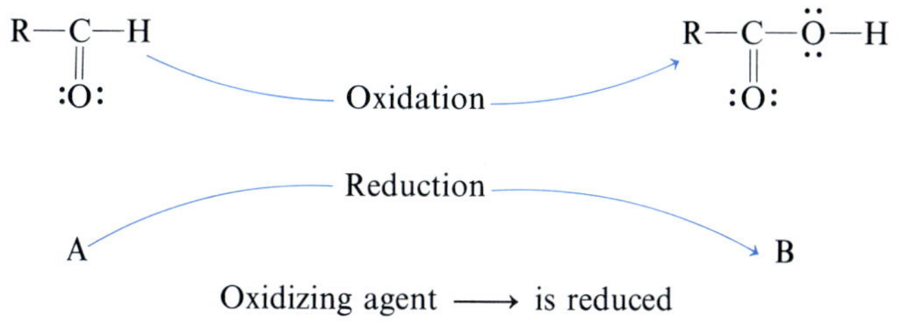
R—C—H
:O:
Oxidation
R—C—O—H
:O:
Reduction
A
B
Oxidizing agent ⟶ is reduced

One example of this type of reaction in biochemistry is the conversion of 3-phosphoglyceraldehyde to 1,3-diphosphoglyceric acid. In this reaction the biological oxidizing agent NAD^+ is reduced to $NADH + H^+$.

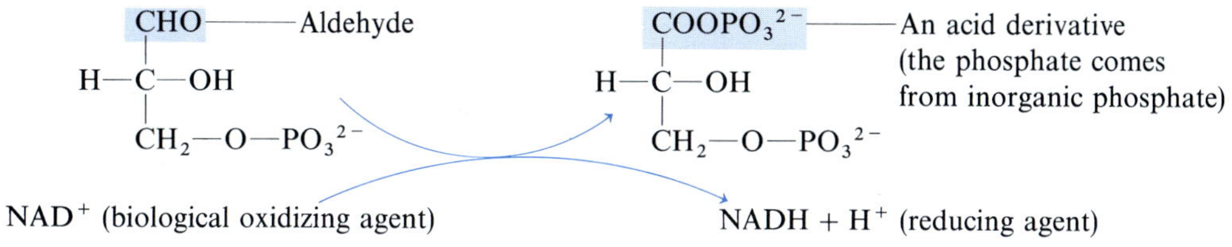
CHO——Aldehyde
H—C—OH
CH2—O—PO3 2−
COOPO3 2−——An acid derivative (the phosphate comes from inorganic phosphate)
H—C—OH
CH2—O—PO3 2−
NAD+ (biological oxidizing agent)
NADH + H+ (reducing agent)

TOLLENS TEST The Tollens test is a simple test for aldehydes. In this procedure, the silver ion in $Ag(NH_3)_2^+$ is the oxidizing agent.

Oxidation

$$R{-}\underset{\underset{O}{\|}}{C}{-}H + AgNO_3 + NaOH + NH_3 \longrightarrow R{-}\underset{\underset{O}{\|}}{C}{-}O^-Na^+ + Ag^0$$

Reduction

Tollens test for aldehydes

A positive result is easily observed for this test because a perfect film of silver coats the inside of the test tube. In this example the Ag^+ has gained electrons (been reduced) to form metallic silver, while the aldehyde has been oxidized to an acid.

BENEDICT'S OR FEHLING'S TEST FOR ALDEHYDES A very similar idea of oxidation–reduction is applied in **Benedict's** or **Fehling's tests**. In these visual tests, the deep blue color of Cu^{2+} disappears as the aldehyde is oxidized to an acid and the copper is reduced to Cu^+. The copper(I) ion gives a very characteristic brick-red color as it forms an insoluble copper(I) oxide. (See Plate 7 in color inserts.) The two tests differ only in the identity of the complex anion that is associated with the Cu^{2+}. Citrate is the anion of citric acid; tartrate is the anion of tartaric acid. Organic acids are discussed in the next chapter.

$$R{-}\underset{\underset{O}{\|}}{C}{-}H \quad + [Cu^{2+}\,(\text{complex})] \longrightarrow R{-}\underset{\underset{O}{\|}}{C}{-}O{-}H + Cu_2O$$

Benedict's reagent	(citrate)	(red)
Fehling's reagent	(tartrate)	

Aldehyde
Acid
Cu^{2+}
Cu^+

Benedict's and Fehling's test for aldehydes

We saw from Table 19.1 that glucose (a common sugar) contains an aldehyde group. One of the simple tests for diabetes (discussed in detail in Chapter 30) is an analysis for glucose in the urine. This is easily done with reagents that give certain color changes. As the concentration of copper(II) ions (blue color) decreases and the color of Cu(I) ion increases (red color), solutions will pass through a color change continuum. This can be used as an indication of the amount of sugar in urine (oxidation of glucose) and can help screen for diabetes.

Color	Yellow-green ⟷ Red
Percentage glucose	0.5% ⟷ 2% (or greater)

The preparation of wine vinegar is simply an example of the oxidation of an alcohol (ethanol) to an aldehyde (acetaldehyde or ethanal) and then further to an acid (acetic or ethanoic acid). The same process occurs in our bodies after alcohol consumption.

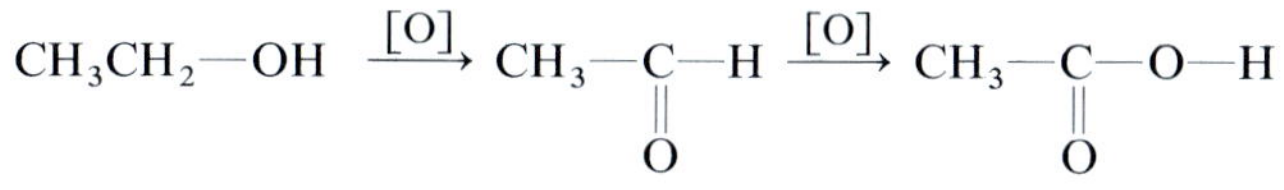

Ethanol (primary alcohol) — **Acetaldehyde** (aldehyde) — **Acetic acid** (acid)

NEW TERMS

oxidation The loss of electrons; in carbonyl chemistry, it relates to the reaction:

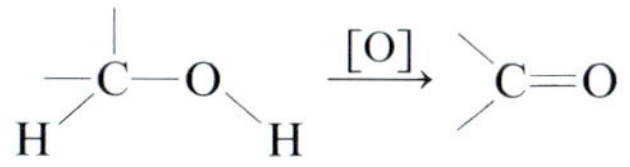

reduction The gaining of electrons; in carbonyl chemistry, it refers to the process:

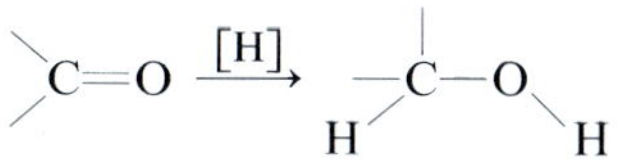

Tollens tests A test for aldehydes using a silver ion as the oxidizing agent. As the aldehyde is oxidized to an acid, the silver ion is reduced to free silver.

Benedict's or Fehling's test A test for aldehydes using copper ion as the oxidizing agent.

TESTING YOURSELF

Reactions of Carbonyl Compounds

1. Give products for each of the following oxidation reactions.

a. 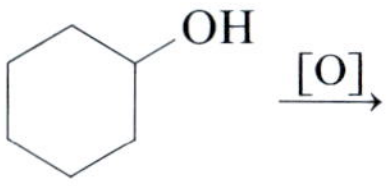

b. 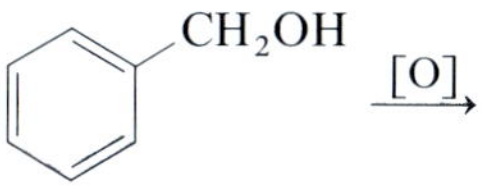

c. 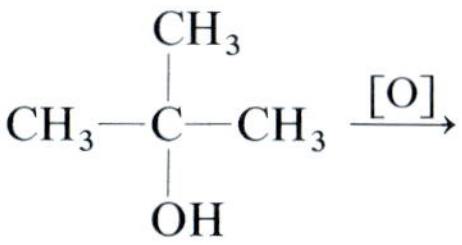

2. Give products for each of the following reduction reactions.

a. 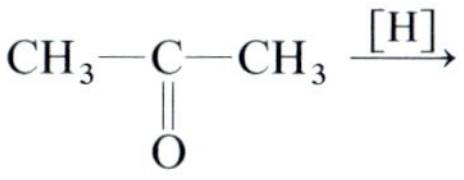

b. 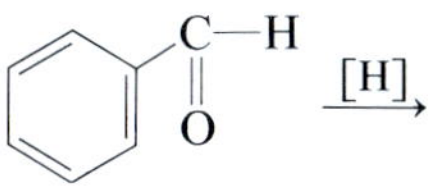

3. Show the acetal or ketal resulting from the following reactions.

a.
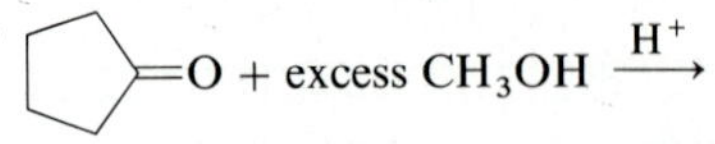

=O + excess CH_3OH
H^+

b.
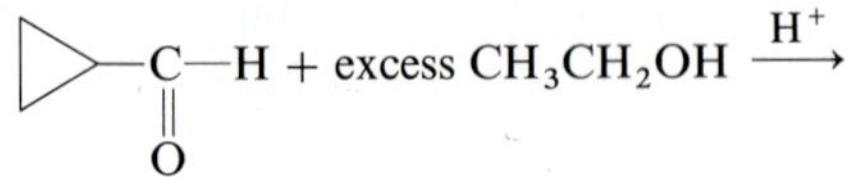

C—H + excess CH_3CH_2OH
H^+
O

Answers **1. a.** cyclohexanone **b.** benzaldehyde **c.** no reaction **2. a.** 2-propanol **b.** benzyl alcohol **3. a.**
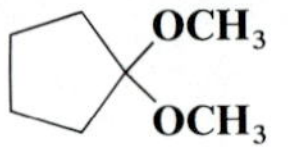

OCH_3
OCH_3

b.
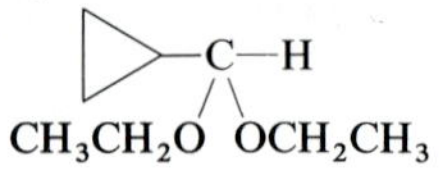

C—H
CH_3CH_2O OCH_2CH_3

19.6 Oxidation of Hydroquinones to Quinones

Quinones are formed from the oxidation of hydroquinones.

Phenols will not readily oxidize; however, *p*-hydroxyphenol, commonly known as **hydroquinone**, can be easily oxidized to **quinone**. Oxidation (and its reverse, reduction) is common to biological systems. Hydroquinone is

ANTIOXIDANTS

Many foods rich in fats and oils become spoiled easily because of oxidation (particularly unsaturated oils because the double bonds are very reactive). Although the reaction processes by which these reactions take place are quite complex, simply stated atmospheric oxygen causes these reactions. To prevent the reactions from taking place, *antioxidants* are sometimes added to food as preservatives. Antioxidants are compounds that inhibit oxidation by molecular oxygen (autoxidation). Phenolic compounds and hydroquinone compounds are especially useful in this regard.

The use of antioxidants to preserve food is not new, only the understanding of their chemical action is. Since ancient times, spices and herbs have been used to help in the safe storage of food. It is interesting to note that many of these natural materials are pleasant smelling—indeed, the *aromatic* spices such as clove, oregano, thyme, rosemary, and sage contain phenolic compounds that were among the very early compounds associated with the aromatic class of hydrocarbons. Even the process of smoking meats provides antioxidants to the food because formaldehyde, as well as phenolic compounds, are found in wood smoke. Antioxidants are also present in animal tissue to prevent oxidation of the natural fats in the body. There is some thought that the "aging process" is accelerated by the oxidation of natural double bonds.

You need not become concerned that the "sea" of oxygen around us is destructive, because the complex reactions involved in autoxidation processes do not use oxygen as it naturally occurs in the atmosphere. However, the pollutants in the environment are capable of initiating some of these reactions, and this is one of the reasons why there is so much concern about atmospheric contamination.

In addition to food spoilage, there is another common autoxidation problem. Most of us drive a car and have probably heard of the fuel delivery system getting "gummed up." This is precisely what happens when the hydrocarbons in fuel undergo autoxidation. Under the high temperature conditions in the combustion chamber, the lubricating oil undergoes this process, accelerating damage to the engine. Natural rubber, synthetic rubber, and most plastics are attacked by oxygen in much the same way. We find that when antioxidants are added to these materials, the structural damage resulting from oxidation is greatly reduced.

often added to commercial products to inhibit their oxidation since it is more reactive to oxidizing agents than most other functional groups. The compound *t*-butylhydroquinone (TBHQ) is added as an antioxidant to many snack foods, fats, and oils.

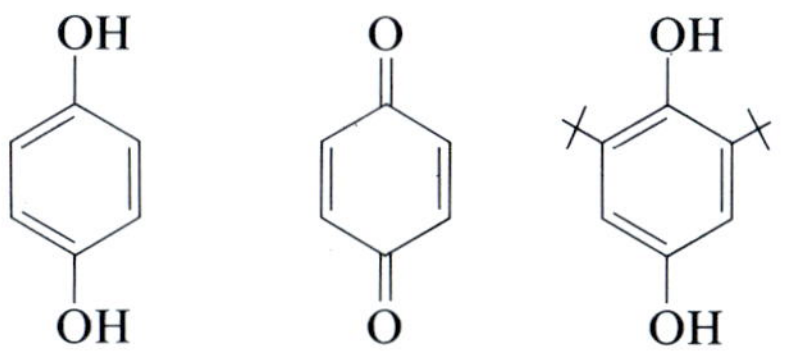
OH
O
OH
OH
O
OH

Hydroquinone **Quinone** **TBHQ**

The reaction of an oxidation of hydroquinone and the reverse reduction of a quinone are shown below.

GENERAL REACTION

Oxidation–Reduction of Hydroquinone and Quinone

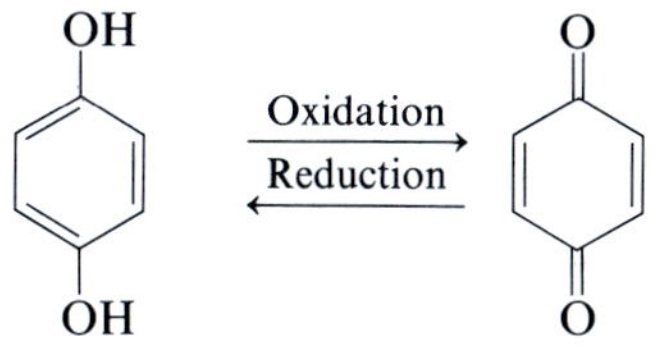
OH
Oxidation
Reduction
O
OH
O

Hydroquinone Quinone

Examples

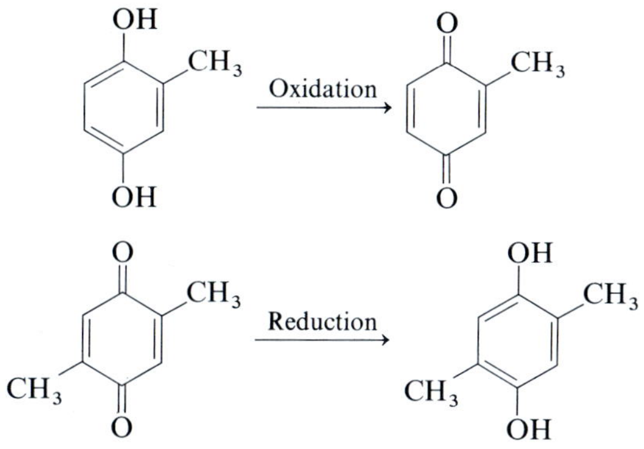
OH
CH_3
Oxidation
O
CH_3
OH
O
O
CH_3
Reduction
OH
CH_3
CH_3
O
CH_3
OH

Quinones in Our Everyday Life

As part of the process of respiration (biological oxidation), oxygen must be converted to water. If we keep track of the electrons involved in this process, we can see that the oxygen has been reduced. This requires that something else has been oxidized. In the biochemistry chapters of this book, it will be noted that carbon compounds are oxidized to carbon dioxide.

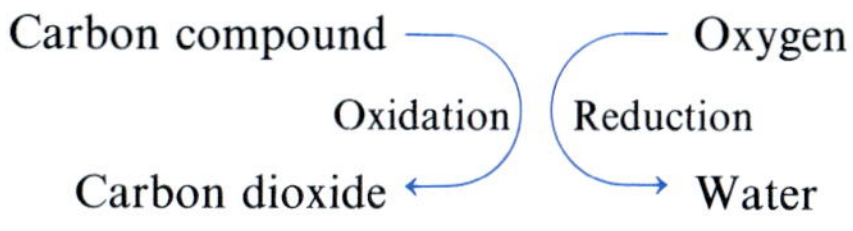
Carbon compound
Oxygen
Oxidation
Reduction
Carbon dioxide
Water

The electrons from the oxidation process get carried down a chain of secondary oxidations and reductions, including a $Fe^{2+} \rightleftharpoons Fe^{3+}$ equilibrium and a hydroquinone ⇌ quinone equilibrium. In a section on photosynthesis in Chapter 29, the same conversion will be seen, but in the reverse sense. Water will be oxidized to oxygen, and the electrons will be used to reduce carbon dioxide into organic compounds.

Vitamin K is important in the maintenance of our blood-clotting ability. Because of its obvious hydrocarbon nature, Vitamin K is a fat-soluble vitamin; it is also a quinone.

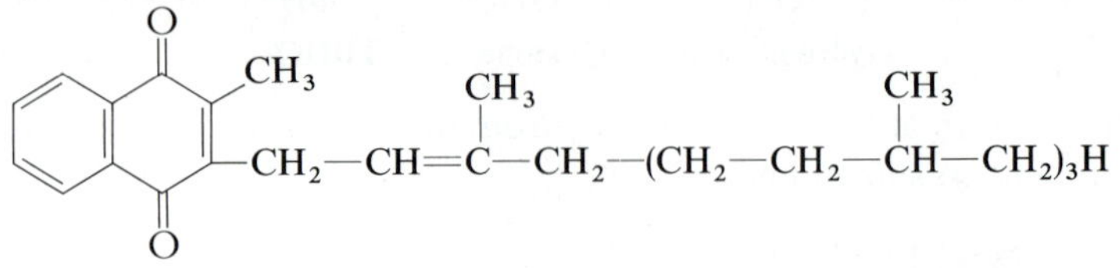

Vitamin K

The Bombardier Beetle Uses Quinones

A particularly interesting example of oxidation of a hydroquinone to a quinone can be found in the defense mechanism of the bombardier beetle (Figure 19.12). This beetle contains separate supplies of hydrogen peroxide and hydroquinone in special storage units in its body. When attacked, the beetle mixes these two chemicals and a powerful exothermic reaction takes place. The oxygen formed from the hydrogen peroxide is then used as a propellant to force out the hot quinone, which has been recorded at a temperature of 100 °C (212 °F)!

FIGURE 19.12
The bombardier beetle has a unique defensive spray that has its origins in the hydrogen peroxide oxidation of hydroquinone to quinone. The exothermic nature of the reaction and the formation of an oxygen-based propellant system makes this a very effective defense.

FIGURE 19.13
The effects of the bombardier beetle's spray on a frog is graphic evidence for the quality of this unique defense system.

The photograph in Figure 19.13 shows the after effects of this spray on a frog that has caught a bombardier beetle. Little comment is needed because the picture speaks for itself. The frog is no longer a threat.

NEW TERMS

hydroquinone A common name for *p*-hydroxyphenol, an effective antioxidant.

quinone A common name for 2,5-cyclohexadiene-1,4-dione, the oxidation product from hydroquinone.

TESTING YOURSELF

Hydroquinones and Quinones

1. Show what happens to TBHQ when it is oxidized.
2. What might be a reasonable structure for the product resulting from reduction of vitamin K?

Answers 1. 2.

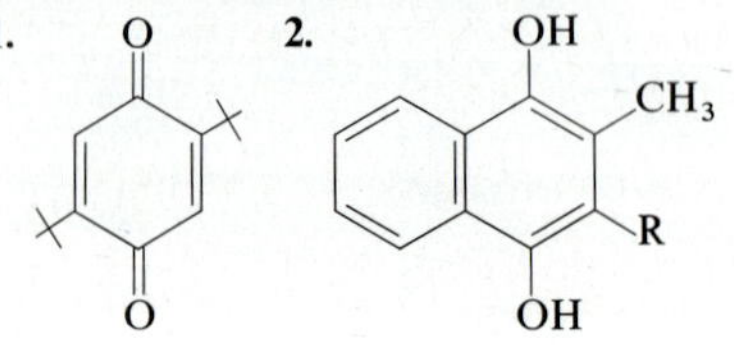

O
O
OH
CH3
R
OH

19.7 Aldol Condensation

The Aldol condensation combines two carbonyl compounds to form a beta-hydroxy carbonyl compound.

Hydrogen atoms on the sigma-bonded carbon of a carbonyl compound are more acidic than one would expect. This hydrogen is readily removed, and the resulting ion is an efficient nucleophile. This results in a very common reaction for carbonyl compounds—the **Aldol condensation**. In a very simple sense, the hydrogen is removed from one molecule of a carbonyl compound, and the resulting ion can attack the carbonyl carbon of a second molecule. The result of this reaction is a beta-hydroxy carbonyl compound (Figure 19.14).

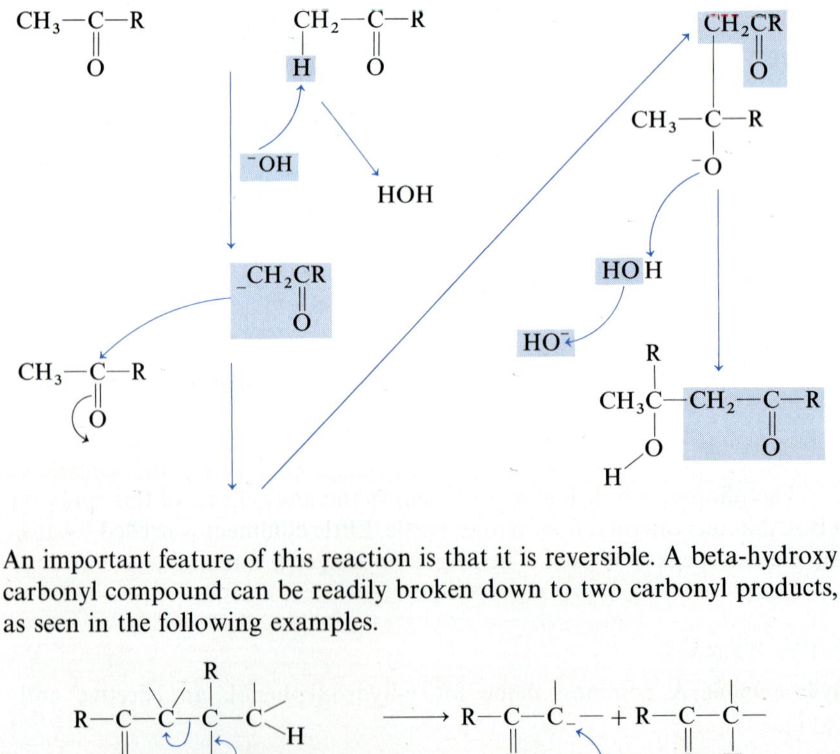

CH3—C—R
O
CH2—C—R
H
O
⁻OH
HOH
CH2CR
O
CH3—C—R
O
CH2CR
O
CH3—C—R
⁻O
HOH
HO⁻
R
CH3C—CH2—C—R
O
H
O

FIGURE 19.14
The Aldol condensation. In this reaction one molecule (highlighted) becomes attached to a second molecule. The acidic hydrogen on an alpha hydrogen is removed by base, and the resulting anion attacks the carbonyl carbon of the second molecule.

An important feature of this reaction is that it is reversible. A beta-hydroxy carbonyl compound can be readily broken down to two carbonyl products, as seen in the following examples.

R
R—C—C—C—C
H
O
O—H
⁻OH
R—C—C⁻ + R—C—C—
O
O H
Can pick up H⁺

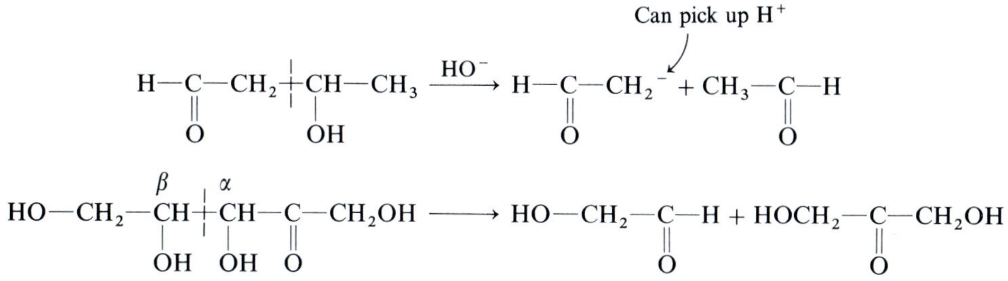

$$\text{H}-\underset{\underset{\text{O}}{\|}}{\text{C}}-\text{CH}_2\!+\!\underset{\underset{\text{OH}}{|}}{\text{CH}}-\text{CH}_3 \xrightarrow{\text{HO}^-} \text{H}-\underset{\underset{\text{O}}{\|}}{\text{C}}-\text{CH}_2^- + \text{CH}_3-\underset{\underset{\text{O}}{\|}}{\text{C}}-\text{H}$$

Can pick up H$^+$

$$\text{HO}-\text{CH}_2-\underset{\underset{\text{OH}}{|}}{\overset{\beta}{\text{CH}}}\!+\!\underset{\underset{\text{OH}}{|}}{\overset{\alpha}{\text{CH}}}-\underset{\underset{\text{O}}{\|}}{\text{C}}-\text{CH}_2\text{OH} \longrightarrow \text{HO}-\text{CH}_2-\underset{\underset{\text{O}}{\|}}{\text{C}}-\text{H} + \text{HOCH}_2-\underset{\underset{\text{O}}{\|}}{\text{C}}-\text{CH}_2\text{OH}$$

The Aldol condensation reaction is the prime method of breaking down six-carbon sugars to three-carbon units (see Chapter 28 for more examples).

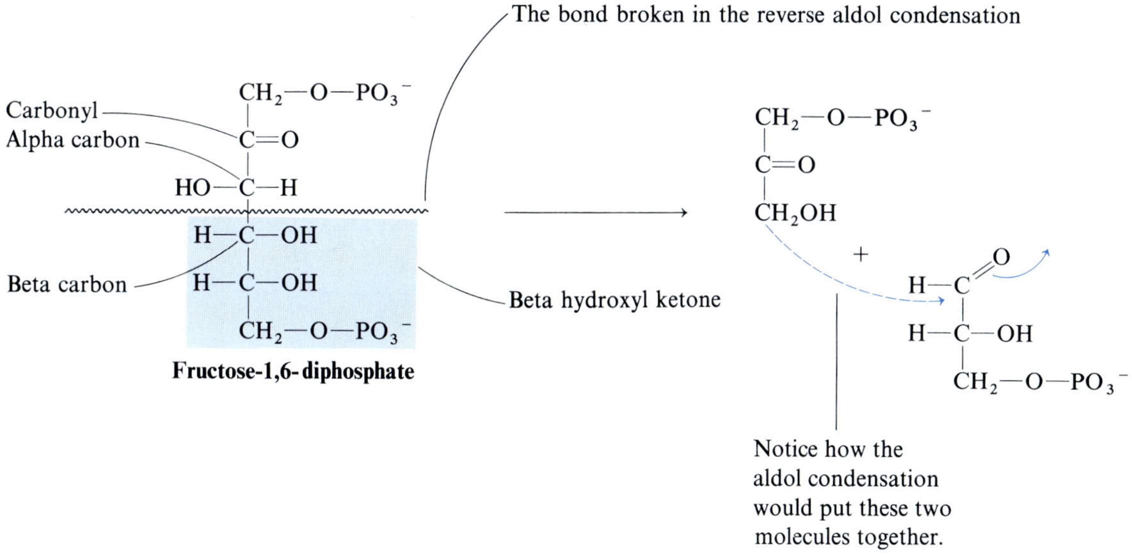

The bond broken in the reverse aldol condensation

Carbonyl
Alpha carbon

$$\begin{array}{c} \text{CH}_2-\text{O}-\text{PO}_3^- \\ | \\ \text{C}=\text{O} \\ | \\ \text{HO}-\text{C}-\text{H} \\ | \\ \text{H}-\text{C}-\text{OH} \\ | \\ \text{H}-\text{C}-\text{OH} \\ | \\ \text{CH}_2-\text{O}-\text{PO}_3^- \end{array}$$

Beta carbon

Fructose-1,6-diphosphate

Beta hydroxyl ketone

$$\begin{array}{c} \text{CH}_2-\text{O}-\text{PO}_3^- \\ | \\ \text{C}=\text{O} \\ | \\ \text{CH}_2\text{OH} \end{array}$$

+

$$\begin{array}{c} \text{H}-\text{C}\overset{\text{O}}{\diagup} \\ | \\ \text{H}-\text{C}-\text{OH} \\ | \\ \text{CH}_2-\text{O}-\text{PO}_3^- \end{array}$$

Notice how the aldol condensation would put these two molecules together.

NEW TERM

Aldol condensation A reaction in which two molecules of a carbonyl containing molecule react to form a beta-hydroxy carbonyl compound.

SUMMARY

The greater electronegativity of oxygen dominates the chemical behavior of the carbon–oxygen double bond of the *carbonyl group*. In reactions, the oxygen atom appears as though it has a negative charge and reacts with electrophiles. The carbon atom of the carbonyl group appears as though it has a positive charge and reacts with nucleophiles. Reactions of the carbonyl double bond include *reduction*, *hydration*, *cyanohydrin* formation, and ketal formation. In general, the addition reactions of the carbonyl group resemble those of alkenes, with an important difference due to the electronegativity differences between the oxygen and the carbon.

Acetals and *ketals* are important derivatives of the carbonyl group and are formed by the reaction of the carbonyl compound with alcohol in the presence of an acid catalyst. With only one equivalent of alcohol, *hemiacetals* and *hemiketals* are formed, respectively. With two equivalents of alcohol, acetals and ketals are produced from the same reaction. These derivatives of carbonyl compounds show up frequently as natural products, particularly sugars.

REACTION SUMMARY

Addition of an Electrophile and Nucleophile to a Carbonyl

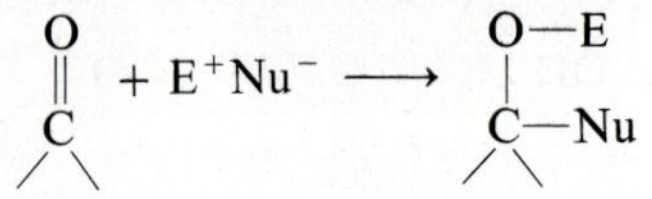
+ E⁺Nu⁻
O—E
C—Nu

Cyanohydrin Formation

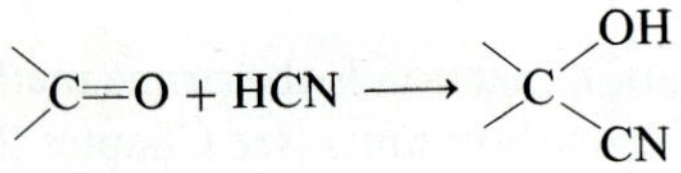
C=O + HCN
OH
CN

Hydrate Formation

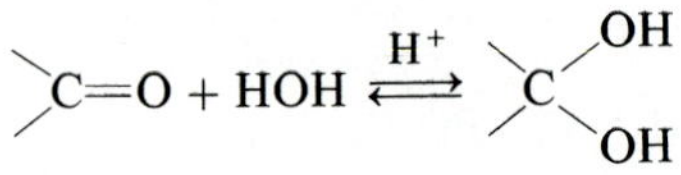
C=O + HOH
H⁺
OH
OH

Carbonyl Reduction

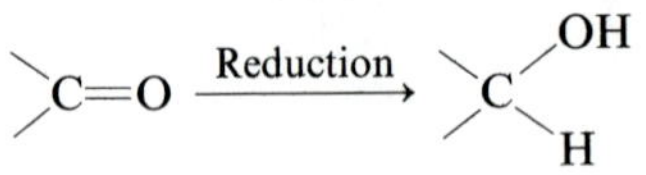
C=O
Reduction
OH
H

Acetal and Ketal Formation

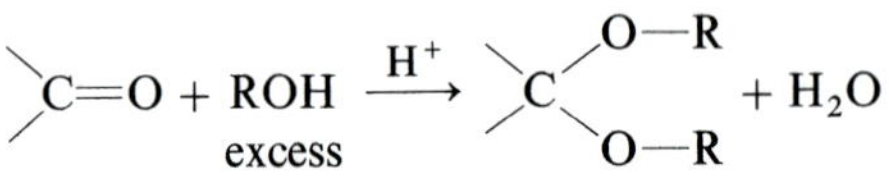
C=O + ROH
excess
H⁺
O—R
O—R
+ H₂O

Alcohol Oxidation

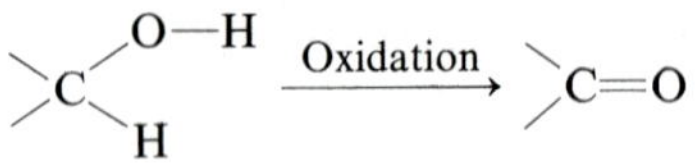
O—H
H
Oxidation
C=O

Aldehyde Oxidation

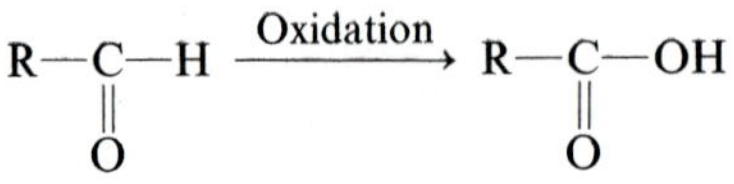
R—C—H
Oxidation
R—C—OH

Oxidation of Hydroquinones

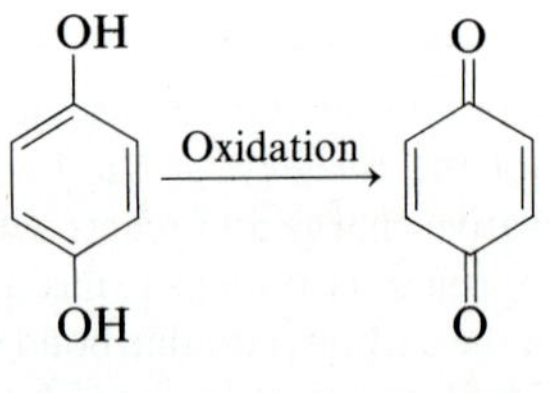
OH
OH
Oxidation
O
O

Reduction of Quinones

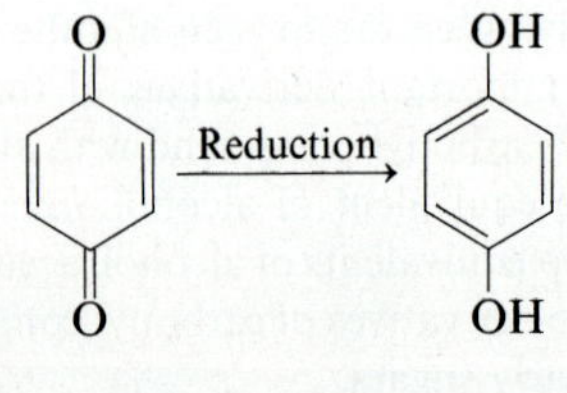
O
O
Reduction
OH
OH

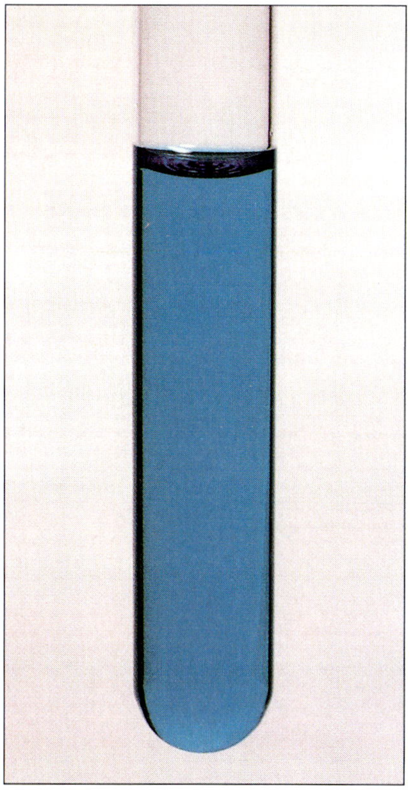

(a)

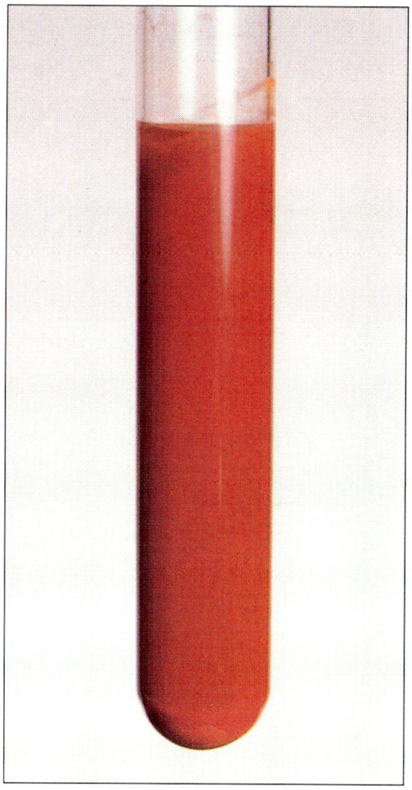

(b)

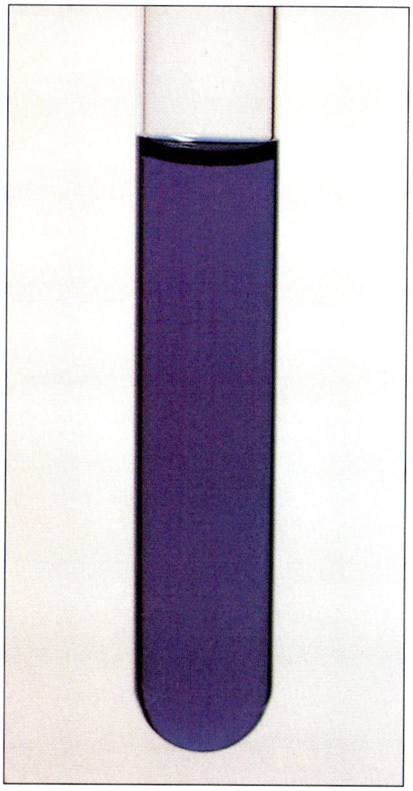

(c)

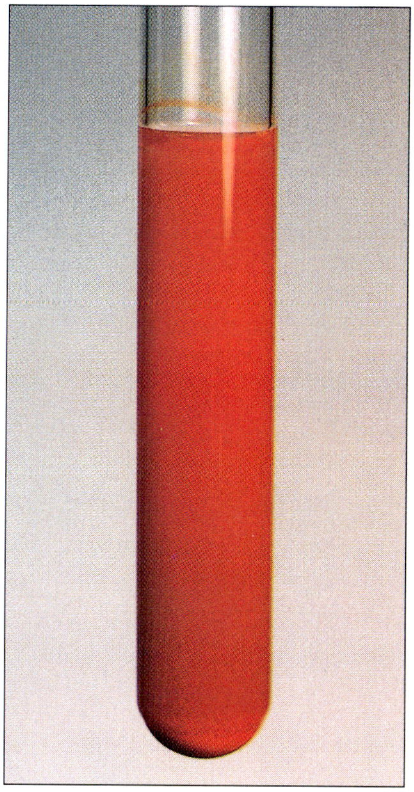

(d)

Plate 7

Blue solutions of Cu(II) ions found in (a) Benedict's and (c) Fehling's solutions turn red (b and d) in the presence of aldehydes, as the copper is reduced to Cu(I). While copper is reduced, the aldehyde is oxidized to a carboxylic acid. (See Chapter 19.)

GROUP CAMPS
BY RESERVATION ON

Plate 8

(Top) Bark beetles, attracted by sex pheromones, cause massive destruction of pine forests in Yellowstone National Park. The beetle carries a fungus into the tree that is the actual cause of tree death. (Bottom) Acres of standing dead wood provided rich and plentiful fuel for the Yellowstone inferno of 1988. (See Chapter 19.)

Plate 9

These Giant Sequoia are a prime example of the use of cellulose to provide a strong structure in plants. When cellulose is embedded in the polymer lignin, the complex is strong, stress resistant, and durable. Cellulose is a polysaccharide, a type of carbohydrate. (See Chapter 23.)

Plate 10

(Below) Honey bees break down sucrose in nectar to yield the fructose and glucose found in honey. They use waxes as building material for combs and as a waterproof layer on their exterior surface. Bee's wax from combs is a mixture of esters, while the surface lipids of bees are primarily hydrocarbons. (See Chapters 23 and 24.)

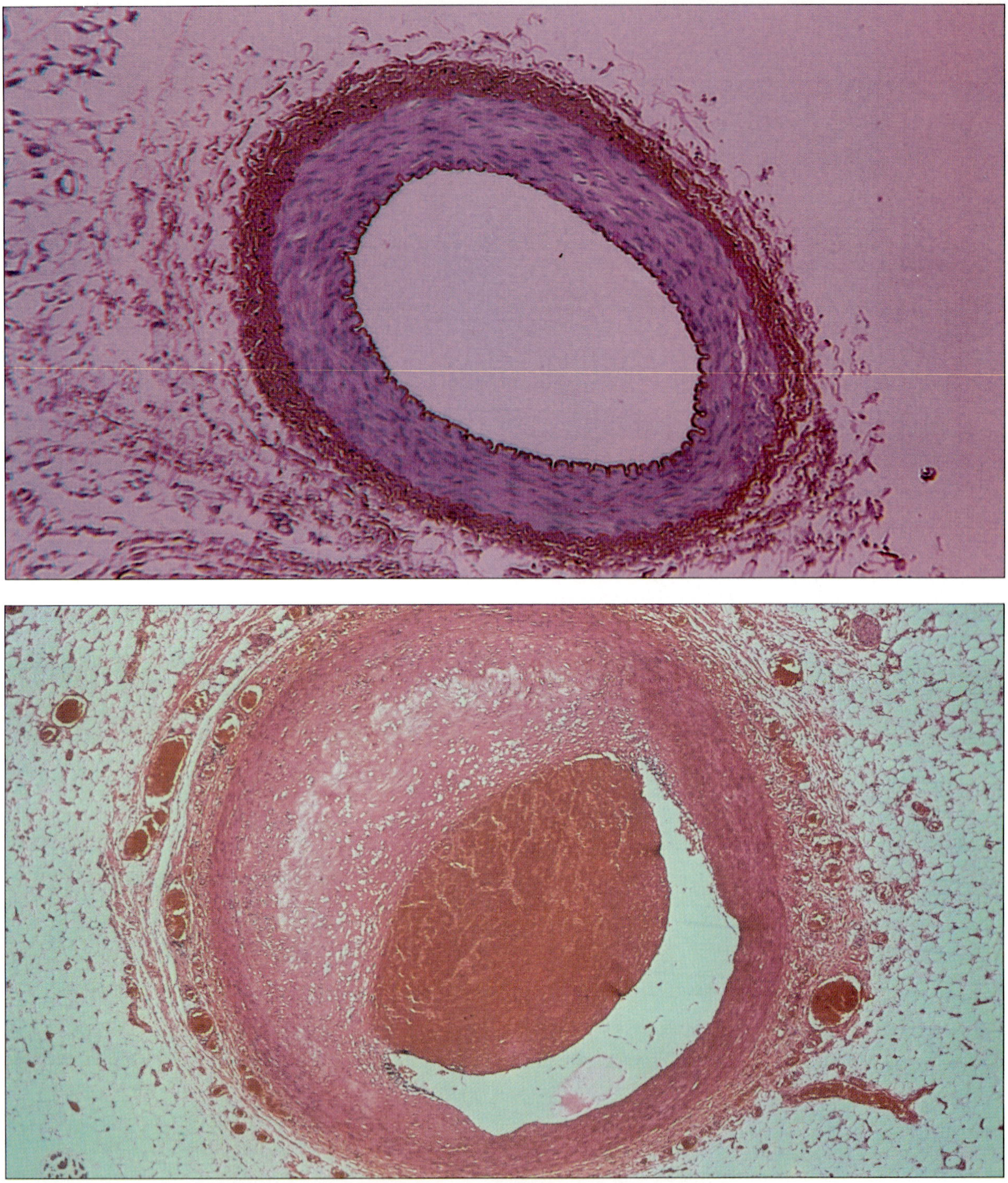

Plate 11

A comparison between a normal artery (top) and one nearly blocked with atherosclerotic plaque and calcified wall (bottom). Blood flow in a normal artery is unrestricted and free flowing. Normal blood flow is significantly reduced in the other artery and may be totally blocked by a blood clot. The atherosclerotic deposits include cholesterol and other lipids. (See Chapter 24.)

TERMS

carbonyl group
aldehyde
ketone
dipole–dipole force
cyanohydrin
hydrate
hemiacetal
hemiketal
acetal
ketal
Le Chatelier's principle
carbohydrates
pheromones
intramolecular reaction
oxidation
reduction
Tollens test
Benedict's or Fehling's test
hydroquinone
quinone
Aldol condensation

CHAPTER REVIEW

QUESTIONS

1. Carbonyl carbon reactivity is best understood by using which representation?
   a. $>C{=}O$ b. $>\overset{+}{C}{-}\overset{-}{O}$ c. $>\overset{-}{C}{-}\overset{+}{O}$ d. none of these
2. Which of these compounds will *not* be easily oxidized?
   a. aldehyde c. secondary alcohol
   b. primary alcohol d. tertiary alcohol
3. The compound $CH_3{-}CH_2{-}CH_2{-}CH_2{-}\underset{\underset{O}{\|}}{C}{-}CH_3$ is best named:
   a. 5-hexanal c. 1-methylpentanal
   b. 5-hexanone d. 2-hexanone
4. The oxidation of cyclopentanol will give
   a. no reaction c. cyclopentene
   b. cyclopentanone d. cyclopentanal
5. The carbon of a carbonyl carbon is ________ hybridized.
   a. *sp* b. sp^2 c. sp^3 d. not
6. Aldehydes are oxidized to
   a. alcohols b. ketones c. acetals d. acids
7. The reaction of ethanal with one equivalent of methanol and a trace of an acid will give
   a. acetal b. ketal c. hemiacetal d. hemiketal
8. The reduction of cyclohexanone with $NaBH_4$ will give
   a. an organic acid b. an alcohol c. an aldehyde d. a hemiketal
9. When the alpha proton is removed from a carbonyl compound, the resulting anion is *not*
   a. a nucleophile c. a Lewis acid
   b. a Lewis base d. an enolate
10. Aldehydes will have boiling points that are ________ the corresponding primary alcohols.
    a. lower than b. about the same as c. higher than

Answers **1.** a **2.** d **3.** d **4.** b **5.** b **6.** d **7.** c **8.** b **9.** c **10.** a

DIAGNOSTIC CHART

Blacken in all of the circles under the number of each question that you missed in the preceding Chapter Review questions. The diagnostic chart will help you identify concept areas that need more study.

	Questions									
Concepts	1	2	3	4	5	6	7	8	9	10
Structure	○				○					
Nomenclature			○	○				○		
Reactions	○	○		○		○	○	○	○	
Physical properties										○
Acetals and ketals							○			

EXERCISES

Structure and Physical Properties

1. Compare the structures of 2-propanone and 2-methylpropene. Which would have the greater dipole? Why?
2. For each of the molecules in question 1, show where an H^+ would become preferentially attached.
3. Rank the following compounds according to boiling point (lowest first).
   a. pentanal c. 1-pentanol
   b. 1-chloropentane d. pentane
4. Why do ethene molecules not interact with each other as well as with molecules of methanal?

Nomenclature

5. Give names to each of the following compounds.

a. 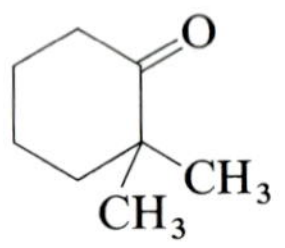

e. 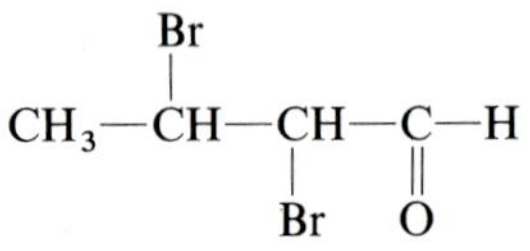

b. 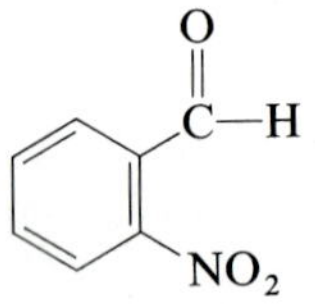

f. 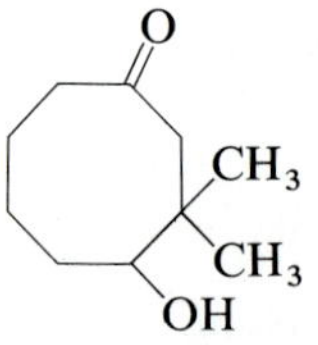

c. 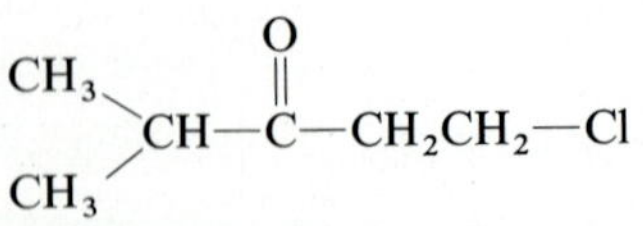

d. 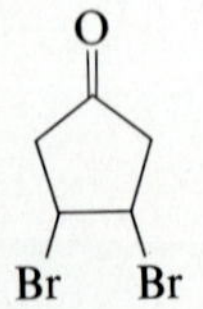

g. 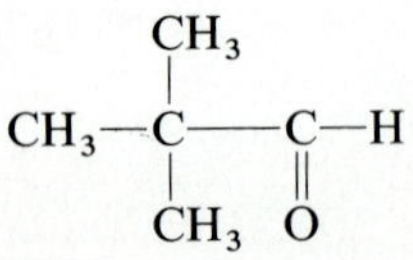

6. Give structures for each of the following compounds.
   a. 5-bromopentanal
   b. 3,4-dimethylcyclohexanone
   c. *p*-chlorobenzaldehyde
   d. 2,2,5,5-tetramethyl-4-heptanone.
7. Give names for each of the following cyclic ketones.

a. 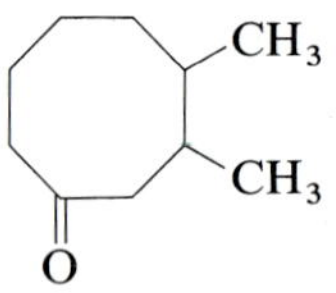

f. 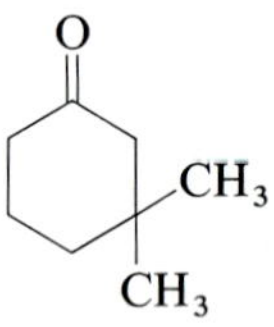

b. 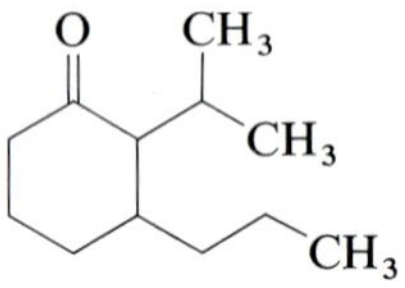

g. 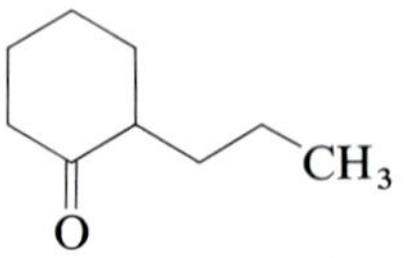

c. 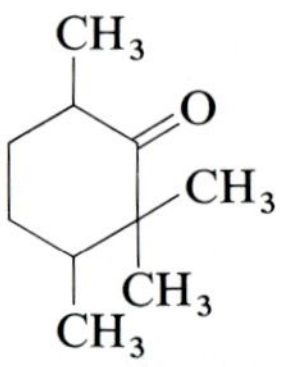

h. 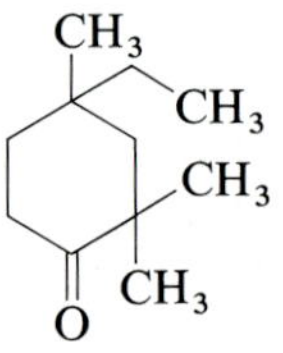

d. 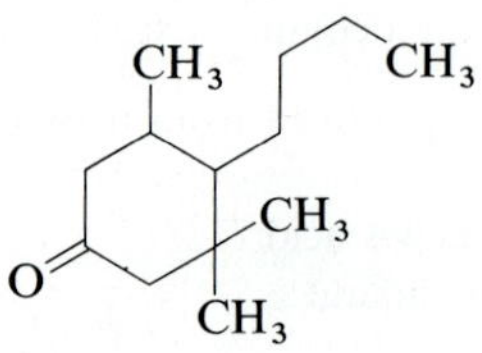

i. 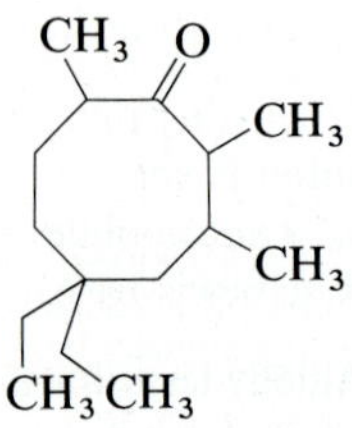

e. 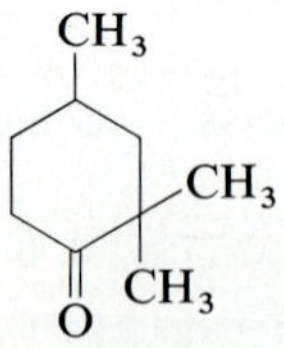

8. Give names for each of the following aldehydes.

a. 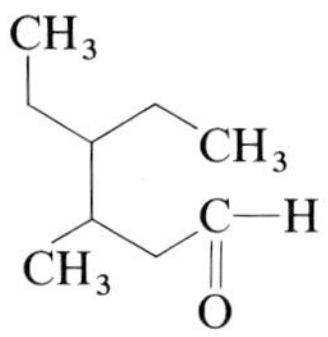

b. 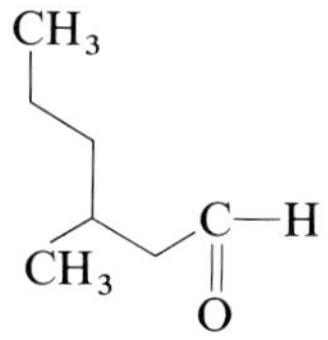

c. 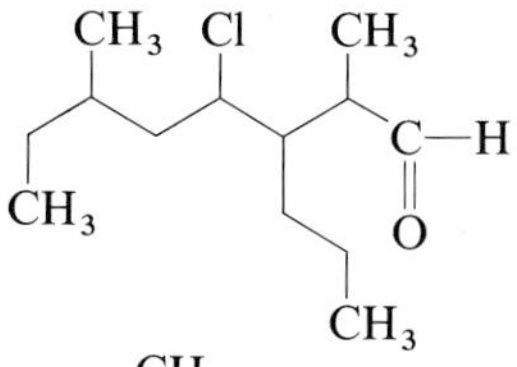

d. 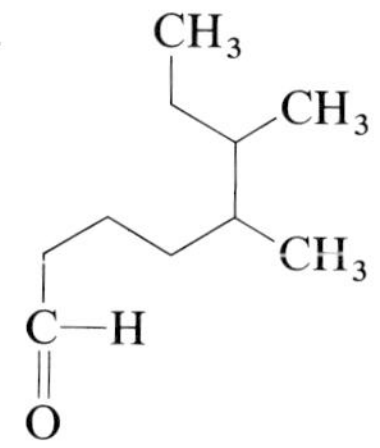

e. 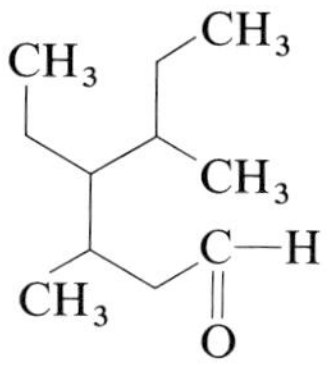

f. 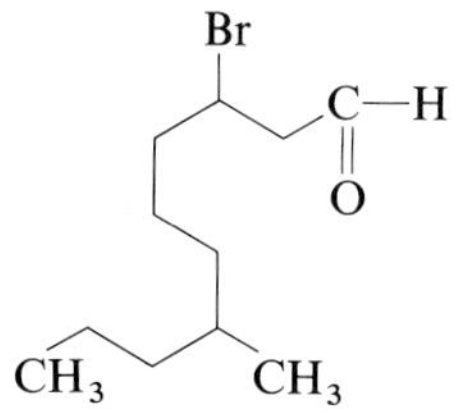

9. Give names for each of the following aldehydes.

a. 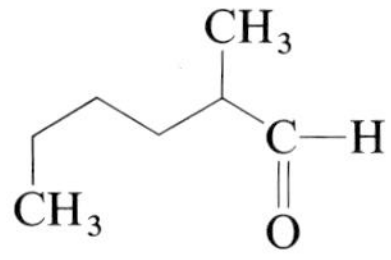

b. 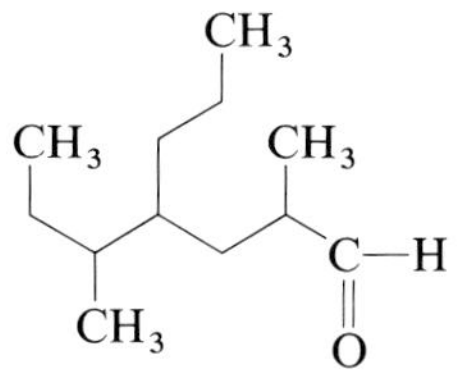

c. 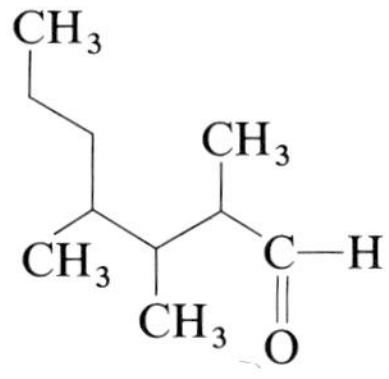

d. 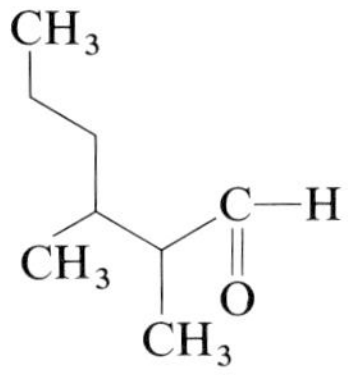

e. 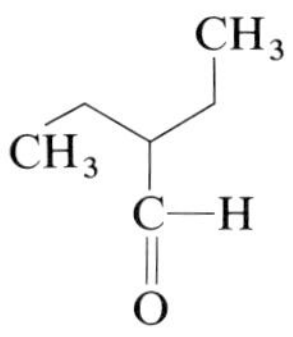

f. 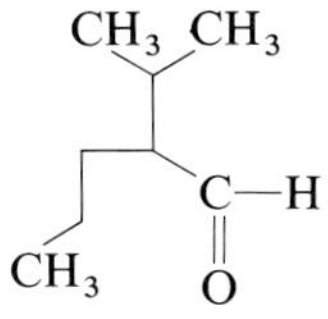

10. Give names for each of the following ketones.

a. 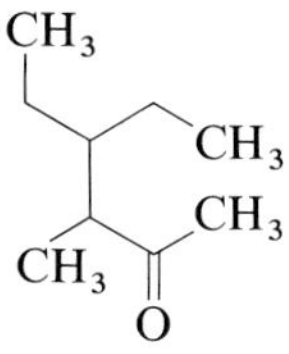

b. 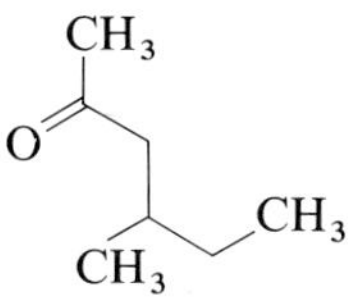

c. 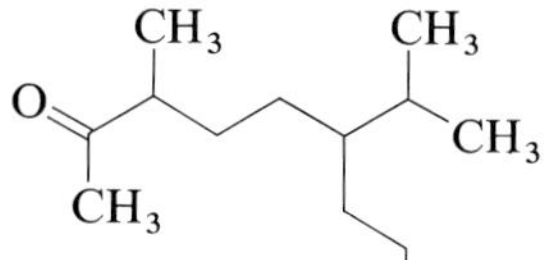

d. 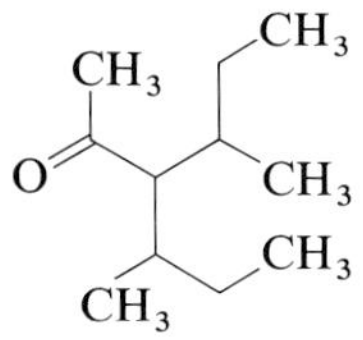

e. 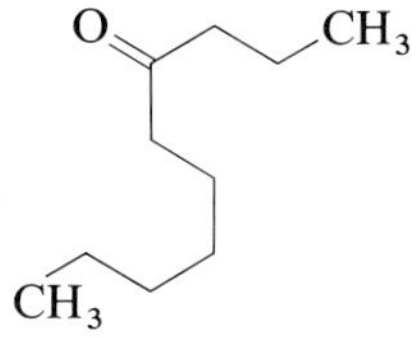

f. 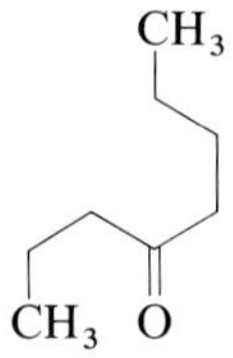

Reactions

11. Give products for the following reduction reactions.

a. 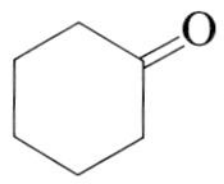 + $LiAlH_4 \longrightarrow$

b. $CH_3-\overset{\overset{O}{\|}}{C}-CH_2CH_3 + LiAlH_4 \longrightarrow$

c. $CH_3-\underset{\underset{O}{\|}}{C}-H + H_2 \xrightarrow{Pt}$

d. 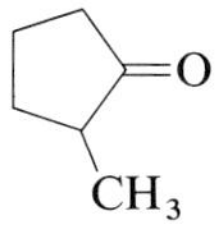 $+ H_2 \xrightarrow{Pt}$

12. Give products for the following acetal and ketal formation reactions

a. 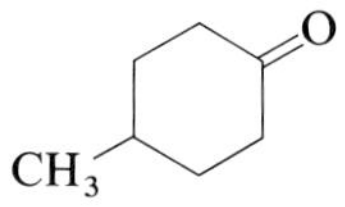 + Excess $CH_3CH_2OH \xrightarrow{H^+}$

b. $H-\overset{\overset{O}{\|}}{C}-CH_2CH_2CH_2-\overset{\overset{OH}{|}}{CH}-CH_3 \xrightarrow{H^+}$

c. 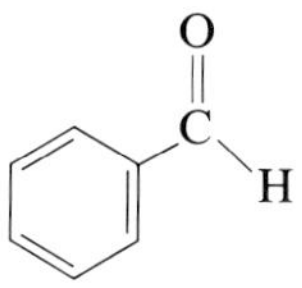 + 1 equiv. $CH_3OH \xrightarrow{H^+}$

d. 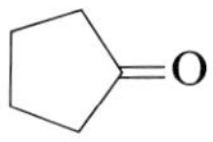 + Excess $CH_3OH \xrightarrow{H^+}$

13. Give products for the following oxidation reactions.

a. 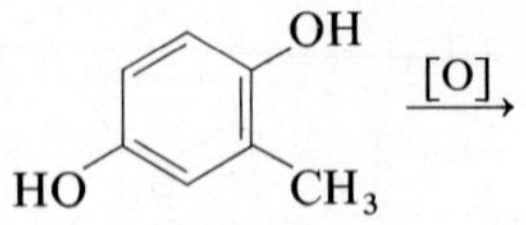

b. 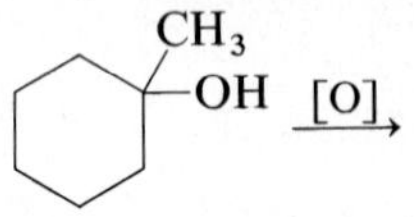

c. 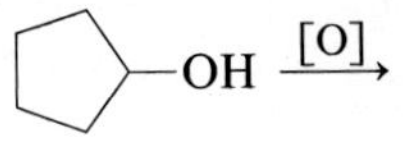

d. $CH_3-CH_2-\underset{\displaystyle O}{\underset{\|}{C}}-H \xrightarrow{[O]}$

e. 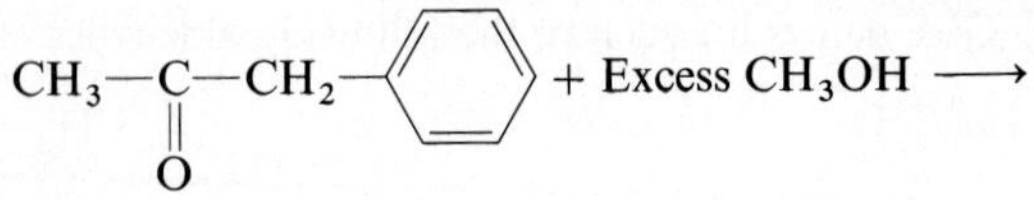

f. $CH_3-\underset{\displaystyle O}{\underset{\|}{C}}-H \xrightarrow{HO^-}$

g. 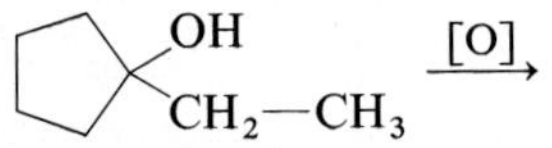

h. $CH_3CH_2CH_2-OH \xrightarrow{[O]}$

i. 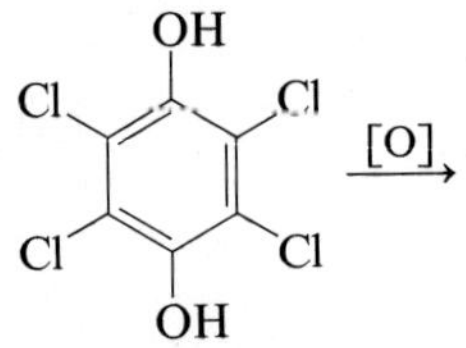

j. 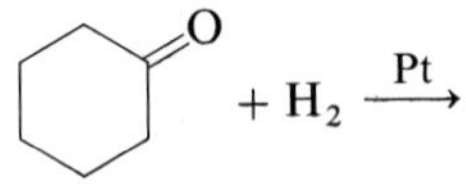

14. Give products for these general carbonyl reactions.

a. $CH_3-\underset{\displaystyle O}{\underset{\|}{C}}-CH_3 + {}^-CN \longrightarrow$

b. 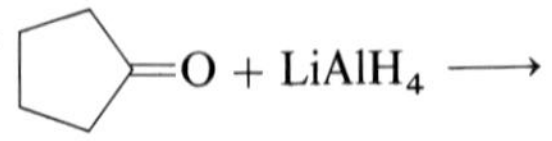

c. 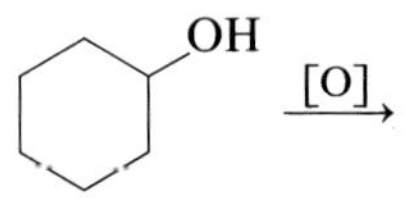

d. 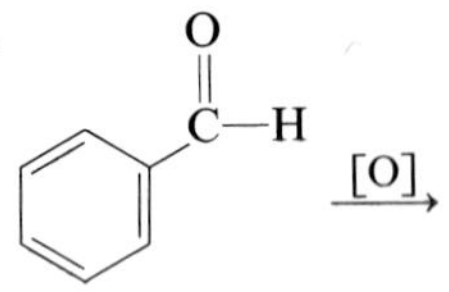

Thought Questions

15. Using the oxidation model introduced in this chapter, show how the oxidation of methane will produce carbon dioxide.

16. Using the same model, show the oxidation of methanol and ethanol.

CHAPTER

20

Carboxylic Acids and Their Derivatives

OBJECTIVES

After completing this chapter, you should be able to

- Explain the reasons for the acidity of carboxylic acids.
- Name and identify representative acids and their derivatives.
- Predict the products of reactions of acids and of their derivatives.
- Describe the process of saponification.
- Explain the chemical behavior of soaps and detergents.

Hydrogen ions are important in living systems. A hydrogen can be readily removed from the O—H bond in a carboxyl group, which is a combination of the carbonyl *and the* hydroxyl *group. Derivatives of carboxylic acids, including esters and anhydrides, have many applications in biological chemistry.*

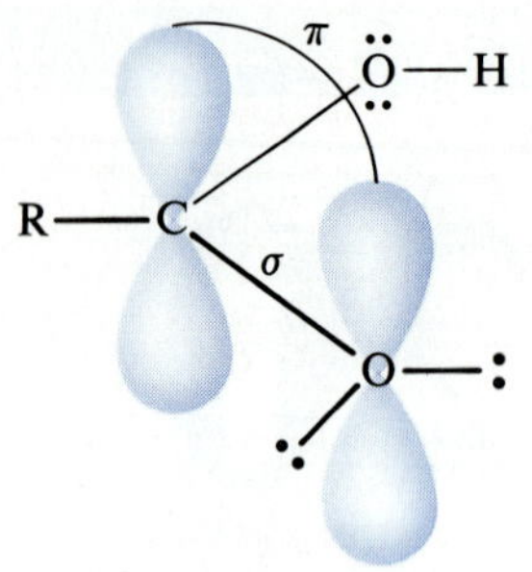

π
O—H
R—C
σ
O—:

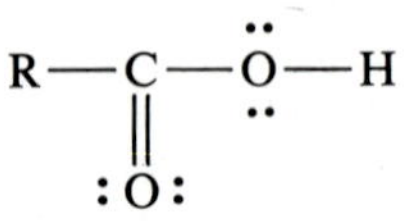

FIGURE 20.1
The electronic structure of carboxylic acid. Notice the carbonyl portion of the carboxyl group.

20.1 Carboxylic Acids: Structure, Properties, and Nomenclature

Structure

From application of the rules for Lewis electron dot structures, an organic acid can be characterized by the following general structural representation:

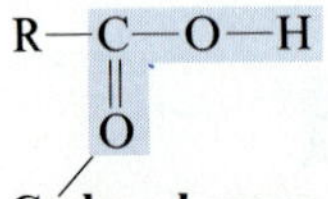

Carboxyl group

The —COOH is the **carboxyl group** (a contraction of the words *carb*onyl and hydr*oxyl*), which is the functional group that characterizes the organic **carboxylic acids**. We see within the carboxyl group a C=O and an —OH group. However, the acid group does *not* react as though it were an alcohol plus a carbonyl—it is a unique functional group. The C=O group, just as in the case of the aldehydes and ketones, has both a sigma and a pi bond (Figure 20.1).

The acid functional group can be designated in several ways, and you should be able to recognize each one.

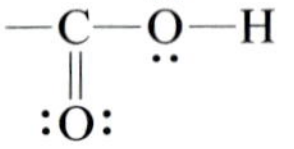　　—CO_2H　　—COOH

Hydrogen Bonding

The electronegative oxygen of the carbonyl participates well in hydrogen bonding with an alcohol.

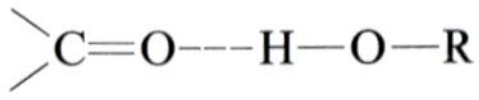

Because the carboxyl group has two different opportunities for hydrogen bonding, it is observed that carboxylic acids form *dimers.* A dimer is a compound made up of two identical pieces. The bonding in the dimer structure is sufficiently strong so that in some methods of determining molecular weights, organic acids seem to be twice as heavy as they really should be. This is due to the formation of the dimer (Figure 20.2).

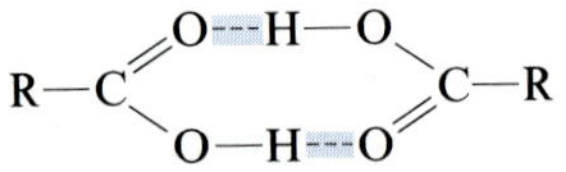

FIGURE 20.2
Strong hydrogen bonding causes the formation of dimers. The carbonyl portion of the carboxyl group of one molecule is hydrogen bonded to the —OH group of the other molecule. Because of the relative positions of the two portions of the functional group, each molecule effectively participates twice in this arrangement, forming a relatively strong attraction of two molecules for one another.

Boiling Points

Because the carboxylic acids are able to form intermolecular dimers and also have the potential for normal, linear hydrogen bonding between adjacent molecules, the boiling points of acids are even higher than the alcohols. Figure 20.3 shows the boiling points of the alcohols, aldehydes, and carboxylic acids. Notice that as the carbon chain gets longer, the influence of the functional group decreases, and the carboxylic acid behaves more and more like a simple hydrocarbon. (Compare this to Figure 18.1 in Chapter 18 where the relationship of the alcohols to the alkanes was shown.)

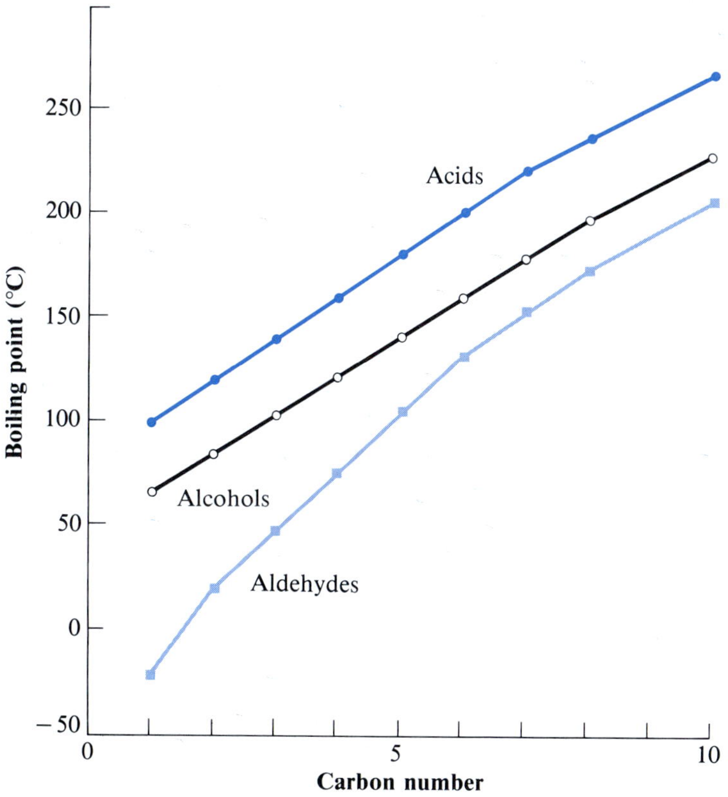
Boiling point (°C)
250
200
150
100
50
0
−50
Acids
Alcohols
Aldehydes
0
5
10
Carbon number

FIGURE 20.3
The stronger hydrogen bonding of acids causes the boiling points to be higher than for the corresponding alcohols. Again notice how the differences become less important as the carbon chain length increases.

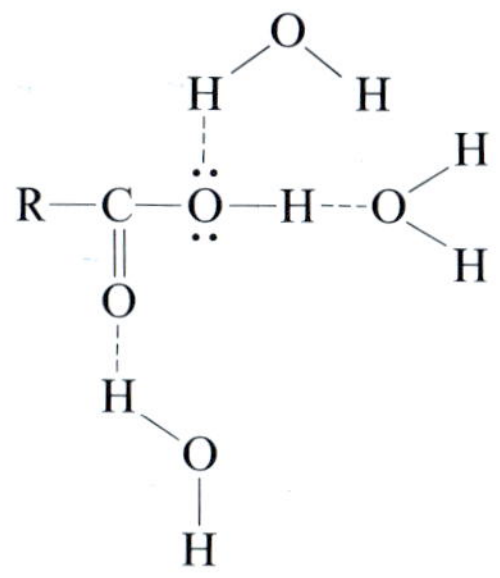

FIGURE 20.4
Extensive hydrogen bonding of the carboxyl group with water gives the acids high water solubility for the shorter chain length acids.

Solubility

The solubility of a carboxylic acid is readily understood by considering the hydrogen bonding of the carboxyl group with water (Figure 20.4). Just as the solubility of alcohols in water rapidly decreases as the carbon number increases, the same trends can be seen for the organic acids. As the carbon chains increase in size, the acids become more and more like a hydrocarbon (Figure 20.5).

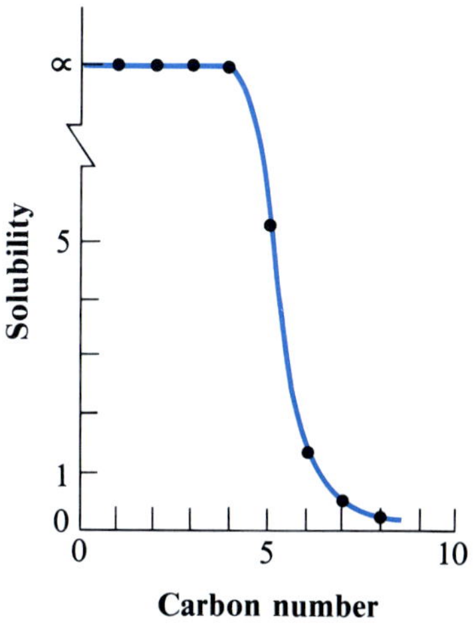
Solubility
∞
5
1
0
0
5
10
Carbon number

FIGURE 20.5
The solubility of acids decreases as the carbon number increases. Again, the dominating role of the functional group rapidly decreases as the carbon chain length increases.

Nomenclature of Carboxylic Acids

GENERAL RULES: THE CARBOXYL GROUP IS NUMBER ONE The nomenclature of carboxylic acids is very similar to that already represented by most of the other functional groups. Here the carboxyl group, —COOH, takes priority, and the carboxyl carbon is assigned the number one position in any numbering scheme. The *-e* ending of alkane, alkene and alkyne is replaced by *-oic acid*. Other substituents on the carbon chain are named and numbered as previously discussed. If there is a carbonyl group somewhere in the molecule (other than that contained in the carboxyl group), it is given a number to show its position and the designation *oxo*. Another way of

ASPIRIN

For over 200 years it has been known that willow bark contains substances useful for reducing pain and fever. In the late 1800s, the sodium salt of salicylic acid was used as a medication; however, this proved to irritate the lining of the stomach. *Salol*, the phenyl ester of salicylic acid, is a milder form of the drug, but the by-product of its hydrolysis in the intestine is phenol, a very toxic compound. In 1899, *aspirin*, or *acetylsalicylic acid*, was introduced by the Baeyer company of Germany. In many countries aspirin means only the Baeyer brand, but in the United States Baeyer no longer holds rights to the name. Thus, aspirin is the generic term used for all brands of salicylic acid. It remains on the market in the same form today.

A powerful boost to the sales of aspirin came in late 1987 with the report that one aspirin tablet per day can reduce the risk of heart attack for middle-aged men by almost 50%. The important activity that appears to be helpful in this respect is the ability of aspirin to reduce clotting of blood. This may have a negative effect for some people because of a slight increase in the probability of a stroke. However, these data are not yet well understood, and much more research needs to be done.

---

naming acids that also contain a ketone carbonyl is to use the prefix *keto* to designate the position and functional group. Often when this is done, Greek symbols are used rather than numbers to show position.

EXAMPLE 20.1

Name these carboxylic acids.

$$\overset{\beta}{\underset{3}{CH_3}}-\overset{\alpha}{\underset{2}{CH}}-\overset{1}{COOH}$$
$$\quad\quad\quad|$$
$$\quad\quad\,Br$$

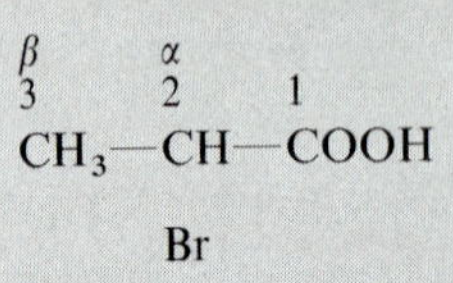

2-Bromopropanoic acid
(or α-bromopropanoic acid)

$$\overset{4}{CH_3}-\overset{3}{CH}=\overset{2}{C}-\overset{1}{COOH}$$
$$\quad\quad\quad\quad|$$
$$\quad\quad\quad CH_3$$

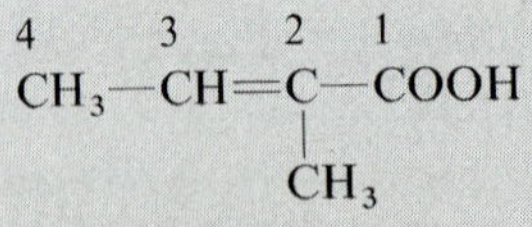

2-Methyl-2-butenoic acid

$$\quad\quad\;Br$$
$$\quad\quad\;|$$
$$\overset{3}{CH_3}-\overset{\alpha\,|\,2}{C}-\overset{1}{COOH}$$
$$\quad\quad\;|$$
$$\quad\quad\;Br$$

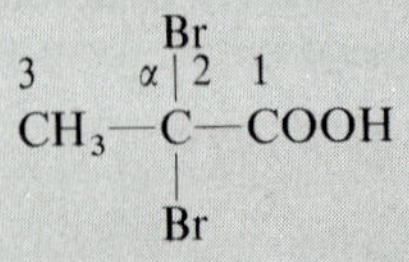

2,2-Dibromopropanoic acid
(or α,α-dibromopropanoic acid)

$$\overset{5}{CH_3}-\overset{4}{CH_2}-\overset{3}{C}-\overset{2}{CH_2}-\overset{1}{COOH}$$
$$\quad\quad\quad\quad\;\|$$
$$\quad\quad\quad\quad\;O$$

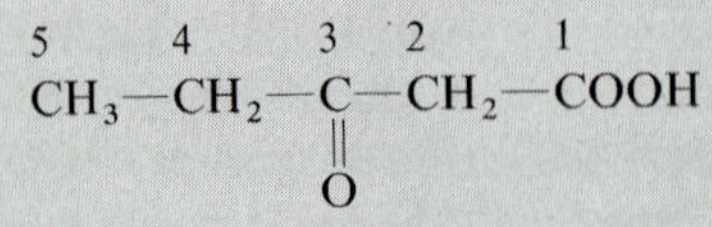

3-Oxopentanoic acid
(or 3-ketopentanoic acid)

COMMON NAMES FOR CARBOXYLIC ACIDS Aromatic carboxylic acids are often named on the basis of the common name, *benzoic acid*, which serves as the parent name. Several other common names are routinely used for organic acids, and several representative examples are shown in Table 20.1.

TABLE 20.1
Representative Naturally Occurring Organic Acids

Name	Description	Structure
Formic acid	Found in ant venom	H—COOH
Acetic acid	Found in vinegar (4 to 5%)	CH_3—COOH
Lactic acid	Formed from pyruvic acid	CH_3—CH—COOH | OH
Salicylic acid	A constituent of aspirin and oil of wintergreen	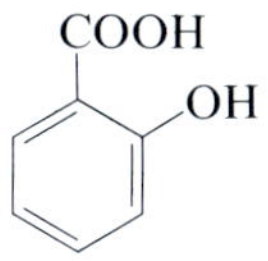
Tartaric acid	Common in grapes	HOOC—CH—CH—COOH | | HO OH
Malic acid	Common in apples	HOOC—CH—CH_2—COOH | OH
Lauric acid (n = 10) Myristic acid (n = 12) Palmitic acid (n = 14) Stearic acid (n = 16)	Saturated fatty acids	CH_3—$(CH_2)_n$—COOH
Oleic acid	Unsaturated fatty acid	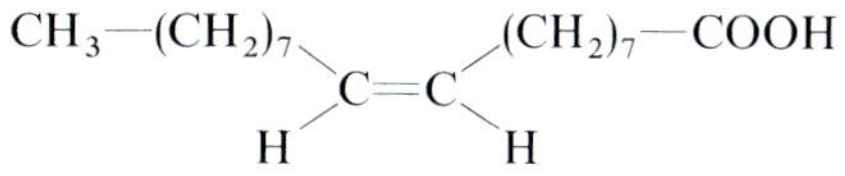

NAMES FOR DICARBOXYLIC ACIDS There are also a number of *di*carboxylic acids, and many of these have common names. The sentence, "*o*h *m*y, *s*uch *g*ood *a*pple *p*ie" is a useful mnemonic device for remembering the list of dicarboxylic acids that have an increasing number of CH_2 groups attached (Table 20.2). Some of the dicarboxylic acids and their derivatives play important roles in biological reactions, and they will be seen again in later chapters.

A number of naturally occurring carboxylic acids are encountered in biochemistry. Long carbon chain acids are called **fatty acids**, and examples of these are found in the biochemistry of *lipids* (Chapter 24).

NEW TERMS

carboxyl group The functional group that characterizes the organic acids:

$$-\overset{\overset{\large O}{\|}}{C}-OH$$

carboxylic acid Organic compounds containing the carboxyl group.

fatty acid A long carbon chain carboxylic acid.

TABLE 20.2
Common Dicarboxylic Acids

Name	Structure
Oxalic acid	HOOC—COOH
Malonic acid	HOOC—CH_2—COOH
Succinic acid	HOOC—CH_2—CH_2—COOH
Glutaric acid	HOOC—CH_2—CH_2—CH_2—COOH
Adipic acid	HOOC—CH_2—CH_2—CH_2—CH_2—COOH
Pimelic acid	HOOC—CH_2—CH_2—CH_2—CH_2—CH_2—COOH

Increasing number of CH_2 groups attached ↓

TESTING YOURSELF

Carboxylic Acid Structure and Nomenclature

1. Name the following.

a. 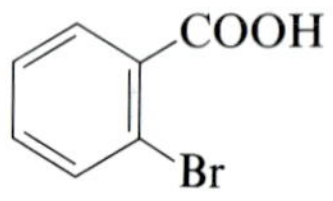

b. F_3C—COOH (F—C(F)(F)—COOH)

c. CH_3—CH(CH_3)—CH_2CH_2—COOH

2. Give structures for each of the following.
   a. 3,5-dinitrobenzoic acid
   b. 2-bromo-3-methylpentanoic acid
   c. 3-hexenoic acid

Answers **1. a.** *o*-bromobenzoic acid **b.** 2, 2, 2-trifluoroethanoic acid (or trifluoroacetic acid) **c.** 4-methylpentanoic acid **2. a.**

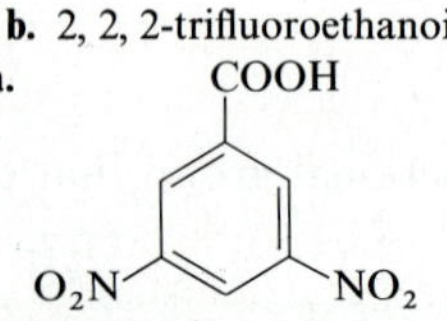

b. CH_3CH_2—CH(CH_3)—CH(Br)—COOH

c. CH_3—CH_2—CH=CH—CH_2—COOH

20.2 Reactions and Preparations of Carboxylic Acids

Organic Acids Are Weaker Than Mineral Acids

GENERAL TRENDS The class of organic compounds called *carboxylic acids* are obviously characterized by a special acidity. What characteristics make these materials so acidic relative to other functional groups already studied? Clearly, we have to ask the question about why the —OH of a carboxylic acid is more able to lose the H as H^+ than the —OH on an alcohol is. First, let us look at the structure of the carboxyl group (Figure 20.6). As we have already seen in the chemistry of the carbonyl group, the polarity of the carbonyl oxygen results in the carbon atom having a partial positive charge. The electrons on the oxygen of the hydroxyl group are attracted toward this positive charge on the carbon. Now the oxygen has a partial positive charge, and because of its high electronegativity, oxygen prefers not to have a positive charge. The electrons attached to the hydrogen are thus pulled toward the oxygen to "neutralize" the positive charge. The net effect of this loosening of the O—H bond is expressed in the ease of the H coming off as H^+.

An acid readily loses a proton and forms resonance-stabilized anions.

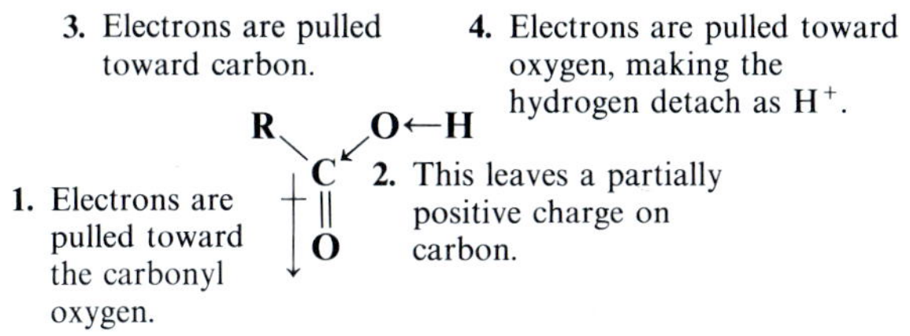

3. Electrons are pulled toward carbon.
4. Electrons are pulled toward oxygen, making the hydrogen detach as H+.
R
O←H
C
O
2. This leaves a partially positive charge on carbon.
1. Electrons are pulled toward the carbonyl oxygen.

FIGURE 20.6
The electronegativity of oxygen in the carbonyl portion of the carboxyl group polarizes electrons in the molecule so that the O—H bond is weakened.

RESONANCE STABILITY OF THE CARBOXYLATE ANION The same generalization already made for the acidity of phenols and the α-hydrogens on carbonyl compounds can be applied to carboxylic acids. After loss of the proton from the carboxylic acid, a **carboxylate anion** remains. By resonance there is charge delocalization of the carboxylate anion. Further, as with the resonance structures of benzene, two *equivalent* resonance structures can be drawn for the carboxylate anion. These resonance structures are shown in Figure 20.7.

QUANTITATIVE LOOK AT THE ACIDITY OF CARBOXYLIC ACIDS Unlike HCl, which we know to be *completely* ionized or dissociated into H^+ and Cl^-, all of the organic carboxylic acids are weak acids. That is, they do *not* ionize completely. As the data in Table 20.3 show, however, the organic acids are more acidic, that is, they have lower pK_a values than most other common functional groups.

It is one thing to use numbers to explain relative acidities and another to realize what the numbers mean. It turns out that a simple acid–base test can show the ranges of acidity. We need only to compare the strength of base required to remove the acidic hydrogen. Comparisons of alcohols, phenols, and acids are given in Table 20.4. In this table we can see that a carboxylic acid has a proton that can be removed even by the weakest bases, while an alcohol shows no reaction. The phenol exhibits acidity between acids and alcohols.

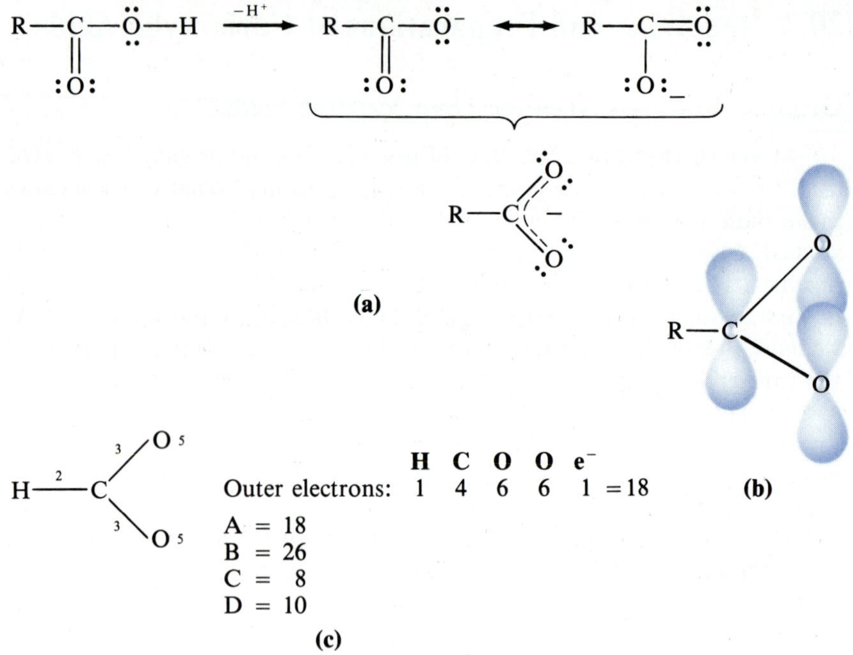

−H⁺
(a)
(b)
H C O O e⁻
Outer electrons: 1 4 6 6 1 = 18
A = 18
B = 26
C = 8
D = 10
(c)

FIGURE 20.7
Structures for the carboxylate anion. (a) Resonance structures are one simple way to examine electron delocalization. (b) An orbital view shows the same features, but it is harder to show chemical behavior with these more complex drawings. (c) The Lewis electron dot approach also shows the delocalization.

HALOGENS CAN INFLUENCE THE ACIDITY OF ORGANIC ACIDS We know that the halogens are electronegative and that the order of *increasing* electronegativity is as follows:

Halogen	I	< Br	< Cl	< F
Electronegativity	2.66	2.96	3.46	3.98

Because of this electron-withdrawing nature of a halogen, we might expect that when a hydrogen atom of the methyl group of acetic acid is replaced by an electron-withdrawing halogen (we do not have to be concerned with how this happens), the acidity will *increase*. Just as we first considered why a carboxylic acid is acidic by examining how the electronegative oxygen

TABLE 20.3
Effects of Functional Groups on Hydrogen Acidity

Functional Group	Approximate K_a	Approximate pK_a	
R—H	10^{-50}	50	Increase in acidity ↓
R—O—H	10^{-18}	18	
H—O—H	10^{-16}	16	
R—S—H	10^{-12}	12	
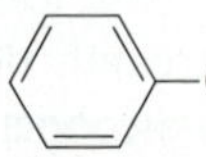—O—H	10^{-10}	10	
R—C(=O)—O—H	10^{-5}	5	

TABLE 20.4
Tests for Acidity

	Functional Groups			
	Alcohol ROH	Phenol  —OH	Acid RCOOH	Base
Products	RO^-	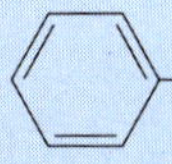 —O^-	$RCOO^-$	Na*
Products	No reaction	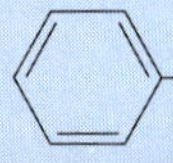 —O^-	$RCOO^-$	NaOH
Products	No reaction	No reaction	$RCOO^-$	$NaHCO_3$

* Formally an oxidation–reduction reaction.

distorts the electrons in the bonds, we can consider the simplifed displacement of electrons shown here.

How electron-withdrawing groups increase acidity

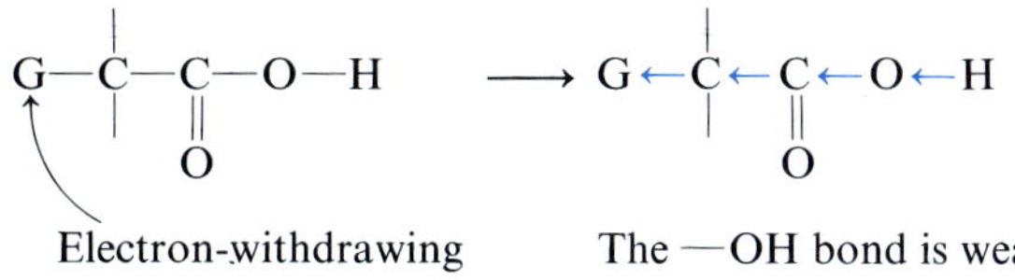
Electron-withdrawing group
The —OH bond is weakened because the electrons are being attracted toward the electronegative group, G

The electron-withdrawing effects of the halogen, coupled with the electron-withdrawing effects of the oxygen atoms of the carboxyl group, make it easier for the hydrogen atom to be given off as an H^+ ion. All of the effects conspire to delocalize negative charge in the carboxylate anion. The effect of halogen substitutions on the strength of carboxylic acids is shown in Figure 20.8.

Fluoroacetic acid, F—CH_2—COOH, the most acidic of the halogenated acetic acids, is a very potent poison. Its sodium salt is known as 1080, a rodent and coyote poison. As we shall see in the chapters on biochemistry, acetic acid is important in the metabolism of almost all important biological molecules. Fluoroacetic acid mimics acetic acid but cannot participate in its reactions. Thus, if fluoroacetic acid is ingested, the chemical processes involving acetic acid in the living cell can no longer continue, and the organism will die.

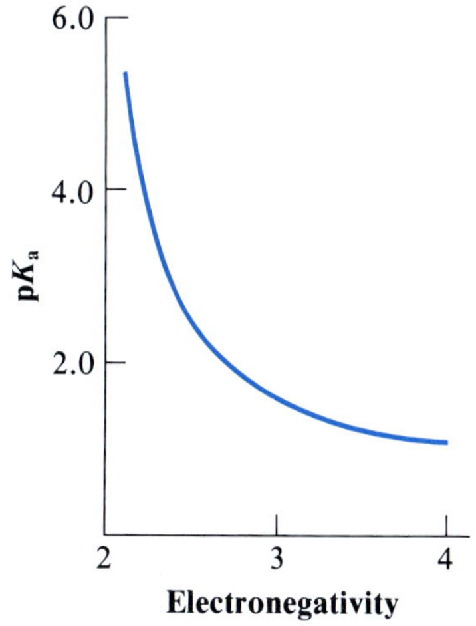
6.0
4.0
2.0
pK_a
2
3
4
Electronegativity

FIGURE 20.8
As the electronegativity of the substituent atom increases, the acidity of the acid increases (pK_a decreases).

Preparation of Carboxylic Acids by Oxidation

Acids are easily prepared by the oxidation of primary alcohols and aldehydes.

The oxidation of aldehydes to organic acids as a test for aldehydes was already discussed in Chapter 19. An aldehyde may serve as a starting material for the preparation of acids. As in the case of alcohol oxidation, the symbol

[O] is employed to denote an oxidizing agent. The usual reagents for oxidizing aldehydes to acids are oxides of chromium or manganese ($KMnO_4$, CrO_3, or $K_2CR_2O_7$). Recall that primary alcohols may be oxidized to aldehydes. However, it is difficult to stop the oxidation at this point, and it is often observed that continuation of the oxidation process provides a synthetic method for preparing acids from alcohols.

GENERAL REACTION

Oxidation of a Primary Alcohol to an Acid

$$R{-}CH_2{-}OH \xrightarrow{[O]} R{-}\underset{\displaystyle \overset{\|}{O}}{C}{-}H \xrightarrow{[O]} R{-}COOH$$

Primary alcohol — Aldehyde (difficult to stop at this stage) — Acid

Examples

$$CH_3{-}OH \xrightarrow{[O]} H{-}\underset{\displaystyle \overset{\|}{O}}{C}{-}H \xrightarrow{[O]} H{-}COOH \xrightarrow{[O]} CO_2$$

$$CH_3{-}CH_2{-}CH_2{-}OH \xrightarrow{[O]} CH_3{-}CH_2{-}\underset{\displaystyle \overset{\|}{O}}{C}{-}H \xrightarrow{[O]} CH_3{-}CH_2{-}COOH$$

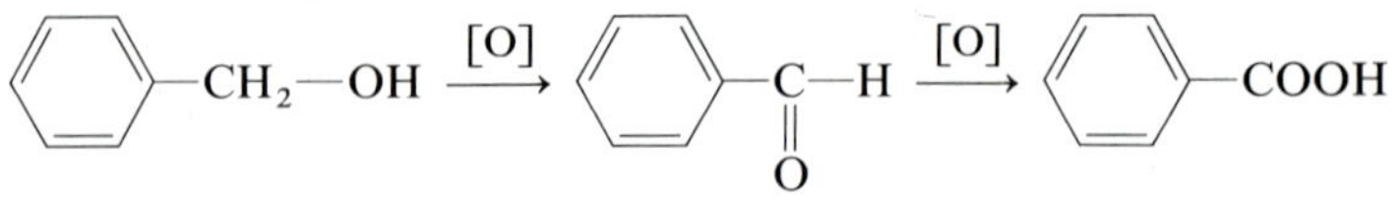

CH2—OH
[O]
C—H
O
[O]
COOH

How can we tell what kind of organic compound can undergo easy oxidation? Is there a way to know when oxidation is likely to continue to another compound, as in the case of primary alcohols to aldehydes and then to acids? These questions can be answered by looking at a general scheme of oxidation (Figure 20.9). Let us *assume* that an oxidation involves the replacement of a C—H bond by a C—OH bond. We see that the C—H bonds on methane are replaced, one at a time, by —OH. In the last chapter we saw that a hydrate of a carbonyl readily lost water to produce the carbonyl. This approach shows the same thing. Thus, we can see why primary alcohols will lead to aldehydes, and they in turn will produce acids. Since a

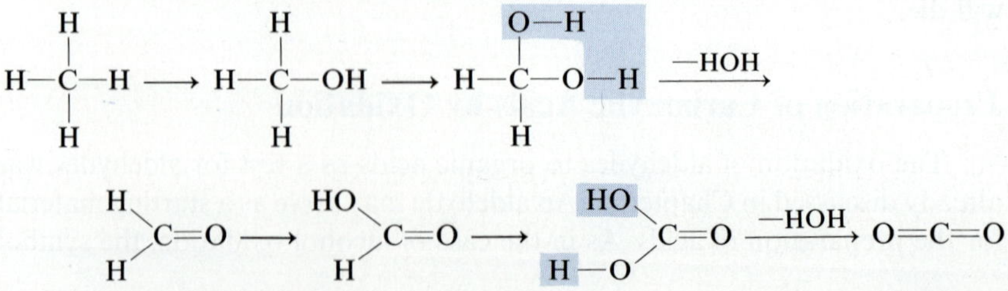

—HOH
—HOH
O=C=O

FIGURE 20.9
A simple model for an oxidation can be considered as replacing a C—H bond with a C—OH bond. Whenever there are two —OH bonds on the same carbon atom, water is removed and a C=O is formed. This model does not represent how the oxidation takes place, but simply shows the increasing oxidation states as H is replaced by O.

ketone has no hydrogens attached to the carbonyl-containing carbon atom, we would correctly predict that ketones were resistant to oxidation. We would also correctly predict that tertiary alcohols would resist chemical oxidation.

In another important oxidation reaction, an alkyl side chain attached to a benzene ring can be converted to a benzoic acid derivative by oxidation. This reaction requires more vigorous oxidation conditions than those previously discussed and heat (Δ) is generally required.

Alkyl side chains on aromatic rings are oxidized to acids.

GENERAL REACTION

Oxidation of an Alkyl Side Chain on a Benzene Ring

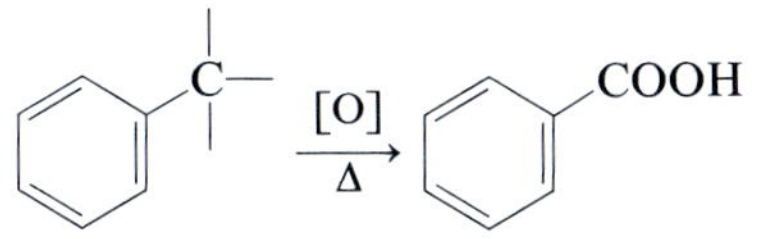
C
[O]
Δ
COOH

Examples

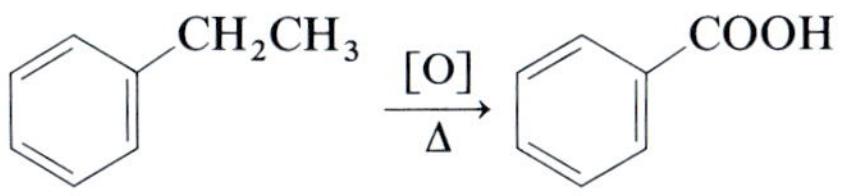
CH_2CH_3
[O]
Δ
COOH

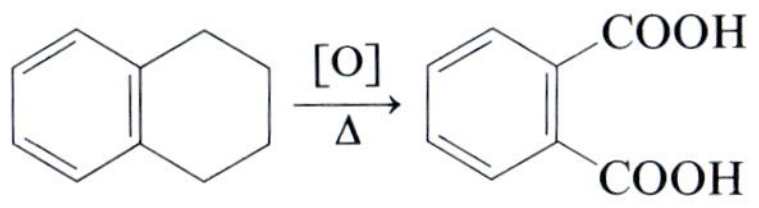
[O]
Δ
COOH
COOH

As a result of this special ability of alkylated benzene rings to become oxidized at the side chain, both in the laboratory as well as under biological conditions, alkyl-substituted benzene derivatives are often easily degraded in the environment. These alkylated compounds are also usually less toxic to living organisms than benzene rings without alkyl groups. Recall from Chapters 17 and 18 that the biological oxidation of benzene gives phenol, a poisonous compound. The oxidation of toluene will form benzoic acid, a relatively harmless compound that can be excreted from the body. It is noteworthy that an aromatic ring is not attacked by the usual oxidizing agents.

PREPARATION OF CARBOXYLIC ACIDS BY NITRILE HYDROLYSIS The cyanide ion, C≡N⁻(or CN⁻), is able to displace a halogen in a nucleophilic substitution reaction. An important class of organic compounds called the **nitriles** results from this reaction. A nitrile is an organocyanide compound.

Nitriles are organocyanide compounds.

$$CH_3{-}CH_2{-}Cl + C{\equiv}N^- \longrightarrow \underset{\textbf{Nitrile}}{CH_3{-}CH_2{-}C{\equiv}N} + Cl^-$$

Note that the negative charge is located on the carbon, making cyanide ion electronically similar to carbon monoxide:

$$:C{\equiv}N: \qquad :C{\equiv}O:$$

GENERAL REACTION

Formation of a Nitrile by a Displacement Reaction

$$R{-}X + CN^- \longrightarrow R{-}C{\equiv}N + X^-$$

Examples

$$CH_3{-}Br + CN^- \longrightarrow CH_3{-}CN + Br^-$$

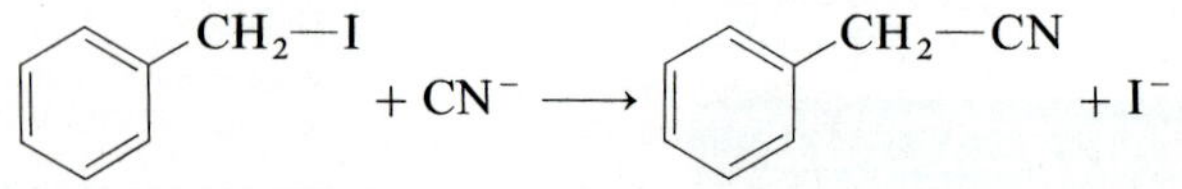

There are not too many natural products containing the nitrile group. We have already seen one example in the millipede defense mechanism (Chapter 19). A common nitrile, known as *ricinine* has been isolated from the caster plant, *Ricinus communis*. The presence of this substance makes the plant poisonous.

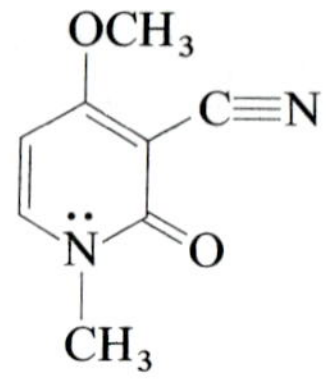

Ricinine

Nitriles can be hydrolyzed to acids.

Although a nitrile is a relatively stable functional group, in the presence of either strong acid or strong base and hot water, it is hydrolyzed to a carboxylic acid. **Hydrolysis** is the reaction of water with a functional group. In the acid-catalyzed hydrolysis of a nitrile, a carboxylic acid is produced and the byproduct is the ammonium ion.

GENERAL REACTION

Acid Hydrolysis of a Nitrile

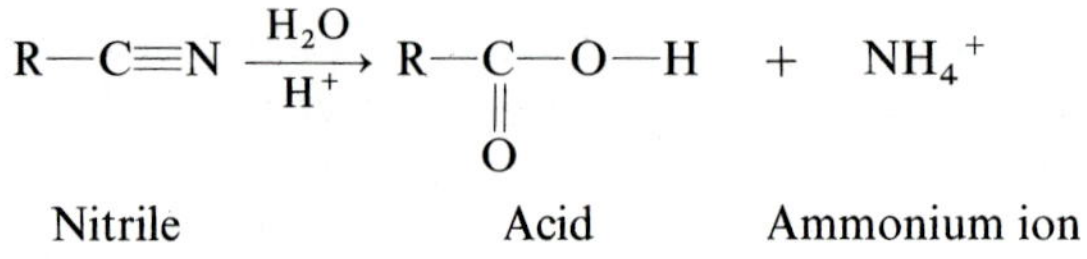

Examples

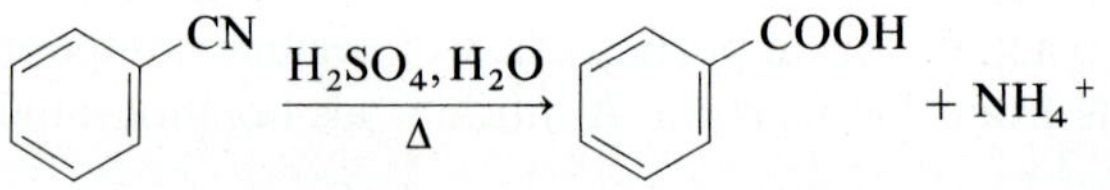

$$CH_3{-}CN \xrightarrow[\Delta]{H_2SO_4,\ H_2O} CH_3{-}COOH + NH_4^+$$

Basic hydrolysis of a nitrile will produce a salt of the carboxylic acid and free ammonia. Addition of acid to the reaction mixture will cause protonation of the carboyxlic acid.

GENERAL REACTION

Basic Hydrolysis of a Nitrile

$$R-C\equiv N \xrightarrow[HO^-]{H_2O} R-\underset{\underset{O}{\|}}{C}-O^- + NH_3$$

Examples

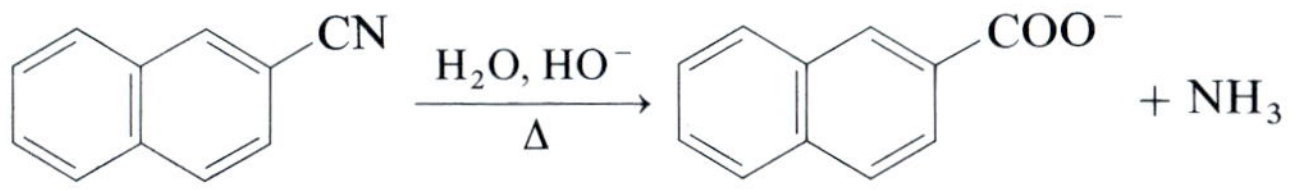

$$CH_3-CH_2-CN \xrightarrow[\Delta]{H_2O,\ HO^-} CH_3-CH_2-COO^- + NH_3$$

Decarboxylation of Beta Keto Acids

Organic acids that also contain a carbonyl group are often given the generic name **keto acid**. When you begin your study of biochemistry, you will find two common kinds of acids that also contain a carbonyl group—the *alpha keto* acids and the *beta keto* acids.

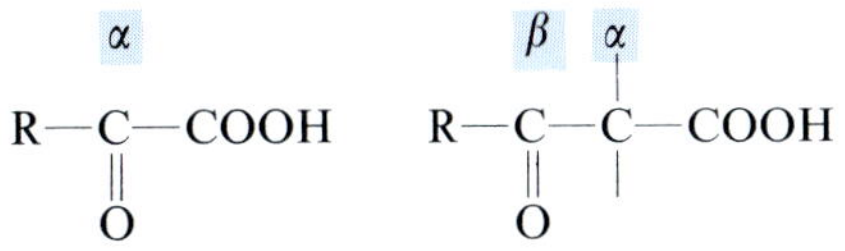

Alpha keto acid **Beta keto acid**

One of the most interesting aspects of the beta keto acid is its ability to undergo easy **decarboxylation**, or loss of carbon dioxide. This decarboxylation provides a useful and simple way to make ketones. In biological systems ketones are not the usual product, but the loss of CO_2 is a very important process.

Beta keto acids are easily decarboxylated (lose carbon dioxide).

GENERAL REACTION

Decarboxylation of a Beta Keto Acid

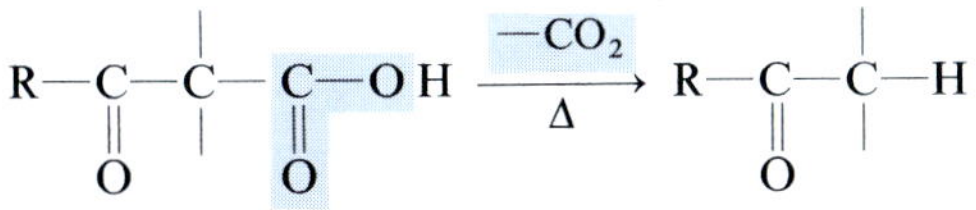

Beta keto acid

Examples

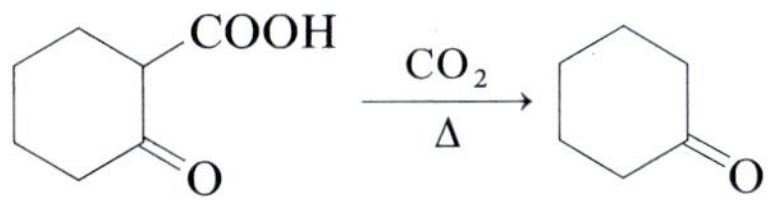

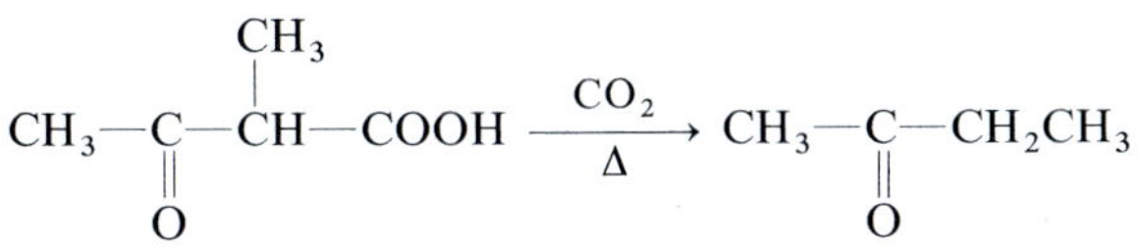

In a later section of this chapter (Section 20.5), a unique preparation of a beta keto acid will be provided. One approach to the beta keto acid would involve the oxidation of a beta hydroxy acid, which can be produced by the hydration (adding of water) of a 2,3-unsaturated acid. This method will be seen as an important feature in biochemical degradation processes. For example, one of the byproducts of oxidation of carbon compounds in our body is carbon dioxide. We are constantly breathing out the carbon dioxide produced as our metabolism breaks down carbon compounds for energy. One of the sources of carbon dioxide is the *citric acid cycle* (or TCA cycle, to be discussed in Chapter 28). (Notice the number of simple organic reactions!)

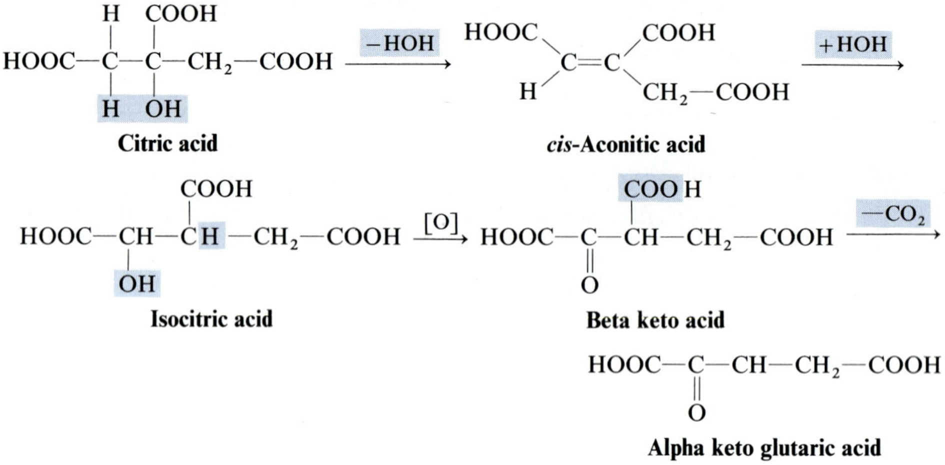

$$HOOC-\underset{\underset{H}{|}}{\overset{\overset{H}{|}}{C}}-\underset{\underset{OH}{|}}{\overset{\overset{COOH}{|}}{C}}-CH_2-COOH \xrightarrow{-HOH} \underset{H}{\overset{HOOC}{>}}C=C\underset{CH_2-COOH}{\overset{COOH}{<}} \xrightarrow{+HOH}$$

Citric acid   *cis*-**Aconitic acid**

$$HOOC-\underset{\underset{OH}{|}}{CH}-\overset{\overset{COOH}{|}}{CH}-CH_2-COOH \xrightarrow{[O]} HOOC-\underset{\underset{O}{\|}}{C}-\overset{\overset{COOH\ H}{|}}{CH}-CH_2-COOH \xrightarrow{-CO_2}$$

Isocitric acid   **Beta keto acid**

$$HOOC-\underset{\underset{O}{\|}}{C}-CH-CH_2-COOH$$

Alpha keto glutaric acid

A general reaction that summarizes these processes is shown below.

GENERAL REACTION

Formation of and Decarboxylation of Beta Keto Acids

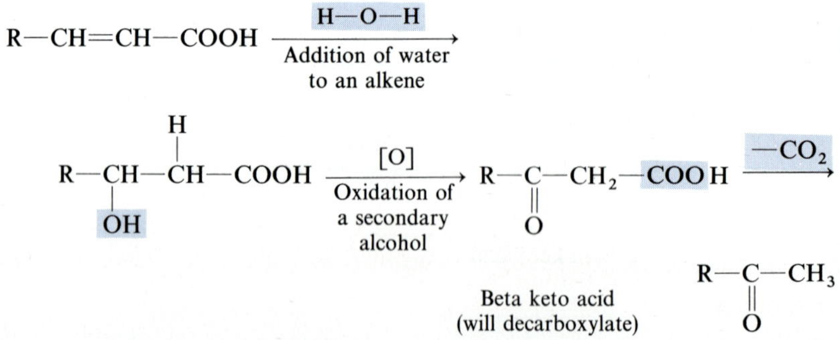

$$R-CH=CH-COOH \xrightarrow[\text{Addition of water to an alkene}]{H-O-H}$$

$$R-\underset{\underset{OH}{|}}{CH}-\overset{\overset{H}{|}}{CH}-COOH \xrightarrow[\text{Oxidation of a secondary alcohol}]{[O]} R-\underset{\underset{O}{\|}}{C}-CH_2-COOH \xrightarrow{-CO_2}$$

Beta keto acid
(will decarboxylate)

$$R-\underset{\underset{O}{\|}}{C}-CH_3$$

NEW TERMS

carboxylate anion The resonance-stabilized anion resulting from removal of the acid proton of an organic acid.

nitrile A functional group generally included within the chemistry of organic acids. Nitriles are organic cyanides and have the general structure R—C≡N.

hydrolysis The reaction of water with a functional group which results in the conversion of the functional group to a different functional group. The reaction is often catalyzed by either acid or base.

decarboxylation The loss of CO_2. This reaction takes place very readily for beta keto acids.

keto acids Organic acids that also contain a carbonyl functional group.

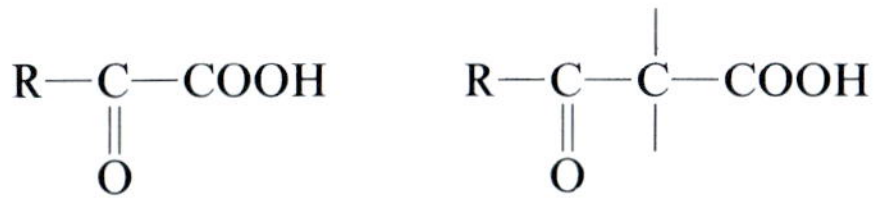

Alpha keto acid   Beta keto acid

TESTING YOURSELF

Reactions of Carboxylic Acids

1. Place the following compounds in order of *increasing* acidity.
   a. CH_3—COOH, F—CH_2—CH_2—COOH, CH_3—CH(F)—COOH
   b. Cl—CH(Cl)—COOH, CH_3—COOH, Cl_3C—COOH
   c. phenol, alcohol, water
2. What is the organic acid formed in each of the following reactions?

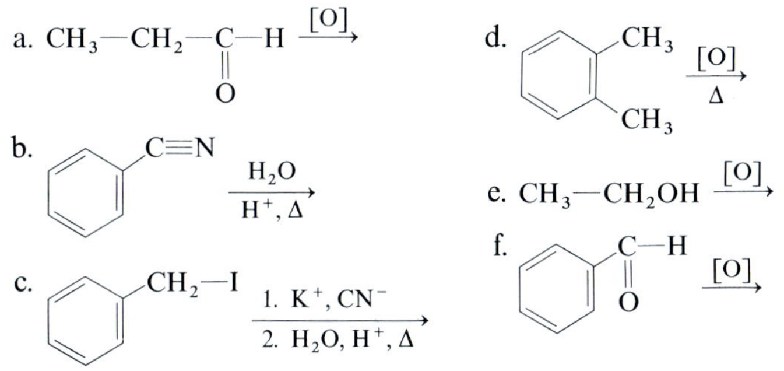

3. Give products for each of the following beta keto acid reactions.

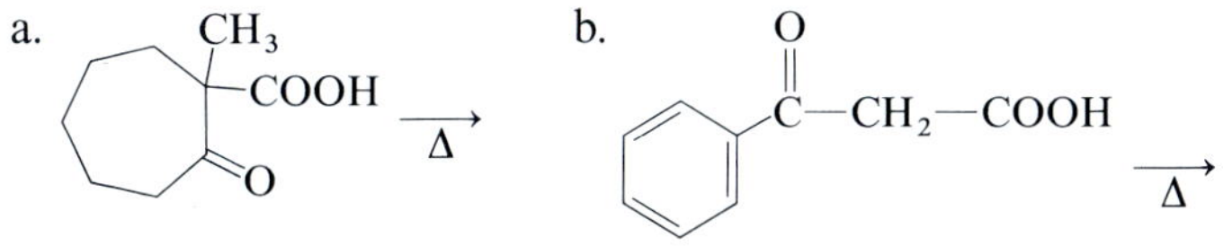

Answers **1. a.** CH_3—COOH, F—CH_2—CH_2—COOH, CH_3—CH(F)—COOH **b.** CH_3—COOH, Cl—CH(Cl)—COOH, Cl_3C—COOH **c.** alcohol, water, phenol **2. a.** propanoic acid **b.** benzoic acid **c.** phenylacetic acid (or 2-phenylethanoic acid) **d.**

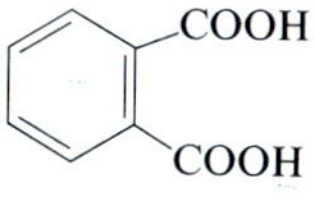

e. acetic acid **f.** benzoic acid **3. a.**

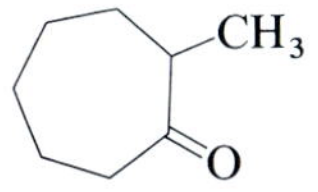

b.

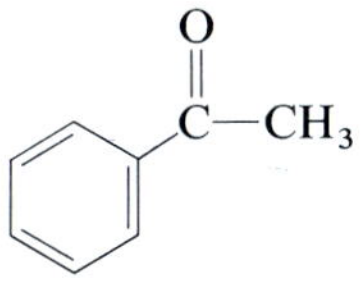

20.3 Esters

Structure and Reactivity

Esters are derivatives of acids and alcohols:

$$R{-}\overset{\overset{\displaystyle O}{\|}}{C}{-}OH + R'{-}OH \longrightarrow R{-}\overset{\overset{\displaystyle O}{\|}}{C}{-}O{-}R'$$

ORGANIC ESTERS One of the most common derivatives of an organic acid is the **ester**. An ester has the structure

$$R{-}\overset{\overset{\displaystyle O}{\|}}{C}{-}O{-}R'$$

Esters are most commonly formed by reaction of an acid and an alcohol. For example, propanoic acid and methanol react in the presence of an acid catalyst to give an ester:

$$\underset{\textbf{Propanoic acid}}{CH_3{-}CH_2{-}\overset{\overset{\displaystyle O}{\|}}{C}{-}O{-}H} + \underset{\textbf{Methanol}}{CH_3{-}O{-}H} \xrightarrow[\text{catalyst}]{\text{Acid}} \underset{\textbf{Ester}}{CH_3{-}CH_2{-}\overset{\overset{\displaystyle O}{\|}}{C}{-}O{-}CH_3} + H_2O$$

Note that in the formation of an ester, the hydroxyl group of the acid (*not* of the alcohol) is lost as water.

GENERAL REACTION

Ester Formation

$$\underset{\text{Acid}}{R{-}\overset{\overset{\displaystyle O}{\|}}{C}{-}O{-}H} + \underset{\text{Alcohol}}{H{-}O{-}R'} \underset{H_2O}{\overset{H^+}{\rightleftarrows}} \underset{\text{Ester}}{R{-}\overset{\overset{\displaystyle O}{\|}}{C}{-}O{-}R'} + \underset{\text{Water}}{H{-}O{-}H}$$

Examples

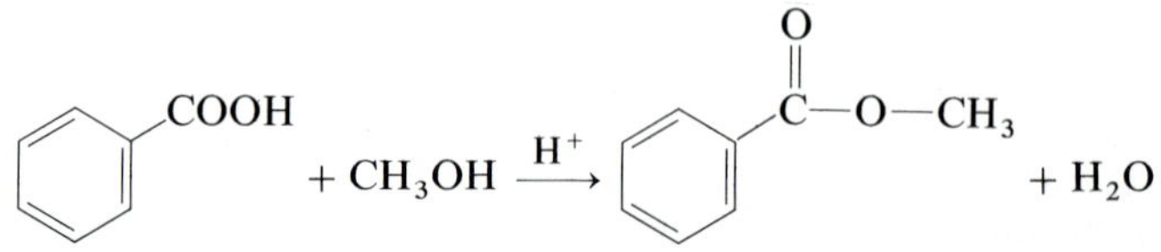
COOH
+ CH3OH
H+
O
C—O—CH3
+ H2O

Methylbenzoate

$$CH_3{-}\overset{\overset{\displaystyle O}{\|}}{C}{-}OH + CH_3{-}CH_2{-}OH \xrightarrow{H^+} \underset{\text{Ethyl acetate}}{CH_3{-}\overset{\overset{\displaystyle O}{\|}}{C}{-}O{-}CH_2CH_3} + H_2O$$

This reaction is an *equilibrium reaction* that is catalyzed by acid. Thus, to obtain good yields of the ester product, the equilibrium must be forced in the direction of the ester. According to Le Chatelier's principle, the easiest way to unbalance the system is to remove the water as it forms. This is easily accomplished in the laboratory (Figure 20.10).

Conversely, one could correctly predict that if water was in excess, the reaction would proceed backward and the ester would react with the water to form the starting acid and alcohol. This is indeed observed. We will discuss this reaction in a later section of this chapter (Section 20.5).

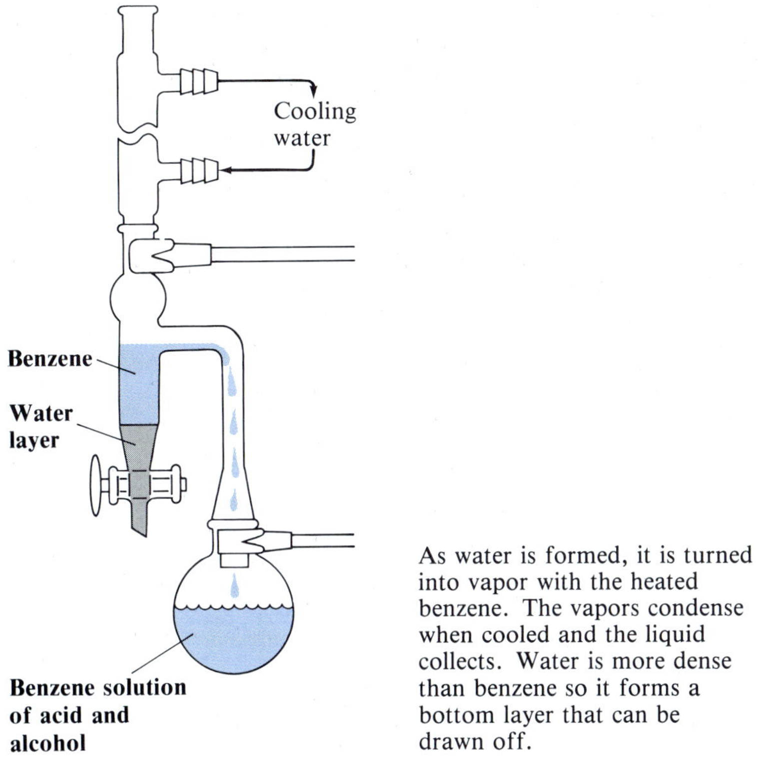
Cooling
water
Benzene
Water
layer
Benzene solution
of acid and
alcohol
As water is formed, it is turned into vapor with the heated benzene. The vapors condense when cooled and the liquid collects. Water is more dense than benzene so it forms a bottom layer that can be drawn off.

FIGURE 20.10
A laboratory method for forcing the equilibrium to favor formation of an ester. Water is removed from the reaction mixture as it is formed.

ESTER FLAVORS AND ODORS Many esters are used as flavorings and fragrances because a common characteristic of many esters is a "fruity" odor. Several common esters and their characteristic odors are shown in Table 20.5. It should be mentioned that the authentic odors and flavors are generally a complex mixture of many compounds. This is one of the reasons that synthetic fragrances are not quite the same as those provided by nature.

Esters are often used in flavorings.

TABLE 20.5
Esters and Their Odors

First Side Group (R) (acid)	Second Side Group (R′) (alcohol)	Fruity Odor
H	Ethyl	Rum
H	Isobutyl	Raspberry
Methyl	Isopentyl	Banana
Methyl	Octyl	Orange
Propyl	Ethyl	Pineapple

THIOESTERS We will find a number of **thioesters** in our study of biological chemistry. They have the same structure as esters but have sulfur replacing oxygen, just as in a thiol.

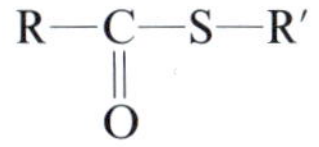

It is appropriate to recognize that their chemistry is not unique; we have already seen that thiols are the sulfur analogs of alcohols. Thus, when an organic acid is reacted with a thiol, a thioester will result. Notice in the following general reaction that if the —OH from the acid was not involved in ester formation, it would not be possible to prepare thioesters from thiols.

GENERAL REACTION

Formation of a Thioester

$$R-\underset{\underset{O}{\|}}{C}-O-H + H-S-R' \xrightarrow{H^+} R-\underset{\underset{O}{\|}}{C}-S-R' + H_2O$$

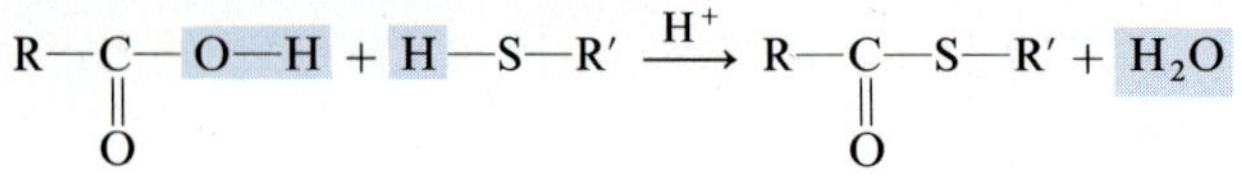

$$\text{Acid} \qquad \text{Thiol} \qquad \text{Thioester}$$

Example

$$CH_3-\underset{\underset{O}{\|}}{C}-OH + CH_3CH_2SH \xrightarrow{H^+} CH_3-\underset{\underset{O}{\|}}{C}-S-CH_2CH_3 + H_2O$$

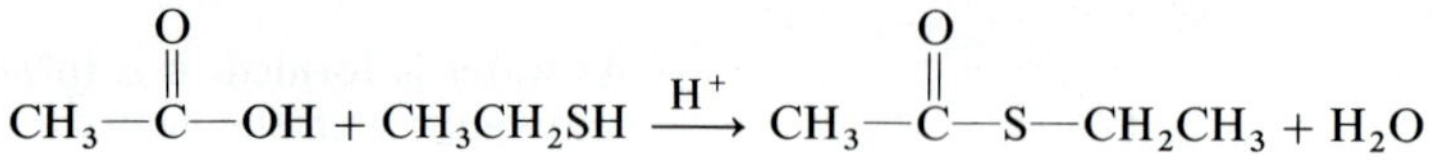

An important thioester found throughout the study of biochemistry is acetylcoenzyme A, which serves as an agent for transferring the acetyl group

$$-\underset{\underset{O}{\|}}{C}-CH_3$$

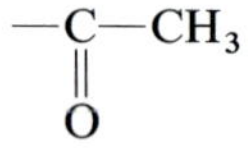

Many biochemical reactions are derived from this acetyl group transfer.

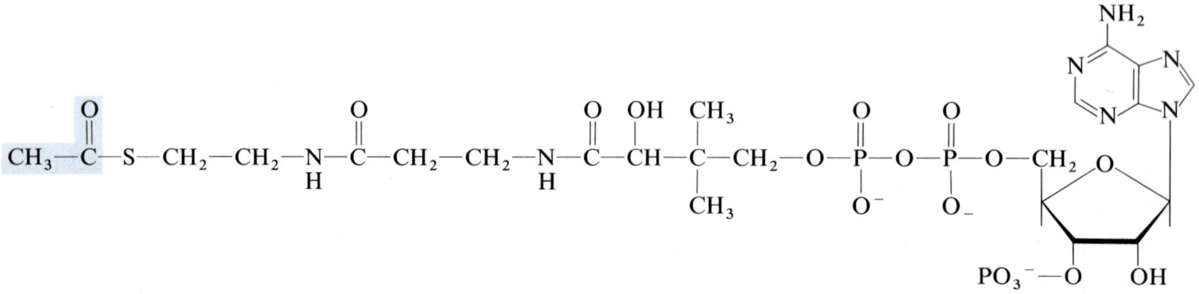

Acetylcoenzyme A

Coenzyme A is often abbreviated Co—A—SH. You can see that it is a thiol. Reaction with acetic acid results in the thioester acetylCoA (CoA—S—$\underset{\underset{O}{\|}}{C}$—CH_3).

INORGANIC ESTERS It is also possible to prepare esters of inorganic acids, particularly sulfuric acid (cold), phosphoric acid, and nitric acid.

$$R-O-H + H-O-SO_3H \longrightarrow R-O-SO_3H + H_2O$$

$$R-O-H + H-O-PO_3H_2 \longrightarrow R-O-PO_3H_2 + H_2O$$

$$R-O-H + H-O-NO_2 \longrightarrow R-O-NO_2 + H_2O$$

Phosphate and sulfate esters are widely used as **detergents** (see pages 568–571). Nitrate esters are often explosive, as best exemplified by *nitroglycerine*, a trinitrate ester of the triol glycerol.

$$\begin{array}{l} CH_2-O-NO_2 \\ | \\ CH-O-NO_2 \\ | \\ CH_2-O-NO_2 \end{array}$$

Nitroglycerine

Alfred Nobel discovered that by mixing nitroglycerine with an inert material (such as diatomaceous earth), the highly explosive and difficult to handle nitroglycerine was rendered relatively safe. This was applied in the early manufacture of dynamite. This was the start of new techniques for preparing "safe" explosives. Nobel endowed the Nobel prizes from the great wealth that he gained from his work.

Nitroglycerine is also used in medicine. Because it has the ability to dilate blood vessels and arteries, it is used for heart pain (angina).

Nomenclature of the Esters

By examining the stucture of an ester we see that there are two R groups, one belonging to the acid and one to the alcohol.

Belongs to acid (R) — Belongs to alcohol (R′)

$$R-\underset{\substack{\|\\O}}{C}-O-R'$$

Esters are named by first assigning the name of the alcohol alkyl group (methyl, ethyl, and so on), followed by the acid in which the *-oic acid* suffix is replaced by *-oate*.

EXAMPLE 20.2

Name these esters.

$$CH_3-\underset{\substack{\|\\O}}{C}-O-CH_3$$

Methyl ethanoate (or methyl acetate)

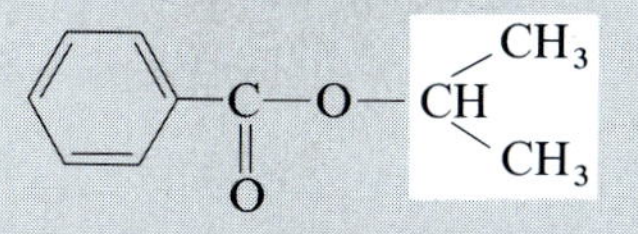

Isopropyl benzoate

$$CH_3-CH_2-O-\underset{\substack{\|\\O}}{C}-CH_2-CH_2-CH_3$$

Ethyl butanoate

Saponification: The Basic Hydrolysis of an Ester

The base-catalyzed hydrolysis of any ester to the salt of the acid and the alcohol is called **saponification**. Here is the general reaction.

GENERAL REACTION

Basic Hydrolysis (Saponification) of an Ester

$$R-\underset{\underset{O}{\|}}{C}-O-R' \xrightarrow[HO^-, \Delta]{H_2O} R-\underset{\underset{O}{\|}}{C}-O^- + R'-OH$$

Ester   Acid salt   Alcohol

Examples

$$Ph-\underset{\underset{\ddot{O}:}{\|}}{C}-\ddot{O}-CH_2CH_3 \longrightarrow Ph-\underset{\underset{\ddot{O}:}{\|}}{C}-\ddot{O}^- + CH_3CH_2\ddot{O}H$$

$$CH_3-\underset{\underset{:\ddot{O}:}{\|}}{C}-\ddot{O}-Ph \longrightarrow CH_3-\underset{\underset{:\ddot{O}:}{\|}}{C}-\ddot{O}^- + Ph-\ddot{O}H$$

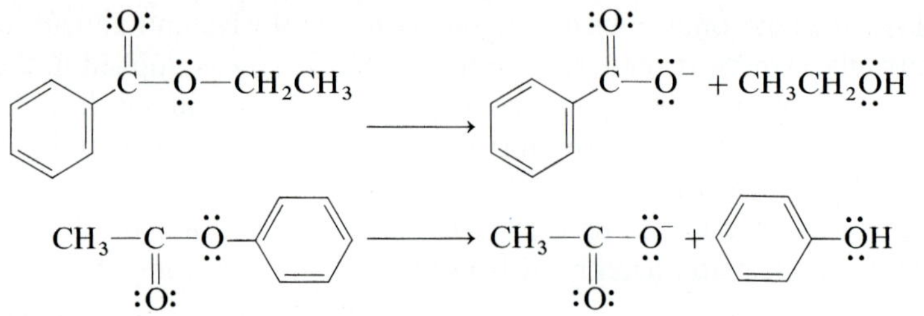

Soap is a salt of a long chain fatty acid.

Fat is a complex organic *tri*ester (that is, it has *three* connected esters) of glycerol and long carbon chain acids called *fatty acids*. Fat is also called a *triglyceride*. It can be hydrolyzed by lye (NaOH) to form the salt of the fatty acid and glycerol. This salt is called **soap**.

$$\begin{array}{l} CH_2-O-\overset{O}{\overset{\|}{C}}-(CH_2)_n-CH_3 \\ CH-O-\overset{O}{\overset{\|}{C}}-(CH_2)_n-CH_3 \\ CH_2-O-\overset{O}{\overset{\|}{C}}-(CH_2)_n-CH_3 \end{array} + 3\,NaOH \longrightarrow \begin{array}{l} CH_2-OH \\ CH-OH \\ CH_2-OH \end{array} + 3\,CH_3-(CH_2)_n-\overset{O}{\overset{\|}{C}}-O^-Na^+$$

Sodium salt of a fatty acid (soap)

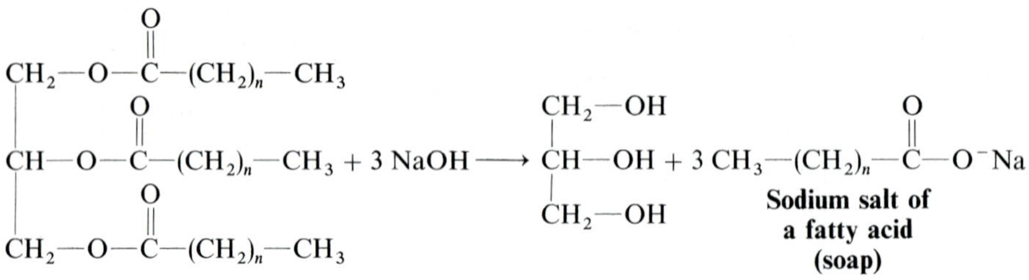

Detergents

SOAP AS ONE KIND OF DETERGENT Although known by the ancient Romans, it was not until the late 1700s that the process of soap making had developed to the craft that allowed almost everyone to have soap. The soap industry didn't start in the United States until the early 1800s. Today, soap is a very common material in our daily world.

But what is soap? It is the salt of a fatty acid. The constitution of the long chain alkyl group (the number of carbon atoms is generally in the 12 to 18 range) determines the specific properties of the soap. Similarly, the way a soap feels depends to some extent on the cation involved. For example, potassium salts are often used for the very fine soft soaps associated with shaving creams.

How does soap work? In our earliest discussions of organic chemistry, we noted that water insolubility was an important feature of hydrocarbons. A particularly important relationship was noted in the solubility of alcohols, where the short chain alcohols are very soluble in water as a result of the ability of the OH group to hydrogen bond with water molecules. As the

The long hydrocarbon chain dissolves "dirt and grime" while the polar end dissolves in water.

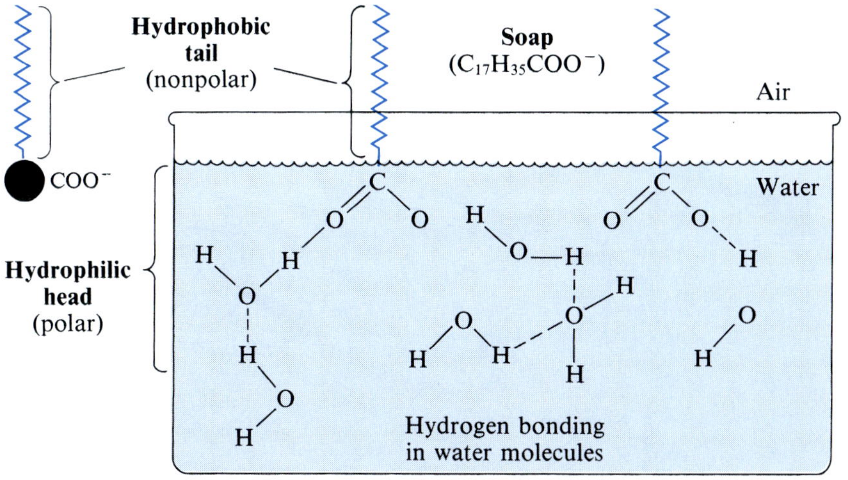
Hydrophobic tail (nonpolar)
Soap ($C_{17}H_{35}COO^-$)
Air
COO^-
Water
Hydrophilic head (polar)
Hydrogen bonding in water molecules

FIGURE 20.11
The chemical and physical behavior of soap is readily understood in terms of the water solubility of the ionic end of the soap and the hydrocarbon solubility of grease, dirt, and oil.

carbon chain gets larger, the solubility rapidly decreases. The alcohol more resembles a hydrocarbon. A similar relationship is found for organic acids. Thus, we expect the hydrocarbon portion of a fatty acid salt to be insoluble in water. However, the ionic end *does* have water solubility, even more than for an alcohol, since it has an ionic group. In a simplified explanation, "dirt," grime, and grease have more hydrocarbon character than water character and thus dissolve in the collective hydrocarbon chains of the soap molecules. The behavior of soap in water is shown in Figure 20.11. This explanation of the action of soap is very similar to the way biological molecules are carried through cell membranes. There is almost always a hydrocarbon (*hydrophobic*) portion of the molecule and a water-soluble (*hydrophilic*) portion.

MICELLES AND SURFACE CHEMISTRY From Figure 20.11, it is apparent that the hydrocarbon "tail" of soap will not be soluble in water and that the soap will align along the surface of the water with the ionic end in the water and the hydrocarbon tail protruding out of the water. But what if there are more soap molecules than can "fit" at the water surface? These molecules have to find a unique orientation that keeps the hydrocarbon portion out of the water. This is achieved by forming clusters of molecules in which the hydrocarbon tails are in the interior of the cluster and the ionic ends are on the surface of the cluster. This formation is called a **micelle**. Soap in the form of a micelle is able to clean since the oily dirt will be collected in the center of the micelle. The micelles stay in solution as a colloid (see Chapter 9) and will not come together to precipitate because of the ion—ion repulsion. Thus, the dirt suspended in the micelles is also easily rinsed away. This process is illustrated in Figure 20.12.

The hydrocarbon part of a soap molecule interacts to form a region of high hydrocarbon density. The ionic ends are on the outside of this region and are in the water phase.

In Chapter 24 the behavior of membranes will be shown to resemble the action of these micelles.

SOAP AND HARD WATER What are the effects of hard water on soap? The term *hard water* simply means that there is a relatively high concentration of divalent metal cations (Fe^{2+}, Mg^{2+}, and Ca^{2+}) in the water. These react with the soap. To balance the charge, however, *two* soap anions must react with one ion. The resulting product is an insoluble salt that forms the

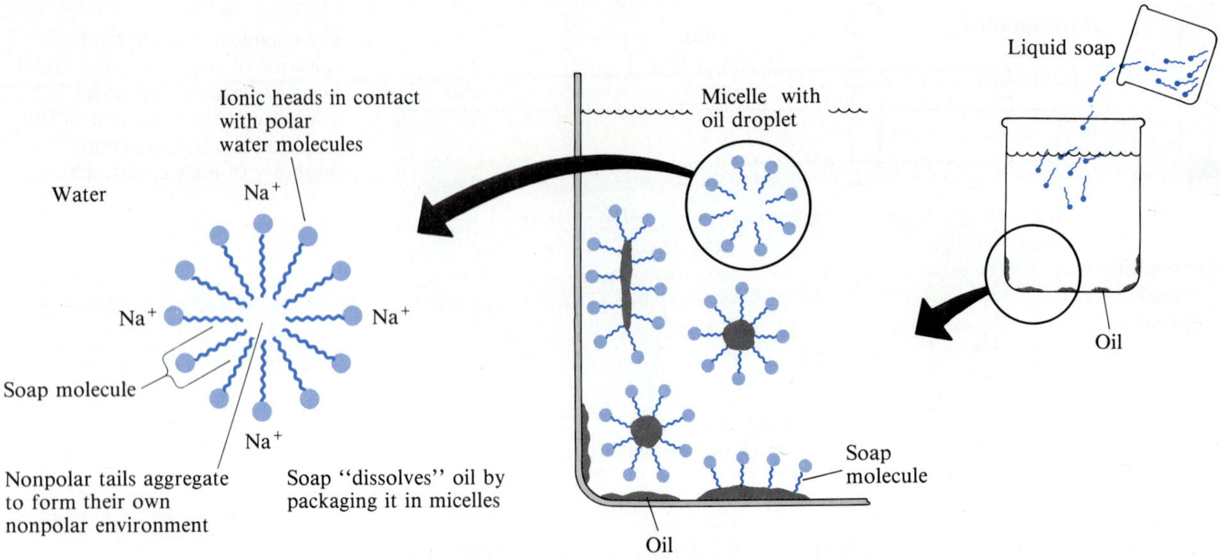

Liquid soap
Ionic heads in contact with polar water molecules
Micelle with oil droplet
Water
Na⁺
Na⁺
Na⁺
Soap molecule
Na⁺
Oil
Soap molecule
Nonpolar tails aggregate to form their own nonpolar environment
Soap "dissolves" oil by packaging it in micelles
Oil

FIGURE 20.12
The chemistry of micelles. This series of magnifications shows what happens to soap in water. The hydrocarbon portion of the soap can dissolve in itself to form a micelle. Grime and grease, being very similar to hydrocarbon, dissolve in the hydrocarbon portion of the soap. We are reminded of the general statement, "like dissolves like."

characteristic scum or "bathtub ring" in a sink or tub. It also serves as a water softener by removing the divalent ions from solution.

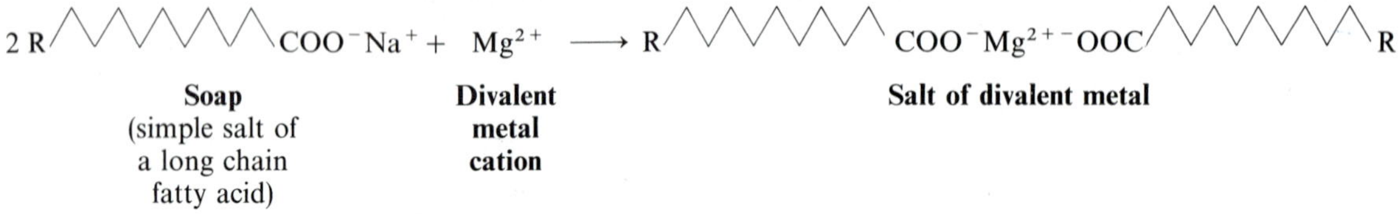

2 R COO⁻Na⁺ + Mg²⁺ ⟶ R COO⁻Mg²⁺⁻OOC R
Soap (simple salt of a long chain fatty acid)
Divalent metal cation
Salt of divalent metal

OTHER DETERGENTS There are two ways to get around the problem of hard water: (1) soften the water or (2) use a different type of detergent. In the former method, the divalent cations are exchanged for sodium ions by using an ion exchange material that has the capacity to "trade" ions. Once the divalent ions in the water are exchanged for sodium ions, the soap works well. The material in the ion exchange apparatus must be exchanged regularly because after a time, all of the sodium ions are used up. The ion exchange column is rinsed with a concentrated solution of sodium salts (often sodium chloride) to replace the divalent ions, and the hard water rinse is discarded.

The second way to get around the problem of hard water is to use a detergent (although soap is classified as a detergent, common use makes a distinction). The detergents have a long hydrocarbon portion, just as a soap. However, the polar ends are different (Table 20.6). Most important, however, is the observation that the polar ends *do not* react with divalent ions in hard water. The charged ends *do not* form insoluble salts with divalent ions, and thus remain effective in hard water. Additionally, most common washing detergents have additives to enhance the cleansing power. These are generally inorganic salts that have the ability to complex with the metal ions in

TABLE 20.6
Examples of Detergents

Type	Name	Structure
Cationic detergent	Trimethylhexadecylammonium chloride	$CH_3-(-CH_2-)_{15}-\overset{+}{N}(CH_3)_3Cl^-$
Anionic detergent	Sodium dodecylbenzene sulfonate	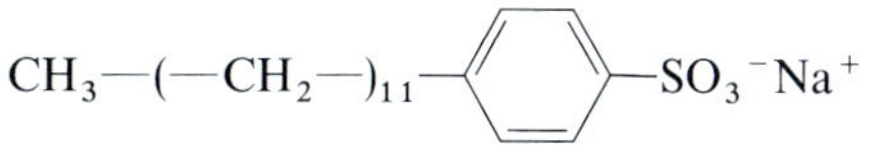
Neutral detergent	Pentaerythrityl palmitate	$CH_3-(-CH_2-)_{14}-C(=O)-O-CH_2-C(CH_2-OH)_3$

hard water, thus removing them from the possibility of reacting with the soap.

NEW TERMS

ester A functional group derived from an acid and an alcohol.

$$R-\underset{\displaystyle O}{\underset{\|}{C}}-O-R'$$

thioester A functional group derived from an acid and a thiol.

$$R-\underset{\displaystyle O}{\underset{\|}{C}}-S-R'$$

detergent A material that is a surface active agent having a long hydrocarbon chain and a polar end. If the polar end is a carboxylate anion, the detergent is called a soap. Most other polar groups are often generically classified as detergents.

saponification The basic hydrolysis of an ester to form the salt of the acid and the alcohol.

soap A detergent consisting of the salt of a long chain fatty acid.

micelle An aggregation of hydrocarbon materials having polar ends in the water phase.

TESTING YOURSELF

Ester Formation and Nomenclature

1. Give the names of the products of these acid-catalyzed ester reactions.

a.

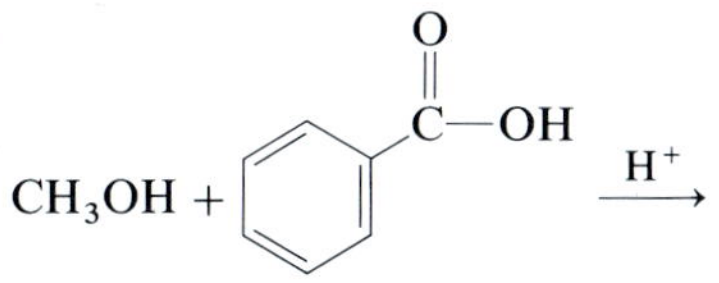

b. $CH_3-CH_2-\underset{\displaystyle O}{\underset{\|}{C}}-OH + CH_3CH_2-OH \xrightarrow{H^+}$

c. $$CH_3-\underset{\overset{|}{}}{\overset{CH_3}{\overset{|}{CH}}}-\underset{\underset{O}{\|}}{C}-OH + CH_3-\overset{CH_3}{\overset{|}{CH}}-OH \xrightarrow{H^+}$$

2. Give the product for the following reaction.

$$2\,CH_3-(CH_2)_8-COO^-Na^+ + Fe^{2+} \longrightarrow$$

Answers **1. a.** methyl benzoate **b.** ethyl propanoate **c.** isopropyl 2-methylpropanoate
2. $CH_3-(CH_2)_8-\underset{\underset{O}{\|}}{C}-O^-Fe^{2+-}O-\underset{\underset{O}{\|}}{C}-(CH_2)_8-CH_3$

20.4 Acid Chlorides and Anhydrides

Preparation of Acid Chlorides

Acid chlorides are prepared from acids by replacing the —OH group with —X.

Just as the hydroxyl group in an alcohol can be replaced by halogen (Chapter 19), this group in an acid can also be replaced by the halogen chlorine.

$$R-OH \longrightarrow R-X$$

$$R-\underset{\underset{O}{\|}}{C}-OH \longrightarrow R-\underset{\underset{O}{\|}}{C}-X$$

The product of this replacement with chlorine is an **acid chloride**. The acid chlorides are convenient to prepare experimentally because one simply has to add a carboxylic acid to *thionyl chloride*, which has the structure

$$Cl-\overset{\overset{O}{\|}}{S}-Cl$$

When evolution of the gases SO_2 and HCl ceases, the excess thionyl chloride is removed by distillation and the acid chloride is collected.

GENERAL REACTION

Preparation of an Acid Chloride

$$\underset{\text{Acid}}{R-\underset{\underset{O}{\|}}{C}-O-H} + \underset{\text{Thionyl chloride}}{SOCl_2} \longrightarrow \underset{\text{Acid chloride}}{R-\underset{\underset{O}{\|}}{C}-Cl} + SO_2 + HCl$$

Examples

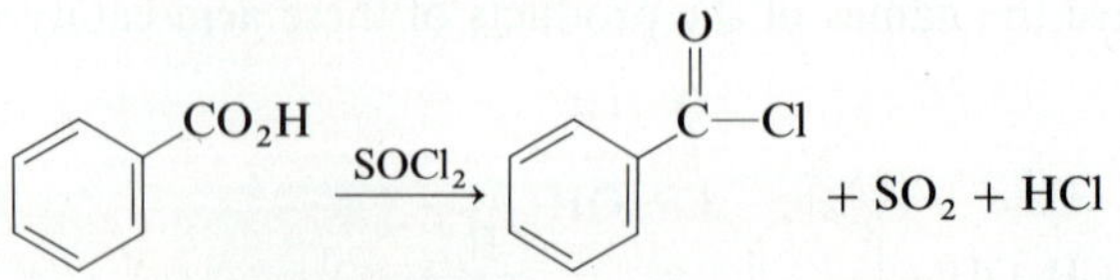

$$CH_3-COOH \xrightarrow{SOCl_2} CH_3-\underset{\underset{O}{\|}}{C}-Cl + SO_2 + HCl$$

Reactions of Acid Chlorides

Recall from the displacement reactions of alkyl halides in Chapter 18 that the chlorine atom is more easily displaced than an —OH group. This tends to make acid chlorides much more reactive than the parent acids. Two reactions to consider include the hydrolysis and the esterification reaction. Notice that the two reactions are almost identical, again showing the similarities of alcohols to water.

Acid chlorides undergo the same reactions as acids, but are generally more reactive.

Reaction of Alkyl Halides	Reaction of Acid Chlorides
$RO^- + R'{-}X \longrightarrow R{-}O{-}R'$	$R{-}O{-}H + R'{-}C(=O){-}Cl \longrightarrow R{-}O{-}C(=O){-}R' + HCl$
$HO^- + R'{-}X \longrightarrow R'{-}OH$	$H{-}O{-}H + R'{-}C(=O){-}Cl \longrightarrow H{-}O{-}C(=O){-}R' + HCl$

GENERAL REACTION

Hydrolysis and Esterification of Acid Chlorides

$$R{-}C(=O){-}Cl + H{-}O{-}Z \longrightarrow R{-}C(=O){-}O{-}Z + HCl$$

where Z = H Hydrolysis
Z = R Esterification

Examples

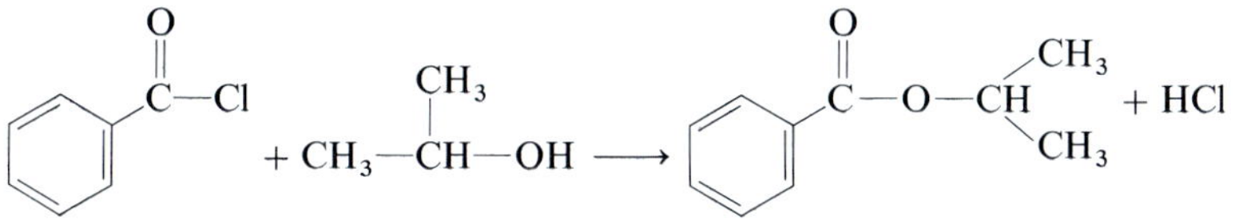

$$CH_3{-}C(=O){-}Cl + H_2O \longrightarrow CH_3{-}C(=O){-}OH + HCl$$

Acid Anhydrides

As the name implies, **acid anhydrides** are anhydrous or dehydrated acids, that is, they have lost water.

Anhydrides can be considered as anhydrous acids and have a reactivity between that of an acid and an acid chloride.

$$R{-}C(=O){-}O{-}H + H{-}O{-}C(=O){-}R \longrightarrow R{-}C(=O){-}O{-}C(=O){-}R + H_2O$$

(The O—H + H shown highlighted leaves as H_2O.)

Anhydride

The anhydride reacts in the same way that an acid chloride reacts. Anhydrides are more reactive than acids but less reactive than acid chlorides.

The anhydride can be hydrolyzed by water or esterified with alcohol, but at a slower rate than observed with an acid chloride.

GENERAL REACTION

Reaction of an Acid Anhydride with an Alcohol

$$\mathrm{R{-}\underset{\underset{\displaystyle O}{\|}}{C}{-}O{-}\underset{\underset{\displaystyle O}{\|}}{C}{-}R + H{-}O{-}R' \longrightarrow R{-}\underset{\underset{\displaystyle O}{\|}}{C}{-}O{-}R' + H{-}O{-}\underset{\underset{\displaystyle O}{\|}}{C}{-}R}$$

Examples

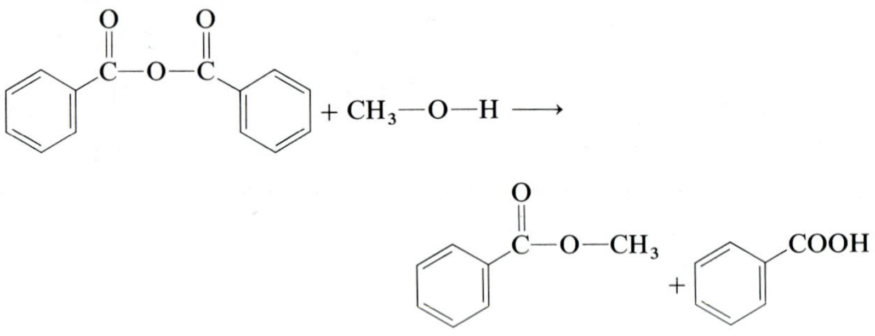

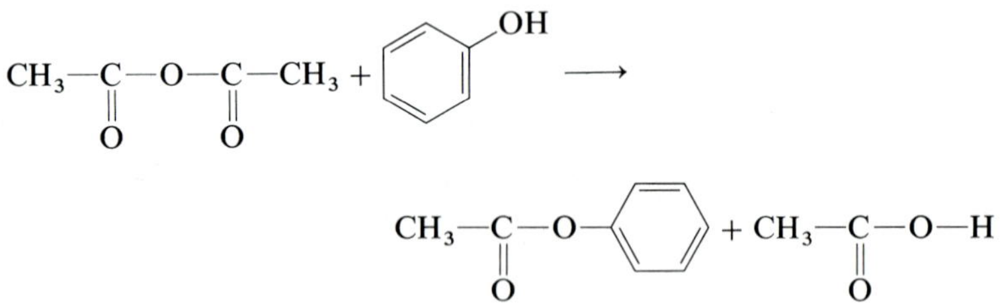

Reactions in organic chemistry have a very consistent pattern. In the reaction with acid chlorides, one of the by-products of the reaction is HCL. In the chemistry of anhydrides, an acid by-product is also obtained, but in these cases, it is an organic acid. The similarities of the chemistry of acid anhydrides and acid chlorides can be illustrated by the following reaction:

$$\mathrm{R{-}\underset{\underset{\displaystyle O}{\|}}{C}{-}G + H{-}O{-}R' \longrightarrow R{-}\underset{\underset{\displaystyle O}{\|}}{C}{-}O{-}R'}$$

where R′ = H or alkyl. If G = —Cl, then it is an acid chloride, and if G = $\mathrm{{-}O{-}\underset{\underset{\displaystyle O}{\|}}{C}{-}R}$, then it is an anhydride.

NEW TERMS

acid chloride A derivative of an organic acid that is more reactive than an acid, having the structure

$$\mathrm{R{-}\underset{\underset{\displaystyle O}{\|}}{C}{-}Cl}$$

acid anhydride A derivative of an organic acid that is dehydrated, having the structure

$$R-\overset{\overset{\large O}{\|}}{C}-O-\overset{\overset{\large O}{\|}}{C}-R$$

TESTING YOURSELF

Reactions of Acid Chlorides and Anhydrides

1. Give products for the following acid chloride reactions.

a.
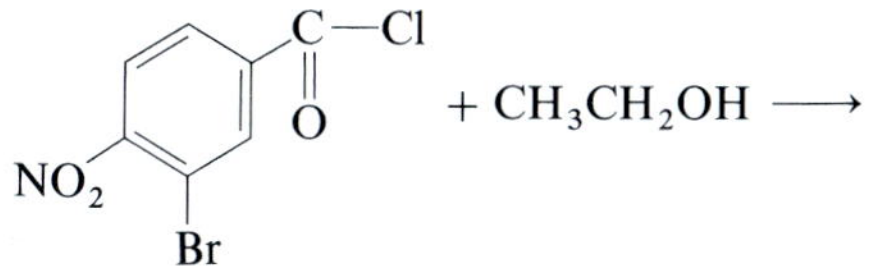

b.
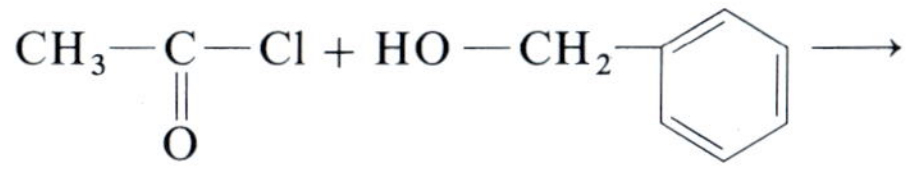

c. $CH_3-\underset{\underset{O}{\|}}{C}-Cl \xrightarrow[HO^-,\ \Delta]{H_2O}$

2. Give products for the following acid anhydride reactions.

a. $CH_3-CH_2-\overset{\overset{O}{\|}}{C}-O-\overset{\overset{O}{\|}}{C}-CH_2CH_3 \xrightarrow{H_2O}$

b.
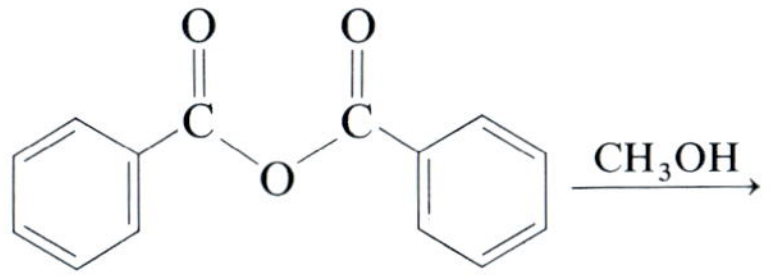

Answers 1. a.
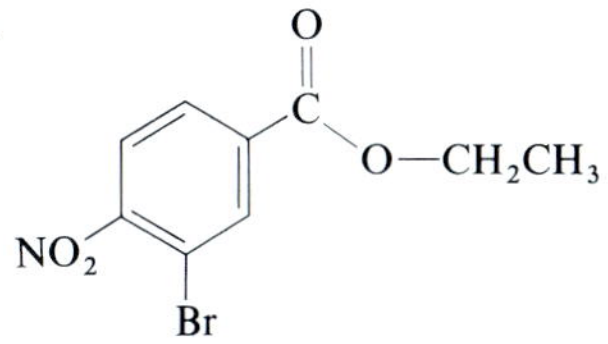

b.
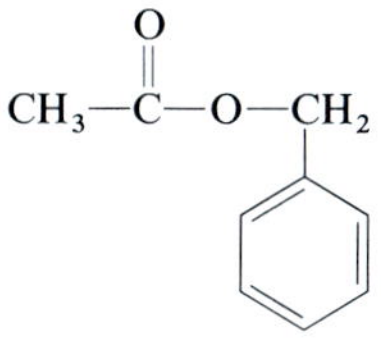

c. $CH_3-\underset{\underset{O}{\|}}{C}-OH$ **2. a.** $CH_3-CH_2-\underset{\underset{O}{\|}}{C}-OH$

b.
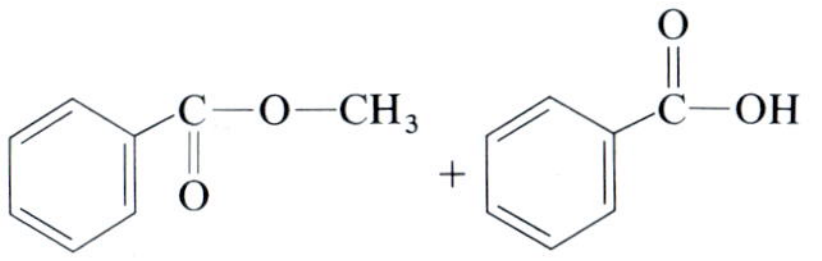

20.5 Claisen Condensation

Just as the Aldol condensation was a method for combining two carbonyl compounds (Chapter 19), the **Claisen condensation** combines two esters. The proton on the carbon atom adjacent to the carboxyl carbon is acidic,

The *Claisen condensation* combines two esters to form a beta keto ester.

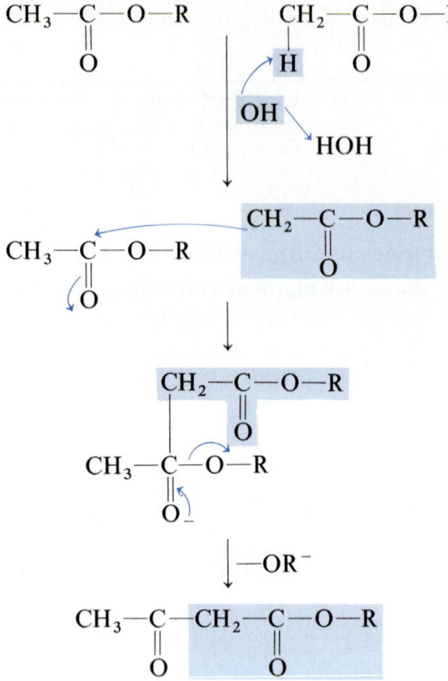

FIGURE 20.13
The Claisen condensation combines two ester molecules and forms a beta keto ester. Two molecules of ester combine (one molecule is highlighted) to give a beta keto ester.

$$CH_3-C(=O)-O-R \quad\quad CH_2-C(=O)-O-R$$
$$\quad\quad\quad\quad\quad\quad\quad\quad\quad\quad |H$$
$$\quad\quad\quad\quad\quad\quad OH \searrow$$
$$\quad\quad\quad\quad\quad\quad\quad\quad\quad HOH$$

$$CH_3-C(=O)-O-R \quad + \quad {}^-CH_2-C(=O)-O-R$$

$$\downarrow$$

$$\begin{array}{c} CH_2-C(=O)-O-R \\ | \\ CH_3-C(-O^-)-O-R \end{array}$$

$$\downarrow -OR^-$$

$$CH_3-C(=O)-CH_2-C(=O)-O-R$$

as is the proton on the carbon atom adjacent to a carbonyl. It can be removed, and the resulting nucleophile can react with another molecule of ester. The product resulting from such a reaction is a beta keto ester. This reaction is shown in Figure 20.13.

We have already mentioned diabetes as a disease caused by improper metabolism of glucose. In serious cases of this disorder, two molecules of acetylCoA undergo Claisen condensation to form a beta keto thioester. Hydrolysis of this ester leads to a beta keto acid, and we have already seen that these readily undergo decarboxylation. The product of this reaction is acetone. One of the symptoms of the disease is the slight odor of acetone on the breath.

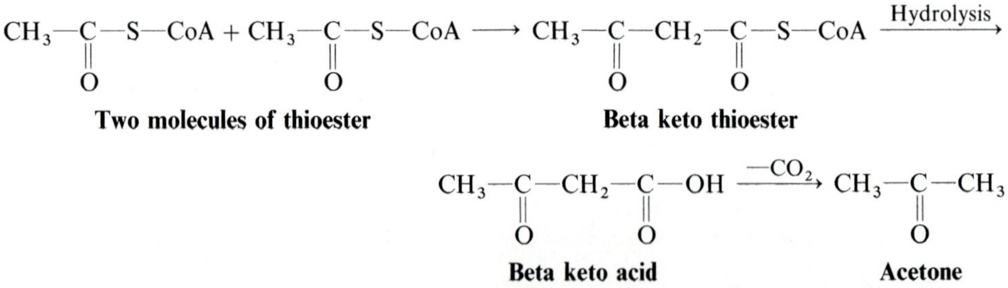

$$CH_3-C(=O)-S-CoA + CH_3-C(=O)-S-CoA \longrightarrow CH_3-C(=O)-CH_2-C(=O)-S-CoA \xrightarrow{Hydrolysis}$$

Two molecules of thioester   **Beta keto thioester**

$$CH_3-C(=O)-CH_2-C(=O)-OH \xrightarrow{-CO_2} CH_3-C(=O)-CH_3$$

Beta keto acid   **Acetone**

Like the products of an aldol condensation, beta keto esters also readily come apart. This reaction becomes very important in biological reactions.

$$R-\cdots-C(=O)-CH_2-C(=O)-OR \xrightarrow{:Nu}$$

$$R-\cdots-C(Nu)(=O) \quad + \quad {}^-CH_2-C(=O)-OR \xrightarrow{H_2O} CH_3-C(=O)-OR$$

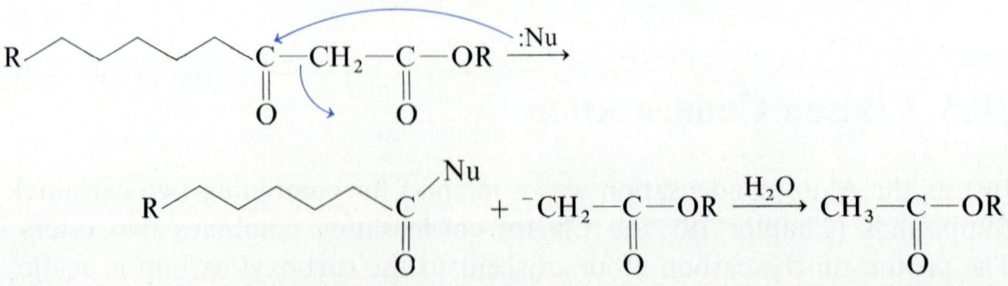

NEW TERM

Claisen condensation A reaction in which two molecules of an ester react to form a beta keto ester.

20.6 Important Organic Acids and Acid Derivatives

We have already discussed how the salt of trifluoroacetic acid is used as a poison. Salts of benzoic acid and propanoic acid are used as preservatives. Sodium benzoate is used as an antimicrobial in beverages, jams and jellies, and a number of other food items. Sodium and/or calcium propionate is used as a preservative in many bread products, where it reduces or inhibits mold growth.

Sorbic acid and its salts have found use as food additives to inhibit mold and yeast growth. They are often the preservatives found in smoked, pickled, and dried foods.

$CH_3CH{=}CH{-}CH_2{-}COOH$

Sorbic acid

A number of interesting esters are found in everyday life. The esters of *p*-hydroxybenzoic acid (the parabens) are used for inhibiting growth of yeasts and molds, but they are not as useful as antibacterials. The polyester fiber *dacron* is a polymeric ester of terphthalic acid and ethylene glycol.

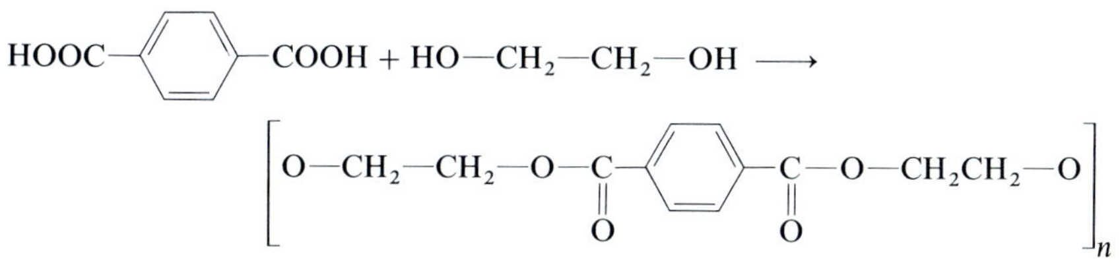

This important polyester has medical applications because it is inert to normal biochemical processes and can be used to repair blood vessels.

SUMMARY

Organic acids are less acidic than the more common mineral acids, but they are more acidic than other organic functional groups. The increased ability of the *carboxyl group* to give up a proton can be attributed to the delocalization of charge on the remaining anion—much the same reasoning used to explain phenol acidity. *Carboxylic acids* are prepared by oxidations of alcohols, aldehydes, and alkyl-substituted aromatic systems. They can also be formed from the *hydrolysis* of *nitriles. Beta keto acids* undergo very easy loss of carbon dioxide. When an acid and an alcohol are mixed under proper conditions, one obtains an *ester. Acid halides* and *anhydrides* are both derivatives of acids, and they are both more reactive than the acids.

REACTION SUMMARY

Oxidation of a Primary Alcohol to an Acid

$$R{-}CH_2{-}OH \xrightarrow{\text{Oxidation}} R{-}\underset{\underset{O}{\|}}{C}{-}H \xrightarrow{\text{Oxidation}} R{-}COOH$$

Oxidation of an Alkyl Side Chain on a Benzene Ring

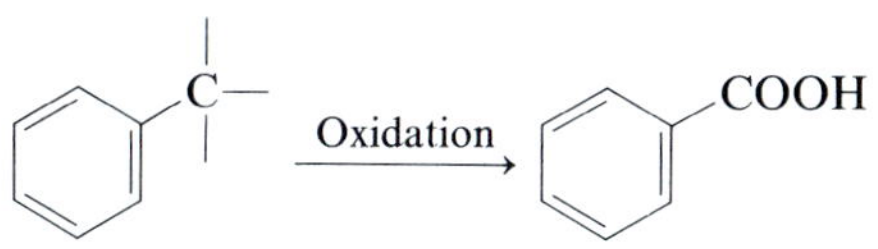

Formation of Nitriles by a Displacement Reaction

$$\mathrm{R{-}X + CN^- \longrightarrow R{-}CN + X}$$

Acid Hydrolysis of a Nitrile

$$\mathrm{R{-}CN \xrightarrow[H^+]{H_2O} R{-}COOH + {}^{+}NH_4}$$

Basic Hydrolysis of a Nitrile

$$\mathrm{R{-}CN \xrightarrow[HO^-]{H_2O} R{-}COO^- + NH_3}$$

Decarboxylation of a Beta Keto Acid

$$\mathrm{R{-}\underset{\overset{\|}{O}}{C}{-}\underset{|}{\overset{|}{C}}{-}COOH \xrightarrow[\Delta]{-CO_2} R{-}\underset{\overset{\|}{O}}{C}{-}\underset{|}{\overset{|}{C}}{-}H}$$

Ester Formation

$$\mathrm{R{-}\underset{\overset{\|}{O}}{C}{-}OH + HOR' \xrightarrow{H^+} R{-}\underset{\overset{\|}{O}}{C}{-}O{-}R' + H{-}O{-}H}$$

Formation of Thioesters

$$\mathrm{R{-}\underset{\overset{\|}{O}}{C}{-}OH + HSR' \xrightarrow{H^+} R{-}\underset{\overset{\|}{O}}{C}{-}S{-}R' + HOH}$$

Basic Hydrolysis (Saponification) of an Ester

$$\mathrm{R{-}\underset{\overset{\|}{O}}{C}{-}O{-}R' \xrightarrow[HO^-,\,\Delta]{H_2O} R{-}\underset{\overset{\|}{O}}{C}{-}O^- + R'OH}$$

Preparation of Acid Chlorides

$$\mathrm{R{-}\underset{\overset{\|}{O}}{C}{-}OH + SOCl_2 \longrightarrow R{-}\underset{\overset{\|}{O}}{C}{-}Cl + SO_2 + HCl}$$

Hydrolysis and Esterification of Acid Chlorides

$$\mathrm{R{-}\underset{\overset{\|}{O}}{C}{-}Cl + HOH \longrightarrow R{-}\underset{\overset{\|}{O}}{C}{-}O{-}H + HCl}$$

$$\mathrm{R{-}\underset{\overset{\|}{O}}{C}{-}Cl + HOR' \longrightarrow R{-}\underset{\overset{\|}{O}}{C}{-}O{-}R' + HCl}$$

Reaction of an Acid Anhydride with an Alcohol

$$\mathrm{R{-}\underset{\overset{\|}{O}}{C}{-}O{-}\underset{\overset{\|}{O}}{C}{-}R + HOR' \longrightarrow R{-}\underset{\overset{\|}{O}}{C}{-}O{-}R' + R{-}\underset{\overset{\|}{O}}{C}{-}OH}$$

TERMS

carboxyl group
carboxylic acid
fatty acid
carboxylate anion
nitrile
hydrolysis
decarboxylation
keto acids
ester
thioester
detergent
saponification
soap
micelle
acid chloride
acid anhydride
Claisen condensation

CHAPTER REVIEW

QUESTIONS

1. 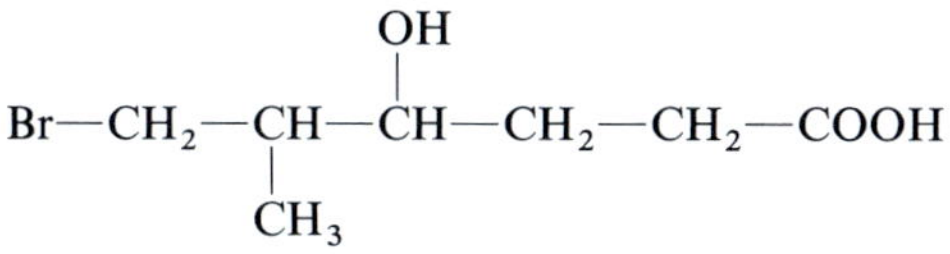 is best named
   a. 1-bromo-2-methyl-3-alcoholhexanoic acid
   b. 1-carboxyl-4-methyl-5-bromo-3-pentanol
   c. 1-bromo-3-hydroxy-2-methylhexanoic acid
   d. 6-bromo-4-hydroxy-5-methylhexanoic acid
2. Which of these is most reactive toward water?
   a. acid chloride b. acid anhydride c. ester d. organic acid
3. A beta keto acid will not
   a. be called a 3-ketoacid
   b. lose carbon dioxide when heated
   c. be called a 2-ketoacid
   d. form a ketone when heated
4. An acid is not formed by
   a. oxidation of a primary alcohol
   b. nitrile hydrolysis
   c. aldehyde oxidation
   d. oxidation of a ketone
5. In an ester, 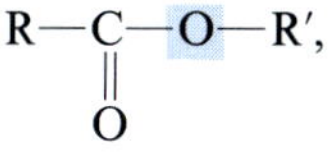 the highlighted oxygen comes from
   a. alcohol b. acid c. water d. acid chloride
6. The salt of a long chain organic acid is called a
   a. carbanion b. cationic detergent c. soap d. saponification
7. A nitrile is sometimes classified as
   a. an organic cyanide
   b. an inorganic ester
   c. a mixed anhydride
   d. a base
8. Oxidation of toluene will give
   a. benzyl alcohol b. benzaldehyde c. benzoic acid d. quinone
9. Basic hydrolysis of a nitrile will *not* give
   a. ammonia b. acid salt c. free acid
10. What is the product of this reaction?

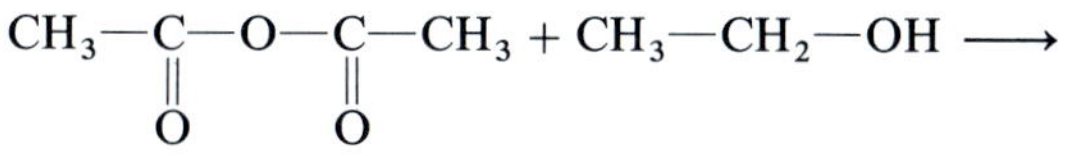

   a. methyl propanoate b. ethyl propanoate c. methyl acetate d. ethyl acetate

Answers **1.** d **2.** a **3.** c **4.** d **5.** a **6.** c **7.** a **8.** c **9.** c **10.** d

DIAGNOSTIC CHART

Blacken in all of the circles under the number of each question that you missed in the preceding Chapter Review questions. The diagnostic chart will help you identify concept areas that need more study.

Concepts	Questions 1	2	3	4	5	6	7	8	9	10
Structure			○		○	○	○			○
Reactions		○		○	○			○	○	○
Function groups	○	○	○	○	○	○	○		○	○
Nomenclature	○		○					○		○

EXERCISES

Structure and Nomenclature

1. Draw electron dot structures for the following.

a. H C O H
O

b. H C N

c. H H
H C C O C H
H O H

2. Name the following compounds.

a. 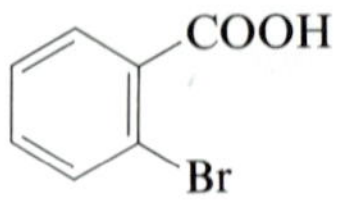

b. $CH_3—CH_2—C(=O)—COOH$

c. $CH_3—C(CH_3)_2—COOH$

d. $CH_3—CBr_2—C(=O)—CH_2—COOH$

e. $(CH_3)_2CH—CH(Cl)—COOH$

f. $CH_3—CH(OH)—CH_2—CH_2—COOH$

3. Draw structures for each of the following.

a. isopropylpropanoate

b. *cis*-2,3-dibromo-2-hexenoic acid

c. beta keto pentanoic acid

d. isopropyl pentanoate

e. 2-bromobutanoic acid

f. *m*-nitrobenzoic acid

g. alpha keto propanoic acid

4. Name the following compounds.

a. 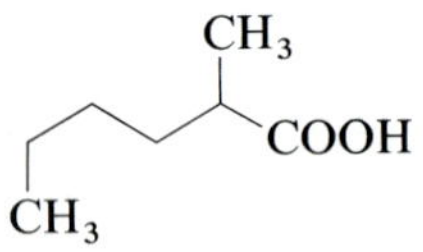

b. 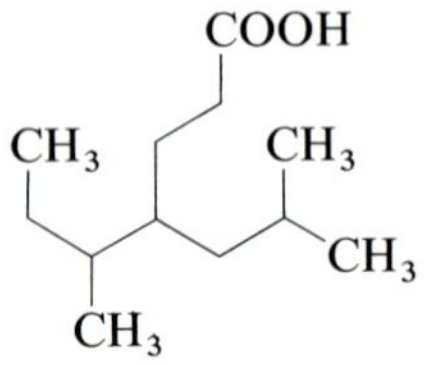

c. 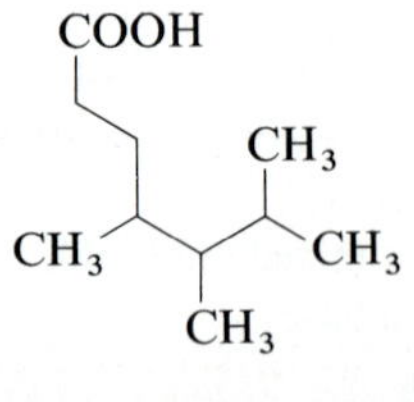

d. 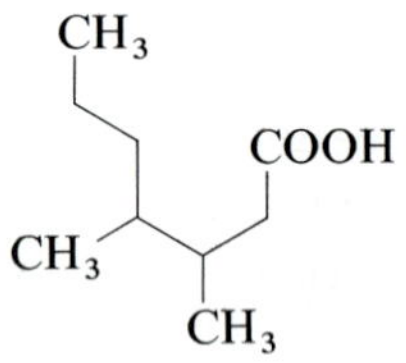

e. 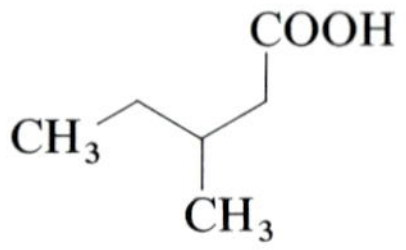

f. 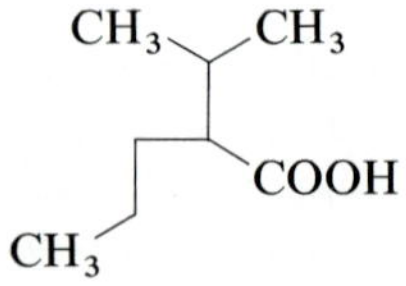

Physical Properties

5. By some measurements, acetic acid seems to have a molecular weight two times larger than it really is. Why?

6. Which would be the better soap, $CH_3—CH_2—COO^-Na^+$ or $CH_3—(CH_2)_{10}—COO^-$? Why?

7. Explain why FCH_2COOH is a stronger acid than ICH_2COOH.

Reactions of Acids

8. Give products for the following reactions.
   a. $CH_3COOH + NaOH \longrightarrow$
   b. $CH_3-CH_2-C(=O)-CH(CH_3)-COOH \xrightarrow{\Delta}$
   c. $(CH_3)_2CH-CH_2-CH_2-OH \xrightarrow{[O]}$
   d. Benzoic acid + Methanol $\xrightarrow{H^+}$
   e. Hexanoic acid + Thionyl chloride $\longrightarrow$
   f. 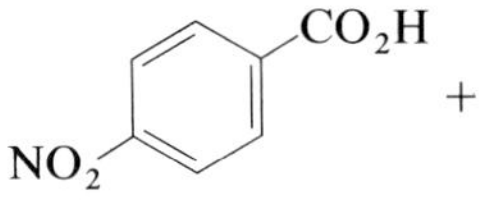 $+ NaOH \longrightarrow$
   g. 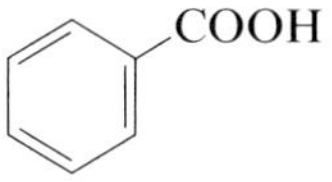 $+ SOCl_2 \longrightarrow$
   h. $CH_3-CH_2-CH_2-C(=O)-CH_2-COOH \xrightarrow{\Delta}$

Reactions of Esters

9. Give structures of the esters formed in the following reactions.
   a. $CH_3-CH_2-C(=O)-Cl$ + *n*-Propanol $\longrightarrow$
   b. 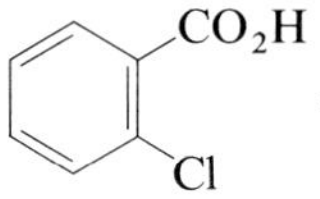 $+ CH_3-CH_2OH \xrightarrow{H^+}$
   c. 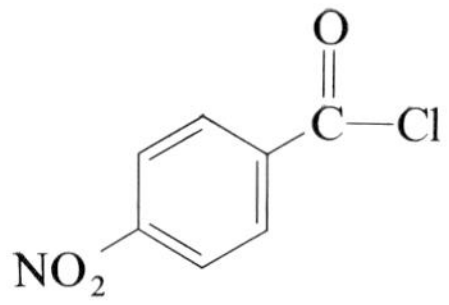 $+ CH_3OH \longrightarrow$
   d. $CH_3-CH_2-C(=O)-Cl +$ 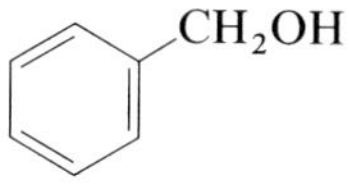 $\longrightarrow$
   e. $CH_3-C(=O)-O-C(=O)-CH_3 + CH_3OH \longrightarrow$
   f. 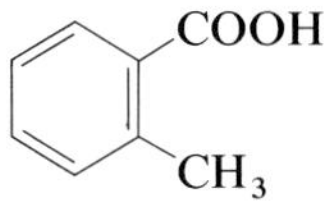 $+ CH_3CH_2OH \xrightarrow{H^+}$
   g. $CH_3-C(=O)-O-C(=O)-CH_3 + CH_3-OH \longrightarrow$
   h. 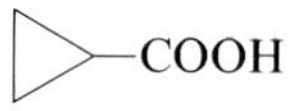 $+ CH_3CH_2OH \xrightarrow{H^+}$
   i. 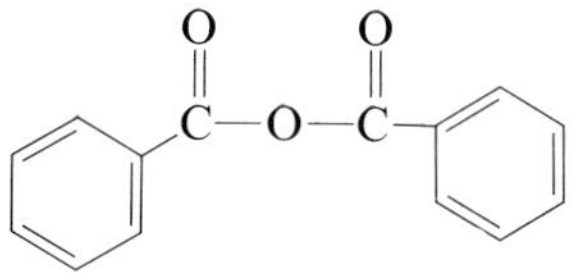 $+ CH_3OH \longrightarrow$

10. Give products for the following reactions of esters.
   a. 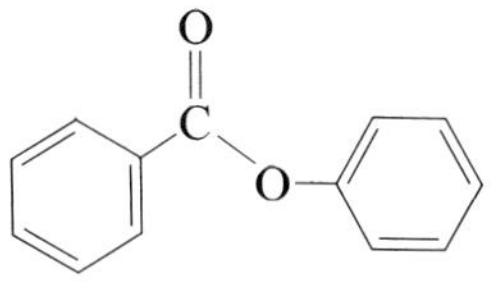 $\xrightarrow[\Delta]{NaOH,\ H_2O}$
   b. $CH_3-(CH_2)_{12}-C(=O)-O-CH_2CH_3 \xrightarrow[\Delta]{NaOH,\ H_2O}$

Other Methods to Form Acids

11. Give products for these reactions.
   a. Butylbenzene $\xrightarrow{[O]}$
   b. 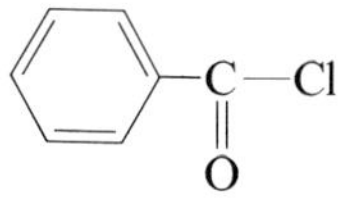 $+ H_2O \longrightarrow$
   c. 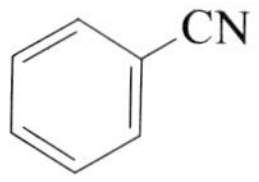 $\xrightarrow[\Delta]{H^+,\ H_2O}$
   d. 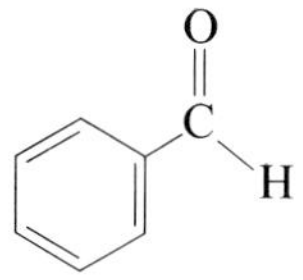 $\xrightarrow{\text{Oxidation}}$
   e. $(CH_3)_2CH-CH_2OH \xrightarrow{\text{Oxidation}}$
   f. $CH_3-C(=O)-O-C(=O)-CH_3 + H_2O \longrightarrow$
   g. $CH_3CH_2CN + H_2O \xrightarrow[\Delta]{H^+}$

Thought Questions

12. Which of the acids in each pair is more acidic? Why?
   a. α-fluoropropanoic acid, β-fluoropropanoic acid
   b. β-chlorohexanoic acid, β-bromohexanoic acid

13. The preparation of aspirin and oil of wintergreen are both esterification reactions, and they both come from the same starting material—salicylic acid. How would you prepare these two molecules?

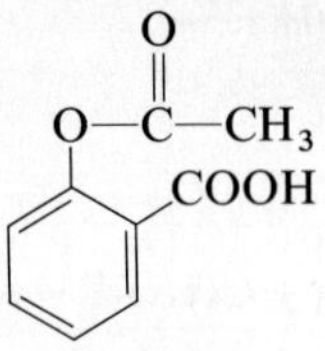

Aspirin

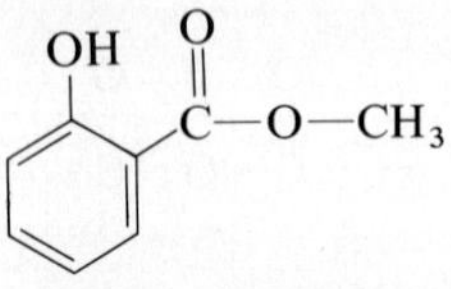

Oil of wintergreen

14. In diabetes, a person may exhibit symptoms of ketosis, and may even have "acetone breath." If one of the metabolic products is beta keto butanoic acid, show how it will form acetone.

15. Explain why a concentrated detergent might be as useful as turpentine for cleaning a paint brush used for oil paints.

CHAPTER

21

Amines and Amides

Amines are organic analogs of ammonia, NH_3, with one or more alkyl groups substituted for the ammonia hydrogens. Because of the pair of nonbonding electrons on the nitrogen atom, amines act as proton acceptors. Amines are the fundamental functional group in a number of important pharmaceutical drugs.

OBJECTIVES

After completing this chapter, you should be able to

- Discuss the origins of basicity of amines.
- Predict the products of amines with acids and carbonyl compounds.
- Name simple amines.
- Discuss how amide bonding controls structure.
- Recognize the amine groups in alkaloids.

Amines are the organic equivalent of ammonia.

Nitrogen is an abundant element on the earth, ranking 14th in elemental abundance in the earth's crust (Table 21.1). Thus, it should be no surprise to see that it is incorporated into a large number of organic compounds. In the biochemistry chapters that follow, nitrogen-containing organic compounds will be seen to have a central role in human physiology. Many biologically important compounds are formed from nitrogen-containing **amines**, which are organic derivatives of ammonia. If one of the hydrogen atoms of ammonia is replaced by an R group, a new compound called a *primary amine* is formed. Replacement of two hydrogens results in a *secondary amine*, and replacement of all three hydrogens gives a *tertiary amine*. The R groups *do not* have to be the same.

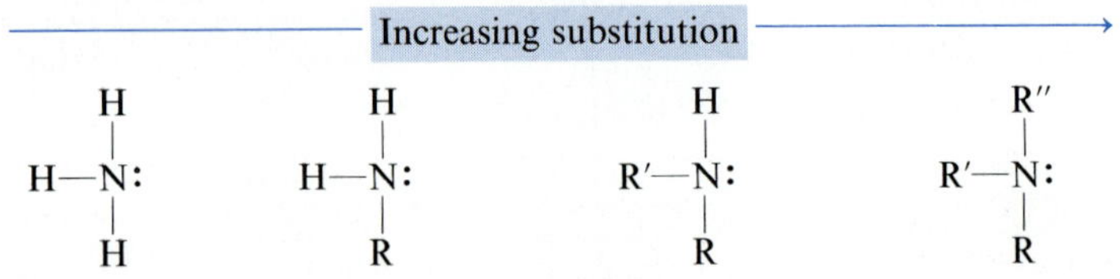
Increasing substitution

Ammonia **Primary amine** **Secondary amine** **Tertiary amine**

21.1 Amines: Structure, Properties, and Nomenclature

Hybridization and Structure

Ammonia is a weak base as a result of its lone pair of electrons (Lewis base). Application of the rules used for Lewis electron dot structures gives a structure in which the nitrogen atom of the amine has three bonds and a lone pair of electrons. This requires that amines have sp^3 hybridization. In general, an N—H bond of ammonia can be substituted by an N—R group to make several different kinds of substituted amines. The lone pair of electrons is responsible for the unique chemical behavior and properties of the amine, just as it provides for the chemical behavior of ammonia. A general structure of an amine is given here, in which R is an alkyl or H.

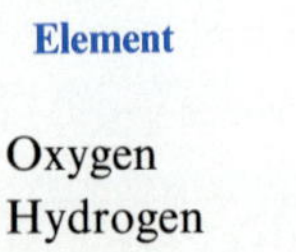

TABLE 21.1
Elemental Abundance in the Earth's Crust

Rank (in Abundance)	Element	Atoms per 1000 Atoms on Earth
1	Oxygen	533
3	Hydrogen	151
5	Carbon	30
12	Chlorine	10
14	**Nitrogen**	**9**

A logical question to ask at this point is why the amines are considered to be basic as a result of having lone pair electrons, while ethers are not, even though they also contain lone pairs of electrons. Recall that an ether *is* a base in the Lewis sense—it is an electron donor. However, amines are better electron donors. This is because nitrogen is less electronegative than oxygen, thus, the electrons on an amine are not held as tightly.

Hydrogen Bonding

Since nitrogen is more electronegative than carbon, polarization of the carbon–nitrogen bond in an amine is expected (Figure 21.1).

The C—N bond is a polar covalent bond with a dipole such that the carbon has a partial positive charge and the nitrogen a partial negative charge. However, the difference in electronegativity is less than for a C—O or C—Cl bond, and less polarization occurs. Similarly, in the N—H bond, the H acts as a partially positive atom. This is similar to the H—O bond of water (but to a lesser extent), so that amines can also hydrogen bond with one another and with other polar molecules.

The hydrogen bonding of amines isn't as strong as that of alcohols or acids.

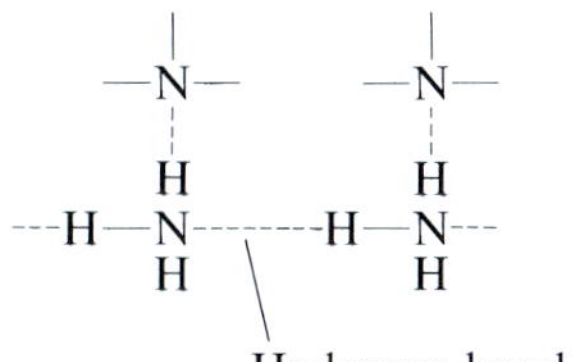
Hydrogen bond

Just as the alcohols had fewer hydrogens for hydrogen bonding than water, the amines decrease in their ability to hydrogen bond as they become more substituted. Since the difference in electronegativity between N and H is less than the difference between O and H in water, hydrogen bonding is less important for the amines. This is illustrated by comparison of boiling points for a series of amines with boiling points of alcohols (Figure 21.2).

Nomenclature

Amines are generally named by identifying all of the alkyl groups on the nitrogen, followed by the suffix *-amine.*

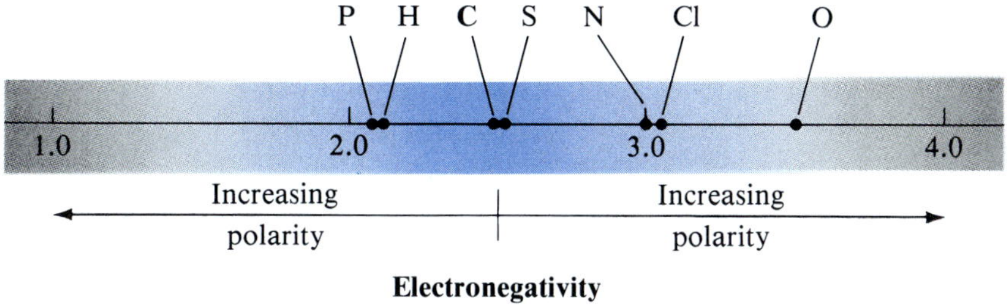
P
H
C
S
N
Cl
O
1.0
2.0
3.0
4.0
Increasing polarity
Increasing polarity
Electronegativity

FIGURE 21.1
The electronegativity difference between carbon and nitrogen suggests that there is a polarized C—N bond.

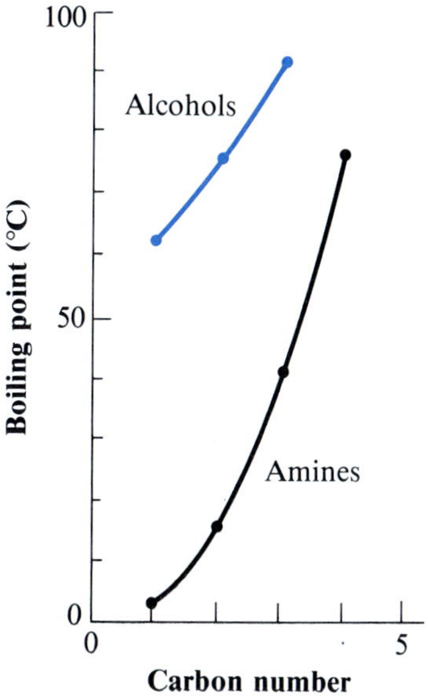
100
50
0
0
5
Alcohols
Amines
Boiling point (°C)
Carbon number

FIGURE 21.2
The boiling points of amines are considerably lower than the boiling points of the corresponding alcohols. This is a result of decreased hydrogen bonding.

EXAMPLE 21.1

Name these amines.

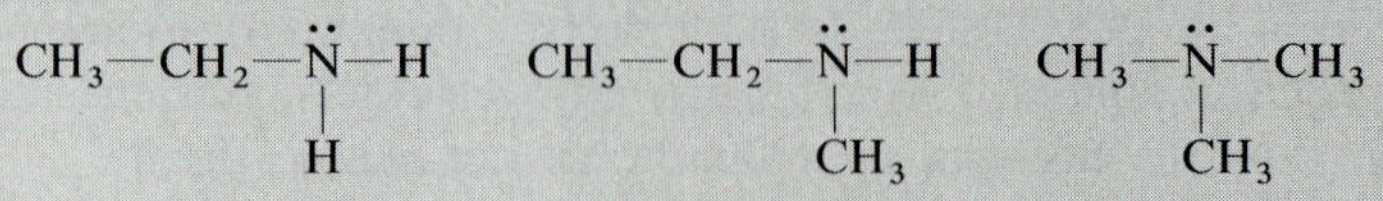

Ethylamine **Ethylmethylamine** **Trimethylamine**

TABLE 21.2
Heterocyclic Amines Having Common Names

Name	Structure
Aniline	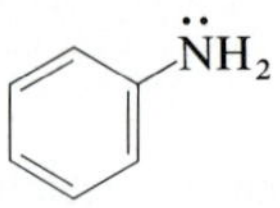
Pyridine	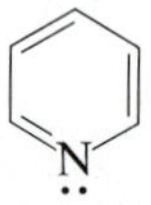
Pyrrolidine	
Piperidine	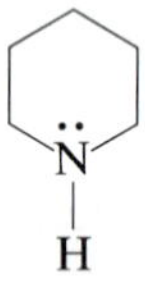
Pyrrole	
Nicotinic acid	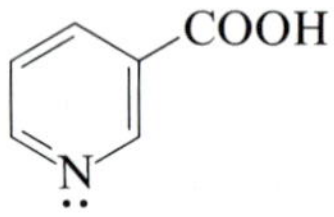
Purine	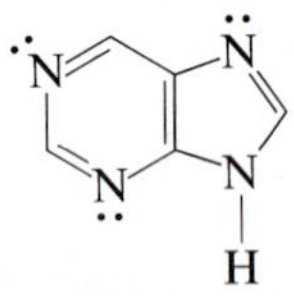
Pyrimidine	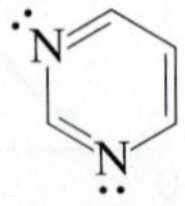
Imidazole	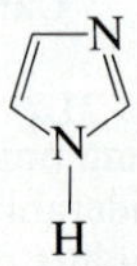

A ring system can also have a nitrogen atom as part of the ring. In general, any ring that contains some atom other than carbon is called a **heterocycle**. Here are three examples of heterocycles:

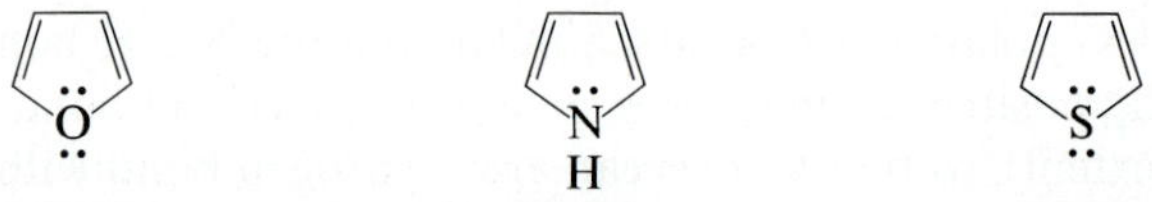

Furan (oxygen heterocycle) **Pyrrole** (nitrogen heterocycle) **Thiophene** (sulfur heterocycle)

Note that the one in the middle (pyrrole) is an amine because it contains nitrogen. Many nitrogen-containing heterocycles have common names, such as those in Table 21.2.

Many amines are characterized by strong, fishy odors; in fact, some have very bad odors. Consider the names of two common diamines, *cadaverine* and *putrescine*; these are produced by microorganisms in rotting meat. Note how these names suggest the odor associated with the amine!

$$H_2N-CH_2-CH_2-CH_2-CH_2-NH_2$$

Putrescine

$$H_2N-CH_2-CH_2-CH_2-CH_2-CH_2-NH_2$$

Cadaverine

One type of derivative of the common amine has the nitrogen atom substituted by an alkyl group. These compounds are called *N*-alkyl derivatives and have the prefix *N*- in their names, which indicates the alkyl group is attached to nitrogen. Thus *N*-methylhexylamine can be differentiated from 2-methylhexylamine.

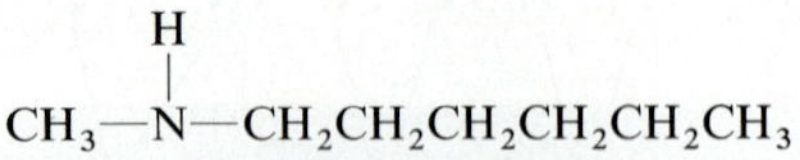

***N*-Methylhexylamine**

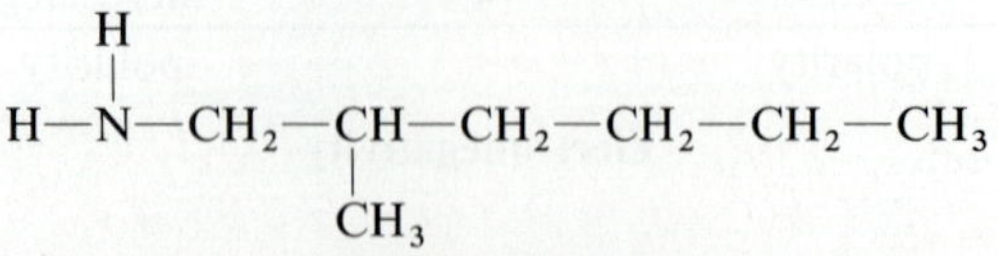

2-Methylhexylamine

EXAMPLE 21.2

Name these *N*-alkyl derivatives of amines.

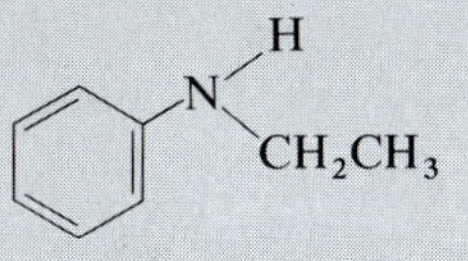
H
N
CH_2CH_3

***N*-Ethylaniline**

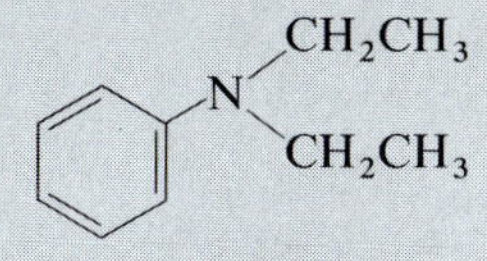
CH_2CH_3
N
CH_2CH_3

***N,N*-Diethylaniline**

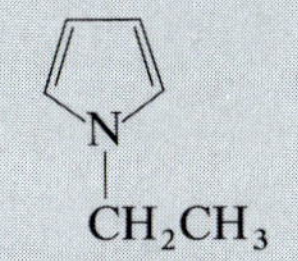
N
CH_2CH_3

***N*-Ethylpyrrole**

NEW TERMS

amine An organic functional group containing a nitrogen.

heterocycle A cyclic system having some atom other than carbon as a structural component of the ring. Some heterocycles are amines.

TESTING YOURSELF

Amine Structure and Nomenclature

1. Name the following amines.

a.

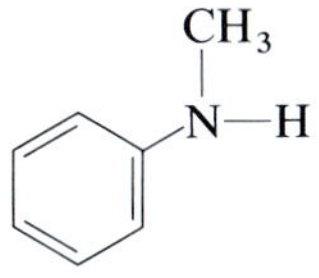
CH_3
N—H

d.

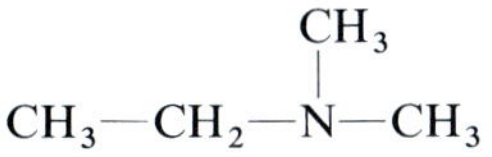
CH_3
CH_3—CH_2—N—CH_3

e. $CH_3—CH_2—CH_2—CH_2—NH_2$

b.

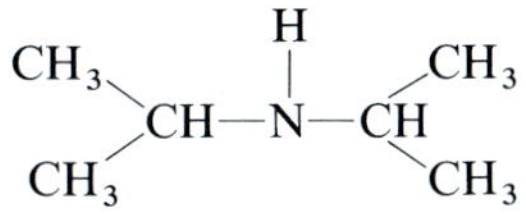
H
CH_3
CH_3
CH—N—CH
CH_3
CH_3

c.

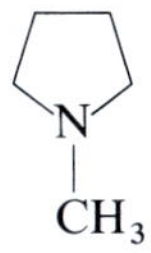
N
CH_3

Answers **1. a.** *N*-methylaniline **b.** diisopropylamine **c.** *N*-methylpyrrolidine **d.** *N,N*-dimethylethylamine or dimethylethylamine **e.** butylamine

21.2 Reactions of Amines

Basicity

The lone pair of electrons on an amine gives it the characteristic basic properties. An amine will react with a proton from an acid according to the following general reaction.

Amines are Lewis bases.

GENERAL REACTION

An Amine as a Base

$$-\ddot{N}: \; + \; H^+X^- \longrightarrow -\overset{|}{\underset{|}{N}}{}^{+}\!-H + X^-$$

Electron donor (Lewis base) | Electron acceptor (Lewis acid) | Ammonium salt

Examples

$$CH_3-NH_2 + HCl \longrightarrow CH_3-\overset{+}{N}H_3 + Cl^-$$

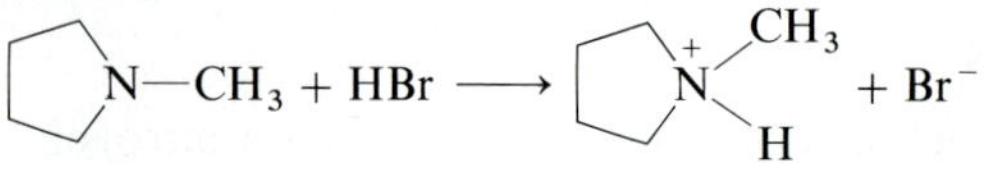

$$CH_3-CH_2-\underset{\displaystyle H}{\underset{|}{N}}-CH_3 + H_2SO_4 \longrightarrow CH_3-CH_2-\overset{\displaystyle H}{\overset{|}{\underset{\displaystyle H}{\underset{|}{N^+}}}}-CH_3 + HSO_4^-$$

Amines are weak bases, much as the organic acids are weak acids. If an amine reacts with water, an equation can be written to express this equilibrium (which also expresses the basicity). This equation relates the reaction of an amine with water and explains why aqueous solutions of amines should be treated as Brønsted–Lowry bases (OH^- donors).

$$R-NH_2 + H-O-H \rightleftharpoons R-\overset{+}{N}H_3 + OH^-$$

$$K_b = \frac{[R-\overset{+}{N}H_3{}^+][OH^-]}{[R-NH_2]}$$

The value $\boldsymbol{K_b}$ is related to basicity, just as K_a is related to acidity of acids. In the reverse reaction,

$$R-NH_3{}^+ + HO^- \rightleftharpoons R-NH_2 + H_2O$$

we can obtain a value for K_a.

Just as K_a (and pK_a) were useful indicators of acidity, K_b (and pK_b) are useful measures of basicity. Several representative amines and their p$K_b$values are given in Table 21.3. As for pK_a, recall that the *smaller* the value of pK_b, the more basic the amine. Note that pK_a + pK_b = 14.0.

Pyridine is a basic aromatic amine. There are six electrons in the pi system, giving pyridine the same kind of stability demonstrated earlier for benzene (Chapter 17). However, since nitrogen needs only three bonds, we see that the bonding electrons are used; thus, a lone pair of electrons is left on nitrogen. This lone pair gives pyridine its observed basicity. A similar argument can be made to account for the basicity of pyrimidine and purine.

Delocalization of electrons from the nitrogen lone pair into the ring is responsible for pyrrole and aniline being weaker bases than other amines.

The amines pyrrole and pyrrolidine look quite similar in structure, but there is a basic difference. Whereas pyrrolidine is a typical amine base, pyrrole is relatively neutral. These differences in basicity are easily explained by examining the lone pair electrons (Figure 21.3). In pyrrolidine (as in other typical bases), the lone pair of electrons occupies an sp^3 orbital and is avail-

TABLE 21.3
Representative Amines and Their pK_b Values

Amine Name	Structure	K_b	pK_b	pK_a
Ammonia	NH_3	1.8×10^{-5}	4.7	9.3
Methylamine	$CH_3—NH_2$	4.4×10^{-4}	3.4	10.6
Diethylamine	$CH_3—CH_2—NH(CH_2—CH_3)$	3.1×10^{-4}	3.3	10.7
Triethylamine	$CH_3—CH_2—N(CH_2—CH_3)—CH_2—CH_3$	1.0×10^{-3}	4.2	9.8
Aniline	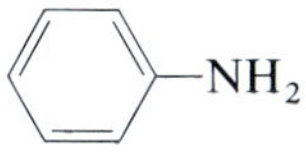	4.2×10^{-10}	9.4	4.6
Pyridine	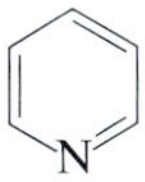	1.8×10^{-9}	8.7	5.3

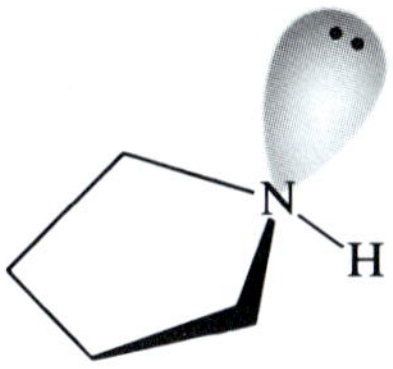
N
H
Pyrrolidine

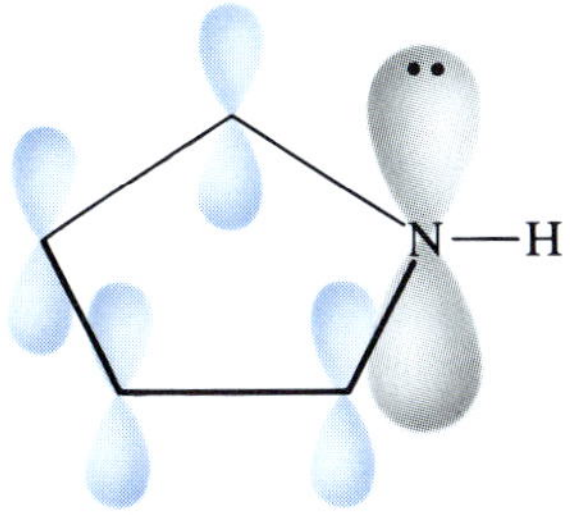
N—H
Pyrrole

FIGURE 21.3
Reasons for the difference in basicity of pyrrolidine and pyrrole are found in the observation that the lone pair electrons of pyrrole are delocalized into the pi system of the ring. These electrons then are not available to act as a base. In pyrrolidine the lone pair of electrons is localized on nitrogen and is available to act as a base.

able to react with a proton. In pyrrole the electrons are in a *p* orbital and are used to complete the delocalized electron network. These electrons are *not* available to react with a proton.

From the examples of amines in Table 21.3, we notice that the range of basicity does not vary too much with structure. A striking exception to this is the relative nonbasic nature of aniline. Why is aniline so weakly basic? Since basicity is related to the ability to donate electrons, the aromatic system electron delocalization essentially removes the electron lone pair from the nitrogen. If the electrons are *not* available on nitrogen, they cannot impart basic character to that site. This idea is well demonstrated by a resonance analysis (Figure 21.4). Note that this rationale is identical to that invoked to explain the pronounced acidity of phenols in Chapter 18.

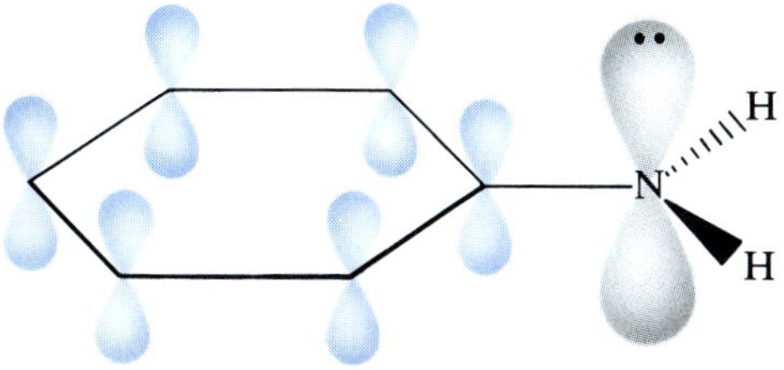
H
N
H
Aniline

FIGURE 21.4
The electron delocalization of the nitrogen electrons into the ring is the reason why aniline is not very basic. This delocalization is very similar to that already discussed for phenol.

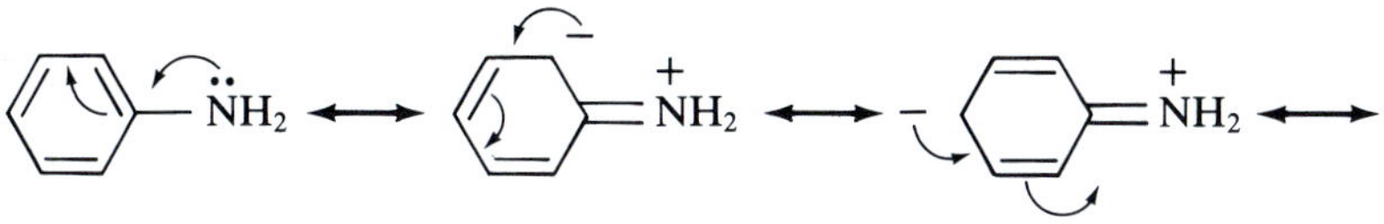
NH₂
+
NH₂
+
NH₂
Continued

Reactions as a Nucleophile

REACTION WITH AN ALKYL HALIDE In the previous section, we saw that the electrons on nitrogen made amines basic. An amine can effectively use its electron lone pair for purposes other than pulling off hydrogens. Some of the more important chemical properties of amines center on their ability to act as nucleophiles. Recall the reaction of a nucleophile with an alkyl halide:

$$Nu: + \ \underset{|}{\overset{|}{C}}-X \longrightarrow Nu-\underset{|}{\overset{|}{C}}- + X^-$$

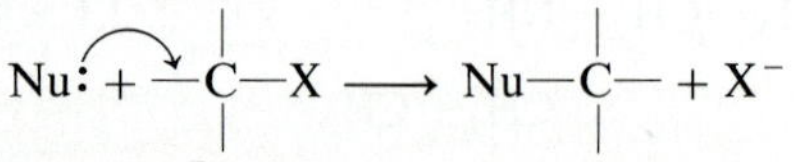

The lone pair of electrons on an amine can participate in the same kind of reaction. Looking at this reaction from another viewpoint, we can also say that alkyl halides can *alkylate* the nitrogen atom of an amine.

GENERAL REACTION

An Amine with an Alkyl Halide

$$R-N: + \ \underset{|}{\overset{|}{C}}-X \longrightarrow R-\overset{+}{N}-\underset{|}{\overset{|}{C}}- + X^-$$

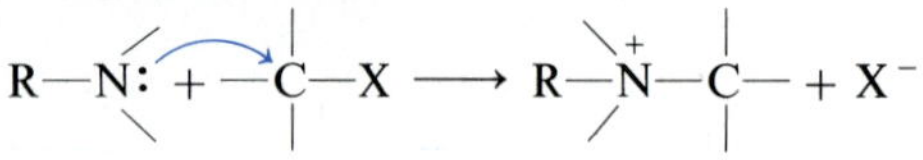

Examples

$$CH_3-\ddot{N}H_2 + CH_3-CH_2-Br \longrightarrow CH_3-\underset{H}{\overset{H^+}{\underset{|}{N}}}-CH_2-CH_3 + Br^-$$

$$\underset{}{\bigcirc}\!\!\!N-H + CH_3-I \longrightarrow \underset{}{\bigcirc}\!\!\!\overset{+}{N}\underset{CH_3}{\overset{H}{\diagup}} + I^-$$

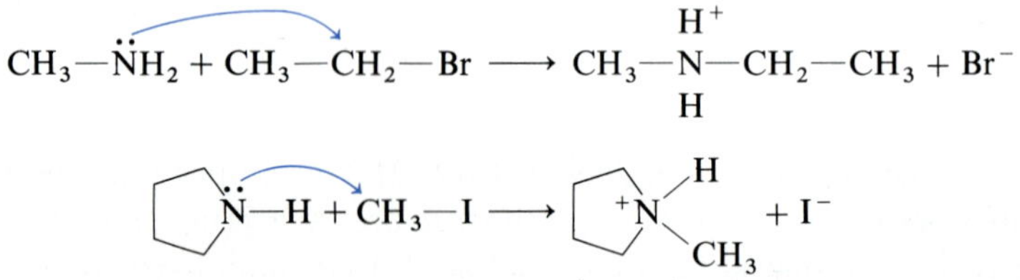

If the alkylation of pyridine is carried out with a long chain alkyl halide (such as *cetyl chloride*, $C_{16}H_{33}Cl$), a detergent is obtained (in this case, cetylpyridinium chloride.)

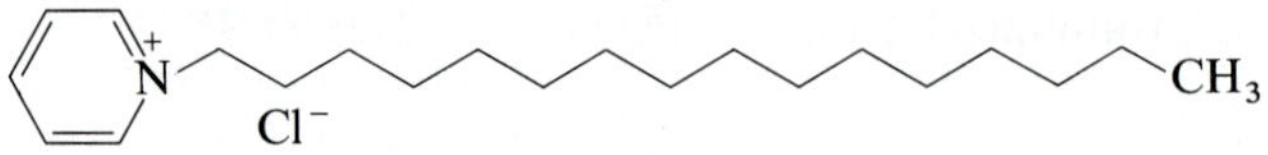

Cetylpyridinium chloride

Notice that this compound has a long hydrocarbon portion and an ionic end. This particular compound is the active ingredient in many mouthwashes. Recall how your mouth bubbles when you gargle. You are gargling with a detergent!

The alkylation reaction is biologically significant because a number of alkylating agents (compounds that readily donate an alkyl group) are considered to be the source of genetic mutations (see Chapter 26). Compounds such as some alkyl halides and dimethylsulfate ($CH_3-O-SO_2-O-CH_3$) alkylate the nitrogens of some of the purines and pyrimidines that are common to DNA.

REACTION WITH AN ACID CHLORIDE TO PRODUCE AN AMIDE Just as alcohols and water react with acid chlorides, so can amines. The only requirement for the following reaction is that the amine cannot be tertiary. The product is an *amide* having unique chemical properties that will be discussed

Amines react with acid chlorides to form amides.

in Section 21.3. In this reaction an acid is given off as a by-product. This will also react with amine. These reactions are conducted with excess amine or in the presence of some other proton trapping agent, such as $CaCO_3$.

GENERAL REACTION

An Amine with an Acid Chloride

$$R-C(=O)-Cl + H-\ddot{N}(R')H(R'') \longrightarrow R-C(=O)-N(R')H(R'') + HCl\ \text{(trapped)}$$

Examples

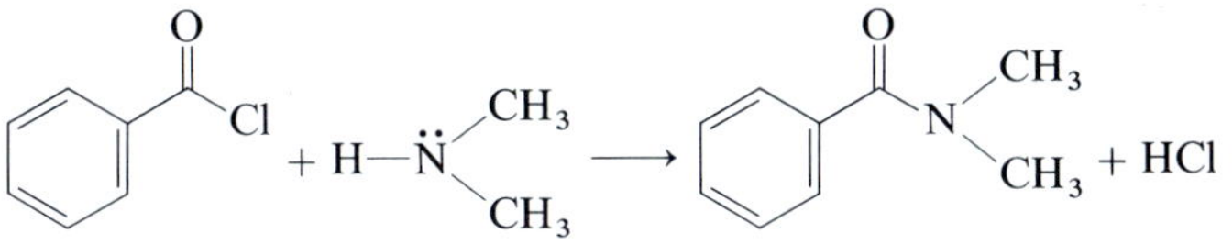

$$CH_3-C(=O)-Cl + H_2N-CH_2CH_3 \longrightarrow CH_3-C(=O)-NH-CH_2CH_3 + HCl$$

REACTION WITH AN ACID ANHYDRIDE The chemical behavior of acid anhydride reactions is similar to that of acid chlorides. When reacted with an amine, both produce an amide. With acid halide the by-product is HX, while acid anhydride gives the by-product $H-O-C(=O)-R$.

Amines react with acid anhydrides to form amides.

GENERAL REACTION

An Amine with an Acid Anhydride

$$R-C(=O)-O-C(=O)-R + H-N(R')R''(H) \longrightarrow \underset{\text{Amide}}{R-C(=O)-N(R')R''(H)} + H-O-C(=O)-R$$

Examples

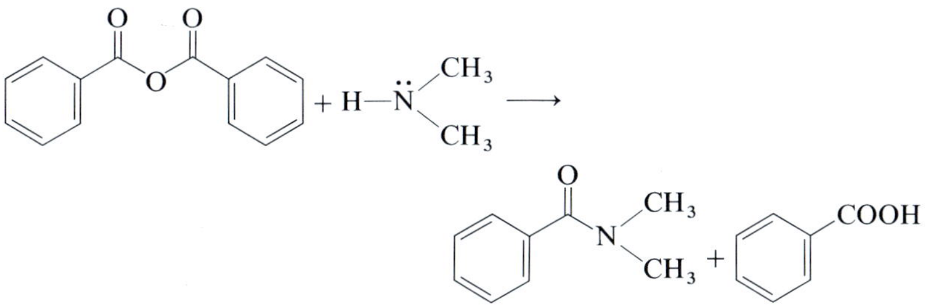

$$CH_3-C(=O)-O-C(=O)-CH_3 + H_2N-CH_2CH_3 \longrightarrow$$

$$CH_3-C(=O)-NH-CH_2-CH_3 + CH_3COOH$$

An amine can react with an ester in a *transamination* reaction to form an amide.

REACTION WITH AN ESTER: TRANSAMINATION Recall from Chapter 20 that the saponification of an ester required attack at the carboxyl carbon by OH^-, with a resulting displacement of an alkoxide group.

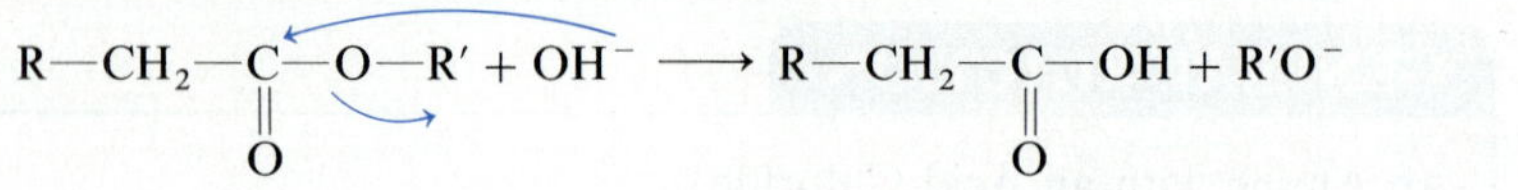
R—CH₂—C—O—R′ + OH⁻ ⟶ R—CH₂—C—OH + R′O⁻

The lone pair of electrons on an amine can also participate in the same kind of reaction. In the process an ester is converted into an amide.

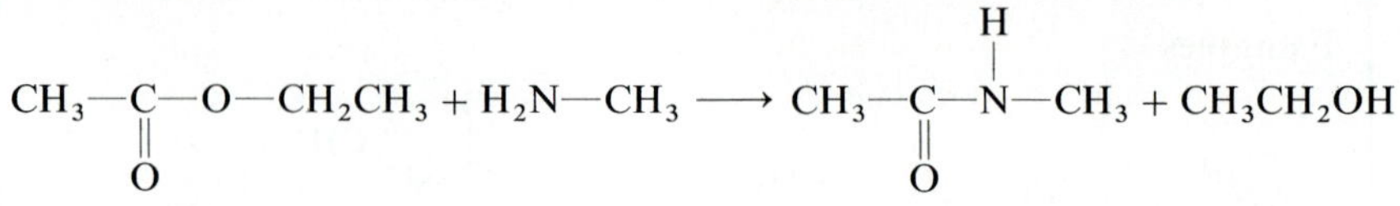
CH₃—C—O—CH₂CH₃ + H₂N—CH₃ ⟶ CH₃—C—N—CH₃ + CH₃CH₂OH

Ester **Amide**

GENERAL REACTION

Conversion of an Ester to an Amide

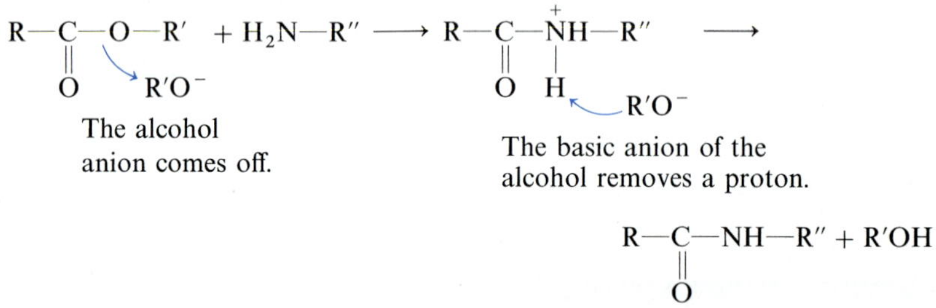
R—C—O—R′ + H₂N—R″ ⟶ R—C—NH—R″ ⟶
R′O⁻
The alcohol anion comes off.
R′O⁻
The basic anion of the alcohol removes a proton.
R—C—NH—R″ + R′OH

Examples

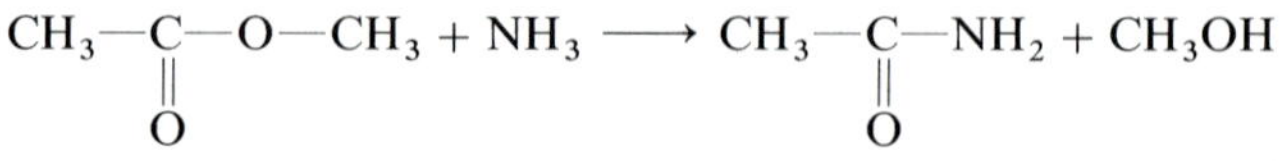
CH₃—C—O—CH₃ + NH₃ ⟶ CH₃—C—NH₂ + CH₃OH

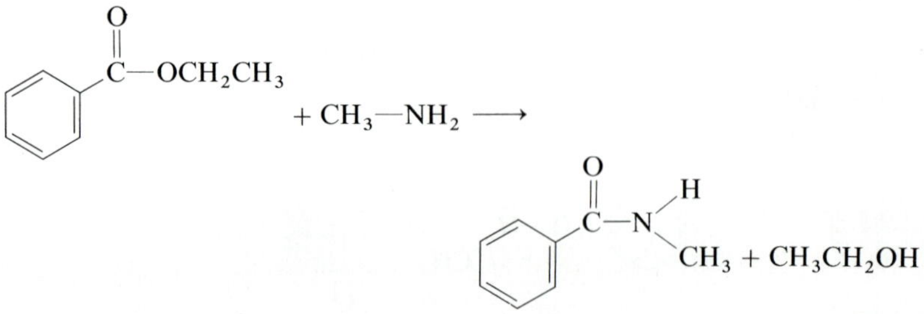
C—OCH₂CH₃
+ CH₃—NH₂ ⟶
C—N
H
CH₃ + CH₃CH₂OH

Amines react with carbonyl compounds to form *imines.*

REACTIONS WITH CARBONYL COMPOUNDS We have already shown that the amines have a lone pair of electrons that make them both basic *and* nucleophilic. These electrons can be used for nucleophilic attack at the carbonyl group of aldehydes and ketones. The product of the reaction of a primary amine with a carbonyl group produces an **imine**, which has a double bond between the carbon and nitrogen.

GENERAL REACTION

An Amine with a Carbonyl Group

$$\rangle C{=}O + H_2N{-}R \xrightarrow{-H_2O} \rangle C{=}N{-}R$$

Carbonyl compound   Amine   Imine

Examples

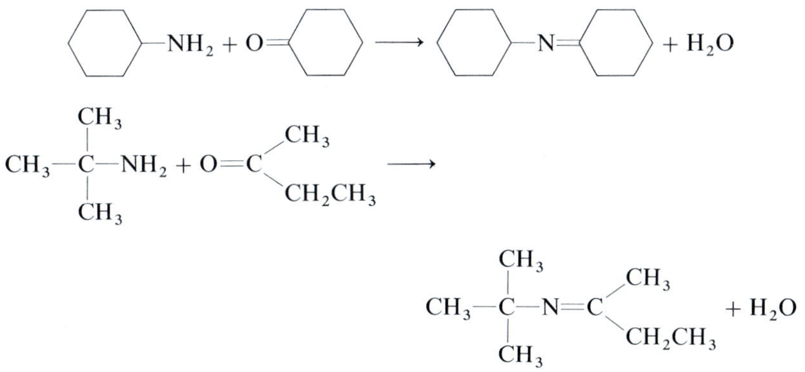

In biological systems the formation of an imine (also called a *Schiff base*) is one of the steps for making amines.

$$\rangle C{=}O + H_2N{-}R \xrightarrow{-H_2O} \rangle C{=}N{-}R \xrightarrow{[H]} \rangle CH{-}NH{-}R$$

Imine

This reaction finds significance in biological systems as a way of converting keto acids to *amino acids* (transamination). Unique to this reaction is the realization that a keto acid and amino acid form a different keto acid and amino acid.

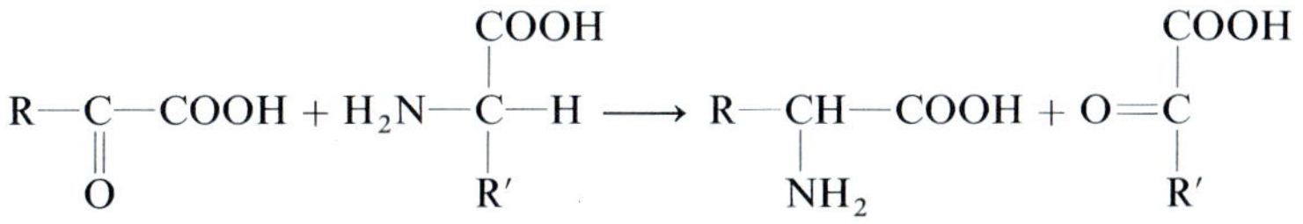

α-Keto acid   **α-Amino acid**   **α-Amino acid**   **α-Keto acid**

Hydrazine is the simplest diamine, $H_2N{-}NH_2$. In terms of their reactions, hydrazines should be considered similar to primary amines, except the R is another NH_2. We now expect that a primary amine will react with a carbonyl to give an imine. We should expect a similar reaction with hydrazine; this product is called a *hydrazone*. Substituted hydrazines react to give the corresponding substituted hydrazones. In particular, 2,4-dinitrophenylhydrazine is useful for making unique derivatives of carbonyl compounds. Most of the hydrazones are easily purified and characterized, and thereby serve as derivatives useful for identifying a particular aldehyde or ketone.

GENERAL REACTION

A Hydrazine with a Carbonyl Compound

$$\rangle C{=}O + H_2N{-}NH_2 \xrightarrow{H^+} \rangle C{=}N{-}NH_2 + H_2O$$

Hydrazine Hydrazone

Examples

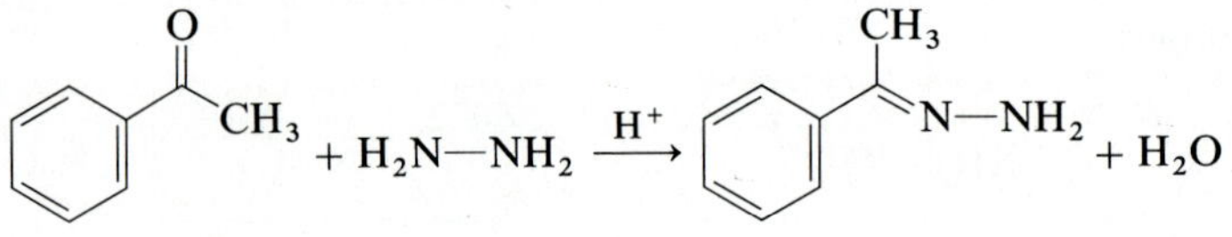

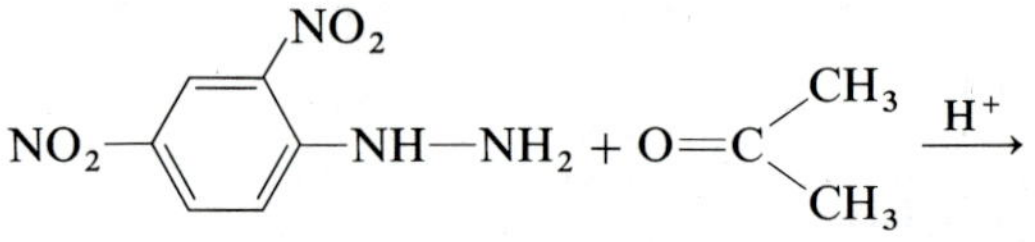

2,4-Dinitrophenylhydrazine

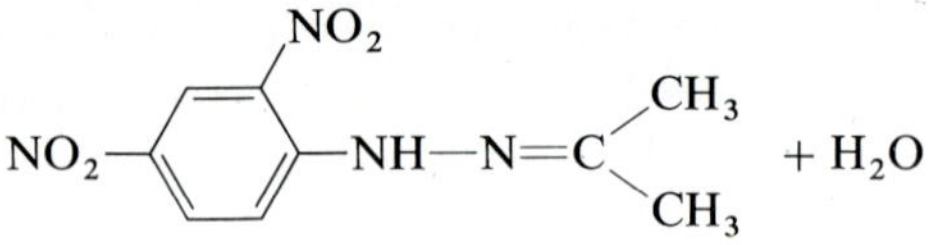

A 2,4-Dinitrophenylhydrazone derivative of a carbonyl group

Oxidation of Secondary Amines to Nitrosamines

Secondary amines are oxidized by nitrous acid to nitrosamines:

$$R{-}\underset{\substack{|\\R}}{N}H + HNO_2^- \longrightarrow R{-}\underset{\substack{|\\R}}{N}{-}N{=}O$$

Nitrous acid **Nitrosamine**

There is some concern that nitrites added as preservatives to certain foods, such as packaged meats, hotdogs, and bacon, are harmful to humans. In the presence of stomach acid, these nitrites form nitrous acid that can then convert amines to nitrosamines. These nitrosamines are known carcinogens. Whether the levels are high enough to be a problem is not yet established.

NEW TERMS

K_b A constant related to the basicity of a compound.

imine A compound formed from the reaction of a carbonyl-containing compound and a primary amine.

TESTING YOURSELF

Amine Reactions

1. Give the products of these reactions of an organic acid derivative with an amine.

a. $CH_3{-}\underset{\substack{\|\\O}}{C}{-}O{-}CH_3 + H{-}N(CH_3)_2 \longrightarrow$

b.

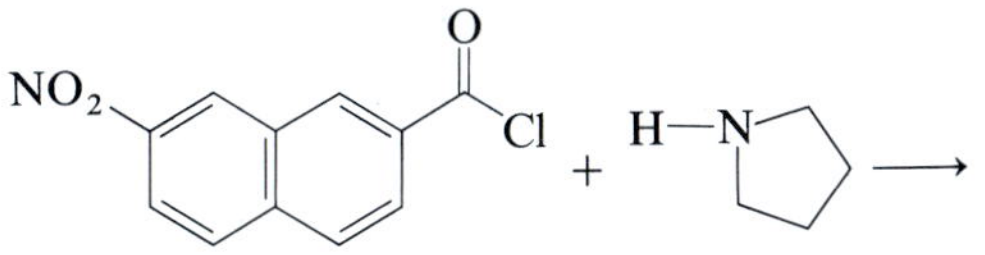

c.

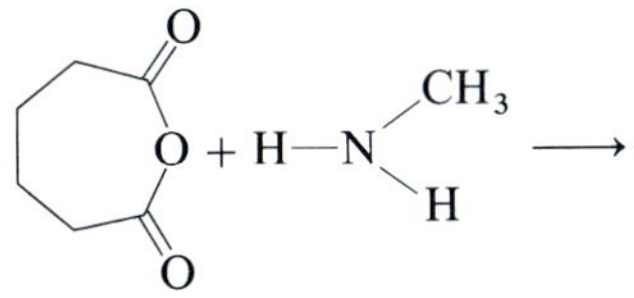

d.

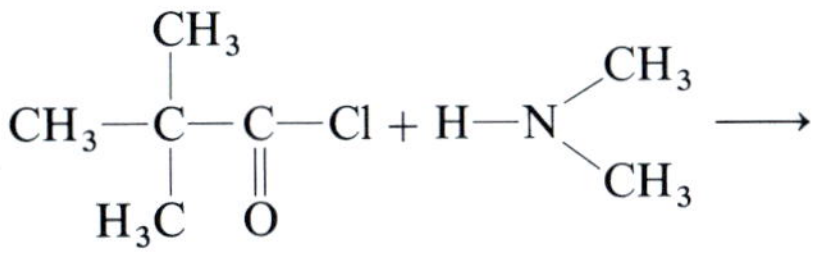

2. Give products for the following amine reactions.

a.

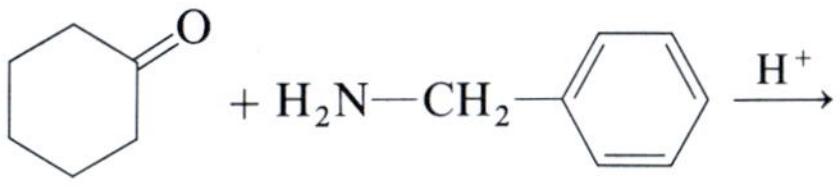

b.

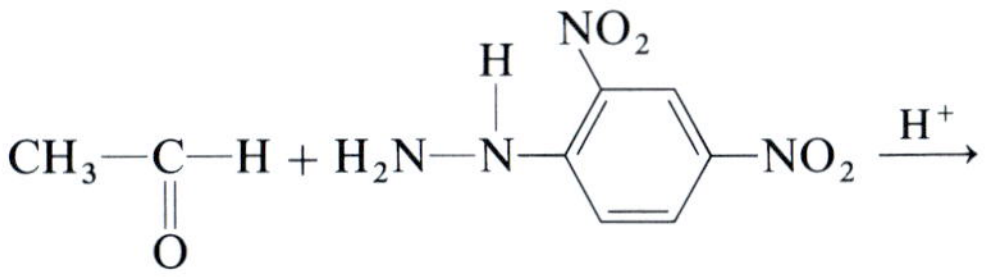

Answers 1. a.

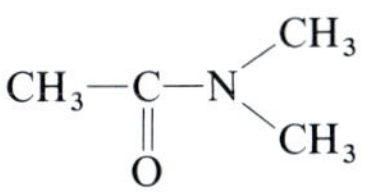

b.

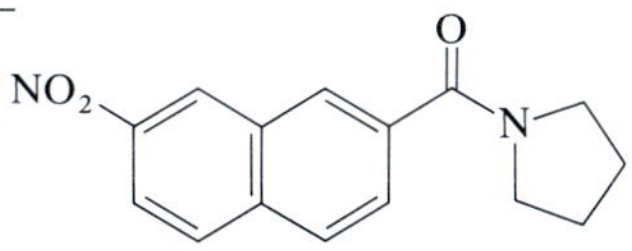

c.

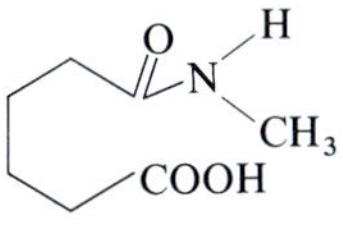

d.

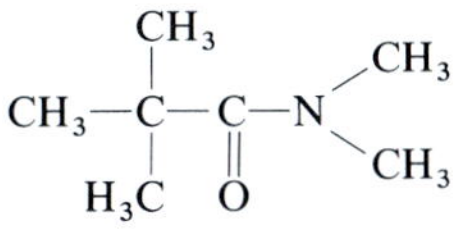

2. a.

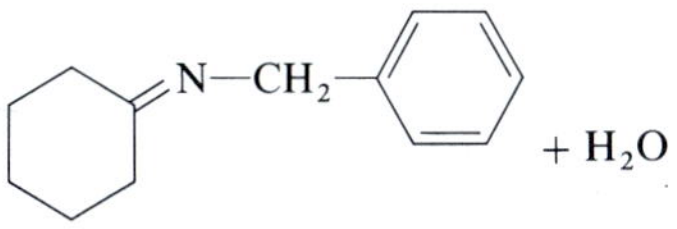

b.

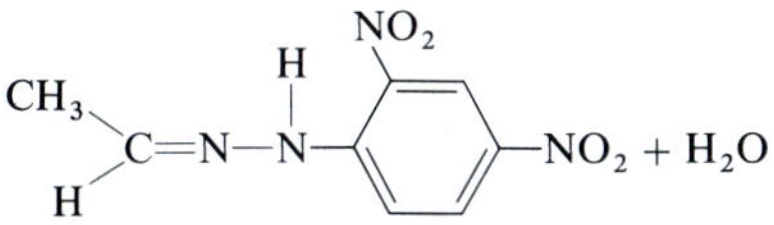

21.3 Amides

Structure

An **amide** is a nitrogen-containing compound with this general structure

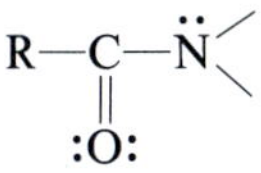

Amides are related to acids and amines in much the same way that esters are related to acids and alcohols. Note the similarity between these two reactions:

$$R-\underset{\underset{O}{\|}}{C}-O-H + H-O-R' \longrightarrow R-\underset{\underset{O}{\|}}{C}-O-R' + H_2O$$

Ester

$$R-\underset{\underset{O}{\|}}{C}-O-H + H-N\langle \longrightarrow R-\underset{\underset{O}{\|}}{C}-N\langle + H_2O$$

Amide

Amides are planar around the nitrogen atom, thus the nitrogen is sp^2 hybridized. As a result of an amide's resonance structure having the nitrogen lone pair delocalized to the carbon (Figure 21.5), amides are not very basic. Also, this causes the amide to be planar around the amide functional group. The planarity of the amide bond plays an important role in the structure of proteins. Because the resonance structure can be drawn to show a positive charge on nitrogen, it is possible to conclude that a N-H bond of an amide is more acidic than an N-H bond on an amine.

Reactions

Amides can be hydrolyzed by reactions similar to the hydrolysis of esters; however, they are generally much more resistant to hydrolysis than esters. This has important biological consequences because the stability of the amide functional group contributes to protein stability.

GENERAL REACTION

Hydrolysis of an Amide

$$R-\underset{\underset{O}{\|}}{C}-N\langle + H_2O \xrightarrow{H^+ \text{ or } HO^-} R-\underset{\underset{O}{\|}}{C}-OH + H-N\langle$$

Examples

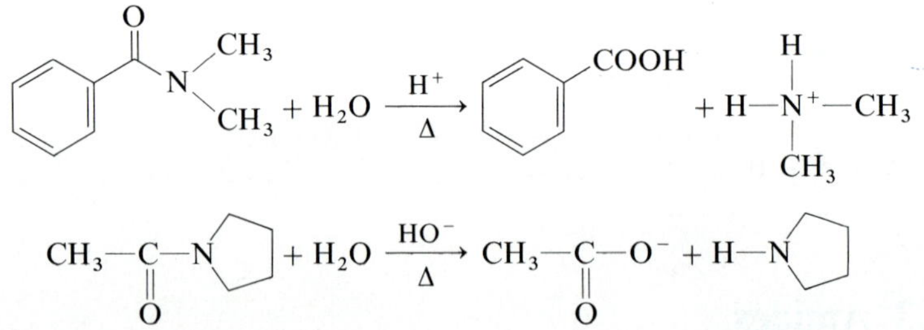
O
CH3
N
CH3
+ H2O
H+
Δ
COOH
H
+ H—N+—CH3
CH3
CH3—C—N
+ H2O
HO−
Δ
CH3—C—O−
+ H—N
O

In basic hydrolysis, the products are the free amine and the salt of the acid, while in acidic hydrolysis the products are the salt of the amine and the free acid. Notice the similarity between the hydrolysis of amides and nitriles (see Chapter 20).

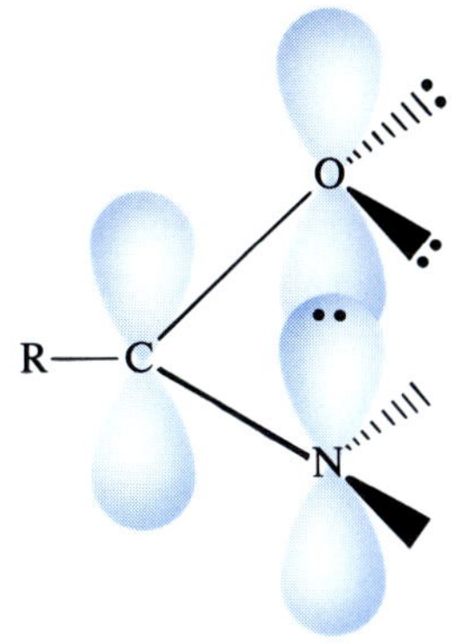

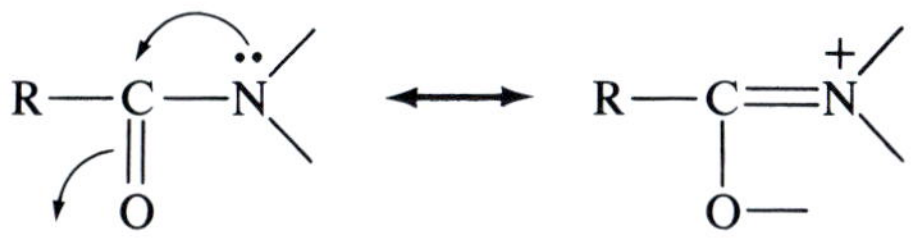

FIGURE 21.5
The resonance structures of an amide can account for the lack of basicity of an amide as compared to an amine. The resonance structure having the "charge on oxygen" contributes greatly to the "real" structure of an amide—particularly the observed lack of rotation about the C—N bond.

Amide

Amides as Polymers

We have mentioned polymers in several preceding chapters. The amide bond is often involved in natural polymeric materials (proteins) and also in synthetic polymers. One of the best known synthetic polymers, *nylon*, is composed of amide linkages. Nylon is formed by reacting the acid chloride of adipic acid (six-carbon dicarboxylic acid) with 1,6-diaminohexane. Nylon is made commercially by heating the acid chloride of adipic acid with 1,6-diaminohexane.

Nylon is a synthetic polymer formed by connected chains of amides.

$$Cl-\underset{\underset{O}{\|}}{C}-CH_2-CH_2-CH_2-CH_2-\underset{\underset{O}{\|}}{C}-Cl + H-\underset{\underset{H}{|}}{N}-CH_2-CH_2-CH_2-CH_2-CH_2-CH_2-\underset{\underset{H}{|}}{N}-H \longrightarrow$$

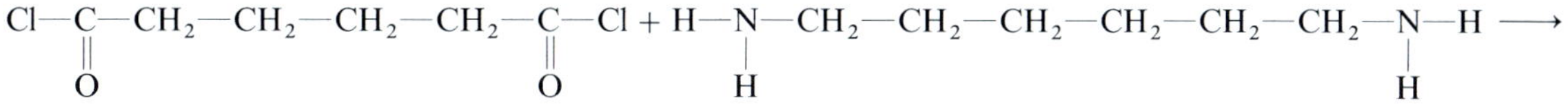

1,6-Diaminohexane

$$\left[\underset{\underset{O}{\|}}{C}-CH_2-CH_2-CH_2-CH_2-\underset{\underset{O}{\|}}{C}-\underset{\underset{H}{|}}{N}-CH_2-CH_2-CH_2-CH_2-CH_2-CH_2-\underset{\underset{H}{|}}{N}-\right]_n$$

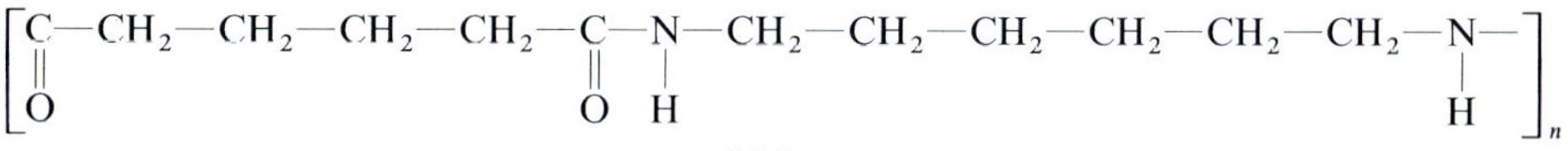

Nylon

The strength of nylon results from hydrogen bonding between chains, much in the same way as we shall see in the chemical behavior of proteins.

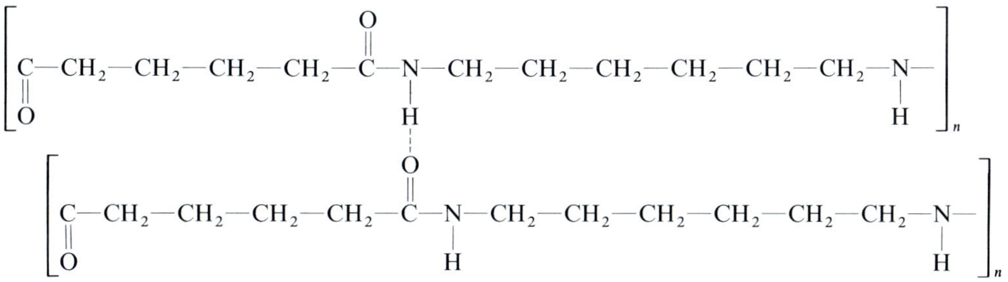

NEW TERMS

amide A nitrogen-containing compound derived from an acid and an amine with the general structure

R—C(=O)—N

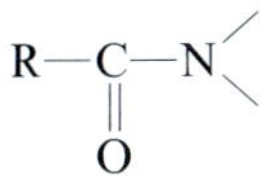

TESTING YOURSELF

Amide Structure and Reactivity

1. Which of the two resonance structures in Figure 21.5 (the one on the left or right) is the more important contributor to the overall structure and reactivity of the amide?
2. Show products from the following hydrolysis reactions of amides.

a. 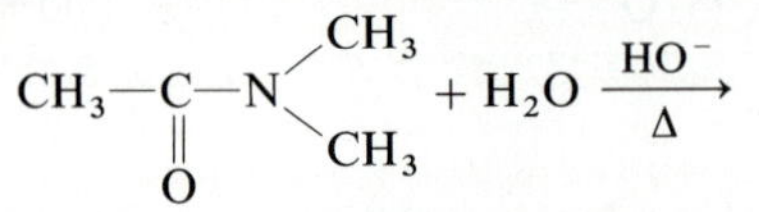

b. 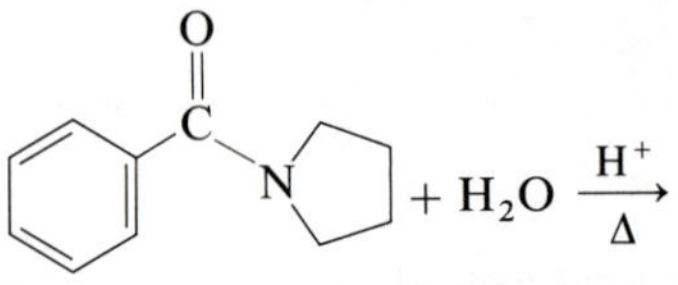

Answers **1.** The one on the left is more important because it is neutral.

2. a. **b.** 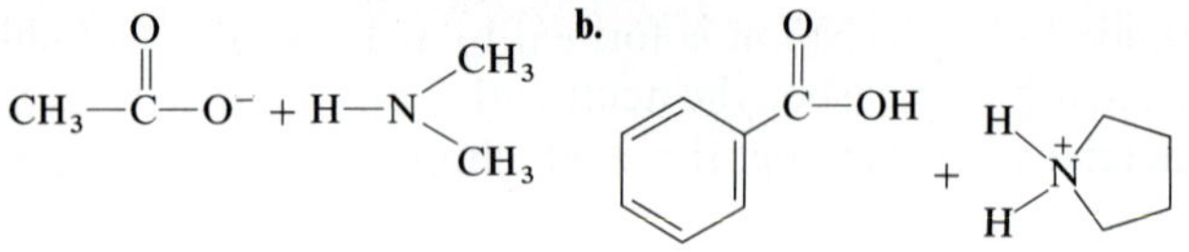

21.4 Special Amines

Amino Acids

As the name implies, these **amino acids** have *both* an amino and a carboxyl group. The chemistry of amino acids will be discussed in detail in Chapter 25. They are introduced at this point to show that they are special amines. Most of the common amino acids are alpha amino acids.

$$\underset{\displaystyle NH_2}{R{-}\underset{|}{CH}{-}COOH}$$

An α-amino acid

Alkaloids

Alkaloids are examples of biologically active basic natural products.

Many of the most well-known natural products can be classified as **alkaloids**, which at one time were defined as basic, nitrogen-containing compounds obtained from plants. There is some debate as to whether this definition is too limited because a number of basic natural products are now known from animal sources. Nevertheless, compounds in this very diverse family have been used as drugs (quinine for malaria), pest poisons (strychnine), analgesics (morphine), and stimulants (caffeine). In the following pages, we discuss many examples of this important group of special amines.

Quinine, obtained from the bark of the cinchona tree, has been used as an effective remedy for malaria for hundreds of years. Long before the structure of quinine was known, quinine bark was used to relieve symptoms of malaria. Who first discovered this use will never be known, but folk medicines have long been an important source in finding active medicinal ingredients. Malaria is the most widespread infectious disease in the world and is caused by a protozoan transmitted through mosquito bite. Quinine effectively destroys the parasite. Today there are many synthetic analogs of quinine.

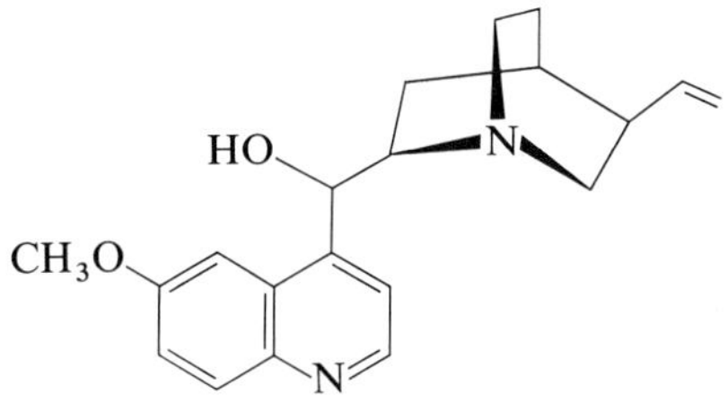
HO
N
CH_3O
N

Quinine

Strychnine has long been known as a deadly killer. It is a complex alkaloid with an extremely bitter taste. It was once common to "tonics" that helped to increase appetites; other less toxic compounds are now used for this purpose. It is a common ingredient in rodent poisons.

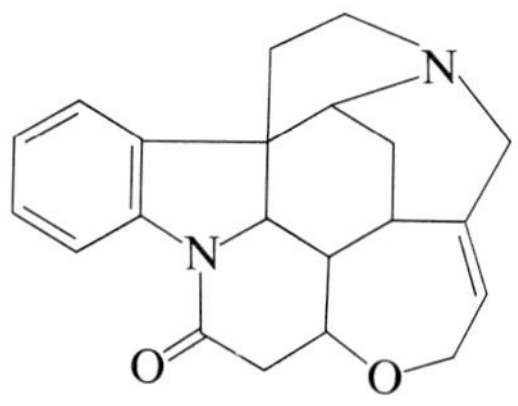
N
N
O
O

Strychnine

Morphine, one of the major alkaloids of the oriental poppy plant, is one of the most effective pain killers known. It is also very addictive. Until the early 1900s, opium (the crude latex that contains morphine and codeine) was found as a common ingredient in many commercial medicines. Indeed, it is estimated that by 1900 there were about 250,000 legally addicted people in the United States.

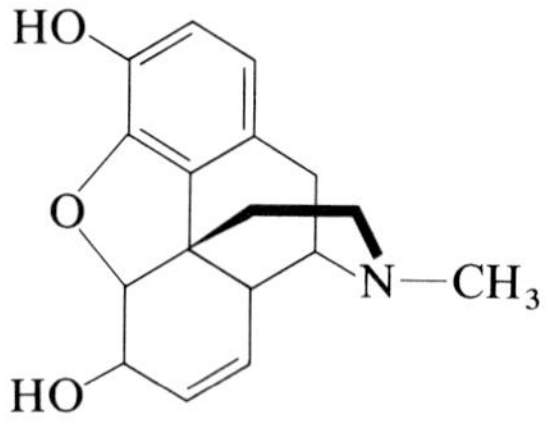
HO
O
N—CH_3
HO

Morphine

Another natural opiate is *codeine*, which is used for cough medicines. It can also be addictive, and its use is only by prescription. Another derivative of morphine is diacetylmorphine or *heroin*, which is much more addictive and is a major problem as a street drug. Chemists have maintained an active interest in preparing compounds that have the pain-killing ability of morphine without its addictive effects. One result of this work is *demerol*. *Methadone* is used as a substitute for heroin in drug treatment programs.

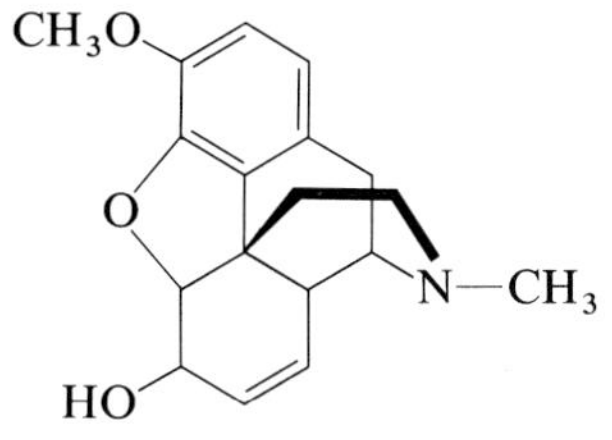
CH_3O
O
N—CH_3
HO

Codeine

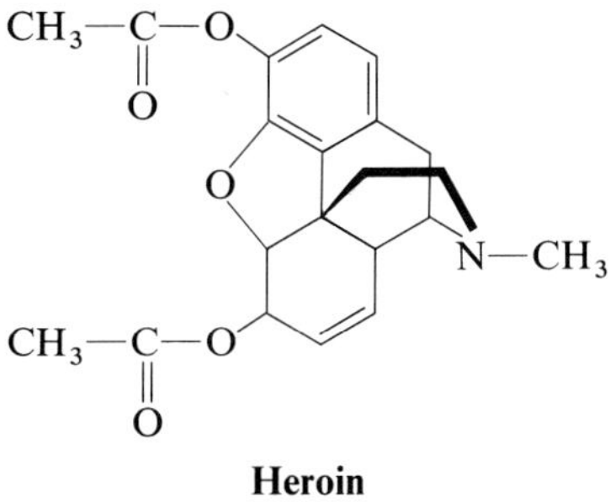
CH_3—C—O
O
O
N—CH_3
CH_3—C—O
O

Heroin

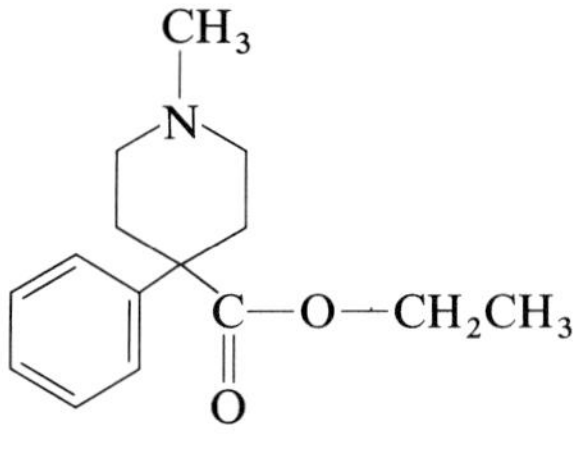
CH_3
N
C—O—CH_2CH_3
O

Demerol

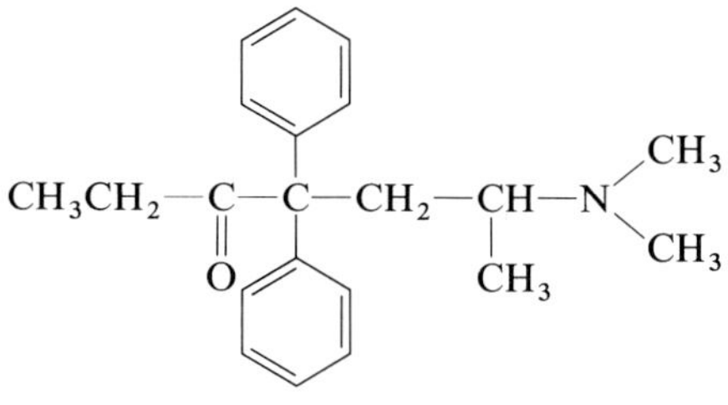
CH_3CH_2—C—C—CH_2—CH—N
O
CH_3
CH_3
CH_3

Methadone

ERGOTISM

The fungus *Claviceps purpurea* parasitizes growing kernels of a number of cereal grains. As the fungus takes over the plant, it produces a number of alkaloids. Ergotism often causes gangrene by causing a decrease in blood circulation in the extremities. Another symptom is epileptic-like convulsions known as St. Anthony's fire. Thousands died of ergot poisoning during the Middle Ages. Infected cereal grains, used by local bake shops, was the reason for many outbreaks of ergotism. In A.D. 994, more than 40,000 people died from a single outbreak.

Hallucinations as a result of ingestion of these alkaloids were common. Ergotism has been used to explain the witchcraft thought to be practiced by a number of young women in New Salem in the late 1690s.

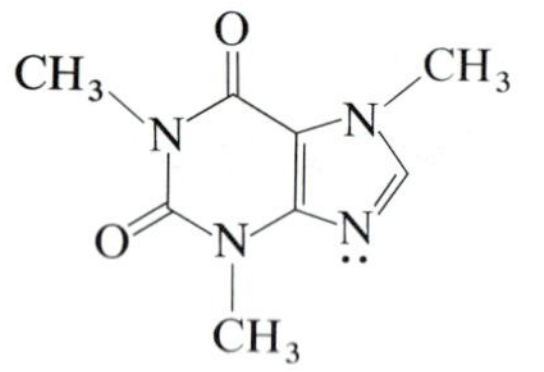

O
CH3
CH3
N
N
N
N
O
CH3

Caffeine

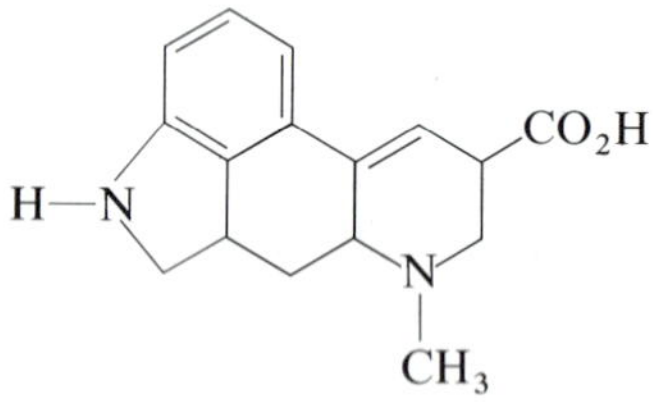

CO2H
H—N
N
CH3

Lysergic acid

Caffeine is a stimulant found in coffee, tea, and many cola drinks. Millions of people begin their days with coffee to "get started." Caffeine affects the central nervous system by stimulating the cerebral cortex. This results in more rapid and clearer thinking and improved coordination. It is also used to prevent some migraine headaches.

Lysergic acid is an alkaloid produced on cereal grains that are infected with a certain fungus called *ergot*. Many derivatives of this alkaloid are produced by this fungus. *LSD* ("acid") is a synthetic derivative of lysergic acid and is a potent hallucinogen (a drug that changes perception). As little as 20 μg (micrograms) can result in noticeable effects. Although the questions of long term health problems as a result of taking this drug are not fully resolved, there is no question that the possibility of a "bad trip" is very real. Under the influence of LSD, serious mental feelings of impending death, insanity, and paralysis are common. Also, users of LSD have reported "flashbacks"—recurrences of the feelings associated with taking the drug even when the drug is not used. These effects have been known to cause serious mental stress.

Phenylcyclidine (PCP or "angel dust") is one of the common street drugs often distributed as LSD. Its effects are quite different, though, more closely resembling alcoholic intoxication. However, feelings of schizophrenia are common, and flashbacks can recur unpredictably for months. Significant numbers of deaths have been reported due to use of PCP. It has also had some use in veterinary medicine.

N

PCP

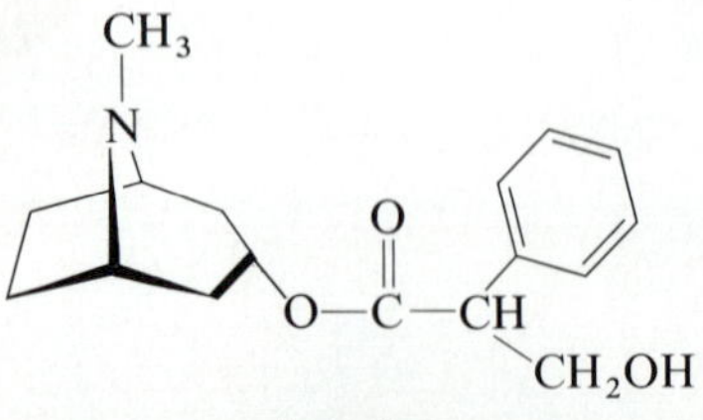

CH3
N
O
O—C—CH
CH2OH

Atropine

Atropine and cocaine are structurally similar, but have markedly different activities. *Atropine* (from the plant Deadly Nightshade) is very toxic, but dilute solutions of this alkaloid cause dilation of the eye pupil and find some use in eye examinations. Atropine has also been used as an antidote for opium poisoning.

CRACK

"Crack" is a solid form of cocaine that has become the major drug problem of the 1980s. Heating this substance and inhaling the vapors is said to cause such an intense response that many users become addicted after their first exposure. Unlike many drugs, the effects of crack are very short-lived, and to maintain the feelings obtained, a person has to have more and more in ever-increasing amounts. A recent essay on the drug sums it up in a most graphic way:

These are the two Americas. No other line you can draw is as trenchant as this. On one side, people of normal human appetites for food and sex and creature comforts; on the other, those who crave only the roar and crackle of their own neurons, whipped into a frenzy of synthetic euphoria.*

* *Newsweek*, Nov. 28, 1988, p. 64.

Cocaine is the active compound in the coca shrub. The leaves of the plant have been chewed by the natives of Peru as a stimulant. Since it was known to have a numbing effect on the lips and gums, there was early interest in its use as an anesthetic. However, the addictive nature of this drug and its resulting adverse side effects, have limited its medical applications. Synthetic cocaine analogs, such as novocaine and xylocaine, have now received much more acceptance.

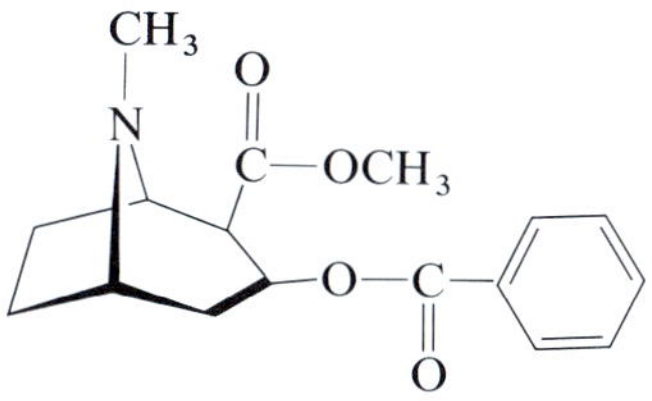

Cocaine

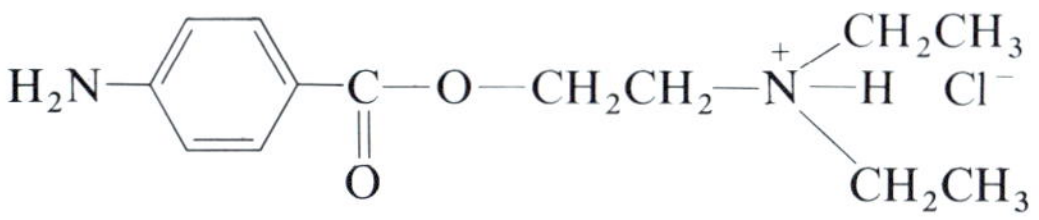

Novocaine

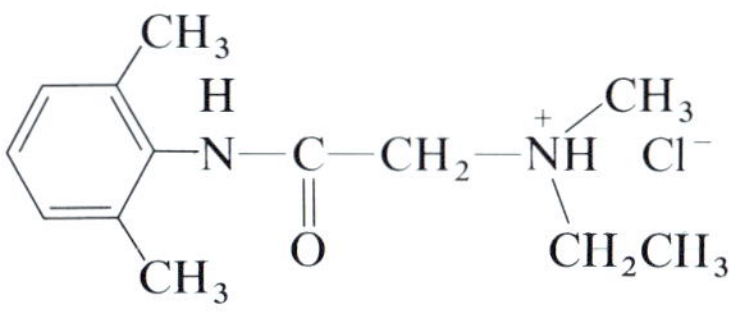

Xylocaine

One of the major alkaloids found in tobacco, *nicotine*, has recently been described by the Surgeon General as an addictive drug.

The most common alkaloid is *nicotine*. Nicotine is the major alkaloid constituent of tobacco. A violent toxin, solutions of nicotine are used as an insecticide. In small doses, nicotine can act as a stimulant and it is addictive. In larger amounts it can cause convulsions, nausea, and death.

Many other diverse structural types are associated with the class of natural products called the alkaloids. Additional examples of interesting alkaloids are given in Table 21.4.

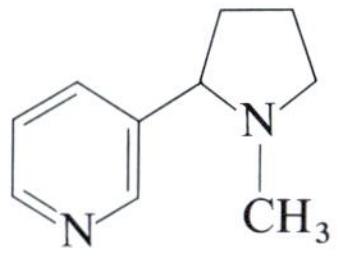

Nicotine

TABLE 21.4
Examples of Important Alkaloids

Name	Description	Structure
Emetine	Active ingredient of ipecac, used as an emetic (to induce vomiting in cases of poisoning)	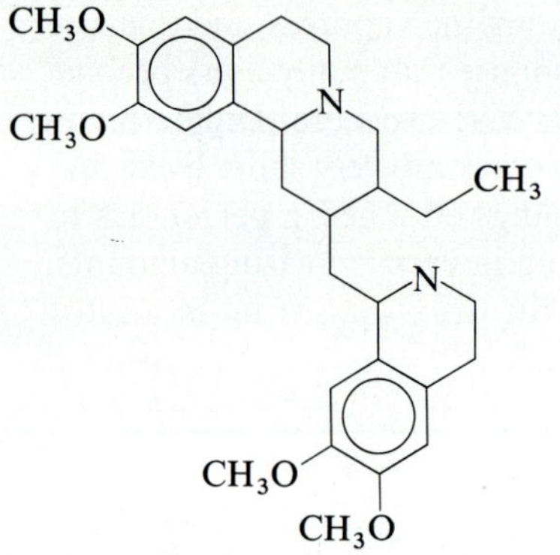 CH₃O CH₃O N CH₃ N CH₃O CH₃O
Ergotamine	Used in treatment of migraine headaches	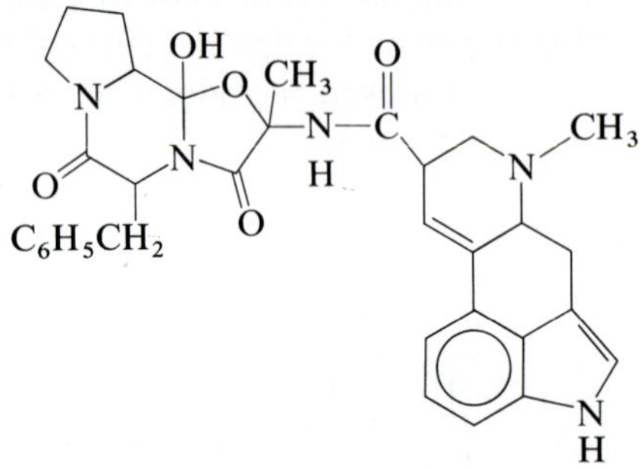 OH O CH₃ O N N N—C H N CH₃ O O C₆H₅CH₂ N H
Tubocurarine	Active ingredient in curare (South American arrow poison); also used in surgery as a muscle relaxant	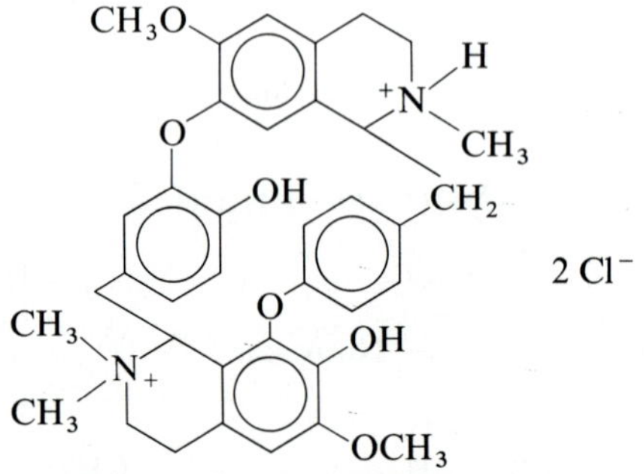 CH₃O H +N CH₃ O OH CH₂ 2 Cl⁻ O CH₃ N+ OH CH₃ OCH₃
Psilocybin	Hallucinogen; active ingredient of the mushroom *Psilocybe mexicana*	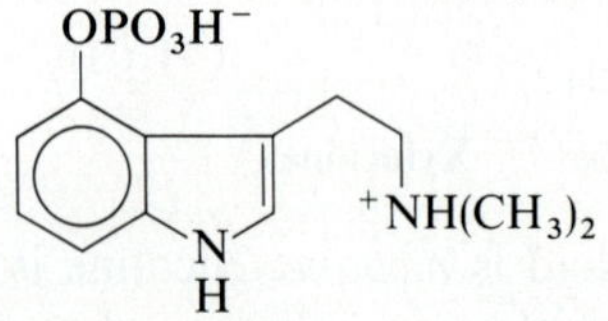 OPO₃H⁻ +NH(CH₃)₂ N H
Dopamine	A neural transmitter; a lack of this compound results in the spastic motions and tremors associated with Parkinson's disease	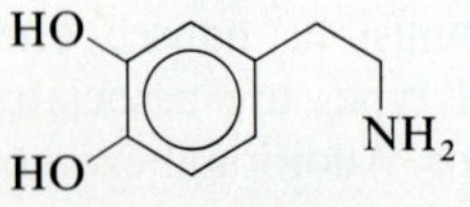 HO NH₂ HO

Name	Description	Structure
Serotonin	A vasoconstrictor present in platelets; also thought to be involved in neural mechanisms of sleep and sensory perception.	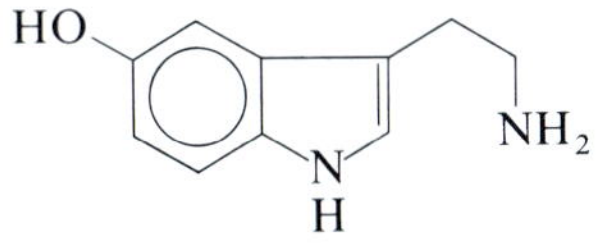
Reserpine	Used to treat hypertension and as a tranquilizer for the emotionally disturbed.	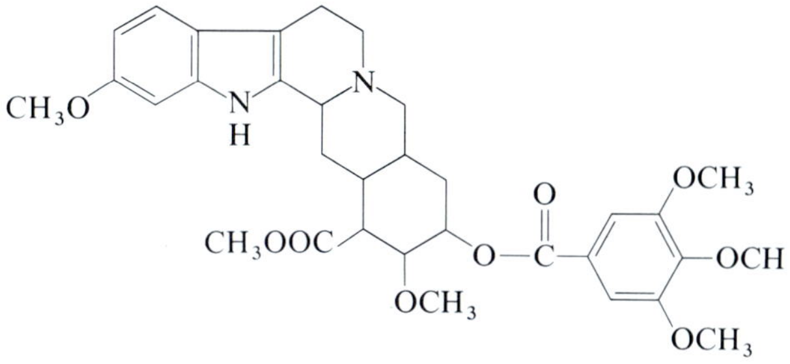
Amphetamine	Synthetic alkaloid: (dexedrine or benzedrine). Used as an appetite suppressant and as a stimulant.	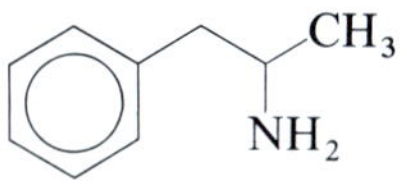
Methamphetamine	Synthetic alkaloid used as an appetite suppressant and as a stimulant.	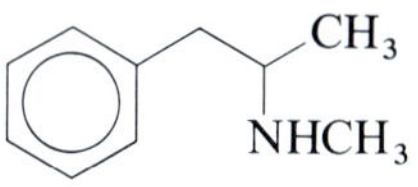

Amines Related to Adrenaline

Amines related to adrenaline are all based on a similar skeleton:

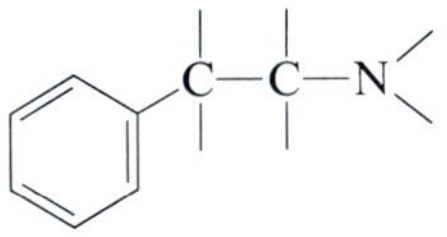

Adrenaline (also known as *epinephrine*) is a hormone produced in the adrenal gland. It is the "fight-or-flight" hormone that prepares the body for instant action. The norepinephrine (*nor* means "loss of methyl") is the hormone that controls blood pressure.

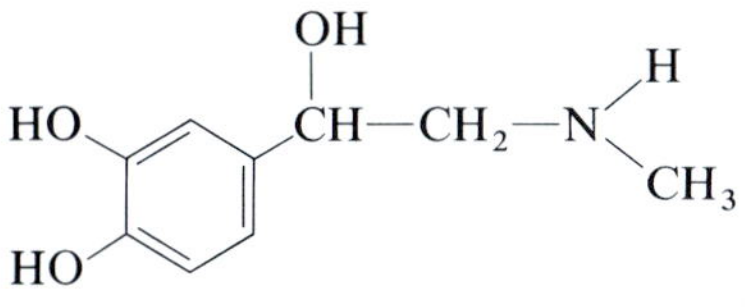
OH
HO
CH—CH₂—N
H
CH₃
HO

Epinephrine

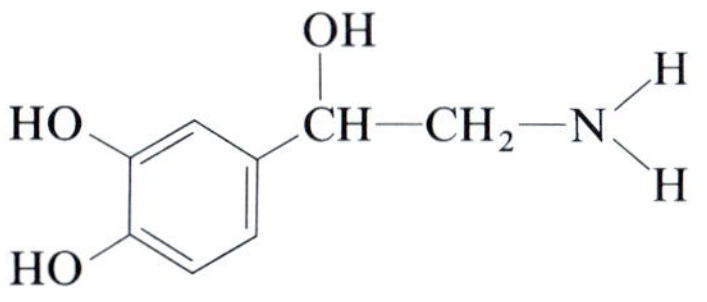
OH
HO
CH—CH₂—N
H
H
HO

Norepinephrine

BARBITURATES

Barbiturates are a group of synthetic molecules that have important application in medicine as sedatives and hypnotics. The basic arrangement of atoms is derived from the reaction of a malonic acid ester with urea.

A number of drugs are related to this general structure. Representative examples are listed in the table below.

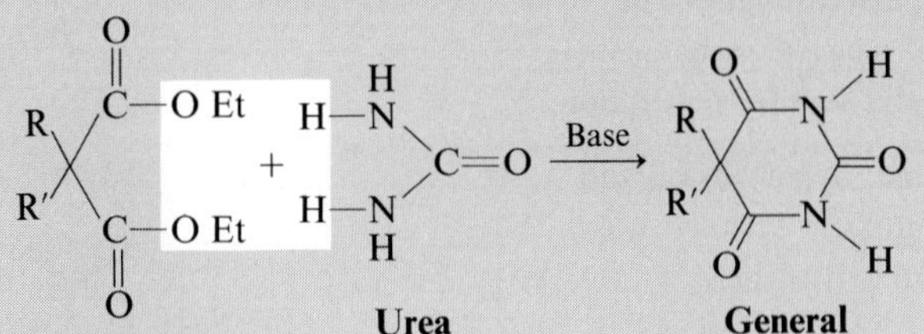

R
R′
C—O Et
C—O Et
O
O
H
H—N
H—N
H
C=O
+
Base
R
R′
O
O
N
N
H
H
=O
Urea
Derivative of malonic acid
General structure of barbituric acid

Name	Drug and Description	R	R′
Barbital	Veranol, a hypnotic and sedative	Et	Et
Phenobarbital	Luminal, an anticonvulsant, hypnotic, and sedative	Et	Ph
Secobarbital	Seconal, a hypnotic	CH_2—CH—CH_2—	CH_3—CH_2CH_2CH—CH_3
Butalbital	Sandloptal, a sedative	CH_2=CH—CH_2—	CH_3—CH(CH_3)—CH_2—
Amobarbital	Amytal, a sedative	Et	CH_3—CH(CH_3)CH_2CH_2—
Pentobarbital	Nembutal, a hypnotic and sedative	Et	CH_3—CH_2CH_2CH(—)—CH_3

The sedative properties of these drugs vary, ranging from the long lasting sedative properties of barbital and phenobarbital to the intermediate sedative properties of secobarbital and pentobarbital. One of the most potent, fast acting drugs is sodium pentothal. This molecule differs slightly in structure from the other barbiturates by the inclusion of a *thiocarbonyl.*

Barbiturates are addictive drugs. Because people can build a tolerance to these drugs, a dosage to effect ratio is difficult to maintain. The continual need to increase the dose to obtain a given response can lead to serious overdoses and even death. In combination with alcohol, the barbiturates are very dangerous compounds.

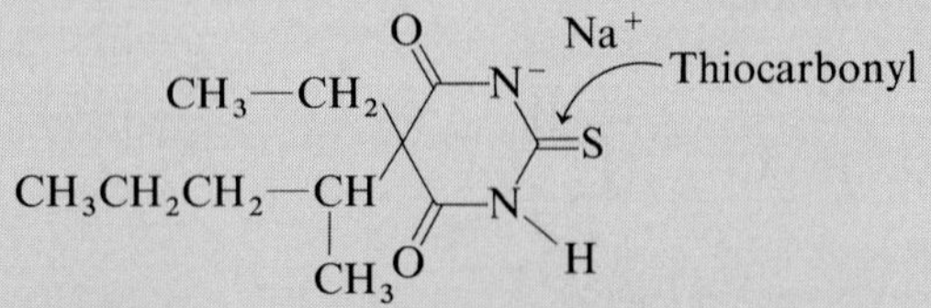

O
Na⁺
Thiocarbonyl
CH₃—CH₂
N⁻
=S
CH₃CH₂CH₂—CH
N
CH₃
O
H

Sodium pentothal

A number of compounds have been prepared that mimic some of the activity of these hormones. These molecules have powerful physiological activities and are illicit drugs. The parent amphetamine ("an upper"), methamphetamine ("speed"), and methoxyamphetamine ("STP") are examples of these hormone mimics.

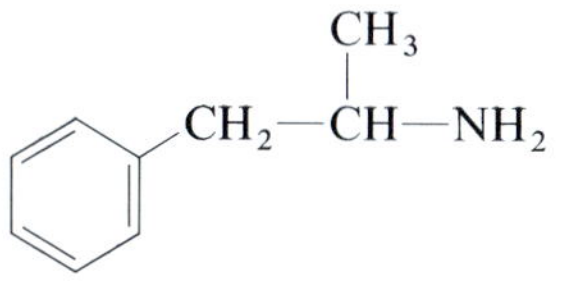

Amphetamine

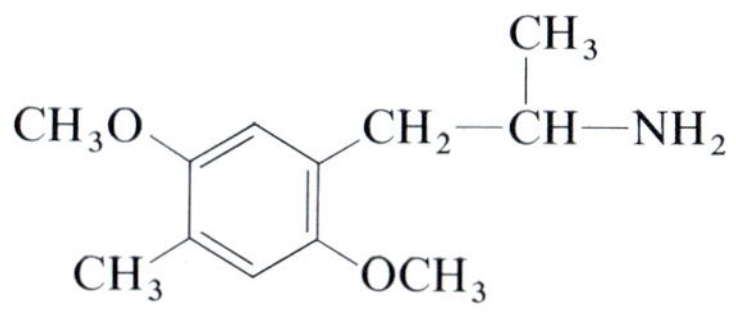

Methoxyamphetamine

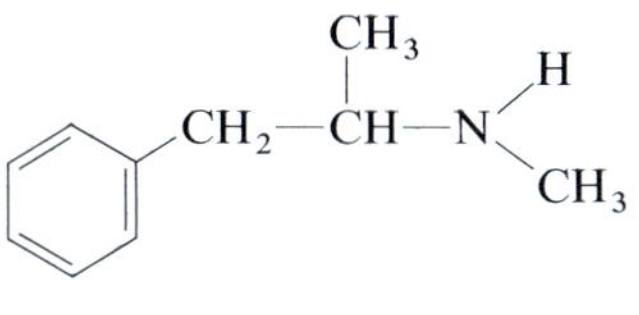

Methamphetamine

Important Heterocyclic Compounds Used in Medicine

We have mentioned that heterocyclic molecules are very important. Many are structurally complex. However, there are a large number of important drugs that are composed of these complex heterocyclic molecules. In Table 21.5 a few of the representative heterocyclic molecules are shown.

TABLE 21.5
Examples of Important Heterocyclic Amines

Name	Description	Structure
Diphenylhydramine	Antihistamine	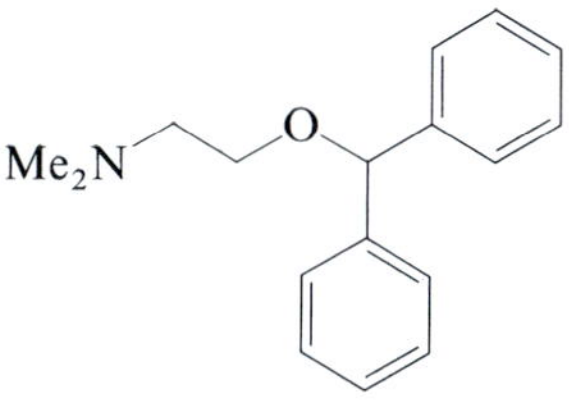
α-Propoxyphene	Analgesic	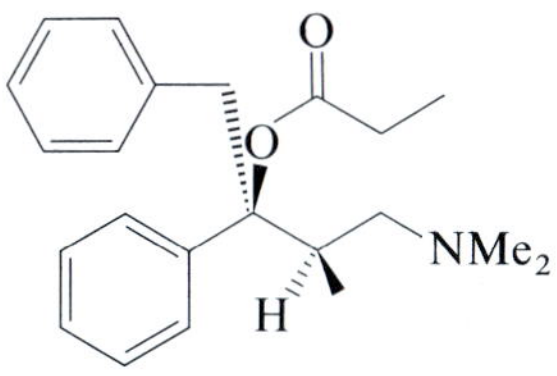
Caramiphene ethanedisulfonate	Anticholinergic and antitussive	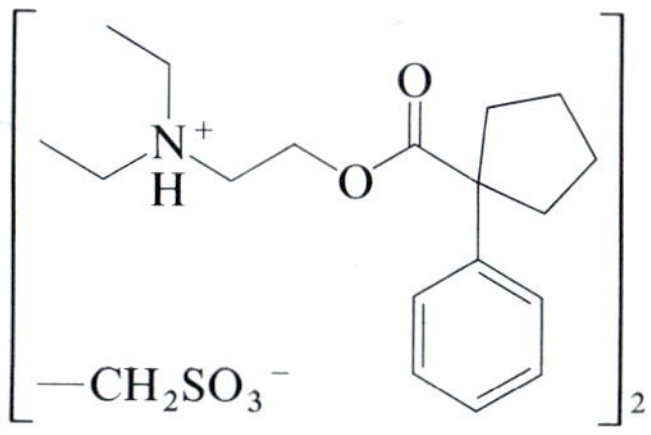

(Continued)

Table 21.5 *(continued)*

Name	Description	Structure
Methyldopa	Antihypertensive	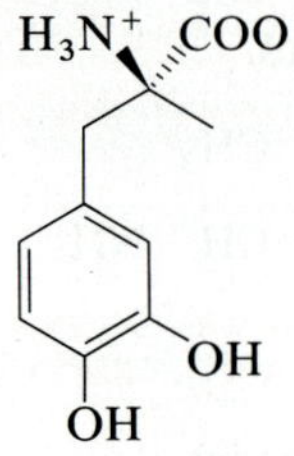
Phenylephrine	Adrenergic	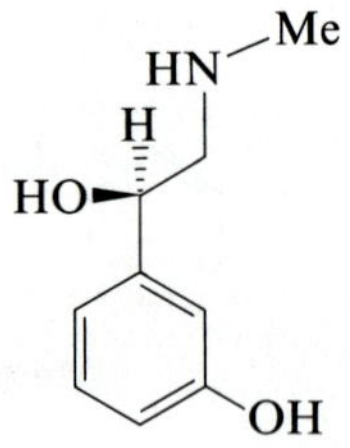
Tolbutamide	Hypoglycemic	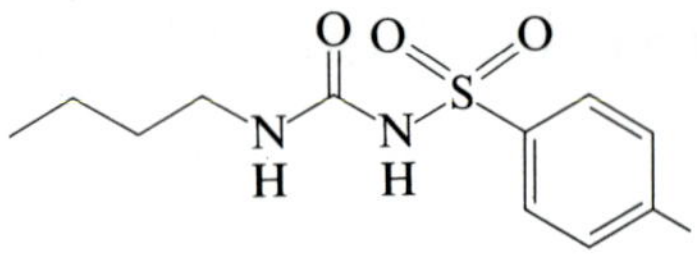
Meprobamate	Tranquilizer and anticonvulsant	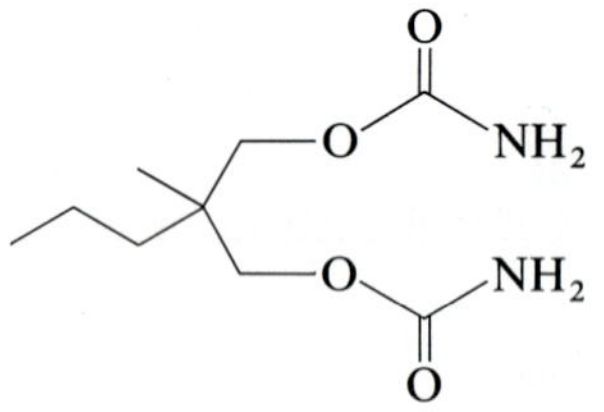
Indomethacin	Antiinflammatory, antipyretic, and analgesic	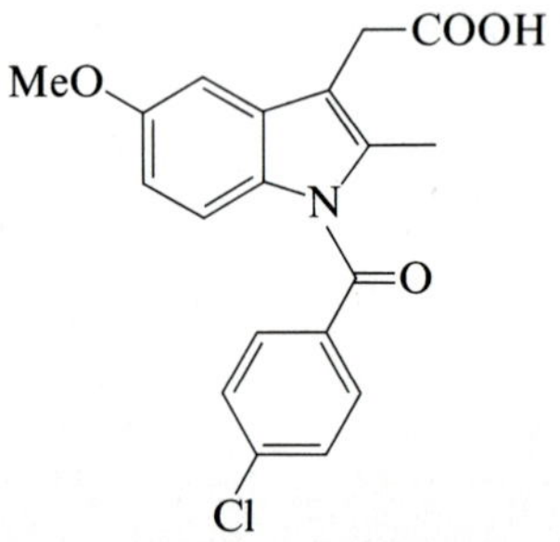
Phenylbutazone	Antiinflammatory	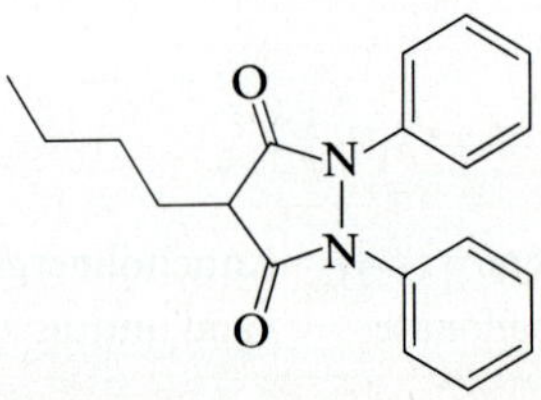
Phenazopyridine	Urinary analgesic	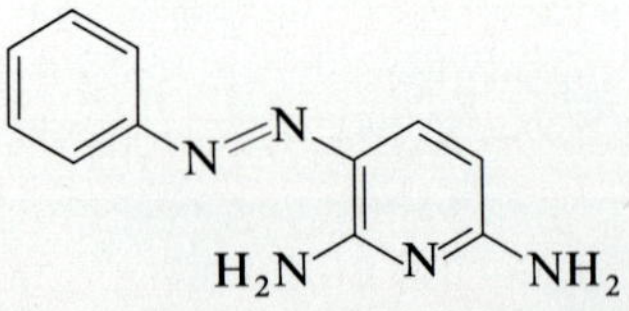

Name	Description	Structure
Diazepam (valium)	Minor tranquilizer	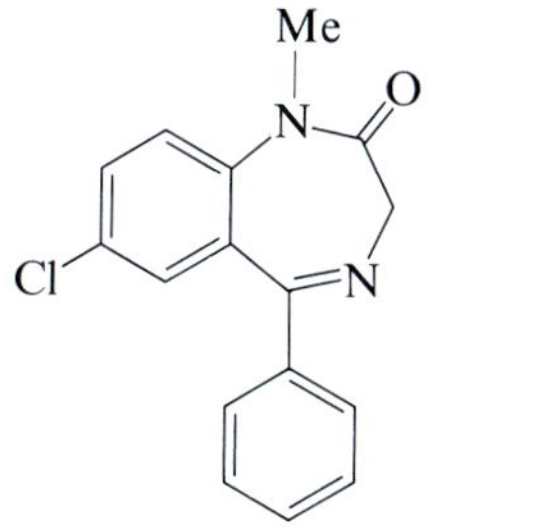
Chlordiazepoxide (librium)	Minor tranquilizer	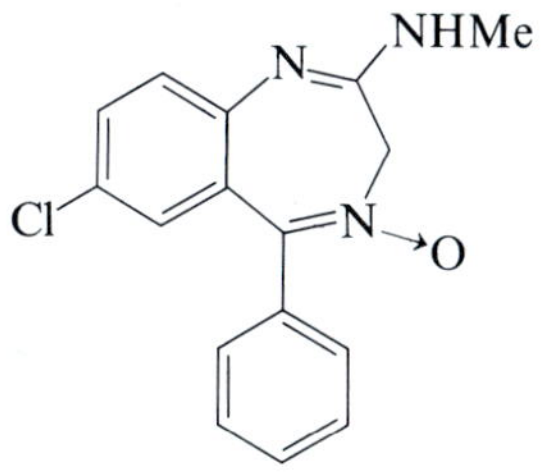
Penicillins	If R = $PhCH_2$: Benzylpenicillin = penicillin G (usual fermentation product) If R = $PhCH_2O$: Phenoxymethyl-penillin = penicillin V If R = 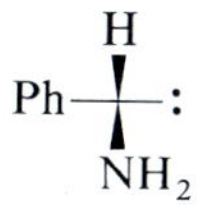 Ampicillin	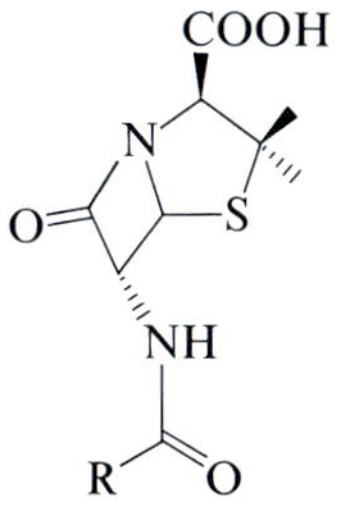

NEW TERMS

amino acid Important molecules containing both an amine functional group and a carboxylic acid functional group. Most of the important natural amino acids are α-amino acids.

alkaloid A naturally occurring, basic, nitrogen-containing compound.

TESTING YOURSELF

Special Amines

1. Based on your understanding of solubility, why would you predict that amino acids are water soluble?
2. You find a rare plant in an isolated region of the world and learn that the local people use this plant as a medicine. How would you get the alkaloids out of the plant.

Answers 1. Since amines are basic and acids are acidic, you expect these two functional groups to undergo acid–base chemistry. An acid and a base react to form a salt. Many salts, being ionic, are water soluble. 2. Since alkaloids are basic, grind the plant up and extract with acid. The basic alkaloids will form water soluble salts.

SUMMARY

The *amines* are an important functional group. As a result of the electron lone pair on nitrogen, amines are basic. The lone pair also makes it possible for an amine to act as a nucleophile. Amines undergo a variety of reactions as a consequence of the nitrogen lone pair. Reactions include acid–base chemistry, nucleophilic substitution with organohalogen compounds, reaction with carbonyl compounds to form *imines*, reactions with hydrazine to form hydrazones, and reactions with acid derivatives to form amides.

Amides are important derivatives of amines, and the amide bond is central to the structure and chemical behavior of proteins. The amide bond is also an important structural feature of nylon.

Alkaloids constitute a diverse and interesting class of basic natural products. Many of these have dramatic pharmacological activity.

REACTION SUMMARY

An Amine as a Base

$$-\overset{|}{\underset{|}{N}} + HX \longrightarrow -\overset{|}{\underset{|}{N^+}}-H + X^-$$

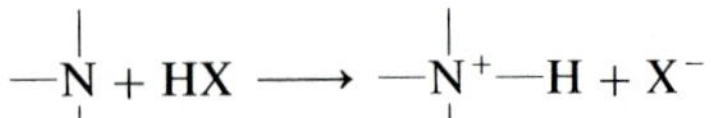

An Amine with an Alkyl Halide

$$R-N{<} + -C-X \longrightarrow R-\overset{+}{N}{<}-C- + X^-$$

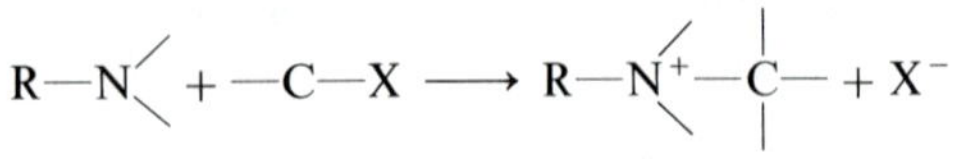

An Amine with an Acid Chloride

$$R-\underset{\underset{O}{\|}}{C}-Cl + H-N{<}^{R'}_{H(R'')} \longrightarrow R-\underset{\underset{O}{\|}}{C}-N{<}^{R'}_{H(R'')} + HCl$$

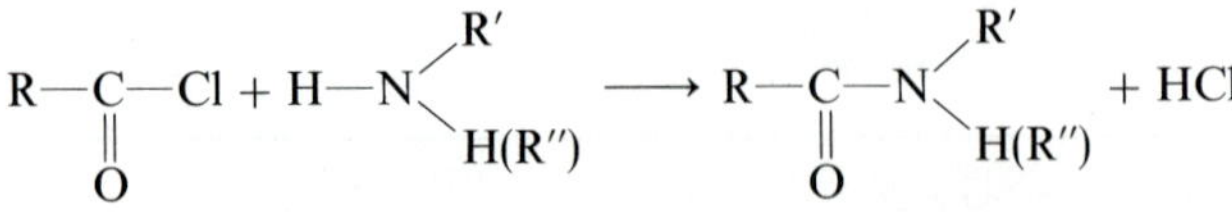

An Amine with an Acid Anhydride

$$R-\underset{\underset{O}{\|}}{C}-O-\underset{\underset{O}{\|}}{C}-R + H-N{<}^{R'}_{R''(H)} \longrightarrow R-\underset{\underset{O}{\|}}{C}-N{<}^{R'}_{R''(H)} + R-\underset{\underset{O}{\|}}{C}-O-H$$

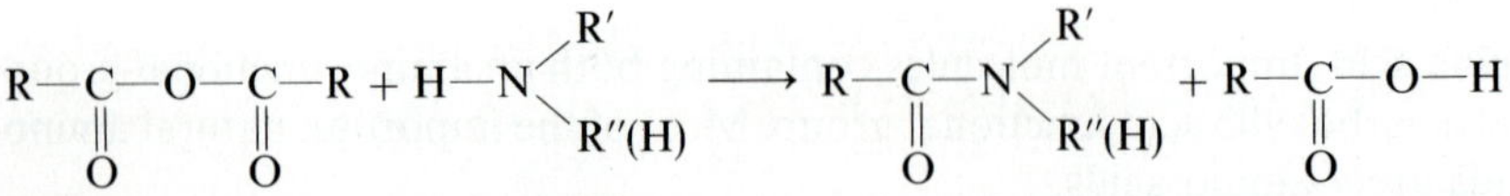

Conversion of an Ester to an Amide

$$R-\underset{\underset{O}{\|}}{C}-O-R' + H_2N-R'' \longrightarrow R-\underset{\underset{O}{\|}}{C}-\overset{H}{\underset{}{N}}-R'' + R'OH$$

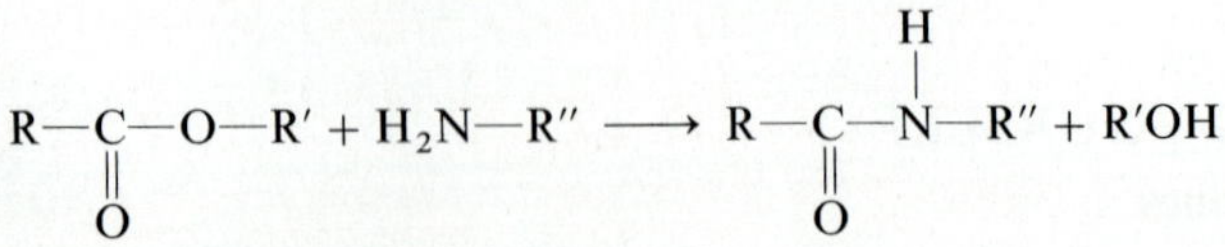

An Amine with a Carbonyl Group

$${>}C{=}O + H_2N-R \xrightarrow{-H_2O} {>}C{=}N-R$$

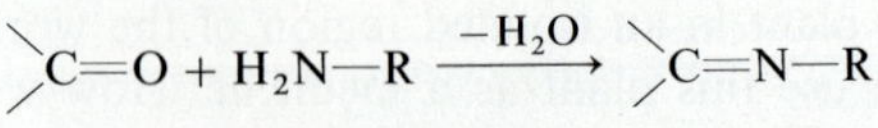

A Hydrazine with a Carbonyl Compound

$$>C{=}O + H_2N{-}NH_2 \xrightarrow{H^+} >C{=}N{-}NH_2 + H_2O$$

Hydrolysis of an Amide

$$R{-}\underset{\displaystyle O}{\underset{\|}{C}}{-}N< + H_2O \xrightarrow{H^+ \text{ or } HO^-} R{-}\underset{\displaystyle O}{\underset{\|}{C}}{-}OH + H{-}N<$$

TERMS

amine
heterocycle
K_b
imine
amide
amino acid
alkaloid

CHAPTER REVIEW

QUESTIONS

1. The best name for $CH_3{-}CH_2{-}\overset{\displaystyle H}{\overset{|}{N}}{-}CH_3$ is
   a. ethylhydromethylamine c. ethylmethylamine
   b. 2-nitrobutane d. 2-aminobutane
2. The hybridization of nitrogen in an amine is
   a. *sp* b. sp^2 c. sp^3 d. sp^4
3. In the reaction, $CH_3{-}NH_2 + CH_3{-}Cl \longrightarrow$, the amine will act as
   a. an electrophile b. a nucleophile c. an alkylating agent d. a base
4. Amines are generally classified as
   a. weak acids b. weak bases c. strong acids d. strong bases
5. The reaction of an acid chloride with methylamine will yield
   a. an amide b. an acid salt c. an imine d. no reaction
6. A hydrazone will result from the reaction of hydrazine with
   a. an acid b. an aldehyde c. an alcohol d. a phenol
7. An amine will exhibit ______ hydrogen bonding than an alcohol.
   a. more b. the same c. less
8. Reduction of an imine will give an
   a. acid b. amide c. amine d. alcohol
9. The reaction of $CH_3{-}\underset{\displaystyle O}{\underset{\|}{C}}{-}O{-}\underset{\displaystyle O}{\underset{\|}{C}}{-}CH_3$ with $CH_3{-}NH_2$ will give
   a. an acid + amine c. an amide + acid
   b. an amide + amine d. no reaction
10. A secondary amine is
    a. methylamine c. ethylmethylpropylamine
    b. ethylmethylamine d. dimethylethylamine

Answers **1.** c **2.** c **3.** b **4.** b **5.** a **6.** b **7.** c **8.** c **9.** c **10.** b

DIAGNOSTIC CHART

Blacken in all of the circles under the number of each question that you missed in the preceding Chapter Review questions. The diagnostic chart will help you identify concept areas that need more study.

	Questions									
Concepts	1	2	3	4	5	6	7	8	9	10
Structure		○					○			○
Physical properties							○			
Nomenclature	○					○				○
Functional groups	○		○	○	○	○		○	○	
Reactions			○	○	○	○		○	○	

EXERCISES

Structure and Nomenclature

1. Give names for the following amines.

a. 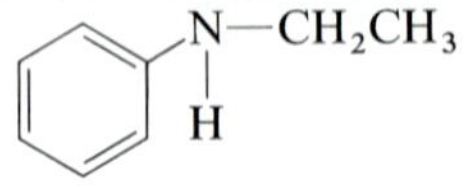

b. 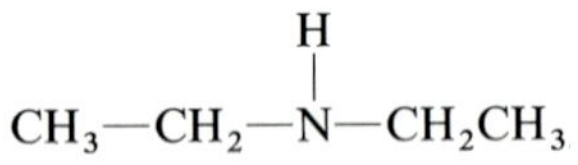

c. 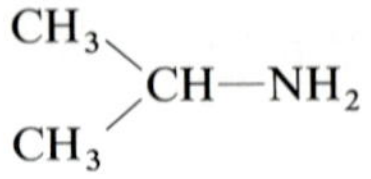

d. 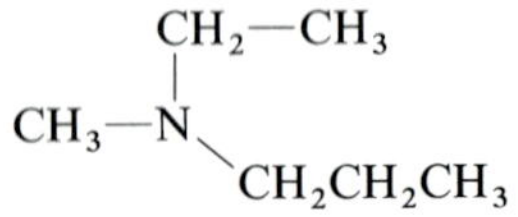

e. 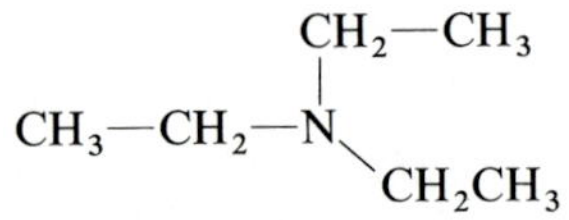

2. Give names for the following compounds.

a. 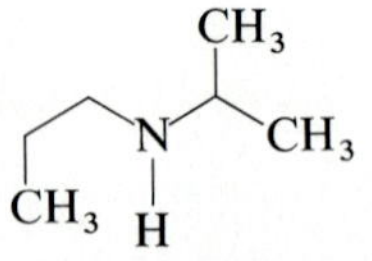

b. 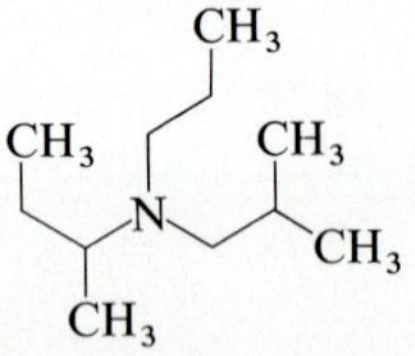

c. 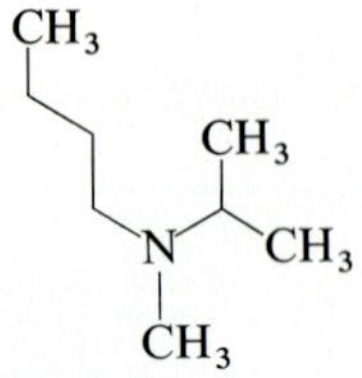

d. 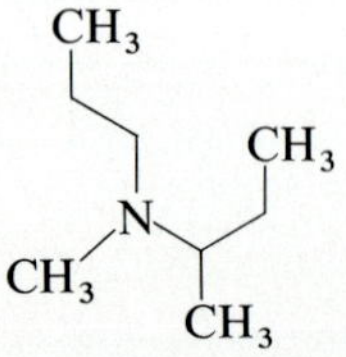

e. 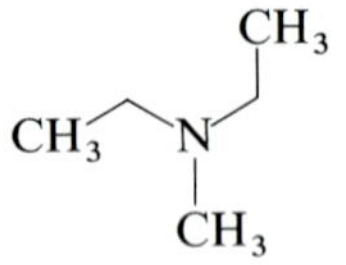

f. 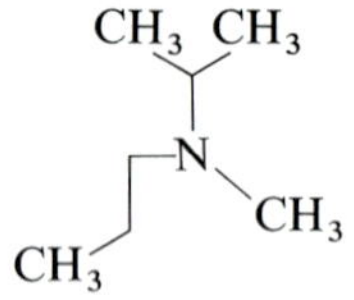

Reactions

3. Give products for the following acid–base reactions.

a. $CH_3-CH_2-NH_2 + HCl \longrightarrow$

b. 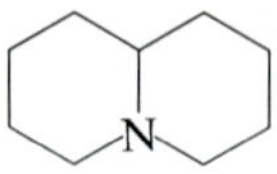 + HBr ⟶

4. Give products for the following reactions of amines with alkyl halides.

a. 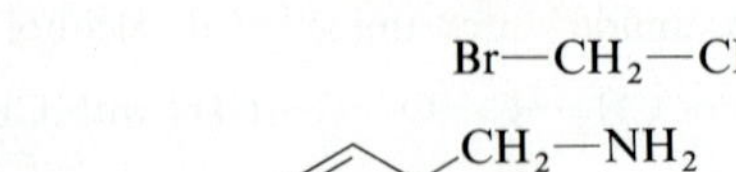

b. 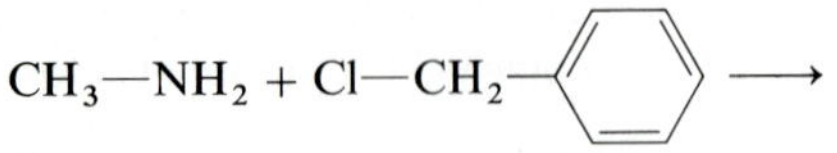

c. 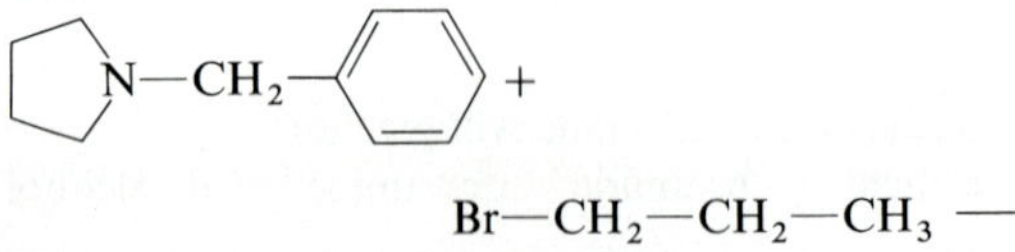

d. 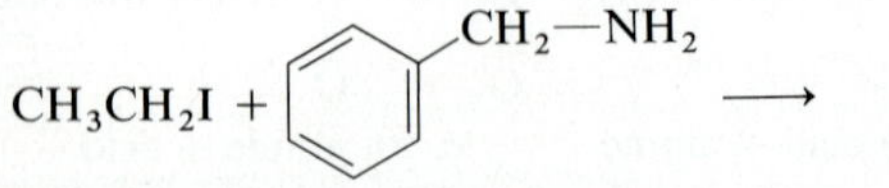

5. Give the structures of the amides resulting from the following reactions.

a. 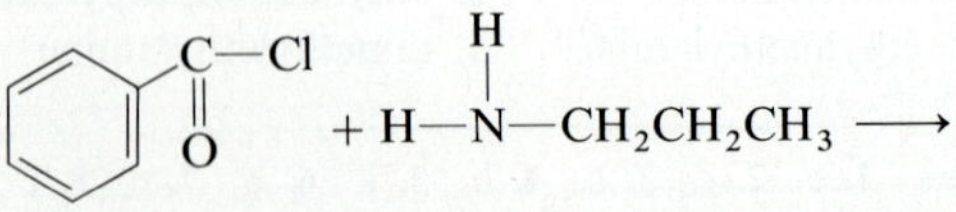

b. 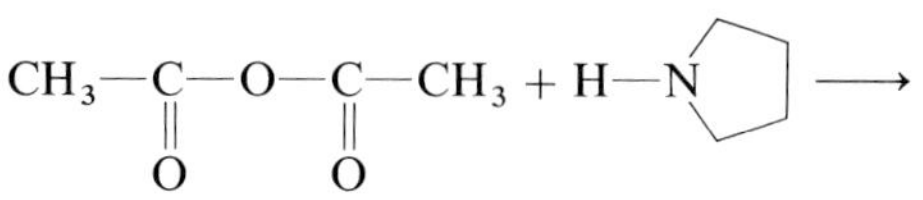

c. 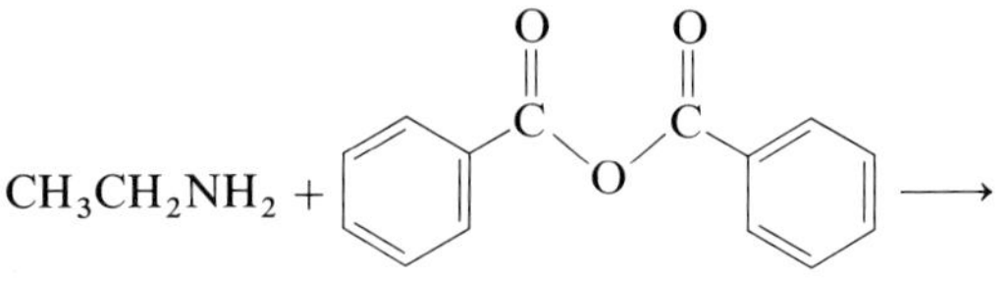

d. 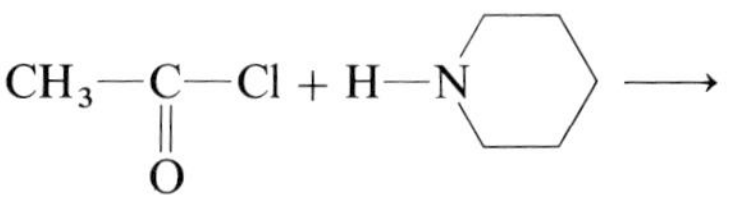

6. Give products for the following carbonyl reactions with amines and hydrazines.

a. 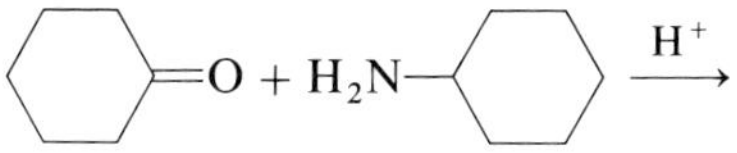

b. 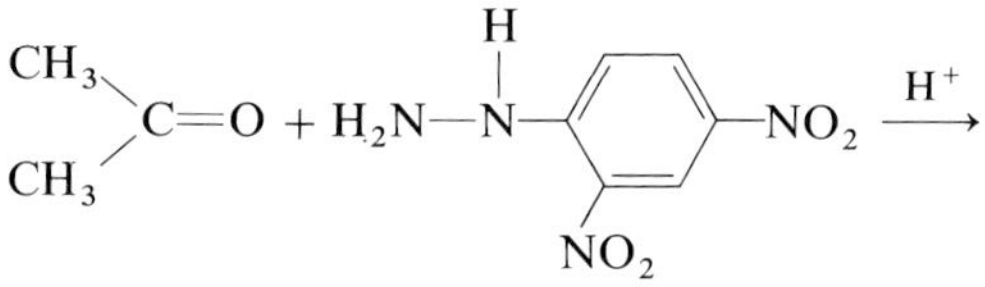

c. 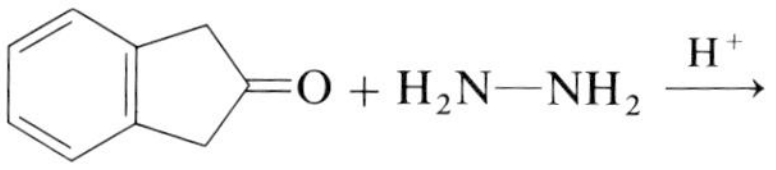

d. 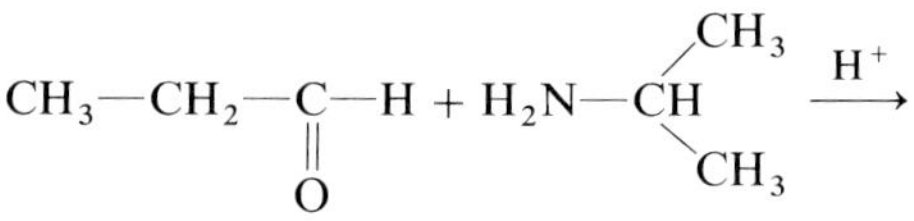

7. Give products for each of the following amide reactions.

a. 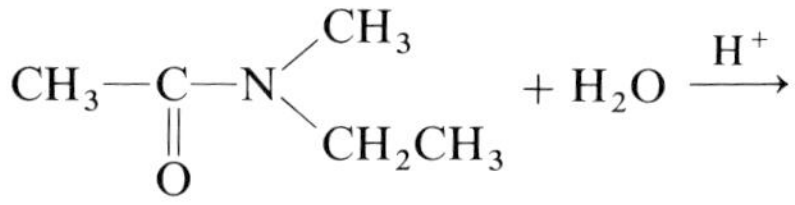

b. 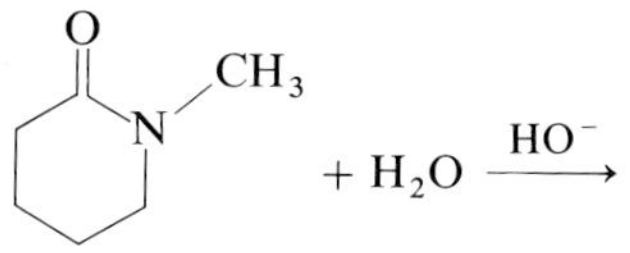

8. Give products for the following miscellaneous reactions.

a. 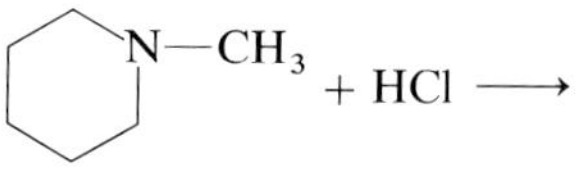

b. 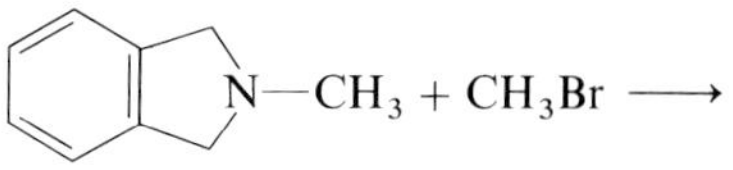

c. 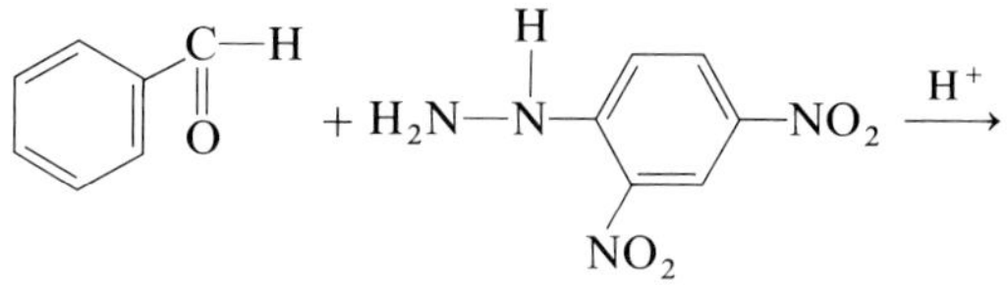

d. 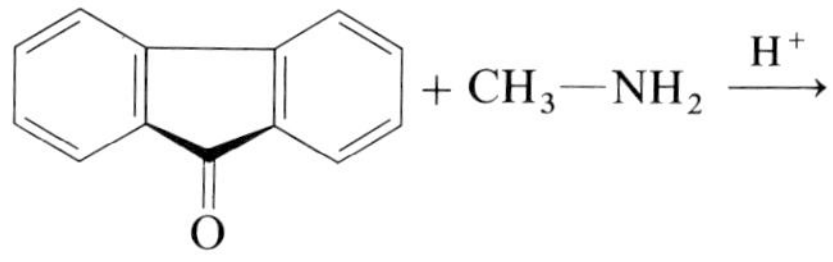

e. 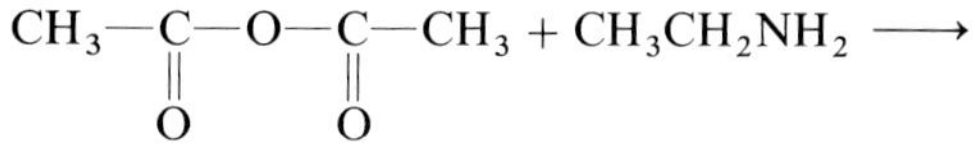

f. 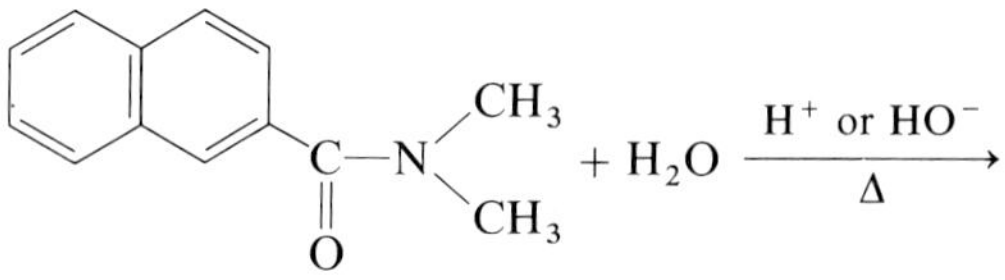

Thought Questions

9. Explain why aniline is considerably less basic than cyclohexylamine.

10. Why would an alcohol have a higher boiling point than an amine with the same number of carbon atoms?

CHAPTER

22

Cells, Biomolecules, and Nutrition

Modern medicine has become more and more molecular in its approach to diagnosis and treatment of disease. Biochemistry, the reactions and molecules that support life, is now an essential part of medicine and health care. But biochemistry cannot be understood without chemical and biological background. The earlier chapters have provided the neccessary chemical background. This chapter introduces the principal biomolecules and provides a survey of cell biology and nutrition that will help you understand biochemistry.

OBJECTIVES

After completing this chapter, you should be able to

- Describe a cell and explain its role in organisms.
- Name the organelles of a procaryotic cell and describe their function.
- Name the organelles of a eucaryotic cell and describe their function.
- Describe the role chirality plays in the interaction of biomolecules.
- Define the term *nutrition* and name the major groups of nutrients.

22.1 Introduction to Cells

From the simplest organism to the most complex, one structural feature remains constant: all living organisms contain one or more cells. The **cell** is the basic unit of life, and every cell, given the proper environment, appears to be capable of living in the absence of other cells. Most of the nutrients absorbed by the human body are used by cells, and much of the chemistry that occurs within the body takes place within cells. Cells contain a wide variety of substances dissolved or dispersed in water, but they also contain specialized structures called **organelles**. Organelles carry out many cellular functions. Some knowledge of cell structure will provide a framework for understanding biochemistry.

The cell is the smallest unit of a living organism that can live independently.

Cell Size

With the exception of some egg cells, most cells are rather small, having dimensions that range from a micrometer to tens of micrometers. There appears to be some lower and upper limits to cell size. Typical cell volume is about 10^{-9} mL (10^{-6} μL) which is a very small reaction volume. This is in marked contrast to a typical laboratory experiment where a vessel or flask may contain from a few microliters to several liters of reactants, or on an industrial scale, where reaction vessels may be as large as houses.

Why are cells not smaller? The most likely reason is the requirement for some minimum number of compounds and some minimum number of copies of these compounds that must be present to sustain life. But why then are cells not larger? What is the constraint that keeps them small? A comparison between the cell as a reaction vessel and a laboratory flask provides some insight into cell size. In a reaction flask, stirring is provided by the experimentalist with a stirring rod or with a mechanical stirrer. This action keeps the solution mixed properly so that reagents can collide and react. A cell has no stirring motor, although cell movement may cause some slight mixing. Movement of cell contents depends primarily upon *diffusion* of the reactants through the cell medium. Diffusion must be fast enough to maintain the proper concentration of the reactants in the various parts of the cell. As a cell gets larger in size, diffusion cannot supply the various reactants fast enough.

Cell size is limited to a certain range by several physical and molecular factors.

The diffusion of oxygen that has just entered a cell can be used as an illustration. When oxygen diffuses into a cell from the environment, it is able to diffuse throughout the whole cell as long as the cell is less than 1 mm in length, width, or depth. The interior of the cell would never receive enough oxygen if the cell were larger than this because the oxygen would be consumed by the outer layer of the cell contents before it got to the core. If this is the case, then it is wasteful to have a cell larger than the maximal size that allows oxygen to diffuse throughout the cell. The size of a cell is limited by the rates of diffusion of molecules within the cell.

Aside from this restriction imposed by diffusion rates, a second factor restricting cell size is the movement of materials into and out of the cell. Nutrients must be able to enter the cell, and waste products must be able to leave. This transport is dependent upon the surface area of the cell. The greater the surface area of the cell, the more rapidly materials can enter or leave the cell.

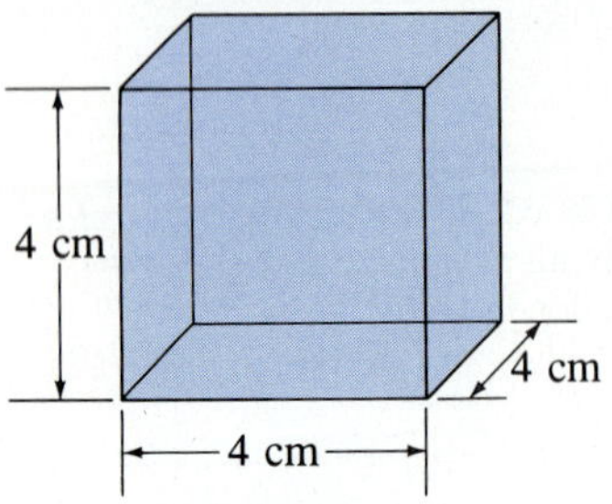
4 cm
4 cm
4 cm

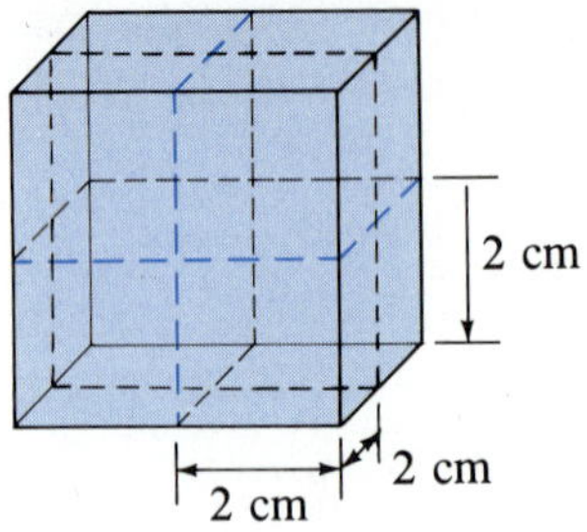
2 cm
2 cm
2 cm

To better understand the relationship between cell surface area and cell volume, consider a cube that is 4 cm on each side. The volume of the cube shown in the top left margin is $4\text{ cm} \times 4\text{ cm} \times 4\text{ cm} = 64\text{ cm}^3$. The surface area is 6 faces $\times$ $(4\text{ cm} \times 4\text{ cm}) = 96\text{ cm}^2$. The ratio of surface area to volume is $96/64 = 1.5$. Now slice the cube into eight equal pieces (below). Each new cube will have a dimension of $2\text{ cm} \times 2\text{ cm} \times 2\text{ cm}$, but the total volume of the 8 small cubes will still equal 64 cm^3. *Each* of the small cubes now has 6 faces that are $2\text{ cm} \times 2\text{ cm}$. This equals 24 cm^2 per cube. There are 8 small cubes, thus the total surface area is $8 \times 24\text{ cm}^2 = 192\text{ cm}^2$. For the *same* volume, the surface area has doubled. The ratio of surface area to volume has changed from 1.5 to $192/64 = 3.0$.

If the small cubes are subdivided again in the same way, the division will yield 64 cubes each with 1-cm dimensions. Again, the total volume will remain the same, but the surface area will have increased to 384 cm^2 and the surface area to volume ratio will be $384/64 = 6.0$. This exercise can be continued over and over again, and each time the surface area to volume ratio increases. But what is the best surface area to volume ratio for a cell? What is the best size for a cell? There is no single correct answer, but a cell must be large enough to hold the needed cell constituents, yet small enough to allow for ready cellular diffusion within the cell and to have a favorable surface area to volume ratio for movement of materials into the cell.

Cells in Multicellular Organisms

The cell is the basic unit of life, but multicellular organisms are not simply a collection of totally independent cells. There are many different kinds of cells in the human body (Figure 22.1), and these cells are arranged into various tissues, organs, and organ systems. There is significant interaction between the cells of the body. Many of these interactions are molecular, thus biochemistry provides insight not only into the workings of cells, but also into how the body functions as a whole.

NEW TERMS

cell The basic unit of life; the smallest piece of a living organism that can carry out life processes by itself.

organelle The structures within a cell that carry out one or more functions of the cell.

22.2 Procaryotic Cells

Bacteria are **procaryotes**, simple one-celled organisms whose cells distinctly differ from the cells of higher organisms, the *eucaryotes*. The procaryotic cell does not contain a true nucleus and is both smaller and structurally simpler than the eucaryotic cell. The structural differences between bacterial cells and animal cells provide a basis for the effectiveness of a number of antimicrobial drugs. These drugs kill or retard growth in bacteria (procaryotic cells) but not in eucaryotic cells because they affect a particular bacterial cell structure or component that is uniquely different from its counterpart in the

Antibiotics used in medicine are effective because they adversely affect bacteria with little or no damage to human cells.

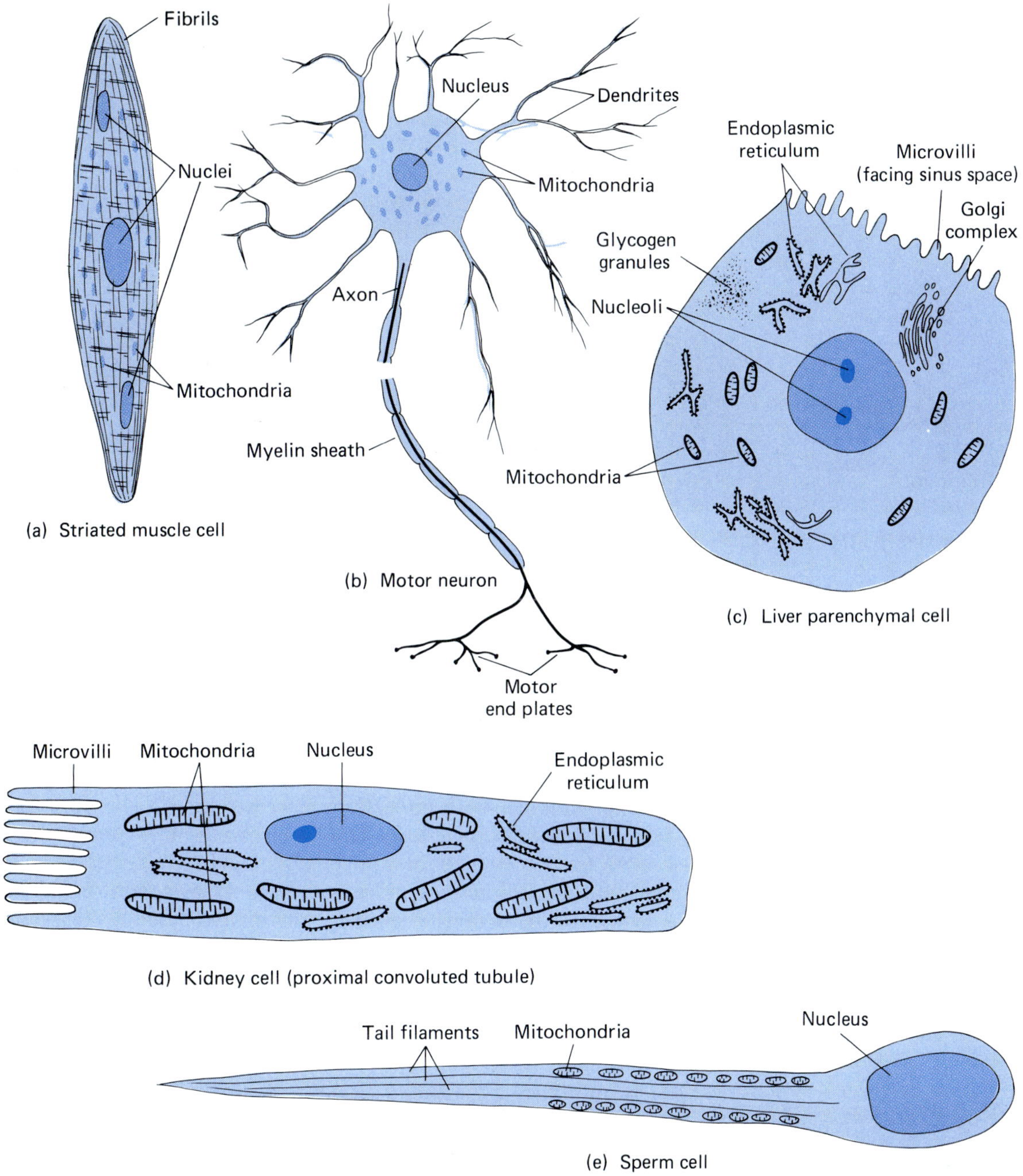
Fibrils
Nuclei
Mitochondria
(a) Striated muscle cell
Nucleus
Dendrites
Mitochondria
Axon
Myelin sheath
Mitochondria
(b) Motor neuron
Motor
end plates
Endoplasmic
reticulum
Microvilli
(facing sinus space)
Golgi
complex
Glycogen
granules
Nucleoli
(c) Liver parenchymal cell
Microvilli
Mitochondria
Nucleus
Endoplasmic
reticulum
(d) Kidney cell (proximal convoluted tubule)
Tail filaments
Mitochondria
Nucleus
(e) Sperm cell

FIGURE 22.1
The cell is the basic unit of life. These are some representative cells of the human body.

animal cell. Several of these differences will be pointed out during the discussion of bacterial cell structure.

The major structural components of the procaryotic cell include the cell membrane, the cytosol, the cell wall, and in some species, an outer layer beyond the cell wall (Figure 22.2). The **cell membrane** is the cell structure that separates the contents of the cell from the external environment. The cell membrane has a large surface area, but it is only a few molecules thick.

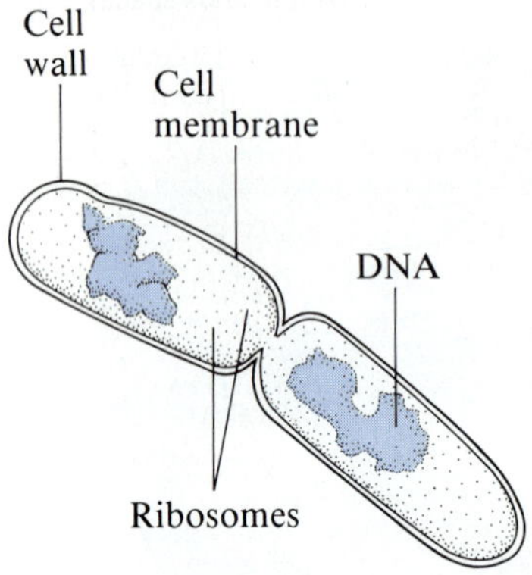
Cell wall
Cell membrane
DNA
Ribosomes

(a)

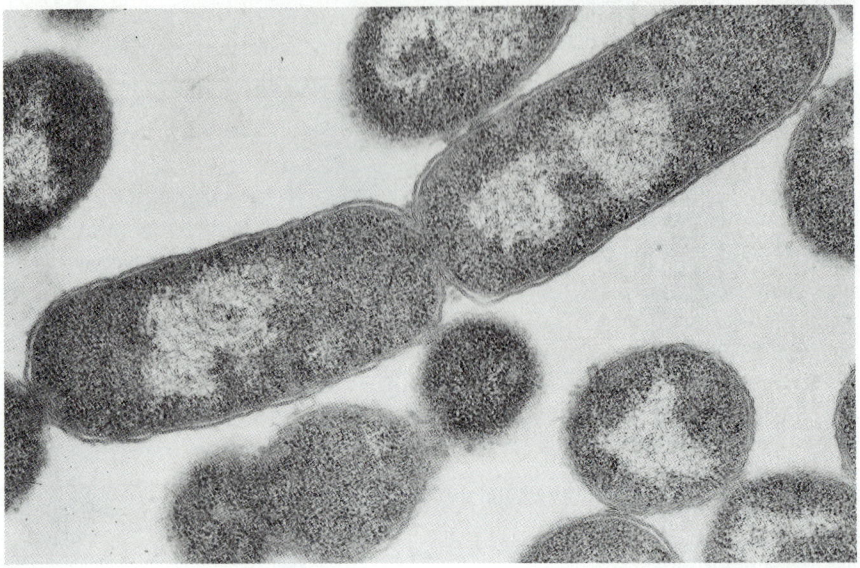

(b)

FIGURE 22.2
This bacterial cell is a typical procaryotic cell. The procaryotic cell is the simplest cell type. (a) Schematic drawing of a dividing cell. (b) Electron micrograph of bacterial cells.

It is not a strong, rigid structure, nor does it just act as a passive barrier between the internal and external environment. The cell membrane is responsible for controlling the entry of nutrients and the departure of waste products. Many transport systems are located in the cell membrane. The structure of this organelle is discussed in more detail in Chapter 24.

In bacteria, the **cytosol** consists of the water, ions, and molecules that are separated from the exterior environment by the cell membrane. This is the internal environment of the cell. The cytosol makes up the bulk of the bacterial cell volume. With an electron microscope, two structures can be seen in the cytosol of procaryotes. One of them is a huge circular molecule called DNA (see Chapter 26 for more detail on DNA). In bacteria, the DNA is sometimes called the *nucleoid* or *nuclear region*. The second structure visible in the cytosol is the organelle called the **ribosome**. These organelles are the site of protein synthesis (Chapter 26), and numerous ribosomes are present in each cell. Procaryotic DNA and ribosomes are structurally different from their eucaryotic counterparts. These differences can be exploited in treatment of bacterial infections. The antibiotic *streptomycin*, for example, affects bacterial ribosomes, resulting in disruption of bacterial protein synthesis (Table 22.1). The ribosomes in the cells of the invaded host organism are not affected.

TABLE 22.1
Some Common Antibiotics and Their Mode of Action

Antibiotic	Mode of Action in Bacteria
Cephalosporins	Disrupt cell wall synthesis
Chloramphenicol	Disrupts protein synthesis
Erythromycin	Disrupts protein synthesis
Novobiocin	Disrupts DNA synthesis
Penicillins	Disrupt cell wall synthesis
Streptomycin	Disrupts protein synthesis

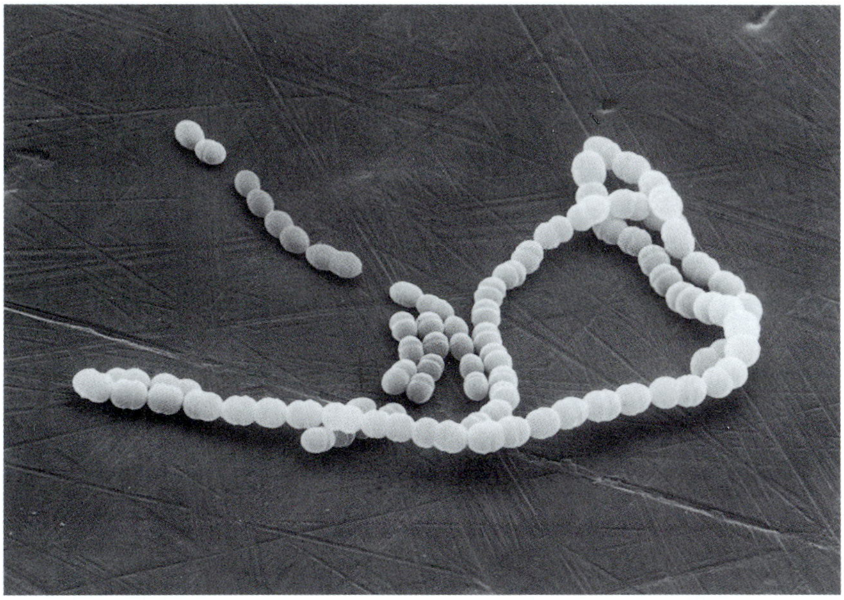

FIGURE 22.3
A scanning electron micrograph of gram positive streptococci. These bacteria are quite vulnerable to penicillin because penicillin disrupts normal cell wall synthesis.

The bacterial **cell wall** lies outside the cell membrane. It is actually one gigantic molecule that forms by bonding numerous smaller molecules together. This structure provides mechanical protection to the cell. If a bacterial cell were placed in a solution hypotonic to it, water would move into the cell via osmosis (see Chapter 9). The relatively weak cell membrane has little strength, so the membrane would expand. In the absence of a cell wall, it would expand indefinitely, then burst. The cell wall prevents cellular expansion that would rupture the cell membrane. Animal cells lack cell walls. They do not need a cell wall, because the composition of the body fluids bathing the cells is maintained within a narrow range of concentrations (Chapter 30).

Cell walls protect bacteria from bursting in hypotonic solutions.

The bacterial cell wall is another feature that can be exploited by antibiotics. Penicillins and cephalosporins are antibiotics that prevent the normal formation of cell walls in some bacterial species (Figure 22.3). Without normal cell walls, the bacteria perish. Except for allergic reactions in some people, penicillins are one of the safest of all antimicrobials because nothing in the host organism appears to be affected by them.

Penicillin was the first antibiotic used in humans. Penicillin and its derivatives are still highly useful and effective.

Some bacterial species possess an outer layer beyond the cell wall. This layer is composed of a material called *lipopolysaccharide* which contains both lipids (nonpolar biomolecules) and carbohydrates (see Chapters 23 and 24). This outer layer is *pyrogenic* in mammals, that is, it causes fever. Even if the bacterium is dead, the lipopolysaccharide can cause adverse reactions in humans. The next time you are around sterile medical equipment such as syringes, note that the wrapper states that the contents are both sterile and nonpyrogenic. This certifies that no living microorganisms are present, and no pyrogens are present that could cause fever in the patient.

NEW TERMS

procaryote The simplest and smallest cell type characterized by the absence of a true nucleus and other membranous organelles.

cell membrane Sheetlike structure that separates the interior of the cell from the external environment.

cytosol The fluid interior of the cell, containing the ions and molecules that support life. Organelles are suspended in it.

ribosome Small organelle that is the site of protein synthesis.

cell wall Rigid structure that surrounds bacterial and plant cells. It protects the cell from rupturing caused by swelling.

TESTING YOURSELF

Cells and Procaryotes

1. Cells fall within a certain size range. (a) Why are they not smaller than this? (b) Why are they not larger?
2. What is the function of the cell membrane of a cell?
3. Compare what would happen to a bacterial cell and an animal cell in a hypotonic solution.

Answers **1. a.** Cells must contain a minimum number of different kinds of biomolecules, and a minimum number of copies of each kind. **b.** Diffusion cannot provide nutrients fast enough if cells are too large. **2.** The cell membrane separates the cell from the external environment and regulates the passage of substances into and out of the cell. **3.** The animal cell would expand and probably burst due to osmotic pressure. The bacterial cell would not be harmed because the cell wall would prevent expansion of the cell membrane.

22.3 Eucaryotic Cells

Eucaryotic cells are larger and more complex than procaryotic cells.

The cells of plants and animals are **eucaryotic** cells. They are generally larger than procaryotic cells and have more structural complexity. All eucaryotes possess cytosol and a cell membrane. The cytosol contains ribosomes, but the ribosomes of eucaryotes are larger and differ somewhat in structure from the ribosomes of procaryotes. Some cells, like those of plants and algae, also possess cell walls and outer layers, but animal cells generally lack these features. Four major organelles are found in eucaryotes that are not found in procaryotes.

1. Eucaryotes possess a true *nucleus*, the organelle in which DNA is located. In bacteria, the DNA is not located in a separate organelle but is simply suspended in the cytosol.
2. Eucaryotic cells also possess an *endoplasmic reticulum*, a complex of membranes that penetrates throughout the cytosol and is involved in synthesis of some biomolecules.
3. Eucaryotes possess *mitochondria*, organelles that are the site of aerobic energy production (see Chapter 28). Most of the useful energy that is produced in animal cells is produced in the mitochondria.
4. Most plants and algae possess *chloroplasts*, organelles that are the site of photosynthesis (see Chapter 29). When light is available, the chloroplast is the major site of energy production in the plant.

Figure 22.4 illustrates the general features of a eucaryotic cell.

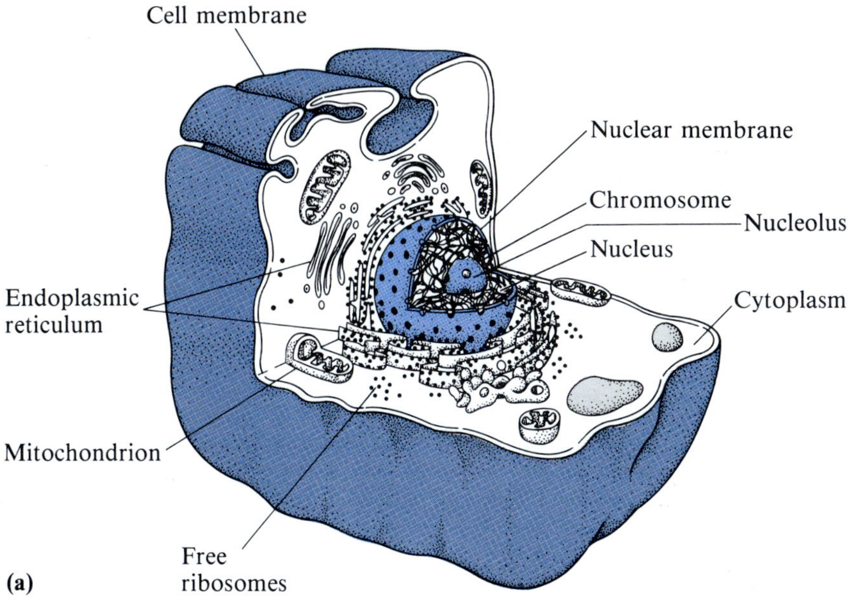
Cell membrane
Nuclear membrane
Chromosome
Nucleolus
Nucleus
Cytoplasm
Endoplasmic reticulum
Mitochondrion
Free ribosomes
(a)

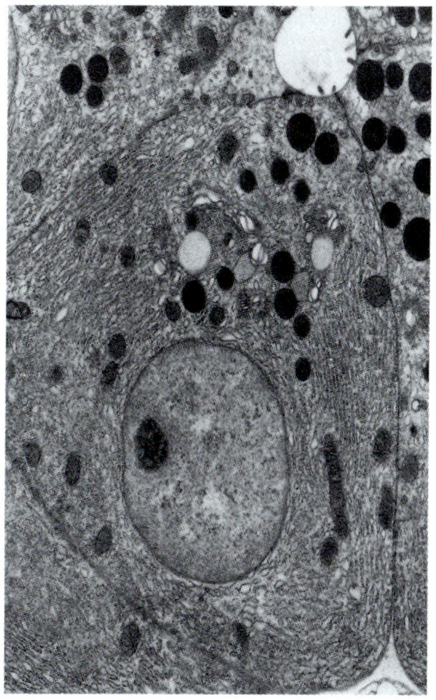
(b)

FIGURE 22.4
(a) This generalized eucaryotic cell illustrates the structures found in an animal cell. Eucaryotic cells are larger and more complex than procaryotic cells.
(b) Electron micrograph of typical mammalian cell.

The **nucleus** is the portion of the cell where information needed for cell growth, division, and maintenance is located. The nucleus is generally visible in cells prepared for viewing with a microscope. It is here that the DNA of the eucaryote is found. The DNA of eucaryotes is present as several pieces instead of the single molecule found in procaryotes. Each of these DNA molecules is associated with proteins in larger structures called *chromosomes*. Chromosomes are complex structures that contain genetic material (see Chapter 26). Also located in the nucleus is the *nucleolus*, which has a major role in the synthesis of ribosomes.

The **endoplasmic reticulum** is a complex network of membranes that permeates and divides the cytosol into various compartments (Figure 22.4). It plays a role in the synthesis and transport of a number of molecules and structures that are used inside and outside the cell (see Chapter 29). Much remains to be learned about this organelle.

The **mitochondrion** (plural, mitochondria) is a bacteria-sized organelle where much of the energy production of the cell takes place (Figure 22.5). Mitochondria are generally located where energy demand is highest in the cell. This organelle has two membranes: a relatively smooth outer membrane that separates the mitochondrion from the cytosol, and a highly folded inner membrane (Figure 22.5). The folds of the inner membrane, which are called *cristae*, greatly increase the surface area of this membrane. Many of the proteins needed for aerobic energy production are located in this membrane. They carry electrons and are involved in the synthesis of ATP (see Chapter 28). The core of the mitochondrion is called the *matrix*. The matrix contains another group of enzymes that are involved in energy production, the enzymes of the tricarboxylic acid cycle (see Chapter 28). Curiously, mitochondria possess DNA and ribosomes that are similar to those of bacteria.

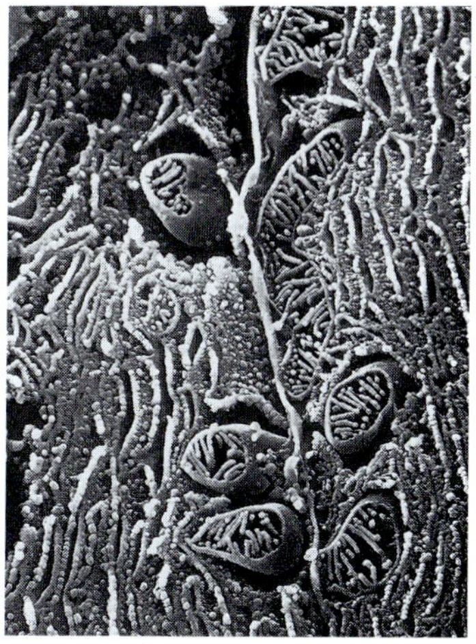

FIGURE 22.5
Mitochrondria surrounded by endoplasmic reticulum. In eucaryotes, mitochondria are the sites of aerobic energy production. (See Figure 28.8 for more on mitochondria.) Structural similarities between mitochondria and bacteria have led many scientists to believe mitochondria evolved from ancestors of bacteria.

TABLE 22.2
A Comparison of the Structural Features of Different Cell Types

	Procaryote	Eucaryote	
Organelle	Bacteria	Animal	Plant
Cell membrane	Present	Present	Present
Cytosol	Present	Present	Present
Cell wall	Present	Absent	Present
Nucleus	Absent	Present	Present
Mitochondrion	Absent	Present	Present
Endoplasmic reticulum	Absent	Present	Present
Chloroplast	Absent	Absent	Present

Table 22.2 summarizes the similarities and differences of bacterial, animal, and plant cells.

NEW TERMS

eucaryote The cells of higher organisms which possess a true nucleus and a number of other membranous organelles that are not present in the procaryotic cell.

nucleus Eucaryotic organelle that contains the DNA of the cell and directs cell division, growth, and maintenance of the cell.

endoplasmic reticulum The membranous organelle within the eucaryotic cell where some of the cellular synthesis and transport of materials occurs.

mitochondrion The organelle of eucaryotes where oxygen-requiring energy production occurs.

TESTING YOURSELF

Eucaryotic Cells

1. Compare the structure of a procaryotic cell with that of a eucaryotic cell.
2. What role does the mitochondrion play in a cell?
3. Compare the location and organization of DNA in procaryotes and eucaryotes.

Answers **1.** Eucaryotes are larger and more complex with various membranous organelles and a true nucleus. Procaryotes are smaller and simpler, lack a nucleus, and have only a cell membrane that is membranous. **2.** Mitochondria are the sites of aerobic energy production and thus provide most of the useful energy in a eucaryotic cell. **3.** DNA in eucaryotes is in the nucleus and is complexed with proteins to form highly organized structures called chromosomes. The DNA of procaryotes is naked and not contained in a nucleus.

22.4 Biomolecules

The compounds found in living organisms do not differ fundamentally from the substances you have studied in the organic chemistry chapters of this book. They possess no vital force nor other mystic properties. Many of them

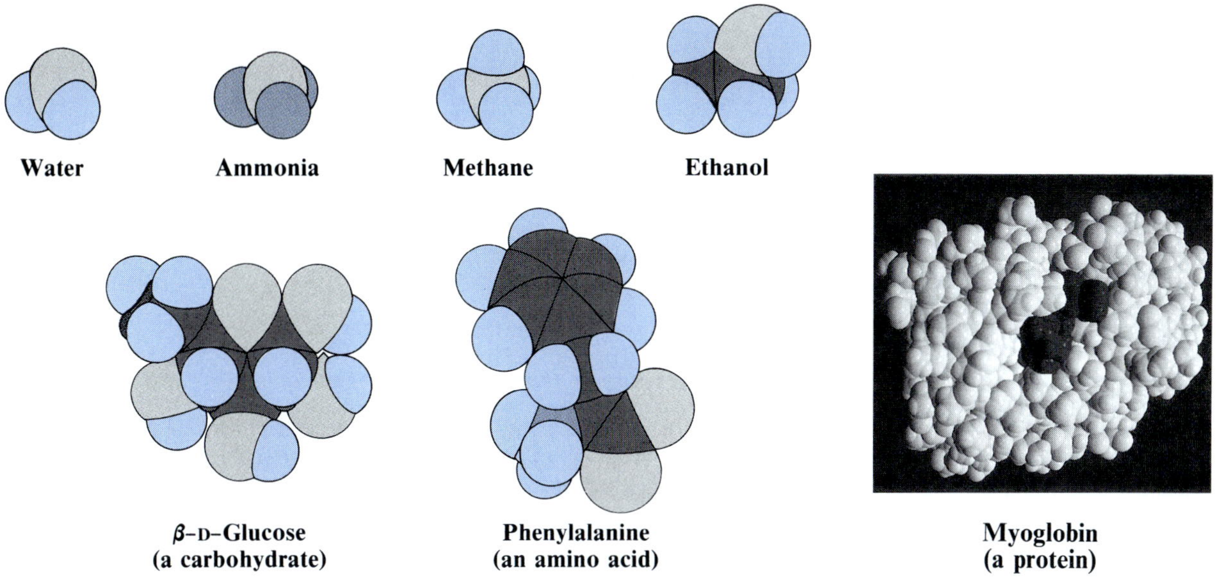
Water
Ammonia
Methane
Ethanol
β-D-Glucose
(a carbohydrate)
Phenylalanine
(an amino acid)
Myoglobin
(a protein)

FIGURE 22.6
These space-filled models illustrate that biomolecules are generally larger and more complex than the compounds you studied in general and organic chemistry. Several simple molecules are shown for comparison along the top. (Relative sizes are not to scale.)

are, however, considerably more complex than the relatively simple organic molecules you have studied so far.

Size and Functional Groups of Biomolecules

The complexity of many biomolecules is due to their relatively large size, the presence of multiple functional groups, and their distinctive three-dimensional structures.

Many biomolecules are quite large. Proteins typically have molecular weights in the tens to hundreds of thousands; DNA molecules often have molecular weights in the millions. Even some of the smaller biomolecules such as simple sugars and fats have molecular weights in the hundreds. To help visualize their relative sizes, Figure 22.6 shows space-filled models of several important biomolecules. Models of the more familiar compounds water, ammonia, methane, and ethanol are included for comparison.

Many biomolecules possess two or more functional groups within the molecule, thus they are a mosaic with respect to chemical and physical properties. Some parts of these molecules are nonpolar while other regions are polar. Some regions react rapidly with certain other molecules while other regions are inert. The size and multiple functionality of biomolecules allows them to interact with other biomolecules in important and unique ways. For example, look at Figure 22.7, which illustrates the amino acid tyrosine. Note that even in this small biomolecule, three different functional groups are present and two polar regions are separated by a nonpolar region within the molecule.

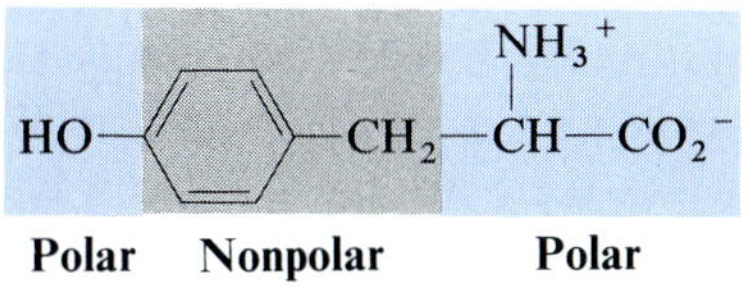
HO
CH_2
NH_3^+
CH
CO_2^-
Polar
Nonpolar
Polar

FIGURE 22.7
This is the amino acid tyrosine. Although this is a relatively small biomolecule, it has several functional groups and three regions of distinctly different polarity.

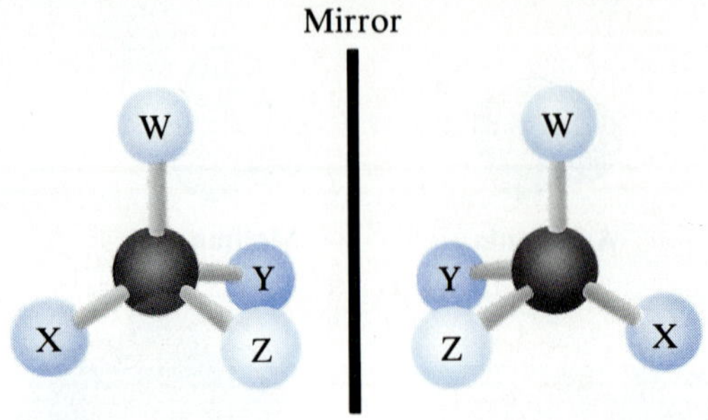
Mirror
W
Y
X
Z
W
Y
Z
X

FIGURE 22.8
These two molecules are stereoisomers. They have the same atoms attached to the same atoms, yet they are not identical structures because their orientation in space is different.

Stereoisomerism in Biomolecules

Stereoisomers differ only in the three-dimensional arrangement of atoms in space.

Many biomolecules show a type of isomerism known as *stereoisomerism.* As with other isomers, **stereoisomers** have the same composition. A set of stereoisomers also has the same atoms attached to the same atoms, but they differ in the way the atoms are oriented in space. You have already seen examples of stereoisomerism in cis-trans isomerism in alkenes and ring compounds. These isomers have the same composition, and they have the same atoms attached to the same atoms, but the atoms have different orientations in space. (Make a pair of cis-trans isomers and examine them for these features.) The stereoisomerism seen in many biomolecules meet these same conditions, but it involves a different structural relationship.

This structural relationship is illustrated with the two models shown in Figure 22.8. Each molecule has the chemical formula CWXYZ, and each has the atoms W, X, Y, and Z attached to the carbon (C) atom. But these two molecules are *not* simply copies of the same molecule. They are uniquely different. They are mirror images of each other.

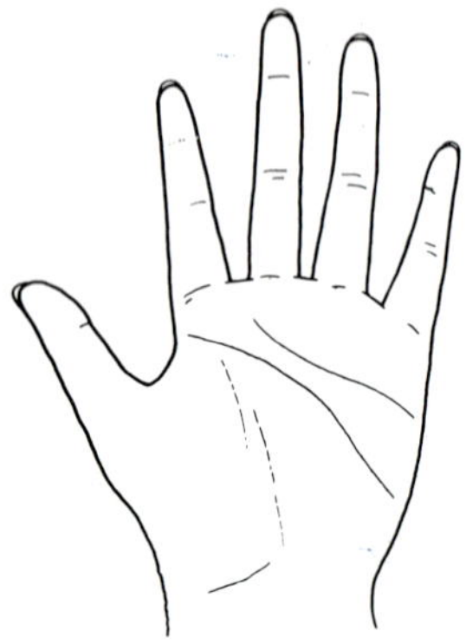

These stereoisomers are related in the same way your hands are related to each other. Are your two hands identical? Look at the hand drawn in the margin. Is it a left hand or a right hand? You have no trouble distinguishing the hand in the margin even though your hands are very similar. If they were identical, you could not have identified the hand correctly except by chance. Each hand has five fingers, a palm, and a back; they have the same composition. The components are attached to the same places on each hand. Identical composition, same attachment, yet your hands are different. Try one more test with your hands. Try to superimpose one of them on the other, keeping both palms up. Superimposition simply means that the two forms can be mentally merged into one as the objects are brought together (Figure 22.9).

A pair of hands display **chirality**, that is, they are *chiral* objects. (The word *chiral* is derived from the Greek word for *hand.*) The pairs of molecules shown in Figure 22.9 are also chiral objects, since they cannot be superimposed upon each other. A pair of molecules are chiral if they are mirror images of each other and are not superimposable. Each molecule in the pair is an **enantiomer** of the other.

A chiral carbon has four different substituents bonded to it.

Why are these molecules chiral? It is due to the presence of an atom with four different substituents. This atom, usually a carbon atom, is called a **chiral center**, once called an *asymmetric carbon.* Whenever an atom has four different substituents, there are always two ways the atoms can be ar-

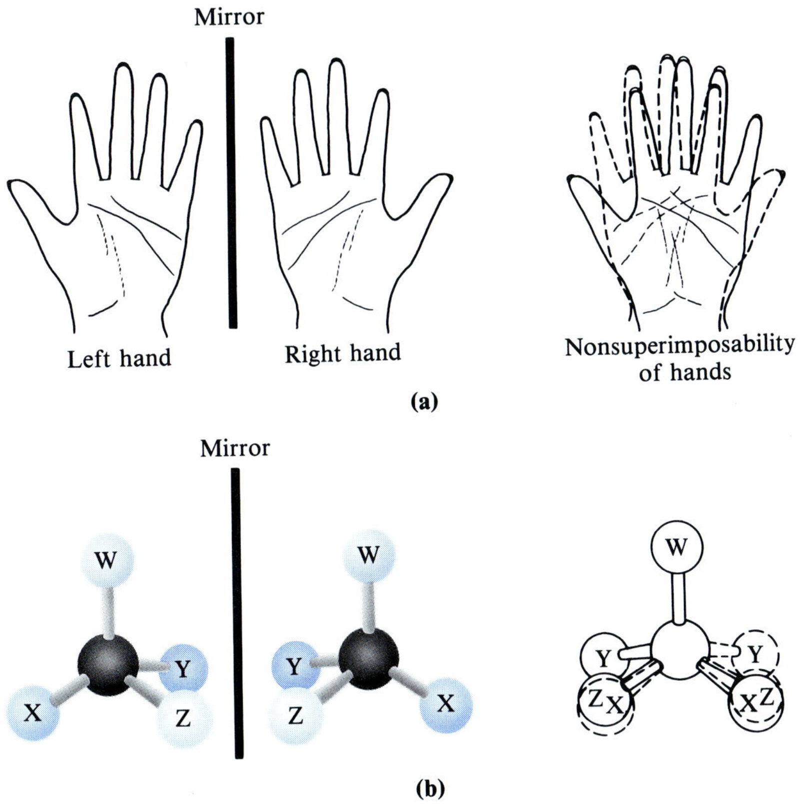
Mirror
Left hand
Right hand
Nonsuperimposability of hands
(a)
Mirror
W
Y
X
Z
W
Y
Z
X
W
Y
Y
Z
X
X
Z
(b)

FIGURE 22.9
(a) Hands are mirror images, but they are not superimposable upon each other. (b) These two molecules are said to be chiral objects because they, like hands, are nonsuperimposable mirror images.

ranged about that atom. There is always the possibility of two isomers. In most biomolecules, however, only one of this possible pair actually exists.

A pair of enantiomers are identical in most physical and chemical properties, but they do differ in two properties. First, they differ in their interaction with plane polarized light and are thus sometimes called *optical isomers.* Second, they differ in their interaction with other chiral compounds. This second difference is of great biological importance; the first difference is useful in studying these compounds.

Enantiomers interact differently with plane polarized light. This interaction can be seen with an instrument called a polarimeter (Figure 22.10). Light that has passed through a polarizing filter is aligned in one plane and is said to be plane polarized (Figure 22.10a). One enantiomer of a pair rotates this plane polarized light clockwise and is designated as the (+) isomer. The other enantiomer rotates the light counterclockwise by the same amount (Figure 22.10b), and is designated as the (−) isomer. Matter that is not chiral does not rotate the light in either direction. A discussion in Chapter 24 illustrates the use of polarimetry in biochemical applications.

Many biomolecules are chiral. Many of the interactions that occur between biomolecules can occur only because of this chirality. Consider for a moment the binding of a small molecule to the surface of a hypothetical macromolecule such as a protein. In Figure 22.11 a pair of enantiomers are shown colliding with the surface of the same protein. The surface has a unique shape due to its chirality, and because of this, one of the enantiomers binds to the surface but the other does not.

One reason enzymes bind small biomolecules with high specificity is because the enzymes and the small molecules are both chiral.

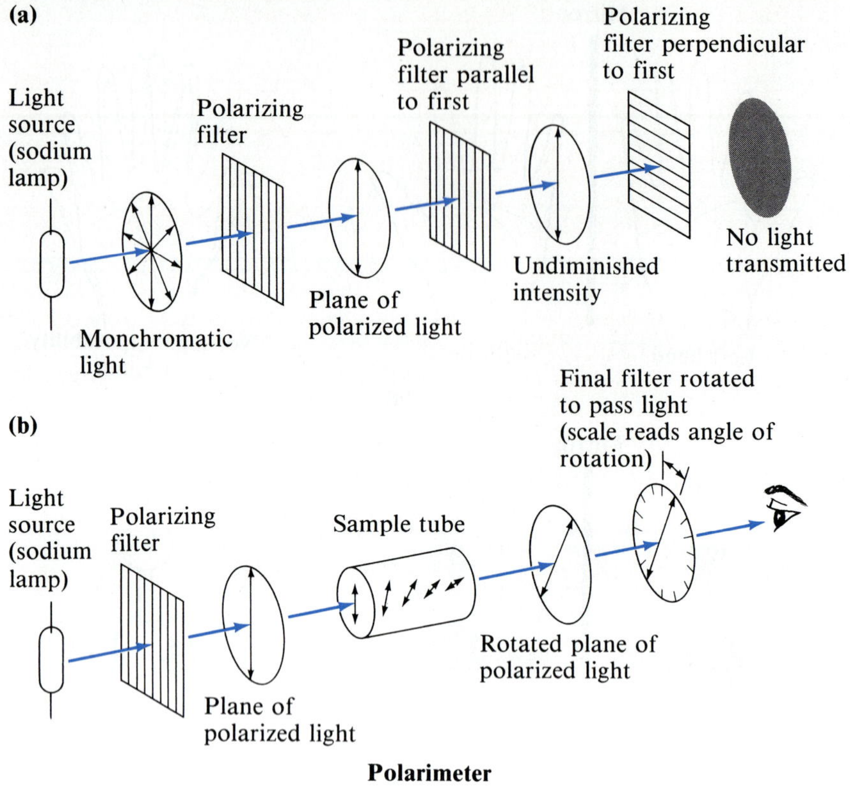
(a)
Light source (sodium lamp)
Polarizing filter
Polarizing filter parallel to first
Polarizing filter perpendicular to first
Monchromatic light
Plane of polarized light
Undiminished intensity
No light transmitted
(b)
Light source (sodium lamp)
Polarizing filter
Sample tube
Final filter rotated to pass light (scale reads angle of rotation)
Plane of polarized light
Rotated plane of polarized light
Polarimeter

FIGURE 22.10
(a) Plane polarized light is produced by passing a beam of light through a polarizing filter. This plane polarized light can pass through any polarizing filter with the same alignment, but it is not transmitted by a polarizing filter of another alignment. (b) A polarimeter measures the rotation of plane polarized light as it passes through a sample. A sample of an enantiomer will rotate light; the observer must rotate the eyepiece containing the final filter to observe the light. The amount of rotation, in degrees, is read directly from the eyepiece. Since the light has been rotated counterclockwise, the (−) isomer was present.

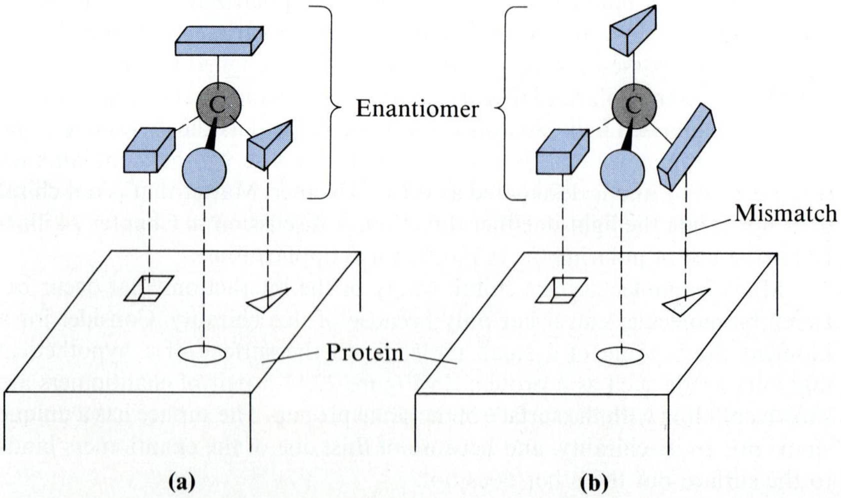
C
C
Enantiomer
Mismatch
Protein
(a)
(b)

FIGURE 22.11
The matching of two enantiomers to the surface of a protein. (a) The three groups of this enantiomer fit perfectly into the three sites on the surface of the protein, thus, this enantiomer will bind with the protein. (b) The groups are oriented in space such that only two groups will fit into the sites on the protein surface. This enantiomer will not bind with the protein.

Your sense of smell is based upon this type of interaction. Consider the two enantiomers shown here.

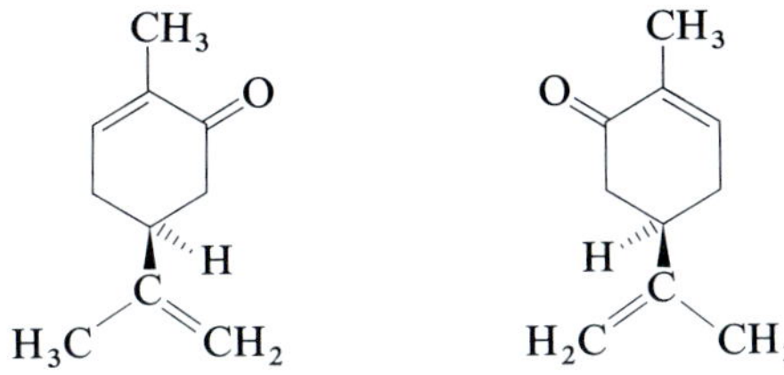
CH3
O
H
C
H3C
CH2
CH3
O
H
C
H2C
CH3
(−)-Carvone (spearmint) (+)-Carvone (caraway)

Your nose possesses a number of receptors for odor, proteins that bind specific molecules and initiate the perception of smell for that compound. The compound (+)-carvone smells like caraway, while (−)-carvone smells like spearmint. One binds to one receptor, and the other to a different one. Thus, they have distinctly different smells. Many of the reactions of the body show equal selectivity. Often one enantiomer is biologically active while the other is totally inactive.

NEW TERMS

stereoisomers Isomers that differ only in the arrangement of atoms in space.

chirality An object is chiral if it cannot be superimposed on its mirror image. This word is derived from the Greek word for hand because hands are chiral.

enantiomers A pair of isomers that are mirror images of each other and are not superimposable; a pair of chiral molecules.

chiral center An atom with four different substituents.

TESTING YOURSELF

Biomolecules

1. Biomolecules differ in three important ways from the organic molecules you studied earlier. What are these three differences?
2. Compare these pairs of objects below and identify them as chiral or nonchiral.
   a. a pair of boots
   b. two basketballs
   c. two tennis rackets
   d.

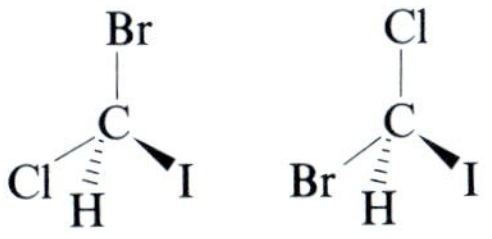
Br
C
Cl
H
I
Cl
C
Br
H
I

Answers **1.** Biomolecules are generally larger, often multifunctional, and are usually chiral. **2. a.** chiral **b.** nonchiral **c.** nonchiral **d.** chiral

22.5 General Principles of Nutrition

The study of **nutrition** is linked to biochemistry. Biochemistry deals with the chemistry of living systems while nutrition is a science that studies several aspects of food. *Food* is anything which, when eaten, provides substances

Nutrients are the substances needed by the body to sustain life.

that the body needs to grow or maintain itself. These substances are referred to as **nutrients**. Food consists of biomolecules. You will find that many of the nutritional terms you have heard on TV commercials or read about on food packaging labels are general classes of biomolecules.

Food provides matter and energy for the body. Through a large number of chemical reactions that are collectively called *metabolism* (see Chapter 27), foods are converted to energy and molecules needed by the body. Food contains energy in the form of potential energy stored in the chemical bonds of molecules. The total amount of potentially available energy in food is measured in **Calories**, a unit of energy used in nutrition that is equal to the kilocalorie of the chemist. (One Calorie equals 1000 "small" calories.) In the body, the bonds of food molecules are broken and other bonds are formed when wastes are produced. The potential energy in the bonds of waste products is less than that in the bonds of food molecules. This difference in energy is available to the body; it is useful energy that a chemist would call *free energy*. Part of it is always lost as heat, but the rest is conserved by the body as the potential energy in certain chemical bonds (see Chapter 27). This energy is used as needed by the body to carry out the numerous tasks that require energy (Figure 22.12).

The Calorie of nutrition is 1000 times larger than the "small" calorie used by chemists.

Food also provides matter needed by the body. During growth, the mass of an organism increases. Since matter cannot be created nor destroyed, the matter must be added to the body from an external source—food. In addition, during normal life maintenance processes, molecules of the body are broken down and excreted as wastes. This matter must be replaced, and again food is the source.

Foods contain nutrients, but the nutritional content of foods varies greatly. Similarly, the needs of the body for various nutrients are different. Nutrients can be classified into six major classes: (1) water, (2) carbohydrates, (3) lipids, (4) proteins, (5) vitamins, and (6) minerals (Table 22.3). These groups collectively provide all of the energy and matter needed by the body. Some classes such as carbohydrates and lipids are most important as sources of energy, while other classes such as vitamins and minerals are important as sources of specific compounds or elements. Water serves yet another and very vital role. It is the solvent in which all the other molecules of the body are either dissolved, dispersed, or suspended. The nutrition of the most important biomolecules is discussed at the end of the appropriate chapters that follow.

FIGURE 22.12
Potential energy stored in chemical bonds can be used for a variety of purposes. Motion results from the mechanical energy produced by muscles. That mechanical energy is derived from energy stored in chemical bonds in adenosine triphosphate (ATP) (see Chapter 27).

NEW TERMS

nutrition The branch of science that deals with the composition, preparation, consumption and processing of food.

nutrients Substances required by the body that are provided by food.

Calorie The unit of energy used in nutrition. It is equal to 1 kcal or 1000 "small" calories.

TESTING YOURSELF

Nutrition

1. a. Convert 7.2 Cal into kilocalories and calories.

TABLE 22.3
Nutrients Required by Humans

Class	Examples and Some Sources	Principle Function
Carbohydrates	Sugars, starches (complex carbohydrates), and fiber from grains, fruits, and vegetables.	Provide energy
Lipids	Fats and oils from vegetable oils, nuts, and nonlean meats	Provide energy and essential fatty acids
Protein	Proteins are widespread in animal and plant tissues; meats are an excellent source of protein	Provide amino acids for protein synthesis and energy
Vitamins	Water soluble vitamins: niacin, riboflavin, thiamine, pyridoxal, pantothenic acid, folic acid, biotin, cobalamin, ascorbic acid; fat soluble vitamins: A, D, E, K	Serve as coenzymes and have a variety of other roles
Minerals	Calcium, chromium, cobalt, copper, iron, magnesium, manganese, molybdenum, potassium, selenium, sodium, zinc, chlorine, fluorine, iodine, nitrogen, phosphorus, sulfur	Many roles—calcium and phosphorus found in bones and teeth; sodium, potassium, and chloride are electrolytes, and so on
Water	Unpolluted, uncontaminated fresh water	Solvent for biomolecules and the reactions of metabolism

b. If this energy were used to increase the temperature of a liter of water (mass = 1000 g), what temperature increase would be seen?

2. List the six major classes of nutrients.

Answers **1. a.** 7.2 kcal, 7200 cal **b.** 7.2 °C **2.** carbohydrates, lipids, proteins, vitamins, minerals, and water

SUMMARY

The *cell* is the basic unit of life. The two basic types of cells are the *procaryote* and the *eucaryote*. The procaryotic cell is smaller and simpler and is typical of bacteria. Procaryotes typically have a *cell membrane*, *cytosol*, *cell wall*, and perhaps an outer layer. Eucaryotic cells are larger, more complex cells found in plants and animals. In addition to a cell membrane, cytosol, and perhaps cell walls, eucaryotes contain a *nucleus*, *mitochondria*, and *endoplasmic reticulum*. The biomolecules in cells possess the characteristics of other organic compounds, but are generally larger, multifunctional, and *chiral*. *Nutrition* is the science that deals with foods and the nutrients they contain. *Nutrients* are substances required by the body for normal growth and maintenance and include water, carbohydrates, lipids, proteins, vitamins, and minerals. Within the cells of the body, the nutrients are processed by the chemical reactions that make up metabolism, to produce the energy and substances needed by the body.

TERMS

cell
organelle
procaryote
cell membrane
cytosol
ribosome
cell wall
eucaryote
nucleus
endoplasmic reticulum
mitochondrion
stereoisomers
chirality
enantiomers
chiral center
nutrition
nutrient
Calorie

CHAPTER REVIEW

QUESTIONS

1. Which of the following organelles is found in both procaryotic and eucaryotic cells?
   a. nucleus b. ribosome c. mitochondrion d. endoplasmic reticulum
2. The endoplasmic reticulum serves what role?
   a. synthesis b. energy production c. protection d. movement
3. The nucleus contains
   a. ribosomes b. mitochondria c. DNA d. the cell membrane
4. What organelle separates the cell from the external environment?
   a. ribosomes b. cell wall c. endoplasmic reticulum d. cell membrane
5. Plane polarized light is affected differently by what kind of molecules?
   a. identical molecules c. macromolecules
   b. chiral molecules d. all kinds of biomolecules
6. Two compounds have the same composition and also have the same atoms attached to the same atoms, although with different orientations in space. These compounds are
   a. identical c. positional isomers
   b. structural isomers d. stereoisomers
7. Food provides energy. How is energy stored in food?
8. Water is a nutrient. What is its biological role?

Answers **1.** b **2.** a **3.** c **4.** d **5.** b **6.** d **7.** Energy is stored in covalent bonds in biomolecules. **8.** Water is the solvent for molecules and ions. It serves as the medium in which metabolic reactions occur.

DIAGNOSTIC CHART

Blacken in all of the circles under the number of each question that you missed in the preceding Chapter Review questions. The diagnostic chart will help you identify concept areas that need more study.

	Questions							
Concepts	**1**	**2**	**3**	**4**	**5**	**6**	**7**	**8**
Procaryotes	○			○				
Eucaryotes	○	○	○	○				
Chirality					○	○		
Nutrition							○	○

EXERCISES

The cell

1. The cell is called the basic unit of life. Why is it given this distinction?
2. a. Calculate the surface area to volume ratio of a cube that is 1 cm on each side.
   b. Cells have a large surface area to volume ratio. What is gained by this large ratio?
3. What physical process is responsible for the movement of substances into and out of cells?
4. a. Are all of the cells of the body identical?
   b. What are tissues?
   c. What are organs?

Procaryotes

5. The cell membrane is more than a passive barrier to movement of substances into and out of the cell. Why is this?
6. What synonyms are used to describe DNA in procaryotes?
7. a. What is the physiological role of ribosomes?
   b. Where are they located in the procaryotic cell?
8. Bacterial cells possess a cell wall that is lacking in animal cells. Why do animal cells lack a cell wall?
9. What is the mode of action of these common antibiotics?
   a. penicillin
   b. erythromycin
   c. chloramphenicol
10. a. What is the composition of lipopolysaccharide?
    b. What effect does lipopolysaccharide have on humans?

Eucaryotes

11. What broad classes of organisms have eucaryotic cells?
12. What is the physiological role of these organelles?
    a. nucleolus c. endoplasmic reticulum
    b. mitochondrion d. chloroplast
13. Make a small table showing which of the organelles in question 12 are found in animals, plants, and bacteria.

Biomolecules

14. Identify the polar and nonpolar regions in a molecule of lysine (a common amino acid).

$$^{+}NH_3CH_2CH_2CH_2CH_2\underset{\displaystyle ^{+}NH_3}{\underset{|}{C}}HCOO^{-}$$

15. Two molecules that are nonsuperimposable mirror images of each other are called ________.
16. Which of the following pairs of objects are superimposable?
    a. a pair of gloves c. your ears
    b. two tennis rackets d. two footballs
17. Which of the pairs of objects in question 16 are chiral?
18. A pair of enantiomers are identical in all but two properties. What are they?

Nutrition

19. Define the term *nutrient*.
20. Give examples of substances that are classified as
    a. carbohydrates c. minerals
    b. lipids d. vitamins
21. Molecules from food are broken down to yield smaller molecules and energy. What happens to this energy?

CHAPTER 23

Carbohydrates

Carbohydrates are one of several major classes of compounds found in living organisms. They serve vital roles in energy storage and transport and are components of cells and extracellular structures in both plants and animals. This chapter discusses the structure, nomenclature, and properties of some important carbohydrates.

OBJECTIVES

After completing this chapter, you should be able to

- Identify and classify some important monosaccharides.
- Describe the glycosidic bond.
- Identify some of the common disaccharides and polysaccharides.
- Recognize some of the chemical properties of carbohydrates and relate them to the chemistry of alcohols and carbonyl compounds.
- Describe the major role of carbohydrates in energy metabolism and structure.

We correctly perceive paper, cotton, crab shells, nectar, and blood as quite different substances. Yet all of these materials are related by composition—they all contain glucose. The glucose in these substances is not in the same form. Blood has glucose (blood sugar) in solution. Glucose provides a significant amount of the body's energy needs. Nectar contains sucrose, common table sugar, which in turn contains glucose. Paper and cotton contain cellulose, a macromolecule containing only glucose. Crab shells are formed from chitin, a macromolecule that is essentially a modified cellulose. Glucose is obviously a widely distributed compound. It is also the single most abundant organic molecule on the earth. Glucose is a carbohydrate.

Glucose is the most abundant biomolecule on earth.

Carbohydrates are polyhydroxy aldehydes, ketones, or derivatives of these compounds. Although *poly-* normally means many, as used here it means two or more groups.

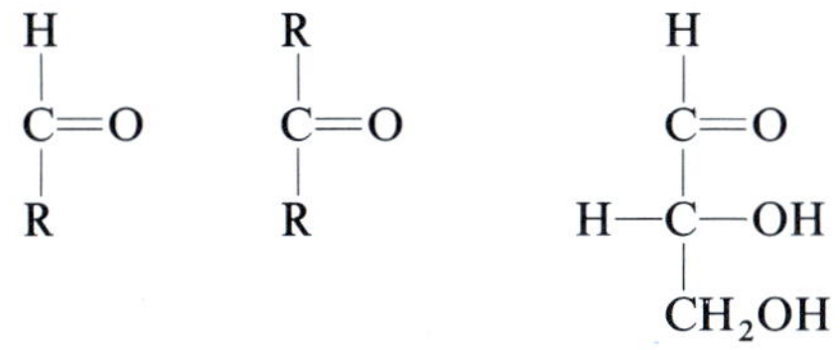

Aldehyde **Ketone** **Dihydroxy aldehyde** (carbohydrate)

Like all organic compounds, the chemistry of carbohydrates can best be explained by considering the functional groups that are present. This chapter introduces the major carbohydrate classes—the monosaccharides, oligosaccharides, and polysaccharides. The structure, nomenclature, and some of the chemical properties of these organic compounds will be presented. In later chapters some of their roles in energy storage and cellular and extracellular structure will be examined.

23.1 Classification of Carbohydrates

You have seen carbohydrates before when they were introduced in the carbonyl chapter as examples of hemiacetals and hemiketals (Chapter 19). The name *carbohydrate* comes from the early observation that some members of this class had formulas that could be written $C_n(H_2O)_n$. They appeared to be "hydrates of carbon." A simple test that seems to confirm this notion is to treat a sugar with a strong mineral acid such as concentrated sulfuric acid. When this acid is added to sugar, the white sugar is converted to a blackish residue that is nearly pure carbon. Water is liberated as steam.

$$C_n(H_2O)_n \xrightarrow[\text{(dehydrating agent)}]{H_2SO_4} n\,C + n\,H_2O$$

Early chemists named these compounds for this property, although today it is clear that this definition is not really accurate nor does it reflect the general chemical properties of these compounds.

The names of many carbohydrates contain the suffix *-ose*. Gluc*ose*, sucr*ose*, and cellul*ose* are clearly identified as carbohydrates. Some carbohydrates do not follow this simple rule of nomenclature—starch, glycogen, and

The suffix *-ose* identifies many compounds as carbohydrates.

chitin are examples. Carbohydrates can be named by a systematic nomenclature similar to the IUPAC names used in organic chemistry, but generally common (trivial) names are used because they are simpler and are universally understood by chemists and biologists alike. The examples cited above are common names.

Carbohydrates are classified by the number of saccharide units found in the molecule.

Carbohydrates come in a wide variety of sizes, but the larger ones contain two or more basic carbohydrate units. This basic unit is called a *saccharide*, derived from the Latin word for sugar. One classification of carbohydrates is based on the number of these units found in the molecule. The smallest carbohydrates, the *monosaccharides*, or simple sugars, are the basic units found in other carbohydrates. These compounds have the general formula $C_nH_{2n}O_n$. The monosaccharides *cannot* be converted to smaller carbohydrates by hydrolysis in dilute acid. Glucose is a monosaccharide.

$$\underset{\text{(Monosaccharide)}}{\text{Glucose}} + H_2O \xrightarrow{\text{dil. } H^+} \text{No reaction}$$

Oligosaccharides are carbohydrates that have two or more monosaccharides covalently linked together. Mild hydrolysis breaks these covalent bonds to yield the individual monosaccharides. As previously mentioned, sucrose is the sugar found in nectar. Mild hydrolysis of sucrose yields two monosaccharides, glucose and fructose.

$$\underset{\text{(Disaccharide)}}{\text{Sucrose}} + H_2O \xrightarrow{H^+} \underset{\text{(Monosaccharides)}}{\text{Glucose} + \text{Fructose}}$$

The prefix *oligo-* means several and will be used in this text to mean two through ten. The oligosaccharides are designated by the actual number of monosaccharides present in the molecule. For example, the *di*saccharide lactose contains *two* monosaccharides, and the *tetra*saccharide maltotetrose contains *four* monosaccharides. Note that the prefixes *di-*, *tri-*, *tetra-*, and so on represent the same numerical value you learned earlier. Most of the oligosaccharides discussed in this text are disaccharides.

Cellulose, starch, and chitin are examples of very large carbohydrates. These substances are classified as **polysaccharides**. They are polymers of monosaccharides that yield many monosaccharide molecules upon hydrolysis.

$$\underset{\text{(Polysaccharide)}}{\text{Cellulose}} + n\,H_2O \xrightarrow{H^+} n + 1 \quad \underset{\text{(Monosaccharide)}}{\text{Glucose}}$$

A carbohydrate is classified as a polysaccharide if it contains more than ten monosaccharide units.

NEW TERMS

carbohydrate A class of compounds consisting of polyhydroxy aldehydes and ketones and derivatives of these compounds.

monosaccharides Simple sugars such as glucose; the smallest compounds that are carbohydrates. These are the basic units from which larger carbohydrates are made.

oligosaccharides Carbohydrates that contain two to ten monosaccharides, such as table sugar and milk sugar (disaccharides).

polysaccharides Carbohydrates that contain many monosaccharide units, such as starch and cellulose.

TESTING YOURSELF

Classification of Carbohydrates

1. How many monosaccharide units are present in a trisaccharide?
2. What name is given to a carbohydrate that contains many monosaccharide units?

Answers **1.** three **2.** polysaccharide

23.2 Monosaccharides

The monosaccharides are biological molecules that are found free in nature and as components of other carbohydrates. Although many monosaccharides are found in nature, only a small number of them will be mentioned here. The ones discussed are those you will see again in later chapters.

Some Common Monosaccharides

The two smallest monosaccharides, *glyceraldehyde* and *dihydroxyacetone*, each contain three carbon atoms. These sugars are not common in the free form, but their derivatives are important in some energy-producing reactions of glycolysis (see Chapter 28).

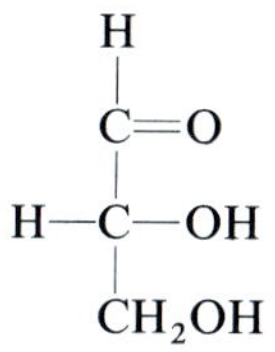

Glyceraldehyde

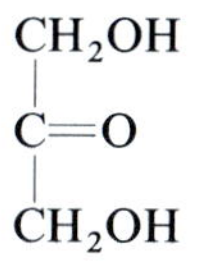

Dihydroxyacetone

Ribose and *2-deoxyribose* are five-carbon sugars. They are components of several biomolecules including nucleotides and nucleic acids that are important in heredity and cell activity (Chapter 26). Note that these sugars and the three that follow are shown in two forms, a straight chain form and a cyclic form. Although the cyclic form predominates, both contribute to the compound's properties.

Ribose and 2-deoxyribose are components of RNA and DNA, respectively.

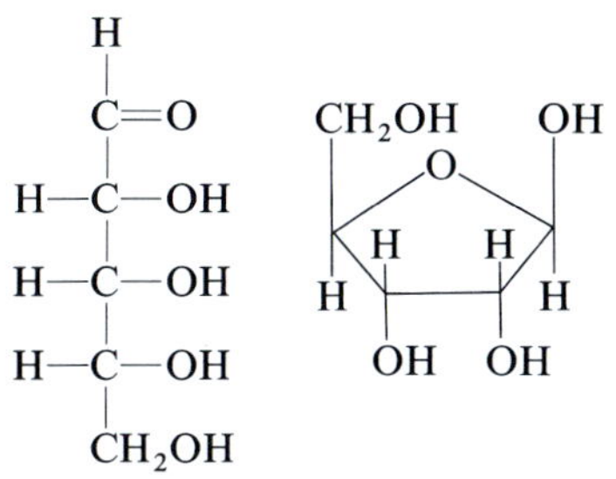

Ribose

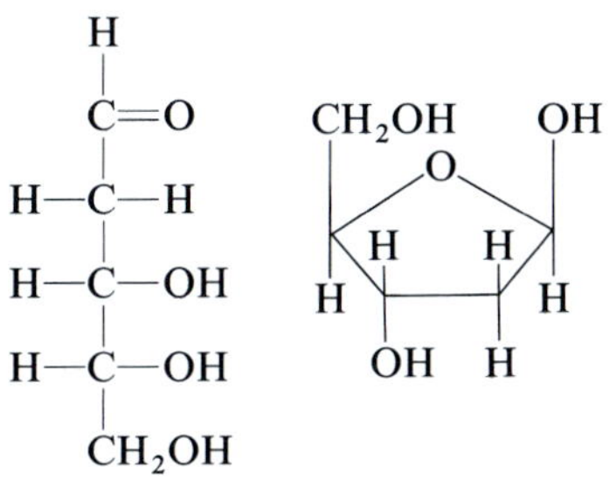

Deoxyribose

Glucose is one of several important monosaccharides that contains six carbon atoms. It is found as a component of many other carbohydrates, but its role as blood sugar makes it of great importance to the health sciences. Blood circulates glucose throughout the body, making it available to all of the cells. The normal range for blood glucose concentration is about 60 to 100 mg per 100 mL. If the concentration is consistently above this normal range, the condition known as hyperglycemia exists. This is seen in the disease diabetes mellitus. If concentrations below the normal range persistently exist, then the person has hypoglycemia (low blood sugar). The regulation of blood sugar concentration is discussed in more detail later in Chapter 30.

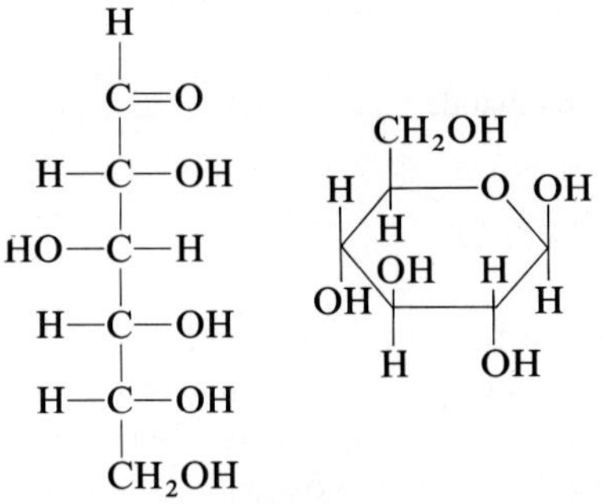

Glucose

Fructose, along with glucose, is found free in honey (see Plate 10 in color insert). Derivatives of fructose are important in energy metabolism. Fructose is now used as a commercial sweetener in the form of high fructose syrups. On a weight basis, fructose is 1.7 times sweeter than table sugar (sucrose), therefore less fructose will yield the same degree of sweetness as a given amount of table sugar. For example, 0.6 g of fructose is as sweet as 1 g of sucrose. Since less is needed for any desired sweet taste, fewer calories are added to the food so it is often used in "diet" foods.

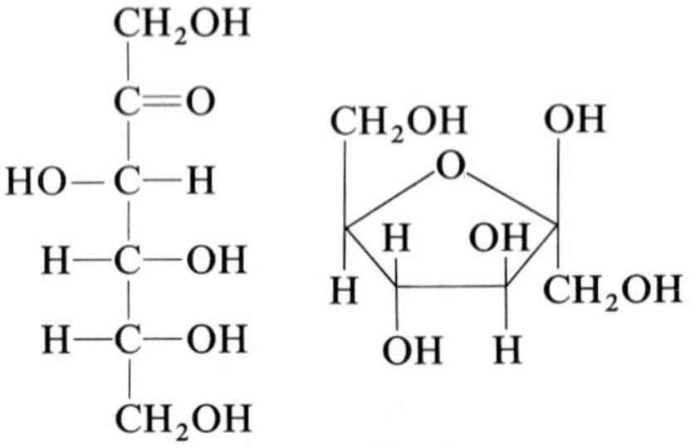

Fructose

Galactose is a third important monosaccharide with six carbon atoms. It is a component of lactose (milk sugar), and it forms part of a variety of larger molecules used by the body. Several genetic diseases involve impaired galactose metabolism.

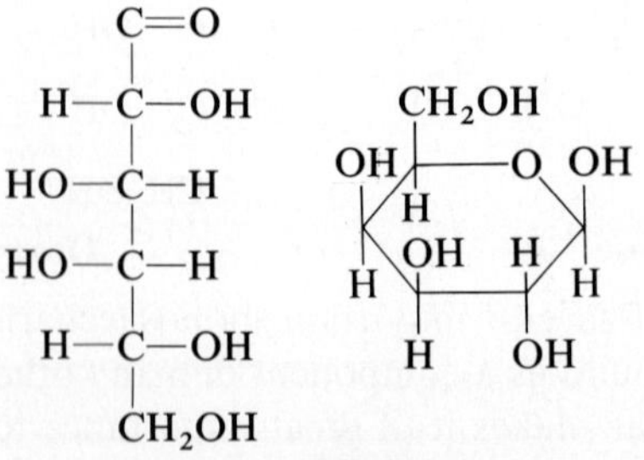

Galactose

Classification of Monosaccharides

Simple sugars can be classified on the basis of several structural features: (1) the number of carbon atoms in the chain, (2) the presence of an aldehyde or ketone group, (3) the size of the ring if the sugar is cyclic, and (4) its stereochemistry. The next four sections discuss the classification of monosaccharides. Because oligosaccharides and polysaccharides are classified in part by the number and kinds of simple sugars they contain, an understanding of monosaccharide structure is necessary for an understanding of the larger carbohydrates.

CLASSIFICATION BY CHAIN LENGTH AND FUNCTIONAL GROUP Monosaccharides can be classified by the number of carbon atoms in the sugar. The number of carbon atoms is indicated by a prefix such as *tri-* (three carbons), *tetr-* (four carbons), *pent-* (five carbons), *hex-* (six carbons), or *hept-* (seven carbons). This numerical prefix is combined with the suffix *-ose*. Thus, a *triose* is a sugar with three carbon atoms, while a *pentose* contains five carbon atoms. Examples of representative sugars are presented in Table 23.1.

Classification by carbon number refers to the number of carbon atoms in the monosaccharide.

The functional group present in the sugar may also be used for classification. For example, sugars having an *ald*ehyde group as part of their structure are classified as **aldoses**. Other sugars have a *keto*ne carbonyl group and are called **ketoses**. A sugar can be classified by both functional group and chain length. The *aldo* or *keto* prefix is placed before the designation for chain length. Thus, in the examples found in Table 23.1 there is an *aldotriose*, an *aldotetrose*, an *aldopentose*, and a *ketohexose*.

All monosaccharides contain hydroxy groups. Classification by function group refers to the presence of an aldehyde group or a ketone group.

CLASSIFICATION OF MONOSACCHARIDES BY RING SIZE In the chapter on carbonyl reactions (Chapter 19), it was shown that a sugar may exist in a cyclic form through the formation of an intramolecular hemiacetal or hemiketal. Hemiacetals and hemiketals are formed when an alcohol reacts with an aldehyde or ketone, respectively.

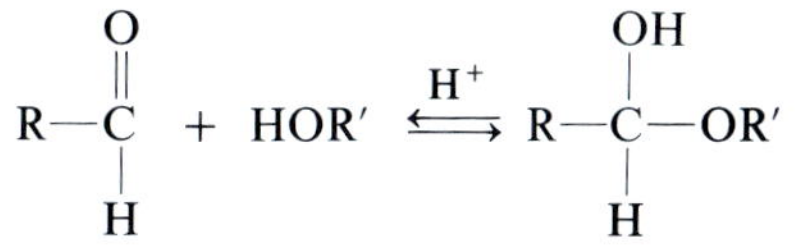

Aldehyde **Alcohol** **Hemiacetal**

The cyclic forms of sugars are hemiacetals or hemiketals.

Because sugars have both hydroxyl groups and an aldehyde or ketone group, intramolecular hemiacetals or hemiketals are certainly possible and, in some sugars, quite likely. Generally, pentoses and hexoses exist as either a five- or six-membered ring. A sugar with fewer than five carbon atoms cannot form a stable ring because the ring would have three or four atoms. Just as for cyclopropane and cyclobutane, these would be unstable.

Cyclic sugars typically have five- or six-membered rings.

Ribose forms a five-membered ring in the following reaction:

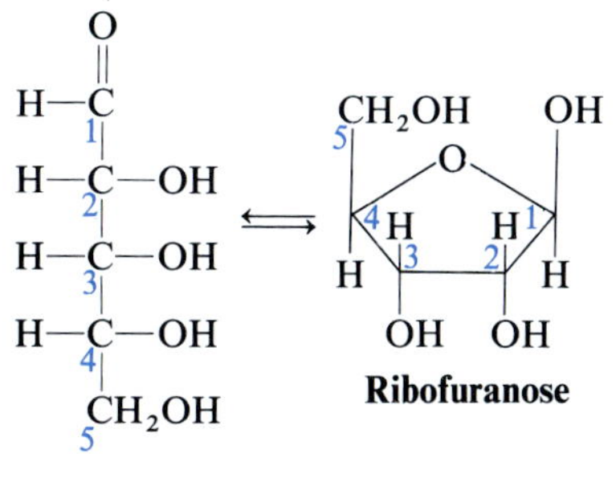

Ribose **Ribofuranose**

TABLE 23.1
Some Representative Sugars

Common Name	Molecular Formula	Structure	Classifications
Glyceraldehyde	$C_3H_6O_3$	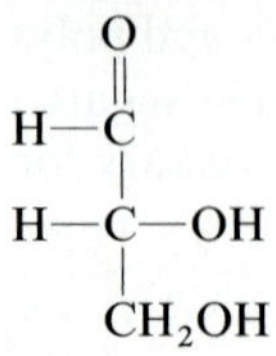	Triose, aldose, aldotriose
Erythrose	$C_4H_8O_4$	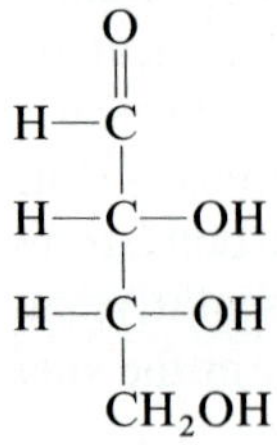	Tetrose, aldose, aldotetrose
Ribose	$C_5H_{10}O_5$	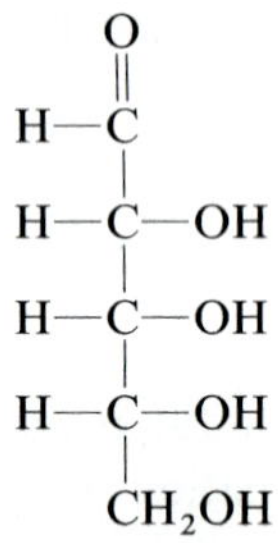	Pentose, aldose, aldopentose
Fructose	$C_6H_{12}O_6$	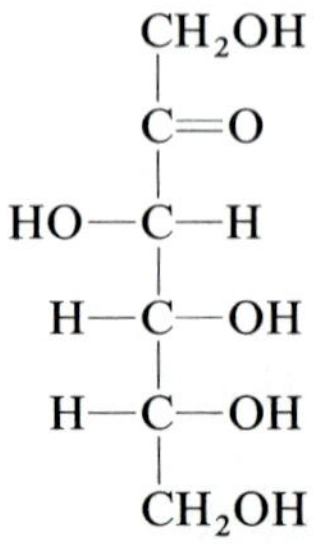	Hexose, ketose, ketohexose

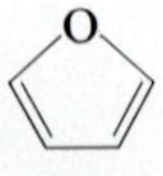

Furan

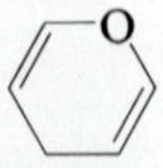

Pyran

Monosaccharides with five-membered rings are called **furanoses**, (note the name ribofuranose in the previous equation) because the cyclic ether *furan* has the same ring structure containing four carbon atoms and a single oxygen atom.

Glucose forms a six-membered ring in the following reaction. A monosaccharide with a six-membered ring is a **pyranose**, because the cyclic ether *pyran* contains five carbon atoms and an oxygen atom. Note that glucose in the cyclic form is called *glucopyranose*.

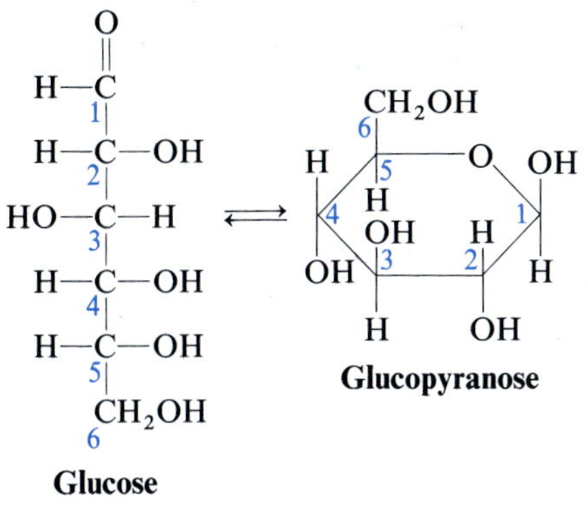
Glucose
Glucopyranose

Let's use glucose and fructose as examples to summarize the classification that has been presented so far. Glucose is an aldose because of its aldehyde functional group, a hexose because it contains six carbon atoms, and a pyranose because its cyclic form resembles the cyclic ether pyran. Fructose is also a hexose, a ketose because it is a ketone, and a furanose because its most stable ring form contains the five-membered ring furan.

CLASSIFICATION BY STEREOCHEMISTRY: ANOMERS Pentoses and hexoses exist as cyclic forms, but the introduction of a ring into the sugar results in the formation of two cyclic isomers. Examine Figure 23.1. Glucose forms two different ring forms. In both cases this occurs through a reversible reaction between the carbonyl group (C-1) and the C-5 hydroxyl group to yield a cyclic hemiacetal. The carbon atoms of the open chain and ring forms of glucose are labeled in Figure 23.1 to assist you in visualizing the formation of the ring. Notice that the reaction yields two different products that differ with respect to the orientation of the —OH group on carbon atom number

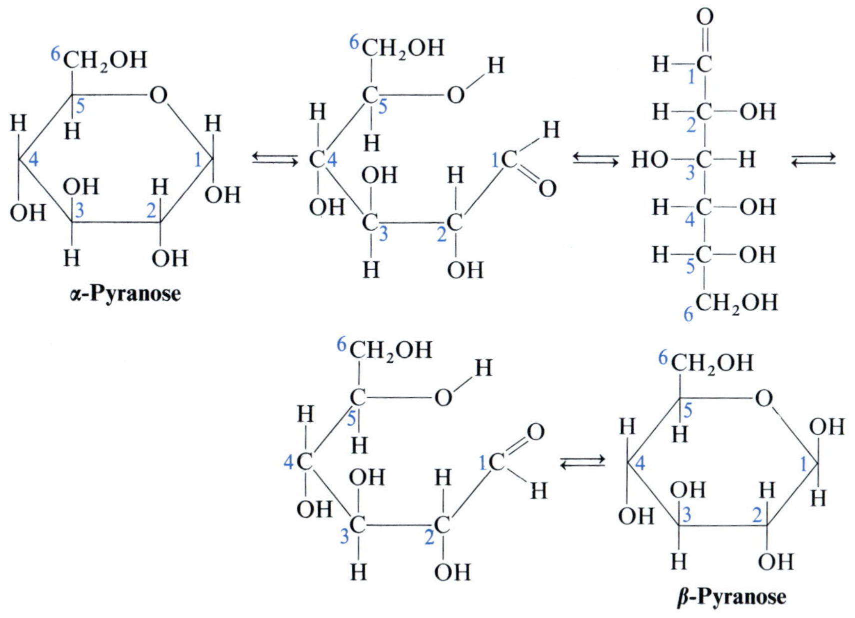
α-Pyranose
β-Pyranose

FIGURE 23.1
The formation of α-D-glucopyranose and β-D-glucopyranose from the open chain form of glucose. Two possible ring forms of glucose exist because the hydroxy group on carbon number five can add to either side of the carbonyl group of carbon number one. α-D-Glucopyranose and β-D-glucopyranose are anomers.

one. The carbonyl group is planar because the carbonyl carbon is sp^2 hybridized. The alcohol can attack either side of the planar carbonyl carbon, yielding two isomers. In sugar chemistry the carbon atom that was the carbonyl carbon before ring formation is called the **anomeric carbon**, and the two products formed are called **anomers**. In the ring form the anomeric carbon has an —OH that can be drawn either above the ring (up), or below the ring (down). If the anomeric —OH is drawn "up" (on the same side as the —CH_2OH), then the anomer is said to be a beta (β) anomer; the other orientation is "down" and is called an alpha (α) anomer. Glucose exists as two different ring forms or anomers, both of which are glucopyranoses. They are distinguished as α-glucopyranose and β-glucopyranose.

The α- and β-glucopyranoses are anomers.

CLASSIFICATION OF STEREOCHEMISTRY: D- AND L- FAMILIES Recall from Chapter 22 that many biological compounds, including most common carbohydrates, possess one or more chiral centers. Recall that a chiral center is an atom with four different substituents attached to it and that two possible stereoisomers called *enantiomers* exist for each chiral center. Each enantiomer is the mirror image of the other. The number of possible stereoisomers depends upon the number (n) of chiral centers in the molecule.

$$\text{Number of isomers} = 2^n$$

Glyceraldehyde is the smallest carbohydrate with a chiral center. Since the middle carbon atom of glyceraldehyde is chiral, glyceraldehyde must exist as a pair of enantiomers (a pair of mirror images). Since $n = 1$ for glyceraldehyde, then $2^1 = 2$.

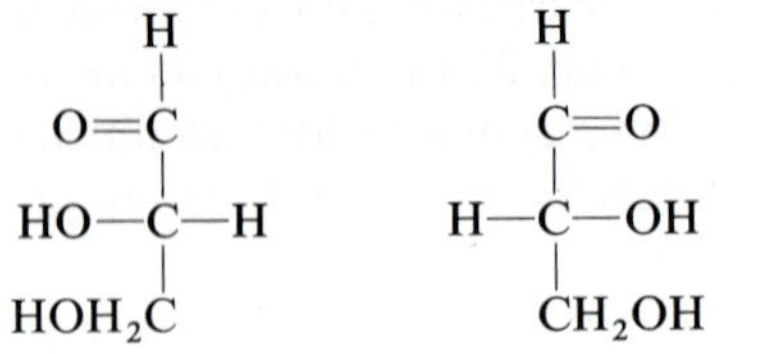
H
O=C
HO—C—H
HOH₂C
H
C=O
H—C—OH
CH₂OH

L-Glyceraldehyde **D-Glyceraldehyde**

When the carbonyl group is placed at the top (as drawn here), the —OH group on the middle carbon of the left molecule is on the left side of the molecule. This enantiomer is designated L-glyceraldehyde. The other enantiomer is designated D-glyceraldehyde. The letters are derived from *levo-* (left) and *dextro-* (right). The stereochemistry of the glyceraldehydes is used to assign all other carbohydrates as D- or L-. For example, D-glucose and L-glucose have the following stereochemistry:

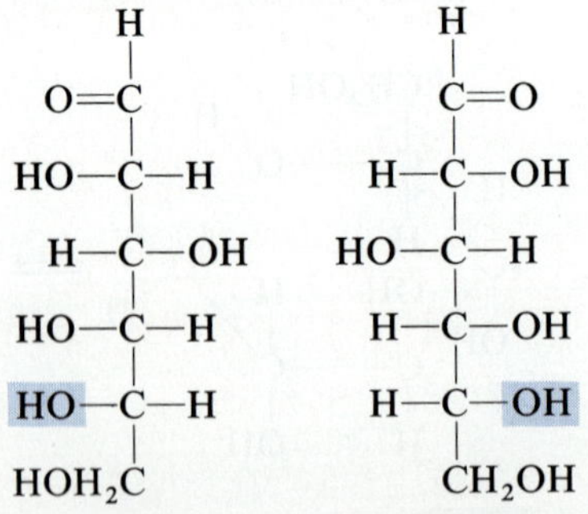
H
O=C
HO—C—H
H—C—OH
HO—C—H
HO—C—H
HOH₂C
H
C=O
H—C—OH
HO—C—H
H—C—OH
H—C—OH
CH₂OH

L-Glucose **D-Glucose**

Note that L-glucose is the mirror image, the enantiomer, of D-glucose. Also note that the hydroxyl group (highlighted) most distant from the carbonyl group has the orientation designated by D- and L-. Since there may be more than one chiral center in the molecule, one particular chiral center must be used to determine whether the molecule is in the D- or L- family. The chiral center most distant from the carbonyl group is used for the determination. Most of the naturally occurring sugars belong to the D- family.

The D- and L- isomers of a compound are enantiomers.

Most naturally occurring carbohydrates belong to the D-stereochemical family.

To fully identify a sugar, the stereochemistry must be included in the name. The name α-D-glucopyranose identifies this sugar as D-glucose in the ring form with the anomeric —OH pointing down.

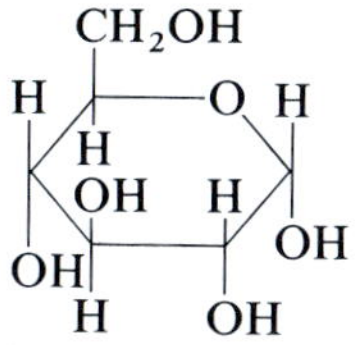

NEW TERMS

aldose A monosaccharide containing an aldehyde group. The prefix *ald-* indicates the presence of an aldehyde group, and the *-ose* suffix signifies a carbohydrate.

ketose A monosaccharide containing a ketone group. Note the prefix *keto-* and the suffix *-ose.*

furanose The five-membered ring form of monosaccharides. The ring resembles the cyclic ether furan.

pyranose The six-membered ring form of monosaccharides. The ring resembles the cyclic ether pyran.

anomeric carbon The carbon atom in the cyclic form of sugars that had been the carbonyl carbon of the open chain form.

anomers The pair of isomers that forms when a monosaccharide forms a ring. The oxygen attached to the anomeric carbon can have either a *beta* (β) ("up") or an *alpha* (α) ("down") orientation.

TESTING YOURSELF

Monosaccharides

1. Describe the following monosaccharides according to their functional group and number of carbon atoms.

a. 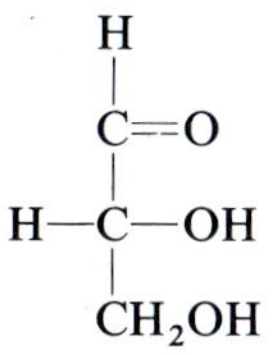

b. 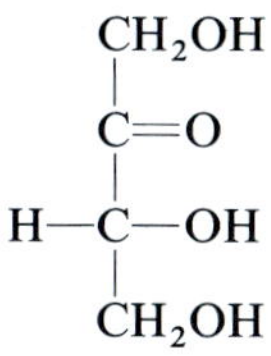

c. 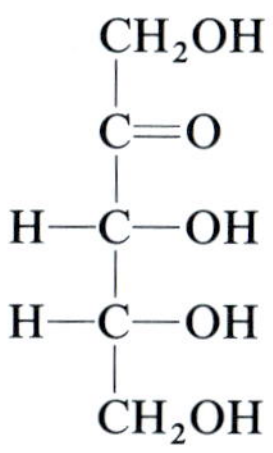

2. Classify glucose according to its carbonyl functional group and the number of carbon atoms it contains.

3. Identify the rings found in deoxyribose, fructose, and galactose as furans or pyrans.
4. Name the ring forms of deoxyribose, fructose, and galactose.
5. An aldohexose such as glucose contains four chiral centers. How many stereoisomers exist?

Answers **1. a** aldotriose **b.** ketotetrose **c.** ketopentose **2.** aldohexose **3.** Deoxyribose and fructose contain a furan ring, while galactose contains a pyran ring. **4.** deoxyribofuranose, fructofuranose, and galactopyranose **5.** $2^n = 2^4 = 16$

23.3 Properties and Reactions of Sugars

Solubility

Sugars are water soluble. This solubility is easily explained because, like the alcohols discussed in Chapter 18, sugars form hydrogen bonds between water and their hydroxyl and carbonyl groups.

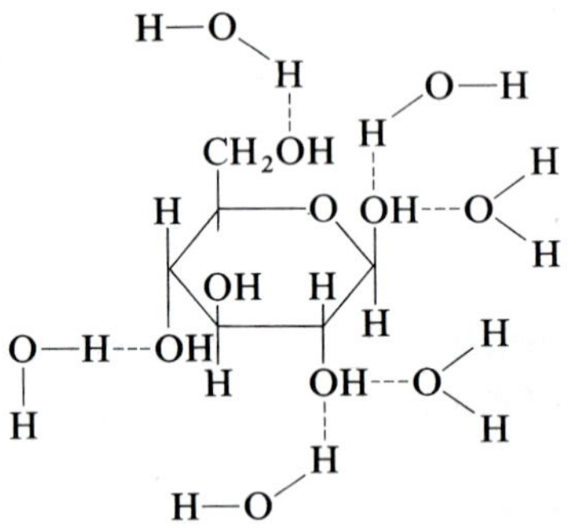

The hydroxy groups in sugars greatly influences their water solubility.

Carbohydrates contain two or more hydroxyl groups that hydrogen bond with water. This accounts for the high water solubility of the smaller carbohydrates. Some of the very large (polymeric) carbohydrates are not water soluble. They form colloidal suspensions or are insoluble.

Optical Properties of Sugars

The direction of rotation of plane polarized light cannot be predicted by stereochemical family.

Sugars contain one or more chiral centers. Compounds with chiral centers are optically active because they rotate plane polarized light (see Chapter 22). One of the pair of enantiomers rotates light clockwise and the other rotates light counterclockwise by the same amount. The clockwise rotation of light is designated by a (+) symbol, and counterclockwise rotation is specified by a (−). D-glyceraldehyde and D-glucose rotate light to the right (+), and L-glyceraldehyde and L-glucose rotate the light to the left (−). It is simply chance that these two D-monosaccharides both rotate light to the right. There is *no* necessary relationship between family and the rotation of light. For example, D-fructose rotates plane polarized light to the left (−). The D- and L- family classification simply relates the arrangement of atoms about a single chiral center in the molecule to the arrangement of atoms about the chiral center in glyceraldehyde. It is not related to optical activity.

Mutarotation

Much of the chemical behavior of a sugar can be related to the form of the carbonyl group (aldehyde or ketone) and the equilibrium between the carbonyl (open chain) form and the cyclic (hemiacetal or hemiketal) forms. Proof of this interconversion is provided by the phenomenon of **mutarotation**, which is a change in the optical activity of sugar solutions.

When one of the anomers of glucose, α-D-glucopyranose, is dissolved in water, the solution initially has a specific rotation of +112°, as measured by a polarimeter.

$$\underset{\text{(solid)}}{\alpha\text{-D-Glucopyranose}} + H_2O \longrightarrow \text{Solution with initial specific rotation of } +112^\circ$$

A similar solution of the other anomer, β-D-glucopyranose, gives a specific rotation of +15°.

$$\underset{\text{(solid)}}{\beta\text{-D-Glucopyranose}} + H_2O \longrightarrow \text{Solution with initial specific rotation of } +15^\circ$$

After either pure anomer is dissolved in water, the specific rotation changes. Eventually, each sugar solution will exhibit the same rotation of +52°.

What has happened? From the previous discussions on hemiacetal formation in this chapter and Chapter 19, the change in specific rotation, or mutarotation, can be explained by a dynamic equilibrium between the cyclic hemiacetal forms and the open chain carbonyl compound (look at Figure 23.1 again). The aldehyde group and the alcohol group of the open chain structure can form a cyclic hemiacetal, and the hemiacetal can form the open chain carbonyl compound. Since the reaction is reversible, the hemiacetal products will eventually isomerize since each can return to the open chain carbonyl compound and then reform either hemiacetal. The alpha anomer and the beta anomer undergo interconversion.

Anomers readily convert to each other through a common intermediate, the free aldehyde or ketone.

Now let's reconsider the behavior of D-glucose. The pure alpha anomer rotates light +112° while the pure beta form rotates light +15°. Since they can interconvert, either one will eventually form an equilibrium mixture of the two anomers. Thus, the experimental optical rotation reflects a *mixture* of about two-thirds β-D-glucopyranose and one-third α-D-glucopyranose. The observed change in optical activity results from the interconversion of anomers. Interconversion of anomers occurs in the cells of the body, thus both anomers are constantly available to the cell.

Oxidation–Reduction of Sugars

The aldehyde group of aldoses is easily oxidized (Chapter 19). In addition, ketoses can be isomerized to aldoses. Thus, all monosaccharides are reducing agents, that is, they give up electrons as they are oxidized. For this reason, the monosacchrides are classified as **reducing sugars**. The carbonyl groups of both ketoses and aldoses can be reduced (Chapter 19). Several derivatives of sugars found in living systems are oxidized or reduced sugars.

The reducing properties of monosaccharides can be used to identify carbohydrates and to determine how much is present. Several convenient analyses for sugars are available, including Benedict's and Fehling's tests

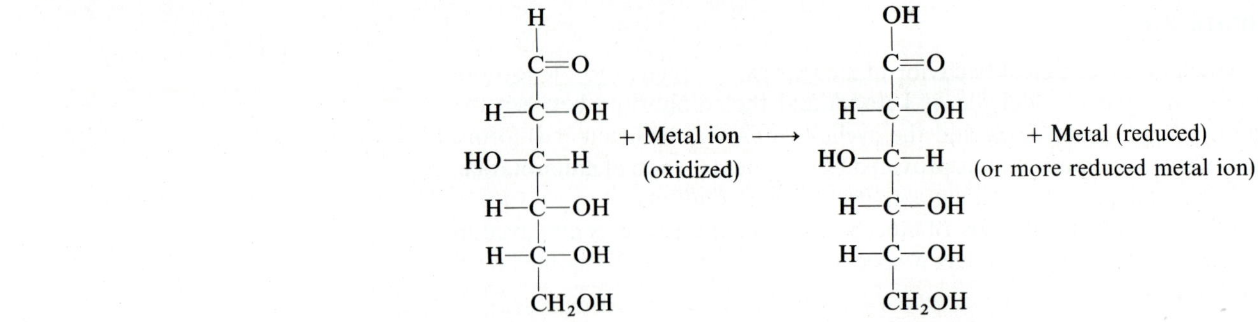
+ Metal ion (oxidized)
+ Metal (reduced) (or more reduced metal ion)

Aldohexose **Carboxylic acid**

FIGURE 23.2
The oxidation of an aldohexose to a carboxylic acid by a metal ion. The ions copper(II) and silver(I) are commonly used for this reaction.

discussed in Chapter 19. The basis for these tests is the reduction of colored metal ions to other products. Although glucose and many other simple sugars exist in ring forms, they still undergo reactions characteristic of the aldehyde group because they are in equilibrium with the open chain form. These sugars can be analyzed by observing the oxidation of the aldehyde group to a carboxylic acid. Since every oxidation requires a reduction, progress of the reaction is observed by following color changes that occur as metal ions are reduced to other products. This process is illustrated in Figure 23.2.

Benedict's and Fehling's tests for reducing sugars use the reduction of Cu^{2+} to Cu^{+} as the basis of the tests. Solutions of Cu^{2+} ions are blue, while the product Cu_2O is brick red. Another useful reaction is the reduction of Ag^{+} to free Ag^{0}. If this reaction is carried out in the presence of clean glass, the free silver will deposit on the glass to form a mirror.

Glycosidic Bonds

The anomeric hydroxyl group of a cyclic sugar is the hydroxyl group of a hemiacetal or hemiketal. This hydroxyl group is somewhat more reactive than the other —OH groups of a sugar molecule. In Chapter 19, it was shown that hemiacetals and hemiketals react with alcohols to form acetals and ketals, respectively.

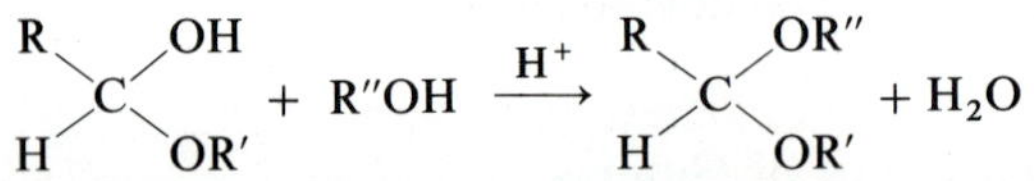

Hemiacetal **Alcohol** **Acetal**

If a sugar is dissolved with an alcohol in the presence of a catalytic amount of acid, similar chemical behavior is predicted and observed.

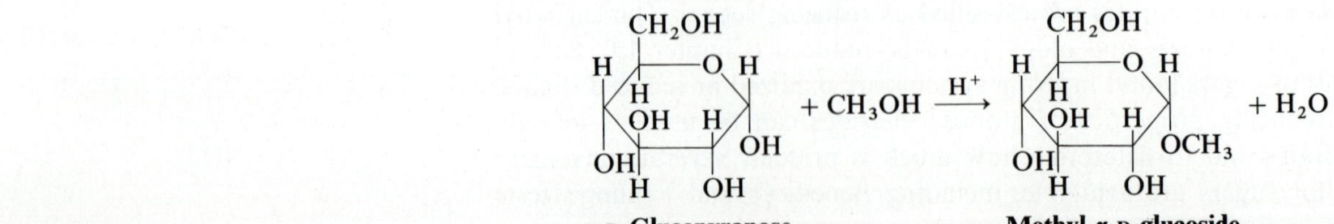

α-D-Glucopyranose **Methyl α-D-glucoside**

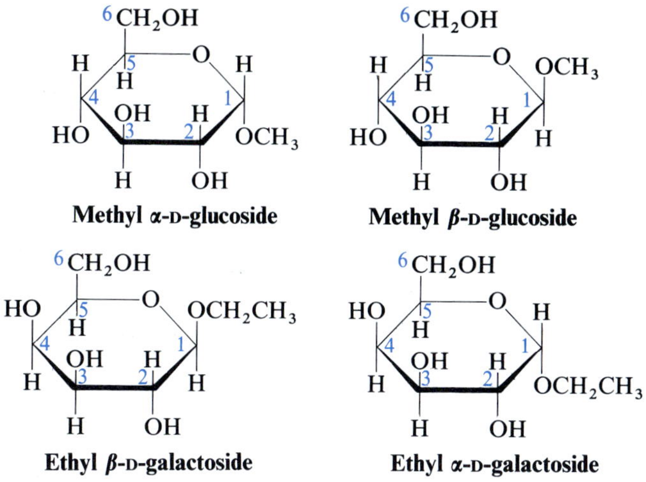
Methyl α-D-glucoside
Methyl β-D-glucoside
Ethyl β-D-galactoside
Ethyl α-D-galactoside

FIGURE 23.3
These glycosides were formed when glucose and galactose reacted with methanol and ethanol. The name indicates which alcohol and sugar formed the glycoside. Note that two possible isomers are formed for each sugar and alcohol because the sugar can exist as either of two anomers.

The product is a carbohydrate equivalent of an acetal or ketal called a **glycoside**. The covalent bond between the hemiacetal or hemiketal and the alcohol is called a **glycosidic bond** (Figure 23.3). The glycoside will have either an alpha or a beta glycosidic bond depending on the orientation of the sugar's anomeric —OH group when the glycosidic bond formed. Once the glycosidic bond has formed, the sugar is locked into the anomeric form it had when the glycosidic bond formed. The ring will not open as long as the glycosidic bond is present.

Glycosidic bonds link the cyclic forms of sugars to alcohols.

The name of the sugar of the glycoside is used to name the glycoside. For example, if the sugar of a glycoside is glucose, the compound is named glucoside. Fructose would be the sugar in a fructoside, and galactose the sugar in a galactoside.

NEW TERMS

mutarotation The change in specific rotation seen in sugar solutions that results from equilibrium between anomeric forms.

reducing sugars Sugars that have a free carbonyl group and are thus reducing agents.

glycoside Compound formed when a cyclic sugar is bonded to an alcohol through a glycosidic bond.

glycosidic bond Bond between the anomeric carbon of a cyclic sugar and the —OH group of another sugar or an alcohol. This bond links sugars together in oligosaccharides and polysaccharides.

TESTING YOURSELF

Properties of Sugars and Glycosidic Bonds

1. a. Classify glucose, lactose, and sucrose as reducing or nonreducing sugars. (Refer to Section 23.4 for additional information.)

b. Why is each reducing or nonreducing?

2. Do the following glycosides have alpha or beta linkage?

a. 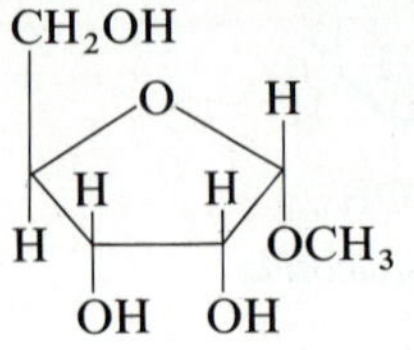

c. 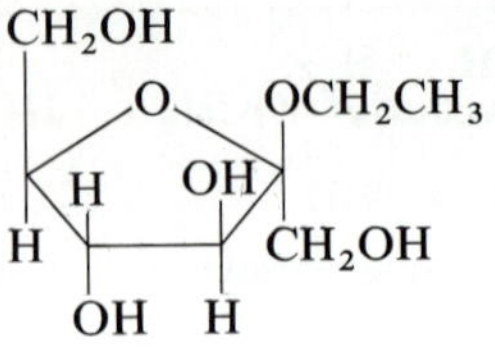

b. 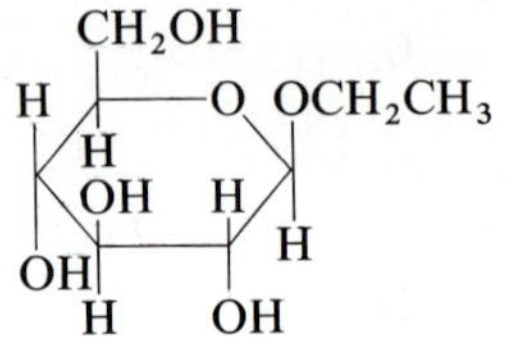

Answers **1. a.** Glucose and lactose are reducing sugars, sucrose is nonreducing. **b.** Glucose and lactose contain a free anomeric carbon that can yield a free carbonyl group. Sucrose has both anomeric carbon atoms tied up in a glycosidic bond. **2. a.** alpha **b.** beta **c.** beta

23.4 Oligosaccharides

Acetals and ketals form when hemiacetals or hemiketals react with an alcohol. Cyclic sugars are hemiacetals and hemiketals with additional hydroxyl groups present. These hydroxyl groups are available to react with the anomeric hydroxyl group of a cyclic sugar. For example, the hydroxyl group on carbon number four of a glucopyranose could react with the anomeric hydroxyl of another glucopyranose:

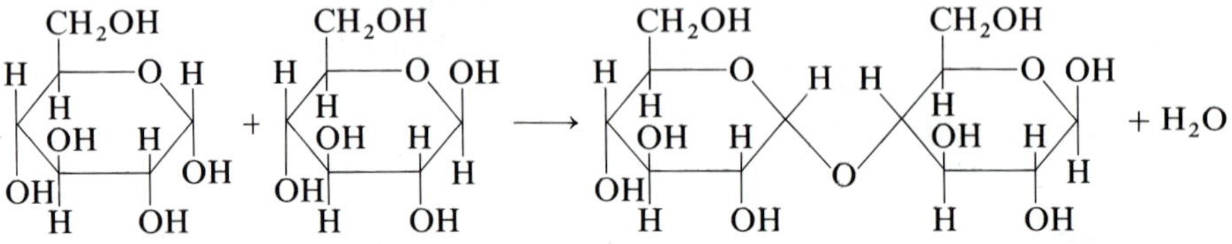

Glucopyranose **Glucopyranose** **Maltose** (a glucoside)

Oligosaccharides are sugars consisting of two to several monosaccharides linked by glycosidic bonds. In biochemistry, the positions within a larger molecule that are occupied by smaller units or building blocks are referred to as **residues**. The residues in an oligosaccharide or polysaccharide are occupied by monosaccharides.

The most important biological oligosaccharides are the disaccharides.

Lactose is milk sugar.

By far the most common oligosaccharides are *disaccharides*. There are several common disaccharides that you should know. Lactose, which is milk sugar, is an important source of energy for the young of mammals. It makes up about 5% of cow's milk, but human milk contains somewhat more. Lactose consists of the monosaccharides galactose and glucose linked by a β 1→4 glycosidic bond (Figure 23.4a). As shown in the figure, the oxygen on the anomeric carbon of galactose is in the beta position and is bonded to the number four carbon of glucose. The designation β 1→4 is used to identify

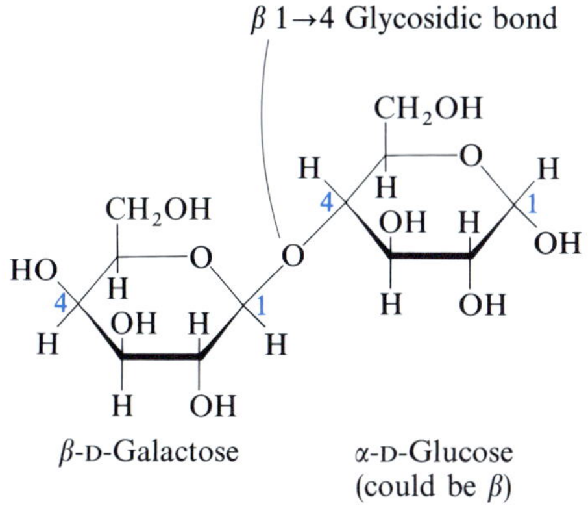
β 1→4 Glycosidic bond
β-D-Galactose
α-D-Glucose
(could be β)

(a) Lactose

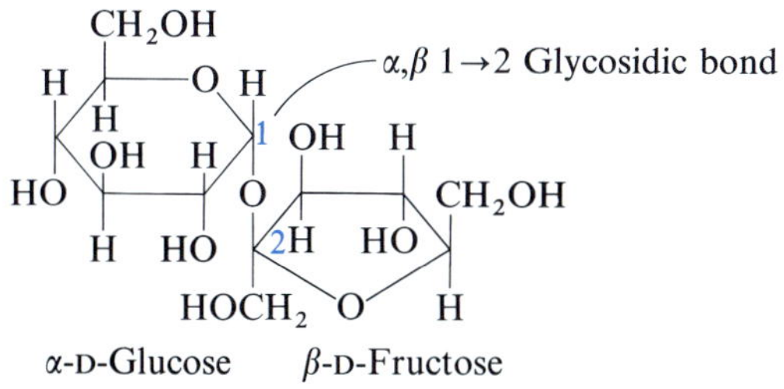
α,β 1→2 Glycosidic bond
α-D-Glucose
β-D-Fructose

(b) Sucrose

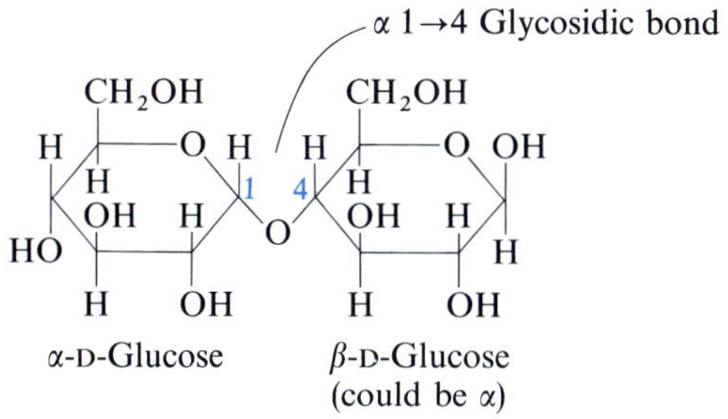
α 1→4 Glycosidic bond
α-D-Glucose
β-D-Glucose
(could be α)

(c) Maltose

FIGURE 23.4
Three important disaccharides that are common components of foods and partially digested starch include lactose, sucrose, and maltose. Take a moment to compare their monosaccharides and glycosidic bonds.

which atoms of the residues are bonded together and indicates the orientation of the anomeric hydroxyl group when the glycosidic bond was formed. The anomeric carbon atom of the glucose residue in lactose does not participate in a glycosidic bond. Like free glucose, this glucose residue exists in an equilibrium between the beta anomer, the alpha anomer, and the open chain form. Since the open chain form has a free carbonyl group that can reduce metal ions, lactose is a reducing sugar.

Sucrose is a disaccharide found in plants where it is used for transportation of sugar residues. In addition, sucrose is found in the nectar of flowers where it serves as an inducement to insects and other animals that in turn pollinate the flowers. Sucrose is a disaccharide of glucose and fructose linked by an α,β 1→2 glycosidic bond (Figure 23.4b). Examine the structure of fructose in the figure to be sure you understand what the designation α,β 1→2 means. Note that in sucrose, the anomeric hydroxyl group of both fructose and glucose are involved in the glycosidic bond. This is stated in the nomenclature of the bond, and since neither has a free anomeric hydroxyl group,

Sucrose obtained from cane is called cane sugar; from beets, it is beet sugar. By any other name, this is still table sugar.

neither can exist as the free carbonyl compound. Unlike the aldoses and lactose, sucrose is not a reducing sugar.

Maltose is a disaccharide made up of two residues of glucose linked by an α 1→4 glycosidic bond (Figure 23.4c). Maltose is obtained from the malt of germinating grain and is present in the digestive tract during the digestion of starch (see Chapter 28). In germinating grain or in the digestive tract, maltose is a degradation product of larger carbohydrates known as polysaccharides. Malt, which is obtained from germinating barley, is used to make beer. The sugars in malt, including maltose, serve as the energy source for the yeast during the fermentation process.

Although there are many oligosaccharides besides the disaccharides that play a role in the normal structure and function of the body, only a few will be mentioned in this text. For example, carbohydrates are bonded to many of the proteins (Chapter 25) and lipids (Chapter 24) of the body. The role of these molecules will be deferred until an understanding of proteins and lipids has been developed.

NEW TERM

residue A position within an oligomer or polymer; the residues of an oligosaccharide are occupied by monosaccharides.

TESTING YOURSELF

Oligosaccharides

1. Name the simple sugars in (a) lactose, (b) maltose, and (c) sucrose.
2. Explain the designation β 1→4 used to describe the glycosidic bond of lactose.

Answers **1. a.** galactose and glucose **b.** glucose **c.** fructose and glucose **2.** The anomeric —OH of galactose is in the beta position and is bonded to an —OH group on carbon number four of glucose.

23.5 Polysaccharides

Polysaccharides are polymers of monosaccharides.

When many monosaccharide units are linked together by glycosidic bonds, the resulting macromolecules are called *polysaccharides.* These are polymers of simple sugars, and some are very abundant in nature. Starch, glycogen, cellulose, and chitin are common examples of polysaccharides. Although each of these polysaccharides is a polymer of glucose or a glucose derivative, they have important differences in properties that result from differences in structure. Among these differences is the suitability of the polysaccharide as a nutrient.

Starch and Glycogen: Storage Polysaccharides

Starch and **glycogen** are similar molecules that contain D-glucose molecules linked by α 1→4 glycosidic bonds and, in some cases, α 1→6 bonds (Figure 23.5). There are two different types of polymers found in starches, *amylose* and *amylopectin.* These polymers differ in size and shape. Amylose

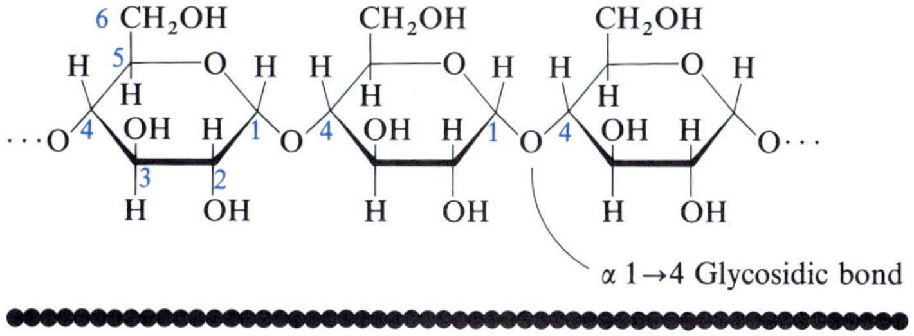
α 1→4 Glycosidic bond

(a) Amylose

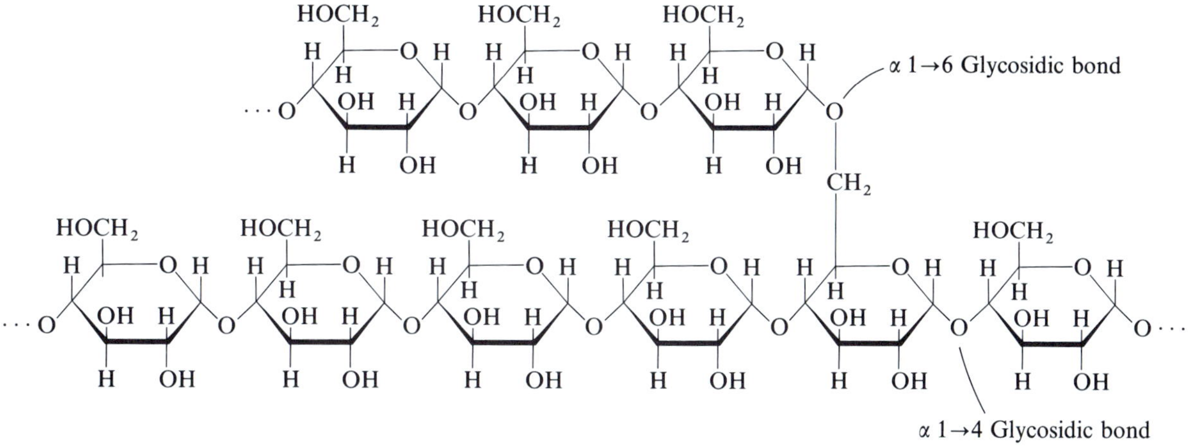
α 1→6 Glycosidic bond
α 1→4 Glycosidic bond

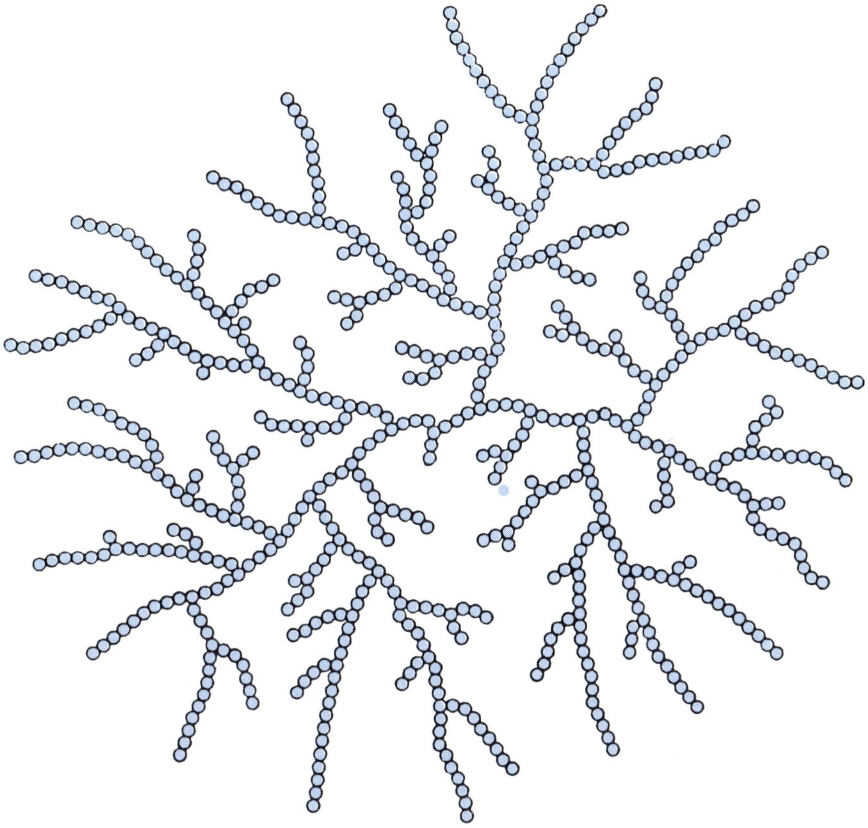

(b) Glycogen

FIGURE 23.5
A comparison of amylose and glycogen. (a) Amylose is a linear polymer composed only of glucose residues linked by α 1→4 bonds. (b) Glycogen (and the similar substance amylopectin) consists of glucose residues linked by α 1→4 and α 1→6 glycosidic bonds. Each α 1→6 bond introduces another branch point that provides an addition terminal sugar residue. These terminal sugars in glycogen can be quickly cleaved, yielding a glucose derivative for quick energy production.

is a linear polymer, that is, it has all of its D-glucopyranose residues connected together into one long chain by α 1→4 glycosidic bonds (Figure 23.5a). Amylopectin is larger than amylose and is branched like glycogen. Some of the D-glucopyranose residues have both α 1→4 and α 1→6 glycosidic bonds, and a branch occurs at each of these residues. The difference in size, from several hundred glucose residues for amylose to several thousand for amylopectin, accounts for the difference in their behavior in water. Neither are truly soluble in water because they are too big to form true solutions. They form colloidal dispersions. Amylose forms these more readily in hot water because it is smaller.

Glycogen, like amylopectin, is also highly branched (Figure 23.5b). The branching in glycogen has known physiological significance. In liver and muscle cells, a specific, highly regulated enzyme named glycogen phosphorylase releases individual phosphorylated glucose molecules at appropriate times from the ends of the chains of glycogen. These molecules can be used by the cell for energy or other needs (Chapter 28), or the phosphate can be removed to form glucose. At any moment, the end of each branch could have a phosphorylase molecule, thus, the highly branched structure of glycogen ensures that rapid formation of glucose or glucose derivatives can occur. A linear polysaccharide would only have one or two ends to act as sites where an enzyme could release individual molecules from the polymer. Clearly the highly branched nature of glycogen facilitates its rapid breakdown when glucose residues are needed.

The highly branched structure of glycogen enhances rapid mobilization of energy reserves.

Cellulose and Chitin: Structural Polysaccharides

Cellulose and **chitin** are structural polymers with glycosidic bonds that are β 1→4. This is in contrast to starches and glycogen (used for energy storage) which have α 1→4 bonds. In general, α 1→4 glycosidic bonds are more easily cleaved than β 1→4 bonds. This makes sense because structural materials should be relatively resistant to breakdown, while energy storage forms should readily be converted to smaller, more easily used energy forms. Cellulose is a polymer of glucose, but in chitin the —OH group of carbon number two in each glucose molecule is replaced by an amide group (Figure 23.6). Chitin is the polysaccharide that gives rigidity to the exoskeletons of crabs, shrimp, insects, and other arthropods. Cellulose is the principal structural polysaccharide of plants, and because plants are so abundant, it is estimated that for every person on earth, the plant world synthesizes about 100 lb of cellulose per day! This biomass has great potential as a source of energy and chemicals.

Cellulose and chitin are structural polysaccharides that help give rigidity to plants and arthropod exoskeletons, respectively.

Wood is primarily cellulose fibers embedded in a highly polymerized substance called lignin. This complex of cellulose and lignin is a strong, enduring substance that is ideal for many structural uses. Steel-reinforced concrete mimics the structure of natural wood because the steel rods are analogous to the cellulose fibers and the concrete is like the lignin. Similarly, fiberglass fishing rods consist of fiberglass embedded in resin.

Starch (α 1→4) and cellulose (β 1→4) differ only in the orientation of the glycosidic linkage between the glucose residues, yet humans efficiently digest starch in food but cannot utilize cellulose. In general, mammals and other higher animals do not possess the necessary enzymes to cleave cellulose to glucose. However, grazing animals such as sheep and cattle (called ungulates)

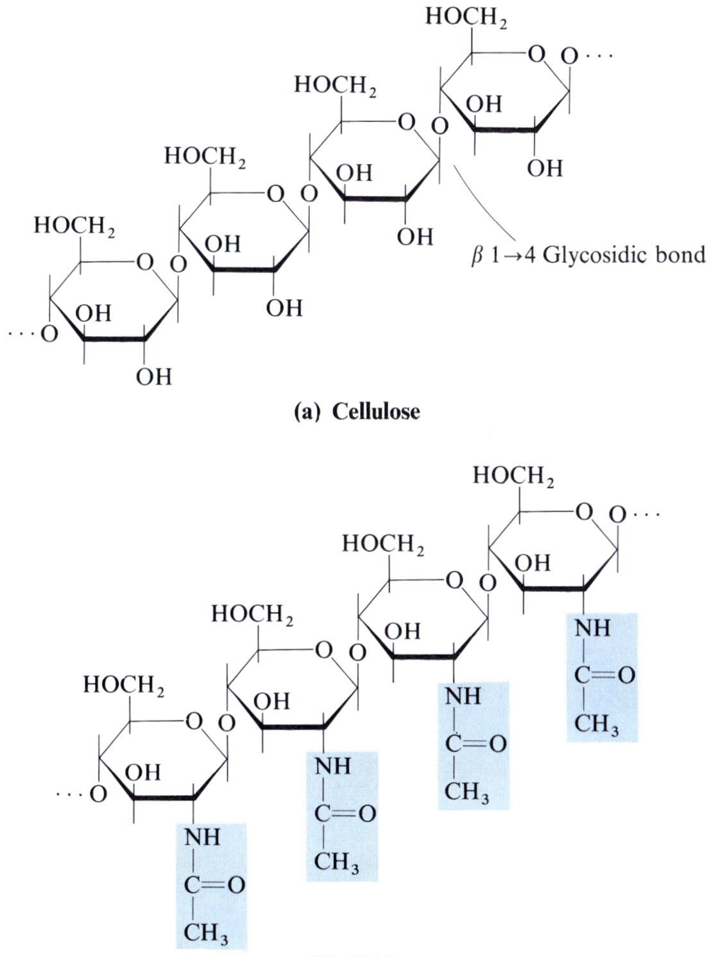
HOCH2
OH
O
β 1→4 Glycosidic bond
(a) Cellulose
NH
C=O
CH3
(b) Chitin

FIGURE 23.6
A comparison of (a) cellulose and (b) chitin. These structural polysaccharides possess beta glycosidic bonds that are not readily broken. Chitin has a different sugar than cellulose; cellulose contains glucose, while chitin has *N*-acetylglucosamine residues (see highlighted portion for structural difference).

possess an intestinal microflora that make these enzymes. Cellulose is thus degraded by the microbial enzymes, and the degradation products are thus available to the microflora and the host animal. Thus, cattle and related organisms can efficiently digest many plants for food that are not suitable for other animals. Dietary cellulose, however, may not be a completely inert, wasted portion of the diet for humans and other animals. Cellulose is a major part of the indigestible, tasteless *fiber* present in our diet as roughage. There appear to be many benefits of a high fiber diet (see next section).

Other Polysaccharides: Hyaluronic Acid and Heparin

The polysaccharides discussed so far have had repeating units consisting of a single sugar or a single sugar derivative. Some important polysaccharides have a more complex repeating unit. **Hyaluronic acid** is found in the connective tissue of higher animals. It is part of the viscous material around bone joints that absorbs shock and acts as a lubricant for the bone surfaces. In Figure 23.7 the unique glycosidic bonding and the *two*-sugar repeating units of hyaluronic acid are shown.

FIGURE 23.7
Hyaluronic acid is a polysaccharide found in connective tissues. (a) This disaccharide is the repeating unit found in hyaluronic acid. (b) An electron micrograph and (c) interpretative drawing of developing cartilage. Polysaccharides found in connective tissue are normally complexed with proteins to form proteoglycans. Hyaluronic acid makes up the central backbone of this aggregate.

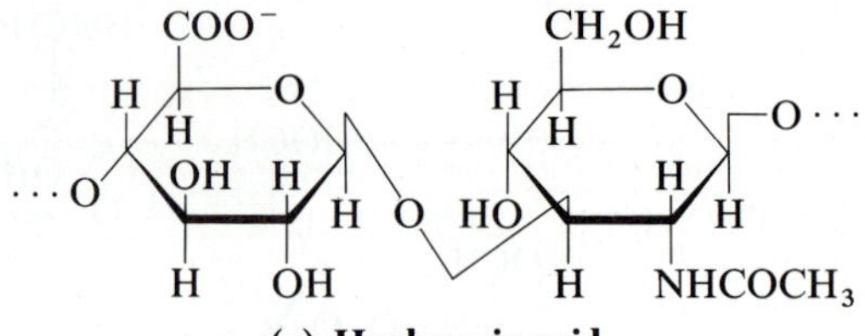
COO⁻
CH₂OH
NHCOCH₃

(a) Hyaluronic acid

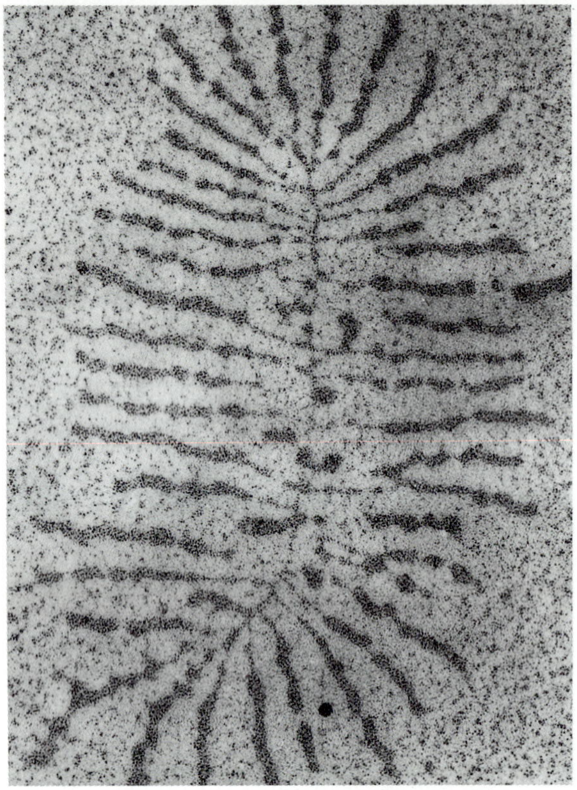

(b)

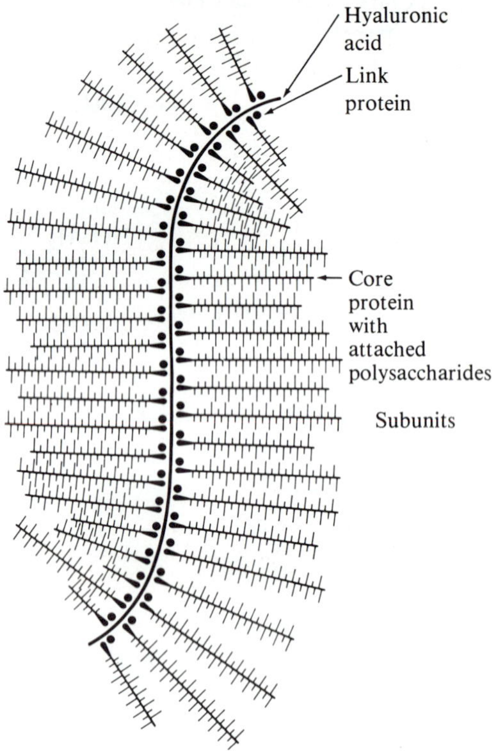
Hyaluronic acid
Link protein
Core protein with attached polysaccharides
Subunits

(c)

Heparin is a polysaccharide produced by some cells of the circulatory system. It is a powerful blood anticoagulant because it inhibits blood clotting. Heparin is used in a number of medical applications where the chance of blood clotting must be reduced or eliminated. Heparin is common to most animals and is an almost linear polysaccharide. It also has a unique two-sugar repeating unit and is characterized by a large number of sulfate groups (Figure 23.8).

FIGURE 23.8
The structure of a short segment of heparin. Note the repeating unit of heparin and the sulfate groups that are present.

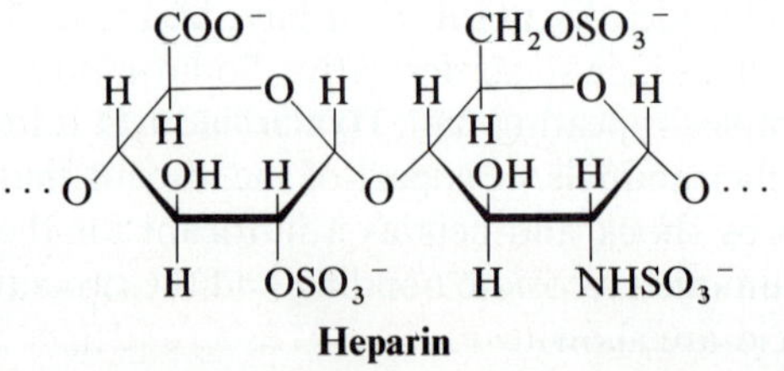
COO⁻
CH₂OSO₃⁻
OSO₃⁻
NHSO₃⁻

Heparin

NEW TERMS

starch A readily digestible plant polysaccharide made of glucose residues bonded by α 1→4 and some α 1→6 glycosidic bonds.

glycogen The animal equivalent of starch; the polysaccharide that is the storage form of glucose in animals.

cellulose A plant polysaccharide made up of β 1→4 linked glucose; the structural material in plants.

chitin A structural polysaccharide found in arthropods similar to cellulose but has an amide in place of an —OH on carbon number two of the glucose residues.

hyaluronic acid A complex polysaccharide found in connective tissue and in bone.

23.6 Carbohydrates and Nutrition

The carbohydrates can be classified into three categories as nutrients—the sugars, the complex carbohydrates, and fiber. Dietary sugars are water-soluble monosaccharides and disaccharides. Sucrose, more commonly known as table sugar, is perhaps the most common dietary sugar, but glucose (blood or grape sugar), fructose, and lactose (milk sugar) are other common sugars of the diet. During digestion, oligosaccharides are broken down to simple sugars (Chapter 28).

Complex carbohydrates and **fiber** are polysaccharides. The complex carbohydrates, which are principally starches, are broken down by digestion to simple sugars that are used by the body (Chapter 28). Fiber consists primarily of polysaccharides and complexes of polysaccharides that cannot be digested. There has been speculation that the higher incidence of cancer of the lower intestinal tract for people of "developed" countries might be related to the relatively low fiber content of their diet. Studies of this nature have resulted in modification by some people of their diets and have led to additional studies that may ultimately clarify the role of fiber in diet.

Simple sugars in the diet and those produced by digestion are a major source of energy for the body. From each gram of digestible dietary carbohydrate 4 Cal (4 kcal) are available. Through a series of chemical reactions, glucose is broken down to carbon dioxide and water (see Chapter 28). The overall process is highly exothermic, and much of the released energy is conserved by the body for various tasks. Simple sugars are also used by the body to make a variety of biomolecules. Sugars are found as components of all of the major classes of biomolecules in the body. Some lipids, some proteins, and all of the nucleotides and nucleic acids contain sugars.

There are many dietary sources of carbohydrates. Generally foods from plants are relatively higher in carbohydrates than foods derived from animals. Roughly 40 to 50% of the calories consumed in the U.S. diet are derived from carbohydrates. In other cultures the percentage is somewhat higher. At the present time, nutritionists believe a larger proportion of dietary energy should come from dietary carbohydrate and that an increase in complex carbohydrates and natural sugars should be accompanied by a decrease in consumption of processed sugars such as table sugar.

Carbohydrates are the main source of energy in the human diet.

ARTIFICIAL SWEETENERS

Many people are overweight and seek to correct this by reducing their caloric intake. Since a significant amount of energy is derived from dietary carbohydrate, many of them reduce their intake of carbohydrate. This can be accomplished by reducing the total intake of food or by reducing or eliminating specific carbohydrate-rich foods from the diet. Many have turned to artificial sweeteners to satisfy their "sweet tooth" while reducing dietary intake of refined sugar and other sweeteners. Today and in the recent past, several artificial sweeteners have been used, including saccharin, cyclamates, and aspartame. The relative sweetness of some of the more common natural and artificial sweeteners are listed in the accompanying table.

Sweeteners	Sweetness Relative to Sucrose
Glucose (blood sugar)	0.5
Fructose (a sugar in honey)	1.7
Sucrose (table sugar)	1.0
Cyclamate	30
Saccharin	450
Aspartame	180

Saccharin was discovered in 1879, and it is the oldest known low calorie sweetener. It was in common use until the early 1960s, when it was supplemented by *cyclamates.* A major reason for the switch to cyclamates was the "aftertaste" associated with saccharin. However, although cyclamates were the preferred sweetener during this time, saccharin was still used in cyclamate preparation because it is so much sweeter. Both of these sweeteners are nonnutritive, that is, the body does not use them for energy or matter.

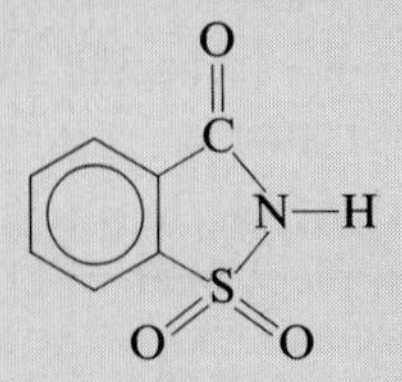
O
C
N—H
S
O
O

Saccharin

Concern over the safety of cyclamates caused their removal from use. Furthermore, since 1978, a warning appears with every product containing saccharin which states that saccharin causes cancer in laboratory animals. In 1981, *aspartame* was approved as a low calorie sweetener. Aspartame lacks the aftertaste of saccharin, and in exhaustive testing, no apparent health concerns were found for this compound, although some individuals still are not convinced it is totally safe. Aspartame, like sugars, is a nutritive sweetener, that is, it can be used by the body. Although aspartame has about the same caloric value per gram as sucrose, it is 180 times sweeter and so much less can be used.

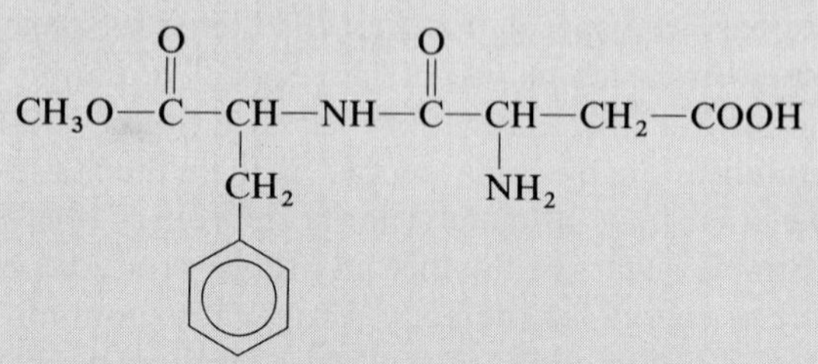
O
O
CH3O—C—CH—NH—C—CH—CH2—COOH
CH2
NH2

Aspartame

NEW TERMS

complex carbohydrates The digestible polysaccharides of the diet; mostly starches.

fiber The indigestible carbohydrates of the diet; mostly polysaccharides such as cellulose and complexes of polysaccharides.

TESTING YOURSELF

Nutrition

1. What are the characteristic glycosidic bonds of (a) amylose, (b) cellulose, and (c) glycogen?

2. A children's breakfast cereal contains 11 g of sugar and 15 g of complex carbohydrate per 1-oz serving. (a) Determine the number of calories provided by these carbohydrates. (b) What percentage is provided in the form of complex carbohydrates?

Answers **1. a.** α 1→4 **b.** β 1→4 **c.** α 1→4 and α 1→6 **2. a.** 11 g + 15 g = 26 g; 26 g × 4 Cal/g = 104 Cal **b.** 15 g/26 g × 100 = 58%

SUMMARY

Carbohydrates are polyhydroxy aldehydes and ketones and derivatives of these compounds. The nomenclature of *monosaccharides* (simple sugars) is closely related to the structure of the sugar. Thus, carbon number, carbonyl type, ring size, and stereochemistry may be included in typical names for monosaccharides. Monosaccharides, particularly glucose, are linked together by *glycosidic bonds* to form *oligosaccharides* and polymers called *polysaccharides*. Most carbohydrates serve as energy reserves, such as *starch* and *glycogen*, or as structural elements in or between cells.

TERMS

carbohydrate	reducing sugar
monosaccharides	glycoside
oligosaccharides	glycosidic bond
polysaccharides	residue
aldose	starch
ketose	glycogen
furanose	cellulose
pyranose	chitin
anomeric carbon	hyaluronic acid
anomers	complex carbohydrates
mutarotation	fiber

CHAPTER REVIEW

QUESTIONS

For questions 1 to 3, use these monosaccharide structures for your answers.

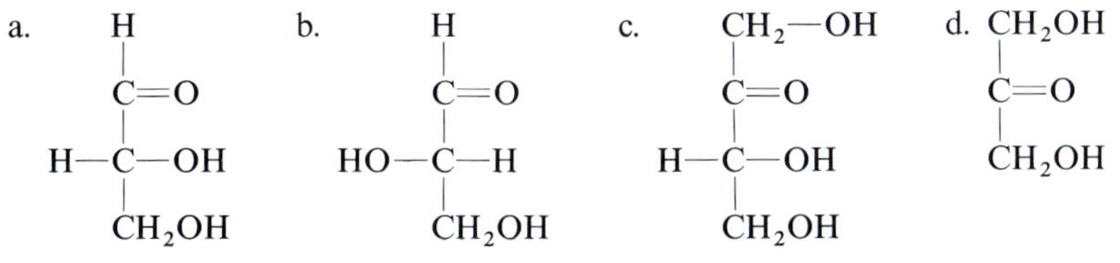
a.
b.
c.
d.

1. Which sugar(s) are aldoses?
2. Which sugar(s) belongs to the L- family?
3. Which sugar(s) would only slowly reduce metal ions?
4. Common table sugar is
   a. glucose b. fructose c. sucrose d. maltose
5. Which is not readily metabolized by humans?
   a. cellulose b. amylopectin c. lactose d. glycogen

6. Mutarotation is a term related to
   a. interconversion of anomers
   b. relationship of D- to L- families
   c. hydrolysis of sucrose
   d. number of simple sugars in a carbohydrate
7. The major difference between cellulose and starch is found in
   a. the sugars
   b. ring size of sugar
   c. linkage between sugars
   d. D- or L- families that the individual sugars belong to
8. Which is a monosaccharide?
   a. sucrose b. galactose c. maltose d. cellulose
9. A reducing sugar will
   a. react with Benedict's reagent c. always be a ketose
   b. have fewer calories d. none of these
10. Which of the following sugars is a monosaccharide and a furanose?
   a. glucose b. galactose c. fructose d. sucrose

Answers **1.** a and b **2.** b **3.** c and d **4.** c **5.** a **6.** a **7.** c **8.** b **9.** a **10.** c

DIAGNOSTIC CHART

Blacken in all of the circles under the number of each question you missed in the preceding Chapter Review questions. The diagnostic chart will help you identify the concept areas that need more study.

	Questions									
Concepts	**1**	**2**	**3**	**4**	**5**	**6**	**7**	**8**	**9**	**10**
Classification								○		○
Monosaccharides	○	○						○		○
Properties			○			○			○	
Glycosidic bond							○			
Oligosaccharides				○						
Polysaccharides					○		○			
Nutrition					○					

EXERCISES

Monosaccharides

1. Define the term *monosaccharide*.
2. Identify each of the following sugars by functional group, carbon length, and D- or L- family.

a.

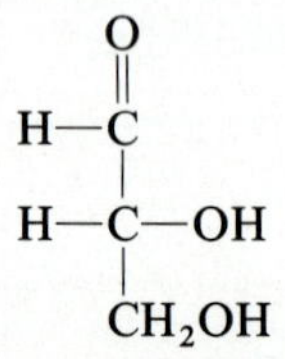

b.

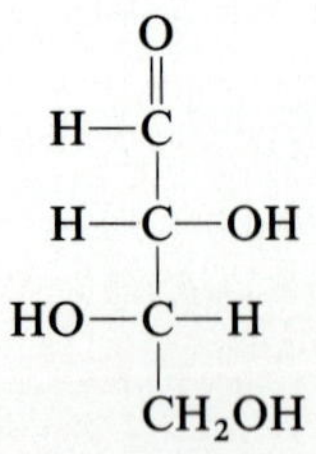

c.

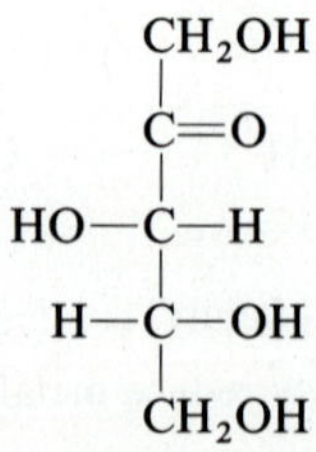

d.

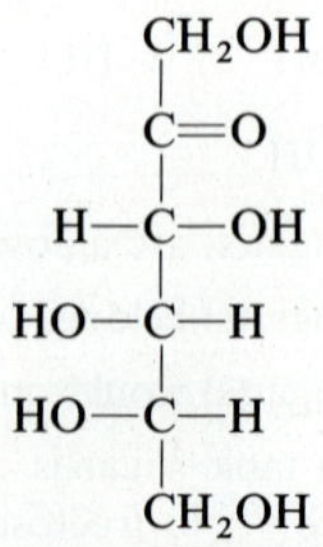

3. What information is contained in the term *ketopentose*?
4. How can an L-sugar be recognized?

5. Identify each of the simple sugars shown here.

a. 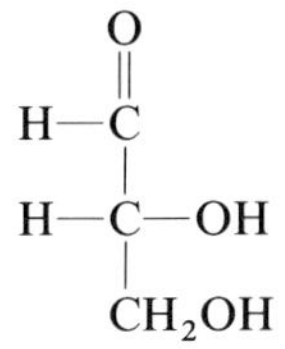

b. 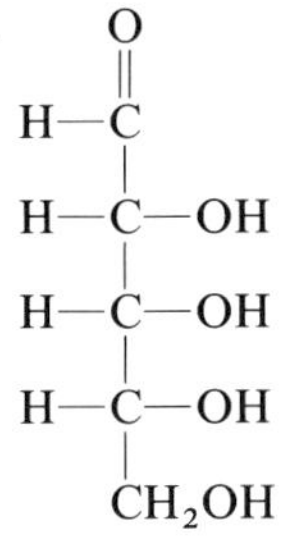

c. 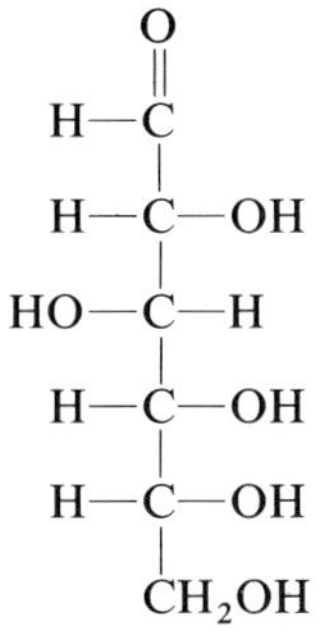

d. 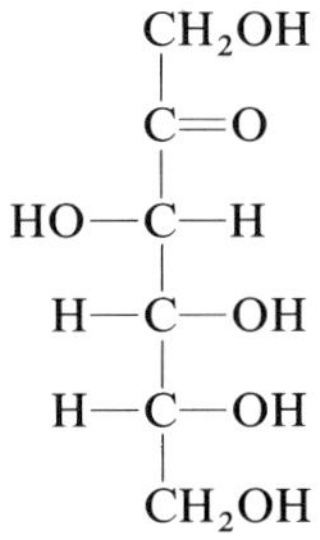

Reactions and Glycosides

6. Identify each of the following glycosidic linkages as alpha or beta.

a. 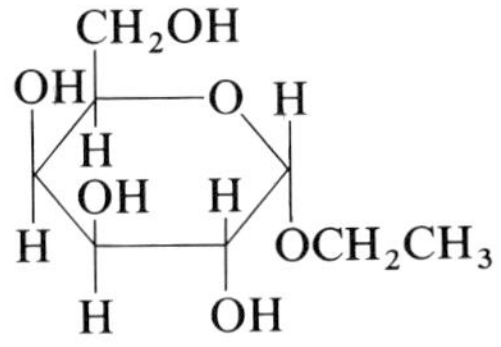

b. 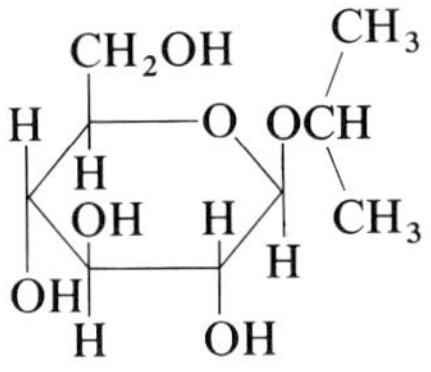

c. 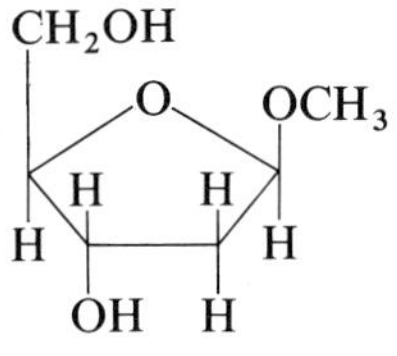

d. 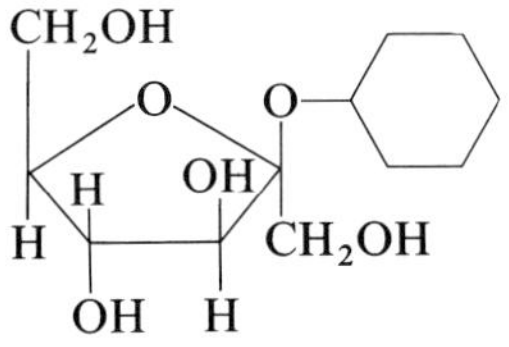

7. How is an alpha anomer recognized?

8. Describe the furanose and pyranose forms of carbohydrates.

9. Describe the process of mutarotation.

10. Show how hemiacetal formation and the cyclic forms of sugars are related.

11. Describe the chemistry involved in the designation "reducing sugar."

12. Identify each of these disaccharides.

a. 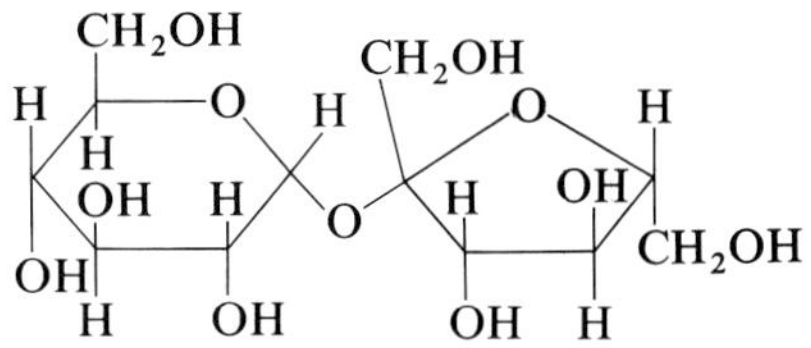

b. 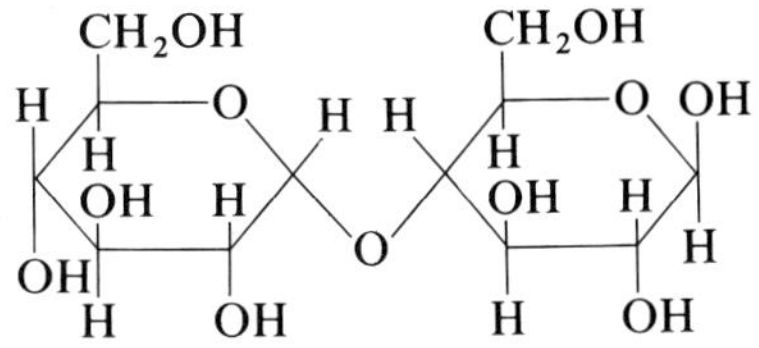

c. 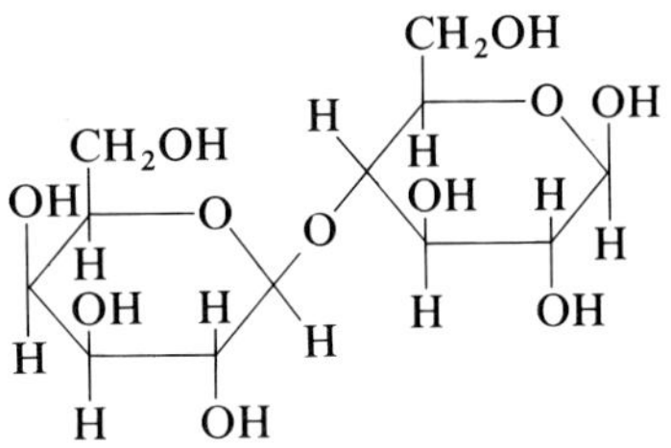

13. The sugar shown here is found in some mushrooms. How does this disaccharide differ from most other disaccharides you have seen? Will it be a reducing sugar?

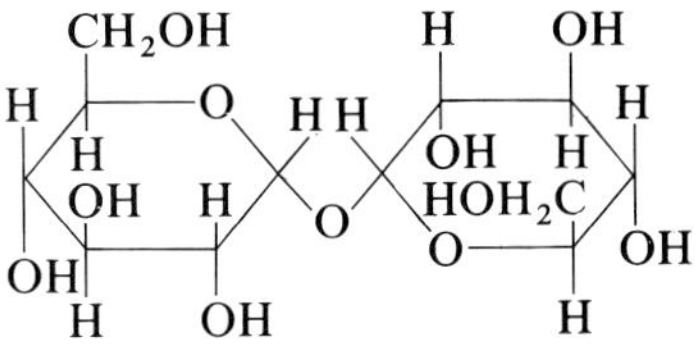

14. Hydrolysis of disaccharides yields two simple sugars. What will be the simple sugars obtained from the hydrolysis of the following?
a. sucrose b. maltose c. lactose

15. What glycosidic linkages are found in the disaccharides in question 12?

Polysaccharides

16. Hydrolysis of cellulose yields what monosaccharide?

17. Describe the similarities and differences between starch and cellulose.

18. Describe the similarities between glycogen and starch.

19. Describe the similarities and differences between cellulose and chitin.

20. Where is hyaluronic acid found and what is its role?

Matching

21. Match the term on the left with the description on the right.

A. amylose	a. ketose
B. glycogen	b. table sugar
C. fructose	c. a polymer of glucose with beta glycosidic linkages
D. sucrose	d. a polymer of glucose stored in the body
E. triose	e. a three carbon sugar
F. cellulose	f. a form of starch

Summary Questions

22. For each of the following words, give an explanation, definition, or example that demonstrates your understanding of the word.

a. carbohydrate	e. ketohexose
b. pentose	f. L-family
c. aldose	g. furanose
d. ketose	h. anomeric carbon
i. anomers	p. cellulose
j. glycosidic linkage	q. amylose
k. glycoside	r. reducing sugar
l. monosaccharide	s. aspartame
m. disaccharide	t. lactose
n. polysaccharide	u. mutarotation
o. glycogen	

23. Draw structures for each of the following.
    a. D-aldopentose
    b. L-glyceraldehyde
    c. α-D-glucopyranose
    d. β-D-galactopyranose
    e. the repeating unit for starch and glycogen
    f. the repeating unit for cellulose
    g. the isomers of "ketohexose"

24. Which of the following carbohydrates would be classified as a reducing sugar?

a. galactose	d. glycogen
b. fructose	e. sucrose
c. starch	f. lactose

CHAPTER

24

Lipids

Lipids are a large class of naturally occurring compounds that are characterized by their insolubility in water and solubility in nonpolar organic solvents. Included in this group of natural products are the fats and oils, water insoluble components of membranes, and steroids. This chapter will describe the structural and chemical characteristics of these compounds, as well as the structure of membranes.

OBJECTIVES

After completing this chapter, you should be able to

- Describe the general structures of the various lipids.
- Classify lipids as saponifiable or nonsaponifiable.
- Describe the structural differences between fats and oils.
- Describe the function of lipids in membranes.
- Identify the common structure of steroids.

Lipids are soluble in nonpolar organic solvents but insoluble in water.

Unlike carbohydrates, the compounds known as lipids are not composed of readily recognizable, chemically related structural units. Instead, lipids represent a class of substances that are related to one another by their solubility properties. **Lipids** are the biological molecules that are insoluble in water and soluble in organic solvents. As the hydrocarbon component (the alkyl group) of an organic compound increases in size, the relative contribution of a functional group to the physical properties of the molecule decreases. For example, low molecular weight acids and alcohols are readily soluble in water because the polar —OH or —COOH groups hydrogen bond with water (see Chapters 18 and 20). High molecular weight acids and alcohols still have the polar functional groups, but they are water insoluble because the large nonpolar alkyl group does not bond to water. It is said to be *hydrophobic* (or "water hating") and therefore water insoluble. As the size of an alkyl group increases, the water solubility of the compound decreases. Lipids are molecules that have a large hydrocarbon component and are largely or entirely nonpolar.

Because lipids are defined in terms of their solubility properties, it should not be surprising to learn that lipids vary greatly in structure and function. Many lipids are involved in energy storage and utilization, but others serve roles in the structure of membranes and other macromolecular complexes. Some are hormones, some play protective roles, and some aid in digestion and nutrient transport. Lipids are an important and diverse group of biomolecules.

24.1 Classification of Lipids

Lipids are often classified as either saponifiable or nonsaponifiable. In Chapter 20 you were shown the reaction for saponification:

$$\underset{\textbf{Ester}}{R{-}\overset{\overset{\displaystyle O}{\|}}{C}{-}OR'} \xrightarrow{H_2O,\ OH^-} \underset{\substack{\textbf{Carboxylic}\\ \textbf{acid}}}{R{-}\overset{\overset{\displaystyle O}{\|}}{C}{-}O^-} + \underset{\textbf{Alcohol}}{R'OH}$$

Saponifiable lipids are hydrolyzed in base to yield fatty acids and alcohols.

Saponifiable lipids are esters that are hydrolyzed by base to give large carboxylic acids called *fatty acids* and an alcohol. Some saponifiable lipids contain one or more other molecules. Saponifiable lipids include the triacylglycerols, glycolipids, sphingolipids, waxes, and a variety of phosphate-containing lipids called phospholipids. **Nonsaponifiable lipids** are those that are *not* hydrolyzed by base, and include the steroids, prostaglandins, leukotrienes, and terpenes. (Terpenes were already discussed in Chapter 16.) Each of these lipid classes are discussed in more detail in the sections that follow. Table 24.1 summarizes these lipid classes.

NEW TERMS

lipid A class of biomolecules characterized by their insolubility in water and their solubility in organic solvents.

saponifiable lipid A lipid that can be hydrolyzed by base to one or more fatty acids and one or more other molecules.

TABLE 24.1
Saponifiable and Nonsaponifiable Lipids

Class of Lipid	Saponification Products
Saponifiable lipids	
Triacylglycerols	Carboxylic acids and alcohol
Glycolipids	Carboxylic acids, alcohol, and sugars
Sphingolipids	Carboxylic acids, sphingosine, and other components
Waxes (some)	Carboxylic acids and alcohols
Phospholipids	Carboxylic acids, alcohol, phosphate, and other components
Nonsaponifiable lipids	
Steroids	These lipids are not cleaved by saponification.
Prostaglandins	
Leukotrienes	
Terpenes	

nonsaponifiable lipid A lipid that does not contain fatty acids and therefore cannot be saponified by base.

TESTING YOURSELF

Lipid Classification

1. An organic compound has been isolated from animal tissue with chloroform and methanol. It does not dissolve in water. To what class of biomolecules does this compound belong?
2. A water insoluble compound is heated in the presence of aqueous base. Several fatty acids and the alcohol glycerol were obtained. How should this lipid be classified?

Answers **1.** lipid **2.** saponifiable lipid

24.2 Fatty Acids

The carboxylic acids found in the saponifiable lipids are rather large carboxylic acids. These acids are called **fatty acids** and typically have 10 to 24 carbon atoms in the carbon chain.

Fatty acids are large, unbranched carboxylic acids containing ten or more carbon atoms.

$$CH_3CH_2CH_2CH_2CH_2CH_2CH_2CH_2CH_2CH_2CH_2CH_2CH_2COOH$$

Myristic acid
(a typical fatty acid)

Fatty acids nearly always contain an even number of carbon atoms and are rarely branched. The only other group commonly found in fatty acids is the carbon–carbon double bond. These structural features are a consequence of their synthesis in the body, as we shall learn later in our discussion of metabolism (Chapter 29). Although fatty acids are a principal component of saponifiable lipids, they are rarely found free in cells or living organisms. In fact, free fatty acids are somewhat toxic to cells. Soaps, which are the salts of fatty acids, have some antibacterial action on the surface of the skin.

Classification

Fatty acids are classified as saturated fatty acids or unsaturated fatty acids. **Saturated fatty acids** contain no carbon—carbon double or triple bonds. The term *saturated* means these fatty acids will not accept any more hydrogen, that is, they are saturated with hydrogen. The most common saturated fatty acids are unbranched, contain an even number of carbon atoms, and have no functional group besides the carboxyl group. Table 24.2 shows the names and structures of some common fatty acids. Fatty acids can be named by the IUPAC rules for carboxylic acids (Chapter 20), but more frequently, common names are used. For example, the unsaturated fatty acid linolenic acid has the complex IUPAC name *cis,cis,cis*-9,12,15-octadecatrienoic acid. Systematic names are descriptive, but they are often too long for frequent usage.

Some of the fatty acids were named for the source from which they were first isolated. Palmitic acid is the saturated fatty acid typical of palm oil. Stearic acid is the principal fatty acid obtained from beef tallow. A convenient shorthand notation for fatty acids consists of two numbers separated by a colon, the first indicating the number of carbon atoms, and the second the number of carbon–carbon double bonds in the molecule. For example, palmitic acid is designated as 16:0 and stearic acid as 18:0.

TABLE 24.2
Some of the Common Fatty Acids Found in Animals and Plants

Number of C Atoms to Number of Double Bonds	Common Name	Structure
Saturated fatty acids		
10:0	Capric acid	 COOH
12:0	Lauric acid	 COOH
14:0	Myristic acid	 COOH
16:0	Palmitic acid	 COOH
18:0	Stearic acid	 COOH
20:0	Arachidic acid	 COOH
Unsaturated fatty acids		
16:1	Palmitoleic acid	 COOH
18:1	Oleic acid	 COOH
18:2	Linoleic acid	 COOH
18:3	Linolenic acid	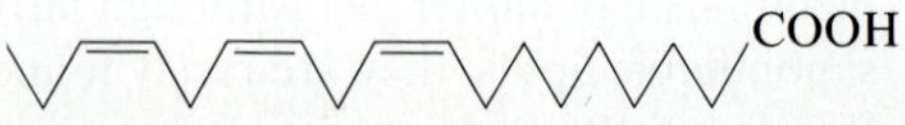 COOH

Unsaturated fatty acids contain one or more carbon–carbon double bonds. If the acid contains just one double bond, it is called a *monounsaturated* fatty acid. If two or more double bonds are present, it is a *polyunsaturated* fatty acid. Table 24.2 lists the more common unsaturated fatty acids found in foods. Since palmitoleic acid (16:1) and oleic acid (18:1) contain one carbon–carbon double bond, they are monounsaturated. Linoleic (18:2) and linolenic (18:3) acids are polyunsaturated fatty acids. In organic chemistry you learned that the orientation and position of the double bond as well as the number of double bonds should be expressed in a name because several isomers of each compound are possible. In nature, the isomers shown in Table 24.2 are by far the most common. The fatty acids in plants and animals nearly always have the cis configuration. In monounsaturated fatty acids, the double bond is usually between carbon atoms 9 and 10. In linoleic acid the double bonds are between carbon atoms 9 and 10 and between carbons 12 and 13. Linolenic acid is similar to linoleic acid, but has a third carbon–carbon double bond between carbons 15 and 16.

The carbon-carbon double bonds of most unsaturated fatty acids are cis.

Physical Properties

The solubility properties of fatty acids depend upon pH. Fatty acids are carboxylic acids with pK_a values between 4 and 5. They can exist in a protonated form as a carboxylic acid or in an unprotonated carboxylate form, which is the conjugate base of the carboxylic acid. These two forms have different solubility properties.

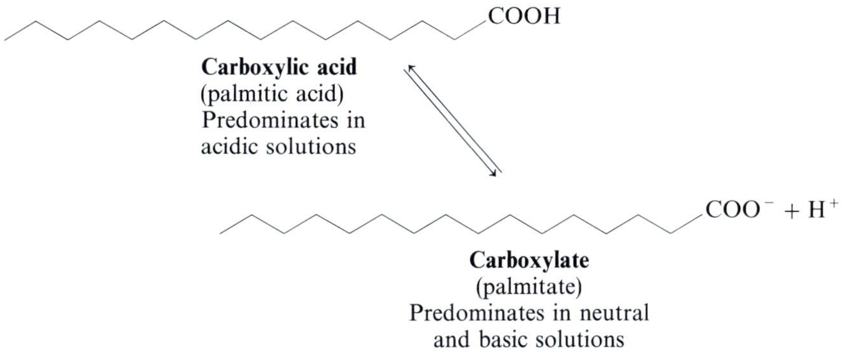
COOH
Carboxylic acid
(palmitic acid)
Predominates in
acidic solutions
COO⁻ + H⁺
Carboxylate
(palmitate)
Predominates in neutral
and basic solutions

Carboxylic acids have a polar head group, the —COOH group, but the large nonpolar alkyl group makes them insoluble in water and soluble in nonpolar organic solvents. The salts of fatty acids are in the carboxylate form, which is ionic. Ionic compounds are generally insoluble in nonpolar organic solvents, but the interactions of water molecules with the ionic group are sufficiently strong that these salts appear to "dissolve or suspend in water." True solutions are not formed, however; micelles form instead (see Figure 20.12).

Fatty acids show increasing melting points and boiling points with increasing size. This increase is also apparent in the other classes of organic compounds that you have studied. The melting points of fatty acids and of the compounds that contain them are also influenced by the presence of cis carbon–carbon double bonds. The cis double bonds in the unsaturated fatty acids impart a definite structure on the acid by, in effect, introducing "kinks" into the molecule (Figure 24.1). These kinks or bends in the molecule reduce

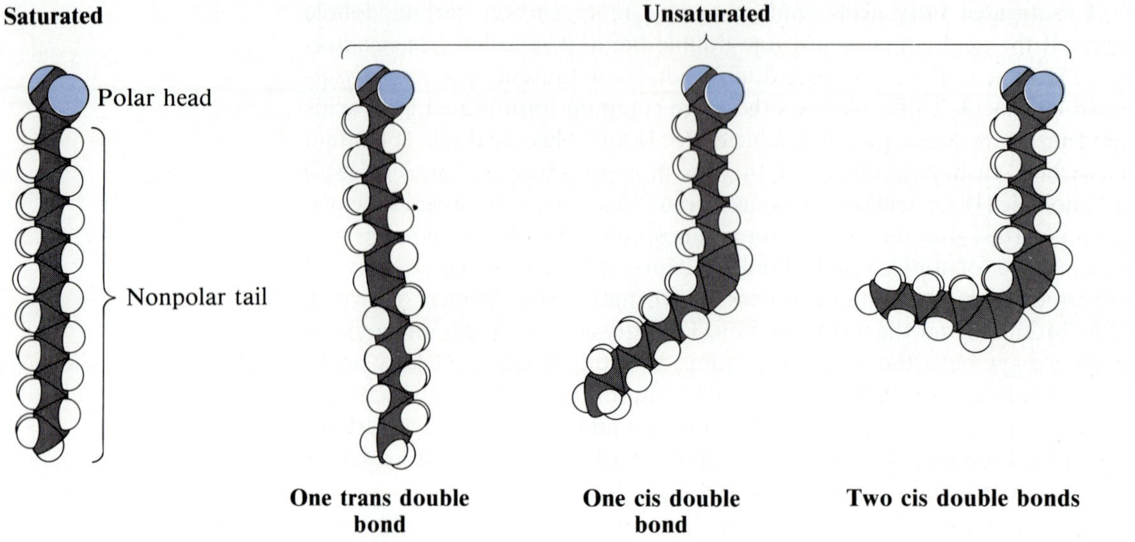
Saturated
Unsaturated
Polar head
Nonpolar tail
One trans double bond
One cis double bond
Two cis double bonds

FIGURE 24.1
These space-filling models illustrate the structure of some saturated (far left) and unsaturated fatty acids. Cis double bonds introduce a "kink" into the molecule that greatly affects the structure.

Compounds containing cis double bonds have lower melting points than the corresponding saturated compound.

the ability of the molecules to "stack" easily in an orderly fashion in the solid state (see Figure 15.6). The net effect is that compounds containing cis double bonds remain liquids at lower temperatures than comparable compounds that lack cis double bonds. Biomolecules that contain appreciable amounts of unsaturated fatty acids are liquids rather than solids at room temperature.

Chemical Properties

The chemical properties of fatty acids are just as would be predicted by the presence of a carboxyl group and carbon–carbon double bonds. The carboxyl group of fatty acids can undergo the same reactions as any carboxyl group, but in biological tissues, the most common derivative is an ester. Ester bonds are very common between fatty acids and a variety of alcohols. The ester bond is seen repeatedly in the sections on saponifiable lipids that follow. Examples of amide bonds will also be seen.

Carbon-carbon double bonds in fatty acids undergo hydration, hydrogenation, halogenation, and other reactions typical of this functional group.

Unsaturated fatty acids are common in most organisms. The carbon–carbon double bonds of these compounds undergo the typical addition reactions of carbon–carbon double bonds (Chapter 16), three of which will be discussed. In the first reaction, hydration, water adds to these double bonds to introduce a hydroxyl group into the molecule. This is important in the metabolic reactions where fatty acids are broken down to produce energy (Chapter 28).

$$\underset{\textbf{Unsaturated fatty acid}}{RCH{=}CH(CH_2)_nCOOH} + H_2O \longrightarrow \underset{\textbf{Hydroxy fatty acid}}{R\overset{OH}{\overset{|}{C}}HCH_2(CH_2)_nCOOH}$$

In a second reaction, hydrogenation, hydrogen adds to these double bonds.

$$RCH{=}CH(CH_2)_nCOOH + H_2 \longrightarrow RCH_2CH_2(CH_2)_nCOOH$$

Again, a number of important reactions of the body involve addition of hydrogen to these bonds, but this reaction is also important in the food industry. When you get a chance, read the label on a number of food packages. Look for the term *partially hydrogenated*, which is used to describe some oils. This term indicates that some of the carbon–carbon double bonds found in the fatty acids of these oils have had hydrogen added to them. A third addition reaction of importance is halogenation, the addition of a halogen to the double bond.

$$RCH{=}CH(CH_2)_nCOOH + X_2 \longrightarrow RCHXCHX(CH_2)_nCOOH$$

The halogens (see Chapter 16) add to the carbon–carbon double bonds of fatty acids.

One other reaction of unsaturated fatty acids is important to an understanding of the properties of fats and oils. Carbon–carbon double bonds can be oxidized by oxygen. This reaction cleaves the double bond, yielding products that contain the aldehyde group or carboxylic acid group. If the product molecules are small enough, they will be volatile.

$$CH_3(CH_2)_nCH{=}CH(CH_2)_nCOOH \xrightarrow{\text{Oxidation}} CH_3(CH_2)_nCHO + CHO(CH_2)_nCOOH$$

NEW TERMS

fatty acids Carboxylic acids from biological sources that generally contain 10 or more carbon atoms.

saturated fatty acid A fatty acid that has no C—C double bonds.

unsaturated fatty acid A fatty acid that has one or more C—C double bonds

TESTING YOURSELF

Fatty Acids

1. Name the following fatty acids.
   a. The saturated fatty acid with 18 carbon atoms.
   b. The unsaturated fatty acid with 16 carbon atoms and one double bond.
2. The double bonds of most fatty acids have what configuration?
3. a. What is the appropriate name for oleic acid at pH 7?
   b. Draw its structure at pH 7.

Answers **1. a.** stearic acid **b.** palmitoleic acid **2.** cis **3. a.** oleate

b. 
COO-

24.3 Triacylglycerols and Waxes

Triacylglycerols and some waxes are lipids that are nonpolar and saponifiable. These properties bring them together in this section, but they have different structures and different biological roles. Triacylglycerols serve as energy reserves, while waxes have structural roles as building and waterproofing materials.

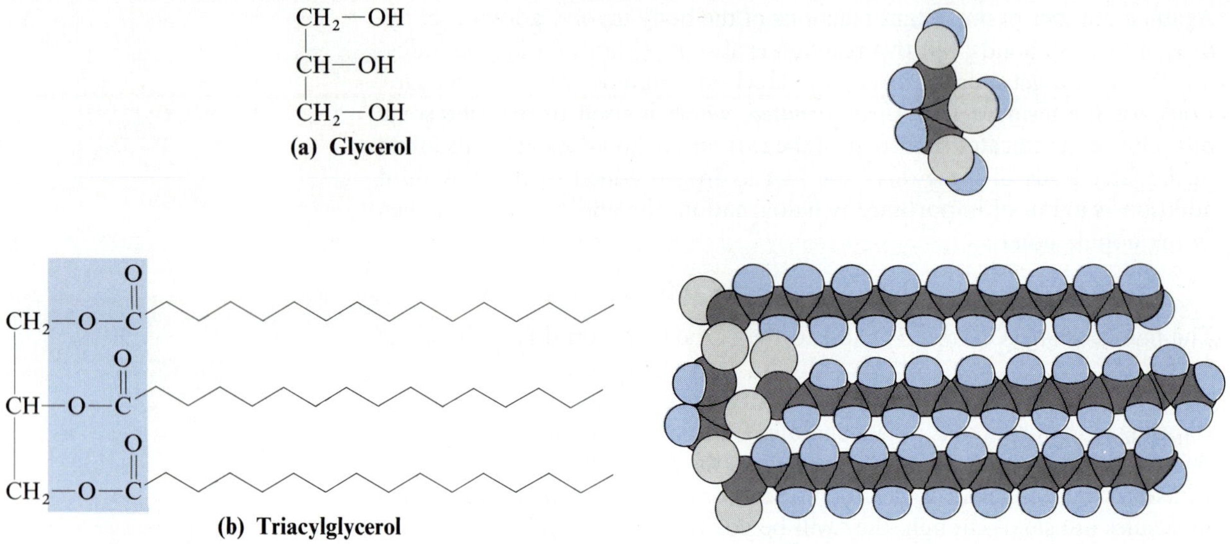
(a) Glycerol
(b) Triacylglycerol

FIGURE 24.2
The structure of glycerol and a triacylglycerol. (a) Glycerol is a trihydroxy alcohol that is a component of many lipids. (b) Triacylglycerols have three fatty acids bound to glycerol through ester bonds (highlighted). Mono- and diacylglycerols are much less common than triacylglycerols.

Triacylglycerols

Triacylglycerols are triesters of three fatty acids and glycerol.

NOMENCLATURE The family of lipids known as **triacylglycerols** are triesters formed from three fatty acids esterified to the three hydroxyl groups of glycerol (Figure 24.2). All triacylglycerols contain glycerol, but they may contain different fatty acids. Some contain three copies of the same fatty acid, but more commonly two or more different fatty acids will be present in a triacylglycerol. During certain reactions in the body, one or two fatty acids may be bonded to a glycerol. These molecules are called *monoacylglycerols* and *diacylglycerols*, respectively. These compounds, when present at all, are not generally found in high concentrations in living systems.

Fats and oils differ primarily in their melting point.

Triacylglycerols are known by several synonyms. Triglyceride is an older name that means the same thing; three fatty acids esterified to glycerol. You may see the term *triglyceride* used frequently, but it is slowly being replaced by triacylglycerol. The terms *fats* and *oils* are common terms used to describe triacylglycerols. A **fat** is a sample of triacylglycerols that is solid at room temperature. Lard from pork and tallow from beef are fats. The triacylglycerols of most animals are fats. The term **oil** is used to describe a sample of triacylglycerols that are liquids at room temperature. Corn oil, cotton seed oil, and sunflower oil are examples. Many plant triacylglycerols are oils. The word oil is also used for other liquids like crude oil (petroleum), mineral or paraffin oil, and motor oil. These materials are hydrocarbons, not triacylglycerols.

PHYSICAL AND CHEMICAL PROPERTIES If fats and oils are both triacylglycerols, why are some triacylglycerols solids at room temperature (fats) while others are liquids (oils)? The basis for this difference is the degree of unsaturation found in the fatty acids in the triacylglycerols. Fats contain fatty acids that are predominantly saturated. The fatty acid molecules fit together relatively well, and thus remain in the solid phase until the temperature is higher. Oils contain a larger percentage of fatty acids that are more highly unsaturated. These fatty acids do not fit together as well because of the "kinks" resulting from the double bonds and thus tend to form a liquid phase at room temperature. The physical state of a triacylglycerol provides some clues to the degree of unsaturation of the fatty acids in it.

There is a convenient test for estimating the amount of unsaturation in a sample of triacylglycerols. This test is the *iodine test*. Iodine as ICl (iodine monochloride) readily adds to carbon–carbon double bonds.

$$\ldots CH_2CH_2CH{=}CHCH_2CH_2 \ldots + ICl \longrightarrow \ldots CH_2CH_2CHI\,CHCl\,CH_2CH_2 \ldots$$

Molecular iodine and ICl are intensely colored substances, but most iodinated organic compounds are colorless. Iodine can be added to a sample of triacylglycerols by titration. When the iodine color persists, all of the carbon–carbon bonds have been iodinated. The more double bonds present in the sample, the more iodine that is required to complete the titration. The **iodine number** of a fat or oil is defined as the grams of iodine, as I_2, required to react with all of the carbon–carbon bonds in 100 g of the sample. A high iodine number indicates a high degree of unsaturation. A low number means the triacylglycerol is relatively saturated. Lard has an iodine number of about 65 to 70, while corn oil has an iodine number of 125 to 130. Iodine numbers vary slightly from sample to sample because environmental factors and diet influence the fatty acid composition of plants and animals. The degree of unsaturation in fats and oils has significant implications in health and nutrition (see Section 24.7).

A large iodine number indicates a high degree of unsaturation in the oil.

The presence of carbon–carbon double bonds makes triacylglycerols susceptible to oxidation. They can readily become **rancid**, a state in which fats and oils smell and taste bad due to the presence of volatile organic acids and aldehydes that have formed via oxidation of carbon–carbon double bonds. Fats and especially oils may become rancid with time after exposure to air. A small amount of oxidative cleavage will make an oil or fat unsuitable for eating. The food industry can do two things to reduce the chance of a food becoming rancid. They can add antioxidants such as BHA (butylated hydroxyanisole) or BHT (butylated hydroxytoluene) (see Chapter 19) that react with the oxidizing species to prevent their reaction with double bonds, or they can partially hydrogenate an oil to reduce the number of double bonds present. This hydrogenation of some of the double bonds does indeed increase shelf life and extend the useful life of a cooking oil, but the degree of saturation in dietary oils and fats is correlated to coronary disease (see Section 24.7). Extended shelf life may be gained at the expense of good nutrition.

Added antioxidants, such as BHA or BHT, or partial hydrogenation of the oil will reduce the chance of rancidity and thus extend shelf life.

BIOLOGICAL ROLES Many animals and plants store triacylglycerols as energy reserves. Triacylglycerols are excellent molecules for energy storage because they are high in energy and are water insoluble. In general, the less oxygen a molecule contains, the more energy it contains. Thus, on a per gram basis, fats and oils have more energy than the sugars, starches, and glycogen studied earlier. These compounds contain numerous oxygen atoms in the hydroxyl groups, and thus the carbon atoms of the molecule are more oxidized than the carbon atoms of the long hydrocarbon chain of fats and oils. The carbon atoms of amino acids in proteins are also more oxidized than the carbon atoms of fats and oils. As a rough guide, fats and oils have about 9 kcal (Cal) of energy per gram of triacylglycerol. Dry carbohydrates and proteins have about 4 kcal of energy per gram. Thus, just on a mass basis, fats and oils are more than twice as efficient for energy storage. Furthermore, fats and oils are stored "dry" in the fat cells of the body, but glycogen is hydrated by numerous water molecules. Thus, 1 g of hydrated glycogen in the body contains far less energy than 1 g of anhydrous body fat.

Triacylglycerols are the most efficient molecular storage form for energy.

Most people have a significant amount of body fat. The average North American has about 10 to 30% body fat. The energy stored in fat can provide usable energy for several months. In contrast, this same person has only enough energy reserves as glycogen for one day or less. Glycogen meets some of our short term energy needs, while fats serve as energy reserves over a longer time frame.

Large amounts of stored fat are common in mammals that hibernate or live in cold climates (Figure 24.3). This fat storage serves two useful purposes: (1) it provides an energy source during hibernation or fast, and (2) it acts as a source of insulation against cold. Animals such as seals use fat as insulation against the cold polar waters and as an energy source during the breeding and pup-rearing seasons. Hibernating animals such as rodents and grizzly bears use it as both insulation and a food source during hibernation. Subcutaneous fat in humans may serve some insulating role. Slender people may feel cold during conditions in which heavier people are comfortable, yet when

FIGURE 24.3
This polar bear relies on stored fat for energy and insulation. Some marine mammals have several inches of blubber for insulation against cold arctic waters.

SPERM WHALES AND LIPIDS

An interesting use of a lipid is found in the sperm whale. Well known for being a rich source of oil, the sperm whale was extensively hunted in the past. The head of the sperm whale contains a large amount (sometimes as much as 4 tons!) of a mixture of oils called *spermaceti oil.* This mixture tends to solidify as it is cooled. Most solids have greater densities than liquids; thus, as the whale feeds at great depths in the cold oceans, the oil freezes, the amount of freezing depending on the temperature. At the colder temperatures, the density of the whale increases because the mass remains constant, but the volume of the whale decreases as the oil solidifies. The buoyancy of the whale decreases, and it is able to stay submerged, without exerting a great amount of energy. The whale is able to control its density and thus its buoyancy by diverting either cold sea water or warm blood through the chambers in its head, thus either freezing or melting the spermaceti oil.

it is hot, the discomfort is seen in the better insulated individual. Fatty tissue also serves a protective role: fatty layers around the internal organs of vertebrates serve as a cushion to protect the organs from injury.

Waxes

Waxes are a heterogeneous group of waxy solids found in nature. Some are hydrocarbons, others are alcohols or ketones, and some are esters and thus saponifiable. These waxes, unlike triacylglycerols, are not used for energy storage. Instead they serve structural roles. Waxes are found as protective coatings on the skins of fruits and leaves and on the exterior surface of the exoskeleton of insects. Waterfowl have preen glands that provide wax the bird uses to coat feathers for water proofing. Repelling water is the role of all of these waxes. Birds use waxes to keep feathers dry, while plants and insects use waxes to prevent water from leaving the organism via transpiration. Bees also use waxes to make combs. These wax cells are marvelous vessels for storing honey and pollen and also serve as growth chambers for the developing young.

Waxes cover and waterproof the exterior surface of many biological surfaces.

Saponifiable waxes are esters of a long chain fatty acid and a long chain alcohol (Table 24.3). The melting points of waxes vary according to their structures in much the same way as fats and oils; thus waxes from different plants and animals have different physical properties. Some of these waxes have been used for commercial uses in the past and present.

NEW TERMS

triacylglycerols Triesters of glycerol and three fatty acids.

fats Triacylglycerols that are solids at room temperature.

oils Triacylglycerols that are liquids at room temperature.

iodine number An index used to indicate the degree of unsaturation present in a fat or oil.

rancid Term used to indicate that a fat or oil is foul smelling and bad tasting due to the presence of volatile acids and aldehydes.

TABLE 24.3
Some Typical Waxes from Plants and Animals

Wax	Structure of an Ester Component	Source	Uses
Beeswax	$CH_3(CH_2)_{14}{-}\overset{O}{\overset{\parallel}{C}}{-}O{-}(CH_2)_{29}CH_3$	Bees	Candles, cosmetics, confections, medicinals, art preservation
Carnauba wax	$HOCH_2(CH_2)_{17}{-}\overset{O}{\overset{\parallel}{C}}{-}O{-}(CH_2)_{31}CH_3$	Brazilian palm trees	Coatings for perishable products; polishing candies and pills; auto and floor waxes
Spermaceti	$CH_3(CH_2)_{14}{-}\overset{O}{\overset{\parallel}{C}}{-}O{-}(CH_2)_{15}CH_3$	Sperm whales	Use in products such as cosmetics, soap, and candles banned in 1976; sperm whales are now protected species
Rice bran wax	$CH_3(CH_2)_{20}{-}\overset{O}{\overset{\parallel}{C}}{-}O(CH_2)_{21}CH_3$	Rice bran	Lipstick base, plastics processing aids
Bayberry wax	$CH_3(CH_2)_{12}{-}\overset{O}{\overset{\parallel}{C}}{-}O{-}(CH_2)_xCH_3$	Berries of myrtle shrubs	Candles
Jojoba oil	$CH_3(CH_2)_7CH{=}CH{-}(CH_2)_9{-}\overset{O}{\overset{\parallel}{C}}{-}O{-}CH_2(CH_2)_9CH{=}CH(CH_2)_7CH_3$	Jojoba beans	Has replaced sperm whale products
Jojoba wax	$CH_3(CH_2)_{18}{-}\overset{O}{\overset{\parallel}{C}}{-}O{-}CH_2(CH_2)_{18}{-}CH_3$	Jojoba beans	Cosmetics and candles

From: Kroschwitz and Winokur, *Chemistry: General, Organic, and Biological*, McGraw-Hill, 1985.

waxes Solid lipids found on the surface of many plants and insects. These materials include hydrocarbons, alcohols, and ketones or esters of fatty acids and fatty alcohols.

TESTING YOURSELF

Triacylglycerols and Waxes

1. A sample of triacylglycerol is found to have an iodine number of 139. Is this sample an oil or fat?
2. How many Calories (kilocalories) are found in 1 oz of margarine? Assume the margarine is pure triacylglycerol.
3. What do triacylglycerols and many waxes have in common?
4. Classify each of these lipids as a fatty acid, wax, or triacylglycerol.

   a. $CH_3(CH_2)_8COOH$

   b. 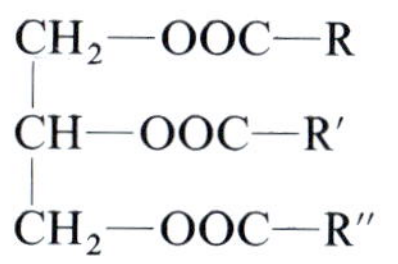

   c. 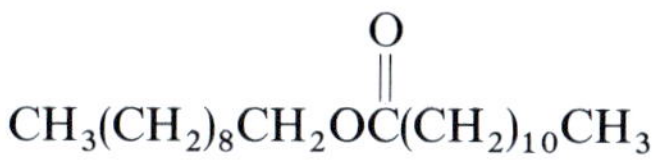

Answers **1.** Corn oil has an iodine number of 125 to 130, and since this sample has a larger iodine number, it is probably an oil. **2.** 1 oz × 28.4 g/oz × 9 kcal/g × 1 Cal/kcal = 256 Cal. (The actual value is slightly smaller because margarines are not pure oil or fat.) **3.** Both are esters. The triacylglycerol is a triester of glycerol and three fatty acids. Many waxes are esters of a fatty acid and a long carbon chain alcohol. **4.** **a.** fatty acid **b.** triacylglycerol **c.** wax

24.4 Saponifiable Lipids of Membranes

Some saponifiable lipids are structural components of cell membranes. Some are also found in the circulatory system where they play a role in normal transport of other lipids. These structural lipids are often referred to as *polar lipids* because one end of the molecule contains one or more polar functional groups, while the other end of the molecule is nonpolar. Molecules that possess a highly polar end and a nonpolar end are **amphipathic molecules**. The carboxylate form of a carboxylic acid is amphipathic.

Amphipathic molecules possess polar and nonpolar regions.

$$\underbrace{CH_3CH_2CH_2CH_2CH_2CH_2CH_2CH_2CH_2CH_2CH_2CH_2CH_2CH_2CH_2}_{\text{Nonpolar}}\underbrace{COO^-}_{\text{Polar}}$$

Some of these lipids are saponifiable, others are not. The saponifiable lipids of membranes and the circulatory system are the topic of this section.

The most abundant class of lipids in the membranes are the lipids that contain phosphorus, the *phospholipids*. Many of these phospholipids are **phosphoacylglycerols**. This important class of compounds is related structurally to the triacylglycerols. Instead of three fatty acids bonded to glycerol through ester bonds, the phosphoacylglycerols have two fatty acids esterified to the first and second hydroxyl groups of glycerol and a phosphoric acid esterified to the third hydroxyl group of the glycerol molecule (Figure 24.4a). Molecules with this structure are called *phosphatidic acids*, but normally an additional molecule is bonded to the phosphate group.

If the compound choline is bonded to the phosphate, then the compound is a phosphatidyl choline, the most abundant phospholipid. It is also known

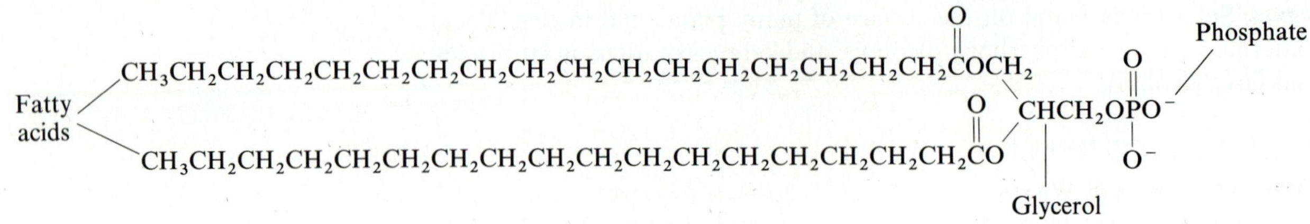
Phosphate
Fatty acids
Glycerol

(a) Phosphatidic acid

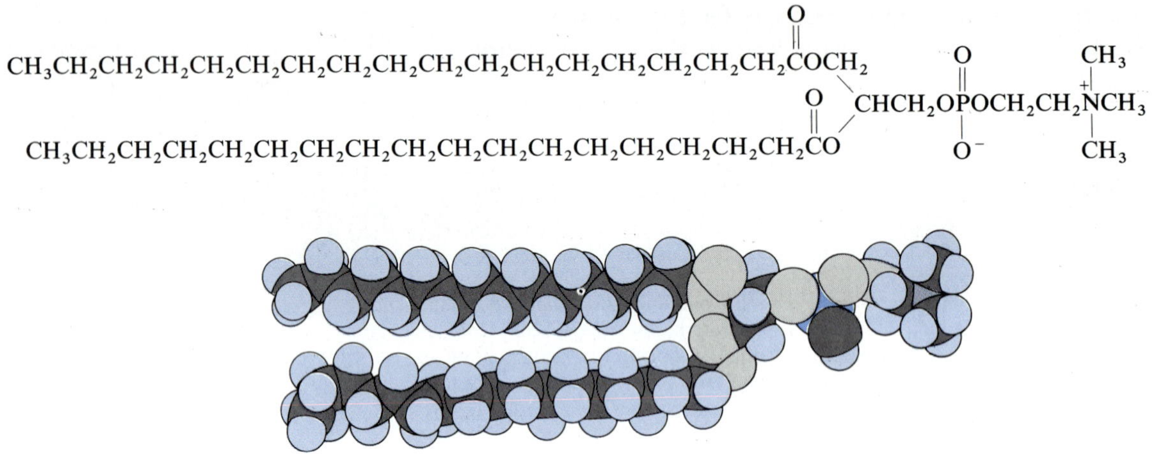

(b) Phosphatidyl choline (lecithin)

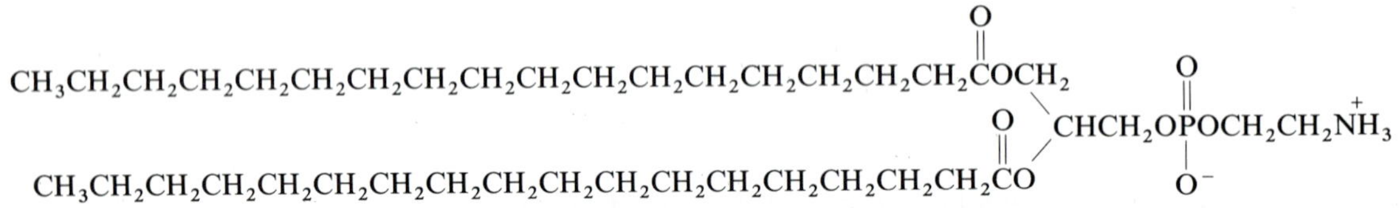

(c) Phosphatidyl ethanolamine (cephalin)

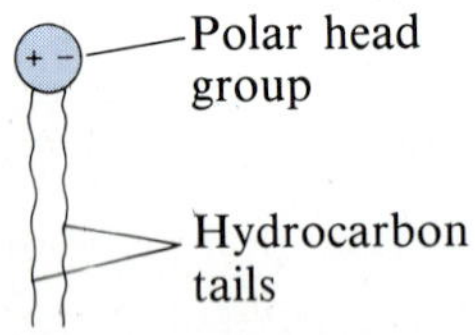
Polar head group
Hydrocarbon tails

(d)

FIGURE 24.4
Phosphoacylglycerols are polar lipids containing phosphatidic acids. (a) Phosphatidic acid contains glycerol, two fatty acids, and a phosphate group. These lipids are the basic unit of phosphoacylglycerols. (b) Phosphatidyl choline (lecithin) is the most common phosphoacylglycerol. It contains a phosphatidic acid and a molecule of choline. Its space-filling model is also shown. (c) Phosphatidyl ethanolamine (cephalin) is another common phosphoacylglycerol. It contains a phosphatidic acid and ethanolamine. (d) Most polar lipids can be represented by a polar head group with two hydrocarbon tails.

as lecithin (Figure 24.4b). If ethanolamine is bonded to the phosphate group, the phospholipid is called phosphatidyl ethanolamine (Figure 24.4c). These compounds are also known as cephalins. Other derivatives of phosphatidic acids are also found in cells including phosphatidyl serine and phosphatidyl inositol. All of these compounds are polar lipids because they contain a polar

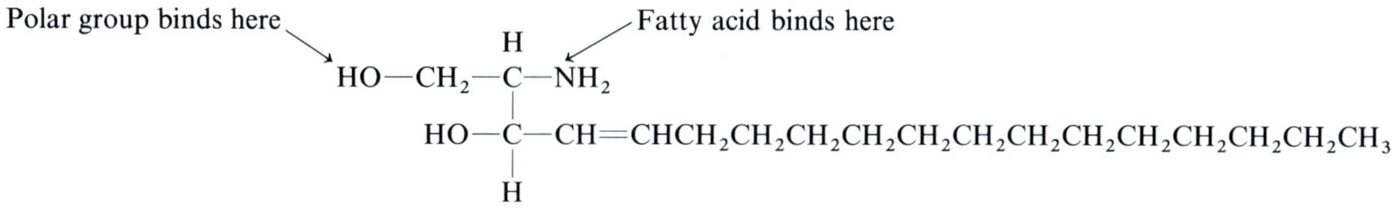
Polar group binds here
Fatty acid binds here

(a) Sphingosine

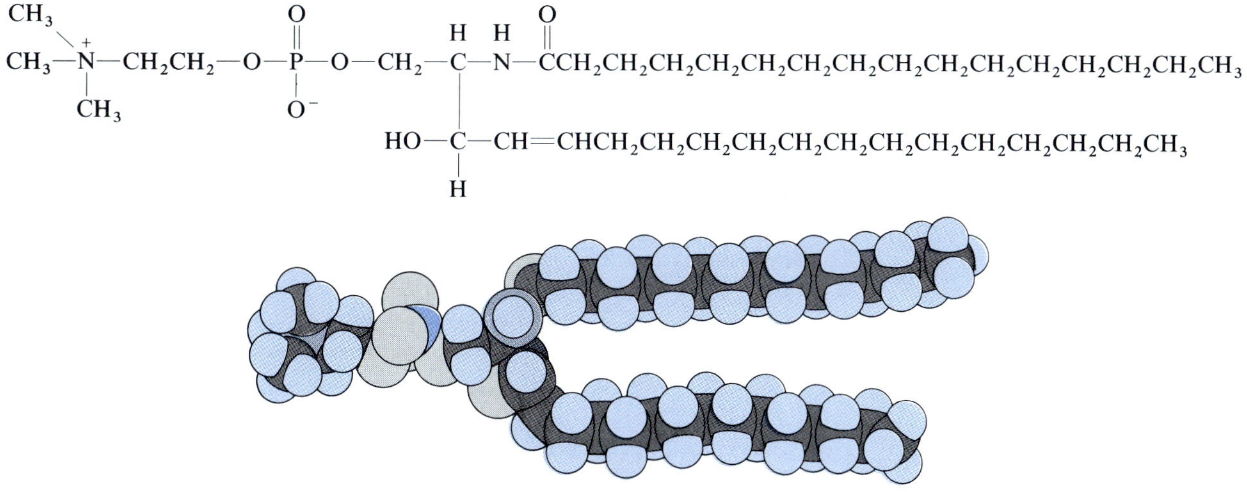

(b) Sphingomyelin

FIGURE 24.5
Sphingosine and sphingomyelin. (a) Sphingosine is found in all sphingolipids. In sphingolipids, a fatty acid is bonded to the amino group through an amide bond, and a polar group is attached to the primary alcohol. (b) Sphingomyelin is a typical sphingolipid. The polar group for this compound is a phosphate group and a molecule of choline. Compare the space-filling model of sphingomyelin with the space-filling model of phosphatidyl choline in Figure 24.4b.

head group, the phosphate and its attached compound, and a nonpolar region commonly referred to as the "tail," which is made up of the two fatty acids.

The **sphingolipids** are another group of saponifiable lipids found in membranes. These lipids contain the base *sphingosine*, a fatty acid, and one or more other molecules (Figure 24.5a). The fatty acid is bonded to sphingosine through an amide bond to the amino group. The other molecules are polar molecules such as sugars or phosphate and choline. These polar groups are bonded to the sphingosine through an ester bond to a hydroxyl group. The sphingolipids are not really similar to phosphoacylglycerols in composition, but they do resemble each other in structure. Sphingolipids are amphipathic because each has a polar end and a nonpolar end consisting of two long hydrocarbon chains. This similarity in structure is why they appear to have similar roles in membranes.

There are two types of sphingolipids, *sphingomyelin* (Figure 24.5b), which contains phosphate and choline, and *cerebroside*, which contains sugars. In your studies you may encounter the term *glycolipid*. This term is used to describe a lipid that has a carbohydrate as a component. Phosphatidyl inositol and the cerebrosides are examples of glycolipids.

NEW TERMS

amphipathic molecules Molecules with both polar and nonpolar regions within it.

phosphoacylglycerols Amphipathic molecules similar to triacylglycerols in structure, but with a substituted phosphoric acid in place of the fatty acid on the third hydroxyl group of glycerol.

sphingolipids A group of polar membrane lipids characterized by the presence of sphingosine.

TESTING YOURSELF

Saponifiable Lipids of Membranes

1. Why would a phospholipid be more likely to be slightly water soluble than a triacylglycerol?
2. Hydrolysis of glycolipids would yield what classes of compounds?
3. What polarity characteristics must a molecule possess if it is amphipathic?

Answers **1.** The polar head group of the phospholipid is water soluble and will increase the overall solubility compared to triacylglycerols. **2.** Carbohydrates and lipids **3.** The molecule must have both polar and nonpolar ends.

24.5 Nonsaponifiable Lipids

The nonsaponifiable lipids do not possess fatty acids bonded to some other molecules. Base hydrolysis, (saponification), has no apparent effect on their structure. These lipids have rather different structures. This section discusses the steroids, prostaglandins, and leukotrienes. Terpenes have already been discussed briefly in Chapter 16.

Steroids

All steroids possess a four membered ring system called the steroid nucleus.

The **steroids** are those compounds that contain a fused ring system commonly known as the *steroid nucleus* (Figure 24.6). Although the steroids are related by structure and synthesis, they possess a great variety of functions. Regardless of the function, however, you can always recognize a steroid by its steroid nucleus.

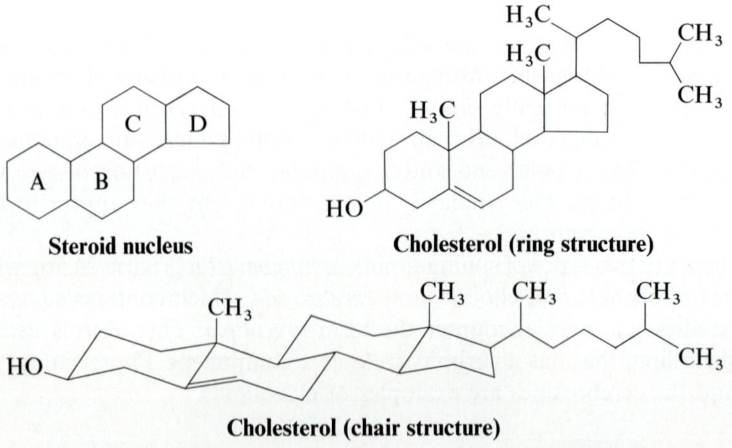
A
B
C
D
Steroid nucleus
H_3C
H_3C
CH_3
CH_3
H_3C
HO
Cholesterol (ring structure)
CH_3
CH_3
CH_3
CH_3
CH_3
HO
Cholesterol (chair structure)

FIGURE 24.6
All steroids possess the steroid nucleus. Cholesterol, the most abundant steroid, is shown here as both a flat ring structure and in the more realistic chair form.

The most abundant of the steroids in animals is the widely distributed compound **cholesterol** (Figure 24.6). Cholesterol is a component of membranes in animals, and like other membrane lipids, it is an amphipathic compound. The steroid nucleus is nonpolar, but the hydroxyl group can hydrogen bond with water or other polar molecules. Many of the other steroids in the body are synthesized from cholesterol.

Most steroids are synthesized from cholesterol, which is the most abundant steroid in animals.

Cholesterol is obtained from the diet and is synthesized by the liver. The circulatory system transports cholesterol to the rest of the body. In some people, deposits containing cholesterol build up on the interior surface of arteries. This condition is known as *atherosclerosis*, the accumulation of fatty substances and growth of abnormal muscle cells in the artery wall (see Plate 11 in color insert). These deposits reduce the diameter of the arteries supplying the heart and other vital organs. This reduced diameter means these vessels are more easily blocked, perhaps by a small clot of blood that would normally pass through. If no blood reaches portions of the heart, that portion of the heart is deprived of oxygen and nutrients and may die. Bypass surgery can be done to circumvent severely constricted arteries, but this is only useful before the artery is totally clogged and while the heart tissue is still alive. Plate 11 shows a normal artery and one clogged by advanced stages of atherosclerosis. Although the specific role of cholesterol in this disease process is not understood, it can be shown that a relationship exists between blood cholesterol levels and atherosclerosis. In general, the more elevated blood cholesterol levels, the higher the probability for heart attack.

Another group of steroids found in the body are the *steroid hormones.* Although never abundant, these compounds have a great influence over body function and activity. Some are involved in salt and water regulation, others in the body's response to stress, still others in sexual function and the development and maintenance of secondary sexual characteristics.

The steroid hormones, although present only in trace quantities, exert powerful effects on the body.

Sex hormones are steroids that determine the secondary sexual characteristics of males and females. Although both sexes carry some of each of the hormone types, a careful balance must be maintained to provide normal growth and development. The male hormones, the androgens, are quite similar in structure to the female hormone progesterone, while the other common female hormones, the estrogens, possess an aromatic ring in the fused ring system (Figure 24.7).

In the human female the concentrations of estrogen and progesterone change during the menstrual cycle. Prior to ovulation, there is a high estrogen level, but after ovulation there is an increase in progesterone. This increase in progesterone helps to prepare the uterine lining for implantation of

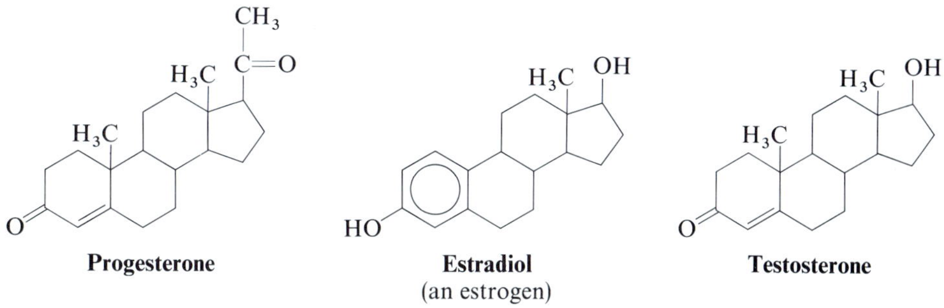

CH3
C=O
H3C
H3C
O
Progesterone
H3C
OH
HO
Estradiol
(an estrogen)
H3C
OH
H3C
O
Testosterone

FIGURE 24.7
The common human sex hormones.

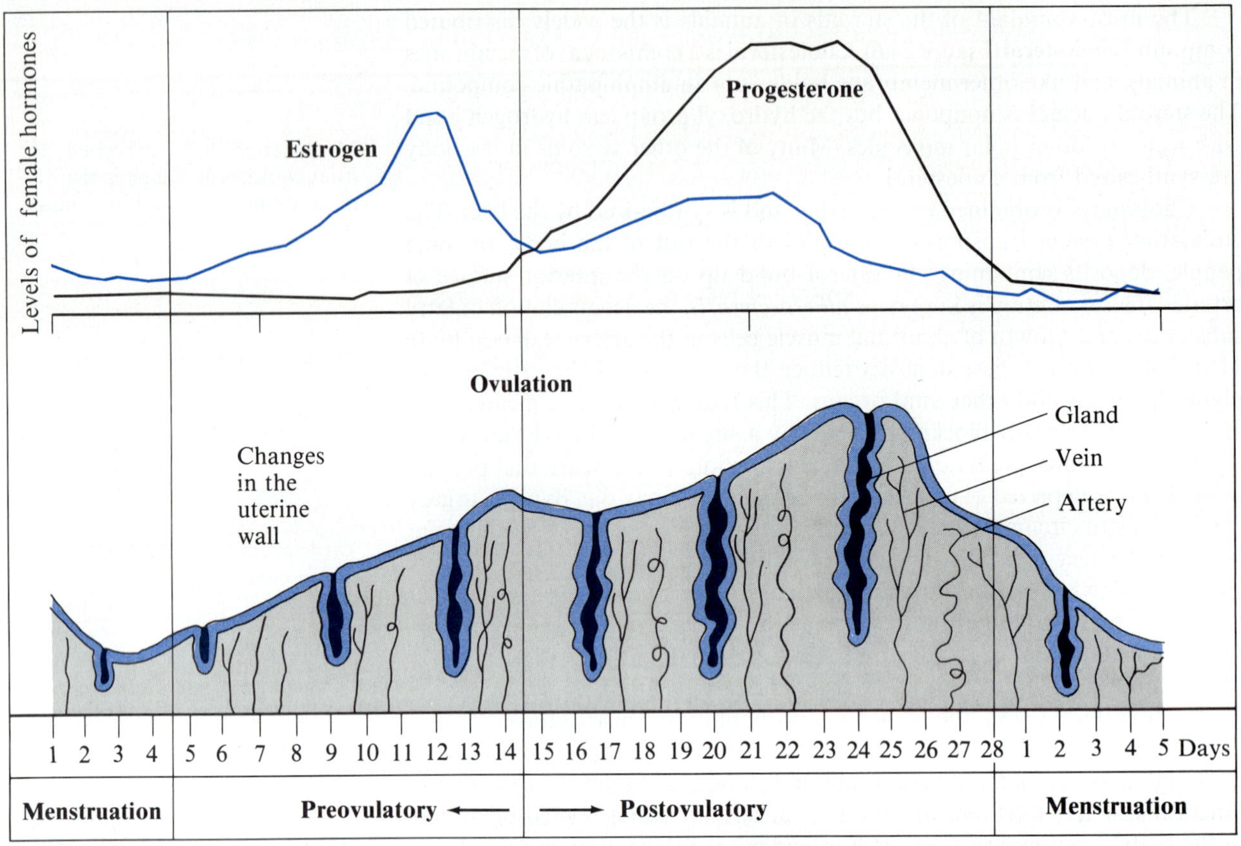
Levels of female hormones
Estrogen
Progesterone
Ovulation
Changes in the uterine wall
Gland
Vein
Artery
1 2 3 4 5 6 7 8 9 10 11 12 13 14 15 16 17 18 19 20 21 22 23 24 25 26 27 28 1 2 3 4 5 Days
Menstruation
Preovulatory
Postovulatory
Menstruation

FIGURE 24.8
The menstrual cycle. The concentrations of estrogen and progesterone vary throughout the cycle (top). These hormones cause significant physiological changes in their target tissues. Changes that occur in the uterine wall during the menstrual cycle are shown along the bottom.

a fertilized egg. If this does not occur, there is a "sloughing off" of the blood cells that had been stored for the implantation, resulting in menstruation. A generalized sequence of this monthly cycle is outlined in Figure 24.8.

An interesting application of structural modification of biomolecules can be found in the steroid analogs used as oral contraceptives. A specific example is *norethindrone*, which closely resembles progesterone (Figure 24.9). By the slight modification in structure shown in the figure, a natural product that cannot be taken orally because of rapid biological degradation is replaced by a synthetic material that has biological activity and is stable enough to be taken orally. This synthetic steroid mimics the action of progesterone and prevents further ovulation. Synthetic estrogens are also used in some contraceptive formulations.

Another important group of steroid hormones are produced by the cortex of the adrenal gland and are known as the *adrenocorticoid* hormones.

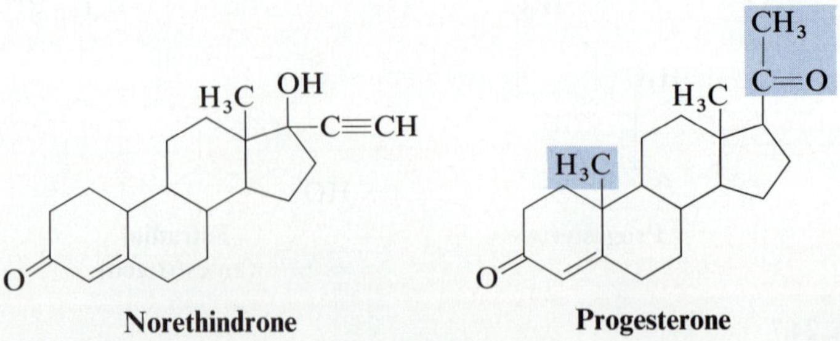
H3C
OH
C≡CH
O
Norethindrone
CH3
C=O
H3C
H3C
O
Progesterone

FIGURE 24.9
A comparison between norethindrone and progesterone with the differences highlighted for emphasis.

ANABOLIC STEROIDS IN ATHLETICS

Humans thrive on competition. Athletics provides an acceptable outlet for the competitive drive. But the internal need to excel or external pressure to succeed puts athletes in an ethical dilemma. Should they win at all cost or play the game by the rules? All serious competitors train hard. Many eat balanced diets, conscientiously obtain enough rest, and abstain from tobacco and other substances that reduce performance. Some take vitamin and mineral supplements to attempt to enhance their performance. But which actions to enhance performance are acceptable and which are not?

Some athletes have overstepped the boundary between acceptable and unacceptable by using performance-enhancing drugs. Within this group of drugs are the *anabolic steroids*. These compounds are natural steroid hormones or, more commonly, synthetic analogs of these steroid hormones. In biochemistry, anabolic refers to reactions that build or synthesize. Anabolic hormones stimulate reactions related to synthesis and growth.

The male sex hormone testosterone is an anabolic hormone. It stimulates muscle development. Thus, at puberty, males develop increased muscle mass, which is a secondary sex characteristic of males. Some athletes use anabolic steroids to stimulate muscle growth. This provides additional strength that is beneficial in many sports, including weightlifting, sprinting, and football. But is this action to enhance performance acceptable? Most people think that it is not. It provides an unfair edge. Furthermore, there are a number of known or suspected side effects that are potentially quite harmful: liver tumors, aggressive behavior, testicular atrophy in males, and the development of masculine traits in females.

These steroids play a central role in the regulation of some aspects of metabolism. Structurally, they are classified according to their high degree of oxygenation. Several alcohol and carbonyl groups within a molecule are not uncommon. The most common member of this class is *cortisol*, a powerful antiinflammatory agent (Figure 24.10). Cortisol and some of its active derivatives, such as cortisone and prednisolone, are used to treat a variety of skin inflammations, arthritis, and asthma.

Although cortisol and its derivatives have highly beneficial properties, prolonged therapeutic use can result in problems, including excessive breakdown of protein (reflected in muscle deterioration, weakness, and excessive excretion of nitrogen-containing waste products), sodium retention, calcium excretion, and increased susceptibility to infection.

Bile salts are another group of steroids that play a role in digestion. These compounds are synthesized from cholesterol in the liver and then stored in the gallbladder. The bile salts are transported to the small intestine during the process of digestion, where they aid in the processes of digestion

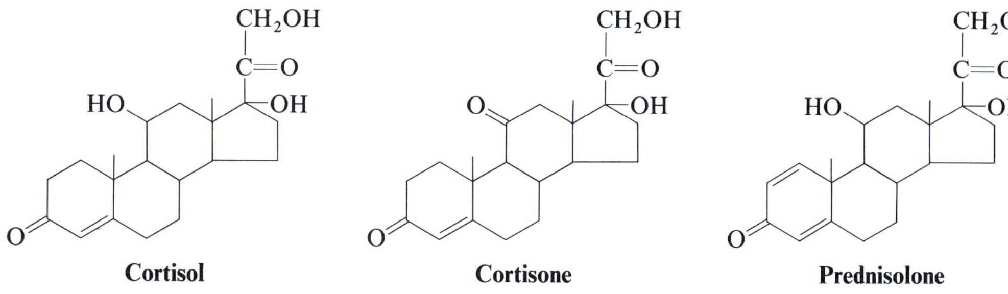

CH$_2$OH
C=O
HO
OH
O
Cortisol
CH$_2$OH
C=O
O
OH
O
Cortisone
CH$_2$OH
C=O
HO
OH
O
Prednisolone

FIGURE 24.10
Cortisol, cortisone, and the synthetic analog prednisolone.

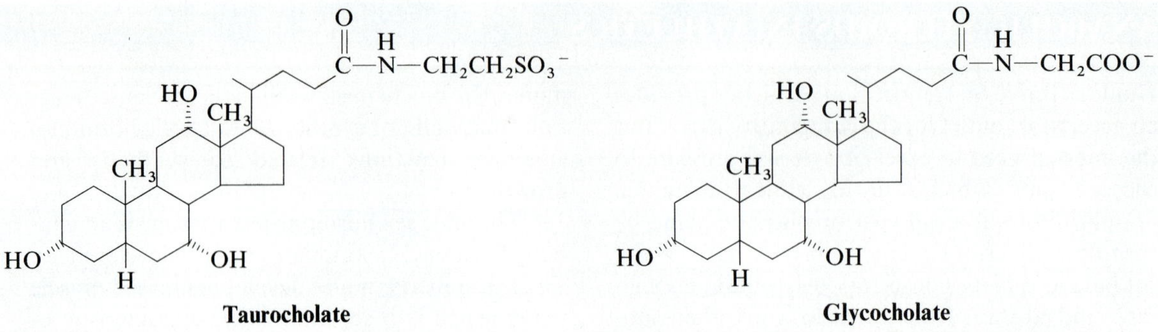
O
H
C—N—CH₂CH₂SO₃⁻
HO
CH3
CH3
HO
OH
H
Taurocholate
O
H
C—N—CH₂COO⁻
HO
CH3
CH3
HO
OH
H
Glycocholate

FIGURE 24.11
The common bile salts of mammals.

Bile salts, which are synthesized from cholesterol, break down fat globules in the small intestine to aid in digestion and absorption.

and absorption of dietary lipids. The specifics of lipid digestion are discussed in Chapter 28, but for now note that these compounds have polar groups projecting from one face of the steroid ring system, while the other face is nonpolar. These compounds are amphipathic, and their role in digestion requires this property. Two typical bile salts, *taurocholate* and *glycocholate*, are shown in Figure 24.11.

Prostaglandins and Leukotrienes

Prostaglandins and leukotrienes are two more classes of nonsaponifiable lipids that play a role in regulation of body function. These compounds are never abundant in the body, but they are very active in trace amounts. These lipids are synthesized from three polyunsaturated fatty acids—linoleic acid, linolenic acid, and arachidonic acid. The **prostaglandins** were first isolated from secretions of the male reproductive tract and were thought to originate in the prostate gland. It is now apparent that they are widespread in both sexes. They are present in extremely small amounts in many tissues of the body and have pronounced regulatory properties. Prostaglandins have been shown to exhibit powerful biological activity in many body functions, including blood pressure regulation, operation and functioning of the lungs, reproductive physiology, and smooth muscle action, including uterine contraction. They also appear to participate in the physiology of fever and pain. Aspirin is the world's most common analgesic and antipyretic. Although aspirin has been used for over 100 years, the details of how it controls pain and reduces inflammation were not understood until recently. It now appears that aspirin interferes with the body's ability to synthesize prostaglandins. This implies that the prostaglandins somehow function in the perception of pain. Laboratory syntheses of many prostaglandins now provide enough material for wide scale biological testing and evaluation. The structures of two common prostaglandins are shown in Figures 24.12a and b. Arachidonic acid, the unsaturated fatty acid from which prostaglandins are synthesized, is shown for comparison.

Prostaglandins and leukotrienes are regulatory molecules.

One of the most recently discovered groups of lipid compounds are the **leukotrienes**. These compounds are also derived from arachidonic acid and appear to play a role in allergic and inflammation responses. Some of these compounds were isolated from leukocytes, and the name reflects this origin as well as the unsaturated nature of the compounds. The structure of a common member of this group is shown in Figure 24.12c.

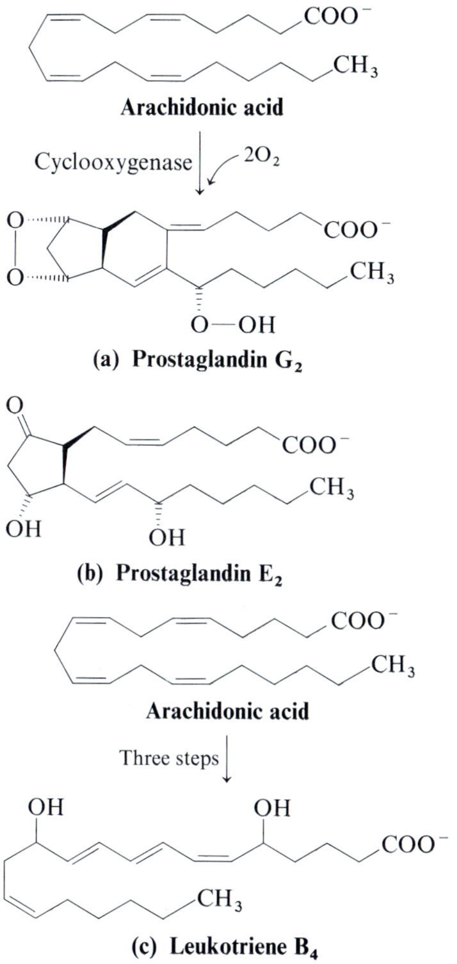
COO⁻
CH₃
Arachidonic acid
Cyclooxygenase
2O₂
O
O
COO⁻
CH₃
O—OH
(a) Prostaglandin G₂
O
COO⁻
CH₃
OH
OH
(b) Prostaglandin E₂
COO⁻
CH₃
Arachidonic acid
Three steps
OH
OH
COO⁻
CH₃
(c) Leukotriene B₄

FIGURE 24.12
Prostaglandins and leukotrienes are made from arachidonic acid and other polyunsaturated fatty acids. (a) Prostaglandin G_2 is made from arachidonic acid in a reaction catalyzed by cyclooxygenase (which is a part of the enzyme prostaglandin synthase). Aspirin inhibits cyclooxygenase activity. Because prostaglandins enhance inflammation, aspirin is an antiinflammatory agent. (b) Prostaglandin E_2 stimulates contraction of uterine muscles. (c) Leukotriene B_4 is made from arachidonic acid.

NEW TERMS

steroids Compounds possessing the steroid nucleus.

cholesterol The most abundant steroid in animals. An important membrane lipid with significant health implications.

prostaglandins Regulatory lipids derived from unsaturated fatty acids. Prostaglandins often have a cyclic portion based on cyclopentane.

leukotrienes Lipids derived from arachidonic acid that are implicated in allergic responses.

TESTING YOURSELF

Nonsaponifiable Lipids

1. Steroids, which are a large and diverse group, share one structural feature. Name and draw this feature.

2. Estrogens and progesterone have what general physiological role?
3. The body uses arachidonic acid to synthesize what two important classes of regulatory molecules?

Answers **1.** the steroid nucleus

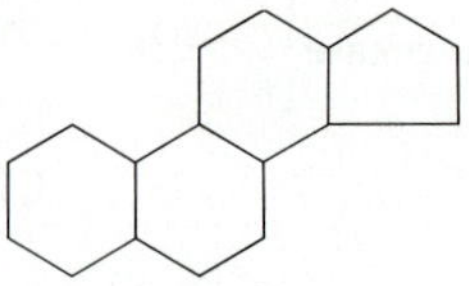

2. female sex hormones that control reproductive activity and influence secondary sex characteristics **3.** prostaglandins and leukotrienes

24.6 Liposomes and Membranes

Micelles, Bilayers, and Liposomes

Recall from Chapter 20 that *micelles* are aggregates of amphipathic molecules that assume a spherical shape in water. Soaps and many other amphipathic molecules will form micelles under certain conditions. In micelles, the long hydrophobic tail of the molecules are clustered together in the core of the micelle, with water excluded from the core (Figure 24.13). Micelles in water provide two distinctly different environments: (1) a nonpolar or hydrophobic environment in the core of the micelle, and (2) a polar, aqueous environment on the surface of the micelle.

The structure of a micelle can be explained in several ways, but the most accurate picture includes water bonding that effectively squeezes out the nonpolar part of the amphipathic molecule. A water molecule normally hydrogen bonds to other water molecules. When a nonpolar entity is present between two water molecules, the strong hydrogen bond between them is broken and replaced by weaker London forces between the water molecules and the nonpolar species (see Figure 15.3). In terms of energy, this is unfavorable because it is less stable. The tendency is thus for the two water molecules to bond together, effectively squeezing the nonpolar molecule out. If several nonpolar entities are in an aqueous solution, they end up clustered together. This is not because they have a particularly strong attraction for each other, but rather because water molecules prefer to bond to other water molecules. Nonpolar molecules or nonpolar parts of amphipathic molecules are squeezed into nonaqueous cavities in a water solution.

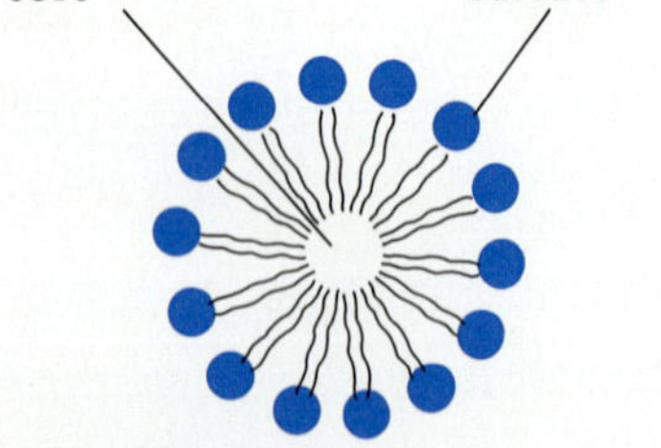
Nonpolar (hydrophobic) core
Polar (aqueous) surface

FIGURE 24.13
The structure of a micelle.

The clustering of nonpolar species in an aqueous solution is referred to as *hydrophobic interactions.* These interactions play a very important role in biological membranes and protein structure. Remember, the term hydrophobic interactions implies that water-hating, nonpolar molecules prefer to be together, but the reality is that they have been squeezed together by polar water molecules.

Another structure that some amphipathic molecules can form when placed into an aqueous solution is a **bilayer** (Figure 24.14a). The amphipathic molecules in a bilayer are arranged with the nonpolar tails projecting into the core of the bilayer and the polar head groups positioned on the exterior surface of the bilayer. Soap bubbles are bilayers of soap molecules with a

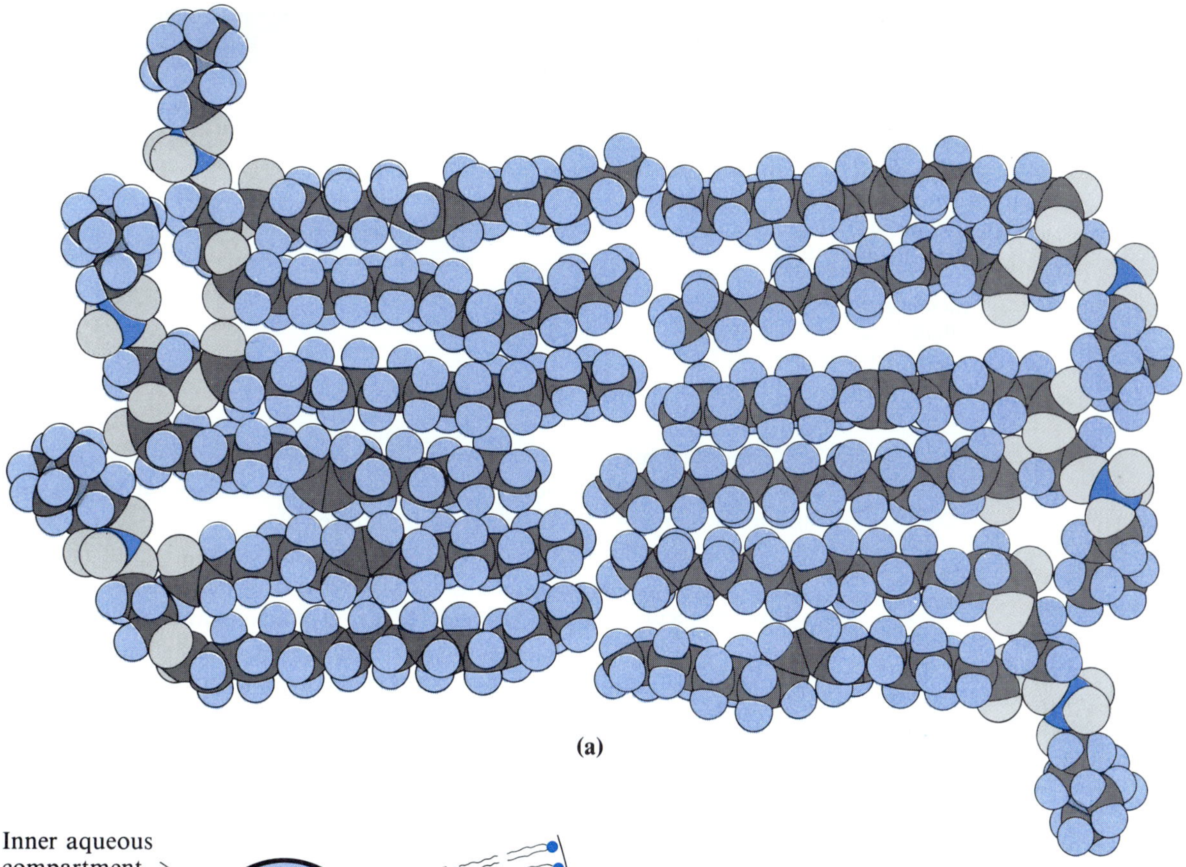

(a)

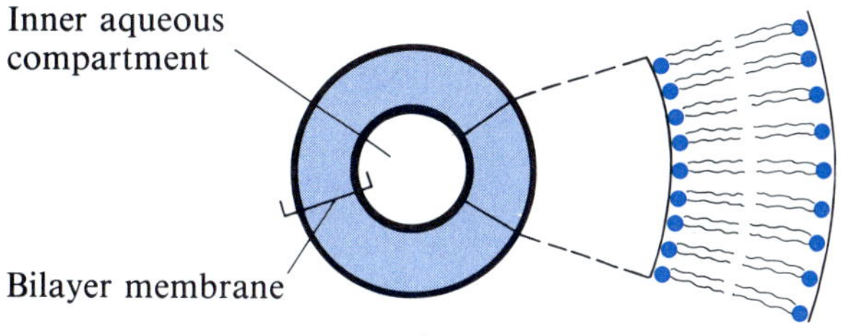
Inner aqueous compartment
Bilayer membrane

(b)

FIGURE 24.14
The structure of bilayers and liposomes. (a) Bilayers consist of polar lipids that have their nonpolar regions aggregated together and their polar head groups facing the aqueous solution. (b) Liposomes are spherical structures that are, in essence, bilayers folded around a core of polar solvent. Researchers are developing drug-filled liposomes as a more efficient drug delivery system.

film of water on either side. Bilayers can close back upon themselves to form a continuous bilayer surrounding a core of water. These structures are called **liposomes** (Figure 24.14b).

Liposomes can be prepared as relatively stable structures. They show some promise in the pharmaceutical industry as "casings" for drugs. The polar surface of the liposome keeps them suspended in blood and other body fluids, while the core carries a solution or suspension of the drug. In the body, the liposomes are slowly broken down, resulting in a gradual, time-released delivery of the drug in appropriate concentrations.

Micelles, bilayers and liposomes are all aggregates of amphipathic molecules that are held together by hydrophobic interactions within the core, and polar interactions at the surface.

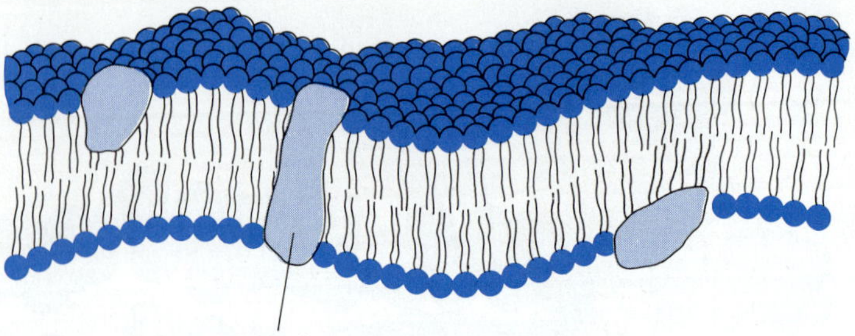
Protein

FIGURE 24.15
The fluid mosaic model of biological membranes. Note the proteins that are imbedded in the lipid matrix. The lipid matrix is made up of polar lipids, primarily phosphoacylglycerols, with some cholesterol, glycolipids, and others. Because covalent bonds do not exist between the components of the membrane, they are free to diffuse throughout the membrane. The membrane is thus "fluid."

Membranes

Micelles, bilayers, and liposomes have served another important role in biology—as models for developing an understanding of biological membranes (Figure 24.15). **Membranes** are bilayer-like structures that separate a cell from the external environment and divide the cell contents into numerous compartments. The most abundant component of membranes are amphipathic molecules, such as phospholipids, glycolipids, and cholesterol. Like micelles and bilayers, membranes have a nonpolar core that is hydrophobic and surfaces that are polar. But polar parts of these molecules include positive and negative charges that attract each other. This attraction significantly stabilizes membranes and liposomes derived from them.

Membranes possess amphipathic lipids and proteins.

Another major component of membranes are proteins. These macromolecules are discussed in detail in Chapter 25, but for now picture them as large molecules that may have both polar or nonpolar regions on their surface. Membrane proteins are imbedded in the lipids of membranes. The nonpolar portions of the protein are in contact with the hydrophobic interior of the membrane, while the polar portions of the protein face out toward the water. Again, hydrophobic interactions hold the nonpolar portion of the protein in the core, while various polar interactions keep the polar part facing the water.

The presently accepted model for cellular membranes is called the *fluid mosaic model* because the proteins form a mosaic in the membrane (Figure 24.15). Both the protein molecules and the lipid molecules are free to diffuse around the surface of the membrane because only weak, noncovalent bonds hold them in the membrane. A membrane's structure and properties provoke the image of a two-dimensional liquid.

Membranes are stable biological entities. The lipid components and some of the proteins are not easily removed from this two-dimensional film. Ionic attractions at the membrane surface must be broken, and a nonpolar tail or surface must be forced between numerous polar molecules and ions.

Membranes separate the cell from the external aqueous environment and divide the cell interior into compartments. This separation is maintained principally by the hydrophobic core of the membrane because polar molecules, including water, cannot easily pass through the nonpolar core. Nonpolar molecules such as molecular oxygen or carbon dioxide do readily cross membranes, and the transport of these molecules appears to be by simple diffusion across the membrane.

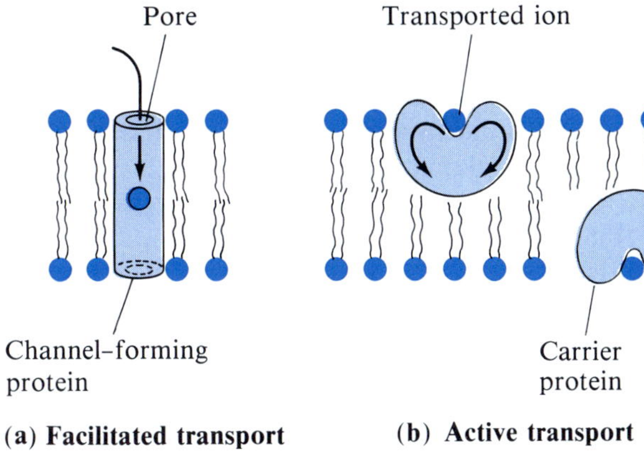
Pore
Transported ion
Channel-forming protein
Carrier protein
(a) Facilitated transport
(b) Active transport

FIGURE 24.16
Models of facilitated transport and active transport in membranes. Membrane proteins function as the channels or carriers. (a) Facilitated transport allows molecules or ions to move from a region of higher concentration on one side of the membrane to the other side where the concentration is lower. It is as though the particles have a channel or pore in the membrane that they can pass through. (b) Active transport requires energy. A carrier accepts a molecule or ion on one side of the membrane and drops it off at the other side. In this model, energy is required to return the carrier protein to its original location.

How do polar molecules and ions cross a membrane? Some of the proteins of the membrane play specific roles in transporting polar molecules. One protein may be involved in the transport of glucose, while others are involved in the transport of sodium ion. You can imagine these transport proteins as acting like channels or pumps. Two common methods for passing materials through a membrane are facilitated diffusion and active transport. **Facilitated diffusion** is simply diffusion that is speeded up by proteins that act as channels (Figure 24.16). Movement of the molecules or ions of the substance is from a region of higher concentration to one of lower concentration, but it can occur quickly because the proteins allow passage across the nonpolar core of the membrane. Facilitated diffusion is independent of externally applied energy.

While lipids prevent movement of polar molecules and ions across the membrane, some membrane proteins act as specific transport systems for these polar species.

Active transport requires external energy for its operation. This energy is provided by ATP or electrochemical gradients (see Chapter 28). Active transport moves molecules against a concentration gradient (from low concentration to high concentration). Picture a protein acting like a pump (Figure 24.16). As in any pump, some kind of energy must be applied to the pump before it can function, but when energy is used, matter is pumped from one region to another even against its gradient.

NEW TERMS

bilayer A sheetlike structure made up of amphipathic molecules. The nonpolar tails form an interior core, and the polar head groups make the surfaces polar.

liposome A structure consisting of an aqueous core separated from the external environment by a bilayer of amphipathic molecules.

membranes The sheetlike structures in cells that separate the cell interior from the external environment and divide the cell interior into compartments.

facilitated diffusion The transport of substances across the membrane of a cell in which proteins speed up the movement. The substances can only flow down their gradient.

active transport The energy-requiring transport of substances across the membrane of a cell.

TESTING YOURSELF

Membranes

1. What are the two principal classes of compounds found in membranes?
2. Consider a protein in a membrane. Describe the interaction (bonding) of the molecule with the molecules surrounding it.
3. What name is given to the energy-requiring process that passes material through a membrane?

Answers **1.** polar lipids and proteins **2.** The protein has hydrophobic interactions with the nonpolar tails of the polar lipids in the interior of the membrane and polar interactions with the polar head groups of the polar lipids and the components of the aqueous environment surrounding the membrane. **3.** active transport

24.7 Lipids and Nutrition

Fats and oils provide a significant amount of our energy needs. In fact, in the developed world, they may be providing too much.

Fats and oils are a primary source of energy in the human diet. The present U.S. diet contains about 40% fats and oils (by caloric intake), while the diets of people in developing countries contain about 15 to 30%. Nutritionists believe 15 to 30% is more appropriate and healthful.

Types and Roles of Dietary Lipids

Linoleic acid and perhaps linolenic acid are essential.

Triacylglycerols are broken down in the body during metabolism and most of the glycerol and fatty acids are used for energy production. However, some of the fatty acids are used as building blocks for other molecules. One of these fatty acids is the polyunsaturated fatty acid, linoleic acid. This is an *essential fatty acid* because the body cannot synthesize it, yet is is needed to synthesize other necessary molecules in the body. There is some evidence that a second polyunsaturated fatty acid, linolenic acid, may also be essential. The human body cannot introduce carbon–carbon double bonds into the proper places in monounsaturated fatty acids to make linoleic or linolenic acids. Many oils are good sources of the polyunsaturated fatty acids. Some marine fish are also good sources of these lipids.

A second lipid found in many foods is cholesterol (see Section 24.5). This compound is more important as a structural element than as a source of energy, and it normally is much less abundant in foods than are fats and oils. Cholesterol in food servings is normally expressed in milligrams rather than grams. It might even be ignored in nutrition except for one thing: there appears to be a correlation between the level of cholesterol in blood and cardiovascular disease. Since this correlation has become apparent, possible links between blood cholesterol and dietary cholesterol have received considerable study. Cholesterol is not an essential nutrient because the body can synthesize it. If the diet lacks or has little cholesterol, the body simply makes what it needs.

Correlations of Lipids and Health

There are several correlations between dietary lipids and good health. The most obvious one is the relationship between good health and obesity.

People who are significantly overweight are at greater risk for several health problems including cardiovascular disease. To reduce weight, caloric intake must be less than the caloric needs of the body. The body must then use stored fat to make up the energy deficiency. Because fats and oils are the most energy rich nutrients, reducing their amounts in the diet is an excellent way to lose weight and ultimately bring energy use and energy intake into balance. However, it is neither easy nor desirable to eliminate fat from the diet. A small amount of the polyunsaturated fatty acids should be eaten regularly. In addition, fat contributes significantly to the taste and texture of foods. A nearly fat free diet is difficult for most people to maintain.

There are several correlations that exist between dietary lipid, blood cholesterol and cardiovascular disease.

A second correlation between dietary lipids and health is related to the degree of saturation in dietary fats and oils. Fats are more saturated than oils. There appears to be a correlation between the amounts of saturated fats and polyunsaturated fats in the diet and cardiovascular disease. In general, if the ratio of polyunsaturated fats to saturated fats is greater than one, then the risk of heart disease is reduced. If the ratio is below one, the risk is increased. Since foods from animal sources are higher in saturated fats, and plants are higher in highly unsaturated fats, a reduction in foods from animal sources and an increase in consumption of fruits and vegetables will normally increase the ratio of these fats in the diet.

There is also a correlation between blood cholesterol levels and cardiovascular disease. Humans obtain cholesterol in the diet and through synthesis. Normally the amount of cholesterol eaten and synthesized is balanced with the amount that is lost from the body or used to synthesize other compounds. Cholesterol and its derivatives play essential roles in the body, thus it is neither necessary nor desirable to rid our systems of cholesterol. Although it can be shown that a correlation exists between blood cholesterol concentrations and heart disease, no similar correlation has as yet been shown between dietary cholesterol and the development of heart disease in all humans. People with known heart conditions should reduce their blood cholesterol levels through medication and diet, but for the general public it has not been shown that reducing dietary cholesterol will by itself reduce the risk of heart attacks. Many factors appear to be correlated to cardiovascular disease—smoking, exercise, obesity, the amount of dietary fats and oils, the degree of saturation in dietary fats and oils, and blood cholesterol levels are all known factors that affect the cardiovascular system. Present evidence does not yet show that a simple reduction in dietary cholesterol, without changes in the other factors that influence cardiovascular disease, is likely to produce a significantly healthier cardiovascular system. Table 24.4 shows the cholesterol content of some common foods.

Fat Soluble Vitamins

The **fat soluble vitamins** are lipids because of their solubility properties. The fat soluble vitamins are vitamins A, D, E, and K. They are stored in the fatty parts of the body, and a diet lacking in any of these vitamins will not normally cause a deficiency until this pool has been depleted.

Vitamins A, D, E, and K are called the fat soluble vitamins because they are nonpolar, hydrophobic compounds that dissolve into body fat.

Vitamin A was the first human vitamin to be discovered. In two of its forms, retinol and retinal, it serves a role in detection of light in the eye (see Figure 16.17). A deficiency may manifest itself in a form of night blindness.

TABLE 24.4
Amount of Cholesterol Found in Some Common Foods

Food	Serving	Amount of Cholesterol (mg)
Egg	1	250
Tuna (canned in oil, drained)	184 g	116
Beef, pork, or turkey (dark meat)	84 g	67
Chicken or turkey (light meat)	84 g	55
Halibut	84 g	55
Salmon	84 g	40
Butter	1 tbsp	35
Hot dog	1	34
Whole milk	1 cup	34
Cheddar cheese	28 g	28
Ice cream	0.5 cup	27
Skim milk	1 cup	5

It is also involved in maintenance of normal skin and other surface tissues, growth, and perhaps membrane function.

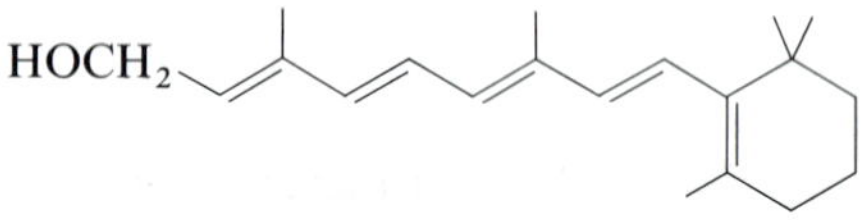
HOCH2

Vitamin A

Toxicity from vitamin A overdose can occur because excess vitamin A is stored in the body and is not excreted. Most overdoses involve dietary supplements or medications; rarely does the diet provide toxic quantities. (Large daily amounts of liver could cause it, but who among you eat large quantities of liver every day?)

Vitamin D, which exists in several forms, is active in the body as dihydroxycholecalciferol. This compound is involved in several ways in calcium uptake and utilization. Proper bone mineralization will not occur unless adequate amounts are present. The disease rickets may result from serious deficiency of vitamin D in children.

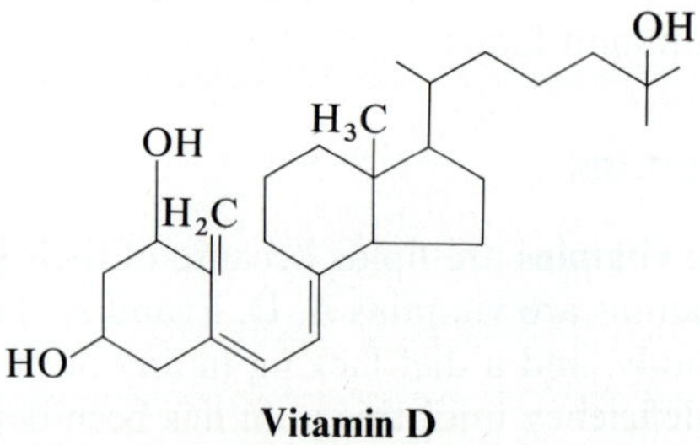
OH
H3C
OH
H2C
HO

Vitamin D

Vitamin D is sometimes called the "sunshine vitamin" because ultraviolet light converts some steroids to this vitamin. These reactions occur in the skin when exposed to sunlight.

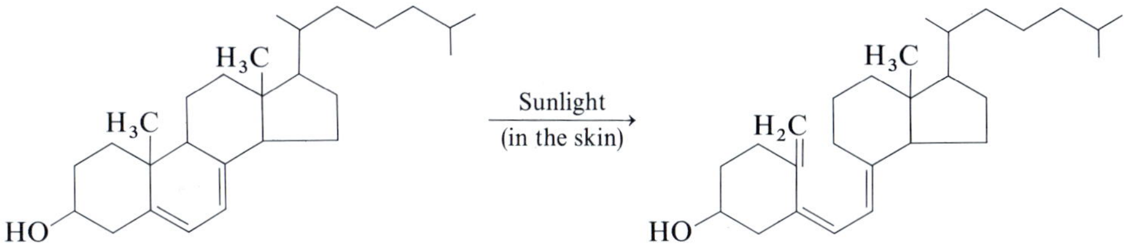
H$_3$C
H$_3$C
HO
Sunlight
(in the skin)
H$_3$C
H$_2$C
HO

7-Dehydrocholesterol **Vitamin D_3** (cholecalciferol)

As long as people are exposed to adequate amounts of sunlight, there should be little or no need for supplementary sources of vitamin D. Few natural sources of this vitamin are available. Liver, fatty fishes, egg yolk, and butter provide some. A more common source is supplementary vitamin D added to dairy products and some other foods. This is the major source of dietary vitamin D in the United States. Strict vegetarians should have no deficiency unless they lack access to sunlight.

Vitamin E is a collection of compounds known as *tocopherols.* The tocopherols are antioxidants, that is, they reduce the rate of oxidation of double bonds, and they appear to function in this role in the membranes of cells. Deficiencies in humans are rare. In animals, several deficiencies have been seen or induced by deficient diets. These symptoms include sterility, muscle atrophy, and nervous disorders.

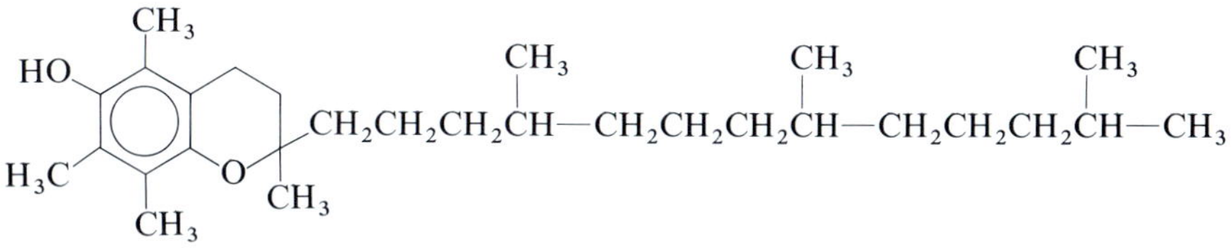
CH$_3$
HO
H$_3$C
O
CH$_3$
CH$_3$
CH$_3$ CH$_3$ CH$_3$
CH$_2$CH$_2$CH$_2$CH—CH$_2$CH$_2$CH$_2$CH—CH$_2$CH$_2$CH$_2$CH—CH$_3$

Vitamin E

Vitamin K has a role in the synthesis of some proteins required for normal blood clotting. Without vitamin K, these proteins are not adequately synthesized, and normal clotting may not occur.

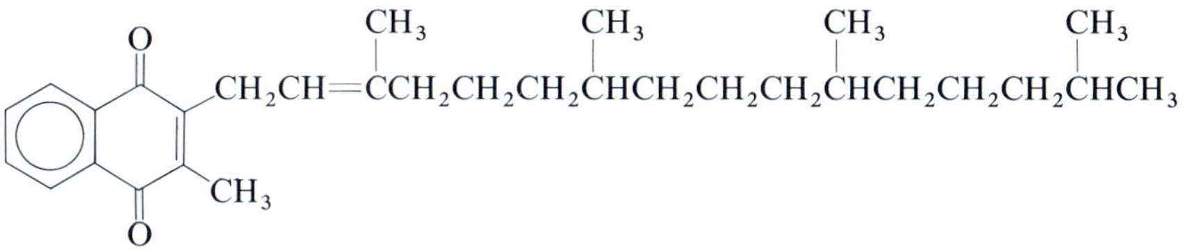
O
CH$_3$ CH$_3$ CH$_3$ CH$_3$
CH$_2$CH=CCH$_2$CH$_2$CH$_2$CHCH$_2$CH$_2$CH$_2$CHCH$_2$CH$_2$CH$_2$CHCH$_3$
CH$_3$
O

Vitamin K

NEW TERMS

fat soluble vitamins Lipids required by the body that cannot be synthesized by it. Vitamins A, D, E, and K are fat soluble.

TESTING YOURSELF

Lipids and Nutrition

1. Name the fat soluble vitamins.
2. Which vitamin is the "sunshine vitamin"?

3. Which vitamin is associated with the chemistry of vision?

Answers **1.** A, D, E, and K **2.** D **3.** A

SUMMARY

Lipids are water insoluble biomolecules that can be extracted from tissues by nonpolar solvents. Lipids can be classified according to their hydrolysis properties. *Saponifiable lipids* are hydrolyzed in aqueous base, while *nonsaponifiable lipids* are not. *Fatty acids* are a component of all saponifiable lipids. These carboxylic acids are long, unbranched molecules containing zero to several cis carbon–carbon double bonds. *Triacylglycerols* (fats and oils) and phospholipids are common saponifiable lipids involved in energy storage and structure, respectively. *Prostaglandins*, *leukotrienes*, and *steroids* are common nonsaponifiable lipids. All three have regulatory roles, but steroids have structural roles as well. Dietary lipids provide a significant amount of our energy needs. Several lipids are essential components of the diet, including linoleic acid and the *fat soluble vitamins*.

TERMS

lipid
saponifiable lipid
nonsaponifiable lipid
fatty acids
saturated fatty acid
unsaturated fatty acid
triacylglycerols
fats
oils
iodine number
rancid
waxes
amphipathic molecules
phosphoacylglycerols
sphingolipids
steroids
cholesterol
prostaglandins
leukotrienes
bilayer
liposome
membranes
facilitated diffusion
active transport
fat soluble vitamins

CHAPTER REVIEW

QUESTIONS

1. The double bond in an unsaturated fatty acid affects or is involved in all but which of the following?
   a. higher melting point
   b. lower melting point
   c. triacylglycerols being more liquid
   d. the formation of prostaglandins
2. A wax is
   a. a nonpolar solid c. a long-chain alcohol
   b. a triacylglycerol d. none of these
3. Which are essential in membrane formation?
   a. waxes b. triacylglycerols c. phospholipids d. terpenes
4. Which of the following is a male sex hormone?
   a. progesterone b. androsterone c. estrone d. cortisone
5. Spermaceti is obtained from
   a. whales b. plants c. insects d. seals
6. The steroid responsible for maintenance of the uterine lining during pregnancy is
   a. progesterone b. testosterone c. estrogen d. cholic acid

7. Lecithin is a phospholipid that also contains
   a. sugar b. ethanolamine c. choline d. sphingosine
8. The precursor to steroids in the human body is
   a. cholesterol b. testosterone c. cortisone d. bile acids
9. Hydrogenation of a vegetable oil will give
   a. soap b. waxes c. spermaceti d. vegetable fat

Answers **1.** a **2.** a **3.** c **4.** b **5.** a **6.** a **7.** c **8.** a **9.** d

DIAGNOSTIC CHART

Blacken in all of the circles under the number of each question that you missed in the Chapter Review questions. The diagnostic chart will help you identify concept areas that need more study.

	Questions								
Concepts	**1**	**2**	**3**	**4**	**5**	**6**	**7**	**8**	**9**
Structure	○	○					○	○	
Nomenclature		○	○	○	○	○	○	○	○
Properties	○		○			○			
Reactions								○	○

EXERCISES

Classification

1. All of the compounds discussed in this chapter are lipids. What do they have in common?
2. What is the difference between a saponifiable and a nonsaponifiable lipid?
3. List the common saponifiable lipids introduced in this chapter.
4. List the nonsaponifiable lipids discussed in this chapter.

Fatty Acids

5. Fatty acids are carboxylic acids. What distinguishes them from other carboxylic acids?
6. Fatty acids are sometimes classified as saturated, monounsaturated, or polyunsaturated. What do these terms mean?
7. What effect does a cis double bond have on the shape of a fatty acid molecule? What effect does this have on melting points of fatty acids and compounds that contain them?

Triacylglycerols

8. A food package label lists an oil as "partially hydrogenated." What does this mean?
9. An oil is said to have an iodine number of 110. Define the term *iodine number*, and describe what it indicates.
10. Fats and oils are classified as triacylglycerols. What structural features do they share?
11. Which of the following two triacylglycerols would most likely be an oil? Why? How would you convert it to a fat?

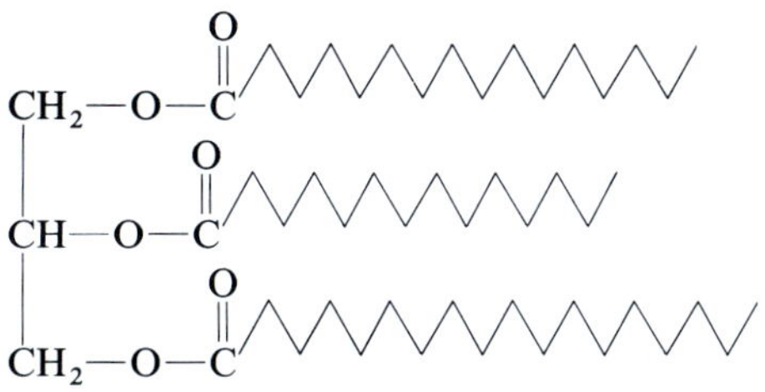

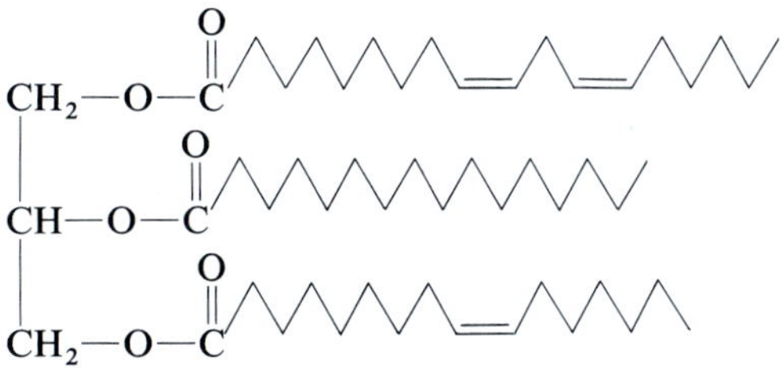

12. The short chain fatty acids have bad odors. Why do some fats and oils become "rancid" when exposed to light, air, and water?

13. What can be done to reduce the chances of rancidity in fats and oils found in food?
14. There are two main factors that make fats more efficient for energy storage than glycogen. What are these factors?
15. A 50-kg woman with 20% body fat begins a fast. Her daily activity consumes 2000 Cal per day during the fast. If we assume that all of the energy used comes from metabolism of fat (not strictly true) and that all 20% of the body fat is available (also not strictly true), how long will her fat sustain her energy needs?
16. Draw a structure that represents a saponifiable wax. What role do waxes have in organisms?

Saponifiable Lipids of Membranes

17. What is an amphipathic molecule? What cell structures contain this type of molecule?
18. The basic unit of the phosphoacylglycerols is phosphatidic acid. Draw this material.
19. Phospholipids are found in membranes, but triacylglycerols are not. Why?
20. What is the structural difference between a cephalin and a lecithin?
21. What is a sphingolipid?
22. Draw the structure of sphingomyelin.

Nonsaponifiable Lipids

23. Draw the structure of cholesterol. What structural feature does it have in common with other steroids?
24. What is atherosclerosis, and how is blood cholesterol related to it?
25. Name the two major female sex hormones and describe their biological function.
26. There has been an increased awareness of athletes taking anabolic steroids (those resembling testosterone). Why would an athlete want to take these?
27. How do the estrogens vary in structure from most other steroids?
28. What is the best known corticoid steroid? What is its biological role?
29. Name the two common bile salts. What is their biological role?
30. What is a prostaglandin, and what role does it play in our well being?
31. How does aspirin reduce inflammation and pain?
32. What synthetic relationship exists between prostaglandins and leukotrienes? Where in the body are leukotrienes found?

Liposomes and Membranes

33. Show the arrangement of amphipathic molecules in a micelle.
34. Compare the polarity of a micelle's core to the polarity of its surface.
35. *Hydrophobic interactions* is a term used to describe the interaction of nonpolar molecules or pieces of molecules when they are in an aqueous environment. Use examples of bonding between these molecules and bonding between water molecules to describe hydrophobic interactions.
36. Draw a short piece of lipid bilayer. Label the parts of the molecules with respect to polarity.
37. Draw a liposome. What potential use might the pharmaceutical industry have for liposomes?
38. What two types of biological molecules are found in biological membranes?
39. Describe the fluid mosaic model for biological membranes.
40. Compare active transport and facilitated diffusion.

Lipids and Nutrition

41. Compare the caloric intake from fats in U.S. diets to the caloric intake recommended by many nutritionists.
42. What are the essential fatty acids?
43. What correlations exist between lipids and cardiovascular disease?
44. Why are some vitamins called fat soluble? List them.

Matching

45. Match the term on the left with the description on the right.

A. triacylglycerols	a. a nonpolar solid
B. wax	b. female sex hormone
C. phospholipid	c. derived from arachidonic acid
D. prostaglandin	d. fats and oils
E. progesterone	e. constituent of membrane

General Questions

46. For each of the following words or phrases, provide a definition, description, or example to show that you understand it.

a. lipid	i. wax
b. triacylglycerol	j. prostaglandin
c. unsaturated fatty acid	k. steroid
d. fat	l. estrogen
e. phosphoacylglycerol	m. testosterone
f. facilitated transport	n. bile salts
g. phosphatidyl choline	o. cholesterol
h. glycolipid	p. cortisol

47. Much of the chemical behavior of compounds can be directly related to the structures of the compounds. For each of the following pairs of compounds, describe the similarities and differences.

a. oil/fat	d. estrogen/progesterone
b. fat/phospholipid	e. prostaglandin/leukotriene
c. fatty acid/wax	

CHAPTER

25

Amino Acids, Peptides, and Proteins

Amino acids are building blocks for two important groups of biomolecules, the peptides and the proteins. Peptides have important biological roles as hormones and antibiotics. Proteins are macromolecules that serve a wide variety of critical functions. This chapter provides the background needed to understand these important biomolecules.

OBJECTIVES

After completing this chapter, you should be able to

- Write the structure of a generalized amino acid.
- Define the electronic properties of amino acids, peptides, and proteins.
- Describe the peptide bond.
- Describe the four levels of protein structure.
- Explain the importance of amino acid sequence in protein structure and function.
- Describe some of the properties of proteins.
- Describe some common functions of proteins.

25.1 Protein Function

Protein is the single most abundant class of biomolecules in cells; protein makes up one-half of the dry mass of a cell. The name *protein* is derived from the Greek word *proteios*, which means of the first rank or importance. Proteins are indeed of the first rank, because they are directly responsible for most of the chemical activity and for much of the physical structure of a cell.

Proteins have many important roles in the body.

One important group of proteins, the *enzymes*, serve as catalysts for the multitude of reactions that occur in the body. Enzymes are discussed in more detail in Chapter 27. Proteins are also involved in transport of materials in the blood. Hemoglobin helps transport oxygen and carbon dioxide, and lipoproteins transport lipids that would otherwise be insoluble (see Chapter 30). Various membrane proteins are involved in movement of substances across membranes via active transport or facilitated diffusion (see Chapter 24).

Proteins serve a wide variety of other critical roles in living organisms. Some proteins have protective roles in the body. Antibodies are proteins that bind to specific foreign substances including viruses and bacteria (see Chapter 30). The antibody–particle complex can then be more easily destroyed by various mechanisms. Lysozyme is an enzyme that cleaves or lyses the cell wall of some bacterial species, thus providing a protective role. Some proteins play a role in movement, such as actin and myosin which are two contractile proteins in muscle. Some proteins serve as storage proteins. Myoglobin stores oxygen in muscle, and casein is a protein in milk that stores amino acids for the nursing young. Some proteins have structural roles. Collagen is the most abundant protein in animals and is a component of skin, bone, teeth, ligaments, tendons, and other extracellular connective structures. Some proteins play a role in regulation of body function. Insulin, glucagon, and growth hormone help control cellular and body activities. Several examples of these assorted functions are presented in Table 25.1.

NEW TERM

protein Polymers of amino acids that have important roles in living systems.

25.2 Alpha Amino Acids

Much of this chapter describes the structural properties of proteins, peptides, and amino acids. The common component of these important biomolecules is the amino acid. Peptides and proteins are oligomers and polymers of amino acids, just as oligosaccharides and polysaccharides are made from monosaccharides. An understanding of protein structure first requires an understanding of amino acids.

Structure of a Generalized Amino Acid

Amino acids contain both an amino group and a carboxylic acid group. Both functional groups are attached to the same carbon atom, which is designated in common nomenclature as the α-carbon (alpha carbon) because it is the first carbon of the main chain. Thus, the amino group is said to be

TABLE 25.1
Common Proteins and Their Functions

Name	Function
Structural	
α-Keratin	Found in skin and hair
Collagen	Fibrous connective tissue
Contractile	
Myosin	Thick muscle filaments
Actin	Thin muscle filaments
Transport	
Hemoglobin	Oxygen transport in blood
Serum albumin	Fatty acid transport in blood
Cytochrome *c*	Transport of electrons
Protective	
Antibodies	React with foreign particles
Fibrinogen and thrombin	Important for blood clotting
Regulation	
Insulin	Involved in regulation of metabolism
Growth hormone	Involved in regulation of metabolism
Storage	
Myoglobin	Stores oxygen in muscle
Ovalbumin	Protein of egg white. Provides amino acids to developing young

an α-amino group because it is bonded to the α-carbon. Figure 25.1 shows a generalized α-amino acid. The α-carbon atom has four substituents. R in this figure represents an organic group or side chain, such as an alkyl group. There are 20 amino acids commonly found in proteins, each with a different side chain. The fourth bond to the α-carbon is always to a hydrogen atom.

The α-carbon of an amino acid has four substituents: a hydrogen, an amino group, a carboxyl group, and a side chain.

The α-carbon atom has four substituents, and unless R represents a hydrogen atom (as it would for the amino acid glycine), all four substituents are different. This makes the α-carbon a chiral center, and a pair of enantiomers (mirror image isomers) exist. Figure 25.1 shows both of these enantiomers, the D and L forms. These designations were used earlier to

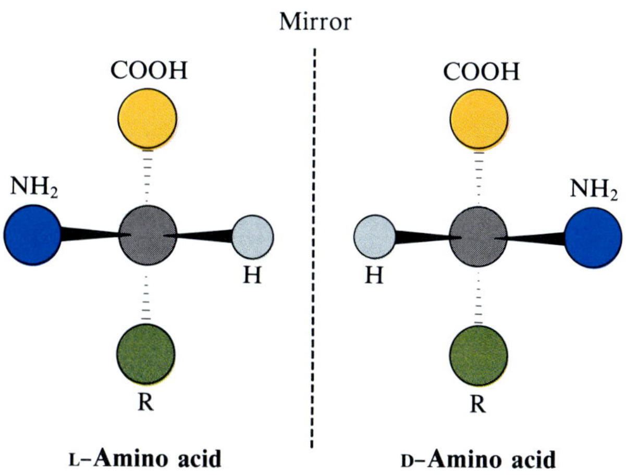
Mirror
COOH
NH$_2$
H
R
L-Amino acid
COOH
NH$_2$
H
R
D-Amino acid

FIGURE 25.1
The generalized α-amino acid possesses a chiral center and thus exists as a pair of enantiomers. Both enantiomers are shown here aligned as the mirror images of each other. The amino acids in proteins are L-amino acids.

TABLE 25.2
The 20 Common Amino Acids

Amino Acid Name	Abbreviations	Structure
Nonpolar R groups		
Alanine	Ala or A	$CH_3-CH(NH_2)-COOH$
Valine	Val or V	$CH_3-CH(CH_3)-CH(NH_2)-COOH$
Leucine	Leu or L	$CH_3-CH(CH_3)-CH_2-CH(NH_2)-COOH$
Isoleucine	Ile or I	$CH_3-CH_2-CH(CH_3)-CH(NH_2)-COOH$
Phenylalanine	Phe or F	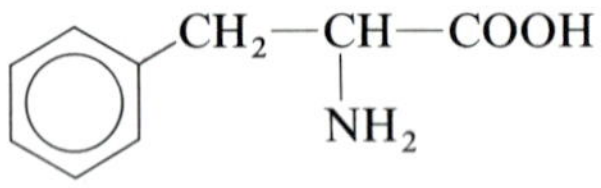
Tryptophan	Trp or W	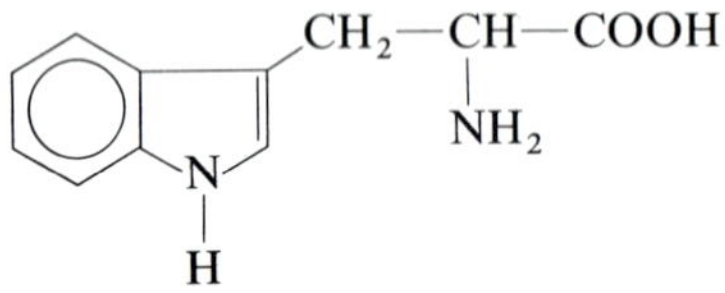
Methionine	Met or M	$CH_3-S-CH_2CH_2-CH(NH_2)-COOH$
Proline	Pro or P	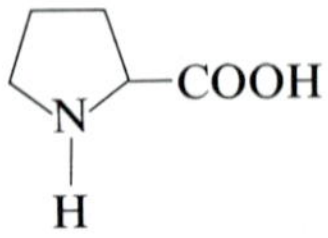
Polar but neutral R groups		
Serine	Ser or S	$HO-CH_2-CH(NH_2)-COOH$
Threonine	Thr or T	$CH_3-CH(OH)-CH(NH_2)-COOH$

The amino acids of the body are L-amino acids.

compare the stereochemistry of monosaccharides to glyceraldehyde. The terms mean the same thing here. When the molecules are arranged with the carboxyl group at the top of the illustration and the side chain down, the L-amino acid has the amino group facing to the left, and the D-amino acid has the amino group pointing to the right. The amino acids found in proteins are L-amino acids. D-Amino acids are uncommon, but they are found in some bacterial cell walls.

With the exception of proline, the α-amino acids are identical except for

Table 25.2 *(continued)*

Amino Acid Name	Abbreviations	Structure
Polar but neutral R groups *(continued)*		
Tyrosine	Tyr or Y	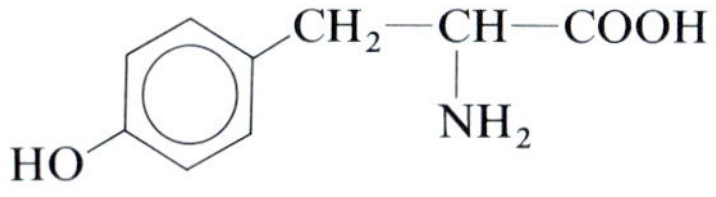
Cysteine	Cys or C	$HS-CH_2-CH(NH_2)-COOH$
Asparagine	Asn or N	$NH_2-C(=O)-CH_2-CH(NH_2)-COOH$
Glutamine	Gln or Q	$NH_2-C(=O)-CH_2CH_2-CH(NH_2)-COOH$
Glycine	Gly or G	$H-CH(NH_2)-COOH$
Acidic R groups		
Glutamic acid	Glu or E	$HO-C(=O)-CH_2CH_2-CH(NH_2)-COOH$
Aspartic acid	Asp or D	$HO-C(=O)-CH_2-CH(NH_2)-COOH$
Basic R groups		
Lysine	Lys or K	$NH_2-CH_2CH_2CH_2CH_2-CH(NH_2)-COOH$
Arginine	Arg or R	$NH_2-C(=NH)-NH-CH_2CH_2CH_2-CH(NH_2)-COOH$
Histidine	His or H	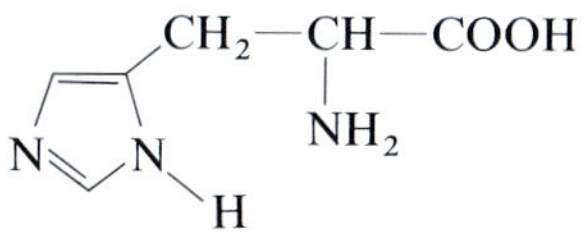

the side chain (Table 25.2). The structure of these side chains distinguishes amino acids from each other physically and chemically. It is worth the time to examine each of them briefly and classify them into general groups. There are several ways to classify the amino acids. The system used in this book is based on the polarity of the side chain. Since the structure of the amino acids in a protein ultimately determines the structure of that protein, an understanding of amino acid structure must be developed to appreciate their contribution to protein structure.

Nonpolar Amino Acids

The nonpolar amino acids all possess a nonpolar *side chain.*

As the amino acids are discussed, refer to Table 25.2 to begin associating names with structures. The *nonpolar amino acids* contain a side chain that is nonpolar. These side chains may be alkyl groups, aromatic rings, or other nonpolar groups. These amino acids play an important role in determining and maintaining protein structure.

Four of these nonpolar amino acids have alkyl groups as side chains. *Alanine* has a methyl group for its side chain. *Valine* has an isopropyl group, *leucine* an isobutyl group, and *isoleucine* a secondary butyl group. A review of alkyl group structure in Chapter 15 may be helpful at this point.

Two nonpolar amino acids have aromatic rings in the side chain. *Phenylalanine* looks like alanine with a phenyl ring replacing one of the hydrogen atoms of the methyl group. The aromatic ring of the other amino acid, *tryptophan,* is fused to a second ring that contains nitrogen. This two-ring system is called *indole* (Chapter 21), and it may be helpful to picture tryptophan as alanine that has an indole replacing one of the methyl hydrogen atoms.

Methionine is a sulfur-containing amino acid that is a thioether (see Chapter 19). Like ethers, thioethers are relatively nonpolar. The final amino acid in this group to be discussed is proline. *Proline* has a ring of five atoms that includes the α-carbon atom and the α-nitrogen atom of the amino group. One end of the side chain is bonded to the α-carbon atom, and the other end is bonded to the nitrogen atom of the α-amino group. This ring structure makes proline unique among the amino acids, and it thus has a special role in protein structure.

Polar Amino Acids

The rest of the amino acids are considered *polar* because the side chain contains one or more polar groups. These amino acids also play vital roles in protein structure, but some of them may play special roles in the function of the protein. The polar amino acids are placed into three different subgroups.

The side chain of the neutral polar amino acids is uncharged but polar.

NEUTRAL POLAR AMINO ACIDS The first subgroup of the polar amino acids is known as the *neutral polar amino acids.* The side chains cf these amino acids are polar, but not ionic. *Serine* has a hydroxyl group in its side chain. For this amino acid, simply picture a hydrogen atom of the methyl group on alanine replaced with a hydroxyl group. A second amino acid with the alcohol functional group is *threonine.* This molecule is like a serine that has had a second methyl hydrogen replaced, this time with a methyl group. The third member of the polar neutral group is *tyrosine,* which also has an —OH group, but tyrosine is a phenol rather than an alcohol. Tyrosine looks like phenylalanine that has a hydroxyl group substituted for a hydrogen in the *para* position of the ring.

Cysteine is a neutral polar amino acid that contains sulfur. It looks like serine that has a sulfur atom replacing the oxygen of the alcohol group. Recall from Chapter 19 that the —SH group is called the thio or sulfhydryl group. *Asparagine* contains an amide group in the side chain as does the similar compound *glutamine.* The final member of this group is *glycine,* which has a hydrogen atom as its side chain. The hydrogen atom is not really polar, but it is too small to add nonpolar character to the amino acid. Glycine is thus sometimes listed with the neutral polar amino acids.

ACIDIC AMINO ACIDS The *acidic amino acids* make up the second group of polar amino acids. There are two of them, *glutamic acid* and *aspartic acid.* Both contain a carboxyl group in the side chain; this second carboxyl group is *in addition* to the one attached to the α-carbon atom. These amino acids differ in size; glutamic acid has one more methylene group, $—CH_2—$. If you look at glutamine and asparagine closely (Table 25.2), you will see that they are the amides of glutamic acid and aspartic acid respectively.

The side chain of the acidic amino acids contains a carboxyl group.

BASIC AMINO ACIDS The third group of polar amino acids is known as the *basic amino acids.* Each contains one or more nitrogen atoms that have an unshared pair of electrons; they are Lewis bases. *Lysine* has an amino group on the end of a chain of four carbon atoms. *Arginine* is more complicated, and has a side chain consisting of four carbon atoms and three nitrogen atoms. The three nitrogen atoms and the carbon atom between them make up the *guanidino group.* This group can accept one hydrogen atom.

The side chain of the three basic amino acids has a nitrogen that can accept a hydrogen ion.

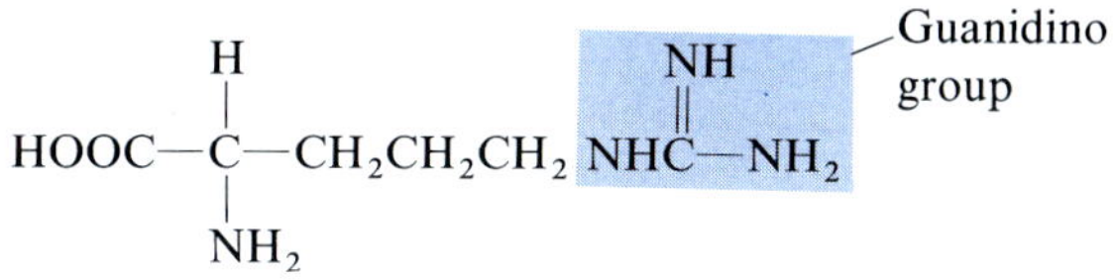
Guanidino group

Arginine

The third basic amino acid is *histidine.* Histidine looks like alanine with a heterocyclic imidazole ring replacing one of the hydrogen atoms of the methyl group. Only one of the ring nitrogen atoms has an unshared pair of electrons that can be donated.

Acid–Base Properties of Amino Acids

The α-amino and α-carboxyl groups of amino acids act as acids and bases. Acids and bases react in neutralization reactions, and the α-amino group and carboxyl group of an amino acid are not exceptions. These groups react in an intramolecular acid–base reaction. The carboxyl group donates a hydrogen ion to the amino group.

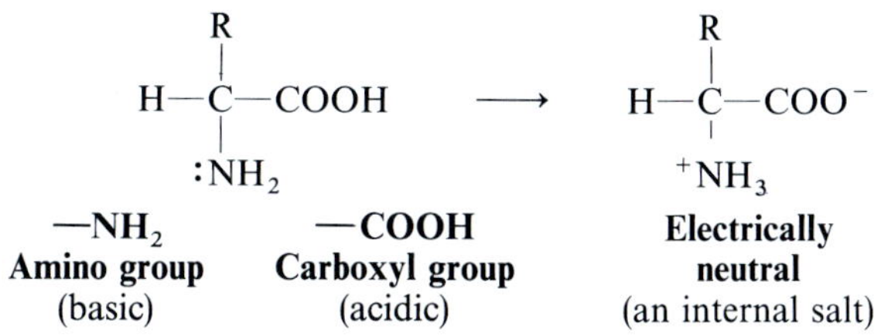
—NH₂
Amino group
(basic)
—COOH
Carboxyl group
(acidic)
Electrically
neutral
(an internal salt)

The product of this reaction is a dipolar ion that is called a **zwitterion**. Dipolar ions are electrically neutral because the opposite charges cancel. A consequence of the ionic nature of the zwitterions is the water solubility and

Amino acids exist as dipolar ions called *zwitterions.*

high melting points of the α-amino acids. These are just the properties you would expect of ionic compounds.

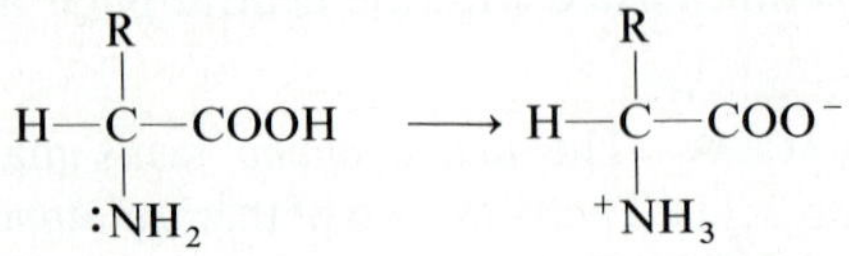

α-Amino acid with free carboxyl group and free amino group — **α-Amino acid in zwitterion form**

A second acid–base property is seen in amino acids. They are **amphoteric**; that is, they can function as *either* acids or bases. In the presence of a base, the amino acid behaves as an acid because it gives up a proton from the ammonium salt portion of the molecule. Likewise, in an acidic solution, the amino acid functions as a base because the carboxylate group accepts a hydrogen ion (Figure 25.2). Although it is important to recognize this dipolar ion character of amino acids, in this text the amino acids are represented in either the dipolar or nonionic form.

In pure water and in aqueous solutions near pH 7, many of the amino acids are neutral dipolar ions. At other pH values, these amino acids are charged. When a hydrogen ion is gained by the molecule, the electrical charge of the amino acid is changed from zero to plus one.

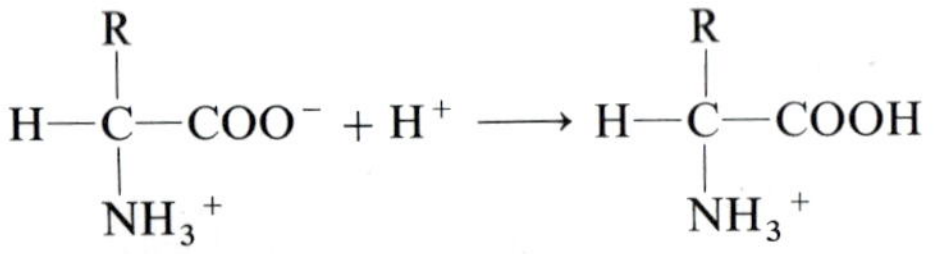

Neutral **+1** **+1**

Similarly, a loss of a hydrogen ion to a base will change the charge on the amino acid from neutral to minus one.

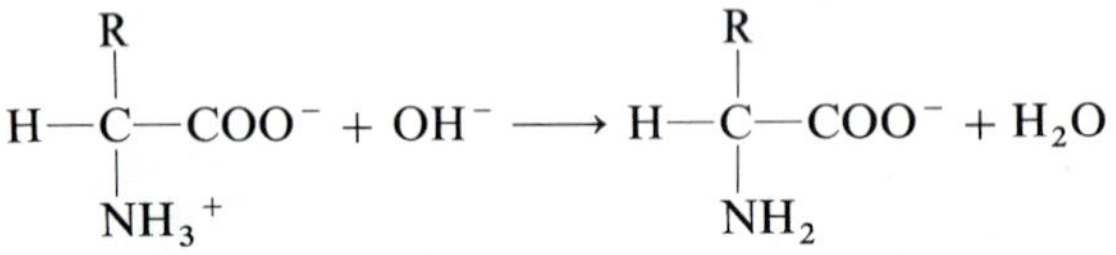

Neutral **−1** **−1**

Amino acids are neutral only at one pH. If the pH is higher or lower, then the amino acid has a negative or positive charge.

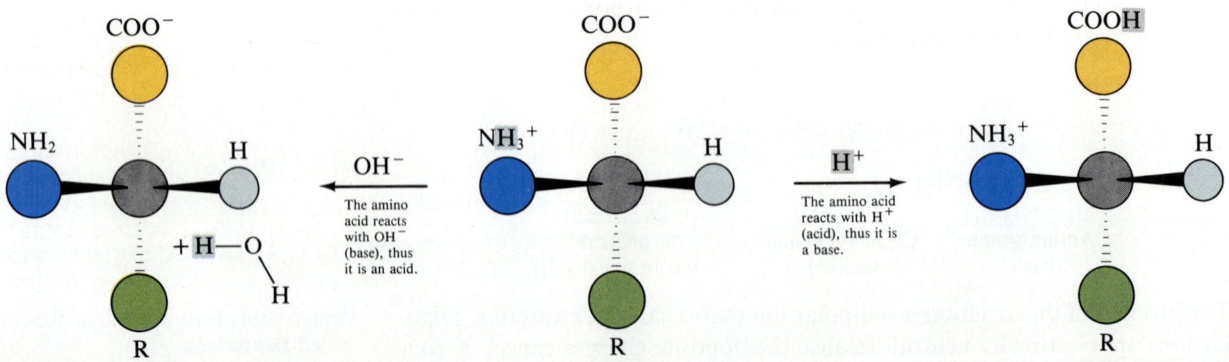

COO⁻
NH₂
H
+H—O
H
R
OH⁻
The amino acid reacts with OH⁻ (base), thus it is an acid.
COO⁻
NH₃⁺
H
R
H⁺
The amino acid reacts with H⁺ (acid), thus it is a base.
COOH
NH₃⁺
H
R

FIGURE 25.2
Amino acids are amphoteric molecules because they can act as both acids and bases. The protonated amino group acts as an acid, and the carboxylate group acts as a base.

The charge on an amino acid varies with pH.

The pH value of a neutral amino acid (that is, with no net charge) is called the **isoelectric point** or **pI**. The pI values of the amino acids are known but can also be calculated from the pK_a values of the amino group and carboxyl group. If the amino acid has a side chain containing an ionizable group, such as the carboxyl group of aspartic acid or the amino group of lysine, then the calculations to determine the pI of the amino acid must include the pK_a of that ionizable group. You will not be asked to do these calculations, but you should know that there is a single pH, the pI, where an amino acid is electrically neutral. At higher pH, the amino acid has a net negative charge, and at lower pH it is positively charged.

None of the amino acids have exactly the same pI value because the side group affects the acid–base properties of the α-amino and α-carboxyl groups. Since the amino acids have different pI values, they will have at least a slightly different magnitude of charge at any given pH. Particles that have different charges are attracted to the poles of an electrical field to different degrees. Thus, amino acids with a net positive charge will be attracted to the negative pole of an electrical field at a velocity determined by the net positive charge on the amino acid. The more positive the charge, the more strongly that amino acid is attracted to the negative pole and the more rapidly the amino acid moves toward the negative pole. Negatively charged amino acids are attracted in a similar way to the positive pole. **Electrophoresis** is a technique used to separate charged particles located in an electrical field. Electrophoresis is an important tool for detection and identification of proteins because, as you shall see, proteins are charged particles due to the presence of amino acids (Figure 25.3).

When amino acids are present in peptides and proteins, the α-amino and carboxyl groups are not free but instead are involved in covalent bonds. The side chain is rarely involved in a covalent bond. It is the side chain of

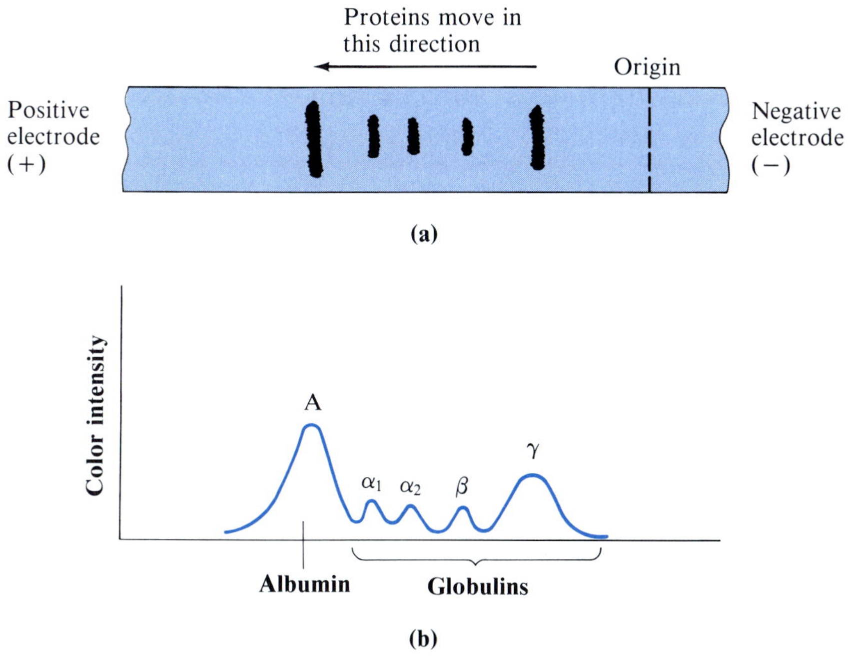
Proteins move in this direction
Origin
Positive electrode (+)
Negative electrode (−)
(a)
Color intensity
A
α1
α2
β
γ
Albumin
Globulins
(b)

FIGURE 25.3
These blood proteins were separated by electrophoresis. (a) Serum was placed onto an electrophoresis strip as a band at the origin. The strip was placed into an electrical field (designated by the positive and negative electrodes). At the pH value of this experiment, blood proteins have a net negative charge and are attracted to the positive electrode. Because the proteins have different sizes and different net charges, they migrate at different rates. (b) After separation, the strip is stained with a protein-specific dye. The intensity of the stain indicates the relative amount of each protein.

an amino acid that contributes most to the acid–base and other properties of a peptide or protein. A brief review of the side chains of the amino acids in Table 25.2 reveals that a variety of functional groups and polarities exist. In Section 25.4 on protein structure, the properties of the amino acid side chains are used repeatedly to explain protein structure.

NEW TERMS

amino acids A class of biological compounds whose members possess both an amino group and a carboxylic acid group. Amino acids are the building blocks of peptides and proteins.

zwitterion A dipolar ionic form of an amino acid that is formed by donation of a H^+ from the α-carboxyl group to the α-amino group. Because both charges are present, the net charge is neutral.

amphoteric molecule A molecule that functions as a base in the presence of an acid and as an acid in the presence of a base.

isoelectric point (pI) The pH at which an amphoteric molecule, such as an amino acid or protein, has no net charge.

electrophoresis Technique that separates charged particles in an electrical field; especially useful for detection of amino acids, peptides, or proteins.

TESTING YOURSELF

Amino Acids

1. Identify each of the following amino acids as neutral, acidic, or basic.

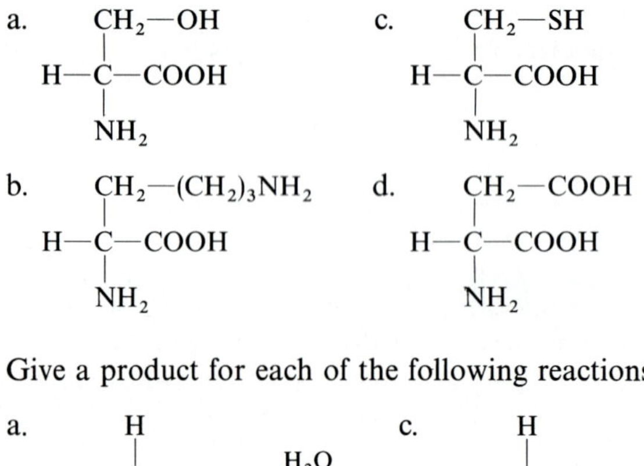

2. Give a product for each of the following reactions.

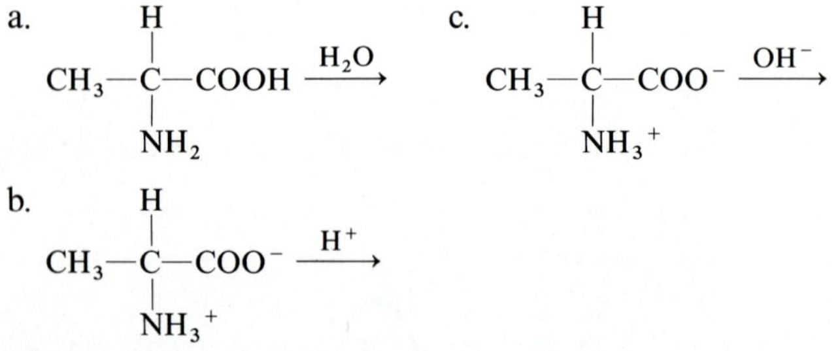

Answers: **1. a.** neutral **b.** basic **c.** neutral **d.** acidic

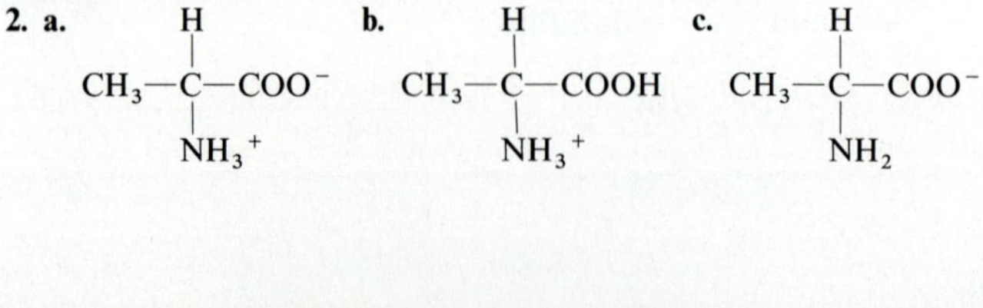

25.3 Peptide Bonds and Peptides

The Peptide Bond

Although amino groups do not readily react with carboxylic acids to form amides, the formation of an amide can be summarized by this equation.

$$R{-}\overset{O}{\overset{\|}{C}}{-}OH + NH_2{-}R' \longrightarrow R\overset{O}{\overset{\|}{C}}{-}NH{-}R' + H_2O$$

In organic chemistry, the bond formed between the carbon atom of the carbonyl group of a carboxylic acid and the nitrogen atom of an amino group is called an amide bond. When this bond involves amino acids, it is called a peptide bond. **Peptide bonds** connect amino acids together in peptides and proteins.

$$H_3\overset{+}{N}{-}\underset{R_1}{\overset{H}{C}}{-}\overset{O}{\overset{\|}{C}}{-}\overset{-}{O} + H_3\overset{+}{N}{-}\underset{R_2}{\overset{H}{C}}{-}\overset{O}{\overset{\|}{C}}{-}\overset{-}{O} \longrightarrow H_3\overset{+}{N}{-}\underset{R_1}{\overset{H}{C}}{-}\overset{O}{\overset{\|}{C}}{-}\underset{H}{N}{-}\underset{R_2}{\overset{H}{C}}{-}\overset{O}{\overset{\|}{C}}{-}\overset{-}{O} + H_2O$$

The reaction that forms an amide or peptide is a readily reversible reaction. This reverse reaction is a hydrolysis reaction, that is, a cleavage by water.

$$\underset{\textbf{Amide (or peptide)}}{R\overset{O}{\overset{\|}{C}}{-}NH{-}R'} + H_2O \longrightarrow \underset{\textbf{Carboxylic acid}}{R\overset{O}{\overset{\|}{C}}{-}OH} + \underset{\textbf{Amine}}{NH_2{-}R'}$$

Both of these reactions occur in the body. The body uses energy to form peptide bonds to build the proteins and peptides needed to carry out the normal functions of the body. Proteins and peptides are broken down by hydrolysis; dietary proteins are cleaved in the digestive tract, and cellular proteins and peptides are hydrolyzed and replaced routinely during the normal course of cellular activity.

The peptide bond has some structural features that are not readily apparent in formulas used to represent it. In the preceding dipeptide reaction, the peptide bond was shown as a single bond between the carbon atom of the carbonyl group and the nitrogen atom, but this bond actually has some double bond character due to delocalization of electrons. A better representation of this bond would be a rigid, platelike structure (Figure 25.4). Because of double bond character, the peptide bond lacks the free rotation typical of a single bond. This results in the two atoms of the carbonyl group and the nitrogen atom with its hydrogen atom lying in the same plane. The atoms of a peptide bond act as a single rigid unit in a peptide or protein.

When a peptide bond is formed between two amino acids, the amino group of one and the carboxyl group of the other participate in the reaction. That leaves one of the amino acids with a free carboxyl group and the other with a free amino group. Each of the amino acids can form two peptide bonds, one with its amino group and one with its carboxyl group. Because amino acids are bifunctional, several to many amino acids can be linked together to form oligomers and polymers called *peptides* or even larger polymers called *proteins.*

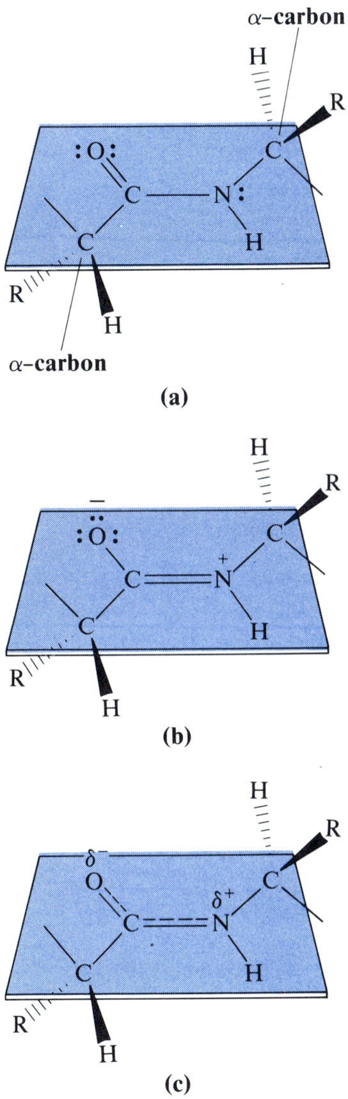
α-carbon
α-carbon
(a)
(b)
(c)

FIGURE 25.4
The structure of the peptide bond. (a) The peptide bond is typically represented as an amide bond with a carbonyl group and a C—N single bond. (b) Because the atoms of the peptide bond have unshared electron pairs and pi electrons, another resonance form for this bond can be drawn. (c) The forms in (a) and (b) can be mentally combined to illustrate the partial double bond character of the C—N bond and the C—O bond. Since free rotation does not occur around double bonds (or partial double bonds), these atoms all lie in a plane.

Structure and Nomenclature of Peptides

Peptides are oligomers and polymers of amino acids. There is no fixed number of amino acids that determines an upper limit for classifying a compound as a peptide. Instead, a compound composed of amino acids with a molecular weight less than 5000 is classified as a peptide. The average molecular weight of an amino acid is somewhat over 100, so on the average, peptides will have fewer than 50 amino acids.

A peptide containing two amino acids is called a *dipeptide*. One with three is a tripeptide and one with four is a tetrapeptide. The prefixes used for the number of amino acids are the same as in organic nomenclature and in the nomenclature of the carbohydrates. Oligopeptides contain several amino acids. For our purposes, ten amino acids is used as the upper limit for an oligopeptide; **polypeptides** contain more than ten amino acids. The boundary between oligopeptides and polypeptides is arbitrary.

Peptides are sometimes named by a systematic nomenclature. For example, the dipeptide in Figure 25.5a is called alanylglycine, because it contains alanine which has a free amino group and glycine with a free carboxyl group. In the systematic nomenclature the N-terminal amino acid (the amino

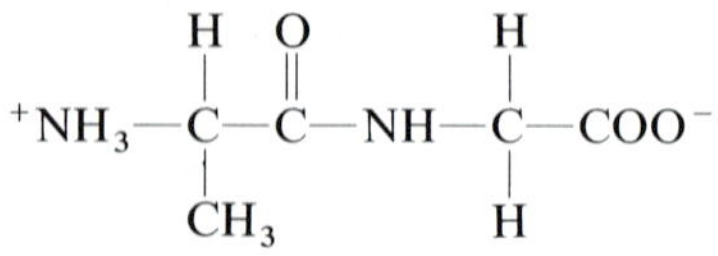

Ala—Gly

(a) Alanylglycine
(a dipeptide)

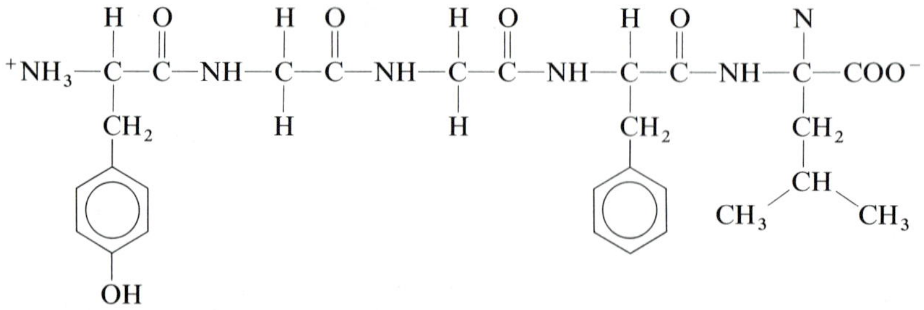

Tyr—Gly—Gly—Phe—Leu

(b) Leu-enkephalin
(a pentapeptide)

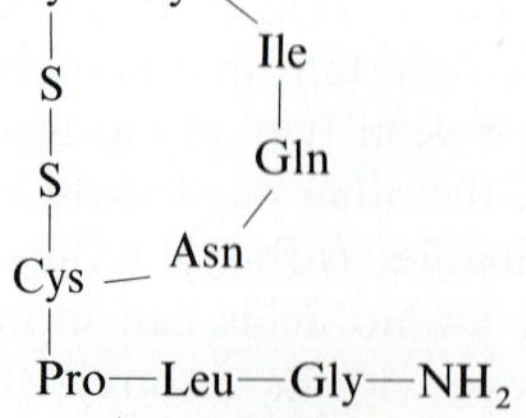
Cys—Tyr
Ile
S
Gln
S
Asn
Cys
Pro—Leu—Gly—NH2

(c) Oxytocin
(a nonapeptide)

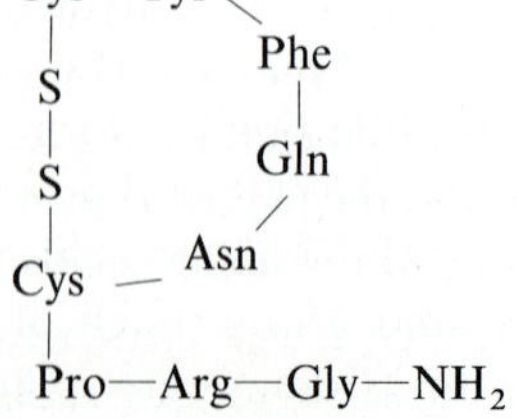
Cys—Tyr
Phe
S
Gln
S
Asn
Cys
Pro—Arg—Gly—NH2

(d) Vasopressin
(a nonapeptide)

FIGURE 25.5
Some common peptides. The entire structures of alanylglycine and leu-enkephalin are shown; only abbreviations are normally used for larger peptides. Amino acid composition and covalent bonding are shown for the other peptides. Note the structural similarity between vasopressin and oxytocin, even though they cause quite different physiological responses. Gly—NH_2 in these two peptides indicates that the carboxyl group of glycine is bonded to ammonia via an amide bond.

acid with the free amino group) is named first, then all others are named in order. Note that only the last amino acid, the C-terminal amino acid, retains its full name. The other names are all shortened to the stem of the original name, and the suffix *-yl* is added. If several amino acids are present in a peptide, the systematic name becomes large and inconvenient. Most peptides obtained from nature have a common name that is used more often than the systematic name. For example, Figure 25.5b shows a peptide that is commonly called leu-enkephalin, but its systematic name is tyrosinylglycylglycylphenylalanylleucine.

Biologically Important Peptides

Many naturally occurring peptides are hormones. Oxytocin (Figure 25.5c) is a nonapeptide that causes contraction of uterine and other smooth muscle. Vasopressin (Figure 25.5d) is a nonapeptide quite similar to oxytocin, but it is involved in water retention and blood pressure regulation. Several peptides, including the endorphins and enkephalins (Figure 25.5b), are neuropeptides that seem to be involved in pain and pleasure sensation. Morphine and heroin (see Chapter 21) appear to mimic some of the effects of endorphins and enkephalins. (The term *endorphin* is a contraction of *endo*genous m*orphin*e.) These peptides may also be the agent responsible for the "runner's high" that follows intense aerobic exercise. Another example is glutathionine, which is a tripeptide that is not a hormone. It appears to serve a role as a reducing agent in cells. The artificial sweetener Aspartame is a modified dipeptide (Chapter 23) that is many times sweeter than sucrose.

NEW TERMS

peptide bond The amide bond between the amino group of an amino acid and the carboxylic acid group of another amino acid.

peptide A compound consisting of amino acids linked by peptide bonds. Often the number of amino acids is indicated by prefixes such as *di-*, *tri-*, or *oligo-*.

polypeptide A macromolecule containing many (ten or more) amino acids. This term is sometimes used to mean the chain in a protein.

TESTING YOURSELF

Peptide Bond and Peptides

1. The peptide bond is essentially what bond of organic chemistry?
2. Examine oxytocin in Figure 25.5c. What amino acid is N-terminal? Also name the C-terminal amino acid of leu-enkephalin (Figure 25.5b).
3. You have isolated a molecule from a human neuron that is made from numerous amino acids connected by peptide bonds. What criterion would you use to determine if it is a peptide or protein?

Answers: **1.** the amide bond **2.** cysteine; leucine **3.** Molecular weight. If the compound had a molecular weight greater than 5000, then it is a protein; if less than 5000, then it is a peptide.

25.4 Protein Structure

A molecular weight of 5000 is the arbitrary boundary between large peptides and proteins.

Proteins are macromolecules containing one or more polypeptides plus, in some cases, one or more other components. These polymers of amino acids have molecular weights greater than 5000. Proteins and the larger peptides are polypeptides and are distinguished arbitrarily by size. Most proteins are very large; many have molecular weights greater than 50,000, and some are more than one million. The amino acids of proteins are linked by peptide bonds. To distinguish each amino acid residue from another in a peptide or polypeptide, each specific residue is numbered. The N-terminal amino acid occupies the first residue in the protein, the next one is the second residue, and the C-terminal amino acid is the last residue in the protein.

If a few reasonable assumptions are made, you can get an idea of how many possible proteins can be formed from amino acids. Consider a protein with 100 residues. This would be a rather small protein. Assume that any of the 20 different amino acids could be in any of the 100 residues. Each of these combinations would be a unique protein. The number of possible combinations that the amino acids can have is equal to 20 times itself 100 times or 20^{100}. This is about 1×10^{130}! And remember, this number was calculated using only 100 amino acid residues. Any number of amino acids from about 50 up to many hundreds could actually be present in a protein. Compare this gigantic number of possible combinations to the estimated number of atoms in the entire universe, which is about 10^{80}, give or take a few orders of magnitude. The number of possible proteins that could be made from the 20 amino acids is truly astronomical. In reality, the number of actual proteins that are found in the living organisms on earth is much smaller. It is estimated that the number of different protein structures present in all organisms is greater than one trillion (10^{12}).

Protein Conformation

Each amino acid sequence in a protein is unique for that protein. Obviously, the number of possible amino acid sequences for proteins is very large, but what about the orientation in space of the atoms of any one of these proteins? Free rotation about single bonds allows for an infinite number of conformations for a molecule of even modest size or complexity. There are large numbers of single bonds in a protein, so an infinite number of possible conformations exist for each protein. Yet biochemists are quite sure that all of the copies of any given functional protein are in one or, at most, a very few conformations. Why does each have only one conformation? Because it is the most stable and lowest energy conformation. There are forces that operate within a protein that cause it to be in that most stable conformation.

The normal conformation of a protein is called its **native conformation.** In this conformation the protein has its native function. If a protein's conformation is changed, it loses its native function or activity. This process is called **denaturation**, and it illustrates that a protein must have its native conformation to be functional. Because protein structure is critical to protein function, much of this chapter deals with protein structure. Protein structure is complex, and to facilitate understanding, protein chemists have described protein structure at four different levels called *primary*, *secondary*, *tertiary*, and *quaternary* structure.

Primary Structure

The **primary structure** of a protein is the amino acid sequence of that protein. The primary structure is known when it has been determined what amino acids are present in the protein, how many copies of each amino acid are present, and which amino acids are bonded to each other. The amino acid sequences of many proteins are now known, and each unique protein appears to have its own combination of amino acids in their own unique sequence. Each protein has its own unique primary structure. The amino acid sequence of the small protein myoglobin is shown in Figure 25.6.

The amino acid sequence of a protein chain is the primary structure of the protein.

Ultimately, the amino acid sequence of a protein, the primary structure, determines the other levels of structure. Since various amino acids are present in different places within proteins, their interactions with water and with each other will differ. This results in unique combinations of interactions within each protein, which in turn produces a unique conformation for the protein. Each protein has its own unique shape because it has its own unique amino acid sequence.

Primary structure determines the shape of a protein.

A single, unique conformation for a molecule should not really be surprising to you. Recall from Chapter 15 that the small hydrocarbon ethane has an infinite number of possible conformations, but only one conformation, the staggered one, is the most stable. It is the bonding of atoms within ethane that ultimately determines its most stable conformation. The hydrogens of ethane get in each others way in other conformations, making them less stable. While more factors than hindrance are involved in protein structure, the same basic argument holds. A protein will have some conformation

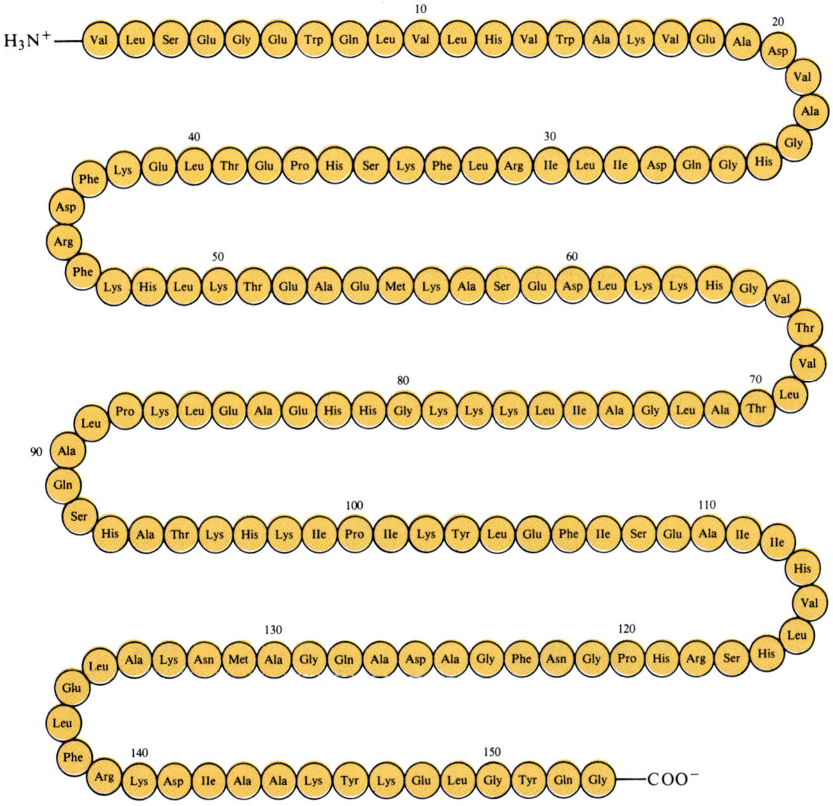
H_3N^+ Val Leu Ser Glu Gly Glu Trp Gln Leu Val Leu His Val Trp Ala Lys Val Glu Ala Asp Val Ala Gly His
10 20
His Gly Gln Asp Ile Leu Ile Arg Leu Phe Lys Ser His Pro Glu Thr Leu Glu Lys Phe Asp Arg Phe
30 40
Lys His Leu Lys Thr Glu Ala Glu Met Lys Ala Ser Glu Asp Leu Lys Lys His Gly Val Thr Val Leu
50 60
Thr Ala Leu Gly Ala Ile Leu Lys Lys Lys Gly His His Glu Ala Glu Leu Lys Pro Leu Ala Gln Ser
70 80 90
His Ala Thr Lys His Lys Ile Pro Ile Lys Tyr Leu Glu Phe Ile Ser Glu Ala Ile Ile His Val Leu
100 110
His Ser Arg His Pro Gly Asn Phe Gly Ala Asp Ala Gln Gly Ala Met Asn Lys Ala Leu Glu Leu Phe
120 130
Arg Lys Asp Ile Ala Ala Lys Tyr Lys Glu Leu Gly Tyr Gln Gly —COO^-
140 150

FIGURE 25.6
The primary structure of myoglobin from a sperm whale. This protein has these 153 amino acids in this specific sequence.

that is more stable, and that particular conformation depends on the atoms in the protein and how they are bonded to each other.

Secondary Structure

The **secondary structure** of a protein involves the arrangement of the atoms of the polypeptide chain or "backbone" of the protein. This chain is the continuous line of atoms that includes the atoms of the peptide bonds plus the α-carbon atom of the amino acids. The atoms that are not part of this chain are the hydrogen atom attached to the α-carbon and the atoms of the side group. Figure 25.7 shows the backbone of a part of a polypeptide chain. Many proteins have portions of the peptide chain or, more rarely, most of the chain arranged in secondary structure. Secondary structure is classified into two categories: (1) α-helix or (2) β-pleated sheet.

In an **α-helix**, the atoms of the backbone are arranged into a spiral shape, a helix. Coiled springs and the toys called slinkies are common examples of helices (Figure 25.8a). The atoms of a peptide bond form a rigid platelike structure, and in an α-helix, these plates are arranged in an orderly, repeating fashion that places them in this spiral arrangement. They are arranged in this way because hydrogen bonds form between some of them (Figure 25.8b and c). When the chain is coiled, an N—H group of one peptide bond is directly over or under a carbonyl group in another loop. They are in position for hydrogen bonding. Each hydrogen bond is a weak dipolar interaction, but the α-helix brings the atoms of the backbone into a position where numerous hydrogen bonds can form between atoms of different coils.

It appears that as long as no other forces are present to prevent an α-helix from forming, the numerous hydrogen bonds will stabilize a protein into an α-helix. Some proteins, such as the hair and skin protein α-keratin, have virtually the entire length of the polypeptide chain arranged in an α-helix. Other proteins have only portions of the chain in this conformation (for example, see myoglobin in Figure 25.10). What prevents the chain from assuming an α-helix conformation? Whenever a proline is present, its unique shape introduces a "kink" in the chain. Just as a cis bond in the fatty acid portion of a triacylglycerol prevents orderly clustering of these lipid molecules, a proline bends the chain and prevents the α-helix conformation. The α-helix can also be prevented by certain side groups. If the size or shape of a side group places a portion of it between atoms that would hydrogen bond, then that large bulky R group acts as a physical barrier that prevents stable formation of hydrogen bonds in that region. Thus, the α-helix does not form.

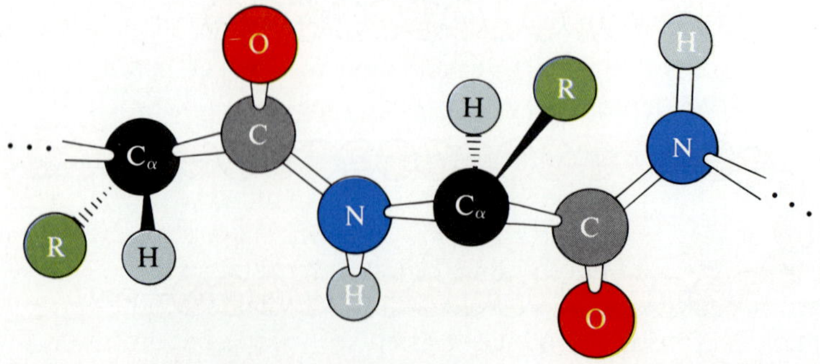
O
H
R
H
Cα
C
N
N
Cα
C
R
H
H
O

FIGURE 25.7
The atoms of part of a polypeptide chain are shown here. Secondary protein structure is defined as the orientation of these atoms with respect to each other.

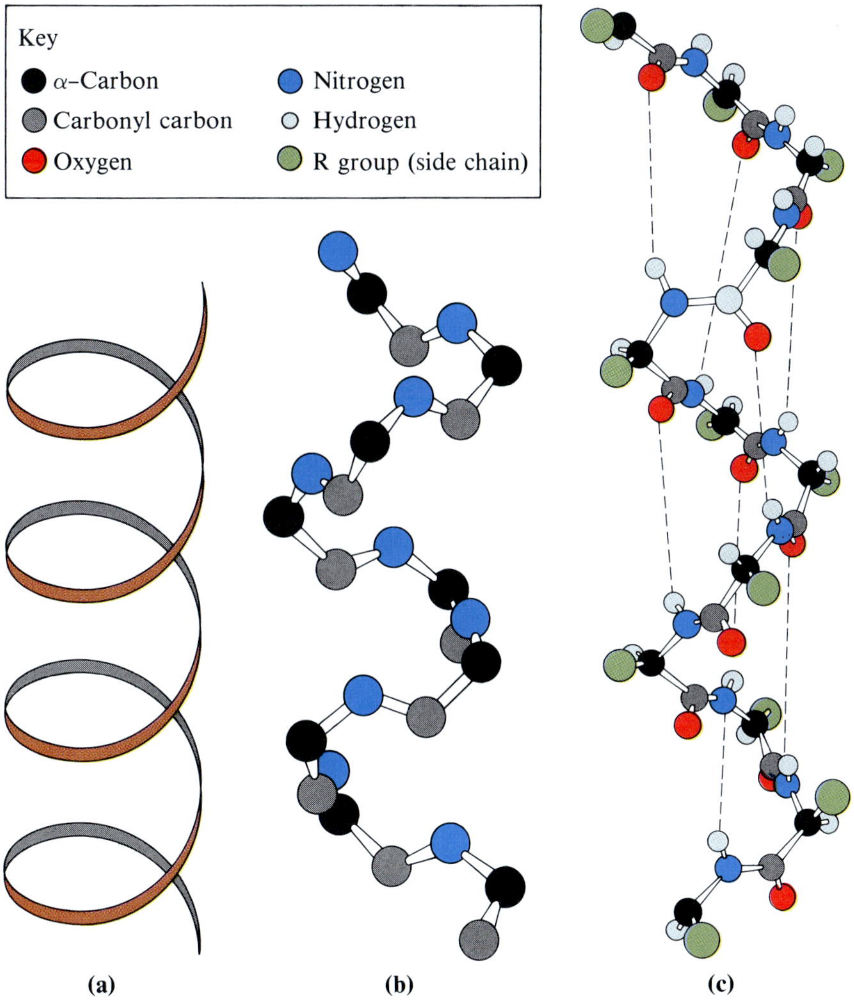

Key
α-Carbon
Carbonyl carbon
Oxygen
Nitrogen
Hydrogen
R group (side chain)
(a)
(b)
(c)

FIGURE 25.8
The arrangement of the polypeptide chain of a protein into a spiral is referred to as an α-helix. (a) Coiled springs and slinkies are common objects with spiral or helical structures. (b) The atoms of this polypeptide chain are oriented into a spiral. (c) Hydrogen bonds (dashed lines) between atoms of the polypeptide chain maintain the α-helix of the protein.

Similarly, side chains with similar charge would repel each other, therefore adjacent or nearby charged side groups may prevent the formation of a stable α-helix.

The second type of secondary structure found in proteins is called the **β-pleated sheet** (Figure 25.9). This structure also results from hydrogen bonding between atoms of the peptide bond, but the bond forms between atoms that are in different polypeptide chains. β-Sheets are found in a number of proteins. The structure of fibroin, the protein of silk, is mostly β-pleated sheets.

Examples of secondary protein structure are α-helix and β-pleated sheet.

If the atoms of a polypeptide backbone are arranged into an α-helix or a β-pleated sheet, then the term *secondary structure* applies, but most proteins are not entirely helical or sheetlike. Higher levels of structure are needed to adequately explain the conformations of these proteins.

Tertiary Structure

If the entire peptide chain of a protein was arranged into an α-helix, the protein would have a long, cylindrical, rodlike shape. If it had only β-sheet structure, then the protein will have a flat, sheetlike structure. Some structural proteins are rodlike or sheetlike in shape, but most soluble proteins

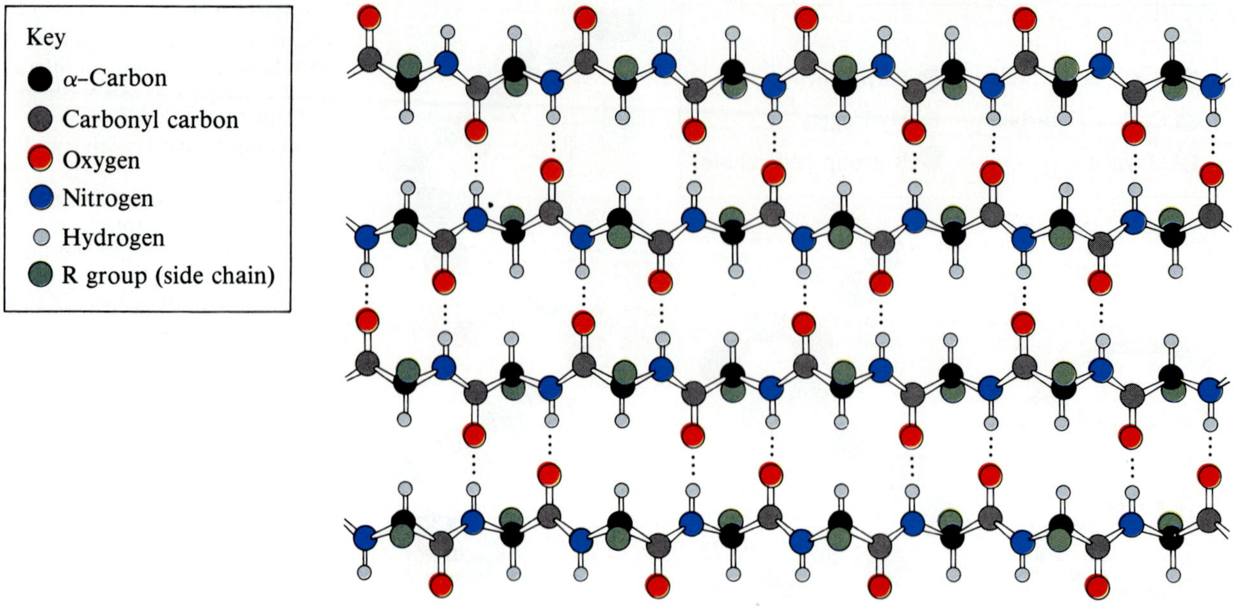
Key
α-Carbon
Carbonyl carbon
Oxygen
Nitrogen
Hydrogen
R group (side chain)

FIGURE 25.9
A representation of a β-pleated sheet in a protein. In this type of secondary protein structure, polypeptide chains are held in place by hydrogen bonding (dots) between atoms of different chains.

are neither. They are often spherical or have a compact globular shape, although nearly all other possible shapes have been found. This bending of the peptide chain to form a compact shape is described as **tertiary structure** (Figure 25.10).

Several forces within a protein are responsible for its assuming and maintaining its characteristic tertiary structure. Those forces include (1) hydrogen bonding, (2) ionic and polar interactions, and (3) Hydrophobic

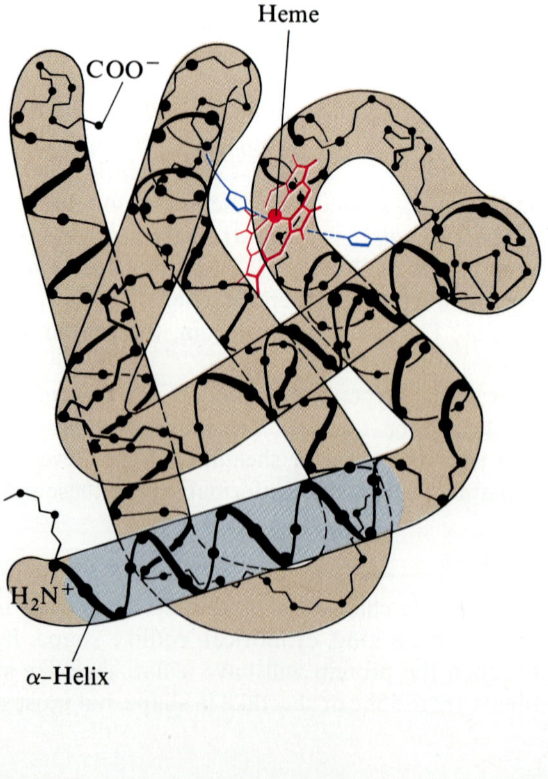
Heme
COO$^-$
H_2N^+
α-Helix

FIGURE 25.10
The tertiary structure of myoglobin. Much of the length of this protein exists as α-helix, one of which is highlighted. Take a moment to identify other regions of α-helix. Heme is the prosthetic group of myoglobin.

interactions. These forces result from interactions between atoms from different portions of the protein, usually between atoms of different side groups. In addition, a covalent bond (called a *disulfide bridge*) may form between two cysteine residues once the protein is in its native conformation. This bond does not play a role in the protein's assumption of native conformation, but it is very important in helping to maintain the native conformation once it has formed.

Disulfide bridges are covalent bonds between cysteine residues that help maintain the tertiary structure of a protein.

In most proteins, hydrogen bonds form between some atoms. These atoms have been brought close together because the protein is folded in a certain way, even though they are atoms in amino acids that may be many residues apart (Figure 25.11). The hydrogen bond may form between atoms of two different side chains or an atom of a side chain and an atom of the backbone. In either case, the formation of hydrogen bonds stabilizes the protein and helps hold the protein in that shape. Again, any one hydrogen bond is weak, but *many* may be present in a globular protein.

Ionic and other polar interactions result whenever a full or partial positive charge is close to a full or partial negative charge. In a compactly folded protein, multiple ionic and polar interactions may be present. Ionic interactions are quite strong, whereas other polar interactions are weaker, but again, these interactions collectively contribute to help make the protein conformation stable (Figure 25.11).

Hydrophobic interactions also contribute to protein shape and stability. Hydrophobic side groups are found within the core of most proteins. In the core, few or no water molecules are found. Most or all of the amino acid side chains within the core of a protein are nonpolar or hydrophobic. This clustering of nonpolar side chains is much like the clustering of the hydrophobic tails of soap molecules in micelles or the clustering of the nonpolar tails of membrane lipids in the core of a membrane. Water-hating nonpolar molecules or parts of molecules are "squeezed out" of water (Chapter 24); they prefer to be together in a hydrophobic environment. Much of the stability of many globular proteins seems to be due to these hydrophobic interactions (Figure 25.11).

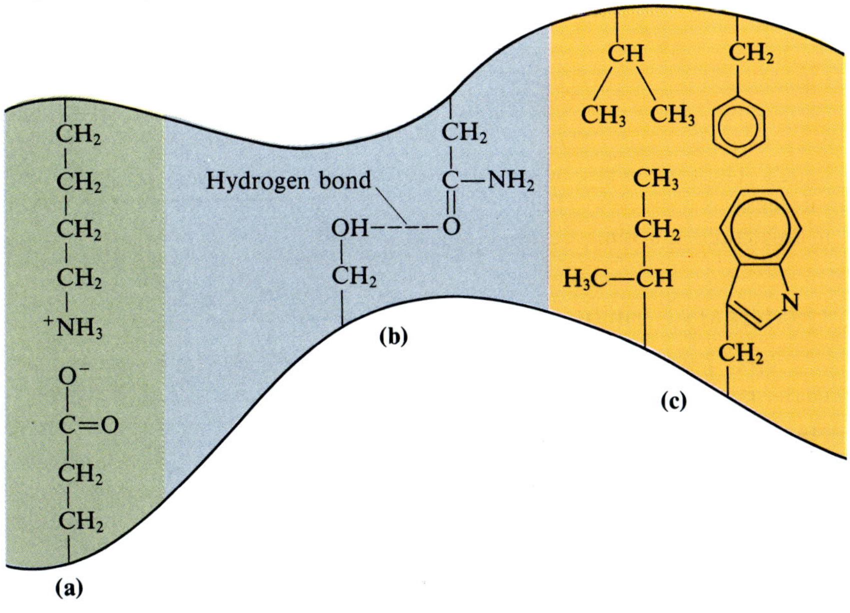
Hydrogen bond
(a)
(b)
(c)

FIGURE 25.11
The forces responsible for tertiary structure of a protein: (a) ionic and polar bonding (green), (b) hydrogen bonding (blue), and (c) hydrophobic interactions (yellow).

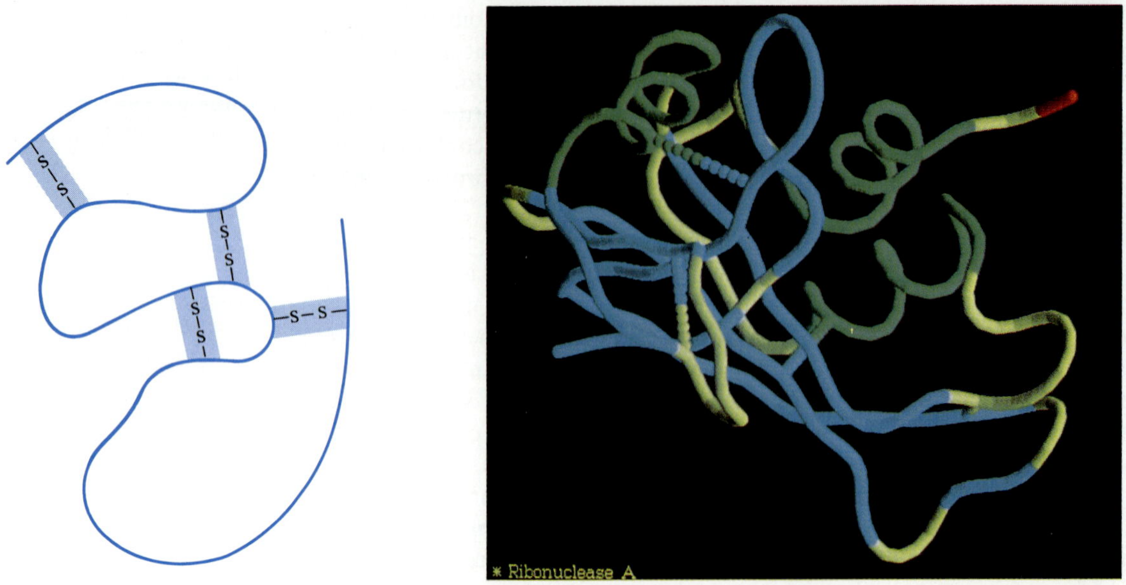
—S—S—
* Ribonuclease A

(a) (b)

FIGURE 25.12
Disulfide bonds are covalent bonds in proteins that help maintain tertiary structure. (a) Schematic drawing of a hypothetical protein with the bonds highlighted. (b) A three-dimensional representation of the enzyme ribonuclease with four disulfide bonds.

In Chapter 19 you were introduced to a covalent bond that forms between two thiols. This bond is called a **disulfide bridge** or bond, and it forms by oxidation. Once formed, the bond can be broken by a reduction reaction.

$$\underset{\textbf{Thiol}}{R{-}SH} + \underset{\textbf{Thiol}}{HS{-}R} \underset{\text{Reduction}}{\overset{\text{Oxidation}}{\rightleftharpoons}} \underset{\textbf{Disulfide}}{R{-}S{-}S{-}R}$$

Once a newly synthesized protein assumes its native conformation, one or more pairs of cysteine side chains (the sulfhydryl groups) may be brought next to each other. These pairs oxidize to form disulfide bonds (Figure 25.12). These covalent bonds are stronger than the individual bonds and interactions previously discussed, and they significantly stabilize the protein in this conformation.

Quaternary Structure

Some proteins possess more than one polypeptide chain, and the term **quaternary structure** is used to describe this level of structure. These proteins are called oligomeric proteins, which means they are oligomers of polypeptide chains. For example, the oxygen transport protein hemoglobin has four *subunits*, two identical subunits designated as α (alpha) and two other identical subunits designated as β (beta). The four subunits in hemoglobin have a very specific arrangement designated $\alpha_2\beta_2$.

The number and arrangement of polypeptide chains in a protein is called its quaternary structure.

What forces hold these subunits together in this particular way? The same forces that are responsible for tertiary structure are responsible for

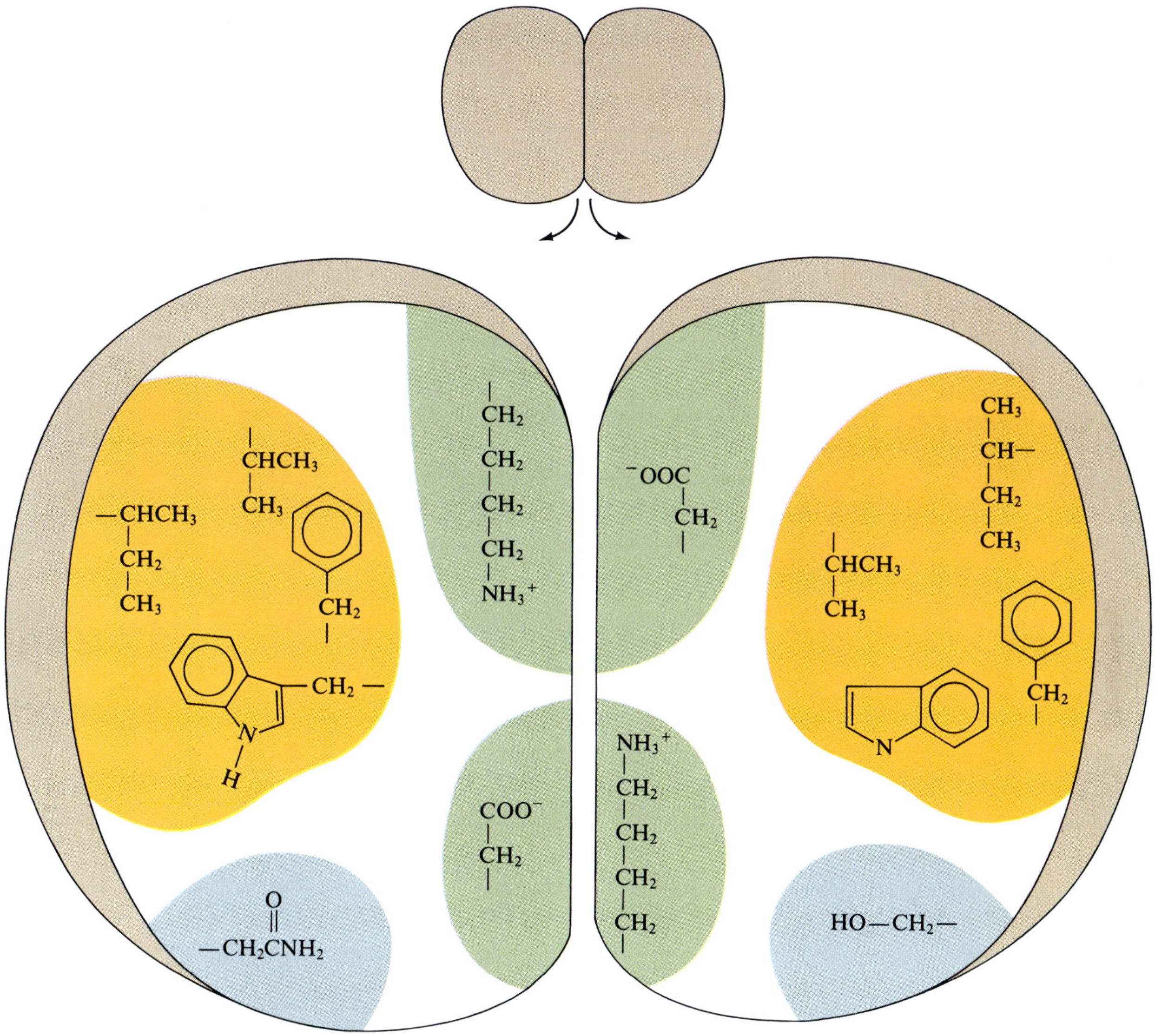
CH2
CH2
CH2
CH2
NH3+
CHCH3
CH3
—CHCH3
CH2
CH3
CH2
CH2 —
N
H
O
—CH2CNH2
COO−
CH2
−OOC
CH2
NH3+
CH2
CH2
CH2
CH2
CH3
CH—
CH2
CH3
CHCH3
CH3
CH2
N
HO—CH2—

FIGURE 25.13
Subunit interaction in a hypothetical dimeric protein. The two faces bind to each other through ionic and polar bonding (green), hydrogen bonding (blue), and hydrophobic interactions (yellow). The forces responsible for quaternary protein structure are essentially the same as those responsible for tertiary structure.

quaternary structure—hydrophobic interactions, hydrogen bonds, and ionic and polar interactions. It may be helpful to picture the two interacting faces of a dimeric protein as being complementary to each other (Figure 25.13). A portion of the face of one of the subunits may have a hydrophobic surface that matches with a hydrophobic surface of the other. One subunit may possess a region with positively charged groups that is attracted to negative groups on the corresponding region of the other subunit. One subunit has just the right groups in just the right orientation to form hydrogen bonds to groups on the face of the other subunit. Collectively, these forces hold the subunits together in a specific orientation.

SICKLE CELL ANEMIA

A look at the human genetic disease *sickle cell anemia* illustrates the importance of protein structure for proper function. Portions of the human population that originated in the malarial belt of the world possess a type of hemoglobin called hemoglobin S. In the United States, some blacks and a smaller proportion of the people of Mediterranean descent possess this type of hemoglobin.

Hemoglobin S differs from normal hemoglobin by one amino acid in the β-chain. In normal hemoglobin, the sixth residue is a glutamic acid, while in hemoglobin S it is a valine. Both hemoglobins bind oxygen properly, but hemoglobin S binds to other hemoglobin S molecules in a way that normal hemoglobin does not. Hemoglobin is tightly packed inside erythrocytes (or red blood cells), and normal hemoglobin molecules do not bind to each other within the erythrocyte. But hemoglobin S molecules can bind to one another if many of them contain no oxygen. When the hemoglobin S molecules bind to each other, they form long filaments. These filaments change the shapes of the red blood cells into sickle-shaped cells (see figure).

These sickle cells are fragile, easily broken, and do not pass as readily through capillaries. Normal blood flow to tissues may be impaired, and anemia (iron deficiency) may result from breakage of the sickled erythrocytes. A single amino acid difference in this protein causes a distinct difference in the solubility behavior of the protein. This affects cell shape and stability, which can lead to impaired cardiovascular function.

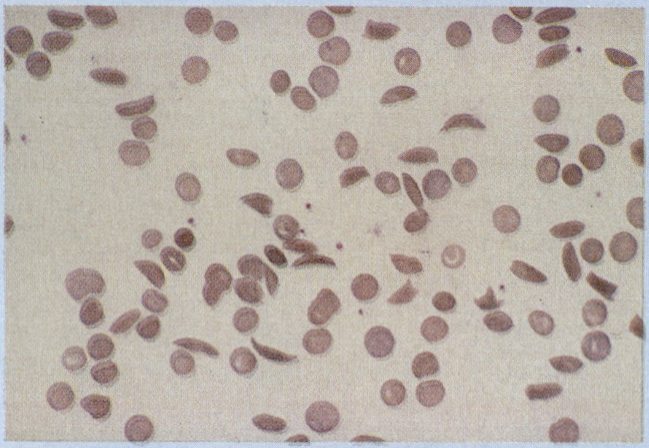

The blood cells of someone with sickle cell anemia.

Other Components in Proteins

The prosthetic group of a conjugated protein is the part that is not an amino acid.

So far we have presented a picture of a protein as one or more polypeptide chains arranged into a specific shape. Proteins possessing only amino acids connected by peptide bonds are called **simple proteins**. Other proteins, called **conjugated proteins**, possess one or more additional components called prosthetic groups. A **prosthetic group** is a non–amino acid component of a protein. For example, the prosthetic group in hemoglobin is *heme*. Hemoglobin contains four copies of heme, one bound to each of the four polypeptides of hemoglobin. A second example is the digestive enzyme carboxypeptidase, which contains zinc ion (Zn^{2+}) as its prosthetic group. Conjugated proteins lose their native activity if the prosthetic group is removed.

Macromolecular Complexes

Proteins have a unique shape because certain forces and interactions occur that cause the protein to assume this most stable conformation. Are these same forces responsible for other molecular complexes and organelles? Most definitely. Hydrophobic interactions, hydrogen bonding, and ionic interactions all play a part in holding many different macromolecular complexes together. For example, ribosomes (organelles involved in protein synthesis) are macromolecular complexes consisting of many different proteins and a few large molecules of ribonucleic acid (RNA). Each of the proteins and RNAs of a ribosome are synthesized separately, yet they bind together to form the subunits of a ribosome. This process is called *self-assembly*. Self-assembly occurs because the components are more stable when they are bound to other components rather than being free in solution. Self-assembly is a characteristic of all macromolecular complexes.

NEW TERMS

native conformation The normal shape or conformation that a protein has in its biological setting performing its normal biological activity.

denaturation Process that results in the loss of a protein's native conformation, and therefore its activity.

primary structure The amino acid sequence in a protein or peptide.

secondary structure The helical or pleated structure of a protein that is due to hydrogen bonding between atoms of the polypeptide chain.

α-helix Secondary structure of a protein in which the polypeptide backbone is arranged into a regular spiral shape.

β-pleated sheet Secondary structure of a protein in which the polypeptide chain or chains are arranged into a sheetlike structure.

tertiary structure The compact, three-dimensional shape of globular proteins.

disulfide bridge A covalent bond between two sulfur atoms.

quaternary structure The structure of an oligomeric protein resulting from specific interactions among the subunits.

simple protein A protein containing only amino acids.

conjugated protein A protein that contains one or more prosthetic groups in addition to amino acids. (The groups may be organic or inorganic.)

prosthetic group A group in a protein that is not an amino acid and which plays a major role in the activity of conjugated proteins.

TESTING YOURSELF

Protein Structure

1. What is the primary structure of a protein?
2. What forces are responsible for maintaining tertiary structure in a protein?
3. A protein is broken down in the laboratory to yield amino acids and an organic molecule called pyridoxal. How would you classify this protein? What is pyridoxal in this case?

Answers **1.** the amino acid sequence in the protein **2.** hydrogen bonding, ionic and other polar interactions, hydrophobic interactions, and disulfide bridges **3.** This is a conjugated protein. It is the prosthetic group of this protein.

25.5 Properties and Classification of Proteins

Properties of Proteins

Proteins are amphoteric molecules, that is, they can act as acids or bases. The amphoteric nature of proteins is due to the presence of acidic and basic side chains in some of the amino acids found in the protein. At the neutral pH typical of a cell, the carboxyl groups in side chains will be deprotonated and will exist as carboxylate ions, $—COO^-$. The basic groups on lysine and arginine will be protonated; for example, the amino group of lysine will be $—NH_3^+$. These ionic groups are generally located on the surface of the protein, so the surface would be a mosaic of charges, some positive and some negative. If the solution were to become more acidic, the carboxyl groups would accept some of the hydrogen ions.

$$—COO^- + H^+ \longrightarrow —COOH$$

A negatively charged carboxylate ion would be changed to a neutral carboxyl group. In addition, if any histidine residues were present, these would also accept a hydrogen ion to become positively charged. Under acidic conditions, the charge on a protein becomes more positive. In basic solutions, the protonated side chains of arginine and lysine will give up hydrogen ions. These positively charged groups become neutral, and the net charge on the protein becomes more negative. The net charge on a protein changes with the pH of the solution.

As the pH of a solution changes, the charge of a protein changes.

What effects will changes in pH have on the properties of a protein? Since pH affects the charge of a protein, then any property that depends on the charge of the protein will be influenced by pH. Two properties that are influenced are solubility and denaturation.

Consider a protein in an aqueous solution. Each of the protein molecules interacts with the water molecules and solutes that may be present (Figure 25.14). Water and water soluble solutes interact with polar and ionic side chains located on the surface of the protein. These polar interactions between the protein and the solvent keep the protein in solution. As long as the interaction of the protein molecules with the solvent are stronger than the interactions between protein molecules, the protein molecules will remain in solution.

Why don't protein molecules bind to each other since their surfaces are polar and therefore should be able to form polar interactions? Each of the protein molecules has some net charge that is a reflection of the ionizable side groups found in the protein (Figure 25.14a). A protein may have more acidic side groups than basic ones, therefore the net charge on the protein will be negative because at normal cellular pH, the carboxyl groups will have lost their hydrogen ion and will have a negative charge. Since each protein molecule has a net negative charge, they tend to repel each other. They remain independent in the solvent rather than aggregating into a complex too large to remain in solution (Figure 25.14b). If the solution becomes more acidic, some of the carboxylate ions will be protonated, which will decrease the negative charge on the protein. At a certain pH value, enough of the carboxylate ions will be protonated that the net charge on the protein will be zero. Just as amino acids are neutral as their isoelectric pH, so are proteins. At the isoelectric point, the proteins are neutral and no longer repel each

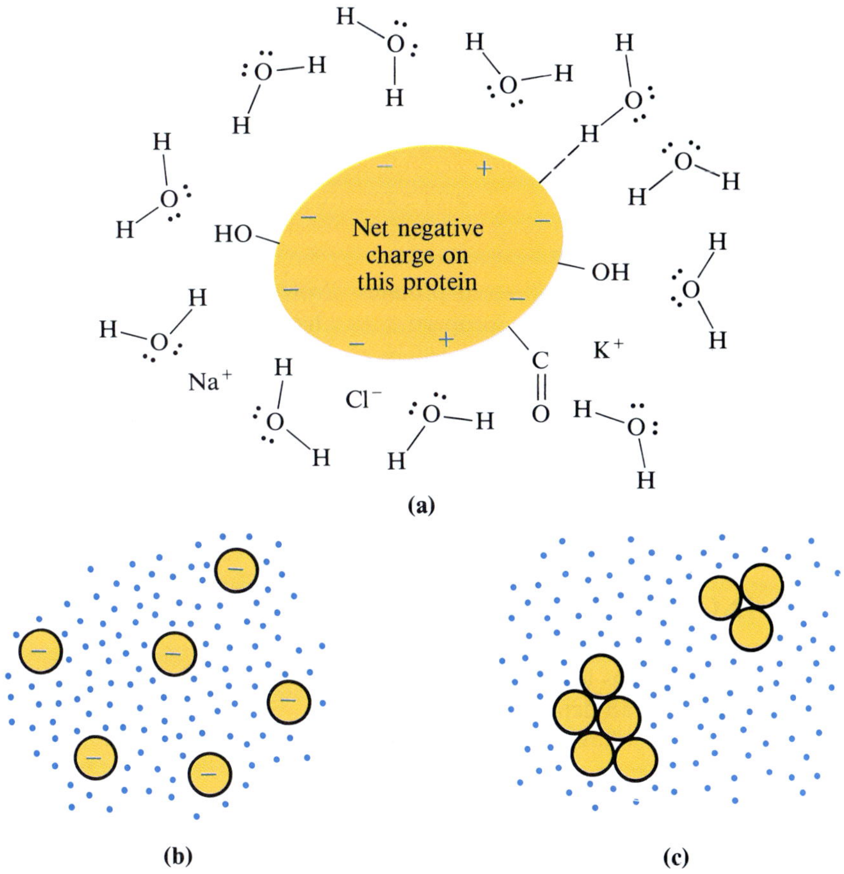
Net negative charge on this protein
HO
OH
Na$^+$
Cl$^-$
K$^+$
(a)
(b)
(c)

FIGURE 25.14
An illustration of protein–solute and protein–protein interactions. (a) The polar and ionic groups on a protein surface interact with water and solutes through non-covalent bonds. The protein is solvated and remains in solution; it has a net negative charge at this pH. (b) The protein molecules will not bind to each other because their net negative charges effectively repel each other. The proteins remain in solution. (c) If the pH of the solution changes, the net charge on the protein changes. At the pI, the protein has no net charge, and the protein molecules no longer repel each other. They aggregate and precipitate from solution.

other. The protein molecules tend to aggregate and will precipitate from solution (Figure 25.14c). In general, proteins are least soluble at their isoelectric point.

Changes in pH can denature a protein. Denaturation occurs whenever a protein changes from its normal or native conformation to some other conformation. Recall that the structure of a protein may depend on specific ionic interactions. If the pH changes, certain ionizable side chains may accept or lose hydrogen ions. Ions may form or disappear, and specific stabilizing ionic interactions may be lost and other destabilizing ionic interactions may occur. Generally, proteins are denatured if the pH of the solution goes beyond a certain range.

Several factors disrupt normal bonding and interactions in proteins and thus lead to denaturation.

The pH of a solution is not the only factor that can affect the native conformation. Heat, solvents, and heavy metals can also change it. Heating a solution increases the thermal energy of the molecules in the solution. As the protein molecules gain thermal energy, the bonds in the molecule vibrate and rotate more strongly. Eventually enough heat is gained that the hydrogen bonds and other interactions maintaining the protein conformation are no longer strong enough to hold the protein in this conformation. The protein assumes other more random shapes, and native function and properties are lost. For example, when you fry an egg, the soluble proteins in the egg white become denatured. These proteins are highly insoluble in the denatured form, thus a white solid forms from the clear, viscous solution that is the egg white.

Classification of Proteins

Classification of proteins by one simple system is not possible because proteins vary greatly in size, shape, function, and properties. As a consequence, proteins are classified by a variety of systems. They are sometimes classified by function, as was shown in Table 25.1. They can be classified by shape or solubility. **Fibrous proteins** are fiberlike and are generally insoluble, although some contractile proteins of muscle and other cells are exceptions. **Globular proteins** are spheroid and tend to be soluble. They can be classified by composition as simple or conjugated proteins. Proteins are indeed numerous, complex, and varied in many properties. In Chapter 27 the enzymes, which are proteins with catalytic activity, will be studied in more detail. Even within this single class of proteins, a great diversity in function and properties is seen.

NEW TERMS

fibrous proteins Fiberlike proteins that are usually insoluble.

globular proteins Spheroid, generally soluble proteins.

TESTING YOURSELF

Proteins

1. Why are proteins least soluble at their pI?
2. What factors can denature proteins?
3. Compare the solubility of fibrous and globular proteins.

Answers **1.** Proteins have no net charge at the isoelectric point, thus they do not repel each other. **2.** heat, pH, change in solvent, and some heavy metals **3.** Most fibrous proteins are insoluble, while many of the globular proteins are soluble.

25.6 Proteins and Amino Acids in Nutrition

Amino acids are needed by the body to synthesize proteins. Some amino acids can be synthesized by animals, others cannot. Those that cannot be synthesized are called **essential amino acids**. They include the following:

histidine	isoleucine
leucine	lysine
methionine	phenylalanine
threonine	tryptophan
valine	

The essential amino acids cannot be synthesized in the body.

These amino acids must be included in the diet to ensure proper growth or function because adequate protein synthesis cannot occur unless all 20 amino acids are present simultaneously in adequate amounts. Two of the nonessential amino acids are synthesized from essential amino acids. Tyrosine is made from phenylalanine, and the sulfur of cysteine comes from methionine. Thus, the diet must provide extra methionine and phenylalanine if tyrosine or cysteine are present in inadequate quantities.

In a nutritional sense, proteins are the sources of amino acids. Dietary proteins are broken down during digestion to amino acids, which are then

PROTEIN–CALORIE MALNUTRITION

Inadequate protein in the diet is most harmful to children. If protein synthesis cannot occur adequately during development, temporary or permanent physical or mental impairment may result. *Protein–calorie malnutrition* (*PCM*) is a disease seen primarily in children of developing nations. This disease appears to be caused by inadequate amounts of protein in the diet, but the total amount of energy available in the diet is also often inadequate. *Kwashiorkor* is a form of the disease in which the child has edema, an enlarged liver, and peeling skin (see figure). This form of the disease may be more common when caloric intake is close to adequate but too little dietary protein is present. Marasmus is another form of the disease in which the body appears more wasted away, but edema tends to be absent. Diets too low in both calories and protein may lead to this form of PCM. Often victims of this disease show some symptoms of both forms.

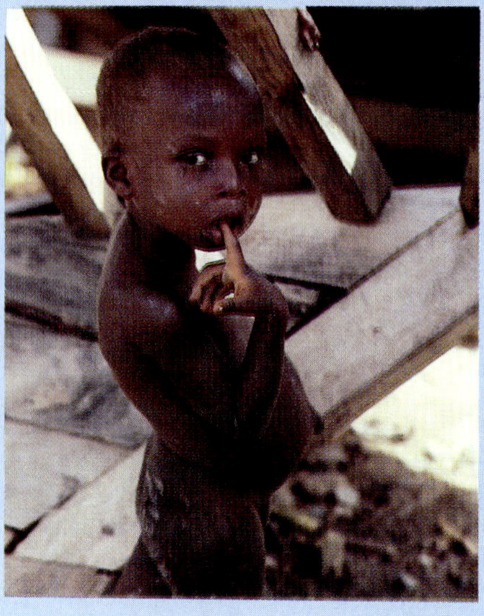

This child has inadequate protein in his diet and is suffering from kwashiorkor, a variety of PCM.

absorbed (see Chapter 28). They are used by cells to make proteins or other compounds. If the amino acids are in excess of body needs for protein synthesis, they are broken down to yield energy. A gram of protein yields 4 Cal (4 kcal) of energy.

Protein yields 4 Cal/g, about the same as carbohydrate and much less than 1 g of fat.

Although many foods contain protein, meat and dairy products are among the best sources; other good sources include foods derived from some seeds and nuts. Generally foods from animal sources are richer in protein and, in a nutritional sense, have better protein than plant sources. Most vegetables and fruits contain relatively little protein.

NEW TERM

essential amino acids The amino acids that cannot be synthesized by the body.

TESTING YOURSELF

Nutrition

1. An ounce of a common breakfast cereal contains 3 grams of protein, 1 gram of fat, and 23 grams of carbohydrate. Estimate the caloric content of one ounce of this cereal. (If necessary, look up the caloric content of these food groups in Chapters 23, 24, and 25.)

Answer (3 g protein) × (4 Cal/g protein) + (1 g fat) × (9 Cal/g fat) + (23 g carbohydrate) × (4 Cal/g carbohydrate) = 113 Cal (The cereal box states 1 oz of the cereal contains 110 Cal.)

SUMMARY

Amino acids are bifunctional molecules that are the building blocks of *proteins* and *peptides.* All amino acids contain an α-hydrogen, α-amino group, and α-carboxyl group, but each of them has a unique side chain. Many of the properties of amino acids are determined or influenced by the side group. Amino acids can be linked together by *peptide bonds* to form peptides or proteins characterized by unique amino acid residues. As a result of the order of amino acids in a protein chain (*primary structure*), these macromolecules have very precise *secondary*, *tertiary*, and in some cases, *quaternary* structure. These unique structural characteristics impart specific biological activity to proteins, including catalytic activity to enzymes. Proteins provide the amino acids needed by the body for protein synthesis and as a source of energy. Proteins control many of the processes that occur within cells. Proteins are, in turn, synthesized by a highly regulated process discussed in Chapter 26, which covers the molecular basis of heredity.

TERMS

protein
amino acids
zwitterion
amphoteric
isoelectric point (pI)
electrophoresis
peptide bond
peptide
polypeptide
native conformation
denaturation
primary structure
secondary structure
α-helix
β-pleated sheet
tertiary structure
disulfide bridge
quaternary structure
simple protein
conjugated protein
prosthetic group
fibrous proteins
globular protein
essential amino acids

CHAPTER REVIEW

QUESTIONS

1. Secondary structure of a protein is shaped almost exclusively by
   a. hydrogen bonds c. ionic bonds
   b. disulfide linkages d. hydrophobic interactions
2. Almost all of the common amino acids
   a. are in the D family c. are split equally between both families
   b. are in the L family
3. An essential amino acid is
   a. incorporated into all proteins
   b. found in all food
   c. cannot be synthesized by the human body
4. A zwitterion is *not*
   a. soluble in hydrocarbons c. electrically neutral
   b. soluble in water d. polar
5. The peptide linkage is an example from organic chemistry of
   a. an ester b. an amide c. an anhydride d. an amine
6. The clustering of the side chain of leucine with the side chain of valine is referred to as
   a. disulfide bonding c. hydrophobic interactions
   b. hydrogen bonding d. ionic bonding

7. The α-helix of a protein is an example of
   a. primary structure b. secondary structure c. tertiary structure
8. Which of the following is a peptide?
   a. hemoglobin b. casein c. myoglobin d. vasopressin
9. Which of the following proteins serves a defensive role?
   a. casein b. myoglobin c. lysozyme d. cytochrome *c*
10. Changes in pH most affect what type of bonding in proteins?
    a. hydrophobic interactions c. hydrogen bonding
    b. ionic interactions d. disulfide bridges

Answers **1.** a **2.** b **3.** c **4.** a **5.** b **6.** c **7.** b **8.** d **9.** c **10.** b

DIAGNOSTIC CHART

Blacken in all of the circles under the number of each question you missed in the preceding Chapter Review questions. The diagnostic chart will help you identify concept areas that need more study.

	Question									
Concepts	**1**	**2**	**3**	**4**	**5**	**6**	**7**	**8**	**9**	**10**
Function									○	
Amino acids		○	○	○		○				
Peptide bond					○					
Peptides								○		
Protein structure	○					○	○			○
Properties										○
Nutrition			○							

EXERCISES

Protein Function

1. List the general functions of proteins discussed in this chapter.
2. Describe the specific function of these proteins.
   a. enzymes
   b. antibodies
   c. blood lipoproteins
   d. hemoglobin
   e. myoglobin
   f. lysozyme
   g. casein
   h. insulin
   i. collagen

Amino Acids

3. Draw the structure of an α-amino acid. Why are these compounds called α-amino acids?
4. Describe zwitterions.
5. The amino acids of the body are related to which glyceraldehyde? Draw the structure of alanine, showing the correct stereochemistry.
6. List the nonpolar amino acids.
7. What is the difference between a basic and an acidic amino acid? Draw the structures of one basic and one acidic amino acid in both their protonated and unprotonated forms.
8. What property must a compound possess to be classed as amphoteric?
9. Explain what is meant by the term *isoelectric point*.
10. Examine Figure 25.3. Predict the migration of these proteins if the pH of the solution were made more acidic.

Peptide Bond and Peptides

11. What is a peptide bond?
12. Draw the tripeptide ala-gly-leu, with alanine as the N-terminal amino acid. HINT: Refer to the list of amino acids.
13. Make a table of the biologically active peptides discussed in this chapter. Include the names, structures, and biological role or effect.

14. Why is there no free rotation about a peptide bond?

Protein Structure

15. Describe what is meant by primary, secondary, tertiary, and quaternary structure of proteins.
16. Describe the forces that are responsible for the maintenance of the protein structures in question 15.
17. Draw a small region of polypeptide in an α-helix. What forces maintain the chain in this conformation?
18. What is a *disulfide linkage* in protein structure?
19. The structure of the enzyme RNA polymerase is sometimes shown as $\alpha_2\beta\beta'\sigma$. What do you think this symbolism means?
20. What does *self-assembly* mean?
21. Compare the composition of a simple protein and a conjugated protein.

Properties of Proteins

22. What is denaturation of a protein?
23. Which amino acids would contribute a positive charge to a protein in solution? Which would contribute a negative charge?
24. A protein has no net charge at pH 6.13. What charge would it have at pH 7? At pH 9? At pH 5.5?
25. At which pH would the protein in question 24 have the lowest solubility?

Nutrition

26. List the essential amino acids.
27. A diet for experimental rats is prepared containing the minimal amounts of the essential amino acids needed for normal growth in this species. The diet is lacking in the amino acid cysteine. Describe the growth of these rats, and explain the basis for the observed growth.
28. What names are given for the forms of protein–calorie malnutrition? What nutritional conditions contribute to these two forms?

General Question

29. A list of some words and phrases introduced in this chapter is presented below. For each, provide a description, definition, or example to show that you understand the word or phrase.
    a. α-amino acid
    b. L-amino acid
    c. essential amino acid
    d. zwitterion
    e. amphoteric
    f. isoelectric point
    g. dipeptide
    h. peptide linkage
    i. protein
    j. primary structure
    k. secondary structure
    l. tertiary structure
    m. hydrophobic interaction
    n. disulfide bridge
    o. denaturation

CHAPTER

26

Molecular Basis of Heredity

Living organisms are unique among the matter found on earth because they alone can reproduce. Reproduction and the day-to-day activities of an organism require organization and information. Nucleic acids are the carriers of the information needed for reproduction and other vital functions. All living organisms use the same kinds of molecules to provide the genetic information that is needed for continuity of life.

OBJECTIVES

After completing this chapter, you should be able to

- Recognize a nucleotide, and identify those that are important in DNA and RNA.
- Describe the structure of DNA, and explain why specific bases are paired.
- Explain the process of DNA replication.
- Describe the structure of RNA, and explain why it is complementary to a strand of DNA.
- Identify the roles of DNA, mRNA, and tRNA in the synthesis of proteins.
- Define the term *codon*, and explain the role of codons in protein synthesis.
- Describe the molecular basis for the introduction of errors into genetic material.
- Explain how induction and repression regulate protein synthesis in bacteria.
- Recognize the overall strategy of genetic engineering for introducing a gene into an organism.

26.1 Search for the Molecular Basis of Heredity

Living organisms are unique among the materials of the earth because they have the ability to replicate themselves. "Like begets like" is a biblical phrase that shows this was known to ancient peoples. Yet variability is present within this constancy—cats beget cats, but kittens differ from each other. We see this in humans, too; while sets of identical twins appear very similar or nearly identical, other people are distinctly different in appearance (Figure 26.1). What is the molecular basis for these similarities and differences? Genetics and molecular biology are disciplines that study how traits are passed from one generation to another. While the study of the biological processes that control heredity began well over a century ago with the pioneering experiments of Gregory Mendel, the molecular basis of heredity was not established for nearly another century.

The search for the molecular basis of heredity has been long and fascinating. It began when scientists studying cells and genetics become convinced that the cell nucleus contained the molecules that controlled heredity. They had observed that nuclear division always preceded the division of

FIGURE 26.1
Individuals always have offspring that are recognizable as members of that species, yet considerable differences are found within a species. This chapter presents the molecular basis for this consistency and variability.

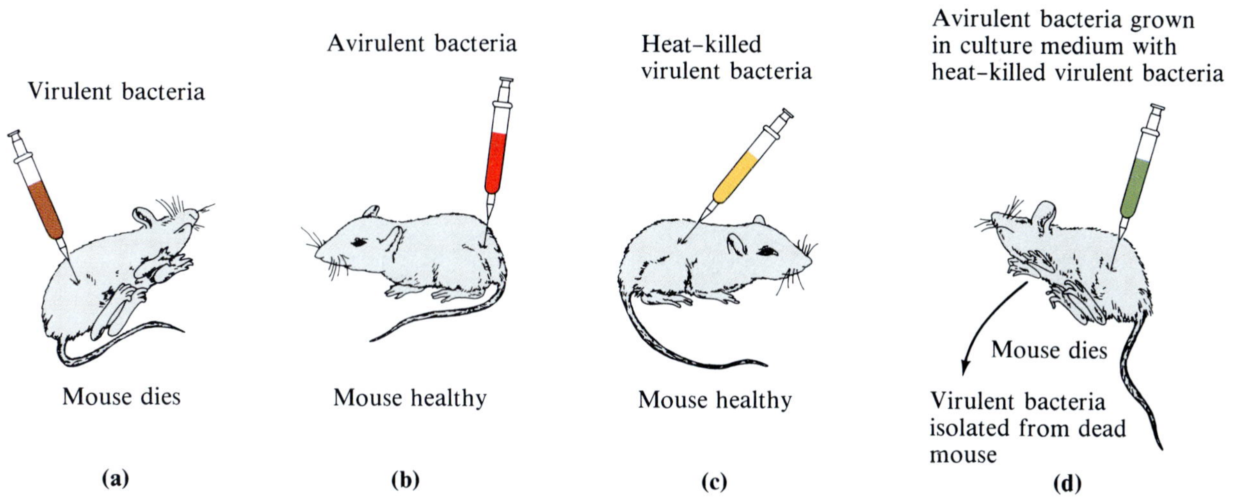
Virulent bacteria
Mouse dies
(a)
Avirulent bacteria
Mouse healthy
(b)
Heat-killed virulent bacteria
Mouse healthy
(c)
Avirulent bacteria grown in culture medium with heat-killed virulent bacteria
Mouse dies
Virulent bacteria isolated from dead mouse
(d)

FIGURE 26.2
Experiment on bacterial transformation. (a) Virulent *Streptococcus pneumoniae* kill mice, (b) but the avirulent strain does not. (c) Heat-killed virulent strain cannot kill mice, (d) but avirulent strain grown with heat-killed virulent strain does kill mice. The avirulent strain becomes virulent because it takes up molecules from the killed virulent bacteria. These molecules transform the avirulent strain into a disease-causing strain.

cells, and this and other data implied that the nucleus was the site where the molecules of heredity were found.

Biochemical studies of the nucleus revealed numerous small molecules, many proteins, and two classes of large acidic molecules called **deoxyribonucleic acid (DNA)** and **ribonucleic acid (RNA)**. At first, it was not known which, if any, of these molecules were the molecules of heredity. With time, scientists began to believe proteins were responsible for heredity because only proteins appeared sufficiently large and complex to contain the vast amount of information that a cell had to have. Small molecules could not possibly contain that much information, and although RNA and DNA were large, they appeared to be composed of only a few basic building blocks. They did not appear complex enough to contain all the needed information. The hypothesis that proteins were carriers of information became generally accepted, but it was supported by little direct evidence.

In the late 1920s some experiments with bacteria provided clues that ultimately lead to the determination of the molecular basis of heredity. These experiments showed that nondisease-producing (*avirulent*) strains of a bacterium could be converted into a disease-producing (*virulent*) strain simply by growing the avirulent strain in the culture medium where a disease-producing strain had been grown (Figure 26.2). In these experiments, all of the virulent bacteria had been killed or removed from the medium, so it was not simply contamination of the avirulent strain with the virulent strain. Apparently, molecules in the medium that came from the virulent strain were able to change the avirulent strain into a virulent one. Most importantly, the transformation was permanent; virulence remained through all future generations of the strain. The heredity of the strain had been altered.

A reasonable hypothesis for these observations was that molecules containing genetic information were released into the culture medium by the virulent strain. The cells of the avirulent strain then took up these molecules and made them a permanent part of the cell. These experiments indicated that molecules are involved in heredity, but they did not establish the identity of the molecules involved.

In the early 1940s, Oswald Avery and his colleagues at the Rockefeller Institute used these bacteria to determine the molecules of heredity. They

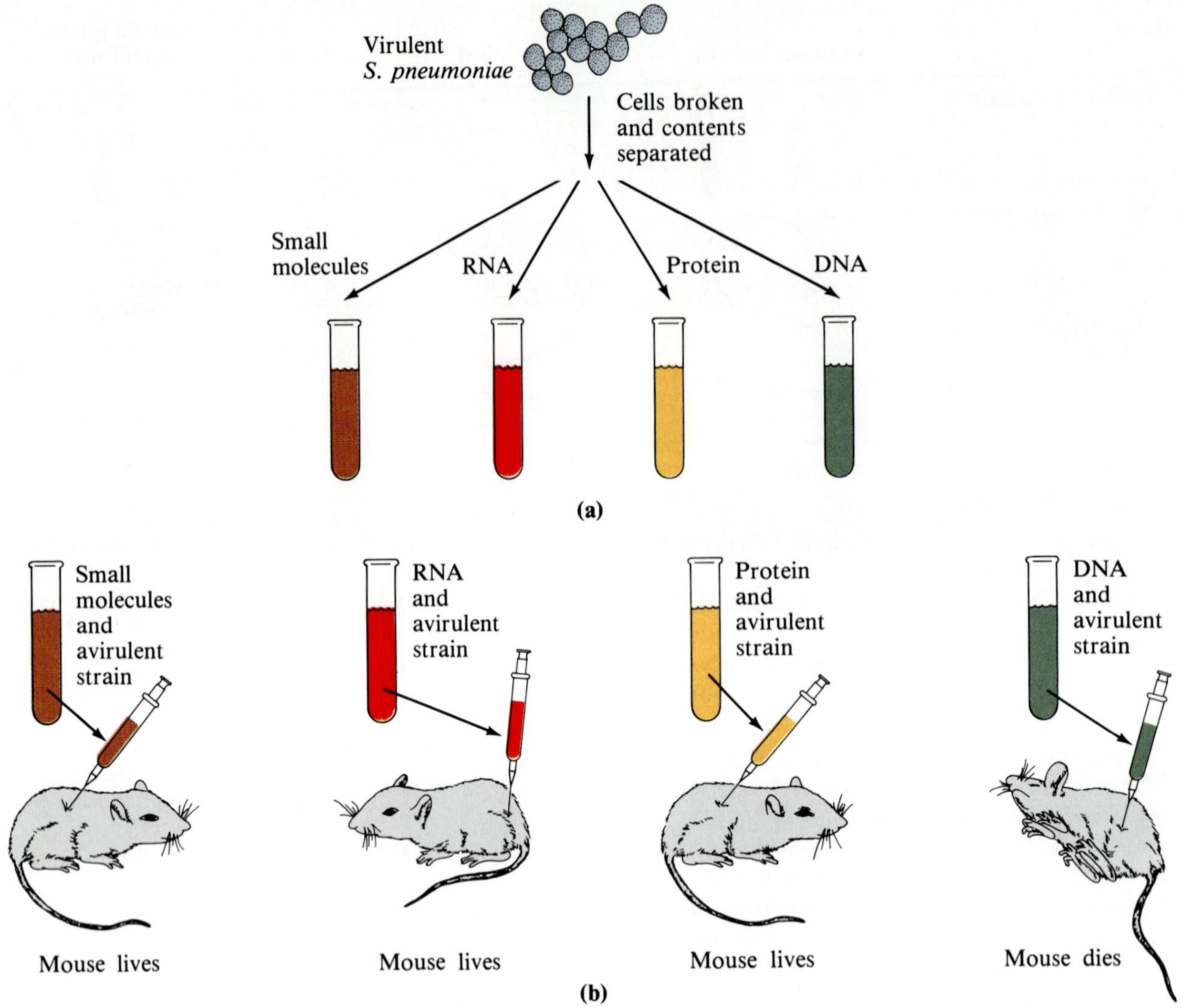
Virulent
S. pneumoniae
Cells broken
and contents
separated
Small
molecules
RNA
Protein
DNA
(a)
Small
molecules
and
avirulent
strain
RNA
and
avirulent
strain
Protein
and
avirulent
strain
DNA
and
avirulent
strain
Mouse lives
Mouse lives
Mouse lives
Mouse dies
(b)

FIGURE 26.3
Avery and his colleagues showed that DNA transforms avirulent bacteria into a virulent strain. (a) Virulent *S. pneumoniae* were killed, broken open, and the cellular contents separated from each other. (b) Avirulent *S. pneumoniae* were transformed into a virulent strain only by the fraction containing DNA.

took virulent cells and removed several kinds of molecules from them. Avirulent cells were mixed with each of these fractions and allowed to grow for a period of time. These cells were then tested for virulence (Figure 26.3). Only the cells grown in the presence of the DNA fraction became virulent, suggesting that DNA was the molecule of heredity. It was possible that the DNA fraction was contaminated with proteins, and the contaminating proteins could have caused the change from avirulent to virulent. They tested this by treating portions of the DNA fraction with either *proteases* (enzymes that destroy proteins) or *DNase* (a DNA-destroying enzyme). The protease-treated DNA fraction caused transformation into virulent strain, but the DNase-treated samples were unable to transform the avirulent bacteria into a virulent strain. Clearly DNA was the molecule of heredity. What remained was to determine how this huge but seemingly monotonous molecule could carry information.

DNA is the molecule that carries genetic information.

NEW TERMS

Deoxyribonucleic acid (DNA) A very large, acidic macromolecule found in the cell nucleus. This molecule is the carrier of genetic information.

Ribonucleic acid (RNA) Several kinds of acidic macromolecules found in the nucleus and other parts of the cell.

TESTING YOURSELF

Experiments with DNA

1. How were the enzymes DNase and protease used to establish that DNA, rather than protein, was the carrier of genetic information?

Answer **1.** They were used to destroy DNA and proteins in solutions that were used in transformation studies. Only solutions containing intact DNA caused transformation.

26.2 Nucleotides

Deoxyribonucleic acid, or DNA, is the molecule of heredity, and it is made up of nucleotides. DNA is a polynucleotide, that is, a polymer of nucleotides, thus an understanding of nucleotide structure must come before a discussion of DNA structure.

Nucleotides contain a sugar, a nitrogenous base, and one or more phosphate groups. The **nitrogenous base** is a heterocyclic base containing nitrogen and is typically bonded to a carbon on one side of the sugar. The phosphate is typically bonded to a carbon on the other side of the sugar. Very roughly, a nucleotide can be represented as

Phosphate \ Sugar / Nitrogenous base

Several different nitrogenous bases are found in nucleotides but only two sugars are common. Generally one to three phosphates are present in nucleotides. Nucleotides are the building blocks of DNA and the other nucleic acid, RNA. They are also present in a number of coenzymes, such as ATP and NAD. Because nucleotides are common in biochemistry, a brief look at their structure and nomenclature is in order.

Although a number of different nitrogenous bases can be found in nucleotides, five are of immediate interest. These five bases are *adenine*, *guanine*, *cytosine*, *thymine*, and *uracil*. These bases are heterocyclic compounds, but adenine and guanine have the fused ring system found in the compound purine (Figure 26.4a), while cytosine, thymine, and uracil resemble the compound pyrimidine (Figure 26.4b). Because of these resemblances, adenine and guanine are called *purines* and the other three bases are called *pyrimidines*. Take a moment to study the ring systems and the functional groups attached to the rings because these features distinguish the purines and pyrimidines from each other.

Adenine and guanine are purines; cytosine, thymine, and uracil are pyrimidines.

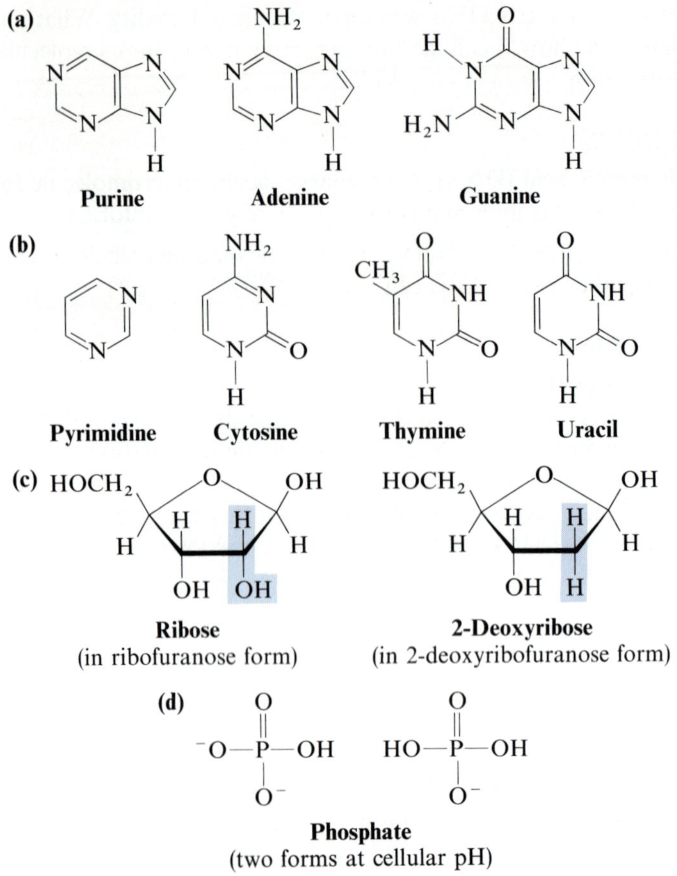
(a)
Purine
Adenine
Guanine
(b)
Pyrimidine
Cytosine
Thymine
Uracil
(c)
Ribose
(in ribofuranose form)
2-Deoxyribose
(in 2-deoxyribofuranose form)
(d)
Phosphate
(two forms at cellular pH)

FIGURE 26.4
The nitrogenous bases, sugars, and phosphate found in common nucleotides. (a) The nitrogenous bases adenine and guanine are shown with purine. The similar ring structure identifies these compounds as purines. (b) The nitrogenous bases cytosine, thymine, and uracil are shown with pyrimidine. Because they contain the same ring system, they are referred to as pyrimidines. (c) Ribose and 2-deoxyribose are the common sugars found in nucleotides. The structural difference between these sugars is highlighted. (d) Phosphate is the common name given to the forms of phosphoric acid found at cellular pH.

The sugars most commonly found in the nucleotides are ribose and deoxyribose (Figure 26.4c). Each is an aldopentose that exists as a furanose in living systems. These two sugars differ only in the substituents on carbon number two (C-2). Ribose has one hydrogen atom and one hydroxyl group as substituents, while deoxyribose lacks the hydroxyl group and thus has two hydrogen atoms as substituents. The prefix *deoxy-* signifies the absence or loss of an oxygen from the molecule. In nucleotides, a bond exists between a specific nitrogen atom of a nitrogenous base and C-1 of the furanose ring. This bond is highlighted in Figure 26.5.

Nucleotides also contain phosphate, which exists as an anion or dianion at cellular pH (see Figure 26.4d). When phosphate is in a nucleotide, it typically has a single negative charge.

Compounds containing only a sugar and a nitrogenous base are called **nucleosides**. Although these compounds are rarely found in biological sam-

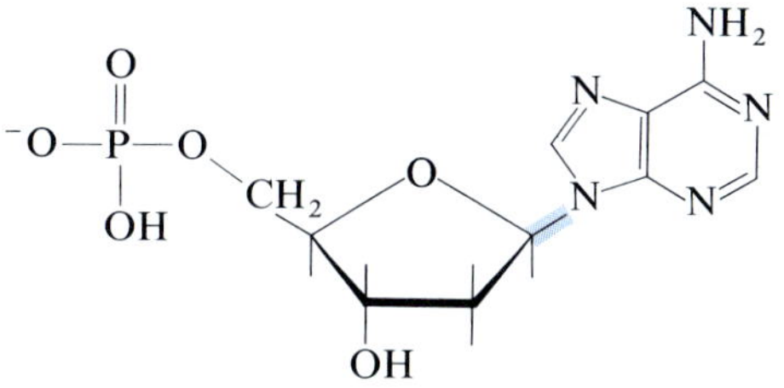

Deoxyadenosine monophosphate (dAMP)

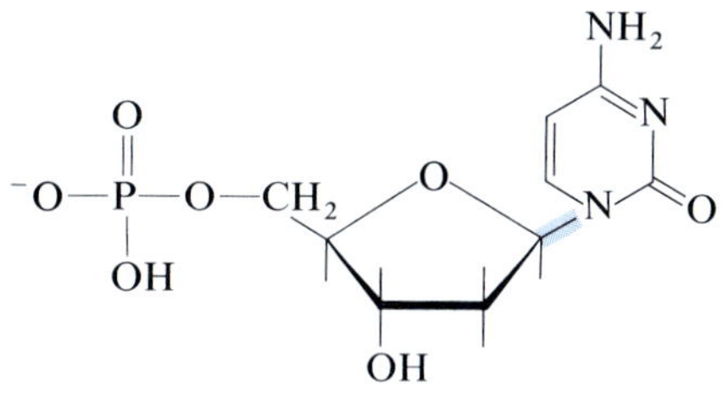

Deoxycytidine monophosphate (dCMP)

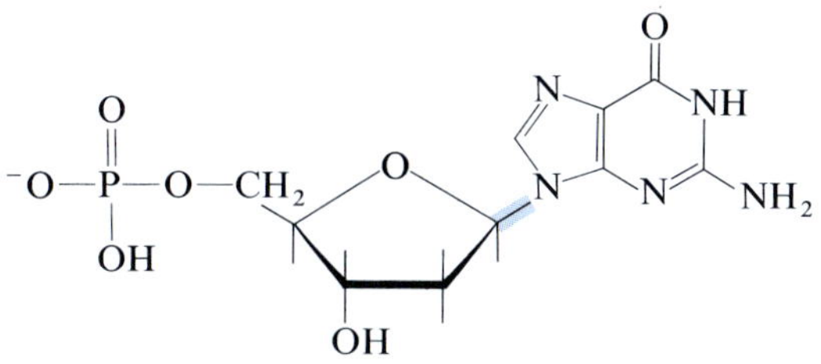

Deoxyguanosine monophosphate (dGMP)

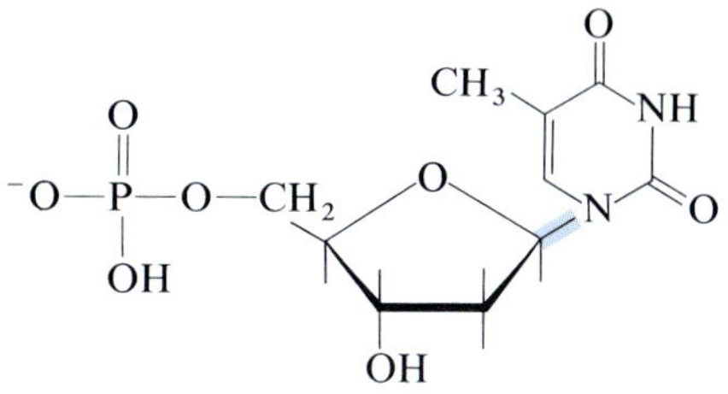

Deoxythymidine monophosphate (dTMP)

FIGURE 26.5 The structures of the deoxyribonucleotides. The highlighted bond between the base and the sugar is a glycosidic bond, and the bond between the sugar and the phosphate is an ester bond.

ples, they are important because the nomenclature of nucleotides is based on nucleoside nomenclature. The name of a nucleoside is derived from the nitrogenous base it contains. For example, a nucleoside containing adenine bonded to deoxyribose is called deoxyadenosine. Note as a general rule that

Base + Sugar = Nucleoside

Base + Sugar + One or more phosphates = Nucleotide

Table 26.1 summarizes the nomenclature of the nucleosides and nucleotides.

TABLE 26.1
Nomenclature of the Nucleosides and Nucleotides of DNA and RNA

Symbol	Base	Sugar	Nucleoside	Nucleotide
A	Adenine	Deoxyribose	Deoxyadenosine	Deoxyadenosine monophosphate (dAMP)
G	Guanine	Deoxyribose	Deoxyguanosine	Deoxyguanosine monophosphate (dGMP)
C	Cytosine	Deoxyribose	Deoxycytidine	Deoxycytidine monophosphate (dCMP)
T	Thymine	Deoxyribose	Deoxythymidine	Deoxythymidine monophosphate (dTMP)
A	Adenine	Ribose	Adenosine	Adenosine monophosphate (AMP)
G	Guanine	Ribose	Guanosine	Guanosine monophosphate (GMP)
C	Cytosine	Ribose	Cytidine	Cytidine monophosphate (CMP)
U	Uracil	Ribose	Uridine	Uridine monophosphate (UMP)

Nucleotides are nucleosides with one or more phosphates added through an ester bond between the phosphate and the hydroxyl group on C-5 of the sugar. An ester bond involving phosphate is often called a *phosphoester bond.*

A nucleotide is named from the nucleoside it contains. For instance, deoxyadenosine monophosphate is the nucleotide containing deoxyadenosine and one phosphate (Table 26.1). Adenosine triphosphate, ATP, is the nucleotide containing the nucleoside adenosine and three phosphates.

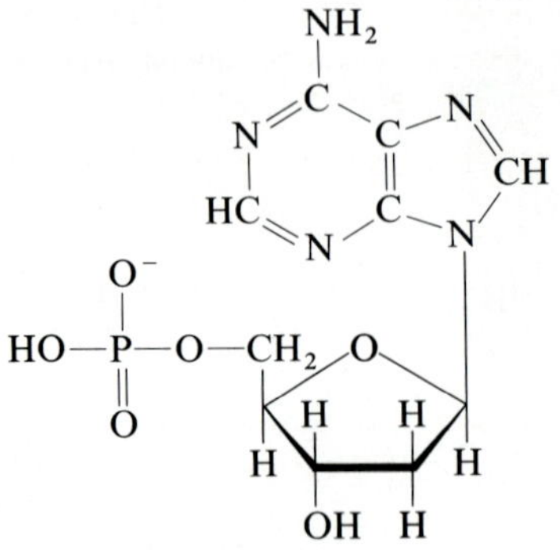

Deoxyadenosine monophosphate (dAMP)

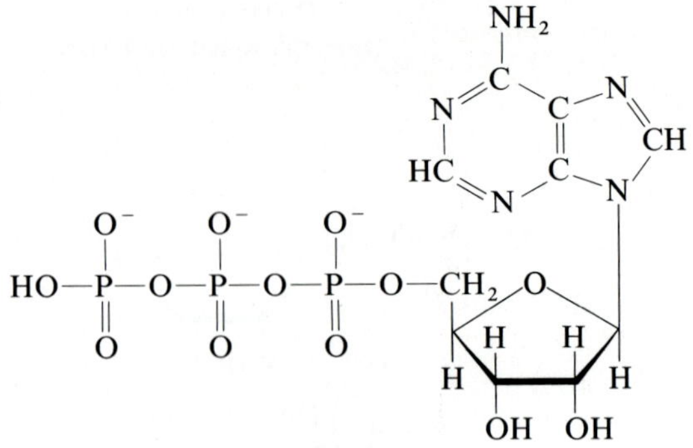

Adenosine triphosphate (ATP)

NEW TERMS

nucleotide A compound consisting of a nitrogenous base, a sugar, and one or more phosphate groups.

nitrogenous base A basic, nitrogen-containing heterocyclic compound, the most common ones being adenine, guanine, cytosine, thymine, and uracil.

nucleoside A compound consisting of a nitrogenous base and a sugar.

TESTING YOURSELF

Nucleotides

1. Name the nitrogenous bases that are (a) purines and (b) pyrimidines.
2. Name the compound containing ribose, one phosphate, and uracil.

Answers **1. a.** adenine and guanine **b.** cytosine, thymine, and uracil **2.** uridine monophosphate (UMP)

26.3 Structure and Replication of DNA

Earlier it was stated that DNA is a polymer of nucleotides. It is perhaps better to say that it is a polymer of deoxyribonucleotides, for DNA is an abbreviation of *deoxyribo*nucleic acid. The name clearly identifies deoxyribose as the sugar found in this polymer. DNA is a polymer of nucleotides, a nucleic acid, but how are the nucleotides arranged in DNA? What is the structure of DNA? Often the structure of a biomolecule provides important clues to its function, and much of DNA's function was predicted once its structure became known.

TABLE 26.2
Percentage Composition of the Bases of DNA in Several Species

Species	% A	% T	% G	% C	% Purines	% Pyrimidines
E. coli	24.7	23.6	26.0	25.7	50.7	49.3
Yeast	31.3	32.9	18.7	17.1	50.0	50.0
Salmon	29.7	29.1	20.8	20.4	50.5	49.5
Human	30.9	29.4	19.9	19.8	50.8	49.2

Determination of DNA Structure: The Double Helix

The determination of DNA structure relied on several lines of evidence. One involved the base composition of DNA. Four bases are found in the deoxynucleotides of DNA; the purines adenine and guanine and the pyrimidines cytosine and thymine. In a sample of DNA, the amount of thymine found is nearly equal to the amount of adenine, and the amount of cytosine is nearly equal to the amount of guanine. This means that the amount of purines equals the amount of pyrimidines. Except for some viruses, this is true for the DNA of all species (Table 26.2). Such similarity could not occur by chance in all species. The structure of DNA must provide an explanation for the existence of these equalities.

A second line of evidence involved the size and shape of DNA. It was known for a long time that DNA was a very long molecule, but in the early 1950s Maurice Wilkins and Rosalind Franklin determined some other dimensions for the molecule and showed that it had repeating units at specific intervals along its length. In 1953, James Watson and Francis Crick proposed a structure for DNA that accounted for the data of Wilkins and Franklin, the equality of bases, and all other known structural data for DNA. They proposed that DNA is a *double helix*, two strands wrapped around each other in a helical fashion. The specific structure they proposed also suggested how DNA could split to provide the exact genetic material for two cells.

Before an understanding of the double helix can be reached, the structure of a single strand of DNA should be examined. The deoxyribonucleotides of a strand are covalently linked to each other through ester bonds between the phosphates and the hydroxyl groups of the sugars (Figure 26.6). Just as

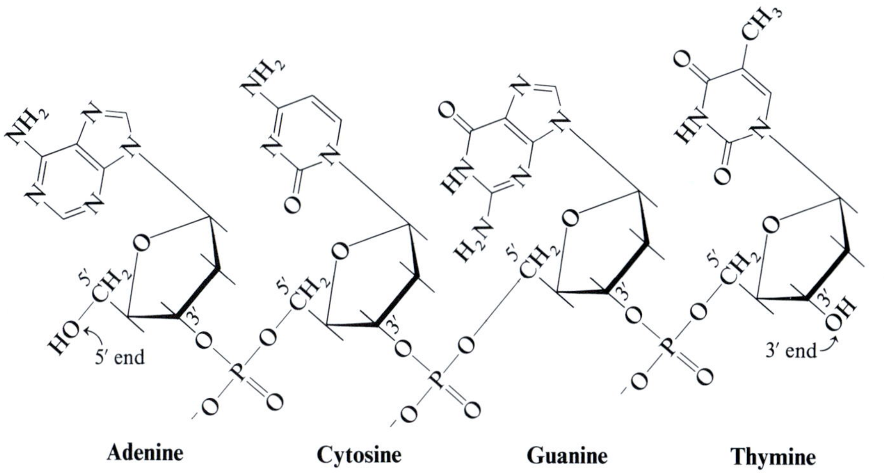
5′ end
3′ end
Adenine
Cytosine
Guanine
Thymine

FIGURE 26.6
The bonding of nucleotides to each other to form a strand of DNA. At one end of the strand, there is a hydroxyl group on the number 5 carbon of ribose that is not bound to another nucleotide. This end of the molecule is called the 5′ end. The other end of the strand has a free hydroxyl group on carbon number 3 of a sugar, thus it is the 3′ end.

proteins are polymers of amino acids bonded through peptide bonds, DNA is a polymer of deoxynucleotides bonded through phosphate ester bonds. If one mentally extends this linear polymer of deoxynucleotides in both directions, a large ribbonlike molecule is formed.

The double helix proposed by Watson and Crick has two strands of DNA present in each molecule. The two strands are arranged with the bases of one strand close to the bases of the other. The bases form the core of the molecule with the sugar–phosphate backbones wound around the core in a helical fashion. The DNA molecule is indeed a double helix (Figure 26.7).

Hydrogen Bonding in DNA

DNA is a double-stranded molecule with the strands held together by hydrophobic interactions and specific hydrogen bonding between complementary bases.

What forces or bonding holds the two strands together? The strands are held together by hydrogen bonds between specific bases and by hydrophobic interactions that are not specific. The force of each hydrogen bond is individually weak, typically only a few kilocalories per mole, but the thousands upon thousands of hydrogen bonds present in a DNA molecule collectively hold the two strands together. Furthermore, hydrophobic interactions involving the bases add more stability. The strands are held tightly together by these interactions, but during certain biological processes they are temporarily separated.

In DNA, adenine is complementary to thymine, and guanine is complementary to cytosine.

In DNA, adenine (A) pairs with thymine (T) through two hydrogen bonds and guanine (G) pairs with cytosine (C) through three hydrogen bonds. Bases that pair with each other in nucleic acids are called **complementary bases**. This accounts for the equality of bases that is noted in Table 26.2, but why is this specific combination of bases seen rather than other ones? This base pairing occurs because of size and shape factors. The core of the double helix is only large enough for one purine and one pyrimidine to fit. Two purines are too large to fit into the core without distorting the shape of the molecule, thus A does not pair with G. Two pyrimidines would not be close enough to each other for hydrogen bonds to effectively form, so C will not pair with T. Furthermore, the shape of A is ideal for hydrogen bonding to T, but in the core of a DNA molecule, A has neither the correct functional groups nor the orientation to form stable hydrogen bonds with C. Similarly, G can hydrogen bond to C but not T (Figure 26.7). Thus, DNA is a double helix held together by hydrogen bonds between specific base pairs.

Higher Levels of DNA Structure

As described so far, DNA is a long, linear molecule with two strands wrapped around each other. But DNA is not found in this form in living cells. In a typical bacterium, the length of the DNA molecule would be greater than 1 mm, yet the entire length of the bacterial cell is only about 10 μm (one one-hundredth as long as the DNA). This long piece of DNA is not just shoved into the volume of the cell like spaghetti into a thimble, but instead the DNA is arranged into a more compact, orderly form. In bacteria, the ends of the DNA molecule are covalently attached to each other to form circular DNA (Figure 26.8a). The circular DNA molecule is supercoiled, that is, the molecule is wrapped around itself to yield a more compact and orderly structure (Figure 26.8a). This arrangement fits into the cell better and allows for the many functions and changes typical of DNA in living cells.

Bacterial DNA is circular and supercoiled.

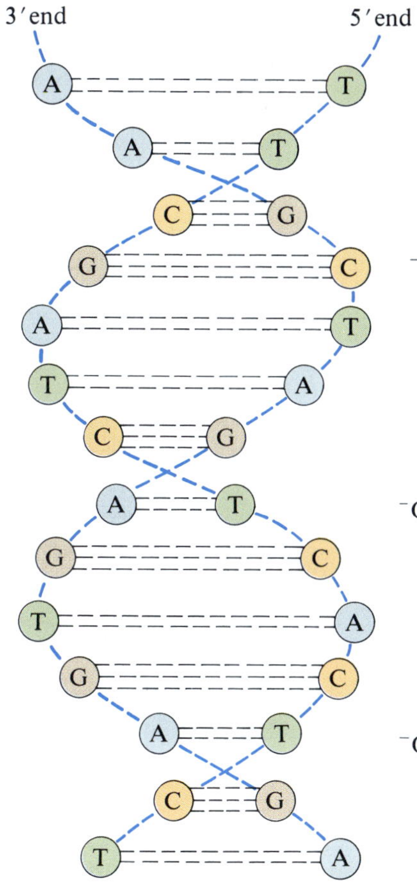

3′end
5′end
A T
A T
C G
G C
A T
T A
C G
A T
G C
T A
G C
A T
C G
T A

(a)

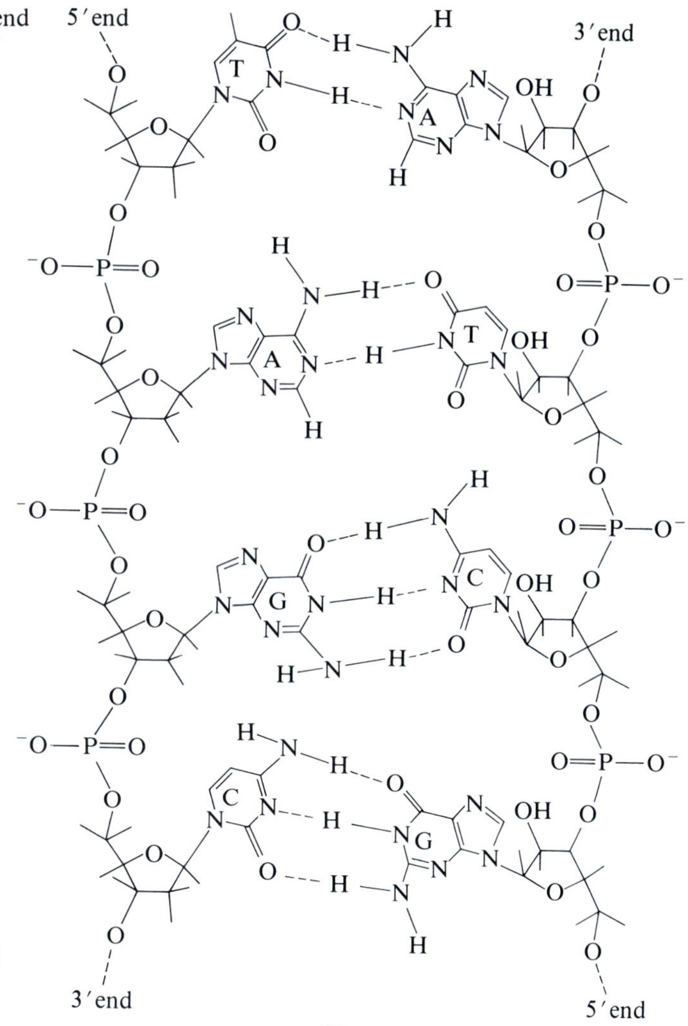

5′end
3′end
T A
A T
G C
C G
3′end
5′end

(b)

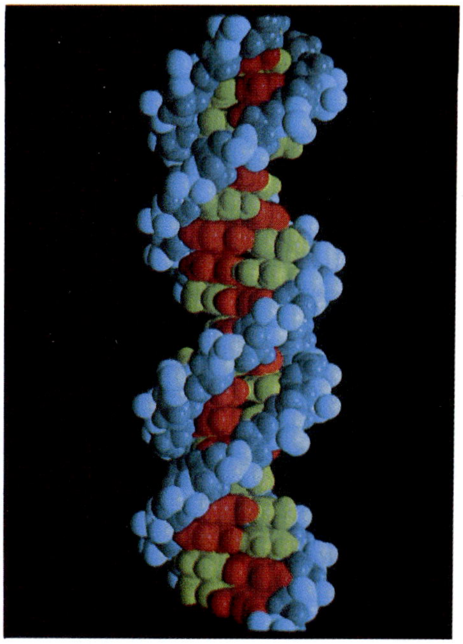

(c)

FIGURE 26.7
Three representations of the structure of a short piece of DNA. (a) In this ribbon model, the two strands of a DNA molecule are held together by hydrogen bonds between bases. Two hydrogen bonds form between adenine (A) and thymine (T) and three hydrogen bonds form between guanine (G) and cytosine (C). (b) This illustration shows the specific hydrogen bonds between groups on pairs of nitrogenous bases. (c) This is a space-filling model of DNA. This model provides the best overview of DNA, but specific interactions between bases are difficult to see.

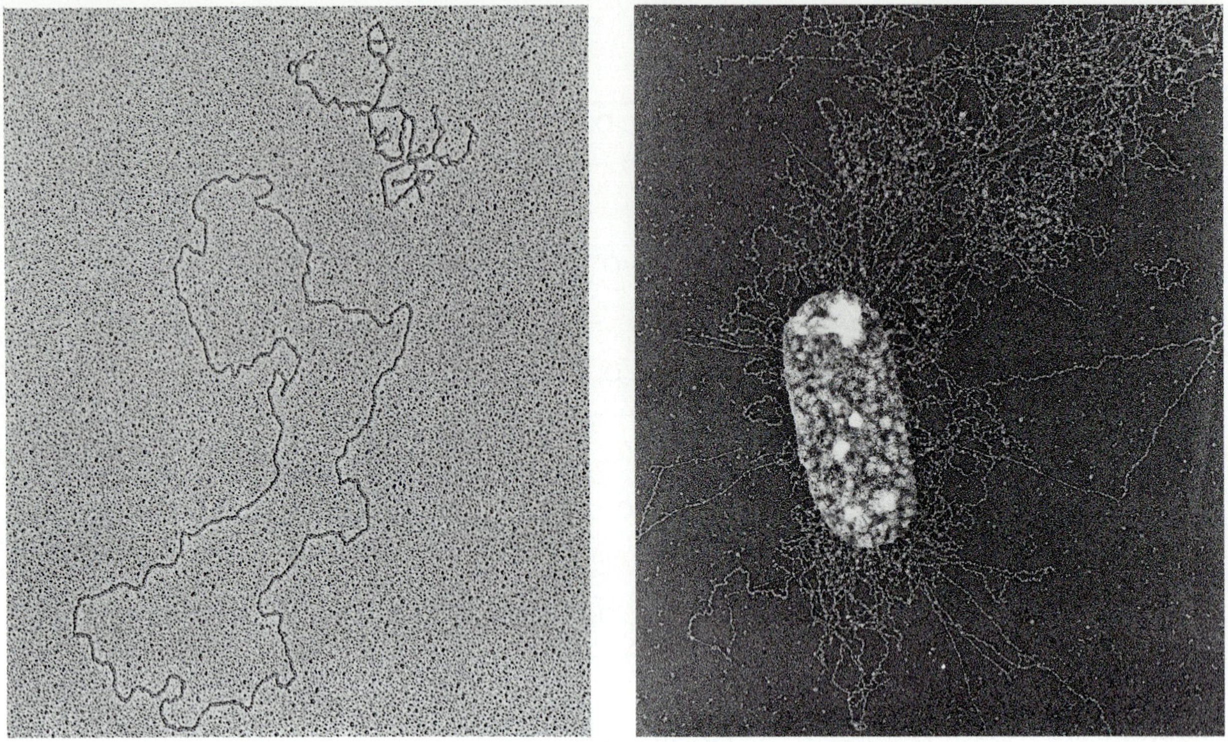

(a)

(b)

FIGURE 26.8
Electromicrographs of DNA. (a) Bacterial DNA is circular (below) and supercoiled (above). (b) DNA is spilling from a ruptured bacterial cell. DNA is much longer than the cell that contains it.

Eucaryotic DNA is complexed with proteins to form chromosomes.

The cells of higher organisms contain several to many long, linear DNA molecules. The DNA molecules of a typical mammalian cell have a combined length of over 1 m, yet the DNA is contained in a nucleus that is much less than 1 mm in diameter. The DNA of these higher cells is arranged into **chromosomes**. The DNA in chromosomes is complexed with a large number of proteins to yield a more compact structure (Figure 26.9). Much of the protein consists of *histones*, which are small positively charged proteins that bind through electrostatic interactions to DNA and, in effect, neutralize the negative charge associated with the phosphates. Protein–DNA interactions make the structures more compact, yet allow for all of the normal functions of DNA.

Replication of DNA

ROLE OF REPLICATION DNA stores genetic information. This information is used to control all of the activities of a cell and, ultimately, those of a multicellular organism. DNA is nearly always passed on exactly to cells arising from cell division. Thus, with rare exceptions, the activities and capabilities of daughter cells are identical to the cell from which they arose.

What structural features of DNA allow genetic information to be passed to daughter cells? The double-stranded nature of the molecule and the com-

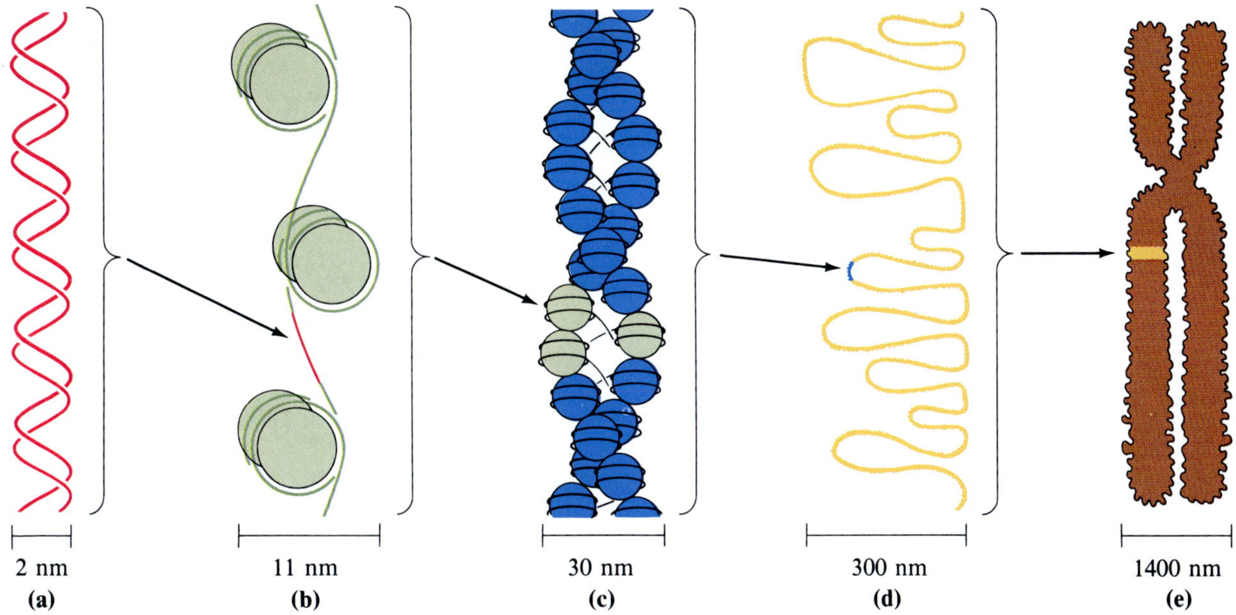
2 nm
(a)
11 nm
(b)
30 nm
(c)
300 nm
(d)
1400 nm
(e)

FIGURE 26.9
The arrangement of DNA in chromosomes. (a) DNA is a double helix of two strands approximately 2 nm in diameter (red). (b) In chromosomes, the DNA molecule is wrapped around an aggregate of proteins called histones to form nucleosomes that have a diameter of about 11 nm (green). (c) The nucleosomes are in turn wound about each other to form orderly clusters of 30 nm diameter (blue). (d) The clusters are part of various size loops that may be 300 nm across (yellow). (e) During certain phases of the cell cycle, chromosomes are visible with a microscope (1400 nm).

plementary nature of the bases allow for it. The two strands of DNA are *complementary.* This means the specific composition and sequence of one strand is related to the other because a specific base of one strand is always paired through hydrogen bonding to a specific base on the other strand. If the two strands are separated and a new strand is made for each of the original strands, then each new strand is complementary to the old strand it is now paired with (Figure 26.10). This process is called **semiconservative replication** because one strand of the original DNA molecule is conserved intact in each of the daughter DNA molecules. Through replication, two new DNA molecules are made that are identical in composition and sequence to the DNA molecule from which they were made. Each daughter cell receives one of the two molecules of DNA that were formed by replication and thus has the same genetic information possessed by the original cell.

DNA replication is semiconservative because each daughter cell gets one of the original strands.

PROCESS OF REPLICATION Although the concept of replication is straightforward, the cellular and molecular mechanics of the process are complex. Two factors contribute greatly to this complexity: (1) the helical nature and size of DNA and its association with proteins complicate the separation of the strands, and (2) the requirement that a specific base must pair with another specific base means that the process must be very accurate. What follows is an overview of replication; no attempt is made here to provide all of the specific details.

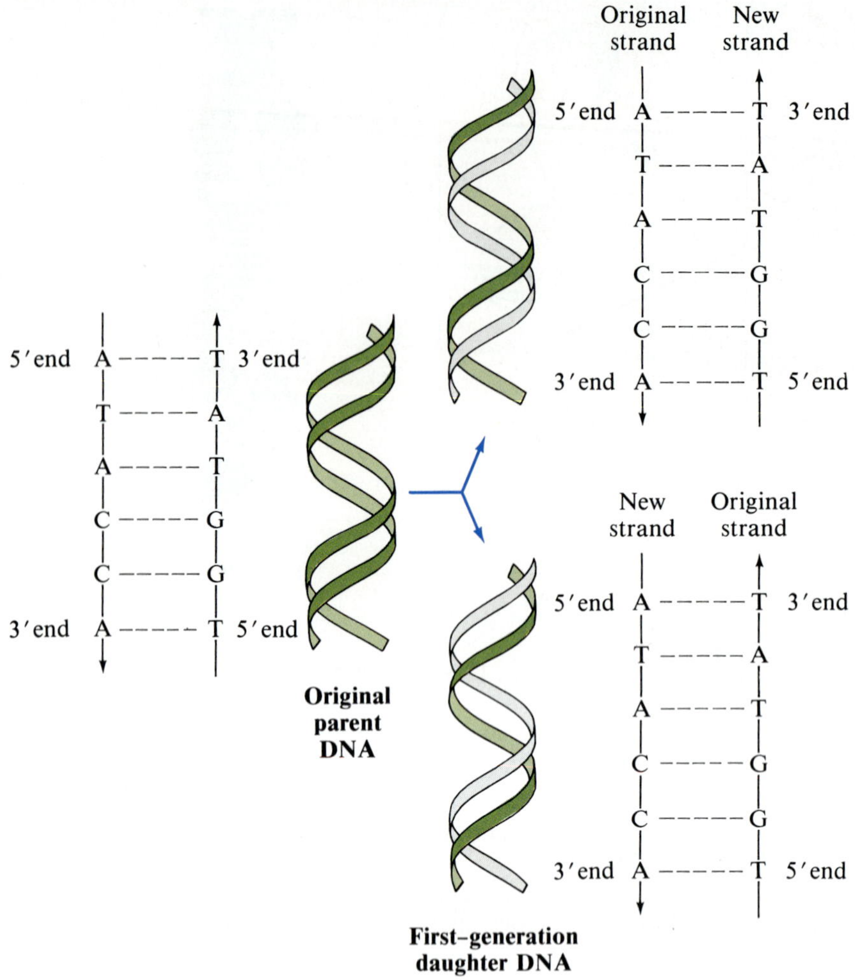

Original strand
New strand
5′end A ----- T 3′end
T ----- A
A ----- T
C ----- G
C ----- G
3′end A ----- T 5′end
5′end A ----- T 3′end
T ----- A
A ----- T
C ----- G
C ----- G
3′end A ----- T 5′end
Original parent DNA
New strand
Original strand
5′end A ----- T 3′end
T ----- A
A ----- T
C ----- G
C ----- G
3′end A ----- T 5′end
First-generation daughter DNA

FIGURE 26.10
Semiconservative replication of DNA. Each of the original strands of the parental DNA molecule is paired with a new strand in the daughter molecules. Note that the bases of the new strands are complimentary to the bases of the original strands.

The increased size and complexity of eucaryotic DNA requires multiple origins of replication.

Replication begins at a specific site called the *origin of replication.* In bacteria, there is only one origin per DNA molecule, but higher organisms have multiple origins per chromosome. Multiple origins are necessary to speed up replication because the chromosomes of higher organisms are much larger and more complex than the bacterial DNA molecule. At the origins of replication, several proteins unravel the DNA molecule, separate the two strands, and maintain them as separated strands (Figure 26.11). This separation is necessary because the enzymes responsible for the synthesis of the new strands must be able to bind to the individual DNA strand and because the nitrogenous bases of the strand must not be bonded to the bases of the other strand.

Enzymes are discussed in detail in Chapter 27. For this overview, consider them simply as efficient catalysts.

Unwinding and separation of the two strands of DNA yield two *replication forks* (Figure 26.11). At each fork, synthesis of both strands will occur simultaneously. Although several enzymes and proteins are involved in replication, the principal enzymes responsible for the synthesis of the new strands

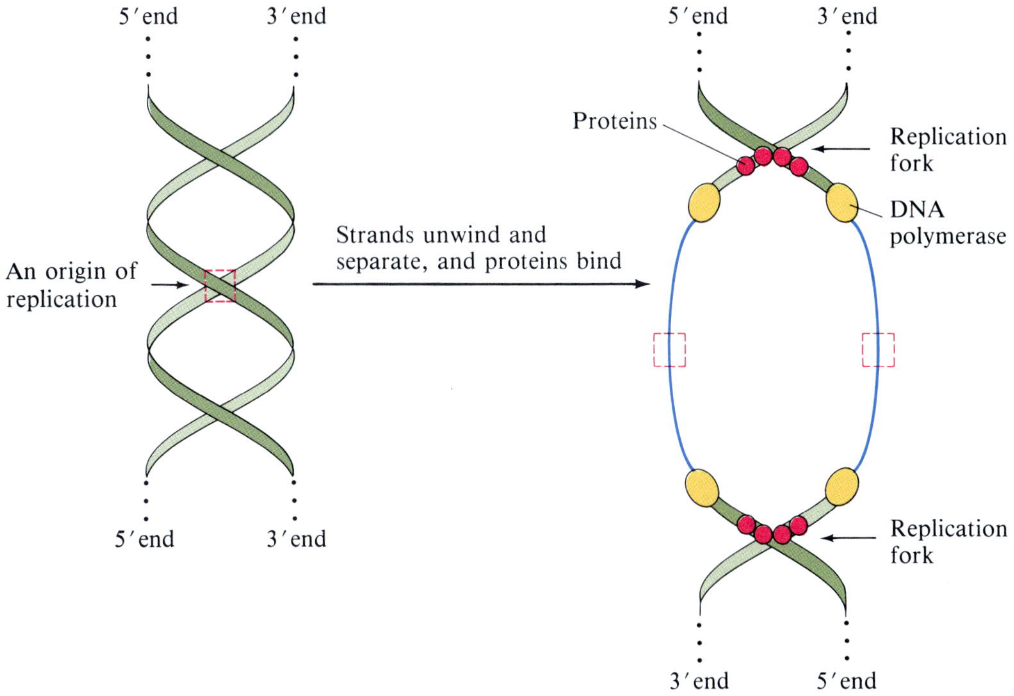
5′end
3′end
An origin of replication
5′end
3′end
Strands unwind and separate, and proteins bind
5′end
3′end
Proteins
Replication fork
DNA polymerase
Replication fork
3′end
5′end

FIGURE 26.11
The origins of replication and replication forks of DNA. Proteins aid in the separation of the DNA strands at the origins of replication. The separation yields two replication forks, which are regions where DNA polymerase can bind to the DNA strand and synthesize DNA complementary to the strand.

are the *DNA polymerases.* DNA polymerase functions in the following way. DNA polymerase binds noncovalently to one of the parent DNA strands. A deoxyribonucleotide triphosphate then binds to the complex (Figure 26.12a). There are four deoxyribonucleotide triphosphates in cells: dATP, dGTP, dCTP, and dTTP. The nucleotide that binds is the one whose base is complementary to the base on the DNA strand at the point where the DNA polymerase is bound. This specific binding occurs because only that base can form hydrogen bonds to the base on the strand. In Figure 26.12a note that the first nucleotide has guanine, which is complementary to cytosine in the strand of DNA. The other three deoxyribonucleotides cannot form these hydrogen bonds with cytosine, so they cannot bind to the complex. Hydrogen bonding between the base of a nucleotide from solution and a base on the strand of DNA correctly places the incoming base needed to synthesize the new strand of DNA.

DNA polymerase can bind two nucleotides (Figure 26.12b). A second nucleotide from solution binds to the complex. Again the nucleotide that binds will be the one whose base is complementary to the base on the strand of DNA. In Figure 26.12b the second nucleotide has adenine, which is the complementary base to thymine on the DNA strand. At this point, the two nucleotides are positioned next to each other, each hydrogen bonded to its complementary base on the parent strand. The DNA polymerase now catalyzes the formation of a phosphoester bond between these two nucleotides (Figure 26.12c). The nucleotides are now covalently attached to each other.

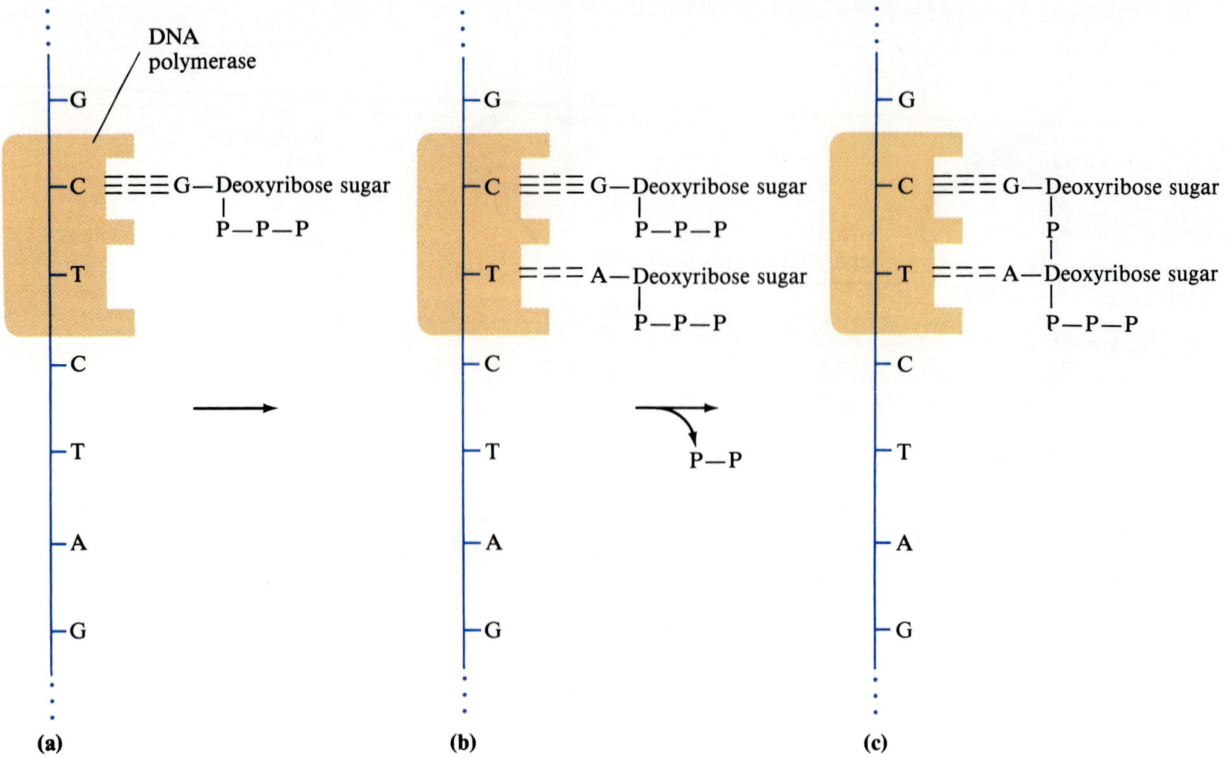
DNA polymerase
G
C
G—Deoxyribose sugar
P—P—P
T
A—Deoxyribose sugar
C
T
A
G
P—P
(a)
(b)
(c)

FIGURE 26.12
The role of DNA polymerase in replication and the formation of phosphoester bonds. (a) Nucleotides bind to the DNA polymerase and DNA strand through hydrogen bonds to the complementary base in the strand of DNA. (b) A second nucleotide binds to the DNA–DNA polymerase complex. (c) DNA polymerase catalyzes the formation of a phosphoester bond between the nucleotides.

The DNA polymerase moves down one more base on the DNA strand, another nucleotide binds (it is complementary to the next base on the strand), and the DNA polymerase catalyzes the formation of an ester bond between this new nucleotide and the previous one. One by one, nucleotides are added to the end of a growing strand of DNA, with the base of each new nucleotide complementary to the base on the parent strand.

DNA polymerase can only move toward the 3′ end of DNA, thus DNA synthesis is said to be 5′ → 3′ (see Figure 26.6 caption). Since DNA synthesis is occurring on both strands simultaneously, one of the new strands is synthesized as a long continuous piece and the other as a series of shorter pieces called *Okazaki fragments.* A ligase connects these pieces into a long, continuous piece.

During replication, a new strand of DNA is synthesized on both parent strands of DNA at both replication forks of each origin of replication. When the ends of these newly synthesized strands come together, they are covalently bonded together by the ligase. Eventually, all of the pieces have been linked together, and a new intact complementary strand is complexed with each of the parent strands, as was shown in Figure 26.10. If no errors have occurred during replication, each of the daughter DNA molecules is identical to the parent DNA molecules.

NEW TERMS

complementary bases Pairs of bases that hydrogen bond to each other in nucleic acids: guanine pairs with cytosine and adenine pairs with thymine (or uracil in RNA).

chromosomes Complexes of DNA and proteins found in the nucleus of eucaryotic cells. These structures carry genetic information.

semiconservative replication The process that produces two DNA molecules from one. The process is semiconservative because each daughter DNA molecule receives one of the strands from the parent molecule.

TESTING YOURSELF

DNA

1. What bond links nucleotides in DNA?
2. A short segment of a DNA strand has the sequence AACTGGC. What is the sequence of the complementary strand?
3. In eucaryotes, DNA is complexed with proteins to form what structures?
4. What enzyme is responsible for the synthesis of the complementary strand of DNA during replication?

Answers **1.** phosphodiester bond **2.** TTGACCG **3.** chromosomes **4.** DNA polymerase

26.4 RNA

Besides DNA, cells contain another kind of nucleic acid called ribonucleic acid or RNA. RNA is also a polymer of nucleotides, but there are several important structural differences between RNA and DNA. First, the sugar found in RNA is ribose, while DNA contains deoxyribose. Second, RNA contains the base *uracil* instead of the thymine found in DNA. Third, RNA is a single stranded molecule while DNA is double stranded. In addition, DNA is a much larger molecule than RNA. The structural differences between DNA and RNA are summarized in Table 26.3.

There are three classes of RNA: messenger, transfer, and ribosomal.

There are three main classes of RNA found in cells. **Messenger RNA (mRNA)** carries the genetic information stored in DNA to the rest of the cell. **Ribosomal RNA (rRNA)** is RNA found in ribosomes, the organelles

TABLE 26.3
Structural Differences Between DNA and RNA

Nucleic Acid	Sugar	Base	Strands	Size
DNA	Deoxyribose	Thymine	Two	Largest molecule in cell
RNA	Ribose	Uracil	One	Macromolecules, but much smaller than DNA

responsible for protein synthesis. **Transfer RNA (tRNA)** is a smaller form of RNA that carries amino acids to the ribosome and helps ensure that the correct amino acids are used at the right time during protein synthesis. Each of these molecules is synthesized by a process called *transcription*. The specific structure and role of each of these RNA types will be discussed more fully in Sections 26.5 and 26.6 on transcription and translation.

NEW TERMS

messenger RNA (mRNA) RNA that carries genetic information from the nucleus to the rest of the cell. The information is used to direct protein synthesis.

ribosomal RNA (rRNA) RNA molecules found in ribosomes, the site of protein synthesis.

transfer RNA (tRNA) RNA molecules that transfer the correct amino acids into the protein that is being synthesized by ribosomes.

TESTING YOURSELF

RNA

1. What are the three classes of RNA?
2. List four differences between DNA and RNA.

Answers **1.** messenger RNA, transfer RNA and ribosomal RNA **2.** DNA has the sugar deoxyribose, RNA has ribose. DNA is double stranded, RNA is single stranded. DNA has the base thymine, RNA has the base uracil. DNA is the largest molecule in cell, RNA is much smaller.

26.5 Transcription

Role of Transcription

Transcription synthesizes an RNA molecule that is complementary to a part of a DNA strand.

Information stored in DNA ultimately controls the activities of cells. These activities are primarily chemical reactions that are catalyzed and controlled by enzymes. Enzymes, and all the other proteins required for life, are synthesized using information contained in DNA. The process called *transcription* synthesizes a molecule of messenger RNA using DNA as a template or guide. This mRNA then carries the information needed to synthesize the protein to the ribosomes. Ribosomes, through the process called *translation*, use the information in the mRNA to synthesize the protein (see Section 26.6). To a large extent, a cell controls its molecular activities by controlling which proteins are present and by controlling how much of each protein is present at any given time.

Process of Transcription

Transcription is a process similar to replication. Replication yields a strand of DNA that is complementary to the parent strand of DNA, while transcription yields an RNA molecule that is complementary to a portion of a strand of DNA. Transcription begins with the binding of the enzyme *RNA polymerase* to one of the strands of DNA at one of many specific sites on the DNA molecule. These sites are called *promoters*, sequences of bases recognized by RNA polymerase as binding sites. Once bound, the RNA polymerase begins to synthesize a strand of RNA that is complementary to the strand of DNA. This strand of DNA is referred to as the *sense strand*, because the information that is needed to synthesize the protein is present in this strand. The other strand is called the *nonsense strand*. Remember, the strands of DNA are complementary, not identical; only one can carry the exact information needed for the synthesis of a polypeptide.

Transcription begins at promoters, which are specific sites on the sense strand of DNA.

Ribonucleotides from solution bind to the RNA polymerase–sense strand complex. The specific ribonucleotide that binds depends upon the base on the sense strand in the complex. Like DNA polymerase, RNA polymerase binds nucleotides from solution that are complementary to the bases of the DNA strand, then it covalently links them. The specifics of RNA synthesis are similar to those shown for DNA polymerase in Figure 26.12, but ribonucleotides bind rather than deoxyribonucleotides. In Figure 26.13b, RNA polymerase is moving along a sense strand, building an RNA molecule by the sequential addition of ribonucleotides to the end of the growing RNA molecule. Each new base is complementary to the base on the sense strand. The RNA polymerase molecule continually moves down one more base, then another nucleotide binds, and a phosphoester bond forms between this base and the previous one. This continues until the RNA polymerase comes to a sequence of bases on the sense strand that signal the termination of transcription. The complex dissociates to give the newly synthesized RNA, the RNA polymerase, and the DNA molecule.

It is worthwhile to compare the accuracy of replication and transcription. Transcription need not be as accurate as replication. Replication yields two daughter DNA molecules that contain all of the genetic information of the organism. If one of them has been synthesized incorrectly, the genetic information of that daughter cell and all the descendants of that cell will be in error. If an error is made in transcription, then some of the copies of a particular protein may be incorrect. Since other correct copies of the protein are also present, the function of that protein will be fulfilled. The cell will

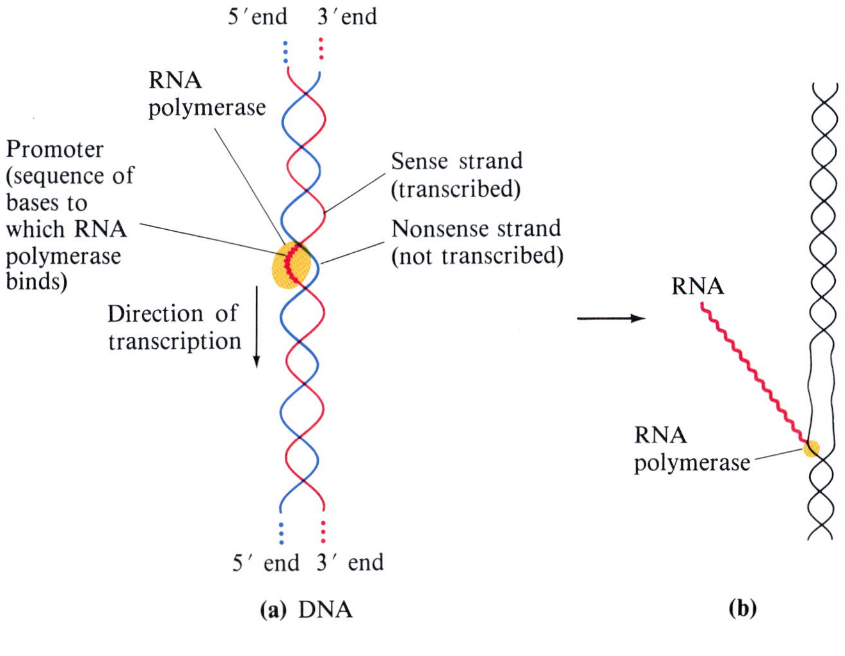
5′end 3′end
RNA polymerase
Promoter (sequence of bases to which RNA polymerase binds)
Sense strand (transcribed)
Nonsense strand (not transcribed)
Direction of transcription
5′ end 3′ end
(a) DNA
RNA
RNA polymerase
(b)

FIGURE 26.13
Some specific steps of transcription. (a) RNA polymerase (red) binds to the sense strand of DNA at a sequence of bases called the promoter (yellow). (b) The RNA polymerase catalyzes the sequential connection of ribonucleotides that are complementary to the bases of the sense strand of DNA. The newly synthesized RNA molecule is thus complementary to a portion of a DNA strand.

waste some material and energy making the defective protein, but that is far less serious than a genetic error.

In eucaryotes, the transcript must be processed to obtain a functional mRNA.

The RNA formed by transcription is called the **transcript**, but the transcript must usually be altered before it can function in the cell. One exception is the transcript in bacteria; this transcript becomes a functional mRNA as soon as transcription is complete. mRNA in higher organisms is formed from transcripts by the action of specific enzymes in the nucleus. These enzymes remove the unneeded segments called *introns* and splice together the remaining segments called *exons*. This splicing plus some other processing converts the transcript into mRNA that diffuses from the nucleus into the rest of the cell. Although some base pairing can occur in mRNA, it is generally regarded as a long linear molecule randomly oriented in solution.

Ribosomal RNA is synthesized from specific transcripts in a slightly different way. A specific transcript is cleaved by enzymes in the nucleus into smaller pieces of RNA. These smaller pieces are the rRNA molecules that join with specific proteins to form functional ribosome subunits. The RNA in a ribosome has a specific shape and is bound to certain proteins in the ribosome.

Transfer RNA is made from specific transcripts that are cleaved by enzymes to yield RNA of the proper size. Other enzymes catalyze covalent changes to some of the nucleotides in the RNA. There are many different tRNA molecules in a cell. Each of them undergoes different processing to become a functional tRNA. All of the tRNAs have the same overall shapes, but each is uniquely different. In two dimensions, tRNAs are often represented as cloverleaf-shaped molecules with specific bases hydrogen bonded to other bases of the same strand. In three dimensions the tRNAs look more L or club shaped (Figure 26.14).

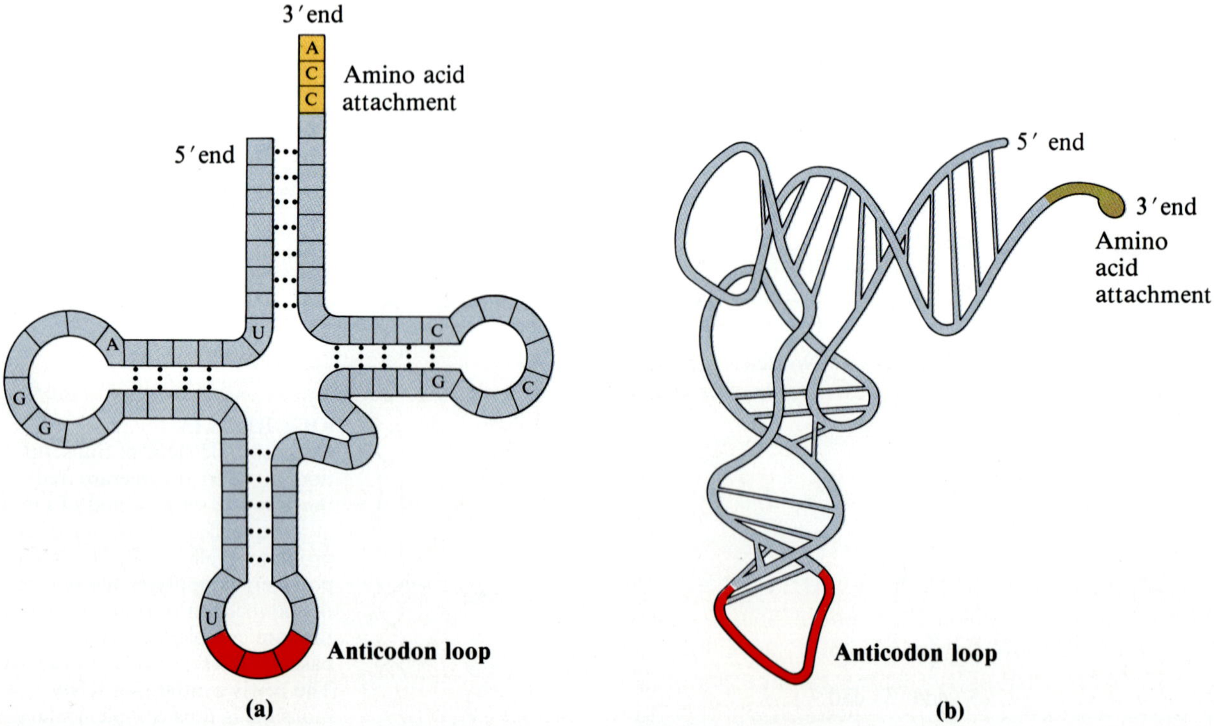
3′end
A
C
C
Amino acid attachment
5′end
U
A
G
G
C
G
C
U
Anticodon loop
(a)
5′ end
3′end
Amino acid attachment
Anticodon loop
(b)

FIGURE 26.14
(a) Two- and (b) three-dimensional representations of tRNA. The anticodon (in red) of a tRNA molecule pairs with the codon of mRNA during translation. The 3′ end of the molecule is the attachment site for the amino acid.

NEW TERMS

transcription The process that synthesizes RNA molecules using a DNA molecule as a template.

transcript The first product of transcription. In higher cells it is modified to yield functional RNA molecules.

TESTING YOURSELF

Transcription

1. What enzyme is responsible for RNA synthesis?
2. What bases in a strand of DNA would be complementary to this segment of RNA?

—A—U—G—C—

—?—?—?—?—

3. Transcription begins where on the sense strand of DNA?

Answers **1.** RNA polymerase **2.** —T—A—C—G— **3.** at a promoter

26.6 Translation

Proteins are synthesized by **translation**, a process that uses information stored in mRNA as instructions for protein synthesis. Because mRNA is synthesized by transcription from DNA, the information comes ultimately from DNA. The term **gene** is often used for that part of a DNA molecule that contains the information needed for the synthesis of a polypeptide or protein. The mRNA transcribed from one gene has the information needed for the synthesis of one particular polypeptide. If a protein has only one kind of polypeptide, then only one mRNA is needed for the synthesis of that protein. If two or more different peptides (subunits) are present in the protein, then two or more kinds of mRNA are needed to synthesize the polypeptides of that protein.

The process called protein synthesis literally translates the information stored in the base sequence of mRNA into a sequence of amino acids in a polypeptide chain.

The Genetic Code

Since mRNA molecules differ from one another only in the sequence of the purine and pyrimidine bases, how does the information stored in a specific set of mRNA molecules convey the correct information for the synthesis of that specific protein? An analogy is helpful. There are 26 letters in the English alphabet. From various combinations of these 26 letters we can make words and sentences that express our thoughts. The writers of philosophy, religion, science, poetry, and mystery novels all use the same letters. The message in each book or article is unique because the letters are arranged into unique combinations. Analogously, the information stored in each mRNA molecule differs from that in all other mRNA molecules because the bases of each mRNA molecule are arranged in a unique order. An "alphabet" of only four "letters"—adenine, guanine, cytosine, and uracil—provides all the information needed to synthesize all of the proteins of all living organisms.

How do four bases code for the twenty amino acids found in proteins? Consider what would be needed to make a code with four letters: A, U, C, and G. If each letter stood for a particular amino acid, only four amino acids would be specified, clearly inadequate for the twenty amino acids found in proteins. If *two* letters were used to code for an amino acid, how many amino acids could be specified? If four rows and columns are filled with each of the four letters and the letters in each column and row are matched, then sixteen combinations of letters are generated. This is still too few to code for twenty amino acids.

	A	C	U	G
A	AA	AC	AU	AG
C	CA	CC	CU	CG
U	UA	UC	UU	UG
G	GA	GC	GU	GG

What would happen if combinations of three letters or bases were used? The possible combinations of three letter words obtained from the four letters A, G, C, and U can be represented by a three-dimensional cube (Figure 26.15). This cube contains 64 combinations of three letters. Each of the small cubes within the larger cube is characterized by three bases. Each cube represents one three-letter word (for example, AUC or GAG) that could code for an amino acid. Now enough code is available to specify all twenty of the amino acids. In fact, there are now more words or codes than amino acids. The spare words or codes provide duplicate code for amino acids as well as words for "starting" and "stopping" protein synthesis.

Three consecutive bases in mRNA make up a codon.

A sequence of three bases in an RNA molecule is called a **codon**. All of the 64 codons have meaning, none are nonsense. Most codons specify an amino acid, although a few codons serve as punctuation for the process of translation. For example, this nine base sequence in an RNA molecule, AAUGCUGGA, corresponds to three codons: AAU, GCU, and GGA. They

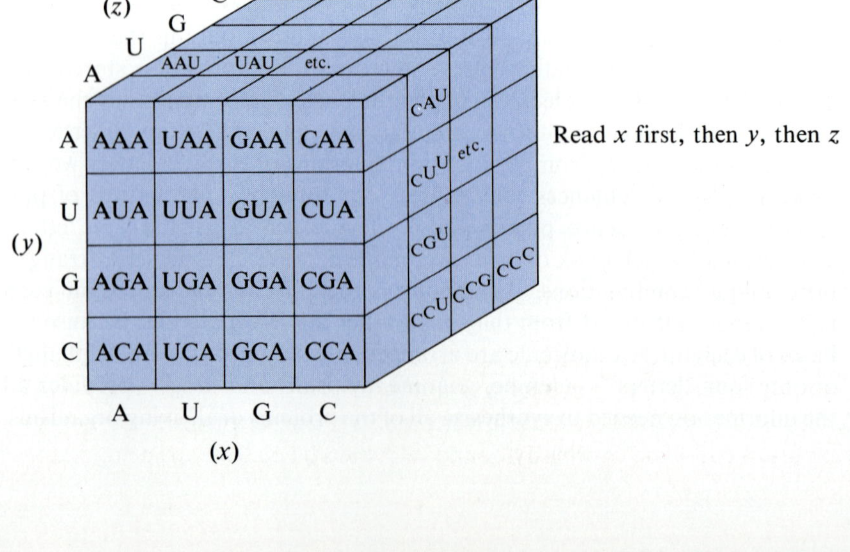
(z)
A
U
G
C
AAU
UAU
etc.
A AAA UAA GAA CAA
U AUA UUA GUA CUA
G AGA UGA GGA CGA
C ACA UCA GCA CCA
(y)
A U G C
(x)
CAU
CUU
etc.
CGU
CCU
CCG
CCC
Read x first, then y, then z

FIGURE 26.15
The genetic code. There are 64 combinations of four bases arranged into sequences of three. These three-base sequences are called *codons*, and codons correspond to a particular amino acid or serve as a stop signal during translation.

Table 26.4
The Genetic Code

Codon	Amino Acid	Codon	Amino Acid	Codon	Amino Acid	Codon	Amino Acid
UUU	Phe	UCU	Ser	UAU	Tyr	UGU	Cys
UUC	Phe	UCC	Ser	UAC	Tyr	UGC	Cys
UUA	Leu	UCA	Ser	UAA	Stop	UGA	Stop
UUG	Leu	UCG	Ser	UAG	Stop	UGG	Trp
CUU	Leu	CCU	Pro	CAU	His	CGU	Arg
CUC	Leu	CCC	Pro	CAC	His	CGC	Arg
CUA	Leu	CCA	Pro	CAA	Gln	CGA	Arg
CUG	Leu	CCG	Pro	CAG	Gln	CGG	Arg
AUU	Ile	ACU	Thr	AAU	Asn	AGU	Ser
AUC	Ile	ACC	Thr	AAC	Asn	AGC	Ser
AUA	Ile	ACA	Thr	AAA	Lys	AGA	Arg
AUG	Met	ACG	Thr	AAG	Lys	AGG	Arg
GUU	Val	GCU	Ala	GAU	Asp	GGU	Gly
GUC	Val	GCC	Ala	GAC	Asp	GGC	Gly
GUA	Val	GCA	Ala	GAA	Glu	GGA	Gly
GUG	Val	GCG	Ala	GAG	Glu	GGG	Gly

code for the amino acids asparagine, alanine, and glycine. Table 26.4 shows that many of the amino acids have two or more codons.

Messenger RNA molecules are longer than the message they contain, that is, there are bases at either end of the molecule that are not used as parts of codons. The message for the amino acid sequence of the protein begins somewhere inside one of the physical ends of the mRNA, and the end of the message occurs before the other end of the mRNA molecule. Certain codons are required to start and stop the process of translation. The stop codons are highlighted in Table 26.4. The codon for start, AUG, is also the codon for the amino acid methionine. How can one codon have two meanings? Wherever AUG occurs in a message, a methionine will be inserted into the protein. But if the AUG codon is preceded by some specific bases in a particular order, then the process of translation will begin with that codon. AUG can thus serve two roles in translation.

Process of Translation

Translation involves numerous molecules including mRNA, ribosomes, many tRNAs with amino acids bound to them, and a variety of proteins. Translation requires energy that is provided by GTP and ATP. The process can be summarized by a series of steps. The steps described here are a simplified version of the actual process.

The molecular interactions that occur during translation are among the most complex yet studied.

In the first step of translation, the *initiation* step, one of the subunits of a ribosome binds to the mRNA at the start codon, AUG. Remember, AUG serves as the start codon only if certain sets of bases precede it. A tRNA that has methionine covalently bound to it then binds by hydrogen bonding

FIGURE 26.16
The process of translation. (a) Initiation. Translation is initiated by the formation of a complex of mRNA, a ribosome, and a tRNA bound to methionine. (b) Elongation. A second tRNA with its amino acid (Phe) binds, then a ribosomal enzyme forms a peptide bond between the amino acids, leaving a free tRNA and a tRNA with a dipeptide. (c) Through translocation, the ribosome and tRNA bearing the peptide move along the mRNA by one codon. The next tRNA with its amino acid then binds, and peptide bond formation and translocation repeat. (d) Termination of translation. When a stop codon (red) is present in the ribosome-mRNA complex, an enzyme cleaves the bond between the tRNA and the polypeptide. The complex then dissociates.

between three of its bases and the three bases of the codon AUG. These three bases on the tRNA are called an **anticodon** because they are complementary to the bases of the codon. Only this tRNA of the many kinds of tRNA in the cell will bind, because only this tRNA has the proper anticodon. Now the other subunit of the ribosome binds to form a complex of a ribosome, mRNA, and the proper tRNA bound to the start codon of the mRNA. Figure 26.16a summarizes this first step.

In the second step of translation, a second tRNA with its amino acid binds to the complex (Figure 26.16b). The tRNA that binds will be the one whose anticodon is complementary to the second codon of the mRNA. This tRNA has the amino acid bound to it that is coded for by the codon on the mRNA. Note in Figure 26.16 that codon AUG is next to codon UUU, and at the other end of the tRNA, the amino acid methionine is next to the amino acid phenylalanine. The tRNA molecules serve as adapters or guides that ensure that the correct amino acids, corresponding to specific codons, are inserted into the growing polypeptide chain. A protein in the ribosome now catalyzes the formation of a peptide bond between the two amino acids leaving an empty tRNA and a tRNA that now has a peptide bound to it (Figure 26.16b).

Next the process of *translocation* occurs, and the ribosome moves one codon along the mRNA. The empty tRNA falls from the complex, and the

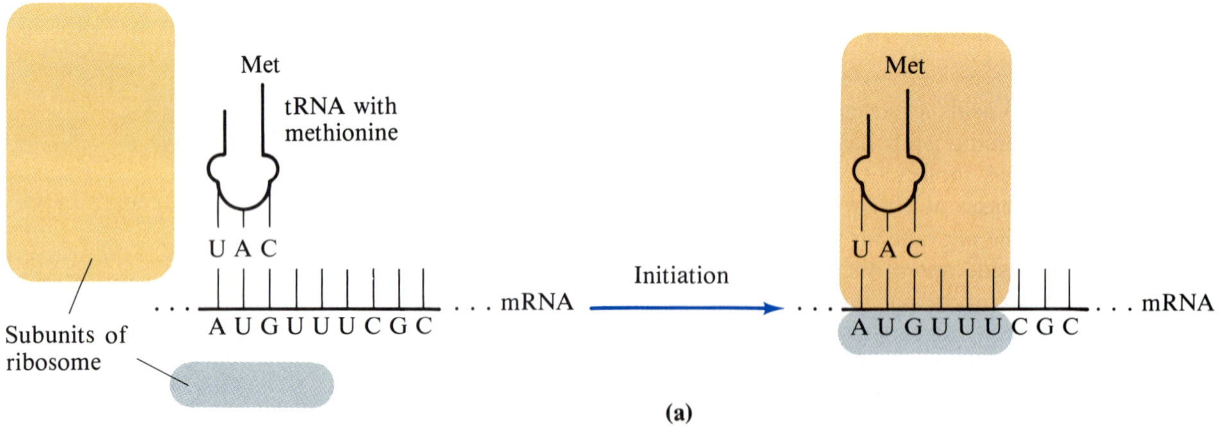
Met
tRNA with methionine
U A C
. . . A U G U U U C G C . . . mRNA
Subunits of ribosome
Initiation
Met
U A C
. . . A U G U U U C G C . . . mRNA

(a)

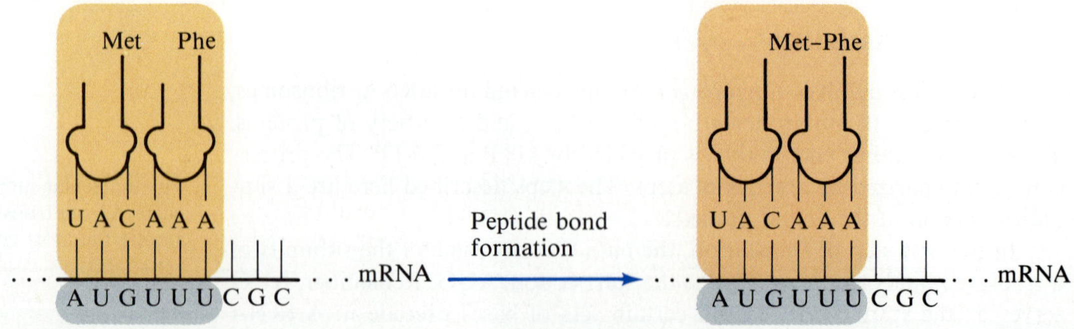
Met Phe
U A C A A A
. . . A U G U U U C G C . . . mRNA
Peptide bond formation
Met–Phe
U A C A A A
. . . A U G U U U C G C . . . mRNA

(b)

peptidyl–tRNA remains with its codon (Figure 26.16c). Another tRNA with its amino acid will bind to the next codon, a peptide bond will form, and translocation will occur again. These events continue as long as the next codon is a codon for an amino acid.

When the next codon is a stop codon, termination occurs. There are no tRNA's for the stop codons. When a stop codon enters the ribosome–mRNA complex, a protein in the ribosome cleaves the polypeptide from the tRNA (Figure 26.16d). The freed polypeptide diffuses from the complex and the two subunits of the ribosome fall from the mRNA. Many times the polypeptide is modified slightly following translation. Often one or more peptide bonds are cleaved, reducing the size of the polypeptide. If the polypeptide requires a prosthetic group or if it is a part of an oligomeric protein, then it will bind to these other parts to become a functional protein.

Most polypeptides are modified slightly or extensively after translation.

NEW TERMS

translation A synonym for protein synthesis. Information stored in mRNA is used to direct the synthesis of protein.

gene That portion of a DNA molecule that codes for a specific transcript. Since most transcripts become mRNA molecules, a gene can be considered the information in a DNA molecule that codes for one polypeptide chain.

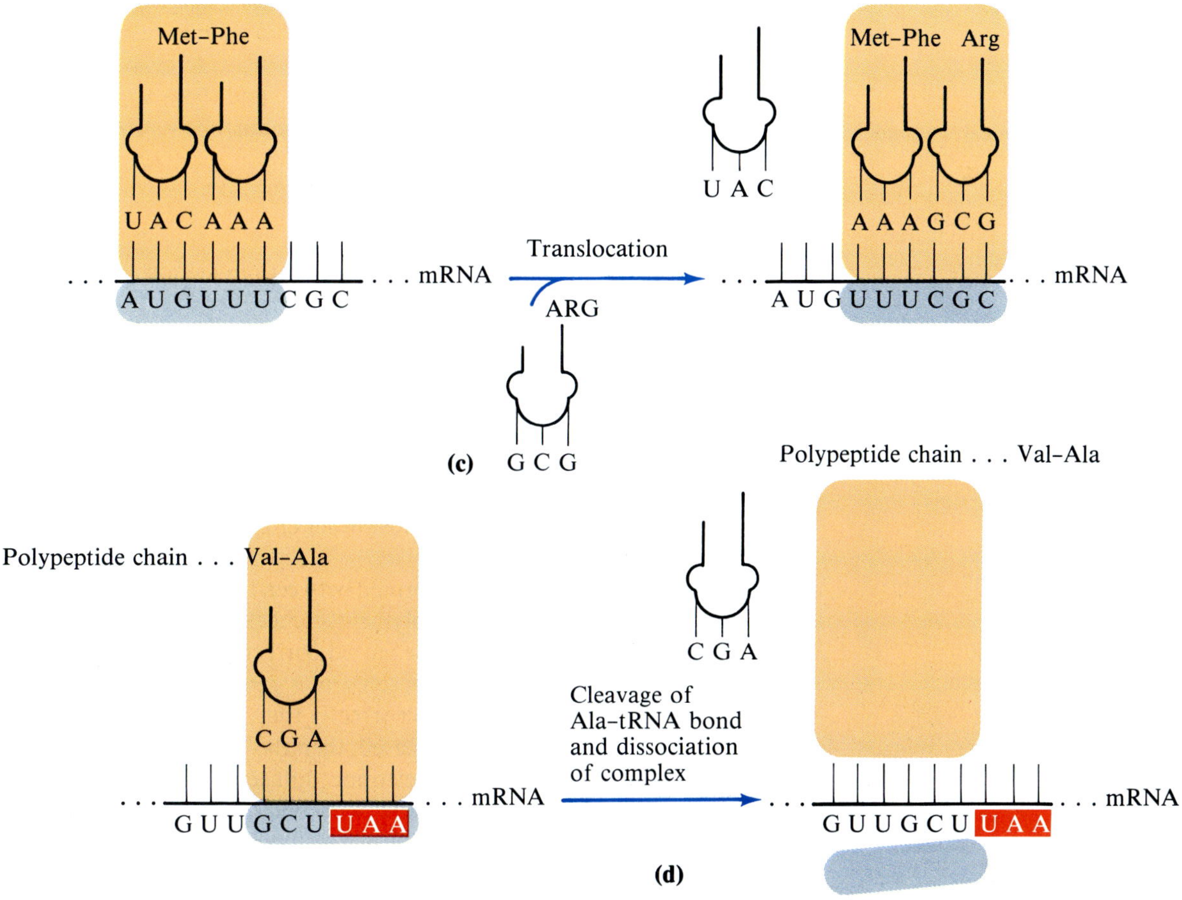
Met–Phe
U A C A A A
. . . mRNA
A U G U U U C G C
Translocation
ARG
G C G
(c)
U A C
Met–Phe Arg
A A A G C G
. . . mRNA
A U G U U U C G C
Polypeptide chain . . . Val–Ala
C G A
. . . mRNA
G U U G C U U A A
Cleavage of
Ala–tRNA bond
and dissociation
of complex
(d)
Polypeptide chain . . . Val–Ala
C G A
. . . mRNA
G U U G C U U A A

codon The three-base sequence in mRNA that determines what amino acid will be inserted into the polypeptide chain. Ultimately the sequence of codons in mRNA determines the primary sequence of proteins.

anticodon The three-base sequence in tRNA that is complementary to a codon on mRNA.

TESTING YOURSELF

Translation

1. What anticodon is complementary to the codon AGC?
2. Use Table 27.4, the genetic code, to determine which amino acid corresponds to the codon CGU.
3. What molecules ensure that the correct amino acid is inserted for a codon on mRNA?
4. What name is given to the process in which a ribosome moves stepwise along a mRNA molecule?

Answers **1.** UCG **2.** arginine **3.** tRNA molecules **4.** translocation

26.7 Mutagenesis

Mutations are permanent changes in genetic material. They arise whenever the structure of DNA is permanently altered. Since the sequence of bases in DNA determines the information that is stored, any change in this sequence changes the information contained in DNA. These changes usually show up as proteins with altered structure and function. The process that leads to the formation of mutations is called *mutagenesis*, and there are several ways a mutation can arise. Mutations can arise spontaneously, that is, without external agents.

Spontaneous Mutations

The small but measurable error rate in replication is one source of mutations.

One form of spontaneous mutation involves replication. Replication is a very accurate process, but it is not perfect. If an error occurs, the information contained in the DNA is altered and a spontaneous mutation has occurred. It is estimated that during replication an incorrect base is added to the growing DNA strand less than one time per one billion base additions. Most enzymes are not this accurate, so why is DNA polymerase an exception? It is believed that the inherent accuracy for base addition by DNA polymerases is about one error per 100,000 base insertions, but these polymerases possess an additional capability not seen in most other enzymes—DNA polymerases have "proof reading" ability.

After the phosphoester bond is formed between the most recent nucleotide and the previous one, the DNA polymerase checks again (proof reads) to see that the newest base is really base paired correctly to the complementary base on the parent strand. It can sense correct geometric alignment, and if two bases are not hydrogen bonded correctly they will not have the correct shape. If the wrong base has been inserted, then the DNA polymerase cleaves the phosphoester bond and the nucleotide diffuses into solution. Another nucleotide, presumably the correct one, will then diffuse in and bind to the complex, and the process is repeated. It is estimated that the error rate of this proof reading operation is also around one error per 100,000 base insertions. The overall error rate is simply the mathematical product of the error rate of each: $(1/100{,}000) \times (1/100{,}000) = 1/10^{10}$. This is one error per ten billion base additions.

What will happen to a base pair in a newly synthesized DNA molecule if the bases are not complementary? Cells possess a set of enzymes that repair DNA. These DNA repair enzymes have a variety of specific functions. The base pair that is not complementary will not hydrogen bond properly, so the strands are not held together at that point. Repair enzymes cut out the incorrect base, insert the proper base, and form phosphoester bonds between this base and the neighboring nucleotides. This operation restores the correct base sequence to the daughter strand. The repair enzymes are generally able to distinguish between the parent and daughter strands. Normally some of the bases of a strand of DNA are modified slightly by addition of methyl groups or other changes. A mature strand of DNA (the parent strand) will have these modifications on it, but a newly synthesized strand will not. Immediately following replication, the two strands are physically different and can be distinguished by the repair enzymes. Normally the incorrect nucleotide is removed from the daughter strand.

If the base is removed from the parent strand and replaced with the base that is complementary to the base in the daughter strand (which happens occasionally), then the DNA molecule will *not* be identical to the DNA molecule from which it was synthesized. It will differ by one base pair (Figure 26.17a). A permanent genetic change involving a replacement of one base for another in DNA is called **substitution**. The change of one base means that one codon has been changed. A change in a codon could result in a whole spectrum of possible consequences. If the codon has been altered to

Normal gene

DNA template strand	AAA	TAG	CGG	TCC ···
mRNA	UUU	AUC	GCC	AGG ···
Normal protein chain	Phe	Ile	Ala	Arg ···

(a) Substitution of G by A on DNA

Template	AAA	TAG	CAG	TCC ···
mRNA	UUU	AUC	GUC	AGG ···
Mutant protein	Phe	Ile	Val	Arg ···

(b) Insertion of a C base in DNA

Template	AAA	TAG	CCG	GTCC ···
mRNA	UUU	AUC	GGC	CAGG ···
Mutant protein	Phe	Ile	Gly	Gln ···

Deletion of an A base from DNA

A

Template	AAT	AGC	GGT ···
mRNA	UUA	UCG	CCA ···
Mutant protein	Leu	Ser	Pro ···

FIGURE 26.17
Introduced changes in DNA structure. (a) Substitution of one base for another in DNA results in one altered codon in mRNA. This means a different amino acid may be inserted into the protein. (b) Frame shift mutations result when a base is inserted or deleted from a DNA strand. This results in altered codons from that corresponding point in mRNA to the end of the gene. Many amino acids will be altered from that point in the protein to the end of the protein.

a different codon but for the same amino acid, then no observable change will occur. The protein produced from this DNA molecule will be identical to the protein produced from the parent DNA molecule. Another possibility is that the new codon may be for a different but similar amino acid. This may have little or no effect on the structure of some proteins, so even though it is different, the protein may still be functional. The codon may also be for a very different amino acid, or perhaps the codon is now a stop codon. The synthesized protein would have the wrong structure in the first case and an incomplete structure in the second case. In either case, the protein would be inactive. Much of the time a mutation results in a defective protein rather than a functional one.

Mutagens Induce Changes in DNA

Mutagens are agents that increase the mutation rate in cells.

Mutations can also be induced by a variety of agents called **mutagens**. An agent is a mutagen if an organism shows a higher rate of mutation in the presence of the agent than in its absence. For example, several forms of electromagnetic radiation cause mutations. Ultraviolet (UV) light and ionizing radiation such as X-rays alter the structure of bases in DNA. Again, repair mechanisms normally repair the damage, but this process is not perfect. If an error is made, a mutation arises. Each time an individual is exposed to sunlight, there is a chance that UV light may cause a change in the DNA of skin cells. The greater the intensity of the sunlight, or the longer the duration of exposure, the greater the chance of a mutation.

There is a reasonable correlation between mutagenesis and *carcinogenesis*, the generation of cancer. The greater the exposure to UV light, the greater the chance of skin cancer, perhaps because one or more mutations in skin cells have resulted in a change in that cell from a normal growth pattern to the growth pattern of a cancerous cell.

A number of substances such as nitrous acid and 2-aminopurine cause mutations. Some of these mutagens chemically modify bases in DNA, and as before, the repair may not be accurate. The result would be a substitution. Some other mutagens become wedged between base pairs in a DNA molecule. Some of these molecules contain three fused rings that roughly resemble a purine–pyrimidine base pair. If this DNA molecule were replicated, a base could be inserted where it should not be, or less frequently, a base may be left out. Again, if this is not corrected, a mutation will result. These mutations are called *insertion* or *deletion* mutations. Collectively they are called **frameshift mutations**, and both change the codon at the point of base insertion or deletion and all of the codons that follow (Figure 26.17b). Nearly all frameshift mutations result in defective proteins since numerous residues in the protein will be affected.

Mutations are the Source of Genetic Variability

Mutations can occur in two types of cells—the somatic (or body) cells and the germ (or sex) cells. The consequences of mutation are quite different in these two cell types. The DNA found in the sperm cells of males and in the eggs (or ova) of females is called *germinal DNA*. The DNA found in all other cells is called *somatic DNA*. An error in somatic DNA results in a mutation of a cell in a specific part of the body. That cell may divide to yield a clone of cells with the same genetic change. The result may eventually be a cancerous tumor, but the results are confined to that individual. The offspring from this individual will not carry the same error because somatic cells are not passed on to offspring.

If a change occurs in the germinal DNA of sperm or ova, it will be transmitted to the offspring as a permanent genetic change. Exposure to mutagens can result in alteration of the germinal DNA, thus directly affecting the offspring. Or instead, the offspring may become a carrier of this genetic trait, which can then be passed on to subsequent offspring.

These genetic changes or errors are the ultimate source of genetic variation seen in organisms. There is no unique sequence of DNA that is the DNA of a species. Instead, the individuals of a species contain very similar, but nonidentical DNA. This is why each kitten of a litter is unique, why each human is an individual distinctly different from all other humans. Unfortunately, some of these genetic differences are the source of genetic diseases or deficiencies that are passed from generation to generation. Some genetic diseases and their causes are listed in Table 26.5.

Mutations are the ultimate source of genetic variation.

Sometimes a greater number of people in specific populations are more prone to certain disorders than others. For example, Tay-Sachs disease, a disorder affecting lipid metabolism, is more common in Eastern European Jewish populations and their descendants than in other populations. About 10% of the Afro-American population carry the sickle cell trait, meaning they are carriers but do not suffer the full effects of the disorder. About 0.4% are afflicted with sickle-cell anemia. Sickle-cell anemia is a well-documented genetic disorder that was discussed in Chapter 25.

TABLE 26.5
Some Human Genetic Diseases

Disorder	Cause
Phenylketonuria (PKU)	An enzyme for converting phenylalanine to tyrosine is defective. The result is that phenylalanine is converted to phenylpyruvate and other phenylketones. The accumulation of these compounds in the developing young causes neurological damage.
Galactosemia	The enzyme needed for changing galactose to glucose is missing. The buildup of galactose is very serious in infants. Mental disorders as well as cataracts of the eye lens result from the buildup of galactose in the body.
Albinism	An enzyme needed for the conversion of tyrosine to a diphenolic compound is missing. This diphenolic compound is needed for formation of normal pigmentation.
Fabrey's disease	An enzyme needed for removal of a galactose from glycolipids is defective. This results in accumulation of glycolipid in a number of tissues. Death is often a result of kidney or cardiac failure due to this buildup.

NEW TERMS

mutation A permanent change in the base sequence of DNA which changes genetic information.

substitution A mutation in which a single base has substituted for another in a DNA strand; a single codon is affected.

mutagens Agents that cause mutations.

frameshift mutation A mutation resulting from insertion or deletion of a base from a DNA strand. The codon at that point plus all that follow are altered.

TESTING YOURSELF

Mutations

Show how substitution, insertion, and deletion mutations can affect protein synthesis. Consider this hypothetical base sequence in mRNA:

··· C—	G—	A—	C—	U—	G—	G—	A—	U—	A—	A—	U—	C—	C ···
1	2	3	4	5	6	7	8	9	10	11	12	13	14

1. What is the corresponding amino acid sequence?
2. Replace ^{4}C (C at number four position) with G. What is the new amino acid sequence?
3. Between ^{3}A and ^{4}C, insert a U. What is the amino acid sequence?
4. Delete ^{5}U. What is the amino acid sequence?

Answers **1.** arg—leu—asp—asn **2.** arg—val—asp—asn **3.** arg—ser—gly—chain termination **4.** arg—arg—ile—ile

26.8 Regulation of Gene Expression

Need for Regulation of Cellular Activity

Information stored in DNA as genes, through the processes of transcription and translation, is used to synthesize the variety of proteins that carry out the numerous and necessary functions of a cell. The activities of all cells of an organism, in turn, contribute to the well being of the organism. All of the somatic cells of a person have essentially the same genetic composition because they have the same DNA. They are derived from a single fertilized cell that marked the beginning of life for that individual. Yet the cells of the body are not identical. Cells of the heart have a different shape and function than liver cells. Neurons do not resemble heart or liver cells in either structure or function. Furthermore, any given cell may vary slightly in capability and function with time. Consider the changes that occur as a fertilized egg develops into an embryo, then a fetus, an infant, a child, and finally an adult.

The structure and function of a cell are a reflection of the proteins that make up the cell. Each cell type is different because it has a different set of proteins. Some proteins, like DNA polymerase, are found in virtually all cells. Other proteins are found in only one cell type, for instance, hemoglobin is found only in red blood cells and the cells that develop into them. Why do different cells have different proteins when they contain the same genes?

Regulation in Bacteria

Cells possess the ability to synthesize specific proteins in specific amounts. Cells regulate the amount and timing of protein synthesis; only some of the genes of the cell are used at any given time. An understanding of the regulation of protein synthesis in higher organisms is just beginning. Because regulation in bacteria is better understood, two examples of regulation in bacteria will be presented here. Some of these principles extend, with modification, to higher organisms.

Cells differ because they contain differing amounts of proteins. This results from regulation of when and how much protein is synthesized.

INDUCTION Many bacteria possess the ability to use several different organic compounds as energy sources and as sources of carbon atoms. For example, the intestinal bacterium *E. coli* grows on glucose, lactose, and a variety of other organic compounds. But if you look at a population of these bacteria that have been grown on glucose, they lack the enzymes needed to consume lactose. Yet if some of these bacteria are placed into a solution that contains only lactose, they will use the lactose and grow well in this solution. Furthermore, if we look for the enzymes needed to consume lactose, they are now present in the bacteria. The presence of lactose *induced* the appearance of these enzymes in the bacteria. This process is called **induction**.

Induction works by altering the rate of formation of mRNA for a specific set of proteins. If the mRNA is not formed, no proteins will be made. This is because mRNA in bacteria exists only a short time; it is broken down to nucleotides by specific enzymes. To continually synthesize proteins, mRNA must be constantly formed by transcription. If transcription is stopped or blocked, no more molecules of that protein will be made.

Induction by lactose works in the following way. The genes for the proteins that are needed to use lactose are adjacent to each other on the DNA molecule. In Figure 26.18a these genes are called structural genes and are labeled Z, Y, and A, symbols chosen for these genes when they were first discovered. In this figure, a promoter is near the Z gene. Remember, promoters are the sites where RNA polymerase binds to begin transcription. If synthesis of the lactose-utilizing proteins were not regulated, RNA polymerase would freely bind here, and transcription would occur to yield mRNA that contains information for the translation of all three genes. But the synthesis of these proteins is regulated. Next to the promoter is a region that is called the operator. The operator is involved in controlling whether transcription will or will not occur. The operator and promoter are labeled control sites in Figure 26.18. The final piece of DNA that is involved in the regulation of lactose use is the regulatory gene. This gene's product is a protein called repressor (see next paragraph). A regulatory gene, a promoter, an operator, and the structural genes are a unit that is called an **operon**. This particular operon is the lactose operon, which provides for the regulated synthesis of lactose-utilizing enzymes.

In the bacterial cell is a *repressor* protein called the lactose repressor. This protein's function is to bind to the operator of the lactose operon (Figure 26.18b). There are other repressors that bind to the operators of the other

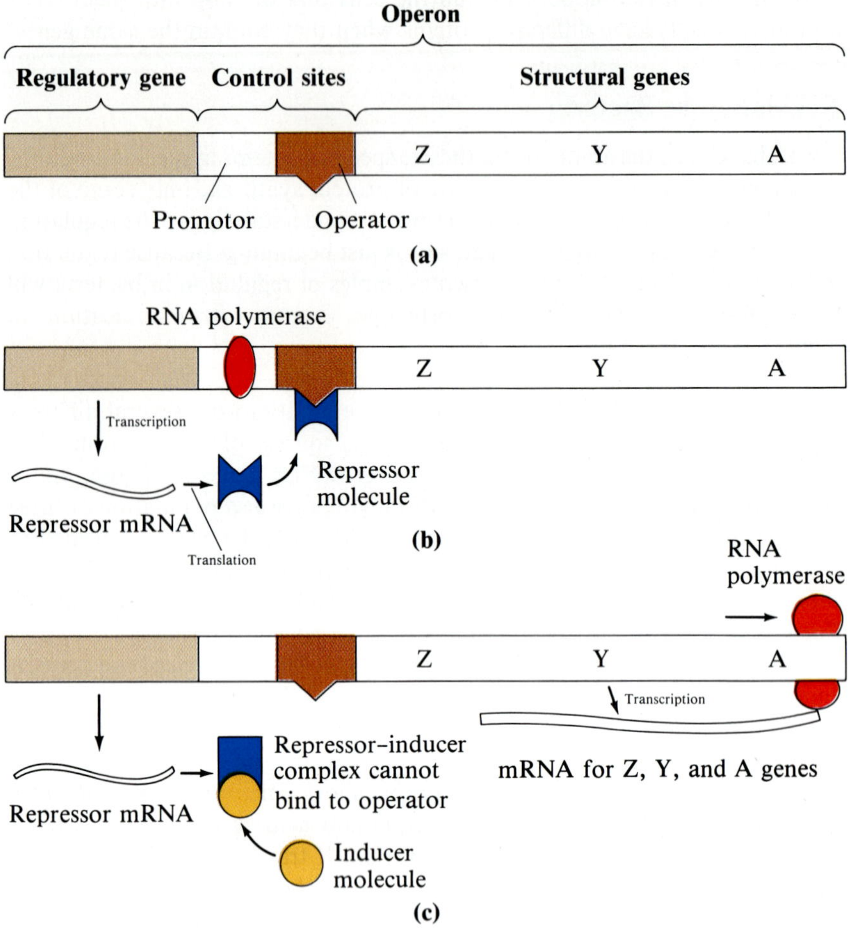
Operon
Regulatory gene
Control sites
Structural genes
Z
Y
A
Promotor
Operator
(a)
RNA polymerase
Z
Y
A
Transcription
Repressor molecule
Repressor mRNA
Translation
(b)
RNA polymerase
Z
Y
A
Transcription
Repressor–inducer complex cannot bind to operator
mRNA for Z, Y, and A genes
Repressor mRNA
Inducer molecule
(c)

FIGURE 26.18
In some bacteria, lactose in the culture medium induces the appearance of the enzymes that break down lactose. (a) The arrangement of the genes in the lactose operon. (b) When lactose is absent, the repressor protein binds to the operator. RNA polymerase cannot transcribe the structural genes because the repressor prevents the polymerase from moving from the promotor to the structural genes. (c) When the inducer (lactose) is present, it binds to the repressor and alters its shape. The complexed repressor can no longer bind to the operator. RNA polymerase can now bind to the promotor and transcribe RNA.

operons. When the repressor is bound to the operator, RNA polymerase *cannot* bind to the promoter, thus transcription cannot occur. The presence of the repressor prevents transcription because it physically blocks the binding of RNA polymerase. In the absence of lactose, the repressor remains on the operator.

If some lactose diffuses into the cell, some of it binds to the repressor molecule. The binding of lactose to the repressor changes the conformation of the repressor. The ability of a protein to bind to a molecule depends on the shape of both the protein and the molecule. If the shape of either changes, the ability to bind changes. In this new conformation the repressor will no longer bind to the operator, so the repressor–lactose complex diffuses from the DNA. Now RNA polymerase molecules are free to bind to the promoter (Figure 26.18c), thus transcription and subsequently translation occurs. The enzymes needed for utilization of lactose are synthesized.

REPRESSION Bacteria possess a second way to regulate protein synthesis. **Repression** is the *turning off* of protein synthesis when the proteins are not needed by the cell. As an example, consider amino acid synthesis in bacteria. Bacteria can produce all 20 of the amino acids used in protein synthesis. If they are given a diet lacking all 20, they synthesize all of them. If

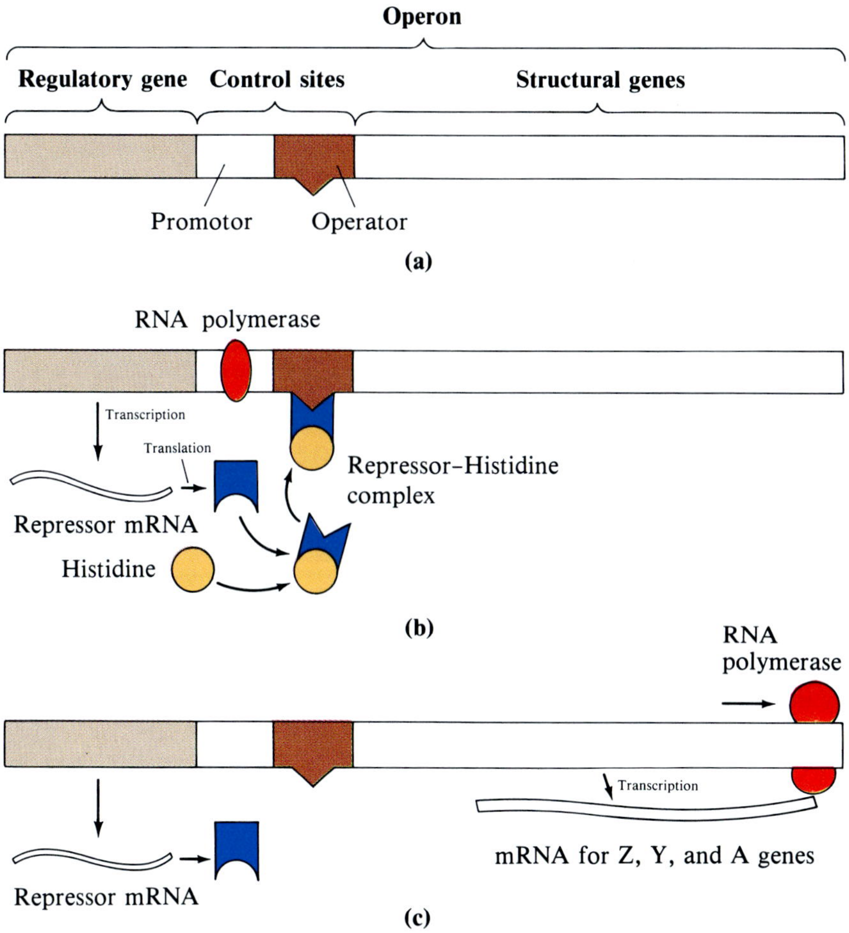

Operon
Regulatory gene
Control sites
Structural genes
Promotor
Operator
(a)
RNA polymerase
Transcription
Translation
Repressor-Histidine complex
Repressor mRNA
Histidine
(b)
RNA polymerase
Transcription
mRNA for Z, Y, and A genes
Repressor mRNA
(c)

FIGURE 26.19
Histidine repression in bacteria. (a) The arrangement of genes in the histidine operon. (b) When histidine is present, it binds to the repressor protein and alters the shape of the repressor. When histidine is bound, the repressor binds to the operator. This prevents RNA polymerase from transcribing the structural genes. (c) When histidine is absent in the cell, the repressor cannot bind to the operator. RNA polymerase is free to transcribe the structural genes.

one of the amino acids, say histidine, is provided to the bacteria, they stop synthesizing the enzymes needed for the synthesis of histidine. The presence of histidine in the diet repressed the synthesis of the histidine-synthesizing enzymes. Repression by histidine also involves an operon with an operator, a promoter, the genes for the proteins, and a repressor. The main difference is in the binding properties of the repressor. The repressor, by itself, *does not* bind to the operator, but if histidine is bound to it, the repressor–histidine complex does bind to the operator.

Consider the initial amount of histidine in the cell to be low because there is none in the solution surrounding the bacterium. When the concentration of histidine is low, it does not bind to the repressor (Figure 26.19). If histidine is not bound to the repressor, then it will not bind to the operator. If the repressor–histidine complex is not bound to the operator, then RNA polymerase can bind to the promoter and transcription can occur. The enzymes needed for histidine will be made by translation of the mRNA that has been transcribed.

But if the concentration of histidine were high, perhaps because it was present in the environment, then histidine binds to the repressor. The repressor with bound histidine has the correct conformation needed to bind to the operator. Once the repressor–histidine complex is bound to the operator,

RNA polymerase can no longer bind to the promoter, transcription no longer occurs, and the proteins are not synthesized. Histidine represses the synthesis of the enzymes needed for its synthesis.

Regulation in Higher Organisms

Specific interactions between proteins and DNA are involved in regulation in eucaryotic cells.

Regulation in higher organisms is not well understood. It is thought that specific proteins may bind to DNA and influence the transcription of genes, somewhat like induction turns on bacterial genes and repression turns them off. At the present time, operons have not been found in higher organisms, but specific interactions between proteins and some portions of DNA molecules do occur. More research is needed to determine the details of regulation in humans and other higher organisms.

NEW TERMS

induction Regulation of gene expression involving the turning on of genes by the presence of a compound (inducer).

operon A set of genes in bacteria that work in concert. Genes within the operon are either turned on or off collectively.

repression Regulation of gene expression involving the turning off of genes by the presence of a compound (repressor).

TESTING YOURSELF

Regulation

1. What is the operator in an operon?
2. How does the regulatory gene contribute to regulation of lactose use?
3. Why does RNA polymerase not transcribe the structural genes constantly?

Answers **1.** The operator is the piece of DNA where the repressor binds. When the repressor is bound, RNA polymerase cannot transcribe the genes. **2.** This gene is transcribed to yield mRNA that is translated to yield the lactose repressor protein. **3.** The repressor–operator complex physically blocks the movement of RNA polymerase to the structural genes.

26.9 Genetic Engineering

Mutations are permanent changes in the genetic material of an organism. Today techniques are available that can be used to permanently alter the DNA of organisms in very specific ways. It is now possible to add whole genes into an organism. This scientific and technical area has several names including **genetic engineering**, *biotechnology*, and *bioengineering*.

Process of Genetic Engineering: An Example

Although the techniques of genetic engineering can be used for a variety of applications, we will use one as an example—the production of synthetic human insulin. Insulin is needed for treatment of patients with the genetic disease diabetes mellitus. In the past, these people were given insulins obtained from the pancreas of slaughtered animals. These insulins were very similar but not identical to human insulin. They did not always function identically to human insulin, and medical complications resulted. In the past no adequate source of human insulin was available. Biotechnology has changed all that. It is now possible to introduce a gene for human insulin (obtained from a human source) into bacteria. These genetically altered bacteria are then grown in large numbers and are able to synthesize usable quantities of human insulin. The insulin is harvested, not unlike antibiotics are harvested from cultured microorganisms, and then processed and provided to the medical community. Quantities of pure human insulin are now available at a reasonable cost.

Although the strategy for obtaining quantities of an unavailable human protein is straightforward, the actual process is much harder. The first complication encountered in this strategy is the phenomenon of split genes in higher organisms. The human gene for our desired protein has introns scattered throughout the gene. This is no problem for a higher cell; the introns are cut out of the transcript and the transcript is otherwise modified to a functional mRNA. But bacteria do not have the enzymes needed to process a transcript to a functional mRNA. They do not need them because the transcript in bacteria *is* a functional mRNA. One cannot simply fragment human DNA and insert pieces of it into the bacterial DNA, because in the bacterium, the gene would not yield a functional mRNA for the protein.

Two alternatives can be used to get around this problem of split genes. First, the gene could be chemically synthesized using sophisticated techniques derived from simpler organic syntheses. If the amino acid sequence of the protein is known, the codons needed for the protein can be determined. A start and stop codon can be added, and perhaps a section at the beginning can be added to act as a promoter. To each end of the synthesized gene is added a short piece of several bases. These pieces are used to connect the gene to bacterial DNA (Figure 26.20).

The second method that could be used involves human mRNA for the desired protein. The mRNA is isolated, then the enzyme *reverse transcriptase* is used to make a single strand of DNA that is complementary to the mRNA. This enzyme is obtained from RNA viruses that make the enzyme during a portion of their life cycle. This single strand of DNA is made into double-stranded DNA by DNA polymerase to yield a DNA molecule that codes for the protein. It is not the natural gene since it has no introns. Instead, it is a copy of the expressed portion of the natural gene. It is again necessary to make modifications to the ends of the gene so it can be inserted into bacterial DNA.

The next step is to insert the gene into bacterial DNA (Figure 26.20). The gene is not put directly into the bacterial chromosome, but is instead placed into a smaller piece of bacterial DNA called a **plasmid**. Plasmids, like bacterial chromosomes, are circular DNA molecules, but plasmids are much smaller. The plasmid DNA is cleaved by one of several enzymes that are called *restriction endonucleases*. These enzymes are used because they cleave DNA, leaving a short piece of single-stranded DNA at each end. These short pieces are complementary to the short pieces that were added to the genes. When the copies of the gene are mixed with the cleaved plasmid DNA, some of the genes are inserted into the plasmids and held by hydrogen bonding between complementary bases. An enzyme is added to form covalent bonds

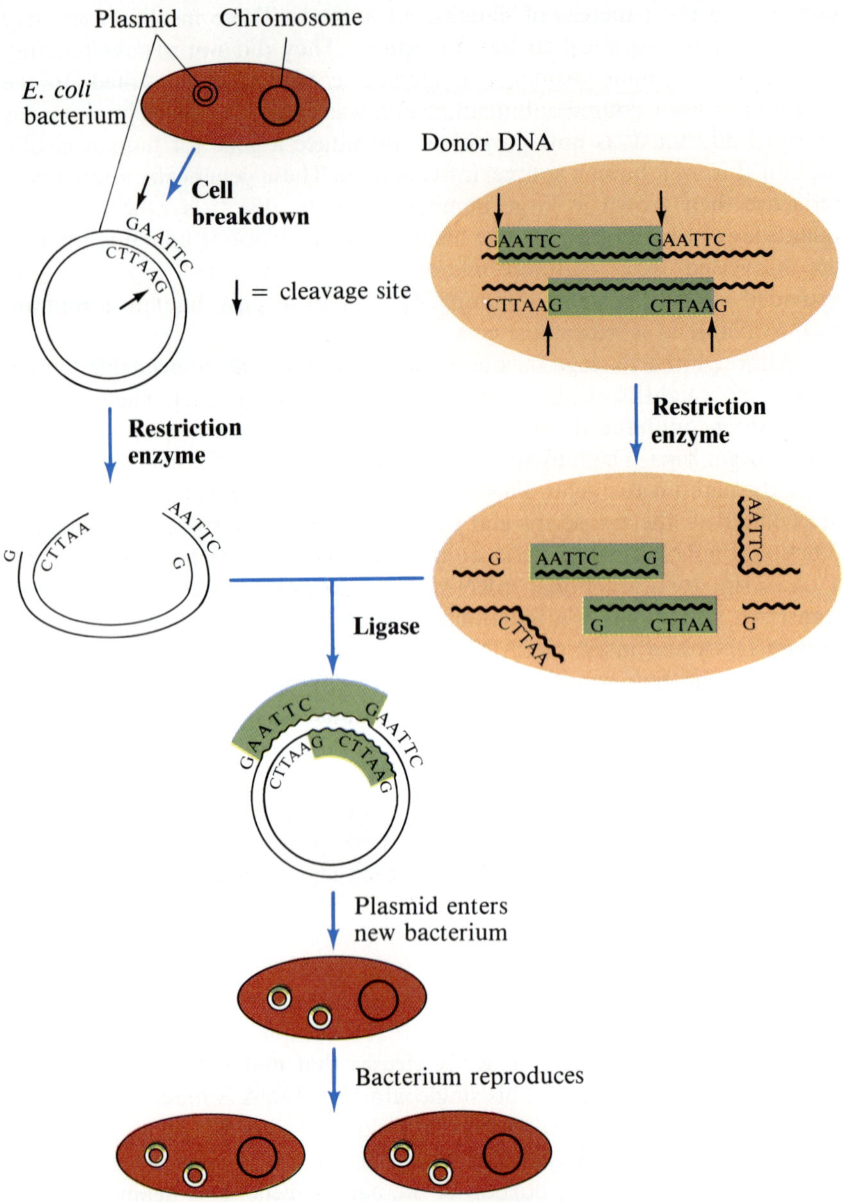
Plasmid
Chromosome
E. coli bacterium
Donor DNA
Cell breakdown
GAATTC
CTTAAG
↓ = cleavage site
Restriction enzyme
Restriction enzyme
Ligase
Plasmid enters new bacterium
Bacterium reproduces

FIGURE 26.20
The insertion of a eucaryotic gene into a bacterium. A plasmid is isolated from bacteria, and the desired eucaryotic DNA is prepared. The plasmid DNA is cleaved with a restriction enzyme. These pieces are mixed, bind together through the complementary bases at the end of the pieces, and are covalently linked by the enzyme ligase. This incorporates the donor DNA into the bacterial plasmid. This plasmid is then inserted into bacteria, which reproduce. Often genes of the donor DNA are expressed in the bacteria.

between the plasmid DNA and the DNA of the inserted gene. The gene is now an integral part of the plasmid.

The plasmid can then be inserted into the bacterial species of choice (Figure 26.20). These bacteria are then cultured. If the process has been successful, the gene will be transcribed to yield mRNA that is in turn translated into functional protein. The protein can then be harvested. This described procedure, while feasible, is very laborious. Much planning, time, effort, and sometimes a little luck go into the successful completion of one of these projects. However, potentially great benefit will come to civilization from these efforts in genetic engineering.

The Future of Genetic Engineering

Bioengineering is already providing us with useful biological products. It may also be used someday to alter the genes of human beings. Already numerous alterations have been made with microorganisms, and some experiments with higher species have produced organisms with portions of their body derived from one species and other parts from a second species. One example of such an organism is part chicken and part quail (Figure 26.21). Although the genetic makeup of microorganisms are readily changed now, it is not yet feasible to alter the genes of higher organisms in a similar manner. There are indications that such manipulations will be available in the future. Perhaps someday human germ cells will be genetically altered by adding a functional gene to replace a nonfunctional gene. A couple who had previously chosen to not have children, because of the risk of a genetic disease, will be able to have children free of the disease. Perhaps someday people with genetic diseases will be cured by inserting a functional gene or set of genes into the individual. A phenylketonuric might metabolize phenylalanine properly, or a diabetic could produce or use adequate amounts of insulin.

FIGURE 26.21
This chick has the head of a chicken and the body of a quail. Early in development, embryonic cells of these species were mixed and joined into one embryo. The embryo developed into a mosaic, part of which is quail-like and part of which is chicken-like.

As the technology progresses, one can imagine that virtually all genetic traits could be altered at the will or whim of society. The potential for altering mankind is most impressive. When and how these changes come about will be determined ultimately by you, by society, not by bioengineers. All of us need to be informed citizens prepared to deal with issues from a wide variety of subjects and areas. Technology opens many doors; society chooses which doors we pass through.

Biotechnology brings not only the promise of great benefits, but the challenge of choice.

NEW TERMS

genetic engineering The branch of technology that manipulates genetic information to produce biological products or organisms with permanently altered abilities.

plasmid Small pieces of DNA found in some bacteria; a common vehicle for introducing new genes into bacteria.

TESTING YOURSELF

Genetic Engineering

1. What type of enzyme is used to cut genes out of larger DNA pieces and to cut plasmids?
2. Why are these enzymes used?
3. What role does the enzyme reverse transcriptase play in the insertion of a human gene into a bacterium?

Answers **1.** Restriction endonucleases. **2.** They cut DNA leaving single stranded ends, sticky ends, which permit pieces of DNA to be rejoined. **3.** This enzyme is used to make single-stranded DNA that is complementary to human mRNA.

SUMMARY

Deoxyribonucleic acid or *DNA* is a polymer of nucleotides whose sequence of purine and pyrimidine bases provides information to living organisms, much like the alphabet is used to provide information for society. The bases adenine, guanine, cytosine, and thymine are combined with deoxyribose and phosphate to form the *nucleotides* that make up DNA. DNA is duplicated with high fidelity by the process of *replication*, and the process of *transcription* produces another nucleic acid, *ribonucleic acid* or *RNA*, the carrier of information from DNA to the cell. This information is needed to synthesize, via *translation*, the proteins that carry out the functions of the cell. The activities of the cell are highly regulated, in part by controlling the rates of transcription. The genetic composition of organisms can now be altered through the process of *genetic engineering*. These potential genetic alterations promise great changes for the future.

TERMS

deoxyribonucleic acid (DNA)
ribonucleic acid (RNA)
nucleotide
nitrogenous base
nucleoside
complementary bases
chromosomes
semiconservative replication
messenger RNA (mRNA)
ribosomal RNA (rRNA)
transfer RNA (tRNA)
transcription
transcript
translation
gene
codon
anticodon
mutation
substitution
mutagens
frameshift mutation
induction
operon
repression
genetic engineering
plasmid

CHAPTER REVIEW

QUESTIONS

1. The base–pair specificity of DNA is
   a. A—G, C—T b. A—T, C—G c. A—U, C—T d. T—U, A—C
2. The sugar associated with RNA is
   a. ribose b. glucose c. deoxyribose d. fructose
3. Codons are part of what molecules?
   a. DNA b. mRNA c. tRNA d. rRNA
4. Purines and pyrimidines are the organic bases found in nucleic acids. The purines are
   a. adenine and thymine c. adenine and guanine
   b. adenine and cytosine d. guanine and cytosine
5. The nucleic acid that brings amino acids to the site of protein synthesis is
   a. mRNA b. tRNA c. rRNA d. DNA
6. What name is given to the turning off of a set of genes in bacteria?
   a. switching b. induction c. repression d. suppression
7. A gene is a segment of
   a. DNA b. RNA c. protein d. codon
8. The backbone of DNA is *not* composed of
   a. sugars b. bases c. phosphoric acid

9. If a gene is to be introduced into a bacterium, what is often used to carry the gene?
   a. transfer RNA c. plasmid
   b. reverse transcriptase d. restriction endonuclease

10. The synthesis of ______ is most directly affected by errors in the DNA code.
   a. proteins b. lipids c. carbohydrates d. vitamins

Answers **1.** b **2.** a **3.** b **4.** c **5.** b **6.** c **7.** a **8.** b **9.** c **10.** a

DIAGNOSTIC CHART

Blacken in all of the circles under the number of each question you missed in the preceding Chapter Review questions. The diagnostic chart will help you identify concept areas that need more study.

	Questions									
Concepts	**1**	**2**	**3**	**4**	**5**	**6**	**7**	**8**	**9**	**10**
Nucleotides	○	○		○						
Nucleic acids	○	○	○	○	○		○	○		
Replication	○									
Transcription	○						○			
Translation			○		○					
Mutations			○				○			○
Regulation						○				
Bioengineering									○	

EXERCISES

Nucleotides and Nucleic Acids

1. What sugars are found in nucleic acids?
2. Which nitrogenous bases are purines? Pyrimidines?
3. What is the difference between ribose and deoxyribose?
4. Describe the base pairing of DNA.
5. How does DNA differ from RNA?
6. What is the difference between a nucleoside and a nucleotide?
7. What are the three major types of RNA? Describe their functions.
8. What two types of bonding hold the two strands of DNA together?

Genetic Code

9. Why are three bases needed to determine the genetic code?
10. If a segment of DNA has the base sequence

—A—C—G—G—T—A—C—T—G—

what will be the corresponding sequence on the mRNA? What amino acids would be coded?

11. Define the terms *codon* and *anticodon.*
12. Why isn't the triplet, T—A—C a codon for protein synthesis?

Protein Synthesis

13. The peptide hormone *oxytocin* is formed by the pituitary gland by cutting it from a larger peptide. The amino acid sequence for oxytocin is

```
HOOC–Cys–Tyr–Ile–Gln–Asn–Cys–Pro–Leu–Gly–NH2
      |                   |
      S-------------------S
```

(Note the disulfide linkage between the two cys residues.) Write the codons that code for oxytocin.

14. A portion of the amino acid sequence for normal hemoglobin is

—Val—His—Leu—Thr—Pro—Glu—Glu—Lys—

The corresponding amino acid sequence for sickle cell hemoglobin (S) is

—Val—His—Leu—Thr—Pro—Val—Glu—Lys—

Write the codon change that has occurred.

15. Briefly describe how a gene on DNA becomes expressed to yield protein.
16. Although DNA carries the "information" or message for all processes in an organism, most of the attention is directed toward the role of DNA in protein synthesis. Discuss why this is reasonable.

DNA Replication

17. Describe the process of DNA replication.
18. Describe how substitution of bases can yield a mutation.
19. What is a frame shift mutation?

General Question

20. The following words or phrases have been used in this chapter. Give an example, definition, or a brief description of each to show that you understand the word or phrase.
    a. deoxyribose
    b. nucleic acid
    c. nucleoside
    d. purine
    e. specific base pairing
    f. t, m, and r prefixes to RNA
    g. codon and anticodon
    h. transcription
    i. sickle cell anemia
    j. mutation

CHAPTER

27

Metabolism, Enzymes, and Bioenergetics

The molecules in the body interact in a complex, highly coordinated set of reactions called metabolism. *These reactions are catalyzed by enzymes, which are proteins with very specific activities. As in all chemical processes, energy plays a critical role in the reactions of the body. This chapter introduces metabolism, discusses the properties of enzymes, and provides insight into the role of energy in metabolism.*

OBJECTIVES

After completing this chapter, you should be able to

- Define the terms *metabolism*, *catabolism*, and *anabolism*.
- Describe the composition and classification of enzymes.
- Discuss how enzymes work and why they are specific.
- List factors that influence the rates of enzymic reactions.
- Describe enzyme inhibition and its importance in medicine.
- Discuss how enzyme activity is regulated.
- Describe the role of energy in maintaining life.

27.1 Metabolism

The chemical reactions of the body make up metabolism.

All of the chemical reactions that take place within the body are collectively called **metabolism**. Some of these reactions produce energy, which is used by the body in a variety of ways. These reactions make up **catabolism**, the subject of Chapter 28. Another set of reactions builds larger, more complex molecules by using energy to connect smaller molecules or pieces of molecules together. These reactions constitute **anabolism**, the topic of Chapter 29. Replication of DNA, transcription, and translation are examples of anabolic processes that you have already seen in Chapter 26.

The body's chemical reactions do not occur in an uncontrolled random fashion, nor do they occur completely independent of each other. Many of the reactions are linked sequentially; the product of one reaction is the reactant of another. A sequence of reactions within the body is called a **metabolic pathway**. A metabolic pathway provides a series of small chemical alterations that changes a substance into something quite different. In the pathway illustrated to the left, reactant A is converted to product B in the first reaction (R_1) of the pathway. B, in turn, is the reactant for the second reaction. For the overall process, substance E is formed from A by a sequence of four chemical reactions. Some of the metabolic pathways of the body have more than a dozen steps.

$$A \xrightarrow[R_1]{} B \xrightarrow[R_2]{} C \xrightarrow[R_3]{} D \xrightarrow[R_4]{} E$$

The chemical reactions of a metabolic pathway are catalyzed by enzymes and occur under the mild conditions that exist within a cell. For example, through many individual steps, cellular glucose is converted to carbon dioxide, water, and energy. Several dozen reactions occur during this transformation. Glucose is also a component of cellulose in wood, and analogously, when glucose burns in a campfire, carbon dioxide, water, and energy are the products. The reactions of the body occur at 37°C, while combustion occurs at temperatures that would destroy living tissues. Special purpose catalysts called *enzymes* permit the occurrence of low temperature metabolic reactions that are essential for life.

Catalysis by enzymes allows metabolic reactions to occur at a reasonable rate in the body.

NEW TERMS

metabolism The sum of all chemical reactions in the body that collectively sustain life.

catabolism The energy-yielding part of metabolism that breaks down larger, more complex molecules into simpler ones.

anabolism The part of metabolism that uses energy to build larger, more complex molecules from simpler ones.

metabolic pathway A series of chemical reactions in the body that converts a substance to another distinctly different substance.

27.2 Enzymes

Definition of Enzymes

The reactions of the body are catalyzed by **enzymes**. Enzymes are proteins, although recently a few RNA molecules have also been shown to possess some catalytic activity. In Chapter 10 catalysts were defined as substances that speed up chemical reactions without themselves being changed during the reaction. Enzymes speed up reactions by lowering the energy of activation of the reaction (see Figure 10.9). A catalyzed reaction requires less time to reach equilibrium and often occurs at lower temperatures and under milder reaction conditions than an uncatalyzed reaction. Enzymes are some of the most efficient catalysts known. Many of the reactions in the laboratory require high temperatures, high concentrations of reagents, and a variety of solvents and reagents that would destroy a living cell or not be available to it. Enzymes permit the body's reactions to occur rapidly and efficiently at body temperature with relatively low reagent concentrations in water, the solvent of the cell. For example, one molecule of carbonic anhydrase, an enzyme of red blood cells, can catalyze the hydration of 600,000 carbon dioxide molecules in 1 second.

Enzymes are more than just efficient catalysts, however. They are often specific for certain substances (see Section 27.3). Cells contain many hundreds of different compounds, yet only one or a few of these compounds are affected by a particular enzyme. Furthermore, the amount and activity of some enzymes varies with the conditions that exist within the cell. These changes in enzyme activity are one important way cells help regulate the body's chemical reactions and activities (see Section 27.5).

Enzyme Nomenclature

Although most organic molecules are named by composition and structure, proteins are too big and complex to be named this way. Enzyme nomenclature is based on the reactions catalyzed by the enzyme and attempts to provide some information about the function of the enzyme. First, the name identifies the substance as an enzyme with the suffix **-ase**. Less commonly, an enzyme is identified by an older suffix *-in*. Ure*ase*, RN*ase*, and hexokin*ase* are clearly enzymes, as are tryps*in*, chymotryps*in*, and peps*in*. Second, the name may describe the reaction catalyzed by the enzyme. For example, an *oxid*ase catalyzes an oxidative reaction, and an *isomer*ase catalyzes an isomerization of the reacting molecule to an isomer. Third, the name may provide information about the reacting molecules, the **substrate** or substrates of the reaction. *Glucose oxid*ase catalyzes the oxidation of glucose, and *hexokin*ase catalyzes reactions involving hexoses.

The suffix *-ase* in a substance name identifies it as an enzyme.

A substrate is the reactant in an enzyme catalyzed reaction.

Many of the common enzyme names contain one or two of the desired features of modern enzyme nomenclature but rarely all three. A more formal nomenclature has been developed by the International Enzyme Commission. This system classifies all enzymes with respect to reaction type and substrates. All enzymes are uniquely identified by a name and number code. Common enzyme names are in general use, but a formal name is usually given at least once in a research paper to identify the enzyme unambiguously. Table 27.1 lists the six major classes of enzymes.

Enzyme Composition

Enzymes are proteins that possess catalytic activity. Some enzymes require no other molecule or ion to be active. These enzymes contain only one or more polypeptide chains and are thus simple proteins. Other enzymes are not catalysts unless one or more inorganic ions or organic molecules are present. The ions and organic molecules required for the activity of these enzymes are called **cofactors**. If the cofactor is an organic compound, it is

Some enzymes require cofactors for catalytic activity.

TABLE 27.1
Major Enzyme Classes

Class of Enzyme	Reaction Type
Oxidoreductase	Oxidation–reduction
Transferase	Transfer one or more atoms from one substance to another
Hydrolase	Hydrolytic cleavage or reverse
Lyase	Cleavage or reverse, but not oxidation–reduction or hydrolysis
Isomerase	Intramolecular rearrangements
Ligase	Energy-requiring bond formation

typically referred to as a **coenzyme**. Some coenzymes are derived from vitamins (see Table 27.2).

An enzyme with its cofactor is called a *holoenzyme*. An enzyme lacking its cofactor is called an *apoenzyme*. If a cofactor is permanently attached to an enzyme, it is called the *prosthetic group* of the enzyme. In other cases, the coenzymes may be attached to the enzyme only part of the time in a manner similar to the substrate.

Minerals and Vitamins in Metabolism

Many coenzymes are synthesized from vitamins.

Macronutrients, including carbohydrates, proteins, and lipids, provide energy and building blocks for the body. Minerals and vitamins serve a different role—they provide a source of cofactors for catalytic activity. For example, the digestive enzyme carboxypeptidase (Chapter 28) requires zinc ion, and many enzymes require a coenzyme called NAD^+ (nicotinamide adenine dinucleotide). The minerals in our diet provide the inorganic cofactors that enzymes require, and many of the vitamins are used to make a variety of coenzymes needed by the body. Table 27.2 provides the names and functions of a few key coenzymes along with the vitamins from which they are formed. Structures of a few representative coenzymes are shown here.

TABLE 27.2
Coenzymes, Their Functions, and Vitamins from Which They Are Derived

Coenzyme	Function	Vitamin
Flavin adenine dinucleotide (FAD)	Redox reactions	Riboflavin
Nicotinamide adenine dinucleotide (NAD^+)	Redox reactions	Niacin
Coenzyme A	Carrier of acyl groups	Pantothenic acid
Thiamine pyrophosphate (and other forms)	Forms covalent intermediates during catalysis	Thiamine
Pyridoxal phosphate (and other forms)	Forms covalent intermediates during catalysis	B_6
Tetrahydrofolate (and other forms)	Transfers groups containing one carbon atom	Folic acid
Cobalamines	Several reaction types	B_{12}
Biotin	Carboxylations	Biotin

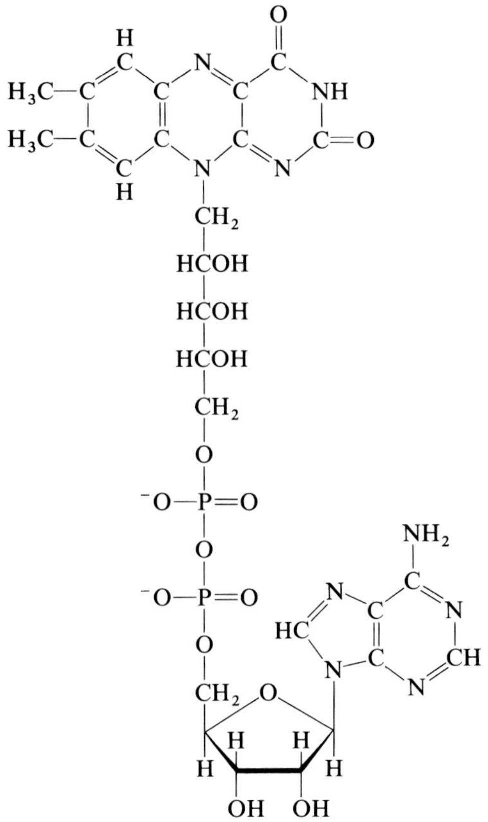

Flavin adenine dinucleotide (FAD)

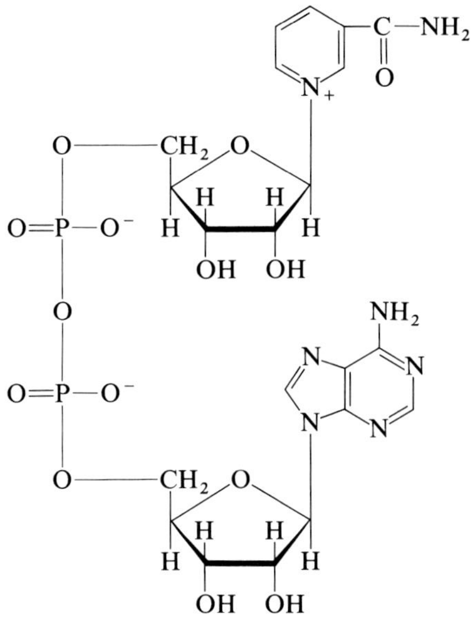

Nicotinamide adenine dinucleotide (NAD^+)

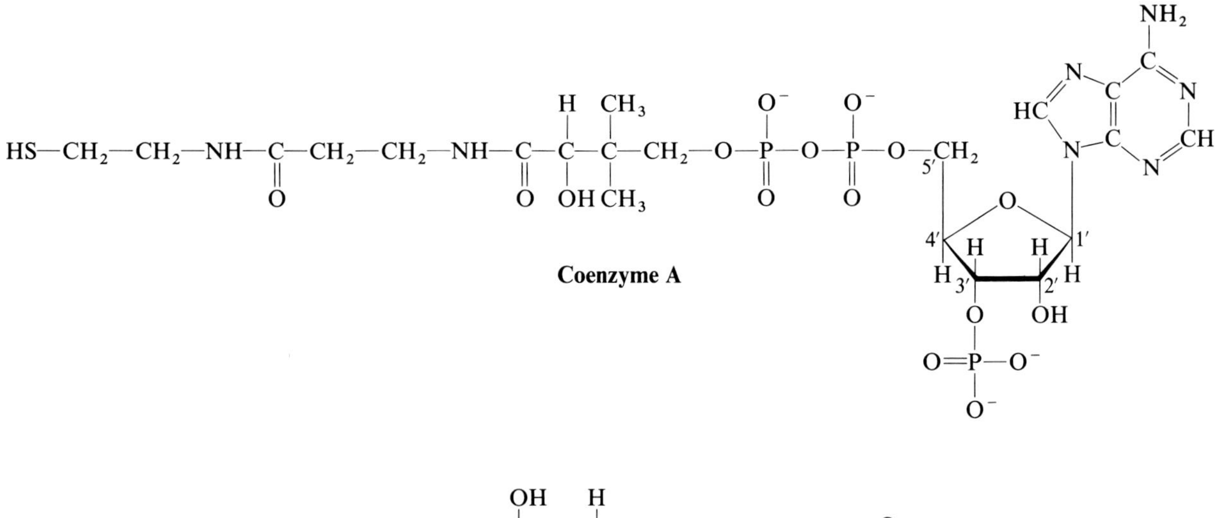

Coenzyme A

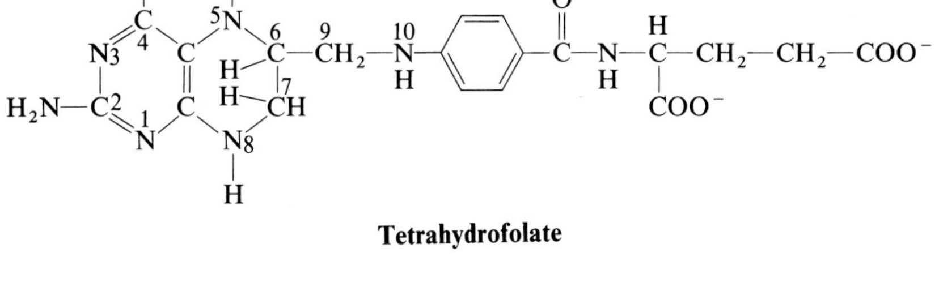

Tetrahydrofolate

NEW TERMS

enzyme A protein that has specific catalytic activity.

-ase A suffix used to indicate that the substance is an enzyme.

substrate The reacting molecule that binds to an enzyme. The enzyme catalyzes its conversion to product.

cofactor The non–amino acid portion of some enzymes.

coenzyme The name given to organic cofactors.

TESTING YOURSELF

Enzymes

1. What name is given to that portion of metabolism that yields useful energy?
2. Use enzyme nomenclature to predict the function of triose phosphate isomerase.
3. What name is given to the non–amino acid part of an enzyme?
4. Nicotinamide adenine dinucleotide (NAD^+) is synthesized from what vitamin?

Answers **1.** catabolism **2.** The enzyme isomerizes triose phosphates. (The actual isomerization is between glyceraldehyde 3-phosphate and dihydroxyacetone phosphate.) **3.** cofactor **4.** niacin (nicotinic acid and nicotinamide)

27.3 Enzyme Specificity and Activity

The place on an enzyme where substrate binds and catalysis occurs is called the *active site.*

Enzymes catalyze the conversion of specific substrate molecules to product molecules. The substrate or substrates bind noncovalently to the **active site** of the enzyme, where they react to yield a product or products. The active site has two distinct functions: the binding of substrate and the catalysis of the reaction.

Enzyme Specificity

Why does the active site bind only one or a few of the many compounds in the cell? Why do enzymes show specificity for substrate? Enzyme specificity is determined by fit, proper alignment of the enzyme and the small molecule (see Figure 22.11), and specific noncovalent bonding between the active site of the enzyme and portions of the substrate molecules. Other molecules in the cell do not fit or form similar bonds, thus they do not bind to the active site. The size and shape of the active site and the specific chemical groups located at the active site contribute to binding specificity.

The active site of an enzyme is on the surface of the enzyme (Figure 27.1). The size and shape of the active site are determined by the amino acid residues at the active site and the prosthetic group, if present. Because each enzyme has different amino acids with different orientations, the active site of each enzyme is different. Similarly, each compound in the cell has its own size and shape so only some of them will fit into the active site of any particular enzyme. The size and shape of an active site influences substrate binding.

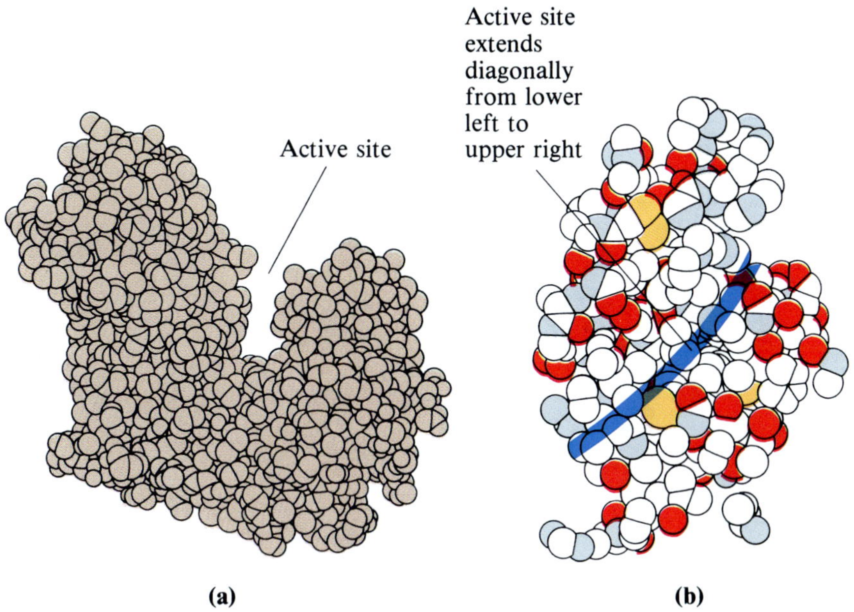
Active site
Active site extends diagonally from lower left to upper right
(a)
(b)

FIGURE 27.1
The active site of two enzymes. (a) The active site of hexokinase is a pocket that can accommodate the substrate hexoses and ATP. (b) The active site of lysozyme is a long cleft that can accommodate the polysaccharide chain that is its substrate.

Consider the enzyme lysozyme shown in Figures 27.1b and 27.2. This enzyme catalyzes the hydrolysis of glycosidic bonds in polysaccharides in bacterial cell walls.

The long cleft of this enzyme's active site accommodates the long polysaccharide molecule quite nicely (Figure 27.2a), but if you think about it, many other molecules are also linear like a polysaccharide. A polypeptide or the chain of a fatty acid of a triacylglycerol might also fit into the cleft. But lysozyme does not catalyze the hydrolysis of these molecules; it is specific for its substrate.

The size and shape of an active site contribute to proper substrate binding, but additional factors are necessary to explain specificity of binding. Look at Figure 27.2b. This close view of binding between lysozyme and its substrate shows specific hydrogen bonds between them. Not only are the sizes and shapes of the substrate and active site complementary, but the orientation of groups on the substrate and active site permit hydrogen bonding between them. Fatty acids and polypeptides may fit, but they do not bond to the active site.

The interactions that occur between an enzyme and its substrate are the same interactions that are responsible for tertiary structure in proteins: hydrogen bonding, hydrophobic interactions, and ionic and polar interactions. A lipid will bind to the active site of an enzyme because the sizes and shapes are complementary and because the active site of the enzyme is lined with nonpolar amino acid residues that form hydrophobic interactions with the lipid. A polysaccharide could not form these hydrophobic interactions because it has no hydrophobic regions. The digestive enzyme trypsin cleaves peptide bonds in proteins at lysine or arginine residues. The active site of trypsin can accommodate the size and shape of the polypeptide chain and

A substrate binds to an enzyme because they are complementary in size, shape, and potential for binding to each other.

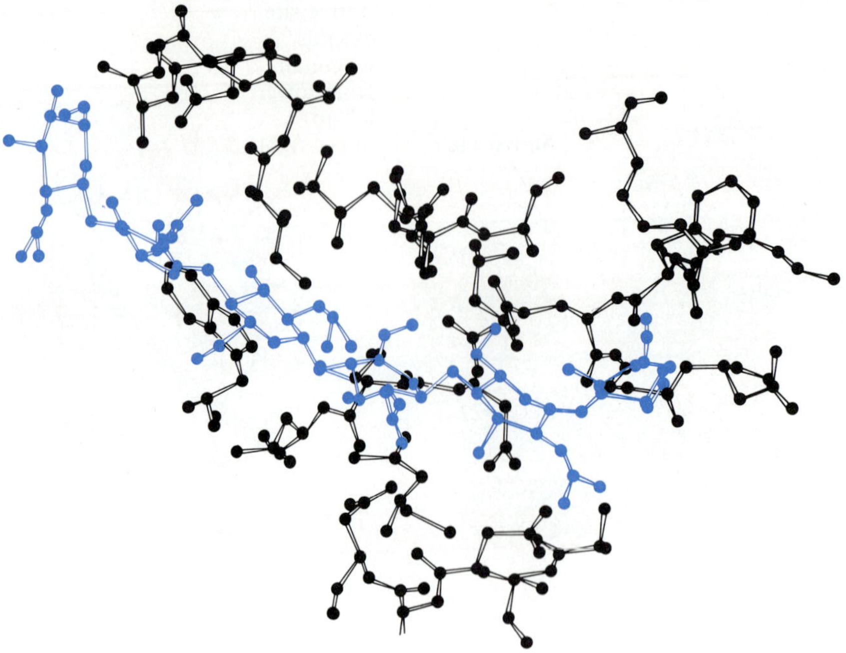

(a)

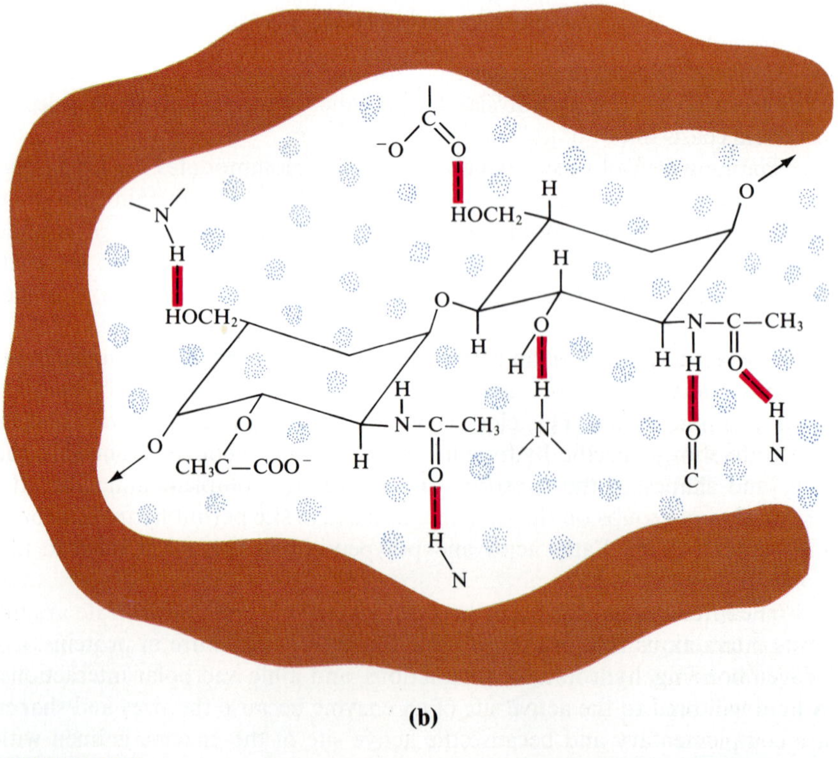

HOCH₂
HOCH₂
N—C—CH₃
N—C—CH₃
CH₃C—COO⁻

(b)

FIGURE 27.2
Binding of a polysaccharide in a bacterial cell wall to the active site of lysozyme. (a) The substrate and active site are complementary in shape, with the long substrate molecule fitting into a cleft (blue) in the enzyme. (b) Noncovalent interactions (red), primarily hydrogen bonds, form between groups on the substrate and enzyme to hold the substrate in the active site.

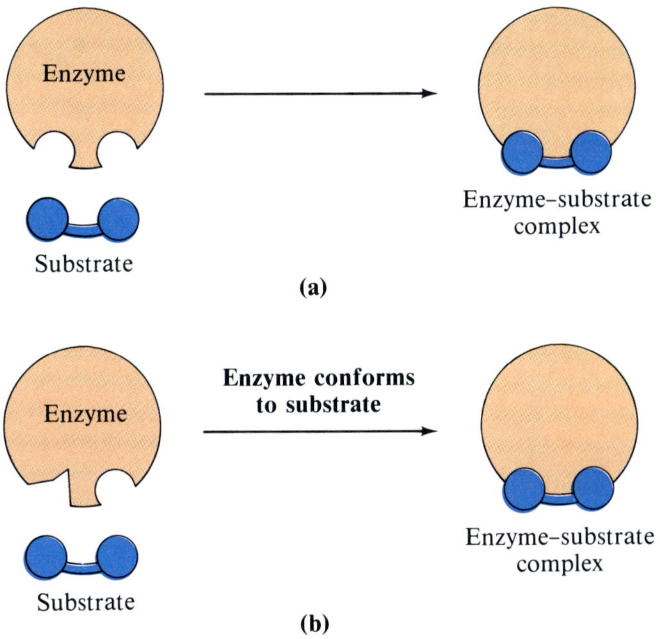
Enzyme
Substrate
Enzyme–substrate complex
(a)
Enzyme
Enzyme conforms to substrate
Substrate
Enzyme–substrate complex
(b)

FIGURE 27.3
Binding of substrate to enzyme. (a) The lock-and-key model describes binding sites on enzymes as complementary to the substrate; the substrate fits into a preexisting site. (b) The induced fit model states that the binding of the substrate to the enzyme causes changes in shape that make the substrate and binding site complementary.

these amino acid residues, but in addition, the active site has a negative charge that forms a strong ionic bond to the positive charge of either lysine or arginine. No other amino acid residue in a polypeptide has the correct shape and charge to bind to the active site. No other peptide bond is normally cleaved by trypsin.

This description of the active site corresponds to the **lock-and-key model** of enzyme substrate interaction (Figure 27.3a). Just as a key is complementary to the keyhole of a lock, the substrate fits into the active site of an enzyme. This model is easily visualized, but for a variety of reasons it is now considered an oversimplification of enzyme substrate interaction. Biochemists today view this interaction instead as an **induced fit model**. In this model, the binding of substrate to enzyme in effect "induces" a change in the shape of one or both of the molecules. When they are bound together, neither has the shape it had when free in solution (Figure 27.3b). Either model can be used to explain enzyme substrate binding, but the induced fit model helps explain enzyme activity better than the lock and key model.

The induced fit model and lock-and-key model both explain substrate-enzyme binding.

Enzyme Activity

Enzymes catalyze the reaction that converts their substrate to product. How can the enzyme's active site catalyze a chemical reaction? No single factor can explain enzyme catalysis; any given enzyme may use any combination of several factors to enhance the reaction rate. Enzymes catalyze reactions through these factors: (1) proximity effects, (2) orientation effects, (3) acid–base catalysis, and (4) strain.

Consider first proximity effects. Some enzymes have two or more substrates. For example, a transferase transfers some group from one molecule to another. One way an enzyme can enhance catalysis is by binding both substrates simultaneously in close proximity at the active site (Figure 27.4a). This effectively increases the concentration of the substrates at the active

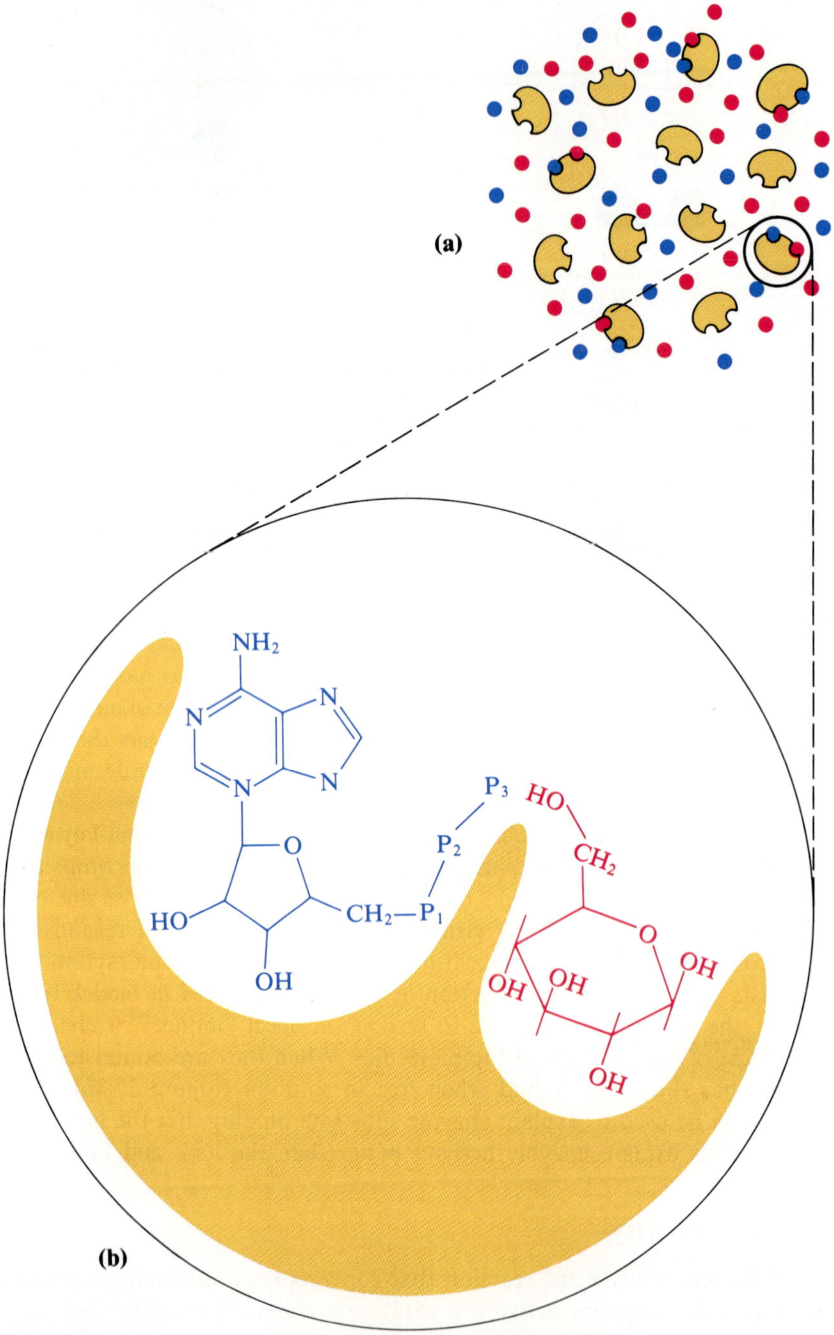
(a)
NH2
N
N
N
N
O
HO
OH
CH2—P1
P2
P3
HO
CH2
O
OH
OH
OH
OH
(b)

FIGURE 27.4
The binding of substrates to an enzyme can result in two effects. (a) Proximity effect. Because these two hypothetical substrates bind to the enzyme simultaneously, they are concentrated at the active site. The chance of them encountering and reacting with each other at the active site is much higher than out in solution. (b) Orientation effect. ATP and a hexose are used to illustrate orientation in a hypothetical enzyme. The third phosphate of ATP is well aligned for transfer to the 6′ hydroxyl group of the hexose.

site, and since an increase in concentration typically increases rate, the reaction occurs faster.

Second, enzymes orient substrates to enhance the rate of a reaction. When a substrate is bound to the active site of an enzyme, it is held in a specific orientation with respect to another substrate molecule or to amino acid side chains in the active site (Figure 27.4b). The atoms of the substrate are in the proper orientation for fast reaction with the other substrate or the group on the enzyme.

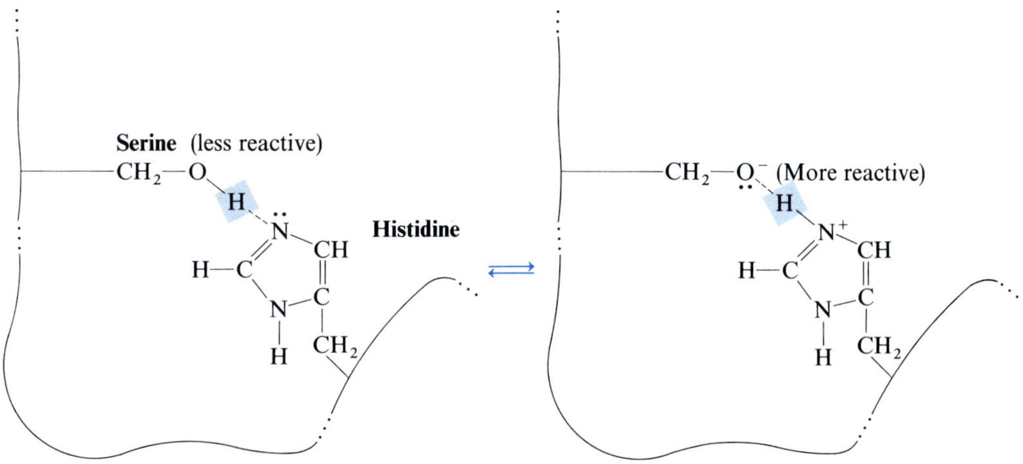
Serine (less reactive)
CH_2—O
H
N
CH
H—C
C
N
H
CH_2
Histidine
CH_2—O^- (More reactive)
H
N^+
CH
H—C
C
N
H
CH_2

FIGURE 27.5
Some of the side chains of amino acids in an enzyme can act as acids and bases that speed up reactions. In this example, a histidine (base), shown on the left side of the diagram, has accepted a hydrogen ion from a serine (acid). The serine on the right side is in the anionic form, and is much more reactive. It reacts more rapidly with substrate than would the —OH form of serine.

Third, enzymes can act as acids and bases during catalysis. This speeds up the rate of a reaction just as acids and bases catalyze many reactions in general and organic chemistry. Recall that certain amino acid side chains may be acidic or basic (Chapter 25)—they can donate or accept hydrogen ions. They can contribute to catalysis in an enzyme by acting as acids or bases (Figure 27.5). The acidic and basic groups at an active site have one enormous advantage over acids and bases in solution—they are properly positioned to quickly give up or accept protons. This orientation allows acid–base catalysis to proceed rapidly and efficiently.

A fourth factor influencing the rate of enzyme reactions is substrate strain. In the induced fit model of substrate–enzyme binding, the enzyme or substrate (or both) undergoes conformational changes when they bind. When a substrate is free in solution, it is generally in the lowest energy (or most stable) conformation. But when it binds to the enzyme, the substrate is forced into a higher energy conformation. This strained conformation of the substrate is more reactive than a free substrate molecule because it more closely resembles the transition state of the reaction, the high energy form between substrate and product (Figure 27.6). Because it is higher in energy, it reacts more quickly. Upon binding, many enzymes turn ordinary substrate molecules into strained, reactive species.

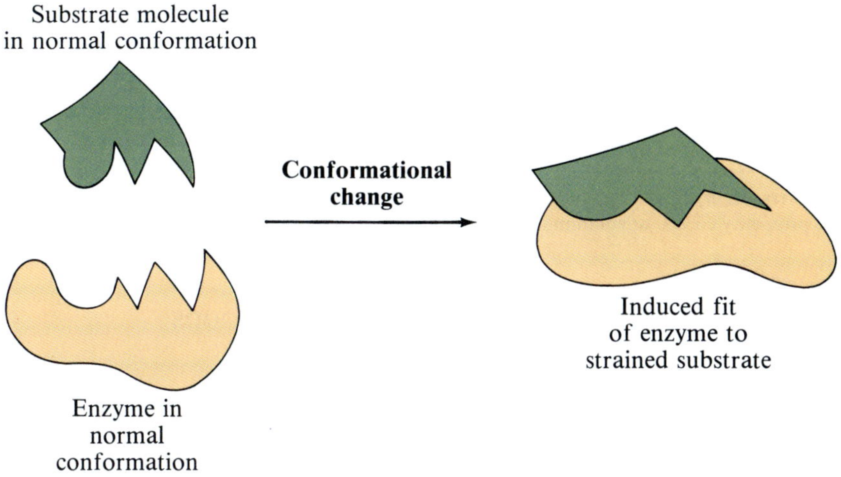
Substrate molecule
in normal conformation
Conformational
change
Induced fit
of enzyme to
strained substrate
Enzyme in
normal
conformation

FIGURE 27.6
According to the induced fit hypothesis, the binding of substrate to enzyme results in conformational changes in these molecules. The substrate in the strained conformation more easily reacts to yield a product than would the normal, unstrained substrate molecule.

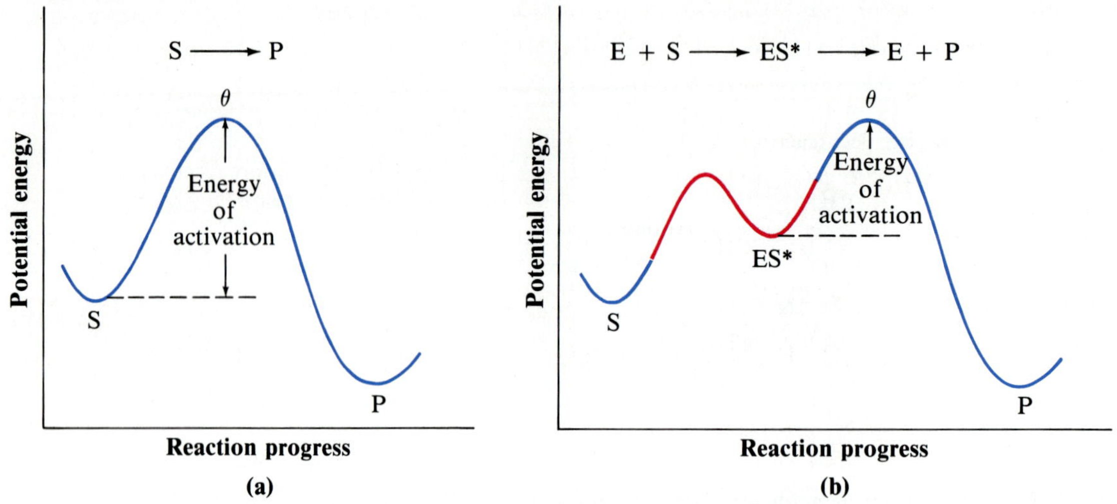
S ⟶ P
θ
Energy
of
activation
S
P
Potential energy
Reaction progress
(a)
E + S ⟶ ES* ⟶ E + P
θ
Energy
of
activation
ES*
S
P
Potential energy
Reaction progress
(b)

FIGURE 27.7
The energy of activation for a reaction in (a) the absence and (b) the presence of an enzyme that induces strain in the substrate. The energy of activation for enzyme-catalyzed reactions is always lower than for the uncatalyzed reaction, but the difference is not always due to strain in the substrate. (a) As S is converted to P, it passes through a high energy state θ. The difference between this state and S is the energy of activation. (b) When S binds to E, conformational changes convert it to a higher energy, less stable form, S*. The difference in energy between the highest energy form and S* is less than the difference shown in part (a).

The strain induced in the substrate upon binding to an enzyme in effect lowers the energy of activation of a reaction. This energy is the difference in energy between the substrate and the highest energy form as the substrate changes to product (Figure 27.7). Binding of the substrate to an enzyme results in a change in the substrate to a form that more nearly has the same energy as the highest energy form. Because they are so similar, even the small amounts of energy available at the temperatures within a cell are enough to get the substrate beyond this energy barrier to form the product.

NEW TERMS

active site The site on an enzyme that binds substrate and catalyzes the reaction to yield product.

lock-and-key model Model for substrate enzyme interaction which states that the two molecules are complementary to each other before binding and fit together like a lock and key.

induced fit model Model for substrate enzyme interaction which states that the binding of substrate to enzyme causes a change in the shapes of one or both of the molecules. When bound to each other, the two are complementary.

TESTING YOURSELF

Enzyme Activity

1. What noncovalent interactions bind substrates to enzymes?
2. List four ways enzymes catalyze reactions.

Answers **1.** hydrogen bonding, ionic interactions, and hydrophobic interactions **2.** proximity effects, orientation effects, acid–base catalysis, and strain

27.4 Rates of Enzyme-Catalyzed Reactions

Reaction Velocities

For enzyme catalyzed reactions, biochemists traditionally but incorrectly use velocity for rate.

Enzyme-catalyzed reactions are always much faster than uncatalyzed reactions occurring under otherwise identical conditions. Whether enzyme catalyzed or not, **reaction rates** or **velocities** are typically expressed in terms of the amount of reactant (substrate) consumed per unit time or the amount of product formed per unit time. The velocity v, of enzyme-catalyzed reactions is affected by a variety of factors, including the concentration of the enzyme. Consider some enzyme, designated as E, with its substrate S. The rate at which S is converted to product P is directly dependent on the amount of enzyme present (Figure 27.8). As the number of molecules (concentration) of the enzyme increases, the rate of the reaction increases. For example, if 4 molecules of E convert 100 molecules of S to product P in 1 sec, 8 molecules of E would convert 200 molecules of S. The reaction rate would double. If all other conditions are kept the same, the rate of the reaction is directly proportional to the amount of enzyme. Living organisms can increase the rate of some reactions by simply making more of the enzyme needed for the reaction (see Section 27.5).

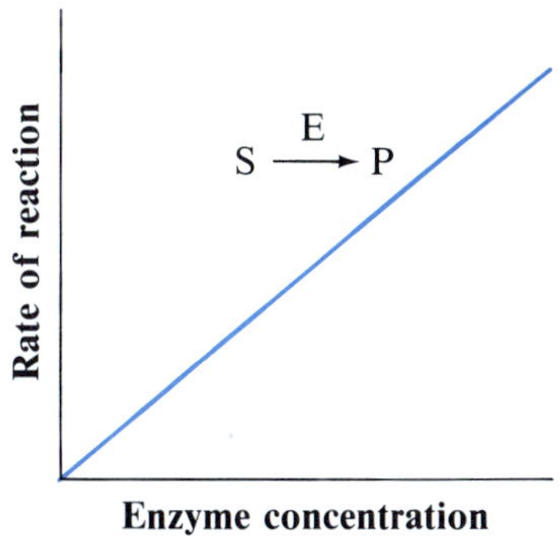
Rate of reaction
S —E→ P
Enzyme concentration

FIGURE 27.8
The effect of increasing concentrations of enzyme on the rate of a reaction. If the enzyme concentration doubles, the probability of the enzyme binding the substrate doubles, and the rate of formation of product doubles.

The rate of enzyme-catalyzed reactions also varies with the concentration of the substrate. But the rate varies with substrate concentration in a way that is substantially different from "ordinary" chemical reactions. In many nonenzymic reactions, the rate is directly proportional to the concentration of the reactant. Doubling the concentration of the reactant doubles the rate of the reaction (Figure 27.9a). Doubling the concentration of S in an enzyme-catalyzed reaction appears to double the rate if the concentrations of S are rather small, but doubling relatively large concentrations of S actually yields only a slight increase in reaction rate (Figure 27.9b). Rather than the usual linear graph seen in Figure 27.9a, a hyperbolic graph is observed. The velocity of the reaction with ever increasing substrate concentration approaches a maximal velocity designated V_{max}. This relationship between substrate concentration and the rate of an enzyme-catalyzed reaction is expressed in an equation developed by L. Michaelis and M. Menten:

$$v = \frac{V_{max}[S]}{[S] + K_m}$$

where K_m is defined as the concentration of S that yields half maximal velocity.

The relationship between rate and substrate concentration expressed by the Michaelis–Menten equation is explained by the formation of an enzyme–substrate complex. In turn, this complex may break down to yield free E and S or product P and free E.

$$E + S \rightleftharpoons ES \longrightarrow E + P$$

At low concentrations of S, a substrate molecule is likely to collide with an unoccupied E molecule to form ES. Doubling S increases the probability of

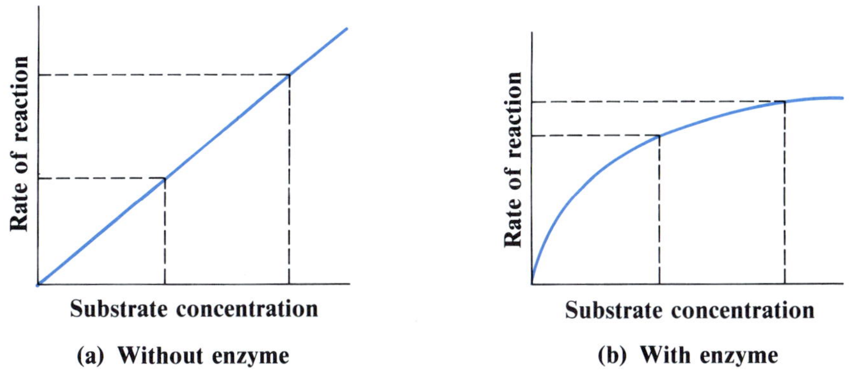
Rate of reaction
Substrate concentration
(a) Without enzyme
Rate of reaction
Substrate concentration
(b) With enzyme

FIGURE 27.9
The effects of substrate concentration on the rate of (a) an uncatalyzed reaction and (b) an enzyme-catalyzed reaction. (The reaction rates are not drawn to scale; the actual rate of the enzyme catalyzed reaction would be much larger.) (a) Doubling the concentration of substrate doubles the rate of the reaction lacking enzyme. (b) Doubling the concentration of substrate in an enzyme-catalyzed reaction increases the rate but does not necessarily double it.

EFFECTS OF TEMPERATURE ON BODY FUNCTION

Humans are homeothermic organisms, that is, we maintain a more or less constant internal temperature. Occasionally, environmental stress or disease causes body temperature to deviate significantly from this normal range. Consider the effects of reduced body temperature. As temperature decreases, life-sustaining biochemical reactions within the body proceed more slowly. For example, normal brain function is dependent on normal metabolic rates in the cells of the central nervous system. As body core temperature decreases, as it does in *hypothermia*, brain chemistry and function slow down. Disorientation, unconsciousness, and death may follow. Conversely, if core temperature increases beyond the normal range, as it can during prolonged high fever or during excessive exposure to very hot weather, some proteins in the body will denature. Loss of certain critical enzymes in the central nervous system results in dysfunction and possibly death.

collision by two. But at higher concentrations of S, many of the enzyme molecules already have S bound to them. Many of the collisions of S with enzyme molecules involve unproductive collisions with ES rather than with E. Because the number of enzyme molecules is normally small, at high concentrations of S most of the E molecules already have S bound to them; they exist as ES, not E. Doubling the concentration of S increases the amount of ES and decreases the amount of E. Ever increasing concentrations of S yield ever smaller increases in rate, which produces the hyperbolic graph shown in Figure 27.9b.

When all enzyme molecules have substrate bound to them, increasing the concentration of substrate has no effect on the reaction rate.

Effects of Temperature on Enzyme-Catalyzed Reactions

The effects of temperature on an enzyme-catalyzed reaction are similar to the effects on other reactions (up to a point). The rates for most chemical reactions increase regularly with increasing temperature (Figure 27.10a). This increase in the rate of a reaction reflects the higher speed and increased energy of molecules. Collisions between reactants are more frequent, and a larger proportion of these collisions possess enough energy to break old bonds and make new ones. The rate of an enzyme-catalyzed reaction also increases with temperature, but the stability of the enzyme is also an impor-

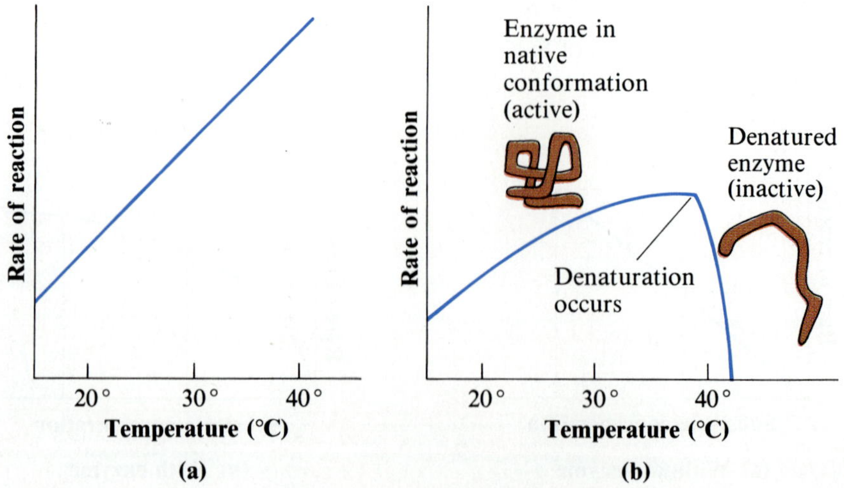
Rate of reaction
20°
30°
40°
Temperature (°C)
(a)
Rate of reaction
Enzyme in native conformation (active)
Denatured enzyme (inactive)
Denaturation occurs
20°
30°
40°
Temperature (°C)
(b)

FIGURE 27.10
The effects of temperature on the rate of (a) uncatalyzed and (b) enzyme-catalyzed reactions. (Rates not to scale.) (a) Near room temperature or body temperature, a 10 °C rise in temperature roughly doubles the rate of a reaction. (b) The rates of enzyme-catalyzed reactions also increase with temperature until the enzyme denatures. Above this temperature, the rate drops.

tant factor. At lower temperatures, enzymes are stable and the rate of the reaction increases with temperature. But there is some upper temperature limit beyond which the protein is unstable. It will be *denatured*, that is, it will lose its normal shape and therefore its normal function (see Chapter 25). When this occurs, the enzyme no longer functions as a catalyst and the reaction rate drops drastically (Figure 27.10b).

Effects of pH on Enzyme-Catalyzed Reactions

Changes in hydronium ion concentration (that is, changes in pH) also alter the rate of enzyme-catalyzed reactions (Figure 27.11). The effect may simply reflect the denaturation of the protein above or below a stable pH range, or it may be more subtle. In our discussion of enzyme activity in Section 27.3, the acid–base properties of some residues were listed as factors that contribute to catalysis in some enzymes. These groups accept and donate protons during the catalytic cycle. If the pH is too high or too low, then these groups will be unprotonated or protonated too much of the time. The enzyme will function much less efficiently. It is only within a certain range of pH that the catalytic groups of the enzyme will be protonated an appropriate amount of time.

Changing the pH will result in the protonation or deprotonation of side chains in certain amino acids in an enzyme.

Changes in pH can also alter the substrate. As the pH changes, a group in the substrate may be protonated or deprotonated. The altered form of the substrate may not bind to the enzyme or may react more slowly.

Inhibition of Enzymes

The rate of enzyme-catalyzed reactions can be decreased or inhibited by a group of substances called *inhibitors*. Inhibitors bind to enzymes, and alter the ability of the enzyme to carry out catalysis. The inhibition may be permanent, that is *irreversible*, or it may be *reversible* if activity returns when the inhibitor is removed.

Inhibitors bind to enzymes and reduce their activity.

Irreversible inhibition occurs whenever an **irreversible inhibitor**, represented by molecule I, binds permanently to an enzyme, E, and the resulting association has little or no activity.

$$\underset{\text{(Active)}}{E} + I \longrightarrow \underset{\text{(Inactive)}}{EI}$$

The association between the inhibitor and the enzyme might be a covalent bond or a very strong noncovalent interaction, but once it is formed, the

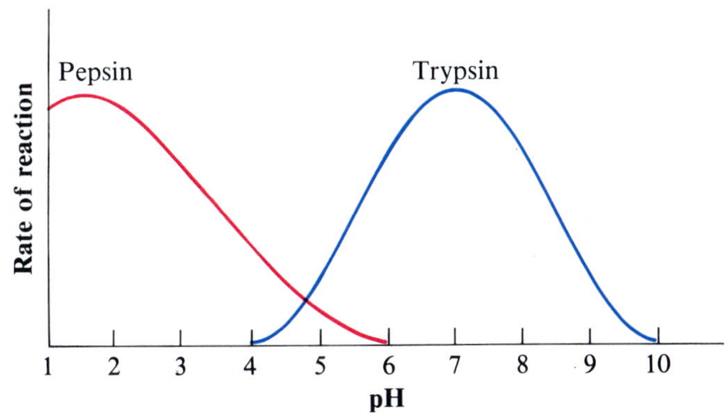

Pepsin
Trypsin
Rate of reaction
1
2
3
4
5
6
7
8
9
10
pH

FIGURE 27.11
The effects of pH on enzyme activity. Pepsin, a digestive enzyme of the stomach, is most active under acidic conditions like those found in the stomach. Trypsin, a digestive enzyme of the small intestine, is most active at neutral to slightly alkaline pH, the pH range of the small intestine.

TABLE 27.3
Nerve Gases and Insecticides that Inactivate Acetylcholinesterase

Compound	Structure	Use	Toxicity
Tabun	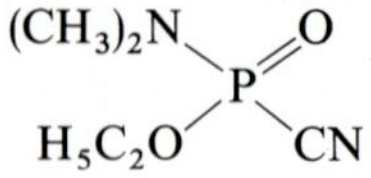	Nerve gas	Highly toxic
Sarin	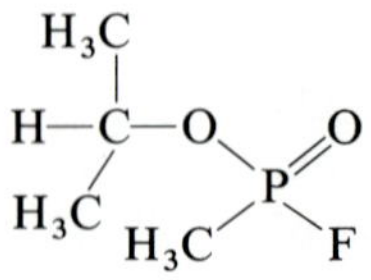	Nerve gas	Highly toxic
Malathion	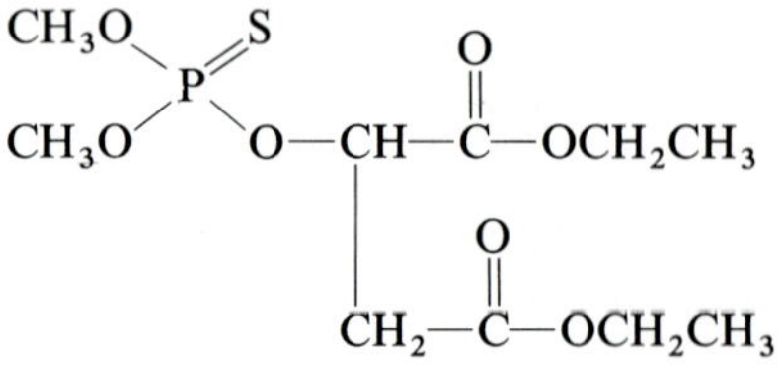	Insecticide	Toxic but can be handled safely with proper precautions
Parathion	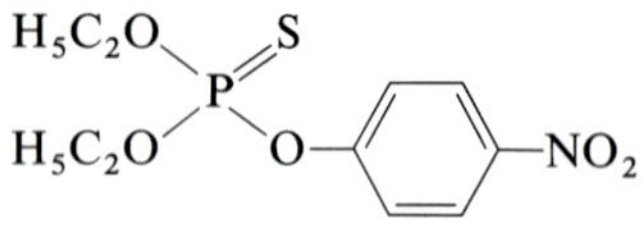	Insecticide	Toxic but can be handled safely with proper precautions

complex remains. The enzyme molecule still exists, but without its catalytic activity it makes no contribution to the cell. The apparent concentration of the enzyme has been reduced.

A group of organophosphates provides an example of irreversible inhibition. This group of compounds includes the nerve gases and certain organophosphate insecticides such as malathion (Table 27.3). They share a common structural feature—each of them possesses a phosphate group that reacts readily with the hydroxyl group of a serine in an enzyme called *acetylcholinesterase* (Figure 27.12). This reaction yields a stable phosphoester bond. Once formed, the complex remains. Because this serine is part of the active site, and is therefore essential for the catalytic activity of this enzyme, the enzyme molecule loses activity once the inhibitor is irreversibly bound.

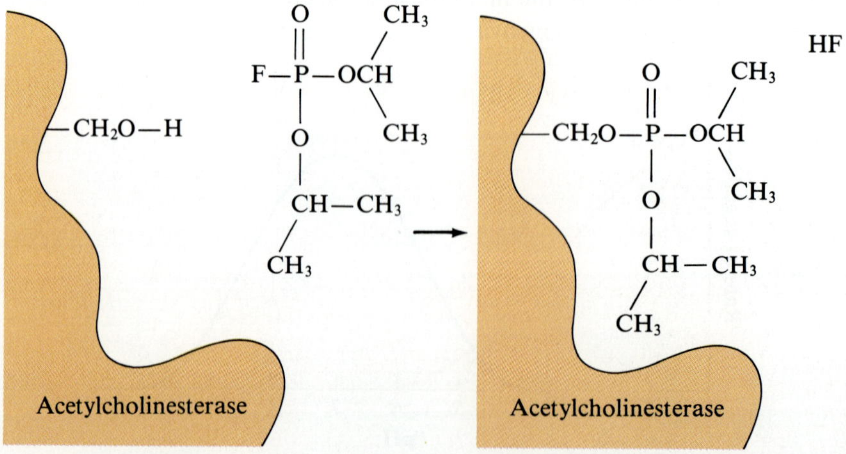
Acetylcholinesterase
Acetylcholinesterase
HF

FIGURE 27.12
The reaction of diisopropylphosphofluoridate (DIPF) with the catalytic serine of acetylcholinesterase. The bond between serine and DIPF is covalent and strong. Once formed, it remains. Without this free serine, the enzyme molecule is inactive.

In some of the junctions of the nervous system, and at neuromuscular junctions, nerve impulses cause the release of acetylcholine into the synapse. In the synapse, acetylcholine stimulates an electrical impulse in a second nerve or in a muscle. Normally acetylcholinesterase hydrolyzes acetylcholine in the synaptic junctions, preparing the synapse to respond to future nerve impulses. If acetylcholine accumulates, however, as it will if enough acetylcholinesterase molecules are inhibited, the muscle or nerve will show increased activity even though the normal stimulus is absent. The result is abnormal nerve transmission and muscle contraction, loss of normal function, and often death.

A second type of inhibition is caused by **reversible inhibitors.** Like irreversible inhibitors, these substances also bind to an enzyme molecule and reduce or eliminate enzymic activity. But these substances bind reversibly to the enzyme; they remain associated with the enzyme for a period of time, then diffuse away. Since a permanent association is not formed, the inhibitor can be removed and activity restored.

$$\underset{\text{(Active)}}{\text{E}} + \text{I} \rightleftharpoons \underset{\text{(Inactive)}}{\text{EI}}$$

Reversible inhibitors may bind to the active site of the enzyme or elsewhere on the enzyme surface. If binding occurs at the active site, the substance is a **competitive inhibitor,** so called because it competes with the substrate for the active site (Figure 27.13). While I is bound to the active site, S cannot bind. For that period of time, that enzyme molecule is catalytically inactive. A reversible inhibitor changes the K_m of the enzyme. A higher concentration of S is needed to obtain half maximal velocity.

Competitive inhibitors compete with substrate for the active site of an enzyme.

The structures of competitive inhibitors resemble the substrate structures. These inhibitors can form many or all of the noncovalent bonds that a sub-

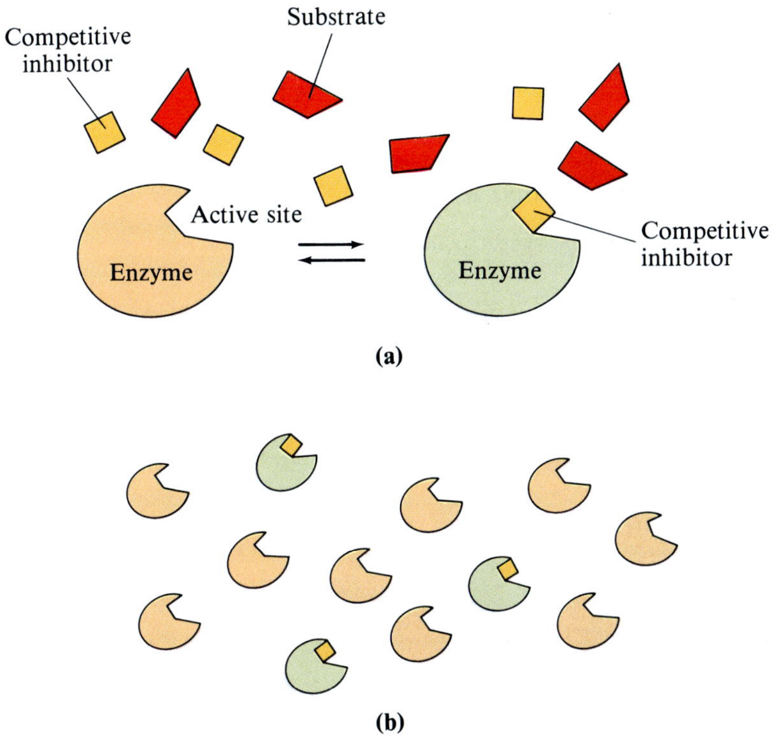
Substrate
Competitive inhibitor
Active site
Enzyme
Enzyme
Competitive inhibitor
(a)
(b)

FIGURE 27.13
The reversible binding of a competitive inhibitor to the active site of an enzyme. (a) A competitive inhibitor binds at the active site and prevents binding of substrate. While it is bound, the enzyme is inactive. (b) A competitive inhibitor effectively reduces the concentration of active enzyme (tan) by temporarily inactivating some of the enzyme molecules (green with yellow inhibitor bound).

strate would form with the enzyme molecule. But competitive inhibitors, unlike substrate, cannot react to yield a product. Competitive inhibitors bind to the active site, but no reaction occurs.

Sometimes competitive inhibitors of enzymes can be used to advantage. Consider the antitumor agent, methotrexate. This compound is structurally similar to dihydrofolate, a coenzyme involved in the synthesis of thymine (dTMP). Thymine is made from uracil and is essential for DNA synthesis. If the nucleotides of thymine are in low concentration or absent, DNA synthesis is slow or halted. Methotrexate inhibits the ability of cells to grow because it binds to the active site of an enzyme needed for thymine synthesis. Rapid cell growth requires DNA synthesis, thus rapidly dividing cancer cells are more adversely affected by methotrexate than are normal cells.

Another example of competitive inhibition is found in the action of the *sulfa drugs.* These drugs are effective against a number of bacteria. Bacteria use *p*-aminobenzoic acid to synthesize folic acid, which the bacteria use to make some coenzymes. The sulfa drugs, including sulfanilamide, resemble *p*-aminobenzoic acid sufficiently well to bind to the active site of dihydropteroate synthetase (Figure 27.14). While sulfanilamide is bound to this enzyme, no folic acid can be produced by the enzyme molecule. If enough of the enzyme is inhibited, insufficient folic acid is produced, and growth of the bacterial population is inhibited. Sulfa drugs do not affect the metabolism of mammals directly because mammals cannot make folic acid; it is obtained from their diets.

Some reversible inhibitors do not bind at the active site of an enzyme. These are **noncompetitive inhibitors** because they do not compete with the substrate. If they do not occupy the active site and prevent S from binding, how do they inhibit the enzyme? Noncompetitive inhibitors bind to the enzyme and cause a conformational change in it (Figure 27.15). In this new

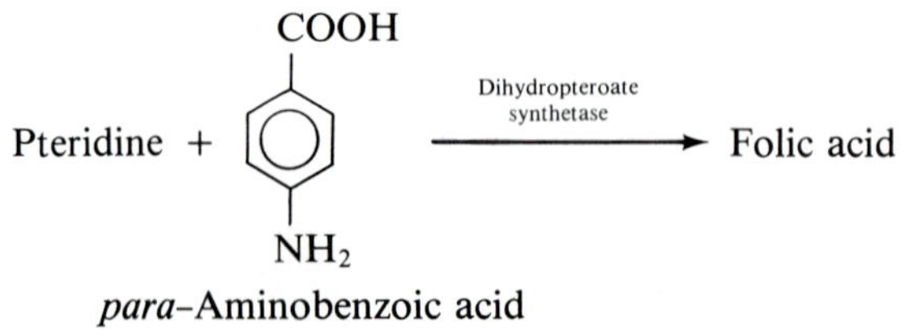
COOH
Dihydropteroate synthetase
Pteridine +
Folic acid
NH2
para-Aminobenzoic acid

Folic acid synthesis in bacteria

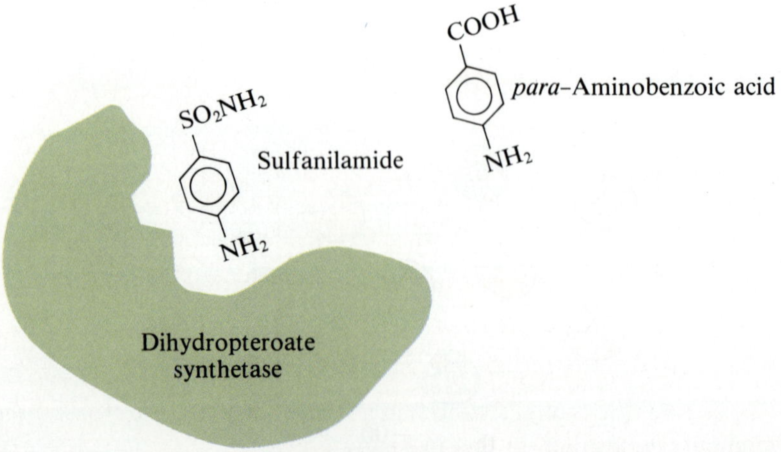
COOH
para-Aminobenzoic acid
NH2
SO2NH2
Sulfanilamide
NH2
Dihydropteroate synthetase

FIGURE 27.14
The effects of sulfanilamide on folic acid synthesis. Sulfanilamide competes with *para*-aminobenzoic acid for the active site of this enzyme. This antimicrobial drug retards bacterial growth because folic acid is essential for proper cell function and growth.

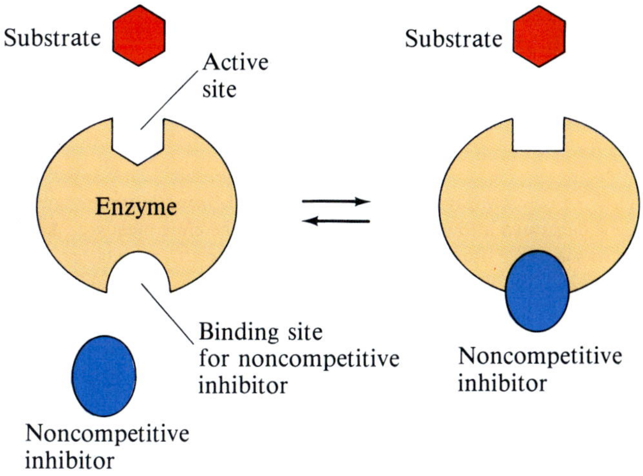
Substrate
Active site
Enzyme
Binding site for noncompetitive inhibitor
Noncompetitive inhibitor
Substrate
Noncompetitive inhibitor

FIGURE 27.15
The interaction of an enzyme and a noncompetitive inhibitor. This type of inhibitor does not bind to the active site, but its binding causes the enzyme to become inactive. In this example, the inhibitor induces conformational changes in the enzyme which prevent the substrate from binding to the active site.

conformation, the enzyme is less active or inactive. It now binds S less efficiently or not at all, or the catalytic groups of the active site are no longer aligned properly for efficient catalysis. Noncompetitive inhibitors reduce V_{max} for the reaction.

NEW TERMS

reaction rate (velocity) The rate at which a reaction occurs, usually expressed in terms of substrate consumed per unit time or product formed per unit time.

irreversible inhibitor Molecule that binds tightly to an enzyme and reduces or eliminates the activity of the enzyme.

reversible inhibitor Molecule that binds temporarily to an enzyme and, while bound, reduces or eliminates the activity of the enzyme.

competitive inhibitor Reversible inhibitor that binds to the active site of an enzyme and thus competes with the substrate.

noncompetitive inhibitor Reversible inhibitor that binds to an enzyme at a location other than the active site and thus does not compete with the substrate.

TESTING YOURSELF

Rates of Enzyme-Catalyzed Reactions

1. Use the Michaelis–Menten equation to calculate velocities for an enzyme-catalyzed reaction at the following substrate concentrations:

$$V_{max} = 2.25\ \mu\text{mol/min}$$
$$K_m = 3.50\ \text{mM}$$
$$[\text{S}] = 1.00, 3.00, 5.00, 10.00, \text{ and } 20.00\ \text{mM}$$

2. For question 1, graph the velocity as a function of substrate concentration. Compare your graph to that in Figure 27.9b.

3. Why is the rate of most enzyme-catalyzed reactions virtually zero at temperatures much above body temperature?
4. Compare the binding of competitive and noncompetitive inhibitors to enzymes.

Answers **1.** velocities = 0.50, 1.04, 1.32, 1.67, and 1.91 μmol/min **2.** A hyperbolic curve drawn through these points yields a curve indistinguishable from that in Figure 27.9b. **3.** The enzyme is denatured at these temperatures and is therefore inactive. **4.** Competitive inhibitors bind to the active site of an enzyme, and noncompetitive inhibitors bind elsewhere on the enzyme.

27.5 Regulation of Enzyme Activity

The activity of enzymes is regulated in several ways. One way is to change the concentration of the enzyme. This can be accomplished by altering the rate at which the enzyme is synthesized or the rate at which it is broken down. Regulation of protein synthesis by control of gene expression was discussed in Chapter 26. Control of gene expression results in increased or decreased rates of enzyme synthesis, which increases or decreases the concentration of the enzyme. The rate of protein breakdown is also regulated. Body proteins are continuously degraded within cells, but the rate of degradation of some enzymes may vary with cellular conditions. Both synthesis and degradation affect the concentration of enzymes; regulation of these processes contributes to the regulation of metabolic activity.

Enzyme activity can be altered by changing enzyme concentration or by altering the activity of the enzyme molecules.

Enzyme activity can also be regulated in several ways that do not change the concentration of the protein. Three examples of this type of enzyme regulation are discussed here: (1) activation of zymogens, (2) reversible covalent modification of an enzyme, and (3) allosteric regulation.

Zymogens

Some enzymes are synthesized by the body in an inactive form called a **zymogen.** Zymogens can be recognized by the suffix *-ogen* or the prefixes *pre-* or *pro-*. Zymogens possess no enzymic activity when synthesized, but

BLOOD CLOTTING

Blood clotting prevents blood loss from injuries, but clotting also contributes to heart attacks. How does this vital process occur? Blood contains several proteins called clotting factors that are zymogens, inactive enzymes. A cut or injury initiates sequential activation of these zymogens, resulting in the formation of an insoluble mass that closes the wound.

Injury stimulates release of clot-activating factors and exposes blood to nonphysiological surfaces that also activate the sequence. At each step another zymogen is activated to an active enzyme. Each step thus amplifies the process, and increasing amounts of zymogen are activated. The activation proceeds through several factors until fibrinogen, a soluble protein, is converted to fibrin, an insoluble fibrous protein. At the end of the sequence, active factor XIII cross-links fibrin molecules to form a massive molecular plug, which along with platelets, blocks blood flow from the wound.

they gain activity when one or more peptide bonds are cleaved at some time following synthesis. Cleavage of these peptide bonds is followed by conformational changes to yield an active form. Examples include some of the digestive enzymes of the gastrointestinal tract and the enzymes responsible for blood clotting (see box).

A specific example is useful here. *Trypsin* is a digestive enzyme that cleaves specific peptide bonds of ingested proteins. Since dietary and body proteins are so similar, what prevents trypsin from breaking down the proteins of the pancreas where it is made? If trypsin were active when synthesized, it would immediately begin to work on the proteins of the pancreas. Instead, trypsin is synthesized as the inactive precursor *trypsinogen*, which is released into the small intestine when partially digested food enters from the stomach.

In the small intestine, trypsinogen is converted to trypsin by trypsin molecules already there or by the enzyme enterokinase. These enzymes cleave a peptide bond near one end of the trypsinogen molecule. When freed of the small peptide, trypsin undergoes a conformational change to the active form (Figure 27.16). The activity of some other digestive enzymes is controlled in a similar manner.

Covalent Modification of Enzymes

The activity of some enzymes can be reversibly increased or decreased by covalent modification of the enzyme. An example is the phosphorylation–dephosphorylation of the enzyme glycogen phosphorylase. This enzyme

The activity of some enzymes can be turned on and off by covalent modification.

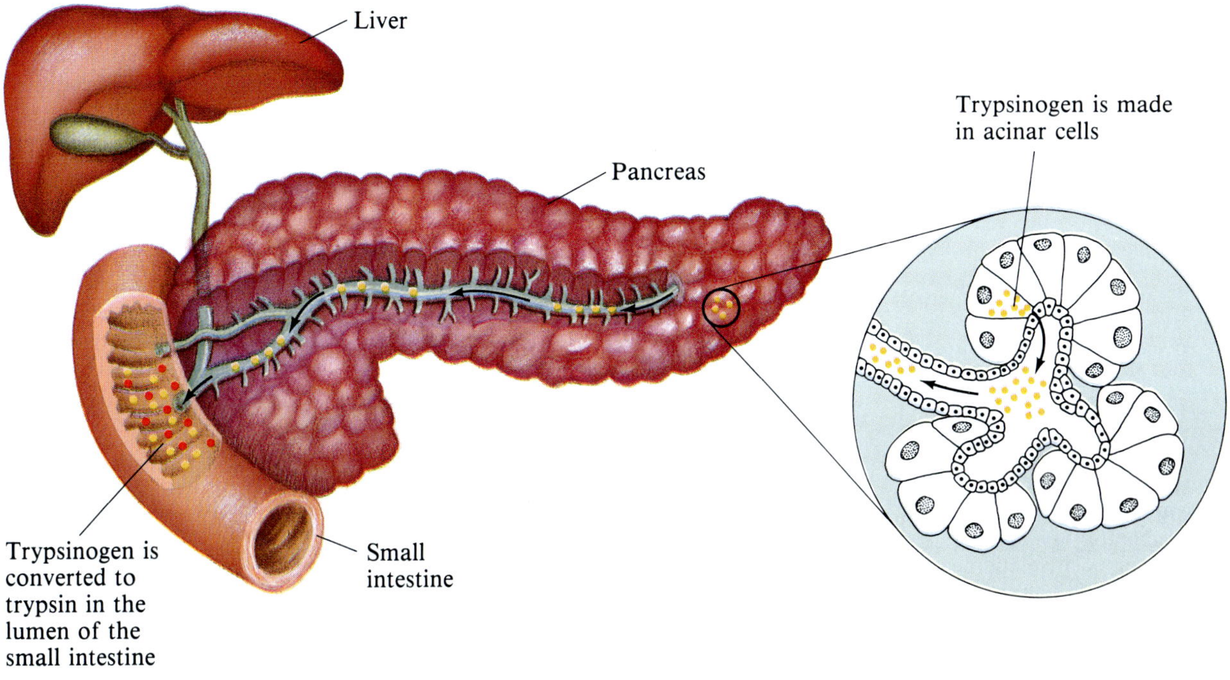
Liver
Pancreas
Trypsinogen is made in acinar cells
Trypsinogen is converted to trypsin in the lumen of the small intestine
Small intestine

FIGURE 27.16
The conversion of trypsinogen molecules (yellow dots) into trypsin (red dots). Trypsinogen is the inactive form of trypsin synthesized in the pancreas. It is released when food enters the small intestine and passes through a system of ducts into the small intestine. Here it is converted to the active form trypsin by the enzyme enterokinase.

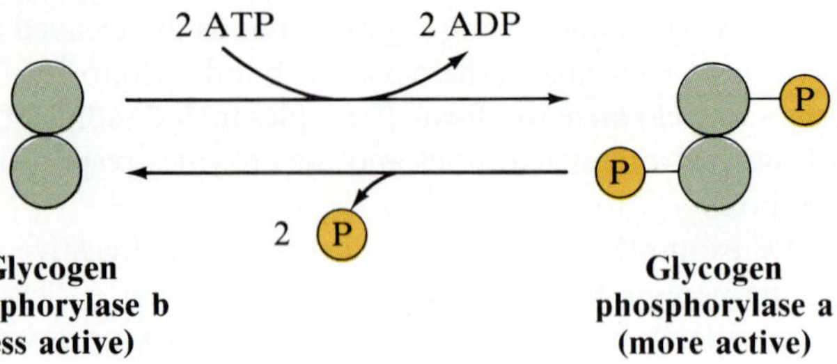
2 ATP
2 ADP
P
P
2 P
Glycogen phosphorylase b (less active)
Glycogen phosphorylase a (more active)

FIGURE 27.17
The reversible covalent modification of glycogen phosphorylase. When increased activity of glycogen phosphorylase is required in muscles, the enzyme is phosphorylated by transfer of phosphate from ATP to each polypeptide chain of this protein. When the activity is not needed, the phosphate is cleaved from the polypeptide chain, returning the enzyme to its less active form.

cleaves glucose molecules from glycogen and attaches a phosphate group to the glucose. This cleavage occurs on demand, that is, whenever more phosphorylated glucose is required by the muscles or liver (see Chapters 28 and 30). When phosphorylase has no phosphate groups covalently attached to certain serine residues in the enzyme, it has little enzymic activity. When the need for phosphorylated glucose increases, a series of events is initiated that results in the attachment of phosphate groups to these serines. When the serines are phosphorylated, glycogen phosphorylase is more active and phosphorylated glucose is rapidly formed from glycogen. The enzyme remains active until dephosphorylation returns it to the less active form (Figure 27.17). The activity of the enzyme is regulated by reversible covalent modification.

Allosteric Regulation of Enzymes

A third mechanism for the regulation of preexisting enzymes is **allosteric regulation**. In addition to the active site, some enzymes possess other sites that can bind certain small molecules found in the cell. These sites are called *allosteric sites*, or "other place" sites. These sites show the same high specificity of binding for their small molecules, called *effectors*, that the active site does for substrates. When an effector binds to the allosteric site, the enzyme undergoes a conformational change. In the new conformation the activity of the enzyme is different. If the effector is a **positive effector**, the enzyme will have greater activity than if the effector had not bound. If the effector is a **negative effector**, then the enzyme is less active in the new conformation (Figure 27.18). Negative effectors are, in essence, normal noncompetitive inhibitors of the enzyme.

Effectors binding to allosteric sites on some enzymes increase or decrease the activity of the enzyme.

Allosteric regulation of the first enzyme of a metabolic pathway allows for the efficient regulation of the entire pathway through a process called *feedback inhibition*. Consider the example shown in Figure 27.19. A pathway

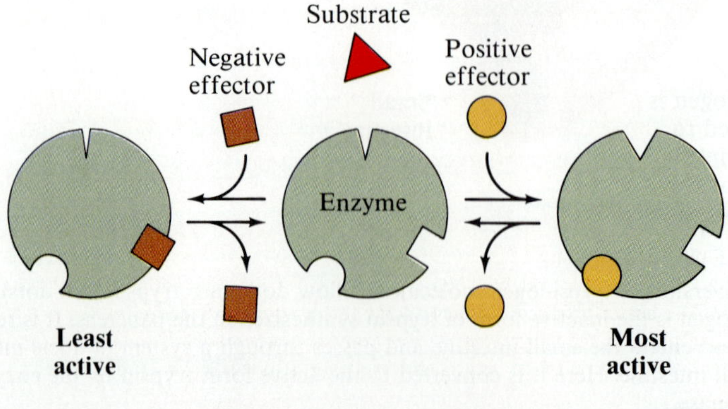
Substrate
Negative effector
Positive effector
Enzyme
Least active
Most active

FIGURE 27.18
The regulation of allosteric enzymes by positive and negative effectors. Binding of a positive effector increases the activity of the enzyme, while binding of a negative effector decreases the activity. In this hypothetical enzyme, the effectors cause conformational changes in the enzyme that influence binding of the substrate.

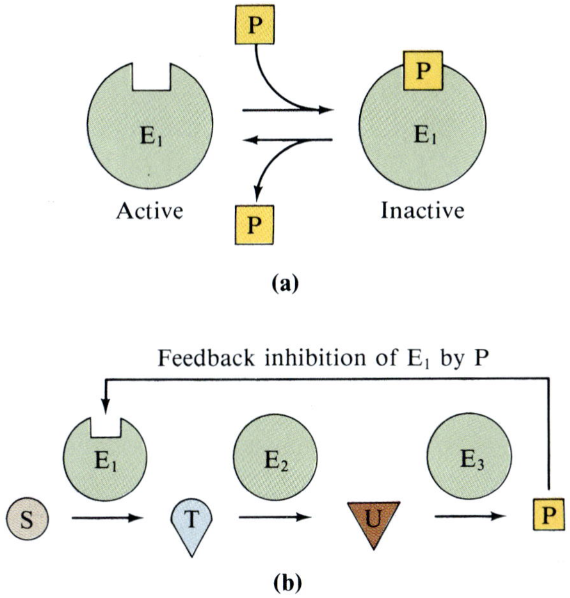
P
P
E1
E1
Active
P
Inactive
(a)
Feedback inhibition of E1 by P
E1
E2
E3
S
T
U
P
(b)

FIGURE 27.19
Feedback inhibition of a metabolic pathway. (a) P is a negative effector of the first enzyme of this pathway, E_1. (b) The activity of E_1 depends on the amount of P in the cell. If the concentration of P is relatively high, E_1 is relatively inactive (P feeds back and controls E_1). If a small amount of P is present, most of the E_1 molecules are active. This allows the synthesis of P to be controlled by the amount of P.

of five enzymes converts substrate S into product P. Product P is a negative effector of the first enzyme (E_1) of the pathway. As S is converted to P, the concentration of P increases and more of it binds to the allosteric site of E_1. Fewer E_1 molecules are now active, so the rate at which P is produced is reduced. The process is readily reversible. If the concentration of P decreases within the cell, some P bound to E will diffuse away. The E_1 enzyme molecules become active, and more P will be formed. Allosteric control provides for rapid and sensitive regulation of metabolic activity.

NEW TERMS

zymogen An inactive form of an enzyme that is activated by cleavage of one or more peptide bonds.

allosteric regulation Regulation of enzyme activity by the binding of small molecules to sites other than the active site.

positive effector A small molecule whose binding to an allosteric enzyme causes an increase in enzymic activity.

negative effector A small molecule whose binding to an allosteric enzyme decreases the activity of the enzyme.

TESTING YOURSELF

Regulation

1. Why are some digestive enzymes synthesized in an inactive form?
2. Explain how a thermostat works by feedback inhibition.

Answers **1.** If the enzymes were synthesized in an active form, they would hydrolyze the proteins, lipids, and carbohydrates of the cells of the pancreas and gastrointestinal tract. **2.** When a room gets too cold, an electrical contact in the thermostat closes and activates the heater. The heater produces thermal energy in the room until the room temperature exceeds a set point in the thermostat. At this point the circuit opens, which shuts down the heater. Heat, the product of the heater, effectively shuts down its own production.

ENZYMES IN MEDICINE

Enzymes are essential to life, but they also play an increasingly important role in medicine. Enzymes are used in diagnosis (the determination of the cause and nature of a disease) and in treatment. The diagnosis and treatment of a heart attack provides an illustration of the role of some specific enzymes in medicine.

The muscle cells of the heart possess some enzymes that are not normally found in blood plasma. When a heart attack occurs, the reduced blood flow to the muscle cells results in the death and rupture of some cells. The contents of these broken cells enter the blood. Enzymes of heart muscle are now in the blood. If a blood sample is analyzed for these enzymes, the sample will be positive in a heart attack victim, but negative in someone suffering from some other ailment (see figure).

Another enzyme is used to treat heart attack victims. When a heart attack occurs, blood flow to a portion of the heart is reduced or stopped. Often this is due to a blood clot blocking an artery that is partially restricted by atherosclerotic deposits (see Plate 11 in color insert). Intense discomfort may occur immediately, and with time more and more permanent damage occurs to the tissues deprived of blood. If the clot can be dissolved before permanent damage results, the probability of death and permanent disability is significantly lessened. Tissue-type plasminogen activator (TPA) is an enzyme that can be administered to heart attack victims. This enzyme converts a zymogen called plasminogen to the enzyme plasmin. Plasmin is the enzyme that normally breaks down blood clots in the body. Prompt treatment with TPA simply activates the normal clot dissolving machinery of the body. If TPA is administered in time, blood flow to the heart muscle is restored before the muscle cells die.

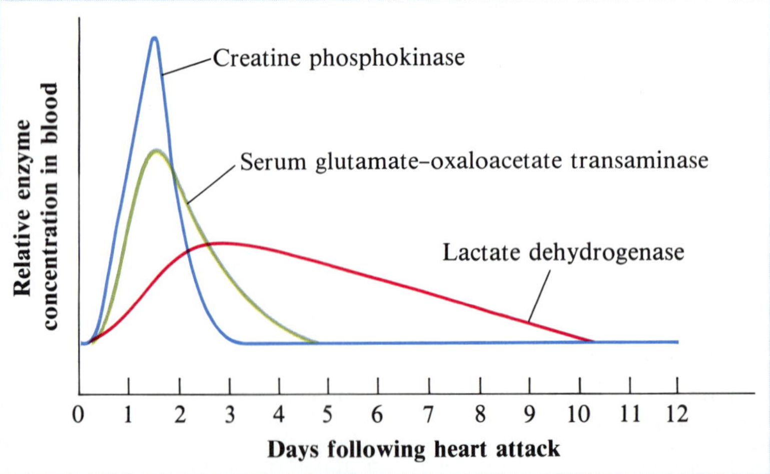
Creatine phosphokinase
Serum glutamate-oxaloacetate transaminase
Lactate dehydrogenase
Relative enzyme concentration in blood
0 1 2 3 4 5 6 7 8 9 10 11 12
Days following heart attack

These three enzymes are normally absent from blood. Heart muscle cells are destroyed during and after a heart attack, and the enzymes of these cells are released into the blood stream. The presence of these enzymes in blood within these concentration ranges is one line of evidence a physician would use to diagnose a heart attack.

27.6 Bioenergetics: Maintaining the State of Life

All too often we forget that chemical reactions involve energy as well as matter. *Thermodynamics* is the study of energy dynamics, and these studies show that energy can be both a reactant and a product of chemical reactions (see Chapter 10).

$$A \longrightarrow B + \text{Energy} \qquad \text{Exothermic reaction}$$
$$C + \text{Energy} \longrightarrow D \qquad \text{Endothermic reaction}$$

The reactions of the body produce or require energy. This section presents some concepts of how energy is involved in metabolism.

The cell is a highly organized system that includes many macromolecules. The general tendency of the universe over time is to move toward more random arrangements of smaller entities. This is formally described as *entropy*. How does the cell resist entropy and maintain its complex state? How does it resist the general tendency toward smaller, more random arrangements without violating the basic laws of the universe? The cell does it through the use of energy. A constant supply of energy is required to keep a living organism from disintegrating into an array of inorganic ions and molecules no different from those found in the soil, water, and rocks of this planet.

Cells constantly expend energy to fight entropy.

Energy is defined as the capacity to do work or the capacity to effect change. Living organisms use energy to bring about three kinds of changes: (1) biosynthetic or chemical work used to build or sustain the molecules needed for life; (2) transport work used to convey molecules and ions into, out of, and around cells; and (3) changes in position, that is, movement or mechanical work. These cellular activities maintain the state of life.

Cells use energy to do work, but where does the energy come from? The energy is produced during the exothermic reactions of catabolism (Chapter 28), but the energy produced by catabolism is first stored in the cell in several forms. The most important form of energy storage is adenosine triphosphate (ATP). ATP is made from adenosine diphosphate (ADP) and inorganic phosphate (P_i) in an endothermic reaction.

ATP is the major storage form for metabolic energy.

$$\text{ADP} + \text{P}_i + 7.3\ \text{kcal/mol} \longrightarrow \text{ATP} + \text{H}_2\text{O}$$

(Nucleotide structures are shown in Section 26.2.) The inorganic pnosphate becomes covalently attached to the terminal phosphate of ADP through an anhydride bond. Note that water is lost, thus the term *anhydride*. Attachment of the phosphate and loss of water requires the input of 7.3 kcal of energy per mole of ATP formed. For practical purposes, it can be said that a mole of ATP has 7.3 kcal more energy than a mole of ADP and P_i. Each mole of ATP formed in the body effectively stores 7.3 kcal of useful energy.

The exothermic reactions of catabolism yield energy used to synthesize ATP, and ATP provides the energy when the cells carry out other endothermic reactions. These paired reactions are examples of coupled reactions; both reactions occur simultaneously with the energy-yielding reaction "driving" or forcing the energy-requiring reaction. Many of the reactions of the body are coupled reactions. For example, the binding of an amino acid to a tRNA (see Chapter 26) is endothermic, and the exothermic hydrolysis of ATP is coupled to it to "drive" the reaction. As long as the exothermic reaction yields more energy than is required by the endothermic reaction, both reactions can occur. Through coupled reactions, the body uses energy to continually carry out the synthetic reactions and transport needed to resist the chaotic tendency of the universe.

TESTING YOURSELF

Bioenergetics

1. What is the source of the energy needed to drive endothermic reactions within the cell?
2. What name is given to the tendency for the universe to go from more order to less order with time?

Answers **1.** The energy released in the exothermic reaction of ATP with water is often coupled to endothermic reactions in the cell. The energy originally comes from potential energy stored in chemical bonds in food molecules. **2.** entropy

SUMMARY

All of the reactions of the body are collectively called *metabolism.* Catabolic reactions yield energy as they break down larger molecules to smaller ones. Anabolic reactions use energy and smaller molecules to build larger molecules. The reactions of the body are catalyzed by *enzymes,* which are either simple or conjugated proteins. Many enzymes can be recognized by the suffix *-ase* in their name. Enzymes bind substrate molecules and catalyze their conversion to products. Many enzymes show considerable specificity for their *substrates.* The catalytic activity of enzymes is influenced by substrate proximity, orientation, and strain. Enzymes may also act as acids or bases during a reaction. The rates of enzyme-catalyzed reactions are affected by the concentrations of substrate and enzyme and are influenced by temperature and pH. Enzymes can also be inhibited by *reversible* or *irreversible inhibitors.* The activity of an enzyme can be modified by changing the concentration of the enzyme or by altering its activity. *Zymogens,* covalent modification, and *allosteric regulation* are examples of enzyme regulation. Medicine uses enzymes in both diagnosis and therapy. The reactions of the body involve energy as well as matter. Catabolic reactions yield energy that is stored in molecules of ATP. This energy is used as needed to drive energy-requiring reactions. The constant input of energy helps maintain the state of life against entropy.

TERMS

metabolism
catabolism
anabolism
metabolic pathway
enzyme
-ase
substrate
cofactor
coenzyme
active site
lock-and-key model
induced fit model
reaction rate (velocity)
irreversible inhibitor
reversible inhibitor
competitive inhibitor
noncompetitive inhibitor
zymogen
allosteric regulation
positive effector
negative effector

CHAPTER REVIEW

QUESTIONS

1. Which reactions of the body are energy yielding?
   a. anabolic reactions
   b. catabolic reactions
   c. reduction reactions
   d. synthesis reactions
2. The enzymes are
   a. carbohydrates
   b. lipids
   c. nucleic acids
   d. proteins

3. Small organic molecules needed for the activity of some enzymes are
   a. coenzymes b. negative effectors c. substrates
4. The binding of a substrate to an enzyme may be influenced by
   a. hydrogen bonding between the substrate and the enzyme
   b. hydrophobic interactions between the substrate and the enzyme
   c. ionic interactions between the substrate and the enzyme
   d. all of the above
5. Which of the following does not influence the activity of an enzyme molecule?
   a. orientation of substrates
   b. proximity of substrates
   c. size of the substrate
   d. acid–base properties of the enzymes
6. For an enzyme-catalyzed reaction over a wide range of substrate concentrations, doubling substrate concentration will
   a. always double the rate of the reaction
   b. generally increase the rate of the reaction
   c. result in no change in the rate of the reaction
   d. decrease the rate of product formation
7. Inhibitors can bind to enzymes
   a. both reversibly and irreversibly
   b. only reversibly
   c. only irreversibly
   d. only at the active site
8. Enzymes that are inactive until one or more peptide bonds are cleaved are called
   a. apoenzymes c. ligases
   b. holoenzymes d. zymogens
9. The inhibition of the first enzyme of a metabolic pathway by the product of the pathway is an example of
   a. antimetabolism c. negative inhibition
   b. feedback inhibition d. irreversible inhibition
10. The presence of this enzyme in blood is evidence for a heart attack.
   a. casein c. creatine phosphokinase
   b. hemoglobin d. carbonic anhydrase

Answers **1.** b **2.** d **3.** a **4.** d **5.** c **6.** b **7.** a **8.** d **9.** b **10.** c

DIAGNOSTIC CHART

Blacken in all of the circles under the number of each question that you missed in the preceding Chapter Review questions. The diagnostic chart will help you identify concept areas that need more study.

	Questions									
Concepts	**1**	**2**	**3**	**4**	**5**	**6**	**7**	**8**	**9**	**10**
Metabolism	○									
Enzymes		○	○	○	○		○			
Specificity and activity				○	○					
Rates						○	○		○	
Regulation								○	○	
Enzymes in medicine										○
Bioenergetics	○									

EXERCISES

Metabolism

1. What are the two main branches of metabolism?
2. What is a metabolic pathway?
3. Why are enzymes needed for metabolic reactions?

Enzymes

4. Which of the following substances are enzymes?
   a. trypsin d. acetylcholinesterase
   b. galactose e. catalase
   c. tryptophan f. sucrase
   What clues do the names of these substances give you?
5. From the name of each of these enzymes, what is its substrate and what type of reaction does it catalyze?
   a. xylulose reductase
   b. glutamine synthetase
   c. xanthine oxidase
6. What is an enzyme's substrate?
7. What is a prosthetic group?
8. What name is given to an enzyme that lacks its prosthetic group?
9. These coenzymes are made from what vitamins?
   a. pyridoxal phosphate
   b. coenzyme A
   c. flavin adenine dinucleotide
   d. thiamine pyrophosphate
10. What is the active site of an enzyme?
11. For substrate–enzyme binding, compare the lock-and-key model and the induced fit model.
12. Make a table that (1) lists the factors affecting the activity of an enzyme and (2) briefly describes these factors.
13. What effect does an enzyme have on the activation energy of a reaction?

Reaction Rates

14. Sketch the rate (velocity) of an enzyme-catalyzed reaction as a function of enzyme concentration. Compare this graph to the one you made for question 2 of the Testing Yourself at the end of Section 27.4. Why does one graph give a straight line but the other does not?
15. What is the definition of K_m?
16. Why does a graph of enzyme activity versus temperature increase in a linear fashion until a precipitous drop is seen at higher temperatures?
17. Compare the activity of pepsin and trypsin in the pH range of 2 to 9. What is the physiological significance of these pH profiles?
18. Compare a reversible and irreversible enzyme inhibitor.
19. What structural relationship exists between an enzyme's substrate and a competitive inhibitor?
20. How do sulfa drugs inhibit bacterial growth?

Regulation

21. Describe two ways cells can control the amount of an enzyme in the cell.
22. From its name, how can you recognize that a substance is a zymogen?
23. What is meant by the term *feedback inhibition*? Does this form of regulation require a positive or negative effector?

Enzymes in Medicine

24. Explain how a study of enzyme concentrations in blood can be used to diagnose a heart attack.
25. What use does the enzyme tissue-type plasminogen activator have in medicine?

Bioenergetics

26. How do cells maintain the highly organized state necessary for life?
27. What is the role of catabolism and anabolism?
28. How is energy stored in ATP?

CHAPTER

28

Catabolism

The reactions of catabolism cleave larger dietary molecules into smaller ones. Catabolic reactions are exothermic, and cells save and use a portion of this energy. The energy and small molecules produced by catabolism are used to carry out the reactions in the body that are needed to sustain life.

OBJECTIVES

After completing this chapter, you should be able to

- Describe how and where carbohydrates, lipids, and proteins are digested.
- Describe the absorption and transport of digestion products.
- Describe the central role of glucose in carbohydrate catabolism.
- Explain the metabolic role of glycolysis and the aerobic and anaerobic fate of pyruvate.
- Describe how the TCA cycle, electron transport, and oxidative phosphorylation provide energy under aerobic conditions.
- Describe the catabolism of fatty acids.
- Describe transamination, deamination, the urea cycle, and catabolism of amino acids.

The reactions of catabolism yield energy and small molecules.

Catabolism is the part of metabolism that breaks down larger, more complex molecules into smaller and generally simpler molecules. Catabolism of very different molecules such as glucose, fatty acids, and some amino acids yields the same products, although by very different reactions. For example, acetyl CoA, which you were introduced to in Chapter 20, is produced from all of the above dietary molecules. These simpler product molecules are used by the body in one of two ways. They may be broken down further through exothermic reactions to yield energy that the cells then use as needed, or these smaller molecules may be used as building blocks for the synthesis of other molecules in the body.

Catabolic reactions are generally exothermic and oxidative. Catabolism yields smaller molecules with more oxidized carbon atoms; carbon dioxide is the final product of a number of catabolic pathways. Carbon in CO_2 has an oxidation number of $+4$, while the oxidation number of carbon in many biological molecules ranges from $+2$ to -3. In general, oxidative reactions are exothermic—as reduced molecules are oxidized, energy is released.

Catabolic reactions occur throughout the body. Although most of these reactions occur inside cells, digestion is extracellular. This chapter will discuss digestion, glycolysis, the tricarboxylic acid cycle, oxidative phosphorylation, electron transport, and several reactions and pathways involving lipids and amino acids.

28.1 Digestion, Absorption, and Transport

In the gastrointestinal tract, large dietary molecules are converted to molecules that can be absorbed and utilized by the cells of the body. This process is called **digestion**, and it involves both mechanical and chemical factors. Chewing is a mechanical process that reduces food particle size, exposing more of the food to digestive enzymes. The chemical aspects of digestion include the hydrolysis of bonds in large molecules to form smaller ones, and the absorption of these smaller molecules into the body. The smaller products are absorbed by the mucosal cells of the small intestine. From the mucosal cells, the molecules enter the cardiovascular system and are distributed throughout the body. The cells of the body then absorb these molecules and use or store them. You can refer to Figure 28.1 as you read about digestion of biomolecules.

Digestion of Carbohydrates

Digestion is an extracellular process that prepares dietary molecules for absorption.

The digestion of carbohydrates begins in the mouth. Food is chewed in the mouth and mixed with saliva which contains the enzyme salivary **amylase**. This digestive enzyme catalyzes the addition of water to glycosidic bonds in the starches amylose and amylopectin. Salivary amylase hydrolyzes α $1 \rightarrow 4$ glycosidic bonds of these starches, yielding smaller polysaccharides, oligosaccharides, and some glucose molecules (Figure 28.2). The effectiveness of salivary amylase is limited in two ways: (1) the amount of contact between the enzyme and the starches is limited because mixing of food and saliva is far from complete, and (2) the amount of time the amylase can work on the starches is limited because the amylase is inactivated by the acidic secretions of the stomach. The digestion of starch that begins in the mouth stops in

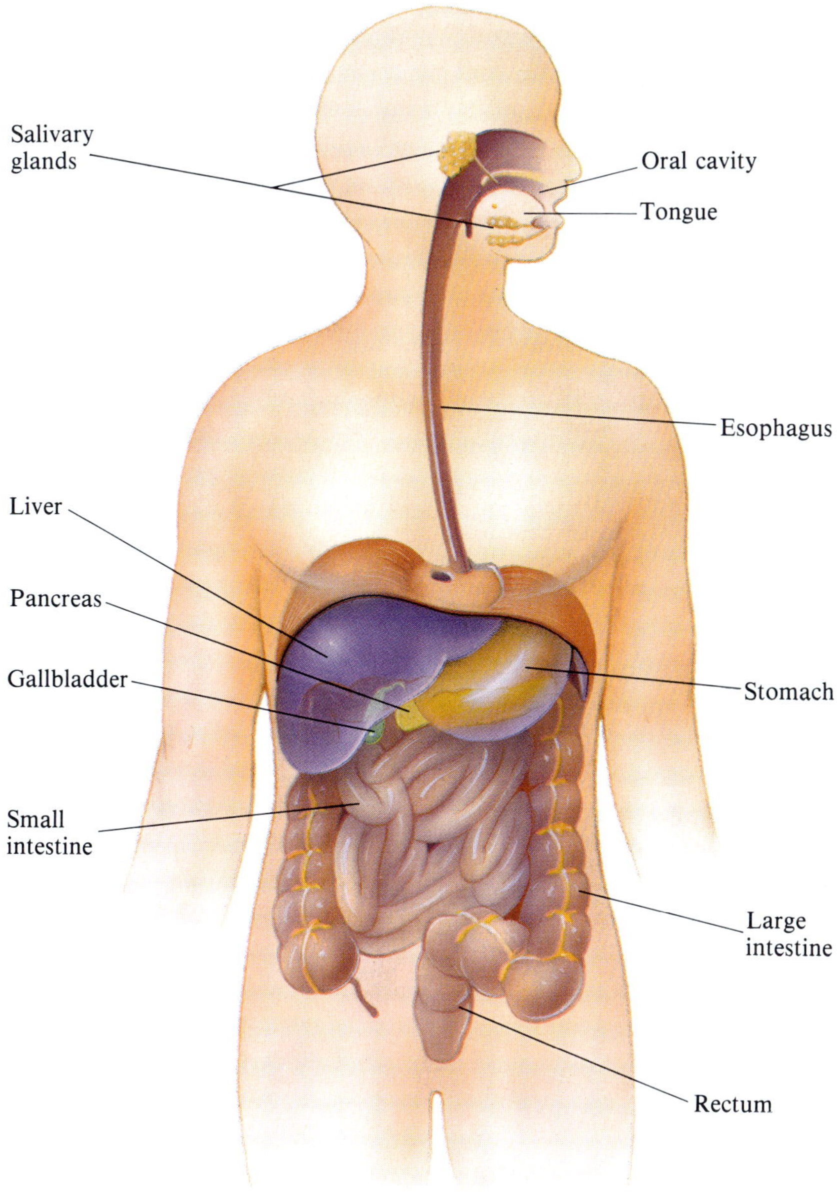
Salivary glands
Oral cavity
Tongue
Esophagus
Liver
Pancreas
Gallbladder
Stomach
Small intestine
Large intestine
Rectum

FIGURE 28.1
The human gastrointestinal tract. This organ system is responsible for the digestion and absorption of food.

the stomach where the amylase activity is lost. Dietary sugars and cellulose are not affected by amylase.

The chewed food that enters the stomach becomes mixed with gastric secretions. These secretions contain hydrochloric acid and are thus quite acidic; the pH of the stomach is less than two. Many proteins, including salivary amylase, are denatured in this acidic environment. No specific enzymes for carbohydrate digestion are found in the stomach. The H^+ found in the stomach may catalyze the hydrolysis of some of the glycosidic bonds of poly- and oligosaccharides, but this makes no significant contribution to the breakdown of carbohydrates.

The partially digested contents of the stomach and small intestine are called **chyme**. Chyme entering the small intestine contains sugars and complex carbohydrates that were not altered in either the mouth or the stomach.

FIGURE 28.2
Digestion of carbohydrates. Dietary carbohydrates, including (a) starches, (b) sugars (shown as sucrose and lactose), and (c) indigestible complex carbohydrates (shown as cellulose), are subjected to a variety of conditions as they pass through the gastrointestinal tract.

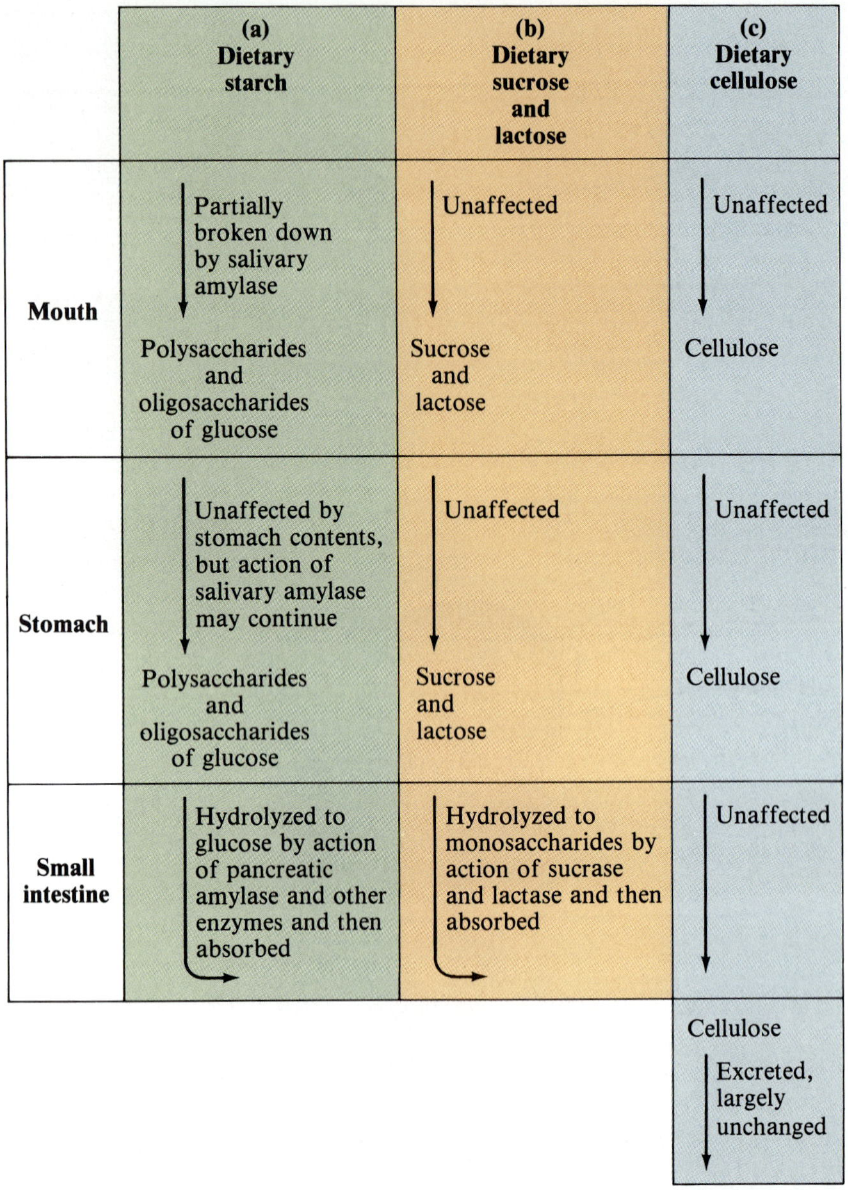

(a) Dietary starch
(b) Dietary sucrose and lactose
(c) Dietary cellulose
Mouth
Partially broken down by salivary amylase
Polysaccharides and oligosaccharides of glucose
Unaffected
Sucrose and lactose
Unaffected
Cellulose
Stomach
Unaffected by stomach contents, but action of salivary amylase may continue
Polysaccharides and oligosaccharides of glucose
Unaffected
Sucrose and lactose
Unaffected
Cellulose
Small intestine
Hydrolyzed to glucose by action of pancreatic amylase and other enzymes and then absorbed
Hydrolyzed to monosaccharides by action of sucrase and lactase and then absorbed
Unaffected
Cellulose
Excreted, largely unchanged

The digestion of starches that began in the mouth is completed in the small intestine.

It also contains the oligosaccharides and polysaccharides formed by the action of amylase on starch. Chyme is acidic, but the small intestine mixes chyme with secretions of the pancreas that contain bicarbonate. This base neutralizes the acid added by the stomach, changing the pH to between 7 and 8. The secretions of the pancreas also contain pancreatic amylase, an enzyme similar to the one in saliva. This enzyme completes the hydrolysis of the partially digested starches. Nearly all of the exposed α 1→4 glycosidic bonds in the complex carbohydrates are hydrolyzed, leaving glucose, maltose, and isomaltose as the products (Figure 28.2). These sugars, like the dietary sugars, are water soluble. Cellulose and related complex carbohydrates are not digested because humans lack enzymes specific for β 1→4 glycosidic bonds in polysaccharides. These substances and anything trapped within them are a major part of the fiber found in foods.

LACTOSE INTOLERANCE

Lactase is found in the young of all mammals, but many adults lack this enzyme and as a result have *lactose intolerance.* If these people eat lactose, the sugar present in milk and other diary products, it is not digested in the small intestine but instead enters the large intestine. The presence of unabsorbed lactose in the large intestine alters normal water absorption. This alteration can lead to diarrhea and dehydration. Furthermore, the large intestine contains numerous bacteria that can absorb and use the lactose. The lactose is an energy-rich addition to the supporting medium for the bacteria. This causes them to grow well and produce gases, which can lead to distress of the lowei intestinal tract for the host.

Dietary disaccharides such as sucrose and lactose and the disaccharides produced by starch digestion are hydrolyzed to monosaccharides by several enzymes in the small intestine (Figure 28.2). The maltose and isomaltose produced by digestion of starches are hydrolyzed by reactions catalyzed by malt*ase* and isomalt*ase*, respectively. The hydrolysis of sucrose is catalyzed by sucr*ase*, and the hydrolysis of lactose is catalyzed by lact*ase*.

The final products of carbohydrate digestion are a variety of monosaccharides. Normally large amounts of glucose are present in the chyme as well as some fructose, galactose, and other sugars. These sugars are absorbed by the mucosal cells of the small intestine. These cells possess a variety of specific transport systems in their cell membranes that move the small molecules produced by digestion into the cells. The absorbed sugars are then secreted by the mucosal cells into the abundant capillaries in the small intestine. Blood distributes the sugars throughout the body making them available to all cells.

Absorbed sugars are made available to all of the cells of the body.

Digestion of Protein

Saliva contains no enzymes that affect proteins. Protein digestion does not begin until food enters the stomach. The acidic environment of the stomach denatures proteins, leaving them in a partially or totally unfolded state. The peptide bonds of a denatured protein are more exposed to digestive enzymes than the peptide bonds of compact proteins. The peptide bonds of dietary proteins are cleaved by hydrolysis in a reaction catalyzed by the enzyme *pepsin.*

The acidic environment of the stomach denatures proteins, which exposes more peptide bonds to pepsin.

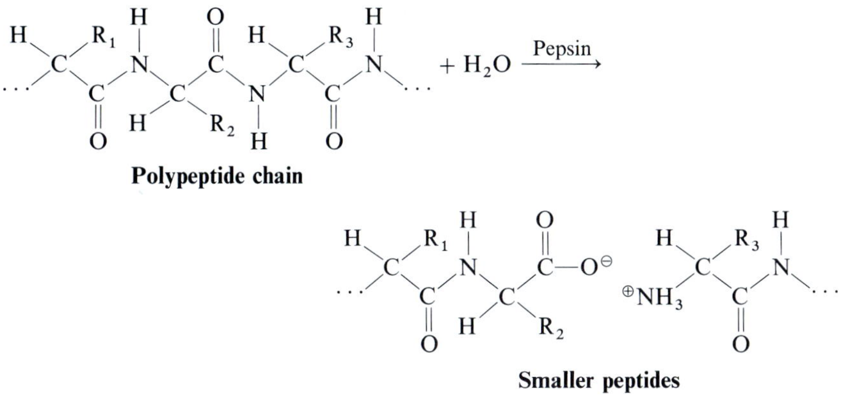

Pepsin
Polypeptide chain
Smaller peptides

The hydrolysis of polypeptides is not completed in the stomach. Portions of the food are not adequately exposed to pepsin in the stomach. Furthermore, pepsin does not work equally well on all peptide bonds; pepsin shows some specificity for substrate. The chyme entering the small intestine contains a mixture of peptides, polypeptides, and free amino acids. Figure 28.3 summarizes the digestion of protein in the stomach and the rest of the gastrointestinal tract.

The secretions of the pancreas entering the small intestine contain several zymogens that are activated to **proteases** and **peptidases**, enzymes that catalyze the hydrolysis of proteins and peptides, respectively. Trypsinogen is converted to the enzyme trypsin by hydrolytic reactions catalyzed by trypsin itself or by the enzyme enterokinase. There is usually a little trypsin or enterokinase in the small intestine, so conversion of trypsinogen to trypsin always occurs. Trypsin catalyzes the activation of other zymogens, including chymotrypsinogen to chymotrypsin and procarboxypeptidase to carboxypeptidase. Collectively, these digestive enzymes and some others convert the peptides and polypeptides to free, water-soluble amino acids. The amino acids are absorbed by mucosal cells, which secrete them into the cardiovascular system for distribution throughout the body. Thus, like monosaccharides, amino acids are made available to all the cells of the body.

Digestion of Lipids

Dietary lipids include triacylglycerols (fats and oils), cholesterol, and polar lipids. We will look at only triacylglycerols and cholesterol here. Digestion of triacylglycerols does not begin until chyme enters the small intestine (Figure 28.4). Saliva and gastric secretions do not contain any significant amounts

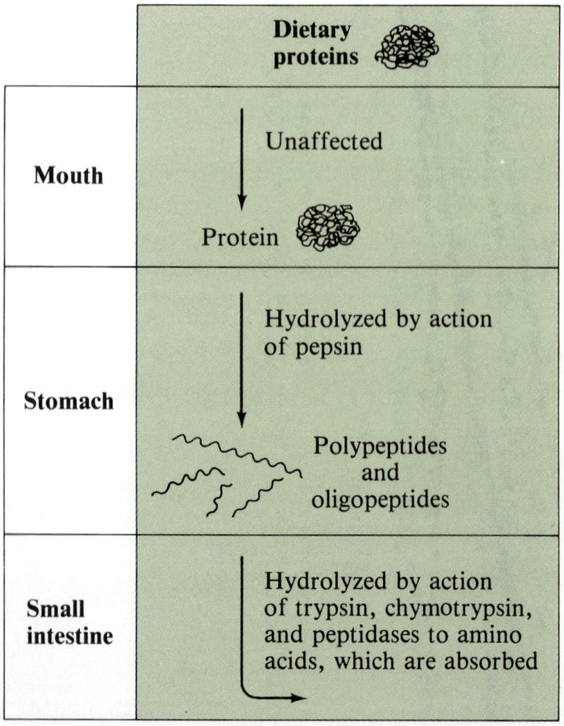

Dietary proteins
Mouth
Unaffected
Protein
Stomach
Hydrolyzed by action of pepsin
Polypeptides and oligopeptides
Small intestine
Hydrolyzed by action of trypsin, chymotrypsin, and peptidases to amino acids, which are absorbed

FIGURE 28.3
Digestion of dietary proteins occurs primarily in the stomach and small intestine.

CHOLESTEROL AND HEART DISEASE

The role played by cholesterol in heart disease is not fully known, but studies of a human genetic disease called *familial hypercholesterolemia* have provided some valuable clues. People with two copies of the defective gene (homozygotes) die at an early age of heart disease. Liver transplants are a treatment for these individuals. People with one copy of the defective gene (heterozygotes) have a much higher risk of heart disease than the normal population. These individuals have elevated blood cholesterol levels because they have elevated levels of low density lipoprotein (LDL). This elevated blood cholesterol is linked to an increased risk of heart attacks. The cause of increased LDL cholesterol appears to be due to a decreased ability of cells to take up LDL from the blood. The cells contain an abnormally low number of receptors for the LDL, thus less cholesterol is taken up by these cells and more cholesterol remains in the blood. These individuals are treated by depriving the body of cholesterol through restricted diet and by treatment with drugs to reduce both the uptake of dietary cholesterol and cholesterol synthesis. Under these conditions, the body makes more of the LDL receptor and uptake of blood cholesterol is stimulated. Whether the normal population benefits from reduced dietary cholesterol has been harder to prove.

of enzymes that affect lipids. Cholesterol requires only absorption, while fats and oils must first be hydrolyzed. Digestion of lipids is complicated by their hydrophobic nature. Carbohydrates, proteins, and peptides are water soluble or suspended in water so digestive enzymes can interact directly with these dietary compounds. But dietary lipids are water insoluble and form droplets and globules in the gastrointestinal tract. Only lipid molecules on the surface of these droplets are exposed to enzymes; the vast majority have little chance

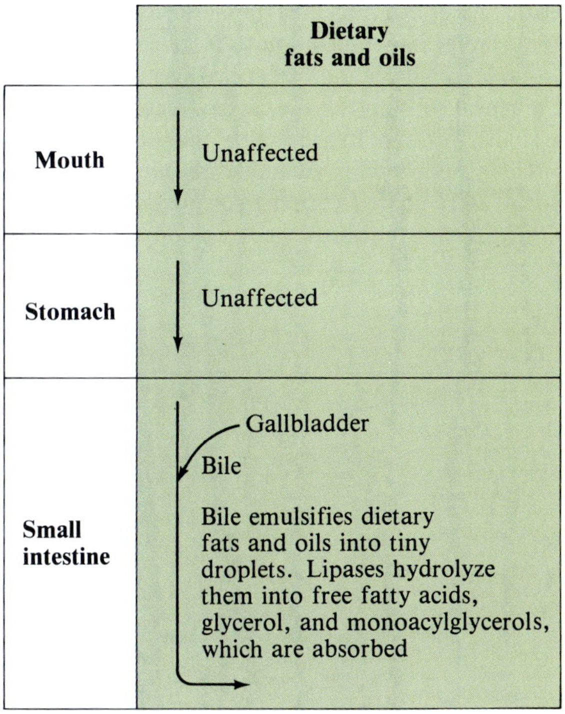
Dietary fats and oils
Mouth
Unaffected
Stomach
Unaffected
Small intestine
Gallbladder
Bile
Bile emulsifies dietary fats and oils into tiny droplets. Lipases hydrolyze them into free fatty acids, glycerol, and monoacylglycerols, which are absorbed

FIGURE 28.4
Digestion of fats and oils occurs in the small intestine through the action of both enzymes and bile.

to come into contact with enzymes because they are buried within the hydrophobic interior of the droplet.

The micelles formed from bile salts and triacylglycerols have large surface areas that are exposed to digestive enzymes.

This complication is eliminated by **bile salts**. These compounds are synthesized by the liver, stored in the gallbladder, and secreted into the small intestine as needed. See Figure 24.11 in Chapter 24 for the structure of two typical bile salts, taurocholate and glycocholate. Bile salts, like soaps, are amphipathic molecules, and they emulsify dietary lipid into many micelles. The micelles collectively have a much greater surface area than the ingested lipid droplets. People who lack a gallbladder should avoid a high fat diet.

The digestion of triacylglycerols involves hydrolysis of two or three of the ester bonds between the fatty acids and the glycerol. This hydrolysis is catalyzed by **lipases** synthesized and secreted by the pancreas.

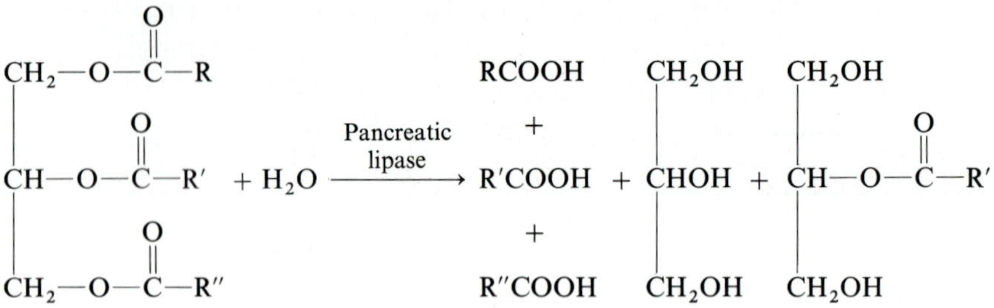
Pancreatic lipase

Triacylglycerols **Fatty acids** **Glycerol** **Monoacylglycerols**

Dietary phosphoacylglycerols are digested by a phospholipase that is synthesized and secreted as a zymogen by the pancreas. This zymogen is activated by trypsin.

Cholesterol, monoacylglycerols, glycerol, and fatty acids are absorbed by the mucosal cells. In these cells, triacylglycerols are reformed from fatty acids and from glycerol or monoacylglycerols. These triacylglycerols along with cholesterol are secreted into the lymphatic vessels of the small intestine. **Lymph** is a fluid similar in composition to blood but lacking red blood cells. Lipids are insoluble in lymph and blood because these fluids are principally water. In the mucosal cells and in the cardiovascular system, the lipids bind to specific proteins through hydrophobic interactions to form complexes called **lipoproteins**. Lipoproteins resemble micelles—they possess a hydrophobic interior of lipid coated with a hydrophilic surface of protein molecules (Table 28.1). Lipids are transported in blood and lymph as lipoproteins.

Because lipids are water insoluble, they must be transported as lipoprotein complexes.

TABLE 28.1
Typical Percentage Composition of Lipoproteins

Lipoprotein	Triacylglycerol (%)	Cholesterol (%)	Phospholipid (%)	Protein (%)	Average Density	Approximate Diameter (nm)
Chylomicrons	87	3–4	8	1.5–2.0	0.92–0.96	100–1000
Very low density lipoprotein (VLDL)	55	15	20	9–10	0.95–1.00	25–75
Low density lipoprotein (LDL)	8–10	45	22	25	1.00–1.06	20–25
High density lipoprotein (HDL)	1–5	18–20	21–30	45–55	1.06–1.21	7–12

The triacylglycerols in lipoproteins are broken down into fatty acids and glycerol, which are then taken up by the cells of the body. Adipose cells take up much of the fatty acids and glycerol, and triacylglycerols are rebuilt and stored. When the body requires more energy, lipases within the adipose cell break down triacylglycerols and release free fatty acids into the bloodstream. These free fatty acids bind to the protein serum albumin by hydrophobic interactions. The fatty acids circulate as part of this protein complex, and they too are absorbed by cells as needed.

NEW TERMS

digestion Process that breaks down food molecules into smaller compounds that are absorbed by the body.

amylases Enzymes that catalyze the hydrolysis of $\alpha\ 1 \rightarrow 4$ glycosidic bonds in starches.

chyme Partially digested food in the stomach and small intestine.

proteases Enzymes that catalyze the hydrolysis of peptide bonds in proteins and polypeptides.

peptidases Enzymes that catalyze the hydrolysis of peptide bonds in peptides. The distinction between a protease and a peptidase is not clear cut.

bile salts Emulsifying agents in bile that break down dietary lipid droplets into micelles.

lipases Enzymes that catalyze the hydrolysis of ester bonds in fats, oils, and similar lipids.

lymph Body fluid similar to blood but lacking red blood cells.

lipoproteins Macromolecules composed of protein and lipid which transport lipids in blood.

TESTING YOURSELF

Digestion

1. Characterize catabolic reactions with respect to (a) size of reactants and products and (b) oxidation state of reactants and products.
2. Where does the digestion of starches occur?
3. How do hydrophobic molecules such as triacylglycerols come in contact with the appropriate water-soluble digestive enzyme?

Answers **1. a.** Catabolic reactions generally convert larger molecules to smaller ones. **b.** The reactants are generally oxidized during catabolism. **2.** Starch digestion begins in the mouth, but most of the digestion occurs in the small intestine. **3.** Dietary lipids are emulsified by bile salts. The micelles that are produced readily contact digestive enzymes in the gut.

28.2 Carbohydrate Catabolism

Carbohydrate digestion and absorption bring sugars into blood from where they are absorbed into cells and used for energy or to build other molecules. For example, some galactose and glucose may be used to synthesize glycolipids or glycoproteins needed by the cell. Glucose in muscle or in liver could also be used to synthesize glycogen, a polymeric storage form for glucose. Most of the dietary sugars, however, are used for energy production. Glucose is the most important monosaccharide and the only one maintained at approximately constant concentrations in the blood.

Glycolysis

Glycolysis is the principal catabolic pathway for sugars.

The principal pathway for catabolism of glucose is **glycolysis**, which occurs in the cytosol of cells. It is a series of ten enzyme-catalyzed reactions that cleaves and oxidizes glucose to two molecules of pyruvic acid. At physiological pH, the pyruvic acid dissociates into pyruvate and H^+.

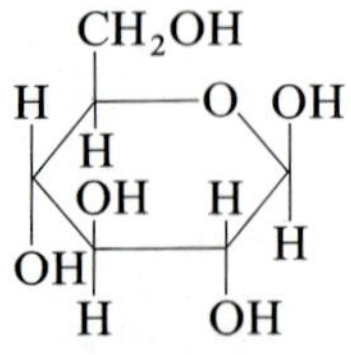

$$\xrightarrow{\text{Glycolysis}} 2\,CH_3\overset{O}{\overset{\|}{C}}COOH \longrightarrow 2\,CH_3\overset{O}{\overset{\|}{C}}COO^- + 2\,H^+$$

Glucose **Pyruvic acid** **Pyruvate**

Glucose catabolism is exothermic. A portion of the energy released is conserved by synthesizing adenosine triphosphate (ATP) from adenosine diphosphate (ADP) and phosphate. Furthermore, the oxidation that occurs during glycolysis is accompanied by the reduction of nicotinamide adenine dinucleotide from its oxidized form (NAD^+) to its reduced form (NADH). A representative equation for glycolysis is

$$\text{Glucose} + 2\,\text{ADP} + 2\,\text{P}_i + 2\,\text{NAD}^+ \xrightarrow{\text{Glycolysis}} 2\,\text{Pyruvate} + 2\,\text{ATP} + 2\,\text{NADH}$$

In the first step of glycolysis, a molecule of glucose is phosphorylated. The enzyme hexokinase catalyzes the transfer of a phosphate from ATP to the hydroxyl group on carbon six of glucose to produce glucose 6-phosphate (Step 1, Figure 28.5). Note that even though glycolysis is a net producer of ATP, two ATPs are consumed in the early stages of glycolysis. Glucose 6-phosphate is a substrate of the next step of glycolysis, but the phosphorylation of glucose accomplishes something else. Glucose 6-phosphate is ionic, and ions do not readily cross cell membranes. Thus, phosphorylation of glucose traps glucose within the cell.

In the second step of glycolysis, glucose 6-phosphate is isomerized to fructose 6-phosphate in a reaction catalyzed by the enzyme phosphoglucose isomerase (Step 2, Figure 28.5). Fructose 6-phosphate is then phosphorylated to fructose 1,6-diphosphate by transfer of a phosphate group from ATP (Step 3, Figure 28.5). This reaction is catalyzed by the enzyme phosphofructokinase. This is the second of two molecules of ATP that are consumed. The changes that occur in the sugars during the first steps of glycolysis are endothermic; ATP hydrolysis provides the energy to drive these reactions.

In the fourth reaction, the enzyme aldolase catalyzes the cleavage of fructose 1,6-diphosphate into two phosphorylated trioses, one molecule of glyceraldehyde 3-phosphate and one molecule of dihydroxyacetone phosphate (Step 4, Figure 28.5). Next, dihydroxyacetone phosphate is isomerized into a molecule of glyceraldehyde 3-phosphate (Step 5, Figure 28.5). This step is catalyzed by the enzyme triose phosphate isomerase. This enzyme and aldolase effectively convert fructose 1,6-diphosphate to two molecules

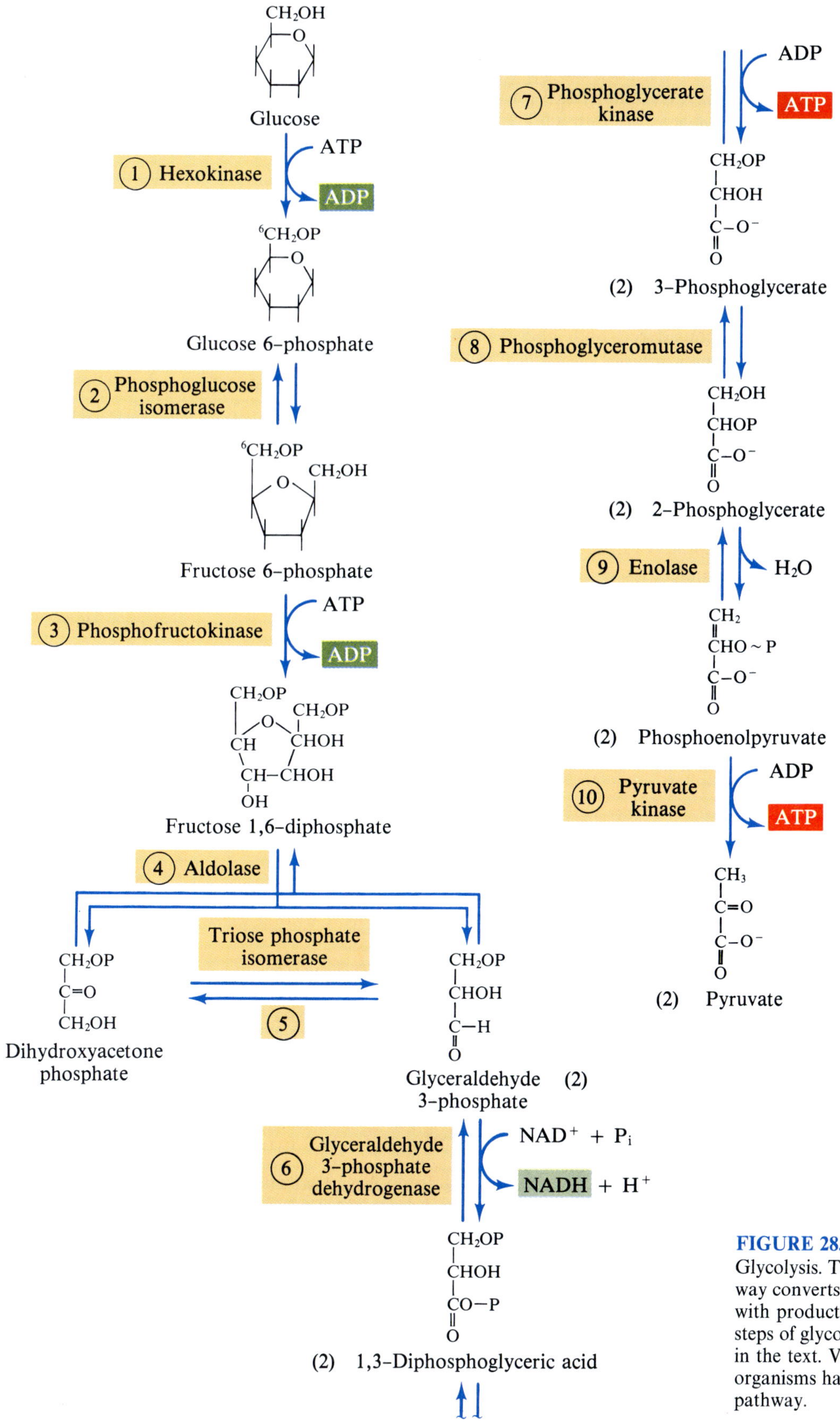

Glucose
ATP
1 Hexokinase
ADP
Glucose 6-phosphate
2 Phosphoglucose isomerase
Fructose 6-phosphate
ATP
3 Phosphofructokinase
ADP
Fructose 1,6-diphosphate
4 Aldolase
Triose phosphate isomerase
5
Dihydroxyacetone phosphate
Glyceraldehyde 3-phosphate (2)
6 Glyceraldehyde 3-phosphate dehydrogenase
NAD⁺ + Pi
NADH + H⁺
(2) 1,3-Diphosphoglyceric acid
ADP
7 Phosphoglycerate kinase
ATP
(2) 3-Phosphoglycerate
8 Phosphoglyceromutase
(2) 2-Phosphoglycerate
9 Enolase
H2O
(2) Phosphoenolpyruvate
ADP
10 Pyruvate kinase
ATP
(2) Pyruvate

FIGURE 28.5
Glycolysis. This metabolic pathway converts glucose to pyruvate with production of energy. The steps of glycolysis are described in the text. Virtually all living organisms have the glycolytic pathway.

$$\textbf{Glucose} + \textbf{2 NAD}^+ + \textbf{2 ADP} + \textbf{2 P}_i \longrightarrow \textbf{2 Pyruvate} + \textbf{2 NADH} + \textbf{2 ATP}$$

of glyceraldehyde 3-phosphate. To balance glycolysis for one mole of glucose, each of the following steps should be multiplied by two.

In the sixth step of glycolysis, glyceraldehyde 3-phosphate is both oxidized and phosphorylated. The enzyme glyceraldehyde 3-phosphate dehydrogenase catalyzes this complex reaction (Step 6, Figure 28.5). The aldehyde group is oxidized to a carboxylic acid with concurrent reduction of NAD^+ to NADH. But the free carboxylic acid is not found because an inorganic phosphate becomes bonded to the carbon atom of the carboxyl group to form an anhydride bond. Just as the anhydride bonds of ATP are considered high energy bonds, this mixed anhydride bond is also a high energy bond. Since two glyceraldehyde 3-phosphates react, two NADH and two molecules of 1,3-diphosphoglycerate are formed for each glucose molecule. Phosphoglycerate kinase catalyzes the seventh step of glycolysis. 1,3-Diphosphoglycerate reacts with ADP to form ATP and 3-phosphoglycerate (Step 7, Figure 28.5). The phosphate group that had been bonded to the carboxyl group is transferred to ADP. The high energy mixed anhydride bond is broken in an exothermic reaction, but a portion of that energy is conserved in the endothermic formation of an anhydride bond in ATP. Two molecules of ATP and 3-phosphoglycerate are formed per glucose. At this point, no net ATP synthesis has occurred because two ATPs were consumed earlier during the phosphorylation of glucose and fructose 6-phosphate.

The compound 3-phosphoglycerate is in turn converted to 2-phosphoglycerate by an isomerization catalyzed by the enzyme phosphoglyceromutase (Step 8, Figure 28.5). The enzyme enolase catalyzes the dehydration of 2-phosphoglycerate into phosphoenolpyruvate (PEP) (Step 9, Figure 28.5). The phosphoester bond in PEP is a high energy bond. Two molecules of PEP are formed from a molecule of glucose. In the final reaction of glycolysis, PEP reacts with ADP in a reaction catalyzed by pyruvate kinase (Step 10, Figure 28.5). Loss of phosphate from PEP to form pyruvate is highly exothermic, and again a portion of this energy is conserved through synthesis of a high energy anhydride bond in ATP. Since two molecules of PEP are formed from each molecule of glucose, two molecules of pyruvate and two molecules of ATP are formed in this last reaction.

Glycolysis has a gross production of four ATPs, but the net ATP production is two.

In summary, glycolysis oxidizes and cleaves one molecule of glucose to two molecules of pyruvate. Two molecules of NADH are formed. Two molecules of ATP are consumed with ADP as product, but later in the pathway four molecules of ATP are formed from four molecules of ADP. The pathway is a net producer of useful energy because two ATPs are now available to the cell.

$$\text{Glucose} + 2\ \text{NAD}^+ + 2\ \text{ADP} + 2\text{P}_i \longrightarrow 2\ \text{Pyruvate} + 2\ \text{NADH} + 2\ \text{ATP}$$

Glucose is the principal compound that enters glycolysis, but it is not the only one. Other monosaccharides also enter glycolysis, though not necessarily at the same place as glucose. Figure 28.6 illustrates the entry point of some common carbohydrates.

Glycogen Catabolism

Glycogen, a polysaccharide found in liver and muscle, serves as a storage form of glucose. When cellular energy is needed in muscle, glycogen reacts with inorganic phosphate in a reaction catalyzed by the enzyme glycogen phosphorylase, to yield glucose 1-phosphate.

$$\text{Glycogen}_n + \text{Phosphate} \xrightarrow[\text{phosphorylase}]{\text{Glycogen}} \text{Glycogen}_{n-1} + \text{Glucose 1-phosphate}$$

This highly regulated process is discussed in more detail in Chapter 30. The glucose 1-phosphate that is produced is converted to glucose 6-phosphate by a reaction catalyzed by the enzyme phosphoglucomutase. This molecule is an intermediate of glycolysis (Figure 28.6).

In liver, the catabolism of glycogen often serves a different purpose. As in muscle, glucose 1-phosphate is produced from glycogen and inorganic phosphate and is isomerized to glucose 6-phosphate. This molecule is then cleaved to glucose and inorganic phosphate in a reaction catalyzed by glucophosphatase. Glucose is secreted from the liver into the blood as needed. Glycogen in muscle thus serves as a storage form of glucose to meet the immediate energy needs of muscle cells, while liver glycogen serves as a storage form of glucose for the entire body.

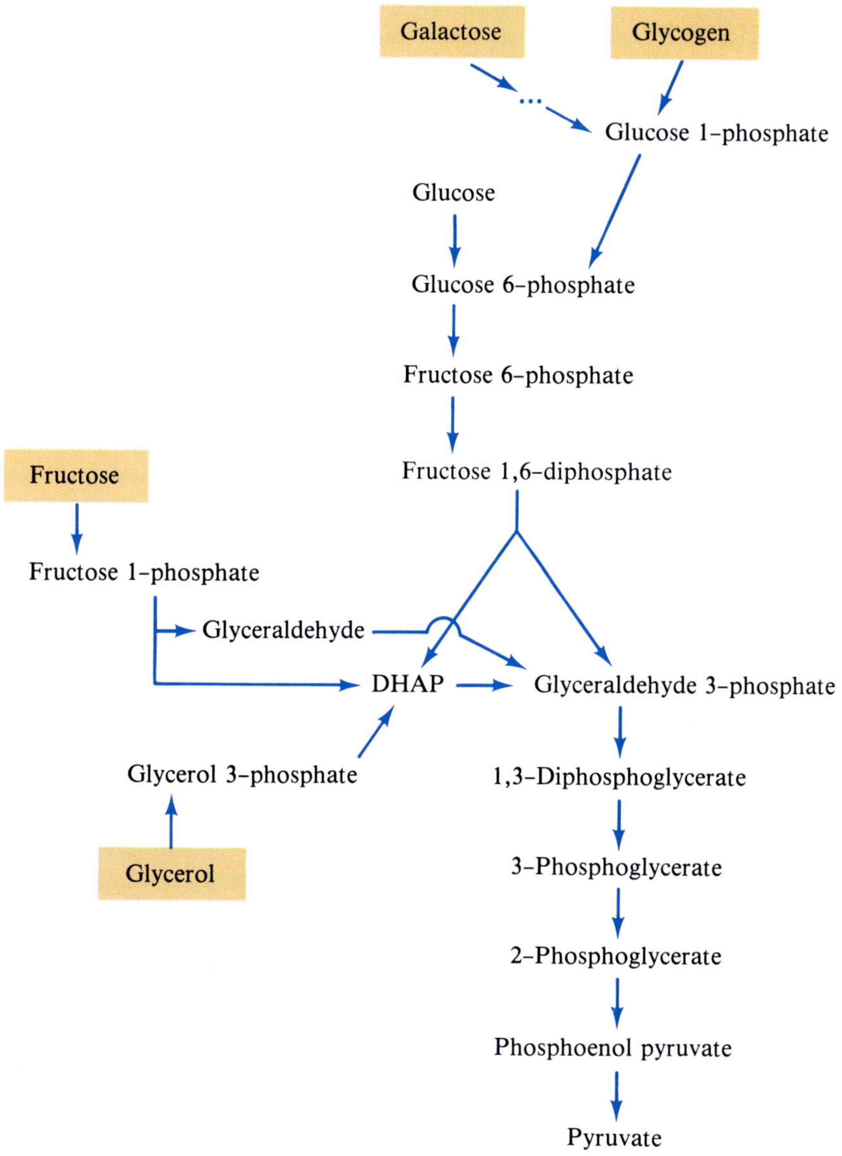
Galactose
Glycogen
Glucose 1–phosphate
Glucose
Glucose 6–phosphate
Fructose 6–phosphate
Fructose 1,6–diphosphate
Fructose
Fructose 1–phosphate
Glyceraldehyde
DHAP
Glyceraldehyde 3–phosphate
Glycerol 3–phosphate
Glycerol
1,3–Diphosphoglycerate
3–Phosphoglycerate
2–Phosphoglycerate
Phosphoenol pyruvate
Pyruvate

FIGURE 28.6
Other carbohydrates enter glycolysis in addition to glucose. This figure shows the entry points for galactose, glycogen, fructose, and glycerol.

Pyruvate Catabolism

The catabolic fate of pyruvate depends on the presence or absence of oxygen.

The pyruvate produced by glycolysis has two possible catabolic fates in mammals. If oxygen is available to the cell, *aerobic* catabolism will result in the oxidative decarboxylation of pyruvate to yield acetyl coenzyme A (acetyl CoA) and CO_2 (Section 28.3). The acetyl CoA, in turn, can be oxidized completely into carbon dioxide and water by the tricarboxylic acid cycle (see Section 28.3) and the electron transport chain (see Section 28.4). These highly exothermic processes occur in the mitochondria. In these reactions, which are examples of **respiration**, the hydrogen atoms of the original substrate, glucose, are passed to an inorganic acceptor, oxygen. The NADH produced by glycolysis is also recycled to NAD^+ in the presence of oxygen (see Section 28.4).

If oxygen is not available to the cell, *anaerobic* catabolism occurs. Pyruvate is reduced to lactate in a reaction using NADH produced by glycolysis.

$$H^+ + CH_3{-}\overset{\overset{\displaystyle O}{\|}}{C}{-}COO^- + NADH \xrightleftharpoons{\text{Lactate dehydrogenase}} CH_3{-}\underset{\underset{\displaystyle H}{|}}{\overset{\overset{\displaystyle OH}{|}}{C}}{-}COO^- + NAD^+$$

Pyruvate **Lactate**

In this reaction, the hydrogen atoms derived from glucose are passed to pyruvate, an organic molecule. This process is a **fermentation**.

If NADH were not recycled to NAD^+ either in the presence or absence of oxygen, glycolysis would quickly come to a halt because cells contain only a small amount of NAD^+ which they continually recycle.

Yeast can also metabolize glucose in the absence of oxygen, but yeast recycle NADH and process pyruvate somewhat differently. In the absence of oxygen, yeast convert pyruvate first to acetaldehyde and carbon dioxide. The acetaldehyde in turn reacts with NADH to yield ethanol.

$$\text{Pyruvate} \xrightarrow{\text{Pyruvate decarboxylase}} CO_2 + \text{Acetaldehyde}$$

$$\text{Acetaldehyde} + NADH \xrightarrow{\text{Alcohol dehydrogenase}} \text{Ethanol} + NAD^+$$

$$\textbf{Net:}\quad \text{Pyruvate} + NADH \longrightarrow \text{Ethanol} + CO_2 + NAD^+$$

NAD^+ is recycled, and ethanol and carbon dioxide are produced. Since the acceptor of hydrogen atoms is an organic molecule, acetaldehyde, this is also a fermentation. This particular fermentation has been used by humans since antiquity—the baking, alcoholic beverage, and chemical industries use alcoholic fermentation as an industrial process.

NEW TERMS

glycolysis Cytosolic process that converts glucose to two molecules of pyruvate with the production of two molecules of ATP and two molecules of NADH.

respiration Oxidation of a compound with transfer of electrons to an inorganic substance. Respiration using oxygen is the principal source of energy in the body.

fermentation Oxidation of a compound with transfer of electrons to an organic molecule. Formation of lactic acid from glucose in anaerobic muscle is a fermentation.

TESTING YOURSELF

Carbohydrate Catabolism

1. Compare the fate of pyruvate in aerobic and anaerobic catabolism.
2. Distinguish between a respiration and a fermentation.
3. Compare the net and gross production of ATP during glycolysis.

Answers **1.** During aerobic conditions pyruvate is converted to acetyl CoA, which is catabolized further. During anaerobic conditions it is converted to lactate. **2.** In respiration, the electrons from the reacting molecule are accepted by an inorganic species, such as oxygen. In a fermentation, an organic molecule accepts the electrons. **3.** For each molecule of glucose, four ATP are produced during glycolysis. This is the gross production. Two ATP are consumed during glycolysis, so the net product is 2 ATP.

28.3 Tricarboxylic Acid Cycle

Anaerobic catabolism of glucose, through glycolysis, yields two net ATPs. The useful energy saved in the synthesis of these two ATPs is about 14.6 kcal per mole of glucose. Complete oxidation of glucose to CO_2 and water yields 686 kcal per mole of glucose. The energy conserved by ATP synthesis in glycolysis is only a little more than 2% of the energy available when glucose is oxidized completely. Of course, glucose is not oxidized completely during glycolysis—much of the original chemical energy of glucose remains in the pyruvate molecules. The tricarboxylic acid cycle and electron transport coupled to oxidative phosphorylation are responsible for the complete oxidation of glucose with the concurrent release of large amounts of energy. This section discusses the role of the tricarboxylic acid cycle in catabolism.

Anaerobic catabolism of glucose is a very inefficient process for energy production.

If glucose is metabolized in the presence of oxygen, the pyruvate produced in glycolysis enters the mitochondrial matrix where it is cleaved and oxidized to carbon dioxide and acetyl CoA.

$$\text{Pyruvate} + \text{CoA} + \text{NAD}^+ \xrightarrow[\text{}]{\text{Pyruvate dehydrogenase}} \text{Acetyl CoA} + \text{CO}_2 + \text{NADH}$$

The acetyl CoA is then oxidized to carbon dioxide by the reactions of the **tricarboxylic acid** or **TCA cycle** (Figure 28.7). These oxidative reactions are highly exothermic. Some of the energy released by these reactions is conserved in the form of reduced and phosphorylated coenzymes. This cycle is also called both the *Krebs cycle*, named for Hans Krebs, the discoverer of many of the reactions in the cycle, and the *citric acid cycle*, named for the first product of the cycle.

In the first reaction of the TCA cycle, acetyl CoA and oxaloacetate react to yield a molecule of citrate and a molecule of CoA (Step 1, Figure 28.7). Citrate is one of two tricarboxylic acids found in this cycle that provide the cycle with its name. Citrate is isomerized by the enzyme aconitase to isocitrate (Step 2).

Isocitrate is the substrate of isocitrate dehydrogenase, an enzyme that catalyzes the oxidation and cleavage of isocitrate to a molecule of CO_2 and

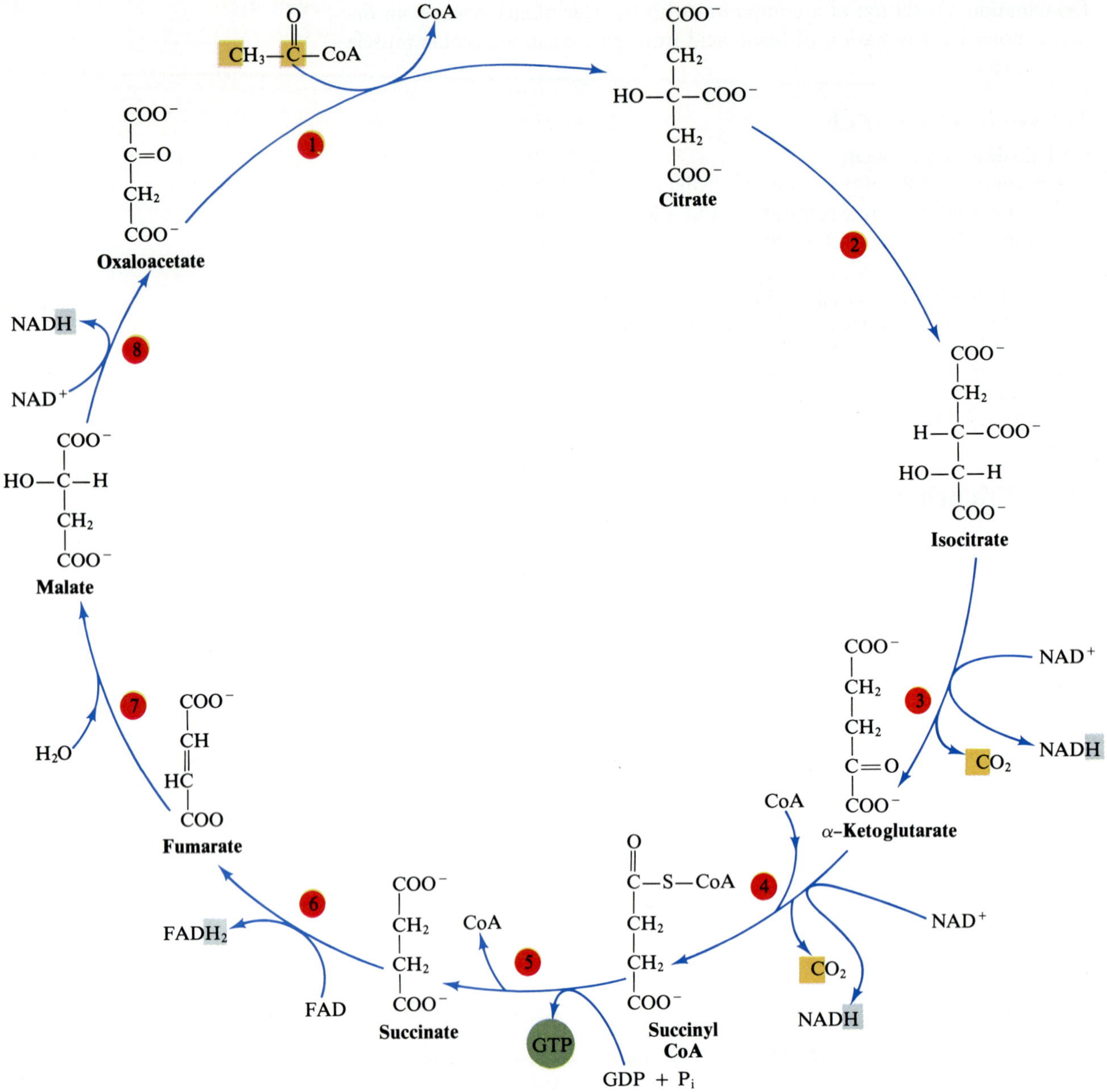
CoA
CH_3—C(=O)—CoA
Citrate
Isocitrate
NAD+
NADH
CO_2
α-Ketoglutarate
CoA
NAD+
CO_2
NADH
Succinyl CoA
GDP + P_i
GTP
CoA
Succinate
FAD
$FADH_2$
Fumarate
H_2O
Malate
NAD+
NADH
Oxaloacetate
1
2
3
4
5
6
7
8

FIGURE 28.7
The tricarboxylic acid cycle. In this cycle, the carbon atoms (yellow) of acetic acid (as acetyl CoA) are oxidized to carbon dioxide. Energy is conserved as reduced NADH and $FADH_2$ (blue) and as GTP (green). Steps 1 through 8 are discussed in the text.

α-ketoglutarate. During this oxidation, a molecule of NAD^+ is reduced to NADH (Step 3). This step is the first of four oxidative steps in the cycle. The next step is an oxidative step that involves several substrates and products. In this reaction, α-ketoglutarate is oxidized, yielding CO_2 and succinyl CoA. NAD^+ is reduced to NADH concurrently (Step 4).

The acetyl group of the acetyl CoA that entered the cycle contained two carbon atoms. So far, the TCA cycle has produced two molecules of CO_2. The rest of the cycle traps additional energy and returns the intermediate molecules to oxaloacetate, the starting compound of the cycle.

Succinyl CoA is the substrate of succinate thiokinase, which catalyzes the hydrolysis of the bond between succinate and CoA. Energy released in this reaction is used to form a bond between guanosine diphosphate (GDP) and inorganic phosphate to form guanosine triphosphate (GTP) (Step 5). A GTP is equivalent in energy to an ATP, so this reaction is equivalent to the synthesis of an ATP.

Succinate is next oxidized to fumarate by succinate dehydrogenase (Step 6). As succinate is oxidized, the coenzyme flavin adenine dinucleotide is reduced from its oxidized form, FAD, to its reduced form, $FADH_2$. Fumarate contains a carbon–carbon double bond that undergoes the reactions typical of this functional group. In the next step, the enzyme fumarase adds a molecule of water to fumarate to yield malate (Step 7). Malate is then oxidized to oxaloacetate by malate dehydrogenase (Step 8). NAD^+ is reduced to NADH concurrently. Oxaloacetate is the starting compound, thus the cycle is complete.

The following equation summarizes the TCA cycle:

As acetyl CoA is oxidized to carbon dioxide, coenzymes are reduced or converted to high energy forms.

$$\text{Acetyl CoA} + 3\ NAD^+ + FAD + GDP + P_i \longrightarrow 2\ CO_2 + CoA + 3\ NADH + FADH_2 + GTP$$

Little available energy comes directly from the TCA cycle because only one ATP equivalent (the GTP) was produced per acetyl CoA. Since one molecule of glucose yields two molecules of pyruvate and thus two molecules of acetyl CoA, two equivalents of ATP as GTP are produced in the TCA cycle per glucose. However, the NADH and $FADH_2$ produced in the TCA cycle and elsewhere contain significant amounts of energy that can be used for the synthesis of ATP. The processcs of clectron transport and oxidative phosphorylation are responsible for this synthesis of ATP.

NEW TERM

tricarboxylic acid or TCA cycle A cyclic pathway that oxidizes acetyl CoA to CO_2, yielding reduced and phosphorylated coenzymes as the other products.

TESTING YOURSELF

TCA Cycle

1. Which steps in the TCA cycle involve oxidation–reduction?

Answer 1. Steps 3 (isocitrate dehydrogenase), 4 (α-ketoglutarate dehydrogenase), 6 (succinate dehydrogenase), and 8 (malate dehydrogenase). This can be recognized because dehydrogenases and the coenzymes NAD^+ and FAD participate in oxidation–reduction reactions.

28.4 Electron Transport and Oxidative Phosphorylation

Aerobic catabolism of glucose yields either 36 or 38 ATPs per glucose molecule, depending upon how the cytosolic NADH is consumed. Two ATPs were produced in glycolysis, and the two GTPs produced in the TCA cycle are equivalent to ATP, so four ATPs are accounted for. The remaining ATPs are synthesized by **oxidative phosphorylation**, a process in which ATP synthesis from ADP and phosphate is coupled to an oxidative process. Transport

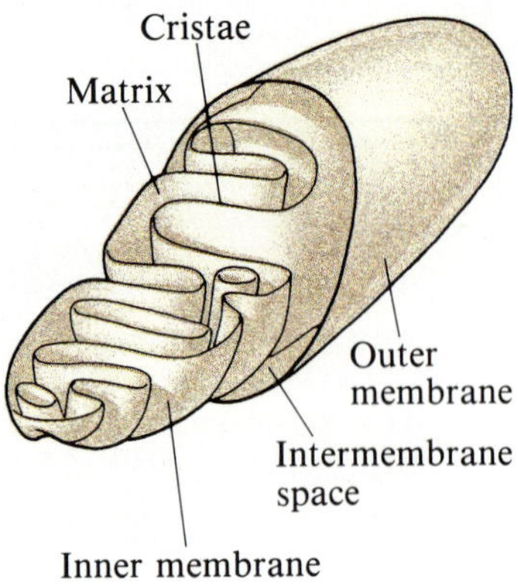
Cristae
Matrix
Outer membrane
Intermembrane space
Inner membrane

FIGURE 28.8
The matrix and inner membrane of a mitochondrion contain the important components of the TCA cycle (most are in the matrix), electron transport (inner membrane), and oxidative phosphorylation (inner membrane). The cristae are invaginations of the inner membrane that greatly increase its surface area.

of electrons via an **electron transport chain** within the mitochondria is intimately involved in this oxidation. This section describes electron transport and oxidative phosphorylation and accounts for the aerobic production of energy in the cell.

Oxidative reactions are generally exothermic, and the oxidation of NADH and $FADH_2$ to NAD^+ and FAD by oxygen are not exceptions.

$$\text{NADH} + \text{H}^+ + 1/2\ \text{O}_2 \longrightarrow \text{NAD}^+ + \text{H}_2\text{O} + \text{Energy}\ (-52.6\ \text{kcal/mol})$$
$$\text{FADH}_2 + 1/2\ \text{O}_2 \longrightarrow \text{FAD} + \text{H}_2\text{O} + \text{Energy}\ (-43.4\ \text{kcal/mol})$$

The formation of ATP from ADP and inorganic phosphate is endothermic; + kcal of energy is required per mole of ATP formed. Clearly enough energy is available during the oxidation of these reduced coenzymes to synthesize some ATP if the two processes are coupled.

Electron Transport Chain

The mechanism by which NADH and $FADH_2$ are oxidized is understood in principle. When NADH is oxidized to NAD^+, two electrons pass from NADH to molecular oxygen through a series of electron carriers that are part of the inner membrane of the mitochondrion (Figure 28.8). Each of the carriers are alternately reduced and oxidized as they first accept then pass the electrons on to the next carrier of the electron transport system. Ultimately, molecular oxygen accepts the electrons along with protons from solution to become water (Figure 28.9).

For each mole of NADH that is oxidized to NAD^+, over 52 kcal of energy are released. The **chemiosmotic theory** provides an explanation for how a portion of this energy is conserved. Energy is released as electrons flow from NADH through the carriers to oxygen. Some of this energy is

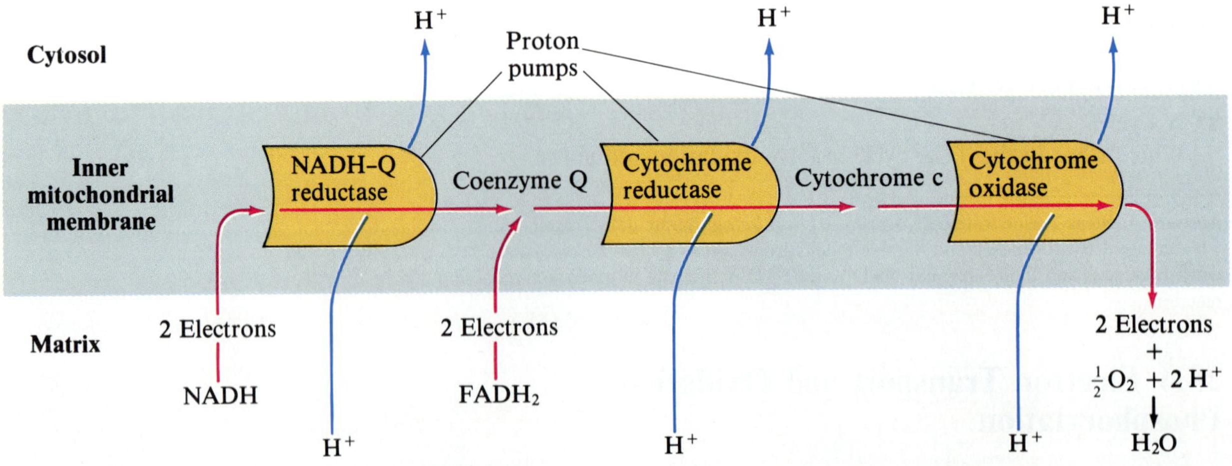
Cytosol
H+
H+
H+
Proton pumps
Inner mitochondrial membrane
NADH–Q reductase
Coenzyme Q
Cytochrome reductase
Cytochrome c
Cytochrome oxidase
Matrix
2 Electrons
NADH
2 Electrons
FADH2
2 Electrons + 1/2 O2 + 2 H+
H2O
H+
H+
H+

FIGURE 28.9
The electron transport chain. As electrons pass from NADH to oxygen (red arrows), they pass through three proton pumps. The pumps are protein complexes that use the energy of this electron flow to pump protons from the mitochondrial matrix into the cytosol (blue arrows). The electrons from $FADH_2$ pass through only two proton pumps, thus fewer protons are transferred. (The number of electrons and protons and the reaction that reduces oxygen to water are not balanced in this illustration.)

used to pump protons (H^+) from the mitochondrial matrix to the outside of the mitochondrion. For each pair of electrons that flows from an NADH to molecular oxygen, three sets of protons are pumped (Figure 28.9). This translocation of protons establishes a gradient that is both chemical and electrical in nature. Whenever a substance has a higher concentration in one region than another, a chemical gradient exists for that substance. If the substance is also electrically charged, as is a proton, then an electrical gradient also exists. Chemical and electrical gradients have potential energy. For example, it is the energy of a gradient that is responsible for the diffusion of particles from a region of higher concentration to a region of lower concentration. The electron transport system of mitochondria is not just a transporter of electrons, it is also a proton pump that derives its energy from that flow of electrons. The pumping of protons in mitochondria is powered by a flow of electrons just as water pumped by an electric pump is powered by a flow of electrons (electricity). The chemical energy stored in the reduced state of NADH is converted to electrical energy that is then used to establish a gradient possessing stored energy.

The energy released when NADH is oxidized is used to establish and maintain an electrochemical gradient.

The oxidation of $FADH_2$ to FAD by oxygen also yields energy, but less than the oxidation of NADH: 43.4 kcal/mol versus 52.6 kcal/mol. This smaller quantity of energy pumps fewer protons. Figure 28.9 shows that the electrons from $FADH_2$ enter the electron transport system at a different point than the electrons from NADH. When $FADH_2$ is oxidized, only two sets of protons are pumped. The energy from the oxidation of NADH drives three proton pumps, while the energy from the oxidation of $FADH_2$ drives two of the pumps.

Oxidative Phosphorylation

ATP is synthesized by the phosphorylation of ADP. When this synthesis is associated with the electron transport system, it is called *oxidative phosphorylation.* It is observed that three ATPs are synthesized for each NADH that is oxidized, and two ATPs are formed for each oxidation of an $FADH_2$.

$$\text{NADH} + \tfrac{1}{2}\,O_2 + H^+ + 3\ \text{ADP} + 3\ P_i \longrightarrow \text{NAD}^+ + H_2O + 3\ \text{ATP}$$
$$\text{FADH}_2 + \tfrac{1}{2}\,O_2 + 2\ \text{ADP} + 2\ P_i \longrightarrow \text{FAD} + H_2O + 2\ \text{ATP}$$

This ratio of three to two is the same as the ratio of protons pumped when NADH and $FADH_2$ are oxidized. The amount of ATP synthesized is a direct result of the proton gradient, which in turn is a reflection of the amount

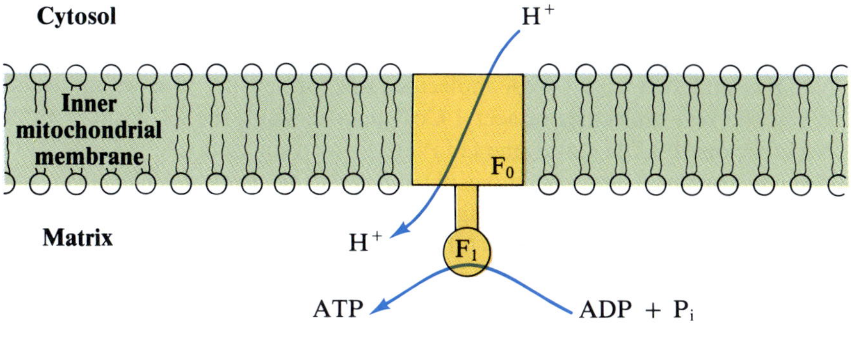
Cytosol
H+
Inner mitochondrial membrane
F0
Matrix
H+
F1
ATP
ADP + Pi

FIGURE 28.10
ATP synthetase consists of protein particles F_1, F_0, and a connecting stalk. As protons pass through F_0, the electrochemical gradient is reduced slightly. A portion of the energy made available from this proton flow is used to phosphorylate ADP to ATP. ATP synthetase is responsible for most of the cellular synthesis of ATP.

of energy that was available to pump protons. The proton gradient is potential energy that can be used to do work. In oxidative phosphorylation, that energy is used to synthesize ATP from ADP and P_i.

Within the inner mitochondrial membrane are multiple copies of a protein complex called ATP synthetase. In Figure 28.10 ATP synthetase is labeled F_0 and F_1. F_0 is imbedded in the membrane and appears to serve as a channel through which protons can pass from the outside of the membrane to the inside. As protons pass in a controlled fashion through this channel, a portion of the potential energy of the gradient is lost. F_1 uses some of that energy to synthesize ATP from ADP and inorganic phosphate (Figure 28.10).

Energy stored in the proton gradient is used to make ATP from ADP and inorganic phosphate.

If the concentration of ATP is high within a cell, the concentration of ADP is necessarily low, because ATP is made from ADP and P_i.

$$ADP + P_i \rightleftharpoons ATP + H_2O$$

The aerobic synthesis of ATP, oxidative phosphorylation, is highly regulated; if little ADP is present in the cell, little oxidative phosphorylation occurs. If the concentration of ATP decreases, its rate of synthesis through oxidative phosphorylation increases. The gradient of protons is maintained in a similar way. If oxidative phosphorylation is occurring rapidly, then protons rapidly enter the mitochondrion. Electron transport must occur rapidly to maintain the proton gradient. Similarly, if NADH and $FADH_2$ are being rapidly consumed by the electron transport system, then the TCA cycle will become very active to reduce the NAD^+ and FAD that is being formed. While the TCA cycle is very active, considerable acetyl CoA is required, and the activity of glycolysis and pyruvate dehydrogenase or some other pathway must provide it in sufficient quantity. For convenience, the parts of metabolism are studied separately, but metabolism actually functions as a highly integrated, highly regulated process.

Although metabolism is broken down into its components for easy study, it is actually a highly integrated process.

Analysis of the Energy from Glucose

Consider now the origin of the 36 or 38 ATPs that are formed from a glucose molecule. Glycolysis produces two ATPs and two NADHs. These NADHs are produced in the cytosol, but NADH cannot enter mitochondria directly. Cells of the body transport the electrons of cytosolic NADH into mitochondria by two different mechanisms: (1) the electrons are passed to NAD^+ within the mitochondria to form NADH or (2) the electrons are passed to an FAD to form $FADH_2$ in the mitochondria (Table 28.2). This is an important difference because $FADH_2$ yields one less ATP than NADH; two fewer ATPs are synthesized per glucose if the second transport mechanism is used.

The amount of ATP formed from aerobic catabolism of glucose depends on how glycolytic NADH is recycled.

When pyruvate is converted to acetyl CoA, an NADH is formed. Two NADHs and two acetyl CoA molecules are formed per glucose molecule. When the two molecules of acetyl CoA pass through the TCA cycle, three NADHs, one $FADH_2$, and one GTP are formed for each acetyl CoA. After electron transport and oxidative phosphorylation, twelve ATPs are made per acetyl CoA that passes through the TCA cycle, twenty-four per glucose molecule. In contrast, only two ATPs are formed per glucose by anaerobic catabolism of glucose. The amount of ATP formed from glucose catabolism is shown in Table 28.2

TABLE 28.2
Two Alternatives for Aerobic Production of ATP from Glucose

Electrons passed to NAD^+ to form NADH

2 ATP	from glycolysis.
6 ATP	2 NADHs from glycolysis. Electrons transported into mitochondria to form NADH. Thus, 2 NADH × 3 ATP/NADH =
6 ATP	2 NADHs from conversion of 2 pyruvate to acetyl CoA. Thus, 2 NADH × 3 ATP/NADH =
24 ATP	2 acetyl CoA molecules through TCA cycle yield 12 ATPs each. Thus, 2 acetyl CoA × 12 ATP/acetyl CoA =
38 ATP	

Electrons passed to FAD to form $FADH_2$

2 ATP	from glycolysis.
4 ATP	2 NADHs from glycolysis. Electrons transported into mitochondria to form $FADH_2$. Thus, 2 $FADH_2$ × 2 ATP/NADH =
6 ATP	2 NADIIs from conversion of 2 pyruvate to acetyl CoA. Thus, 2 NADH × 3 ATP/NADH =
24 ATP	2 acetyl CoA molecules through TCA cycle yield 12 ATPs each. Thus, 2 acetyl CoA × 12 ATP/acetyl CoA =
36 ATP	

NEW TERMS

oxidative phosphorylation A process that synthesizes ATP from ADP and inorganic phosphate using energy stored in a proton gradient.

electron transport chain A series of electron carriers that transport electrons from the reduced coenzymes NADH and $FADH_2$ to oxygen. Energy released during this process is used to pump protons.

chemiosmotic theory A theory proposed by Peter Mitchell stating that the energy released by the flow of electrons along the electron transport chain is used to establish a proton gradient. This high energy gradient can be used to do work, including the synthesis of ATP.

TESTING YOURSELF

Electron Transport and Oxidative Phosphorylation

1. Twelve ATPs are obtained from each acetyl CoA molecule that passes through the TCA cycle, yet no ATP is produced directly. Account for the twelve ATPs.

2. When 1 mol of NADH is oxidized to NAD^+, -52.6 kcal of energy is released. When 1 mol of ATP is synthesized from ADP and P_i, $+7.3$ kcal of energy must be added. If three ATPs are synthesized for each NADH oxidized, what is the efficiency of this coupling?
3. Proteins in the inner membrane of a mitochondrion pump protons from inside the mitochondrion to the outside to maintain a proton gradient. What is the source of the energy for this pumping?

Answers **1.** One GTP is produced and is equivalent to one ATP in energy. Three NADHs are produced, and when one NADH is oxidized by oxygen, enough energy is released to synthesize three ATPs. One $FADH_2$ is produced, and oxidation of this molecule by oxygen yields two ATPs. **2.** $[(3 \times 7.3 \text{ kcal/mol})/52.6 \text{ kcal/mol}] \times 100 = 42\%$ **3.** When NADH or $FADH_2$ is oxidized, electrons pass from it to oxygen. The energy released during this oxidation is used to pump protons.

28.5 Lipid Catabolism

Adipose tissue plays an important role in storing and releasing lipids, in much the same way as the liver stores glucose as glycogen and releases it as needed. Earlier it was stated that triacylglycerols are hydrolyzed in adipose cells to glycerol and fatty acids (see Section 28.1). The fatty acid molecules are released into the blood, where they bind to serum albumin, which is a transport protein for these hydrophobic molecules. The cells of the body absorb these free fatty acids as needed. The concentration of fatty acids in the blood is rather low, only a few milligrams per 100 mL, but they are constantly being absorbed by the cells of the body and replaced by new fatty acids from the adipose cells. Serum fatty acids are in rapid dynamic equilibrium with fatty acids in body and adipose cells. A significant amount of the energy needs of a human are met by catabolism of fatty acids.

Free fatty acids bound to serum albumin are constantly available to the cells of the body.

The glycerol produced by hydrolysis of triacylglycerols is readily catabolized. Glycerol is a carbohydrate, not a lipid. Within cells, glycerol is readily converted to glycerol 3-phosphate. This is oxidized to dihydroxyacetone phosphate, which is an intermediate of glycolysis (see Figure 28.6).

Once inside the cell, the water insoluble fatty acids are bound to a coenzyme A molecule which serves as a carrier. This reaction is an energy-requiring reaction in which concurrent ATP hydrolysis to adenosine monophosphate (AMP) and pyrophosphate (PP_i) provides the energy. Because two P_i molecules must be added to AMP to get back to ATP, hydrolysis of ATP to AMP is equivalent in energy to two high energy bonds.

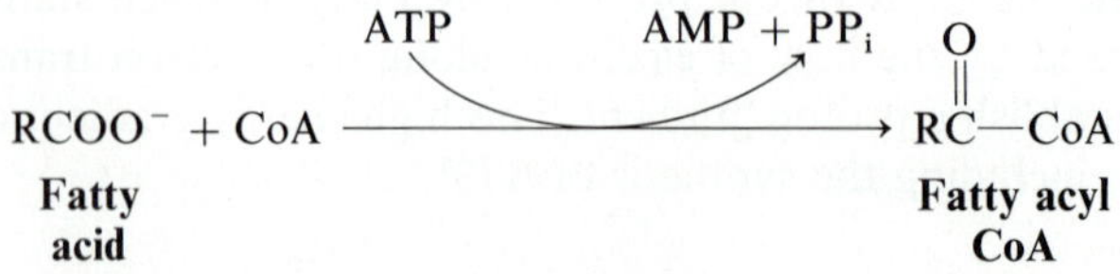
ATP
AMP + PPi
RCOO⁻ + CoA
RC—CoA
O
Fatty acid
Fatty acyl CoA

Fatty acyl CoA molecules diffuse through the cytosol to mitochondria where they are split. The fatty-acid part is transported into the mitochondrion bound to carnitine, which is $^+N(CH_3)_3CH_2CHOHCH_2COO^-$. Inside the mitochondrion the fatty acid is rebound to CoA.

$$\underset{\text{(Cytosol)}}{\text{Fatty acyl CoA + Carnitine}} \longrightarrow \underset{\text{(Cytosol)}}{\text{Fatty acyl carnitine + CoA}}$$

$$\underset{\text{(Cytosol)}}{\text{Fatty acyl carnitine}} \longrightarrow \underset{\text{(Mitochondrion)}}{\text{Fatty acyl carnitine}}$$

$$\underset{\text{(Mitochondrion)}}{\text{Fatty acyl carnitine + CoA}} \longrightarrow \underset{\text{(Mitochondrion)}}{\text{Fatty acyl CoA + Carnitine}}$$

Beta Oxidation of Fatty Acids

Complete lipid catabolism is an aerobic process that occurs within mitochondria. The series of reactions that catabolize fatty acids constitute a cycle or spiral called **beta oxidation**. In each cycle, the fatty acid is oxidized and cleaved, yielding a smaller fatty acid, an acetyl CoA, an NADH, and a $FADH_2$. In essence, beta oxidation catabolizes fatty acids just as glycolysis catabolizes sugars.

Beta oxidation is the principal catabolic pathway for fatty acids.

The compound palmitoyl CoA (Figure 28.11) will be used to illustrate the steps of beta oxidation. In the first step, palmitoyl CoA is oxidized to an unsaturated compound by removal of two hydrogen atoms. This oxidation is accompanied by the reduction of FAD to $FADH_2$ (Figure 28.11). This oxidation is similar to the oxidation of succinate to fumarate in the TCA cycle (see Section 28.3). The name *beta oxidation* signifies that the oxidation occurs between the β- and γ-carbons of the acyl CoA.

The unsaturated product of the first step is hydrated by the addition of water to the double bond. This addition of water yields an alcohol, just as water was added to fumarate to make malate in the TCA cycle. The specificity of the enzyme ensures that the hydroxyl group is added to the β-carbon of the intermediate (Step 2, Figure 28.11). In Step 3, the β-hydroxy group is oxidized to a β-keto group. This oxidation is accompanied by the reduction of NAD^+ to NADH. This intermediate has the same number of carbon atoms as the starting acyl CoA, but the β-carbon atom has been oxidized

TAY-SACHS DISEASE

One of the better known genetic disorders associated with lipid metabolism is *Tay-Sachs disease.* Gangliosides, which are glycolipids derived from sphingosine, are common in healthy brain tissue. Gangliosides are constantly being made and broken down in tissues. But in individuals with Tay-Sachs disease, the enzyme hexosaminidase A is missing. This enzyme is responsible for the cleavage of a sugar derivative from a specific ganglioside.

$$\text{Tay-Sachs ganglioside} \xrightarrow{\text{Hexosaminidase A}} \text{Hematoside} + N\text{-Acetylneuraminate}$$

Because this enzyme is not present, the concentration of Tay-Sachs ganglioside builds up in the brain. This accumulation results in paralysis, blindness, and early death, typically by the age of four. Symptoms appear in the first several months of life. The patients typically have enlarged heads with doll-like faces and a characteristic alteration of the retina. No cure or treatment has been found for this disease at this time. However, the condition can be detected in embryos by testing for hexosaminidase A activity in amniotic fluid.

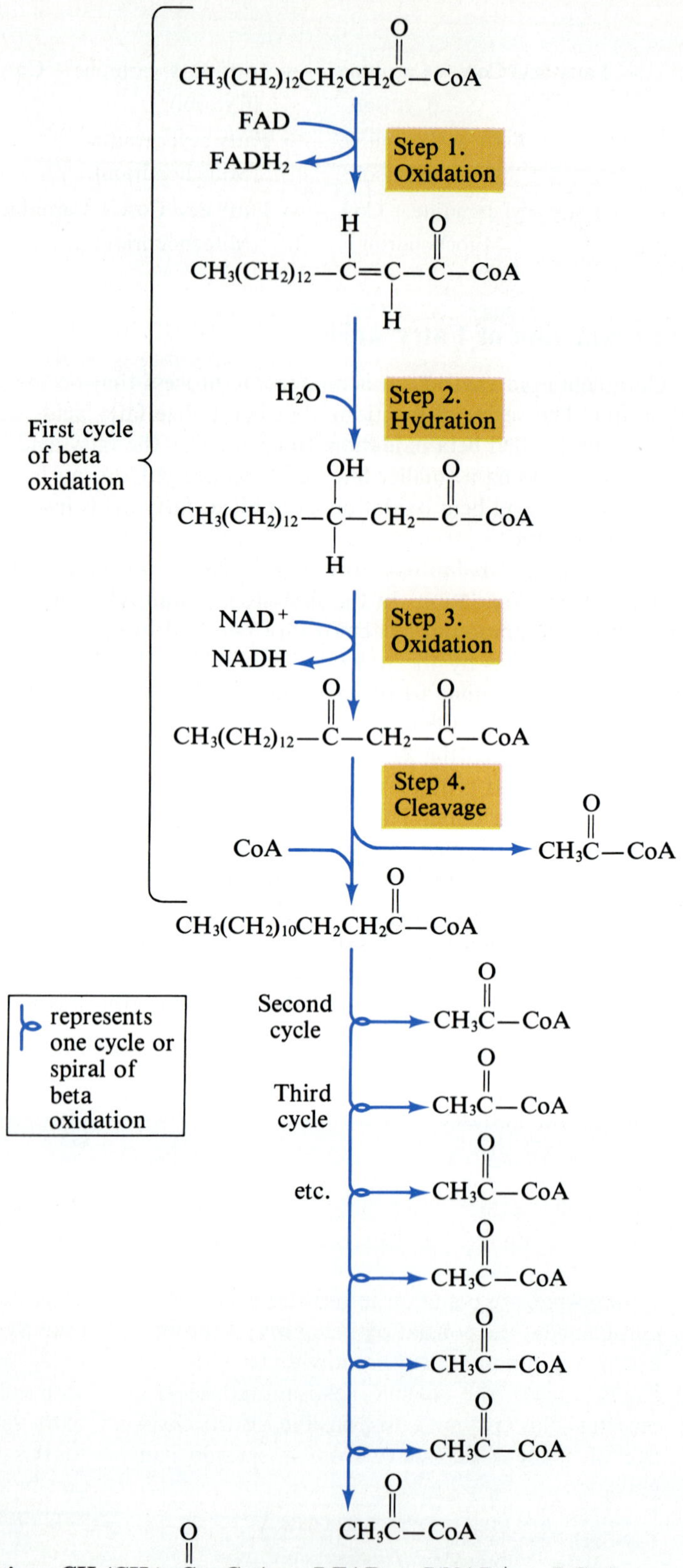

First cycle of beta oxidation
FAD
FADH₂
Step 1. Oxidation
H₂O
Step 2. Hydration
NAD⁺
NADH
Step 3. Oxidation
Step 4. Cleavage
CoA
Second cycle
Third cycle
etc.
represents one cycle or spiral of beta oxidation

FIGURE 28.11
Beta oxidation of fatty acids. The specific steps are noted in the text, but notice that one more acetyl CoA (eight) is produced than $FADH_2$ and NADH (seven each).

Net reaction: $$CH_3(CH_2)_{14}\overset{O}{\overset{\|}{C}}—CoA + 7\,FAD + 7\,NAD^+ + 7\,CoA \longrightarrow 8\,CH_3\overset{O}{\overset{\|}{C}}CoA + 7\,FADH_2 + 7\,NADH$$

from the oxidation number of an alkane to that of a carbon atom of a carbonyl group.

In the fourth step of beta oxidation, the intermediate is cleaved into two smaller molecules. The bond between the α- and β-carbon atoms is broken, and the β-carbon atom is transferred to a molecule of coenzyme A to form myristoyl CoA, the CoA derivative of myristic acid. The rest of the intermediate leaves the reaction as acetyl CoA. The product, myristoyl CoA, is a compound identical to the starting material except that it has two fewer carbon atoms. These four steps of beta oxidation can be repeated on myristoyl CoA to yield lauroyl CoA, which in turn is beta oxidized. Each subsequent acyl CoA is oxidized and cleaved until the final step cleaves butyryl CoA, yielding two molecules of acetyl CoA. From palmitoyl CoA, eight molecules of acetyl CoA are produced along with seven molecules each of $FADH_2$ and NADH. Acetyl CoA and NADH are also produced when glucose is catabolized aerobically. This is an important feature of catabolic pathways. Although the starting compounds may be very different, the products are generally one of a few small, common molecules found in the cell.

Beta oxidation of fatty acids yields acetyl CoA, NADH, and $FADH_2$.

Recall that the common fatty acids found in nature have an even number of carbon atoms. Beta oxidation of these compounds always yields acetyl CoA. The unsaturated fatty acids are also oxidized by beta oxidation, but the product yield is different and a few additional enzymes are needed. Each double bond in a fatty acid reduces the yield of $FADH_2$ by one. An $FADH_2$ is produced whenever a carbon–carbon double bond is formed during beta oxidation; if the bond already exists in the molecule, then the reaction does not occur.

Energy Production from Fatty Acids

Consider the energy production from a molecule of palmitic acid that passes through beta oxidation as palmitoyl CoA. This sixteen-carbon fatty acid is cleaved into eight acetyl CoA molecules, each of which yields twelve ATP when it is metabolized in the TCA cycle (Table 28.3). Beta oxidation also yields NADH and $FADH_2$ directly, but there is one fewer of each of these products than acetyl CoA because the final step of beta oxidation yields two acetyl CoA molecules. These molecules also yield ATP when they are

TABLE 28.3
Production of ATP from Palmitic Acid

96 ATP	8 acetyl CoA × 12 ATP/acetyl CoA =
21 ATP	7 NADH × 3 ATP/NADH =
14 ATP	7 $FADH_2$ × 2 ATP/$FADH_2$ =
131 ATP	Gross production of ATP from palmitic acid
−2 ATP	Activation of fatty acid for entry into beta oxidation =
129 ATP	Net production of ATP from palmitic acid

oxidized (Table 28.3). But there is a cost associated with beta oxidation; an ATP was broken down into an AMP when the fatty acid is attached to coenzyme A. The hydrolysis of ATP to AMP reduces the net yield of high energy bonds by two.

Fatty acids can also be oxidized at other atoms besides the β-carbon. Alpha oxidation yields fatty α-keto acids, and omega oxidation yields dicarboxylic acids. These pathways are quantitatively less important than beta oxidation, but they do yield essential products.

Other Lipid Catabolism

Triacylglycerols are by far the most abundant class of lipids in the diet. Other saponifiable lipids are hydrolyzed to fatty acids that are catabolized by beta oxidation. There are specific catabolic pathways for many of the other components, but they are quantitatively less important and too numerous for a detailed study. The catabolism of cholesterol deserves comment because cholesterol has such an apparent role in health. Small amounts of cholesterol are converted to steroid hormones, and some more is lost in the digestive tract. But the principal route for catabolism of cholesterol, about three-fourths of it, is its conversion to bile salts. These compounds emulsify dietary lipid in the small intestine and are then reabsorbed from the small intestine. The reabsorption is not 100% efficient, however; amounts of cholesterol must be continually converted to bile salts.

NEW TERMS

beta oxidation Metabolic pathway that oxidizes fatty acids to acetyl CoA and reduced coenzymes.

TESTING YOURSELF

Lipid Catabolism

1. Use Table 28.2 as a guide to determine the production of ATP by beta oxidation of stearic acid.
2. Triacylglycerols are composed of three fatty acids and a glycerol molecule. These components are catabolized in what pathways?

Answers **1.** Stearoyl CoA will yield 9 acetyl CoA, 8 NADH, and 8 $FADH_2$. These yield 108, 24, and 16 ATP, respectively, for a total gross production of 148 ATP. Two ATP equivalents are required to activate the fatty acid, thus the net production is 146 ATP. **2.** The fatty acids are catabolized by beta oxidation. The glycerol is converted to dihydroxyacetone phosphate, which is an intermediate of glycolysis.

28.6 Amino Acid Catabolism

Amino acids derived from dietary protein are absorbed by the blood from the small intestines and join a pool of amino acids that circulate in the blood. A portion of these amino acids are used to synthesize proteins within cells of the body, but most diets provide more amino acids than are needed for protein synthesis. Excess amino acids are metabolized for energy or converted to other molecules.

Transamination of Amino Acids

The catabolism of amino acids ultimately requires twenty different pathways because the structures of the amino acids are so diverse. There are, however, several features of amino acid catabolism that are common to all of the amino acids. The first of these features is **transamination**. In this reaction, the α-amino group of an amino acid is transferred to an acceptor molecule, α-ketoglutarate (α-KG).

Catabolism of amino acids begins with the removal of amino groups via transamination.

$$\text{Amino acid} + \alpha\text{-KG} \xrightarrow{\text{Transaminase}} \text{Keto acid} + \text{Glutamate}$$

In the cells of the liver, several transaminases are present. These enzymes collectively transfer the amino groups of all of the amino acids to α-KG. These transaminations effectively collect all of the nitrogen of the α-amino groups into a single molecule, glutamate, and generate keto acids that are the substrates for the specific catabolic pathways. The role of transamination in amino acid catabolism is shown in Figure 28.12.

Oxidative Deamination

The process of transamination collects nitrogen atoms as amino groups in glutamate. The nitrogen of these amino groups is not needed by the body and must be eliminated safely and efficiently. The next step leading to nitrogen excretion is **deamination**. In this reaction, the enzyme glutamate dehydrogenase catalyzes the deamination of glutamate to α-KG.

$$\text{Glutamate} + NAD^+ \xrightleftharpoons{\text{Glutamate dehydrogenase}} \alpha\text{-KG} + NADH + NH_3$$

This reaction recycles glutamate to α-KG and effectively converts what had been the amino groups of amino acids to ammonia (Figure 28.12). This does pose a potential problem because ammonia is toxic. Note that the deamination reaction is reversible. If the concentration of ammonia gets high

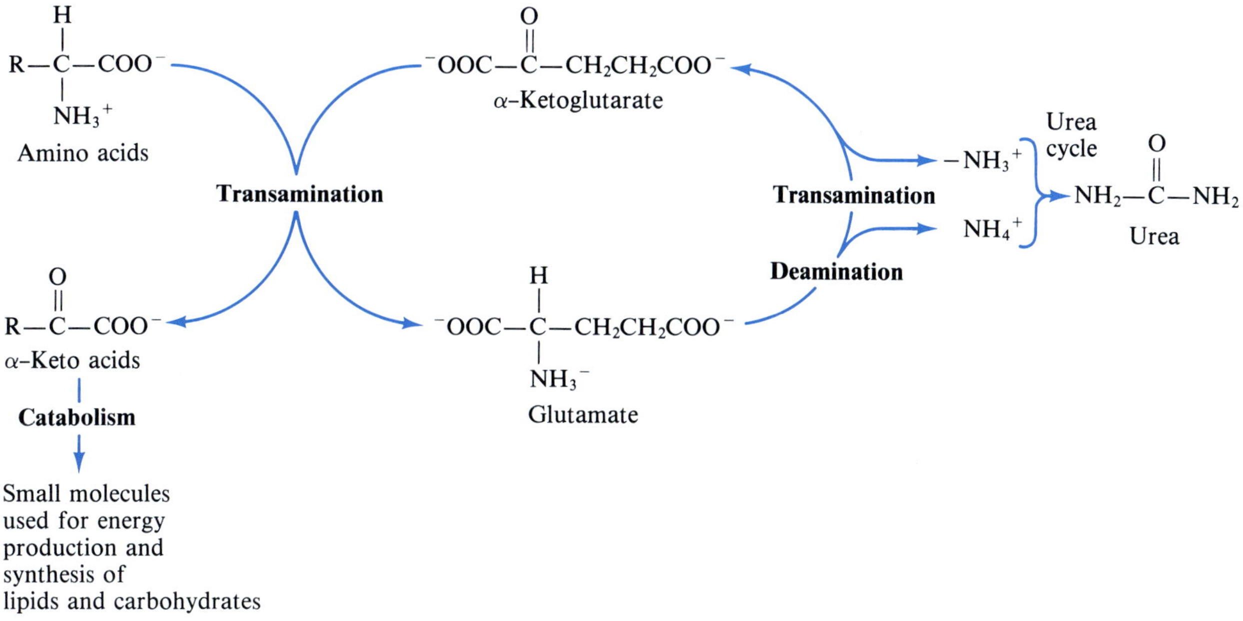

Amino acids
α-Ketoglutarate
Transamination
Transamination
Deamination
Urea cycle
Urea
α-Keto acids
Catabolism
Small molecules used for energy production and synthesis of lipids and carbohydrates
Glutamate

FIGURE 28.12
An overview of amino acid catabolism. Transamination yields α-keto acids that are metabolized to smaller molecules. The α-amino group from the amino acids combines with α-ketoglutarate to form glutamate. The amino groups of glutamate molecules are removed via transamination and deamination reactions. The nitrogenous products of these reactions are used to make urea.

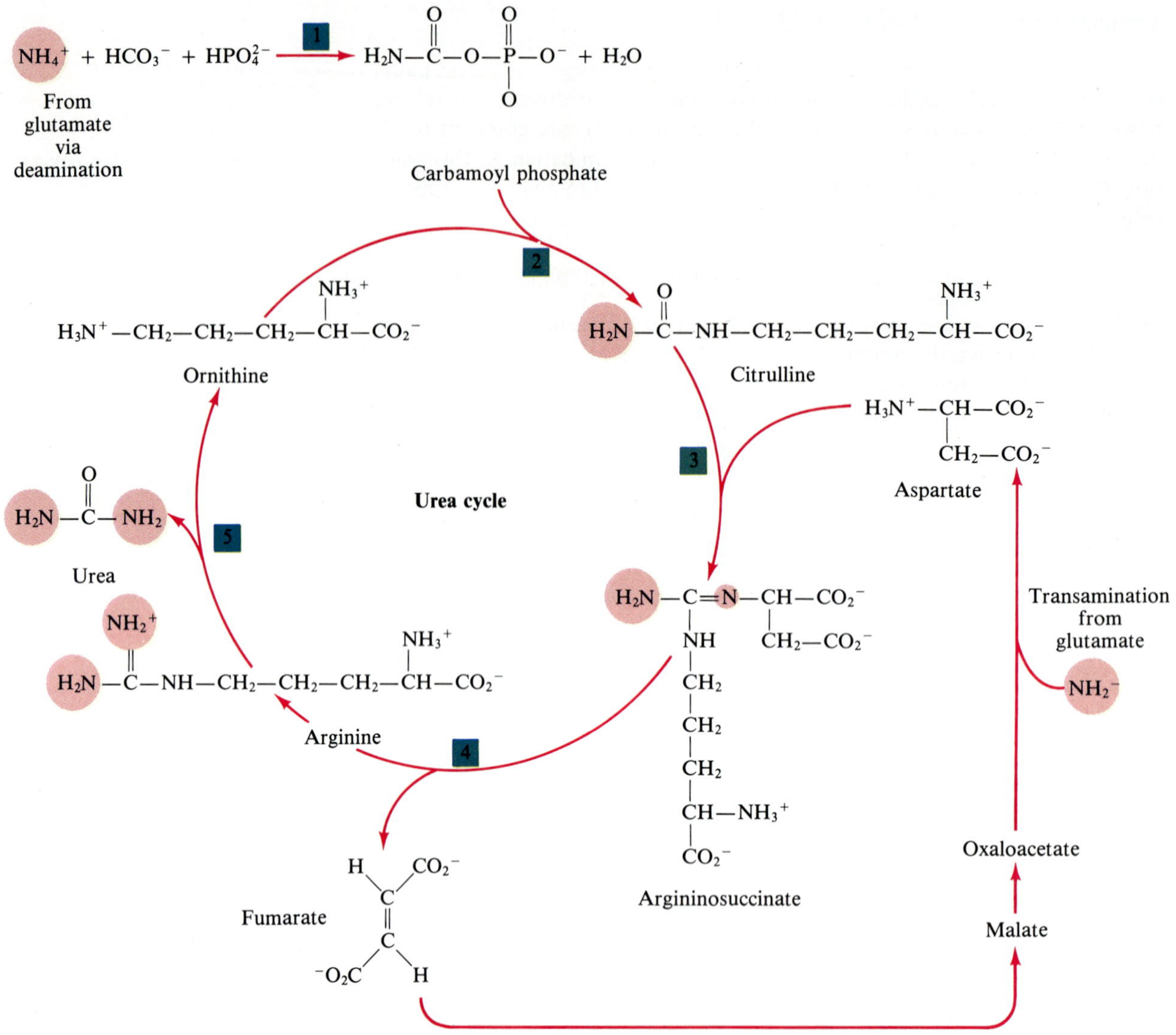

NH4+ + HCO3− + HPO4 2− → H2N—C—O—P—O− + H2O
From glutamate via deamination
Carbamoyl phosphate
1
2
3
4
5
Urea cycle
Ornithine
Citrulline
Aspartate
Urea
Arginine
Argininosuccinate
Fumarate
Transamination from glutamate
Oxaloacetate
Malate

FIGURE 28.13
The urea cycle. The two nitrogen atoms of urea both come from glutamate, one via deamination to yield ammonium ion, the other via transamination to oxaloacetate which yields aspartate.

enough, the reverse reaction will occur, and cellular α-KG will be converted to glutamate. α-KG is an intermediate of the TCA cycle, and if it is not available, the cell cannot carry out the aerobic metabolism that is responsible for most of the available energy in cells. Some fish and other aquatic species that have abundant environmental water available simply excrete ammonia directly into the environment before the concentration builds up to toxic levels. Other species, including mammals, convert the ammonia and other excess metabolic nitrogen to less toxic nitrogenous compounds that can be safely excreted.

Urea Cycle

The relatively nontoxic nitrogenous waste urea is synthesized in the urea cycle.

The nitrogenous excretion product of mammals and some other groups is urea, which is synthesized in the **urea cycle** (Figure 28.13). The ammonia released from glutamate is combined with carbon dioxide to form *carbamoyl phosphate*. This endothermic reaction is coupled to the hydrolysis of ATP.

$$NH_3 + CO_2 + 2\,ATP \longrightarrow H_2N\overset{O}{\overset{\|}{C}}O\overset{O}{\overset{\|}{\underset{\underset{OH}{|}}{P}}}{-}O^- + 2\,ADP + P_i$$

Carbamoyl phosphate

Carbamoyl phosphate enters the urea cycle by reacting with ornithine to form citrulline. Citrulline reacts with aspartate to form a molecule of arginosuccinate. Note in Figure 28.13 that citrulline is bonded to aspartate through the amino group of aspartate. The nitrogen of this amino group and the nitrogen of ammonia are ultimately the nitrogen atoms that are found in urea. Arginosuccinate is cleaved to yield arginine and fumarate. The fumarate is recycled to aspartate, with the amino group of aspartate coming from glutamate through a transamination reaction. Both the nitrogen atoms of urea ultimately come from glutamate.

$$\text{Oxaloacetate} + \text{Glutamate} \xrightarrow{\text{Transaminase}} \text{Aspartate} + \alpha\text{-Ketoglutarate}$$

The arginine that is formed is cleaved to yield urea, the product of the cycle, and ornithine, which can combine with carbamoyl phosphate to repeat the cycle.

Catabolism of Keto Acids

The specific reactions that break down the keto acids formed from the twenty amino acids are too detailed to pursue. However, the products of these pathways are compounds you have seen before. Figure 28.14 shows

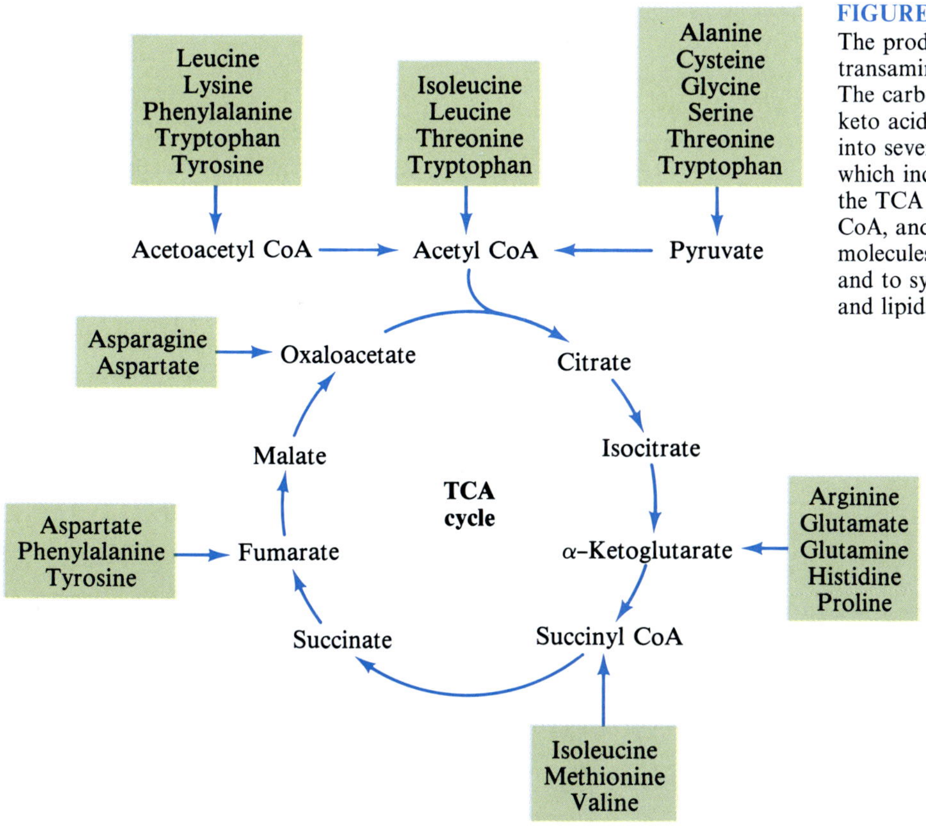
Leucine
Lysine
Phenylalanine
Tryptophan
Tyrosine
Isoleucine
Leucine
Threonine
Tryptophan
Alanine
Cysteine
Glycine
Serine
Threonine
Tryptophan
Acetoacetyl CoA
Acetyl CoA
Pyruvate
Asparagine
Aspartate
Oxaloacetate
Citrate
Malate
Isocitrate
TCA cycle
Aspartate
Phenylalanine
Tyrosine
Fumarate
α-Ketoglutarate
Arginine
Glutamate
Glutamine
Histidine
Proline
Succinate
Succinyl CoA
Isoleucine
Methionine
Valine

FIGURE 28.14
The products of amino acid transamination are keto acids. The carbon skeletons of the keto acids are broken down into several smaller metabolites, which include intermediates of the TCA cycle, pyruvate, acetyl CoA, and acetoacetyl CoA. These molecules can be used for energy and to synthesize carbohydrates and lipids (see Chapter 29).

The various carbon skeletons found in the twenty amino acids are broken down into seven common biomolecules.

these compounds and the amino acids they came from. Note that the amino acids are degraded to intermediates of the TCA cycle, pyruvate, acetyl CoA, and acetoacetyl CoA. The dozens of different carbon atoms of twenty amino acids appear in seven common metabolites.

Pyruvate and the TCA cycle intermediates can be consumed for energy, but they can also be used to synthesize glucose and other carbohydrates through gluconeogenesis. This topic is discussed in the next chapter.

NEW TERMS

transamination Reaction that transfers amino groups from amino acids to α-ketoglutarate to yield keto acids and glutamate.

deamination Reaction that removes the α-amino group from glutamate yielding free ammonia and α-ketoglutarate.

urea cycle Pathway that converts excess metabolic nitrogen to relatively nontoxic urea.

TESTING YOURSELF

Amino Acid Catabolism

1. Urea contains two nitrogen atoms. What is the direct source of these atoms in the urea cycle? What is the ultimate source of these atoms?

Answer 1. One comes from carbamoyl phosphate and the other comes from aspartate. They ultimately come from the α-amino groups of amino acids via glutamate through transamination reactions.

SUMMARY

The reactions of *catabolism* convert many large, complex molecules to a smaller set of simpler ones. These reactions are generally oxidative and energy yielding. Catabolism occurs within cells, but digestion prepares dietary molecules for cellular absorption by hydrolyzing food molecules into smaller molecules. *Digestion* yields monosaccharides from carbohydrates, amino acids from proteins, and fatty acids and glycerol from triacylglycerols. These products of digestion are absorbed by mucosal cells and released into the circulatory system. Cells take in these nutrients as needed. *Glycolysis* is the principal catabolic pathway for sugars and yields pyruvate as the product. Glycolysis occurs in the presence or absence of oxygen.

Lactate is produced from pyruvate in the absence of oxygen. When oxygen is available, pyruvate is converted to acetyl CoA, which is consumed in the *TCA cycle*. This cycle produces carbon dioxide, GTP, and the reduced coenzymes NADH and $FADH_2$. These reduced compounds are oxidized back to NAD^+ and FAD via the *electron transport chain*. The energy produced by this oxidation is used to pump protons from the mitochondria to maintain an electrochemical gradient. The energy stored in this gradient is used by ATP synthetase to make ATP from ADP and P_i. Fatty acid catabolism yields acetyl CoA, $FADH_2$, and NADH. These products are consumed in the TCA cycle and electron transport chain. Amino acid catabolism begins with removal of amino groups via *transamination*. The nitrogen from these amino groups is used to make urea, and the carbon skeletons of the amino acids are catabolized to several intermediates of metabolism including acetyl CoA. Catabolism of sugars, lipids, and amino acids yields the same small set of metabolites.

TERMS

digestion
amylases
chyme
proteases
peptidases
bile salts
lipases
lymph
lipoproteins
gylcolysis
respiration
fermentation
tricarboxylic acid or TCA cycle
oxidative phosphorylation
electron transport chain
chemiosmotic theory
beta oxidation
transamination
deamination
urea cycle

CHAPTER REVIEW

QUESTIONS

1. After digestion occurs, dietary carbohydrates are in what form?
   a. free amino acids c. monosaccharides
   b. disaccharides d. polysaccharides
2. In the absence of oxygen, the net energy yield from glycolysis in ATP/glucose is
   a. 1 b. 2 c. 4 d. 7
3. If no oxygen is present, the pyruvate produced by glycolysis is converted to
   a. acetyl CoA c. glycerol
   b. fatty acids d. lactate
4. Acetyl CoA that enters the TCA cycle comes from what source?
   a. amino acids c. glucose
   b. fatty acids d. all of the above
5. The catabolism of acetyl CoA by the TCA cycle yields enough energy to make how many ATPs?
   a. 2 b. 4 c. 12 d. 36
6. The electrons lost by NADH during oxidation via the electron transport chain are accepted by
   a. oxygen c. pyruvate
   b. lactate d. acyl CoA
7. The storage of energy as an electrochemical gradient is a feature of
   a. the TCA cycle c. the urea cycle
   b. the chemiosmotic theory d. pyruvate oxidation
8. Beta oxidation refers to oxidation of
   a. beta amino acids c. cholesterol
   b. glucose d. fatty acids
9. During amino acid catabolism, the amino groups of amino acids are transferred directly to what acceptor molecule?
   a. aspartate c. succinate
   b. α-ketoglutarate d. urea
10. The catabolism of sugars, fatty acids, and amino acids yield
    a. polymers and macromolecules c. small molecules and energy
    b. cholesterol and other lipids d. enzymes

Answers **1.** c **2.** b **3.** d **4.** d **5.** c **6.** a **7.** b **8.** d **9.** b **10.** c

DIAGNOSTIC CHART

Blacken in all of the circles under the number of each question that you missed in the Chapter Review questions. The diagnostic chart will help you identify concept areas that need more study.

	Questions									
Concepts	**1**	**2**	**3**	**4**	**5**	**6**	**7**	**8**	**9**	**10**
Digestion	○									
Carbohydrate catabolism		○	○	○						○
TCA cycle				○	○					
Electron transport and oxidative phosphorylation						○	○			
Lipid catabolism				○				○		○
Amino acid catabolism				○					○	○

EXERCISES

Digestion

1. How does chewing aid digestion?
2. Only a small amount of starch digestion occurs before food enters the small intestine. Why?
3. What are the roles of the enzymes maltase, sucrase, and lactase?
4. Why do digestive proteases not break down the stomach and pancreas cells that synthesize them?
5. What is the function of the highly acidic environment of the stomach?
6. What is gained by the emulsification of dietary lipids? What compounds serve as emulsifying agents?
7. Describe why lipoproteins are needed for lipid transport in the blood.
8. List the major lipoproteins and the types of lipids they transport.
9. Complete the following table.

Food Class	Enzymes	Sites of Digestion	Products
Carbohydrates		Mouth and small intestine	
Proteins	Proteases and peptidases		
Lipids			Fatty acids and glycerol

10. Describe where and how monosaccharides are absorbed.

Glycolysis

11. What steps of glycolysis are energy requiring? What is the source of energy for these reactions?
12. What steps of glycolysis are energy yielding? What molecule stores this energy? During glycolysis, how many of these molecules are formed per mole of glucose?
13. Rapid anaerobic catabolism of glucose may lower cellular and blood pH. Why?
14. Other monosaccharides besides glucose enter glycolysis. Where do these other monosaccharides enter into glycolysis?
15. Write an equation for the conversion of glycogen to an intermediate of glycolysis.
16. What fermentation occurs in human muscle cells?

TCA Cycle

17. How is pyruvate converted to acetyl CoA?
18. Compare acetyl CoA, which is the reactant in the TCA cycle, to the carbon-containing products of the TCA cycle. From an inspection of these compounds, what do you think is the role of the TCA cycle?
19. What roles do NAD^+ and FAD serve in the TCA cycle and other metabolic pathways? How many of each coenzyme are involved in one cycle of the TCA cycle?

Electron Transport and Oxidative Phosphorylation

20. What is meant by the expression "oxidative phosphorylation"?
21. How are NADH and $FADH_2$ oxidized in mitochondria?
22. Why does oxidation of NADH yield 3 ATPs but oxidation of $FADH_2$ yield only two?

23. The acceptance and donation of electrons is only one function of the electron transport chain components. What is the other one?
24. Describe the chemiosmotic theory.
25. What enzyme uses the energy stored in a proton gradient to drive the reaction that makes ATP? Describe the location and spatial arrangement of this enzyme.
26. What effects would a high concentration of ATP have on the activity of catabolic pathways?

Lipid Catabolism

27. What role does serum albumin play in fatty acid metabolism?
28. Fatty acids are not water soluble. How are they transported in cells and mitochondria?
29. Compare the ATP yield from palmitic acid (16:0) and palmitoleic acid (16:1).

Amino Acid Catabolism

30. What are the two most common fates for dietary amino acids?
31. From the perspective of waste disposal, what is accomplished by transamination during catabolism of amino acids?
32. Name the reactants, products, and enzyme for deamination. Since NAD^+ is converted to NADH, is this process oxidative or reductive?

General Questions

33. Where does respiration occur in human cells?
34. The catabolism of glucose is often shown as

$$C_6H_{12}O_6 + 6\,O_2 \longrightarrow 6\,CO_2 + 6\,H_2O$$

Where in metabolism are the carbon dioxide molecules and water molecules produced?

35. Some cells obtain 38 moles of ATP per mole of glucose oxidized, while others obtain 36. What is the basis for this difference?
36. Where does oxidative catabolism of fatty acids, amino acids, and the products of glycolysis occur? What are the products of these reactions?

CHAPTER

29

Anabolism

Energy provided by catabolism is only one aspect of metabolism. Cells must constantly build biomolecules for growth and to replace cells that are lost or degraded. This chapter discusses how the body uses small molecules, energy, and electrons for reduction reactions to build the molecules required to sustain life.

OBJECTIVES

After completing this chapter, you should be able to

- Define the terms *autotrophic* and *heterotrophic* organisms.
- Describe the light and dark reactions of photosynthesis.
- Describe gluconeogenesis and its metabolic role.
- Summarize the synthesis of glycogen.
- Describe the biosynthesis of fatty acids.
- Describe the synthesis of cholesterol and other steroids.
- Determine the dietary or metabolic origins of the amino acids.

29.1 Introduction to Anabolism

Digestion of foods yields a variety of small molecules that the body can then use for energy production (see Chapter 28) or synthesis of larger molecules. **Anabolism** is that part of metabolism that involves the synthesis of larger molecules needed by the body. To carry out synthetic activity, a cell must have the right enzymes (Chapter 27), energy, small molecules that serve as building blocks, and a source of electrons that can be used for reductions.

Anabolic reactions are those involved in biosynthesis of molecules.

Many of the small molecules that are used for synthesis of larger molecules are acquired through diet and digestion or are produced by catabolism. You have already learned that dietary protein provides the amino acids needed for protein synthesis and that dietary glucose is needed to synthesize glycogen. Intermediates of metabolism may also serve as building blocks. For example, acetyl CoA is used to synthesize fatty acids. Anabolism requires a pool of appropriate building blocks.

Energy for synthesis is provided primarily by adenosine triphosphate (ATP). Energy is provided by coupling energy-requiring synthetic reactions with energy-yielding reactions, such as ATP hydrolysis. Now is a good time to review Section 27.6 in Chapter 27 if you have forgotten the concept of coupled reactions.

ATP provides the energy for many anabolic reactions.

The principal source of electrons and hydrogen ions (reducing power) is the coenzyme **nicotinamide adenine dinucleotide phosphate (NADPH)**. This molecule is identical to NADH except for the presence of an additional phosphate group (highlighted).

NADPH provides the electrons needed for reductive anabolic reactions.

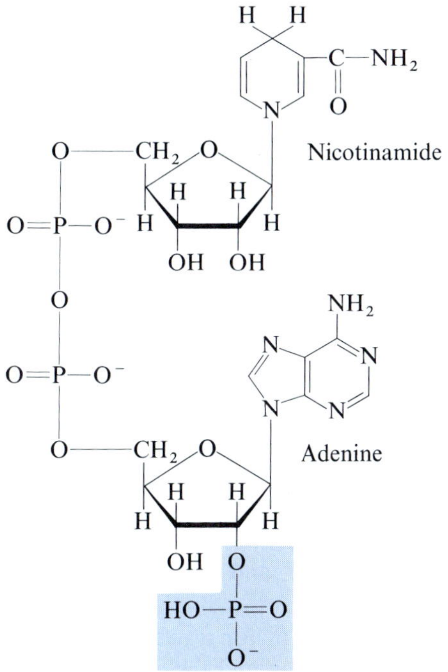
Nicotinamide
Adenine

Like the very similar coenzyme NAD^+, $NADP^+$ alternately accepts and donates a pair of electrons and a hydrogen ion.

$$NADP^+ + :H^- \rightleftharpoons NADPH$$

NADPH must be in its reduced form to be used in biosynthesis. The formation of NADPH is discussed in Sections 29.2 and 29.4.

Catabolism uses NAD^+–NADH, while anabolism uses $NADP^+$–NADPH.

The enzymes involved in the oxidation–reduction reactions of catabolism use NAD^+–NADH, while those involved in anabolism use $NADP^+$–NADPH. This is another example of the order and regulation of metabolism. Each branch of metabolism has its own oxidation–reduction coenzyme that allows for efficient regulation of both anabolism and catabolism. Catabolism works best when the concentration of NAD^+ is high and the concentration of NADH is low. Anabolism works best when the concentration of NADPH is high. By maintaining two pools of oxidation–reduction coenzymes, both anabolism and catabolism can operate simultaneously and efficiently.

Anabolism is a vast and complex field of study. This chapter concentrates on those examples that are of fundamental importance or are involved in human health.

NEW TERMS

anabolism The reactions of the body that synthesize larger, more complex molecules from smaller, simpler ones.

nicotinamide adenine dinucleotide phosphate (NADPH) A coenzyme involved in reduction reactions in anabolism.

29.2 Photosynthesis

Photosynthesis is the ultimate source of all biomolecules on earth.

Heterotrophic organisms are those that feed on other creatures because they cannot obtain energy and building blocks in any other way. Humans, other animals, fungi, and many bacteria are heterotrophic. **Autotrophic organisms**, or "self-feeders," can take simple inorganic molecules such as water and carbon dioxide and ions such as ammonium ion and use energy to convert them to whatever they need. Plants, algae, and some bacteria are autotrophic. Plants and algae use light as their energy source through a chemical process called **photosynthesis**. All life on earth ultimately depends upon photosynthesis; without photosynthetic organisms, all the heterotrophic organisms, including humans, would perish.

Photosynthesis in plants and higher algae occurs in chloroplasts (see Chapter 22) and can be summarized with a single equation:

$$6\,CO_2 + 6\,H_2O \xrightarrow{\text{Light}} \underset{\text{Glucose}}{C_6H_{12}O_6} + 6\,O_2$$

Carbon dioxide is reduced to yield glucose (or some other carbohydrate), and water is oxidized to oxygen. As in Chapter 19, we see that oxidation reactions are coupled with reduction reactions. The photosynthetic reaction requires a great deal of energy, and light provides the needed energy. Although a single equation serves as a summary, the actual process is really many reactions. However, two major reactions are usually used to describe photosynthesis: the light reaction and the dark reaction.

The Light Reaction of Photosynthesis

The synthesis of glucose from carbon dioxide requires energy, provided as ATP, and the reduction of carbon dioxide. The reduction of carbon dioxide, like all reductions, requires electrons. In photosynthesis, water is the

SYNTHESIS OF VITAMIN D

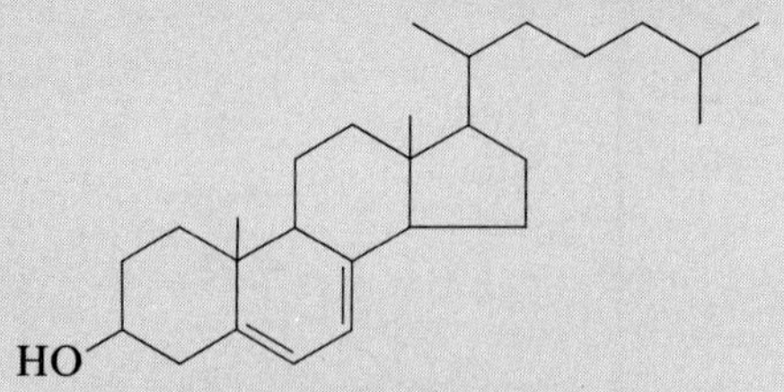

7-Dehydrocholesterol

Ultraviolet light

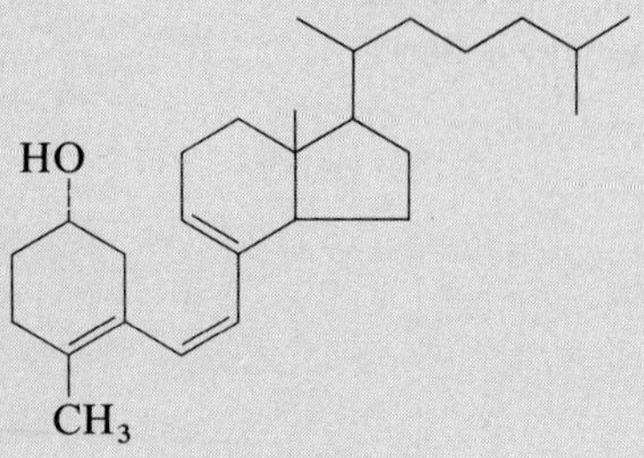

Previtamin D_3

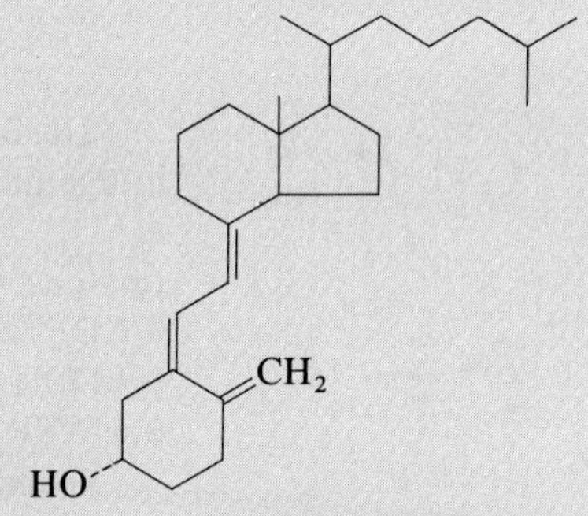

Vitamin D_3

Animals cannot carry out photosynthesis, but they can do some photochemistry. That is, they can carry out some light-dependent reactions. The synthesis of vitamin D is an example. Vitamin D is made from 7-dehydrocholesterol, which is in turn made from cholesterol. (The synthesis of cholesterol is described in Section 29.4.) 7-Dehydrocholesterol is circulated throughout the body in the blood. When blood passes through capillaries in the skin, ultraviolet light is absorbed by 7-dehydrocholesterol. This absorbed energy cleaves a bond in the molecule, yielding previtamin D_3, which isomerizes to vitamin D_3.

The role of sunlight in the synthesis of vitamin D is why it is sometimes referred to as the "sunshine vitamin." The diet of people in polar and cold temperate zones must be fortified with vitamin D, since they may not be exposed to enough sunshine during the winter to make adequate amounts of vitamin D.

ultimate source of these electrons. But water is a very poor reducing agent because it does not readily donate electrons. The **light reaction** uses light energy to strip electrons from certain chlorophyll molecules, and then uses the electrons to convert $NADP^+$ to NADPH. The electron-deficient chlorophyll molecules regain electrons from water.

The light reaction of photosynthesis provides energy (as ATP) and electrons (as NADPH).

The actual path of the electrons can be followed most easily by going backward through the light reaction (Figure 29.1). $NADP^+$ and H^+ gain electrons from a series of electron carriers that obtain electrons from a chlorophyll molecule in photosystem I. The transfer of electrons from photosystem I to the first of these carriers is endothermic. Energy from two photons of light is needed to strip two electrons from photosystem I. This leaves photosystem I electron deficient. It in turn gains electrons from a series of

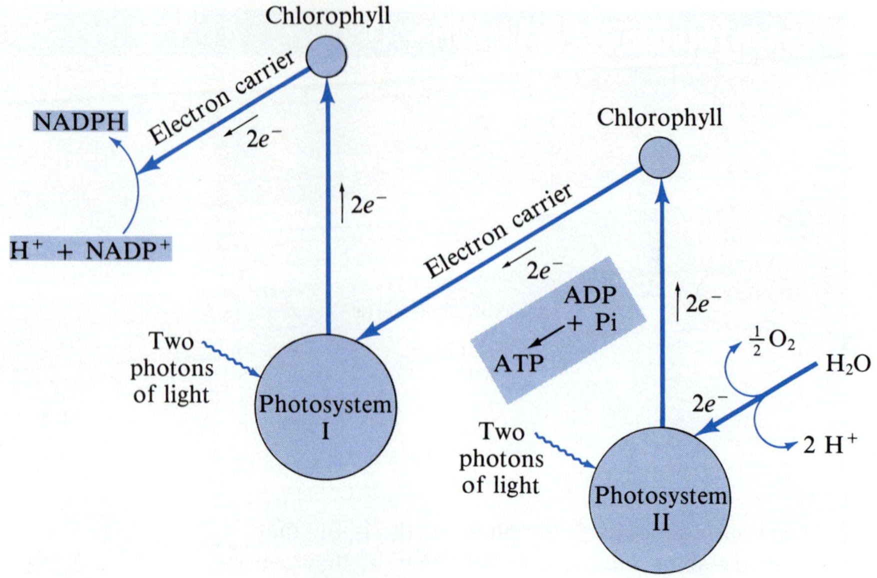
Chlorophyll
Electron carrier
2e⁻
NADPH
H⁺ + NADP⁺
2e⁻
Chlorophyll
Electron carrier
2e⁻
ADP
+ Pi
ATP
2e⁻
½O₂
H₂O
2e⁻
2 H⁺
Two
photons
of light
Photosystem
I
Two
photons
of light
Photosystem
II

FIGURE 29.1
The light reaction of photosynthesis. Electrons from water pass through a series of electron carriers to $NADP^+$. Light provides the energy for this reaction. As electrons pass from photosystem II to photosystem I, protons are pumped to establish a proton gradient, which is stored energy. This energy can be used to synthesize ATP in a manner analogous to ATP synthesis in mitochondria.

electron carriers that get electrons from a chlorophyll molecule in photosystem II. Again, two photons of light are needed to strip two electrons from photosystem II. Photosystem II in turn strips electrons from water, leaving molecular oxygen. The net light reaction can be summarized by this equation:

$$H_2O + NADP^+ \xrightarrow[\text{}]{\text{Light energy}} NADPH + \tfrac{1}{2}O_2 + H^+$$

During this reduction of $NADP^+$, ATP is also made from ADP and inorganic phosphate. In the light reaction, electrons flow from water to $NADP^+$ through a series of electron carriers (Figure 29.1). Note that four photons of light are needed to move the pair of electrons from water to $NADP^+$. There is more than enough energy in four photons to reduce NADPH, however. A part of the energy is used to make some ATP. At least one ATP is made for each NADPH formed.

A proton gradient across the chloroplast membrane is the source of energy needed for ATP synthesis.

During the light reaction, ATP is formed in chloroplasts in a manner similar to oxidative phosphorylation in mitochondria (see Chapter 28). As electrons pass from photosystem II to photosystem I, protons are pumped from the cytosol into the matrix of the chloroplasts. This establishes an electrochemical gradient similar to the one in mitochondria. When protons pass from the matrix to the cytosol, a portion of the energy is used to make ATP from ADP and inorganic phosphate. The chemiosmotic theory thus plays a role in oxidative phosphorylation in mitochondria and in light-dependent ATP synthesis in chloroplasts.

The Dark Reaction of Photosynthesis

The light reaction produces NADPH (the reducing source) and ATP (the energy), but it does not make any glucose directly. The synthesis of glucose occurs in the **dark reaction**. This reaction uses carbon dioxide and the products of the light reaction. The dark reaction, also called the *Calvin cycle*, involves several enzymes of glycolysis, some enzymes for the synthesis

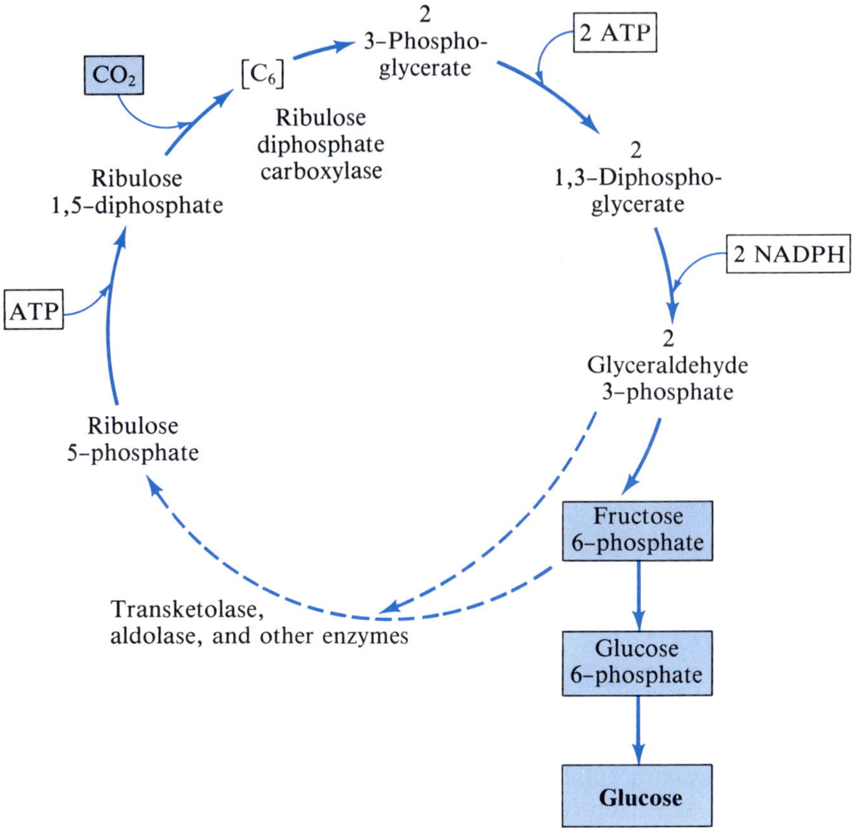

2
3-Phospho-
glycerate
2 ATP
CO_2
[C_6]
Ribulose
diphosphate
carboxylase
2
1,3-Diphospho-
glycerate
Ribulose
1,5-diphosphate
2 NADPH
ATP
2
Glyceraldehyde
3-phosphate
Ribulose
5-phosphate
Fructose
6-phosphate
Transketolase,
aldolase, and other enzymes
Glucose
6-phosphate
Glucose

FIGURE 29.2
The dark reaction of photosynthesis. This reaction uses ATP and NADPH from the light reaction to make carbohydrate (glucose) from CO_2. A portion of the 3-phosphoglycerate that is formed is used to make glucose, and the rest is recycled to ribulose 1,5-diphosphate.

and isomerization of carbohydrates, and the single most abundant protein on earth: ribulose diphosphate carboxylase.

Ribulose diphosphate carboxylase catalyzes the attachment of CO_2, an inorganic molecule, to ribulose diphosphate, an organic molecule.

The first step in the dark reaction is the attachment of carbon dioxide to the organic molecule ribulose diphosphate (Figure 29.2). The product is an unstable one that breaks into two molecules of 3-phosphoglycerate, an intermediate of glycolysis. Through a series of reactions that is partly a reversal of glycolysis, 3-phosphoglycerate is converted to glucose. Note that NADPH and ATP produced by the light reaction are needed for this synthesis.

In the equation for photosynthesis, glucose was shown to be made from carbon dioxide. Glucose is produced by the reactions shown in Figure 29.2, but only one of the six carbon atoms came from carbon dioxide; the remaining five already existed as ribulose diphosphate. Where do the other five CO_2 molecules enter into the picture? How is ribulose diphosphate formed? The whole process is balanced by multiplying the first step by six and adding several steps that synthesize sugars.

$$6\,CO_2 + 6\text{ Ribulose diphosphate} \xrightarrow[\text{carboxylase}]{\text{Ribulose diphosphate}} 12\text{ 3-Phosphoglycerate}$$

$$\underset{\text{(inorganic)}}{6\text{ C atoms}} + \underset{\text{(organic)}}{30\text{ C atoms}} = \underset{\text{(organic)}}{36\text{ C atoms}}$$

Twelve molecules of the product are formed in the preceding equation. Two of these product molecules, each containing three carbon atoms, are used to make glucose, the product of photosynthesis. Ten of them, collectively containing thirty carbon atoms, are rearranged and phosphorylated through the action of several enzymes to yield six ribulose diphosphate molecules, which can then participate in the dark reaction (Figure 29.2).

NEW TERMS

heterotrophic organism An organism that must consume organic matter to obtain energy and carbon atoms.

autotrophic organism An organism that obtains carbon from nonliving sources such as carbon dioxide.

photosynthesis A process by which plants and algae make organic molecules from carbon dioxide using light as the energy source.

light reaction The part of photosynthesis that uses light energy to make ATP and NADPH.

dark reaction The part of photosynthesis that uses ATP, NADPH, and carbon dioxide to make glucose.

TESTING YOURSELF

Photosynthesis

1. What is the site of photosynthesis in a plant? What common organisms carry out photosynthesis?
2. Eighteen ATPs and twelve NADPHs are needed for the synthesis of one glucose molecule in photosynthesis. If an NADPH is equivalent to three ATP, compare the number of ATP required to synthesize glucose to the number produced during aerobic catabolism of glucose.

Answers **1.** The chloroplast. Plants, algae, and some photosynthetic bacteria (cyanobacteria, the blue-green algae). **2.** 54 ATP needed for synthesis, 36 or 38 produced during aerobic catabolism.

29.3 Biosynthesis of Carbohydrates

Digestion of carbohydrates yields large amounts of glucose and smaller amounts of other monosaccharides. These molecules are used for energy production (see Chapter 28), but they are also used to make oligosaccharides, polysaccharides, glycolipids, and glycoproteins within the body. Furthermore, the monosaccharides can be converted to other monosaccharides, and under certain physiological conditions, they can be made from other compounds within the body.

Gluconeogenesis

Glucose can be made from several small molecules that are produced from the catabolism of sugars, amino acids, and the glycerol of fats and oils.

Although the diet provides large amounts of glucose to the body, there are circumstances that require the body to make glucose. This is accomplished by the process of **gluconeogenesis**, which literally means "new synthesis of glucose." The term does not refer to the conversion of other

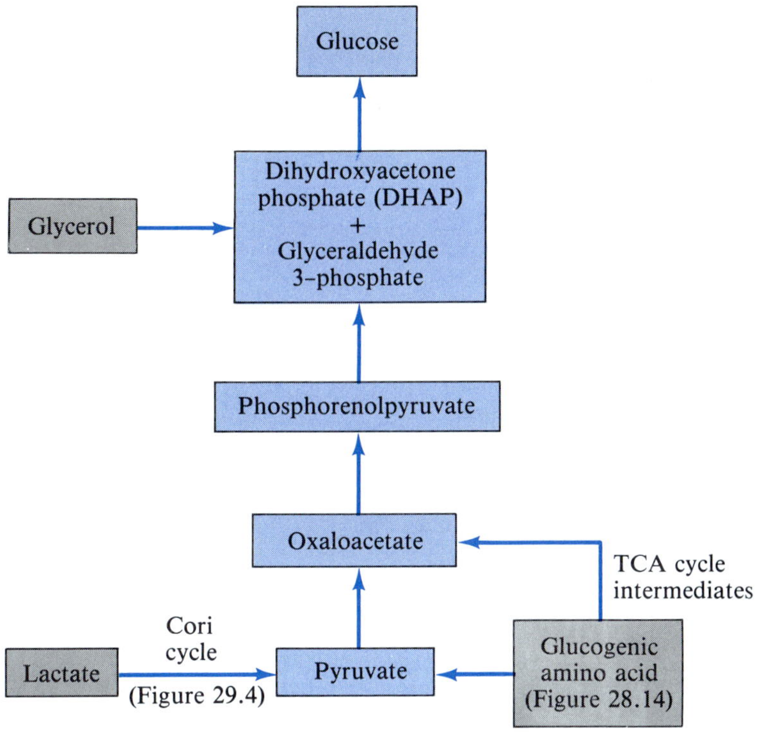
Glucose
Dihydroxyacetone phosphate (DHAP) + Glyceraldehyde 3-phosphate
Glycerol
Phosphorenolpyruvate
Oxaloacetate
TCA cycle intermediates
Cori cycle
Lactate
(Figure 29.4)
Pyruvate
Glucogenic amino acid (Figure 28.14)

FIGURE 29.3
Glucose can be synthesized from several metabolites. Glycerol is produced by catabolism of triacylglycerols; lactate is produced by anaerobic glycolysis. Catabolism of glucogenic amino acids, which is all of them except leucine, yields TCA cycle intermediates and pyruvate.

monosaccharides into glucose; instead, it refers to synthesis from smaller metabolic intermediates produced by catabolism. A summary of the molecules that can be used to make glucose is given in Figure 29.3. Note that the carbon atoms for glucose come from three sources: (1) lactate, (2) the glucogenic amino acids, and (3) glycerol.

Lactate is produced in oxygen deficient muscle via glycolysis and reduction of pyruvate. Lactate diffuses into blood and is taken up by the liver. When sufficient oxygen is available to the liver, the lactate is oxidized to pyruvate. This reaction is simply the reverse of the reaction that led to the production of lactate.

$$\text{Pyruvate} + \text{NADH} + \text{H}^+ \xrightleftharpoons[]{\text{Lactate dehydrogenase}} \text{Lactate} + \text{NAD}^+$$

Since this reaction is reversible, the concentrations of the four species determines the direction of the reaction (Le Chatelier's principle). When lactate and NAD^+ concentrations are high, NADH and pyruvate are produced. There are several fates that are possible for the pyruvate: lipid synthesis, oxidation via the TCA cycle to produce energy, or conversion by gluconeogenesis to produce glucose. Much of this glucose is released into the blood. Exercising muscle converts glucose to lactate and is thus depleting its glucose stores. Under these conditions, muscle readily takes up glucose from blood.

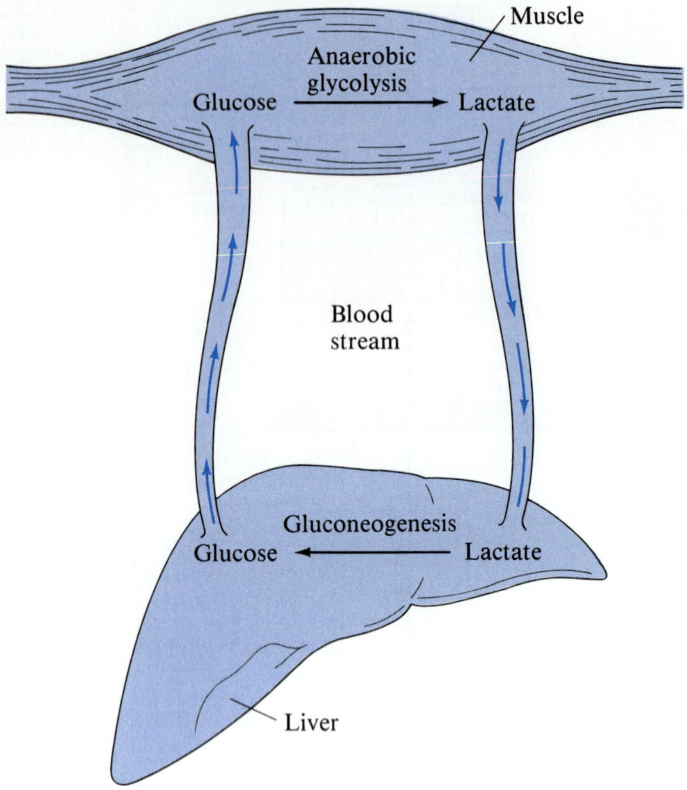
Muscle
Anaerobic glycolysis
Glucose
Lactate
Blood stream
Gluconeogenesis
Glucose
Lactate
Liver

FIGURE 29.4
Cori cycle. Lactate is produced from glucose by glycolysis under anaerobic conditions. The lactate leaves muscles, and is carried to the liver by blood where it is converted back to glucose.

This anaerobic catabolism of glucose to lactate in muscle, resynthesis of glucose from lactate via gluconeogenesis in liver, and transport of glucose back to muscle constitutes a cycle called the **Cori cycle** (Figure 29.4).

Muscle cells synthesize two ATP during the synthesis of lactate. The Cori cycle removes lactate from the body by using it to make glucose. But in terms of energy, there is no such thing as a "free lunch" or breaking even. The synthesis of glucose from lactate requires four ATP and two GTP. The Cori cycle requires four moles of ATP equivalents per mole of glucose.

Most amino acids can be used, in whole or part, to make glucose. These amino acids are called *glucogenic amino acids*. Two conditions lead to significant gluconeogenesis from amino acids: (1) fasting and (2) a protein-rich diet. During fasts, the body breaks down body protein into amino acids. Small amounts of these amino acids are used to remake essential proteins, but the rest are used for energy and glucose synthesis. It may seem odd that under these conditions the body would bother to make glucose. Why not just use the amino acids for energy production? Because the brain and central nervous system use glucose as their principal energy source. Blood glucose must be maintained or dysfunction of the central nervous system will occur. During fasting, amino acids, through gluconeogenesis, provide much of the blood glucose. When much of the caloric intake of the diet is in the form of protein, it is also necessary to make glucose from amino acids to maintain adequate blood glucose levels.

Glycerol is the third source of glucose via gluconeogenesis. Glycerol is produced by catabolism of triacylglycerols. In animals, the fatty acids of fats and oils cannot be used to make glucose. During fasting, the amount of glycerol available to the body increases through increased catabolism of

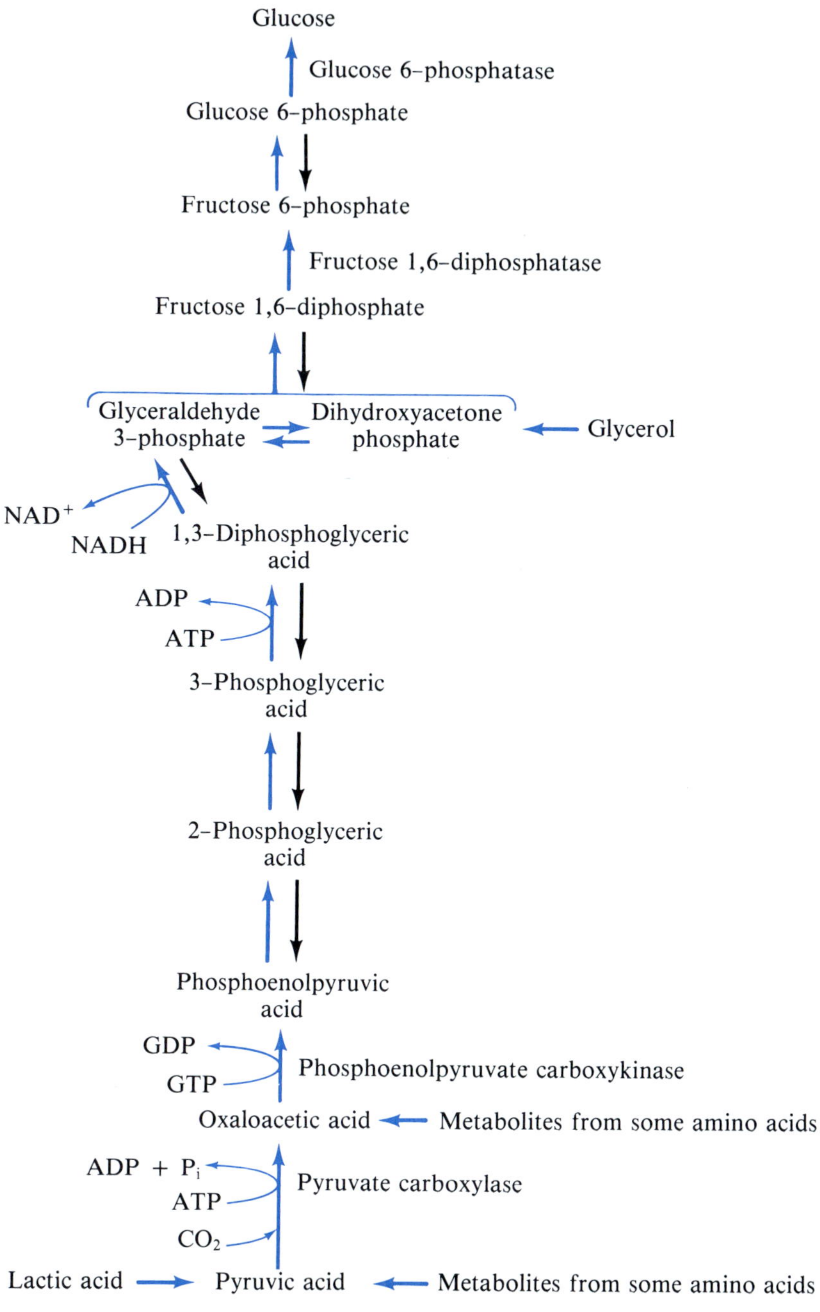

Glucose
Glucose 6-phosphatase
Glucose 6-phosphate
Fructose 6-phosphate
Fructose 1,6-diphosphatase
Fructose 1,6-diphosphate
Glyceraldehyde 3-phosphate
Dihydroxyacetone phosphate
Glycerol
NAD+
NADH
1,3-Diphosphoglyceric acid
ADP
ATP
3-Phosphoglyceric acid
2-Phosphoglyceric acid
Phosphoenolpyruvic acid
GDP
GTP
Phosphoenolpyruvate carboxykinase
Oxaloacetic acid
Metabolites from some amino acids
ADP + Pi
ATP
CO2
Pyruvate carboxylase
Lactic acid
Pyruvic acid
Metabolites from some amino acids

Equation for synthesis from two pyruvates:

$$\textbf{2 Pyruvate} + \textbf{4 ATP} + \textbf{2 GTP} + \textbf{2 NADH} \longrightarrow \textbf{Glucose} + \textbf{4 ADP} + \textbf{2 GDP} + \textbf{6 P}_i + \textbf{2 NAD}^+$$

FIGURE 29.5
Gluconeogenesis includes many of the steps of glycolysis (blue arrows), but some of the steps of glycolysis are irreversible. Pyruvate carboxylase, phosphoenolpyruvate carboxykinase, fructose 1,6-diphosphatase, and glucose 6-phosphatase are enzymes that catalyze reactions that are not shared by gluconeogenesis and glycolysis. Six high energy bonds in ATP and GTP are cleaved per glucose synthesized by gluconeogenesis.

triacylglycerols. Amino acids and glycerol provide the carbon atoms for glucose during fasting.

Some of the specific reactions of gluconeogenesis are already familiar to you because they are reverse reactions of glycolysis. Figure 29.5 shows gluconeogenesis, with the reactions of glycolysis indicated with blue arrows.

The body can make all of the simple sugars that it needs.

The body can make all of the monosaccharides that it requires. Unlike lipids and amino acids, there are no essential sugars. Glucose is available from the diet or through gluconeogenesis. The other monosaccharides are obtained from the diet or made from glucose. The body possesses the necessary enzymes to make the required amounts of oligosaccharides and polysaccharides. Except for lactose produced during milk production, free oligosaccharides are uncommon; however, they are made as components of glycoproteins and some glycolipids.

Glycogenesis

Glucose and other monosaccharides are used as building blocks for the polysaccharides of the body. Each polysaccharide has its own pathway for its synthesis, but we will consider here only **glycogenesis**, the synthesis of glycogen. Glycogen is the principal storage form for glucose. Glycogen is synthesized via this reaction (added residue is highlighted):

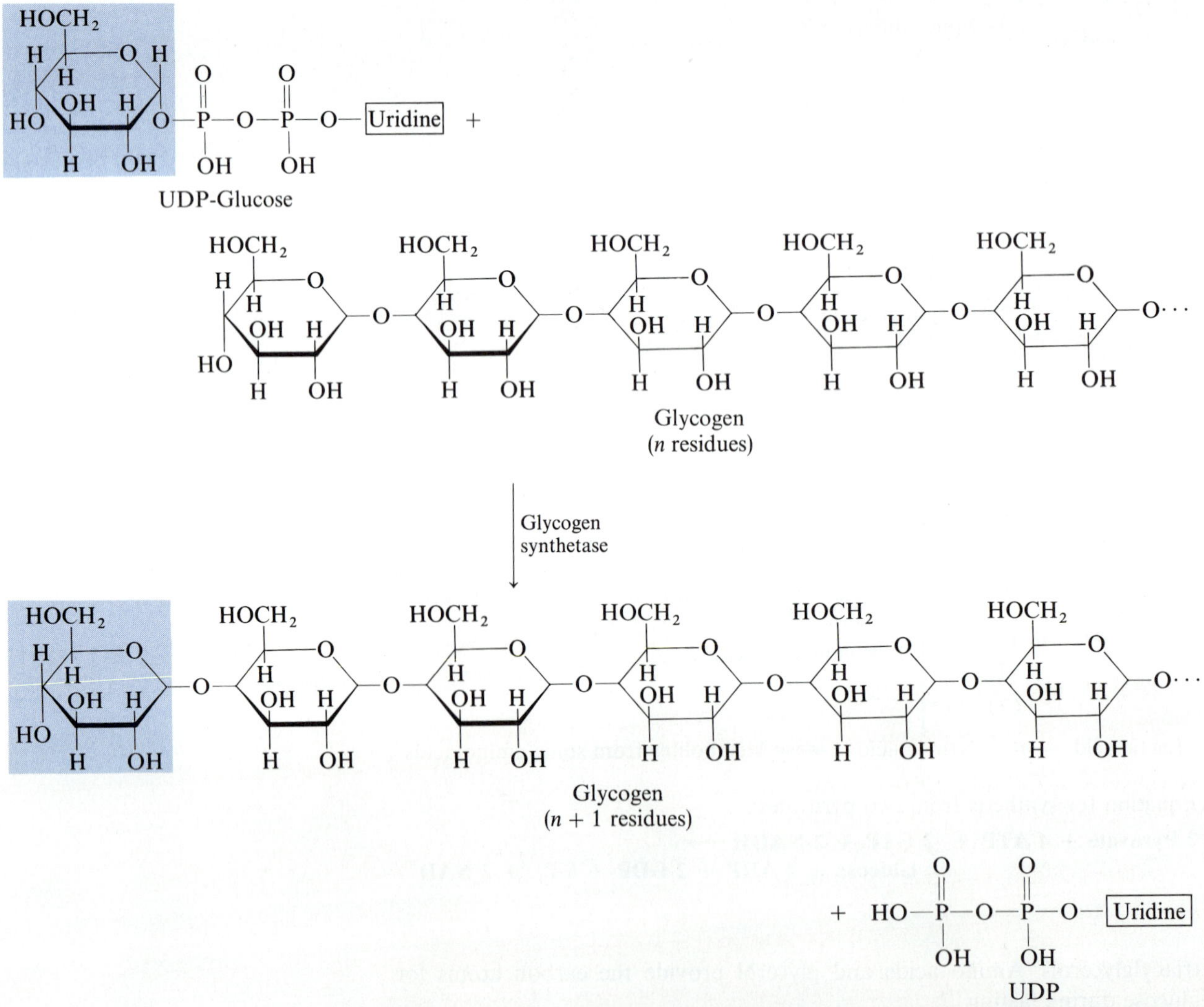
UDP-Glucose
Glycogen
(n residues)
Glycogen
synthetase
Glycogen
(n + 1 residues)
UDP

Uridine diphosphate-glucose (UDP–glucose) is glucose bound through a high energy phosphate ester bond to UDP. The glucose part of this activated

ENZYME MODIFICATION LEADS TO LACTOSE SYNTHESIS

Lactose is synthesized only by the mammary glands following birth. The control of the enzyme that synthesizes lactose is rather interesting. The cells of the body possess the enzyme galactosyl transferase, which transfers galactose residues onto appropriate acceptor molecules. Thus, a general reaction that occurs within all body cells is

Galactose (activated) + Acceptor molecule $\xrightarrow{\text{Galactosyl transferase}}$ Galactosylated product

This enzyme alone cannot make lactose because glucose does not act as an acceptor molecule. But after a female gives birth, the concentration of the hormone prolactin increases in her body. In the mammary glands this hormone stimulates the production of another protein, α-lactalbumin, a protein that binds to galactosyl transferase. This complex has a different specificity than galactosyl transferase alone; it is a lactose synthetase. The complex catalyzes the attachment of galactose to glucose, yielding lactose.

Galactose (activated) + Glucose (an acceptor molecule) $\xrightarrow{\text{Lactose synthetase}}$ Lactose (galactosylated product)

A common enzyme of the body is thus altered through allosteric regulation to yield a special duty enzyme.

molecule is transferred onto one of the branches of an existing glycogen molecule, increasing the number of glucose residues by one. Glucose cannot be added directly. The synthesis of the glycosidic bonds between glucose residues in glycogen requires energy. The energy needed for formation of this bond is found in the cleavage of the high energy bond between the glucose and the UDP of UDP–glucose. The synthesis and breakdown of glycogen are highly regulated processes that are discussed in Section 30.6.

NEW TERMS

gluconeogenesis The synthesis of glucose from several small common molecules of metabolism.

Cori cycle A cycle in which lactic acid produced in muscle is converted back to glucose in the liver.

glycogenesis The synthesis of glycogen within the muscles and liver.

TESTING YOURSELF

Biosynthesis of Carbohydrates

1. Describe the Cori cycle and its metabolic role.
2. The central nervous system requires glucose as an energy source. During a prolonged fast, how does the body provide glucose to the brain?

Answers **1.** This cycle includes the production of lactate from glucose in anaerobic muscle, the transport of lactate to the liver in blood, the synthesis of glucose from lactate via gluconeogenesis in the liver, and transport of the glucose back to muscle. Lactate is recycled, rather than excreted. This is useful because lactate contains considerable energy that would be lost if it were excreted. **2.** The body breaks down cellular protein and uses many of the amino acids to make glucose via gluconeogenesis.

29.4 Biosynthesis of Lipids

As we saw in Chapter 24, the lipids are a diverse group of natural products that can be characterized as lipophilic (nonpolar) and hydrophobic (water hating). In this chapter we will discuss the origins of some of these materials.

In general, separate pathways are used for anabolism and catabolism. This is also true for the lipids. Moreover, for fatty acids, the two processes are located in different parts of the cell.

Lipogenesis

Acetyl CoA is the source of the carbon atoms for fatty acid synthesis.

Lipogenesis is a term that literally means "the synthesis of lipids," but it is more often used to mean the synthesis of fatty acids. Many fatty acids can be made by the body. Linoleic and linolenic acids are exceptions. In humans, the synthesis of fatty acids takes place in the cytoplasm of liver cells. Acetyl CoA is the major precursor to the fatty acids and supplies two carbon fragments for assembling these acids. Any excess acetyl CoA that does not pass through the TCA cycle for energy production may be used for fatty acid synthesis. This is the basic problem with weight control. If we do not use the acetyl CoA for energy, it becomes stored as fat. Citrate formed from acetyl CoA passes from the mitochondria to the cytosol where it reacts with CoA to yield acetyl CoA and oxaloacetate. The acetyl CoA is then available for fatty acid synthesis in the cytosol.

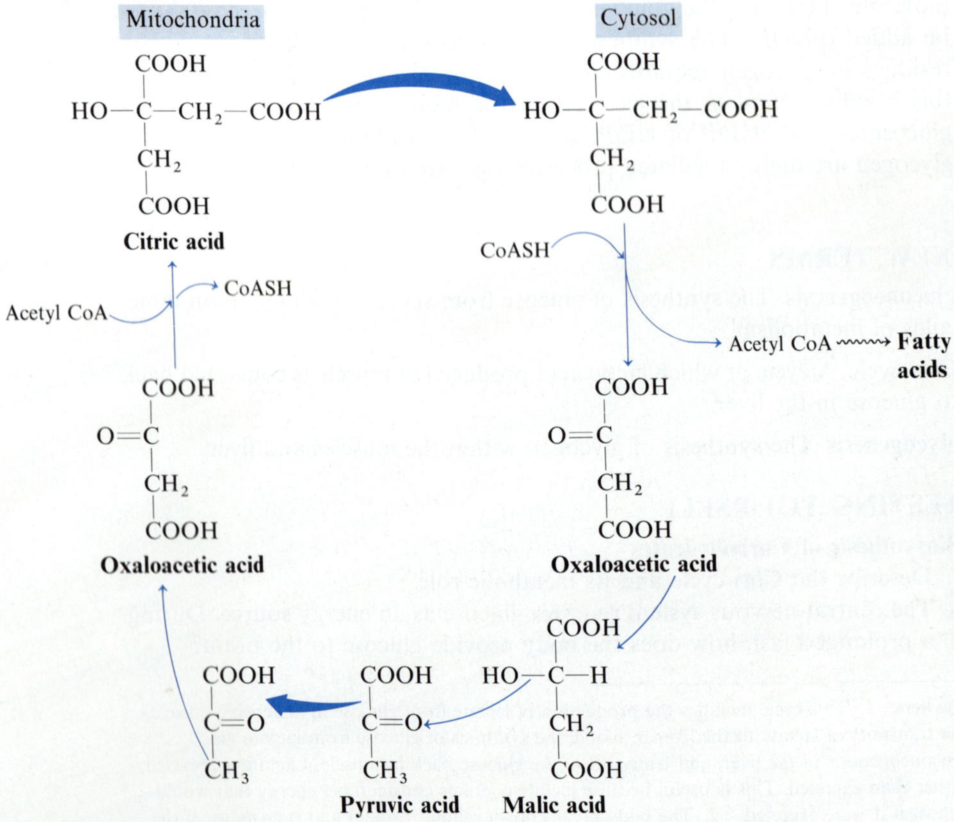
Mitochondria
Cytosol
Citric acid
Acetyl CoA
CoASH
CoASH
Acetyl CoA
Fatty acids
Oxaloacetic acid
Oxaloacetic acid
Pyruvic acid
Malic acid

The oxaloacetate passes through intermediates and returns to the mitochondria. When cytosolic oxaloacetic acid is converted to malic acid, electrons are donated by NADH. When cytosolic malic acid is converted to pyruvic acid, electrons are transferred to $NADP^+$ to form NADPH. This sequence effectively makes NADPH from NADH. Thus, the cycle shown above not only transfers acetyl groups from mitochondria to the cytosol, it also transfers electrons from NADH to $NADP^+$ forming NADPH.

How does the two-carbon acetate unit of acetyl CoA become incorporated into a fatty acid? Fatty acids are synthesized by sequentially adding two carbon pieces to a two-carbon precursor molecule. The carbon atoms come from acetyl CoA, but as you shall learn shortly, lipogenesis does *not* directly use acetyl CoA. Acetyl CoA is converted to malonyl CoA and acetyl ACP which participate directly in fatty acid synthesis.

Let's examine the synthesis of palmitic acid, $CH_3(CH_2)_{14}COOH$, the saturated fatty acid with sixteen carbon atoms. This synthesis requires eight acetyl CoA molecules to provide the sixteen carbon atoms. The acetyl group of one of these acetyl CoA molecules is attached to the **acyl carrier protein (ACP)** to form acetyl ACP:

$$CH_3{-}\underset{\underset{\displaystyle O}{\|}}{C}{-}CoA + HS{-}ACP \longrightarrow CH_3{-}\underset{\underset{\displaystyle O}{\|}}{C}{-}S{-}ACP + CoA$$

ACP is part of a large enzyme called *fatty acid synthase*. ACP serves as a carrier for the acetyl group and the growing intermediates during the synthesis.

The other acetyl CoA molecules are converted to another intermediate, malonyl CoA. This reaction is a carboxylation of acetyl CoA with bicarbonate and requires energy provided by ATP hydrolysis.

The carboxylation of acetyl CoA is an endothermic reaction.

$$CH_3{-}\underset{\underset{\displaystyle O}{\|}}{C}{-}CoA + HCO_3^- \xrightarrow[\text{ATP} + H_2O \;\to\; \text{ADP} + P_i]{\text{Acetyl CoA carboxylase}} {}^-O\underset{\underset{\displaystyle O}{\|}}{C}CH_2{-}\underset{\underset{\displaystyle O}{\|}}{C}{-}CoA$$

The malonyl group of malonyl CoA is then transferred to ACP to yield malonyl ACP.

The subsequent steps in the synthesis of palmitic acid are catalyzed by fatty acid synthase, a protein with several enzymic activities and which serves as the acyl carrier protein. A summary of this synthesis is shown in Figure 29.6. It is interesting to note that these reactions are biological examples of familiar organic reactions: reduction of a ketone to a secondary alcohol, dehydration of an alcohol, and hydrogenation of an alkene.

Humans synthesize other fatty acids in addition to palmitic acid. An additional enzyme system elongates some of the palmitic acid, two carbon atoms at a time, again using acetyl CoA to provide the carbon atoms. The steps are very similar to those of the fatty acid synthase. Still another enzyme system introduces carbon–carbon double bonds to yield unsaturated fatty acids. This system cannot introduce double bonds beyond carbon atoms nine and ten of a chain. That is why linoleic acid and linolenic acid are provided only by the diet.

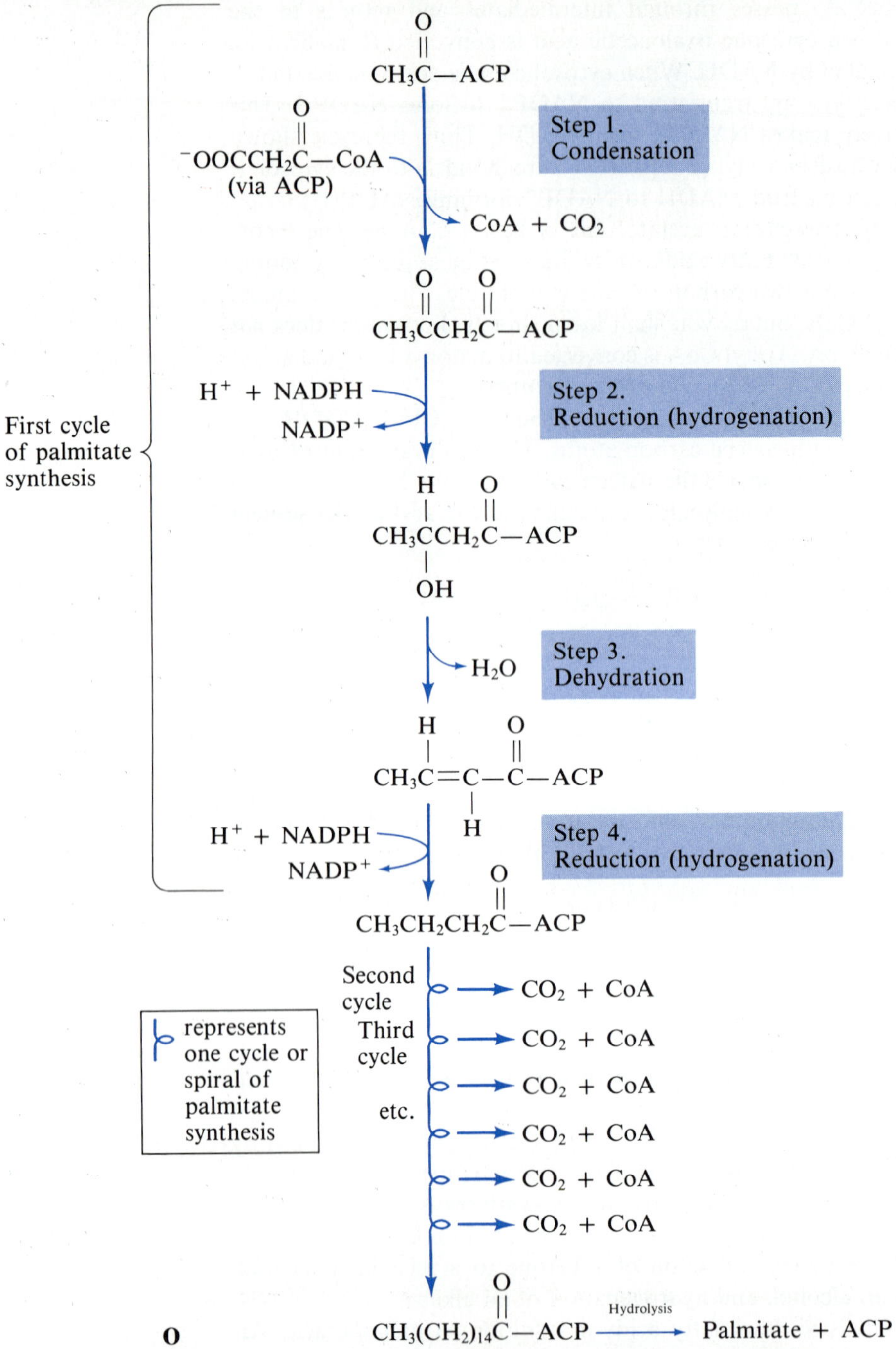

First cycle of palmitate synthesis
Step 1. Condensation
(via ACP)
CoA + CO2
Step 2. Reduction (hydrogenation)
H+ + NADPH
NADP+
Step 3. Dehydration
H2O
Step 4. Reduction (hydrogenation)
Second cycle
Third cycle
etc.
CO2 + CoA
represents one cycle or spiral of palmitate synthesis
Hydrolysis
Palmitate + ACP

Net reaction:

$$\underset{\text{(via ACP)}}{CH_3\overset{O}{\overset{\|}{C}}-CoA} + 7\ \underset{\text{(via ACP)}}{{}^{-}OOCCH_2\overset{O}{\overset{\|}{C}}-CoA} + 14\ NADPH + 14\ H^+ \longrightarrow \text{Palmitate} + 14\ NADP^+ + 7\ CO_2 + 8\ CoA$$

FIGURE 29.6
Fatty acid biosynthesis. Note that both acetate and malonate are attached to a carrier protein during the synthesis.

Note that NADPH was used in the synthesis of fatty acids just as it is used whenever reducing power is needed in synthesis. NADPH is produced during the cyclic transfer of acetyl CoA from mitochondria to the cytosol, and by the **pentose phosphate pathway** (Figure 29.7). This pathway also yields pentoses that can be used by the body or recycled to hexoses.

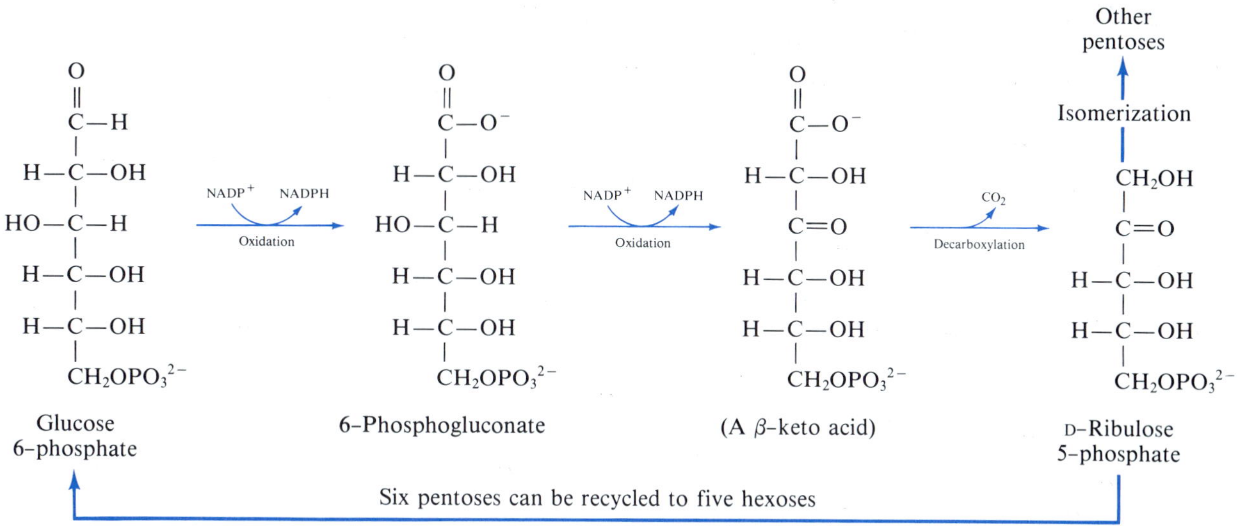
Other pentoses
Isomerization
NADP$^+$
NADPH
Oxidation
NADP$^+$
NADPH
Oxidation
CO_2
Decarboxylation
Glucose 6-phosphate
6-Phosphogluconate
(A β-keto acid)
D-Ribulose 5-phosphate
Six pentoses can be recycled to five hexoses

FIGURE 29.7
Pentose phosphate pathway. Glucose 6-phosphate is oxidized to a phosphorylated pentose, yielding two NADPHs per mole. Pentoses can be made from these phosphorylated pentoses, or they can be recycled to glucose 6-phosphate.

Synthesis of Triacylglycerols and Phosphoacylglycerols

Fatty acids are components of the saponifiable lipids, notably the triacylglycerols and the phosphoacylglycerols. These compounds are esters that contain not only fatty acids but also glycerol. Triacylglycerols are formed in adipose cells by transferring the fatty acid (the acyl group) from CoA to yield the ester.

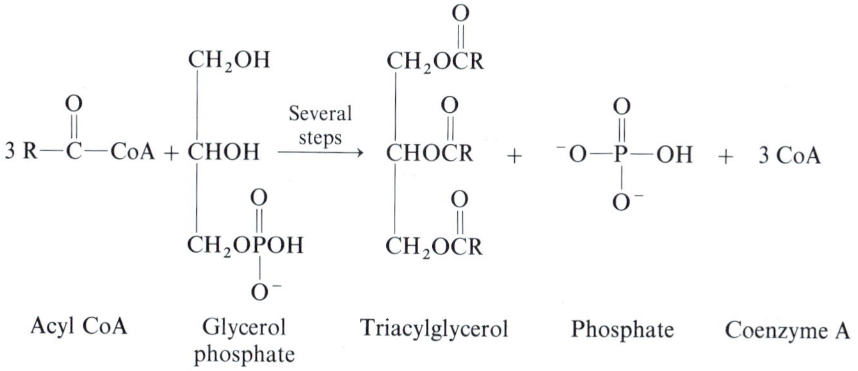
Several steps
Acyl CoA
Glycerol phosphate
Triacylglycerol
Phosphate
Coenzyme A

Ester bond formation is endothermic, but the cleavage of the bond between the fatty acid and the CoA yields energy.

The synthesis of phosphoacylglycerols has some of the same features as the synthesis of triacylglycerols. Two acyl groups are transferred from acyl CoA to glycerol phosphate to form a diacylglycerol phosphate (a phosphatidic acid) (see Chapter 24, Section 24.4). The phosphate is then cleaved from the molecule to yield a diacylglycerol.

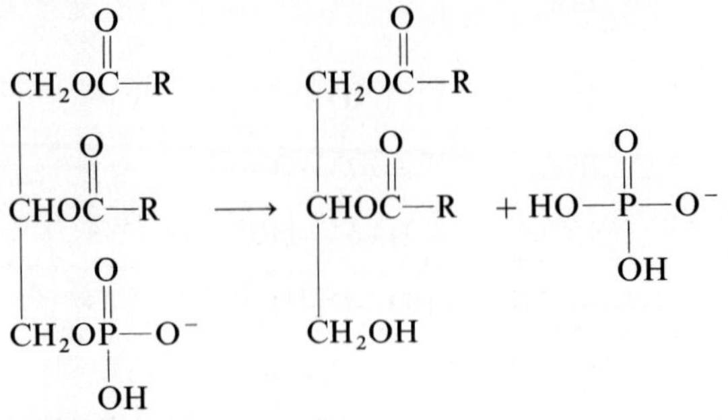

Diacylglycerol phosphate — Diacylglycerol — Phosphate

A molecule of phosphorylated choline or ethanolamine is then transferred from a carrier molecule (cytidine diphosphate, or CDP) onto the diacylglycerol to yield the phosphoacylglycerol. The synthesis of phosphatidyl choline is summarized in the following equation.

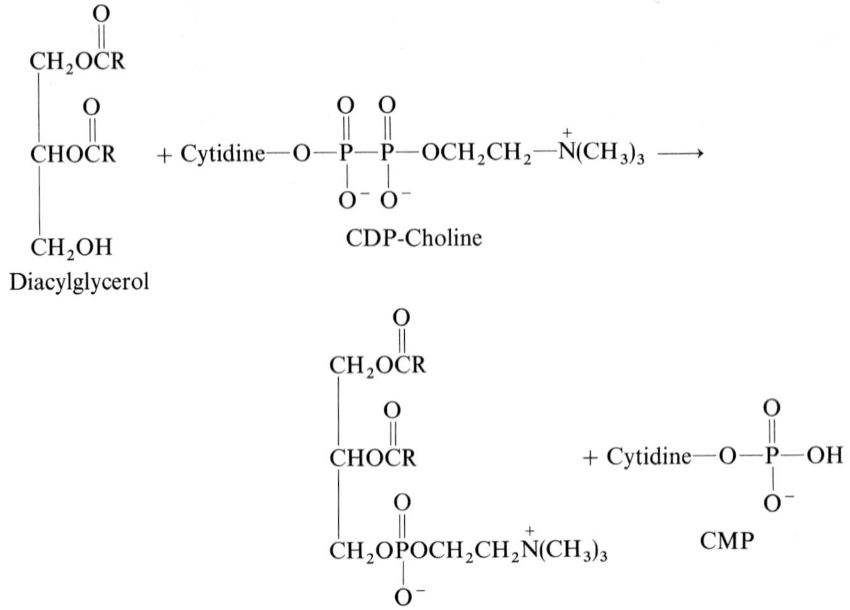

Diacylglycerol — CDP-Choline — Phosphatidyl chloline — CMP

Glycerol phosphate is needed for the synthesis of both triacylglycerols and phosphoacylglycerols. This compound is made by reducing one of the intermediates of glycolysis, dihydroxyacetone phosphate (DHAP).

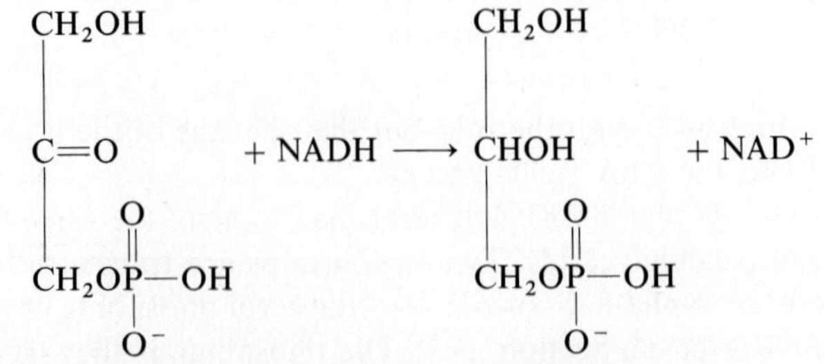

DHAP — Glycerol phosphate

Biosynthesis of Cholesterol

Cholesterol biosynthesis has recently received much attention because blood cholesterol levels have been linked to heart disease. This synthesis occurs primarily in the liver, and the rate of synthesis is normally linked to the amount of dietary cholesterol. All of the carbon atoms of cholesterol are derived from the carbons of acetate in acetyl CoA.

In two steps, three acetyl CoA molecules are condensed into hydroxymethylglutaryl CoA (HMG CoA).

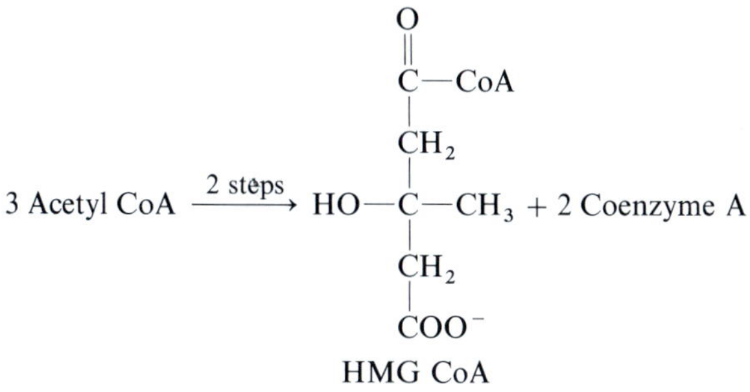
3 Acetyl CoA
2 steps
HO—C—CH$_3$ + 2 Coenzyme A
C—CoA
CH$_2$
COO$^-$
HMG CoA

This compound is a key intermediate in metabolism because it can be used to make sterols or ketone bodies. If cholesterol is needed, some HMG CoA is converted to mevalonic acid in a reaction catalyzed by the enzyme HMG CoA reductase.

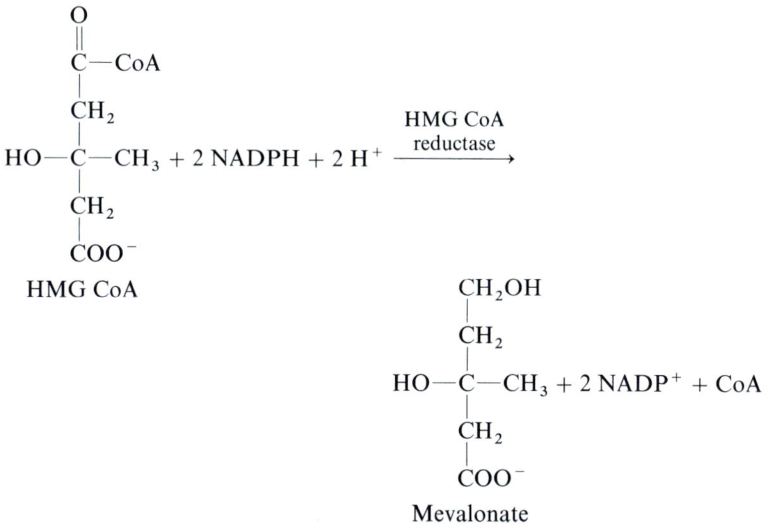
HO—C—CH$_3$ + 2 NADPH + 2 H$^+$
HMG CoA reductase
HMG CoA
CH$_2$OH
HO—C—CH$_3$ + 2 NADP$^+$ + CoA
COO$^-$
Mevalonate

This enzyme is involved in the regulation of cholesterol biosynthesis and is thus a target for drug treatment of individuals affected by high blood cholesterol. HMG CoA reductase is competitively inhibited by mevinolin. This compound is used as a drug to help reduce blood cholesterol. The structures of HMG CoA and mevinolin are shown in the margin, with their structural similarities highlighted.

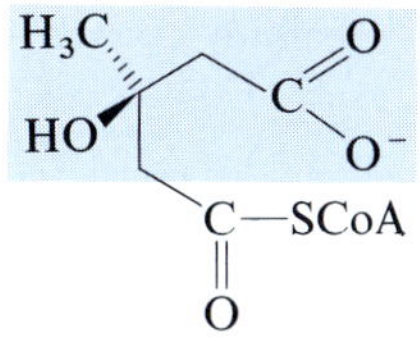
H$_3$C
HO
O
O$^-$
C—SCoA

HMG CoA

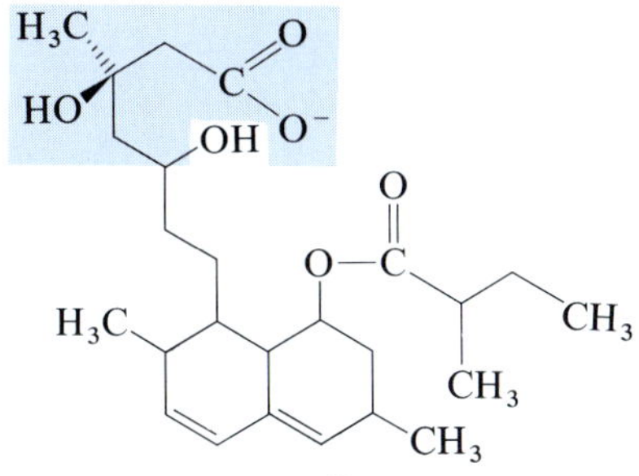
H$_3$C
HO
OH
O$^-$
CH$_3$

Mevinolin

Mevalonate is used by a variety of organisms to make such diverse compounds as terpenes and sterols. The terpenes are natural products that include a number of fragrances and oils. Mevalonate can be converted to isopentyl pyrophosphate.

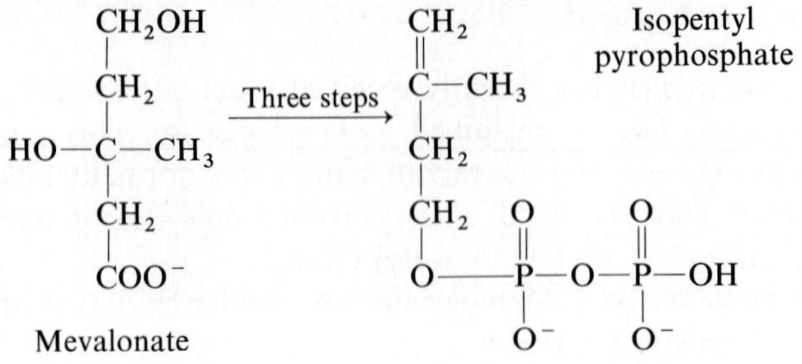
Mevalonate
Three steps
Isopentyl pyrophosphate

Isopentyl pyrophosphate is used to make a host of terpenes, but in mammals, six of these molecules condense to form squalene, an oil found in the secretions on the surface of skin.

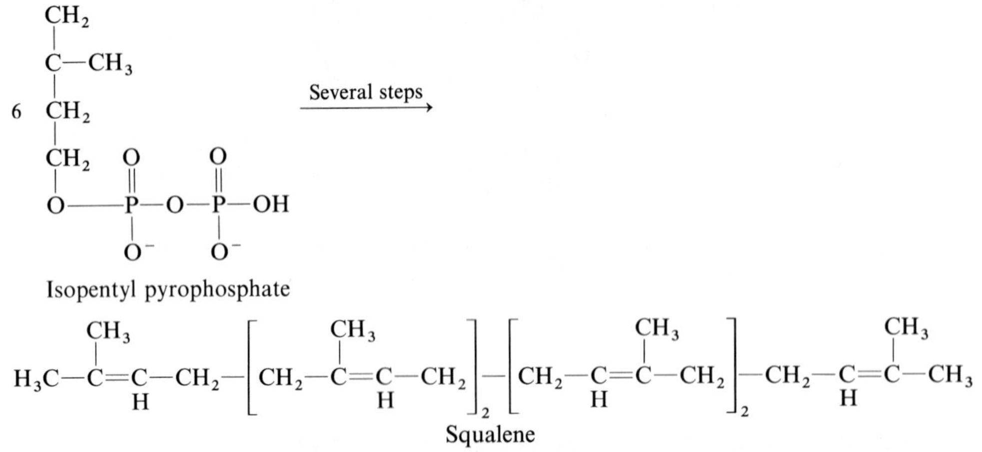
6
Several steps
Isopentyl pyrophosphate
Squalene

Squalene is the precursor to some large terpenes and all of the steroids. Squalene is converted to a steroid ring system to form the sterol lanosterol, which is then converted to cholesterol. (Only carbon skeletons are shown here.) The other steroids of the body are derived from cholesterol.

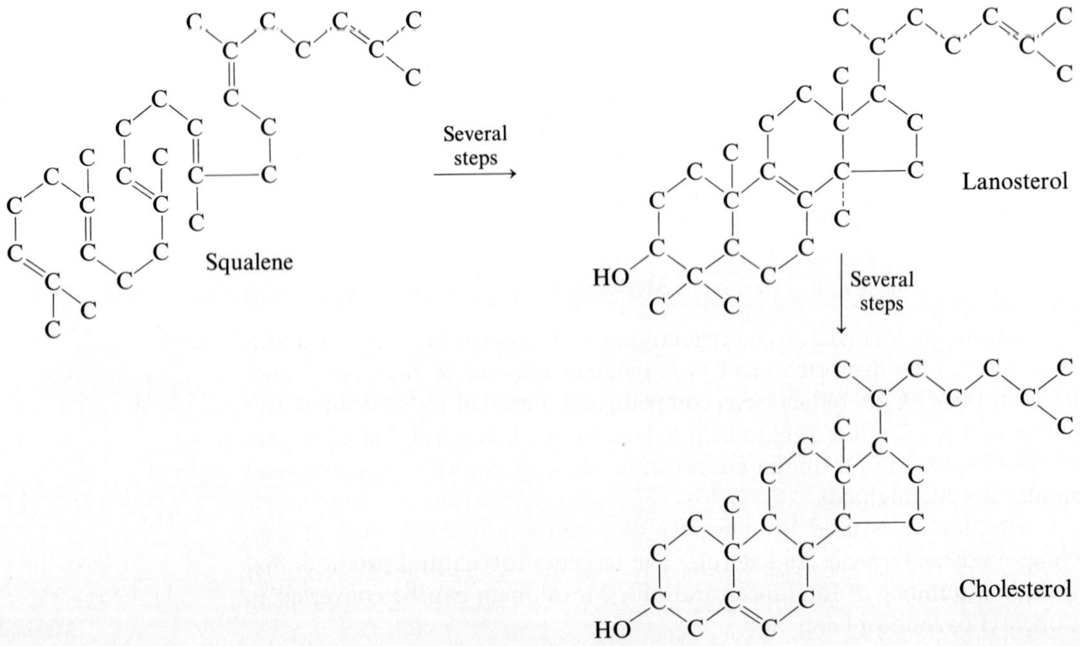
Squalene
Several steps
Lanosterol
Several steps
Cholesterol

Synthesis of Ketone Bodies

HMG CoA is also used to make the ketone bodies, acetoacetate, 3-hydroxybutyrate, and acetone, which provide a portion of the body's energy needs. In a single step HMG CoA is converted to acetyl CoA and acetoacetate.

HMG CoA is used for the synthesis of cholesterol and ketone bodies.

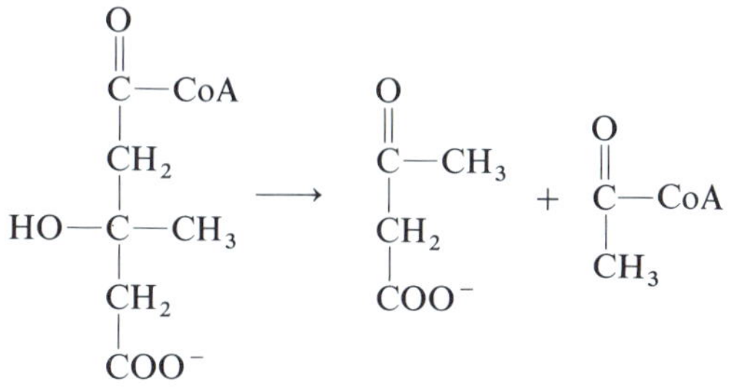

HMG CoA   Acetoacetate   Acetyl CoA

Acetoacetate is reduced to form 3-hydroxybutyrate.

$$CH_3\overset{\overset{\displaystyle O}{\|}}{C}CH_2COO^- + NADH + H^+ \longrightarrow CH_3\overset{\overset{\displaystyle OH}{|}}{C}HCH_2COO^- + NAD^+$$

Small quantities of acetoacetate are decarboxylated to yield acetone.

$$CH_3\overset{\overset{\displaystyle O}{\|}}{C}CH_2COOH \longrightarrow CH_3\overset{\overset{\displaystyle O}{\|}}{C}CH_3 + CO_2$$

The ketone bodies are synthesized in the liver and circulate throughout the body at low concentrations in blood. Many organs use ketone bodies for fuel, and during fasting, their concentrations increase markedly.

NEW TERMS

lipogenesis The biosynthesis of fatty acids.

acyl carrier protein (ACP) Protein involved in lipogenesis that carries the growing fatty acid and the molecules that condense to form the fatty acid.

pentose phosphate pathway Series of reactions that effectively converts glucose to pentoses and NADPH.

TESTING YOURSELF

Biosynthesis of Lipids

1. The carbon atoms of fatty acids come from what molecule?
2. What sources yield the NADPH needed to synthesize fatty acids and other reduced compounds?
3. Why does mevinolin decrease the biosynthesis of cholesterol?

Answers **1.** acetyl group of acetyl CoA **2.** acetyl CoA transport and the pentose phosphate pathway **3.** Mevinolin binds to the active site of HMG CoA reductase, preventing HMG CoA from binding. If HMG CoA is not reduced to mevalonate, then cholesterol synthesis cannot proceed.

29.5 Biosynthesis of Amino Acids

The twenty common amino acids are required for protein synthesis. Diet normally provides these amino acids unless the amount or quality of dietary protein is too low. Some of the amino acids can be synthesized by the body,

An essential amino acid cannot be made by the body.

but others cannot. Recall that an amino acid that cannot be synthesized in adequate quantities is called an essential amino acid. The essential amino acids (as previously discussed in Chapter 25) include the following:

Histidine	Tryptophan
Lysine	Leucine
Threonine	Phenylalanine
Isoleucine	Valine
Methionine	

Plants and many bacteria synthesize all of the amino acids. It appears that humans, along with many other heterotrophic organisms, have lost the capability to make some of these amino acids. When a compound is readily available in the diet, an inability to synthesize the compound is not necessarily a disadvantage.

Transamination of α-keto acids provides some of the amino acids needed for protein synthesis.

Although the specific pathways through which humans make some of the nonessential amino acids are too long and complicated to present here, there are some common features that can be described. Some of the amino acids are made through transamination of the appropriate keto acid, a reaction that is essentially the reverse of a key step in catabolism of amino acids (see Chapter 28). Here are three keto acids and the amino acids produced by transamination.

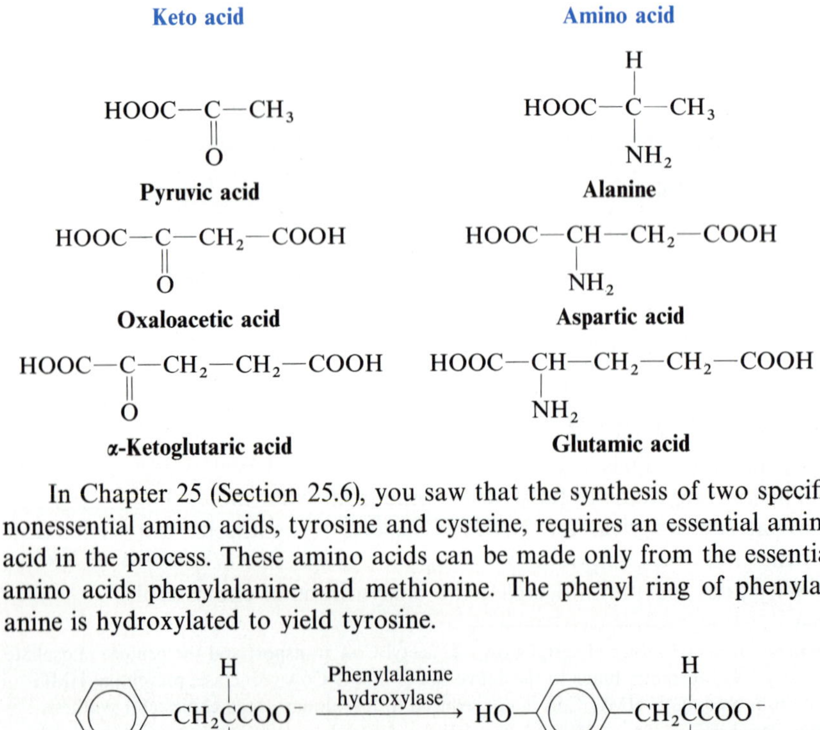
Keto acid
Amino acid
HOOC—C—CH3
O
H
HOOC—C—CH3
NH2
Pyruvic acid
Alanine
HOOC—C—CH2—COOH
O
HOOC—CH—CH2—COOH
NH2
Oxaloacetic acid
Aspartic acid
HOOC—C—CH2—CH2—COOH
O
HOOC—CH—CH2—CH2—COOH
NH2
α-Ketoglutaric acid
Glutamic acid

In Chapter 25 (Section 25.6), you saw that the synthesis of two specific nonessential amino acids, tyrosine and cysteine, requires an essential amino acid in the process. These amino acids can be made only from the essential amino acids phenylalanine and methionine. The phenyl ring of phenylalanine is hydroxylated to yield tyrosine.

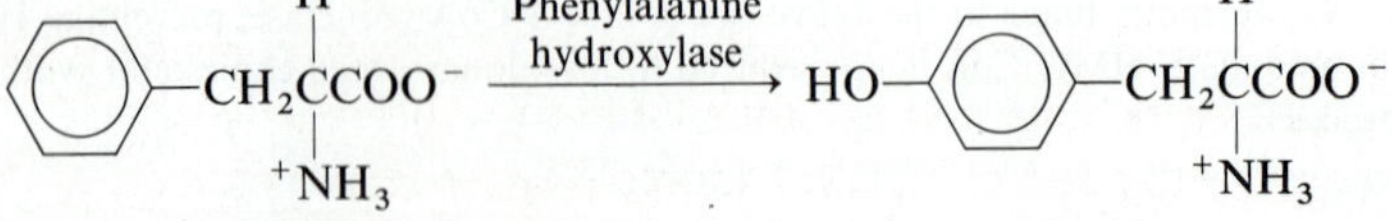
H
CH2CCOO−
+NH3
Phenylalanine
hydroxylase
HO
H
CH2CCOO−
+NH3

Phenylalanine Tyrosine

During the synthesis of cysteine, the sulfur atom is donated by methionine. The synthesis of an amino acid from an essential amino acid places greater dietary demand for those essential amino acids if the other amino acid is

PHENYLKETONURIA (PKU)

Have you ever noticed the warning to phenylketonurics on the label of foods containing Nutrasweet? Nutrasweet® is the brand name for aspartame, an artificial sweetener that contains phenylalanine. Phenylketonuria (PKU) is a genetic disease in which the patients are unable to properly metabolize phenylalanine to tyrosine because they lack the enzyme phenylalanine hydroxylase. If the diet contains more phenylalanine than is needed for protein synthesis, it will accumulate in cells and blood. If the concentration gets high enough, phenylalanine will convert to several phenylketo acids, including phenylpyruvic acid, which are then excreted in the urine. These compounds disrupt normal nervous system development in newborns. If the disease is untreated in a young baby, several severe developmental problems, including mental retardation, can result. Most newborns are tested for PKU. If they have this condition, they are given a diet with just enough phenylalanine for protein synthesis. Under these conditions, phenylpyruvate will not accumulate. Adult phenylketonurics should avoid excess amounts of dietary phenylalanine. The warning label on food packages and beverages containing Nutrasweet helps them to monitor their intake of this amino acid.

absent. For example, the amount of phenylalanine required in the diet is less if tyrosine is present than if it is absent. The same relationship exists between methionine and cysteine.

TESTING YOURSELF

Biosynthesis of Amino Acids

1. Why do mammals lack the ability to synthesize nearly half of the amino acids needed for protein synthesis? Is this a significant handicap for these organisms?
2. Show the reaction through which aspartate is synthesized in the body.

Answers **1.** They lack enzymes for the synthesis of these compounds. No. The diet of mammals normally includes all of the amino acids in sufficient amounts.

2.

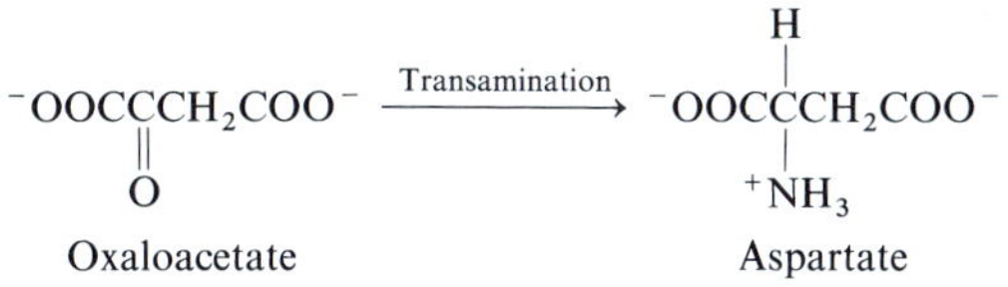
Transamination
Oxaloacetate
Aspartate

29.6 Biosynthesis of Nucleotides

Nucleotides are involved in the transfer of hereditary information and numerous other reactions and are thus crucial to life. The nucleotides are all synthesized by the body; a dietary source is not needed. Again, each nucleotide has its own biosynthetic pathway that is too detailed for our purposes here. However, it is worthwhile to examine a part of one of these pathways because some anticancer agents are involved in this metabolism. Deoxythymidine monophosphate (dTMP) is formed from deoxyuridine monophos-

All needed nucleotides can be made by the body.

phate (dUMP) by the addition of a methyl group donated by a coenzyme derived from folic acid (methylene THF).

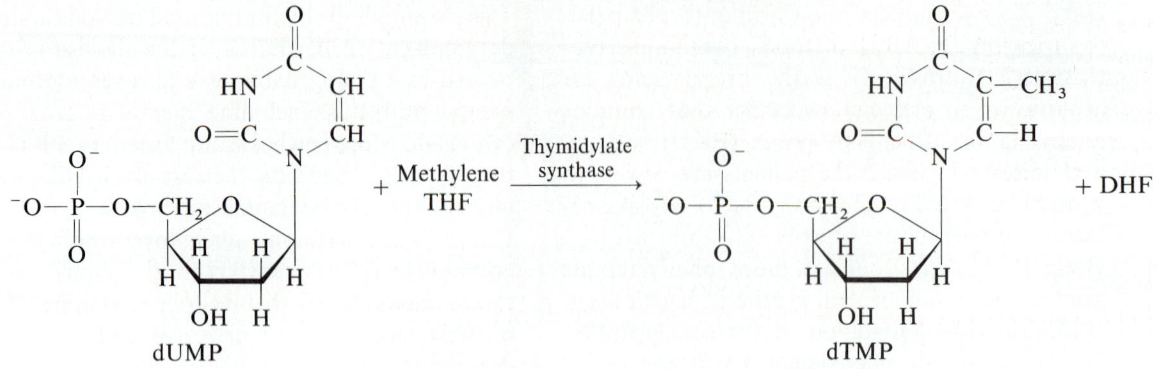
dUMP
Methylene THF
Thymidylate synthase
+ DHF
dTMP

In the body, the anticancer drug fluorouracil is altered to form an irreversible inhibitor of thymidylate synthetase. When the appropriate amounts of fluorouracil have been administered, only a small amount of dTMP is formed. Other compounds, such as methotrexate, prevent the recycling of the coenzyme DHF to methylene THF (see Chapter 27, Section 27.4). If the concentration of methylene THF is low, only a very little dTMP is formed.

Why do these agents work as anticancer drugs? Cancer cells divide more rapidly than most normal cells. Cell division requires DNA replication, which in turn requires the deoxynucleotides. Inhibition of dTMP synthesis kills rapidly dividing cells. Many of the side effects of cancer chemotherapy can be traced to the loss of normal rapidly dividing cells. Hair cells die, and cells lining the gastrointestinal tract are affected as are the cells that divide to form blood cells. Hair loss, gastrointestinal problems, and suppressed immunity are all side effects of chemotherapy.

TESTING YOURSELF

Biosynthesis of Nucleotides

1. How does fluorouracil kill rapidly dividing cells such as cancer cells?

Answer 1. This compound is metabolized to an inhibitor of thymidylate synthase, a key enzyme in the synthesis of deoxytymidylate. This nucleotide is needed for the synthesis of DNA, which must be synthesized *before* cell division can occur. If there is little or no DNA synthesis, there will be little or no cell division.

SUMMARY

Anabolism is that part of metabolism that is involved in the synthesis of larger biomolecules. ATP provides much of the energy for these endothermic reactions and *NADPH* provides the electrons needed for reductions. *Photosynthesis* is the ultimate anabolic process because photosynthetic organisms provide the biomolecules needed by all other organisms. The *light reaction* of photosynthesis provides ATP and NADPH, while the *dark reaction* uses these compounds and carbon dioxide to make glucose. Glucose can also be made through *gluconeogenesis* from lactate, glycerol, and glucogenic amino acids, but it cannot be made from fatty acids. Glucose is used to make the other monosaccharides, disaccharides, and glycogen that are needed by the body.

Lipogenesis is the process that synthesizes fatty acids in the body. ATP provides the energy, and NADPH produced in part by the *pentose phosphate pathway* provides electrons. Fatty acids combine with glycerol to make triacylglycerols, and with glycerol, phosphate, and several small molecules to make phosphoacylglycerols. Cholesterol is synthesized from acetyl CoA. This synthesis can be inhibited by drugs in some people to help reduce high blood cholesterol levels. The body also synthesizes ketone bodies from acetyl CoA. Only some of the 20 amino acids needed for protein synthesis are synthesized by the body. The rest are essential and are required in the diet. All of the nucleotides required by the body can be synthesized by the body. Rapidly dividing cells require more nucleotides for DNA synthesis, thus several inhibitors of nucleotide synthesis are used as anticancer drugs.

TERMS

anabolism
nicotinamide adenine dinucleotide phosphate (NADPH)
heterotrophic organism
autotrophic organism
photosynthesis
light reaction
dark reaction
gluconeogenesis
Cori cycle
glycogenesis
lipogenesis
acyl carrier protein (ACP)
pentose phosphate pathway

CHAPTER REVIEW

QUESTIONS

1. The molecule that provides energy to drive many of the endothermic reactions of anabolism is
   a. ATP b. NAD^+ c. $FADH_2$ d. NADPH
2. What molecule provides electrons for reduction reactions in the synthesis of palmitic acid?
   a. ATP b. NADH c. $FADH_2$ d. NADPH
3. In photosynthesis, what compound is the ultimate source of electrons?
   a. acetyl CoA b. chlorophyll c. glucose d. water
4. What molecule accepts carbon dioxide during the dark reaction of photosynthesis?
   a. NADPH b. glucose c. ribulose diphosphate d. water
5. Which of these processes does not yield glucose as a product?
   a. Cori cycle b. gluconeogenesis c. glycolysis d. photosynthesis
6. Which of the following cannot be used to synthesize glucose?
   a. fatty acids b. glycerol c. lactate d. alanine
7. The synthesis of which of these compounds does not require acetyl CoA?
   a. acetoacetic acid b. cholesterol c. glucose d. palmitic acid
8. Glycerol is used to synthesize which of these compounds?
   a. phosphatidyl choline c. glucose
   b. triacylglycerols d. all of the above
9. Oxaloacetate can be used to make which amino acid?
   a. aspartate b. α-ketoglutarate c. glutamate d. urea
10. What is the role of methylene THF in thymidine biosynthesis?
    a. donate the phosphate group
    b. donate a methyl group
    c. donate a pentose group
    d. donate a nitrogenous base

Answers **1.** a **2.** d **3.** d **4.** c **5.** c **6.** a **7.** c **8.** d **9.** a **10.** b

DIAGNOSTIC CHART

Blacken in all of the circles under the number of each question you missed in the Chapter Review questions. The diagnostic chart will help you identify concept areas that need more study.

	Questions									
Concepts	1	2	3	4	5	6	7	8	9	10
Anabolism	○									○
Photosynthesis			○	○						
Carbohydrate biosynthesis					○	○	○	○		
Lipid biosynthesis		○					○	○		
Amino acid biosynthesis									○	
Nucleotide biosynthesis										○

EXERCISES

Anabolism

1. Compare anabolism to catabolism with respect to: (a) size of reactants and products, (b) whether the process is oxidative or reductive, and (c) whether the process is endothermic or exothermic.
2. Compare the roles and structures of NAD^+–NADH and $NADP^+$–NADPH. Why are there two coenzymes with such similar structures yet different biological roles?

Photosynthesis

3. How do heterotrophic organisms obtain organic compounds? How do autotrophic organisms obtain organic compounds?
4. Write a balanced equation for the synthesis of 1 mol of glucose by the process of photosynthesis.
5. Make a table showing the products and reactants of the light reaction of photosynthesis.
6. Is the light reaction exothermic or endothermic? Where does the energy go to or come from?
7. Explain how the chemiosmotic theory is involved in the light reaction.
8. Compare the roles of the chemiosmotic theory in photosynthesis and oxidative phosphorylation.
9. How are ATP and NADPH used in the dark reaction of photosynthesis?
10. What enzyme is involved in "fixing" inorganic carbon dioxide into organic molecules?

Biosynthesis of Carbohydrates

11. What is the role of gluconeogenesis in carbohydrate metabolism?
12. What molecules serve as precursors to gluconeogenesis? What is the product?
13. What is the normal fate for lactate produced in oxygen deficient muscles? Where do these reactions occur?
14. What is meant by glucogenic amino acids?
15. Write a balanced word equation for glycogen synthesis.
16. Formation of the glycosidic bonds in glycogen is endothermic. What is the source of energy for this reaction?

Biosynthesis of Lipids

17. Explain the transport of acetyl CoA from the mitochondrion to the cytosol.
18. Compare the number of moles of acetyl CoA, ATP, $FADH_2$, and NADH (NADPH) produced in the catabolism of palmitate to those consumed in the anabolism of palmitate. If acetyl CoA and reduced coenzymes were used to make ATP, how much more ATP is needed to make palmitate than is produced from it?
19. What is the product of fatty acid synthase? How do humans make the other fatty acids?
20. How is glycerol phosphate synthesized in the body?
21. Describe the synthesis of phosphatidyl ethanolamine.
22. What is squalene? What important class of biomolecules is synthesized from squalene?

Biosynthesis of Amino Acids

23. You have made a controlled diet for a population of mice. This diet is complete except for the lack of cysteine, a nonessential amino acid. The other 19 amino acids are present in the minimal amount needed by mice. The young mice grow slowly and show signs of malnutrition. What is wrong with your diet?
24. How can transamination serve a role in both catabolism and anabolism of amino acids?

Biosynthesis of Nucleotides

25. Chemotherapy for cancer patients has several common side effects. What is the cause of these side effects?

CHAPTER

30

Blood: The Constant Internal Environment

The body maintains a constant internal environment through self-regulated feedback mechanisms. Blood plays a key role in this control through transport of gases and wastes and by maintenance of nutrients and pH within a normal range. These processes are highly regulated, in part by hormones that circulate in the blood. Blood also provides protection from disease by circulating components of the immune system.

OBJECTIVES

After completing this chapter, you should be able to

- Define the term *homeostasis.*
- Explain the role of blood in gas exchange and transport.
- Describe how filtration and reabsorption are involved in elimination of wastes.
- Describe the factors that influence and regulate blood pH.
- Explain the mechanism of hormone action.
- Describe the role of glucagon and insulin in the maintenance of nutrients in blood.
- Explain how antibodies contribute to the immune system.

Warm-blooded organisms maintain a more or less constant internal temperature. But temperature is only one of several factors that the body maintains within certain limits. Concentrations of nutrients, waste products, and oxygen are also closely regulated. **Homeostasis** is a state of dynamic equilibrium that maintains a constant internal environment in the body. This is accomplished through self-regulating processes that rely on feedback. An example of a homeostatic process is the regulation of body temperature. When the body is too warm, blood flow increases in the skin, which increases the rate of heat loss to air. When the body is too cold, blood flow to the skin is reduced. Internal temperature is self-regulating since body temperature influences the rate of heat loss.

Blood plays an important role in the maintenance of a constant internal environment.

Blood plays a major role in homeostasis because it is in contact with the interstitial fluids that surround all cells. Nutrient concentrations in the blood are maintained within a narrow range; the nutrients are thus available for ready absorption by cells. Cellular wastes diffuse from cells into blood and are transported to excretory organs. Blood circulates hormones that help control and coordinate body activity. Blood also contains a variety of cells and proteins that neutralize or destroy foreign cells or chemicals. Many of these processes involve chemistry that you have already learned in this book. This chapter discusses the chemistry of blood.

30.1 Gas Exchange

Gas Exchange in the Lungs

The cells of the body consume oxygen in the electron transport chain and produce carbon dioxide, principally in the TCA cycle (see Chapter 28). Oxygen must be absorbed from the atmosphere by the lungs, transported by the blood to the tissues, and moved by diffusion from the blood into the tissues. Carbon dioxide produced by aerobic catabolism diffuses from the cells through the interstitial fluid into the blood. It is then transported by the blood to the lungs, from where it leaves the body.

Gas exchange occurs across the alveoli of the lungs.

Inhaled air that is relatively rich in oxygen passes through increasingly smaller passageways in the lungs to enter microscopic sacs called **alveoli** (Figure 30.1). Oxygen within the alveoli dissolves into the moist alveolar membranes. The concentration of oxygen within the alveolar membranes is higher than the concentration of oxygen in the blood that is flowing through the capillaries of the lungs. Because materials generally move from high concentration to low concentration, oxygen thus diffuses into the blood.

Gas Transport in the Blood

The solubility of oxygen in plasma (the liquid part of blood) is rather low (0.3 mL per 100 mL), and the body cannot transport enough dissolved oxygen in blood plasma to sustain life. This limitation is overcome by the presence of an additional oxygen transport system. The red blood cells contain **hemoglobin**, an oxygen-binding protein. Hemoglobin greatly increases the capacity of the blood to transport oxygen. As the concentration of dis-

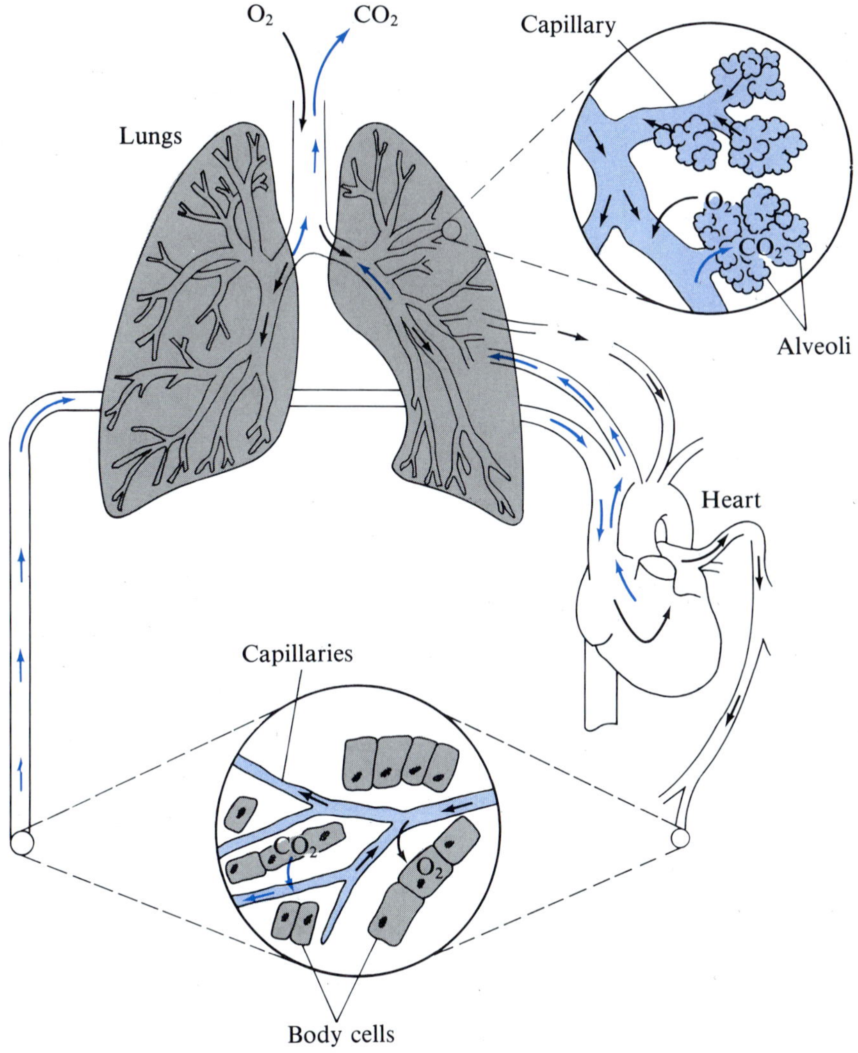
O_2
CO_2
Capillary
Lungs
O_2
CO_2
Alveoli
Heart
Capillaries
CO_2
O_2
Body cells

FIGURE 30.1
The movement of oxygen and carbon dioxide within the respiratory and circulatory systems. Oxygen (black arrows) moves from the lungs through the circulatory system to the tissues of the body. Carbon dioxide (blue arrows) flows in the opposite direction. These gases pass from one phase or system to another by diffusion.

solved oxygen increases within the red blood cells, oxygen binds to hemoglobin molecules within the cells. Each hemoglobin molecule can bind four oxygen molecules, one to each of the four subunits present in the protein (see Figure 25.10). Oxygen binds to hemoglobin *cooperatively*, that is, the binding of the first oxygen molecule to hemoglobin makes the binding of other oxygen molecules easier. Thus, other oxygen molecules rapidly bind to the hemoglobin molecule. When blood leaves the capillaries of the lungs, most hemoglobin molecules have four oxygen molecules bound to them (Figure 30.2). The amount of oxygen in blood leaving the lungs is 20 mL per 100 mL, mostly as oxygen bound to hemoglobin. This is far more than can be transported by a simple aqueous solution of oxygen.

Hemoglobin greatly increases the oxygen carrying capacity of blood.

Oxygen rich blood is pumped to the tissues of the body. In these tissues, the concentration of oxygen is appreciably lower than the concentration in blood. Oxygen diffuses from the blood into the surrounding fluids. This diffusion decreases the amount of dissolved oxygen in the blood and promotes the dissociation of oxygen molecules from hemoglobin. Like binding, the

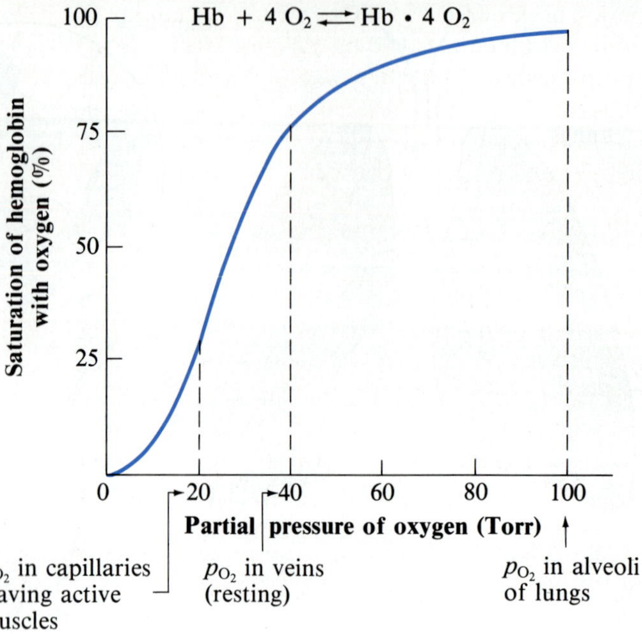
Hb + 4 O2 ⇌ Hb • 4 O2
Saturation of hemoglobin with oxygen (%)
100
75
50
25
0
20
40
60
80
100
Partial pressure of oxygen (Torr)
pO2 in capillaries leaving active muscles
pO2 in veins (resting)
pO2 in alveoli of lungs

FIGURE 30.2
Hemoglobin (Hb) binds oxygen in the lungs and releases it to the tissues of the body. Blood leaving active tissues has less oxygen (p_{O_2} = 20 Torr), and hemoglobin is only 25% saturated with oxygen. In a resting individual, veins have more oxygen (p_{O_2} = 40 Torr), and hemoglobin is around 75% saturated. In lungs, there is more oxygen (p_{O_2} = 100 Torr), and hemoglobin is nearly 97% saturated with oxygen. Hemoglobin loads up with oxygen in the lungs and delivers it to tissues of the body.

dissociation of oxygen from hemoglobin is cooperative; the first dissociation is quickly followed by others. This unloading of oxygen from hemoglobin ensures that the concentration of dissolved oxygen in the blood exceeds the concentration in the interstitial fluids. Adequate amounts of oxygen diffuse into the tissues as the blood passes through the capillaries. In a resting individual about 25% of the oxygen in blood is delivered to the tissues as the blood passes through. Considerably more is delivered to tissues that are metabolically active.

Gas Utilization and Production by Cells

The oxygen that diffuses into the cells of the body is used within the mitochondria for aerobic metabolism. Carbon dioxide is a product of aerobic metabolism and is only slightly soluble in body fluids. However, it reacts with water to form carbonic acid, which readily dissociates to yield bicarbonate ion.

$$\underset{\textbf{Carbon dioxide}}{CO_2} + H_2O \xrightleftharpoons{\text{Carbonic anhydrase}} \underset{\textbf{Carbonic acid}}{H_2CO_3} \rightleftharpoons \underset{\textbf{Bicarbonate ion}}{HCO_3^-} + H^+$$

The formation of carbonic acid is catalyzed by the enzyme *carbonic anhydrase* found within red blood cells. Some carbon dioxide also reacts with certain amino groups within hemoglobin molecules. Carbon dioxide in blood is thus transported in several forms: (1) as dissolved carbon dioxide, (2) as a complex with hemoglobin, and (3) as bicarbonate ion. Of the three, bicarbonate is the principal transport form. When venous blood arrives at the lungs, some of the dissolved carbon dioxide diffuses into the alveoli because the alveolar concentration of carbon dioxide is lower due to exchange with inhaled air. As the concentration of carbon dioxide decreases in the blood, carbonic acid dissociates to yield more carbon dioxide and water. This decreases the amount of dissolved carbonic acid, so bicarbonate combines with H^+ to form more carbonic acid, which yields even more carbon dioxide and water. This conversion of carbonic acid and bicarbonate into carbon dioxide is an excellent example of Le Chatelier's principle. In addition, as the concentration of carbon dioxide in the blood decreases, it is released from the amino groups of hemoglobin. By the time the blood has passed through the capillaries of the lungs, a significant amount of carbon dioxide has diffused into the lungs (Figure 30.1). Simultaneously, oxygen has been taken up by the blood.

Bohr Effect

The description given so far of gas movement between lungs and tissues implies that the entire process is diffusion driven—an increase in oxygen utilization in tissues would increase the concentration gradient of oxygen between the blood and the tissues. More oxygen would diffuse more rapidly from blood to tissues because of the steeper concentration gradient. A similar effect would occur at the lungs. This description is true, but hydrogen ion concentration (pH) and carbon dioxide concentration also influence the amount of oxygen delivered to tissues. The increased delivery of oxygen due to increased concentrations of hydrogen ion and carbon dioxide is called the **Bohr effect**.

Hemoglobin binds oxygen reversibly. At the relatively high concentrations of oxygen in the lungs, hemoglobin readily binds oxygen, while at the relatively low concentrations characteristic of tissues, hemoglobin releases oxygen. But hemoglobin also binds hydrogen ions (H^+) and carbon dioxide molecules. Carbon dioxide and hydrogen ions work like allosteric effectors to produce the Bohr effect. When they bind to hemoglobin, they induce a conformational change in the hemoglobin molecule to a form that binds oxygen molecules less tightly. Increased concentrations of these substances result in their increased binding to hemoglobin, which decreases the ability of hemoglobin to bind oxygen (Figure 30.3).

Increased concentrations of H^+ and CO_2 in blood result in greater oxygen unloading from hemoglobin.

The physiological consequences of the Bohr effect are twofold: (1) increased delivery of oxygen upon demand and (2) increased binding of oxygen to hemoglobin in the lungs. When the metabolism of muscles increases during exercise, carbon dioxide is produced by aerobic metabolism and lactic acid is produced during anaerobic exercise. Both are acids and will yield hydrogen ion in water. Thus, the pH of metabolically active tissue decreases during exercise. The hydrogen ion and carbon dioxide diffuse into blood where these molecules bind to hemoglobin. This binding promotes rapid unloading of

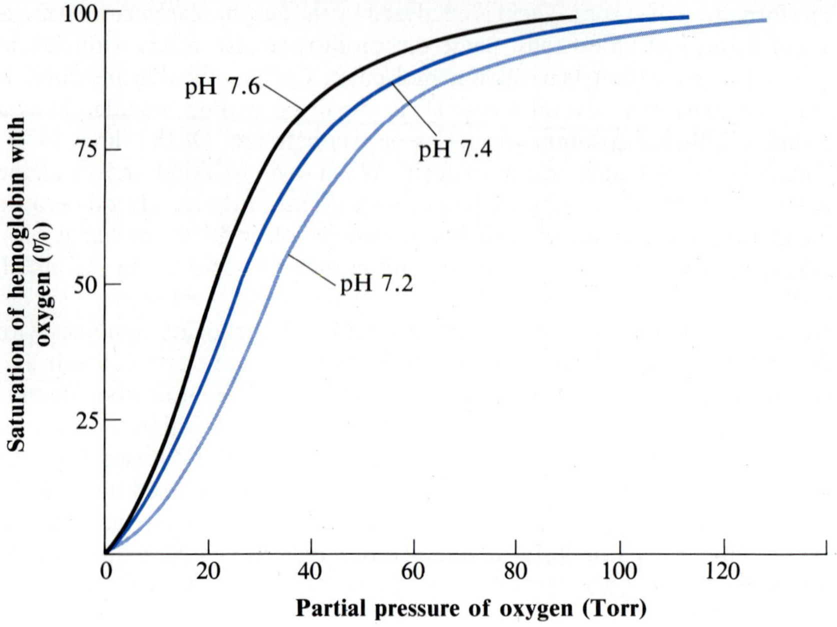
100
75
50
25
Saturation of hemoglobin with oxygen (%)
pH 7.6
pH 7.4
pH 7.2
0
20
40
60
80
100
120
Partial pressure of oxygen (Torr)

FIGURE 30.3
The effects of pH on oxygen binding to hemoglobin. Increased concentration of hydrogen ion (lower pH) reduces the binding of oxygen to hemoglobin. Since active tissue is more acidic (pH = 7.2) than less active tissue (pH = 7.4), more oxygen is delivered to active tissue than to less active tissue. Increased concentrations of carbon dioxide also reduce oxygen binding to hemoglobin.

TABLE 30.1
Effects of pH on the Amount of Oxygen Delivered to Tissues

Tissue pH	mL of Oxygen Delivered per 100 mL of Blood
7.2 (active tissue)	6.6
7.4 (normal)	5.0
7.6	3.0

oxygen. Thus, more oxygen is delivered to the locations where it is most needed, that is, to metabolically active tissues (Table 30.1). The liver removes lactic acid from the blood (see Chapter 29), and carbon dioxide is expelled in the lungs. Loss of these metabolites increases blood pH. Under these conditions, hemoglobin binds oxygen more tightly. The decrease in carbon dioxide concentration and the increased pH promote the oxygenation of hemoglobin in the lungs.

NEW TERMS

homeostasis Maintenance of a constant internal environment.

alveoli Tiny sacs in the lungs where gas exchange occurs.

hemoglobin Protein in red blood cells that binds oxygen cooperatively. The presence of hemoglobin in blood greatly increases the capacity of blood for oxygen transport.

Bohr effect Increases in the concentration of carbon dioxide and hydrogen ions increase dissociation of oxygen from hemoglobin. Decreases in concentration have the opposite effect.

TESTING YOURSELF

Gas Exchange

1. Where is the site of gas exchange between the body and air?
2. Compare the oxygen carrying capacity of whole blood and plasma.
3. What effect would alkalosis (increased blood pH) have on oxygen delivery to tissues?

Answers **1.** in the alveoli of the lungs **2.** Plasma carries about 0.3 mL per 100 mL, whole blood about 20.0 mL per 100 mL. **3.** Less oxygen would be delivered (Bohr effect).

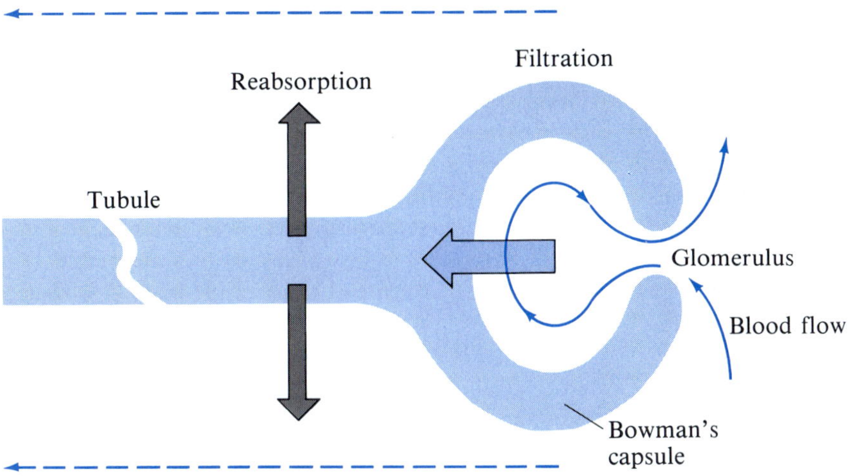

Filtration
Reabsorption
Tubule
Glomerulus
Blood flow
Bowman's capsule

FIGURE 30.4
Filtration and reabsorption in a nephron of the kidney. Blood enters the glomerulus of the kidney and is filtered. The filtrate is channeled by the Bowman's capsule into the tubule (thick blue arrow). Cells lining the tubule reabsorb water and nutrients (dark gray arrows) and return them to the blood.

30.2 Removal of Wastes from the Blood

Carbon dioxide is a waste product that is circulated by blood to the lungs, from where it leaves the body. Some water is also lost from the lungs. Sweat glands may also excrete significant amounts of water and salt under certain conditions. Most other wastes carried by blood are excreted by the kidneys. The physiological function of the kidneys is beyond the scope of this book, but several of the processes have chemical components that you have learned and are thus worth reviewing here.

The walls of the capillaries in parts of the kidneys are quite porous. As blood passes through the glomerulus, which is a cluster of capillaries, blood pressure forces water and solutes through the walls into Bowman's capsule, which funnels them into tubules. Cells and macromolecules such as proteins are too large to pass through the pores in the capillaries, thus the blood is filtered. Through **filtration**, large particles and cells are retained in the circulatory system and water and solutes pass out of it (Figure 30.4). Water, nutrients, and waste products are thus all filtered from the blood into the tubules of the kidney.

Nutrients do not remain in the kidney filtrate, however. The cells lining the tubules have active transport systems to reabsorb nutrients such as glucose and amino acids. Water and ions may also be reabsorbed, but the actual amount reabsorbed is highly regulated. Normal kidney function can be lost through infection or environmental stress. Artificial kidney dialysis can be used to remove wastes from the blood; this process was discussed in Chapter 9.

The ability to reabsorb nutrients in the kidneys is limited. If the concentration of a particular solute exceeds a concentration known as the **renal threshold**, then reabsorption will be incomplete and some of the solute will appear in the urine. This occurs with untreated diabetics. Following a meal,

Diabetics sometimes excrete glucose because the concentration of blood glucose exceeds its renal threshold.

KIDNEY DIALYSIS

Kidneys perform the vital function of clearing wastes and excess water and salts from blood. If the kidneys fail, the constant internal chemical environment changes, and life-threatening problems arise. *Kidney* or *renal dialysis* is a treatment for kidney failure. In one form of dialysis, the patient's blood is passed through fine, semipermeable tubules that are bathed in a solution called the *dialysate*. This solution contains glucose, electrolytes, and other needed substances. Since both blood and the dialysate contain about the same concentration of these nutrients, there is no net diffusion between them. The dialysate does not contain waste products, and these materials diffuse from blood into the dialysate. Excess electrolytes can be removed by using a dialysate with the appropriate concentration of electrolytes. Excess water can be removed by making the dialysate hypertonic to the blood. A simpler form of renal dialysis takes advantage of the semipermeable nature of the membranes of the body. In this method, a dialysate is introduced into the abdominal cavity, where diffusion between the dialysate and blood can occur across the lining of the abdominal cavity. Both methods use simple physical properties—diffusion and osmosis—to accomplish what the kidneys can no longer do.

glucose is absorbed from the small intestine into the blood. The cells of diabetics do not take up glucose as rapidly as normal cells (Section 30.4). Glucose accumulates in the blood and may exceed the renal threshold for glucose, which is 180–200 mg per 100 mL of blood.

NEW TERMS

filtration Process that separates large molecules from small ones by exclusion. Small molecules pass through pores, but large molecules and particles are retained.

renal threshold The maximum concentration of a substance that can be effectively reabsorbed by the kidneys. If the concentration exceeds this threshold, the substance will appear in the urine.

30.3 Regulation of pH

You learned in Chapter 25 that proteins have a native conformation that can be lost if the pH varies. Control of pH within a cell is critical, but because intracellular pH is influenced by extracellular fluids, control of the pH of these fluids is equally critical. If blood pH changes, the proteins of blood and cells may be denatured.

Buffering Systems in Blood

The best way to control pH is with *buffers*. As previously defined in Chapter 11, buffers are solutions of a weak acid and its conjugate base. If hydrogen ion is lost from the system, the acid will give up hydrogen ions to replace those lost by the system.

$$HA \longrightarrow H^+ + A^-$$

This hydrogen ion is released from the conjugate acid to replace those lost from the system.

If hydrogen ion is added to the system, then the conjugate base accepts the H^+ to become the conjugate acid.

$$HA \longleftarrow H^+ + A^-$$

This hydrogen ion added to the system is taken up by the conjugate base.

Collectively these two reactions provide protection against addition or loss of hydrogen ion.

Buffers resist changes in pH because they can accept or donate hydrogen ions.

$$HA \rightleftharpoons H^+ + A^-$$

The conjugate acid and base of this system take up or give up hydrogen ions to prevent changes in pH.

Blood contains three important buffering systems: (1) a bicarbonate system, (2) a phosphate system, and (3) a protein system. Recall that carbon dioxide produced during respiration combines with water to form carbonic acid which dissociates to yield bicarbonate.

Bicarbonate, phosphate, and proteins buffer blood.

$$\underset{\textbf{Carbon dioxide}}{CO_2} + H_2O \rightleftharpoons \underset{\textbf{Carbonic acid}}{H_2CO_3} \rightleftharpoons \underset{\textbf{Bicarbonate}}{HCO_3^-} + H^+$$

Because the concentration of bicarbonate is relatively high in blood, carbonic acid and bicarbonate are a conjugate acid–base pair that makes a significant contribution to the buffering of blood.

Phosphoric acid is a weak triprotic acid, that is, it can donate three hydrogen ions. Near physiological pH (pH 7), this acid exists mainly as either $H_2PO_4^-$ or HPO_4^{2-}. These two ions are an acid–base conjugate pair that contributes to blood buffering.

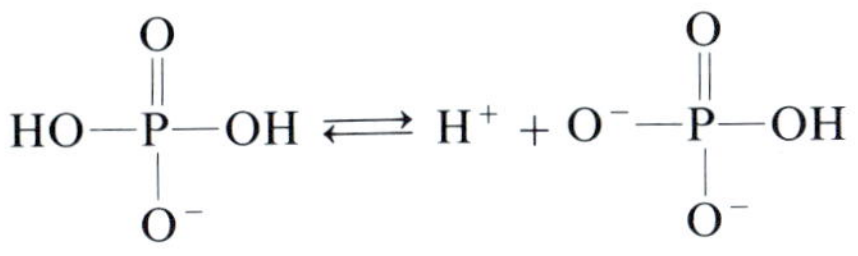

$H_2PO_4^-$ $\qquad$ HPO_4^{2-}

The third buffering system of the blood involves proteins. All blood proteins contribute to the buffering of blood, but hemoglobin, the most abundant blood protein, makes the largest contribution. All proteins possess some amino acids with basic or acidic side groups, and they also possess N-terminal amino groups and C-terminal carboxyl groups. These groups are potential hydrogen ion donors and acceptors that moderate blood pH.

Acidosis and Alkalosis

The body normally maintains blood pH within a narrow range from pH 7.35 to 7.45, but certain physiological states may push the pH beyond this range. If the pH drops below this range, the condition known as **acidosis** exists. If blood pH rises above this range, then **alkalosis** exists. Both of these conditions can be described as respiratory or metabolic, depending upon the cause of the pH imbalance.

Respiratory acidosis occurs when carbon dioxide accumulates in the blood. This can occur through *hypoventilation*, which is abnormally low breathing rates and shallow breathing resulting in reduced loss of carbon dioxide. Whenever the rate of carbon dioxide loss is below the rate of production, carbon dioxide accumulates. The increased carbon dioxide is accompanied by an increase in carbonic acid, which yields an increase in hydrogen ion. *Hyperventilation* (breathing that is too rapid and deep) can cause *respiratory alkalosis* because carbon dioxide is lost at a rate greater than its production. Less carbon dioxide means less carbonic acid and less hydrogen ion. Loss of this acid means the blood will become more basic, that is, more alkaline.

Alterations in normal metabolism can cause *metabolic acidosis* and alkalosis. A principal cause of metabolic acidosis is *ketosis*, the presence of an excess of ketone bodies in the blood. The principal ketone bodies, acetoacetic acid and β-hydroxybutyric acid, ionize to yield hydrogen ion. If the concentration of ketone bodies is high enough, the buffering capacity of the blood is overwhelmed and the blood pH drops. *Metabolic alkalosis* is relatively uncommon, perhaps because the common body metabolites are not alkaline, but excess loss of stomach acid can yield metabolic alkalosis. Ingestion or intravenous feeding of excess bases can also raise blood pH.

NEW TERMS

acidosis Condition that exists whenever the blood pH drops below the normal range; can be metabolic or respiratory in origin.

alkalosis Condition that exists when blood pH exceeds the normal range; like acidosis, it can result from respiratory or metabolic abnormalities.

TESTING YOURSELF

Blood Buffers

1. What are the three principal buffering systems in blood?
2. a. What is respiratory acidosis?
   b. What is the actual source of the hydrogen ion?

Answers **1.** bicarbonate, phosphate, and blood protein **2. a.** increased hydrogen ion concentration in blood due to an increase in blood carbon dioxide **b.** carbonic acid that forms from carbon dioxide and water

30.4 Circulation of Hormones and Their Role in Nutrient Maintenance

The chemical messengers circulated in blood are hormones.

Hormones are chemical messengers that regulate and coordinate the activity of the many different kinds of cells. Certain cells in endocrine glands produce hormones and secrete them into the blood where they are distributed throughout the body. The hormone molecules are then taken up by specific cells called *target cells*. The interaction between hormones and specific proteins of the cells results in an alteration of the cellular metabolism. Although there is considerable variation in the structures of hormones, they are classified into two main categories according to how they initiate a cellular response. Hormones of the first group actually enter the cell and exert their effect within the cell. This group includes the steroid hormones derived from cho-

TABLE 30.2
Some Important Hormones of the Body

Hormone	Endocrine Gland	Function
Cortisol	Adrenal cortex	Controls a variety of metabolic processes
Estrogen	Ovary	Involved in secondary sex traits and female reproductive cycle
Progesterone	Corpus luteum	Involved in female reproductive cycle
Testosterone	Testes	Principal male sex hormone
Insulin	Pancreas	Regulates concentration of nutrients in blood
Glucagon	Pancreas	Stimulates release of glucose and lipid into blood
Epinephrin	Adrenal	Induces a variety of effects in response to environmental stimuli

lesterol, the thyroid hormones, and some protein hormones such as epidermal growth factor. Members of the second group bind to the surface of the cell and exert their influence by initiating a chain of events that results in metabolic changes. This group includes such proteins as glucagon and insulin, peptides, and some amino acid derivatives such as epinephrin. Table 30.2 lists some important hormones in the body.

Hormones That Work within Cells

The hormones that enter a cell, such as the steroid hormones, diffuse into the cell and bind to specific receptor proteins. These complexes enter the nucleus where they bind at specific locations to the DNA–protein complexes located there (Figure 30.5). The binding of the hormone alters the expression

Some hormones enter cells and bind to DNA, thus affecting gene expression.

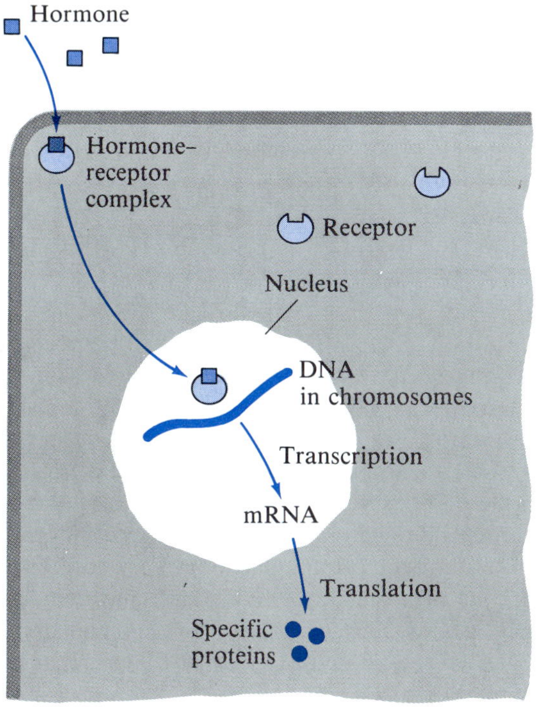
Hormone
Hormone–
receptor
complex
Receptor
Nucleus
DNA
in chromosomes
Transcription
mRNA
Translation
Specific
proteins

FIGURE 30.5
The entry and effects of a steroid hormone on a target cell. Hormone molecules in the blood stream bind to a cytosolic receptor protein at the cell membrane of the target cell. The hormone–receptor complex diffuses into the nucleus and binds at specific sites on DNA in chromosomes. This binding stimulates transcription of the DNA in that region, resulting in production of specific proteins.

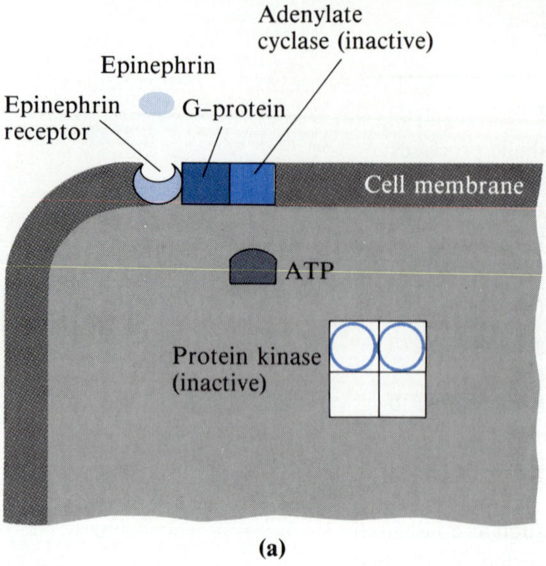
Adenylate
cyclase (inactive)
Epinephrin
Epinephrin
receptor
G-protein
Cell membrane
ATP
Protein kinase
(inactive)
(a)
(active)
Cell membrane
ATP
cAMP
Protein
kinase
(active)
(b)

FIGURE 30.6
The interaction between a target cell and epinephrin. (a) Target cells possess receptor molecules that bind the hormone on the outside surface of the cell. The membrane also has G-protein, and an enzyme (adenylate kinase) that is inactive if hormone is not bound. (b) When epinephrin binds to its receptor, the receptor conformation changes, which activates G-protein, which in turn activates adenylate cyclase. Adenylate cyclase catalyzes the formation of cAMP from ATP. cAMP binds to the enzyme protein kinase, which activates some of the subunits of this enzyme. (The role of protein kinase is shown in Figure 30.8.)

of genes in that region of the DNA. Often the activity of the cell is altered for an indefinite period of time, and the cell requires several hours or more to respond to the hormone.

Hormones That Work at Cell Surfaces

Members of the other group of hormones do not enter the cell. Instead, they bind to a specific receptor molecule in the cell membrane. The hormone–receptor complex serves as a signal to alter parts of the cell's metabolism. For example, the complex formed between the hormone epinephrin and its receptor triggers the activation of the enzyme adenylate cyclase to form **cyclic AMP** (3′,5′-adenosine monophosphate) within the cell. Cyclic AMP (or cAMP) activates the enzyme *protein kinase*, which in turn alters the activity of some other key proteins within the cell (Figure 30.6). Thus, the activity of the cell is changed. This process occurs rapidly, with a time frame measured in seconds, and is readily reversible. The hormone glucagon functions through this mechanism (see the next section).

Maintenance of Nutrients in Blood

The concentration of nutrients in the blood is normally maintained within a narrow range. This ensures that cells have these nutrients available whenever they need them. Humans do not eat continuously. Nutrients in the blood do not simply reflect the concentration of nutrients in the gastrointestinal tract. This section provides an introduction to the mechanisms through which the body uses and regulates nutrients in blood.

Insulin signals that an abundance of nutrients is present in blood.

Nutrient concentrations in blood are influenced by hormones. Although there are other hormones involved, the two primary hormones that influence blood nutrient concentrations are insulin and glucagon. **Insulin**, a "feast" hormone, signals an abundance of nutrients in the bloodstream. It is released when nutrients are available and need to be taken up, stored, or utilized.

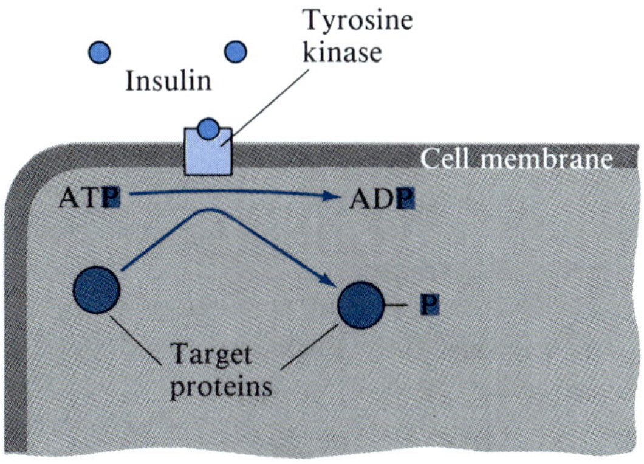
Tyrosine
kinase
Insulin
Cell membrane
ATP
ADP
P
Target
proteins

FIGURE 30.7
The activation of tyrosine kinase by insulin. Insulin binds to its receptor molecule, the enzyme tyrosine kinase, in the cell membrane of the target cell. The binding of insulin activates tyrosine kinase. This enzyme catalyzes the transfer of a phosphate (P) from ATP to target proteins in the cell. Phosphorylation of these proteins changes their activity and thus alters the metabolism of the cell.

Glucagon, a "famine" hormone, serves as a signal that nutrients are scarce and should be released from storage and mobilized into blood to make them available to body cells. Let's first consider the feast hormone insulin.

Glucagon signals that the concentration of fuel molecules is low and must be increased.

INSULIN: THE FEAST HORMONE Shortly after a meal is eaten, glucose, lipids, and amino acids begin to enter the bloodstream through absorption from the small intestine. Depending upon the size and composition of the meal, absorption can continue for several hours. This absorption increases the concentration of glucose, lipids, and amino acids in the blood. There is an ample supply of nutrients available to cells.

Under these conditions, insulin is secreted by the beta cells of the pancreas, and secretion of glucagon is inhibited. Insulin stimulates anabolic processes. Cellular uptake of sugars and amino acids from blood is stimulated. Cellular synthesis of fatty acids, glycogen, and proteins increases. Insulin signals that nutrients are abundant; the cellular equivalent of "eat, drink, and be merry" can now occur. The cells make needed proteins, and lipid and carbohydrate energy stores are replenished.

Insulin binds to insulin receptor proteins in target cells. This binding activates the receptor, which is an enzyme called *tyrosine kinase*. This enzyme, in turn, phosphorylates some of its own tyrosines and some tyrosines in several proteins within the target cells (Figure 30.7). Although specific details are not yet clear, insulin binding helps regulate anabolic processes in target cells, apparently by modifying the activity of one or more proteins.

GLUCAGON: THE FAMINE HORMONE Glucagon becomes important when the concentration of glucose in blood drops toward the lower end of the normal range. This can be thought of as a "famine" condition—nutrients for energy are becoming limited. At lower blood glucose concentrations, glucagon secretion from the pancreas is stimulated, and secretion of insulin is suppressed. The principal target cells for glucagon are liver cells and adipose cells. The ultimate effect of glucagon on these cells is the liberation of fuel molecules into the blood.

Consider the effects of glucagon on a liver cell. Glucagon binds to a specific receptor protein on the surface of the liver cell (Figure 30.8). This binding initiates a series of events that results in the activation of the membrane-bound enzyme adenylate cyclase. This enzyme converts ATP to cyclic AMP (cAMP). cAMP is released into the cell where it acts as a positive

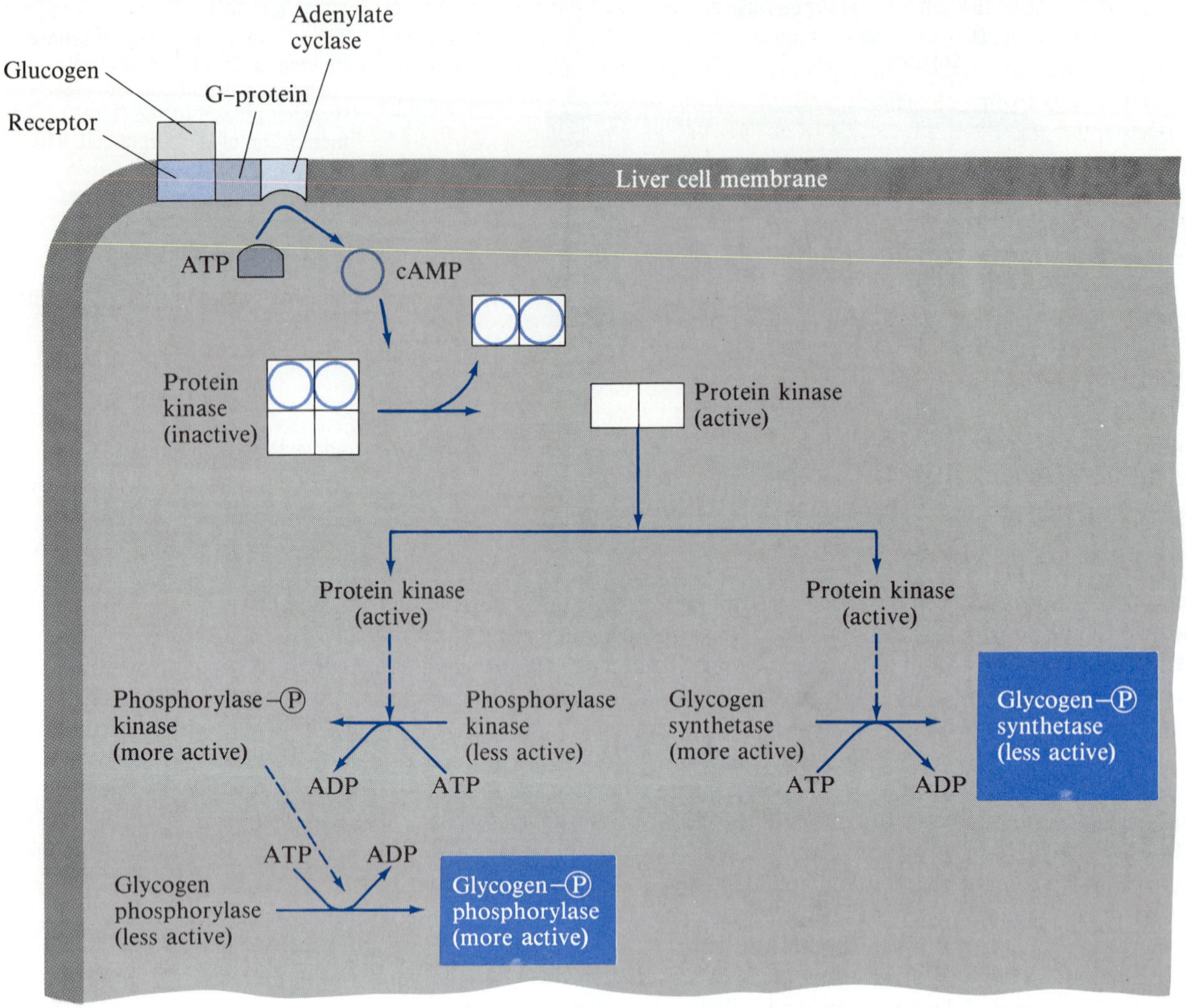
Adenylate cyclase
Glucogen
G-protein
Receptor
Liver cell membrane
ATP
cAMP
Protein kinase (inactive)
Protein kinase (active)
Protein kinase (active)
Protein kinase (active)
Phosphorylase–Ⓟ kinase (more active)
ADP
ATP
Phosphorylase kinase (less active)
Glycogen synthetase (more active)
ATP
ADP
Glycogen–Ⓟ synthetase (less active)
ATP
ADP
Glycogen phosphorylase (less active)
Glycogen–Ⓟ phosphorylase (more active)

FIGURE 30.8
The effects of glucagon on glycogen metabolism in liver cells. Glucagon, like epinephrin, activates protein kinase through a hormone receptor, G-protein, adenylate cyclase, and cAMP (see Figure 30.6). Protein kinase phosphorylates glycogen synthetase, which decreases glycogen synthesis, and it phosphorylates a second kinase (phosphorylase kinase), activating it. This enzyme catalyzes the phosphorylation of glycogen phosphorylase, activating it, which increases the rate of glycogen breakdown. In liver cells, protein kinase coordinates glycogen metabolism by simultaneously decreasing glycogen synthesis and increasing glycogen breakdown.

allosteric effector of an enzyme called protein kinase. cAMP is often referred to as a second messenger. The hormone serves as the first messenger, an extracellular messenger. Cyclic AMP is the second intracellular messenger.

Activated protein kinase attaches a phosphate group to several proteins in the cell. The activity of these proteins is thus increased or decreased by covalent modification. In liver, the enzyme glycogen phosphorylase is activated by covalent modification (Figure 30.8). This enzyme catalyzes the cleavage of glucose from glycogen with attachment of phosphate to the glucose to yield glucose 1-phosphate. The equation for this reaction (which we saw in Chapter 28) is

$$\text{Glycogen } (n \text{ residues}) + \text{Phosphate} \xrightarrow[\text{phosphorylase}]{\text{Glycogen}} \text{Glycogen } (n-1 \text{ residues}) + \text{Glucose 1-phosphate}$$

The phosphate is then cleaved from the glucose, which is then released into the blood. Glucagon thus causes an increase in blood glucose concentration by stimulating the release of glucose from liver glycogen.

Consider now the enzyme glycogen synthetase. This enzyme is responsible for the synthesis of glycogen from phosphorylated glucose. A moment's thought brings the realization that glycogen phosphorylase and glycogen synthetase have opposite effects. One is catabolic—it converts the macromolecule glycogen to a phosphorylated form of glucose. The other is anabolic—it takes a phosphorylated form of glucose and makes glycogen. The two should not have significant activity at the same time because the action of one, in essence, negates the action of the other. To work properly, one must be active while the other is relatively inactive.

The activities of glycogen phosphorylase and glycogen synthetase are regulated in concert. Protein kinases attach phosphate groups to both of these enzymes. But the effect on the catalytic activity of the enzymes is exactly opposite. Phosphorylation of the phosphorylase increases its activity; phosphorylation of the synthetase decreases its activity. One signal, the hormone, initiates a series of events that speeds up the release of energy molecules, a part of catabolism, and reduces the formation of glycogen, an anabolic process.

Glucagon simultaneously stimulates glycogen breakdown and inhibits glycogen synthesis.

LIPID MAINTENANCE Glucagon also affects fat cells. It binds to a receptor, and adenylate cyclase is activated to form cAMP, which activates protein kinase. Protein kinase, in turn, activates several proteins in fat cells including a lipase. The lipase hydrolyzes triacylglycerols to glycerol and fatty acids that are released into the blood. Glucagon, through its effects on the liver and adipose tissue, mobilizes glucose and lipid.

Glucagon mobilizes glucose and lipids when they become low in blood.

NEW TERMS

hormone Messenger molecule produced in one part of the body that is transported throughout the body in blood and that binds to target cells triggering events that alter the metabolism of the cell.

cyclic AMP Molecule produced in some cells in response to hormone binding to the cell membrane; cAMP is a second messenger.

insulin A hormone that signals that nutrients are abundant and stimulates anabolic activity in target cells.

glucagon A hormone that stimulates the release of fuel molecules into blood as needed.

TESTING YOURSELF

Hormones

1. Hormones must interact with their target cells to exert their effect. Describe the two types of initial interaction.
2. Why is cyclic AMP called a second messenger?
3. Describe how glucagon exerts opposite effects on glycogen breakdown and glycogen synthesis simultaneously.

Answers **1.** Some hormones bind to the target cell surface to exert their effects, others must enter the cell to exert their effects. **2.** Some hormones (first messenger) bind to the target cell surface and stimulate the synthesis of cAMP; cAMP then serves as a second, intracellular messenger. **3.** Glucagon stimulates the production of cAMP, which activates the enzyme protein kinase. This enzyme phosphorylates glycogen synthetase (which decreases its activity) and another kinase that phosphorylates glycogen phosphorylase (which increases its activity).

ABO BLOOD GROUP

Blood group involves the immune system and the concept of self and nonself. The ABO blood system has four different blood groups: A, B, AB, and O (Table 30.3). Individuals with blood group A have the A antigen protein on the surface of their red blood cells. Those with group B have B antigen and those with group AB have both. Individuals with Group O have neither. In an individual possessing the A protein or the B protein, the protein is not an antigen because it is a self-protein. If the A or B antigen is absent from red blood cells, the blood plasma contains antibodies (called *agglutinins*) to A or B (Table 30.3).

What would happen if group A blood were given to a group B individual? Since the individual with group B blood lacks the A protein, A is an antigen to that individual. The agglutinin (anti-A) in the recipient's blood plasma would bind to the A antigen on the donor red blood cells causing clumping (agglutination), which would initiate a response that could have grave consequences.

Because the antigens of the ABO blood group can elicit agglutination in other individuals, it is necessary to match donor blood to recipient blood for transfusions. Group O blood is considered to be the *universal donor* because type O blood contains no AB antigens on the red blood cells and thus can cause no agglutination in the recipient. Type AB has both A and B antigens, thus anti-A and anti-B are absent. People with AB blood are thus *universal recipients.*

TABLE 30.3
The ABO Blood Group System

	Blood Type			
Characteristic	A	B	AB	O
Antigen present on the red blood cells	A	B	Both A and B	Neither A nor B
Agglutinin normally present in the plasma	anti-B	anti-A	Neither anti-A nor anti-B	Both anti-A and anti-B
Incompatible with donor blood group	B, AB	A, AB	None	A, B, AB
Percent in a mixed Caucasian population	41	10	4	45
Percent in a mixed Black population	27	20	7	46

30.5 Blood and Immunity

The immune system protects against bacteria, viruses, parasites, and other foreign matter that enter the body. It is able to distinguish between self and nonself—molecules of the body (self) do not normally cause an immune response, but foreign (nonself) molecules do. The immune system has a chemical component and a cellular component. The chemical component includes numerous proteins that bind foreign molecules, and the cellular component includes a variety of cells such as white blood cells that are responsible for defending the body against invasion. There is considerable overlap in function and much interaction between them (Figure 30.9).

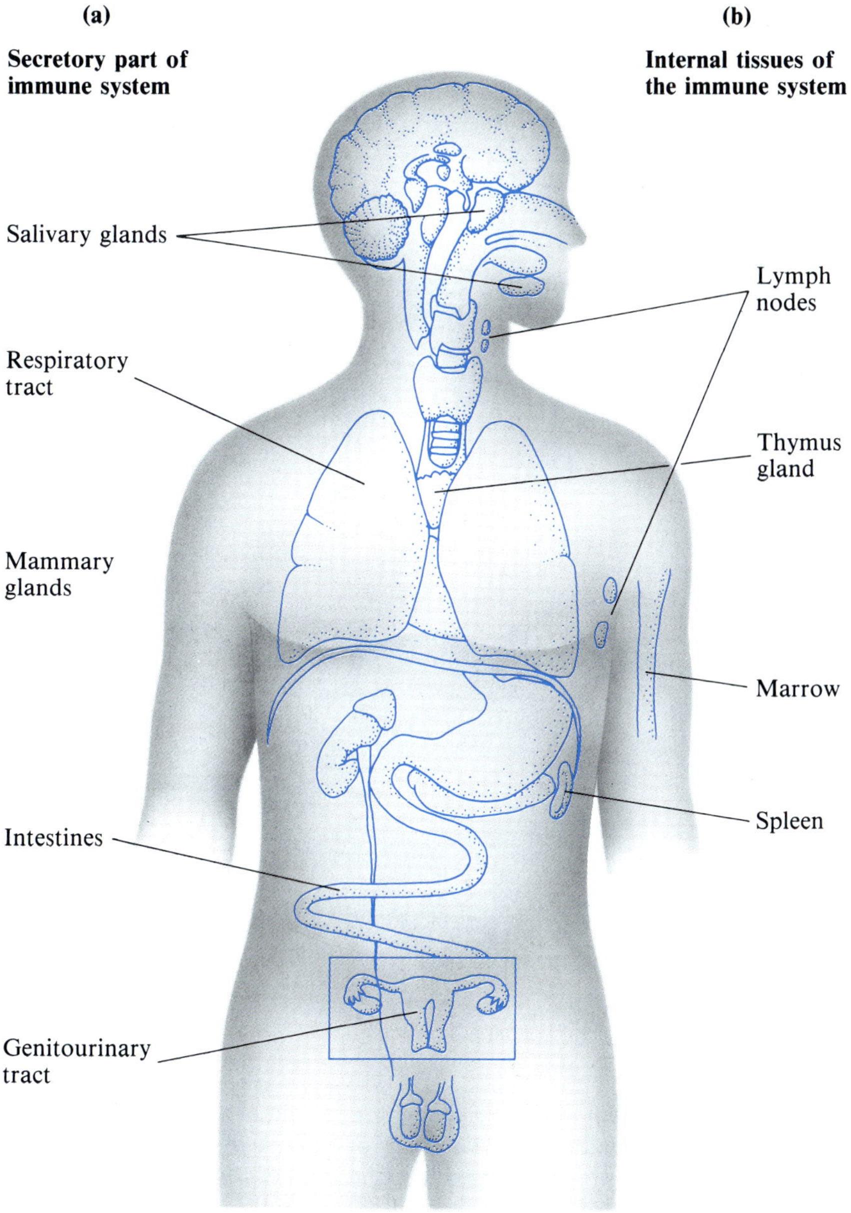

(a)
Secretory part of immune system
Salivary glands
Respiratory tract
Mammary glands
Intestines
Genitourinary tract
(b)
Internal tissues of the immune system
Lymph nodes
Thymus gland
Marrow
Spleen

FIGURE 30.9
The immune system produces proteins and cells that protect the body from foreign matter. (a) Several secretions (listed on the left) contain proteins (antibodies), which reduce the chances of infection. (b) The internal tissues (listed on the right) produce, modify, and distribute the cells and proteins of the immune system.

Antibodies and Antigens

The principal components of chemical defense are the **antibodies**, which are also called *immunoglobulins*. These macromolecules are proteins that bind specifically to large foreign molecules called **antigens**. Antibodies bind antigens like enzymes bind substrates, but antibodies normally have no catalytic activity. Antigens may be free molecules, but they are often portions of the cell surface of an invading cell or part of the surface of a viral particle that has entered the body. Complexes formed between antibodies and antigens are readily recognized by cells of the immune system. The cells ingest the complex and destroy the foreign cell or molecule. Antibodies aid immune cells in clearing the body of foreign bodies.

Antibodies are proteins that bind foreign molecules called antigens.

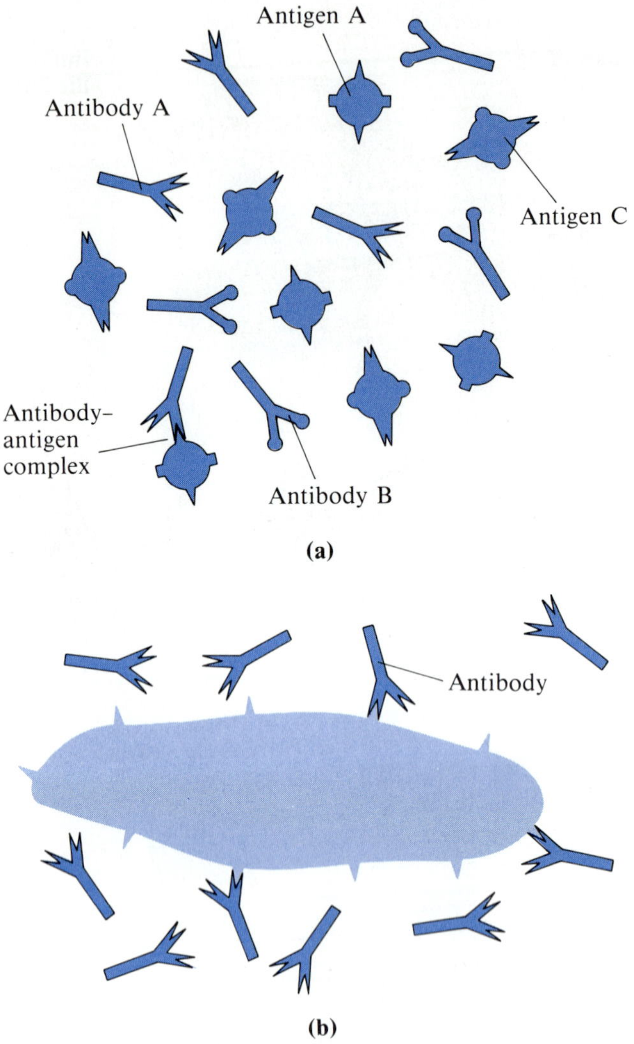
Antigen A
Antibody A
Antigen C
Antibody–
antigen
complex
Antibody B
(a)
Antibody
(b)

FIGURE 30.10
The binding of antibodies to antigens. (a) Antibodies bind to soluble antigens, usually proteins or polysaccharides, that are present in the body. Antibodies can only bind to antigen surfaces that are complementary to their binding site. Antibody A is binding to antigen A. (b) Antibodies can bind to the surfaces of bacteria, viruses, and parasites. Again, the surface must have a region that is complementary to the binding site of the antibody.

Each antibody can bind specifically to one kind of antigen. But because the body can potentially make many tens of millions of different antibodies that differ in their ability to bind specific antigens, virtually all foreign molecules will be an antigen to one or more antibodies (Figure 30.10). This ensures that antigen–antibody complexes will form to aid in ingestion by various immune cells.

All possible antibodies are not present in the body at any one time. Instead they are formed by an immune response whenever a new antigen enters the body. The antigen binds to the surface of an immune cell called a *B lymphocyte*. This cell possesses surface proteins that can bind specific parts of the antigen. The binding of the antigen to the B cell causes two major changes in the cell: (1) it begins to divide very rapidly, yielding many copies of the cell, each of which is a clone of the original cell; and (2) it changes into a new cell type called a *plasma cell*. Once developed, the plasma cells secrete antibodies that bind specifically to the antigen. The binding properties of the new antibodies are identical to the binding properties of the proteins on the B cell surface. This ensures that only populations of cells

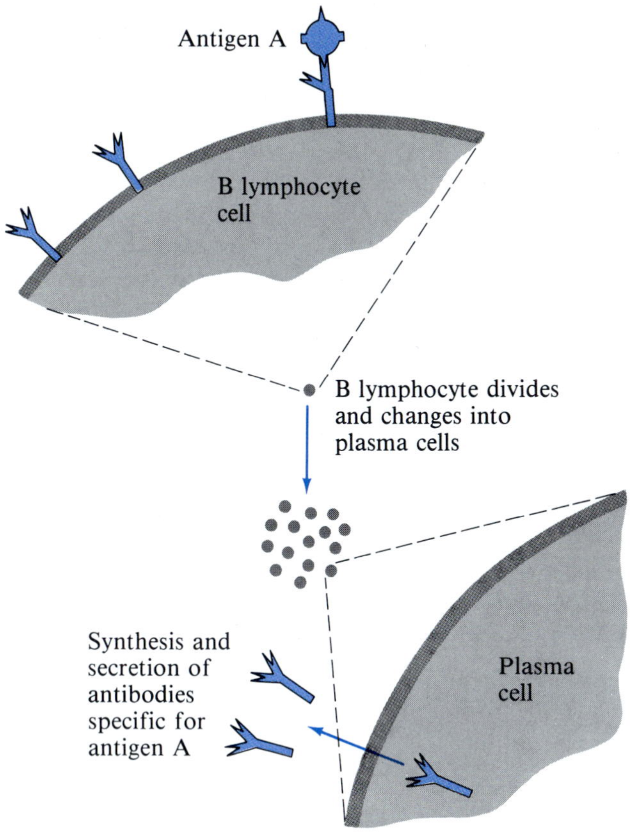
Antigen A
B lymphocyte cell
B lymphocyte divides and changes into plasma cells
Synthesis and secretion of antibodies specific for antigen A
Plasma cell

FIGURE 30.11
The immune response to an antigen. Antigens bind to complementary binding proteins on the surface of B lymphocytes. This binding stimulates the B lymphocyte to divide and initiates changes in the dividing cells that transform them into plasma cells. These plasma cells make and secrete antibodies specific for the antigen that bound to the B lymphocyte.

that will produce the proper antibodies are stimulated by the antigen (Figure 30.11).

The immune response just described takes seven to fourteen days to develop. The next time the organism is exposed to the same antigen, the response is much more rapid and massive. In a sense, the immune system has "memory" and thus responds more quickly to antigens it has previously encountered.

Interferons

Interferons are a group of proteins that interfere with the replication of viruses in cells; thus, they can also be considered as part of the chemical defense system. Interferons are produced by cells that are undergoing attack by viruses. The activity of viruses within body cells stimulates the cells to make and secrete interferons. Like some hormones, interferons bind to specific receptor molecules on the surface of other body cells. This binding stimulates the production of antiviral proteins within the cell that appear to protect

Interferons reduce viral replication in body cells.

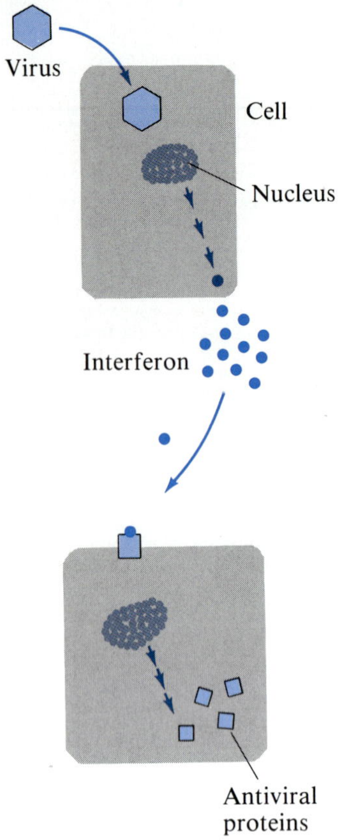
Virus
Cell
Nucleus
Interferon
Antiviral proteins

FIGURE 30.12
The antiviral action of interferons. Viral invasion of cells stimulates the synthesis and secretion of interferons. Even though the invaded cell may die, these proteins bind to noninfected cells and stimulate the production of antiviral proteins. If this cell is invaded by a virus, these antiviral proteins protect the cell by disrupting normal viral reproduction.

the cell from viral attack (Figure 30.12). Interferons have recently attracted considerable attention in medical research because they also appear to affect the growth of some kinds of cancer cells.

Cellular Immunity and AIDS

The cellular component of the immune system involves elements of biology that have not been covered in this book. Immune cells include such cells as lymphocytes that function through molecular interactions with antigens and other cells and molecules of the body. They possess surface proteins that, through binding, recognize antigens, antibodies, and some surface proteins of other cells. Binding stimulates cellular activity that contributes to the defense of the body.

Acquired immune deficiency syndrome, or AIDS, results from the infection of a specific type of T lymphocyte by a virus called HIV. The virus may lie dormant within the lymphocytes for years, but eventually, it multiplies within the lymphocyte and kills the cell by emerging from it. The virus infects other T cells and the cycle repeats. With time, the virus destroys most of these T lymphocytes. Since these cells play a vital role in controlling the immune response, the immune system functions poorly when these cells are lost.

When the immune system fails, the AIDS patient is vulnerable to a host of organisms that would normally be destroyed without injury or harm to the individual. These opportunistic infections include *Pneumocystis carinii*, a normally harmless organism that causes pneumonia in many AIDS patients. The immune system also normally destroys cancer cells, so it is not surprising that cancers such as Kaposi's sarcoma, an otherwise rare skin malignancy, appear in AIDS patients. The Center for Disease Control now predicts that AIDS infection will always result in death, unless a cure is developed.

How can AIDS be cured or an HIV vaccine be developed? The interactions between the HIV virus and the T lymphocytes are all molecular interactions. Specific binding between viral proteins and T cell proteins must occur before infection can occur. The replication of the virus within the cell requires specific viral molecules interacting with host molecules. The search for a cure and a vaccine for AIDS requires a basic understanding of these molecular interactions. Many research projects studying AIDS are of a molecular nature. With ever-increasing frequency, biochemistry and molecular biology are being used to improve health care.

NEW TERMS

antibodies Proteins produced by the body that bind foreign molecules and particles. These complexes are then readily taken up and destroyed by cells of the immune system.

antigen Large foreign molecules within the body. These molecules cause an immune response that results in production of antibodies to the antigen.

interferons Proteins produced by viral infected cells. They bind to other cells and stimulate an antiviral state in them.

TESTING YOURSELF

Immunity

1. Molecules that are normally absent from the body which cause an immune response are called:
2. What cells produce antibodies during an immune response?
3. What group of immune cells is infected by the AIDS virus?

Answers **1.** antigens **2.** plasma cells **3.** T lymphocytes

SUMMARY

The body maintains a constant internal environment through *homeostasis*, which is a set of self-regulating mechanisms. The composition of blood is an important part of homeostasis. Oxygen and carbon dioxide are exchanged between the body and air in the *alveoli* of the lungs. These gases are transported by blood. Their movement into and out of blood is driven by diffusion. *Hemoglobin* greatly enhances the capacity of blood to carry oxygen; this molecule is adapted to readily take up oxygen in the lungs and deliver it to tissues. Wastes are removed from blood by filtration in the kidneys, while nutrients are reabsorbed.

Maintenance of a narrow pH range is critical in blood because denaturation of proteins occurs outside this range. Bicarbonate, phosphate, and proteins are the main buffers in blood. If the pH extends beyond this narrow range, *acidosis* or *alkalosis* will result. Hormones are chemical messengers in blood. *Insulin* serves as a signal that nutrients are abundant, so cells take in more nutrients. *Glucagon* signals that fuel molecules are scarce, and the liver and adipose cells respond by releasing more fuel molecules.

Blood also carries many components of the immune system. *Antibodies* are proteins that bind to foreign molecules to help with their removal. *Interferons* affect viral replication, thus they reduce the chance of viral infections. AIDS is a viral disease that destroys T lymphocytes involved in regulation of the immune response. When these cells are destroyed, the immune system fails.

TERMS

homeostasis
alveoli
hemoglobin
Bohr effect
filtration
renal threshold
acidosis
alkalosis
hormone
cyclic AMP
insulin
glucagon
antibodies
antigen
interferons

CHAPTER REVIEW

QUESTIONS

1. What protein binds oxygen and transports it in blood?
   a. albumin c. antibody
   b. alveolus d. hemoglobin
2. In the buffering of blood, bicarbonate (HCO_3^-) acts as a(n)
   a. acid b. base

3. The nutrients present in the filtrate in kidney tubules are
   a. excreted c. reabsorbed
   b. complexed to urea d. metabolized
4. If a person has ketosis, he or she may also have
   a. metabolic acidosis c. respiratory acidosis
   b. metabolic alkalosis d. respiratory alkalosis
5. Which of the following hormones does not enter the target cell?
   a. estrogen c. epidermal growth factor
   b. testosterone d. insulin
6. The activation of genes within the nucleus is signaled by which hormone?
   a. glucagon c. insulin
   b. estrogen d. epinephrin
7. The proteins that bind foreign molecules in blood are called
   a. antibodies c. lymphocytes
   b. antigens d. interferons
8. The proteins that help inhibit viral replication are called
   a. antibodies c. lymphocytes
   b. antigens d. interferons
9. AIDS is caused by destruction of
   a. antibodies c. plasma cells
   b. B lymphocytes d. T lymphocytes
10. The maintenance of a constant internal environment is called
   a. immunity c. gene expression
   b. homeostasis d. respiration

Answers **1.** d **2.** b **3.** c **4.** a **5.** d **6.** b **7.** a **8.** d **9.** d **10.** b

DIAGNOSTIC CHART

Blacken in all of the circles under the number of each question that you missed in the Chapter Review questions. The diagnostic chart will help you identify concept areas that need more study.

	Questions									
Concepts	**1**	**2**	**3**	**4**	**5**	**6**	**7**	**8**	**9**	**10**
Blood gases	○	○								○
Wastes			○							○
Regulation of pH		○		○						○
Hormones					○	○				○
Immunity							○	○	○	○

EXERCISES

Gas Exchange

1. What role does hemoglobin play in transporting oxygen? In transporting carbon dioxide?
2. The binding of oxygen to hemoglobin is said to be cooperative. What does this mean?
3. What is the role of carbonic anhydrase?
4. Describe the effects of increased acidity and increased carbon dioxide concentration on the oxygen-carrying capacity of hemoglobin. What is this effect called?

Removal of Wastes

5. What organs are responsible for removal of carbon dioxide? Urea? Water? Salt?

6. Describe the process of filtration and reabsorption in the kidney.
7. What is meant by renal threshold? What is the renal threshold for glucose?

Regulation of pH

8. Describe how bicarbonate and carbonic acid contribute to the buffering of blood.
9. If the blood contained bicarbonate but no carbonic acid, would it be buffered?
10. What effects can hypoventilation have on blood pH? What is this condition called?
11. An individual has an infection of the gastrointestinal tract that causes vomiting and diarrhea. What effect could this have on blood pH?

Hormones

12. Hormones circulate throughout the body in blood, and thus come in contact with all cells. Why are only some cells affected by any given hormone?
13. Describe how hormones that enter cells cause changes in the cell.
14. How does protein kinase alter the activity of a cell? What activates protein kinase?
15. How is cyclic AMP formed in a cell?
16. Why is insulin considered a feast hormone?
17. Why is glucagon considered a famine hormone?
18. What role does tyrosine kinase play in the effect of insulin on cells?
19. What effect does glucagon have on adipose cells? How does glucagon cause this effect?

Immunity

20. What happens to a B lymphocyte when an antigen binds to its surface?
21. Describe how an interferon affects replication of a virus in body cells.
22. How can an otherwise harmless organism kill an AIDS patient?
23. What agglutinins would be found in the blood of an individual with group B blood?
24. Who could receive blood from a group B donor?
25. An individual with group B blood could receive blood from whom?

APPENDIX

I

Supplemental Exercises

Numbering is continuous with end-of-chapter exercises.

CHAPTER 1

Length

42. Prior to the development of the metric system, what standards were commonly used for units of length?
43. What is the conversion factor relating yards and inches?
44. The original standard for the meter was one ten-millionth the measured distance from the equator to the north pole. What is today's definition for the SI meter?

Metric Prefixes

45. Convert a distance of 105 cm into the following units.
    a. millimeters b. meters c. micrometers
46. Convert a distance of 2050 m into the following units.
    a. meters b. centimeters c. millimeters

Dimensional Analysis

47. Convert a distance of 16 ft to inches.
48. Convert a distance of 24 ft to yards.
49. Convert a distance of 76 in. to feet.
50. Convert a distance of 6.0 cm to inches.
51. Convert a distance of 25 m to feet.
52. Convert a distance of 4000 ft to meters.
53. Convert a distance of 25,000 ft to miles.
54. Convert a distance of 2.60 km to meters.
55. Convert a distance of 6.2 ft to centimeters.
56. Convert a distance of 140 cm to yards.

Volume

57. How is the SI volume unit liter related to the SI distance unit?
58. Convert a volume of 5.0 qt to gallons.
59. Convert a volume of 30 qt to fluid ounces.
60. Convert a volume of 1450 mL to liters.
61. Convert a volume of 12.0 L to quarts.
62. Convert a volume of 250 mL to cups.
63. Convert a volume of 4.2 L to cups.
64. Convert a volume of 2.0 L to fluid ounces.

Mass and Weight

65. A liter of water has a mass of how many kilograms?
66. Differentiate between the terms *mass* and *weight*.
67. Convert a mass of 1.5 kg to grams.
68. Convert a mass of 500 g to kilograms.
69. Convert a mass of 2.2 kg to ounces.
70. Convert a mass of 1.0 kg to pounds.
71. Express a weight of 150 lb in kilograms.
72. Express a weight of 24 lb in ounces.

Density

73. The density of metallic copper is 8.9 g/cm^3. What will be the volume of 100 g of copper?
74. The density of metallic gold is 19.8 g/cm^3. What will be the volume of 100 g of gold?
75. The density of metallic lead is 11.0 g/cm^3. What will be the volume of 100 g of lead?
76. 250.0 mL of isopropyl alcohol (rubbing alcohol) had a mass of 195 g. What was the density of this sample?
77. A block of walnut $10.0 \times 14.1 \times 20.0$ cm^3 has a mass of 1579 g. What is the density of the walnut sample?
78. Suppose that a 100-g cube of aluminum were dropped in a graduated cylinder that had exactly 50.0 mL of water in it. To what volume would the water rise?
79. Suppose that a 100-g cube of iron were dropped in a graduated cylinder that had exactly 50.0 mL of water in it. To what volume would the water rise?

Specific Gravity

80. A cubic sample $20.0 \times 20.00 \times 20.0\ cm^3$ has a mass of 7900 g. Will it float in water?
81. A 30.0 mL urine sample has a mass of 31.50 g. Is the urine in the normal range of specific gravity?

Significant Figures

82. The density of white pine is $0.37\ g/cm^3$. What will be the mass, expressed to the proper number of significant figures, of a block of white pine with dimensions $3.21 \times 24.0 \times 10.11\ cm^3$?
83. A sample of alcohol was measured to have a volume of 25.0 mL and a mass of 19.512 g. Express the density of this sample to the proper number of significant figures.
84. A cylinder of douglas fir is 6.0 cm in diameter and has a length of 10.10 cm. It has a mass of 145.641 g. What is the density of this sample, expressed to the proper number of significant figures?
85. Which measurement in question 83 limits the accuracy with which the result may be expressed?
86. Which measurement in exercise 84 limits the accuracy with which the result may be expressed?

Scientific Notation

87. Please express the following values in scientific notation, indicating the proper number of significant figures in each case.

1490 L	5225 g
490 cm	55045 kg
1.60 m	2946 km
0.045 g	0.000658 m
0.00000623 L	0.9700 mL
0.0000000500 g	0.067 cm

CHAPTER 2

Temperature Scales

27. To what standard is 0 °C referenced?
28. What is the definition of zero Kelvin?
29. Please make the following Celsius to Kelvin temperature conversions.

Celsius	Kelvin
20	?
100	?
−40	?
?	273
?	373
?	73

30. Please make the following Celsius to Fahrenheit conversions:

Celsius	Fahrenheit
0	?
20	?
37	?
60	?
?	100
?	−40
?	250
?	68

Temperature and Heat

31. After running hard, a person evaporates 25.0 mL of perspiration from her body. The person feels cold because her body is giving up heat to evaporate the water. Presume that this water was heated from 37 °C to 100 °C and then evaporated. How many calories will the person's body supply?
32. How many calories of heat must a person's hand supply if a 25-g ice cube at 0 °C is dropped in his hand and held until the melted water reaches body temperature?

Change of State

33. Change of physical state involves a change in the amount of order in the material. Addition of energy causes the atoms and molecules to move faster and breaks down this order. Atoms and molecules are well ordered in solids, slide across each other in liquids, and float independently in gases. Which change of state requires the most energy? Solid to liquid, or liquid to gas?
34. You can make ice cubes by taking heat away from water. Suppose that you start with an ice cube tray that has 500 g of water at room temperature, 20 °C.
    a. How many calories of heat must be removed to cool the water to 0 °C?
    b. How many calories of heat must be removed to change water from liquid to solid?
    c. How many calories, in total, must be removed to cool the water and convert it to ice?
35. Suppose that you wish to evaporate 25 g of water, which is currently at room temperature (20 °C).
    a. How many calories of heat must be added to heat the water to 100 °C?
    b. How many calories of heat must be added to evaporate the water sample?
    c. How many calories, in total, must be added to heat and evaporate the water?
36. Suppose that a 100-g snowball falls on the sidewalk. The sun eventually melts the snowball and evaporates

the water. How many calories are required to accomplish this feat? Presume that the snowball was at 0 °C at the beginning of the problem.

37. Suppose that the body temperature of a 70-kg person fell to 32 °C. How many calories of heat must be added, externally or by the person's metabolism, to bring body temperature back to normal?

38. Thunderstorms are powered by the heat liberated as water condenses from the gaseous to the liquid state. How many calories of heat would be released if a gallon of water (about 1816 g) condensed?

Chemical and Physical Change

39. Classify the following occurrences as physical or chemical change.

    burning wood in a campfire
    melting candle wax
    using gasoline to run a lawnmower
    nail polish drying
    ice cube melting
    rain
    mixing sugar in coffee

40. Differentiate between chemical and physical change in terms of the following.
    a. amount of energy change involved
    b. ease of reversibility
    c. chemical identity of beginning and ending products

Mixtures, Compounds, and Elements

41. Which of the following classes of materials has a variable composition?
    a. mixtures b. compounds c. elements

42. A certain liquid material always measures 11.1% hydrogen and 88.9% oxygen whenever a sample is analyzed. What is this material?
    a. mixture b. compound c. element

43. Classify the following materials as mixtures, compounds, or elements.

seawater	distilled water
lime, $Ca(OH)_2$	silver spoon
stew	salt, NaCl
milk	copper wire
sulfur	

Atomic Weight

44. Acetone has the molecular composition C_3H_6O. What is the mass of this molecule expressed in terms of the mass of a hydrogen atom?

45. Acetic acid, the active ingredient in vinegar, has the molecular composition $C_2H_4O_2$. What is the mass of this molecule expressed in terms of the mass of a hydrogen atom?

46. A compound of hydrogen and sulfur is 5.92% by weight hydrogen and 94.08% by weight sulfur. Two hydrogen atoms combine with one sulfur atom to form this compound, H_2S. What is the relative weight of a sulfur atom as compared to a hydrogen atom?

47. Hydrogen combines with chlorine to form the compound hydrogen fluoride, HF. This compound is 94.96% by weight fluorine and 5.04% by weight hydrogen. What is the relative weight of a fluorine atom as compared to a hydrogen atom?

Chemical Periodicity

48. Which of the following elements will behave chemically in a manner most similar to magnesium?
    a. calcium (Ca) d. sodium (Na)
    b. potassium (K) e. sulfur (S)
    c. nickel (Ni)

49. Which of the following elements will behave chemically in a manner most similar to sulfur?
    a. rubidium (Rb) d. barium (Ba)
    b. oxygen (O) e. gallium (Ga)
    c. chlorine (Cl)

50. Hydrogen forms compounds with the oxygen family of the periodic table as shown here: H_2O, H_2S, and H_2Se. How many hydrogens would you predict would combine with the element tellurium, Te?

51. Based on the behavior of hydrogen indicated in exercise 50, what do you think the formula of a compound of sodium and oxygen would be?
    a. NaO b. Na_2O c. Na_3O d. NaO_2 e. Na_2O_3

52. How many sodium atoms would you expect to combine with one atom of neon?

53. How many argon atoms would you expect to combine with one krypton atom?

54. The physical properties of each family of elements change smoothly as one moves up or down in the periodic table. Given the following data, what would be a reasonable estimate of the density of the element gallium, Ga?

Element	Density (g/mL)
Boron	2.34
Aluminum	2.70
Indium	7.31
Thallium	11.85

    a. 2.0 g/mL d. 6.0 g/mL
    b. 2.9 g/mL e. 9.0 g/mL
    c. 3.2 g/mL

CHAPTER 3

Cathode Rays

14. Cathode rays carry what charge?
15. Do cathode rays move from plus to minus or minus to plus charge?
16. Will a positively charged plate placed near the path of a beam of cathode rays deflect the beam toward the plate or away from the plate?
17. Are cathode rays larger or smaller than air molecules?

Electrons

18. Electrons are
    a. larger than ordinary atoms
    b. about the same size as hydrogen atoms
    c. much smaller than the smallest atom
19. Electrons carry a
    a. + charge
    b. – charge
    c. zero charge
20. Electrons appear to be located
    a. in the outer part of atoms
    b. in the center of atoms

The Plum Pudding Atom

21. What is the charge carried by canal rays?
22. Why were canal rays *not* observed unless some air was present in the discharge tube?
23. Which part of the atom appears to be heavier—the positive part or the negative part?
24. Describe the "plum pudding atom" proposed by J. J. Thomson.

Natural Radioactivity

25. Which of the following types of radiation (alpha, beta, or gamma) carries
    a. a negative charge
    b. a positive charge
    c. no charge
26. Which of the types of radiation listed in exercise 25 is a wave?
27. Which type of nuclear radiation is the *least* penetrating?

The Nuclear Atom

28. Why was alpha radiation (instead of beta radiation) chosen to probe the distribution of positive charge in the "plum pudding atom"?
29. Describe the nuclear atom model proposed to explain the scattering of alpha particles observed by Geiger, Marsen, and Rutherford.
30. Where are the electrons located in the nuclear atom model?
31. Where is the positive charge located in the nuclear atom model?
32. Approximately what fraction of the total volume of the atom is occupied by the nucleus?
    a. 1/2 d. 1/100
    b. 1/4 e. less than 1/1000.
    c. 1/10

CHAPTER 4

Waves and Light

17. Which of the following characteristics of a wave changes when its frequency is increased?
    a. amplitude b. wavelength c. velocity
18. The energy of a wave increases as its ________ increases.
    a. wavelength b. frequency
19. Which of the following colors of light has the lowest frequency?
    a. red b. yellow c. green d. blue e. violet
20. Which of the colors of light listed in question 19 has the longest wavelength?
21. Which of the colors of light listed in question 19 carries the greatest energy?
22. Which of the colors of light listed in question 19 has the shortest wavelength?

Band and Line Spectra

23. Solids and gases give off different types of spectra when heated. Which type of material emits a band spectrum, comprised of all colors of light?
24. In which physical state must a material be to emit a "fingerprint" spectrum that can be used to identify the element(s) being heated.

The Bohr Atom

25. When an electron jumps from an outer orbit to an inner orbit, light is given off. Consider electron jumps from orbit 3 to orbit 2 and from orbit 4 to orbit 2. Which would give off the shortest wavelength light?
26. The visible region emission lines of hydrogen are shown in the chart below. Why does gaseous hydrogen emit only certain colors of the spectrum, rather than all colors (a band spectrum)?

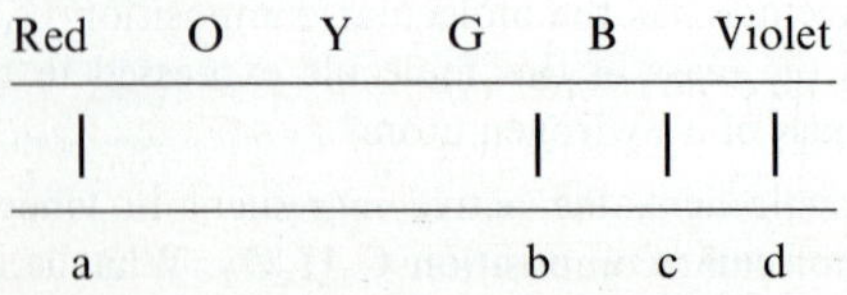
Red
O
Y
G
B
Violet
a
b
c
d

27. The spectral lines shown in problem 26 are caused by electron jumps that end in Bohr orbit 2, (3→2, 4→2, 5→2, and 6→2). Please assign the correct spectral line to each of these electron jumps.

Electron Filling Series

28. What orbital will the outermost electron of a neutral atom of sodium occupy?
29. The electron distribution of which neutral atom is represented by this orbital notation: $1s^2 2s^2 2p^6 3s^2 3p^4$.
30. The electron distribution of the outermost electron group of a neutral atom of the element chlorine is $3s^2 3p^5$.
    a. What is the orbital notation for the electron distribution of a neutral atom of fluorine?
    b. What is the orbital notation for the electron distribution of a neutral atom of bromine?
31. Use the electron filling series to predict the electron distribution of the outermost electron group of neutral atoms of these elements. What do these elements have in common?
    a. oxygen b. sulfur c. selenium
32. Use the electron filling series to predict the electron distribution of the outermost electron group of neutral atoms of these elements. What do these elements have in common?
    a. magnesium b. calcium c. strontium
33. Potassium chloride is often used as a salt substitute for people who should decrease their intake of sodium chloride. Use the electron filling series to predict the electron distribution of the outermost electron group of neutral atoms of sodium and potassium. What do these two elements have in common?

CHAPTER 5

Stability Rules

20. Predict the orbital notation for the outermost electron groups of neutral atoms of these elements. What do these elements have in common?
    a. neon b. argon c. krypton
21. All of the elements listed in problem 20 are chemically stable. They are called the inert or noble gases and are thought to be stable because of their unique electron configuration. Using these three elements as examples, state a rule for chemical stability.
22. Predict the orbital notation of the outermost electron group for a neutral atom of helium. Helium is also chemically stable. Modify the rule for chemical stability you developed in problem 21 to include helium.

Ionic Charge and Compound Formulas

23. Using the electron filling series and your rule for chemical stability developed in problems 21 and 22, predict the charge of stable ions of these elements,

Li	Be	O	F
Na	Al	S	Cl
K	Ga	Se	Br

    Write the symbols for these elements as a table on a worksheet just as shown above. Show the predicted ionic charge in parenthesis with the element. For example, Li (+1).
24. The elements listed in problem 23 are arranged in vertical columns as they are found in the periodic table. What generalization can be stated relating position in the periodic table to the charge of the stable ion of the element?
25. Predict the formula of a compound composed of stable ions of lithium and chlorine.
26. Predict the formula of a compound composed of stable ions of sodium and oxygen.
27. Predict the formula of a compound composed of stable ions of aluminum and fluorine.
28. Predict the formula of a compound composed of stable ions of gallium and bromine.
29. Predict the formula of a compound composed of stable ions of aluminum and oxygen.
30. Predict the formula of a compound composed of stable ions of gallium and sulfur.

Properties of Ionic Compounds

31. The following examples of ionic compounds all have the same charge differential (+1/−1), but differ in bond strength as indicated by their melting points. Melting point is an indication of bond strength; the temperature at which the compound melts indicates the strength with which individual ions are held in the crystal lattice. Please order these compounds from lowest to highest melting point. (HINT: Can you predict the relative size of the four group 1 ions?)

    NaCl KCl RbCl CsCl
32. The melting points of the four compounds listed in problem 31 are 646 °C, 715 °C, 776 °C, and 801 °C. Please assign melting points to each of the four compounds.

Electron Attraction

33. The attraction an atom exerts on its electrons is determined largely by the charge of the nucleus and the distance from the nucleus to the electron. If the positive charge of the nucleus is doubled, all other factors remaining constant, what would happen to the atom's

attraction for its electrons? Would it become greater or smaller and by what amount?

34. Consider now the opposite hypothetical situation where all other factors are held constant and the distance from the nucleus to the outermost electron is doubled. What would happen to the force by which the atom holds this electron? Will it increase or decrease and by what amount?

35. Which causes the greatest change in the force by which an atom holds its outermost electrons: changing the nuclear charge, or changing the nucleus to electron distance by the same factor?

36. In the periodic table, elements are arranged horizontally in terms of increasing atomic number and vertically in families with the same outermost electron configuration. Consider the halogen family: fluorine, chlorine, bromine, iodine, and astatine.
   a. What happens to the nucleus to outermost electron distance as one moves up the table in this column? What would this do to the electron-attracting ability of the atom?
   b. What happens to the charge of the nucleus as one moves up the table from At to F? What would this do to the electron-attracting ability of the atom?
   c. Which of these two effects will be greater?
   d. Which way will electron-attracting ability change as you move up the table from At to F? Increase or decrease?

37. Consider the horizontal row of the periodic table that begins with sodium (Na) and ends with chlorine (Cl). Since the outermost electrons of all of the elements in this row are in the third major electron group, there is little difference in the nucleus to electron distance.
   a. What happens to the nuclear charge as one moves from left to right in this row? Does it increase or decrease?
   b. What does your response to problem 36a suggest will happen to the electron-attracting ability of the elements as one moves along a row, from left to right in the periodic table? Will it increase or decrease?
   c. Since increased nuclear charge will exert greater attraction on an atom's outer electrons, what will happen to the diameter of the neutral atom as one moves left to right in this row, from sodium to chlorine?
   d. Four elements from the third row of the period table are listed below, together with four values for atomic radii. Please match the elements with the appropriate atomic radii.

sodium	2.33 Å
magnesium	0.97 Å
silicon	1.46 Å
chlorine	1.72 Å

38. State a generalization relating position of the element in the periodic table with its electron-attracting ability. Which part of the table has elements with the greatest electron-attracting ability? Which has elements with the least electron-attracting ability?

39. Consider a negative ion, such as Cl^-. One extra electron has been added to this atom to achieve a stable outer group electron configuration, s^2p^6.
   a. The chlorine nucleus has 17 protons. In the neutral atom, these positive charges provided the attractive force for 17 electrons. Now that an 18th electron has been added, is there more or less attractive force available to hold each of the chlorine's electrons?
   b. Will the chloride ion (Cl^-) be larger or smaller than the neutral chlorine atom?

40. Consider a positive ion, such as Na^+. One electron has been removed from this atom to achieve a stable outer electron group configuration, s^2p^6.
   a. The sodium nucleus has 11 protons. In the neutral atom, these positive charges provided the attractive force for 11 electrons. Now that one electron has been removed, is there more or less attractive force available to hold each of the sodium's electrons?
   b. Will the positive sodium ion be larger or smaller than its neutral sodium counterpart?

41. State a generalization relating the size of positive and negative ions and their neutral atom counterparts. Does the same generalization hold for positive ions and negative ions?

Electron Sharing and Covalent Bonding

42. Use your knowledge of the factors that determine an atom's electron-attracting ability to decide the following, given potassium, nitrogen, and oxygen.
   a. Which two elements are closest together in electron-attracting ability and thus which will form a bond of the greatest covalent character?
   b. Which two elements would form a bond of greatest covalent or electron-sharing character?

43. Methyl alcohol is a covalently bonded compound with the structural formula

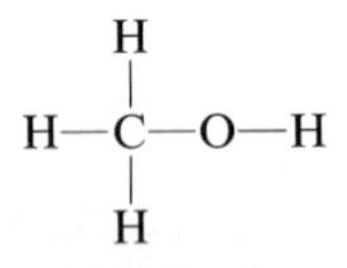

Sketch an electron dot model for this molecule.

44. Diethyl ether, once used as an anesthetic, has the structural formula

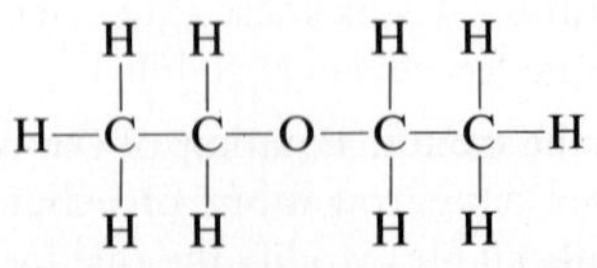

Sketch an electron dot model of this molecule.

45. Acetone, a common solvent, has the structural formula

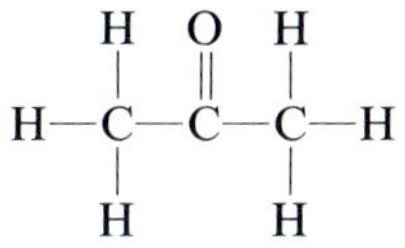

Sketch an electron dot model of this molecule.

Electronegativity and Bond Type Prediction

46. Use your knowledge of the electron filling series and electronegativity to predict the formulas and bond types for compounds of the following elements.

	Bond Type	Formula
Aluminum and fluorine		
Sodium and oxygen		
Calcium and oxygen		
Magnesium and chlorine		
Strontium and oxygen		
Lithium and bromine		
Sodium and iodine		
Aluminum and oxygen		
Carbon and hydrogen		
Carbon and oxygen		
Nitrogen and oxygen		
Nitrogen and hydrogen		

CHAPTER 6

Bond Types

13. Using your knowledge of electronegativity, predict the type of bond that will form between carbon and each of the following elements.

Elements Bonding to Carbon	Bond Type
Hydrogen	
Nitrogen	
Oxygen	
Fluorine	
Sulfur	
Chlorine	
Bromine	

14. Carbon has a unique position in the electronegativity table. Find the element that is farthest above carbon in electronegativity. What type of bond will carbon make with this element?
15. Find the element that is farthest below carbon in electronegativity. What type of bond will carbon form with this element?
16. Can carbon be expected to participate in ionic bonding?

Electron Pair Repulsion and Molecular Polarity

17. Consider the compound ammonia, NH_3.
    a. What type(s) of bonds exist in this molecule?
    b. Draw an electron dot model of this molecule.
    c. How many pairs of bonding electrons are there surrounding the nitrogen?
    d. How many pairs of nonbonding electrons are there surrounding the nitrogen?
    e. What will be the shape of the nitrogen atom?
    f. Will the molecule have a dipole?
18. Consider the compound glycerol, commonly known as glycerine. Its molecular structure is shown here.

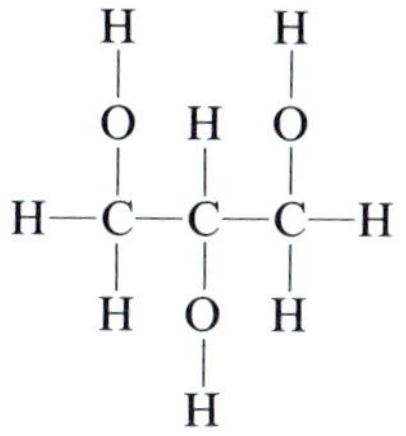

    a. What type(s) of bonds are found in this molecule?
    b. How many pairs of electrons surround each carbon atom?
    c. What will be the shape of each carbon atom?
    d. How many pairs of electrons surround each oxygen atom?
    e. What will be the shape of each oxygen atom?
    f. Does the molecule have one or more polar regions? If so, where are they located?

Orbital Hybridization

19. Refer to the drawing of ammonia in problem 6 (p. 156). What is the hybridization of the outer electron group in the nitrogen atom in this compound?
    a. *s* b. *sp* c. sp^2 d. sp^3
20. What is the hybridization of the outer electron group of a carbon atom in glycerine?
21. What is the hybridization of the outer electron group of an oxygen atom in glycerine?

Molecular Polarity and Solubility

Insert the electron dots in the molecules below and classify these molecules as polar or nonpolar. If they are polar, show the location of their dipole.

22. Propane

```
      H  H  H
   H  C  C  C  H
      H  H  H
```

23. Methyl alcohol

```
      H
   H  C  O  H
      H
```

24. Ethyl alcohol

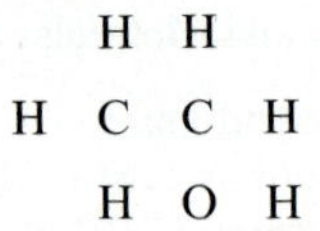

25. Methyl ethyl ketone (finger nail polish remover)

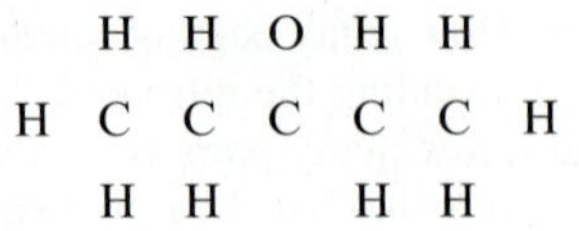

26. Which of the compounds in exercises 22–25 would be least soluble in water?
27. Which of these compounds would be most effective in dissolving salt?
28. Which of these compounds have both polar and non-polar parts and would therefore be good solvents to dissolve nonpolar materials in water?

Hydrogen Bonding

29. What effect does hydrogen bonding have on the effective atomic weight of a molecule: increase, decrease, or no change?
30. In general, what is the relationship between the molecular weight of a molecular and the energy required to make it boil?
31. Do molecules that hydrogen bond show an increase, a decrease, or no change in boiling point as compared to molecules of similar molecular weight that do not hydrogen bond?
32. Compute the electronegativity difference across each bond in the following molecules, and decide which have the greatest polarity and therefore the greatest tendency for their molecules to hydrogen bond.

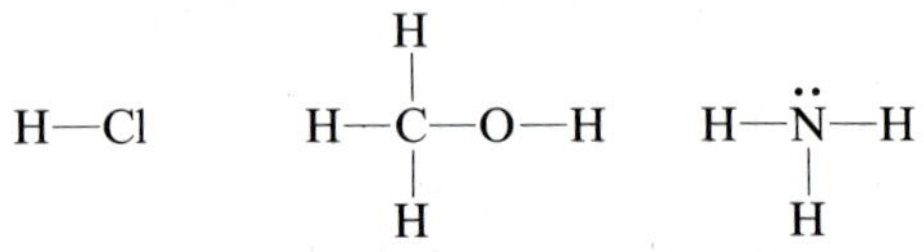

Hydrochloric acid   Methyl alcohol   Ammonia

CHAPTER 7

Naming Simple Compounds

Name the following compounds.

9. HBr	12. LiF	15. $CaBr_2$
10. KCl	13. H_2S	16. HF
11. MgO	14. CaS	17. KI

Balanced Formulas

Write formulas for the following compounds.

18. calcium chloride
19. potassium sulfide
20. strontium bromide
21. carbon dioxide
22. aluminum chloride
23. aluminum oxide
24. magnesium sulfide
25. barium oxide
26. rubidium oxide
27. gallium fluoride

Oxidation Numbers

Determine the oxidation number of the following.

28. sulfur in sulfuric acid, H_2SO_4
29. sulfur in hydrogen sulfide, H_2S
30. nitrogen in ammonia, NH_3
31. nitrogen in nitric acid, HNO_3
32. copper in cupric oxide, CuO
33. iron in ferric oxide, Fe_2O_3
34. phosphorus in phosphoric acid, H_3PO_4
35. tin in stannous fluoride, SnF_2
36. oxygen in hydrogen peroxide, H_2O_2

Polyatomic Ions

Name the following compounds.

37. KNO_3	40. $CaCO_3$	43. Na_2SO_4
38. NaOH	41. $Ag(OH)_2$	44. $Pb(SCN)_2$
39. NH_4Br	42. Ag_2CrO_4	

Write formulas for the following compounds.

45. lead nitrate
46. potassium cyanide
47. magnesium sulfate
48. nickel acetate
49. sodium bicarbonate
50. ammonium hydroxide
51. ammonium thiocyanate
52. potassium permanganate
53. potassium sulfite
54. lead carbonate
55. aluminum hydroxide
56. zinc chromate

Balancing Equations and Predicting Solubility

Write balanced equations for the following chemical reactions.

57. Hydrogen chloride + Sodium hydroxide →
58. Hydrogen chloride + Aluminum hydroxide →
59. Silver nitrate + Ammonium sulfate →
60. Silver nitrate + Ammonium hydroxide →
61. Lead nitrate + Ammonium hydroxide →
62. Lead nitrate + Ammonium sulfate →
63. Potassium nitrate + Sodium chloride →
64. Sodium carbonate + Iron(III) nitrate →
65. Lithium hydroxide + Potassium nitrate →
66. Aluminum nitrate + Potassium carbonate →
67. Copper(II) nitrate + Hydrogen sulfide →

68. Copper(II) nitrate + Hydrogen chloride →
69. Iron(III) nitrate + Ammonium hydroxide →
70. Sodium sulfide + Nickel nitrate →
71. Review the products of reactions 57–70. Three of these reactions do not produce insoluble products. Which reactions are these?

Reaction Classification

Classify the following reactions as synthesis, decomposition, single replacement, or double replacement. Also, balance the equations describing the reactions.

72. Hydrogen chloride + Sodium bicarbonate ⟶ Carbon dioxide + Water + Salt

$$HCl + NaHClO_3 \longrightarrow CO_2 + H_2O + NaCl$$

73. Determine which element, if any, was oxidized during the reaction of question 72 and which, if any, was reduced.

74. Methane (natural gas) + Oxygen ⟶ Water + Carbon dioxide

$$CH_4 + O_2 \longrightarrow H_2O + CO_2$$

75. Determine which element, if any, was oxidized during the reaction of problem 74, and which, if any, was reduced.

76. Magnesium hydroxide + Hydrogen chloride ⟶ Magnesium chloride + Water

$$Mg(OH)_2 + HCl \longrightarrow MgCl_2 + H_2O$$

77. Determine which element, if any, was oxidized during the reaction of problem 76 and which, if any, was reduced.

Electron Transfer Reactions

Which of the following electron transfer reactions will take place spontaneously? (NOTE: The equations are not balanced. Just decide if they will proceed in the direction indicated.)

78. Metallic zinc placed in a solution containing iron(III) ions.

$$Zn^0 + Fe^{3+}(NO_3)_3 \longrightarrow Zn(NO_3)_2 + Fe^0$$

79. If the reaction of exercise 78 proceeds as written, which element is oxidized and which is reduced? Write half-reactions for the oxidation and reduction reactions.
80. Metallic silver placed in a solution containing copper(II) ions.

$$Ag^0 + Cu(NO_3)_2 \longrightarrow AgNO_3 + Cu^0$$

81. If the reaction of problem 80 proceeds as written, which element is oxidized and which is reduced? Write half-reactions for the oxidation and reduction reactions.
82. Metallic lead is placed in a solution containing copper(II) ions:

$$Pb^0 + Cu(NO_3) \longrightarrow Pb(NO_3)_2 + Cu^0$$

83. If the reaction of problem 82 proceeds as written, which element is oxidized and which is reduced? Write half-reactions for the oxidation and reduction reactions.

CHAPTER 8

Formula and Molecular Weight

Compute the mass (in grams) of one mole of each of the following compounds.

45. sodium hydroxide, $NaOH$
46. potassium chloride, KCl
47. lead sulfide, PbS
48. silver chromate, Ag_2CrO_4
49. sodium sulfate, Na_2SO_4
50. calcium carbonate, $CaCO_3$
51. lithium fluoride, LiF
52. iron(III) nitrate, $Fe(NO_3)_3$
53. carbon tetrachloride, CCl_4
54. zinc chromate, $ZnCrO_4$

Percent Composition

Determine the elemental percentage composition of each of the following compounds.

55. propane, C_3H_8
56. water, H_2O
57. salt, $NaCl$
58. stannous fluoride, SnF_2
59. sodium hydroxide, $NaOH$
60. ethylene glycol, $C_2H_6O_2$
61. sulfuric acid, H_2SO_4
62. potassium nitrate, KNO_3

Moles

Compute the number of moles of material represented by each of the following samples.

63. 2.0 g of sodium, Na
64. 14 g of nitrogen gas, N_2
65. 50 g of aluminum, Al
66. 150 g of sucrose, $C_{12}H_{22}O_{11}$
67. 100 g of calcium carbonate, $CaCO_3$
68. 250 g of sodium hydroxide, $NaOH$

Compute the number of grams represented by each of the following samples.

69. 0.5 mol of lithium
70. 5.0 mol of silver
71. 25.0 mol of iron
72. 0.25 mol of calcium
73. 1.5 mol of lithium fluoride
74. 2.5 mol of methane, CH_4
75. 55.5 mol of water, H_2O
76. 0.20 mol of sodium bicarbonate, $NaHCO_3$

Avogadro's Number

How many atoms are contained in each of the following?

77. 8.0 g of oxygen, O.
78. 21.0 g of nitrogen, N.
79. 8.0 g of helium, He.
80. 8.0 g of neon, Ne.

Empirical Formulas

Compute empirical formulas for each of the following compounds, and name each compound.

81. Sn = 62.60%, Cl = 37.40%
82. U = 67.62%, F = 32.38%
83. C = 15.77%, S = 84.23%
84. C = 13.65%, F = 86.35%

Weight Relationships

85. When burned, propane reacts with oxygen producing carbon dioxide, water vapor, and heat. The balanced equation describing this reaction is shown here.

$$C_3H_8 + 5\ O_2 \longrightarrow 3\ CO_2 + 4\ H_2O$$

Suppose that one burns 10.0 g of propane to cook dinner with a camp stove.
a. How many moles of propane were used?
b. How many moles of oxygen are required?
c. How many grams of oxygen are required?
d. How many grams of carbon dioxide are produced?
e. How many grams of water vapor are produced?

86. Excess stomach acid is often neutralized by taking an antacid tablet, which usually contains aluminum hydroxide. The aluminum hydroxide reacts with the stomach acid (HCl) to form water and aluminum chloride.

$$3\ HCl + Al(OH)_3 \longrightarrow 2\ H_2O + AlCl_3$$

Suppose that one swallowed 15.0 g of aluminum hydroxide.
a. How many moles of aluminum hydroxide are present to react?
b. How many moles of hydrochloric acid will be neutralized?
c. How many grams of hydrochloric acid will be neutralized?
d. How many grams of water are produced?
e. How many grams of aluminum chloride are produced?

87. Silver bromide is used in some photographic processes. To synthesize silver bromide, one may react silver nitrate with potassium bromide.

$$AgNO_3 + KBr \longrightarrow AgBr + KNO_3$$

Suppose that you have 100 g of silver nitrate.
a. How many moles of silver nitrate are available to react?
b. How many moles of potassium bromide will be required?
c. How many grams of potassium bromide will be required?
d. How many grams of silver bromide will be produced?
e. How many grams of byproduct potassium nitrate will be produced?

CHAPTER 9

Solutions, Colloids, and Suspensions

18. Define what solutions, colloids, and suspensions are and how they differ from one another.
19. Solutions and colloids are both mixtures of a solute and a liquid. Neither mixture will settle out when allowed to stand. How does a solution differ from a colloidal mixture?
20. Suspensions, on the other hand, will ultimately settle out if allowed to stand for a long enough time. How do the particles of a suspension differ from those of a colloid?
21. How can you experimentally differentiate between a true solution and a colloidal mixture?
22. Explain what is meant by the Tyndall effect.

Separation Techniques

23. Salt dissolves in water to make a true solution. Can you separate the salt from the water by passing it through a filter paper?
24. Propose a method of separating the salt from water.
25. Powdered milk will form a colloidal mixture when added to water. Propose a way of separating the milk solids from the water.
26. The suspended colloidal silt carried by rivers settles out when the river reaches salt water. As a result, most large rivers have sand bars across their entrance. What effect causes the colloidal silt to settle out?

27. Dialysis is a separation technique that works when the colloidal particles are larger than the solvent molecules. Describe briefly how a kidney dialysis machine works.

Solution Concentration

28. A potassium chloride (KCl) solution is 12% by weight KCl. What is the composition of 1000 g of this solution?
29. A solution of NaCl was prepared by mixing 10.0 g of NaCl in 100 g of water. What is the concentration of this solution, in weight percent?
30. Suppose that you wished to prepare 1000 mL of a solution that was 14% by volume ethyl alcohol. You would place the alcohol in a 1000-mL flask and then add water to bring the total volume up to the 1000-mL mark. How many milliliters of alcohol would you use?
31. Suppose that you wished to prepare 250 mL of a solution that was 25% volume ethylene glycol. How many milliliters of ethylene glycol will be required?
32. A solution is prepared by dissolving 20.0 g of sodium hydroxide in water and diluting to a final volume of 1000 mL. What is the concentration of this solution, in moles NaOH per liter of solution?
33. A solution is prepared by dissolving 160 g of lead nitrate in water and diluting to a final volume of 1000 mL. What is the concentration of lead nitrate in the solution, expressed in moles per liter?
34. How many grams of silver nitrate will be required to prepare 100 mL of a solution that is 0.10 *M* $AgNO_3$?
35. How many grams of acetic acid will be required to prepare 500 mL of a solution that is 0.2 *M* acetic acid, $HC_2H_3O_2$.

Colligative Properties

36. The freezing point depression of water is −1.86 °C for each mole of particles that are dissolved in 1000 g of water. For example, if one dissolves 32 g of methyl alcohol, (CH_3OH, MW = 32) in 1000 g of water, the resulting solution will have a volume of about 1032 mL and will have a freezing point of −1.86 °C.
    a. What would be the freezing point of a solution prepared by adding 96 g of methyl alcohol to 1000 g of water?
    b. What would be the freezing point of a solution prepared by adding 96 g of ethyl alcohol (CH_3CH_2OH, MW = 46) to 1000 g of water?
    c. What would be the freezing point of a solution prepared by adding 96 g of rubbing alcohol (isopropyl alcohol) to 1000 g of water?
37. Salt ionizes when dissolved in water, yielding two particles for each NaCl unit added to the solution. Suppose that you dissolved 1 mol of NaCl in 1000 g of water. What would be the freezing point of the resulting solution?
38. Calcium chloride is sometimes used for melting salt on sidewalks. What advantage would this have over NaCl? What would be the freezing point of a solution prepared by dissolving 1 mol of $CaCl_2$ in 1000 g of water?
39. Compare the expected freezing points of solutions containing the following.
    a. 58.45 g of NaCl in 1000 g of water
    b. 58.45 g of $CaCl_2$ in 1000 g of water
    c. 58.45 g of ethyl alcohol in 1000 g of water

CHAPTER 10

Energy Relationships in Chemical Reactions

13. Is the combustion of a match an endothermic or exothermic reaction?
14. The process of baking a cake also involves one or more chemical reactions. Is the overall process endothermic or exothermic?
15. What does the term *activation energy* mean?
16. A gasoline lawnmower motor provides activation energy for its combustion of gasoline with what?

Reaction Rate

17. In hospitals, patients are often given pure oxygen. Ordinary air is only about 20% oxygen. What will this increased availability of oxygen do to the chemical reactions that use oxygen for cell metabolism?
18. If you paint a car in the sun, the paint will dry too rapidly and a smooth job is difficult to achieve. The increased temperature results in a more rapid collision rate for the reacting molecules. Which two of the three factors in the molecular collision model play a part in this problem?
19. The body temperature of surgical patients is sometimes lowered by ice bath prior to heart surgery, when the circulation of oxygenated blood will be stopped temporarily. What will this lowered body temperature do to the patient's metabolism rate, and which two factors of the molecular collision model are involved?
20. A catalyzed reaction proceeds more rapidly than an uncatalyzed reaction. Which of the three factors in the molecular collision model are affected by the presence of a catalyst?
21. Write a general reaction for a catalyzed reaction. By use of this reaction, explain why only a small number of catalyst molecules are required.

CHAPTER 11

Definitions

34. The Arrhenius model proposed that acids were molecules that ionized to give up a hydrogen ion.

$$HCl \longrightarrow H^+ + Cl^-$$

A basic principle of chemical stability unknown to Arrhenius makes this impossible. Arrhenius was also unaware of the role played by water in the solution containing the acid. What is the problem with the Arrhenius proposal?

35. The Brønsted–Lowry model predicted the existence of a hydronium ion. What is a hydronium ion, and how is it formed when acids are placed in water solutions?

36. Give a general definition for a base.

Classification of Molecules as Acids or Bases

37. Many molecules may behave either as acids or bases, depending upon the molecule they are paired with. Which of these molecules could *never* act as an acid?

$$HCl \quad KCl \quad H_2O \quad H_2CO_3$$

38. Suppose that HCl were added to a solution of HCO_3^-. By reference to Table 11.2 (page 274), decide whether or not the following reaction will take place.

$$HCl + HCO_3^- \longrightarrow H_2CO_3 + Cl^-$$

39. Suppose that water solutions of acetic, nitric, and hydrochloric acids were prepared. Each solution contains 0.1 mol of the acid in a total volume of 1000 mL. In each solution, the acid is reacting with water to form hydronium ion.

$$HC_2H_3O_2 + H_2O \rightleftharpoons H_3O^+ + C_2H_3O_2^-$$
$$HNO_3 + H_2O \rightleftharpoons H_3O^+ + NO_3$$
$$HCl + H_2O \rightleftharpoons H_3O^+ + Cl^-$$

a. Which solution will have the greatest concentration of hydronium ion?
b. Which solution will be considered the "weakest" acid?
c. Which solution will be considered the "strongest" acid?

40. Phosphoric acid, H_3PO_4 is found in some soft drinks. When dissolved in water, phosphoric acid ionizes as shown here.

$$H_3PO_4 + H_2O \rightleftharpoons H_3O^+ + H_2PO_4$$

a. By reference to the information in Table 11.2, which will be the predominant direction for this reaction?
b. After deciding which direction the reaction will proceed, identify the compound acting as an acid and the compound acting as a base.
c. Is phosphoric acid a strong or weak acid?

Hydronium Ion and pH

41. The term pH is used to describe the concentration of hydronium ion in a solution. Classify solutions with the following pH values as acidic, neutral, or basic.

lemon juice	pH 2.5
aspirin	pH 3.5
apple juice	pH 4.5
black coffee	pH 5.0
saliva	pH 7.0
soap	pH 8.0
milk of magnesia	pH 9.5
household ammonia	pH 11
oven cleaner	pH 13

42. Which of the pH values listed in exercise 41 represents the highest concentration of hydronium ion?

43. What is the concentration of hydronium ion in the black coffee listed in question 41, expressed in moles per liter?

44. By what factor must one increase the hydronium ion concentration in the soap solution presented in question 41 to change its pH to 7.0?

45. By what factor must one increase the hydronium ion concentration of the ammonia solution listed in question 41 to make the solution neutral?

46. Based on the pH data in question 41, how many times more concentrated is hydronium ion in lemon juice than in apple juice?

Chemical Indicators

47. A chemical indicator is itself a weak acid. What will be the predominant form of the indicator when in water solution: HIn or In^-?

48. When tested with four indicators, a solution displays the following behavior (refer to Figure 11.11). What is a good estimate of the pH of this solution?

with methyl red	red
with bromthymol blue	blue
with thymol blue	blue
with phenolphthalein	red

49. When tested with four indicators, a solution displays the following behavior. What is a good estimate of the pH of the solution?

with methyl orange	red
with methyl red	red
with phenol red	yellow
with cresol red	yellow

50. By reference to Figure 11.11, estimate the K_a for the indicator congo red.

Acid–Base Reactions

51. Compute the number of grams of solute required to prepare 1 L of each of the following solutions.
a. 0.04 *M* NaCl solution
b. 0.20 *M* $AgNO_3$ solution
c. 0.60 *M* $HC_2H_3O_2$ (acetic acid) solution
d. 0.50 *M* LiCl solution
e. 0.3 *M* NaOH solution

52. Compute the number of grams of solute required to prepare each of the following solutions.
    a. 500 mL 0.20 *M* NaOH solution
    b. 200 mL 0.40 *M* LiF solution
    c. 100 mL 0.80 *M* H_2SO_4 solution

53. What would be the concentration, in moles per liter, of each of the following solutions?
    a. 1.00 L of solution containing 2.0 g of NaOH
    b. 1.00 L of solution containing 9.8 g of H_3PO_4
    c. 500 mL of solution containing 6.3 g of HNO_3
    d. 200 mL of solution containing 3.38 g of $AgNO_3$

54. Predict the products of each of the following acid–base reactions, name the salt produced, and balance the equation describing the reaction.
    a. Hydrochloric acid + Lithium hydroxide

$$HCl + LiOH \longrightarrow$$

    b. Nitric acid + Aluminum hydroxide

$$HNO_3 + Al(OH)_3 \longrightarrow$$

    c. Boric acid + sodium hydroxide

$$H_3BO_3 + NaOH \longrightarrow$$

    d. Sulfuric acid + magnesium hydroxide

$$H_2SO_4 + Mg(OH)_2 \longrightarrow$$

    e. Acetic acid + ammonium hydroxide

$$HC_2H_3O_2 + NH_4OH \longrightarrow$$

55. Suppose that 25.0 mL of a HCl solution of unknown concentration was completely neutralized by 75.0 mL of a 0.10 *M* KOH solution.

$$HCl + KOH \longrightarrow KCl + H_2O$$

    a. How many moles of KOH were used to complete the reaction?
    b. How many moles of HCl were present in the 25.0-mL sample?
    c. How many moles of HCl would 1.0 L of this solution contain?

56. A common antacid tablet contains magnesium hydroxide. Suppose that a 2.0 g tablet of this material is found to be completely neutralized by 64.0 mL of 0.20 *M* HCl.
    a. Write the balanced equation for this reaction.
    b. How many moles of HCl were used to complete the reaction?
    c. How many moles of $Mg(OH)_2$ were present in the tablet?
    d. How many grams of magnesium hydroxide were present in the tablet?
    e. What was the percent by weight of magnesium hydroxide in the 2.0-g tablet?

57. Suppose that the phosphoric acid present in a 20.0-mL sample of a soft drink could be neutralized by 70.0 mL of 0.10 *M* NaOH solution. What is the concentration of phosphoric acid in the soft drink, expressed in moles per liter?

$$H_3PO_4 + 3\ NaOH \longrightarrow$$

58. Suppose that a 40.0-mL sample of sulfuric acid (H_2SO_4) can be exactly neutralized by addition of 90.0 mL of 0.20 *M* ammonium hydroxide. What is the concentration of the sulfuric acid sample in moles per liter?

59. A common drain cleaner contains sodium hydroxide. One tablet of this cleaner weighs 3.0 g. If it is dissolved in water and titrated with 0.10 *M* hydrochloric acid, 300 mL of HCl solution are required to neutralize the NaOH. What is the percent by weight of NaOH in the tablet?

Weak Acids and Buffers

60. What would be the hydronium ion concentration in a solution of ascorbic acid of concentration 0.0126 *M*? Assume that a negligible amount of the acid ionizes. (You may wish to refer to Table 11.4 for K_a values.)

61. What is a reasonable estimate of the pH of this solution?

62. What would be the hydronium ion concentration in a solution of lactic acid of concentration 0.0714 *M*? Again, assume that a negligible amount of the acid ionizes.

63. What is a reasonable estimate of the pH of this solution?

64. What would be the hydronium ion concentration in a solution of oxalic acid of concentration 0.00153 *M*? Assume that a negligible amount of the acid ionizes.

65. What is a reasonable estimate of the pH of this solution?

66. Compare now the concentration of hydronium ion in the solution of question 64 with the initial concentration of the un-ionized oxalic acid. Is the assumption that a negligible amount of acid ionizes valid when the K_a value approaches the concentration of the un-ionized acid?

67. Consider the buffer formed by a mixture of acetic acid and potassium acetate. Suppose that the acetic acid and potassium acetate were both present at a concentration of 0.20 *M*.

Acetic acid (0.20 *M*) $HC_2H_3O_2$

$\downarrow$

Buffer system equilibrium $HC_2H_3O_2 + H_2O \rightleftharpoons H_3O^+ + C_2H_3O_2^-$

$\uparrow$

Potassium acetate (0.20 *M*) $KC_2H_3O_2 \longrightarrow K^+ + C_2H_3O_2^-$

    a. By reference to the K_a value shown for acetic acid in Table 11.4, compute the hydronium ion concentration of this buffer solution.
    b. By use of a pocket calculator or logarithm table, compute the pH of this buffer solution.

68. Suppose that one prepared an identical buffer solution, except that the concentrations of acetic acid and potassium acetate were 0.40 *M*.
   a. What would be the hydronium ion concentration of this solution?
   b. What would be the pH of this solution?
   c. How would the buffer capacity of this solution differ from that of problem 67?

69. Consider the buffer formed by a mixture of oxalic acid and potassium hydrogen oxalate. Suppose that the oxalic acid and potassium hydrogen oxalate were both present at a concentration of 0.10 *M*.

Oxalic acid (0.10 *M*) $H_2C_2O_4$

$\downarrow$

Buffer system equilibrium $H_2C_2O_2 + H_2O \rightleftharpoons H_3O^+ + HC_2O_2^-$

$\uparrow$

Potassium hydrogen oxalate (0.10 *M*) $KHC_2O_2 \longrightarrow K^+ + HC_2O_2^-$

   a. By reference to the K_a value shown for oxalic acid in Table 11.4, compute the hydronium ion concentration of this buffer solution.
   b. By use of a pocket calculator or logarithm table, compute the pH of this buffer solution.

CHAPTER 12

Celsius and Kelvin Temperature

17. Convert the following Celsius temperatures to Kelvin.

    27 °C   37 °C   100 °C   −40 °C

18. Convert the following Kelvin temperatures to Celsius.

    250 K   293 K   350 K   393 K

Volume–Temperature Relationships

19. Consider a gas sample of volume 400 mL and temperature 0 °C. What will be the volume of this sample if the temperature is increased to 25 °C and the pressure remains constant?

20. Suppose that you inhaled 1.5 L of air at −4 °C and held it long enough to warm it to body temperature, 37 °C. What would be the new volume of this air, assuming constant pressure?

Volume–Pressure Relationships

21. Express the following pressures in Torr.

    0.5 atm   1.6 atm   2.5 atm

22. Express the following pressures in atmospheres.

    500 Torr   640 Torr   1200 Torr

23. Consider a gas sample of volume 600 mL at a pressure of 640 Torr. What would be the volume of this sample if the pressure were 760 Torr?

24. A diver inhaled 1.4 L of air from his scuba tank while underwater at a depth where the pressure was 2.0 atm. He then swam for the surface with his mouthpiece out. If he were able to hold his breath all the way to the surface, what volume would the 1.4 L of air now occupy at 1 atm pressure? Why does a diver normally exhale all the way to the surface?

Combined Gas Law

25. Consider an automobile tire filled to 3.0 atm pressure at a temperature of 20 °C. The automobile is then driven on a hot day, and the temperature of the air in the tire reaches 30 °C. The volume of the tire changes very little. What is the new pressure in the tire?

26. A gas sample has a volume of 1.2 L when collected at 20 °C and 1.0 atm pressure. What will be the new volume of this sample if the pressure falls to 0 °C and the pressure falls to 0.80 atm?

27. A gas sample has a volume of 3.4 L at 15 °C and a pressure of 640 Torr. What will be the volume of this sample at STP?

Ideal Gas Law

28. What volume would be occupied by 1 mol of carbon dioxide gas (CO_2) at 20 °C and 1 atm pressure?

29. Consider the gas sample of question 28.
   a. What would be the mass of this gas sample?
   b. What would be the density (in g/mL) of carbon dioxide gas at this temperature and pressure?
   c. The density of air at 20 °C and 1 atm is about 1.2 g/mL. Is carbon dioxide lighter or heavier than air?

30. A small propane cylinder holds 1.0 L of propane. The pressure inside the propane tank is 3000 Torr and the temperature is 10 °C.
   a. How many moles of propane are in the cylinder?
   b. Propane reacts with oxygen gas to form carbon dioxide and water vapor.

$$C_3H_8 + 5\,O_2 \longrightarrow 3\,CO_2 + 4\,H_2O$$

   How many moles of oxygen will be required to react with the propane in this cylinder?
   c. At 10 °C and 760 Torr, how many liters of oxygen will be required to react with the propane in this cylinder?
   d. Since air is only 21% oxygen by volume, what volume of air will be required to burn all of the propane in this cylinder?

31. Dry ice is frozen carbon dioxide. Suppose that a 20-g sample of dry ice is placed in a 1.0-L flask, which is then evacuated. The flask and dry ice are then allowed to warm to room temperature, 20 °C., and the dry ice turns into a gas. What will be the pressure in this 1.0-L flask after the gas has reached room temperature?
32. A 2.0-g sample of a liquid hydrocarbon is placed in a 1.0-L flask, which is evacuated. The flask is then placed in a water bath and warmed to 75 °C. The liquid evaporates. The pressure in the flask is measured to be 500 Torr. What is the molecular weight of the hydrocarbon?

Partial Pressure

33. Gases are often collected in the laboratory by displacing water from a collecting flask. Consider, for example, a burette full of water held upside down with its bottom in a beaker of water. The gas to be collected is then passed through a tube into the bottom of the burette. As the gas bubbles in, water is displaced from the burette into the beaker and the gas collects at the top of the burette. Since the pressure exerted on the surface of the water is atmospheric pressure, one can equalize the pressure inside and outside of the burette by simply moving it up or down until the liquid surfaces are at the same level inside and outside. At this point, the pressure inside the burette is the same as atmospheric pressure. However, this pressure is due to two factors—the pressure exerted by the collected gas, and the pressure exerted by water vapor that evaporates from the surface of the water in the burette. The vapor pressure of water at several temperatures is shown in the following table.

    Suppose that a 45.0-mL sample of gas were collected over water as discussed above. The temperature in the laboratory is 20 °C; the water is the same temperature as the laboratory. The pressure in the laboratory is 710 Torr.
    a. What is the partial pressure of water vapor in the sample?
    b. What is the partial pressure of the unknown gas?
    c. How many moles of unknown gas were collected?

Temperature (°C)	Vapor Pressure (Torr)
0	4.58
10	9.20
20	17.54
30	31.82
40	55.32
50	92.51
60	149.38
70	233.7
80	355.1
90	525.76
100	760.00

34. Water may be evaporated from a substance by simply pumping the air from the container and allowing the water to "boil" away. Fruits, for example, can be dehydrated in this manner without heating and thereby destroying them. The technique is called "freeze drying." To what value must the pressure be reduced to make water boil at 20 °C?
35. The concept of partial pressure can be used to solve problems involving combustion of fuels such as gasoline. Consider, for example, the combustion of octane.

$$2\,C_8H_{18} + 25\,O_2 \longrightarrow 16\,CO_2 + 18\,H_2O$$

    Octane has a density of about 0.70 g/mL. Thus, 1.0 mL will have a mass of 0.70 g.
    a. How many moles of octane are represented by the 1.0-mL sample described above?
    b. How many moles of oxygen will be required to react with this amount of octane?
    c. The partial pressure of oxygen in air at STP is about 159 Torr. How many liters of air will be required to completely burn 1.0 mL of octane?
    d. Consider the size of the fuel line and the size of the air intake on an automobile engine. Does your result for part (c) explain this difference?

CHAPTER 13

Nuclear Decay

21. Predict the daughter products of the following nuclear decay reactions.
    a. $^{60}_{27}Co \longrightarrow B^- +$ ______
    b. $^{131}_{53}I \longrightarrow B^- +$ ______
    c. $^{26}_{13}Al \longrightarrow B^+ +$ ______
    d. $^{232}_{90}Th \longrightarrow$ Alpha + ______
    e. $^{238}_{92}U \longrightarrow$ Alpha + ______
22. The table here presents time and count rate data describing the decay of a radioisotope. Plot these data and estimate the half-life of the isotope.

Time (Minutes)	Counts per Minute
0	12000
2	9500
4	7600
8	4800
10	3800

23. Estimate the activity of the sample described in question 12 after 18 minutes has passed.

Interaction of Radiation and Matter

24. Which type of nuclear radiation carries a positive charge?

25. Which type of nuclear radiation carries a zero charge?
26. Which type of nuclear radiation carries a negative charge?
27. Explain how alpha particles interact with the atoms they pass.
28. Why will beta radiation penetrate farther into a substance than alpha radiation?

Detection of Nuclear Radiation

29. Explain how a scintillation detector detects ionizing radiation.
30. Explain how a Geiger counter detects beta particles.
31. A Geiger counter is commonly used to detect beta radiation, but seldom to detect alpha radiation. Why?

Radiation Biology

32. What radiation source provides the largest component of our annual radiation exposure?
    a. natural sources such as cosmic radiation and radioactive elements in the earth
    b. medical exposure to radiation
    c. the nuclear power industry
33. Radioisotope tracers used in medicine are chosen for their chemical ability to concentrate in one organ, for example, iodine in the thyroid gland. Tracers are also chosen by the type of radiation they emit. To be of most value in medicine, should a radioisotope emit alpha, beta, or gamma radiation?

Energy

34. Which fossil fuel is in greatest supply: coal, oil, or natural gas?
35. About what length of time is required for the radioactivity of spent nuclear fuel elements to fall to the level of the original uranium ore?
36. Define the term *critical mass.*
37. Define the term *chain reaction.*
38. Define the term *fission fragment.*
39. Define the term *fallout*, as applied to nuclear weapons.

CHAPTER 14

Electron Dot Structures

7. Give an electron dot structure for formaldehyde.

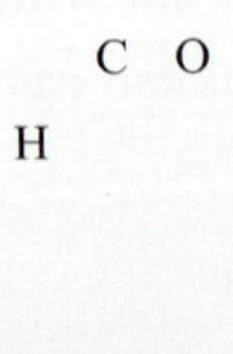

8. Give an electron dot structure for diethylether.

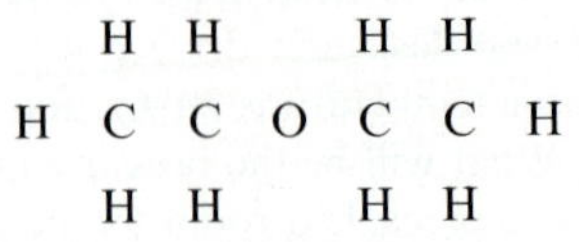

9. Give an electron dot structure for acetonitrile.

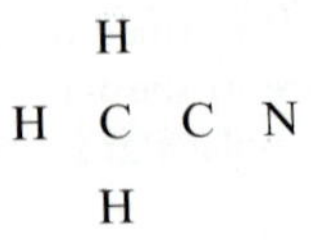

10. Give the hybridization of each atom (other than hydrogen) in the examples in the previous three exercises.
11. What is a functional group?
12. What is the functional group in the following compounds?
    a. alcohol c. ether
    b. ketone d. amine
13. Classify each of the following general structures according to functional group.

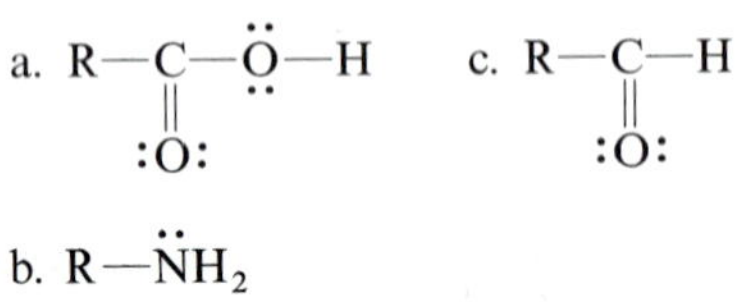

14. Classify each of the following general structures according to functional group.

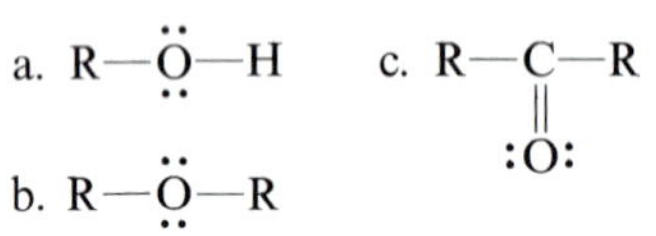

15. How does an alkyne differ from an alkane?
16. Give electron dot structures for each of the following.
    a. H C N b. O C c. H Br
17. What is the hybridization of each highlighted atom?

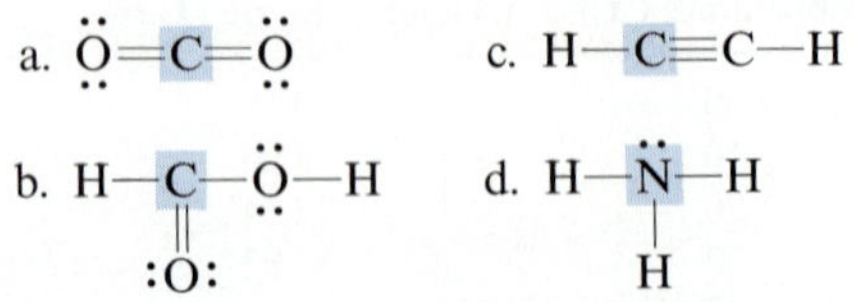

18. What is the geometry of an atom having sp^2 hybridization?
19. What is a Lewis base?
20. Why would you expect that alkanes are not good Lewis bases?

CHAPTER 15

Structure and Physical Properties

26. What is the hybridization of the carbon atoms found in alkanes?
27. What is the name of the attractive forces that attract hydrocarbon molecules to one another?
28. Why do the "straight chain" hydrocarbons have higher boiling points than the branched isomers?
29. Why is the term *straight chain* hydrocarbon somewhat misleading?

Nomenclature and Alkyl Groups

30. What is a homologous series?
31. What is an isomer?
32. Draw and name the isomers of butane.
33. Draw and name the alkyl groups derived from the butane isomers.
34. Why aren't alkyl groups stable compounds?
35. Name each of the following compounds.

a.

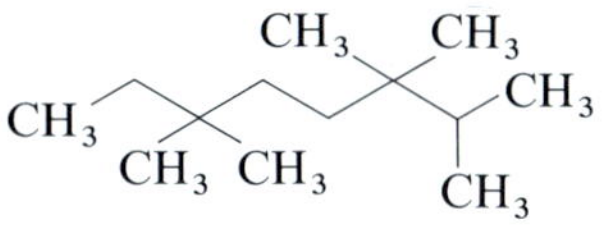

b

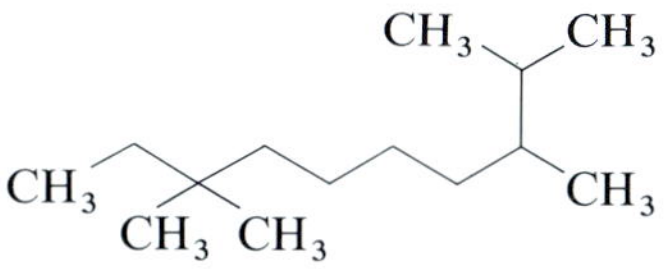

c.

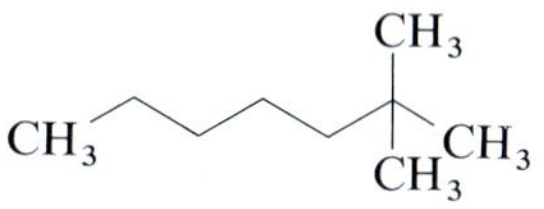

d

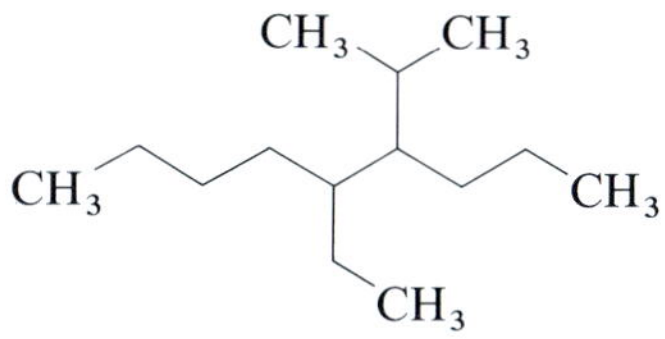

Cycloalkanes

36. Name each of the following compounds.

a

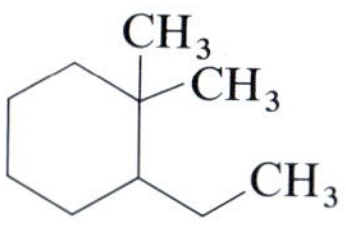

c

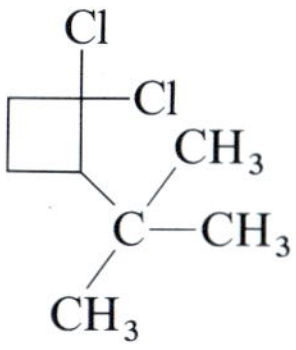

b

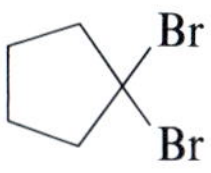

d

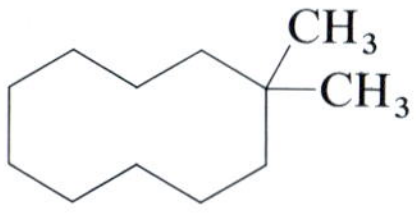

37. Draw structures corresponding to each of the following names.
    a. *cis*-1,2-dimethylcyclopentane
    b. *trans*-1-isobutyl-3-ethylcyclooctane
    c. 3-isopropyloctane
    d. 1,5-dibromo-2,2-dimethylhexane
    e. 1,1-dimethylcyclopropane

Conformations

38. What do we mean by the conformation of an alkane?
39. What is the preferred orientation of a methyl group attached to a cyclohexane ring?

Reactions

40. Why are alkanes and cycloalkanes so unreactive?
41. What are the common products to complete combustion of any hydrocarbon?

CHAPTER 16

Structure and Physical Properties

17. What do we mean by an "unsaturated" hydrocarbon?
18. What is the hybridization of the carbon atoms involved in the alkene bond?
19. Explain why there is no rotation about the C—C double bond.
20. What is the hybridization of carbon atoms involved in the alkyne bond?
21. What do we mean by cis and trans isomers for alkenes?

Nomenclature

22. Give structures for the following compounds.
    a. cyclopentene
    b. *cis*-3-heptene
    c. 4-octyne
    d. *trans*-2-butene
    e. 2,5-octadiene
    f. *trans*-3,4-dimethylcyclohexene
23. Give names for each of the following compounds.

a.

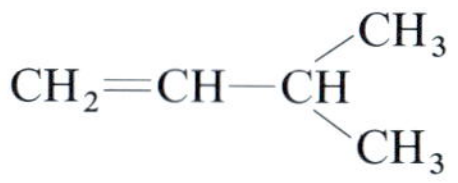

b

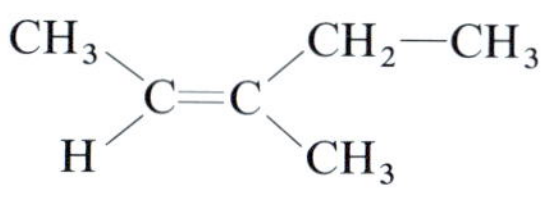

c

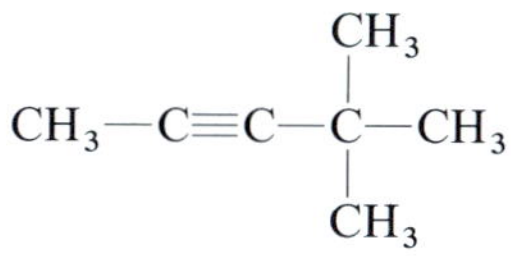

d

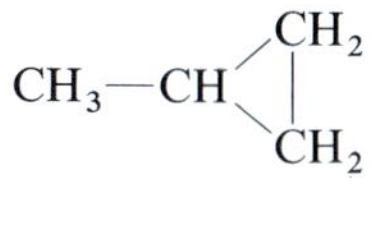

24. Give names for each of the following compounds.

a. 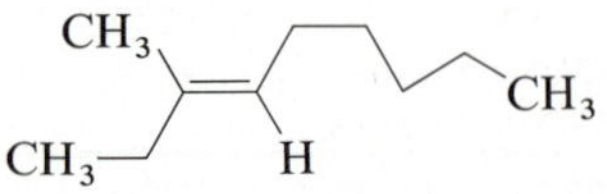

b. 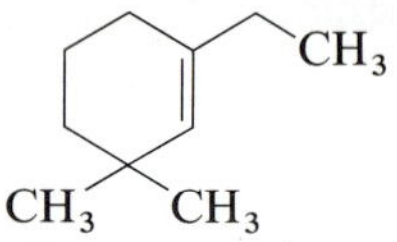

c. 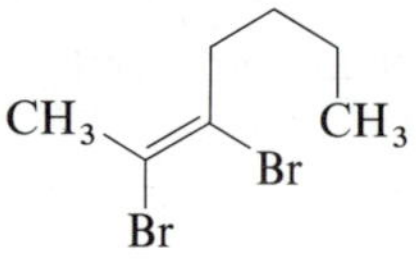

d. 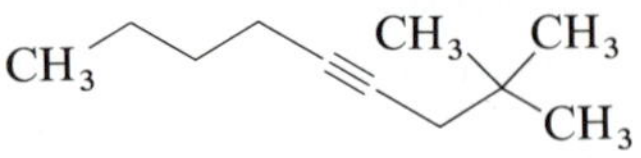

Reactions

25. Why are alkenes and alkynes more reactive than alkanes?

26. What are the products of a reaction of an alkene with hydrogen and a catalyst?

27. What are the products of a reaction of an alkene with bromine?

28. What is the significance of Markovnikov's rule?

29. What will be the site of attachment of an electrophile (E^+) to 2-methyl-2-pentene?

30. Any alkene will react with dilute permanganate. What is the net reaction at the site of the double bond?

31. In the addition of HBr to an alkene, what is the electrophile?

32. When hydrogen adds to an alkene, the hybridization of the carbon atoms in the double bond changes to what?

33. What are the products found from the following reactions.

a. 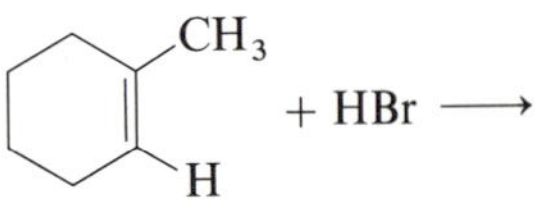

b. 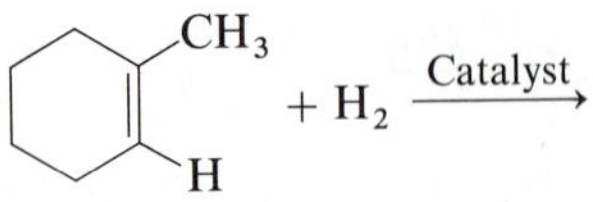

c. 

d. 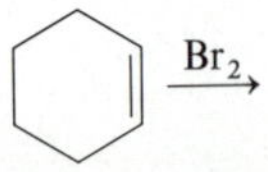

e. 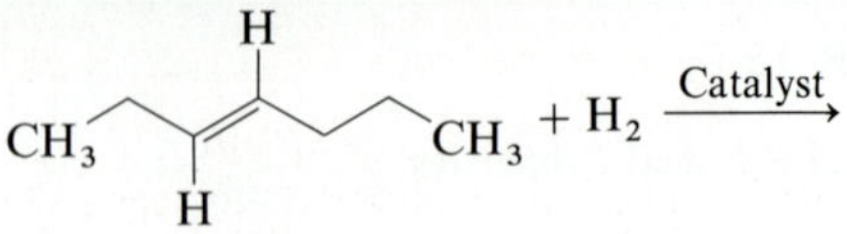

f. 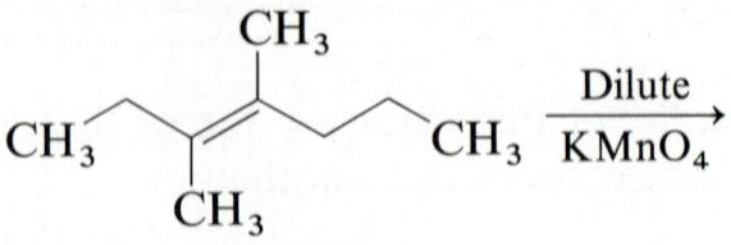

34. What is a polymer?

35. If a polymer had the general structure —$(CCl_2—CCl_2)_n$—, what would be the basic building block alkene?

CHAPTER 17

Structure and Stability

11. What is meant by "delocalized electrons" in benzene?

12. What is the $4n + 2$ rule?

13. What is meant by "resonance structures"?

14. Why does benzene *not* undergo the addition reactions typical of alkenes?

15. Why is (a) a better representation of benzene than (b)?

a.  b. 

Nomenclature

16. Draw structures for the following common names.
 a. toluene b. benzene c. cumene

17. What is the ortho position of toluene?

18. What is the para position of nitrobenzene?

19. What is the meta position of ethylbenzene?

20. Draw structures for the following compounds.
 a. *o*-ethyltoluene c. *m*-bromocumene
 b. *p*-ethyltoluene d. 2,3-dichlorotoluene

21. Draw structures for the following compounds.
 a. *m*-iodoethylbenzene c. *m*-chlorotoluene
 b. *p*-bromochlorobenzene d. 2,4,6-trinitrotoluene

22. Give names for each of the following aromatic compounds.

a. 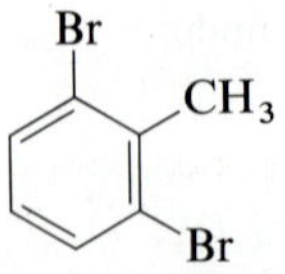

c. 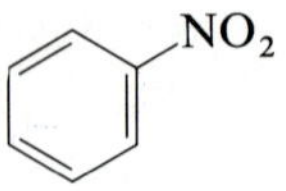

b. 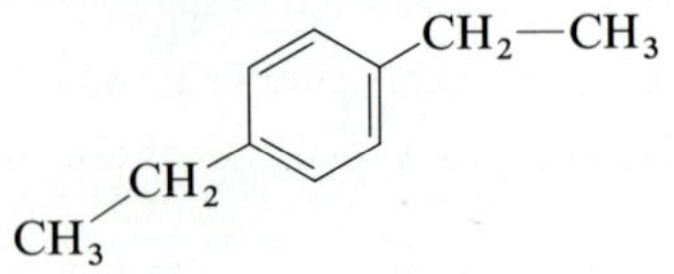

d. 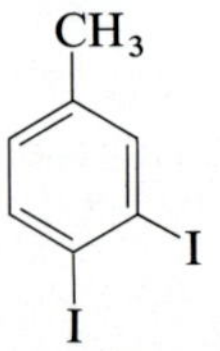

Reactions

23. Benzene derivatives undergo substitution reactions rather than the addition reactions of alkenes. Why?
24. What is meant by "ortho-para director" in reference to reactions of substituted benzenes?
25. List the following groups as *o*-, *p*-, or *m*-directors.
    a. methyl d. bromo
    b. nitro e. ethyl
    c. hydroxy f. sulfonic acid
26. What is meant by "side-chain oxidation"?
27. Give products for the following aromatic reactions.

a. 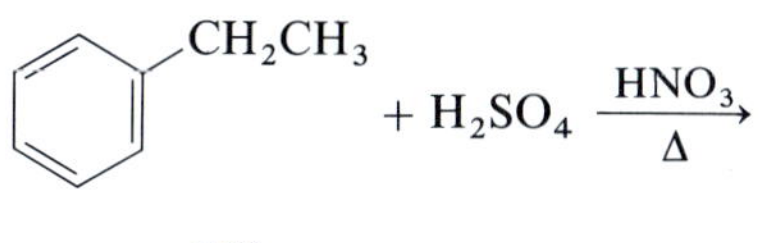

b. 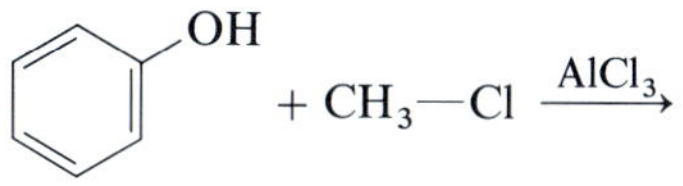

c. 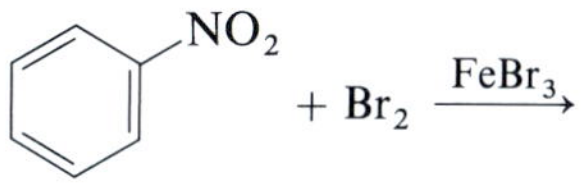

28. Give products for the following aromatic reactions.

a. 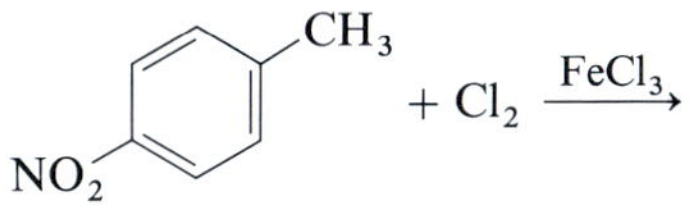

b. 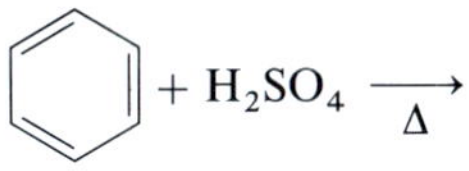

c. 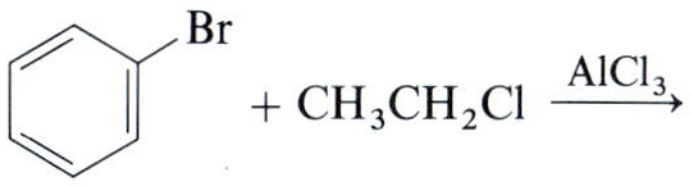

CHAPTER 18

Alcohols

20. What is the hybridization of the oxygen in alcohols?
21. Why does methanol have a lower boiling point than water?
22. Account for the water solubility of lower carbon number (less than five) alcohols.
23. Why is butanol water soluble while butane is not?
24. Why is decanol not appreciably soluble in water?
25. Give names for each of the following alcohols.

a. 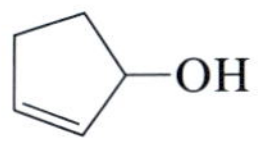

b. 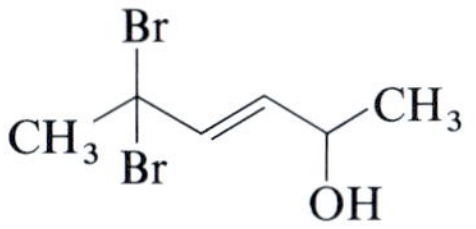

c. 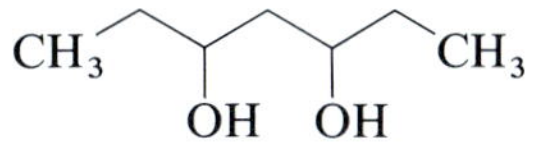

d. 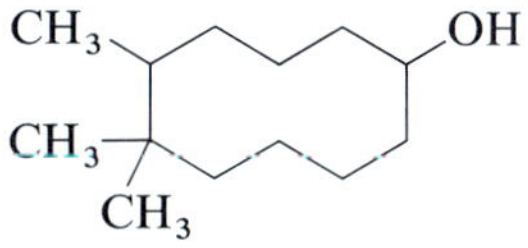

26. Draw structures for each of the following alcohols.
    a. 2-methylcyclohexanol
    b. 2,3-dimethyl-2-octanol
    c. 4-isopropyl-3-decanol
    d. *cis*-1,2-dimethylcyclohexanol
    e. 3-hexen-2-ol
    f. 2-pentyn-1-ol
27. Why can an alcohol be considered a Lewis base?
28. An alcohol can also be considered an acid. Explain.
29. Give products for the following reactions of alcohols.

a. 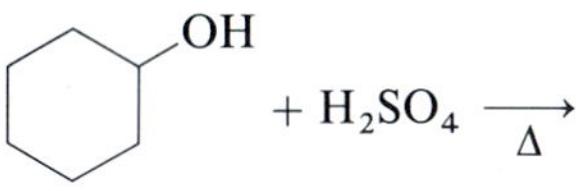

b. 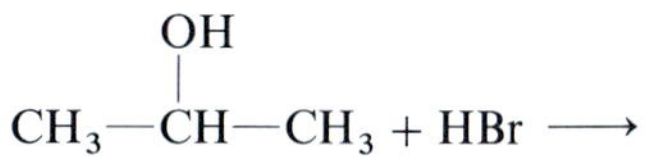

c. 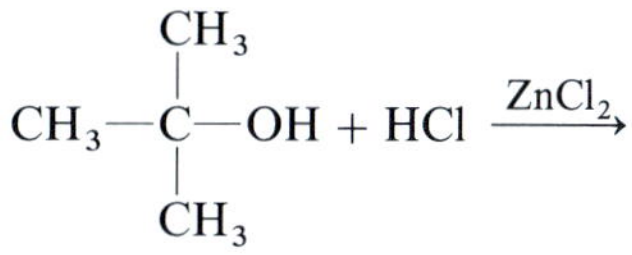

d. 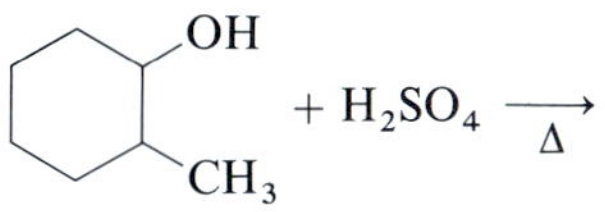

30. What is a polyhydroxylic compound?

Phenols

31. Give names for the following phenols.

a. 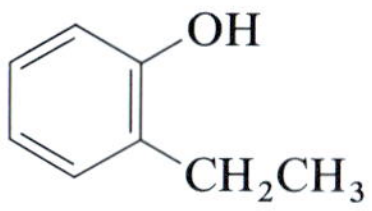

c. 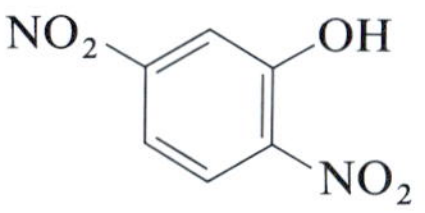

b. 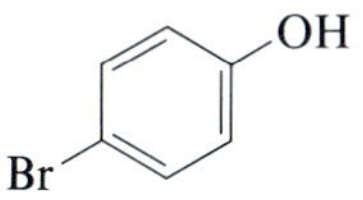

d. 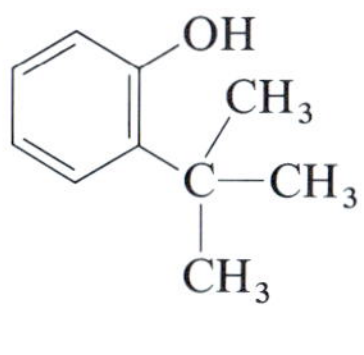

32. Explain why phenols are more acidic than alcohols.
33. Give products for the following reactions.

a. 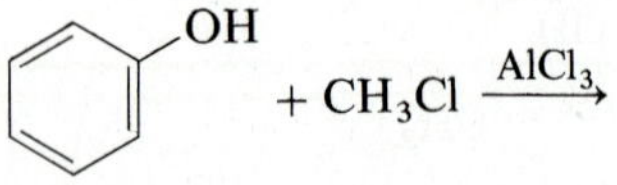

b. 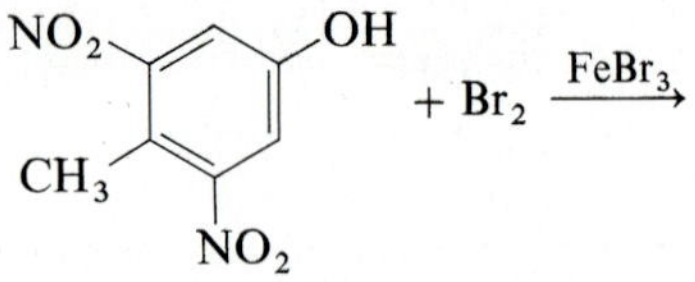

Ethers

34. What is the hydridization of the oxygen in an ether?
35. Ethanol (C_2H_6O) has a higher boiling point than dimethylether, even though both have the same empirical formula. Explain.
36. What is an epoxide?
37. Ethanol is very water soluble while dimethyl ether is not. Explain.
38. Give a product for the following reaction.

$$CH_3{-}O^- + CH_3CH_2Br \longrightarrow$$

Thiols

39. Thiols have lower boiling points than alcohols of the same carbon number. Why?
40. Give names for the following compounds:

a. 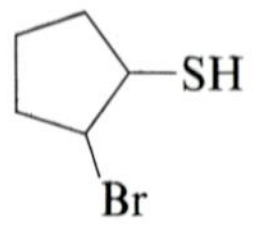

b. 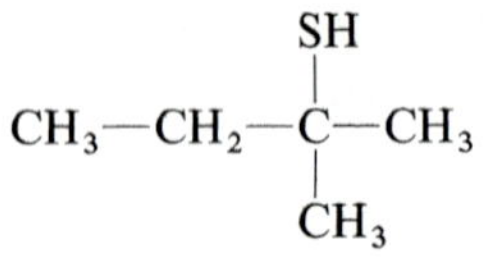

41. Give products for the following reactions.

a. $CH_3{-}SH \xrightarrow{[O]}$

b. 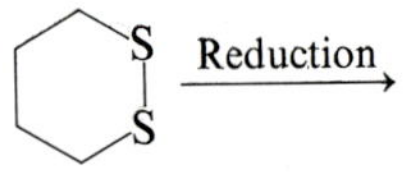

c. $CH_3CH_2CH_2SH + Hg^{2+} \longrightarrow$

CHAPTER 19

Structure and Properties

17. Describe the bonding between carbon and oxygen in the carbonyl group.
18. What is the hybridization of carbon in the carbonyl group?
19. Describe dipole attractive forces for carbonyl compounds.
20. How are the electrons polarized on the carbon and oxygen atoms of the carbonyl group?
21. Describe how the carbonyl group resembles an alkene. What is the major difference?
22. How does the bonding of the carbonyl group influence the addition reactions to the carbonyl double bond?

Nomenclature

23. Give names for the following compounds.

a. 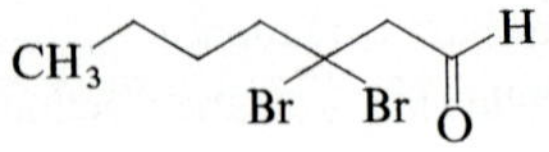

b. 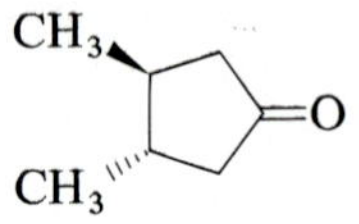

c. 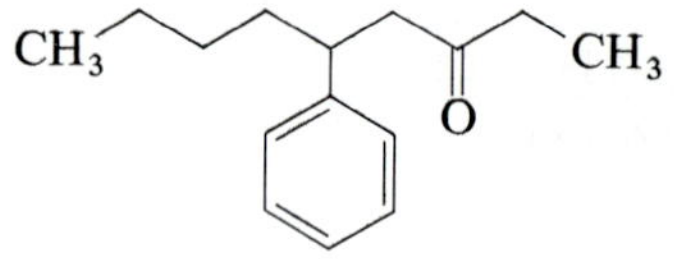

d. 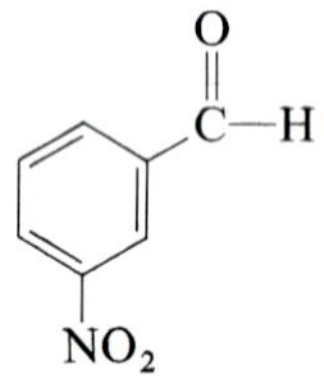

24. Draw structures for the following compounds.
   a. *cis*-2,3-dimethylcyclohexanone
   b. 3-isopropylheptanal
   c. 2-phenylcyclobutanone
   d. 3-hexenal

Reactions

25. Describe why a nucleophile will react at the carbonyl carbon of a carbonyl group.
26. What is a cyanohydrin?
27. What is a hydrate of a carbonyl compound?
28. A secondary alcohol will be formed by the reduction of what?
29. What is an acetal?
30. Give products for each of the following reactions.

a. 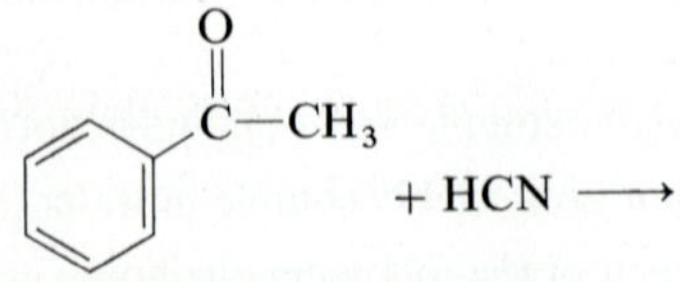

b. 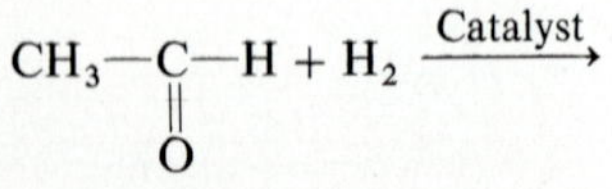

c. 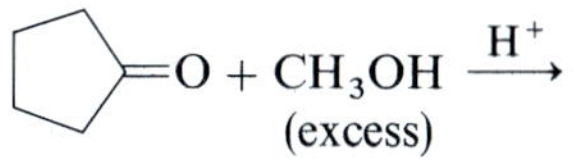

d. 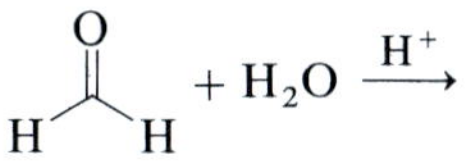

31. In a Tollens test for aldehydes, what is reduced?
32. What will oxidation of a tertiary alcohol produce?
33. What will oxidation of a hydroquinone produce?
34. Give *general* products from the following general reactions.
    a. An aldehyde + oxidation →
    b. An aldehyde + reduction →
    c. A ketone + oxidation →
35. Give products for the following reactions.

a. 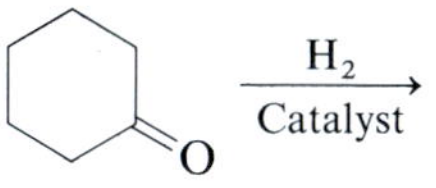

b. 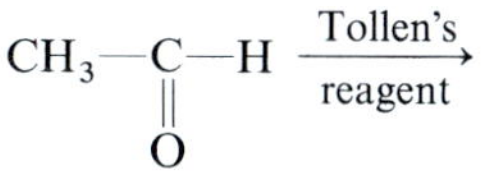

c. 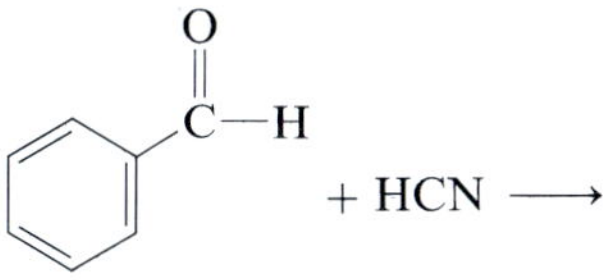

CHAPTER 20

Structure and Physical Properties

16. What is the carboxyl group?
17. What is the hybridization of the carbon of the carboxyl group?
18. Why do acids exhibit "higher than expected" boiling points?
19. Why is chloroacetic acid more acidic than acetic acid?
20. Why are acids of lower carbon number water soluble?
21. Why are long chain carboxylic acids *not* water soluble?

Nomenclature

22. Give names for each of the following compounds.

a. 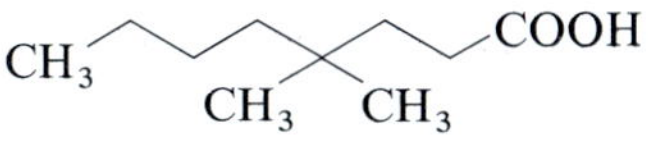

b. 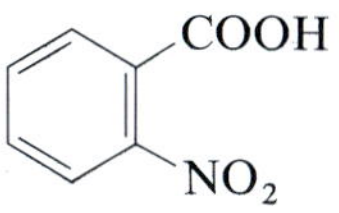

c. 

d. 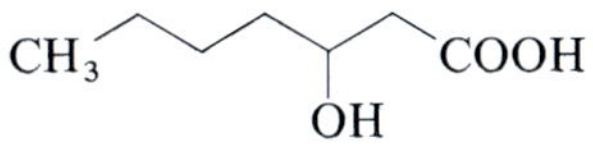

23. Give structures for the following compounds.
    a. *o*-chlorobenzoic acid
    b. 3-isobutyloctanoic acid
    c. *cis*-3-pentenoic acid
    d. 2-hydroxybutanoic acid
    e. 3-oxoheptanoic acid

Reactions

24. Show the resonance structures for the carboxylate anion formed from reaction of acetic acid with sodium hydroxide.
25. How is an acid prepared from an aldehyde?
26. What is a nitrile and how would it be converted to an acid?
27. What is the product formed from heating 3-oxooctanoic acid?
28. What is the product of the reaction of acetic acid with sodium hydroxide?
29. If you mix acetic acid with sodium bicarbonate, a gas is produced. What is the gas? Write an equation to show this reaction.

Esters

30. What is an ester, and how is one prepared?
31. Draw a structure for methyl pentanoate.
32. What is saponification?
33. Describe the action of a detergent in water.
34. Why wouldn't the salt of propanoic acid serve as a soap?

Acid Chlorides and Anhydrides

35. Draw general structures for an acid chloride and an acid anhydride, and describe the similarities of their reactions with an alcohol.
36. Give products for the following reactions.

a. 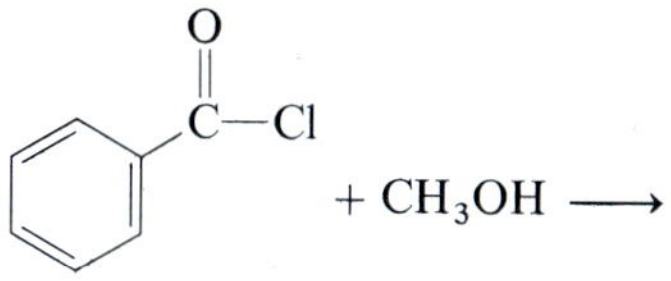

b. 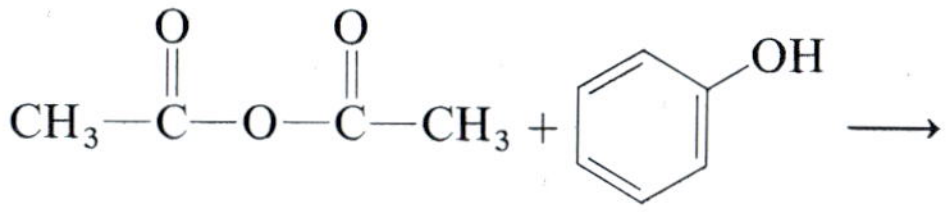

c. 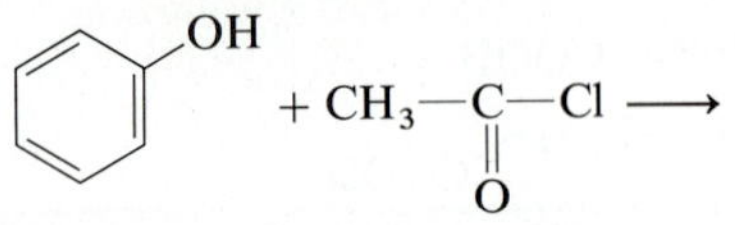

d. 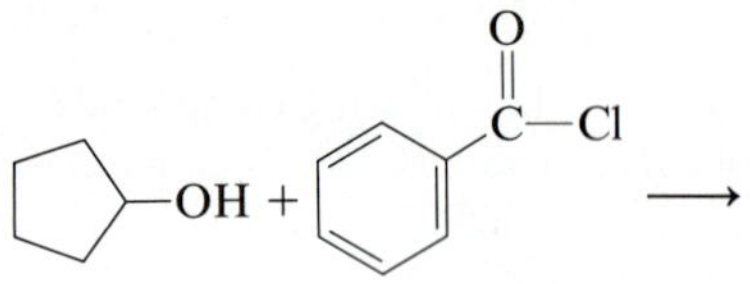

CHAPTER 21

Structure and Properties

11. Why is methyl amine considered as a Lewis base?
12. Give a Lewis electron dot structure for methyl amine.
13. Why does methyl amine have a lower boiling point than methanol?
14. What are the characteristics that make a molecule a Lewis base?

Nomenclature

15. Give names for each of the following amines.

a. $CH_3-CH_2-N(H)-CH_3$

b. 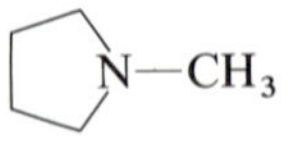

c. $CH_3-N(CH_2-CH_3)(CH_2-CH_3)$

16. Draw structures for each of the following amines.
    a. diisopropylamine c. benzylamine
    b. isobutylmethylamine d. triethylamine
17. Draw structures for the following amines.
    a. *N*-methylaniline b. pyrrole

Reactions of Amines

18. Show the reaction of triethylamine with HCl.
19. Why would you expect that $Cl^{-}-Me_3N^{+}-CH_2-(CH_2)_{12}-CH_3$ would behave like a soap in water?
20. Show the reaction of dimethyl amine with the acid chloride of acetic acid.
21. Give products for the following reactions of amines.

a. $CH_3-NH_2 + HBr \longrightarrow$

b. 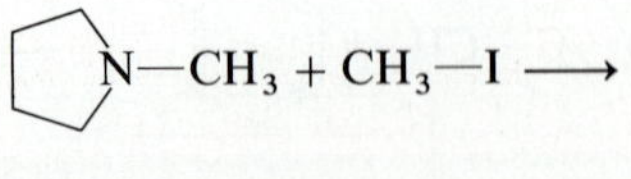

c. $CH_3-C(=O)-Cl + CH_3-NH_2 \longrightarrow$

d. $CH_3-CH_2-CH_2-CH_2-N(CH_3)(CH_3) + CH_3CH_2Br \longrightarrow$

Amides

22. Why is an amide less basic than an amine?
23. Give products for the following reactions of amides.

a. $CH_3-C(=O)-NH_2 + HO^{-} \xrightarrow[\Delta]{H_2O}$

b. 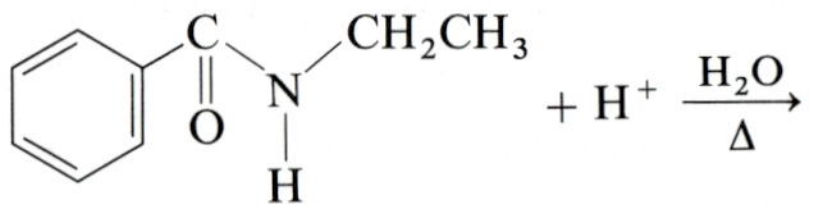

Special Amines

24. Give the general structure for an alpha amino acid.
25. What is an alkaloid?

CHAPTER 22

22. Calculate the volume in μm^3 of a typical bacterium and a eucaryotic cell. Consider the bacterium to be a cylinder 10 μm in height and 2 μm in diameter. Consider the cell to be a cube with 100-μm sides. (HINT: The volume of a cylinder is $\pi r^2 h$.)
23. Compare the surface area to volume ratio of the bacterium and eucaryotic cell in exercise 22.
24. Does the cell wall protect a bacterial cell in hypertonic solutions? Why or why not?
25. What macromolecule is found in the cytosol of bacteria but is found only in the nucleus of eucaryotic cells?
26. Streptomycin disrupts protein synthesis. It is given to humans as an antibiotic to treat certain infections. Why are the human host cells not killed?
27. What is the meaning of the word *nonpyrogenic*?
28. Where is the nucleolus found in eucaryotes? Where is it found in procaryotes?
29. You have isolated two liquid samples that are pure compounds. They have the same composition. What simple tests could you use to determine whether they are the same compound, enantiomers, or some other type of isomer?
30. Use a molecular model kit to make (−)- and (+)-carvone. Note that the ring can have several conforma-

tions since it is not aromatic. Are these models identical? Are they mirror images? Are they superimposable?

31. Identify the chiral center in the carvones in exercise 30.

CHAPTER 23

25. Compare the appearance of a dilute glucose solution and a more concentrated glucose solution when they are mixed with Benedict's solution.
26. Compare the similarities and differences between glucose and fructose.
27. Compare the similarities and differences between ribose and 2-deoxyribose.
28. Compare the composition and bonding in amylose, amylopectin, and glycogen.
29. D-Ribose is levorotatory but D-glucose is dextrorotatory. How can this be?
30. Draw the structure of a disaccharide discussed in this chapter that is a reducing sugar and yields only glucose upon hydrolysis. Can you draw the structure of another disaccharide that meets these conditions?
31. One of your classmates says α-D-glucopyranose and β-D-glucopyranose are enantiomers, another says D-glucose and L-glucose are enantiomers. Which is correct? Justify your choice.
32. Examine the formula of ribose and glucose and use them to explain the origin of the term *carbohydrate*.
33. Fructose is not an aldose, yet it is classified as a reducing sugar. How can this be?
34. Dextrose is a common name for what sugar?
35. Humans cannot digest cellulose. Why?
36. Cellobiose is a disaccharide obtained by partial hydrolysis of cellulose. What is the composition of this disaccharide? What glycosidic bond is present?
37. Explain why glycogen can be broken down rapidly in liver or muscle cells.
38. If 0.03 mol of a carbohydrate yield 0.12 mol of glucose upon hydrolysis, what kind of carbohydrate is it?
39. Name the smallest aldose and ketose.
40. What functional groups are found in carbohydrates?
41. A starch sample is estimated to have an average molecular weight of 140,000. How many glucose residues are in an average starch molecule of this sample? (HINT: first calculate the molecular weight of glucose, then subtract the molecular weight of water from it because a water molecule is lost during glycosidic bond formation.)
42. Many distance runners practice "carbo loading" (eating carbohydrate rich foods) for one to three days preceding a race. What do they hope to accomplish?

CHAPTER 24

48. Is petroleum a lipid?
49. Which of the following solvents may be suitable for extracting (dissolving) lipids from biological tissues?
    a. chloroform d. toluene
    b. ethanol e. water
    c. saturated NaCl (aq.)
50. Consider linolenic acid (18:3). All three double bonds are cis in this molecule. How many isomers exist if both cis and trans isomers are considered for each carbon–carbon double bond?
51. Which of the following groups are likely to be found in a fatty acid?
    a. C═C d. $—CH(CH_3)_2$
    b. C≡C e. —COOH
    c. —OH
52. How many different triacylglycerol molecules can be made from one glycerol molecule, one molecule of oleic acid, and two molecules of palmitic acid?
53. How many of the molecules formed in exercise 52 could be optically active?
54. In Chapter 27 you will learn that enzymes are generally specific for substrate and product. For example, usually only one of two possible enantiomers acts as a reactant or is produced by the reaction. Given this information, how many of the possible optically active molecules in exercise 52 would actually be formed?
55. Write a balanced reaction for the hydrogenation of the following.
    a. oleic acid b. palmitic acid c. linolenic acid
56. Write a balanced reaction for the iodination of the three acids in exercise 55.
57. Vitamin D is synthesized from what precursor molecule?

CHAPTER 25

30. Examine the structure of threonine. How many chiral centers are present in this molecule? Which is compared to the glyceraldehydes to determine stereochemical family?
31. Draw the structure of lysine showing the correct charge on the ionizable groups at pH 1, pH 7, and pH 13. If necessary, review the pK values for these groups in the organic section.
32. Draw a molecule of aspartic acid showing the correct charge on the ionizable groups at pH 1, pH 7, and pH 13.
33. Draw the structure of the tripeptide glutamylphenylalanylleucine.

34. Name this tripeptide.

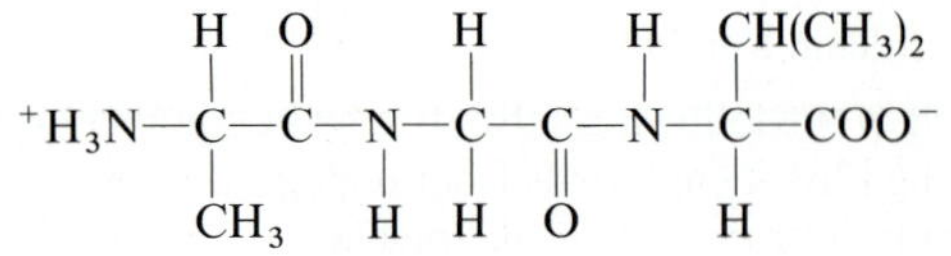

35. Peptide bonds do not show free rotation. Why?
36. Consider a globular protein that is water soluble. Which of these amino acids will probably be on the exterior surface and which are more likely to be buried within the core of the protein?
    a. alanine e. threonine
    b. glutamate f. valine
    c. leucine g. phenylalanine
    d. lysine
37. Why are the amino acid residues in exercise 36 arranged this way?

Use the amino acids listed in exercise 36 to answer exercises 38 through 41.

38. Which of these amino acids could form ionic interactions?
39. Which could form hydrogen bonds?
40. Which could form hydrophobic interactions?
41. How many pentapeptides can you make from these amino acids? (You may use any of them as often as you wish.)
42. Explain the molecular basis of sickle cell anemia.
43. Two protein samples have been shown to have the same amino acid composition. Does it follow that they must have the same structure?
44. Proteins that have been isolated from a biological source are typically stored in buffered solutions at cold temperatures. Why?

CHAPTER 26

21. Where in cells are DNA and RNA found?
22. Draw the structure of deoxyadenosine monophosphate and identify the three components of this nucleotide.
23. Draw the nucleoside consisting of uracil and ribose.
24. Name the nucleotide in exercise 23.
25. If three phosphates were attached to the nucleoside in exercise 23, what would it be named?
26. To what class of compounds would the compound in exercise 25 belong?
27. You have isolated a nucleic acid from a virus and have found the bases adenine, guanine, uracil, and cytosine. Which nucleic acid is present in this virus?
28. You have isolated a nucleic acid from a second virus. The bases are adenine, guanine, thymine, and cytosine, but the percentage of adenine does not equal that of thymine, nor does the percentage of guanine equal that of cytosine. Explain these curious observations.
29. Compare the number of possible trinucleotides from the four bases in DNA to the number of possible tripeptides from the twenty amino acids.
30. A classmate asks you which amino acid corresponds to the codon TTT. What is your response?
31. Write a sequence of bases in mRNA that would code for the tripeptide phenylalanylprolylleucine.
32. Is the base sequence in exercise 31 the only base sequence that will code for this tripeptide? How many base sequences are possible for it?
33. Synthetic DNA could be made for a gene if the amino acid sequence of its protein were known. Would the base sequence of the synthetic gene resemble the actual base sequence of the gene? Why or why not?
34. A graduate student wants to examine a protein involved in the synthesis of the amino acid tryptophan in a bacterium. She provides the bacterium with a complete medium including all twenty amino acids. After substantial growth, she isolates the protein from the bacteria, but is unable to isolate any of the particular protein that she is interested in. What went wrong?
35. What is an intron?
36. What is an exon?

CHAPTER 27

29. What part of metabolism consumes energy for synthesis of larger molecules?
30. Describe the changes in the rate of an enzyme-catalyzed reaction as substrate concentration is changed from zero to a high level.
31. What mathematical equation fits the change in rate observed in exercise 30?
32. Write the equation described in exercise 31.
33. Explain why a graph of the activity of an enzyme versus pH yields a bell-shaped curve.
34. The side-chain of some amino acids can act as general acids or bases during catalysis. List the amino acids that can donate H^+ and those that can accept H^+ during catalysis.
35. An enzyme has optimal activity at 42 °C. Why is the activity lower at both higher and lower temperatures?
36. Give three examples of proteins that are synthesized as zymogens.
37. You have isolated an enzyme from bacteria through a series of steps that removed all other proteins and small

molecules. Your enzyme was active during the first steps of the isolation, but appeared less and less active as its purity increased. In its pure form, it has no activity. Curiously, full activity is restored if a small amount of the ground-up bacteria is added. Explain these observations.

38. You have discovered an enzyme that is involved in the oxidation of crude oil. What coenzyme(s) might be involved in this reaction?

39. Enzymes show varying rates with changes in pH and temperature. Are these factors normally important in the activity of the enzymes within the body?

40. Succinate, $^-OOCCH_2CH_2COO^-$, is an intermediate in aerobic metabolism (see the TCA cycle in Chapter 29). The enzyme succinate dehydrogenase oxidizes succinate to fumarate, $^-OOCCH{=}CHCOO^-$. This enzyme is inhibited by malonate, $^-OOCCH_2COO^-$. What kind of inhibition is this?

41. Compare the structure of the substrate and inhibitor in exercise 40. What similarities and differences may account for the type of inhibition that is observed?

42. Classify the following reactions and pathways as either catabolic or anabolic. (Some of these pathways are covered in later chapters.)

a. digestion	d. transcription
b. DNA replication	e. glycolysis
c. translation	f. glycogen synthesis

43. Briefly describe the concept of coupled reactions in metabolism.

CHAPTER 28

37. What coenzymes are needed in glycolysis?

38. What coenzymes are needed for beta oxidation of fatty acids?

39. How many moles of nitrogen, as ammonia and amino groups, are removed for each mole of urea that is synthesized?

40. Less energy is produced from the oxidation of oleic acid than stearic acid. Why is this?

41. A long, hard hike up a particular mountain path requires 200 kcal of energy as ATP. Hydrolysis of 1 mol of ATP yields 7.3 kcal of energy. Calculate the number of grams of glucose needed to produce 200 kcal under aerobic conditions.

42. Solve exercise 41 for anaerobic conditions. Do you think the body could generate this amount of energy using only anaerobic metabolism?

43. Compare the structures of GTP and ATP. Why are they comparable in energy?

44. An ounce of a breakfast cereal provides 25 g of carbohydrate. If this were all glucose (which is a reasonable approximation since the cereal is primarily wheat starch and sucrose), how many moles of ATP can be formed during aerobic metabolism? (HINT: Assume 36 mol ATP per mol glucose.)

45. Dietary carbohydrate has about 4 kcal/g. A mole of ATP is equivalent to 7.3 kcal when it is hydrolyzed. Calculate the percent efficiency of ATP synthesis from this dietary carbohydrate.

46. You have learned that enzyme names often provide information about the enzyme substrate and the type of reaction that the enzyme catalyzes. Look at the first three reactions and enzymes of glycolysis (Figure 28.5, p. 797), then translate the enzyme name into a brief sentence or phrase that summarizes the reaction and substrate.

CHAPTER 29

26. In Chapter 26 you learned about the interaction of enzymes and their substrates. Use this type of interaction to explain why both NADPH and NADH are not substrates for the same enzymes.

27. Describe the flow of electrons from water to NADPH during the light reaction of photosynthesis.

28. Indicate how and where light is involved in the flow of electrons described in exercise 27.

29. How many moles of photons are used to make the NADPH needed for the synthesis of a mole of glucose?

30. What is the most abundant protein on earth?

31. Briefly explain how a glucose molecule is formed as a product of the dark reaction and how ribulose diphosphate is recycled.

32. Name the intermediates formed as lactate is converted to glucose during gluconeogenesis.

33. Where does gluconeogenesis take place?

34. What relationship exists between galactosyl transferase and lactose synthetase?

35. In Chapter 24 you learned that fatty acids nearly always have an even number of carbon atoms. Can you now explain why this is so?

36. What is the role of acyl carrier protein in fatty acid synthesis? What molecule serves as a carrier of acyl groups in other parts of the cell?

37. What six-carbon molecule serves as a precursor to both cholesterol and the ketone bodies?

38. Name the ketone bodies and describe how two of them are synthesized from the other one.

39. Name the essential amino acids of humans.

40. Name a competitive and noncompetitive enzyme inhibitor that is used in cancer therapy.

CHAPTER 30

26. Carbon dioxide is transported in the blood in what forms?
27. How large might an organism be if an oxygen-transporting protein such as hemoglobin were not available to it? (HINT: review Section 22.1 before answering this question.)
28. Examine the equation in Section 30.1 (p. 848) that deals with CO_2 in equilibrium with H_2CO_3 and HCO_3^-. Describe the effects of an increase in CO_2 on this equilibrium.
29. What term is used to describe the effects seen in exercise 28?
30. The reabsorption of bicarbonate from the tubules in the kidneys is regulated. What effect would a decrease in the reabsorption of bicarbonate have on blood pH?
31. What form of phosphate ion will absorb an H^+ at the conditions found in blood?
32. Identify the conjugate acid and the conjugate base for the bicarbonate buffer system in blood.
33. Which blood protein makes the greatest contribution to the buffering of blood?
34. Compare the structures of cyclic AMP and AMP. What structural difference results in binding of cAMP to protein kinase but no binding of AMP?
35. Name three hormones that are steroids.
36. Name two hormones that are proteins.
37. A third group of hormones are derived from amino acids. Give an example.
38. Would you expect to find relatively high concentrations of insulin and glucagon in blood at the same time? Why or why not?
39. The immune system is said to have memory. What does this mean?

APPENDIX

II

Answers to Odd-Numbered Exercises

Included here are answers to both the end-of-chapter exercises and the supplemental exercises in Appendix I.

CHAPTER 1

1. The meter was defined by French scientists as one ten-millionth the distance from the equator to the North Pole. The best estimate of this distance was scribed on a platinum–irridium bar in the Bureau of Standards in Paris, which became the standard meter.
3. 50 m (100 cm/m) = 5000 cm
5. 453.6 g (1000 mg/g) = 453,600 mg
7. 1 fathom (6 ft/fathom) (12 in./ft) = 72 in.
9. 21 gal (4 qt/gal) (946 mL/qt) (1 L/1000 mL) = 79.5 L
11. a. 36 yd (3 ft/yd) (12 in./ft) = 1296 in.
    b. 6 miles (5280 ft/mile) = 31,680 ft
    c. 184 in. (1 ft/12 in.) = 15.3 ft
    d. 50 m (100 cm/m) = 5000 cm
    e. 60 km (1000 m/km) = 60,000 m
    f. 3540 cm (1 m/100 cm) = 35.40 m
    g. 420 cm (1 in./2.54 cm) = 165.35 in.
    h. 4 ft (12 in./ft) = 48 in.
    i. 30 cm (10 mm/cm) = 300 mm
    j. 12 m (100 cm/m) (1 in./2.54 cm) (1 ft/12 in.) = 39.37 ft
13. 1.0×10^3 L (corrected for significant figures)
15. 10 oz (1 qt/32 oz) (946 mL/qt) = 295.625 mL (3.0×10^2 mL corrected for significant figures)
17. $1 \text{ U.S. gal}\left(\frac{1 \text{ Imp. gal}}{1.2 \text{ U.S. gal}}\right)\left(\frac{10 \text{ lb water}}{1 \text{ Imp. gal}}\right) = 8.33 \text{ lb}$
19. 150 lb (1 kg/2.2 lb) (1.5 mg/kg) = 102 mg
21. $20 \text{ gal}\left(\frac{4 \text{ qt}}{\text{gal}}\right)\left(\frac{946 \text{ mL}}{\text{qt}}\right)\left(\frac{0.7 \text{ g}}{\text{mL}}\right)\left(\frac{1 \text{ kg}}{1000 \text{ g}}\right)\left(\frac{2.2 \text{ lb}}{\text{kg}}\right) = 116.5 \text{ lb}$
23. Octane will float on water because it is less dense.
25. Water has a density of about 1 g/mL, so,

$$1 \text{ qt}\left(\frac{946 \text{ mL}}{\text{qt}}\right)\left(\frac{1 \text{ g}}{\text{mL}}\right)\left(\frac{1 \text{ lb}}{453.6 \text{ g}}\right)\left(\frac{16 \text{ oz}}{\text{lb}}\right) = 33.4 \text{ oz}$$

27. $v = l \times w \times h = 2.00 \text{ cm} \times 5.00 \text{ cm} \times 1000 \text{ cm}$

$$= 10{,}000 \text{ cm}^3\left(\frac{1 \text{ mL}}{\text{cm}^3}\right)\left(\frac{2.702 \text{ g}}{\text{mL}}\right)\left(\frac{1 \text{ kg}}{1000 \text{ g}}\right)$$

$$= 27.0 \text{ kg}$$

29. a. $4.5 \times 10^4 = 45{,}000$
    b. $4.5 \times 10^{-4} = 0.00045$
    c. $4.5 \times 10^{-1} = 0.45$
    d. $3.541 \times 10^3 = 3541$
    e. $3.541 \times 10^{-1} = 0.3541$
    f. $3.541 \times 10^6 = 3{,}451{,}000$
    g. $7.59 \times 10^2 = 759$
    h. $7.59 \times 10^{-4} = 0.000759$
    i. $7.59 \times 10^{-2} = 0.0759$
31. $40.1 \text{ g}/6.02 \times 10^{23} \text{ atoms} = 6.66 \times 10^{-22} \text{ g}$
33. 3 km (30 °C/km) = 90 °C increase
    20 °C + 90 °C = 110 °C
    110 °C (9/5) + 32 = 230 °F
35. 95 °C(9/5) + 32 = 203 °F
37. a. 38 °C (9/5) + 32 = 100.4 °F
    b. (32 °F − 32 °F)(5/9) = 0.0 °C
    c. 63 °C + 273 = 336 K
    d. (27 °F − 32 °F)(5/9) = −2.78 °C
       −2.78 °C + 273 = 270.22 K

e. 300 K − 273 = 27 °C
f. 84 °C (9/5) + 32 = 183.2 °F

39. $\left(\frac{1 \text{ cal}}{\text{g °C}}\right)(50{,}000 \text{ g})(7.5 \text{ °C}) = 375{,}000 \text{ cal}$

41. $(0.1 \text{ g Al})(63 \text{ °C})\left(\frac{0.214 \text{ cal}}{\text{g °C}}\right) = 1.35 \text{ cal}$

43. $\frac{1 \text{ yd}}{36 \text{ in.}} = 1$

45. a. 105 cm (10 mm/cm) = 1,050 mm
b. 105 cm (0.01 m/cm) = 1.05 m
c. 105 cm (10,000 μm/cm) = 1,050,000 μm

47. 16 ft (12 in./ft) = 192 in.

49. 76 in. (1 ft/12 in.) = 6.33 ft

51. $25 \text{ m}\left(\frac{100 \text{ cm}}{\text{m}}\right)\left(\frac{1 \text{ in.}}{2.54 \text{ cm}}\right)\left(\frac{1 \text{ ft}}{12 \text{ in.}}\right) = 82 \text{ ft}$

53. 25,000 ft (1 mile/5,280 ft) = 4.74 miles

55. 6.2 ft (12 in./ft) (2.54 cm/in.) = 189 cm

57. A cube 10 cm on each side will contain a volume of 1 L.

59. 30 qt (32 oz/qt) = 960 oz

61. 12.0 L (1.057 qt/L) = 12.7 qt

63. 4.2 L (1.057 qt/L) (4 cups/qt) = 17.8 cups

65. 1 L of water has a mass of 1 kg.

67. 1.5 kg (1000 g/kg) = 1,500 g

69. 2.2 kg (2.2 lb/kg) (16 oz/lb) = 77.44 oz

71. 150 lb (1 kg/2.2 lb) = 68.1 kg

73. 100 g Cu (1 cc/8.9 g) = 11.2 cc Cu

75. 100 g Pb (1 cc/11.0 g) = 9.10 cc Pb

77. $v = 10.0 \text{ cm} \times 14.1 \text{ cm} \times 20.0 \text{ cm} = 2{,}820 \text{ cm}^3$
$d = m/V = 1579 \text{ g}/2820 \text{ cm}^3 = 0.560 \text{ g/cm}^3$

79. 100 g (1 mL/7.8 g) = 12.82 mL of water would be displaced, so the water would rise to 50.00 + 12.82 = 62.82 mL.

81. 31.50 g/30.0 mL = 1.05 g/mL. No, the normal range is from 1.018 to 1.025 g/mL.

83. $d = m/v$ = 19.512 g/25.0 mL = 0.781 g/mL

85. The volume has only three significant figures.

87.

1.490×10^3 L	5.225×10^3 g
4.90×10^2 cm	5.5045×10^4 kg
1.60×10^0 m	2.946×10^3 km
4.5×10^{-2} g	6.58×10^{-4} m
6.23×10^{-6} L	9.700×10^{-1} mL
5.00×10^{-8} g	6.7×10^{-2} cm

CHAPTER 2

1. 1 cal/g °C (1 g/mL) (1000 mL/L) (20 °C) = 20,000 cal

3. 80 cal/g °C (946 g) = 75,680 cal

5. Evaporating 10 g of snow requires (1) melting the snow (heat of fusion), (2) warming the snow (heat), and (3) evaporating the water (heat of vaporization).

$$10 \text{ g}\left(\frac{80 \text{ cal}}{\text{g °C}}\right) + 10 \text{ g}\left(\frac{1 \text{ cal}}{\text{g °C}}\right)(100 \text{ °C}) + 10 \text{ g}\left(\frac{540 \text{ cal}}{\text{g °C}}\right)$$
$$= 800 \text{ cal} + 1000 \text{ cal} + 5400 \text{ cal} = 7200 \text{ cal}$$

7. Follow the same procedure as in exercise 5.

9.

Occurrence	Type of Change
Burning leaves	Chemical
Melting chocolate	Physical
Drying dishes	Physical
Dissolving sugar	Physical
Burning gas	Chemical
Melting snow	Physical
Eating candy	Chemical
Burning candles	Chemical
Making toast	Chemical

11. Ice + heat → water is a readily reversible, physical change.

13. Compounds have a fixed proportion of atoms.

15.

Material	Classification
Coffee with sugar	Mixture
Dried cocoa mix	Mixture
Salt	Compound
Sugar	Compound
Milk	Mixture
Gold ring	Element
Aluminum cookware	Element
Lemonade	Mixture
Silver dish	Element
Baking soda	Compound

17. If H = 1 g/mol, then C_3H_7OH = 60 g/mol.

19. 97.23/2.76 = 35.23 times the mass of H.
No, chlorine does not appear to be made up of a whole number of hydrogen-sized building blocks.

21. One, to make the compound HI.

23. None; krypton is a noble gas.

25. PH_3

27. The freezing point of water is 0 °C.

29.

Celsius	Kelvin
20 + 273 =	**293**
100	**373**
−40	**233**
0	273
100	373
−200	73

31. You must account for the amount of heat required to warm the water from 37 to 100 °C and the heat required to evaporate the water (the heat of vaporization).

$$(25\text{ g } H_2O)(1\text{ cal/g °C})(63) + (25\text{g})(540\text{ cal/g})$$
$$= 1{,}575\text{ cal} + 13{,}500\text{ cal} = 15{,}075\text{ cal}$$

33. liquid to gas
35. a. Heating the water: (1 cal/g °C) (25 g) (80 °C) = 2,000 cal
    b. Evaporation of water: (540 cal/g) (25 g) = 13,500 cal
    c. Total heat required to evaporate the water: 2,000 + 13,500 = 15,500 cal
37. Use 1 cal/g °C for the specific heat of the person: 70 kg (1000 g/kg)(1 cal/g °C) (5 °C) = 350 kcal

39.

Occurrence	Type of Change
Burning wood	Chemical
Melting wax	Physical
Burning gasoline	Chemical
Drying nail polish	Physical
Melting ice	Physical
Rain	Physical
Mixing sugar	Physical

41. Mixtures have variable compositions.

43.

Material	Classification
Seawater	Mixture
Lime	Compound
Stew	Mixture
Milk	Mixture
Sulfur	Element
Distilled Water	Compound
Silver spoon	Element
Salt (NaCl)	Compound
Copper	Element

45. If H = 1 g/mol, then $C_2H_4O_2$ = 60 g/mol
47. Since H and F combine in a 1:1 ratio, 94.96/5.04 = 18.8 or about 19 times that of hydrogen.
49. Oxygen would behave most similarly to sulfur.
51. Na_2O
53. None; both are noble gases.

CHAPTER 3

1. It will repel the beam.
3. Electrons carry a charge (a) equal in magnitude but opposite in sign to a proton.
5. c. the canal ray experiment
7. c. Gamma particles have no mass.
9. c. Gamma particles are most penetrating.
11. See your instructor.

13.

	^{63}Cu	^{65}Cu
a.	29	29
b.	34	36

c. 0.6909(63) + 0.3091(65) = 63.6182 g/mol

15. Cathode rays move from (−) to (+).
17. Cathode rays are smaller than air molecules.
19. Electrons carry a (−) charge.
21. Canal rays have a positive charge.
23. The positive part is more massive than the negative part.
25. a. beta (−) radiation
    b. Alpha particles have a +2 charge.
    c. Gamma rays have no charge.
27. Alpha particles are the least penetrating.
29. A nucleus where all of the positively charged particles were located in a minute volume at the center of the atom. The nucleus occupied less than 0.01% of the volume of the atom, while the electrons occupied more than 99.99% of the atomic volume.
31. at the very center

CHAPTER 4

1. a. red, orange, yellow, green, blue, violet
3. Ultraviolet light, which because of its short wavelength and high frequency, carries high energy and causes sunburn by breaking chemical bonds in skin molecules. This increases the chance of mutation and thus increases the probability of skin cancer.
5. Calcium flares emit red-orange light.
7. Fireworks containing strontium would emit red-orange light.
9. spectral line C
11. Ca has 20 electrons.

13.

a.	Na: $1s^22s^22p^63s^1$	11 electrons
b.	B: $1s^22s^22p^1$	5
c.	C: $1s^22s^22p^2$	6
d.	N: $1s^22s^22p^3$	7
e.	S: $1s^22s^22p^63s^23p^4$	16
f.	Cl: $1s^22s^22p^63s^23p^5$	17
g.	F: $1s^22s^22p^5$	9
h.	As: $1s^22s^22p^63s^23p^64s^23d^{10}3p^4$	34
i.	Ca: $1s^22s^22p^63s^23p^64s^2$	20
j.	Be: $1s^22s^2$	4
k.	Ne: $1s^22s^22p^6$	10

15. *d*; s^2p^6 is the noble gas configuration

17. Wavelength is inversely related to frequency.
19. Red light has the longest wavelength and thus the lowest frequency.
21. Violet light has the greatest energy.
23. Solids emit a band spectrum.
25. Orbit jump 4→2 would emit the highest energy and thus the shortest wavelength of light.
27. a. a 3→2 jump (lowest energy)
    b. a 4→2 jump
    c. a 5→2 jump
    d. a 6→2 jump (highest energy)
29. Element 16 is S (sulfur)
31. a. oxygen: $2s^2p^4$ b. sulfur: $3s^2p^4$ c. selenium: $4s^2p^4$
    All of these elements have the same outermost electron group configuration of s^2p^4.
33. Na is $3s^1$ and K is is $4s^1$; they share the same outermost electron group configuration.

CHAPTER 5

1. because a full complement of *p* electrons in group 1 is zero
3.

Li^{+1}	K^{+1}
Ca^{+2}	Mg^{+2}
O^{-2}	S^{-2}
F^{-1}	Cl^{-1}

5. LiF Na_2O Al_2O_3 CaO
7. The distance separating the atomic centers is more significant because ionic strength varies by the square of the atomic distance.
9. MgS would have a stronger bond because the ions carry +2 and −2 charges. In NaCl, the ions carry +1 and −1 charges.
11. Al^{+3} would be the smallest due to its nuclear charge of +3.
13. Cesium would have the least amount of hold on its outermost electron because that electron has the greatest nucleus to electron distance; Cs is in the lowest period (period 6).
15. Carbon and oxygen have the most similar electron-attracting ability.
17. N (5 electrons), H (1 electron each)
19.

CO_2	O::C::O	$4 + (6 \times 2) = 16$ electrons
C_2H_2	H:C:::C:H	$(4 \times 2) + (1 \times 2) = 10$
N_2	:N:::N:	$5 \times 2 = 10$
O_2	O::O	$6 \times 2 = 12$
Cl_2	:Cl:Cl:	$7 \times 2 = 14$
H_2CO	H:C:H :: O	$(1 \times 2) + 4 + 6 = 12$

21. An atom must have a full complement of outer group *s* and *p* electrons to be chemically stable.
23. Li^+, Na^+, K^+, Be^{+2}, Al^{+3}, Ga^{+3}, O^{-2}, S^{-2}, Se^{-2}, F^-, Cl^-, Br^-
25. LiCl 27. AlF_3 29. Al_2O_3
31. $Na^+ < K^+ < Rb^+ < Cs^+$
    Smallest Largest
33. The attraction becomes greater.
35. The nucleus to electron distance varies by the square of the difference and therefore has a greater effect on the atom's ability to hold its outermost electrons.
37. a. It increases.
    b. It increases as the nuclear charge increases.
    c. The diameter decreases as the increased nuclear charge draws the electron cloud in tighter.
    d. Na (2.33 Å); Mg (1.72 Å); Si (1.46 Å); Cl (0.97 Å)
39. Less; the number of electrons exceeds the number of protons.
41. Positive ions (cations) are smaller than the parent atom due to increased attraction of the electrons to the nucleus by increasing the charge on the nucleus. Negative ions (anions) are larger than the parent atom due to increased repulsion among the electrons by increasing the number of electrons surrounding the nucleus.
43. 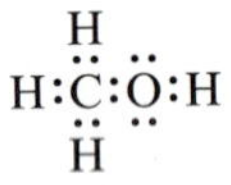
45. 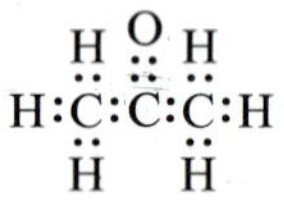

CHAPTER 6

1. e. carbon
3. O (oxygen) = 3.5; H (hydrogen) = 2.1; 3.5 − 2.1 = 1.4; this would be a polar covalent bond.
7.

acetone:	Polar (along the C—O bond) Nonpolar (along the C—H bonds)
isopropyl alcohol:	Polar (along the C—O bond) Nonpolar (along the C—H bonds)
octane:	Nonpolar
formaldehyde:	Polar

9. octane
11. H_2O since it is more polar than H_2S or HCl
13.

	Bond Type
Hydrogen	Covalent
Nitrogen	Covalent
Oxygen	Polar–covalent
Fluorine	Polar–covalent
Sulfur	Covalent
Chlorine	Covalent
Bromine	Covalent

15. Francium is the least electronegative element; it would form a polar–covalent bond with carbon.

17. a. Polar covalent
 b. 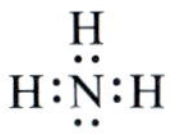
 c. There are 6 bonding electrons (one pair between N and each H).
 d. There are two nonbonding electrons.
 e. pyrimidal
 f. yes, with the negative pole at nitrogen

19. d. sp^3

21. sp^3

23. polar 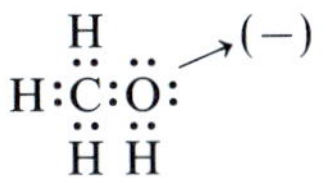

25. polar 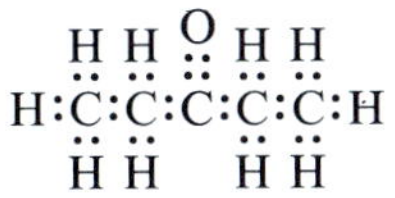

27. methyl alcohol

29. Atomic weight would have no change on hydrogen bonding.

31. Molecules that hydrogen bond show an increase in boiling point.

CHAPTER 7

1. First row (H^+):

$H^+ + Br^- \longrightarrow HBr$
$H^+ + HCO_3^- \longrightarrow H_2CO_3$
$H^+ + ClO_3^- \longrightarrow HClO_3$
$H^+ + F^- \longrightarrow HF$
$H^+ + OH^- \longrightarrow H_2O$
$H^+ + NO_3^- \longrightarrow HNO_3$
$2H^+ + CO_3^{-2} \longrightarrow H_2CO_3$
$2H^+ + CrO_4^{-2} \longrightarrow H_2CrO_4$
$2H^+ + S^{-2} \longrightarrow H_2S$
$2H^+ + SO_4^{-2} \longrightarrow H_2SO_4$
$2H^+ + MnO_4^{-2} \longrightarrow H_2MnO_4$
$3H^+ + PO_4^{-3} \longrightarrow H_3PO_4$

Second through fifth rows: For the rest of the +1 ions (NH_4^+, Li^+, K^+, Ag^+, and Na^+) simply replace H^+ in the formula with the appropriate ion.

Sixth row (Ca^{+2}):

$Ca^{+2} + 2\ Br^- \longrightarrow CaBr_2$
$Ca^{+2} + 2\ HCO_3^- \longrightarrow Ca(HCO_3)_2$
$Ca^{+2} + 2\ ClO_3^- \longrightarrow Ca(ClO_3)_2$
$Ca^{+2} + 2\ F^- \longrightarrow CaF_2$
$Ca^{+2} + 2\ OH^- \longrightarrow Ca(OH)_2$
$Ca^{+2} + 2\ NO_3^- \longrightarrow Ca(NO_3)_2$
$Ca^{+2} + CO_3^{-2} \longrightarrow CaCO_3$
$Ca^{+2} + CrO_4^{-2} \longrightarrow CaCrO_4$
$Ca^{+2} + S^{-2} \longrightarrow CaS$
$Ca^{+2} + SO_4^{-2} \longrightarrow CaSO_4$
$Ca^{+2} + MnO_4^{-2} \longrightarrow CaMnO_4$
$3\ Ca^{+2} + 2\ PO_4^{-3} \longrightarrow Ca_3(PO_4)_2$

Seventh through twelfth rows: For the other +2 ions, simply replace Ca^{+2} with the appropriate +2 ion.

Thirteenth row (Al^{+3}):

$Al^{+3} + 3\ Br^- \longrightarrow AlBr_3$
$Al^{+3} + 3\ HCO_3^- \longrightarrow Al(HCO_3)_3$
$Al^{+3} + 3\ ClO_3^- \longrightarrow Al(ClO_3)_3$
$Al^{+3} + 3\ F^- \longrightarrow AlF_3$
$Al^{+3} + 3\ OH^- \longrightarrow Al(OH)_3$
$Al^{+3} + 3\ NO_3^- \longrightarrow Al(NO_3)_3$
$2Al^{+3} + 3\ CO_3^{-2} \longrightarrow Al_2(CO_3)_3$
$2Al^{+3} + 3\ CrO_4^{-2} \longrightarrow Al_2(CrO_4)_3$
$2Al^{+3} + 3\ S^{-2} \longrightarrow Al_2S_3$
$2Al^{+3} + 3\ SO_4^{-2} \longrightarrow Al_2(SO_4)_3$
$2Al^{+3} + 3\ MnO_4^{-2} \longrightarrow Al_2(MnO_4)_3$
$Al^{+3} + PO_4^{-3} \longrightarrow AlPO_4$

Fourteenth row: For the other +3 ion, substitute Fe^{+3} for Al^{+3}.

3. 133 of the compounds are soluble; 2 are slightly soluble [$Ca(OH)_2$, $Sr(OH)_2$]; and 45 of them are insoluble and are listed below:

Ag:	$AgBr$, AgF, $AgOH$, Ag_2CO_3, Ag_2S, Ag_2SO_4, Ag_3PO_4
Pb:	$PbBr_2$, PbF_2, $Pb(OH)_2$, $PbCO_3$, PbS, $PbSO_4$, $Pb_3(PO_4)_2$
Ca:	$CaCO_3$, CaS, $CaSO_4$, $Ca_3(PO_4)_2$
Co:	$Co(OH)_2$, $CoCO_3$, CoS, $Co_3(PO_4)_2$
Cu:	$Cu(OH)_2$, $CuCO_3$, CuS, $Cu_3(PO_4)_2$
Fe(II):	$Fe(OH)_2$, $FeCO_3$, FeS, $Fe_3(PO_4)_2$
Mg:	$Mg(OH)_2$, $MgCO_3$, MgS, $Mg_3(PO_4)_2$
Sr:	$SrCO_3$, SrS, $Sr_3(PO_4)_2$
Al:	$Al(OH)_3$, $Al_2(CO_3)_3$, Al_2S_3, $AlPO_4$
Fe(III):	$Fe(OH)_3$, $Fe_2(CO_3)_3$, Fe_2S_3, $FePO_4$

5. $AgNO_3 + HCl \longrightarrow AgCl + HNO_3$
$AgNO_3 + NH_4OH \longrightarrow AgOH + NH_4NO_3$
$2\ AgNO_3 + (NH_4)_2S \longrightarrow Ag_2S + 2\ NH_4NO_3$
$2\ AgNO_3 + K_2SO_4 \longrightarrow Ag_2SO_4 + 2\ KNO_3$
$3\ AgNO_3 + Na_3PO_4 \longrightarrow Ag_3PO_4 + 3\ NaNO_3$

$Pb(NO_3)_2 + 2\ HCl \longrightarrow PbCl_2 + 2\ HNO_3$
$Pb(NO_3)_2 + 2\ NH_4OH \longrightarrow Pb(OH)_2 + 2\ NH_4NO_3$
$Pb(NO_3)_2 + (NH_4)_2S \longrightarrow PbS + 2\ NH_4NO_3$
$Pb(NO_3)_2 + K_2SO_4 \longrightarrow PbSO_4 + 2\ KNO_3$
$3\ Pb(NO_3)_2 + 2\ Na_3PO_4 \longrightarrow Pb_3(PO_4)_2 + 6\ NaNO_3$

$NaNO_3 + HCl \longrightarrow NaCl + HNO_3$
$NaNO_3 + NH_4OH \longrightarrow NaOH + NH_4NO_3$
$2\ NaNO_3 + (NH_4)_2S \longrightarrow Na_2S + 2\ NH_4NO_3$
$2\ NaNO_3 + K_2SO_4 \longrightarrow Na_2SO_4 + 2\ KNO_3$
$3\ NaNO_3 + Na_3PO_4 \longrightarrow Na_3PO_4 + 3\ NaNO_3$

$Cu(NO_3)_2 + 2\ HCl \longrightarrow CuCl_2 + 2HNO_3$
$Cu(NO_3)_2 + 2\ NH_4OH \longrightarrow Cu(OH)_2 + 2\ NH_4NO_3$
$Cu(NO_3)_2 + (NH_4)_2S \longrightarrow CuS + 2\ NH_4NO_3$
$Cu(NO_3)_2 + K_2SO_4 \longrightarrow CuSO_4 + 2\ KNO_3$
$3\ Cu(NO_3)_2 + 2\ Na_3PO_4 \longrightarrow Cu_3(PO_4)_2 + 6\ NaNO_3$

$Ni(NO_3)_2 + 2\ HCl \longrightarrow NiCl_2 + 2\ HNO_3$
$Ni(NO_3)_2 + 2\ NH_4OH \longrightarrow Ni(OH)_2 + 2\ NH_4NO_3$
$Ni(NO_3)_2 + (NH_4)_2S \longrightarrow NiS + 2\ NH_4NO_3$
$Ni(NO_3)_2 + K_2SO_4 \longrightarrow NiSO_4 + 2\ KNO_3$
$3\ Ni(NO_3)_2 + 2\ Na_3PO_4 \longrightarrow Ni_3(PO_4)_2 + 6\ NaNO_3$

$LiNO_3 + HCl \longrightarrow LiCl + HNO_3$
$LiNO_3 + NH_4OH \longrightarrow LiOH + NH_4NO_3$
$2\ LiNO_3 + (NH_4)_2S \longrightarrow Li_2S + 2\ NH_4NO_3$
$2\ LiNO_3 + K_2SO_4 \longrightarrow Li_2SO_4 + 2\ KNO_3$
$3\ LiNO_3 + Na_3PO_4 \longrightarrow Li_3PO_4 + 3\ NaNO_3$

$Co(NO_3)_2 + 2\ HCl \longrightarrow CoCl_2 + 2\ HNO_3$
$Co(NO_3)_2 + 2\ NH_4OH \longrightarrow Co(OH)_2 + 2\ NH_4NO_3$
$Co(NO_3)_2 + (NH_4)_2S \longrightarrow CoS + 2\ NH_4NO_3$
$Co(NO_3)_2 + K_2SO_4 \longrightarrow CoSO_4 + 2\ KNO_3$
$3\ Co(NO_3)_2 + 2\ Na_3PO_4 \longrightarrow Co_3(PO_4)_2 + 6NaNO_3$

$Al(NO_3)_3 + 3\ HCl \longrightarrow AlCl_3 + 3\ HNO_3$
$Al(NO_3)_3 + 3\ NH_4OH \longrightarrow Al(OH)_2 + 3\ NH_4NO_3$
$2\ Al(NO_3)_3 + 3\ (NH_4)_2S \longrightarrow Al_2S_3 + 6\ NH_4NO_3$
$2Al(NO_3)_3 + 3K_2SO_4 \longrightarrow Al_2(SO_4)_3 + 6\ KNO_3$
$Al(NO_3)_3 + Na_3PO_4 \longrightarrow AlPO_4 + 6\ NaNO_3$

7. a. All four are single replacement reactions.
   b. Zn, Cu, Pb, and Ag are oxidized; Fe and H are reduced.
   c. $FeCl_2 + Zn^0 \longrightarrow$ proceeds spontaneously
      $Pb^0 + 2\ HCl \longrightarrow$ proceeds spontaneously

9. hydrogen bromide
11. magnesium oxide
13. hydrogen sulfide
15. calcium bromide
17. potassium iodide
19. K_2S
21. CO_2
23. Al_2O_3
25. BaO
27. GaF_3
29. (+2)(?)
    $H_2\ S \quad S = -2$
31. (+1)(?)(−6)
    $H\ N\ O_3 \quad N = +5$
33. (?) (−6)
    $Fe_2\ O_3 \quad Fe = +3$
35. (?)(−2)
    $Sn\ F_2 \quad Sn = +2$
37. potassium nitrate
39. ammonium bromide
41. silver(II) hydroxide
43. sodium sulfate
45. $Pb(NO_3)_2$
47. $MgSO_4$
49. $NaHCO_3$
51. NH_4SCN
53. K_2SO_3
55. $Al(OH)_3$
57. $HCl + NaOH \longrightarrow H_2O + NaCl$
59. $2\ AgNO_3 + (NH_4)_2SO_4 \longrightarrow Ag_2SO_4 + 2\ NH_4NO_3$
61. $Pb(NO_3)_2 + NH_4OH \longrightarrow Pb(OH)_2 + 2\ NH_4NO_3$
63. $KNO_3 + NaCl \longrightarrow KCl + NaNO_3$
65. $LiOH + KNO_3 \longrightarrow LiNO_3 + KOH$
67. $Cu(NO_3)_2 + H_2S \longrightarrow CuS + 2\ HNO_3$
69. $Fe(NO_3)_3 + 3\ NH_4OH \longrightarrow Fe(OH)_3 + 3\ NH_4NO_3$
71. Reactions 57, 63, and 65 do not produce insoluble products.
73. Cl is oxidized from −1 (in HCl) to +4 (in $NaHClO_3$).
75. Carbon was oxidized from −4 (in CH_4) to +4 (in CO_2).
77. There is no change in oxidation state.
79. Zn is oxidized from zero to +2; iron is reduced from +2 to zero.

$$Zn \longrightarrow Zn^{+2} + 2\,e^-$$
$$Fe^{+3} + 3\,e^- \longrightarrow Fe$$

81. Ag is oxidized from zero to +1; copper is reduced from +2 to zero.

$$Ag^0 \longrightarrow Ag^+ + e^-$$
$$Cu^{+2} + 2\,e^- \longrightarrow Cu^0$$

83. Pb is oxidized from zero to +2; copper is reduced from +1 to zero.

$$Pb^0 \longrightarrow Pb^{+2} + 2\,e^-$$
$$Cu^+ + e^- \longrightarrow Cu^0$$

CHAPTER 8

1. NaCl = 58.44 g/mol
   Na = 22.99 g/mol
   Cl = 35.45 g/mol
3. $Pb(OH)_2$ = 241.21 g/mol
   Pb = 207.19 g/mol
   2 O = 32.00 g/mol
   2 H = 2.01 g/mol
5. $AgNO_3$ = 169.88 g/mol
   Ag = 107.87 g/mol
   N = 14.01 g/mol
   3 O = 48.00 g/mol
7. $ZnSO_4$ = 161.43 g/mol
   Zn = 65.37 g/mol
   S = 32. 06 g/mol
   4 O = 64.00 g/mol
9. CH_4, MW = 16.05 g/mol

$$\%C = \frac{12.01}{16.05} = 74.83\% \qquad \%H = \frac{4.04}{16.05} = 25.17\%$$

11. $C_{12}H_{22}O_{11}$, MW = 342.34 g/mol

$$\%C = \frac{(12.01)12}{342.34} = 42.10\% \qquad \%H = \frac{(1.01)22}{342.34} = 6.49\%$$

$$\%O = \frac{(16.00)11}{342.34} = 51.41\%$$

13. NaOH, MW = 40.00 g/mol

$$\%Na = \frac{22.99}{40.00} = 57.48\% \qquad \%O = \frac{16.00}{40.00} = 40.00\%$$

$$\%H = \frac{1.01}{40.00} = 2.52\%$$

15. 16.0 g S (1 mol S/32 g S) = 0.500 mol S
17. 67.45 g Al (1 mol Al/26.98 g Al) = 2.500 mol Al
19. 53.55 g $AlCl_3$ (1 mol $AlCl_3$/133.33 g) = 0.4016 mol $AlCl_3$
21. 366.5 g $Ni(NO_3)_3$ (1 mol/244.74 g $Ni(NO_3)_3$ = 1.498 mol $Ni(NO_3)_3$
23. 2.5 mol Co (58.93 g/mol) = 147.3 g Co
25. 0.50 mol Hg (200.59 g/mol) = 100.30 g Hg
27. 12.0 mol C (12.01 g/mol) = 144.1 g C
29. 0.41 mol $CaCO_3$ (100.09 g/mol) = 41.037 g $CaCO_3$
31. 0.15 mol $Al_2(SO_4)_3$ (342.14 g/mol) = 51.321 g $Al_2(SO_4)_3$
33. $12.7 \text{ g Cu}\left(\frac{1 \text{ mol Cu}}{63.54 \text{ g}}\right)\left(\frac{6.02 \times 10^{23} \text{ atoms}}{\text{mol}}\right)$

    $= 1.20 \times 10^{23}$ atoms Cu
35. $50.76 \text{ g I}\left(\frac{1 \text{ mol I}}{126.90 \text{ g}}\right)\left(\frac{6.02 \times 10^{23} \text{ atoms}}{\text{mol}}\right)$

    $= 2.40 \times 10^{23}$ atoms I
37. For a 100 g of compound,

    33.88% Cu × 100 g = 33.88 g Cu
    14.94% N × 100 g = 14.94 g N
    51.18% O × 100 g = 51.18 g O

    Conversion to moles gives

    33.88 g Cu (1 mol Cu/63.54 g) = 0.5332 mol Cu
    14.94 g N (1 mol N/14.01 g) = 1.066 mol N
    51.18 g O (1 mol O/16.00 g) = 3.198 mol O

    The mole ratios of the atoms are

$$\text{Cu: } \frac{0.5332}{0.5332} = 1 \qquad \text{N: } \frac{1.066}{0.5332} = 2 \qquad \text{O: } \frac{3.198}{0.5332} = 6$$

    The empirical formula is CuN_2O_6, and the name is copper (II) nitrate, $Cu(NO_3)_2$.
39. Conversion to moles for 100 g of compound gives

    28.36 g Na (1 mol/22.99 g) = 1.234 mol Na
    14.81 g C (1 mol/12.01 g) = 1.233 mol C
    39.55 g S (1 mol/32.06 g) = 1.234 mol S
    17.28 g N (1 mol/14.01 g) = 1.233 mol N

    This is a 1:1:1:1 ratio, thus the empirical formula is NaSCN, and the name is sodium thiocyanate.
41. a. 0.200 mol of $ZnCrO_4$ are produced
    b. 0.200 mol $ZnCrO_4$ (181.37 g/mol) = 36.3 g $ZnCrO_4$
    c. 0.200 mol $Zn(NO_3)_2$
    (2 mol $NaNO_3$/mol $Zn(NO_3)_2$ = 0.400 mol
    d. 0.400 mol $NaNO_3$ (85.00 g/mol) = 34.0 g $NaNO_3$
43. a. 151.3 g C
    b. 12.60 mol Zn produced (65.37 g/mol) = 823.7 g Zn
    c. 12.60 mol CO (28.01 g/mol) = 352.9 g CO
45. 40.00 g/mol
47. 239.3 g/mol
49. 141.04 g/mol
51. 25.94 g/mol
53. 153.81 g/mol
55. %C = 81.8; %H = 18.2
57. %Na = 39.3; %Cl = 60.7
59. %Na = 57.4
    %O = 40.0
    %H = 2.6
61. %S = 32.7
    %O = 65.2
    %H = 2.1
63. 2.0 g (1 mol/22.99 g) = 0.087 mol
65. 1.9 mol
67. 1.0 mol
69. 0.5 mol (6.94 g/mol) = 3.47 g Li
71. 1.40 kg Fe
73. 38.92 g LiF
75. 1000 g H_2O
77. 8 g O (1 mol/16 g) (6.02×10^{23} atoms/mole)
    $= 3.01 \times 10^{23}$ atoms
79. 8 g He (1 mol He/4 g) (6.02×10^{23} atoms/mol)
    $= 1.20 \times 10^{24}$ atoms
81. $SnCl_2$
83. CS_2
85. a. 0.227 mol propane
    b. 1.136 mol O_2
    c. 36.4 g O_2
    d. 30.0 g CO_2
    e. 16.3 g H_2O
87. a. 0.589 mol $AgNO_3$
    b. 0.589 mol KBr
    c. 70.0 g KBr
    d. 110.6 g AgBr
    e. 59.5 g KNO_3

CHAPTER 9

1. $AgNO_3$, $Pb(NO_3)_2$, NH_4Cl, NH_4NO_3, $(NH_4)_2SO_4$, $(NH_4)_2S$, NH_4OH, NaCl, $NaNO_3$, Na_2SO_4, Na_2S, NaOH, KCl, KNO_3, K_2SO_4, K_2S, KOH, $Fe_2(SO_4)_3$, $FeCl_3$, and $Fe(NO_3)_3$ are water soluble.
   AgCl, Ag_2SO_4, Ag_2S, AgOH, $PbCl_2$, $PbSO_4$, PbS, $Pb(OH)_2$, Fe_2S_3 and $Fe(OH)_3$ are precipitates (solids) in water.
3. By heating the solution to 56 °C, the acetone would boil and leave the solution as a vapor, leaving the water behind. This process is called distillation.

5. a. From 0 to 4 minutes the sample is heated.
   b. From 4 to 5 minutes the ethyl alcohol distills.
   c. From 5 to 6 minutes the remaining water is heated.
   d. At 6 minutes the water begins to boil.

7. $\frac{0.083 \text{ mol}}{0.100 \text{ L}} = 0.083\ M$

9. Plain water boils at much lower temperatures than a water–antifreeze mixture and would boil dry in the summer.

11. First row: −2.76 °C; −0.27 °C; −1.43 °C
    Second row: −5.28 °C; −0.49 °C; −2.72 °C
    Third row: −29.0 °C; −15.0 °C

13. The concentration increases because the number of sugars and biochemical molecules stays the same but the total volume decreases.

15. The blood cells will absorb water to equalize the concentrations.

17. The presence of sugar forces the bacteria to dehydrate and die.

19. Suspended particles in a colloidal mixture are much larger than those in a solution.

21. Particles in a colloidal mixture will scatter light, those in a solution will not.

23. no

25. Filtration will remove colloidal particles.

27. Dialysis involves the use of a semipermeable membrane that allows the solvent and dissolved solutes to pass through, but prevents the passage of colloidal particles.

29. 10 g NaCl/100 g H_2O = 10% by weight

31. 62.5 mL of ethylene glycol added to 187.5 mL of H_2O

33. 0.483 *M*

35. (0.200 mol/L)(0.500 L)(60.05 g/mol) = 6.005 g acetic acid

37. −3.72 °C

39. a. −37.2 °C b. −2.94 °C c. −2.36 °C

CHAPTER 10

1. The activation energy required to begin combusting logs is much greater than that for small sticks. Also, more oxygen would be available around the surface of the sticks than the logs, so they can collide with sufficient energy.

3. The molecules must collide.

5. They must collide with sufficient energy to break chemical bonds.

7. They must collide with sufficient energy to break chemical bonds.

9. It lowers the activation energy of the chemical reaction.

11. a. To the left
    b. To the left; concentration of hydronium ion would decrease.

13. exothermic (gives off heat)

15. the energy required to cause a reaction to proceed

17. It should increase the metabolic rates due to the increase in the concentration of O_2 in the bloodstream.

19. Fewer collisions per second and lower collision energy will lower the metabolic rate.

21. $A + Cat \longrightarrow ACat$
    $ACat + B \longrightarrow AB + Cat$

CHAPTER 11

1. Arrhenius proposed that acids are molecules that when ionized give up hydrogen ions.

3. Bases are proton acceptors.

5. a. to the left
   b. NH_4^+ acts as an acid (donates H^+)

7. a. to the left
   b. H_3O^+ acts as an acid (donates H^+).
      HCO_3^- acts as a base (accepts H^+).
   c. weak acid

9. pH = 1.5 represents the most acidic concentration given.

11. a factor of 10 less concentrated (from 10^{-1} to 10^{-2})

13. A chemical indicator is a compound that is dependent upon the pH of the solution in which it resides.

15. $K_a \approx 1 \times 10^{-9}$

17. a. 0.050 mol KOH; 2.81 g KOH
    b. 0.040 mol NaF; 1.68 g NaF
    c. 0.100 mol LiCl; 4.24 g LiCl

19. a. $HCl + NaOH \longrightarrow H_2O + NaCl$
       Sodium Chloride
    b. $2\ HNO_3 + Mg(OH)_2 \longrightarrow 2\ H_2O + Mg(NO_3)_2$
       Magnesium nitrate
    c. $H_3BO_3 + 3\ NH_4OH \longrightarrow 3\ H_2O + (NH_4)_3BO_3$
       Ammonium borate
    d. $H_2SO_4 + Ca(OH)_2 \longrightarrow 2\ H_2O + CaSO_4$
       Calcium sulfate
    e. $HC_6H_7O_7 + NaOH \longrightarrow H_2O + NaC_6H_7O_7$
       Sodium citrate

21. a. $3\ HCl + Al(OH)_3 \longrightarrow 3\ H_2O + AlCl_3$
    b. 0.20 mol HCl/L (0.064 L) = 0.128 mol HCl
    c. 0.0128 mol HCl (1 mol $Al(OH)_3$/3 mol HCl) = 0.00427 mol $Al(OH)_3$
    d. 0.00427 mol $Al(OH)_3$ (78.0 g/mol) = 0.333 g $Al(OH)_3$
    e. %$Al(OH)_3$ = 0.333/(2.000 × 100%) = 16.7%

23. 0.40 *M* H_2SO_4

25. hydronium ion concentration = 0.03 *M*

27. hydronium ion concentration = 0.01 *M*

29. A buffer is made of a weak acid and a salt of the acid.
31. Buffer capacity is determined by the concentration of the weak acid and the concentration of its salt.
33. a. $K_a = \frac{[H_3O^+][A^-]}{[HA]}$
    b. $[HA]$ = concentration of citric acid = 0.20 M
    $[A^-]$ = concentration of sodium citrate = 1.68 M
    $[H_3O^+] = 1.00 \times 10^{-5}$ M; pH = 5
35. Hydronium ion, H_3O^+, is formed when an acid (HA) donates a proton to water: $H_2O + HA \rightarrow H_3O^+ + A^-$
37. KCl
39. a. $HCl + H_2O \longrightarrow H_3O^+ + Cl^-$
    b. $HC_2H_3O_2 + H_2O \rightleftharpoons H_3O^+ + C_2H_3O_2^-$
    c. $HCl + H_2O \longrightarrow H_3O^+ + Cl^-$
41. pH < 7.0 is acidic, pH = 7.0 is neutral, and pH > 7.0 is basic.
43. 1.0×10^{-5} M
45. A factor of 10^4 or 10,000 times more H_3O^+ is needed.
47. HIn (only a small amount dissociates to form In^-)
49. a pH value between 1 and 3
51. a. 2.338 g NaCl
    b. 33.98 g $AgNO_3$
    c. 36.03 g $HC_2H_3O_2$
    d. 20.70 g LiCl
    e. 12.00 g NaOH
53. a. 0.050 M NaOH
    b. 0.10 M H_3PO_4
    c. 0.20 M HNO_3
    d. 0.020 M $AgNO_3$
55. a. 0.0075 mol KOH
    b. 0.0075 mol HCl
    c. $\frac{0.0075 \text{ mol HCl}}{0.025\text{L}} = 0.300$ mol HCl or 0.300 M HCl
57. $$\left(\frac{0.10 \text{ mol NaOH}}{\text{L}}\right)(0.070 \text{ L NaOH})\left(\frac{1 \text{ mol } H_3PO_4}{3 \text{ mol NaOH}}\right) \times \left(\frac{1}{.020 \text{ L}}\right) = 0.117 \text{ mol } H_3PO_4/\text{L}$$
59. $$\left(\frac{0.10 \text{ mol HCl}}{1.00 \text{ L}}\right)(0.300 \text{ L HCl})\left(\frac{1 \text{ mol NaOH}}{1 \text{ mol HCl}}\right) \times \left(\frac{40.00 \text{ g}}{1 \text{ mol NaOH}}\right) = 1.20 \text{ g NaOH}$$
    1.20 g NaOH/3.00 g tablet × 100 % = 40.00%
61. pH = 3
63. pH = 3
65. pH = 2
67. a. $[H_3O^+] = 1.76 \times 10^{-5}$ M
    b. pH = 4.76
69. a. $[H_3O^+] = 6.5 \times 10^{-2}$ M
    b. pH = 1.19

CHAPTER 12

1. 0°C = 273 K   50°C = 323 K
   20°C = 293 K   70°C = 343 K
3. $v_2 = v_1(t_2/t_1)$; 250 mL (100/37) = 675 mL
5. 1 atm = 760 Torr
   1.5 atm (760 Torr/atm) = 1140 Torr
   0.30 atm (760 Torr/atm) = 228 Torr
   0.90 atm (760 Torr/atm) = 684 Torr
7. $v_2 = v_1\ (p_2/p_1)$; 550 mL (690/760) = 500 mL
9. $v_1p_1/t_1 = v_2p_2/t_2$
   $$v_2 = \frac{v_1p_1t_2}{t_1p_1} = (500 \text{ L})\left(\frac{750 \text{ Torr}}{299 \text{ K}}\right)\left(\frac{230 \text{ K}}{210 \text{ Torr}}\right) = 1374 \text{ L}$$
11. $pv = nrt$;
    $v = nrt/p$ = (0.4)(62.4)(293 K)/760 Torr = 9.62 L
13. a. $pv = nrt$
    $n = pv/rt$ = (8 atm)(8 L)/(0.0821 L atm/mol K) × (293 K) = 2.66 mol CH_4
    b. 2.66 mol CH_4 (2 mol O_2/mol CH_4) = 5.32 mol O_2
    c. $v = nrt/p$ = (5.32 mol)(0.0821 L atm/mol K) × (293 K)/1 atm = 128 L O_2
    d. 128 L O_2 (100% air/21% O_2) = 610 L of air are required.
15. a. Moisture is added to air as it passes through the nose and throat.
    b. because air is mostly nitrogen (78%)
    c. 28/700 × 100% = 3.68%
17. 27 °C = 300 K   100 °C = 373 K
    37 °C = 310 K   −40 °C = 233 K
19. $v_2 = (t_2/t_1)v_1$ = (373/273)(400 mL) = 548 mL
21. 0.5 atm = 380 Torr
    1.6 atm = 1216 Torr
    2.5 atm = 1900 Torr
23. $(p_1/p_2)v_1 = v_2$ = (640/760)600 mL = 505 mL
25. $$P_2 = \frac{p_1v_1t_2}{v_2t_1} = \frac{(3 \text{ atm})(1 \text{ volume unit})(303 \text{ K})}{(1 \text{ volume unit})(293 \text{ K})}$$
    = 3.10 atm
27. $$v_2 = \frac{p_1v_1t_2}{t_1p_2} = \frac{(640 \text{ Torr})(3.4 \text{ L})(273 \text{ K})}{(288 \text{ K})(760 \text{ Torr})} = 2.7 \text{ L}$$
29. a. 1 mol CO_2 (44 g/mol) = 44 g CO_2
    b. $d = m/v$; v = 24,060 mL;
    44 g/24,060 mL = 0.00183 g/mL
    c. much lighter
31. $p = nrt/v$ = (0.45 mol CO_2)(0.0821 L atm/mol K) × (293 K)/1 L = 10.82 atm
33. a. 17.54 Torr
    b. 692.46 Torr
    c. $n = pv/rt$ = (692.46 Torr)(0.045 L)/(0.0821 L atm/mol K)(293 K) = 1.30 mol
35. a. 1 mL (0.70 g/mL)(1 mol/114.09 g) = 0.00614 mol
    b. (0.00614 mol C_8H_{18})(25 mol O_2/2 mol C_8H_{18}) = .0768 mol O_2
    c. $v = nrt/p$ = (0.0768 mol)(0.0821 L atm/mol K) × (273 K)/(0.209 atm) = 8.24 L air

d. You need only a small tube to allow a few milliliters of gasoline to flow through at a time. However, you need about 10,000 mL of air to mix with it to combust.

CHAPTER 13

1. a. $^{32}_{15}P \longrightarrow$ Beta(−) + $^{32}_{16}O$
   b. $^{45}_{20}Ca \longrightarrow$ Beta(+) + $^{45}_{19}K$
   c. $^{96}_{38}Sr \longrightarrow$ Beta(+) + $^{96}_{37}Rb$
   d. $^{65}_{30}Zn \longrightarrow$ Beta(+) + $^{65}_{29}Cu$
   e. $^{35}_{92}U \longrightarrow$ Alpha(+) + $^{231}_{90}Th$
   f. $^{234}_{94}Pu \longrightarrow$ Alpha(+) + $^{230}_{92}U$
   g. $^{212}_{83}Bi \longrightarrow$ Alpha(+) + $^{208}_{81}Tl$
   h. $^{22}_{11}Na \longrightarrow$ Beta(+) + $^{22}_{10}Ne$
3. approximately 37,500 disintegrations per minute
5. Gamma radiation carries neither charge nor mass.
7. A scintillation crystal is more dense than the gas in a G-M detector, thus interacting better with gamma radiation.
9. Alpha radiation cannot penetrate the film's protective cover.
11. alpha
13. gamma
15. A curie is a measure of the number of atoms that disintegrate per second, whereas the rem is a measure of the biological damage caused by exposure to radiation.
17. The fission product atoms capture neutrons and stop the chain reaction.
19. 4%; 90%
21. a. $^{60}_{27}Co \longrightarrow$ Beta(−) + $^{60}_{26}Fe$
    b. $^{131}_{53}I \longrightarrow$ Beta(−) + $^{131}_{52}Te$
    c. $^{26}_{13}Al \longrightarrow$ Beta(+) + $^{26}_{14}Si$
    d. $^{232}_{90}Th \longrightarrow$ Alpha + $^{228}_{88}Ra$
    e. $^{238}_{92}U \longrightarrow$ Alpha + $^{234}_{90}Th$
23. The half-life of the sample in exercise 12 is about 3 min. A period of 18 min would thus be six half-lives. During a period of six half-lives, the activity would fall from 1000 counts to about 156 counts.
25. Gamma radiation carries a zero charge.
27. Alpha particles cause electrons to be ejected from atoms as they pass.
29. A scintillation detector detects the flash of light caused when an electron ejected from an atom by gamma radiation falls back into its orbital.
31. Alpha particles cannot pass through the window of most G-M detectors.
33. Radioisotopes used in medicine emit gamma radiation so that they may be detected outside of the body.
35. The radioactivity of the fission products in nuclear fuel elements will fall to the level of the initial uranium ore if stored for about 1000 years.
37. A chain reaction occurs when neutrons emitted from one fission event trigger one or more subsequent fissions.
39. The term *fallout* describes the dispersion of radioactive fission products caused by a nuclear explosion.

CHAPTER 14

1

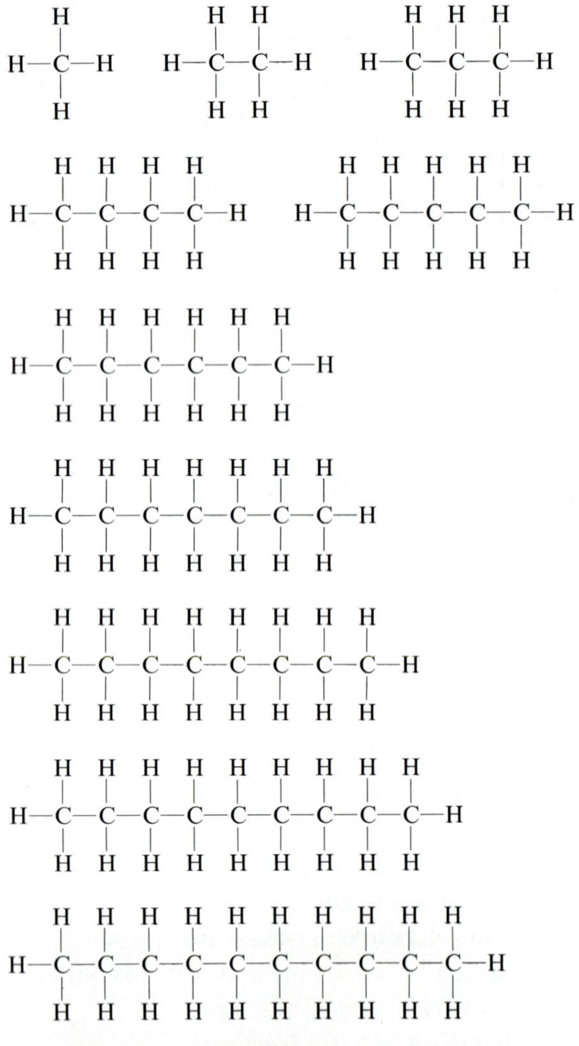

3.

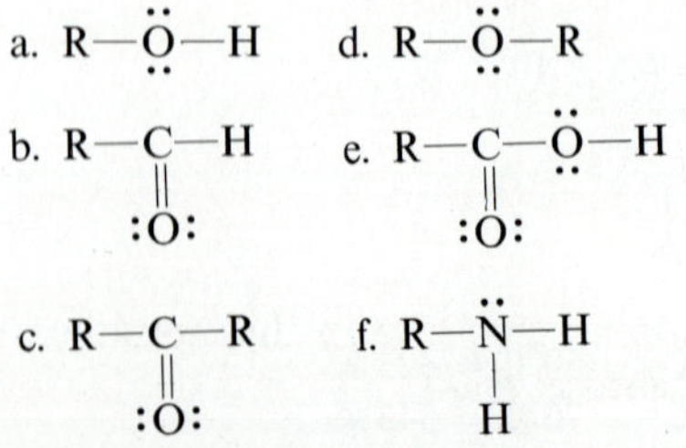
a. R—Ö—H
d. R—Ö—R
b. R—C—H
:O:
e. R—C—Ö—H
:O:
c. R—C—R
:O:
f. R—N—H
H

5.

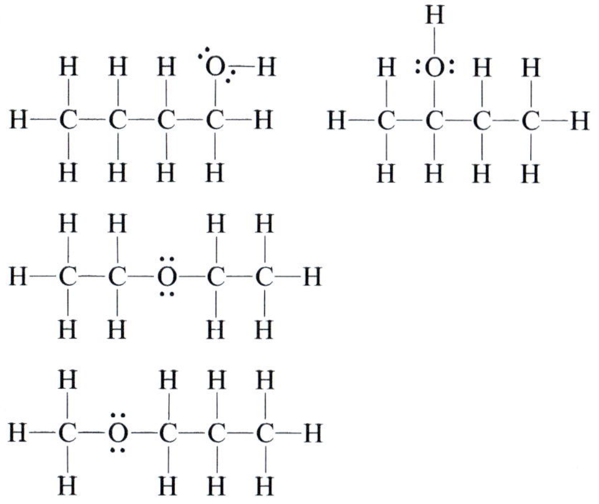

Notice that the only two alcohols that can be made with a "continuous chain" of carbon atoms are the two shown. One with an —OH on the "end" carbon atom and one with an —OH on the "second" carbon. There is, however, another way to attach the carbon atoms:

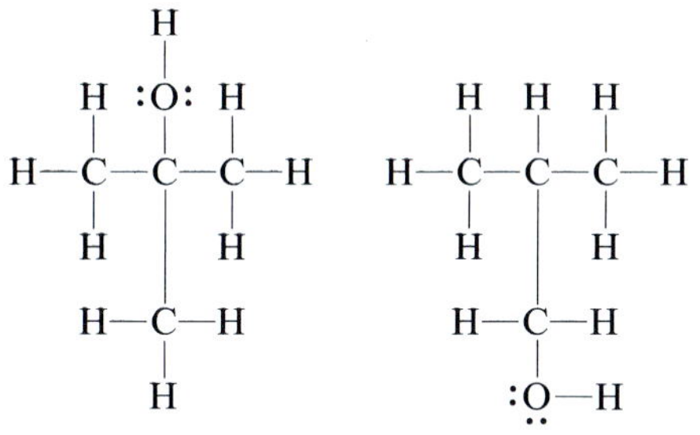

7.

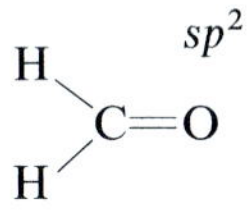

9.

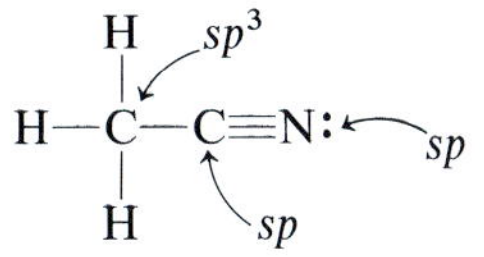

11. A particular arrangement of atoms that function as a single unit. The group of atoms exhibits particular and predictable chemical behavior.

13. a. carboxylic acid b. amine c. aldehyde

15. An alkyne has a triple bond, while an alkane has only single bonds.

17. a. *sp* b. sp^2 c. *sp* d. sp^3

19. an electron donor

CHAPTER 15

1. Carbon atoms in alkanes have sp^3 hybridization; this results in a tetrahedral shape.

3. The conformation with the equatorial isopropyl group will be more stable.

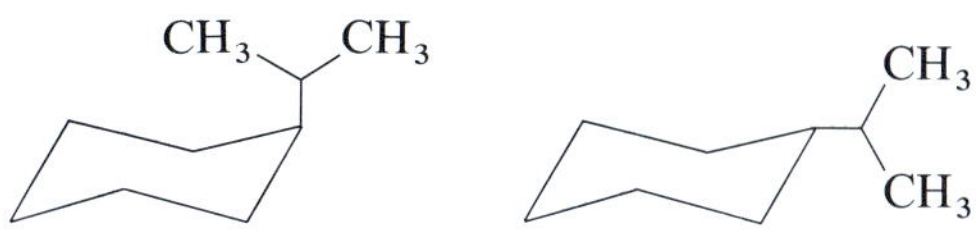

Axial Equatorial

5. a. 2-methylhexane
   b. 4-*sec*-butyl-2-methylheptane
   c. 2,3,4-trimethylheptane
   d. 3,4-dimethylheptane
   e. 3-methylpentane
   f. 2,3-dimethylhexane

7. a. 1,2-dimethylcyclooctane
   b. 1-isopropyl-2-propylcyclohexane
   c. 1,1,2,5-tetramethylcyclohexane
   d. 2-butyl-1,1,3-trimethylcyclohexane
   e. 1,1,3-trimethylcyclohexane
   f. 1,1-dimethylcyclohexane

9. a. 2,2-dibromo-6-methyloctane
   b. 5-bromo-5-methylundecane (Note: Obviously, we can expect hydrocarbons to have chain lengths greater than 10. There is a systematic nomenclature for these also: C_{11} is an undecane, C_{12} is a duodecane, and so on.)

11. a. 1,1-dibromo-3-isopropylcyclohexane.
    b. 1-*tert*-butyl-2,2-dichlorocyclopentane.
    c. 1-isobutyl-1-iododcycloheptane.
    d. 1-bromo-1-chloro-5-*sec*-butylcyclooctane.

13. $CH_3—CH_2—CH_2—CH_2—CH_2—CH_3$

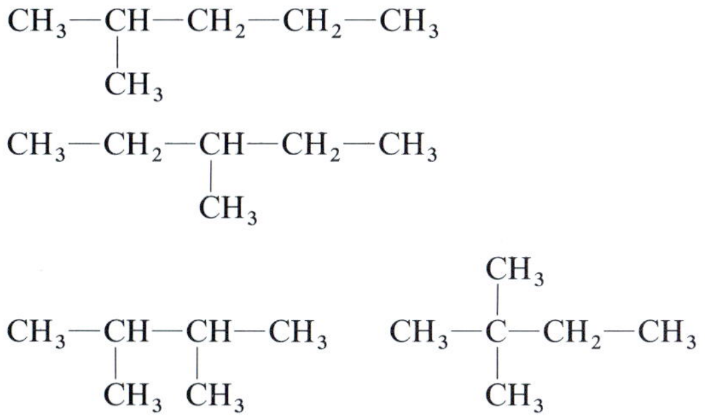

15.

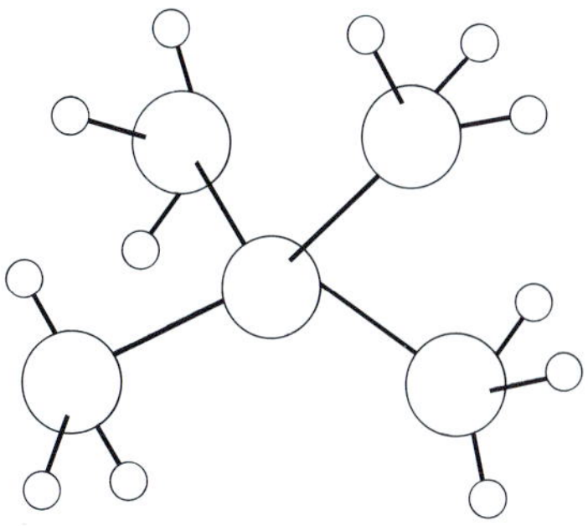

17. a. 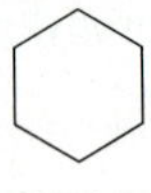

b. 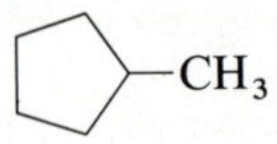

c. 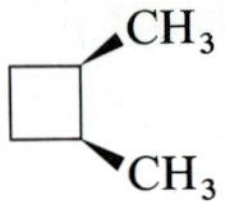

d. 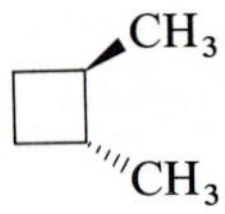

e. 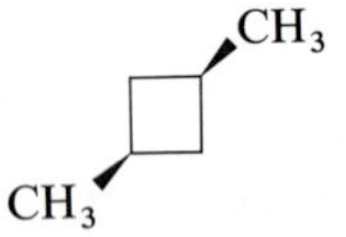

f. 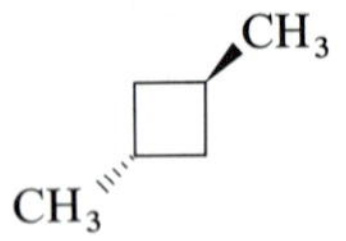

g. 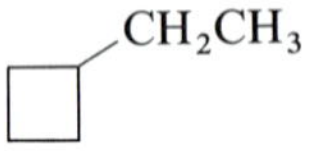

h. 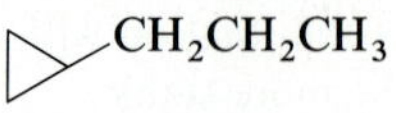

i. 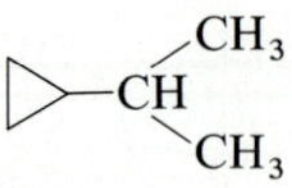

j. 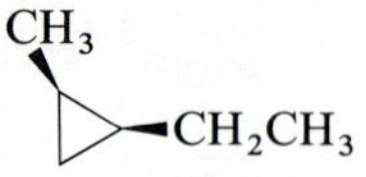

k. 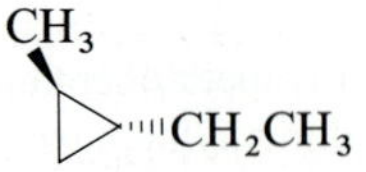

l. 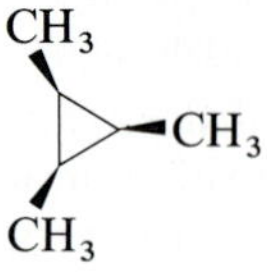

m. 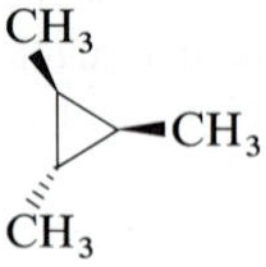

n. 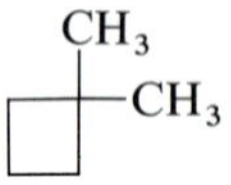

19. a. $3\ CO_2 + 4\ H_2O$ d. $7\ CO_2 + 2\frac{1}{2}\ H_2O$
 b. $4\ CO_2 + 5\ H_2O$ e. chlorocyclopentane
 c. $6\ CO + 7\ H_2O$ f. bromoethane

21. a. should be 3,5-dimethylheptane
 b. *cis*-1,2-dimethylcyclohexane
 c. 2,3-diemthylpentane
 d. 1,3-diethylcyclopentane
 e. 5,5-dibromo-2,2,3,3-tetramethylhexane

23. 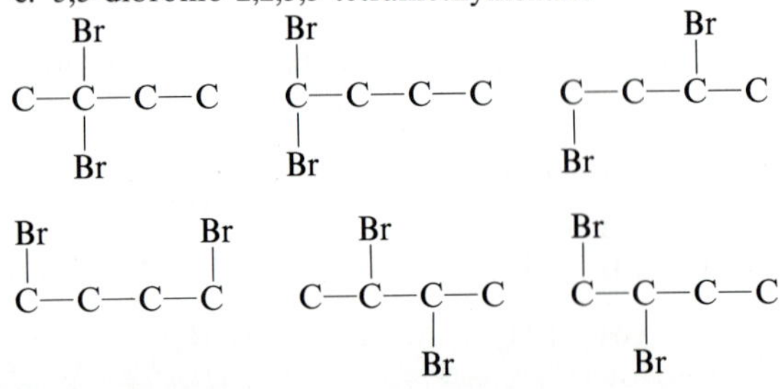

25. b > a > c

27. London forces

29. The sp^3 hybridization of carbon requires bond angles of 109°.

31. molecules having the same empirical formula but different structures

33. 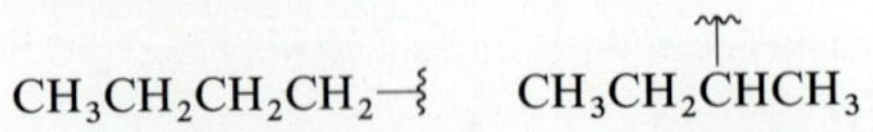

n-Butyl *sec*-Butyl

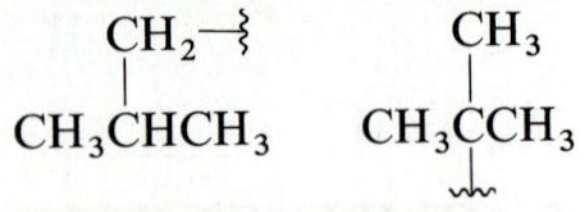

Isobutyl *tert*-Butyl

35. a. 2,3,3,6,6-pentamethyloctane
 b. 3,3,8,9-tetramethyldecane
 c. 2,2-dimethylheptane
 d. 4-isopropyl-5-ethylnonane

37. a. 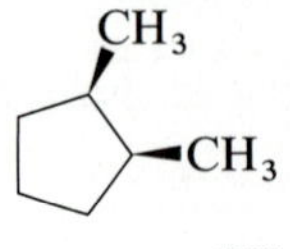

b. 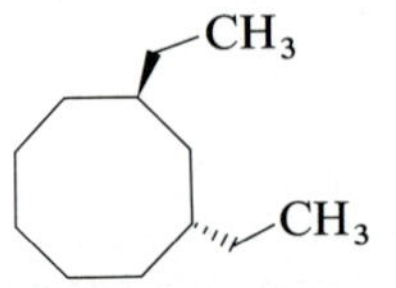

c. 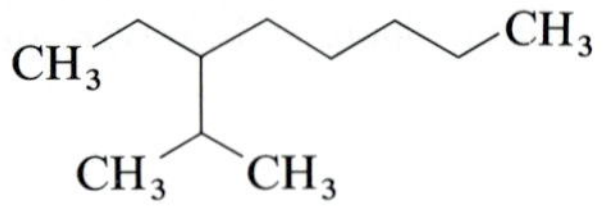

d. 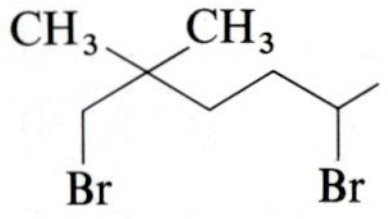

e. 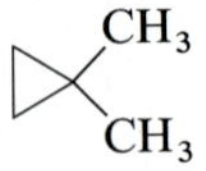

39. equatorial

41. $CO_2 + H_2O$

CHAPTER 16

1. sp^2

3. It would be very difficult to have the planar, trans double bond in a seven-membered ring. At least four carbon atoms would have to be in a plane because of the bonding type; it would be most difficult to bridge these atoms with a three-carbon unit.

5. two pi bonds and one sigma bond

7. An alkene has a double bond (planar, with bond angles of 120°), while an alkyne has a triple bond (linear).

9. Note: Although you can draw a structure for this name, the name is *not* correct. It should be 5-methyl-*cis*-3-heptene.

a. 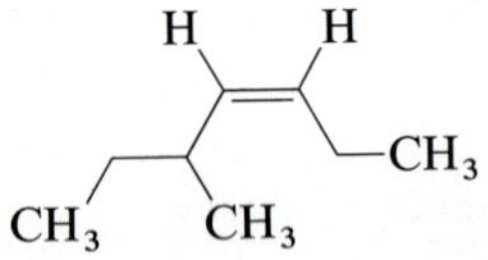

b. 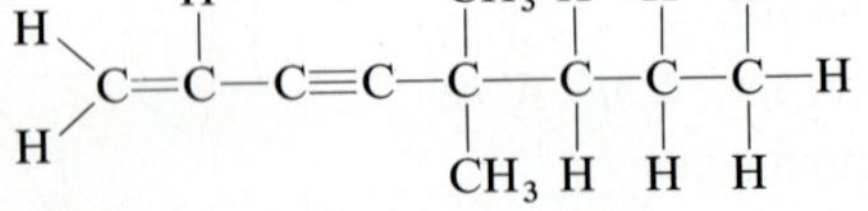

c. 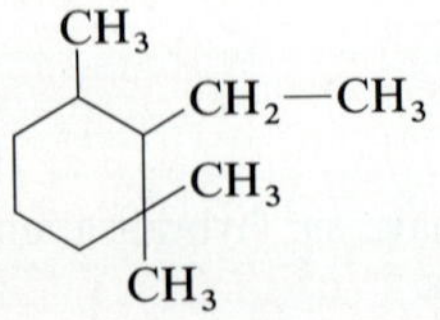

d.
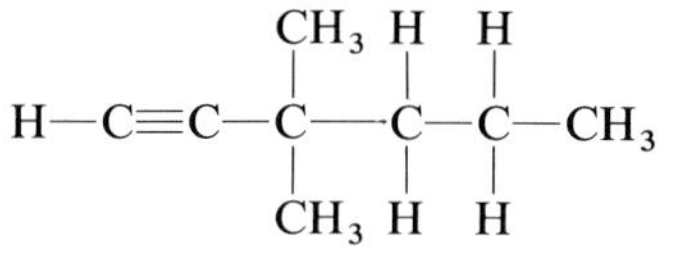

e.
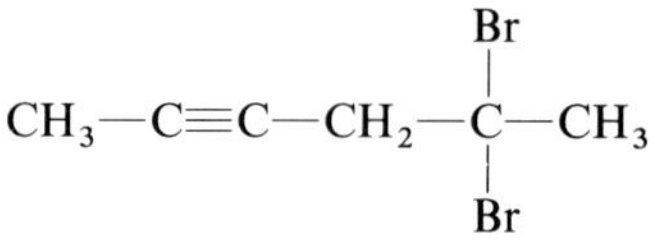

f.
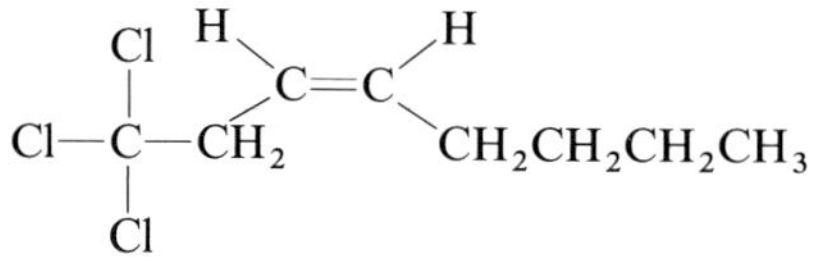

g.
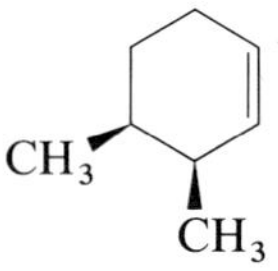

h.
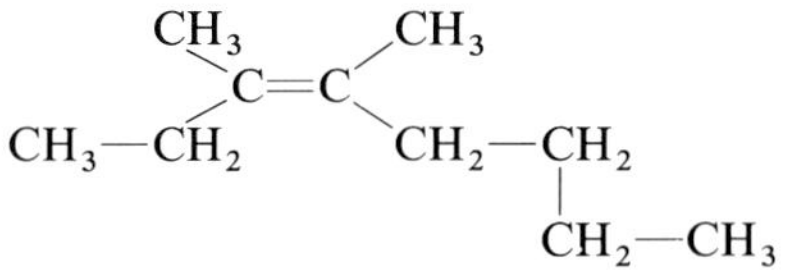

i.
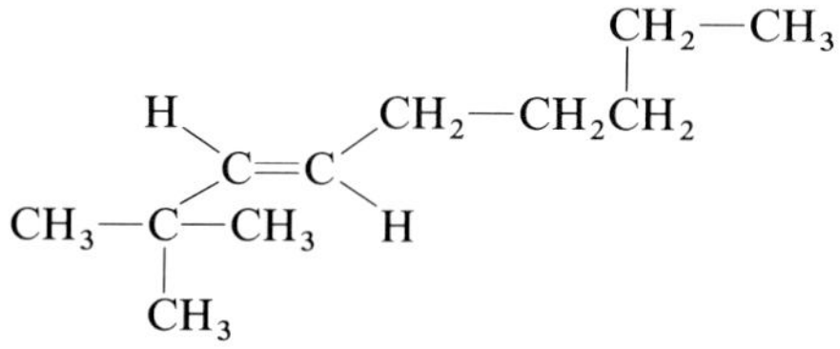

j.
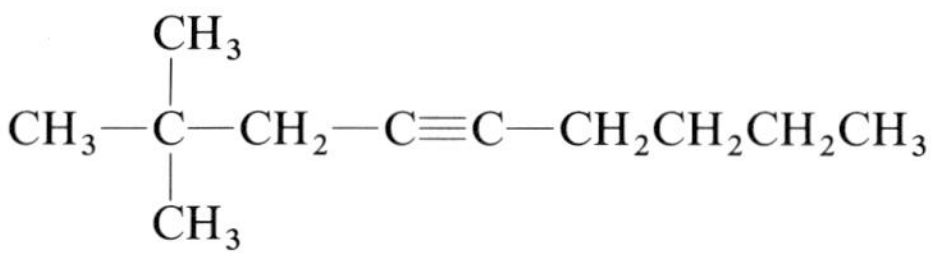

k.
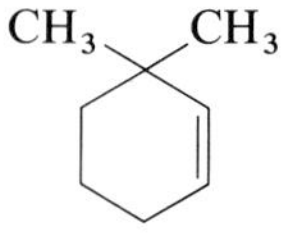

l.
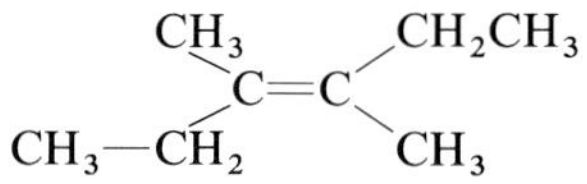

m.
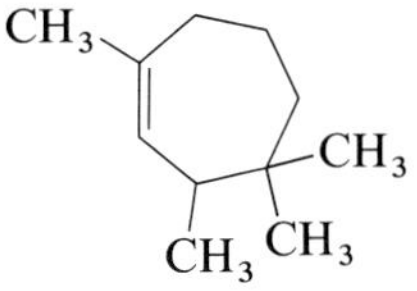

11. carbon dioxide and water

13. a. 4-ethyl-4-bromodecane

b. 1,2-dichloro-1,2-dimethylcyclopentane,
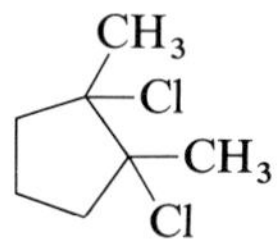

c. butane

d.
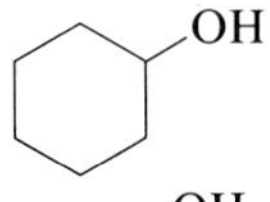
is 1-hydroxycyclohexane (cyclohexanol). The —OH group can be called the hydroxy group.

e.
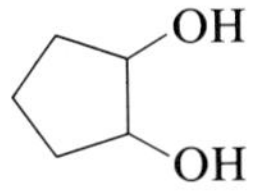
is 1,2-dihydroxycyclopentane.

f. 1,2-dimethylcyclohexane
g. 1-bromo-1-methylcyclohexane
h. 1,2-dibromo-5-methylcycloheptane
i. CO_2 and water
j. Two concepts can come up here. The cyclohexane ring (or any ring) can be used as an alkyl group (or cycloalkyl group). Thus, the group is cyclohexyl. If we assume one mole of HBr adds per one mole of alkyne, we get

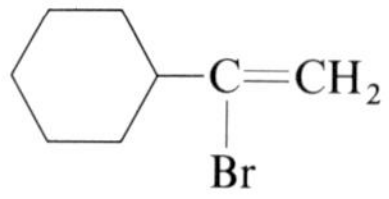

1-bromo-1-cyclohexylethene

If we add excess HBr, we get

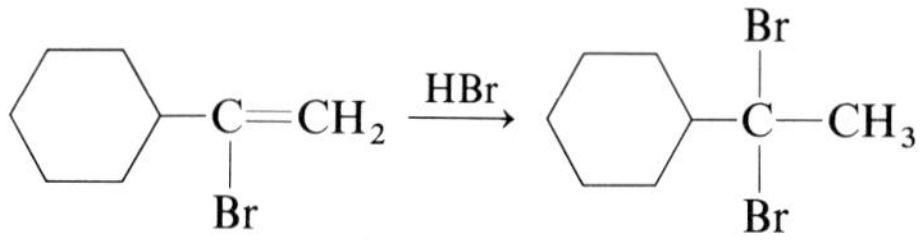

1-cyclohexyl-1, 1-dibromoethane

15. Cl^+ would act as the electrophile because chlorine is more electronegative than bromine.

17. a hydrocarbon containing a double or triple bond

19. The overlap of electrons from the adjacent *p*-orbitals making up the pi-bond cannot be easily broken.

21.
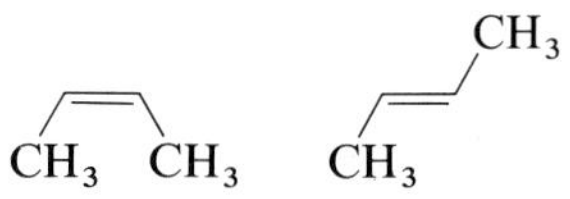

cis trans

23. a. 3-methyl-2-butene
b. *cis*-3-methyl-2-pentene
c. 4,4-dimethyl-2-pentyne
d. methylcyclopropane

25. The pi-electrons can function as a Lewis base.

27. dibromide

29. carbon-3

31. H^+

33. a. 1-bromo-1-methylcyclohexene
    b. 1-methylcyclohexane
    c. 1,2-dihydroxycyclopentane
    d. 1,2-dibromocyclohexane
    e. heptane
    f. 3,4-dimethyl-3,4-dihydroxyheptane

35. 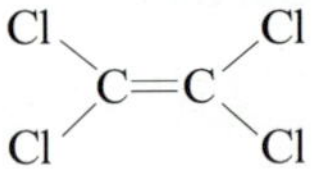

CHAPTER 17

1. a. *m*-bromoanisole (3-bromoanisole) or *m*-bromomethoxybenzene
   b. *o*-bromochlorobenzene or *o*-chlorobromobenzene
   c. *O*-nitrobenzenesulfonic acid
   d. *O*-ethylcumene
   e. 1,4-dimethyl-1,4-cyclohexadiene
   f. *o*-xylene (or *o*-methyltoluene)
   g. 2,3-dibromotololuene
   h. 7,7-dimethyl-2,5-octadiene (7,7-dimethylocta-2,4-diene)
   i. *cis*-2-hexen-5-yne
   j. 3-methyl-2,4-heptadiene (3-methylhepta-2,4-diene)

3. a. *o*-benzyltoluene
   b. *m*-bromocumene
   c. *o*-dibromobenzene
   d. *p*-iodoethylbenzene (*p*-ethyliodobenzene)
   e. *p*-chlorotoluene
   f. 3,3-dimethyl-7-phenylheptane

5. These are structures differing only in the localization of electrons. Relative atomic positions must remain constant.

7. a. Br^+ b. $^+SO_3H$ c. $^+NO_2$ d. $CH_3—CH_2{}^+$

9. a. 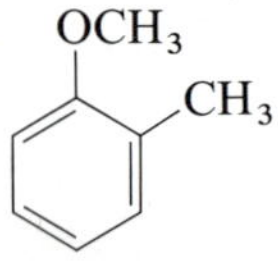 or 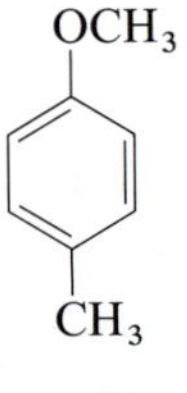

   b. 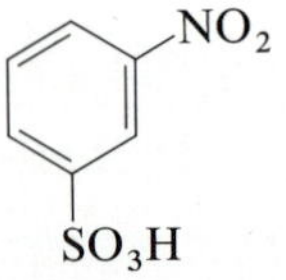

   c. 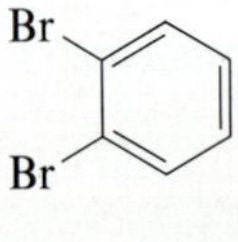 or 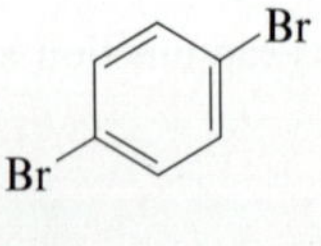

   d. 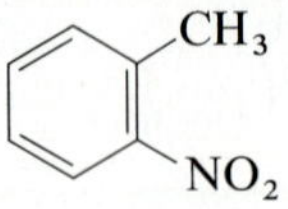 or 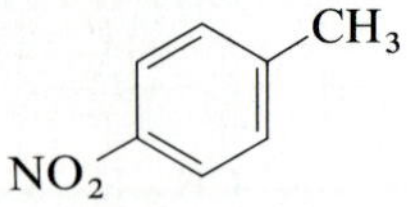

   e. 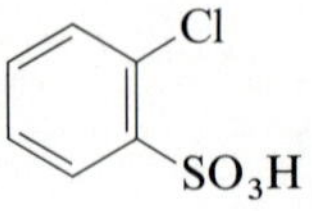 or 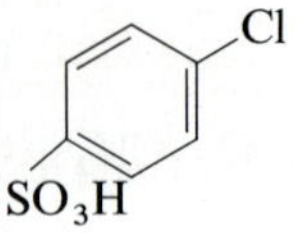

   f. 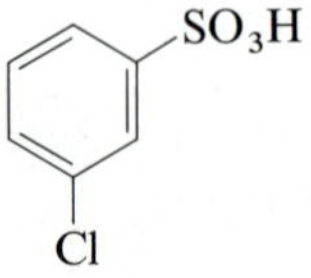

11. Electrons are *not* localized in alternating double and single bonds.

13. These are "equivalent" structures with the same arrangement of atoms but different arrangements of electrons between the atoms.

15. (b) requires delocalized electrons.

17. 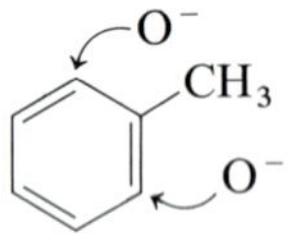

19. 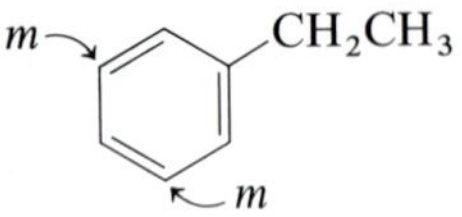

21. a. 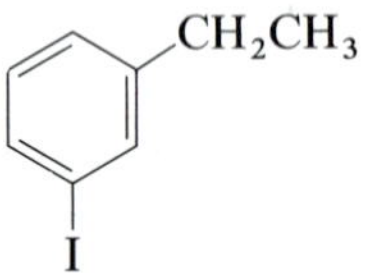

    b. 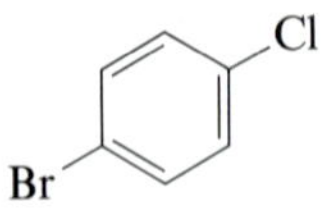

    c. 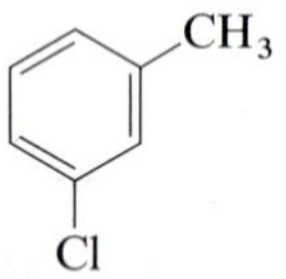

    d. 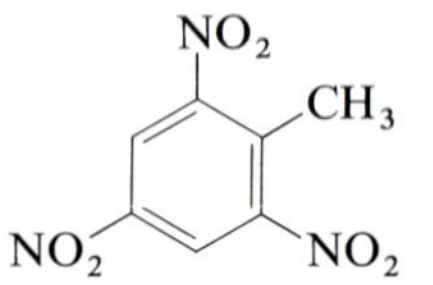

23. Substitution allows for the ring to maintain its electron delocalization.

25. a. *o, p* b. *m* c. *o, p* d. *o, p* e. *o, p* f. *m*

27. a. *o*- and/or *p*-nitroethylbenzene
    b. *o*- and/or *p*-methylphenol
    c. *m*-bromonitrobenzene

CHAPTER 18

1. a. 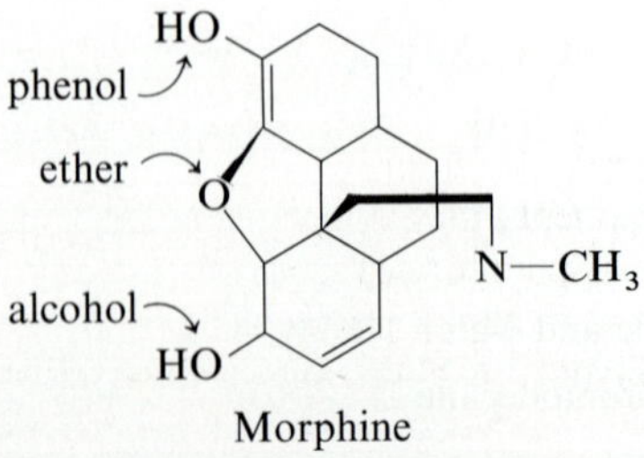
HO
phenol
ether
alcohol
HO
N—CH₃

Morphine

b.

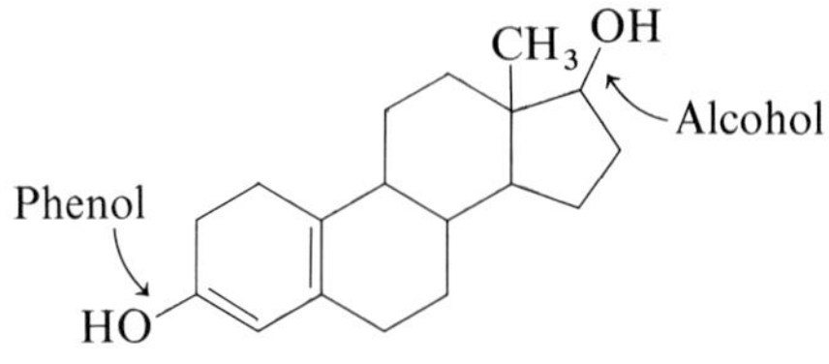
CH3
OH
Alcohol
Phenol
HO

Estradiol

c.

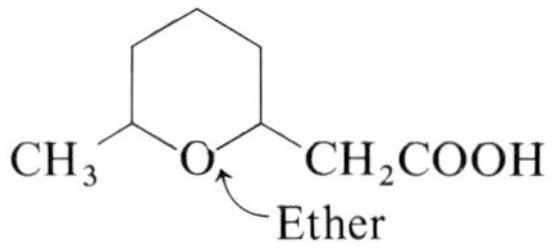
CH3
O
CH2COOH
Ether

Civet constituent

d.

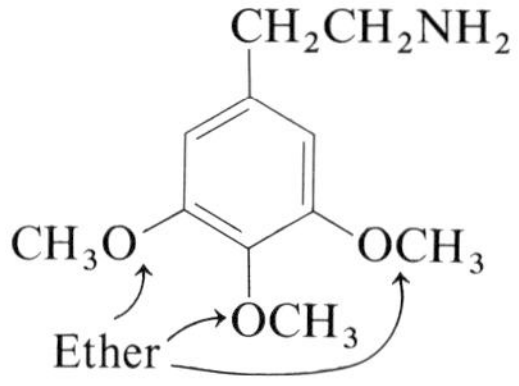
CH2CH2NH2
CH3O
OCH3
OCH3
Ether

Mescaline

3. a.

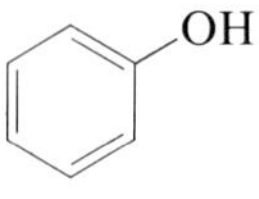

b.

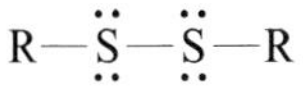

c.

d.

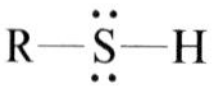

e.

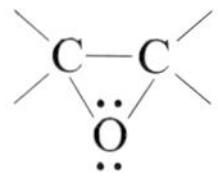

f.

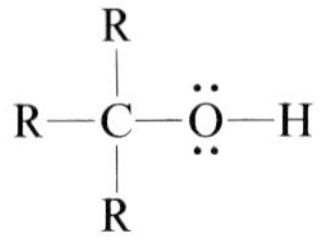

g.

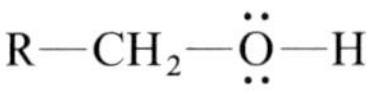

h.

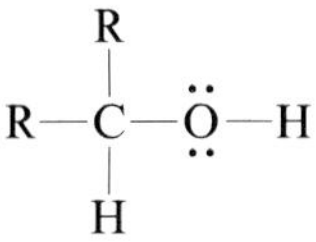

5. The hydroxyl group of an alcohol can participate in hydrogen bonding with water.

7.

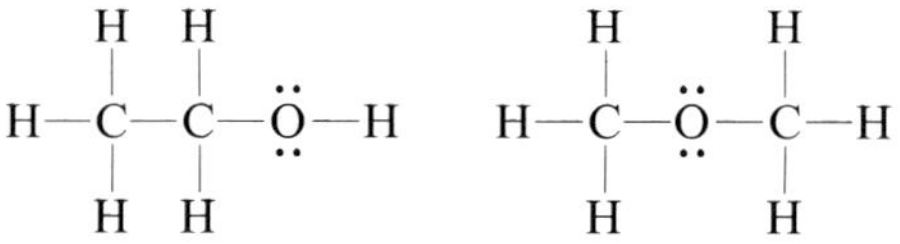

Can participate in hydrogen bonding

9. a. 3-methyl-2-cyclohexen-1-o1
   b. 2,4,5-trichlorophenol
   c. 2-isopropylcyclopentanol
   d. 2-chlorocycloheptanthiol
   e. 2,2-dibromo-3-pentyn-1-o1
   f. 4,4-dibromo-2-methyl-2-pentanol
   g. *trans*-1,2-dihydroxycyclohexane (*trans*-cyclohexan-1, 2-diol)
   h. *o*-nitrophenol
   i. 1-phenyl-1-pentanol
   j. cyclohexan-1,2-dithiol
   k. 2-methyl-1,2-butandiol (2-methylbutan-1,2-diol)
   l. 2,2-dimethylcyclopentanol
   m. 3-butyn-1-thiol
   n. ethylpropyl ether
   o. *tert*-butylisopropyl ether

11. a. 2,3-dimethylcyclooctanol
   b. 2-propyl-3-isopropylcyclohexanol
   c. 1,3,3,4-tetramethylcyclohexanol
   d. 1-butyl-2,2,6-trimethylcyclohexanol
   e. 2,2,4-trimethylcyclohexanol
   f. 2,2-dimethylcyclohexanol
   g. 1-propylcyclohexanol
   h. 4-ethyl-2,2,4-trimethylcyclohexanol
   i. 2,2-diethyl-5,7,8-trimethylcyclooctanol

13. a.

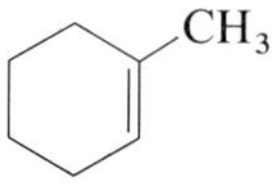

d.

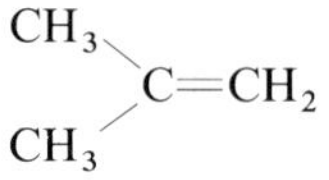

b.

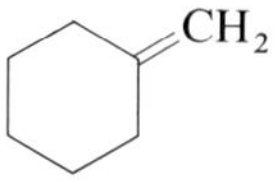

e.

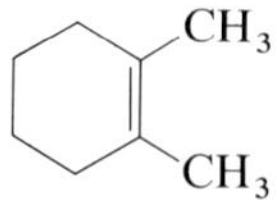

c.

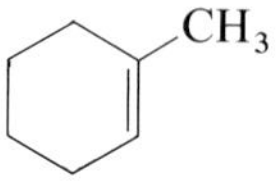

f.

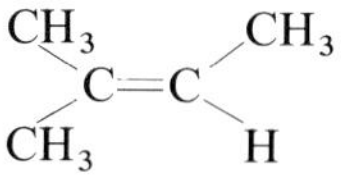

15. a.

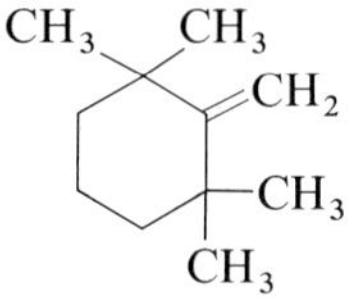

d.

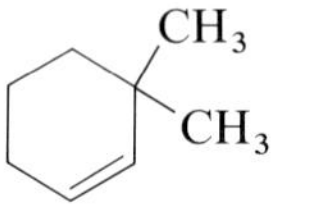

b.

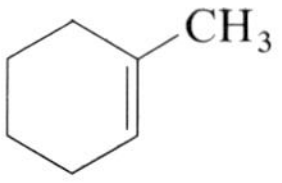

e.

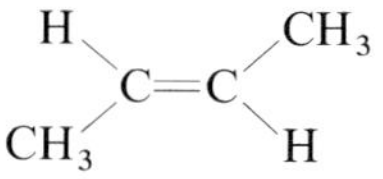

c.

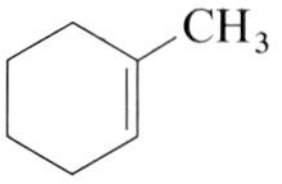

f.

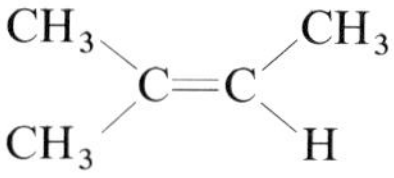

17. a.

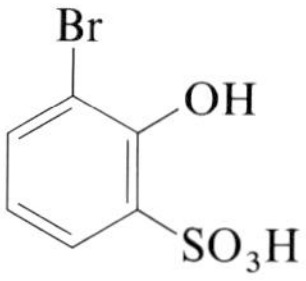

or

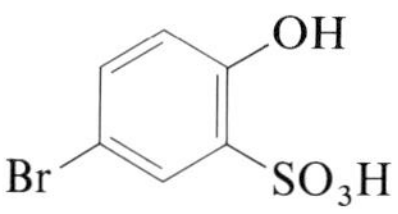

b.

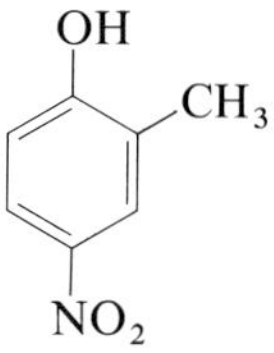

c. 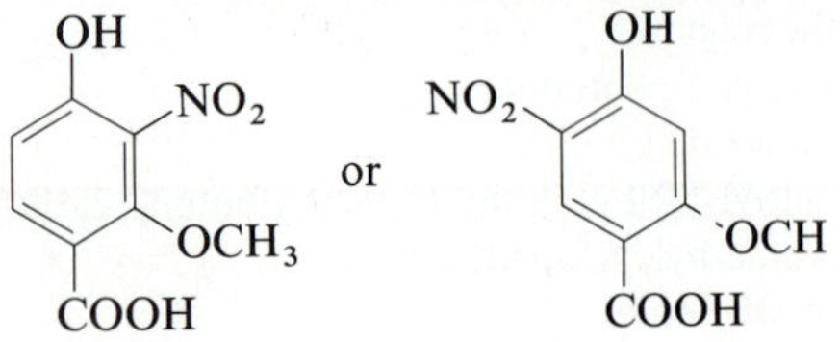

d. 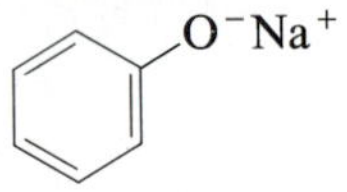

e. 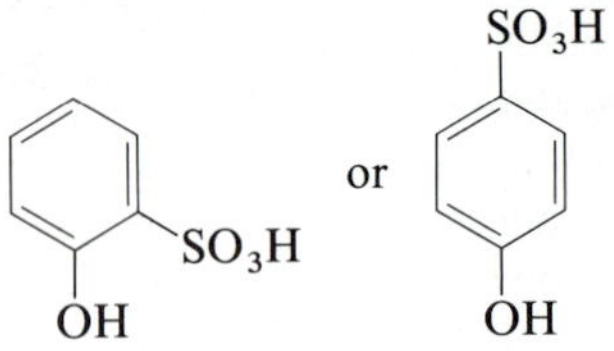

f. No reaction

19. a. 3-ethyl-4-methyl-2-hexanol
b. 4-methyl-3-hexanol
c. 3-methyl-6-isopropyl-5-nonanol
d. 4-octanol
e. 4-ethyl-3,5-dimethyl-4-heptanol
f. 3-decanol

21. less hydrogen bonding

23. The —OH can participate in hydrogen bonding.

25. a. 2-cyclopentenol
b. 5,5-dibromo-*trans*-3-hexen-2-o1
c. 3,5-heptandiol
d. 5,6,6-trimethylcyclodecanol

27. The oxygen atom has nonbonding electrons that can be donated.

29. a. cyclohexene
b. 2-bromopropane
c. 2-chloro-2-methylpropane
d. 1-methylcyclohexene

31. a. *o*-ethylphenol
b. *b*-bromophenol
c. 2,5-dinitrophenol
d. *o*-*tert*-butylphenol

33. a. *o*- and/or *p*-methylphenol
b. 2-bromo-3,5-dinitro-4-methylphenol

35. Alcohols actively participate in hydrogen bonding.

37. Hydrogen bonding is the key to increased solubility of alcohols.

39. Sulfur, being less electronegative than oxygen, does not participate as effectively in hydrogen bonding.

41. a. $CH_3—S—S—CH_3$
b. $HS—CH_2—(CH_2)_4—CH_2—SH$
c. $CH_3—CH_2—CH_2—S—Hg—S—CH_2—CH_2—CH_3$

CHAPTER 19

1. 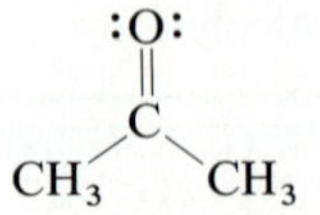 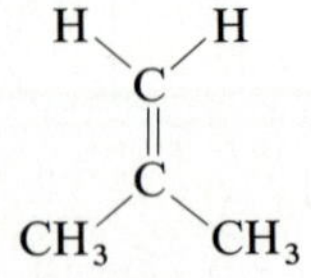

Greater dipole

There is a greater difference in electronegativity for the C—O bond.

3. d (1); b (2); a (3); c (4)

5. a. 2,2-dimethylcyclohexanone
b. 2,3-dibromobutanal
c. *o*-nitrobenzaldehyde
d. 4-hydroxy-3,3-dimethylcyclooctanone
e. 1-chloro-4-methyl-3-pentanone
f. 3,4-dibromocyclopentanone
g. 2,2-dimethylpropanal

7. a. 3,4-dimethylcyclooctanone
b. 2-isopropyl-3-propylcyclohexanone
c. 2,2,3,6-tetramethylcyclohexanone
d. 4-butyl-3,3,5-trimethylcyclohexanone
e. 2,2,4-trimethylcyclohexanone
f. 3,3-dimethylcyclohexanone
g. 2-propylcyclohexanone
h. 4-ethyl-2,2,4-trimethylcyclohexanone
i. 5,5-diethyl-2,7,8-trimethylcyclooctanone

9. a. 4-ethyl-3-methylhexanal
b. 3-methylhexanal
c. 4-chloro-2,6-dimethyl-3-propyloctanal
d. 5,6-dimethyloctanal
e. 4-ethyl-3,5-dimethylheptanal
f. 3-bromo-7-methyldecanal

11. a. cyclohexanol
b. 2-butanol
c. ethanol
d. 2-methylcyclopentanol

13. a. 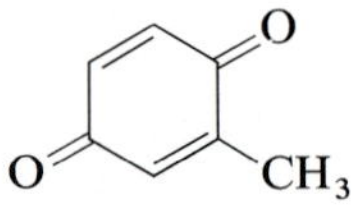
c. 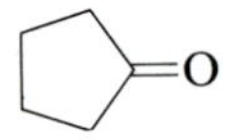

b. No reaction. Tertiary alcohols are resistant to oxidation.

d. $CH_3—CH_2—COOH$

15. 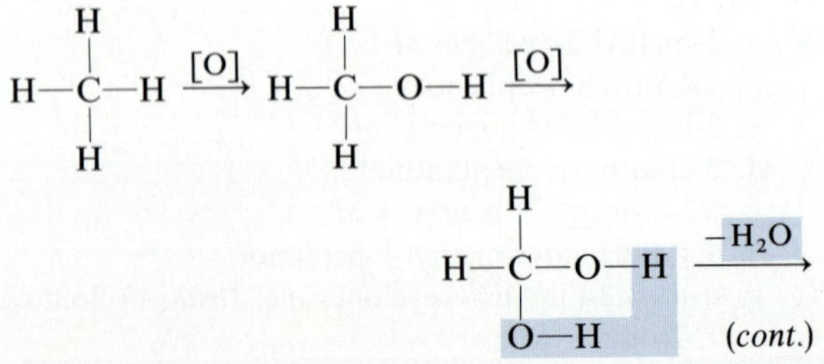

(*cont.*)

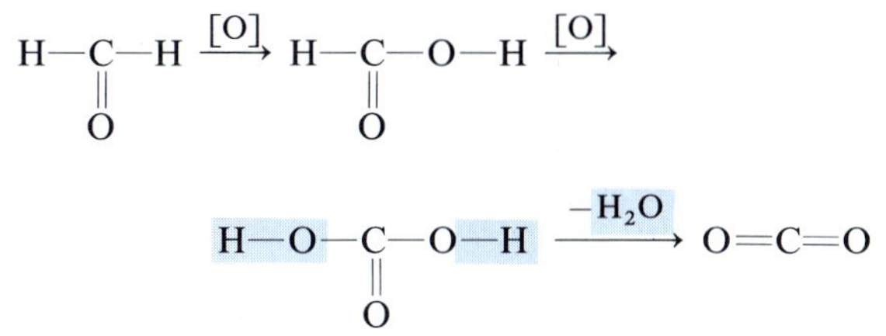

17. The bond is polarized so that the oxygen has a partial negative charge and the carbon a partial positive charge.

19. 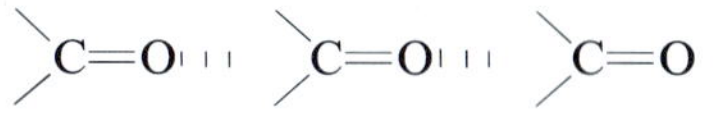

21. Both the carbonyl and the alkene are characterized by a double bond. The double bond of the carbonyl group is polarized.

23. a. 3,3-dibromoheptanal
    b. *trans*-3,4-dimethylcyclopentanone
    c. 5-phenyl-2-nonanone
    d. *m*-nitrobenzaldehyde

25. The polarization of the carbonyl group results in the carbon bearing a partially positive charge.

27. 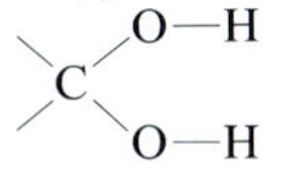

29. 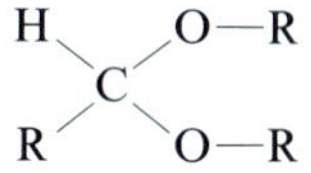

31. Ag^+

33. quinone

35. a. cyclohexanol
    b. $CH_3—CH_2—COOH$ + free Ag
    c. 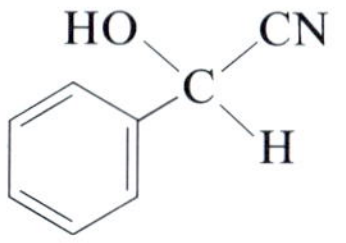

CHAPTER 20

1. a. 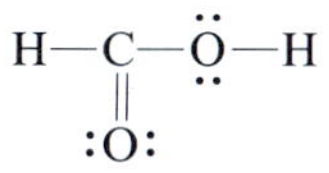
   b. H—C≡N:
   c. 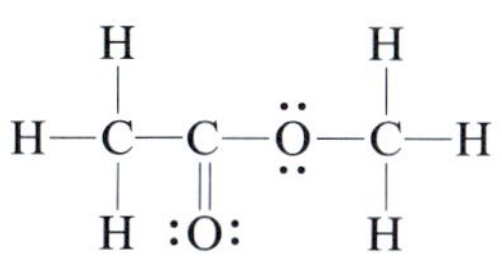

3. a. 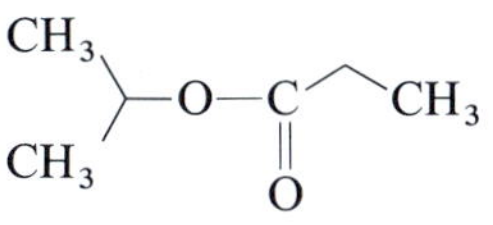
   b. 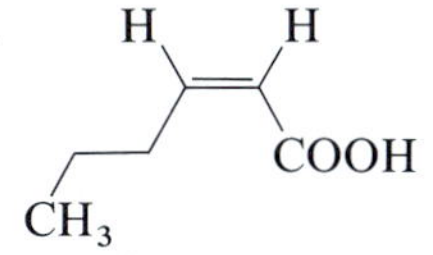
   c. 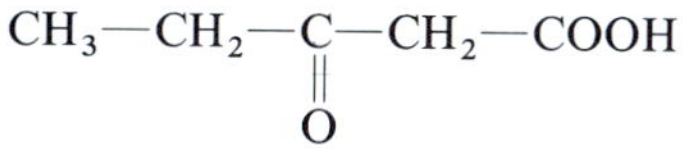
   d. 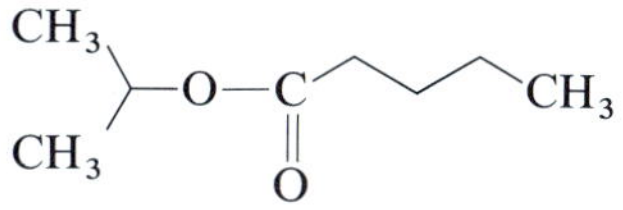
   e. 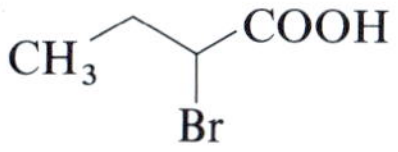
   f. 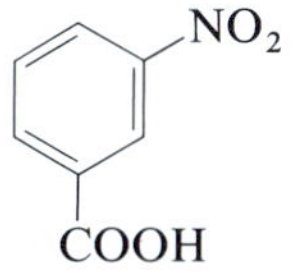
   g. 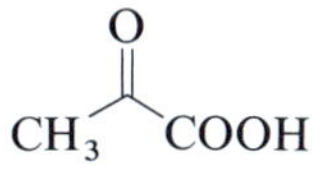

5. Because of strong ability to hydrogen bond, carboxylic acids can dimerize. Thus, there appears to be two molecules rather than one.

7. F is more electronegative.

9. a. 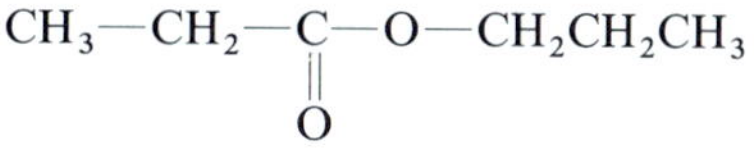
   b. 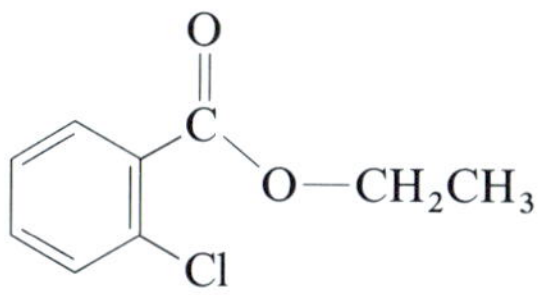
   c. 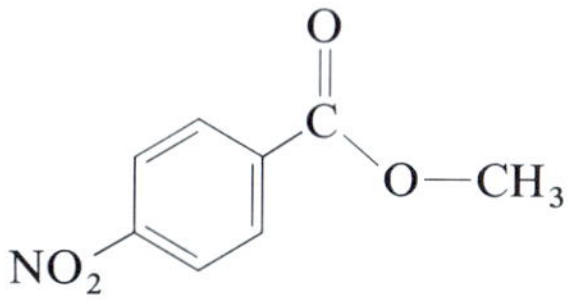
   d. 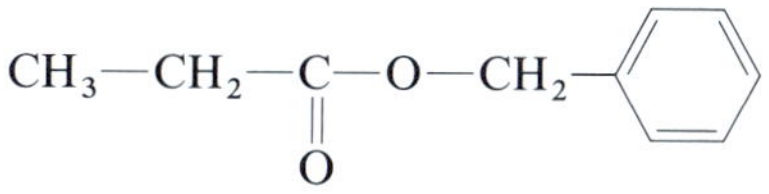
   e. 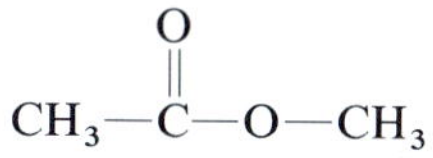
   f. 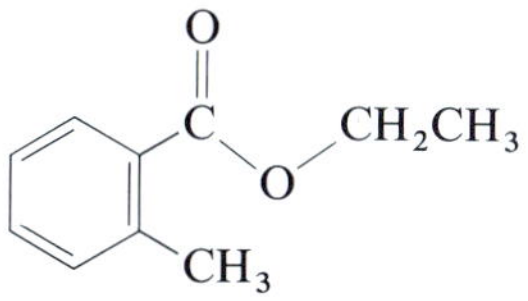
   g. 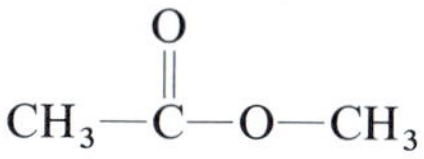

h. 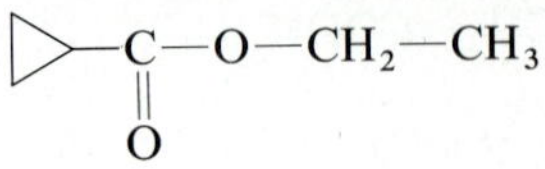
C—O—CH₂—CH₃
O

i. 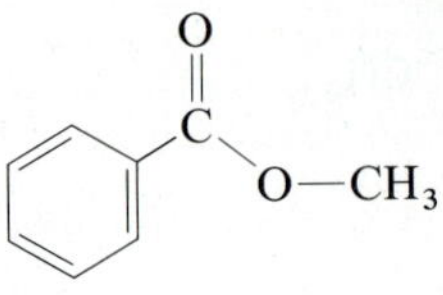
O
C
O—CH₃

11. a. 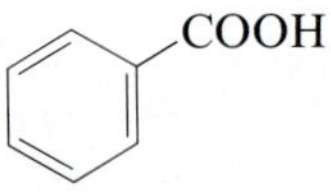
COOH

d. 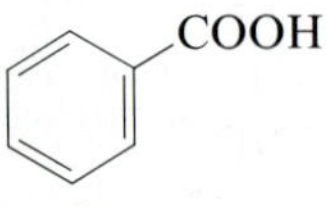
COOH

b. 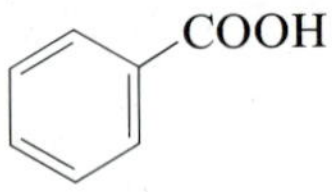
COOH

e. 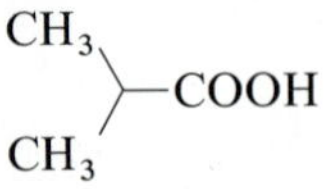
CH₃
COOH
CH₃

c. 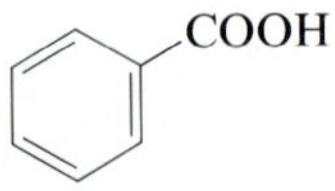
COOH

f. $CH_3—COOH$

g. CH_3CH_2COOH

13. Aspirin: start with salicylic acid and esterify the phenolic group with an acetic acid derivative (experimentally, acetic anhydride is often used). Oil of wintergreen: start with salicylic acid and esterify the carboxylic acid portion with methanol and acid.

15. A concentrated detergent (in the absence of water) very much resembles a hydrocarbon: "like dissolves like."

17. sp^2

19. The chlorine is electronegative and a strong electron-withdrawing group.

21. The nonsoluble hydrocarbon chain becomes more important than the carboxyl group.

23. a. 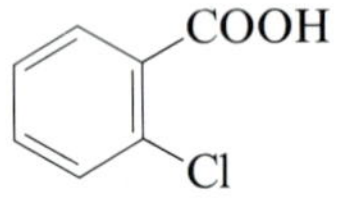
COOH
Cl

b. 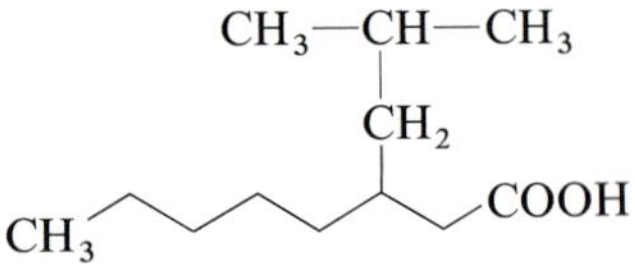
CH₃—CH—CH₃
CH₂
COOH
CH₃

c. 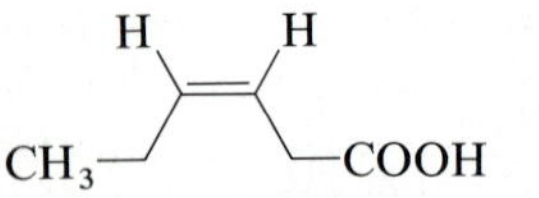
H
H
CH₃
COOH

d. 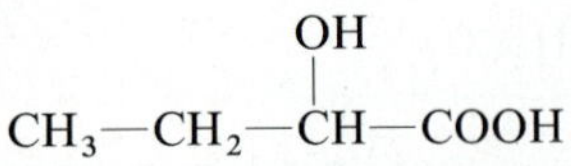
OH
CH₃—CH₂—CH—COOH

e. 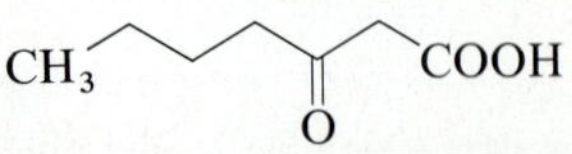
CH₃
COOH
O

25. oxidation of the aldehyde

27. 2-heptanone via decarboxylation of the beta-keto acid

29. carbon dioxide

$$CH_3—COOH + NaHCO_3 \longrightarrow CH_3—COO^- \; Na^+ + H_2O + CO_2$$

31. 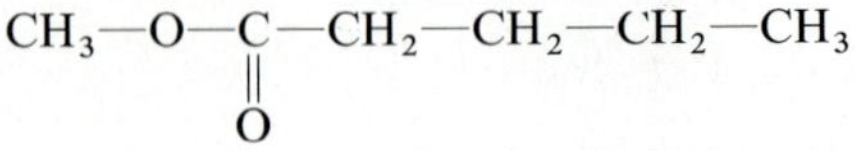
CH₃—O—C—CH₂—CH₂—CH₂—CH₃
O

33. The polar end is water soluble, while the hydrocarbon end is able to mix with other hydrocarbon (oily) materials.

35. 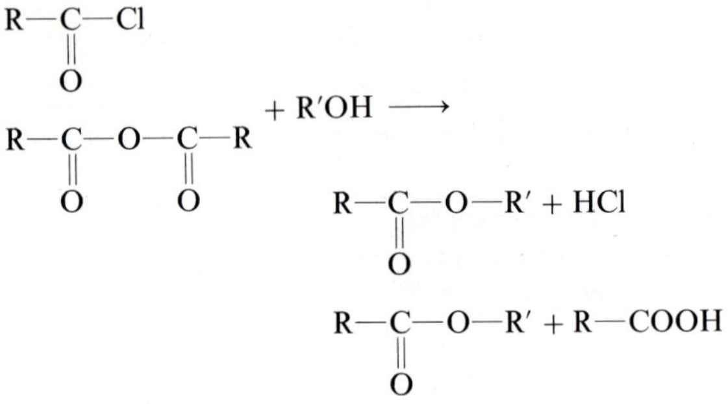
R—C—Cl
O
+ R′OH ⟶
R—C—O—C—R
O O
R—C—O—R′ + HCl
O
R—C—O—R′ + R—COOH
O

CHAPTER 21

1. a. ethylphenyl amine (*N*-ethylaniline)
   b. diethylamine
   c. isopropyl amine
   d. ethylmethylpropyl amine
   e. triethylamine

3. a. $CH_3—CH_2—\overset{+}{N}H_3 \; Cl^-$

b. 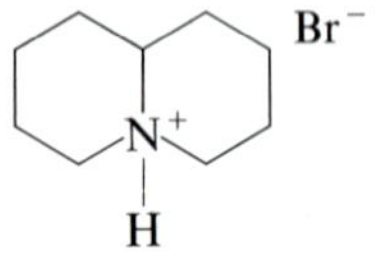
N⁺
H
Br⁻

5. a. 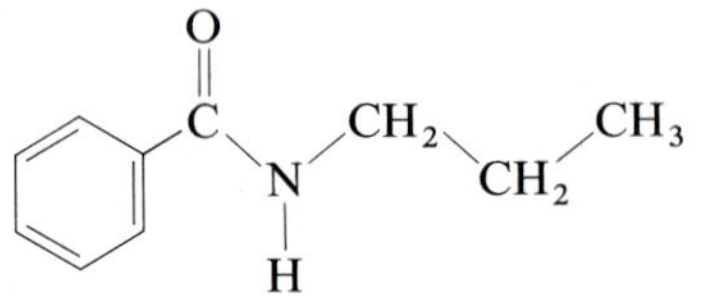
O
C
N
H
CH₂
CH₂
CH₃

b. 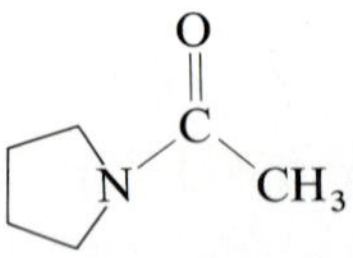
O
C
N
CH₃

c. 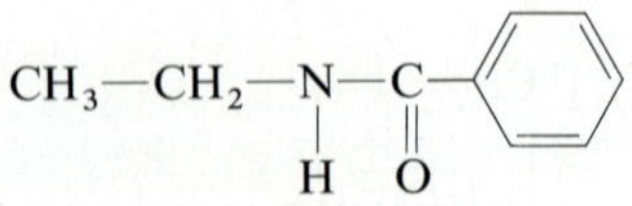
CH₃—CH₂—N—C
H O

d. 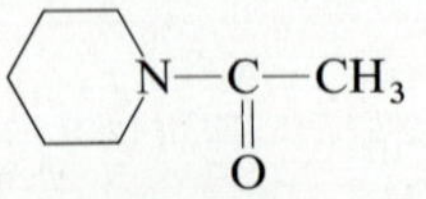
N—C—CH₃
O

7. a. 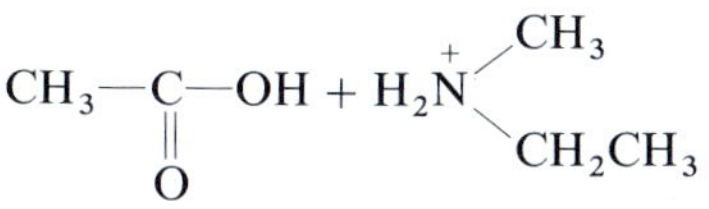

b. 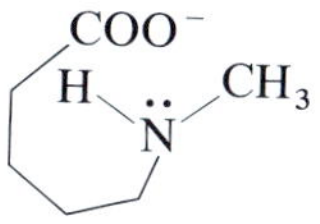

9. The lone pair of electrons on the nitrogen of aniline is delocalized into the benzene ring and is thus not as available to react as a base.

11. The lone pair of electrons on the nitrogen can be donated to a Lewis acid.

13. Nitrogen is less electronegative than oxygen and thus does not participate as effectively in hydrogen bonding.

15. a. ethylmethyl amine
b. *N*-methylpyrrolidine
c. diethylmethyl amine

17. a. 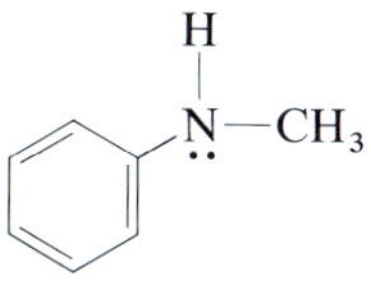 b. 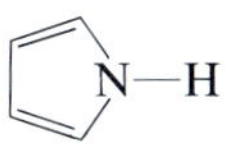

19. The long hydrocarbon portion would be water insoluble, while the polar end (amine salt) would be water soluble.

21. a. $CH_3—\overset{+}{N}H_3\ Br^-$

b. 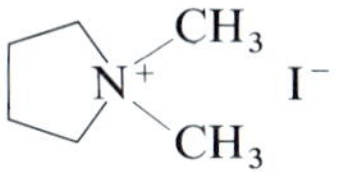

c. 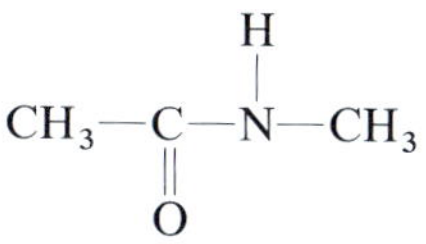

d. 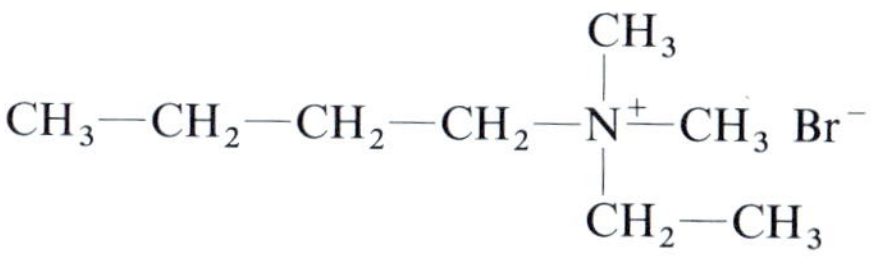

23. a. 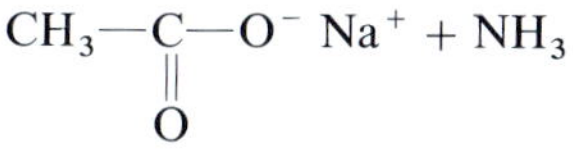

b. 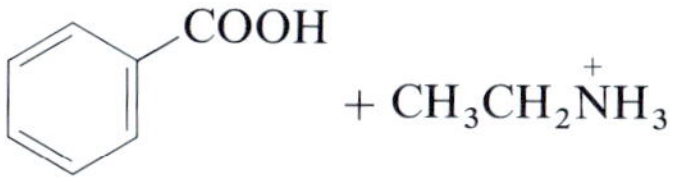

25. a naturally occurring, nitrogen-containing (basic) organic compound

CHAPTER 22

1. The cell is the smallest unit of life that is capable of sustaining itself.
3. Diffusion; some substances involve active transport.
5. The cell membrane possesses transport systems that selectively move some substances across the membrane. Thus, this membrane helps control the contents of the cell.
7. a. They are the site of protein synthesis.
b. cytosol
9. a. block cell wall synthesis
b. disrupt protein synthesis
c. disrupt protein synthesis
11. animals, plants, fungi, and true algae

13.

Organelle	Animals	Plants	Bacteria
Nucleolus	Yes	Yes	No
Mitochondrion	Yes	Yes	No
Endoplasmic reticulum	Yes	Yes	No
Chloroplast	No	Yes	No

15. enantiomers
17. a, c
19. It is a substance provided by food that the body must have for growth or maintenance.
21. The body conserves some of this energy as potential energy stored in chemical bonds of some molecules. The rest is lost as heat.
23. bacterium: surface area $= 2 \times \pi r^2 + 2\pi r h = 69\ \mu m^2$
volume $= \pi r^2 h = 31\ \mu m^3$
surface area/volume $= 2.2\ \mu m^2/\mu m^3$
cell: surface area $= 6 \times \text{length}^2 = 6.0 \times 10^4 \mu m^2$
volume $= \text{length}^3 = 1.0 \times 10^6 \mu m^3$
surface area/volume $= 0.06\ \mu m^2/\mu m^3$
The bacterium has 37 times the surface area per unit volume.
25. DNA
27. does not cause fever
29. Measure several physical properties such as boiling point. Most isomers will have different values for these properties. If they are the same compound or enantiomers, they will have the same values. Then check optical activity with a polarimeter. If the samples are the same compound, the value will be identical. If they are enantiomers, they will rotate light in opposite directions by the same magnitude.
31. Chiral centers have four different substituents, so the chiral center in these compounds is the ring carbon atom

with the dashed bond to the hydrogen. It is bound to the hydrogen, two different ring carbon atoms, and another carbon atom through the wedge bond.

CHAPTER 23

1. A carbohydrate that cannot be broken down into a smaller unit (sugar) by simple hydrolysis.
3. This sugar contains a ketone functional group and has five carbon atoms.
5. a. D-glyceraldehyde c. D-glucose
   b. D-ribose d. D-fructose
7. In the ring structure that is typically illustrated, the hydroxyl group on carbon-1 points down, below the ring.
9. This is the change in rotation of plane polarized light that is seen in a solution of a pure anomer as some of the anomer isomerizes to the other anomer.
11. A sugar that can reduce a metal ion to one of smaller oxidation number is called a reducing sugar. An aldehyde group must be present, or a keto group that can be isomerized to an aldehyde must be present.
13. The glycosidic bond involves the anomeric hydroxyl group of both sugars. Because of this, it is not a reducing sugar.
15. a. $\alpha,\beta\ 1 \rightarrow 2$ b. $\alpha\ 1 \rightarrow 4$ c. $\beta\ 1 \rightarrow 4$
17. Similarities: both are polymers containing only glucose.
    Differences: the glycosidic bonds in cellulose are $\beta\ 1 \rightarrow 4$, those in starch are primarily $\alpha\ 1 \rightarrow 4$.
19. Similarities: both are unbranched polymers of monosaccharides connected by $\beta\ 1 \rightarrow 4$ bonds.
    Differences: cellulose contains glucose; chitin contains a modified glucose, *N*-acetylglucosamine.
21. A. f D. b
    B. d E. e
    C. a F. c
23. a. There are eight aldopentoses; ribose is shown here:

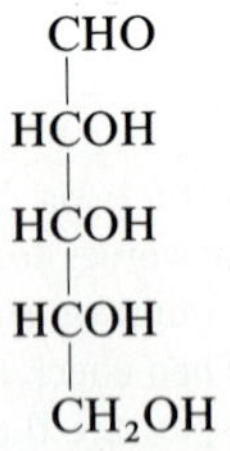

c.

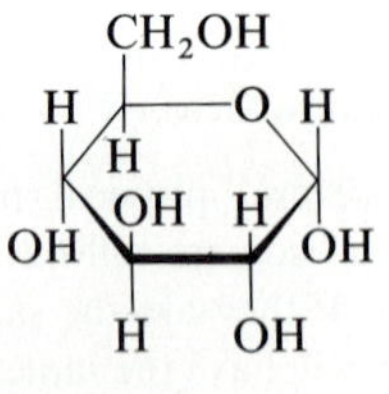

b

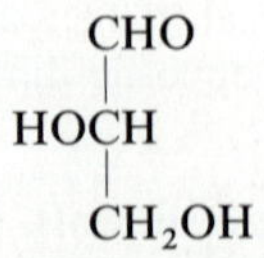

d.

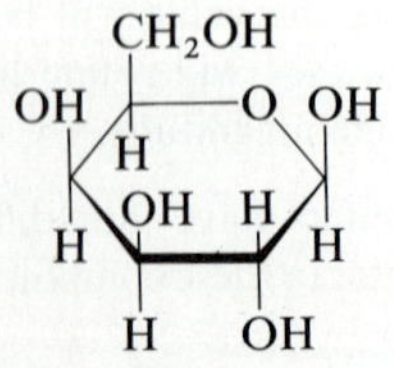

e.

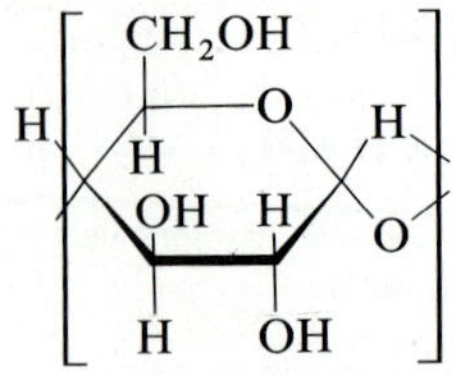

f.

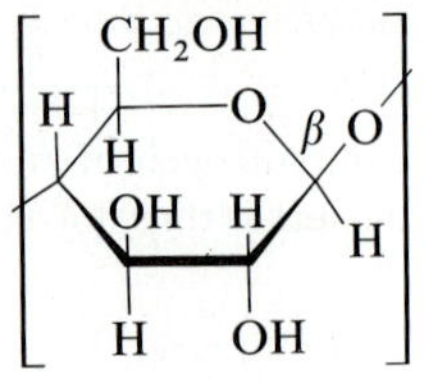

g.

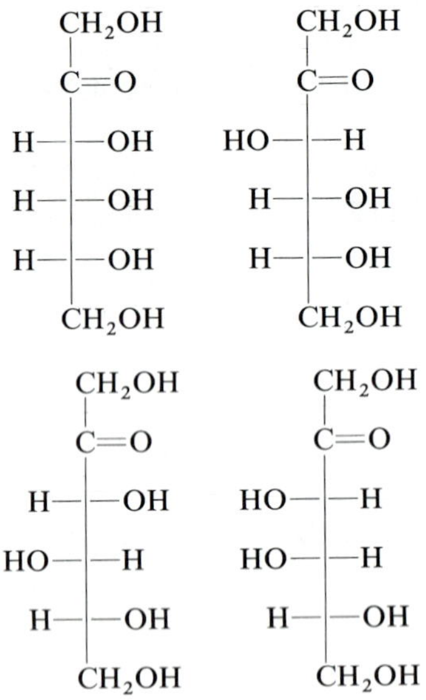

These four ketohexoses are the D-stereoisomers. Their mirror images are the other four ketohexoses.

25. The appearance would be concentration dependent. A very dilute solution would yield a slightly murky solution that is slightly green. A fairly concentrated solution would yield a more obvious reddish precipitate leaving little blue color in the mixture.
27. 2-Deoxyribose has two hydrogen atoms as the substituents on ring carbon-2. Ribose has one hydrogen atom and a hydroxyl group.
29. There is no relationship between the D- and L-stereochemical families and the direction of rotation of plane polarized light.
31. D- and L-glucose are enantiomers because they are non-superimposable mirror images. The other two are anomers.
33. Fructose is a ketose that can isomerize to an aldose. While it is an aldose, it is a reducing sugar.
35. Humans lack enzymes for the hydrolysis of most $\beta\ 1 \rightarrow 4$ glycosidic bonds.

37. Glycogen is a highly branched polymer of glucose. The ends of these numerous branches are sites where enzymes can catalyze the release of glucose residues from the polymer.
39. glyceraldehyde; dihydroxyacetone
41. 140,000 amu × 1 glucose/162 amu = 864 glucose residues. (If your answer is 778 residues, you forgot to correct for the loss of water during glycosidic bond formation.)

CHAPTER 24

1. They are generally soluble in nonpolar organic solvents and insoluble in water.
3. triacylglycerols, phosphoacylglycerols, sphingolipids, glycolipids, and some waxes
5. Their size; they have a long hydrocarbon-like tail.
7. It introduces a bend or kink in the nonpolar tail. It lowers the melting point.
9. This is the amount of iodine, in grams, that will react with the carbon–carbon double bonds in 100 g of the fat or oil. The iodine number is an indication of the amount of unsaturation present in the fat or oil.
11. The bottom one; it possesses double bonds that lower its melting point. Hydrogenate it.
13. Antioxidants such as BHA and BHT can be added, or the oils used in the food can be partially hydrogenated.
15. 45 days
17. It is a molecule that has both polar and nonpolar regions; membranes.
19. Phospholipids are amphipathic molecules that can form the polar and nonpolar interactions needed for membrane structure; triacylglycerols are not amphipathic and thus cannot form these interactions.
21. A saponifiable lipid containing the base sphingosine.
23. The structure of cholesterol is shown in Figure 24.6. The fused ring system is known as the steroid nucleus.
25. Estrogen and progesterone. They play major roles in reproductive physiology, including the stimulation of ovulation by estrogen and stimulation and maintenance of the uterine lining by progesterone.
27. They contain an aromatic ring.
29. Glycocholate and taurocholate; they emulsify dietary lipids in the digestive tract.
31. Aspirin inhibits the synthesis of prostaglandins that are involved in inflammation and pain.
33. Figure 24.13 illustrates a micelle. The polar head groups face out toward water, while the nonpolar tails are clustered in the center.
35. For nonpolar molecules to dissolve in water, they must get between water molecules. This requires the breaking of relatively strong hydrogen bonds between the water molecules. The new bonds between nonpolar molecules and water are weak London forces. The replacement of stronger bonds by weaker ones is unfavorable and does not occur. Thus, the nonpolar molecules aggregate together.
37. The structure of a liposome is shown in Figure 24.14. They may gain use as a delivery system for drugs.
39. Amphipathic lipids and proteins are aggregated into a sheetlike structure held together by noncovalent bonds. The components are free to diffuse, thus the membrane is fluid.
41. The average North American diet gets about 40% of its calories from lipid. A diet with 30% or less from lipid is recommended.
43. The more obese an individual, the greater the risk of heart disease. The larger the dietary intake of fats and oils, the greater the risk. The more saturated the dietary triacylglycerols, the greater the risk. The greater the amount of blood cholesterol, the greater the risk.
45. A. d C. e E. b
    B. a D. c
47. a. Both are triacylglycerols. Oils have lower melting points due to more carbon–carbon double bonds or, less commonly, due to shorter chain fatty acids.
    b. Both are saponifiable lipids derived from glycerol. Phospholipids (phosphoacylglycerols) are amphipathic, but fats are not.
    c. Both are essentially insoluble in water and are linear molecules. Fatty acids contain a carboxyl group that is ionized at neutral pH; thus, they are amphiphatic and form micelles. Waxes are not amphipathic and do not form micelles.
    d. Both are steroid female sex hormones, but they possess different physiological roles.
    e. Both are nonsaponifiable lipids derived from unsaturated fatty acids. Both serve as signals or mediators, but the specific roles are different.
49. The nonpolar solvents, a, b, d
51. a, e
53. Two; they are enantiomers.
55. a. $CH_3(CH_2)_7CH{=}CH(CH_2)_7COOH + H_2 \longrightarrow CH_3(CH_2)_{16}COOH$
    b. no reaction
    c. $CH_3CH_2CH{=}CHCH_2CH{=}CHCH_2CH{=}CH(CH_2)_7COOH + 3\ H_2 \longrightarrow CH_3(CH_2)_{16}COOH$
57. 7-dehydrocholesterol

CHAPTER 25

1. structural, contractile, transport, protective, regulatory, and storage

3. 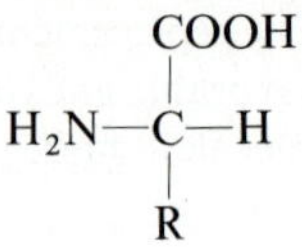 Because the amino group is attached to the α carbon of the carboxylic acid.

5. L-glyceraldehyde 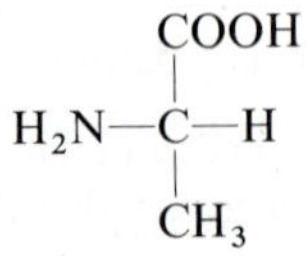

7. The basic amino acids have a basic group in the side chain; an acidic amino acid has a carboxyl group in the side chain. Lysine and aspartic acid are illustrated here.

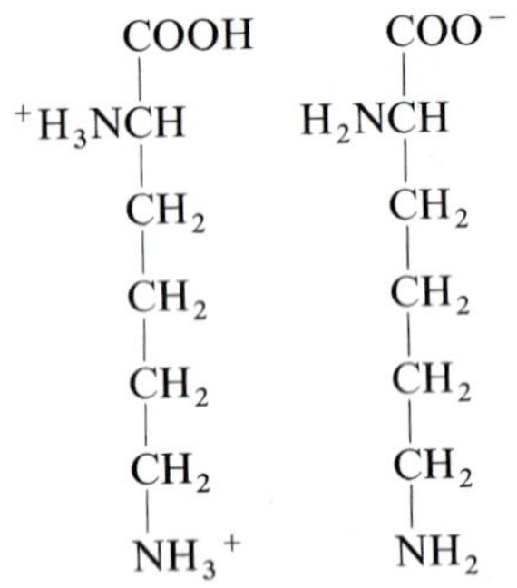

Lysine

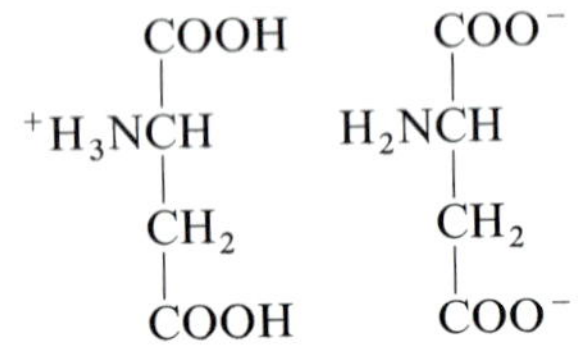

Aspartic acid

9. This is the pH at which an amino acid (or any amphoteric molecule) has no net charge.

11. The amide bond is between the amino group of one amino acid and the carboxyl group of another one. The peptide bond links amino acids in proteins and peptides.

13.

Name	Structure	Biological role
Oxytocin		Contraction of smooth muscle in uterus and other target organs
Vasopressin	See Figure 25.5	Contraction of smooth muscle in arterioles and other target organs
Enkephalin		Pain and pleasure preception

15. primary: the amino acid sequence
secondary: the arrangement of nearby residues due to hydrogen bonding between atoms of the peptide bonds
tertiary: the compact folding of the polypeptide chain back upon itself
quaternary: the arrangements of subunits in an oligomeric protein

17. Compare your structure to the α-helix in Figure 25.8. There are hydrogen bonds between atoms of the peptide bonds.

19. This oligomeric protein contains five subunits: two α subunits, two very similar subunits designated β and β′, and one σ subunit.

21. A simple protein contains only one or more polypeptide chains. A conjugated protein has, in addition, a prosthetic group.

23. Positive: lysine, arginine, and histidine. Negative: aspartate and glutamate. This applies to solutions near pH 7.

25. pH 6.13

27. The rats would fare poorly on this diet. Because cysteine is lacking, methionine is needed for cysteine synthesis. Since a minimal amount of methionine is provided, there is not enough to meet a rat's needs for both methionine and cysteine made from methionine.

29. a. A carboxylic acid with an amino group on the α carbon; the basic building blocks of proteins; any of the common twenty amino acids serve as an example.
b. Amino acids that are in the same stereochemical family as L-glyceraldehyde.
c. Amino acids that the body cannot synthesize.
d. A molecule possessing both a positive and negative charge. The 20 common amino acids exist as zwitterions.
e. Possessing both acidic and basic properties simultaneously. The amino acids are amphoteric.
f. The pH at which a charged molecule has no net charge. Each amino acid has its own unique pI.
g. A molecule composed of two amino acids linked by a peptide bond.
h. The amide bond linking two amino acids.
i. A macromolecule of more than 5000 MW that is made of amino acids linked by peptide bonds. Some have an additional group called a prosthetic group. Many examples can be given.
j. The amino acid sequence of a peptide or protein.
k. The short term arrangement of the polypeptide chain in a protein. Can be arranged into a spiral (α helix) or into a sheet (β-pleated sheet).
l. The compact shape of most proteins resulting from the folding of the polypeptide chain.
m. The clustering together of nonpolar parts of a molecule in an aqueous solution.
n. Covalent bond between the sulfur atoms of two cysteine residues in a protein. Contributes stability to the tertiary structure of a protein.

o. Loss of native conformation (shape) of a protein or other macromolecule.

31. 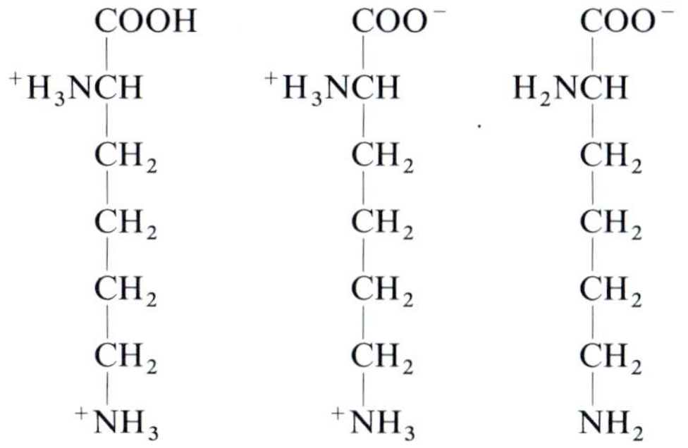

pH 1   pH 7   pH 13

33. 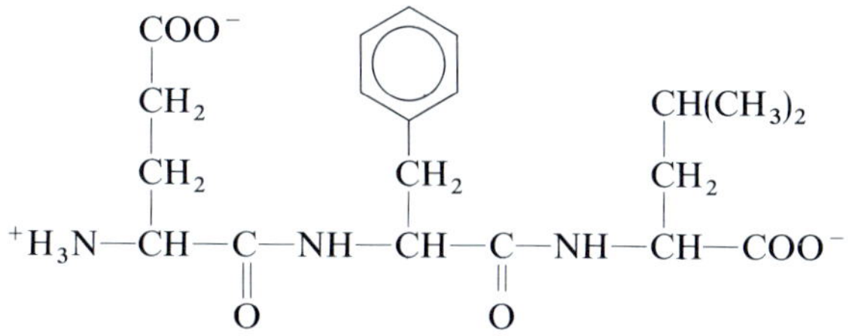

35. The C—N single bond actually has partial double bond character due to resonance. Thus, rotation about that bond is restricted.

37. The hydrophilic amino acids are on the exterior surface where their side chain can interact with the solvent water and polar and ionic solutes. The hydrophobic amino acids are in the interior where they form hydrophobic interactions.

39. b, d, e

41. $7^5 = 16807$

43. No. The amino acids could be bonded to each other in different orders, thus they could have different primary structures.

CHAPTER 26

1. Ribose is found in RNA; deoxyribose is found in DNA.

3. Deoxyribose has two hydrogens on carbon two; ribose has one hydrogen and one hydroxyl group.

5. DNA has deoxyribose; RNA has ribose. DNA has the base thymine; RNA has uracil. DNA is double stranded; RNA is single stranded. DNA is the largest molecule in the cell; RNA is much smaller.

7. Messenger RNA (mRNA): carries genetic information from DNA to the cell for protein synthesis.
Ribosomal RNA (rRNA): component of ribosomes, which are the site of protein synthesis.
Transfer RNA (tRNA): carries the proper amino acid to ribosomes for insertion during protein synthesis.

9. At least 20 codons are needed because there are 20 amino acids. Codons of one or two bases only have 4 or 16 combinations of bases, respectively.

11. A codon is a sequence of three bases in mRNA that codes for a particular amino acid or that serves as a stop signal.
An anticodon is a sequence of three bases in tRNA that pair with a codon in mRNA. This pairing ensures that the proper amino acid will be inserted into the growing polypeptide chain.

13. There is no unique set of codons for oxytocin; any of several sequences will be correct. This is due to multiple codons for most amino acids. Your sequence should begin with either UGU or UGC, since these are the codons for cysteine.

15. An RNA primary transcript is made that is complementary to the gene on DNA. The transcript is modified to yield mRNA, which provides instructions for the synthesis of the protein during translation.

17. The process of replication yields two new strands that are complementary to each of the original strands. The enzyme DNA polymerase synthesizes the new strand by sequentially attaching a new nucleotide whose base is complementary to the corresponding base on the original strand. Specific details are described in Section 26.3.

19. A mutation that results from the insertion or deletion of a base into or from the base sequence of a gene. All codons from that point on in the product mRNA are shifted out of the correct reading frame.

21. DNA is found only in the nucleus of eucaryotic cells and in the cytosol of procaryotic cells. RNA is found in both the nucleus and cytosol of eucaryotes and in the cytosol of procaryotes.

23. 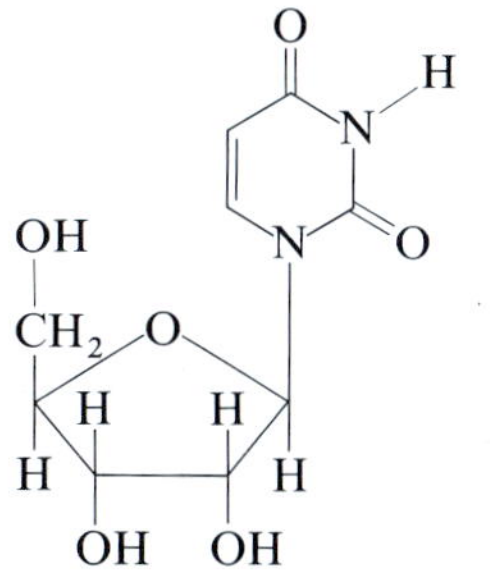

25. uridine triphosphate

27. RNA; the presence of uracil and absence of thymine are the important clues.

29. Trinucleotides: $4^3 = 64$. Tripeptides: $20^3 = 8000$.

31. Several sequences of codons will work. Since there are two codons for phe, four for pro, and six for leu, there are $2 \times 4 \times 6 = 48$ possible answers. This is one of them: UUUCAUUUA

33. No. Since most amino acids have several corresponding codons, it is highly unlikely that a synthetic gene would have exactly the same sequence as the real gene. Furthermore, introns would be a factor if this were a eucaryotic gene.
35. An intron is a part of a eucaryotic gene that is not expressed. There are no codons in the RNA that correspond to these bases in the DNA.

CHAPTER 27

1. catabolism and anabolism
3. They catalyze metabolic reactions, allowing them to proceed rapidly and efficiently at the temperature and conditions that exist within a cell.
5.

Substrate	Reaction Type
a. Xylulose	Oxidation–reduction
b. Glutamine	Synthesis (cannot tell by name which of six classes this enzyme belongs to)
c. Xanthine	Oxidation–reduction

7. the nonamino acid part of an enzyme or other protein
9. a. B_6 c. riboflavin
   b. pantothenic acid d. thiamine
11. Lock-and-key: the substrate and enzyme possess complementary shapes that fit together directly. No change in shape of either is required.
    Induced fit: the substrate and enzyme are not exactly complementary. When they bind, one, the other, or both change shape. They are only complementary after binding occurs.
13. It lowers (decreases) the energy of activation.
15. It is the concentration of substrate in an enzyme-catalyzed reaction that will yield a rate that is one-half the maximal rate of the reaction.
17. Pepsin is active around pH 2, trypsin is active around pH 7 to 8. Each is most active in the pH range where it is found: pepsin in the stomach (pH 2) and trypsin in the small intestine (pH 7–8).
19. A competitive inhibitor resembles the substrate well enough to bind to the active site, but it lacks the bond or group that actually participates in the reaction.
21. Cells regulate the rates at which enzymes are (1) synthesized and (2) broken down.
23. The product of a process influences its own production by decreasing its rate of synthesis when the amount of the product exceeds some threshold value. Negative effector.
25. It can be used to break down blood clots during a heart attack.
27. Catabolism breaks down larger molecules to smaller ones with energy releasing reactions. Anabolism uses smaller molecules and energy to make larger molecules.
29. anabolism
31. Michaelis–Menten equation
33. For activity, the side chains of certain amino acids must be protonated or deprotonated. At pH values outside the normal activity range, these groups become deprotonated or protonated. The enzyme is inactive under these conditions.
35. At higher temperatures, the protein denatures and loses its activity. Like other chemical reactions, the rates of enzymic reactions decrease with temperature, due to decreased translational (thermal) energy.
37. The enzyme may require a prosthetic group or coenzyme that is removed during the isolation procedure.
39. No. Temperature and pH are normally maintained within a narrow range.
41. The substrate and inhibitor both possess two carboxylate groups separated by a region of nonpolarity. The inhibitor cannot be oxidized to a carbon–carbon double bond because it only has one methylene group.
43. An endothermic reaction can be coupled to an exothermic one as long as the amount of energy released by the exothermic reaction exceeds the amount required by the endothermic one. The two collectively are thus exothermic.

CHAPTER 28

1. It reduces food particle size which increases surface area; it mixes food with saliva which contains the enzyme amylase.
3. These enzymes catalyze the hydrolysis of the disaccharides maltose, sucrose, and lactose, respectively.
5. It denatures proteins which exposes the peptide bonds to digestive enzymes.
7. Lipids are insoluble in the aqueous environment of blood.
9.

Food Class	Enzymes	Site of Digestion	Products
Carbohydrates	Amylases, hydrolases for sugars	Primarily mouth and small intestine	Simple sugars
Proteins	Proteases and peptidases	Stomach and small intestine	Amino acids
Lipids	Lipases	Small intestine	Fatty acids and glycerol

11. The steps that involve hexokinase and phosphofructokinase; hydrolysis of ATP.

13. Lactic acid is the product of anaerobic metabolism and may accumulate under these conditions. This acid will donate its hydrogen ion under physiological conditions, thus making blood or the cells more acidic.

15. glucose$_n$ + P_i
$\longrightarrow$ glucose$_{n-1}$ + glucose 1-phosphate
glucose 1-phosphate
$\longrightarrow$ glucose 6-phosphate

17. It is oxidatively decarboxylated in a reaction catalyzed by pyruvate dehydrogenase.

19. These coenzymes accept a pair of electrons and hydrogen ions during oxidation–reduction reactions; 3 NAD^+ and 1 FAD.

21. They are oxidized by the electron transport chain, with oxygen acting as the ultimate acceptor of the electrons.

23. They serve as proton pumps that use energy derived from the oxidation of NADH and $FADH_2$.

25. ATP synthetase; a portion of it (F_0) is embedded in the inner mitochondrial membrane; a stalk and the F_1 part project into the mitochondrial matrix.

27. It transports fatty acids in blood.

29. Palmitoleic acid will yield two fewer ATP because it already has a carbon–carbon double bond. One fewer $FADH_2$ will be produced from palmitoleic acid because of that double bond.

31. The α-amino groups of amino acids are collected into a single molecule, glutamate.

33. in the mitochondria

35. The difference is due to the way the electrons of cytosolic NADH are brought into the mitochondrion of that cell. If mitochondrial FAD accepts them, then one fewer ATP is formed per cytosolic NADH than if mitochondrial NAD^+ accepts them.

37. ATP, NAD^+, ADP

39. two

41. (200 kcal) × (1 mol ATP/7.3 kcal) ×
(1 mol glucose/36 mol ATP)
(180 g glucose/1 mol glucose) = 137 g glucose

43. Both are nucleotide triphosphates; they possess identical phosphate–phosphate anhydride bonds. They are equivalent in energy because hydrolysis of an anhydride bond yields very similar products: a nucleotide diphosphate and inorganic phosphate.

45. 1 mol of glucose yields 36 mol ATP. 36 ATP times 7.3 kcal/mol = 263 kcal of stored energy. 1 mol of glucose = 180 g. 180 g times 4 kcal/g = 720 kcal in 180 g of dietary glucose. 263/720 × 100 = 36.5% efficient

CHAPTER 29

1. (a) Anabolism has small reactants and larger products; catabolism is the opposite. (b) Anabolic reactions tend to be reductive; catabolic are generally oxidative. (c) Anabolism is endothermic; catabolism is exothermic.

3. From their environment; they make them from carbon dioxide.

5.

Reactants	Products
H_2O	O_2
ADP	ATP
P_i	—
$NADP^+$	NADPH

7. As electrons flow from photosystem II to photosystem I through an electron transport chain, protons are pumped. This proton gradient is stored energy and can be used to do work.

9. ATP is used to phosphorylate intermediates of the dark reaction, and NADPH is used to reduce 1,3-diphosphoglycerate to glyceraldehyde 3-phosphate.

11. This pathway provides for the synthesis of glucose from several precursor molecules.

13. It is recycled into glucose via gluconeogenesis. Lactate is formed in muscle, transported to the liver, used to make glucose, and transported back into the blood. This is called the Cori cycle.

15. glucose$_n$ + UDP—glucose → glucose$_{n+1}$ + UDP

17. Citrate leaves the mitochondrial matrix and enters the cytosol. It is cleaved to acetyl CoA and oxaloacetate, which is recycled back into the mitochondrion.

19. Palmitate; palmitate can be elongated and double bonds added to it.

21. A diacylglycerol reacts with CDP-ethanolamine to yield phosphatidyl ethanolamine and CMP.

23. The synthesis of cysteine requires methionine. If only a minimal amount of methionine is provided, there is not enough methionine in the diet to meet both the need for methionine and the need for synthesis of cysteine.

25. The chemotherapy kills cancer cells and other rapidly dividing cells of the body.

27. Electrons flow from water molecules to a chlorophyll in photosystem II, through electron carriers to a chlorophyll in photosystem I, through electron carriers to $NADP^+$

29. 12 NADPH are required, and 4 photons are needed per NADPH. Thus, 48 mol of photons are needed.

31. 6 carbon dioxide molecules combine with 6 ribulose diphosphate molecules to form 12 3-phosphoglycerate molecules; 2 of these are used to make glucose, the remaining 10 are used to make ribulose 1,5-diphosphate.

33. primarily in the liver
35. They are synthesized by sequential addition of acetate (as acetyl CoA). Since acetate contains an even number of carbon atoms, the product must contain an even number of carbon atoms.
37. hydroxymethylglutaryl CoA (HMG CoA)
39. histidine, isoleucine, leucine, lysine, methionine, phenylalanine, threonine, tryptophan, and valine

CHAPTER 30

1. Hemoglobin binds oxygen, which greatly increases the oxygen carrying capacity of blood. Hemoglobin also binds some carbon dioxide, facilitating its transport to the lungs.
3. This enzyme speeds up the reversible hydration of carbon dioxide to form carbonic acid.
5. lungs; kidneys; kidneys (also lungs and skin); kidneys (also skin)
7. If the concentration of a substance in blood exceeds this value, a portion of the substance will not be reabsorbed in the kidneys and will thus be present in urine; approximately 180–200 mg per 100 mL blood.
9. Disregarding the other buffering systems, this blood sample would be buffered only to acid. Without carbonic acid, any additions of base would increase the pH.
11. Loss of stomach acid (HCl) could lead to metabolic alkalosis.
13. Steroid hormones bind to a receptor molecule within the cell. The complex diffuses into the nucleus and binds to specific protein–DNA complexes. This binding alters the rate at which the DNA is transcribed at that region.
15. The enzyme adenylate cyclase catalyzes its formation from ATP.
17. It is released from the pancreas when the blood concentration of certain molecules (glucose) fall below some minimum concentration. Glucagon stimulates the release of glucose and lipid to prevent further decreases in their concentration.
19. It stimulates the hydrolysis of triacylglycerols to fatty acids and glycerol and the release of these compounds to the blood. Glucagon binds to adipose cells and initiates a series of events that lead to the activation of lipases.
21. Interferons bind to cells and stimulate the production of antiviral proteins within the cell. These proteins prevent normal viral replication within that cell.
23. anti-A
25. individuals with B or O blood group
27. Less than 1 mm in diameter, unless a mechanism is provided to deliver air to within a millimeter of all cells.
29. These effects are described by Le Chatelier's principle.
31. HPO_4^{2-}
33. hemoglobin
35. Progesterone, estrogen, testosterone, and cortisol are all examples.
37. epinephrin
39. The second time an organism is exposed to an antigen, the immune response is more rapid and more massive.

Illustration Credits

Page 2 Courtesy of NASA; **6** (*top*) © 1989 Tony Freeman/Photoedit; **6** (*bottom*) © 1989 Barnett/Custom Medical Stock Photo; **14** Montana State University Photo Service; **17** John Amend; **21** Montana State University Photo Service; **22** Montana State University Photo Service; **29** Alan Oddie/Photoedit; **45** John Amend; **47** © Henry Lansford 1983; **50** Jean-Claude Lejeune/Stock Boston; **90** John Amend; **116** Montana State University Photo Service; **150** Bullaty-Lomeo/Image Bank; **151** Peter Menzel/Stock Boston; **188** John Amend; **189** Montana State University Photo Service; **201** John Amend; **223** Montana State University Photo Service; **224** Montana State University Photo Service; **226** Courtesy of NASA; **244** Brian Cristopher; **333** C. Herbert Wagner/Phototake; **334** Montana State University; **348** Courtesy of the Department of Energy; **354** L. J. Conrad; **356** (*top*) Courtesy of W. L. Gore & Associates, Inc.; **356** (*bottom*) Courtesy of W. L. Gore & Associates, Inc.; **357** (*both*) © Tony Freeman 1989/Photoedit; **407** (*left*) © Mickey Gibson/Animals Animals; **407** (*right*) © Kathie Atkinson-Oxford Scientific Films/Animals Animals; **415** Courtesy of Leland Clark; **448** Giraudon/Art Resource, NY; **441** Brad Mundy; **449** (*top*) © 1988 Tony Freeman/Photoedit; **449** (*bottom*) © Leslye Borden/Photoedit; **500** (*top*) Courtesy of the Boston Medical Library in the Francis A. Countway Library of Medicine; **500** (*bottom*) From "Anesthesiology" by Peter M. Winter and John N. Miller. Copyright © 1985 by Scientific American Inc. All rights reserved; **506** (*all*) Terry Pagos; **524** © Alastair Shay-Oxford Scientific Films/Animals Animals; **532** Brad Mundy; **540** TIME/Thomas Eisner with Daniel Aneshansley; **541** (*all*) Thomas Eisner, Charles Walcott; **616** Courtesy of Antoinette Ryter/Pasteur Institute; **617** (*top*) From D. L. Shungo, J. B. Cornett, & G. D. Shockman *Journal of Bacteriology*, **138**: 598–608, 1979. American Society for Microbiology. Reproduced by permission; **617** (*bottom*) © Lennart Nilsson, *The Body Victorious*, Delacorte Press, Bonnier Fakta; **619** Photo Researchers; **621** Courtesy of Richard J. Feldmann and Neil Patterson Publishers; **626** Co Rentemeester/FPG International; **650** (*b* and *c*) From Buckwalter, J. A., and Rosenberg, L. *Collagen and Related Research*, **3**, 489–504 (1983). Courtesy of Dr. Lawrence Rosenberg; **666** David C. Fritts/Animals Animals; **708** Courtesy of Richard J. Feldmann and Neil Patterson Publishers; **710** Photo Researchers; **715** Dagmar Fabricius/Stock Boston; **720** (*top*) Walter Chandoha; **720** (*bottom*) Elyse Lewin/Image Bank; **729** From R. T. Morrison and R. N. Boyd, *Organic Chemistry*, Fifth Edition, copyright © 1987 by Allyn & Bacon, Inc. Reprinted by permission; **730** (*left*) Courtesy of Dr. David Clayton; **730** (*right*) Dr. Jack Griffith and *Scientific American;* **755** Courtesy of Claude Martin and Francoise Dieterlen-Lievre. **Color Plate 2** © 1989 Scott Camazine/Photo Researchers; **3** Dr. K. Sikora/Photo Researchers; **4** (*both*) SAGE II Images courtesy of M. Patrick McCormick and J. C. Larson, NASA Langley Research Center, Hampton, Virginia; **5** (*both*) © 1989 Tony Freeman/Photoedit; **6** Jeff Dunn/Stock Boston; **7** (*all*) © 1989 M. Davidson/Custom Medical Stock Photo; **8** (*top*) George Wuerthner; **8** (*bottom*) Courtesy of Jim Peako; **9** Jeff Albertson/Stock Boston; **10** Oxford Scientific Films/Animals Animals; **11** (*top*) © Chuck Brown 1983/Photo Researchers; **11** (*bottom*) © Bio Photo Associates/Photo Researchers.

Index

Page numbers set in boldface indicate pages where terms are defined in New Terms lists.

C

F

U

V

W